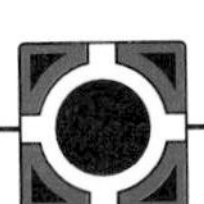

Contemporary Precalculus

A Graphing Approach

Second Edition

Thomas W. Hungerford
CLEVELAND STATE UNIVERSITY

Saunders College Publishing

HARCOURT BRACE COLLEGE PUBLISHERS

Philadelphia Fort Worth San Diego New York Orlando
Austin San Antonio Toronto Montreal London Sydney Tokyo

Text Typeface: Times Roman
Compositor: Progressive Information Technologies
Developmental Editor: Marc Sherman
Managing Editor: Carol Field
Project Editor: Bonnie Boehme
Copy Editor: Charlotte Nelson
Manager of Art and Design: Carol Bleistine
Art Director: Joan Wendt
Illustration Supervisor: Sue Kinney
Art and Design Coordinator: Kathleen Flanagan
Text Designer: Kathyrn Needle
Cover Designer: Joan Wendt
Text Artwork: ST Associates
Director of EDP: Tim Frelick
Manager of Production: Joanne Cassetti
Production Manager: Alicia Jackson
Marketing Manager: Nick Agnew

Cover Credit: Neal Lavey/Phototake NYC

Printed in the United States of America

CONTEMPORARY PRECALCULUS: A GRAPHING APPROACH
Second Edition

ISBN 0-03-018544-0

Library of Congress Catalog Card Number: 95-072986

9012345 039 10 987

Dedicated to the Parks sisters,

whose presence in my life has greatly enriched it:

To my aunt,

Irene Parks Mills

And to the memory of my mother,

Grace Parks Hungerford

and my aunt,

Florence M. Parks

Preface

Contemporary Precalculus: A Graphing Approach, Second Edition, is intended to provide the essential mathematical background needed in calculus for students who have had two or three years of high school mathematics. It integrates graphing technology into the course without losing sight of the fact that the underlying mathematics is the crucial issue. The book enables students to become active participants in developing their understanding of mathematics, but avoids the very real danger of making technology an end in itself. We have done our best to present sound mathematics in an informal manner that stresses meaningful motivation, careful explanations, and numerous examples, with an ongoing focus on real-world problem solving.

As in the first edition, the emphasis is on developing the concepts that play a central role in calculus by exploring these ideas from graphical, numerical, and algebraic perspectives. With the assistance of technology, the interplay between these viewpoints can be fully exploited to give students insight into what is going on and the confidence to work with it on their own. Instructors can spend less time on ''mechanics'' and more on the underlying concepts. They can focus on the essentials, without getting bogged down in calculations that obscure the key ideas.

Changes in the Second Edition

When the first edition was written, the graphing calculator of choice was the TI-81 or comparable Casio and Sharp models.* By the time the book was published, the next generation of calculators, such as the TI-82, TI-85, and Sharp 9300, were just coming into widespread use. Now these calculators, together with the TI-83, HP-38 and Casio 9800, are the precalculus standard, with the even more powerful TI-92 just coming into wider use. Although these improvements in technology do not affect the mathematical concepts involved, they do make available a wider range of tools and change the procedures used in many cases, and thus the need for a technologically updated edition.

*Various HP models, such as the HP-48, were much more powerful, but were not as widely used at the precalculus level because they were perceived to be less user-friendly.

Users of the first edition have suggested a number of other changes that are now incorporated in the text.

Technology Assistance Instructors were happy that the first edition was not tied to a particular brand of graphing calculator, but also reported that many students failed to make full use of the capabilities of their calculators, often because they didn't realize what the calculators could do. Consequently, we have kept the discussions of technology in the body of the text as generic as possible, while adding two new features. *Technology Tips* in the margin offer advice on how to use specific graphing calculators. *Calculator Investigations* precede many of the exercise sets in the early sections of the book to encourage and enable students to become familiar with the capabilities and limitations of graphing and other technology. Both new features can be easily omitted by those who don't need them, but should provide significant assistance to others.

Program Appendix A small collection of useful programs has been included. Most of them are not needed by students who have state-of-the-art calculators, but will provide others with features that are not built into their calculators.

Exercises The majority of the exercises are retained, but ones that "didn't work" as teaching devices have been deleted or altered. Many new exercises, particularly applications, have been added.

Limits and Continuity Because very few instructors who adopted the first edition used this chapter, it is now published separately to save space. It is available at nominal cost to those who request it.

Instructor's Manual and Student Solutions Manual A new author has reworked all solutions, which were then accuracy reviewed, in order to ensure the highest quality solutions manual.

In addition to these global changes, mathematical and pedagogical improvements have been made in various parts of the text, including the following.

Technology and Equations The discussion of equations in Section 1.5 has been expanded to include numerical as well as graphical solution methods. The applications that follow are now in two sections: Applications of Equations and Optimization Applications.

Secant Method An optional excursion that explains the secant method for solving equations has been added to Chapter 3.

Exponential Functions Section 5.2 now has a more complete presentation of exponential growth and decay.

Logarithmic Functions The emphasis on natural logarithms is retained, but common logarithms are introduced simultaneously with them in Section 5.3 because students seem to find them easier to understand.

Trigonometry The number and variety of applied exercises in Chapter 7 have been increased and approximately two dozen new figures have been added.

Systems of Equations The Gauss-Jordan method is now included in Section 11.2.

Mathematical and Pedagogical Features

The mathematical approaches to important topics used in the first edition have been retained.

Functional notation and its uses are thoroughly treated.

The natural exponential and logarithmic functions are emphasized because of their central role in calculus.

Trigonometric functions of real numbers—the ones most widely used in calculus—are introduced first, with traditional triangle trigonometry treated later.

Parametric graphing is introduced early and used thereafter to illustrate such concepts as inverse functions, the definition of trigonometric functions, and the graphs of conic sections.

Average rates of change—a crucial concept for calculus—are fully treated and the calculator is used to explore the intuitive connections between average and instantaneous rates of change.

All of the student-oriented pedagogical features of the first edition are included here.

Graphing Explorations Students are expected to participate actively in the development of concepts and examples by using graphing technology to complete many discussions in the text.

Warnings Students are alerted to common errors and misconceptions (both mathematical and technological) by clearly marked warning boxes.

Exercises Exercise sets proceed from routine calculation and drill to exercises requiring some thought, including graph interpretation and word problems. Some sets include problems labeled *Thinkers,* most of which are not difficult, but simply different from what students may have seen before; a few of the *Thinkers* are quite challenging. Answers for all odd-numbered problems are given in the back of the book, and solutions for these problems are in the Student Solutions Manual.

Chapter Reviews Each chapter concludes with a list of important concepts (referenced by section and page number), a summary of important facts and formulas, and a set of review questions.

Algebra Review Basic algebra is reviewed in an appendix, which can be omitted by well-prepared students or covered as an introductory chapter if necessary.

Geometry Review Frequently used facts from plane geometry are summarized, with examples, in an appendix.

Supplements

The following chart summarizes the print and software supplements available to users of this text. A full description of each is given in the "To the Instructor" section on page xiv or the "To the Student" section on page xvii.

	Print Ancillaries	***Software***
Instructors	Instructor's Manual Test Bank Graphing Calculator Manual Transparency Masters	F/C/P Graph (DOS, Mac, and Windows) EXAMaster+ Computerized Test Bank
Students	Student Solutions Manual Graphing Calculator Manual	F/C/P Graph (DOS, Mac, and Windows)

Acknowledgments

My sincere thanks go to the following reviewers who provided many helpful suggestions for improving the text:

Deborah Adams, *Jacksonville University*
Kelly Bach, *University of Kansas*
Bettyann Daley, *University of Delaware*
Betty Givan, *Eastern Kentucky University*
William Grimes, *Central Missouri State University*
Charles Laws, *Cleveland State Community College*
Martha Lisle, *Prince George's Community College*
Ruth Meyering, *Grand Valley State University*
Philip Montgomery, *University of Kansas*
Roger Nelsen, *Lewis & Clark College*
Ann Steen, *Santa Fe Community College*
Hugo Sun, *California State University at Fresno*
Bettie Truitt, *Black Hawk College*

Thanks also go to the following respondents to a graphing calculator survey:

Anne Brown, *Indiana University at South Bend*
Bruce Hoelter, *Raritan Valley Community College*
Marian Hukle, *University of Kansas*
Sandra Johnson, *St. Cloud State University*
John Khalilian, *University of Alabama*
Jack Porter, *University of Kansas*
Jan Rizzuti, *Central Washington University*
Howard Rolf, *Baylor University*

I am grateful to the accuracy reviewers

Joan McCarter, *Arizona State University*
Flauren Ricketts, *Normandale Community College*

who examined (and corrected where necessary) the examples and exercises; as well as checking the Student Solutions Manual. Their work has greatly improved the final product.

The text maintains its high level of accuracy due to the accuracy reviewers of the first edition:

John Khalilian, *University of Alabama*
Lynn Kotrous, *Central Community College–Platt Campus*
Seyed Pedram, *University of Alabama*

It is a pleasure to acknowledge the invaluable assistance of the Saunders staff, particularly Marc Sherman, Senior Developmental Editor, and Bonnie Boehme, Project Editor. Their fine work has made a difficult project seem easy.

The last word goes to my wife, Mary Alice, whose patience has been sorely tried during the past year. She has provided understanding and support when it was most needed, and I love her dearly.

Thomas W. Hungerford
Cleveland, Ohio

Contents

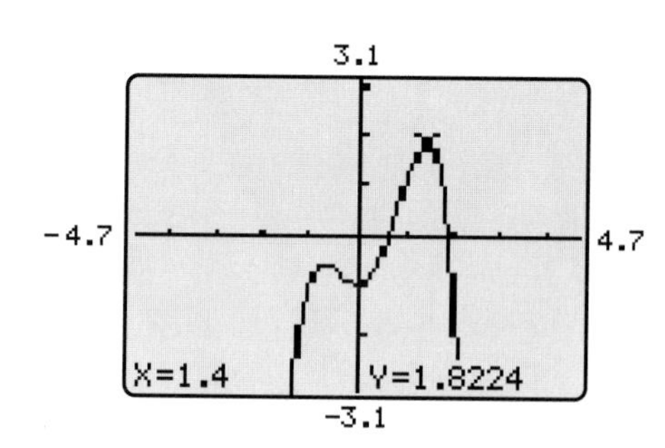

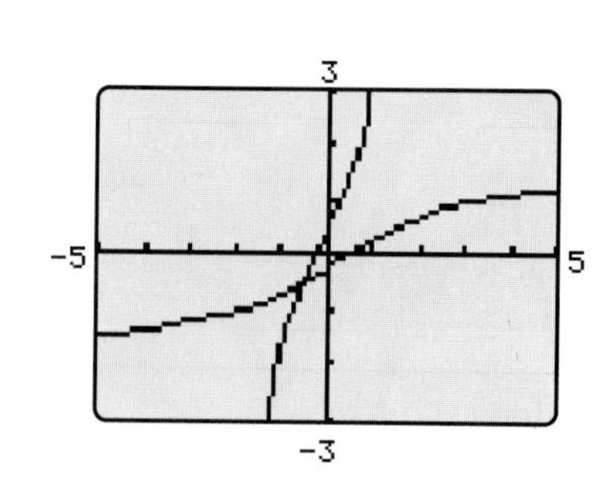

CHAPTER 3

Linear and Quadratic Functions 165

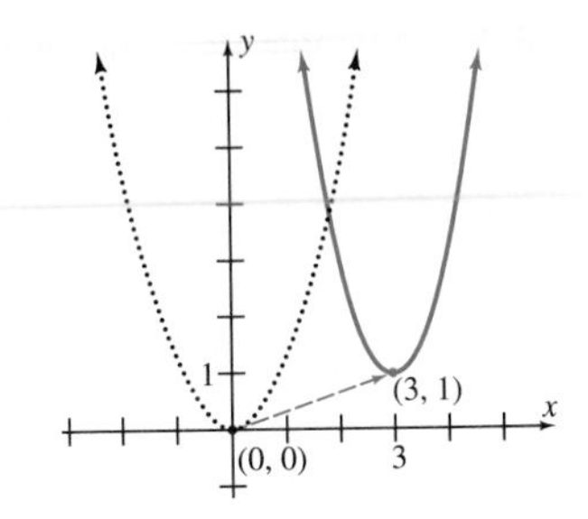

CHAPTER 4

Polynomial and Rational Functions 209

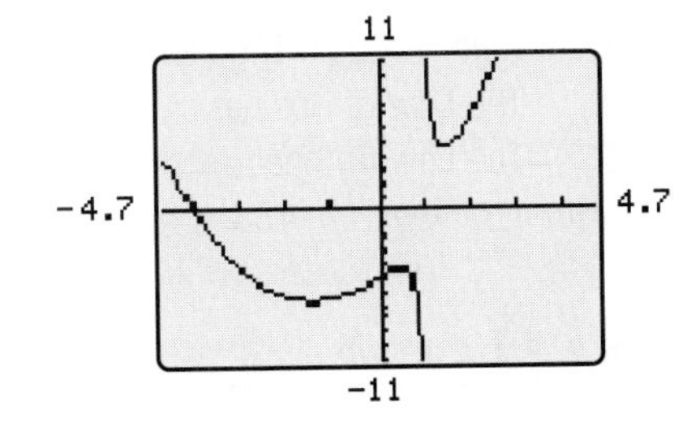

CHAPTER 5

Exponential and Logarithmic Functions 299

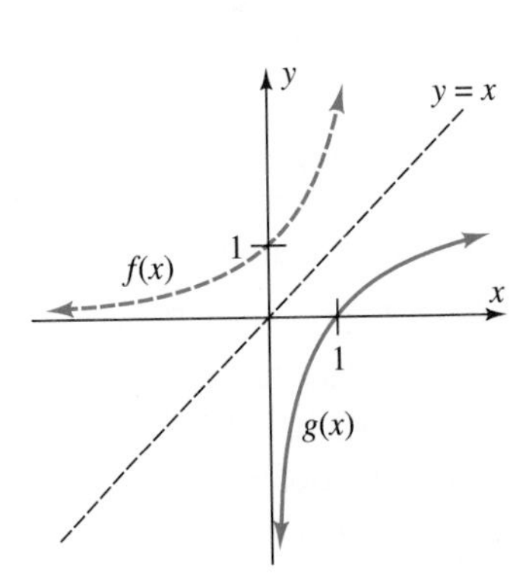

CHAPTER 6

Trigonometric Functions 365

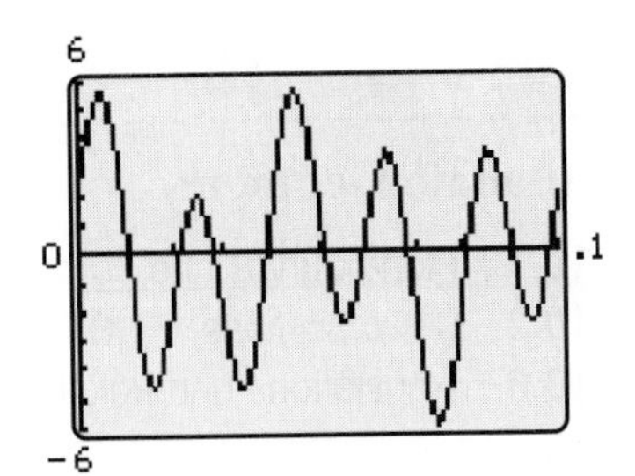

CHAPTER 7

Triangle Trigonometry 451

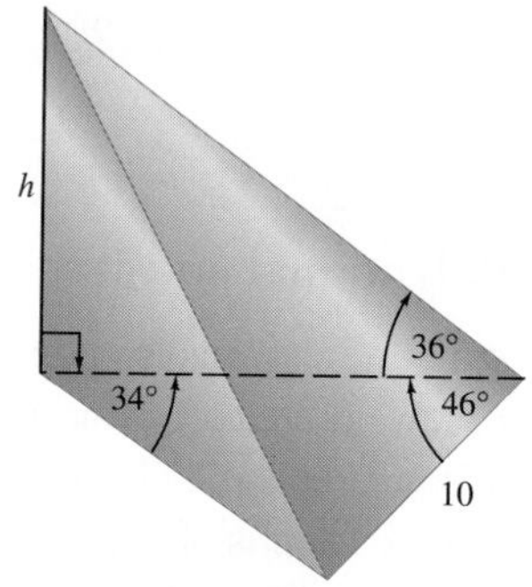

CHAPTER 8

Trigonometric Identities and Equations 486

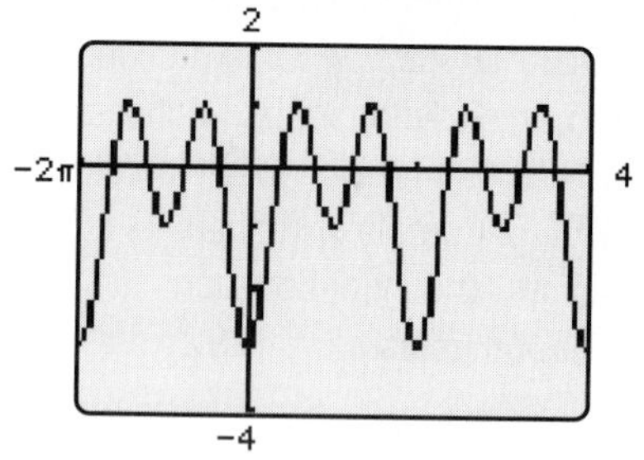

CHAPTER 9

Applications of Trigonometry 542

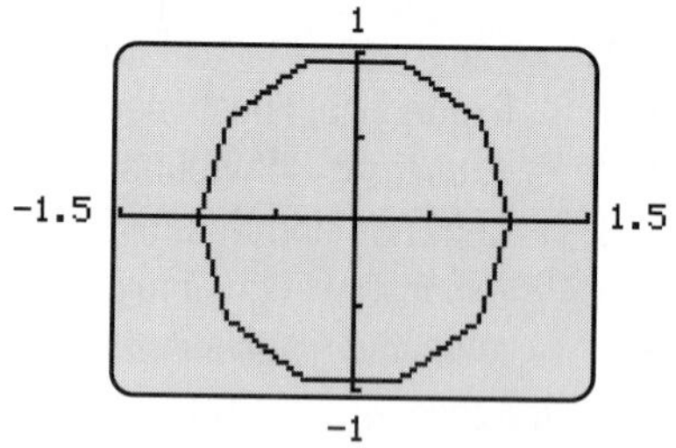

CHAPTER 10

Analytic Geometry 587

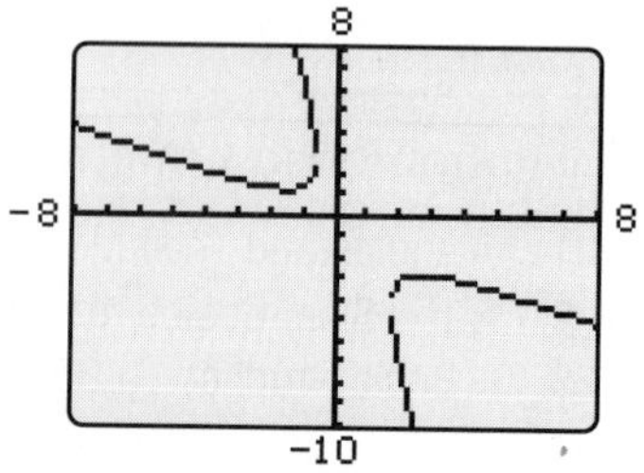

CHAPTER 11

Systems of Equations 652

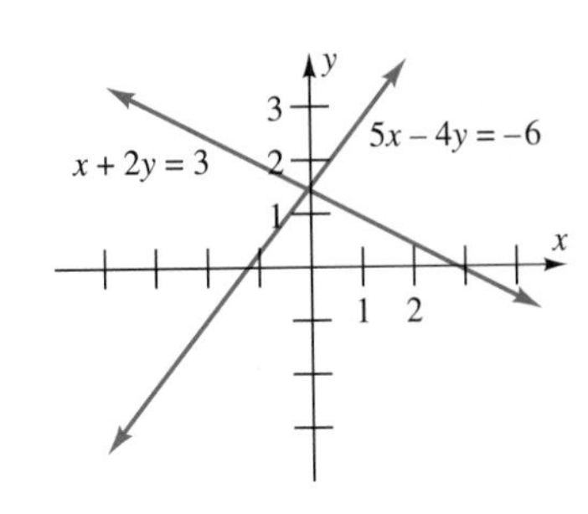

CHAPTER 12

Discrete Algebra 698

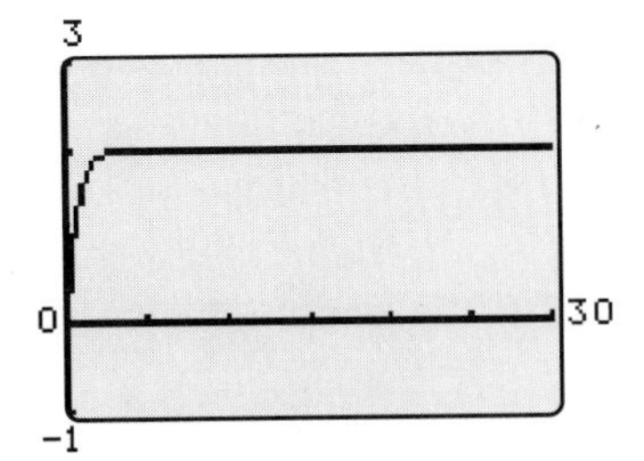

CHAPTER 13

Limits and Continuity

(available as a separate supplement)

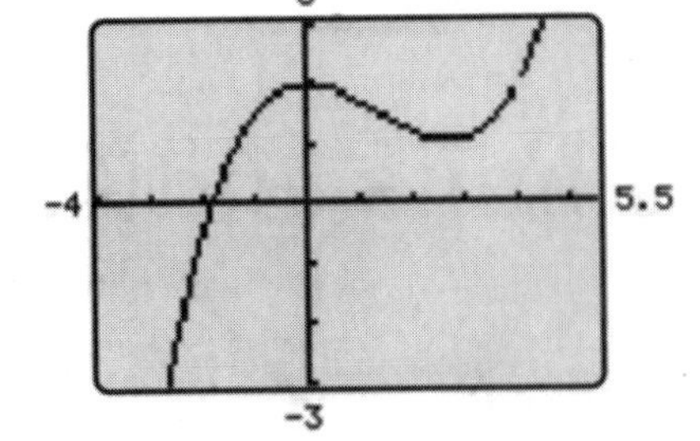

To the Instructor

Every effort has been made to make this text as flexible as possible. With minor exceptions (usually exercises or occasional examples), the interdependence of chapters is given by the chart on the facing page. Each chapter begins with a **Roadmap** that indicates the interdependence of sections within the chapter.

As noted in the preface, the standard review of basic algebra is in the Algebra Review Appendix. This material, which is a prerequisite for the entire book, may be covered as Chapter 0 if necessary, or omitted by well-prepared classes.

A few sections are labeled as **Excursions.** Each Excursion is closely related to the section that precedes it and usually has that section as a prerequisite. No Excursion is a prerequisite for any other section of the text. With rare exceptions, each Excursion is a complete discussion with a full set of exercises. The ''Excursion'' label is designed solely to make syllabus planning easier and is *not* intended as any kind of value judgment on the topic in question.

The use of the **Graphing Explorations** in the text are discussed in the ''To the Student'' section on page xvii.

An optional final chapter on **Limits and Continuity** is published separately and available at nominal cost to schools that adopt the text. If you wish to include this material in your course, please contact your local sales representative.

Supplements

Instructors who adopt this text may receive, free of charge, the following items:

Instructor's Manual with Transparency Masters Written by Matt Foss of North Hennepin Community College, this manual contains detailed solutions to all the exercises and end-of-chapter Review Questions to assist the instructor in the classroom and in grading assignments. Additionally, more than 60 transparency masters of important figures, theorems, and charts from the text are provided.

Test Bank Written by Bruce Hoelter of Raritan Valley Community College, this manual provides 2000 multiple choice and open-ended questions arranged in five forms per chapter, each form containing about 30 questions. Master answer sheets and a complete answer section are included.

Interdependence of Chapters

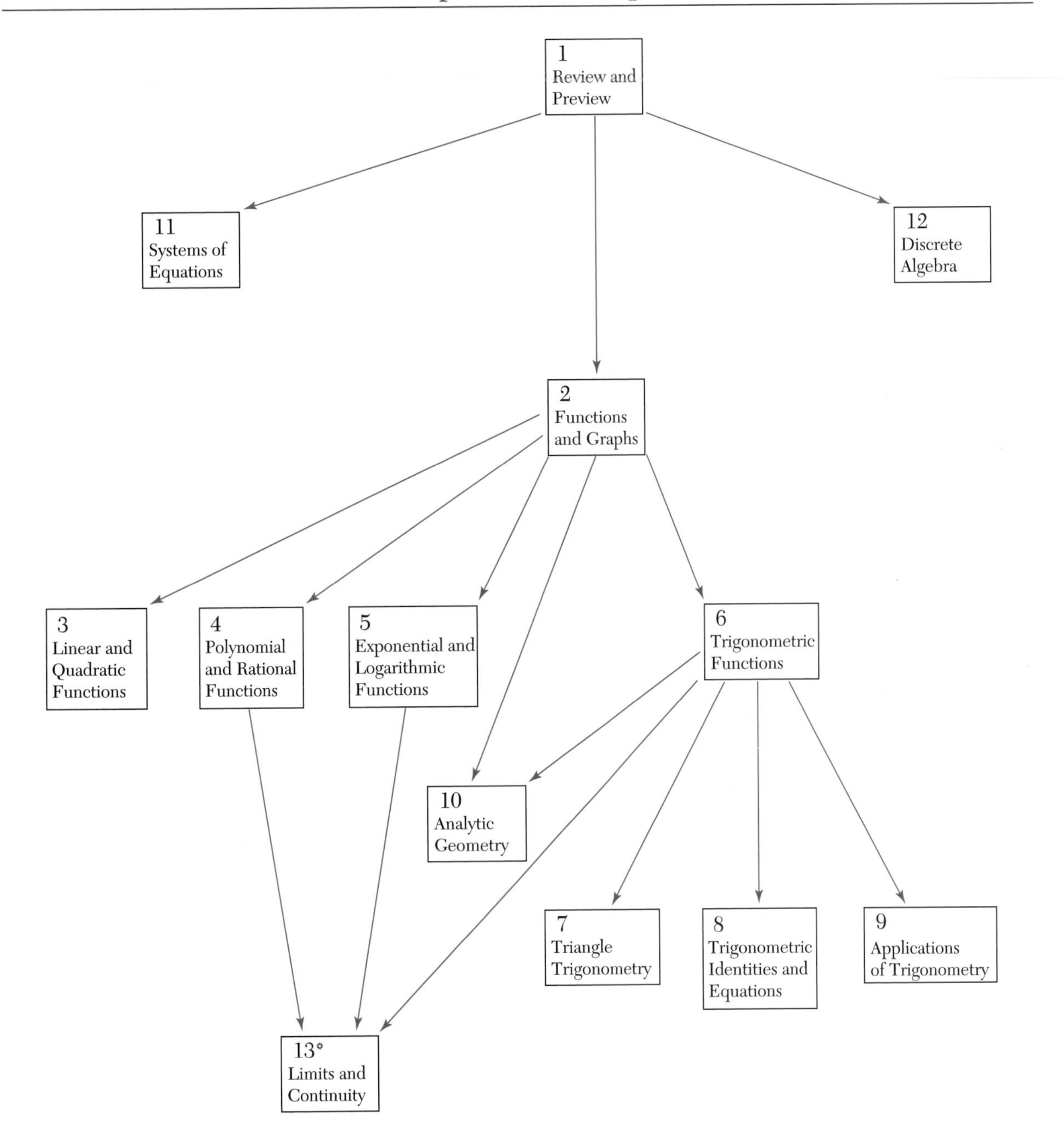

* Chapter 13 is published separately and is available at nominal cost to schools that adopt this text.

EXAMaster+™ Computerized Test Bank (IBM, Mac, and Windows versions) The computerized test bank contains all the test bank questions and allows instructors to prepare quizzes and examinations quickly and easily. Features include: (1) ability to convert multiple choice questions into short answer questions, (2) ability to enter and edit instructor's questions, (3) ability to create and administer a test via computer over a network or with floppies, whereby answers are scored and grades are transferred to an electronic gradebook, and (4) user-friendly printing capability to accommodate all printing platforms.

Graphing Calculator Supplement This supplement covers several major brands of the latest calculator models for problem solving in precalculus. It is written by Ron Marshall and Nicholas Norgaard of Western Carolina University.

F/C/P Graph software (IBM and Mac versions) Offered free to instructors and copyable upon adoption. This user-friendly, interactive software on Functions, Conics, and Parametric equations, written by George W. Bergeman of Northern Virginia Community College, allows students to build their graphing skills and explore precalculus topics in a self-paced, colorized format, and can be used in most instances in place of a graphing calculator. Computer lab exercises accompany the software. Special features include: a root-finder, a maxima-minima finder, ability to graph up to four functions simultaneously, and printing support.

The **Windows™** version of F/C/P Graph includes two features especially helpful to instructors who create handouts and exams, and for students doing projects or homework:

Users can print graphs and scale them to their size requirements.
Users can save graphs to bitmap files for import into their word processing programs.

To the Student

In order to use this text effectively you must have a graphing calculator or a computer equipped with appropriate graphing software (or both). (Free software is available from your instructor, as explained below.) In the text, the terms ''calculator'' and ''graphing calculator'' are used interchangeably. It is understood that, with obvious modifications, all discussions of calculators apply equally well to graphing software for computers.

The text is not written for a specific calculator, so some of the illustrations of calculator screens may not look exactly like yours. Furthermore, you may have to figure out the keystrokes needed to carry out particular procedures. To assist you in doing this, there are frequent **Technology Tips** in the margin. These Tips describe the proper menus or keys to be used on specific calculators that use standard algebraic notation (such as TI-81/82/83/85/92, HP-38, Casio 7700/8700/9800, and Sharp 9300).* When the Tips are not sufficient, consult the instruction manual for your calculator.

Since many students are not aware of the full capabilities of their calculators, many of the early sections in this book contain **Calculator Investigations** (just before the exercise sets). These investigations will help you to become familiar with your calculator and to maximize the mathematical power it provides. Even if your instructor does not assign any of these investigations, you might want to look through them to be sure you are getting the most you can from your calculator.

The key to succeeding in this course is to remember that *mathematics is not a spectator sport.* You can't expect to learn mathematics without *doing* mathematics, any more than you could learn to swim without getting wet. You have to take an active role, making use of all the resources at your disposal: your instructor, your fellow students, your calculator (and its instruction manual), and this book.

It's no secret that many students use their math books only to find out what the homework problems are. If you are one of these, we strongly suggest that you change your ways. There is no way that your instructor can possibly cover the essential topics, clarify ambiguities, explain the fine points, and answer all your questions during class time. You simply will not develop the level of understanding you need to succeed in this course and calculus unless you read the text fully

*HP-48 users should consult the Graphing Calculator Supplement, which is described below, to adapt these tips to the Reverse Polish Notation used on that calculator.

and carefully. In particular, you should read the appropriate section of the text *before* beginning the exercises in that section.

You can't read a math book the way you read a novel or even a history book. You need pencil, paper, and your calculator at hand to work out the statements that you don't understand and to make notes on things to ask your fellow students and/or your instructor. One feature of this book will assist you to become such an interactive reader. The label **Graphing Exploration** indicates that you are to use your calculator as directed in order to complete the discussion. Typically, this will involve graphing one or more equations and answering some questions about the graphs. Doing these explorations as they arise will improve your understanding and clarify issues that might otherwise cause difficulties.

Finally, remember the words of the great Hillel: ''The bashful do not learn.'' There is no such thing as a ''dumb question'' (assuming, of course, that you have read the book and your notes). Your instructor will welcome questions that arise from a serious effort on your part. In any case, your instructor is being paid (with your tuition money) to answer questions. So do yourself a favor and get your money's worth—ask questions.

Supplements

Students using *Contemporary Precalculus* may obtain the following software at no cost.

F/C/P Graph software (IBM, Windows, and Mac versions) is user-friendly, interactive software that can be used in most instances in place of a graphing calculator. Special features include: a root-finder, a maxima-minima finder, ability to graph up to four functions simultaneously, and printing support (so you can save your work). Your instructor has (or can obtain) this software, which you are free to copy.

Students may purchase the following supplements.

The **Student Solutions Manual,** written by Matt Foss of North Hennepin Community College, contains detailed solutions to all odd-numbered Exercises and end-of-chapter Review Questions. Specific instructions for solving graphing calculator problems are included, as are accurate representations of graphing calculator screens.

Graphing Calculator Supplement This supplement covers several major brands of the latest calculator models for problem solving in precalculus. It is written by Ron Marshall and Nicholas Norgaard of Western Carolina University.

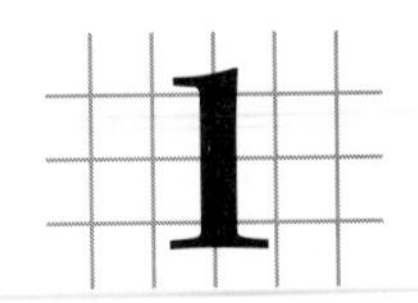

Review and Technology Preview

The first half of this chapter reviews the essential facts about real numbers, equations, and the coordinate plane that are needed in this course and in calculus. The Algebra Review Appendix at the end of the book is a prerequisite for this material.

The graphing calculator is introduced in the second half of the chapter. Graphing techniques are presented that enable you to solve complicated equations easily and to deal effectively with real-world problems. However, a graphing calculator cannot be used with maximum efficiency unless you have a sound knowledge of a variety of different types of functions. These functions are studied in later chapters, at which time we shall return to this material and explore it in greater depth.

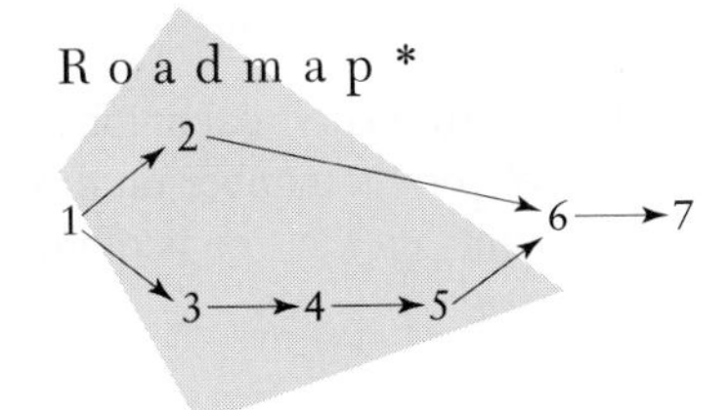

1.1 THE REAL NUMBER SYSTEM

You have been using **real numbers** most of your life. They include:

Natural numbers (or **positive integers**): 1, 2, 3, 4, ...

Integers: 0, 1, −1, 2, −2, 3, −3, 4, −4, ...

Rational numbers: every number that can be expressed as a fraction r/s, with r and s integers and $s \neq 0$; for instance, $\frac{1}{2}$, $-.983 = \frac{-983}{1000}$, $47 = \frac{47}{1}$, and $8\frac{3}{5} = \frac{43}{5}$. Alternatively, rational numbers may be described as numbers that can be expressed as terminating or repeating decimals; for instance, $.5 = \frac{1}{2}$ or $.3333\cdots = \frac{1}{3}$. See Excursion 1.1.A for details.

*The Roadmap at the beginning of each chapter shows the interdependence of sections in the chapter. An arrow, such as 1 ⟶ 2, means that Section 1 is a prerequisite for Section 2.

Irrational numbers:* every real number that *cannot* be expressed as a fraction, or alternatively, real numbers that can be expressed as infinite *nonrepeating* decimals (see Excursion 1.1.A). For example, the number π, which is used to calculate the area of a circle, is irrational.**

We assume that you are familiar with the basic properties of real number arithmetic and with the geometric representation of real numbers as points on a **number line** as in Figure 1–1.

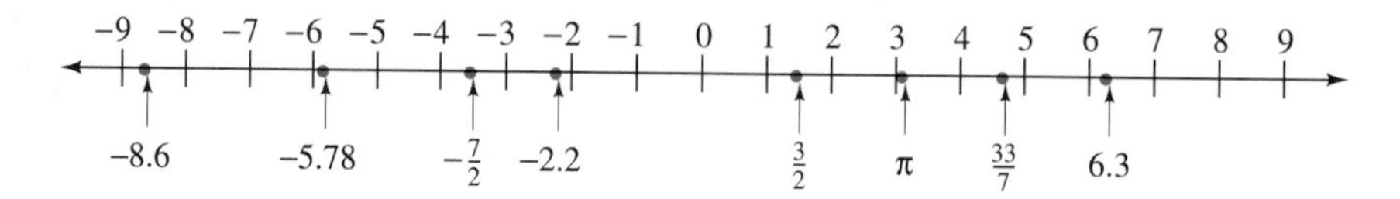

Figure 1-1

We shall assume that there is exactly one point on the line for every real number (and vice versa) and use phrases such as "the point 3.6" or "a number on the line." This mental identification of real numbers and points on the line is often extremely helpful.

Order

The statement **"*c* is less than *d*"** (written $c < d$) and the statement **"*d* is greater than *c*"** (written $d > c$) mean exactly the same thing:

c lies to the *left* of d on the number line.

For example, Figure 1–1 shows that $-5.78 < -2.2$ and $\pi < 4$.

The statement $c \leq d$ (or $d \geq c$) means ***c* is less than or equal to *d*** (or ***d* is greater than or equal to *c***). The statement $b \leq c \leq d$ means

$$b \leq c \qquad \textit{and} \qquad c \leq d.$$

Geometrically, this means that c lies between b and d on the number line (and may possibly be equal to one or both of them). Similarly, $b \leq c < d$ means $b \leq c$ *and* $c < d$, and so on.

Certain sets of numbers, defined in terms of the order relation, appear frequently enough to merit special notation. Let c and d be real numbers with $c < d$:

*Ir-rational = not rational = not a ratio.

**This fact is difficult to prove. In grade school you may have used 22/7 as π; a calculator might display π as 3.141592654. However, these numbers are just *approximations* of π (*close* to but *not quite equal* to π).

Interval Notation ▶

$[c, d]$ denotes the set of all real numbers x such that $c \le x \le d$.
(c, d) denotes the set of all real numbers x such that $c < x < d$.
$[c, d)$ denotes the set of all real numbers x such that $c \le x < d$.
$(c, d]$ denotes the set of all real numbers x such that $c < x \le d$.

All four of these sets are called **intervals** from c to d. The numbers c and d are the **endpoints** of the interval. $[c, d]$ is called the **closed interval** from c to d (both endpoints included and *square* brackets) and (c, d) is called the **open interval** from c to d (neither endpoint included and *round* brackets). For example,*

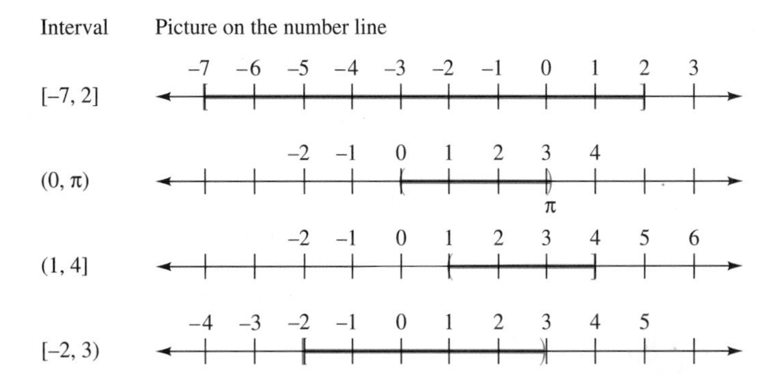

Figure 1-2

If b is a real number, then the half-line extending to the right or left of b is also called an interval. Depending on whether or not b is included, there are four possibilities:

Interval Notation ▶

$[b, \infty)$ denotes the set of all real numbers x such that $x \ge b$.
(b, ∞) denotes the set of all real numbers x such that $x > b$.
$(-\infty, b]$ denotes the set of all real numbers x such that $x \le b$.
$(-\infty, b)$ denotes the set of all real numbers x such that $x < b$.

*In Figures 1–2 and 1–3 a round bracket such as) or (indicates that the endpoint is *not* included, whereas a square bracket such as] or [indicates that the endpoint is included.

For example,

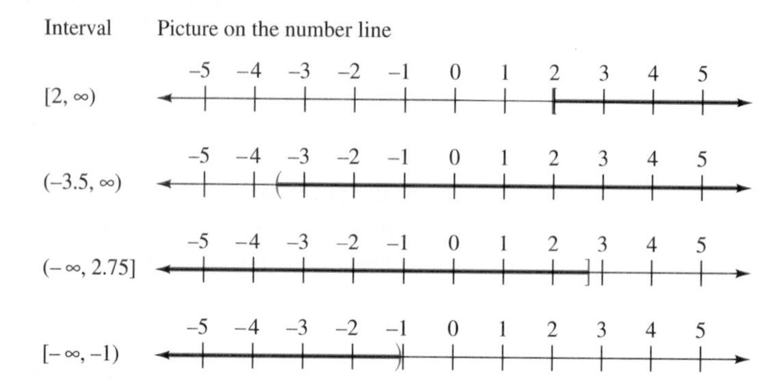

Figure 1-3

In a similar vein, **$(-\infty, \infty)$ denotes the set of all real numbers.**

Note The symbol ∞ is read "infinity" and we call the set $[b, \infty)$ "the interval from b to infinity." The symbol ∞ does *not* denote a real number; it is simply part of the notation used to label the first two sets of numbers defined in the box above. Analogous remarks apply to the symbol $-\infty$, which is read "negative infinity."

> **▶ TECHNOLOGY TIP**
>
> To enter a negative number, such as -5, on most calculators, you must use the negation key: [(−)] [5]. If you use the subtraction key on such calculators and enter [−] [5], the display will read
>
> ANS − 5
>
> which tells the calculator to subtract 5 from the previous answer.

Negative Numbers and Negatives of Numbers

The **positive numbers** are those to the right of 0 on the number line, that is,

all numbers c with $c > 0$.

The **negative numbers** are those to the left of 0, that is,

all numbers c with $c < 0$.

The **nonnegative** numbers are the numbers c with $c \geq 0$.

The word "negative" has a second meaning in mathematics. The **negative *of a* number *c*** is the number $-c$. For example, the negative of 5 is -5 and the negative of -3 is $-(-3) = 3$. Thus the negative of a negative number is a positive number. Zero is its own negative since $-0 = 0$. In summary,

Negatives ▶

The negative of the number c is $-c$.
If c is a positive number, then $-c$ is a negative number.
If c is a negative number, then $-c$ is a positive number.

Square Roots

The **square root** of a nonnegative real number d is defined to be the *nonnegative* number whose square is d; it is denoted $\sqrt{d}$. For instance,

$$\sqrt{25} = 5 \quad \text{because} \quad 5^2 = 25 \qquad \text{and} \qquad \sqrt{1.21} = 1.1 \quad \text{because} \quad (1.1)^2 = 1.21.$$

In the past you may have said $\sqrt{25} = \pm 5$ since $(-5)^2$ is also 25. It is preferable, however, to have a single unambiguous meaning for the symbol $\sqrt{25}$. The definition given above is one way to do this: In the real number system, the term "square root" and the radical symbol $\sqrt{}$ always denote a *nonnegative* number. To express -5 in terms of square roots, we write $-5 = -\sqrt{25}$.

Although $-\sqrt{25}$ is the real number -5, the expression $\sqrt{-25}$ is *not defined* in the real numbers because there is no real number whose square is -25. In fact, since the square of every real number is nonnegative,

No negative number has a square root in the real numbers.*

You can easily convince yourself that $\sqrt{cd} = \sqrt{c}\sqrt{d}$ for any nonnegative real numbers c and d. On the other hand, heed this

▶ TECHNOLOGY TIP

To compute an expression such as $\sqrt{7^2 + 51}$ on a calculator you must use parentheses:

$$\sqrt{}\ (7^2 + 51).$$

Without the parentheses the calculator will compute

$$\sqrt{7^2} + 51 = 7 + 51 = 58$$

instead of the correct answer

$$\sqrt{7^2 + 51} = \sqrt{49 + 51} = \sqrt{100} = 10.$$

WARNING If c and d are positive real numbers, then

$$\sqrt{c + d} \neq \sqrt{c} + \sqrt{d}.$$

For example,

$$\sqrt{9 + 16} = \sqrt{25} = 5, \qquad \text{but} \qquad \sqrt{9} + \sqrt{16} = 3 + 4 = 7.$$

Absolute Value

On an informal level most students think of absolute value like this:

> The absolute value of a nonnegative number is the number itself.
>
> The absolute value of a negative number is found by "erasing the minus sign."

If $|c|$ denotes the absolute value of c, then, for example, $|5| = 5$ and $|-4| = 4$.

This informal approach is inadequate, however, for finding the absolute value of a number such as $\pi - 6$. It doesn't make sense to "erase the minus sign" here. So we must develop a more precise definition. The statement $|5| = 5$ suggests that the absolute value of a positive number ought to be the number itself. For negative numbers, such as -4, note that $|-4| = 4 = -(-4)$, that is, the absolute value of the negative number -4 is *the negative* of -4. These facts are the basis of the formal definition:

*In Section 4.6, we shall study the complex numbers, a larger system in which negative numbers *do* have square roots. These square roots, however, are not real numbers.

Absolute Value ▶

The *absolute value* of a real number c is denoted $|c|$ and is defined as follows:

$$\text{If } c \geq 0, \text{ then } |c| = c.$$

$$\text{If } c < 0, \text{ then } |c| = -c.$$

EXAMPLE 1

(a) $|3.5| = 3.5$ and $|-7/2| = -(-7/2) = 7/2$.

(b) To find $|\pi - 6|$ note that $\pi \approx 3.14$, so that $\pi - 6 < 0$. Hence, $|\pi - 6|$ is defined to be the *negative* of $\pi - 6$, that is,

$$|\pi - 6| = -(\pi - 6) = -\pi + 6.$$

(c) $|5 - \sqrt{2}| = 5 - \sqrt{2}$ because $5 - \sqrt{2} \geq 0$. ■

Here are the important facts about absolute value:

Properties of Absolute Value ▶

1. $|c| \geq 0$ **and** $|c| > 0$ **when** $c \neq 0$.
2. $c \leq |c|$ **and** $-c \leq |c|$.
3. $|c| = |-c|$.
4. $|c^2| = c^2 = |c|^2$.
5. $\sqrt{c^2} = |c|$.
6. $|cd| = |c||d|$ **and if** $d \neq 0$, $\left|\dfrac{c}{d}\right| = \dfrac{|c|}{|d|}$.

▶ **TECHNOLOGY TIP**

To compute absolute values on a calculator use the ABS key and parentheses. To find $|9 - 3\pi|$, for example, key in

ABS(9 − 3π).

If the ABS key is not on the keyboard, look in the MATH menu or its NUM submenu.

EXAMPLE 2 Here are examples of some of the properties listed in the box.

2. $|6| = 6$ and hence $6 \leq |6|$ and $-6 \leq |6|$. Similarly, $|-4| = 4$ and $-(-4) = 4$, so that $-4 \leq |-4|$ and $-(-4) \leq |-4|$.

3. $|3| = 3$ and $|-3| = 3$, so that $|3| = |-3|$.

4. If $c = -5$, then

$$|c^2| = |(-5)^2| = |25| = 25,$$

$$c^2 = (-5)^2 = 25, \qquad \text{and} \qquad |c|^2 = |-5|^2 = 5^2 = 25$$

so that $|(-5)^2| = (-5)^2 = |-5|^2$.

5. When c is negative, it is *not* true that $\sqrt{c^2} = c$, because square roots are always *nonnegative.* For instance, $\sqrt{(-3)^2} = \sqrt{9} = 3$. Hence, $\sqrt{(-3)^2} = |-3|$, *not* -3.
6. If $c = 6$ and $d = -2$, then

$$|cd| = |6(-2)| = |-12| = 12 \quad \text{and}$$

$$|c||d| = |6||-2| = 6 \cdot 2 = 12$$

so that $|6(-2)| = |6||-2|$. ■

When dealing with long expressions inside absolute value bars, do the computations inside first, and then take the absolute value.

EXAMPLE 3

(a) $|5(2 - 4) + 7| = |5(-2) + 7| = |-10 + 7| = |-3| = 3.$

(b) $4 - |3 - 9| = 4 - |-6| = 4 - 6 = -2.$ ■

WARNING When c and d have opposite signs, $|c + d|$ is *not equal* to $|c| + |d|$. For example, when $c = -3$ and $d = 5$, then

$$|c + d| = |-3 + 5| = 2,$$

but $$|c| + |d| = |-3| + |5| = 3 + 5 = 8.$$

The warning shows that $|c + d| < |c| + |d|$ when $c = -3$ and $d = 5$. In the general case, we have the following fact (which is proved in Exercise 111).

The Triangle Inequality ▶ $|c + d| \le |c| + |d|$ **for any real numbers c and d.**

Distance on the Number Line

Observe that the distance from -5 to 3 on the number line is 8 units:

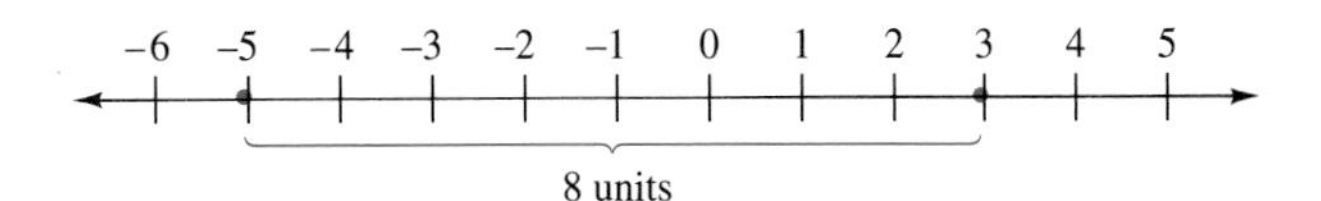

Figure 1-4

This distance can be expressed in terms of absolute value by noting that $|(-5) - 3| = 8$. That is, the distance is the *absolute value of the difference* of the two numbers. Furthermore, the order in which you take the difference doesn't matter; $|3 - (-5)|$ is also 8. This reflects the geometric fact that the distance from -5 to 3 is the same as the distance from 3 to -5. The same thing is true in the general case:

Distance on the Number Line ▶

The distance between c and d on the number line is the number

$$|c - d| = |d - c|.$$

EXAMPLE 4 The distance from 4.2 to 9 is $|4.2 - 9| = |-4.8| = 4.8$ and the distance from 6 to $\sqrt{2}$ is $|6 - \sqrt{2}|$. ■

In the special case when $d = 0$, the distance formula shows that $|c - 0| = |c|$. Hence,

Distance to Zero ▶

$|c|$ is the distance between c and 0 on the number line.

EXAMPLE 5 How many real numbers have absolute value 3? In geometric terms this question is: How many real numbers are 3 units from 0 on the number line? Figure 1–5 shows that 3 and -3 are the only numbers whose distance to 0 is 3 units. ■

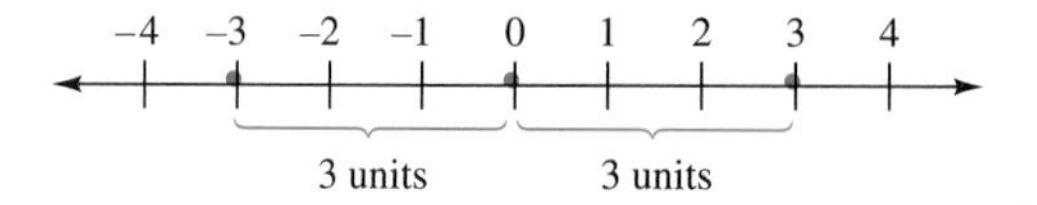

Figure 1-5

The procedure in Example 5 works in the general case:

If k is a positive number, then the only numbers with absolute value k are k and $-k$.

Example 5 shows that algebraic problems can sometimes be solved by translating them into equivalent geometric problems. The key is to interpret statements involving absolute value as statements about distance on the number line.

EXAMPLE 6 To solve the equation $|x + 5| = 3$, we rewrite it as $|x - (-5)| = 3$. In this form it states that

*the distance between x and -5 is 3 units.**

*It's necessary to rewrite the equation first because the distance formula involves the *difference* of two numbers, not their sum.

Figure 1–6 shows that -8 and -2 are the only two numbers whose distance to -5 is 3 units:

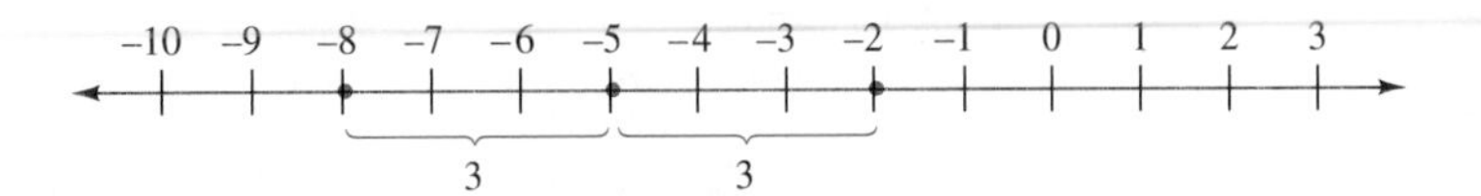

Figure 1-6

Thus $x = -8$ and $x = -2$ are the solutions of $|x + 5| = 3$. ■

EXAMPLE 7 The solutions of $|x - 1| \geq 2$ are all numbers x such that

the distance between x and 1 is greater than or equal to 2.

Figure 1–7 shows that the numbers 2 or more units away from 1 are the numbers x such that

$$x \leq -1 \qquad \text{or} \qquad x \geq 3.$$

So these numbers are the solutions of the inequality. ■

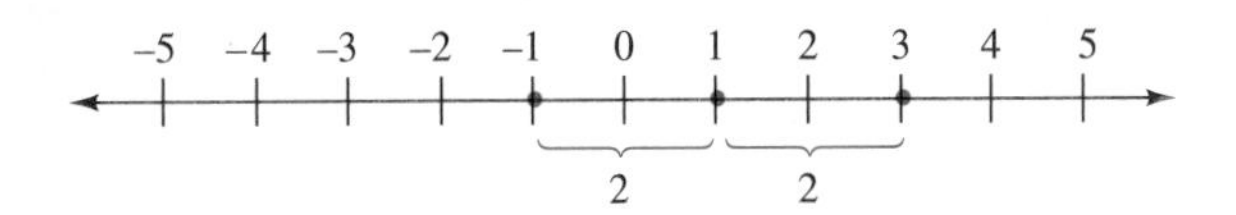

Figure 1-7

EXAMPLE 8 The solutions of $|x - 7| < 2.5$ are all numbers x such that

the distance between x and 7 is less than 2.5.

Figure 1–8 shows that the solutions of the inequality, that is, the numbers within 2.5 units of 7, are the numbers x such that $4.5 < x < 9.5$, that is, the interval $(4.5, 9.5)$. ■

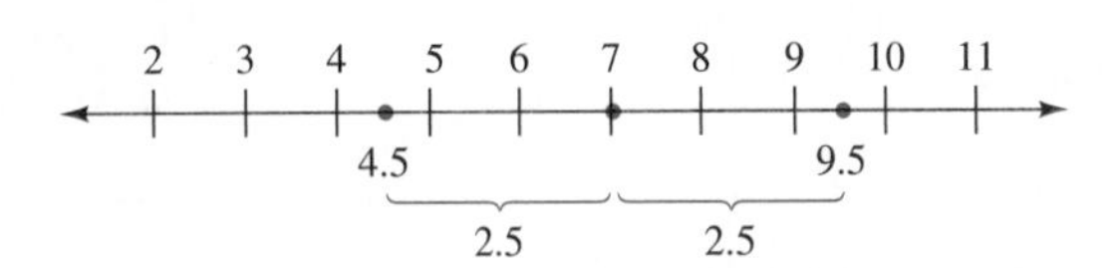

Figure 1-8

CALCULATOR INVESTIGATIONS 1.1

1. MATH Menu

(a) Find the ABS key on your calculator and make these computations:

ABS (−) 9 + 2 ENTER and ABS ((−) 9 + 2) ENTER.

Explain why the answers are different.

(b) Find the INT key on the MATH menu or one of its submenus [on TI-92 type in "int("]. Find out what this key does. Answers will vary depending on the brand of calculator.

(c) Find a command on the MATH menu that you haven't seen before. Find out how to use it.

2. TEST Menu If your calculator has a TEST menu (on the keyboard or in the MATH menu), find out what happens when you key in each of these statements and press ENTER: $8 < 9$, $8 < 5$, $9 > 2$, $4 > 10$.

3. Symbolic Calculations Learn how to store numbers in memories A, B, C of your calculator. Then

(a) Store 2 in memory A, 5 in memory B, and -10 in memory C. Using the ALPHA keys, display this expression on the screen: $B + C/A$. If you press ENTER, what happens? Explain what the calculator is doing.

(b) Experiment with other expressions, such as $B^2 - 4AC$.

EXERCISES 1.1

1. Draw a number line and mark the location of each of these numbers: 0, -7, $8/3$, 10, -1, -4.75, $1/2$, -5, and 2.25.

In Exercises 2–14, express the given statement in symbols.

2. 5 is less than 7.

3. -4 is greater than -8.

4. -17 is less than 14.

5. π is less than 100.

6. x is nonnegative.

7. y is less than or equal to 7.5.

8. z is greater than or equal to -4.

9. t is positive.

10. d is not greater than 2.

11. c is at most 3.

12. z is at least -17.

13. c is less than 4 and d is at least 4.

14. x is greater than -6 and y is at most -6.

In Exercises 15–19, fill the blank with = or <, or >, so that the resulting statement is true.

15. -6 ____ -2 **16.** 5 ____ -3

17. 3/4 ____ .75 **18.** 3.1 ____ π

19. 1/3 ____ .33

In Exercises 20–26, fill the blank so as to produce two equivalent statements. For example, the arithmetic statement "a is negative" is equivalent to the geometric statement "the point a lies to the left of the point 0."

	Arithmetic Statement	Geometric Statement
20.	$a \geq b$	________
21.	________	a lies c units to the right of b
22.	________	a lies between b and c
23.	$a - b > 0$	________
24.	a is positive	________
25.	________	a lies to the left of b
26.	$a + b > c \quad (b > 0)$	________

In Exercises 27–38, simplify and write the given number without using absolute values.

27. $|3 - 14|$ **28.** $|(-2)3|$ **29.** $3 - |2 - 5|$

30. $-2 - |-2|$ **31.** $|6 - 4| + |-3 - 5|$

32. $|-6| - |6|$ **33.** $|(-13)^2|$ **34.** $-|-5|^2$

35. $|\pi - \sqrt{2}|$ **36.** $|\sqrt{2} - 2|$ **37.** $|3 - \pi| + 3$

38. $|4 - \sqrt{2}| - 5$

In Exercises 39–44, fill the blank with <, =, or > so that the resulting statement is true.

39. $|-2|$ ____ $|-5|$ **40.** 5 ____ $|-2|$

41. $|3|$ ____ $-|4|$ **42.** $|-3|$ ____ 0

43. -7 ____ $|-1|$ **44.** $-|-4|$ ____ 0

In Exercises 45–50, draw a picture on the number line of the given interval.

45. $(0, 8]$ **46.** $(0, \infty)$ **47.** $[-2, 1]$ **48.** $(-1, 1)$

49. $(-\infty, 0]$ **50.** $[-2, 7)$

In Exercises 51–58, use interval notation to denote the set of all real numbers x that satisfy the given inequality.

51. $5 \leq x \leq 8$ **52.** $-2 \leq x \leq 7$

53. $-3 < x < 14$ **54.** $7 < x < 135$

55. $x \geq -8$ **56.** $x \geq 12$

57. $x \leq 15$ **58.** $x < \sqrt{2}$

In Exercises 59–62, find two pairs of numbers that make the given statement true and two pairs that make it false. For example, $|x| < |y|$ is true for $x = 1$, $y = 2$ and $x = -1$, $y = 7$, but it is false for $x = 3$, $y = 1$ and $x = -3$, $y = -2$.

59. $|x| + |y| = 1$ **60.** $|y| - |x| < 0$

61. $|x + y| = |x| + |y|$ **62.** $|x + y| < |x| + |y|$

In Exercises 63–70, find the distance between the given numbers.

63. -3 and 4 **64.** 7 and 107

65. -7 and $15/2$ **66.** $-3/4$ and -10

67. π and 3 **68.** π and -3

69. $\sqrt{2}$ and $\sqrt{3}$ **70.** π and $\sqrt{2}$

In Exercises 71–80, write the given expression without using absolute values.

71. $|t^2|$ **72.** $|u^2 + 2|$ **73.** $|(-3 - y)^2|$

74. $|-2 - y^2|$ **75.** $|b - 3|$ if $b \geq 3$

76. $|a - 5|$ if $a < 5$ **77.** $|c - d|$ if $c < d$

78. $|c - d|$ if $c \geq d$ **79.** $|u - v| - |v - u|$

80. $\dfrac{|u - v|}{|v - u|}$ if $u \neq v$, $u \neq 0$, $v \neq 0$

In Exercises 81 and 82, explain why the given statement is true for any numbers c and d. [Hint: *Look at the properties of absolute value on page 6.*]

81. $|(c - d)^2| = c^2 - 2cd + d^2$

82. $\sqrt{9c^2 - 18cd + 9d^2} = 3|c - d|$

In Exercises 83–89, express the given geometric statement about numbers on the number line algebraically, using absolute values.

83. The distance from x to 5 is less than 4.

84. x is more than 6 units from c.

85. x is at most 17 units from -4.

86. x is within 3 units of 7.

87. c is closer to 0 than b is.

88. x is closer to 1 than to 4.

89. x is farther from 0 than from -6.

90. For what real numbers c is it true that $|-c| = -c$?

In Exercises 91–96, translate the given algebraic statement into a geometric statement about numbers on the number line.

91. $|x - 3| < 2$ **92.** $|x - c| > 6$

93. $|x + 7| \leq 3$ **94.** $|u + v| \geq 2$

95. $|b| < |c - 3|$ **96.** $|b - 2| \geq |b - 5|$

97. Explain geometrically why this statement is always false:

$$|c - 1| < 2 \text{ and simultaneously } |c - 12| < 3.$$

98. What is $|-c|$ when $c < 0$? What is $|-c|$ when $c > 0$?

In Exercises 99–108, use the geometric approach explained in the text to solve the given equation or inequality.

99. $|x| = 1$ **100.** $|x| = 3/2$

101. $|x - 2| = 1$ **102.** $|x + 3| = 2$

103. $|x + \pi| = 4$ **104.** $|x - \frac{3}{2}| = 5$

105. $|x| < 7$ **106.** $|x - 5| < 2$

107. $|x| \geq 5$ **108.** $|x - 6| > 2$

Thinkers

109. Explain why the statement $|a| + |b| + |c| > 0$ is algebraic shorthand for "at least one of the numbers a, b, c is different from zero."

110. Find an algebraic shorthand version of the statement "the numbers a, b, c are all different from zero."

111. Prove the triangle inequality as follows:
(a) The box on page 6 shows that $c \leq |c|$ and $d \leq |d|$, and also that $-c \leq |c|$ and $-d \leq |d|$. Use these facts to show that $c + d \leq |c| + |d|$ and $-(c + d) \leq |c| + |d|$.
(b) Use part (a) to show that $|c + d| \leq |c| + |d|$. [*Hint*: What is $|c + d|$ when $c + d$ is positive? When $c + d$ is negative?]

1.1.A *Excursion* DECIMAL REPRESENTATION OF REAL NUMBERS

Every rational number can be expressed as a terminating or repeating decimal. For instance, $3/4 = .75$. To express $15/11$ as a decimal, divide the numerator by the denominator:

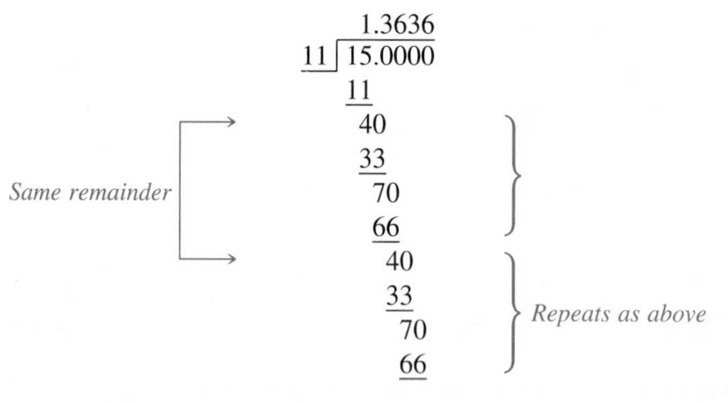

Since the remainder at the first step (namely 4) occurs again at the third step, it is clear that the division process goes on forever with the two-digit block "36" repeating over and over in the quotient $\frac{15}{11} = 1.3636363636\cdots$.

The method used in the preceding example can be used to express any rational number as a decimal. During the division process some remainder *neces-*

sarily repeats.* If the remainder at which this repetition starts is 0, the result is a repeating decimal ending in zeros—that is, a terminating decimal (for instance, $.75000\cdots = .75$). If the remainder at which the repetition starts is nonzero, then the result is a nonterminating repeating decimal, as in the example above.

The long division needed to express a rational number as a repeating decimal is usually done with a calculator or computer. However, a typical calculator displays only the first ten digits of a number in decimal form (although it uses several additional digits in its internal computations). For example, a calculator might display the decimal expansion of 1/17 as .0588235294, although the actual expansion has a repeating block of 16 digits:

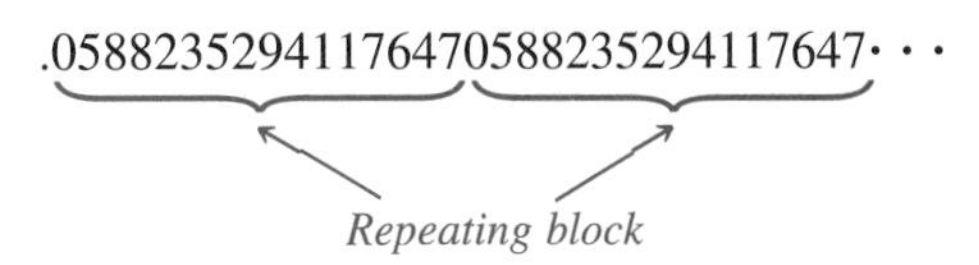

Thus, a calculator can contain the *exact* value of a rational number only if its decimal expansion terminates after approximately ten decimal places. Techniques for obtaining the full decimal expansion of a rational number from a calculator are discussed in Exercise 25.

Conversely, there is a simple method for converting any repeating decimal into a rational number.

EXAMPLE 1 Write $d = .272727\cdots$ as a rational number.

Solution Assuming that the usual rules of arithmetic hold, we see that

$$100d = 27.272727\cdots \qquad \text{and} \qquad d = .272727\cdots$$

Now subtract d from $100d$:

$$\begin{aligned} 100d &= 27.272727\cdots \\ -d &= -.272727\cdots \\ \hline 99d &= 27 \end{aligned}$$

Dividing both sides of this last equation by 99 shows that $d = \frac{27}{99} = \frac{3}{11}$. ■

> ▶ **TECHNOLOGY TIP**
> The FRAC key on TI-82/83/85 calculators automatically converts decimals to fractions (subject to some limitations). The same thing can be done on HP-38 by choosing FRACTION number format in the MODE menu and then entering the decimal. Conversion programs for TI-81 and Casio are in the Program Appendix.

Irrational Numbers

There are many nonterminating decimals that are *nonrepeating* (that is, no block of digits repeats forever), such as $.202002000200002\cdots$ (where after each 2 there is one more zero than before). Although the proof is too long to give here, it is in fact true that every nonterminating and nonrepeating decimal represents an *irrational* real number.

*For instance, if you divide a number by 11 as in the example above, the only possible remainders at each step are the 11 numbers 0, 1, 2, 3, 4, 5, 6, 7, 8, 9, and 10. Hence after *at most* 11 steps, some remainder must occur for a second time (it happened at the third step in the example). At this point (if there are no new digits in the dividend) the division process and hence the quotient begin to repeat.

Conversely every irrational number can be expressed as a nonterminating and nonrepeating decimal (no proof to be given here). For instance, the decimal expansion of the irrational number π begins $3.1415926535897 \cdots$. This computation has actually been carried out to over a billion decimal places by computer.

Since a calculator can deal with only the first 12–15 digits of a number in decimal form, *a calculator cannot contain the exact value of any irrational number.* Furthermore, there are no easy calculator techniques for carrying out the decimal expansion of an arbitrary irrational number to a specified number of decimal places, as there are for rational numbers.

Since every real number is either a rational number or an irrational one, the preceding discussion can be summarized as follows.

Decimal Representation ▶

1. **Every real number can be expressed as a decimal.**
2. **Every decimal represents a real number.**
3. **The terminating decimals and the nonterminating repeating decimals are precisely the rational numbers.**
4. **The nonterminating, nonrepeating decimals are precisely the irrational numbers.**

CALCULATOR INVESTIGATION 1.1A

1. FRAC Key If your calculator has a FRAC key or program (see the Program Appendix), test its limitations by entering each of the following numbers and then pressing the FRAC key.

   (a) .058823529411 **(b)** .0588235294117

   (c) .058823529411724 **(d)** .0588235294117985

   Which of your answers are correct? [*Hint:* Look at the decimal expansion of 1/17 on page 13.]

EXERCISES 1.1.A

In Exercises 1–6, express the given rational number as a repeating decimal.

1. 7/9 **2.** 2/13 **3.** 23/14 **4.** 19/88

5. 1/19 (long) **6.** 9/11

In Exercises 7–14, state whether a calculator can express the given number exactly.

7. 2/3 **8.** 7/16 **9.** 1/64 **10.** 1/22

11. $3\pi/2$ **12.** $\pi - 3$ **13.** 1/.625 **14.** 1/.16

In Exercises 15–21, express the given repeating decimal as a fraction.

15. $.373737 \cdots$ **16.** $.929292 \cdots$

17. $76.63424242 \cdots$ [*Hint*: Consider $10{,}000d - 100d$, where $d = 76.63424242 \cdots$.]

18. $13.513513 \cdots$ [*Hint*: Consider $1000d - d$, where $d = 13.513513 \cdots$.]

19. $.135135135 \cdots$ [*Hint*: See Exercise 18.]

20. .33030303 · · · **21.** 52.31272727 · · ·

22. If two real numbers have the same decimal expansion through three decimal places, how far apart can they be on the number line?

Thinkers

23. Use the methods in Exercises 15–21 to show that both .74999 · · · and .75000 · · · are decimal expansions of 3/4. [Every terminating decimal can also be expressed as a decimal ending in repeated 9's. It can be proved that these are the only real numbers with more than one decimal expansion.]

24. *Finding remainders with a calculator* If you use long division to divide 369 by 7, you obtain:

$$\begin{array}{rrl} & 52 & \longleftarrow \textit{Quotient} \\ \textit{Divisor} \longrightarrow 7 & \overline{)369} & \longleftarrow \textit{Dividend} \\ & \underline{35} & \\ & 19 & \\ & \underline{14} & \\ & 5 & \longleftarrow \textit{Remainder} \end{array}$$

If you use a calculator to find $369 \div 7$, the answer is displayed as 52.71428571. Observe that the integer part of this calculator answer, 52, is the quotient when you do the problem by long division. The usual "checking procedure" for long division shows that

$$7 \cdot 52 + 5 = 369 \quad \text{or equivalently,} \quad 369 - 7 \cdot 52 = 5.$$

Thus the remainder is

dividend − (divisor)(integer part of calculator answer).

Use this method to find the quotient and remainder in these problems:

(a) $5683 \div 9$ **(b)** $1{,}000{,}000 \div 19$
(c) $53{,}000{,}000 \div 37$

In Exercises 25–30, find the decimal expansion of the given rational number. All these expansions are too long to fit in a calculator, but can be readily found by using the hint in Exercise 25.

25. 6/17 [*Hint*: The first part of dividing 6 by 17 involves working this division problem: $6{,}000{,}000 \div 17$. The method of Exercise 24 shows that the quotient is 352,941 and the remainder is 3. Thus the decimal expansion of 6/17 begins .352941 and the next block of digits in the expansion will be the quotient in the problem $3{,}000{,}000 \div 17$. The remainder when 3,000,000 is divided by 17 is 10, so the next block of digits in the expansion of 6/17 is the quotient in the problem $10{,}000{,}000 \div 17$. Continue in this way until the decimal expansion repeats.]

26. 3/19 **27.** 1/29 **28.** 3/43 **29.** 283/47
30. 768/59

31. (a) Show that there are at least as many irrational numbers (nonrepeating decimals) as there are terminating decimals. [*Hint*: With each terminating decimal associate a nonrepeating decimal.]
(b) Show that there are at least as many irrational numbers as there are repeating decimals. [*Hint*: With each repeating decimal associate a nonrepeating decimal by inserting longer and longer strings of zeros: for instance, with .11111111 · · · associate the number .101001000100001 · · ·.]

1.2 SOLVING EQUATIONS ALGEBRAICALLY

This section deals with equations such as

$$3x - 6 = 7x + 4, \qquad x^2 - 5x + 6 = 0, \qquad 2x^4 - 13x^2 = 3.$$

A **solution** of an equation is a number that, when substituted for the variable x, produces a true statement.* For example, 5 is a solution of $3x + 2 = 17$ because $3 \cdot 5 + 2 = 17$ is a true statement. To **solve** an equation means to find all its solutions. Throughout this chapter we shall deal only with **real solutions,** that is, solutions that are real numbers.**

*Any letter may be used for the variable.

**Solutions that are not real numbers are considered in Sections 4.6 and 4.7.

Two equations are said to be **equivalent** if they have the same solutions. For example, $3x + 2 = 17$ and $x - 2 = 3$ are equivalent because 5 is the only solution of each one.

Basic Principles for Solving Equations ▶

Performing any of the following operations on an equation produces an equivalent equation:

1. **Add or subtract the same quantity from both sides of the equation.**
2. **Multiply or divide both sides of the equation by the same *nonzero* quantity.**

The usual strategy in equation solving is to use these Basic Principles to transform a given equation into an equivalent one whose solutions are known.

A **first-degree** or **linear equation** is an equation that involves only constants and the first power of the variable. Every first-degree equation has exactly one solution, which is easily found.

EXAMPLE 1 To solve $3x - 6 = 7x + 4$ we use the Basic Principles to transform this equation into an equivalent one whose solution is obvious:

$$3x - 6 = 7x + 4$$

Add 6 to both sides:
$$3x = 7x + 10$$

Subtract 7x from both sides:
$$-4x = 10$$

Divide both sides by −4:
$$x = \frac{10}{-4} = -\frac{5}{2}$$

Since $-5/2$ is the only solution of this last equation, $-5/2$ is the only solution of the original equation, $3x - 6 = 7x + 4$. ■

Quadratic Equations

A **second-degree** or **quadratic equation** is an equation that can be written in the form

$$ax^2 + bx + c = 0$$

for some constants a, b, c with $a \neq 0$. There are several techniques for solving such equations. We begin with the **factoring method,** which makes use of this property of the real numbers:

Zero Products ▶

If a product of real numbers is zero, then at least one of the factors is zero; in other words,

If $cd = 0$, then $c = 0$ or $d = 0$ (or both).

EXAMPLE 2 To solve $3x^2 - x = 10$, we first rearrange the terms to make one side 0 and then factor:

Subtract 10 from each side: $3x^2 - x - 10 = 0$

Factor left side: $(3x + 5)(x - 2) = 0$

If a product of real numbers is 0, then at least one of the factors must be 0. So this equation is equivalent to:

$$3x + 5 = 0 \quad \text{or} \quad x - 2 = 0$$
$$3x = -5 \qquad\qquad x = 2$$
$$x = -5/3$$

Therefore the solutions are $-5/3$ and 2. ■

WARNING To guard against mistakes, check your solutions by substituting each one in the *original* equation to make sure it really *is* a solution.

The solutions of $x^2 = 7$ are the numbers whose square is 7. Although 7 has just *one* square root, there are *two* numbers whose square is 7, namely, $\sqrt{7}$ and its negative, $-\sqrt{7}$. So the solutions of $x^2 = 7$ are $\sqrt{7}$ and $-\sqrt{7}$, or in abbreviated form, $\pm\sqrt{7}$. The same argument works for any positive real number d:

The solutions of $x^2 = d$ are $\sqrt{d}$ and $-\sqrt{d}$.

We now use a slight variation of this idea to develop a method for solving quadratic equations that don't readily factor. The method depends on this fact: A polynomial of the form $x^2 + bx$ can be changed into a perfect square by adding a suitable constant.* For example, $x^2 + 6x$ can be changed into a perfect square by adding 9:

$$x^2 + 6x + 9 = (x + 3)^2.$$

This process is called **completing the square.**

To complete the square in $x^2 + 6x$, we added $9 = 3^2$. Note that 3 is one-half the coefficient of x in the original polynomial $x^2 + 6x$. The same idea works in the general case. The multiplication pattern for perfect squares shows that for any real number b:

$$\left(x + \frac{b}{2}\right)^2 = x^2 + 2\left(\frac{b}{2}\right)x + \left(\frac{b}{2}\right)^2 = x^2 + bx + \left(\frac{b}{2}\right)^2.$$

Therefore,

Completing the Square ▶

To complete the square in $x^2 + bx$, add the square of one-half the coefficient of x, namely, $\left(\frac{b}{2}\right)^2$. This produces a perfect square:

$$x^2 + bx + \left(\frac{b}{2}\right)^2 = \left(x + \frac{b}{2}\right)^2$$

*A quadratic polynomial $x^2 + bx + c$ is a **perfect square** if it factors as $(x + d)^2$ for some constant d.

The following example shows how completing the square can be used to solve quadratic equations.

EXAMPLE 3 To solve $x^2 + 6x + 1 = 0$, we first rewrite the equation as $x^2 + 6x = -1$. Next we complete the square on the left side by adding the square of half the coefficient of x, namely, $(\frac{6}{2})^2 = 9$. In order to have an equivalent equation, we must add 9 to *both* sides:

$$x^2 + 6x + 9 = -1 + 9$$

Factor left side:
$$(x + 3)^2 = 8$$

Thus $x + 3$ is a number whose square is 8. The only numbers whose squares equal 8 are $\sqrt{8}$ and $-\sqrt{8}$. So we must have:

$$x + 3 = \sqrt{8} \quad \text{or} \quad x + 3 = -\sqrt{8}$$
$$x = \sqrt{8} - 3 \quad \text{or} \quad x = -\sqrt{8} - 3.$$

Therefore the solutions of the original equation are $\sqrt{8} - 3$ and $-\sqrt{8} - 3$, or in more compact notation, $\pm\sqrt{8} - 3$. ■

WARNING Completing the square only works when the coefficient of x^2 is 1. In an equation such as

$$5x^2 - x + 2 = 0$$

you must first divide both sides by 5, and *then* complete the square.

We can use the completing-the-square method to solve *any* quadratic equation:*

$$ax^2 + bx + c = 0$$

Divide both sides by a:
$$x^2 + \frac{b}{a}x + \frac{c}{a} = 0$$

Subtract $\frac{c}{a}$ from both sides:
$$x^2 + \frac{b}{a}x = -\frac{c}{a}$$

*Add $\left(\frac{b}{2a}\right)^2$ to both sides:***
$$x^2 + \frac{b}{a}x + \left(\frac{b}{2a}\right)^2 = \left(\frac{b}{2a}\right)^2 - \frac{c}{a}$$

Factor left side:
$$\left(x + \frac{b}{2a}\right)^2 = \left(\frac{b}{2a}\right)^2 - \frac{c}{a}$$

Find common denominator for right side:
$$\left(x + \frac{b}{2a}\right)^2 = \frac{b^2}{4a^2} - \frac{c}{a} = \frac{b^2 - 4ac}{4a^2}$$

Since the square of $x + \frac{b}{2a}$ is $\frac{b^2 - 4ac}{4a^2}$, we must have:

$$x + \frac{b}{2a} = \pm\sqrt{\frac{b^2 - 4ac}{4a^2}} = \pm\frac{\sqrt{b^2 - 4ac}}{2a}$$

Subtract $\frac{b}{2a}$ from both sides:
$$x = \frac{-b}{2a} \pm \frac{\sqrt{b^2 - 4ac}}{2a} = \frac{-b \pm \sqrt{b^2 - 4ac}}{2a}$$

*If you have trouble following any step here, do it for a numerical example, such as the case when $a = 3, b = 11, c = 5$.

**This is the square of half the coefficient of x.

We have proved:

The Quadratic Formula ▶

The solutions of the quadratic equation $ax^2 + bx + c = 0$ are:

$$x = \frac{-b \pm \sqrt{b^2 - 4ac}}{2a}$$

You should memorize the quadratic formula.

EXAMPLE 4 To solve $x^2 + 3 = -8x$, rewrite the equation as $x^2 + 8x + 3 = 0$ and apply the quadratic formula with $a = 1$, $b = 8$, and $c = 3$:

$$\begin{aligned} x &= \frac{-b \pm \sqrt{b^2 - 4ac}}{2a} = \frac{-8 \pm \sqrt{8^2 - 4 \cdot 1 \cdot 3}}{2 \cdot 1} \\ &= \frac{-8 \pm \sqrt{52}}{2} = \frac{-8 \pm \sqrt{4 \cdot 13}}{2} \\ &= \frac{-8 \pm 2\sqrt{13}}{2} = -4 \pm \sqrt{13} \end{aligned}$$

Therefore the equation has two distinct real solutions, $-4 + \sqrt{13}$ and $-4 - \sqrt{13}$. ■

EXAMPLE 5 To solve $x^2 - 194x + 9409 = 0$, use a calculator and the quadratic formula with $a = 1$, $b = -194$, and $c = 9409$:

$$\begin{aligned} x = \frac{-b \pm \sqrt{b^2 - 4ac}}{2a} &= \frac{-(-194) \pm \sqrt{(-194)^2 - 4 \cdot 1 \cdot 9409}}{2 \cdot 1} \\ &= \frac{194 \pm \sqrt{37636 - 37636}}{2} = \frac{194 \pm 0}{2} = 97 \end{aligned}$$

Thus 97 is the only solution of the equation. ■

EXAMPLE 6 The solutions of $2x^2 + x + 3 = 0$ are given by the quadratic formula with $a = 2$, $b = 1$, and $c = 3$:

$$\begin{aligned} x = \frac{-b \pm \sqrt{b^2 - 4ac}}{2a} = \frac{-1 \pm \sqrt{1^2 - 4 \cdot 2 \cdot 3}}{2 \cdot 2} &= \frac{-1 \pm \sqrt{1 - 24}}{4} \\ &= \frac{-1 \pm \sqrt{-23}}{4} \end{aligned}$$

Since $\sqrt{-23}$ is not a real number, this equation has *no real solutions* (that is, no solutions in the real number system). ■

The expression $b^2 - 4ac$ in the quadratic formula is called the **discriminant.** As the last three examples demonstrate, the discriminant determines the *number* of real solutions of the equation $ax^2 + bx + c = 0$.

Real Solutions of a Quadratic Equation ▶

Discriminant $b^2 - 4ac$	*Number of Real Solutions of* $ax^2 + bx + c = 0$	*Example*
> 0	**Two distinct real solutions**	$x^2 + 8x + 3 = 0$
$= 0$	**One real solution**	$x^2 - 194x + 9409 = 0$
< 0	**No real solutions**	$2x^2 + x + 3 = 0$

The quadratic formula and a calculator can be used to solve any quadratic equation with nonnegative discriminant. Experiment with your calculator to find the most efficient sequence of key strokes for doing this.

EXAMPLE 7 Use a calculator to solve $3.2x^2 + 15.93x - 7.1 = 0$.

Solution First compute $\sqrt{b^2 - 4ac} = \sqrt{15.93^2 - 4(3.2)(-7.1)}$ and store the result in memory D. Then the solutions of the equation are given by

$$\frac{-15.93 + D}{2(3.2)} \approx .411658467 \quad \text{and} \quad \frac{-15.93 - D}{2(3.2)} \approx -5.389783347.$$

Remember that these answers are *approximations,* so they may not check exactly when substituted in the original equation. ■

▶ **TECHNOLOGY TIP**

Any calculator can be programmed to solve quadratic equations, so that you need only enter the coefficients a, b, c to obtain the (approximate) solutions; see the Program Appendix. This is not necessary on calculators that have a built-in quadratic equation solver; check your instruction manual.

Higher Degree Equations

A **polynomial equation of degree *n*** is one that can be written in the form

$$a_n x^n + \cdots + a_3x^3 + a_2x^2 + a_1x + a_0 = 0,$$

where n is a positive integer, each a_i is a constant, and $a_n \neq 0$. For instance, $4x^6 - 3x^5 + x^4 + 7x^3 - 8x^2 + 4x + 9 = 0$ is a polynomial equation of degree 6. As a general rule, polynomial equations of degree 3 and above are best solved by the numerical or graphical methods presented in Section 1.5. However, there are exceptions. If a higher degree equation can be transformed into a quadratic equation by making a suitable substitution for the variable, then it can be solved by factoring or the quadratic formula.

EXAMPLE 8 To solve $4x^4 - 13x^2 + 3 = 0$, substitute u for x^2 and solve the resulting quadratic equation:

$$4x^4 - 13x^2 + 3 = 0$$
$$4(x^2)^2 - 13x^2 + 3 = 0$$
$$4u^2 - 13u + 3 = 0$$
$$(u - 3)(4u - 1) = 0$$
$$u - 3 = 0 \quad \text{or} \quad 4u - 1 = 0$$
$$u = 3 \qquad\qquad 4u = 1$$
$$u = \frac{1}{4}$$

Since $u = x^2$ we see that

$$x^2 = 3 \quad \text{or} \quad x^2 = \frac{1}{4}$$
$$x = \pm\sqrt{3} \qquad x = \pm\frac{1}{2}$$

Hence the original equation has four solutions: $-\sqrt{3}, \sqrt{3}, -1/2, 1/2$. ■

EXAMPLE 9 To solve $x^4 - 4x^2 + 1 = 0$, let $u = x^2$:

$$x^4 - 4x^2 + 1 = 0$$
$$u^2 - 4u + 1 = 0$$

The quadratic formula shows that

$$u = \frac{-(-4) \pm \sqrt{(-4)^2 - 4 \cdot 1 \cdot 1}}{2 \cdot 1} = \frac{4 \pm \sqrt{12}}{2}$$
$$= \frac{4 \pm \sqrt{4 \cdot 3}}{2} = \frac{4 \pm 2\sqrt{3}}{2} = 2 \pm \sqrt{3}$$

Since $u = x^2$, we have the equivalent statements:

$$x^2 = 2 + \sqrt{3} \quad \text{or} \quad x^2 = 2 - \sqrt{3}$$
$$x = \pm\sqrt{2 + \sqrt{3}} \qquad x = \pm\sqrt{2 - \sqrt{3}}$$

Therefore, the original equation has four solutions. ■

EXERCISES 1.2

In Exercises 1–12, solve the equation by factoring.

1. $x^2 - 8x + 15 = 0$
2. $x^2 + 5x + 6 = 0$
3. $x^2 - 5x = 14$
4. $x^2 + x = 20$
5. $2y^2 + 5y - 3 = 0$
6. $3t^2 - t - 2 = 0$
7. $4t^2 + 9t + 2 = 0$
8. $9t^2 + 2 = 11t$
9. $3u^2 + u = 4$
10. $5x^2 + 26x = -5$
11. $12x^2 + 13x = 4$
12. $18x^2 = 23x + 6$

In Exercises 13–16, solve the equation by completing the square.

13. $x^2 - 2x = 15$
14. $x^2 - 4x - 32 = 0$
15. $x^2 - x - 1 = 0$
16. $x^2 + 3x - 2 = 0$

In Exercises 17–28, use the quadratic formula to solve the equation.

17. $x^2 - 4x + 1 = 0$
18. $x^2 - 2x - 1 = 0$
19. $x^2 + 6x + 7 = 0$
20. $x^2 + 4x - 3 = 0$
21. $x^2 + 6 = 2x$
22. $x^2 + 11 = 6x$
23. $4x^2 - 4x = 7$
24. $4x^2 - 4x = 11$
25. $4x^2 - 8x + 1 = 0$
26. $2t^2 + 4t + 1 = 0$
27. $5u^2 + 8u = -2$
28. $4x^2 = 3x + 5$

In Exercises 29–34, find the number *of real solutions of the equation by computing the discriminant.*

29. $x^2 + 4x + 1 = 0$
30. $4x^2 - 4x - 3 = 0$
31. $9x^2 = 12x + 1$
32. $9t^2 + 15 = 30t$
33. $25t^2 + 49 = 70t$
34. $49t^2 + 5 = 42t$

In Exercises 35–38, solve the equation and check your answers. [Hint: *First, eliminate fractions by multiplying both sides by a common denominator.*]

35. $1 - \dfrac{3}{x} = \dfrac{40}{x^2}$
36. $\dfrac{4x^2 + 5}{3x^2 + 5x - 2} = \dfrac{4}{3x - 1} - \dfrac{3}{x + 2}$
37. $\dfrac{2}{x^2} - \dfrac{5}{x} = 4$
38. $\dfrac{x}{x - 1} + \dfrac{x + 2}{x} = 3$

In Exercises 39–48, solve the equation by any method.

39. $x^2 + 9x + 18 = 0$
40. $3t^2 - 11t - 20 = 0$
41. $4x(x + 1) = 1$
42. $25y^2 = 20y + 1$
43. $2x^2 = 7x + 15$
44. $2x^2 = 6x + 3$
45. $t^2 + 4t + 13 = 0$
46. $5x^2 + 2x = -2$
47. $\dfrac{7x^2}{3} = \dfrac{2x}{3} - 1$
48. $25x + \dfrac{4}{x} = 20$

In Exercises 49–52, use a calculator and the quadratic formula to find approximate solutions of the equation.

49. $4.42x^2 - 10.14x + 3.79 = 0$
50. $8.06x^2 + 25.8726x - 25.047256 = 0$
51. $3x^2 - 82.74x + 570.4923 = 0$
52. $7.63x^2 + 2.79x = 5.32$

In Exercises 53–60, find all real solutions of the equation.

53. $y^4 - 7y^2 + 6 = 0$
54. $x^4 - 2x^2 + 1 = 0$
55. $x^4 - 2x^2 - 35 = 0$
56. $x^4 - 2x^2 - 24 = 0$
57. $2y^4 - 9y^2 + 4 = 0$
58. $6z^4 - 7z^2 + 2 = 0$
59. $10x^4 + 3x^2 = 1$
60. $6x^4 - 7x^2 = 3$

In Exercises 61–64, find a number k such that the given equation has exactly one real solution.

61. $x^2 + kx + 25 = 0$
62. $x^2 - kx + 49 = 0$
63. $kx^2 + 8x + 1 = 0$
64. $kx^2 + 24x + 16 = 0$

65. Find a number k such that 4 and 1 are the solutions of $x^2 - 5x + k = 0$.

66. Suppose a, b, c are fixed real numbers such that $b^2 - 4ac \geq 0$. Let r and s be the solutions of $ax^2 + bx + c = 0$.
    - **(a)** Use the quadratic formula to show that $r + s = -b/a$ and $rs = c/a$.
    - **(b)** Use part (a) to verify that $ax^2 + bx + c = a(x - r)(x - s)$.
    - **(c)** Use part (b) to factor $x^2 - 2x - 1$ and $5x^2 + 8x + 2$.

1.2.A *Excursion* ABSOLUTE VALUE EQUATIONS

As we saw on page 8, 3 and -3 are the only numbers with absolute value 3, that is, the only solutions of $|x| = 3$. A similar procedure can be used to solve other absolute value equations.

EXAMPLE 1 To solve $|3x - 4| = 8$, use the fact that there are exactly two numbers whose absolute value is 8, namely, 8 and -8. Since $3x - 4$ is a number with absolute value 8, there are only two possibilities:

$$\begin{array}{rcr} 3x - 4 = 8 & \text{or} & 3x - 4 = -8 \\ 3x = 12 & & 3x = -4 \\ x = 4 & & x = -\dfrac{4}{3} \end{array}$$

You can readily verify that both 4 and $-4/3$ are solutions of the original equation $|3x - 4| = 8$. ■

EXAMPLE 2 Solve the equation $|x^2 + 4x - 3| = 2$.

Solution Since 2 and -2 are the only numbers with absolute value 2, we must have:

$$\begin{array}{rcr} x^2 + 4x - 3 = 2 & \text{or} & x^2 + 4x - 3 = -2 \\ x^2 + 4x - 5 = 0 & \text{or} & x^2 + 4x - 1 = 0 \end{array}$$

The first of these equations can be solved by factoring and the second by the quadratic formula:

$$\begin{array}{rcl} (x + 5)(x - 1) = 0 & \text{or} & x = \dfrac{-4 \pm \sqrt{4^2 - 4 \cdot 1 \cdot (-1)}}{2 \cdot 1} \\ x = -5 \quad \text{or} \quad x = 1 & \text{or} & x = \dfrac{-4 \pm \sqrt{20}}{2} = \dfrac{-4 \pm 2\sqrt{5}}{2} \\ & & x = -2 + \sqrt{5} \quad \text{or} \quad x = -2 - \sqrt{5} \end{array}$$

Therefore the original equation has four real solutions. ■

EXERCISES 1.2.A

In Exercises 1–8, find all real solutions of each equation.

1. $|2x + 3| = 9$

2. $|3x - 5| = 7$

3. $|6x - 9| = 0$

4. $|4x - 5| = -9$

5. $|x^2 + 4x - 1| = 4$

6. $|x^2 + 2x - 9| = 6$

7. $|x^2 - 5x + 1| = 3$

8. $|2x^2 + 5x - 7| = 4$

1.3 THE COORDINATE PLANE

Just as the number line associates the points on a line with real numbers, a similar construction in two dimensions associates points in the plane with ordered *pairs* of real numbers. Draw two number lines in the plane, one horizontal with the positive direction to the right, and one vertical with the positive direction upward, as shown in Figure 1–9 below. The lines are called the **coordinate axes** and their point of intersection the **origin.** The horizontal axis is usually called the ***x*-axis,** and the vertical axis is called the ***y*-axis.*** The coordinate axes divide the plane into four regions, called **quadrants,** that are numbered as in Figure 1–9. A plane equipped with coordinate axes is called a **coordinate plane** and is said to have a **rectangular** (or **Cartesian**) **coordinate system.**

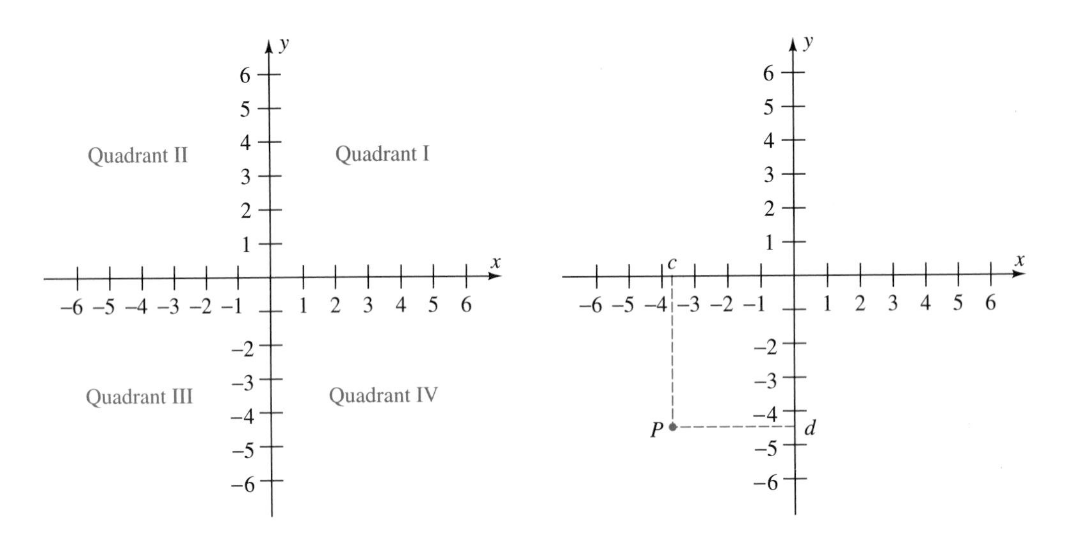

Figure 1-9

Figure 1-10

Given a point P in the plane, draw a vertical line from P to the x-axis and a horizontal line from P to the y-axis. These lines intersect the axes at some numbers c and d, as shown in Figure 1–10. The point P is now associated with the ordered pair of numbers (c, d). The number c is called the ***x*-coordinate** of P and d is called the ***y*-coordinate** of P. Here are some examples of points and their coordinates:

*Although x and y are the traditional labels for the coordinate axes, there is nothing sacred about them. Other letters may be used when convenient.

WARNING The coordinates of a point are an *ordered* pair. Figure 1–11 shows that the point P with coordinates $(-5, 2)$ is quite different from the point Q with coordinates $(2, -5)$. The same numbers (2 and -5) occur in both cases, but in *different order.*

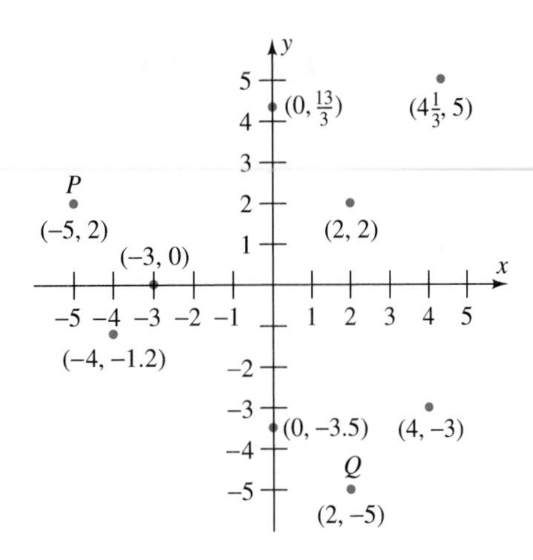

Figure 1-11

We shall often identify a point with its coordinates and refer, for example, to the point (2, 3). When dealing with several points simultaneously, it is customary to label the coordinates of the first point (x_1, y_1), the second point (x_2, y_2), the third point (x_3, y_3), and so on.* Once the plane is coordinatized, it's easy to compute the distance between any two points:

The Distance Formula ▶

The distance between points (x_1, y_1) and (x_2, y_2) is

$$\sqrt{(x_1 - x_2)^2 + (y_1 - y_2)^2}$$

Before proving the distance formula, we shall see how it is used.

EXAMPLE 1 To find the distance between the points $(-1, -3)$ and $(2, -4)$ in Figure 1–12, substitute $(-1, -3)$ for (x_1, y_1) and $(2, -4)$ for (x_2, y_2) in the distance formula:

$$\sqrt{(x_1 - x_2)^2 + (y_1 - y_2)^2} = \sqrt{(-1 - 2)^2 + (-3 - (-4))^2}$$
$$= \sqrt{9 + (-3 + 4)^2} = \sqrt{9 + 1} = \sqrt{10}$$

The order in which the points are used in the distance formula doesn't make a difference. If we substitute $(2, -4)$ for (x_1, y_1) and $(-1, -3)$ for (x_2, y_2), we get the same answer:

$$\sqrt{(2 - (-1))^2 + (-4 - (-3))^2} = \sqrt{3^2 + (-1)^2} = \sqrt{10}. \quad \blacksquare$$

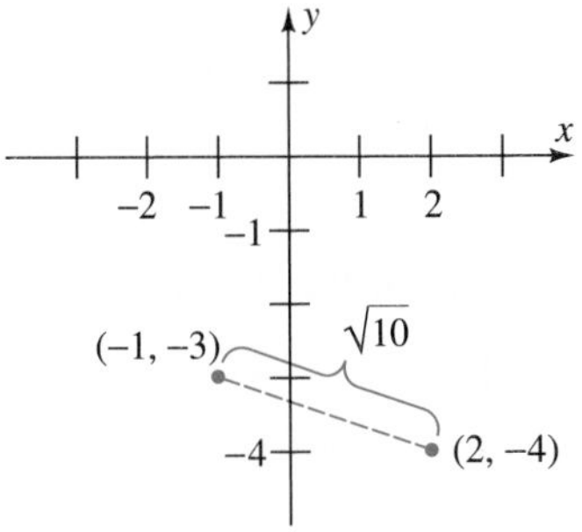

Figure 1-12

*"x_1" is read "x-one" or "x-sub-one"; it is a *single symbol* denoting the first coordinate of the first point, just as c denotes the first coordinate of (c, d). Analogous remarks apply to y_1, x_2, etc.

EXAMPLE 2 To find the distance from (a, b) to $(2a, -b)$, where a and b are fixed real numbers, substitute a for x_1, b for y_1, $2a$ for x_2, and $-b$ for y_2 in the distance formula:

$$\begin{aligned}\sqrt{(x_1 - x_2)^2 + (y_1 - y_2)^2} &= \sqrt{(a - 2a)^2 + (b - (-b))^2} \\ &= \sqrt{(-a)^2 + (b + b)^2} = \sqrt{a^2 + (2b)^2} \\ &= \sqrt{a^2 + 4b^2}. \quad \blacksquare\end{aligned}$$

Proof of the Distance Formula Let P and Q be points that don't lie on the same vertical or horizontal line, as illustrated in Figure 1–13.

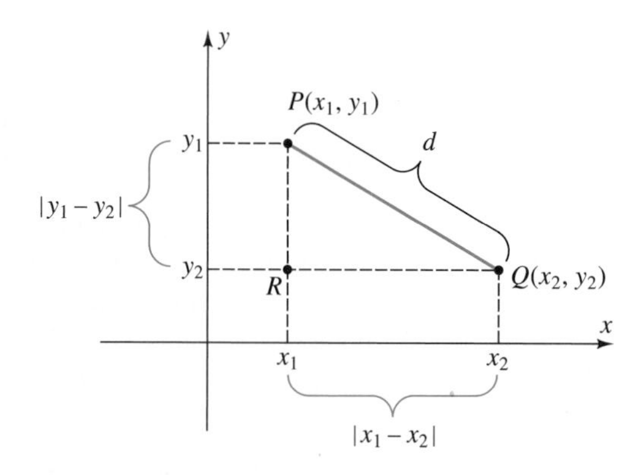

Figure 1-13

As shown in Figure 1–13, the length of RQ is the same as the distance from x_1 to x_2 on the x-axis (number line), namely, $|x_1 - x_2|$. Similarly, the length of PR is the same as the distance from y_1 to y_2 on the y-axis, namely, $|y_1 - y_2|$. According to the Pythagorean Theorem* the length d of PQ is given by:

$$(\text{length } PQ)^2 = (\text{length } RQ)^2 + (\text{length } PR)^2$$

$$d^2 = |x_1 - x_2|^2 + |y_1 - y_2|^2$$

Since $|c|^2 = c^2$ for any real number c (see page 6), this equation becomes:

$$d^2 = (x_1 - x_2)^2 + (y_1 - y_2)^2$$

Since the length d is nonnegative we must have

$$d = \sqrt{(x_1 - x_2)^2 + (y_1 - y_2)^2}$$

This completes the proof in this case. The proof of the cases when P and Q lie on the same vertical or horizontal line is outlined in Exercise 64. ❑

*See the Geometry Review Appendix.

The distance formula can be used to prove this useful fact:

The Midpoint Formula ▶

The midpoint of the line segment from (x_1, y_1) to (x_2, y_2) is

$$\left(\frac{x_1 + x_2}{2}, \frac{y_1 + y_2}{2}\right)$$

EXAMPLE 3 To find the midpoint of the segment joining $(-1, 4)$ and $(3, 1)$, use the formula in the box with $x_1 = -1$, $y_1 = 4$, $x_2 = 3$, and $y_2 = 1$. The midpoint is

$$\left(\frac{x_1 + x_2}{2}, \frac{y_1 + y_2}{2}\right) = \left(\frac{-1 + 3}{2}, \frac{4 + 1}{2}\right) = \left(1, \frac{5}{2}\right)$$

as shown in Figure 1–14. ■

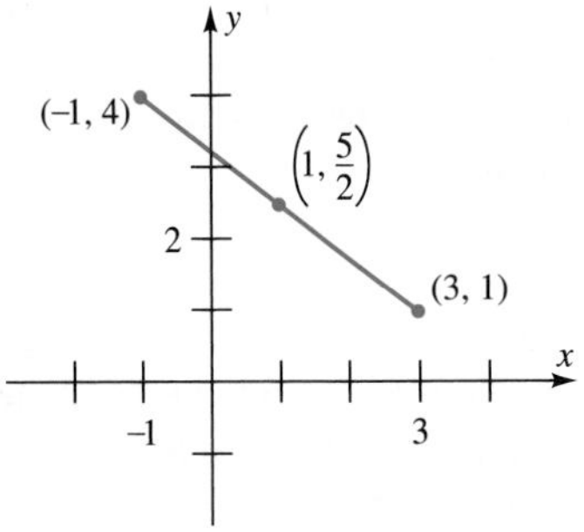

Figure 1-14

Proof of the Midpoint Formula The situation is shown in Figure 1–15.

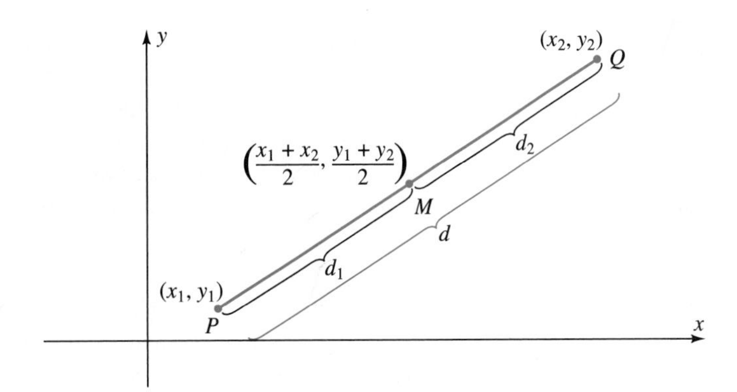

Figure 1-15

In order to show that M is the midpoint of segment PQ, we must show that

$$d_1 = d_2 \quad \text{and} \quad d_1 + d_2 = d.$$

Using the distance formula with the points P and M, we have

$$\begin{aligned}
d_1 &= \sqrt{\left(\frac{x_1 + x_2}{2} - x_1\right)^2 + \left(\frac{y_1 + y_2}{2} - y_1\right)^2} \\
&= \sqrt{\left(\frac{x_1 + x_2}{2} - \frac{2x_1}{2}\right)^2 + \left(\frac{y_1 + y_2}{2} - \frac{2y_1}{2}\right)^2} \\
&= \sqrt{\left(\frac{x_2 - x_1}{2}\right)^2 + \left(\frac{y_2 - y_1}{2}\right)^2} = \sqrt{\frac{(x_2 - x_1)^2}{4} + \frac{(y_2 - y_1)^2}{4}} \\
&= \sqrt{\frac{1}{4}((x_2 - x_1)^2 + (y_2 - y_1)^2)} = \frac{1}{2}\sqrt{(x_2 - x_1)^2 + (y_2 - y_1)^2} = \frac{1}{2}d
\end{aligned}$$

A similar calculation for M and Q (Exercise 66) shows that $d_2 = \frac{1}{2}d$. Hence

$$d_1 = \frac{1}{2}d = d_2 \qquad \text{and} \qquad d_1 + d_2 = \frac{1}{2}d + \frac{1}{2}d = d$$

and the proof is complete. ❑

Graphs of Equations

A **solution** of an equation in two variables, x and y, is a pair of numbers such that the substitution of the first number for x and the second for y produces a true statement. In this case, we say that the pair of numbers **satisfies the equation.** For example, $(3, -2)$ is a solution of $5x + 7y = 1$ because

$$5 \cdot 3 + 7(-2) = 1$$

and $(-2, 3)$ is *not* a solution because $5(-2) + 7 \cdot 3 \neq 1$.

The **graph of an equation** in two variables is the set of points in the plane whose coordinates are solutions of the given equation. Thus the graph is a *geometric picture of the solutions.* For example, the graph of $y = x^2 - 2x - 1$ is shown in Figure 1–16. It includes, among other points, $(-1, 2)$, $(0, -1)$, and $(1.5, -1.75)$ each of which represents a solution of the equation.

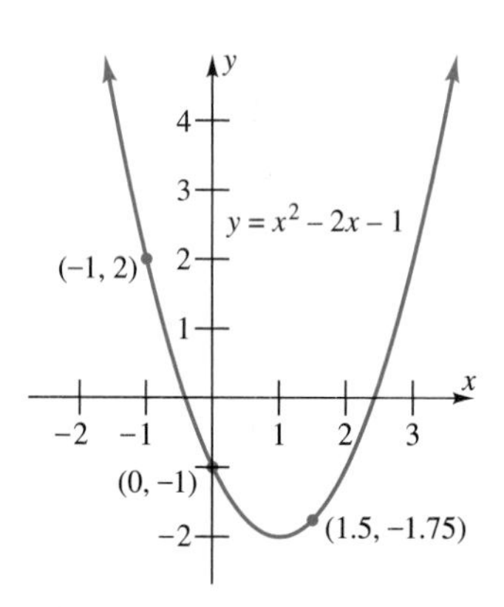

Figure 1-16

Although a principal theme of this book is the use of graphing calculators or other graphing technology to determine the graphs of equations, technology is not always the best approach. So we first consider a situation in which accurate graphs can best be determined by means of the distance formula and algebra.

Circles

If (c, d) is a point in the plane and r a positive number, then the **circle with center (c, d) and radius r** consists of all points (x, y) that lie r units from (c, d), as shown in Figure 1–17. According to the distance formula, the statement that "the distance from (x, y) to (c, d) is r units" is equivalent to:

$$\sqrt{(x - c)^2 + (y - d)^2} = r$$

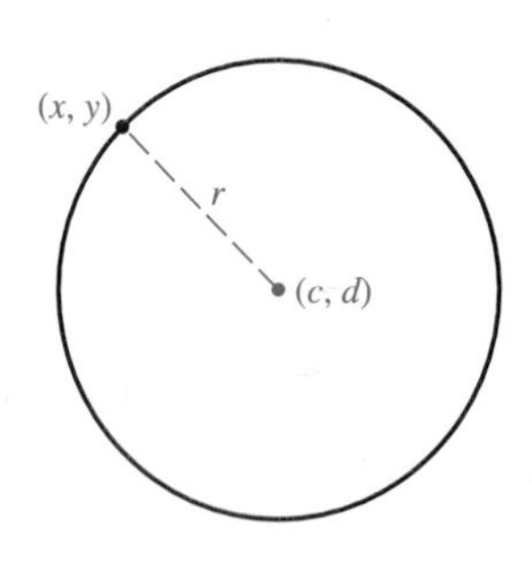

Figure 1-17

Squaring both sides shows that (x, y) satisfies this equation:

$$(x - c)^2 + (y - d)^2 = r^2$$

Reversing the procedure shows that any solution (x, y) of this equation is a point on the circle. Therefore,

Equation of the Circle ▶

> **The circle with center (c, d) and radius r is the graph of**
>
> $$(x - c)^2 + (y - d)^2 = r^2$$

We say that $(x - c)^2 + (y - d)^2 = r^2$ is the **equation of the circle** with center (c, d) and radius r. If the center is at the origin, then $(c, d) = (0, 0)$ and the equation has a simpler form:

Circle with Center at the Origin ▶

The circle with center (0, 0) and radius r is the graph of

$$x^2 + y^2 = r^2$$

EXAMPLE 4

(a) Letting $r = 1$ shows that the graph of $x^2 + y^2 = 1$ is the circle of radius 1 centered at the origin, as shown in Figure 1–18. This circle is called the **unit circle.**

(b) The circle with center $(-3, 2)$ and radius 2, shown in Figure 1–19, is the graph of the equation

$$(x - (-3))^2 + (y - 2)^2 = 2^2$$

or equivalently,

$$(x + 3)^2 + (y - 2)^2 = 4. \quad \blacksquare$$

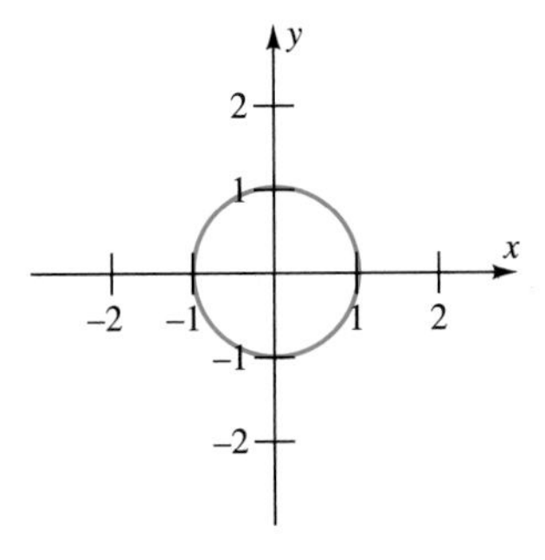

Figure 1-18

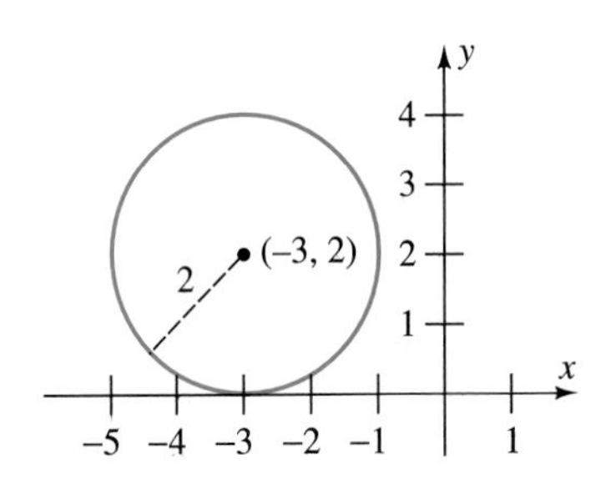

Figure 1-19

EXAMPLE 5 Find the equation of the circle with center $(3, -1)$ that passes through $(2, 4)$.

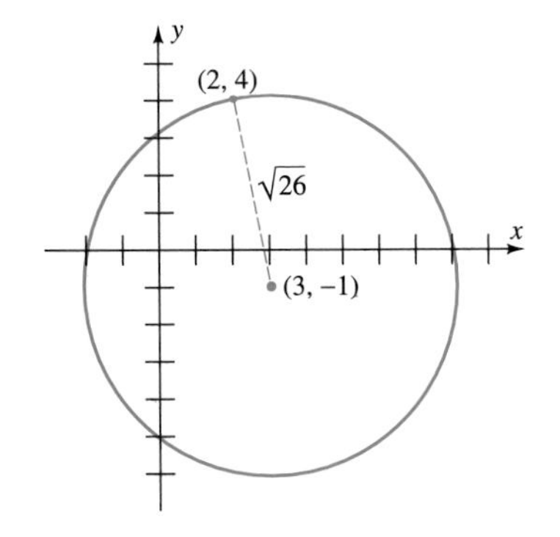

Figure 1-20

Solution We must first find the radius. Since $(2, 4)$ is on the circle, the radius is the distance from $(2, 4)$ to $(3, -1)$ as shown in Figure 1–20, namely,

$$\sqrt{(2 - 3)^2 + (4 - (-1))^2} = \sqrt{1 + 25} = \sqrt{26}.$$

The equation of the circle with center at $(3, -1)$ and radius $\sqrt{26}$ is

$$(x - 3)^2 + (y - (-1))^2 = (\sqrt{26})^2$$

$$(x - 3)^2 + (y + 1)^2 = 26$$

$$x^2 - 6x + 9 + y^2 + 2y + 1 = 26$$

$$x^2 + y^2 - 6x + 2y - 16 = 0. \quad \blacksquare$$

The equation of any circle can always be written in the form

$$x^2 + y^2 + Bx + Cy + D = 0$$

for some constants B, C, D, as in Example 5 (where $B = -6, C = 2, D = -16$). Conversely, the graph of such an equation can always be determined.

EXAMPLE 6 To find the graph of $3x^2 + 3y^2 - 12x - 30y + 45 = 0$, we divide both sides by 3 and rewrite the equation as

$$(x^2 - 4x) + (y^2 - 10y) = -15.$$

Next we complete the square in both expressions in parentheses (see page 17). To complete the square in $x^2 - 4x$ we add 4 (the square of half the coefficient of x) and to complete the square in $y^2 - 10y$ we add 25 (why?). In order to have an equivalent equation we must add these numbers to *both* sides:

$$(x^2 - 4x + 4) + (y^2 - 10y + 25) = -15 + 4 + 25$$

$$(x - 2)^2 + (y - 5)^2 = 14$$

Since $14 = (\sqrt{14})^2$, this is the equation of the circle with center (2, 5) and radius $\sqrt{14}$. ■

EXERCISES 1.3

1. Find the coordinates of the points in Figure 1–21.

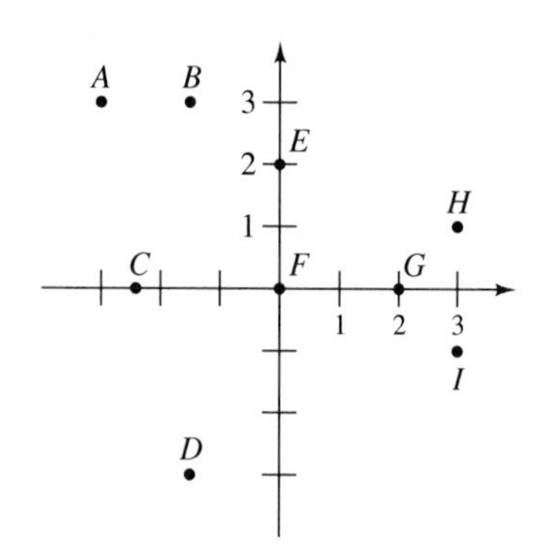

Figure 1-21

2. Draw coordinate axes and plot these points: (0, 0), (−3, 2.1), (2.1, −3), (−4/3, 1) $(5, \pi)$, $(2, \sqrt{2})$, $(-3, \pi)$, (4, 6), $(-\sqrt{3}, \sqrt{3})$, $(\sqrt{3}, -\sqrt{3})$, (5/2, 17/3).

3. **(a)** Plot the points (3, 2), (4, −1), (−2, 3), and (−5, −4).
(b) Change the sign of the y-coordinate in each of the points in part (a) and plot these new points.
(c) Explain how the points (a, b) and $(a, -b)$ are related graphically. [*Hint:* What are their relative positions with respect to the x-axis?]

4. **(a)** Plot the points (5, 3), (4, −2), (−1, 4), and (−3, −5).
(b) Change the sign of the x-coordinate in each of the points in part (a) and plot these new points.
(c) Explain how the points (a, b) and $(-a, b)$ are related graphically. [*Hint:* What are their relative positions with respect to the y-axis?]

In Exercises 5–12, find the distance between the two points and the midpoint of the segment joining them.

5. (−3, 5), (2, −7)
6. (2, 4), (1, 5)
7. (1, −5), (2, −1)
8. (−2, 3), (−3, 2)
9. $(\sqrt{2}, 1)$, $(\sqrt{3}, 2)$
10. $(-1, \sqrt{5})$, $(\sqrt{2}, -\sqrt{3})$
11. (a, b), (b, a)
12. (s, t), (0, 0)

In Exercises 13–18, determine whether the point is on the graph of the given equation.

13. $(1, -2)$; $3x - y - 5 = 0$

14. $(2, -1)$; $x^2 + y^2 - 6x + 8y = -15$

15. $(6, 2)$; $3y + x = 12$ **16.** $(1, -2)$; $3x + y = 12$

17. $(3, 4)$; $(x - 2)^2 + (y + 5)^2 = 4$

18. $(1, -1)$; $\dfrac{x^2}{2} + \dfrac{y^2}{3} = 1$

In Exercises 19–22, find the equation of the circle with given center and radius r.

19. $(-3, 4)$; $r = 2$ **20.** $(-2, -1)$; $r = 3$

21. $(0, 0)$; $r = \sqrt{2}$ **22.** $(5, -2)$; $r = 1$

In Exercises 23–26, sketch the graph of the equation.

23. $(x - 2)^2 + (y - 4)^2 = 1$

24. $(x + 1)^2 + (y - 3)^2 = 9$

25. $(x - 5)^2 + (y + 2)^2 = 5$

26. $(x + 6)^2 + y^2 = 4$

In Exercises 27–32, find the center and radius of the circle whose equation is given.

27. $x^2 + y^2 + 8x - 6y - 15 = 0$

28. $15x^2 + 15y^2 = 10$

29. $x^2 + y^2 + 6x - 4y - 15 = 0$

30. $x^2 + y^2 + 10x - 75 = 0$

31. $x^2 + y^2 + 25x + 10y = -12$

32. $3x^2 + 3y^2 + 12x + 12 = 18y$

In Exercises 33–35, show that the three points are the vertices of a right triangle and state the length of the hypotenuse. [You may assume that a triangle with sides of lengths a, b, c is a right triangle with hypotenuse c provided that $a^2 + b^2 = c^2$.]

33. $(0, 0)$, $(1, 1)$, $(2, -2)$

34. $\left(\dfrac{\sqrt{2}}{2}, 0\right)$, $\left(\dfrac{\sqrt{2}}{2}, \dfrac{\sqrt{2}}{2}\right)$, $(0, 0)$

35. $(3, -2)$, $(0, 4)$, $(-2, 3)$

36. What is the perimeter of the triangle with vertices $(1, 1)$, $(5, 4)$, and $(-2, 5)$?

In Exercises 37–42, describe the set of all points (x, y) that satisfy the given condition.

37. $y = 5$ **38.** $x = -5$

39. $xy = 0$ [*Hint*: When is a product zero?]

40. $xy > 0$ **41.** $|x| < 1$ **42.** $|y| \leq 2$

In Exercises 43–48, find the equation of the circle.

43. Center $(2, 2)$; passes through the origin.

44. Center $(-1, -3)$; passes through $(-4, -2)$.

45. Center $(1, 2)$; intersects x-axis at -1 and 3.

46. Center $(3, 1)$; diameter 2.

47. Center $(-5, 4)$; tangent (touching at one point) to the x-axis.

48. Center $(2, -6)$; tangent to the y-axis.

49. One diagonal of a square has endpoints $(-3, 1)$ and $(2, -4)$. Find the endpoints of the other diagonal.

50. Find the vertices of all possible squares with this property: Two of the vertices are $(2, 1)$ and $(2, 5)$. [*Hint*: There are three such squares.]

51. Do Exercise 50 with (c, d) and (c, k) in place of $(2, 1)$ and $(2, 5)$.

52. Find the three points that divide the line segment from $(-4, 7)$ to $(10, -9)$ into four parts of equal length.

53. Find all points P on the x-axis that are 5 units from $(3, 4)$. [*Hint*: P must have coordinates $(x, 0)$ for some x and the distance from P to $(3, 4)$ is 5.]

54. Find all points on the y-axis that are 8 units from $(-2, 4)$.

55. Find all points with first coordinate 3 that are 6 units from $(-2, -5)$.

56. Find all points with second coordinate -1 that are 4 units from $(2, 3)$.

57. Find a number x such that $(0, 0)$, $(3, 2)$, and $(x, 0)$ are the vertices of an isosceles triangle, neither of whose two equal sides lies on the x-axis.

58. Do Exercise 57 if one of the two equal sides lies on the positive x-axis.

59. Show that the midpoint M of the hypotenuse of a right triangle is equidistant from the vertices of the triangle. [*Hint:* Place the triangle in the first quadrant of the plane, with right angle at the origin so that the situation looks like Figure 1–22.]

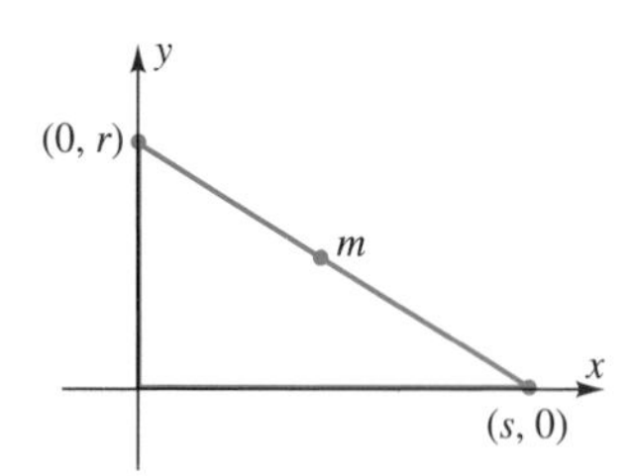

Figure 1-22

60. Show that the diagonals of a parallelogram bisect each other. [*Hint:* Place the parallelogram in the first quadrant with a vertex at the origin and one side along the x-axis, so that the situation looks like Figure 1–23.]

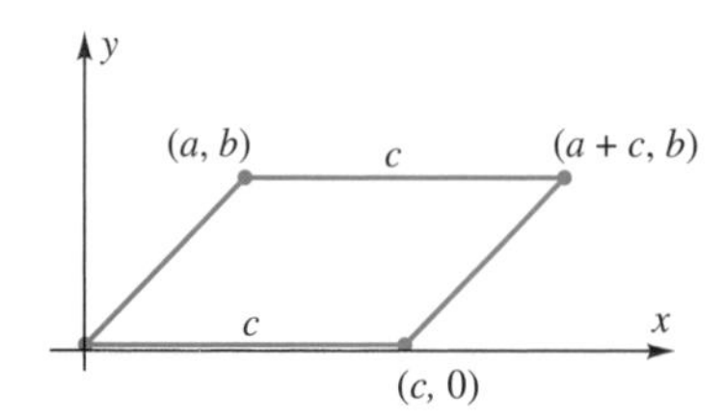

Figure 1-23

61. Show that the diagonals of a rectangle have the same length. [*Hint:* Place the rectangle in the first quadrant of the plane and label its vertices appropriately, as in Exercises 59–60.]

62. If the diagonals of a parallelogram have the same length, show that the parallelogram is actually a rectangle. [*Hint:* See Exercise 60.]

Thinkers

63. For each nonzero real number k, the graph of $(x - k)^2 + y^2 = k^2$ is a circle. Describe all possible such circles.

64. **(a)** Prove that the distance formula is valid for two points on the same vertical line. [*Hint*: The points must have the same first coordinate (why?), say (x_1, y_1) and (x_1, y_2). The distance between them is $|y_1 - y_2|$ (why?). See the box on page 6.]
(b) Prove that the distance formula is valid for two points on the same horizontal line.

65. Suppose every point in the coordinate plane is moved 5 units straight up.
(a) To what point does each of these points go: $(0, -5)$, $(2, 2)$, $(5, 0)$, $(5, 5)$, $(4, 1)$?
(b) Which points go to each of the points in part (a)?
(c) To what point does (a, b) go?
(d) To what point does $(a, b - 5)$ go?
(e) What point goes to $(-4a, b)$?
(f) What points go to themselves?

66. Complete the proof of the Midpoint Formula by showing that $d_2 = \frac{1}{2}d$.

67. Let (c, d) be any point in the plane with $c \neq 0$. Prove that (c, d) and $(-c, -d)$ lie on the same straight line through the origin, on opposite sides of the origin, the same distance from the origin. [*Hint:* Find the midpoint of the line segment joining (c, d) and $(-c, -d)$.]

1.4 GRAPHS AND GRAPHING CALCULATORS*

A **graph** is a set of points in the coordinate plane. Graphs in this general sense include graphs of equations in two variables, such as circles or the curves in Figures 1–24 and 1–25, as well as the curve in Figure 1–26.

*Although we generally mention only graphing calculators, virtually the entire discussion applies (with obvious changes) to graphing software for a computer.

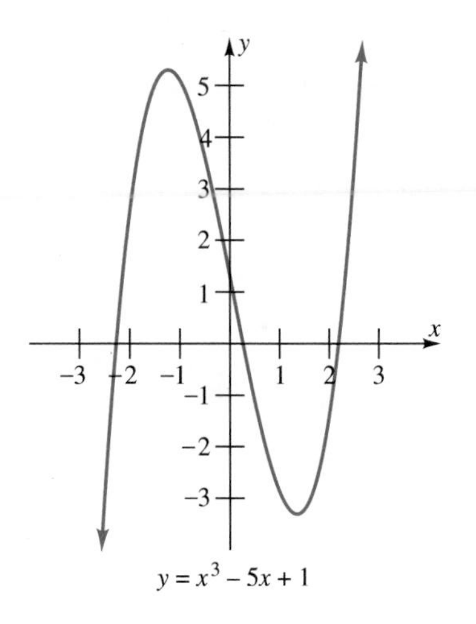

Figure 1-24

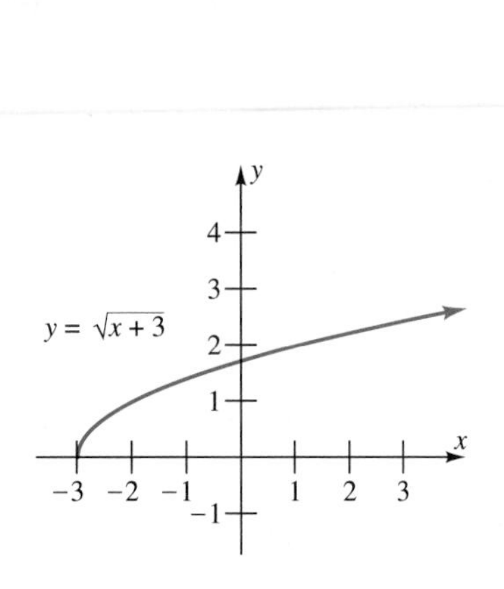

Figure 1-25

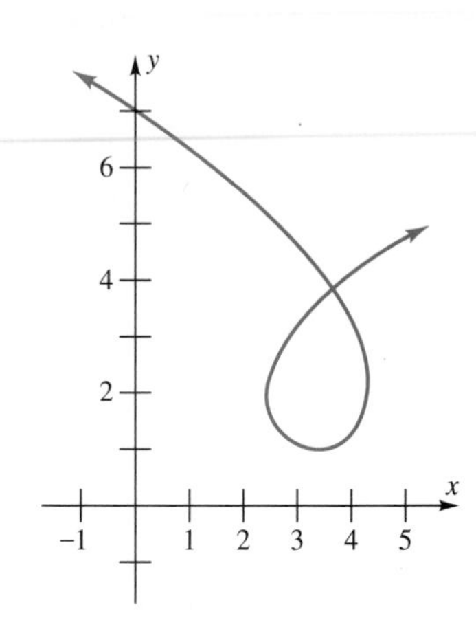

Figure 1-26

Graphs such as the one in Figure 1–26 are considered briefly in Section 2.3 and in more detail in Chapter 10. In the rest of the book, unless noted otherwise, the discussion is limited to graphs of equations such as $y = x^3 - 5x + 1$ and $y = \sqrt{x + 3}$ in Figures 1–24 and 1–25, that is, equations that can be written in the form

$$y = \text{expression in } x.$$

Viewing Windows

When a graph extends infinitely far in some direction, an accurate physical representation of the *entire* graph is not possible. Any figure or calculator screen represents some region of the coordinate plane and includes only the portion of the graph that is located in that region. The rectangular region represented on a screen or in a figure is called the **viewing window** or **viewing rectangle.**

EXAMPLE 1 The viewing window for the color graph in Figure 1–27 on the next page is the rectangular region indicated by the dashed gray lines, whose corner points are $(-4, -3)$, $(-4, 6)$, $(5, 6)$, $(5, -3)$.

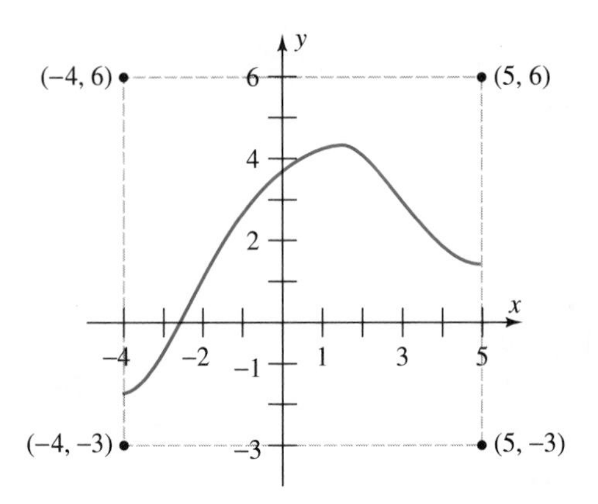

Figure 1-27

This viewing window may also be described as the set of all points whose x-coordinates satisfy $-4 \leq x \leq 5$ and whose y-coordinates satisfy $-3 \leq y \leq 6$. ■

To set up a viewing window on a calculator, you must specify the minimum and maximum values for the x-coordinates and y-coordinates and the spacing of the tic marks on the axes. For instance, to display the viewing window shown in Figure 1–27 you press the RANGE (or WINDOW or EQTN or PLOT SETUP) key and enter the appropriate numbers, as shown here for a TI-82 (other calculators are similar).*

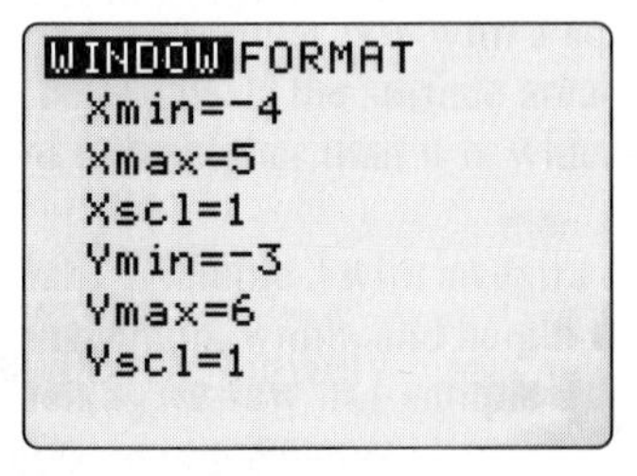

Figure 1-28

The settings Xscl $= 1$ and Yscl $= 1$ put the tic marks 1 unit apart on each axis. This is usually the best setting for smaller viewing windows, but not for large ones.

*On HP-38 ''*X*scl'' and ''*Y*scl'' are called ''*X*tick'' and ''*Y*tick''. On TI-81, 83, and 92, and on HP-38 there is also an ''*X*res'' or ''res'' setting. Unless directed otherwise, set *X*res = 1 on TI and res = detail on HP.

Note The symbol  GRAPHING EXPLORATION indicates that you are to use your graphing calculator or computer software as directed to complete the discussion.

GRAPHING EXPLORATION Set up a viewing window with $-200 < x < 200$ and $-30 < y < 30$ and $X\text{scl} = 1$, $Y\text{scl} = 1$. Then press GRAPH (or PLOT) to display this window on the screen. (If a graph also appears, ignore it; just look at the axes). Can you distinguish the tic marks on each axis? Now press RANGE (or WINDOW or EQTN or PLOT SETUP) and change the scl settings to $X\text{scl} = 20$ and $Y\text{scl} = 5$, so that adjacent tic marks will be 20 units apart on the x-axis and 5 units apart on the y-axis, and press GRAPH (or PLOT) again. Can you distinguish the tic marks now?

Each illustration of a calculator screen in this book has an outside border. The numbers outside the border indicate the minimum and maximum values of x and y. For instance, the viewing window in Figure 1–29 has $-8 \leq x \leq 6$ and $-4 \leq y \leq 5$. The Xscl and Yscl settings are not displayed in Figure 1–29, but are easily determined. Since a 14-unit long segment of the x-axis is shown, the tic marks on the x-axis are two units apart. Similarly, a 9-unit long segment of the y-axis is shown and the tic marks are one unit apart.

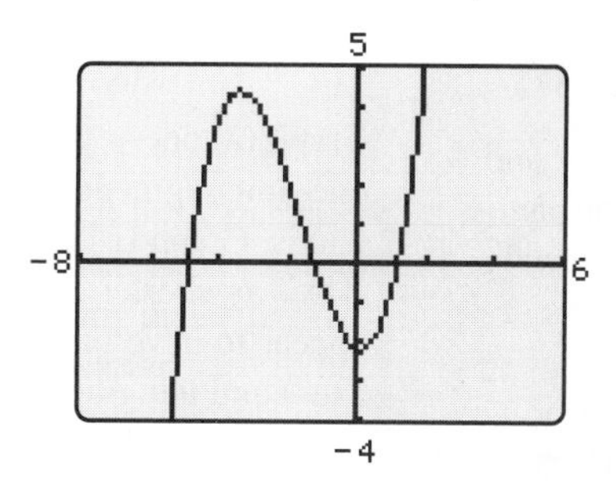

Figure 1-29

▶ **TECHNOLOGY TIP**

Some calculators let you select certain viewing windows, such as the standard window, with a single keystroke. Look in the ZOOM menu on TI, HP-38, and Casio 9800, and in the RANGE MENU (two keys) on Sharp 9300.

Throughout this book the viewing window with $-10 \leq x \leq 10$ ($X\text{scl} = 1$) and $-10 \leq y \leq 10$ ($Y\text{scl} = 1$) is called the **standard viewing window.**

Square Viewing Windows

Specifying the maximum and minimum values for x amounts to choosing the **scale** for the x-axis, that is, determining the size of a line segment representing 1 unit of length; and similarly for y. When different scales are used on the two axes the perceived shape of the graph may be affected.

EXAMPLE 2 The graph of the equation $x^2 + y^2 = 4$ is a circle of radius 2 with center at the origin (see page 29). Solving this equation for y shows that

$$y = \sqrt{4 - x^2} \qquad \text{or} \qquad y = -\sqrt{4 - x^2}.$$

Consequently the graph of $y = \sqrt{4 - x^2}$ is part of a circle (the top half, actually) and should look round. Figure 1–30 shows how it actually looks in different viewing windows on a typical calculator (with $Xscl = 1 = Yscl$ in each case):

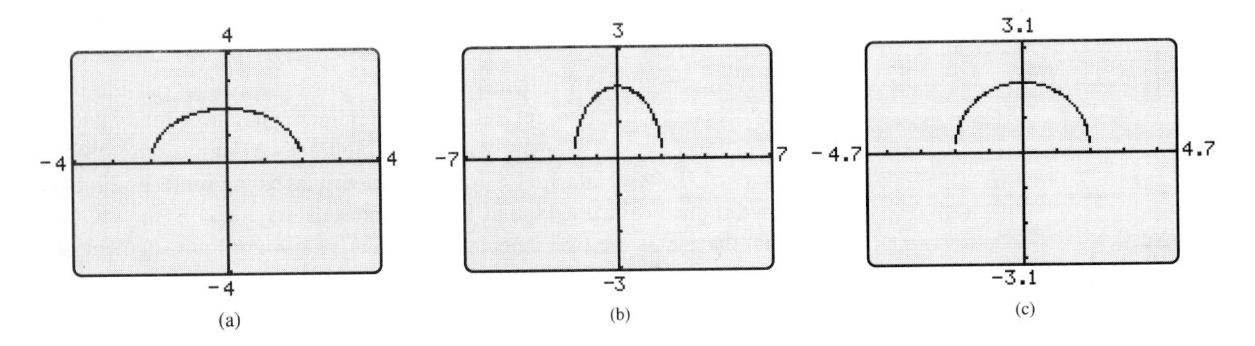

Figure 1-30

Graph (a) doesn't look like a half-circle and the two axes have different scales; a one-unit segment on the x-axis is a bit longer than a one-unit segment on the y-axis. Graph (b) is also oval rather than circular; here a one-unit segment on the x-axis is shorter than a one-unit segment on the y-axis. The axes in graph (c) appear to have the same scale (one-unit segments have the same length on both axes) and the graph does look like half of a circle. ■

> **▶ TECHNOLOGY TIP**
>
> On HP-38, Casio 9800 and all TI calculators you can change the current viewing window to a square one with SQUARE (or ZSQUARE or ZSQR or SQR or ZOOMSQR) in the ZOOM menu.
>
> The following are square windows: the ZDECIMAL window on TI-82/83 and 92 (but not TI-85), the ZOOMDEC window on TI-92, the DECIMAL window on HP-38 (in the VIEWS menu), the DEFAULT window on Sharp 9300 (in the RANGE menu), and the INIT window on Casio (in the RANGE menu).

A calculator viewing window in which a one-unit segment on the x-axis has the same length as a one-unit segment on the y-axis, such as Figure 1–30(c), is called **a square viewing window.** You should use square viewing windows when you want circles to look round, perpendicular straight lines to look perpendicular, etc., but use any convenient window for the rest of your graphs.

Since the typical calculator screen is wider than it is high, setting the same range of values for both x and y (such as $-10 \le x \le 10$ and $-10 \le y \le 10$) does *not* produce a square window because a one-unit segment on the y-axis is shorter than a one-unit segment on the x-axis. Most calculator screens are almost two-thirds as high as they are wide, so a square window can be obtained on such a calculator by displaying a y-axis that is about 2/3 the length of the x-axis (for instance, $-15 \le x \le 15$ and $-10 \le y \le 10$).*

*To obtain a square window on wide-screen calculators (TI-85 and 92, HP-38), the y-axis should be approximately half as long as the x-axis (3/5 as long on TI-85).

Hand-Drawn and Calculator-Drawn Graphs

The traditional method of graphing an equation "by hand" is as follows: Construct a table of values with a reasonable number of entries, plot the corresponding points, and use whatever algebraic or other information is available to make an "educated guess" about the rest.

EXAMPLE 3 The graph of $y = x^2$ consists of all points (x, x^2), where x is a real number. You can easily construct a table of values and plot the corresponding points, as in Figure 1–31. These points suggest that the graph looks like the one in Figure 1–32, which is obtained by connecting the plotted points and extending the graph upward. ■

x	$y = x^2$
–2.5	6.25
–2	4
–1.5	2.25
–1	1
–.5	.25
0	0
.5	.25
1	1
1.5	2.25
2	4
2.5	6.25

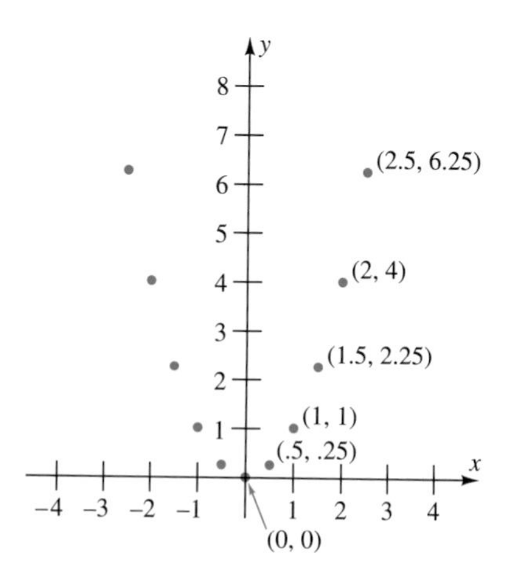

Figure 1-31

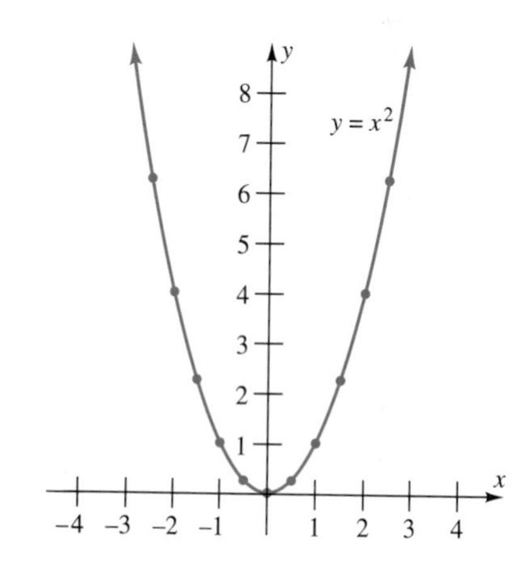

Figure 1-32

A graphing calculator or computer graphing program follows essentially the same procedure. Once the viewing window is chosen and the equation entered, the calculator selects a large number of x values (95 or more) equally spaced along the x-axis and plots the corresponding points, simultaneously connecting them with line segments.

▶ TECHNOLOGY TIP

Calculators can also be set to plot points without connecting them (producing graphs like Figure 1–31). Look in the index of your instruction manual for DOT (TI calculators), or CONNECT (HP-38), or PLOT1 (Sharp 9300), or PLT or PLOT (Casio).

EXAMPLE 4 Graph $y = x^2$ on a calculator and compare the result with Figure 1–32.

Solution First, choose a viewing window. Since we want to compare the result with Figure 1–32 (in which one-unit segments have the same length on both axes), we choose a square window that shows that same portion of the graph: $-7.5 \leq x \leq 7.5$ and $-1 \leq y \leq 9$.

▶ TECHNOLOGY TIP

When entering an equation for graphing, use the "variable" key rather than the ALPHA X key. The variable key has a label like X,T,θ or x-VAR or X|T. There is no such key on TI-92; use the x key on the keyboard.

Next, call up the **equation memory** (press *Y*= on TI, SYMB on HP-38, GRAPH EXE on Casio 9700/9800, EQTN on Sharp 9300, and FMEM on Casio 7700/8700) and enter the equation. Figure 1–33 shows how it looks on a TI-82 (other calculators are similar). Now press GRAPH (or PLOT or DRW) and obtain Figure 1–34.*

Figure 1-33

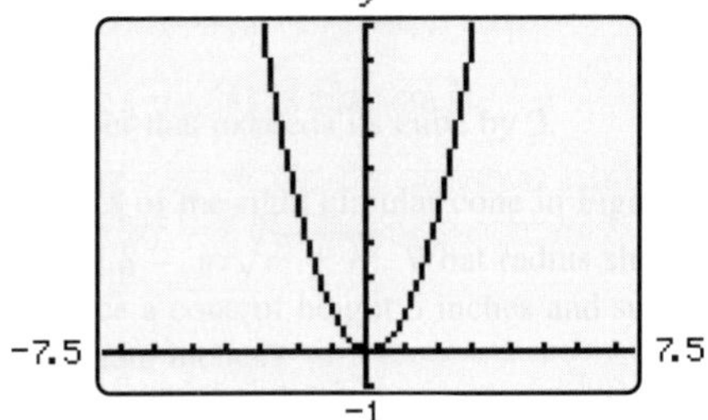

Figure 1-34

How does Figure 1–34 compare to the hand-drawn graph in Figure 1–32? The two graphs have the same general shape, but the hand-drawn graph is a smooth, connected curve, while the graph in Figure 1–34 is a sequence of disconnected dots and short vertical line segments. ■

To see just which points your calculator plots when it draws a graph, press TRACE and a flashing cursor appears on the graph.** Use the *left* and *right* arrow keys to move the cursor along the graph. As you do this the coordinates of the cursor point are displayed at the bottom of the screen, as in Figure 1–35 (the graph of $y = x^2$ from Figure 1–34).

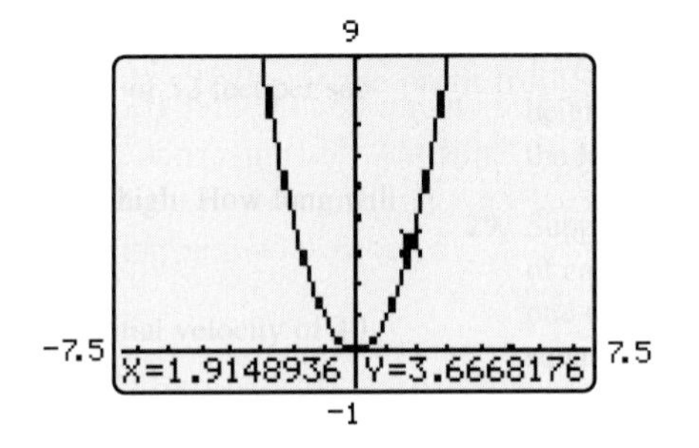

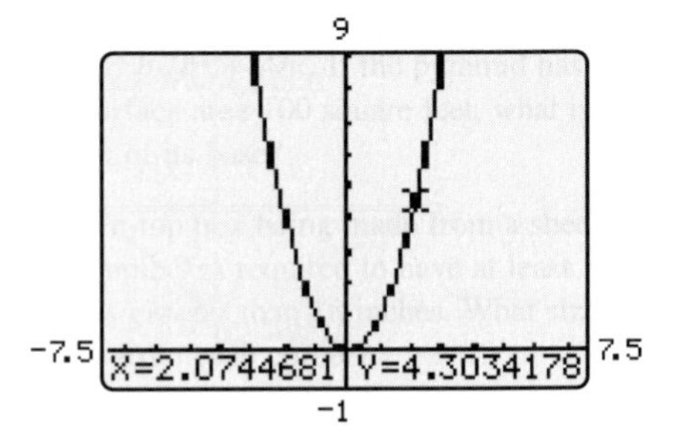

Figure 1-35

*On Casio models 7700-8700, graph equation 1 in the memory by pressing GRAPH FMEM RCL 1 EXE.

**On Sharp calculators, press the left or right arrow key to start the trace feature; on HP-38, the trace is automatically on after graphing.

The trace feature moves the cursor only to those points whose coordinates the calculator computed when drawing the graph. So, in Figure 1–35 the cursor does not land on (2, 4), but it does land on the two nearby points whose coordinates *were* computed.

WARNING On TI, Casio, and HP-38 calculators, you can move the cursor around the screen by using the arrow keys, without using TRACE (on Casio, press PLOT). When this is done, the cursor may appear to be on the graph when it is only *near* a point on the graph, so the displayed coordinates may not be coordinates of a point on the graph as they would be if TRACE were used.

Calculator Quirks*

Although a calculator greatly improves speed, convenience, and accuracy in graphing, it does not produce perfect graphs, as we saw in Example 4. One reason is that a calculator screen is actually a grid of tiny rectangles, called **pixels.** The calculator displays a graph by darkening certain pixels. Since a single pixel appears to be a large dot, we tend to think of it as a point in the plane. This is not the case, however. In the standard viewing window on a typical calculator, for example, a single pixel represents a rectangular region measuring approximately .2 by .3 units. A calculator does not plot a point (x, y) exactly, but darkens the pixel in which this point lies.

Because of the way a calculator plots points, a straight line segment may appear jagged and a smooth connected curve like the graph of $y = x^2$ may look rough and include vertical line segments, even though the actual graph has no such segments.** In order to use a calculator effectively you must be able to interpret the imperfect screen image, that is, visualize what the actual graph should look like. This is easily done in most cases if you understand how a calculator graphs and you know some algebra.

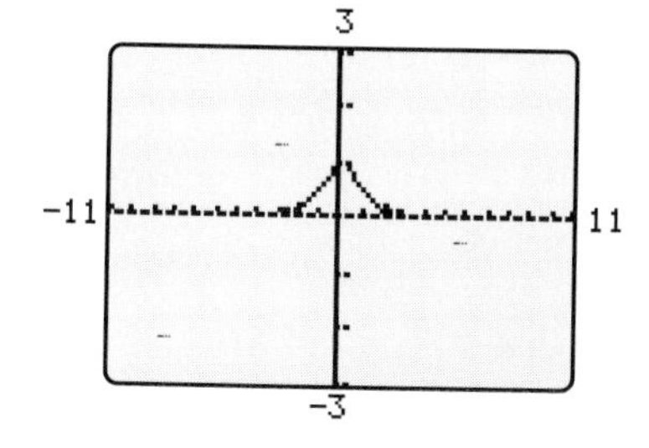

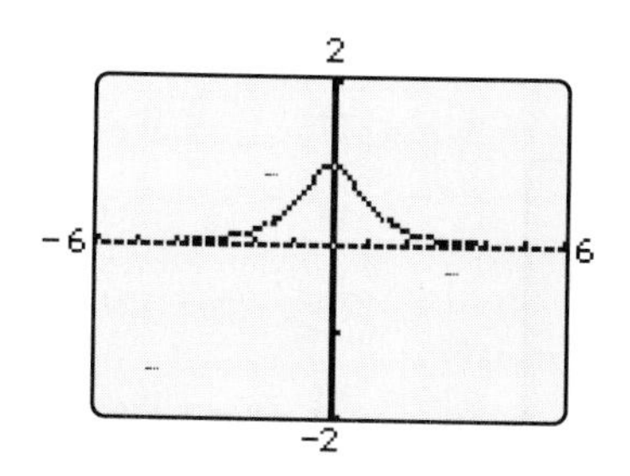

Figure 1-36

EXAMPLE 5 The equation $y = \dfrac{1}{x^2 + 1}$ is graphed in two viewing windows in Figure 1–36 and appears to stop abruptly part way along the x-axis, even though y is defined for every real number x.

*Computers use the same graphing method as do calculators, but their larger screens generally have better resolution, so many (but not necessarily all) of the problems discussed here may not occur with computer graphing.

**In an equation such as $y = x^2$, substituting a number for x produces exactly one value of y, so two points on the graph cannot have the same x-coordinate and different y-coordinates. Since all points on a vertical line segment have the same x-coordinate, no such segment can be part of the graph of $y = x^2$. Similar remarks apply to any equation that can be graphed on a calculator.

GRAPHING EXPLORATION Graph the equation in the right-hand window in Figure 1–36. In this window, each pixel represents a rectangle that is approximately .06 units high, so points within .06 of the x-axis will be in the same pixel as points on the x-axis. Use TRACE to move to the right along the curve past the point where the graph appears to end. As the trace cursor moves from point to point along the curve, watch the y-coordinates. Explain why the ends of the graph appear to coincide with the x-axis. ■

Other anomalies of calculator-drawn graphs will be discussed as the need arises. Once you are aware of the various ways that technology can misrepresent a graph, it is usually not too difficult to determine how the graph should actually look and to avoid erroneous conclusions.

Graphing Several Equations

There are many occasions when you want to graph two or more equations on the same coordinate plane and graphing technology makes this easy.

EXAMPLE 6

(a) Graph the circle $x^2 + y^2 = 4$ on a calculator.

(b) Graph the lower half of this circle on a calculator.

Solution **(a)** Solving the equation for y shows that

$$y = \sqrt{4 - x^2} \qquad \text{or} \qquad y = -\sqrt{4 - x^2}.$$

The graph of each equation is part of the circle. Graphing them both on the same screen will produce the graph of the circle. Enter each of these equations in the equation memory (Figure 1–37); note that the equal sign is shaded in both equa-

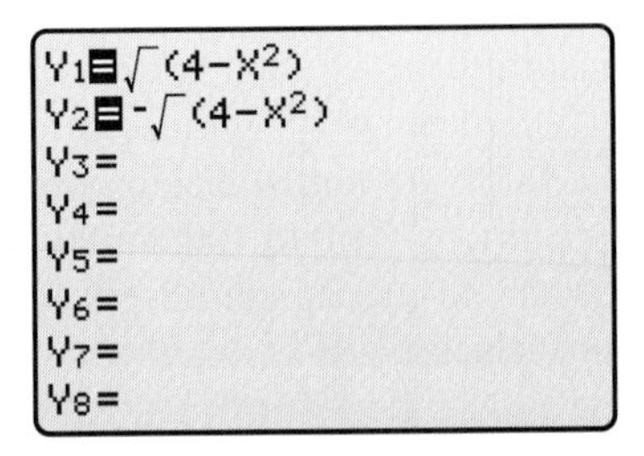

Figure 1-37

tions, indicating that both will be graphed. Press GRAPH (or PLOT or DRW) to obtain Figure 1–38.*

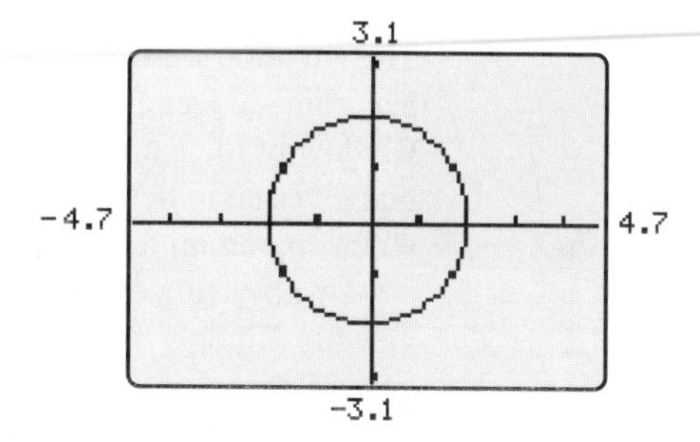

Figure 1-38

We choose the square window in Figure 1–38 so that the circle would look round (if your calculator has a wider screen than ours, your picture may not look round).

> GRAPHING EXPLORATION Enter these two equations in your equation memory. Change to the standard viewing window and regraph these two equations (see the first Tip in the margin). Does the circle look round?

(b) Every point on the lower half of the circle has a negative y-coordinate, so the lower half of the circle must be the graph of $y = -\sqrt{4 - x^2}$. To graph just this equation, call up the equation memory and "turn off" the equation $y_1 = \sqrt{4 - x^2}$ as in Figure 1–39 (see the second Tip in the margin). The equal sign is now shaded only on equation y_2, meaning that only this equation will be graphed. Press GRAPH to obtain Figure 1–40. ■

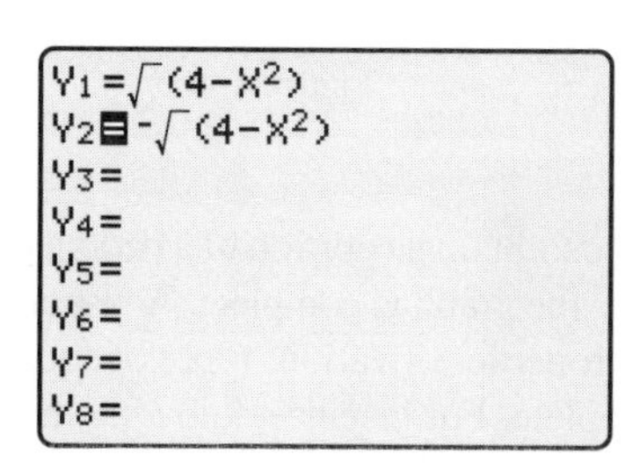

Figure 1-39

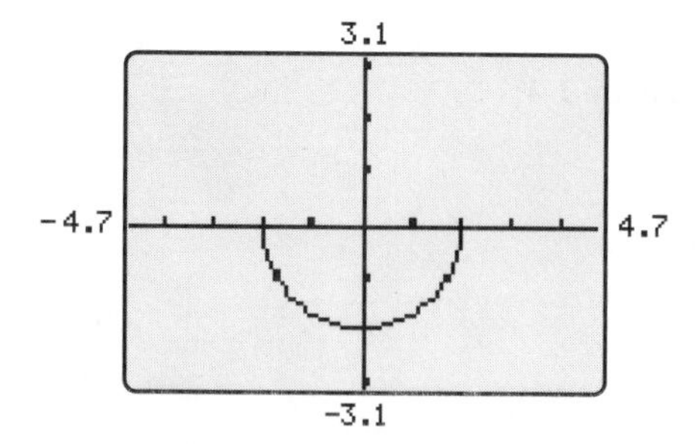

Figure 1-40

*On Casio calculators other than 9800, graph each equation in succession, without clearing the screen, and they will all appear on the same screen.

▶ **TECHNOLOGY TIP**

Once an equation has been graphed, you can change viewing windows without re-entering the equation: just press RANGE (or WINDOW or PLOT SETUP), make the changes, and press GRAPH (on TI and Sharp), PLOT (on HP-38), or EXIT DRW (on Casio 9800), or EXE (on other Casios).

Changing to the standard window on TI calculators is even easier: press ZSTANDARD in the ZOOM menu.

▶ **TECHNOLOGY TIP**

On most calculators an equation is "on" if its equal sign is shaded and "off" if its equal sign is clear. On TI-92 and HP-38 an equation that is "on" has a check mark ✓ next to it.

Only the equations that are "on" will be graphed when you press GRAPH (or PLOT). To turn an equation "on" or "off" place the cursor on its equal sign and press ENTER (TI-81/82/83, Sharp 9300) or move the cursor to the equation and press SELECT (TI-85, Casio 9800) or CHECK (TI-92, HP-38).

Complete Graphs

A key question is whether or not the viewing window shows enough of the graph so that the reader has a reasonably good idea of what the entire graph looks like. The situation is somewhat analogous to photographing a building with a zoom lens camera. Depending on the zoom lens setting (which determines the viewing window of the camera), the resulting photograph may only show a small portion that gives no indication of what the rest of the building looks like, or it may show all the important features and give a good idea what the entire building looks like.

A physical representation of a graph is said to be a **complete graph** if its viewing window shows all the important features of the graph and suggests the general shape of any portions of the graph that aren't shown. In any particular case, many different viewing windows may show a complete graph. It's usually best to use a window that is small enough to show as much detail as possible.

For the color figures in this book, several conventions are used to indicate a complete graph. An arrowhead at the end of a curve, as in Figure 1–41, means that the graph continues on forever in the general direction of the arrowhead, without any twists or breaks. However, when a graph repeats a pattern, as in Figure 1–42, an arrowhead means that this same pattern continues on forever.

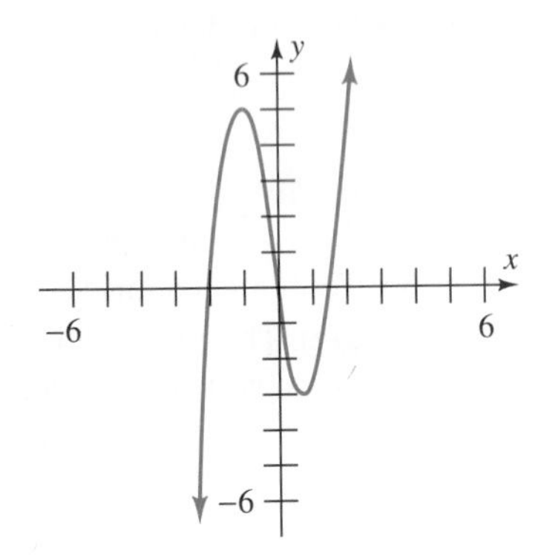

Figure 1-41

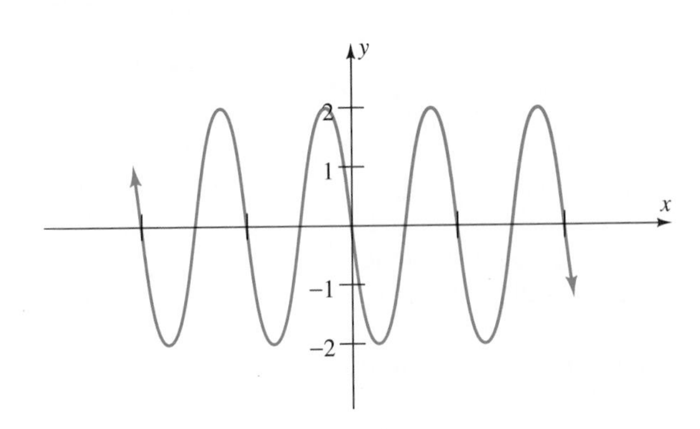

Figure 1-42

When you obtain a (portion of a) graph on a calculator screen, it isn't always clear whether the graph is complete. A principal theme of this book is that knowledge of the properties of various types of equations can be used to determine that a graph is complete. For instance, knowing the equation of the circle enabled us to identify the graph in Figure 1–30 and to know that it was complete. Without similar knowledge in other cases, it may not be possible to say with absolute certainty that a graph is complete. Nevertheless, trying various viewing windows often provides evidence that a particular graph is probably complete. For now, this is usually the best you can do.

EXAMPLE 7 Sketch a complete graph of

$$y = .007x^5 - .2x^4 + 1.332x^3 - .004x^2 + 10.$$

Solution The graph of the equation in the standard viewing window (Figure 1–43) is certainly not complete since it shows no points to the right of the y-axis. Extending the range of y values in Figure 1–44 produces a graph that *might* be complete.

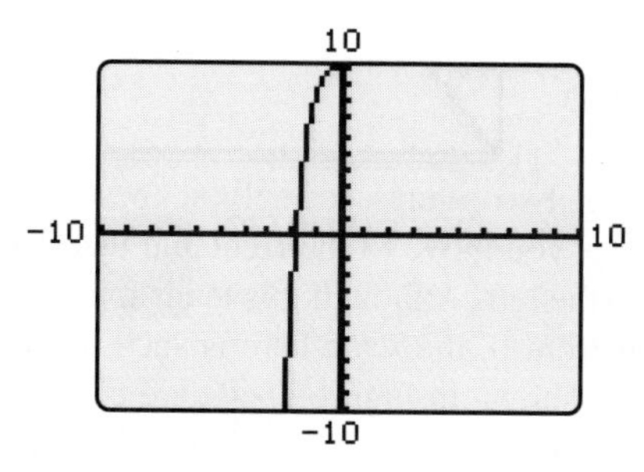

Figure 1-43

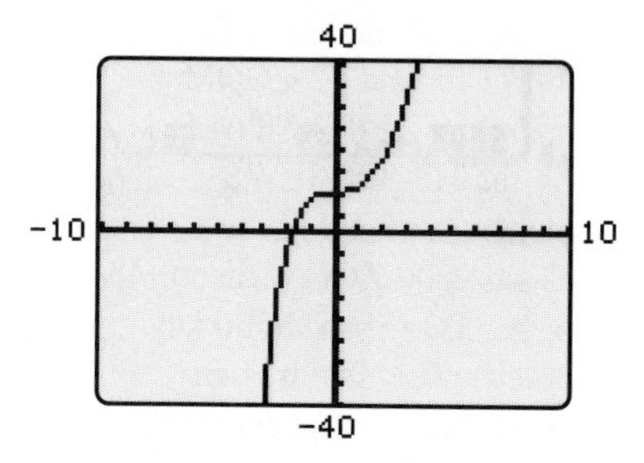

Figure 1-44

Just to be sure, however, we enlarge the range of x values as well, and obtain Figure 1–45, which suggests that parts of the graph lie outside this viewing window. Enlarging the range of y values again, we obtain Figure 1–46, which looks as if it might be a complete graph.

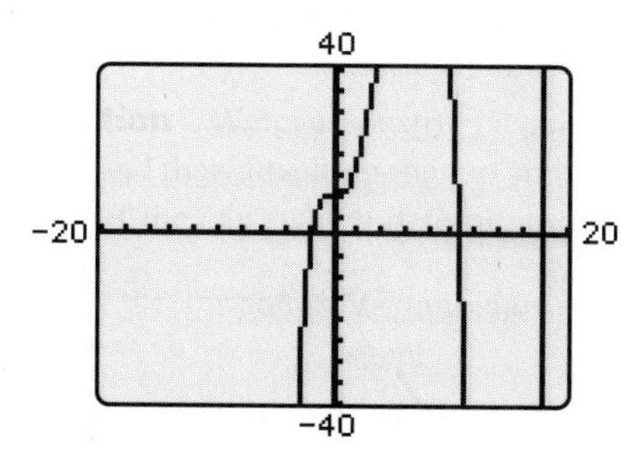

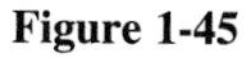

Figure 1-45

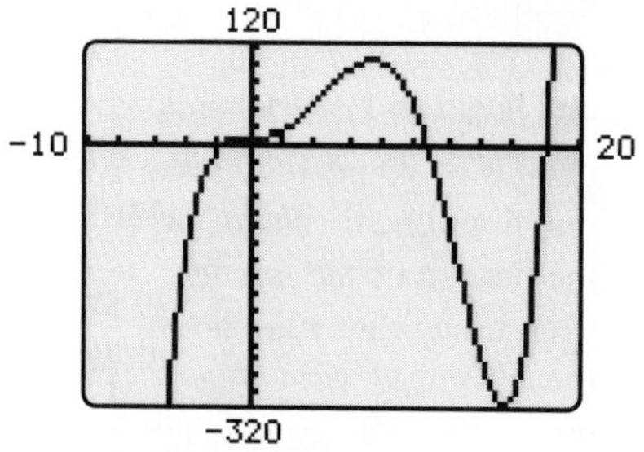

120
-10
20
-320

Figure 1-46

To see if the graph turns or bends again farther out, we try two very large viewing windows in Figures 1–47 and 1–48.

> **▶ TECHNOLOGY TIP**
> Some calculators have an "auto scaling" feature. Once the range of x values has been set, the calculator selects a viewing window that includes all points on the graph whose x-coordinates are in the chosen range. It's labeled ZFIT (in the TI-85 and 92 ZOOM menu) or AUTO-SCALE (in the HP-38 VIEWS menus); or AUT in the Casio 9800 ZOOM menu; or AUTO (Sharp 9300 keyboard). This feature can eliminate some guesswork, but may produce a large window that hides some features of the graph.

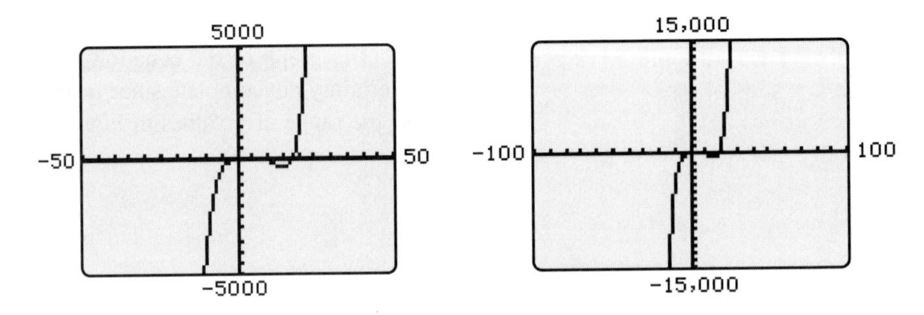

Figure 1-47 **Figure 1-48**

These last two figures strongly suggest that the graph keeps climbing sharply forever as you move to the right and that it keeps falling sharply forever as you move to the left. Although this tends to confirm the shape of the graph for values of x far from 0, the scale here is such that the features of the graph around the origin, as shown in Figure 1–45, are completely obscured.

Consequently, we conclude that Figure 1–45 is probably a complete graph since it shows the important features (twists and turns) near the origin as well as suggesting the appropriate shape farther out. In Chapter 4, we shall see that this graph is in fact complete. ■

Example 7 shows that too small a viewing window may miss twists and turns of the graph and too large a window may hide these features. So don't be afraid to experiment with your calculator until you are reasonably sure that you have a complete graph. In some cases, several different viewing windows may be needed for a complete graph.

> **▶ TECHNOLOGY TIP**
> If you get a blank screen instead of a graph on TI, HP-38 or Casio 9800, press TRACE and use the left/right arrow keys. The coordinates of points on the graph will be displayed at the bottom of the screen, even though they are not in the current viewing window. Use these coordinates as a guide for selecting a viewing window in which the graph does appear.

EXAMPLE 8 If you graph $y = -2x^3 + 26x^2 + 18x + 50$ in the standard viewing window, you obtain Figure 1–49. A blank screen is no cause for panic; it just means that no part of the graph is in this particular window. In such

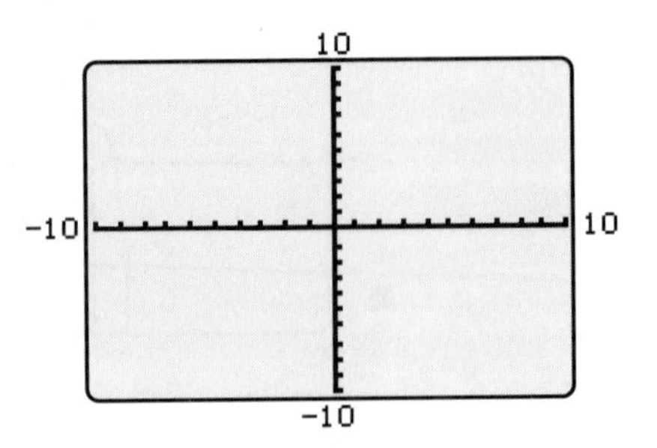

Figure 1-49

cases you can usually find at least one point on the graph by setting $x = 0$ and determining the corresponding y (which is called the **y-intercept** of the graph because the graph touches the y-axis at this point). If $x = 0$ here, then $y = 50$, so the point (0, 50) is on the graph. Enlarging the viewing window to include (0, 50) leads to Figure 1–50.

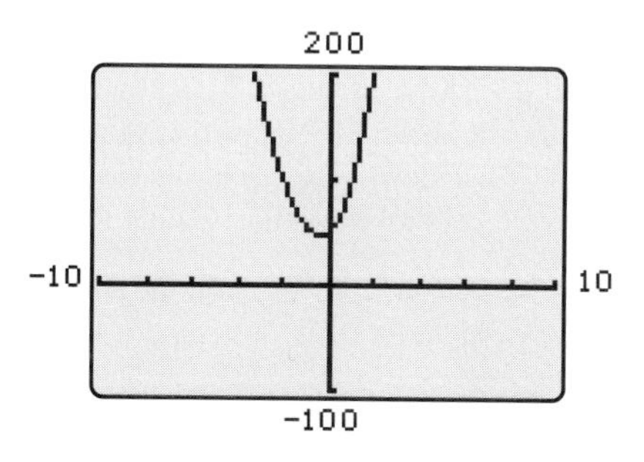

Figure 1-50

GRAPHING EXPLORATION Is the graph in Figure 1–50 complete? If not, find a complete graph. [*Hint:* The graph crosses the x-axis once, has one "peak" and one "valley."] ■

As a general rule, you should follow the directions below when graphing equations.

Graphing Convention ▶

1. **Unless directed otherwise, use a calculator for graphing.**
2. **Complete graphs are required unless a viewing window is specified or the context of a problem indicates that a partial graph is acceptable.**
3. **If the directions say "obtain the graph," "find the graph," or "graph the equation," you need not actually draw the graph on paper. For review purposes, however, it may be helpful to record the viewing window used.**
4. **The directions "sketch the graph" mean "draw the graph on paper, indicating the scale on each axis." This may involve simply copying the display on the calculator screen, or it may require more work if the calculator display is misleading.**

CALCULATOR INVESTIGATIONS 1.4

1. Tic Marks
   (a) Set Xscl = 1 so that adjacent tic marks on the x-axis are one unit apart. Find the largest range of x values such that the tic marks on the x-axis are clearly distinguishable and appear to be equally spaced.
   (b) Do part (a) with y in place of x.
2. Viewing Windows Look in the ZOOM menu on TI, the VIEWS menu on HP-38, or in the RANGE MENU (two keys) on Sharp calculators to find out how many "built-in" viewing windows your calculator has (that is, what viewing windows can be obtained by one or two keystrokes, without setting RANGE (or WINDOW) values manually). Take a look at each one.
3. Screen Size Find out the size of your calculator screen: how many pixels wide and how many high.
4. Square Windows Find a square viewing window on your calculator that has $-10 \le x \le 10$.
5. Trace Graph the two equations that produce the entire circle in Figure 1–38. Press TRACE and move the cursor along the circle. What happens when you use the left and right arrow keys? What happens when you use the up and down arrow keys?
6. Basic Graphs Using the square window found in Investigation 4, graph each of the following equations separately.
   (a) $y = x$ (b) $y = x^2$ (c) $y = x^3$
   (d) $y = |x|$ (e) $y = \sqrt{x}$ (f) $y = \sqrt[3]{x}$

   You should know the shapes of these basic graphs well enough that you can quickly sketch them by hand without using a calculator.

EXERCISES 1.4

Exercises 1–4 are representations of calculator screens. State the approximate coordinates of the points P and Q.

1.

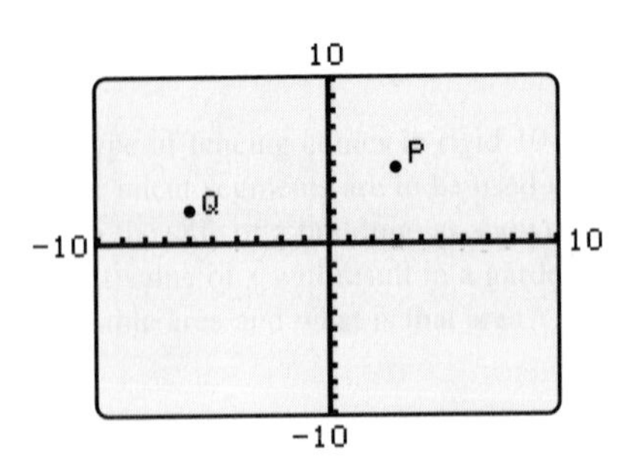

2.

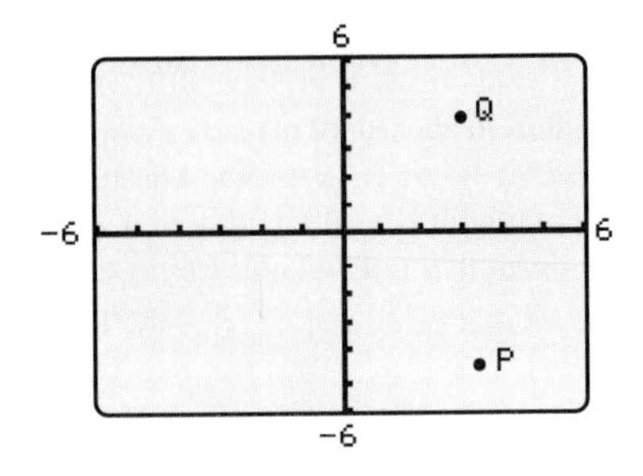

3.

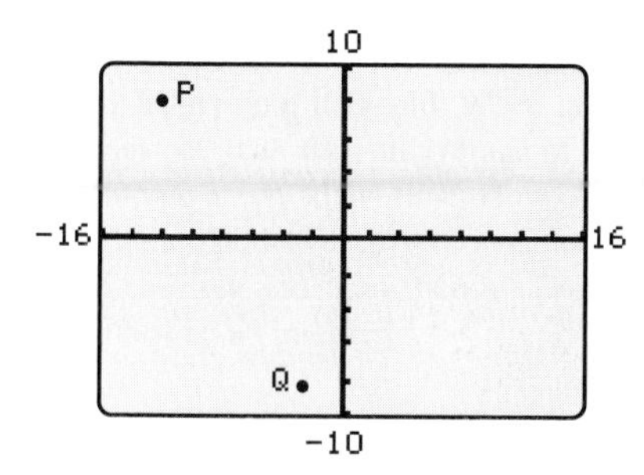

4.

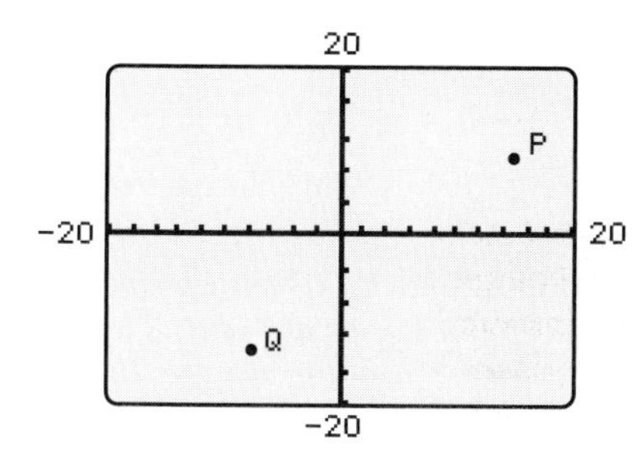

In Exercises 5–10, find at most 6 points on the graph of the equation and plot them. Without using your calculator, use these points to sketch a complete graph of the equation as best you can. Then use the calculator to graph the equation and compare the results.

5. $y = x^2 + x + 1$

6. $y = 2 - x - x^2$

7. $y = (x - 3)^3$

8. $y = x^3 + 2x^2 + 2$

9. $y = x^3 - 3x^2 + x - 1$

10. $y = x^4 - 2x^3 + 2x$

In Exercises 11–16, determine which of the following viewing windows gives the best complete graph of the given equation:

a. $-10 \le x \le 10$; $-10 \le y \le 10$

b. $-5 \le x \le 25$; $0 \le y \le 20$

c. $-10 \le x \le 10$; $-100 \le y \le 100$

d. $-20 \le x \le 15$; $-60 \le y \le 250$

e. None of a, b, c, d gives a complete graph.

11. $y = 18x - 3x^2$

12. $y = 4x^2 + 80x + 350$

13. $y = \frac{1}{3}x^3 - 25x + 100$

14. $y = x^4 + x - 5$

15. $y = x^2 + 50x + 625$

16. $y = .01(x - 15)^4$

17. A toy rocket is shot straight up from ground level and then falls back to earth; wind resistance is negligible. Use your calculator to determine which of the following equations has a graph whose portion above the x-axis provides the most plausible model of the path of the rocket.

(a) $y = .1(x - 3)^3 - .1x^2 + 5$
(b) $y = -x^4 + 16x^3 - 88x^2 + 192x$
(c) $y = -16x^2 + 117x$
(d) $y = .16x^2 - 3.2x + 16$
(e) $y = -(.1x - 3)^6 + 600$

18. Monthly profits at DayGlo Tee Shirt Company appear to be given by the equation

$$y = -.00027(x - 15{,}000)^2 + 60{,}000,$$

where x is the number of shirts sold that month and y is the profit. DayGlo's maximum production capacity is 15,000 shirts per month.

(a) If you plan to graph the profit equation, what range of x values should you use? [*Hint:* You can't make a negative number of shirts.]

(b) The president of DayGlo wants to motivate the sales force (who are all in the profit sharing plan) so he asks you to prepare a graph that shows DayGlo's profits increasing *dramatically* as sales increase. Using the profit equation and the x range from part (a), what viewing window would be suitable?

(c) The City Council is talking about imposing more taxes. The president asks you to prepare a graph showing that DayGlo's profits are essentially flat. Using the profit equation and the x range from part (a), what viewing window would be suitable?

In each of the applied situations in Exercises 19–22, find an appropriate viewing window for the equation (that is, a window that includes all the points relevant to the problem, but does not include large regions that are not relevant to the problem, and has easily readable tic marks on the axes). Explain why you chose this window. See the Hint in Exercise 18(a).

19. A cardiac test measures the concentration y of a dye x seconds after a known amount is injected into a vein near the heart. In a normal heart $y = -.006x^4 + .14x^3 - .053x^2 + 179x$.

20. Beginning in 1905 the deer population in a region of Arizona rapidly increased because of a lack of natural predators. Eventually food resources were depleted to such a degree that the deer population completely died out. In the equation $y = -.125x^5 + 3.125x^4 + 4000$, y is the number of deer in year x, where $x = 0$ corresponds to 1905.

21. A winery can produce x barrels of red wine and y barrels of white wine, where

$$y = \frac{200{,}000 - 50x}{2000 + x}.$$

22. The concentration of a certain medication in the bloodstream at time x hours is approximated by the equation $y = \dfrac{375x}{.1x^3 + 50}$, where y is measured in milligrams per liter. After two days the medication has no effect.

23. **(a)** Confirm the accuracy of the factorization $x^2 - 5x + 6 = (x - 2)(x - 3)$ graphically. [*Hint:* Graph $y = x^2 - 5x + 6$ and $y = (x - 2)(x - 3)$ on the same screen. If the factorization is correct the graphs will be identical (which means that you will see only a single graph on the screen).]
(b) Show graphically that $(x + 5)^2 \neq x^2 + 5$. [*Hint:* Graph $y = (x + 5)^2$ and $y = x^2 + 5^2$ on the same screen. If the graphs are different, then the two expressions cannot be equal.]

True or False *In Exercises 24–28 use the techniques of Exercise 23 to determine graphically whether the given statement is possibly true or definitely false. [We say "possibly true" because two graphs that appear identical on a calculator screen may actually differ by very small amounts or at places not shown in the window.]*

24. $\dfrac{1}{x^2 + 2} = \dfrac{1}{x^2} + \dfrac{1}{2}$ **25.** $\sqrt{x + 9} = \sqrt{x} + 3$

26. $\left(\dfrac{1}{\sqrt{x}} + \dfrac{1}{\sqrt{5}}\right)^2 = \dfrac{1}{x} + \dfrac{1}{5}$

27. $(1 - x)^6 = 1 - 6x + 15x^2 - 20x^3 + 15x^4 - 6x^5 + x^6$

28. $x^5 - 8x^4 + 16x^3 - 5x^2 + 4x - 20 = (x - 2)^2(x - 5)(x^2 + x + 1)$

In Exercises 29–34 use the techniques of Example 6 to graph the equation in a suitable square viewing window.

29. $x^2 + y^2 = 9$ **30.** $y^2 = x + 2$

31. $3x^2 + 2y^2 = 48$

32. $25(x - 5)^2 + 36(y + 4)^2 = 900$

33. $(x - 4)^2 + (y + 2)^2 = 25$

34. $2y^2 - x^2 = 2$

In Exercises 35–40, obtain a complete graph of the equation by trying various viewing windows. List a viewing window that produces this complete graph. [Many correct answers are possible; consider your answer as correct if your window shows all the features in the window given in the answer section.]

35. $y = 7x^3 + 35x + 10$ **36.** $y = x^3 - 5x^2 + 5x - 6$

37. $y = \sqrt{x^2} - x$ **38.** $y = 1/x^2$

39. $y = -.1x^4 + x^3 + x^2 + x + 50$

40. $y = .002x^5 + .06x^4 - .001x^3 + .04x^2 - .2x + 15$

In Exercises 41–44, sketch a complete graph of the equation. None of these graphs contain any vertical line segments, although your calculator may display some. The hints for Exercises 41 and 42 may be helpful.

41. $y = \dfrac{1}{x - 3}$ [*Hint:* If $x = 3$ what can be said about y and what does this imply about the graph?]

42. $y = \dfrac{x}{x - 3}$ [*Hint:* When x is very large in absolute value what is the approximate value of y?]

43. $y = \dfrac{3x + 1}{x - 4}$ **44.** $y = \dfrac{1}{(x - 3)(x + 2)}$

In Exercises 45–48, graph all four equations on the same screen, using a sufficiently large square viewing window, and answer this question: What is the geometric relationship of graphs (b), (c), and (d) to graph (a)?

45. **(a)** $y = x^2$ **(b)** $y = x^2 + 5$
(c) $y = x^2 - 5$ **(d)** $y = x^2 - 2$

46. **(a)** $y = \sqrt{x}$ **(b)** $y = \sqrt{x - 3}$
(c) $y = \sqrt{x + 3}$ **(d)** $y = \sqrt{x - 6}$

47. **(a)** $y = \sqrt{x}$ **(b)** $y = 2\sqrt{x}$
(c) $y = 3\sqrt{x}$ **(d)** $y = \frac{1}{2}\sqrt{x}$

48. **(a)** $y = x^2$ **(b)** $y = -x^2$
(c) $y = -\frac{1}{2}x^2$ **(d)** $y = -2x^2$

In Exercises 49–53, graph the two given equations and the equation $y = x$ on the same screen, using a sufficiently large *square viewing window and answer this question: What is the geometric relationship between graphs (a) and (b)?*

49. **(a)** $y = x^3$ **(b)** $y = \sqrt[3]{x}$

50. **(a)** $y = \frac{1}{2}x^3 - 4$ **(b)** $y = \sqrt[3]{2x + 8}$

51. **(a)** $y = e^x$ **(b)** $y = \ln x$*

52. **(a)** $y = x + 1$ **(b)** $y = x - 1$

53. **(a)** $y = 5x - 15$ **(b)** $y = .2x + 3$

54. Put your calculator in *radian* mode and use the viewing window given by $0 \le x \le 6.28$ and $-2 \le y \le 2$.**

(a) Graph $y = \sin x$

(b) Graph $y = \sin(2x)$

(c) Graph $y = \sin(3x)$

(d) Based on parts (a)–(c), what do you think the graphs of $y = \sin(4x)$, $y = \sin(5x)$, $y = \sin(6x)$, etc., will look like? Use the calculator to verify your answer.

(e) Based on part (d), what do you think the graphs of $y = \sin(50x)$ and $y = \sin(100x)$ will look like? What does a calculator display instead? What might explain the graphs of the calculator?

1.5 SOLVING EQUATIONS NUMERICALLY AND GRAPHICALLY

There are three basic ways to solve equations:

algebraic methods,

numerical approximation, and

graphical approximation.

Algebraic techniques for solving linear and quadratic equations were considered in Section 1.2, where we saw that the quadratic formula provides exact solutions for every second-degree polynomial equation. There are analogous, but more complicated formulas for solving third- and fourth-degree polynomial equations. However, there are no such formulas for the solution of all polynomial equations of degree five or larger.† For these and many other types of equations approximation methods are the only practical alternative. In this section we present some numerical and graphical approximation methods for solving equations, all of which require technology for effective implementation. These methods depend on the connection between equations and graphs, so we begin with that.

In the coordinate plane all points on the x-axis have second coordinate 0. Consequently, when a graph intersects the x-axis, the intersection point has coordinates of the form $(a, 0)$ for some real number a. The number a is called an **x-intercept** of the graph. For example, the graph of $y = x^2 + x - 2$ in Figure 1-51 has x-intercepts at -2 and 1 because it intersects the x-axis at $(-2, 0)$ and $(1, 0)$. To say that $(-2, 0)$ is on the graph of the equation $y = x^2 + x - 2$ means that when -2 is substituted for x, then $y = 0$. In other words, *the x-intercept, -2,*

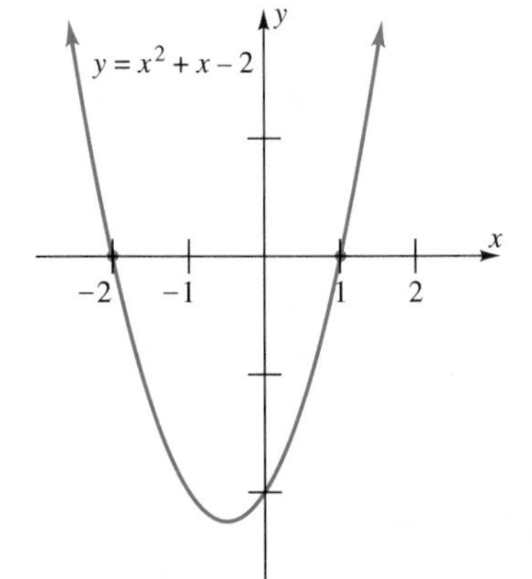

Figure 1-51

*You don't have to know what e^x or $\ln x$ means—just use the calculator keys with these labels to obtain the graphs.

**You don't need to know what radian mode is, or what ''sin'' means. Just use the calculator key with this label.

†This fact was proved by the Norwegian mathematician Niels Henrik Abel in 1824, when he was 21 years old.

of the graph of $y = x^2 + x - 2$ is a solution of the equation $x^2 + x - 2 = 0$. Similarly, the other x-intercept, 1, is also a solution of $x^2 + x - 2 = 0$, as you can easily verify. The same argument works in the general case:

Solutions and Intercepts ▶

The real solutions of a one-variable equation of the form

$$\textbf{expression in } x = 0$$

are the x-intercepts of the graph of the two-variable equation

$$y = \textbf{expression in } x.$$

In other words, to solve an equation, you need only find the x-intercepts of its graph.

Numerical Approximation

Four common numerical methods of approximating the solutions of an equation are

- the bisection method,
- the secant method,
- Newton's method, and
- computer/calculator solver programs.

The secant method requires some information about straight lines and will be considered in Chapter 3. Newton's method is based on concepts from calculus and will not be covered here. The bisection method and solver programs are discussed in the next two examples.

EXAMPLE 1 Use the bisection method to find a solution of the equation $x^5 - 2x^4 - x^3 + 3x + 1 = 0$.

Solution The solutions of the equation are the x-intercepts of the graph of $y = x^5 - 2x^4 - x^3 + 3x + 1$. The following table of values shows that y is positive when $x = 1$ and negative when $x = 2$.

x	0	1	2
$y = x^5 - 2x^4 - x^3 + 3x + 1$	1	2	−1

Even without seeing the graph we know that it must cross the x-axis between 1 and 2 as y changes from positive to negative.* So the equation has a solution

*Here and in the examples below we assume that the graph is an unbroken curve. The justification of this assumption is considered in later chapters.

between 1 and 2. As a first approximation we choose the midpoint of the interval (1, 2), namely, $x = 1.5$. Since the solution (x-intercept) lies between 1 and 2, it must lie with .5 units of 1.5. In other words, the maximum error in our approximation is .5 (half the length of the interval (1, 2)).

Now when $x = 1.5$, then

$$y = 1.5^5 - 2(1.5)^4 - 1.5^3 + 3(1.5) + 1 = -.40625.$$

Since y is positive at $x = 1$ and negative at $x = 1.5$, the graph actually crosses the axis between 1 and 1.5. So our second approximation is the midpoint of the interval (1, 1.5), namely, $x = 1.25$. When $x = 1.25$, then

$$y = 1.25^5 - 2(1.25)^4 - 1.25^3 + 3(1.25) + 1 \approx .96582.$$

Thus y is positive at $x = 1.25$ and negative at $x = 1.5$, so the x-intercept (solution) lies between 1.25 and 1.5 and we use the midpoint of this interval as our third approximation; $x = 1.375$. Every point in the interval (1.25, 1.5) is within .125 units (half the length of the interval) of the midpoint, so .125 is the maximum possible error in this approximation.

This process can be continued indefinitely. At each stage y has opposite signs at the endpoints of an interval and the midpoint of that interval is an approximation of the solution, with a maximum error of half the length of the interval. The midpoint becomes an endpoint of the interval used in the next stage. The results of the first ten iterations of this process are shown in the table below.*

Step	*Interval (c, d)*	*Midpoint m*	*Sign of y when* $x = c$	$x = m$	$x = d$	*Maximum Error*
1	(1, 2)	1.5	+	−	−	.5
2	(1, 1.5)	1.25	+	+	−	.25
3	(1.25, 1.5)	1.375	+	+	−	.125
4	(1.375, 1.5)	1.4375	+	−	−	.0625
5	(1.375, 1.4375)	1.4063	+	+	−	.0313
6	(1.4063, 1.4375)	1.4219	+	+	−	.0156
7	(1.4219, 1.4375)	1.4297	+	−	−	.0078
8	(1.4219, 1.4297)	1.4258	+	+	−	.0039
9	(1.4258, 1.4297)	1.4278	+	−	−	.0020
10	(1.4258, 1.4278)	1.4268				.0010

*The numbers in the table were rounded to four decimal places *after* all calculations were made.

The tenth approximation, $x = 1.4268$, is within .0010 of the actual solution. Greater accuracy could be obtained by further iterations. ■

> **TECHNOLOGY TIP**
> A program for implementing the bisection method is given in the Program Appendix.

The bisection method is based on facts about graphs, but as Example 1 illustrates, you don't need the graph in order to use it. The bisection method can be started as soon as you find two values of x at which y has opposite signs, which guarantees that a solution is between these two values of x. Of course, graphing technology makes it easy to find the approximate location of the x-intercepts (solutions) and some x-values with which to start the process.

Many calculators have a built-in equation solver (that typically uses Newton's method).

EXAMPLE 2 In Example 1 we found a solution of

$$x^5 - 2x^4 - x^3 + 3x + 1 = 0$$

between 1 and 2. Use a solver to find another solution.

> **TECHNOLOGY TIP**
> If your calculator has an equation solver, learn how to use it. Look for SOLVER (on the TI-85 and Sharp 9300 keyboards) or SOLVE (in the TI-82/83 MATH menu or TI-92 ALGEBRA menu or the HP-38 LIB menu).
> The TI-92 solver finds all real solutions simultaneously; the others find one solution at a time. However, you can use POLY on TI-85 or POLYROOT in the POLYNOMIAL submenu of the HP-38 MATH menu to find all solutions of a polynomial equation in one step. In equation mode Casio 9800 can solve second and third degree polynomial equations.

Solution The graph of $y = x^5 - 2x^4 - x^3 + 3x + 1$ shows that there is another x-intercept (solution) between -1 and 0 (this fact can be discovered without the graph by noting that y is negative when $x = -1$ and positive with $x = 0$). We shall use the TI-82 solver; the syntax and procedures for other calculators vary, so check your instruction manual. The TI-82 solver requires you to specify an interval and/or a first approximation (guess) of a solution in that interval. So we use the interval $(-1, 0)$ with $-.5$ as an initial guess. Display "solve(" from the MATH menu and key in

$$\text{solve}\,(\underbrace{x^5 - 2x^4 - x^3 + 3x + 1}_{\textit{equation}},\quad \underbrace{x}_{\textit{variable}},\quad \underbrace{-.5}_{\textit{guess}},\quad \underbrace{\{-1, 0\}}_{\textit{interval}})$$

Press ENTER and the calculator finds the approximate solution $x \approx -.3360526828$. ■

Graphical Approximation

The basic idea of graphical solution approximation is illustrated in the following example.

EXAMPLE 3 Solve the equation $x^4 + x^2 = x^3 + x + 5$ graphically.

Solution We first rewrite the equation so that one side is zero:

$$x^4 - x^3 + x^2 - x - 5 = 0.$$

Now we graph $y = x^4 - x^3 + x^2 - x - 5$ in the standard viewing window:

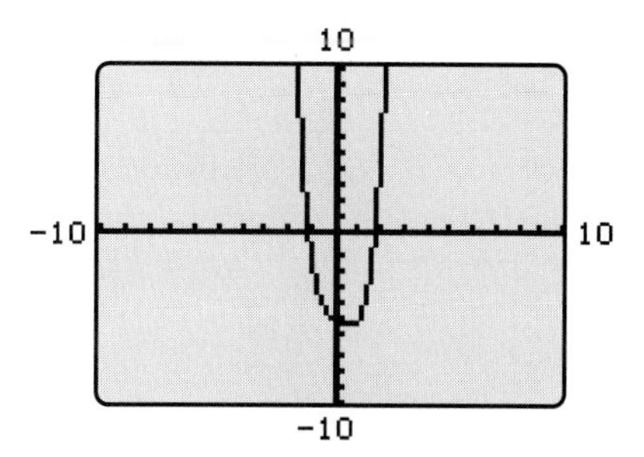

Figure 1-52

Since the tic marks on the x-axis are one unit apart, we see that there is one x-intercept (solution of the equation) between -2 and -1 and another one between 1 and 2. Assuming that Figure 1–52 includes all x-intercepts of the graph, this means that the original equation has exactly two real solutions.*

To get a better approximation of the solution between 1 and 2, we change the viewing window to show only the portion of the graph with $1 \leq x \leq 2$ and $-1 \leq y \leq 1$ (Figure 1–53).** The *X*scl setting is also changed in Figure 1–53 in order to make the tic marks on the x-axis easily readable. The *Y*scl setting is not changed because there is no need to read the tic marks on the y-axis.

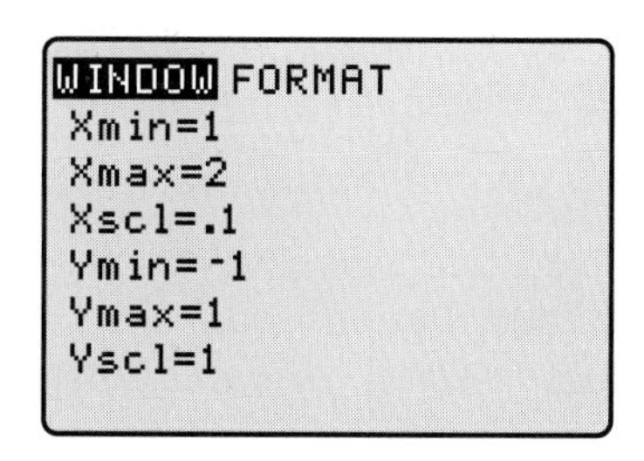

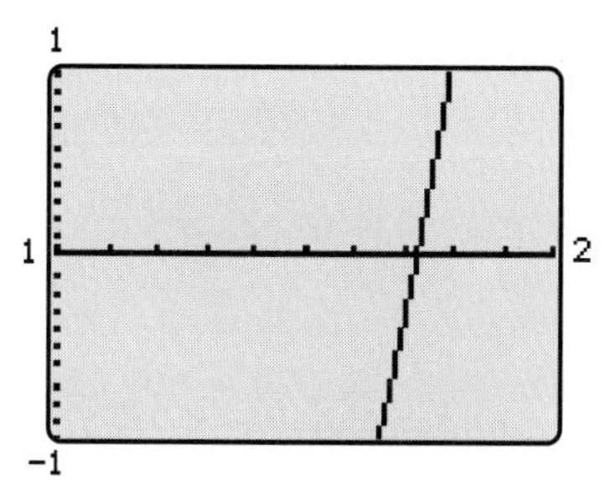

Figure 1-53

*Unless stated otherwise in the examples and exercises of this section, you may assume that the standard viewing window includes all x-intercepts of a graph.

**In a very small viewing window such as this one and those below, the displayed portion of any graph is greatly magnified and consequently looks like a straight line. Since the y-axis is not in the viewing window, the range of y-values is indicated on the edge of the window.

Since the tic marks on the x-axis are now .1 units apart, we see that the solution (x-intercept) lies between 1.7 and 1.8, approximately at 1.72. Greater accuracy can be achieved by changing the viewing window (and the Xscl setting) again:

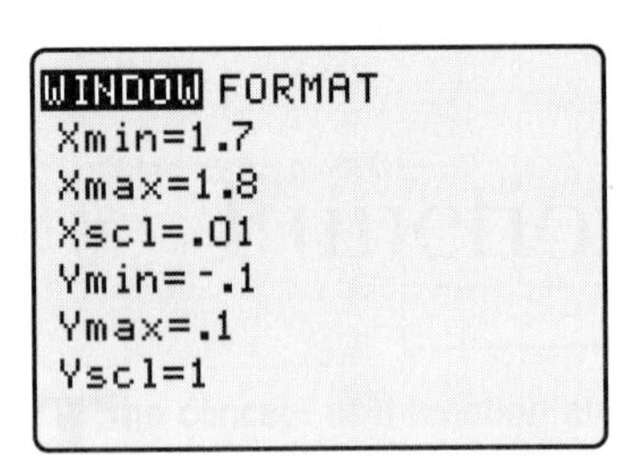

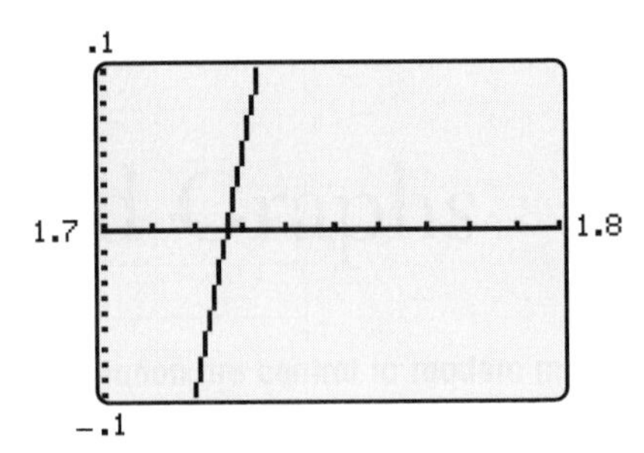

Figure 1-54

Now the tic marks on the x-axis are .01 units apart and the x-intercept (solution) is between 1.72 and 1.73. So any number between 1.72 and 1.73 will approximate the solution with an error of at most .01 (the distance between adjacent tic marks). Figure 1–54 suggests that 1.727 or 1.728 would be reasonable approximations.

Continuing in the same manner with the viewing window in Figure 1–55, which has $1.72 \leq x \leq 1.73$ and tic marks that are .001 units apart, we see that the solution is very close to 1.727 with an error of at most .001. In the viewing window with $1.727 \leq x \leq 1.728$, the graph does not cross the x-axis (try it), so we use the viewing window in Figure 1–56, which has $1.726 \leq x \leq 1.727$ and tic marks .0001 units apart.

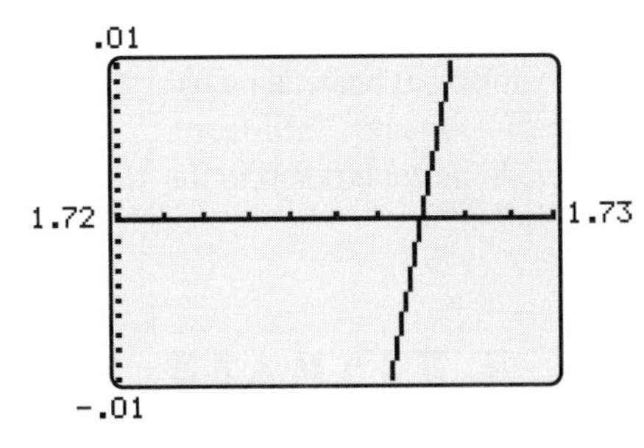

Figure 1-55

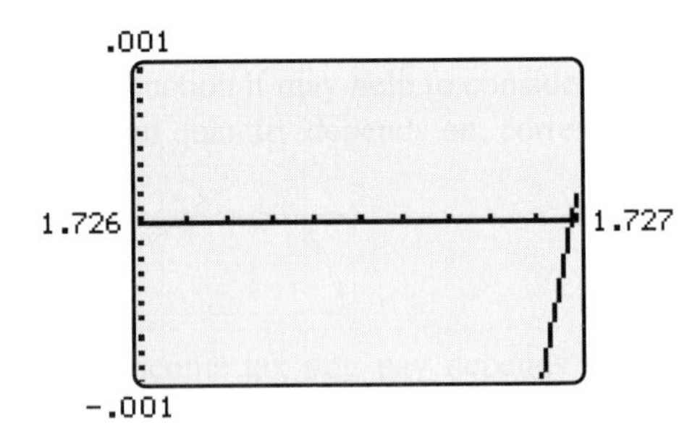

Figure 1-56

Figure 1–56 shows that 1.72698 approximates the solution with a maximum error of .0001. Continuing this process will lead to more accurate approximations, up to the limits of the calculator's precision. Approximations of the other solution of the equation can be obtained in a similar fashion (Exercise 37). ■

The procedure used in Example 3, in which smaller and smaller viewing windows "trap" the x-intercept (solution), is called **(manual) zoom-in.** It may be summarized as follows:

1. Write the equation with 0 on the right side and graph the left side; the graph need not be complete but should include all the x-intercepts. Choose an x-intercept (solution) to approximate.
2. Select a one-unit-long interval that contains the chosen x-intercept and graph the equation over that interval (with Xscl = .1 and $-1 \le y \le 1$). For each graph thereafter use an interval containing the x-intercept that is 1/10 as long as the previous one and decrease the Xscl, Ymin, and Ymax settings to 1/10 of the previous ones.*
3. Repeat step 2 for each of the other x-intercepts (solutions).

Example 3 illustrates the basic method for error estimation when solving equations graphically. When the x-axis has equally spaced tic marks that are k units apart and a graph has an x-intercept (solution) between adjacent tic marks r and s, as in Figure 1–57, then *any* number between r and s is within k units of the x-intercept.

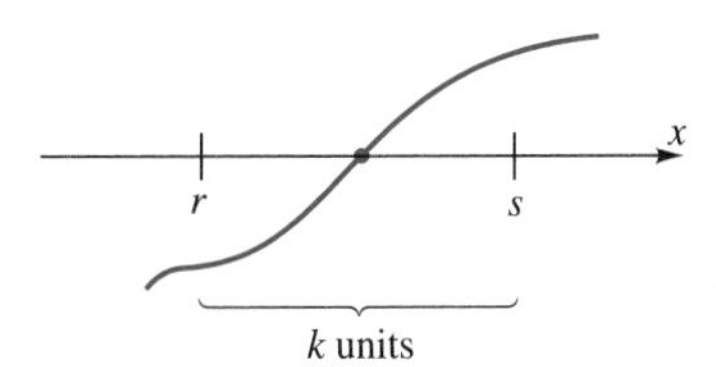

Figure 1-57

So any number between the adjacent tic marks r and s can be used as an approximation of the x-intercept with a maximum error (distance from the actual x-intercept) of at most k units. For instance, if the x-intercept is between adjacent tic marks that are .01 units apart, then the maximum error in using any number between these same tic marks as an approximation is .01. In this book we shall follow this:

Accuracy Convention ▶

Unless stated otherwise, "solve an equation graphically" means approximate the real solutions with a maximum error of at most .01.

*The Yscl need not be changed and it may not be necessary to change the range of y values at every step. However, not changing the y-range may result in a graph that lies so close to the x-axis that its x-intercepts are impossible to read (see Exercise 48).

Automated Graphical Methods

The zoom-in procedure can be automated on most calculators by using either of the following variations:

> **▶ TECHNOLOGY TIP**
> To set the zoom factors, use ZOOM MEMORY SET-FACTORS on TI-82/83, GRAPH ZOOM ZFACT on TI-85, ZOOM SET-FACTORS on TI-92 and HP-38, ZOOM FACTOR on Sharp 9300, and ZOOM FCT on Casio.

Auto-zoom Set each of the zoom factors to 10 (see Tip in the margin). Graph the equation and use the arrow keys to move the cursor as close to the intercept as possible (on Casio use TRACE to do this). Now bring up the ZOOM menu and select ZOOM IN (labeled ZIN on TI-85 and $\times f$ on Casio). The calculator regraphs the equation in a window centered around the x-intercept, in which both axes are reduced to 1/10 of the former lengths. Repeat as often as necessary and use the trace or arrow keys to approximate the x-intercept. Although the calculator automatically changes the Xmin, Xmax, Ymin, and Ymax settings at each stage, the Xscl setting is *not* changed, so after one or two steps no tic marks show on the x-axis. In order to estimate the accuracy of your approximation, you must press RANGE and change the Xscl appropriately.

Box-zoom Graph the equation. Then bring up the ZOOM menu and select BOX (labeled ZBOX on TI-82/83 and ZOOMBOX on TI-92). Following the directions in your instruction manual, draw a small rectangle around the x-intercept (it helps to make it wider than it is high). Press ENTER and the calculator regraphs the equation, using the box you drew as the window. Repeat as often as necessary and use the trace or arrow keys to approximate the x-intercept. Once again, the Xscl setting must be done manually in order to have tic marks visible on the x-axis.

EXAMPLE 4 Use auto-zoom to solve $x^3 + 15x^2 + 70x + 95 = 0$ with a maximum error of .001.

Solution The graph of $y = x^3 + 15x^2 + 70x + 95$ in the standard viewing window (Figure 1–58) is complete (for reasons to be discussed in Chapter 4) and shows that there is a single real solution located between -3 and -2.

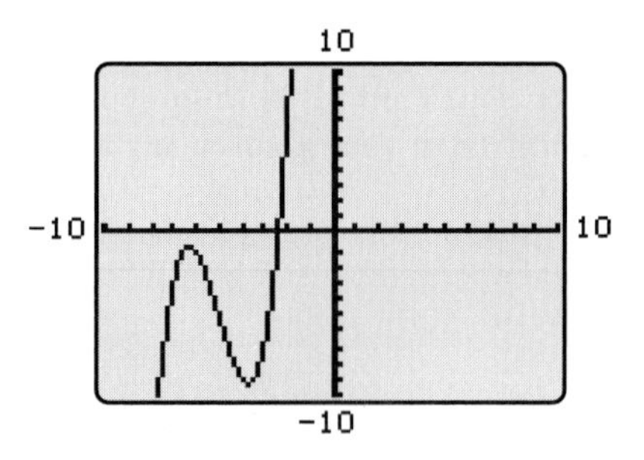

Figure 1-58

Using the ZOOM IN on the ZOOM menu three times in succession, we obtain Figures 1–59, 1–60, and 1–61.

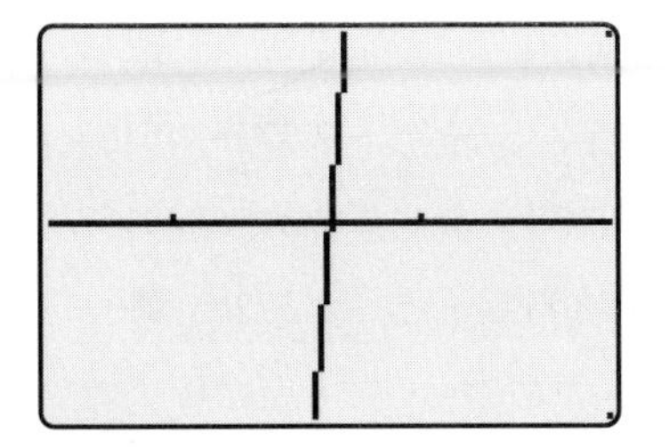

Figure 1-59

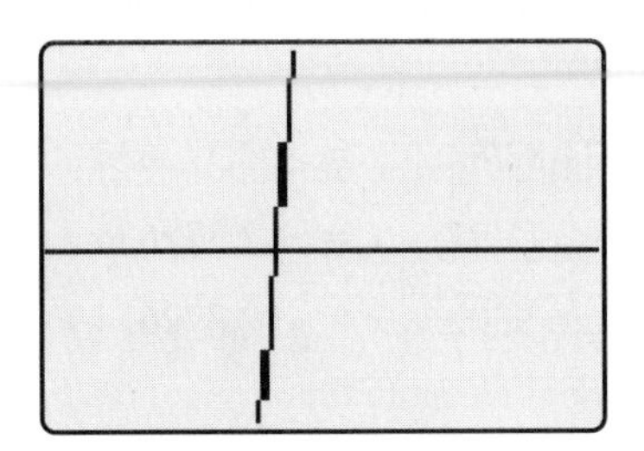

Figure 1-60

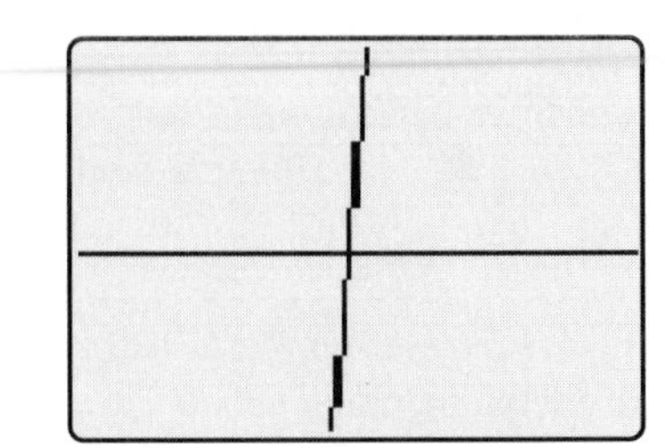

Figure 1-61

Since there are no tic marks on the x-axis, we press RANGE, set Xscl = .001, and regraph. This gives Figure 1–62 in which the tics on the x-axis are .001 units apart (if no tic marks had appeared on the axis we would have made Xscl smaller and graphed again). By using TRACE (Figures 1–63 and 1–64) we see that the intercept is between $x = -2.3727$ and $x = -2.3725$.*

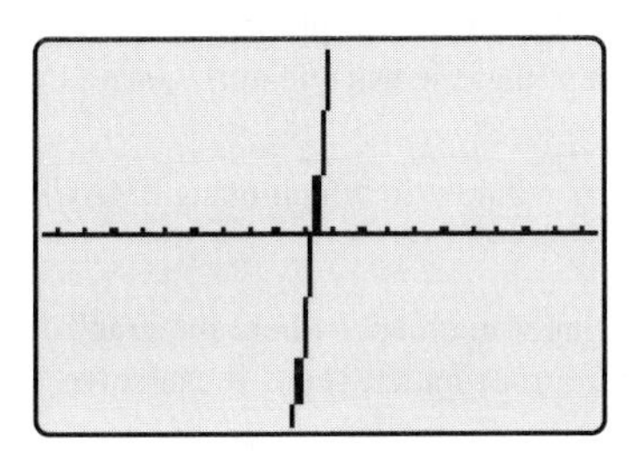

Figure 1-62

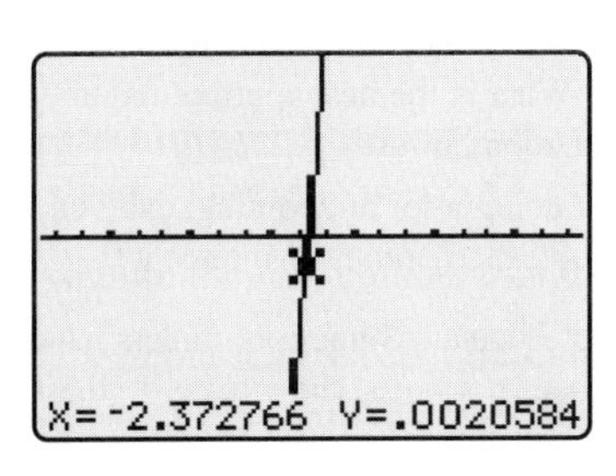

Figure 1-63

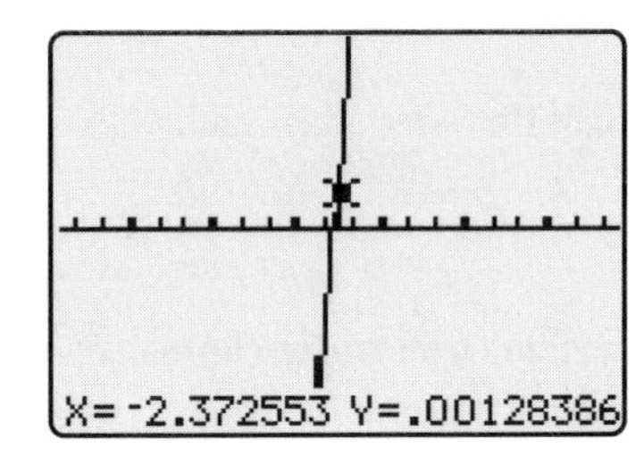

Figure 1-64

▶ TECHNOLOGY TIP

The graphical root finder is labeled ROOT in the TI-82 CALC menu, the MATH submenu of the TI-85 GRAPH menu, and the HP-38 PLOT FCN menu. It is labeled ZERO in the TI-83 CALC menu, the MATH submenu of the TI-92 GRAPH menu, and X-INTERCEPT in the Casio 9800 G-SOLVE menu and Sharp 9300 JUMP menu.

So any number between these two, in particular $x = -2.3726$, is an approximation of the solution with a maximum error of .001 (the distance between two adjacent tic marks on the x-axis). ■

The zoom-in procedure can be skipped entirely on those calculators that have a ''root finder'' (which is a graphical version of an equation solver). This feature, which can approximate an x-intercept with very few keystrokes, typically requires the user to make an initial guess (and possibly to specify an interval containing the x-intercept) by ''marking'' points on the graph. The accuracy of these root finders is generally quite good. If your calculator has one, learn how to use it.

*Alternatively, you can check the RANGE settings and count tic marks to determine the approximate location of the x-intercept. On TI calculators you can also use the arrow keys to approximate the x-intercept.

▶ TECHNOLOGY TIP

A root finder program for TI-81 is in the Program Appendix.

EXAMPLE 5 In Example 4 we saw that $x = -2.3726$ was an approximate solution of the equation $x^3 + 15x^2 + 70x + 95 = 0$ with a maximum error of .001. The root finders on various calculators give the following approximations:

TI-82/83: $x = -2.372635$ TI-85: $x = -2.372634915$

TI-92: $x = -2.372635$ HP-38: $x = -2.3726349153$

Sharp 9300: $x = -2.372634$ Casio 9800: $x = -2.3726$ ■

CALCULATOR INVESTIGATIONS 1.5

1. TRACE and Solutions In theory, you can solve an equation graphically by using TRACE to move along the graph until you land on an x-intercept.
   (a) Try this with the equation $x - 3 = 0$ in the standard viewing window. As you move from left to right on the graph of $y = x - 3$ toward the x-intercept 3, watch the displayed coordinates carefully. Do you land on the point with $x = 3$, $y = 0$? If not, why not? [*Hint:* What is the connection between TRACE and the points the calculator plotted to draw the graph?] What is the best approximation of the x-intercept that the trace feature gives (without zooming-in)?
   (b) Without doing a lot of zooming, can you find a viewing window in which using TRACE lands you exactly on the x-intercept?
2. Graphical Root Finders Some root finders have trouble finding x-intercepts where the graph touches, but does not cross, the x-axis. If your calculator has a root finder, see if it can solve the following equations.
   (a) $\sqrt{x - 2} = 0$ [If asked to specify a starting point or to make an initial guess, use a number other than 2.]
   (b) $x^2 = 0$ [Don't make an initial guess of 0.]

EXERCISES 1.5

In Exercises 1–4, assume the graph of an equation has an x-intercept in the given interval and answer these questions: (a) If you choose a number in this interval as an approximation of the x-intercept, what is the maximum possible error? (b) If you choose the midpoint of the interval, what is the maximum possible error?

1. $[-1.6, -1.5]$
2. $[4.61, 4.65]$
3. $[3.1245, 3.1247]$
4. $[-0.5555, -0.5554]$

In Exercises 5–10, determine graphically the number *of solutions of the equation, but do not solve the equation. You may need a viewing window other than the standard one to find all the x-intercepts.*

5. $x^5 + 5 = 3x^4 + x$
6. $x^3 + 5 = 3x^2 + 24x$
7. $x^7 - 10x^5 + 15x + 10 = 0$
8. $x^5 + 36x + 25 = 13x^3$

9. $x^4 + 500x^2 - 8000x = 16x^3 - 32{,}000$

10. $6x^5 + 80x^3 + 45x^2 + 30 = 45x^4 + 86x$

In Exercises 11–14, use the bisection method to find a solution of the equation in the given interval, with an error of at most .01.

11. $12x^3 - 28x^2 - 7x = 10$; $(2, 3)$

12. $x^5 + x^2 = 7$; $(1, 2)$

13. $x^3 + 4x^2 + 10x + 15 = 0$; $(-3, -2)$

14. $x^3 - 3x^2 - 6x + 9 = 0$; $(1, 2)$

In Exercise 15–18, use graphical approximation to find a solution in the given interval, with an error of at most .01.

15. $x^4 + x - 3 = 0$; $(0, \infty)$

16. $x^3 - 3x^2 + 1 = 0$; $(1, \infty)$

17. $x^5 - 3x^4 - x + 5 = 0$; $(-\infty, 1)$

18. $x^3 - 3x + 1 = 0$; $(-\infty, -1)$

In Exercises 19–30, solve the equation graphically.

19. $x^3 - 3x^2 + 5 = 0$

20. $x^3 + x - 1 = 0$

21. $2x^3 - 4x^2 + x - 3 = 0$

22. $6x^3 - 5x^2 + 3x - 2 = 0$

23. $x^5 - 6x + 6 = 0$

24. $x^3 - 3x^2 + x - 1 = 0$

25. $10x^5 - 3x^2 + x - 6 = 0$

26. $\frac{1}{4}x^4 - x - 4 = 0$

27. $2x - \frac{1}{2}x^2 - \frac{1}{12}x^4 = 0$

28. $\frac{1}{4}x^4 + \frac{1}{3}x^2 + 3x - 1 = 0$

29. $x^6 + x^2 - 2x - 1 = 0$

30. $x^4 + x - 3 = 0$

In Exercise 31–36, use graphical approximation to find a solution in the given interval, with an error of at most .01. [You can avoid any problems with hard to read graphs by using the hints in Exercises 31 and 34.]

31. $\sqrt{x^4 + x^3 - x - 3} = 0$; $(0, \infty)$ [*Hint*: $\sqrt{c} = 0$ exactly when $c = 0$.]

32. $\sqrt{8x^4 - 14x^3 - 9x^2 + 11x - 1} = 0$; $(-\infty, 0)$

33. $\sqrt{\frac{2}{5}x^5 + x^2 - 2x} = 0$; $(0, \infty)$

34. $\dfrac{x^3 - 4x + 1}{x^2 + x - 6} = 0$; $(1, \infty)$ [*Hint*: A fraction c/d is zero exactly when $c = 0$ and $d \neq 0$.]

35. $\dfrac{2x^5 - 10x + 5}{x^3 + x^2 - 12x} = 0$; $(-\infty, 0)$

36. $\dfrac{3x^5 - 15x + 5}{x^7 - 8x^5 + 2x^2 - 5} = 0$; $(1, \infty)$

In Exercises 37–38, find all real solutions of the equation, with an error in x of at most .0001.

37. $x^4 - x^3 + x^2 - x - 5 = 0$ [*Hint*: One solution was found in Example 3.]

38. $3x^3 - 8x^2 + x + 2 = 0$

In Exercises 39–44, find an exact solution of the equation in the given interval. [For example, if the graphical approximation of a solution begins .3333, check to see if 1/3 is the exact solution. Similarly, $\sqrt{2} \approx 1.414$; so if your approximation begins 1.414, check to see if $\sqrt{2}$ is a solution.]

39. $3x^3 - 2x^2 + 3x - 2 = 0$; $(0, 1)$

40. $4x^3 - 3x^2 - 3x - 7 = 0$; $(1, 2)$

41. $12x^4 - x^3 - 12x^2 + 25x - 2 = 0$; $(0, 1)$

42. $8x^5 + 7x^4 - x^3 + 16x - 2 = 0$; $(0, 1)$

43. $4x^4 - 13x^2 + 3 = 0$; $(1, 2)$

44. $x^3 + x^2 - 2x - 2 = 0$; $(1, 2)$

Exercises 45–47 deal with exponential, logarithmic, and trigonometric equations, all of which will be dealt with in detail in later chapters. If you are familiar with these concepts, solve each equation graphically.

45. $10^x - \frac{1}{4}x = 28$

46. $x + \sin\left(\frac{1}{2}x\right) = 4$

47. $\ln x - x^2 + 3 = 0$

48. **(a)** Use the standard viewing window to verify that the graph of $y = x^3 - 3x^2 + 5x - 8$ has an x-intercept between 2 and 3.

(b) Change the range of x values to $2 \leq x \leq 3$ and make the scale marks on the x-axis .1 units apart. Without changing the range of y values, graph the equation again. If you can, locate the x-intercept with a maximum error of .1.

(c) Change the range of x values to $2.3 \leq x \leq 2.4$ and make the scale marks on the x-axis .01 units apart. Without changing the range of y values, graph the equation again. Can you locate the x-intercept with a maximum error of .01?

(d) Now adjust the range of y values appropriately and locate the x-intercept with a maximum error of .01.

1.6 APPLICATIONS OF EQUATIONS

Actual problem situations are usually described verbally. In order to solve such problems you must interpret this verbal information and express it as an equivalent mathematical problem. The following guidelines may be helpful.

Setting Up Applied Problems ▶

1. ***Read*** **the problem carefully and determine what is asked for.**
2. ***Label*** **the unknown quantities by letters (variables) and, if appropriate, draw a picture of the situation.**
3. ***Translate*** **the verbal statements in the problem and the relationships between the known and unknown quantities into mathematical language.**
4. ***Consolidate*** **the mathematical information into an equation in one variable that can be solved or an equation in two variables that can be graphed in order to determine at least one of the unknown quantities.**

Here are some examples of how these guidelines are applied.

EXAMPLE 1 Set up the following problem: The average of two real numbers is 41.125 and their product is 1683. What are the numbers?

Solution *Read:* We are asked for two numbers. *Label:* Call the numbers x and y. *Translate:*

English Language	*Mathematical Language*
Two numbers	x and y
Their average is 41.125	$\frac{x+y}{2} = 41.125$
Their product is 1683.	$xy = 1863$

Consolidate: One technique when you have two unknowns is to express one in terms of the other and use this to obtain an equation in one variable. In this case we can do that by solving the second equation for y:

$$xy = 1683$$

$$y = \frac{1683}{x}$$

and substituting the result in the first equation:

$$\frac{x + y}{2} = 41.125$$

$$\frac{x + \frac{1683}{x}}{2} = 41.125$$

$$x + \frac{1683}{x} = 82.25$$

The solution of this equation is one of the numbers and $1683/x$ is the other. ■

EXAMPLE 2 Set up the following problem: A rectangle is twice as long as it is wide. If it has an area of 24.5 square inches, what are its dimensions?

Solution *Read:* We are asked to find the length and width. *Label:* Let x denote the width and y the length and draw a picture of the situation, as in Figure 1–65. *Translate:* Use the fact that the area of a rectangle is length × width.

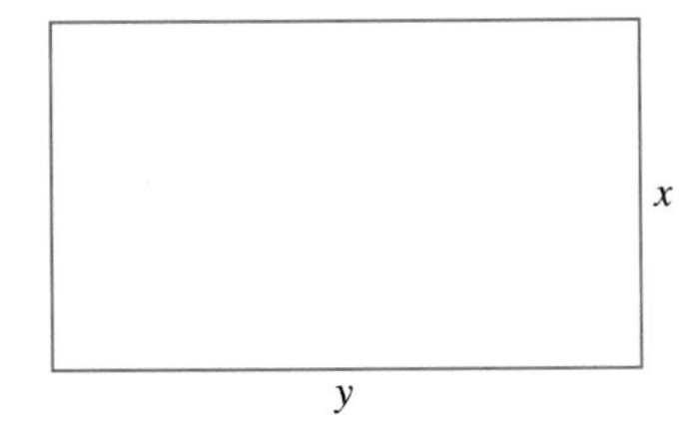

Figure 1-65

English Language	*Mathematical Language*
The width and length of the rectangle	x and y
The length is twice the width.	$y = 2x$
The area is 24.5 square inches.	$xy = 24.5$

Consolidate: Substitute $y = 2x$ in the area equation:

$$xy = 24.5$$

$$x(2x) = 24.5$$

So the equation to be solved is $2x^2 = 24.5$. ■

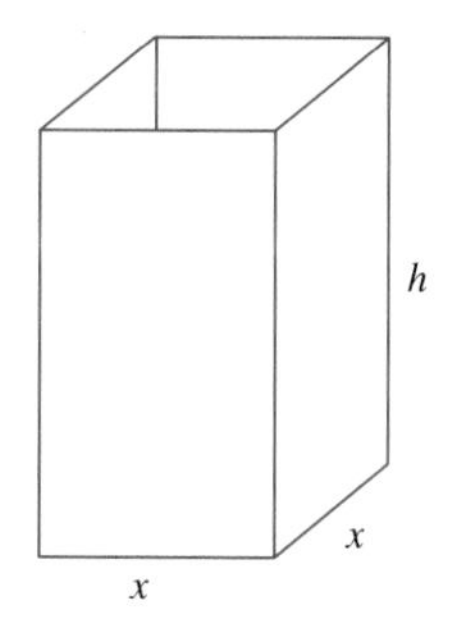

Figure 1-66

EXAMPLE 3 Set up this problem: A rectangular box with a square base and no top is to have a volume of 30,000 cubic cm. If the surface area of the box is 6000 square cm, what are its dimensions?

Solution *Read:* We must find the length, width, and height of the box. *Label:* Let x denote the length. Since the base is square, the length and width are the same. Let h denote the height, as in Figure 1–66. *Translate:* Recall that the volume of a box is given by the product length × width × height and that the surface area is the sum of the area of the base and the area of the four sides of the box. Then we have these translations:

English Language	*Mathematical Language*
The length, width, and height	x, x, and h
The volume is 30,000 cu cm.	$x^2h = 30{,}000$
The surface area is 6000 sq cm.	$x^2 + 4xh = 6000$

Consolidate: We have two equations in two variables, so we solve the first equation for h

$$h = \frac{30{,}000}{x^2}$$

and substitute this result in the second equation:

$$x^2 + 4x\left(\frac{30{,}000}{x^2}\right) = 6000$$

$$x^2 + \frac{120{,}000}{x} = 6000$$

The solution of this last equation will provide the solution of the problem. ■

Once you are comfortable with the setup process illustrated in the preceding examples, you can often do it mentally instead of writing out translation tables and making obvious substitutions. For instance, in Example 2 most people would simply say, "Let x be the width; then $2x$ is the length and the area equation is $2x^2 = 24.5$." Similarly, in the examples below the setup process will be simplified whenever this won't cause any difficulty.

Of course, setting up a problem is only half the job. You must then solve the equation or read the graph you have constructed. When you get an answer, it is extremely important that you

> **Interpret your answers in terms of the original problem. Do they make sense? Do they satisfy the required conditions?**

In particular, an equation may have several solutions, but not all of them may make sense in the context of the problem (for instance, distance can't be negative, the number of people in a room cannot be a proper fraction, etc.).

Applications

We begin with some problems involving interest. Recall that 8% means .08 and that "8% *of* 227" means ".08 *times* 227," that is, $.08(227) = 18.16$. The basic rule of annual simple interest is:

$$\textbf{Interest} = \textbf{rate} \times \textbf{amount.}$$

EXAMPLE 4 A high-risk stock pays dividends at a rate of 12% per year, while a savings account pays 6% interest per year. How much of a \$9000 investment should be put in the stock and how much in the savings account in order to obtain a return of 8% per year on the total investment?

Solution *Label*: Let x be the amount invested in stock. Then the rest of the \$9000, namely, $(9000 - x)$ dollars, goes in the savings account. *Translate*: We want the total return on \$9000 to be 8%, so we have:

$$\begin{pmatrix}\text{Return on } x \text{ dollars}\\ \text{of stock at } 12\%\end{pmatrix} + \begin{pmatrix}\text{Return on } (9000 - x)\\ \text{dollars of savings at } 6\%\end{pmatrix} = 8\% \text{ of } \$9000$$

$$(12\% \text{ of } x \text{ dollars}) + (6\% \text{ of } (9000 - x) \text{ dollars}) = 8\% \text{ of } \$9000$$

$$\begin{aligned}
.12x + .06(9000 - x) &= .08(9000)\\
.12x + .06(9000) - .06x &= .08(9000)\\
.12x + 540 - .06x &= 720\\
.12x - .06x &= 720 - 540\\
.06x &= 180\\
x &= \frac{180}{.06} = 3000.
\end{aligned}$$

Therefore, \$3000 should be invested in stock and $(9000 - 3000) = \$6000$ in the savings account. If this is done, the total return will be 12% of \$3000 (\$360) plus 6% of \$6000 (\$360), a total of \$720, which is precisely 8% of \$9000. ■

Read

EXAMPLE 5 A car radiator contains 12 quarts of fluid, 20% of which is antifreeze. How much fluid should be drained and replaced with pure antifreeze in order that the resulting mixture be 50% antifreeze?

Solution Let x be the number of quarts of fluid to be replaced by pure antifreeze.* When x quarts are drained, there are $12 - x$ quarts of fluid left in the radiator, 20% of which is antifreeze. So we have:

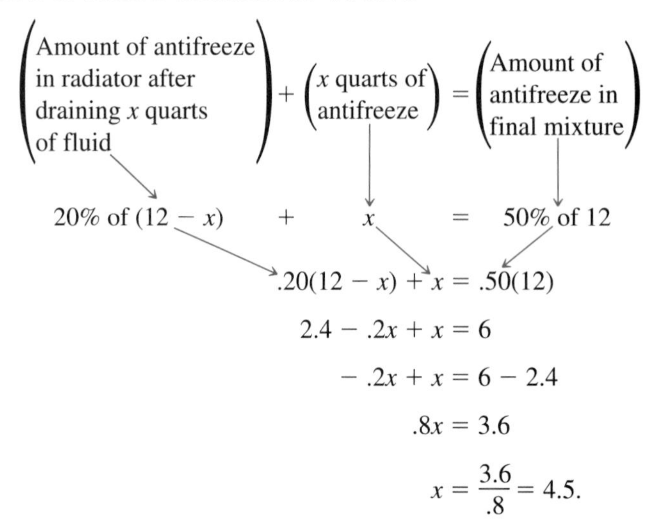

$$.20(12 - x) + x = .50(12)$$

$$2.4 - .2x + x = 6$$

$$-.2x + x = 6 - 2.4$$

$$.8x = 3.6$$

$$x = \frac{3.6}{.8} = 4.5.$$

Therefore, 4.5 quarts should be drained and replaced with pure antifreeze. ■

The basic formula for problems involving distance and a uniform rate of speed is:

Distance = rate × time.

For instance, if you drive at a rate of 55 mph for 2 hours, you travel a distance of $55 \cdot 2 = 110$ miles.

EXAMPLE 6 A pilot wants to make an 840-mile trip from Cleveland to Peoria and back in 5 hours flying time. There will be a headwind of 30 mph going to Peoria and it is estimated that there will be a 40-mph tailwind returning to Cleveland. At what constant air speed should the plane be flown?

Solution Let x be the engine speed of the plane. On the trip to Peoria, a distance of 420 miles, the actual speed will be $x - 30$ because of the headwind. Since rate · time = distance, the time to Peoria will be $\frac{\text{distance}}{\text{rate}} = \frac{420}{x - 30}$. On the return trip the actual speed will be $x + 40$ because of the tailwind and the time

*Hereafter we omit the headings Label, Translate, etc.

will be $\dfrac{\text{distance}}{\text{rate}} = \dfrac{420}{x + 40}$. Therefore:

$$5 = \begin{pmatrix}\text{Time from Cleveland} \\ \text{to Peoria}\end{pmatrix} + \begin{pmatrix}\text{Time from Peoria} \\ \text{to Cleveland}\end{pmatrix}$$

$$5 = \frac{420}{x - 30} + \frac{420}{x + 40}$$

Multiplying both sides by the common denominator $(x - 30)(x + 40)$ and simplifying, we have:

$$5(x - 30)(x + 40) = \frac{420}{x - 30} \cdot (x - 30)(x + 40) + \frac{420}{x + 40} \cdot (x - 30)(x + 40)$$

$$5(x - 30)(x + 40) = 420(x + 40) + 420(x - 30)$$

$$(x - 30)(x + 40) = 84(x + 40) + 84(x - 30)$$

$$x^2 + 10x - 1200 = 84x + 3360 + 84x - 2520$$

$$x^2 - 158x - 2040 = 0$$

$$(x - 170)(x + 12) = 0$$

$$x - 170 = 0 \qquad \text{or} \qquad x + 12 = 0$$

$$x = 170 \qquad\qquad x = -12.$$

Obviously, the negative solution doesn't apply. Since we multiplied both sides by a quantity involving the variable, we must check that 170 actually is a solution of the original equation. It is; so the plane should be flown at a speed of 170 mph. ■

The preceding examples used only algebraic models (equations in one variable). Sometimes a diagram (geometric model) is helpful in visualizing the situation and setting up an appropriate equation.

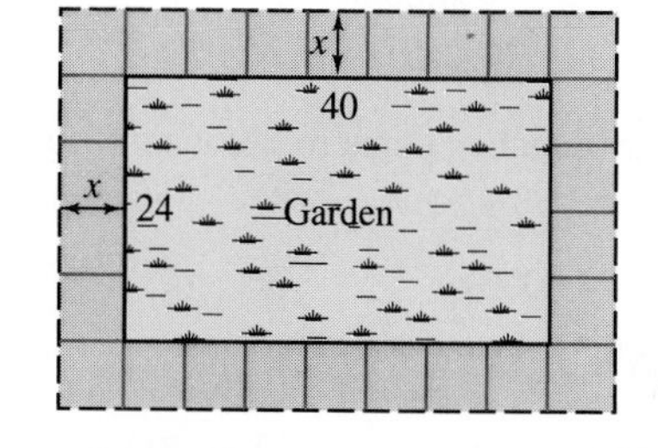

Figure 1-67

EXAMPLE 7 A landscaper wants to put a cement walk of uniform width around a rectangular garden that measures 24 by 40 feet. She has enough cement to cover 660 square feet. How wide should the walk be in order to use up all the cement?

Solution Let x denote the width of the walk (in feet) and draw a picture of the situation (Figure 1–67).

The length of the outer rectangle is $40 + 2x$ (the garden length plus walks on each end) and its width is $24 + 2x$.

$$\begin{pmatrix}\text{Area of outer}\\ \text{rectangle}\end{pmatrix} - \begin{pmatrix}\text{Area of}\\ \text{garden}\end{pmatrix} = \text{Area of walk}$$

$$\text{length} \cdot \text{width} - \text{length} \cdot \text{width} = 660$$

$$(40 + 2x)(24 + 2x) - 40 \cdot 24 = 660$$

$$960 + 128x + 4x^2 - 960 = 660$$

$$4x^2 + 128x - 660 = 0$$

Dividing both sides by 4 and applying the quadratic formula yields

$$x^2 + 32x - 165 = 0$$

$$x = \frac{-32 \pm \sqrt{(32)^2 - 4 \cdot 1 \cdot (-165)}}{2 \cdot 1}$$

$$x = \frac{-32 \pm \sqrt{1684}}{2} \approx \begin{cases} 4.5183 \\ \text{or} \\ -36.5183 \end{cases}$$

Only the positive solution makes sense in the context of this problem. The walk should be approximately 4.5 feet wide. ■

EXAMPLE 8 A rectangular box with a square base and no top is to have a volume of 30,000 cubic cm. If the surface area of the box is 6000 square cm and the box is required to be higher than it is wide, what are its dimensions?

Solution This is essentially Example 3 with an extra condition (the box must be higher than it is wide). If the length, width, and height of the box are x, x, and h as shown in Figure 1–68, then as we saw in Example 3,

$$x^2 + \frac{120{,}000}{x} = 6000 \qquad \text{and} \qquad h = \frac{30{,}000}{x^2}.$$

Multiplying both sides of the first equation by x (which is nonzero because it is a length), we obtain

$$x^3 + 120{,}000 = 6000x$$

$$x^3 - 6000x + 120{,}000 = 0.$$

This equation can be solved graphically. The graph of $y = x^3 - 6000x + 120{,}000$ in Figure 1–69 has one negative x-intercept (solution) and two positive ones.

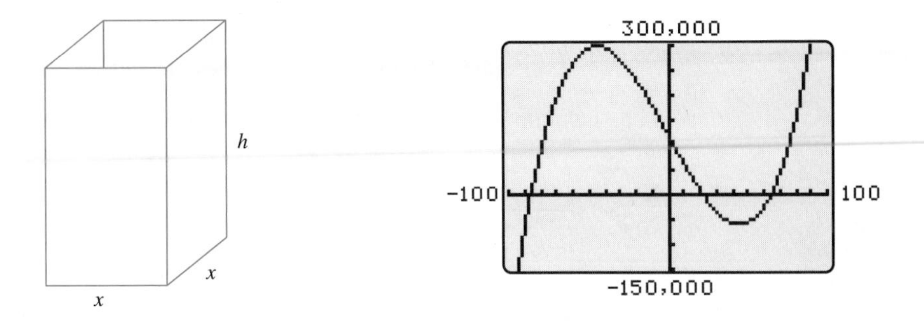

Figure 1-68 **Figure 1-69**

The negative solution does not apply here. The tic marks on the x-axis in Figure 1–69 are ten units apart, so one positive solution is near $x = 20$ and the other between $x = 60$ and $x = 70$. If $60 < x < 70$, however, then the height $h = 30{,}000/x^2$ will be smaller than the width x (for instance, if $x = 64$, then $h = 30{,}000/64^2 \approx 7.3$). So the solution near $x = 20$ is the only one that applies to this situation.

GRAPHING EXPLORATION Use zoom-in or a graphical root finder to obtain an approximation of this solution. Then find the corresponding value of h. ■

EXAMPLE 9 A box (with no top) of volume 1000 cubic inches is to be made from a 22 by 30 inch sheet of cardboard by cutting squares of equal size from each corner and bending up the flaps, as shown in Figure 1–70. If the length, width, and height of the box are each to be less than 18 inches, what size square should be cut from each corner?

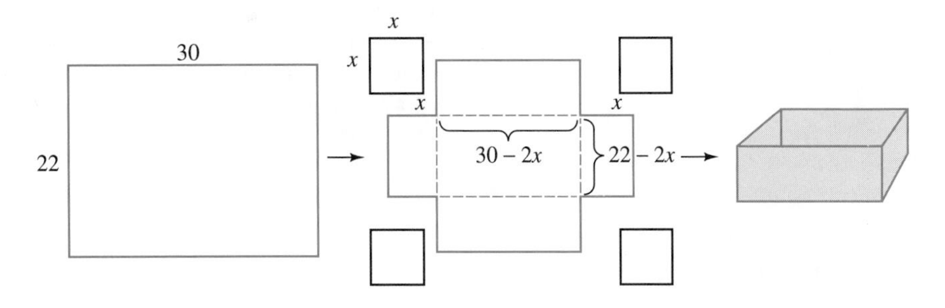

Figure 1-70

Solution Let x denote the length of the side of the square to be cut from each corner. The dashed rectangle in Figure 1–70 is the bottom of the box. Its length is

$30 - 2x$ as shown in the figure. Similarly, the width of the box will be $22 - 2x$ and its height will be x inches. Therefore,

$$\text{length} \times \text{width} \times \text{height} = \text{volume of box}$$

$$(30 - 2x) \cdot (22 - 2x) \cdot x = 1000$$

$$(660 - 104x + 4x^2)x = 1000$$

$$4x^3 - 104x^2 + 660x - 1000 = 0$$

Since the cardboard is 22 inches wide, x must be less than 11 (otherwise you can't cut out two squares of length x). Since x is a length, it is positive. So we need only find solutions of the equation between 0 and 11, which we shall do graphically. The viewing window in Figure 1–71 shows that the graph of $y = 4x^3 - 104x^2 + 660x - 1000$ has x-intercepts between 2 and 3 and between 6 and 7, roughly at 2.2 and 6.4.

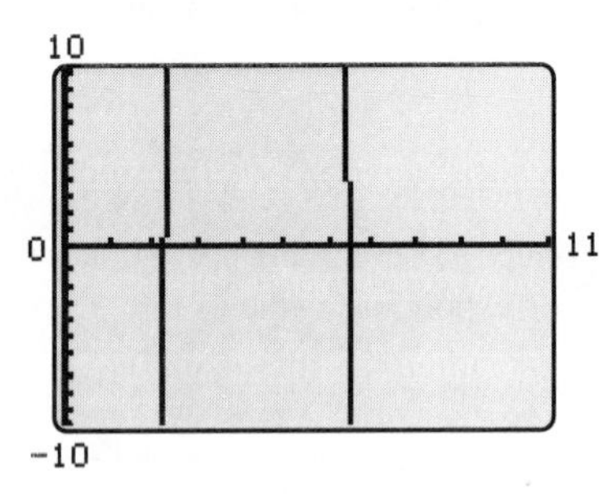

Figure 1-71

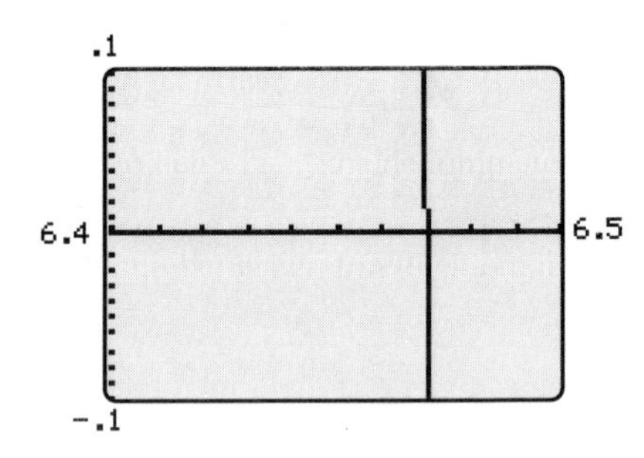

Figure 1-72

When $2 < x < 3$, however, the length $30 - 2x$ of the box will be more than 18 inches (for instance, when $x = 2.2$, the length is $30 - 2(2.2) = 25.6$). Verify that the solution between 6 and 7 will result in a box with each dimension less than 18 inches. Using zoom-in we obtain Figure 1–72 and conclude that 6.47 approximates the solution with an error of at most .01. Therefore a 6.47 by 6.47 inch square should be cut from each corner, resulting in a box that measures 17.06 by 9.06 by 6.47 inches. The actual volume of this box is approximately 1000.03 cubic inches. ■

EXERCISES 1.6

In Exercises 1–4, a problem situation is given.

(a) *Decide what is being asked for and label the unknown quantities.*

(b) *Translate the verbal statements in the problem and the relationships between the known and unknown quantities into mathematical language, using a table as in Examples 1–3. The table is provided in Exercises 1and 2. You need not find an equation to be solved.*

1. The sum of two numbers is 15 and the difference of their squares is 5. What are the numbers?

English Language	*Mathematical Language*
The two numbers	
Their sum is 15.	
The difference of their squares is 5.	

2. The sum of the squares of two consecutive integers is 4513. What are the integers?

English Language	*Mathematical Language*
The two integers	
The integers are consecutive.	
The sum of their squares is 4513.	

3. A rectangle has perimeter of 45 cm and an area of 112.5 sq cm. What are its dimensions?

4. A triangle has area 96 sq in. and its height is two thirds of its base. What are the base and height of the triangle?

In Exercises 5–8, set up the problem by labeling the unknowns, translating the given information into mathematical language, and finding an equation that will produce the solution to the problem. You need not solve this equation.

5. A worker gets an 8% pay raise and now makes $1600 per month. What was the worker's old salary?

6. A merchant has 5 pounds of mixed nuts that cost $30. He wants to add peanuts that cost $1.50 per pound and cashews that cost $4.50 per pound to obtain 50 pounds of a mixture that costs $2.90 per pound. How many pounds of peanuts are needed?

7. The diameter of a circle is 16 cm. By what amount must the radius be decreased in order to decrease the area by 48π square centimeters?

8. A corner lot has dimensions 25 by 40 yards. The city plans to take a strip of uniform width along the two sides bordering the streets in order to widen these roads. How wide should the strip be if the remainder of the lot is to have an area of 844 square yards?

In the remaining exercises, solve the applied problem.

9. You have already invested $550 in a stock with an annual return of 11%. How much of an additional $1100 should be invested at 12% and how much at 6% so that the total return on the entire $1650 is 9%?

10. If you borrow $500 from a credit union at 12% annual interest and $250 from a bank at 18% annual interest, what is the *effective annual interest rate* (that is, what single rate of interest on $750 would result in the same total amount of interest)?

11. A radiator contains 8 quarts of fluid, 40% of which is antifreeze. How much fluid should be drained and replaced with pure antifreeze so that the new mixture is 60% antifreeze?

12. A radiator contains 10 quarts of fluid, 30% of which is antifreeze. How much fluid should be drained and replaced with pure antifreeze so that the new mixture is 40% antifreeze?

13. Two cars leave a gas station at the same time, one traveling north and the other south. The northbound car travels at 50 mph. After 3 hours the cars are 345 miles apart. How fast is the southbound car traveling?

14. An airplane flew with the wind for 2.5 hours and returned the same distance against the wind in 3.5 hours. If the cruising speed of the plane was a constant 360 mph in air, how fast was the wind blowing? [*Hint*: If the wind speed is r miles per hour, then the plane travels at $(360 + r)$ mph with the wind and at $(360 - r)$ mph against the wind.]

15. The average of two real numbers is 41.125 and their product is 1683. What are the numbers? [*Hint:* See Example 1.]

16. A rectangle is twice as long as it is wide. If it has an area of 24.5 inches, what are its dimensions? [*Hint:* See Example 2.]

17. A 13-foot-long ladder leans on a wall. The bottom of the ladder is 5 feet from the wall. If the bottom is pulled out 3 feet farther from the wall, how far does the top of the ladder move down the wall? [*Hint*: The ladder, ground, and wall form a right triangle. Draw pictures of this triangle before and after the ladder is moved. Use the Pythagorean Theorem to set up an equation.]

18. A rectangular theater seats 1620 people. If each row had six more seats in it, the number of rows would be reduced by nine. How many seats would *then* be in each row?

19. Red Riding Hood drives the 432 miles to Grandmother's house in 1 hour less than it takes the Wolf to drive the same route. Her average speed is 6 mph faster than the Wolf's average speed. How fast does each drive?

20. To get to work Sam jogs 3 kilometers to the train, then rides the remaining 5 kilometers. If the train goes 14 kilometers per hour faster than Sam's constant rate of jogging and the entire trip takes 45 minutes, how fast does Sam jog?

Background for Exercises 21–24: *If an object is thrown upward, dropped, or thrown downward and travels in a vertical line subject only to gravity (with wind resistance ignored), then the height h of the object above the ground (in feet) after t seconds is given by:*

$$h = -16t^2 + v_0t + h_0$$

where h_0 is the initial height of the object at starting time $t = 0$ and v_0 is the initial velocity (speed) of the object at time $t = 0$. The value of v_0 is taken as positive if the object starts moving upward at time $t = 0$ and negative if the object starts moving downward at $t = 0$. An object that is dropped (rather than thrown downward) has initial velocity $v_0 = 0$.

21. How long does it take an object to reach the ground if
- **(a)** it is dropped from the top of a 640-foot-high building?
- **(b)** it is thrown downward from the top of the same building, with an initial velocity of 52 feet per second?

22. You are standing on a cliff 200 feet high. How long will it take a rock to reach the ground if
- **(a)** you drop it?
- **(b)** you throw it downward at an initial velocity of 40 feet per second?
- **(c)** How far does the rock fall in 2 seconds if you throw it downward with an initial velocity of 40 feet per second?

23. A rocket is fired straight up from ground level with an initial velocity of 800 feet per second.
- **(a)** How long does it take the rocket to rise 3200 feet?
- **(b)** When will the rocket hit the ground?

24. A rocket loaded with fireworks is to be shot vertically upward from ground level with an initial velocity of 200 feet per second. When the rocket reaches a height of 400 feet on its upward trip the fireworks will be detonated. How many seconds after lift-off will this take place?

25. The dimensions of a rectangular box are consecutive integers. If the box has volume 13,800 cu cm, what are its dimensions?

26. Find a real number that exceeds its cube by 2.

27. The surface area S of the right circular cone in Figure 1–73 is given by $S = \pi r\sqrt{r^2 + h^2}$. What radius should be used to produce a cone of height 5 inches and surface area 100 square inches?

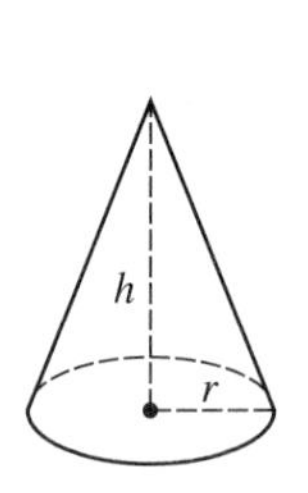

Figure 1-73

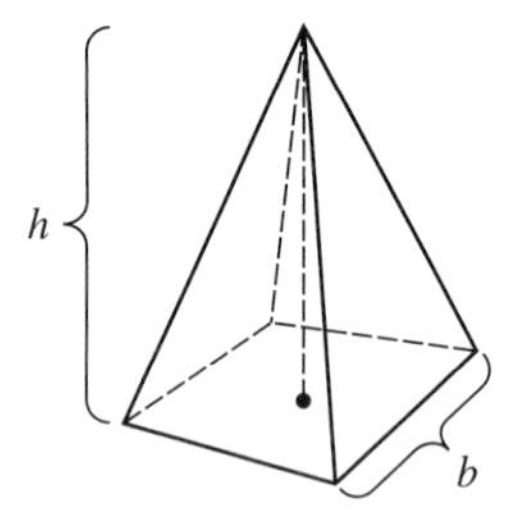

Figure 1-74

28. The surface area of the right square pyramid in Figure 1–74 is given by $S = b\sqrt{b^2 + 4h^2}$. If the pyramid has height 10 feet and surface area 100 square feet, what is the length of a side b of its base?

29. Suppose that the open-top box being made from a sheet of cardboard in Example 9 is required to have at least one of its dimensions *greater* than 18 inches. What size square should be cut from each corner?

30. A homemade loaf of bread turns out to be a perfect cube. Five slices of bread, each .6 in. thick, are cut from one end of the loaf. The remainder of the loaf now has a volume of 235 cu in. What were the dimensions of the original loaf?

31. A rectangular bin with an open top and volume of 38.72 cubic feet is to be built. The length of its base must be twice the width and the bin must be at least 3 feet high. Material for the base of the bin costs \$12 per square foot and material for the sides costs \$8 per square foot. If it costs \$538.56 to build the bin, what are its dimensions?

32. One corner of an 8.5 by 11 inch piece of paper is folded over to the opposite side, as shown in Figure 1–75. If the area of the shaded triangle is 6 square inches, what is x?

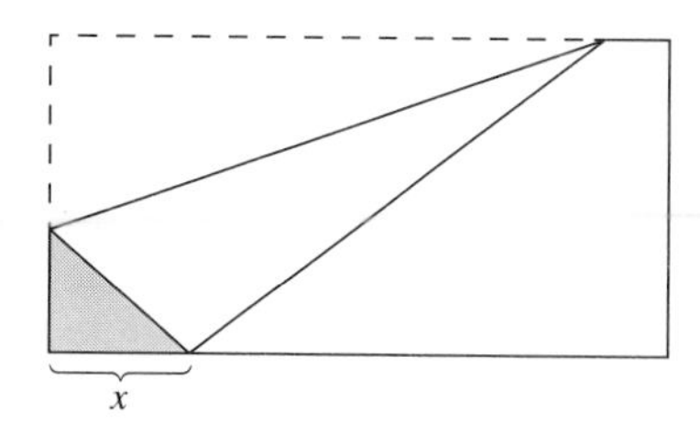

Figure 1-75

1.7 OPTIMIZATION APPLICATIONS

Many real-life situations require you to find the largest or smallest quantity satisfying certain conditions. For instance, automotive engineers want to design engines with maximum fuel efficiency. Similarly, a cereal manufacturer who needs a box of volume 300 cubic inches might want to know the dimensions of the box that requires the least amount of cardboard (and hence is cheapest). The exact solutions of such minimum/maximum problems require calculus. However, graphing technology can provide very accurate approximate solutions as we shall see below. To explain the technique needed to solve such problems, we begin with a purely mathematical example.

EXAMPLE 1 The graph of $y = -x^4 + x^3 + 2x^2 - 1$ is shown in Figure 1–76.

(a) Find the approximate coordinates of the highest point on the graph.

(b) Estimate the accuracy of this approximation.

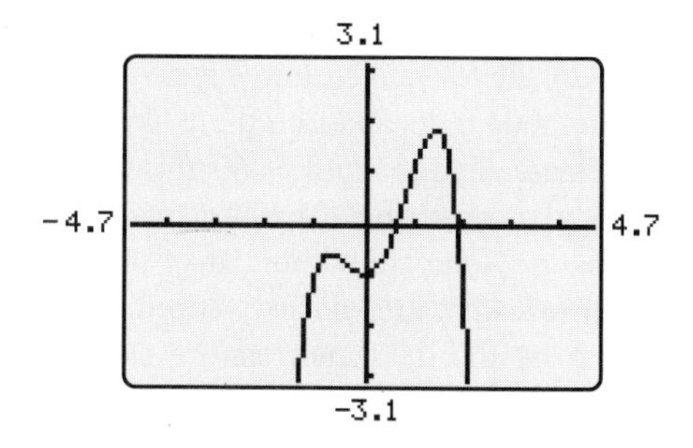

Figure 1-76

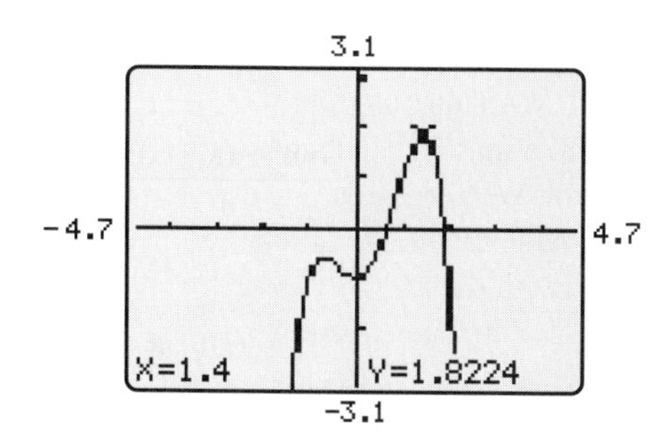

Figure 1-77

Solution **(a)** The highest point on the graph is the one with the largest y-coordinate. You can get a first estimate by using TRACE to move along the curve until you find the largest y-coordinate. As shown in Figure 1–77, this occurs at (1.4, 1.8224). This is not necessarily the highest point, but only the highest one that the calculator actually plotted when drawing the graph in this particular window. We can get a better approximation by zooming in near the highest point (by manual zoom, auto-zoom, or box-zoom) to obtain Figure 1–78. Using TRACE then gives (1.4426, 1.8334) as an approximation of the highest point (Figure 1–79).

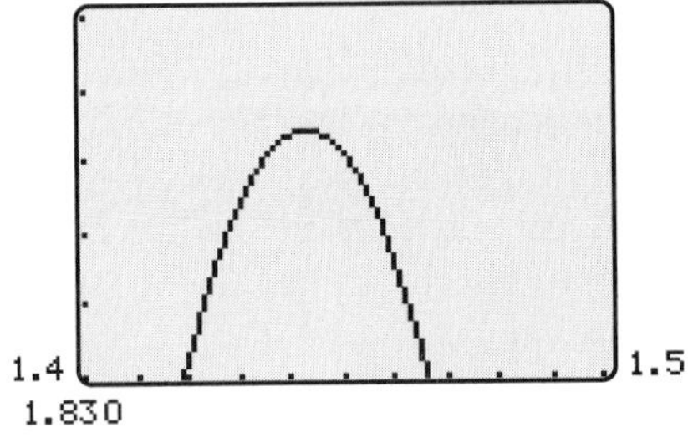
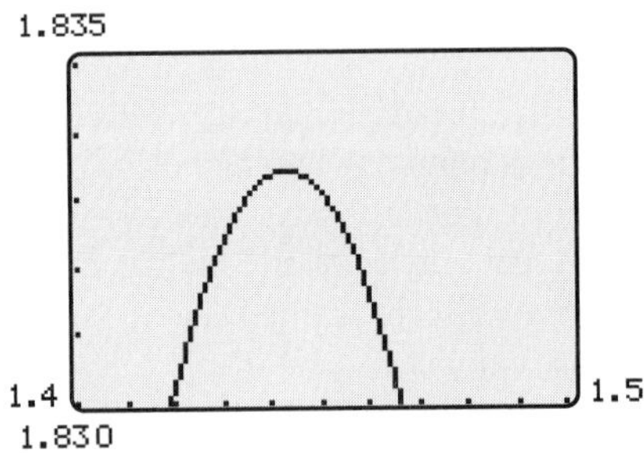

Figure 1-78

Figure 1-79

Note Calculators with different screen widths (in pixels) may not plot the same points on the graph of an equation, even though identical viewing windows are chosen. Consequently, the trace feature on your calculator may produce slightly different results than those shown in the figures here.

▶ TECHNOLOGY TIP

To display the grid determined by the tic marks on the axes, as in Figure 1–81, choose "Grid On" in the FORMAT menu on TI (which may be a submenu of the WINDOW or GRAPH menu), or check GRID in the HP-38 PLOT SETUP menu, or choose PLT in the G-TYPE submenu of the Casio 9800 graphing mode set-up menu.

(b) Although the axes are not shown in Figure 1–78, the location of the axis tic marks is indicated on the left and bottom edges of the screen; they are .01 units apart on the x-axis and .001 units apart on the y-axis (that is, $X\text{scl} = .01$, $Y\text{scl} = .001$). As shown schematically in Figure 1–80, this means that the highest point is contained in a rectangle measuring .01 wide by .001 high. Any point in this rectangle will have an x-coordinate within .01 of the x-coordinate of the highest point and a y-coordinate within .001 of the y-coordinate of the highest point. The rectangles determined by the tic marks can be shown on the screen on most calculators (see the Tip in the margin). In particular, our approximation is in the rectangle containing the highest point (Figure 1–81), so its maximum error is .01 in the first coordinate and .001 in the second. ■

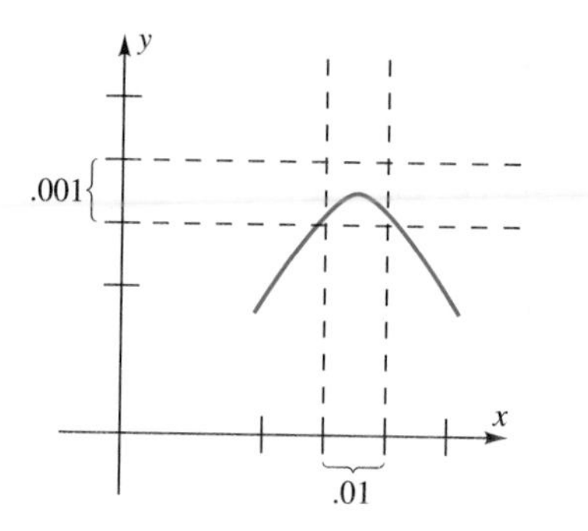

Figure 1-80

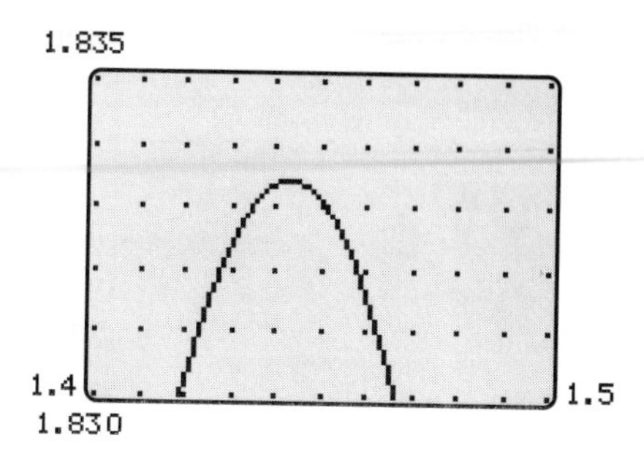

Figure 1-81

> ▶ **TECHNOLOGY TIP**
> The graphical min/max finder is in the CALC menu on TI-82/83, in the MATH submenu of the GRAPH menu on TI-85 and 92, in the G-SOLVE menu on Casio 9800, and in the JUMP menu on Sharp 9300. It is labeled EXTREMUM in the HP-38 PLOT FCN menu. A min/max finder program for TI-81 is in the Program Appendix.

Some calculators have a ''minimum/maximum finder,'' which finds highest and lowest points automatically (see the Tip in the margin). The TI-82 min/max finder says the highest point on the graph in Example 1 is (1.4430021, 1.8334224).

Applications

Before reading the following examples, you should review the guidelines for setting up applied problems (page 60).

EXAMPLE 2 Find two negative numbers whose product is 50 and whose sum is as large as possible.

Solution Let x and z be the two negative numbers and let y be their sum. Then $xz = 50$ and $y = x + z$. Solving $xz = 50$ for z we have $z = 50/x$ so that

$$y = x + z = x + 50/x.$$

We must find the value of x that makes y as large as possible. Since x must be negative we graph $y = x + 50/x$ in a window with $-20 \leq x \leq 0$ (Figure 1–82).

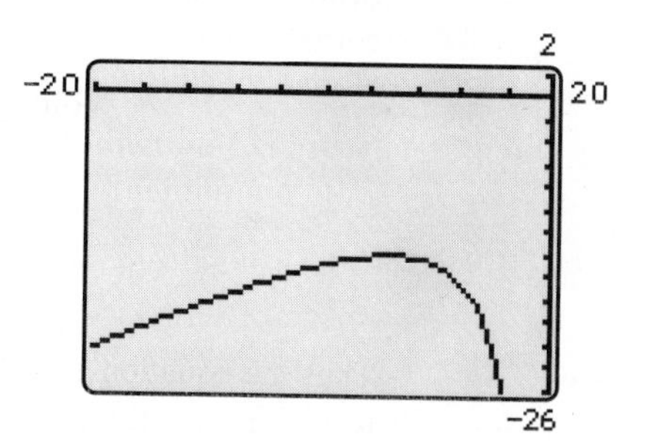

Figure 1-82

Each point (x, y) on this graph represents one of the possibilities:

x represents one of the two negative numbers ($50/x$ is the other);

y is the sum of the two negative numbers.

Since we want y to be as large as possible, we must find the point on the graph (with $x < 0$) with largest y-coordinate, that is, the highest point on this part of the graph.

> ▶ **TECHNOLOGY TIP**
> Zooming-in may lead to a "flat" graph in which the highest or lowest point cannot be easily read. You can avoid this problem by choosing a viewing window that is much wider than it is high.

GRAPHING EXPLORATION Use zoom-in or a min/max finder to approximate the highest point. Its first coordinate x is one of the two numbers and $50/x$ is the other. ■

EXAMPLE 3 A box with no top is to be made from a 22 by 30 inch sheet of cardboard by cutting squares of equal size from each corner and bending up the flaps, as shown in Figure 1–83. To the nearest hundredth of an inch, what size square should be cut from each corner in order to obtain a box with the largest possible volume and what is the volume of this box?

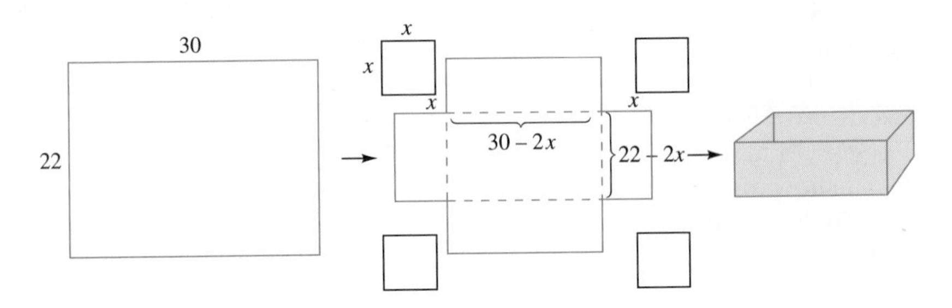

Figure 1-83

Solution Let x denote the length of the side of the square to be cut from each corner. Then,

$$\begin{aligned}\text{Volume of box} &= \text{length} \times \text{width} \times \text{height}\\ &= (30 - 2x) \cdot (22 - 2x) \cdot x\\ &= 4x^3 - 104x^2 + 660x\end{aligned}$$

Thus the equation $y = 4x^3 - 104x^2 + 600x$ gives the volume y of the box that results from cutting a square of side x from each corner. Since the shortest side of the cardboard is 22 inches, the length x of the side of the cut-out square must be less than 11 (why?).

Each point on the graph of the volume equation $y = 4x^3 - 104x^2 + 600x$ $(0 < x < 11)$ in Figure 1–84 represents one of the possibilities:

The x-coordinate is the size of the square to be cut from each corner;
The y-coordinate is the volume of the resulting box.

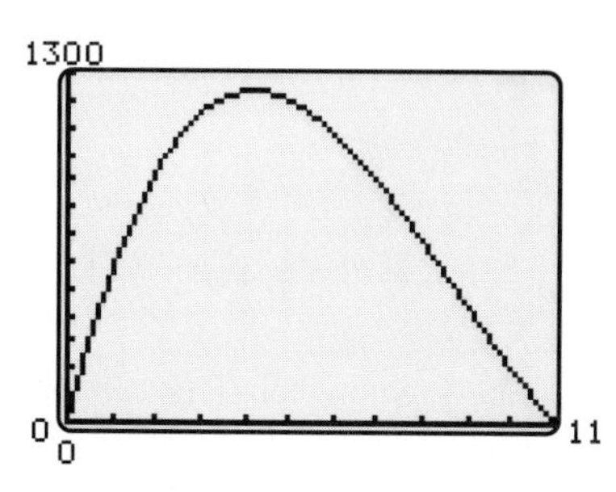

Figure 1-84

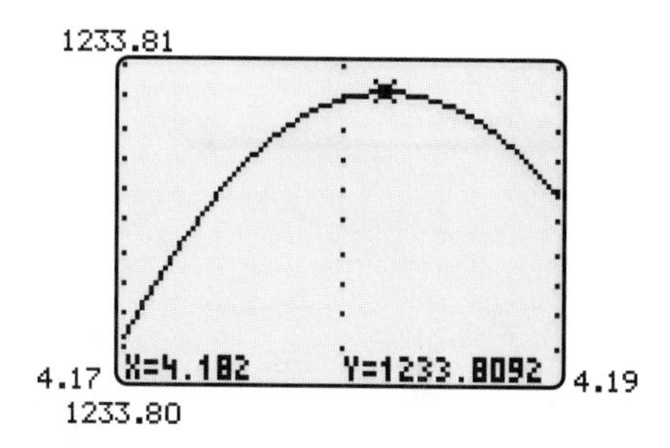

Figure 1-85

The box with the largest volume corresponds to the point on the graph with the largest y-coordinate, that is, the point in this viewing window that is highest above the x-axis. Using zoom-in and trace (Figure 1–85) we find that the highest point has approximate coordinates $x = 4.182$ and $y = 1233.8092$. The tic mark grid in Figure 1–85 shows that this approximation has an error of at most .01 in the x-coordinate and .001 in the y-coordinate.

Therefore, a square measuring approximately 4.18 by 4.18 should be cut from each corner, producing a box of volume approximately 1233.81 cubic inches. ■

EXAMPLE 4 A cylindrical can of volume 58 cubic inches (approximately 1 quart) is to be designed. For convenient handling, it must be at least 1 inch high and 2 inches in diameter. What dimensions will use the least amount of material?

Solution We can construct a can by rolling a rectangular sheet of metal into a tube and then attaching the top and bottom, as shown in Figure 1–86. The surface area of the can (which determines the amount of material) is

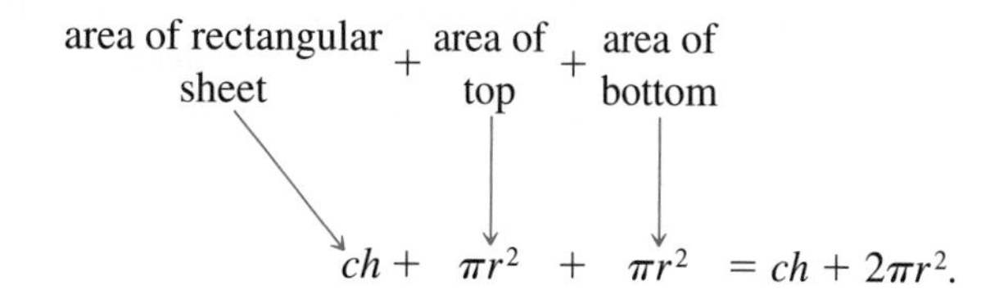

$$ch + \pi r^2 + \pi r^2 = ch + 2\pi r^2.$$

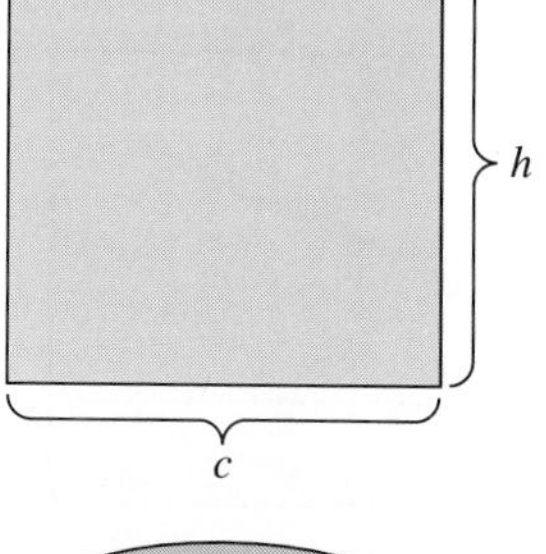

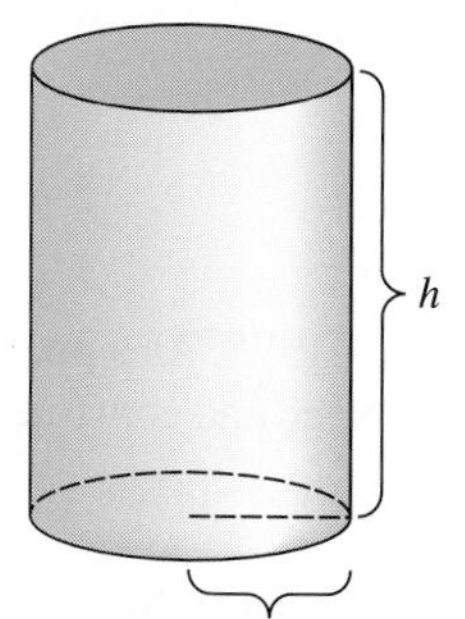

Figure 1-86

When the sheet is rolled into a tube, the width c of the sheet is the circumference of the end of the can, so that $c = 2\pi r$ and, hence,

$$\text{surface area} = ch + 2\pi r^2 = 2\pi rh + 2\pi r^2.$$

The volume of a cylinder of radius r and height h is $\pi r^2 h$. Since the can is to have volume 58 cubic inches, we have

$$\pi r^2 h = 58, \qquad \text{or equivalently,} \qquad h = \frac{58}{\pi r^2}.$$

Therefore,

$$\text{surface area} = 2\pi rh + 2\pi r^2 = 2\pi r\left(\frac{58}{\pi r^2}\right) + 2\pi r^2 = \frac{116}{r} + 2\pi r^2.$$

Note that r must be greater than 1 (since the diameter $2r$ must be at least 2). Furthermore, r cannot be more than 5 (if $r > 5$ and $h \geq 1$, then the volume $\pi r^2 h$ would be at least $\pi \cdot 25 \cdot 1$, which is greater than 58).

The situation can be represented by the graph of the equation $y = 116/x + 2\pi x^2$, with $1 \leq x \leq 5$: The x-coordinate of each point represents a possible radius and the y-coordinate the surface area of the corresponding can. We must find the point with the smallest y-coordinate, that is, the lowest point on the graph. The viewing window in Figure 1–87 and the trace feature suggest the point (2.106383, 82.948253). A graphical min/max finder (Figure 1–88) shows that a better approximation of this point is (2.09773, 82.946845).

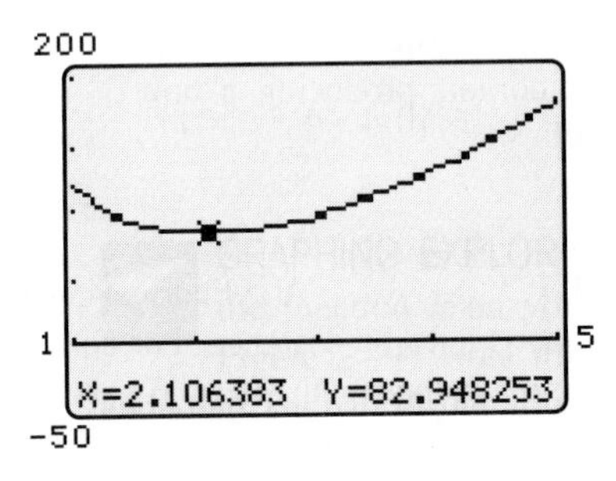

Figure 1-87

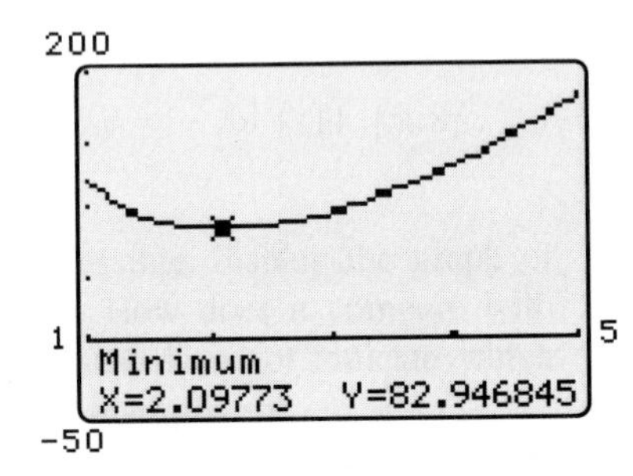

Figure 1-88

If the radius is 2.09773, then the height is $58/\pi(2.09773^2) \approx 4.1995$. As a practical matter, it would probably be best to round to one decimal place and construct a can of radius 2.1 and height 4.2 inches. ■

As illustrated in the preceding examples, the solution of an optimization problem is typically represented by the coordinates of a point, each of which provides relevant information. When a graphical min/max finder is used to find the point, the results are generally accurate to at least four decimal places (often more) in each coordinate. When zoom-in and trace are used, you should estimate *each* coordinate with an error of at most .01.*

CALCULATOR INVESTIGATION 1.7

1. Min/Max Finder If your calculator has a min/max finder (or program), learn how to use it. Then use it to find the x-coordinate of the lowest point on the graph of $y = x^3 - 2x + 5$ in the window with $0 \leq x \leq 5$ and $-3 \leq y \leq 8$. The correct answer is $x = \sqrt{\frac{2}{3}} \approx$ .816496580928. How good is your approximation?

EXERCISES 1.7

In Exercises 1–6, find the coordinates of the highest or lowest point on the part of graph of the equation in the given viewing window. Only the range of x-coordinates for the window are given; you must choose an appropriate range of y-coordinates.

1. $y = 2x^3 - 3x^2 - 12x + 1$; highest point when $-3 \leq x \leq 3$

2. $y = 2x^6 + 3x^5 + 3x^3 - 2x^2$; lowest point when $-3 \leq x \leq 3$

3. $y = 4/x^2 - 7/x + 1$; lowest point when $-10 \leq x \leq 10$

4. $y = 1/(x^2 + 2x + 2)$; highest point when $-5 \leq x \leq 5$

5. $y = \dfrac{x^2(x + 1)^3}{(x - 2)^2(x - 4)^2}$; highest point when $-1 \leq x \leq 0$ [*Hint:* Think small.]

6. $y = \dfrac{x^2(x + 1)^3}{(x - 2)^2(x - 4)^2}$; lowest point when $-10 \leq x \leq -1$

7. Find the highest point on the part of the graph of $y = x^3 - 3x + 2$ that is shown in the given window. The answers are not all the same.
(a) $-2 \leq x \leq 0$ **(b)** $-2 \leq x \leq 2$
(c) $-2 \leq x \leq 3$

8. Find the lowest point on the part of the graph of $y = x^3 - 3x + 2$ that is shown in the given window.
(a) $0 \leq x \leq 2$ **(b)** $-2 \leq x \leq 2$
(c) $-3 \leq x \leq 2$

9. Find the dimensions of the rectangular box with a square base and no top that has volume 30,000 cubic cm and the smallest possible surface area. [*Hint:* See Example 3 in Section 1.6.]

10. An open-top box with a square base is to be constructed from 120 square centimeters of material. What dimensions will produce a box
(a) of volume 100 cubic centimeters?
(b) with largest possible volume?

11. A 20-inch square piece of metal is to be used to make an open-top box by cutting equal-sized squares from each corner and folding up the sides (as in Example 3). The length, width, and height of the box are each to be less than 12 inches. What size squares should be cut out to produce a box with
(a) volume 550 cubic inches?
(b) largest possible volume?

12. A cylindrical waste container with no top, a diameter of at least 2 feet, and a volume of 25 cubic feet is to be constructed. What should its radius be if
(a) 65 square feet of material are to be used to construct it?

*It may sometimes be necessary to use an even smaller maximum error in one coordinate in order to ensure that both errors are at most .01, as was the case in Example 3.

(b) the smallest possible amount of material is to be used to construct it? In this case, how much material is needed?

13. If $c(x)$ is the cost of producing x units, then $c(x)/x$ is the *average cost* per unit.* Suppose the cost of producing x units is given by $c(x) = .13x^3 - 70x^2 + 10{,}000x$ and that no more than 300 units can be produced per week.

(a) If the average cost is \$1100 per unit, how many units are being produced?

(b) What production level should be used in order to minimize the average cost per unit? What is the minimum average cost?

14. A manufacturer's revenue (in cents) from selling x items per week is given by $200x - .02x^2$. It costs $60x + 30{,}000$ cents to make x items.

(a) Approximately how many items should be made each week in order to make a profit of \$1100? [Don't confuse cents and dollars.]

(b) How many items should be made each week in order to have the largest possible profit? What is that profit?

15. (a) A company makes novelty bookmarks that sell for \$142 per hundred. The cost (in dollars) of making x hundred bookmarks is $x^3 - 8x^2 + 20x + 40$. Because of other projects, a maximum of 600 bookmarks per day can be manufactured. Assuming that the company can sell all the bookmarks it makes, how many should it make each day to maximize profits?

(b) Due to a change in other orders, as many as 1600 bookmarks can now be manufactured each day. How many should be made in order to maximize profits?

16. If the cost of material to make the can in Example 4 is 5 cents per square inch for the top and bottom and 3 cents per square inch for the sides, what dimensions should be used to minimize the cost of making the can? [The answer is not the same as in Example 4.]

17. A certain type of fencing comes in rigid 10-ft-long segments. Four uncut segments are to be used to fence in a garden on the side of a building, as shown in Figure 1–89. What value of x will result in a garden of the largest possible area and what is that area?

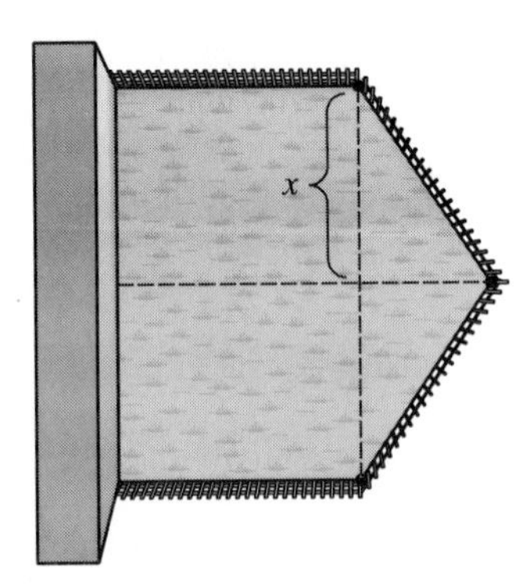

Figure 1-89

18. A rectangle is to be inscribed in a semicircle of radius 2, as shown in Figure 1–90. What is the largest possible area of such a rectangle? [*Hint*: The width of the rectangle is the second coordinate of the point P (why?) and P is on the top half of the circle $x^2 + y^2 = 4$.]

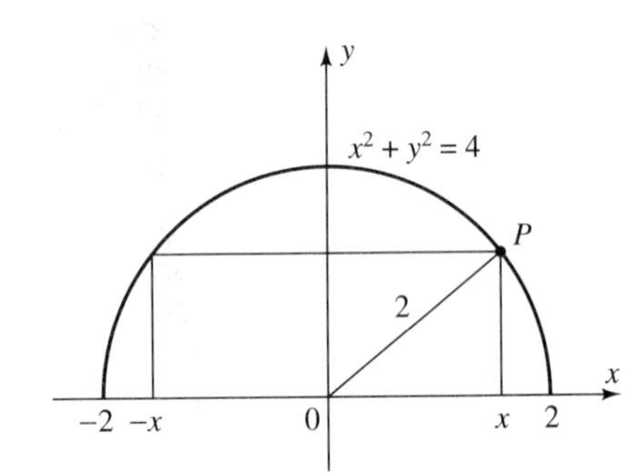

Figure 1-90

19. Find the point on the graph of $y = 5 - x^2$ that is closest to the point (0, 1) and has positive coordinates. [*Hint*: The distance from the point (x, y) on the graph to (0, 1) is $\sqrt{(x-0)^2 + (y-1)^2}$; express y in terms of x.]

20. A manufacturer's cost (in thousands of dollars) of producing x thousand units is $x^3 - 6x^2 + 15x$ dollars and the revenue (in thousands) from x thousand units is $9x$ dollars. What production level(s) will result in the largest possible profit?

*Depending on the situation, a unit of production might consist of a single item or several thousand items. Similarly, the cost of x units might be measured in thousands of dollars.

21. A hardware store sells ladders throughout the year. It costs $20 every time an order for ladders is placed and $10 to store a ladder until it is sold. When ladders are ordered x times per year, then an average of $300/x$ ladders are in storage at any given time. How often should the company order ladders each year in order to minimize its total ordering and storage costs? [*Be careful*: The answer must be an integer.]

22. A mathematics book has 36 square inches of print per page. Each page has a left side margin of 1.5 inches and top, bottom, and right side margins of .5 inch. If a page cannot be wider than 7.5 inches, what should its length and width be in order to use the least amount of paper?

CHAPTER 1 *Review*

Important Concepts ▶

Important Facts and Formulas

- $|c - d|$ = distance from c to d on the number line
- *Quadratic Formula:* If $a \neq 0$, then the solutions of $ax^2 + bx + c = 0$ are $x = \dfrac{-b \pm \sqrt{b^2 - 4ac}}{2a}$.
- If $a \neq 0$, then the number of real solutions of $ax^2 + bx + c = 0$ is 0, 1, or 2, depending on whether the discriminant $b^2 - 4ac$ is negative, zero, or positive.
- *Distance Formula:* The distance from (x_1, y_1) to (x_2, y_2) is
$$\sqrt{(x_1 - x_2)^2 + (y_1 - y_2)^2}$$
- *Midpoint Formula:* The midpoint of the line segment from (x_1, y_1) to (x_2, y_2) is $\left(\dfrac{x_1 + x_2}{2}, \dfrac{y_1 + y_2}{2}\right)$.
- Equation of the circle with center (c, d) and radius r:
$$(x - c)^2 + (y - d)^2 = r^2$$

Review Questions

1. Fill the blanks with one of the symbols $<$, $=$, or $>$ so that the resulting statement is true.
(a) 142 ______ $|-51|$ **(b)** $\sqrt{2}$ ______ $|-2|$
(c) -1000 ______ $\dfrac{1}{10}$ **(d)** $|-2|$ ______ $-|6|$
(e) $|u - v|$ ______ $|v - u|$ where u and v are fixed real numbers.

2. List two real numbers that are *not* rational numbers.

3. Express $0.282828 \cdots$ as a fraction.

4. Express $0.362362362 \cdots$ as a fraction.

5. Express in symbols:
(a) y is negative, but greater than -10;
(b) x is nonnegative and not greater than 10.

6. Express in symbols:
(a) $c - 7$ is nonnegative; **(b)** .6 is greater than $|5x - 2|$.

7. Express in symbols:
(a) x is less than 3 units from -7 on the number line;
(b) y is farther from 0 than x is from 3 on the number line.

8. Simplify: $|b^2 - 2b + 1|$

9. Solve: $|x - 5| = 3$

10. Solve: $|x + 2| = 4$

11. Solve: $|x + 3| = \dfrac{5}{2}$

12. Solve: $|x - 5| \leq 2$

13. Solve: $|x + 2| \leq 2$

14. Solve: $|x - 1| > 4$

15. **(a)** $|\pi - 7| =$ ______ **(b)** $|\sqrt{23} - \sqrt{3}| =$ ______

16. If c and d are real numbers with $c \neq d$ what are the possible values of $\dfrac{c - d}{|c - d|}$?

17. Express in interval notation:
(a) the set of all real numbers that are strictly greater than -8;
(b) the set of all real numbers that are less than or equal to 5.

18. Express in interval notation:
(a) the set of all real numbers that are strictly between -6 and 9;
(b) the set of all real numbers that are greater than or equal to 5, but strictly less than 14.

19. Solve for x: $2\left(\frac{x}{5}+7\right)-3x=\frac{x+2}{5}-4$

20. Solve for x in terms of y: $xy+3=x-2y$

21. Solve for x: $3x^2-2x+5=0$

22. Solve for y: $3y^2-2y=5$

23. Solve for z: $5z^2+6z=7$

24. Solve for x: $325x^2+17x-127=0$

25. Find the *number* of real solutions of the equation $20x^2+12=31x$.

26. For what value of k does the equation $kt^2+5t+2=0$ have exactly one real solution for t?

In Questions 27–30, find all real solutions of the equation. Do not approximate.

27. $x^4-11x^2+18=0$

28. $x^6-4x^3+4=0$

29. $|3x-1|=4$

30. $|x^2+2x-3|=4$

31. A jeweler wants to make a 1-ounce ring consisting of gold and silver, using \$200 worth of metal. If gold costs \$600 per ounce and silver \$50 per ounce, how much of each metal should she use?

32. A calculator is on sale for 15% less than the list price. The sale price, plus a 5% shipping charge, totals \$210. What is the list price?

33. Jack can do a job in 5 hours and Walter can do the same job in 4 hours. How long will it take them to do the job together?

34. A car leaves the city traveling at 54 mph. One-half hour later, a second car leaves from the same place and travels at 63 mph along the same road. How long will it take for the second car to catch up with the first?

35. A 12-foot-long rectangular board is cut in two pieces so that one piece is four times as long as the other. How long is the bigger piece?

36. George owns 200 shares of stock, 40% of which are in the computer industry. How many more shares must he buy in order to have 50% of his total shares in computers?

37. A square region is changed into a rectangular one by making it 2 feet longer and twice as wide. If the area of the rectangular region is three times larger than the area of the original square region, what was the length of a side of the square before it was changed?

38. The radius of a circle is 10 inches. By how many inches should the radius be increased so that the area increases by 5π square inches?

39. Find the distance from $(1, -2)$ to $(4, 5)$.

40. Find the distance from $(3/2, 4)$ to $(3, 5/2)$.

41. Find the distance from (c, d) to $(c - d, c + d)$.

42. Find the midpoint of the line segment from $(-4, 7)$ to $(9, 5)$.

43. Find the midpoint of the line segment from (c, d) to $(2d - c, c + d)$.

44. Find the equation of the circle with center $(-3, 4)$ that passes through the origin.

45. **(a)** If $(1, 1)$ is on a circle with center $(2, -3)$, what is the radius of the circle?
(b) Find the equation of the circle in part (a).

46. Sketch the graph of $3x^2 + 3y^2 = 12$.

47. Sketch the graph of $(x - 5)^2 + y^2 - 9 = 0$.

48. Find the center and radius of the circle whose equation is
$x^2 + y^2 - 2x + 6y + 1 = 0$.

49. Which of statements (a)–(d) are descriptions of the circle with center $(0, -2)$ and radius 5?
(a) The set of points (x, y) that satisfy $|x| + |y + 2| = 5$.
(b) The set of all points whose distance from $(0, -2)$ is 5.
(c) The set of all points (x, y) such that $x^2 + (y + 2)^2 = 5$.
(d) The set of all points (x, y) such that $\sqrt{x^2 + (y + 2)^5} = 5$.

50. If the equation of a circle is $3x^2 + 3(y - 2)^2 = 12$, which of the following statements is true?
(a) The circle has diameter 3.
(b) The center of the circle is $(2, 0)$.
(c) The point $(0, 0)$ is on the circle.
(d) The circle has radius $\sqrt{12}$.
(e) The point $(1, 1)$ is on the circle.

51. The graph of one of the equations below is *not* a circle. Which one?
(a) $x^2 + (y + 5)^2 = \pi$
(b) $7x^2 + 4y^2 - 14x + 3y^2 - 2 = 0$
(c) $3x^2 + 6x + 3 = 3y^2 + 15$
(d) $2(x - 1)^2 - 8 = -2(y + 3)^2$
(e) $\dfrac{x^2}{4} + \dfrac{y^2}{4} = 1$

52. The point $(7, -2)$ is on the circle whose center is on the midpoint of the segment joining $(3, 5)$ and $(-5, -1)$. Find the equation of this circle.

In Questions 53–58,

(a) Determine which of the viewing windows a–e shows a complete graph of the equation.

(b) For each viewing window that does not show a complete graph, explain why.

(c) Find a viewing window that gives a "better" complete graph than windows a–e (meaning that the window is small enough to show as much detail as possible, yet large enough to show a complete graph).

a. standard viewing window
b. $-10 \le x \le 10, -200 \le y \le 200$
c. $-20 \le x \le 20, -500 \le y \le 500$
d. $-50 \le x \le 50, -50 \le y \le 50$
e. $-1000 \le x \le 1000, -1000 \le y \le 1000$

53. $y = .2x^3 - .8x^2 - 2.2x + 6$ **54.** $y = x^3 - 11x^2 - 25x + 275$

55. $y = x^4 - 7x^3 - 48x^2 + 180x + 200$

56. $y = x^3 - 6x^2 - 4x + 24$ **57.** $y = .03x^5 - 3x^3 + 69.12x$

58. $y = .00000002x^6 - .0000014x^5 - .00017x^4 + .0107x^3 + .2568x^2 - 12.096x$

In Questions 59–62, sketch a complete graph of the equation and give reasons why it is complete.

59. $y = x^2 - 10$ **60.** $y = x^3 + x + 4$

61. $y = \sqrt{x - 5}$ **62.** $y = x^4 + x^2 - 6$

In Questions 63–66, sketch a complete graph of the equation.

63. $y = x^2 - 13x + 43$ **64.** $y = |x|$

65. $y = |x + 5|$ **66.** $y = 1/x$

In Questions 67–74, solve the equation graphically. You need only find solutions in the given interval.

67. $x^3 + 2x^2 - 11x = 6;\quad [0, \infty)$

68. $x^3 + 2x^2 - 11x = 6;\quad (-\infty, 0)$

69. $x^4 + x^3 - 10x^2 - 8x + 16 = 0;\quad [0, \infty)$

70. $2x^4 + x^3 - 2x^2 + 6x + 2 = 0;\quad (-\infty, -1)$

71. $\dfrac{x^3 + 2x^2 - 3x + 4}{x^2 + 2x - 15} = 0;\quad (-10, \infty)$

72. $\dfrac{3x^4 + x^3 - 6x^2 - 2x}{x^5 + x^3 + 2} = 0;\quad [0, \infty)$

73. $\sqrt{x^3 + 2x^2 - 3x - 5} = 0;\quad [0, \infty)$

74. $\sqrt{1 + 2x - 3x^2 + 4x^3 - x^4} = 0;\quad (-5, 5)$

75. The cost of manufacturing x caseloads of ball point pens is $\dfrac{600x^2 + 600x}{x^2 + 1}$ dollars. How many caseloads should be manufactured in order to have an *average cost* of \$25? [Average cost was defined in Exercise 13 of Section 1.7.]

76. An open-top box with a rectangular base is to be constructed. The box is to be at least 2 inches wide, twice as long as it is wide and have a volume of 150 cubic inches. What should the dimensions of the box be if the surface area is to be
(a) 90 square inches? **(b)** as small as possible?

77. A farmer has 120 yards of fencing and wants to construct a rectangular pen, divided in two parts by an interior fence, as shown in Figure 1–91. What should the dimensions of the pen be in order to enclose the maximum possible area?

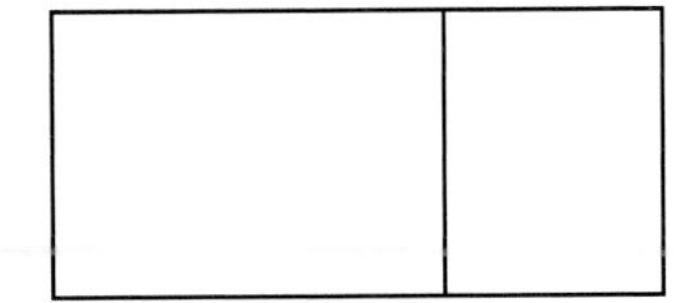

Figure 1-91

78. The top and bottom margins of a rectangular poster are each 5 inches and each side margin is 3 inches. The printed material on the poster occupies an area of 400 square inches. Find the dimensions that will use the least possible amount of posterboard.

79. A rectangle has one side on the x-axis and its other two corners sit on the graph of $y = 9 - x^2$, as shown in Figure 1–92. What value of x gives a rectangle of maximum area?

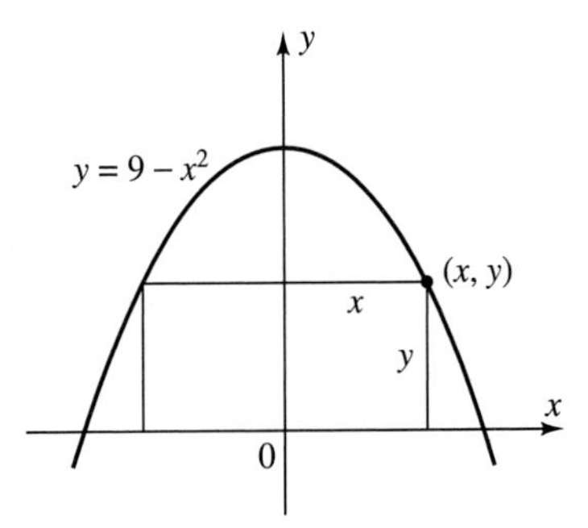

Figure 1-92

80. The window in Figure 1–93 has a rectangular bottom, with a semicircle of radius r lying on top of it, and a perimeter of 40 feet. In order that the window have the maximum possible area, what should r and h be?

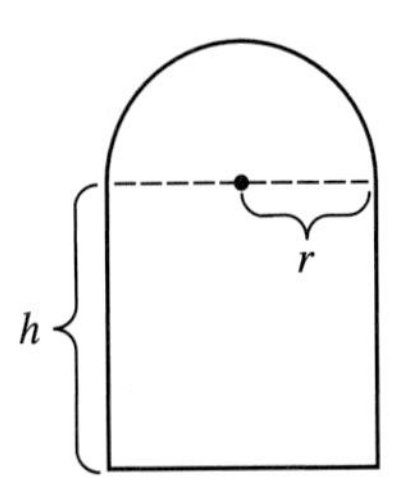

Figure 1-93

2

Functions and Graphs

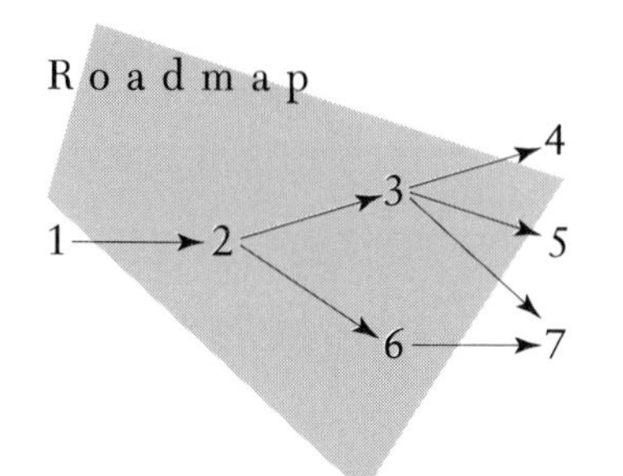

The concept of a function and functional notation are central to modern mathematics. In this chapter you will be introduced to functions and operations on functions, learn how to use functional notation, and develop skill in constructing and interpreting graphs of functions.

2.1 FUNCTIONS

To understand the origin of the concept of function it may help to consider some "real-life" situations in which one numerical quantity depends on, corresponds to, or determines another.

EXAMPLE 1 The amount of income tax you pay depends on the amount of your income. The way in which the income determines the tax is given by the tax law. ■

EXAMPLE 2 The weather bureau records the temperature over a 24-hour period in the form of a graph (Figure 2–1). The graph shows the temperature that corresponds to each given time. ■

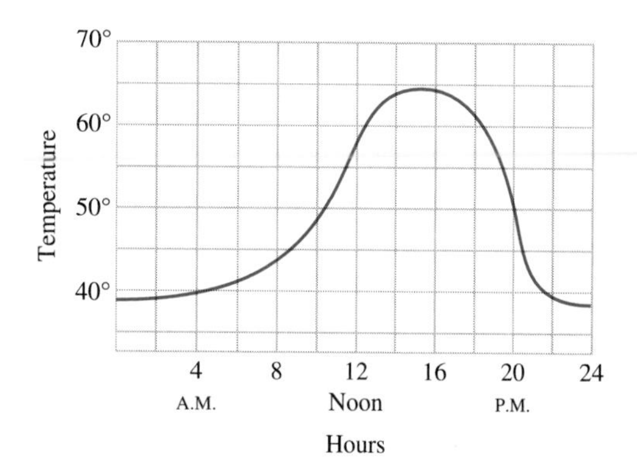

Figure 2-1

EXAMPLE 3 Suppose a rock is dropped straight down from a high place. Physics tells us that the distance traveled by the rock in t seconds is $16t^2$ feet. So the distance depends on the time. ■

The first common feature shared by these examples is that each involves two sets of numbers, which we can think of as a set of inputs and a set of outputs:

	Set of Inputs	*Set of Outputs*
Example 1	All incomes	All tax amounts
Example 2	Hours since midnight	Temperatures during the day
Example 3	Seconds elapsed after dropping the rock	Distances rock travels

The second common feature is that in each example there is a definite *rule* by which each input determines an output. In Example 1 the rule is given by the tax law, which specifies how each income (input) determines a tax amount (output). Similarly, the rule is given by the time/temperature graph in Example 2 and by the formula (distance $= 16t^2$) in Example 3.

Each of these examples may be mentally represented by an idealized calculator that has a single operation key and is capable of receiving or displaying any real number: A number is entered [*input*], the ''rule key'' is pressed, and an answer is displayed [*output*]:

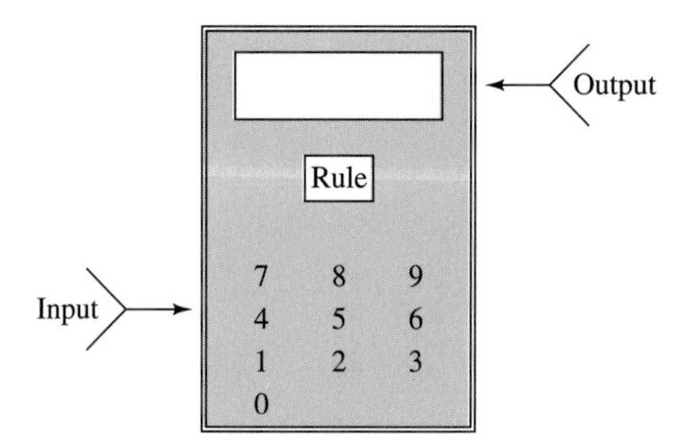

The formal definition of function incorporates these common features (input/rule/output) and a slight change in terminology. The set of all possible inputs is called the *domain* and the set of corresponding outputs is called the *range.*

Definition of Function ▶

A *function* consists of:

A set of numbers (called the *domain*);

A *rule* by which each number in the domain determines one and only one number.

The set of all numbers determined by applying the rule to the numbers in the domain is called the *range* of the function.

The domain in Example 1 consists of all possible income amounts and the range consists of all possible tax amounts. The domain in Example 2 is the set of hours in the day (that is, all real numbers from 0 to 24). The rule is given by the time/temperature graph, which shows the temperature at each time. The graph also shows that the range (the temperatures that actually occur during the day) consists of the numbers from 38 to 64.9.

In Example 2, for each time of day (number in the domain) there is one and only one temperature (number in the range). Notice, however, that it is possible to have the same temperature (number in the range) corresponding to two different times (numbers in the domain). In general,

> **By the rule of a function, each number in the domain determines *one and only one* number in the range. However, several different numbers in the domain may determine the same number in the range.**

In other words, exactly one output is produced for each input, but different inputs may produce the same output.

Although real-world situations, such as Examples 1–3, are the motivation for studying functions, much of the emphasis in mathematics courses is on the functions themselves, independent of possible interpretations in specific situations. Here are some examples of functions in a purely mathematical context.

EXAMPLE 4 The domain of the **absolute value function** is the set of all real numbers. The rule is ''take the absolute value.'' In other words, for each input x, the corresponding output is $|x|$. For instance, applying the rule to the number -6 in the domain produces the number $|-6| = 6$ in the range. Since $|x| \geq 0$ for every x, the range of the function (all absolute values of real numbers) consists of all nonnegative real numbers, that is, the interval $[0, \infty)$. ■

EXAMPLE 5 For each real number s that is not an integer, let $[s]$ denote the *integer* that is closest to s on the *left* side of s on the number line; if s is itself an integer, we define $[s] = s$. Here are some examples:

$$[-4.7] = -5, \quad [-3] = -3, \quad [-1.5] = -2, \quad [0] = 0, \quad \left[\frac{5}{3}\right] = 1, \quad [\pi] = 3.$$

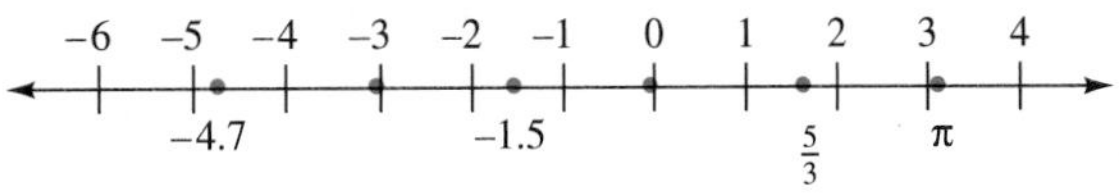

> **▶ TECHNOLOGY TIP**
> The greatest integer function is denoted INT on TI-81/82/83/85 and Sharp 9300, but is denoted FLOOR on TI-92 and HP-38, and INTG on Casio (where ''INT'' denotes a different function). Look in the MATH menu or its NUM or REAL submenu. Also see Exercise 39.

Figure 2-2

The **greatest integer function** is the function whose domain is the set of all real numbers and whose rule is

assign to the real number x in the domain the integer $[x]$,

that is, the input x produces the output $[x]$. The range of the greatest integer function is the set of all integers. ■

Functions Defined by Equations and Graphs

Equations in two variables are not the same thing as functions. However, many equations can be used to define functions in a natural way.

EXAMPLE 6 If a number is substituted for x in the equation

$$y = x^3 + 6x^2 - 5,$$

then exactly one value of y is produced. We can use this equation to define a function whose domain is the set of all real numbers as follows. Think of x as the input and y as the output, so that the rule of the function is

Assign to each number x in the domain the unique number y such that $y = x^3 + 6x^2 - 5$.

We describe this situation by saying that the equation **defines y as a function of x.** ■

Graphing calculators are designed to deal with equations that define y as a function of x, and you should learn how to use your calculator to evaluate such functions quickly and easily. In particular, it is not necessary to key in the entire equation each time you want to find its value at a value of x; see Calculator Investigations 1–5 at the end of the section for details. Although most calculators use only x and y as variables, the letters used for equation variables don't matter.

EXAMPLE 7 The equation $u = v^2 + 1$ defines u as a function of v because for each value of v (input) there is exactly one corresponding value of u (output). ■

EXAMPLE 8 The equation $4x - 2y + 5 = 0$ can be solved uniquely for y:

$$2y = 4x + 5$$

$$y = 2x + \frac{5}{2}$$

For each value of x (input) there is exactly one corresponding value of y (output). Hence, the equation defines y as a function of x. The equation can also be solved uniquely for x in terms of y:

$$4x = 2y - 5$$

$$x = \frac{2y - 5}{4}$$

For each value of y there is exactly one corresponding value of x. So this equation also defines x as a function of y (that is, we consider y as the input and x as the output). ■

Some equations do *not* define one variable as a function of the other as the following example demonstrates.

EXAMPLE 9 The equation $y^2 = 4x^2$ does not define y as a function of x because for every nonzero x (input), there are two values of y (outputs) that satisfy the equation. For instance, when $x = 3$, then $y^2 = 4 \cdot 3^2 = 36$ and both $y = 6$ and $y = -6$ satisfy $y^2 = 36$. Similarly, this equation does not define x as a function of y. ■

Some graphs can also be used to define functions.

EXAMPLE 10 The graph in Figure 2–3 defines a function whose domain is the interval $[-4, 5]$ and whose rule is:

> Assign to the number x in the domain the unique number y such that (x, y) lies on the graph.

WARNING Not every graph defines a function in this way; see Exercise 36.

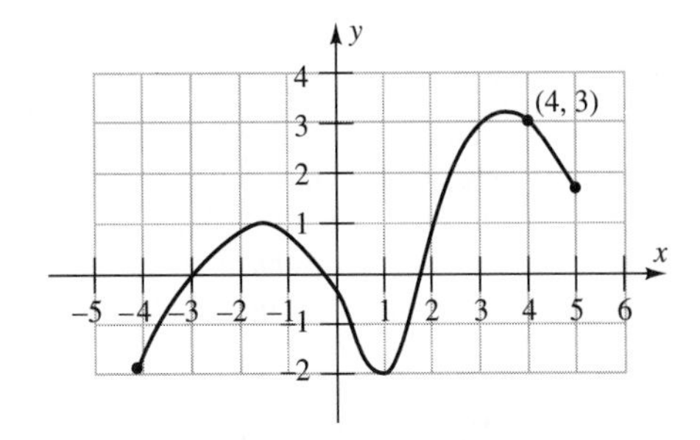

Figure 2-3

For instance, the rule assigns to the number 4 in the domain the number 3 because $(4, 3)$ is on the graph. Similarly, -3 in the domain determines the number 0 in the range since $(-3, 0)$ is on the graph. Since all the second coordinates of points on the graph lie between -2 and 3.3,* the range of this function is the interval $[-2, 3.3]$ ■

CALCULATOR INVESTIGATIONS 2.1

The equation $y = x^3 + 6x^2 - 5$ defines y as a function of x (Example 5). Investigations 1–5 examine various ways of using a calculator to evaluate this function. Some methods will not work with certain calculators.

1. Edit and Replay Find the value of y when $x = -7$ by keying in

(*) $$(-7)^3 + 6(-7)^2 - 5$$

and pressing ENTER to obtain the answer.

(a) Find the value of y when $x = 9$ by editing your previous calculation. Press SECOND ENTER (on TI) or the left or right arrow key (on TI-92, Casio and Sharp) or the up arrow and COPY (on HP-38). This activates the **replay** feature: Your screen will return to the calculation (*). Now use the arrow and DELETE keys to move through the equation and replace -7 by 9. When you are finished press ENTER and the calculator will display the value of y when $x = 9$.

(b) Use the replay feature again to find the value of y when $x = 108$. [You may need to use the INSERT key to avoid unnecessary retyping.]

*We are assuming here that we can read the graph with perfect accuracy. In actual practice, your results will only be as accurate as your measuring ability; but the basic idea should be clear. Also, the color printing process may produce small differences between graphs in two different copies of this book.

2. Equation Memory Enter the equation in the equation memory as $y_1 = x^3 + 6x^2 - 5$ (or $f_1 = \cdot\cdot$ on HP-38 and some Casios). Exit the equation memory. To find the value of y when $x = 12$, store 12 in memory X (use the X-variable key that you use for entering equations). Then call up y_1. Type in y_1 on TI-85, $y_1(x)$ on TI-92, $f_1(x)$ on HP-38; use y_1 on the TI-81/82/83 Y-VARS menu (or its FUNCTION submenu); use Y on the GPH submenu of the Casio 9800 VAR menu and FMEM *fn*1 on Casio 77/8700. Press ENTER and the calculator will display the value of y when $x = 12$. Now evaluate the function when $x = -4.7$ and when $x = 6.43$.

3. Automatic Evaluation Enter the equation as y_1 in the equation memory. Exit the equation memory. Evaluate the function when $x = 29$ as follows. On TI-82/83, TI-92, key in $y_1(29)$ ENTER (see Investigation 2 for how to do this on TI-82/83). On HP-38 key in $f_1(29)$ ENTER. On TI-85, key in EVAL 12 ENTER (you will find EVAL on the MISC submenu of the MATH menu). Use this method to evaluate the function when $x = -13.8$ and when $x = 15.74$. (On TI-82/83 and 92 this can be done simultaneously by using $\{-13.8, 15.74\}$.)

4. Graphical Evaluation Enter the equation as y_1 in the equation memory. To evaluate the function at $x = 9.8$, graph the equation in a window that includes 9.8 on the x-axis (you need not have a complete graph). Then press VALUE (on the TI-82/83 CALC menu or TI-92 graph MATH menu) or EVAL (on the TI-85 GRAPH menu) or Y-CA (on the Casio 9800 G-SOLVE menu) and enter $X = 9.8$. Use this method to evaluate the function when $x = -8$ and when $x = 22.5$.

5. TABLE You can display a table of values of the function on TI-82/83 and 92, HP-38, and Casio 9800. (Table Programs for TI-81 and TI-85, which work similarly, are in the Program Appendix.) First set up the table by specifying the starting value of x and the increment (start at 0 with increment 1 for now) and choose a table type. On TI-82/92 use TBLSET (the increment is denoted "ΔTbl") and set both "Indpnt" and "Depend" at AUTO. On HP-38, use NUM SETUP and set "Numtype" at AUTOMATIC. On Casio 9800 use RNG in the RANGE FUNCTION menu of TABLE mode (increment is denoted "pitch" and you must also enter a maximum value for x; use 12 for now). When the setup is complete, press TABLE (or NUM on HP-38).
   - **(a)** Use the up and down arrow keys to find the value of the function when $x = -4$ and when $x = 11$. [On Casio you can only display values between the minimum and maximum selected in setup, so it won't go to $x = -4$ here.]
   - **(b)** To find the value of the function when $x = 11.4$ and $x = 12.5$, press TBLSET and change the increment to .1. To avoid a lot of scrolling, change the starting value to 11. Then press TABLE or NUM.
   - **(c)** Display a table that shows the value of the function when $x = 35.97$ and $x = 36.03$. Compare the y_1 entries in the table to the values you get by using automatic evaluation (see Investigation 3). Are they different? Why?

EXERCISES 2.1

The notation $[s]$ *used in Exercises 1–8 is explained in Example 5.*

1. $[6.75] = ?$ **2.** $[.75] = ?$ **3.** $[-4/3] = ?$

4. $[-10/3] = ?$ **5.** $[2/3] = ?$ **6.** $[-2/3] = ?$

7. $[-16\frac{1}{2}] = ?$ **8.** $[16.0001] = ?$

In Exercises 9 and 10, give examples of two numbers, u and v, for which the given statement is true and two numbers for which it is false.

9. $[u + v] = [u]$ **10.** $[u] + [v] < [u + v]$

In Exercises 11–14, a functional situation is described that involves at least one function. Verbally describe the domain, range, and rule of each such function, providing there is sufficient information to do so. For example,

Situation: Harry Hamburger owns a professional baseball team. He decides that the only factor to be considered in determining a player's salary for this year is his batting average last season. *Function:* Salaries are a function of batting averages. *Domain:* All team members' batting averages. *Range:* All team members' salaries. *Rule:* There isn't enough information given to determine the rule.

11. The area of a circle depends on its radius.

12. If you drive your car at a constant rate of 55 mph, then your distance from the starting point varies with the time elapsed.

13. When more is spent on advertising at Grump's Department Store, sales increase. If less is spent on advertising, sales drop.

14. A manufacturer has a contract to produce flower pots at a price of $1 each. His profit per pot is determined by the cost of manufacturing the pot. (Note that if it costs more than $1 to make a pot, he loses money; interpret such a loss as "negative profit.") Find an algebraic formula for the rule of this function.

Exercises 15–18 refer to Example 1 in the text. Assume that the state income tax law reads as follows:

Annual Income	*Amount of Tax*
Less than $2000	0
$2000–$6000	2% of income over $2000
More than $6000	$80 plus 5% of income over $6000

15. Find the number in the range (tax amount) that is assigned to each of the following numbers in the domain (incomes):

$500, $1509, $3754,
$6783, $12,500, $55,342.

16. Find four different numbers in the domain of this function that are associated with the same number in the range.

17. Explain why your answer in Exercise 16 does *not* contradict the definition of a function (in the box on page 88).

18. Is it possible to do Exercise 16 if all four numbers in the domain are required to be greater than 2000? Why or why not?

19. The amount of postage required to mail a first-class letter is determined by its weight. In this situation, is weight a function of postage? Or vice versa? Or both?

20. Could the following statement ever be the rule of a function?

Assign to a number x in the domain the number in the range whose square is x.

Why or why not? If there is a function with this rule, what is its domain and range?

In Exercises 21–28, determine whether the given equations defines y as a function of x or defines x as a function of y.

21. $y = 3x^2 - 12$ **22.** $y = 2x^4 + 3x^2 - 2$

23. $y^2 = 4x + 1$ **24.** $5x - 4y^4 + 64 = 0$

25. $3x + 2y = 12$ **26.** $y - 4x^3 - 14 = 0$

27. $x^2 + y^2 = 9$ **28.** $y^2 - 3x^4 + 8 = 0$

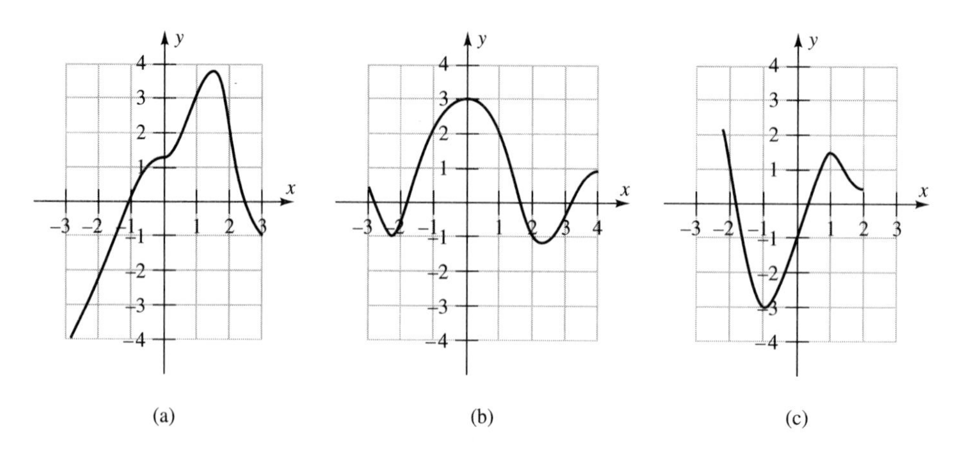

Figure 2-4

Use Figure 2–4 for Exercises 29–35. Each of the graphs in the figure defines a function as in Example 10.

29. State the domain and range of the function defined by graph (a).

30. State the number in the range that the function of Exercise 29 assigns to the following numbers in the domain: $-2, -1, 0, 1$.

31. Do Exercise 30 for these numbers in the domain: $1/2$, $5/2$, $-5/2$.

32. State the domain and range of the function defined by graph (b).

33. State the number in the range that the function of Exercise 32 assigns to the following numbers in the domain: $-2, 0, 1, 2.5, -1.5$.

34. State the domain and range of the function defined by graph (c).

35. State the number in the range that the function of Exercise 34 assigns to the following numbers in the domain: $-2, -1, 0, 1/2, 1$.

36. Explain why none of the graphs in Figure 2–5 defines a function according to the procedure in Example 10. What goes wrong?

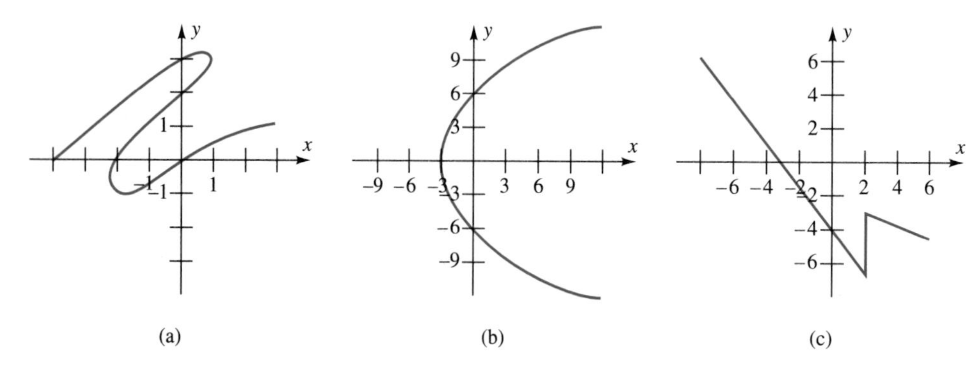

Figure 2-5

Thinkers

37. Consider the function whose rule uses a calculator as follows: ''press COS and then press LN ; then enter a number in the domain and press ENTER or EXE .''* Experiment with this function, then answer the following questions. You may not be able to prove your answers—just make the best estimate you can based on the evidence from your experiments.

(a) What is the largest set of real numbers that could be used for the domain of this function? [If applying the rule to a number produces an error message or a complex number (on TI-85, TI-92, or HP-38), that number cannot be in the domain.]

(b) Using the domain in part (a), what is the range of this function?

38. Do Exercise 37 for the function whose rule is ''press 10^x and then press TAN ;then enter a number in the domain and press ENTER .''

39. The *integer part* function has the set of all real numbers (written as decimals) as its domain. The rule is ''assign to a number in the domain the part of the number to the left of the decimal point.'' For instance, the input 37.986 produces the output 37 and the input -1.5 produces the output -1. On most calculators, the integer part function is denoted ''iPart.'' On calculators that use ''Intg'' or ''Floor'' for the greatest integer function, the integer part function is denoted by ''INT.''

(a) For each nonnegative real number input, explain why both the integer part function and the greatest integer function [Example 5] produce the same output.

(b) For which negative numbers do the two functions produce the same output?

(c) For which negative numbers do the two functions produce different outputs?

2.2 FUNCTIONAL NOTATION

Functional notation is a convenient shorthand language that facilitates the analysis of mathematical problems involving functions. Here are Examples 1 and 3 from the last section.

EXAMPLE 1 ***(Income Tax)*** One state's income tax law reads:

Income	*Amount of Tax*
\$5000 or less	0
More than \$5000	10% of income over \$5000

Let I denote income and write $T(I)$ (read ''T of I'') to denote the amount of tax on income I. In this shorthand language $T(7500)$ denotes ''the tax on an income of \$7500.'' The sentence ''The tax on an income of \$7500 is \$250'' is abbreviated as $T(7500) = 250$. There is nothing that forces us to use the letters T and I here:

Any choice of letters or symbols will do, provided we make clear what is meant by the letters or symbols chosen. ■

*You don't need to know what these keys mean in order to do this exercise. On a standard nongraphing scientific calculator the rule would be ''enter a number, press the ln key and then the COS key.''

EXAMPLE 3 *(Falling Rock)* Let t denote the time (in seconds) and $d(t)$ "the distance (in feet) that the rock has traveled after t seconds." The fact that "the distance traveled after t seconds is $16t^2$ feet" can then be abbreviated as $d(t) = 16t^2$. For instance,

$$d(1) = 16 \cdot 1^2 = 16$$

means "the distance the rock has traveled after 1 second is 16 feet" and

$$d(4) = 16 \cdot 4^2 = 256$$

means "the distance the rock has traveled after 4 seconds is 256 feet." ■

WARNING The parentheses in $d(t)$ do *not* denote multiplication as in the algebraic equation $3(a + b) = 3a + 3b$. The entire symbol $d(t)$ is part of a *shorthand language.* In particular,

$$d(1 + 4) \text{ is } \textit{not} \text{ equal to } d(1) + d(4).$$

For we saw above that $d(1) = 16$ and $d(4) = 256$, so that $d(1) + d(4) = 16 + 256 = 272$. But $d(1 + 4)$ is "the distance traveled after $1 + 4$ seconds," that is, the distance after 5 seconds, namely, $16 \cdot 5^2 = 400$. In general,

Functional notation is a convenient shorthand for phrases and sentences in the English language. It is *not* the same as ordinary algebraic notation.

Functional notation is easily adapted to the usual mathematical setting where the particulars of time, distance, etc., are eliminated. Suppose a function is given. Denote the function by f and let x denote a number in the domain. Then

$f(x)$ denotes the number in the range determined by the number x according to the rule of the function f.

In other words,

$f(x)$ is the output produced by input x.

For example $f(6)$ is the output produced by input 6, that is, $f(6)$ is the number in the range determined by the number 6 in the domain. The sentence

"y is the number in the range determined by the number x in the domain according to the rule of the function f"

is then abbreviated as

$$y = f(x)$$

which is read "y equals f of x."

Each of the following sentences means *exactly the same thing* as $y = f(x)$:

The **value of the function** f at x is y.
The function f **maps** x to y.
y is the **image** of x under (the function) f.

Similarly, the number $f(x)$ is sometimes called

the **value** of (the function) f at x, or
the **image** of x (under the function f).

> **▶ TECHNOLOGY TIP**
>
> Functional notation can be used directly on TI-82/83, TI-92, and HP-38. For example, if the function is entered in the equation memory as y_1, then keying in $y_1(5)$ ENTER evaluates the function at $x = 5$.
>
> See Calculator Investigation 3 in Section 2.1 for details.

In actual practice, functions are seldom presented in the style of domain, rule, range, as they have been here. Usually the rule is given by an equation or formula, such as $f(x) = \sqrt{x^2 + 1}$. This formula can be thought of as a set of directions:

Name of function — *Input number*

$$f(x) = \sqrt{x^2 + 1}$$

Output number — *Directions that tell you what to do with input x in order to produce the corresponding output f(x), namely, "square it, add 1, and take the square root of the result."*

For example, to find $f(3)$, simply replace x by 3 in the formula:

$$f(3) = \sqrt{3^2 + 1} = \sqrt{10}.$$

Similarly, replacing x by -5 and 0 shows that

$$f(-5) = \sqrt{(-5)^2 + 1} = \sqrt{26} \quad \text{and} \quad f(0) = \sqrt{0^2 + 1} = 1.$$

These directions can be applied to any quantities, such as $a + b$ or c^4 (where a, b, c are real numbers). Thus, to compute $f(a + b)$, the output corresponding to input $a + b$, we square the input [obtaining $(a + b)^2$], add 1 [obtaining $(a + b)^2 + 1$] and take the square root of the result:

$$f(a + b) = \sqrt{(a + b)^2 + 1} = \sqrt{a^2 + 2ab + b^2 + 1}.$$

Similarly, the output $f(c^4)$ corresponding to the input c^4 is computed by squaring the input $[(c^4)^2]$, adding 1 $[(c^4)^2 + 1]$, and taking the square root of the result:

$$f(c^4) = \sqrt{(c^4)^2 + 1} = \sqrt{c^8 + 1}.$$

> **▶ TECHNOLOGY TIP**
> Learn how to use your calculator to evaluate functions quickly and easily, with a minimum of keystrokes.
>
> In particular, TI, HP-38, and Casio 9800 users should know how to display a table of values for a function.
>
> See Calculator Investigations 1–5 in Section 2.1 for details.

EXAMPLE 1 The expression $h(x) = \dfrac{x^2 + 5}{x - 1}$ defines the function h whose rule is: Assign to x the number $\dfrac{x^2 + 5}{x - 1}$. Find each of the following:

$$h(\sqrt{3}), \qquad h(-2), \qquad h(-a), \qquad h(r^2 + 3), \qquad h(\sqrt{c + 2}).$$

Solution To find $h(\sqrt{3})$ and $h(-2)$, replace x by $\sqrt{3}$ and -2, respectively, in the rule of h:

$$h(\sqrt{3}) = \frac{(\sqrt{3})^2 + 5}{\sqrt{3} - 1} = \frac{8}{\sqrt{3} - 1} \quad \text{and} \quad h(-2) = \frac{(-2)^2 + 5}{-2 - 1} = -3.$$

The value of the function h at any quantity, such as $-a$, $r^2 + 3$, etc., can be found by using the same procedure: *replace x in the formula for $h(x)$ by that quantity:*

$$h(-a) = \frac{(-a)^2 + 5}{-a - 1} = \frac{a^2 + 5}{-a - 1}$$

$$h(r^2 + 3) = \frac{(r^2 + 3)^2 + 5}{(r^2 + 3) - 1} = \frac{r^4 + 6r^2 + 9 + 5}{r^2 + 2} = \frac{r^4 + 6r^2 + 14}{r^2 + 2}$$

$$h(\sqrt{c + 2}) = \frac{(\sqrt{c + 2})^2 + 5}{\sqrt{c + 2} - 1} = \frac{c + 2 + 5}{\sqrt{c + 2} - 1} = \frac{c + 7}{\sqrt{c + 2} - 1}. \quad ■$$

When functional notation is used in expressions such as $f(-x)$ or $f(x + h)$, the same basic rule applies: Replace x in the formula by the *entire* expression in parentheses.

EXAMPLE 2 If $f(x) = x^2 + x - 2$, then

$$f(-x) = (-x)^2 + (-x) - 2 = x^2 - x - 2$$

Note that in this case $f(-x)$ is *not* the same as $-f(x)$, because $-f(x)$ is the negative of the number $f(x)$, that is,

$$-f(x) = -(x^2 + x - 2) = -x^2 - x + 2. \quad ■$$

If f is any function and h is a nonzero number, then the **difference quotient** of f is the quantity

$$\frac{f(x + h) - f(x)}{h}.$$

Difference quotients, whose significance is explained in Section 3.2, play an important role in calculus. You must be able to use functional notation correctly in order to compute and simplify them.

EXAMPLE 3 Find the difference quotient of $f(x) = x^2 - x + 2$.

Solution Use the definition of difference quotient and algebra:

$$\frac{f(x+h)-f(x)}{h} = \frac{\overbrace{[(x+h)^2-(x+h)+2]}^{f(x+h)} - \overbrace{[x^2-x+2]}^{f(x)}}{h}$$

$$= \frac{(x^2+2xh+h^2)-(x+h)+2-(x^2-x+2)}{h}$$

$$= \frac{x^2+2xh+h^2-x-h+2-x^2+x-2}{h}$$

$$= \frac{2xh+h^2-h}{h} = \frac{h(2x+h-1)}{h} = 2x-1+h \quad ■$$

There are a number of common mistakes in using functional notation, most of which arise from treating it as ordinary algebra rather than a specialized shorthand language.

WARNING ***Common Mistakes with Functional Notation***

Each of the following statements may be FALSE:

1. $f(a+b) = f(a) + f(b)$
2. $f(a-b) = f(a) - f(b)$
3. $f(ab) = f(a)f(b)$
4. $f(ab) = af(b)$
5. $f(ab) = f(a)b$

EXAMPLE 4 Here are examples of three of the errors listed above.

1. If $f(x) = x^2$, then $f(3+2) = f(5) = 5^2 = 25$. But $f(3) + f(2) = 3^2 + 2^2 = 9 + 4 = 13$. So, $f(3+2) \neq f(3) + f(2)$.
3. If $f(x) = x + 7$, then $f(3 \cdot 4) = f(12) = 12 + 7 = 19$. But $f(3)f(4) = (3+7)(4+7) = 10 \cdot 11 = 110$. So $f(3 \cdot 4) \neq f(3)f(4)$.
5. If $f(x) = x^2 + 1$, then $f(2 \cdot 3) = (2 \cdot 3)^2 + 1 = 36 + 1 = 37$. But $f(2) \cdot 3 = (2^2 + 1)3 = 5 \cdot 3 = 15$. So $f(2 \cdot 3) \neq f(2) \cdot 3$. ■

Domains

When the rule is given by a formula, as above, the domain is determined by this convention.

Domain Convention ▶

Unless specific information to the contrary is given, the domain of a function f is taken to be the set consisting of every real number for which the rule of f produces a real number.

EXAMPLE 5 The expression $f(x) = \sqrt{x^2 + 1}$ defines the function f whose rule is: Assign to x the number $\sqrt{x^2 + 1}$. Since $x^2 \geq 0$ always, $\sqrt{x^2 + 1}$ is a real number for every x. So the domain of the function f consists of all real numbers. ■

There are two common situations in which the domain of a function may not consist of all real numbers. These occur when applying the rule of a function leads to division by zero or to the square root of a negative number.

EXAMPLE 6 The expression $k(x) = \dfrac{x^2 - 6x}{x - 1}$ defines the function k whose rule is: Assign to x the number $\dfrac{x^2 - 6x}{x - 1}$. When $x = 1$, the denominator of $\dfrac{x^2 - 6x}{x - 1}$ is 0, but when $x \neq 1$, $\dfrac{x^2 - 6x}{x - 1}$ is a real number. Therefore the domain of k consists of all real numbers *except* 1. ■

EXAMPLE 7 Find the domain of the function given by

$$f(u) = \sqrt{u + 2}.$$

Solution $\sqrt{u + 2}$ is a real number only when the expression under the radical is nonnegative, that is, when $u + 2 \geq 0$. However,

$$u + 2 \geq 0 \qquad \text{is equivalent to} \qquad u \geq -2.$$

Therefore, the domain of f consists of all real numbers greater than or equal to -2, that is, the interval $[-2, \infty)$. ■

EXAMPLE 8 A **piecewise-defined function** is one whose rule involves several formulas, such as:

$$f(x) = \begin{cases} 2x + 3 & \text{if } x < 4 \\ x^2 - 1 & \text{if } 4 \leq x \leq 10 \end{cases}$$

The rule of f gives no directions for numbers x with $x > 10$. So the domain of f is the interval $(-\infty, 10]$. If c is a real number, you cannot find $f(c)$ unless you know whether $c < 4$ or $4 \leq c \leq 10$.

To find $f(-5)$, note that $-5 < 4$, so the first part of the rule applies: $f(-5) = 2(-5) + 3 = -7$. To find $f(8)$, however, use the second part of the rule (because 8 is between 4 and 10): $f(8) = 8^2 - 1 = 63$. Also use the second part of the rule to find $f(4) = 4^2 - 1 = 15$. ■

Applications

Most applications require one quantity to be expressed as a function of another. In such situations, the domain convention presented above may not be applicable. The rule of the distance function for falling objects, for example, is $d(t) = 16t^2$ (see Example 3, p. 96).* According to the domain convention, its domain is the set of all real numbers (since $16t^2$ is a real number for any t). Such a domain, however, doesn't make sense in the real world.

Since t represents time, only nonnegative values of t are meaningful here. Furthermore, if the object were dropped from the top of a fifty-story building, it would hit the ground in approximately 6 seconds. In this case, therefore, the approximate domain of the distance function would be the interval $[0, 6]$. Similarly, the range of the distance function would consist of all real numbers between 0 and the height of the building. Analogous comments apply to other applications:

A real-life situation may lead to a function whose domain does not include all the numbers for which the rule of the function is defined.

EXAMPLE 9 A glassware factory has fixed expenses (mortgage, machinery, etc.) of \$2000 per week. It costs 50¢ to manufacture one cup and at most 17,000 cups can be manufactured each week. Experience has shown that no cups can be sold at a price of \$2 and 1000 cups per week can be sold at a price of \$1.90 each. For each 10¢ decrease in price, 1000 more cups will be sold each week (that is, 2000 per week at \$1.80 each, 3000 at \$1.70 each, and so on). Express each of the following as a function of the number of cups sold each week:

(a) price per cup;

(b) weekly income of the factory;

(c) weekly costs of the factory;

(d) weekly profit of the factory.

Finally, what is the domain of each of these functions?

*Wind resistance is ignored in this example.

Solution

(a) Let x be the number of cups sold each week. Then $x/1000$ is the number of *thousands* of cups sold. No cups can be sold at \$2 each. For each thousand cups to be sold, the price must be decreased by 10¢. Consequently, if $p(x)$ is the price per cup when x cups are sold per week, then

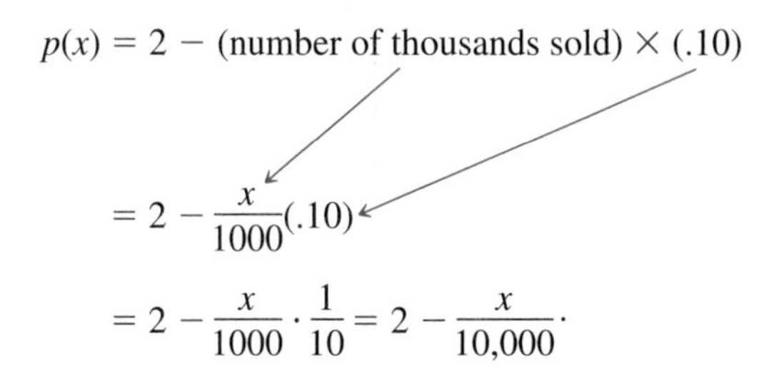

$$p(x) = 2 - (\text{number of thousands sold}) \times (.10)$$

$$= 2 - \frac{x}{1000}(.10)$$

$$= 2 - \frac{x}{1000} \cdot \frac{1}{10} = 2 - \frac{x}{10{,}000}.$$

(b) If $I(x)$ is the income from selling x cups per week, then

$$I(x) = (\text{price per cup}) \times (\text{number sold})$$

$$= \left(2 - \frac{x}{10{,}000}\right) x$$

$$= 2x - \frac{x^2}{10{,}000}.$$

(c) If $C(x)$ is the cost of manufacturing x cups per week, then

$$C(x) = (\text{cost per cup}) \times (\text{number sold}) + (\text{fixed expenses})$$

$$= .50x + 2000.$$

(d) If $P(x)$ is the profit from selling x cups per week, then

$$P(x) = \text{Income} - \text{Cost} = I(x) - C(x)$$

$$= \left(2x - \frac{x^2}{10{,}000}\right) - (.50x + 2000)$$

$$= -\frac{x^2}{10{,}000} + 1.5x - 2000.$$

The rules of each of the functions $p(x)$, $I(x)$, $C(x)$, $P(x)$ are defined for all real numbers. Each of their domains, however, consists of the possible number of cups that can be manufactured each week. Furthermore, only whole cups are made. Hence the domain consists of all integers in the interval $[0, 17{,}000]$. ■

CALCULATOR INVESTIGATIONS 2.2

1. Function Evaluation Find the easiest way to evaluate a function on your calculator. See Calculator Investigations 1–5 in Section 2.1 for some of the possibilities.
2. Custom Menu The CUSTOM key on TI-85 brings up a menu selected by the user. It is convenient to have frequently used operations on the Custom Menu so that they can be performed with a minimum of keystrokes (and no searching in submenus). Learn how to put "eval" (from the MISC submenu of the MATH menu) onto your Custom Menu so that you can quickly and easily evaluate functions.

EXERCISES 2.2

Exercises 1–24 refer to these three functions:

$$f(x) = \sqrt{x+3} - x + 1 \qquad g(t) = t^2 - 1$$

$$h(x) = x^2 + \frac{1}{x} + 2$$

In each case find the indicated value of the function.

1. $f(0)$ **2.** $f(1)$

3. $f(5/2)$ **4.** $f(\pi)$

5. $f(\sqrt{2})$ **6.** $f(\sqrt{2} - 1)$

7. $f(-2)$ **8.** $f(-3/2)$

9. $h(3)$ **10.** $h(-4)$

11. $h(3/2)$ **12.** $h(\pi + 1)$

13. $h(a + k)$ **14.** $h(-x)$

15. $h(2 - x)$ **16.** $h(x - 3)$

17. $g(3)$ **18.** $g(-2)$

19. $g(0)$ **20.** $g(x)$

21. $g(s + 1)$ **22.** $g(1 - r)$

23. $g(-t)$ **24.** $g(t + h)$

In Exercises 25–32, compute:

(a) $f(r)$ **(b)** $f(r) - f(x)$ **(c)** $\dfrac{f(r) - f(x)}{r - x}$

In part (c) assume $r \neq x$ and simplify your answer. Example: If $f(x) = x^2$, then

$$\frac{f(r) - f(x)}{r - x} = \frac{r^2 - x^2}{r - x} = \frac{(r + x)(r - x)}{r - x} = r + x.$$

25. $f(x) = x$ **26.** $f(x) = -10x$

27. $f(x) = 3x + 7$ **28.** $f(x) = x^3$

29. $f(x) = x - x^2$ **30.** $f(x) = x^2 + 1$

31. $f(x) = \sqrt{x}$ **32.** $f(x) = 1/x$

In Exercises 33–40, assume $h \neq 0$. Compute and simplify the difference quotient

$$\frac{f(x + h) - f(x)}{h}.$$

33. $f(x) = x + 1$ **34.** $f(x) = -10x$

35. $f(x) = 3x + 7$ **36.** $f(x) = x^2$

37. $f(x) = x - x^2$ **38.** $f(x) = x^3$

39. $f(x) = \sqrt{x}$ **40.** $f(x) = 1/x$

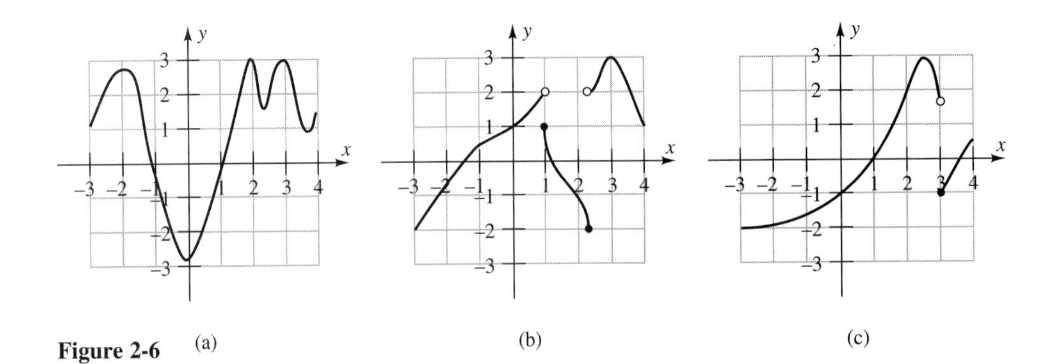

Figure 2-6

Exercises 41–44 refer to the graphs in Figure 2–6, each of which defines a function with domain $[-3, 4]$, as in Example 10 on page 91.

41. In graph (a), find $f(-3), f(-3/2), f(0), f(1), f(5/2), f(4)$. Careful approximate answers are acceptable.

42. In graph (a), find $f(0) - f(2)$; $f(5/2) - f(3)$; $f(4) + 3f(-2)$.

43. In graph (b), find $f(-5/2), f(-3/2), f(0), f(3), f(4)$.

44. In graph (c), find $f(-3) - f(3)$; $f(0) + f(1)$; $3f(2) + 2$.

45. In a certain state the sales tax $T(p)$ on an item of price p dollars is 5% of p. Which of the following formulas give the correct sales tax in all cases?
 (i) $T(p) = p + 5$
 (ii) $T(p) = 1 + 5p$
 (iii) $T(p) = p/20$
 (iv) $T(p) = p + (5/100)p = p + .05p$
 (v) $T(p) = (5/100)p = .05p$

46. Let T be the sales tax function of Exercise 45 and find $T(3.60)$, $T(4.80)$, $T(.60)$, and $T(0)$.

In Exercises 47–64, determine the domain of the function according to the usual convention.

47. $f(x) = x^2$ **48.** $g(x) = (1/x^2) + 2$

49. $h(t) = |t| - 1$ **50.** $k(u) = \sqrt{u}$

51. $f(x) = [x]^2$ [*Hint:* See Example 5 in Section 2.1.]

52. $g(t) = |t - 1|$ **53.** $k(x) = |x| + \sqrt{x} - 1$

54. $h(x) = \sqrt{(x + 1)^2}$

55. $g(u) = \dfrac{|u|}{u}$ **56.** $h(x) = \dfrac{\sqrt{x - 1}}{x^2 - 1}$

57. $g(t) = \sqrt{t^2}$ **58.** $f(x) = x^3 + 2$

59. $g(y) = [-y]$ **60.** $f(t) = \sqrt{-t}$

61. $g(u) = \dfrac{u^2 + 1}{u^2 - u - 6}$ **62.** $f(t) = \sqrt{4 - t^2}$

63. $f(x) = -\sqrt{9 - (x - 9)^2}$ **64.** $f(x) = \sqrt{-x} + \dfrac{2}{x + 1}$

65. Give an example of two different functions f and g that have all of the following properties:

$$f(-1) = 1 = g(-1) \quad \text{and} \quad f(0) = 0 = g(0)$$
$$\text{and} \quad f(1) = 1 = g(1).$$

66. Give an example of a function h that has the property that $h(u) = h(2u)$ for every real number u.

67. Give an example of a function f that has the property that $f(x) = 2f(x)$ for every real number x.

68. Give an example of a function g with the property that $g(x) = g(-x)$ for every real number x.

In Exercises 69–72, a function f is given. Determine whether each of the following statements about the function f is true for every real number x. If a statement is not true, give a numerical example to demonstrate this fact.
 (i) $f(x^2) = (f(x))^2$ [*Hint:* Use the rule of f to compute $f(x^2)$; do you get the same expression as $f(x)\cdot f(x)$?]
 (ii) $f(-x) = f(x)$ [*Hint:* Use the rule of f to compute $f(-x)$; is this expression the same as $f(x)$?]
 (iii) $f(|x|) = |f(x)|$
 (iv) $f(3x) = 3f(x)$?

69. $f(x) = 5x$ **70.** $f(x) = -x$ **71.** $f(x) = x^2$

72. $f(x) = |x|$

73. Define a function that expresses the circumference of a circle as a function of the radius.

74. Define a function that expresses the area of a circle as a function of the diameter.

75. Define a function that expresses the area of a square as a function of the length of a side of the square.

76. Define a function that expresses the area of a square as a function of the length of a diagonal of the square.

77. Suppose a car travels at a constant rate of 55 mph for 2 hours and travels at 45 mph thereafter. Show that distance traveled is a function of time and find the rule of the function.

78. A man walks for 45 minutes at a rate of 3 mph, then jogs for 75 minutes at a rate of 5 mph, then sits and rests for 30 minutes, and finally walks for $1\frac{1}{2}$ hours. Find the rule of the function that expresses his distance traveled as a function of time. [*Warning:* Don't mix up the units of time; use either minutes or hours, not both.]

79. The distance between city C and city S is 2000 miles. A plane flying directly to S passes over C at noon. If the plane travels at 475 mph, express the distance of the plane from city S as a function of time.

80. Do Exercise 79 for a plane that travels at 325 mph.

81. The list price of a workbook is \$12. But if 10 or more copies are purchased, then the price per copy is reduced by 25¢ for every copy above 10. (That is, \$11.75 per copy for 11 copies, \$11.50 per copy for 12 copies, and so on.)

(a) The price per copy is a function of the number of copies purchased. Find the rule of this function.

(b) The total cost of a quantity purchase is

$$(\text{number of copies}) \times (\text{price per copy}).$$

Show that the total cost is a function of the number of copies and find the rule of the function.

82. A potato chip factory has a daily overhead from salaries and building costs of \$1800. The cost of ingredients and packaging to produce a pound of potato chips is 50¢. A pound of potato chips sells for \$1.20. Show that the factory's daily profit is a function of the number of pounds of potato chips sold and find the rule of this function. (Assume that the factory sells all the potato chips it produces each day.)

83. A rectangular region of 6000 sq ft is to be fenced in on three sides with fencing costing \$3.75 per ft and on the fourth side with fencing costing \$2.00 per ft. Express the cost of the fence as a function of the length x of the fourth side.

84. A box with a square base measuring $t \times t$ ft is to be made of three kinds of wood. The cost of the wood for the base is 85¢ per sq ft; the wood for the sides costs 50¢ per sq ft and the wood for the top \$1.15 per sq ft. The volume of the box is to be 10 cu ft. Express the total cost of the box as a function of the length t.

2.3 GRAPHS OF FUNCTIONS AND EQUATIONS

The graph of a function f is the graph of the *equation* $y = f(x)$. So, the point (x, y) is on the graph of f precisely when x is a number in the domain of f and y is the number $f(x)$, the value of the function at x. In other words,

The graph of the function f consists of all points $(x, f(x))$, where x is any number in the domain of f.

Graphs of functions can be distinguished from other graphs by:

The Vertical Line Test ▶

The graph of a function $y = f(x)$ has this property:

No vertical line intersects the graph more than once.

Conversely, any graph with this property is the graph of a function.

To see why this statement is true, note that a point with first coordinate $x = c$ on the graph of a function f must have second coordinate $f(c)$. Since a function cannot have two different values at c, there can't be two different points on the graph of f with first coordinate c. But distinct points on a vertical line necessarily have the same first coordinate and different second coordinates. So the graph of a function cannot contain more than one point from any vertical line. For the converse statement, see Exercise 70.

Technology makes it relatively easy to graph most functions, but it is not appropriate in every situation. For instance, there are some things that you should just *know*—like you know your multiplication tables without using a calculator. In particular, you should know the shapes of the graphs of the following basic functions and be able to sketch these graphs without a calculator:

$$f(x) = x, \qquad g(x) = x^2, \qquad h(x) = x^3,$$
$$k(x) = |x|, \qquad p(x) = \sqrt{x}, \qquad q(x) = \sqrt[3]{x}.$$

If you can't do that now, see Exercises 1–6.

The emphasis in this book is on analyzing various algebraic facts in order to "get a feel" for a graph, so that you will know what viewing window to use to obtain a complete graph and how to interpret the graphs produced by the calculator or computer. Occasionally such analysis leads directly to the graph, without the need for any technology.

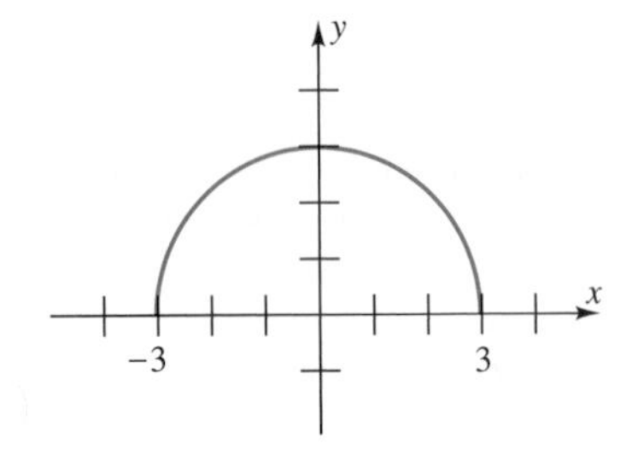

Figure 2-7

EXAMPLE 1 To graph $f(x) = \sqrt{9 - x^2}$, we must graph the equation $y = \sqrt{9 - x^2}$, in which y is always nonnegative. Squaring both sides of this equation shows that $y^2 = 9 - x^2$, so that $x^2 + y^2 = 9$. The graph of this last equation is a circle with center (0, 0) and radius 3 (see Section 1.3). Consequently, the graph of the function f will be the upper half of this circle (the points with nonnegative second coordinates), as shown in Figure 2–7. ■

We now consider two types of functions whose calculator-generated graphs may be misleading and require interpretation: linear functions and piecewise-defined functions. A **linear function** is one whose rule is of the form $f(x) = ax + b$, with a and b constants, such as:

$$f(x) = 3x - 4, \qquad g(x) = -2x + 2, \qquad h(x) = -6.$$

We shall assume here a fact that will be proved in Section 3.1: *The graph of every linear function is a straight line.*

EXAMPLE 2 Figure 2–8 shows the graphs of the three linear functions mentioned in the preceding paragraph. Each graph was obtained by plotting two points and drawing the straight line determined by them.

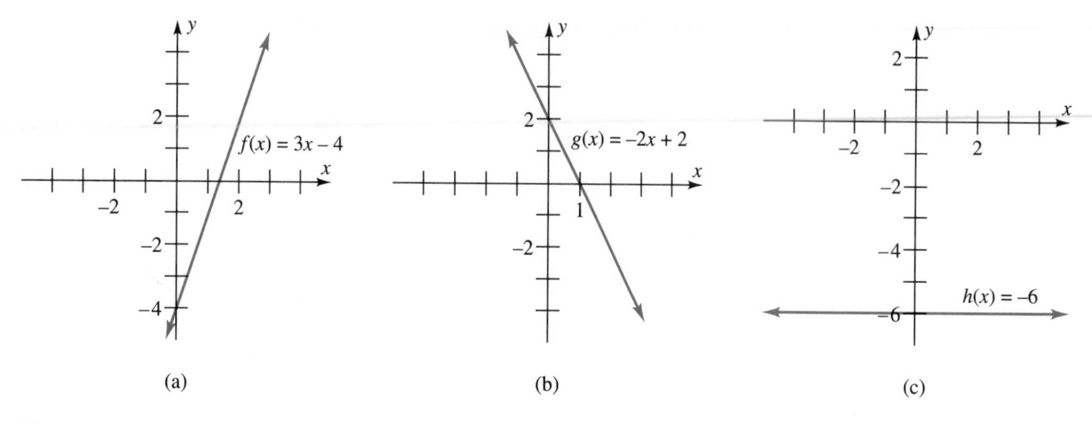

Figure 2-8

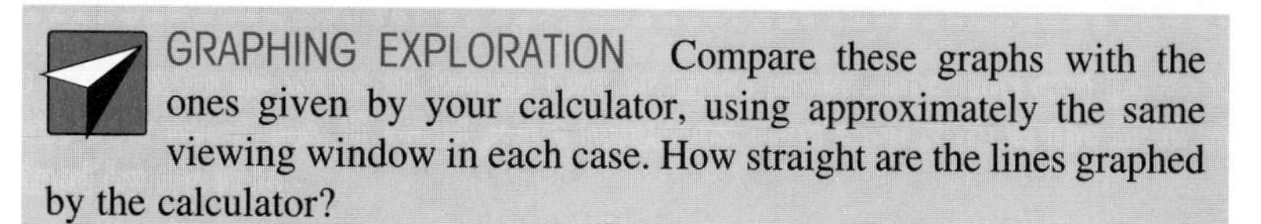
GRAPHING EXPLORATION Compare these graphs with the ones given by your calculator, using approximately the same viewing window in each case. How straight are the lines graphed by the calculator? ■

It is often easier to graph simple piecewise-defined functions by hand than to use a calculator.* Even when a calculator is used, its display does not indicate which endpoints are part of the graph.

EXAMPLE 3 The graph of the piecewise-defined function

$$h(x) = \begin{cases} x^2 & \text{if } x < 0 \\ x & \text{if } 0 \le x < 5 \\ -2x + 11 & \text{if } x \ge 5 \end{cases}$$

is made up of *parts* of the graphs of three functions:

$x < 0$ For these values of x, the graph of h coincides with the graph of $f(x) = x^2$, which is sketched in Figure 2–11(a) on page 109.

$0 \le x < 5$ For these values, the graph of h coincides with the graph of $g(x) = x$, which is a straight line.

$x \ge 5$ For these values of x, the graph of h coincides with the graph of $k(x) = -2x + 11$, which is also a straight line.

Therefore we must graph:

$f(x) = x^2$ when $x < 0$;

$g(x) = x$ when $0 \le x < 5$;

$k(x) = -2x + 11$ when $x \ge 5$.

*Piecewise-defined functions were introduced in Example 8 on page 100.

Combining these partial graphs produces the graph of h (Figure 2-9).

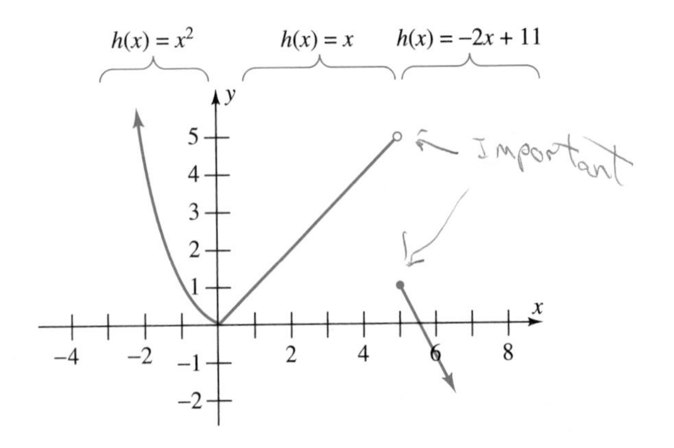

Figure 2-9

The open circle at (5, 5) indicates that this point is *not* on the graph of h (although it is on the graph of $y = x$). The closed circle at (5, 1) indicates that this point *is* on the graph of h.

The best way to graph h on HP-38 or TI calculators (except TI-92) is to use the Tip in the margin to graph three separate equations on the same screen:

$$y = x^2/(x < 0), \quad y = x/((x \geq 0)(x < 5)), \quad y = (-2x + 11)/(x \geq 5).^*$$

> **▶ TECHNOLOGY TIP**
> Inequality symbols are in the TEST menu or the TEST submenu of the MATH menu on TI and HP-38 calculators.

TI-92 uses a different syntax; check your instruction manual. To graph h on Casio, in a viewing window with $-4 \leq x \leq 8$, graph these three equations (including commas and square brackets) on the same screen:

$$y = x^2, [-4, 0] \qquad y = x, [0, 5] \qquad y = -2x + 11, [5, 8].$$

> GRAPHING EXPLORATION If possible, display the graph of the function h on your calculator. How does it compare with Figure 2–9? Note that the calculator does not indicate which endpoints are part of the graph and which ones are not.

■

EXAMPLE 4 The greatest integer function was defined on page 89. When dealing with a function whose rule involves the greatest integer function, such as $g(x) = x - [x]$, consider what happens between each two consecutive integers. For instance,

*The method suggested by the instruction manuals for these calculators, namely, to graph the single equation

$$y = (x^2)(x < 0) + (x)(x \geq 0)(x < 5) + (-2x + 11)(x \geq 5),$$

also works, but may produce erroneous vertical line segments between pieces of the graph that should not be connected.

If	*Then* $[x] =$	*So That* $x - [x] =$
$-2 \le x < -1$	-2	$x - (-2) = x + 2$
$-1 \le x < 0$	-1	$x - (-1) = x + 1$
$0 \le x < 1$	0	x
$1 \le x < 2$	1	$x - 1$
$2 \le x < 3$	2	$x - 2$

Thus, g is actually a piecewise-defined function and each part of its graph is a straight line segment.

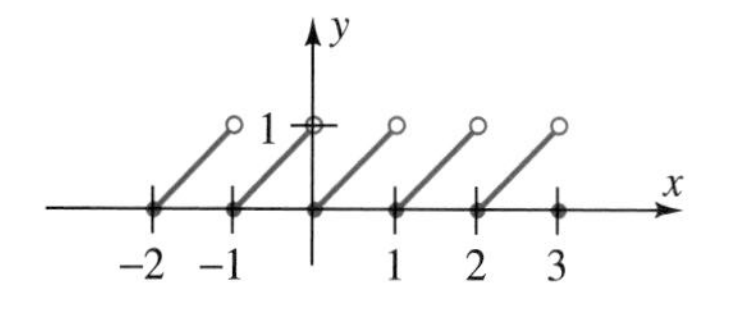

Figure 2-10

GRAPHING EXPLORATION If possible, graph $g(x)$ on your calculator. Compare the result with the part of the graph of g in Figure 2–10, which was obtained by using the preceding table to plot points.

Increasing and Decreasing Functions

A function is said to be **increasing on an interval** if its graph always *rises* as you move from left to right over the interval. The functions whose graphs are pictured in Figure 2–11 are increasing on the stated intervals.

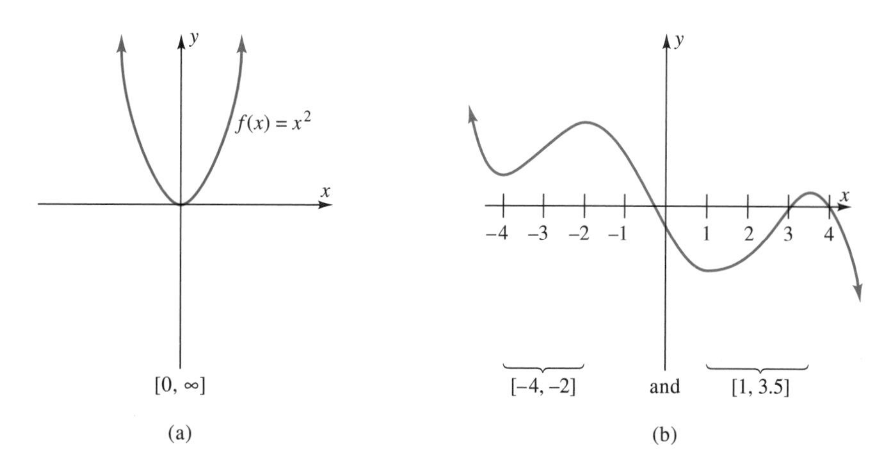

Figure 2-11

A function is said to be **decreasing on an interval** if its graph always *falls* as you move from left to right over the interval. For example, $f(x) = x^2$, whose graph is shown in Figure 2–11(a) is decreasing on the interval $(-\infty, 0]$. The function in Figure 2–11(b) is decreasing on the intervals $(-\infty, -4]$, $[-2, 1]$, and $[3.5, \infty)$.

By examining Figure 2–12 we obtain an algebraic description of increasing and decreasing functions:

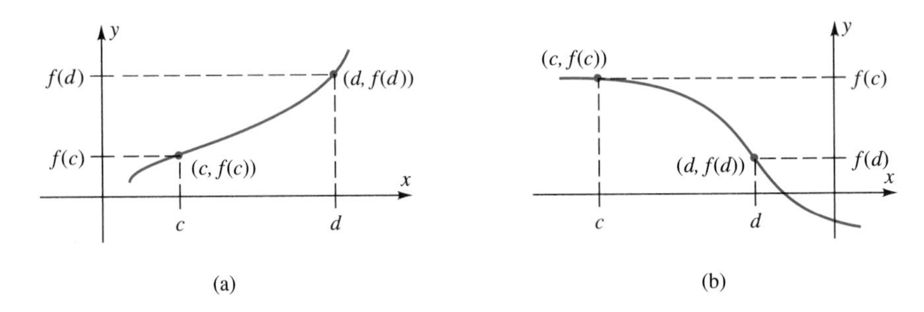

Figure 2-12

Suppose $c < d$. If the graph moves upward as you go from $x = c$ to $x = d$ (Figure 2–12(a)), then the second coordinate at d must be *larger* than the second coordinate at c. Thus,

> **The function f is increasing on an interval provided that for any numbers c, d in the interval,**
>
> **whenever $c < d$, then $f(c) < f(d)$.**

Similarly, if the graph moves downward (Figure 2–12(b)), the second coordinate at d must be *smaller* than the second coordinate at c. Hence,

> **The function f is decreasing on an interval provided that for any numbers c and d in the interval,**
>
> **whenever $c < d$, then $f(c) > f(d)$.**

EXAMPLE 5 On what (approximate) intervals is the function $f(x) = .5x^3 - 3x$ increasing and on what intervals is it decreasing?

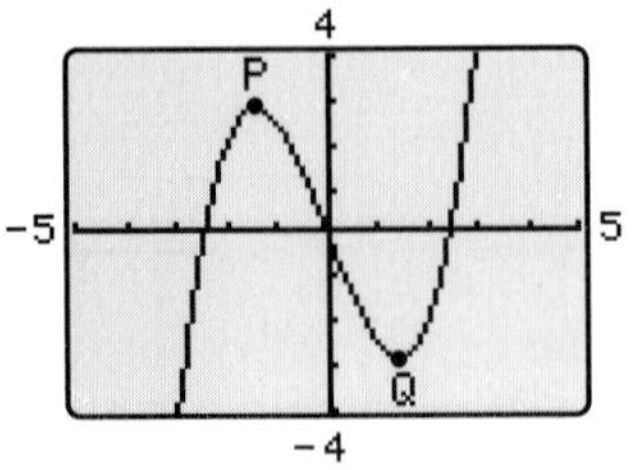

Solution The (complete) graph of f in Figure 2–13 rises to P, falls from P to Q, and then rises again. By using zoom-in or a min/max finder we determine that the approximate coordinates of P and Q are

$$P = (-1.4142, 2.8284) \qquad \text{and} \qquad Q = (1.4142, -3.8284).$$

Hence, the function is increasing on the approximate intervals $(-\infty, -1.4142)$ and $(1.4142, \infty)$ and decreasing on the interval $(-1.4142, 1.4142)$. ■

Figure 2-13

A function whose graph appears ''flat'' over an interval may be increasing or decreasing there.

> GRAPHING EXPLORATION Graph the function $g(x) = 2x^3 + x^2 + 1$ over the interval $(-.5, .1)$. Use the trace feature to determine whether g is increasing or decreasing on this interval. [As you move from left to right the x-coordinates increase; what happens to the corresponding y-coordinates?]

Equation Graphing

Most graphing calculators have several graphing modes. Up to this point we have used ''function'' graphing mode (called ''xy-coordinate'' or ''rectangular mode'' on some calculators), which is available on every graphing calculator. In this mode a calculator can graph only equations that define y as a function of x, that is, equations whose graphs pass the vertical line test. However, a calculator in function graphing mode can be used to graph equations whose graphs don't pass the vertical line test, provided that the graph of the equation is made up of graphs of functions of the form $y = f(x)$.

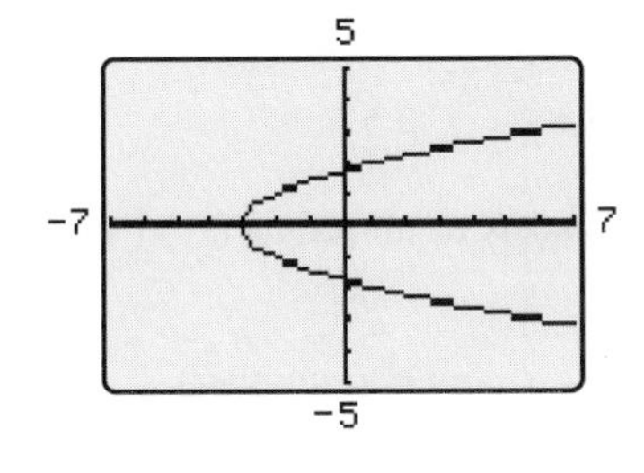

Figure 2-14

EXAMPLE 6 The equation $y^2 = x + 3$ does not define y as a function of x (why?). To graph this equation on a calculator, we first solve for y:

$$y = \sqrt{x+3} \qquad \text{or} \qquad y = -\sqrt{x+3}.$$

Each of these last two equations defines y as a function of x (since in each case there is a unique value of y for each x in the domain). By graphing both $y = \sqrt{x+3}$ and $y = -\sqrt{x+3}$ on the same screen, we obtain the graph of the original equation in Figure 2–14. ■

EXAMPLE 7 To graph the equation $12x^2 - 4y^2 + 16x + 12 = 0$, solve the equation for y:

$$4y^2 = 12x^2 + 16x + 12$$

$$y^2 = 3x^2 + 4x + 3$$

$$y = \pm\sqrt{3x^2 + 4x + 3}$$

Therefore, every point on the graph of the equation is also on the graph of either

$$y = \sqrt{3x^2 + 4x + 3} \qquad \text{or} \qquad y = -\sqrt{3x^2 + 4x + 3}$$

Each of these last two equations defines y as a function of x and can be graphed by a calculator in function graphing mode.

▶ **TECHNOLOGY TIP**

On TI calculators (except TI-81) you can simultaneously graph both functions in the Exploration by keying in

$y = \{1, -1\}\sqrt{3x^2 + 4x + 3}$.

GRAPHING EXPLORATION Use your calculator to graph both of these functions on the same screen. The result will be the graph of the original equation $12x^2 - 4y^2 + 16x + 12 = 0$. Note that one function gives the top half of the graph and the other the bottom half. ■

Parametric Equations

The techniques of Examples 6 and 7 cannot readily be used to graph

$$x = y^3 - 3y^2 - 4y + 7$$

because it's difficult to solve the equation for y. Unlike previous examples, however, this equation defines x as a function of y; if you consider y as the input and x as the output, then for each input y, the equation produces a unique output x. This fact can be used to graph the equation on any calculator that has a ''parametric'' graphing mode.

EXAMPLE 8 To graph $x = y^3 - 3y^2 - 4y + 7$, let t be any real number. If $y = t$, then

$$x = y^3 - 3y^2 - 4y + 7 = t^3 - 3t^2 - 4t + 7.$$

Since the same thing is true for any t, the graph consists of all points (x, y) such that

$$x = t^3 - 3t^2 - 4t + 7 \qquad \text{and} \qquad y = t \qquad (t \text{ any real number}).$$

When written in this form the equation can be graphed as follows.

Use the ''mode,'' ''set up,'' or ''lib'' key on your calculator to change the graphing mode to ''parametric'' and enter the equations $x = t^3 - 3t^2 - 4t + 7$ and $y = t$.* Set the viewing window by entering appropriate maximum and minimum values for x and y. You must also set minimum and maximum values for t and, on some calculators, specify the ''t step'' (or ''t pitch''), which determines how much t changes before the next point is plotted.** Since $y = t$ here, we use the same range settings for y and t. Finally, press GRAPH (or PLOT or DRW). For each number t in the prescribed range, the calculator then plots the point (x, y) determined by t and produces the graph in Figure 2–15. ■

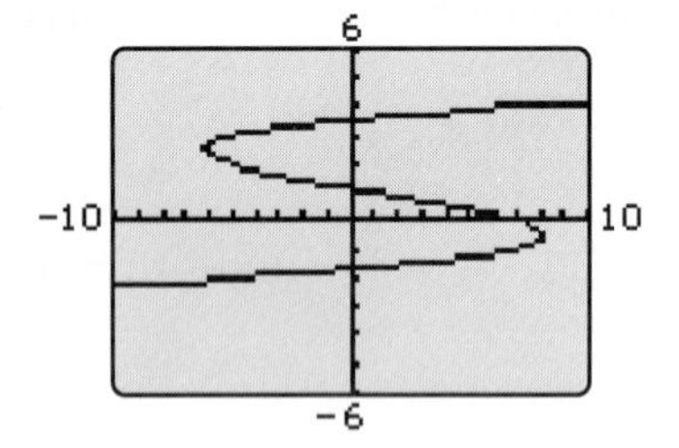

Figure 2-15

*Procedures for doing this vary. Check your instruction manual under the heading ''parametric graphing'' or ''parametric equations'' or ''parametric mode.''

**A few calculators automatically determine the t step when the range of t values is entered. As a general rule, a t step between .05 and .15 usually produces a relatively smooth graph in a reasonable amount of time.

> ▶ **TECHNOLOGY TIP**
> For TI-82/83/92, and HP-38 users: If you have trouble finding appropriate ranges for t, x, and y, it may help to use the TABLE feature to display a table of $t-x-y$ values produced by the parametric equations.

As illustrated in Example 8, the underlying idea of **parametric graphing** is to express both x and y as functions of a third variable t. The equations that define x and y are called **parametric equations** and the variable t is called the **parameter.** Example 8 illustrates just one of the many applications of parametric graphing. It can also be used to graph curves that are not graphs of a single equation in x and y.

EXAMPLE 9 Graph the curve given by

$$x = t^2 - t - 1 \quad \text{and} \quad y = t^3 - 4t - 6 \quad (-2 \le t \le 3).$$

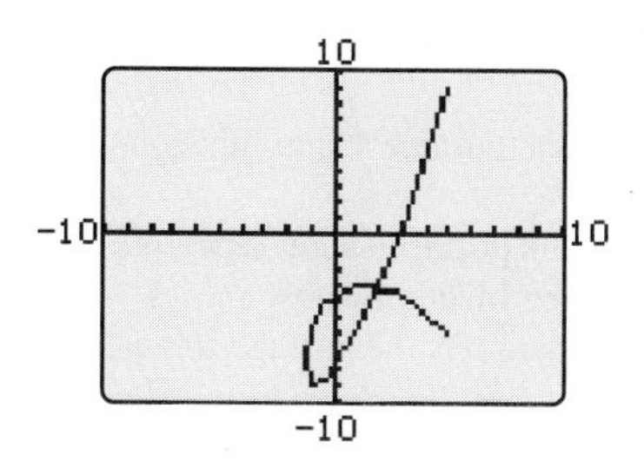

Figure 2-16

Solution Using the standard viewing window we obtain the graph in Figure 2–16. Note that the graph crosses over itself at one point and that it does not extend forever to the left and right, but has "endpoints."

GRAPHING EXPLORATION Graph these same parametric equations, but set the range of t values so that $-4 \le t \le 4$. What happens to the graph? Now change the range of t values so that $-10 \le t \le 10$. Find a viewing window large enough to show the entire graph, including endpoints. ■

Any function of the form $y = f(x)$ can be expressed in terms of parametric equations and graphed that way. For instance, to graph $f(x) = x^2 + 1$, let $x = t$ and $y = f(t) = t^2 + 1$. Parametric graphing will be used hereafter whenever it is convenient and will be studied more thoroughly in Section 10.1.

EXERCISES 2.3

In Exercises 1–6, sketch the graph of the function. You should be able to do this without using a calculator. Regardless of how you obtain these graphs, memorize their shapes.

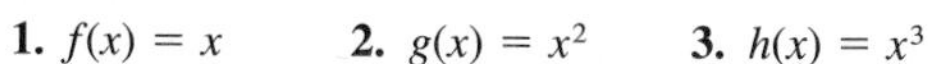

1. $f(x) = x$ **2.** $g(x) = x^2$ **3.** $h(x) = x^3$

4. $p(x) = \sqrt{x}$ **5.** $q(x) = \sqrt[3]{x}$

6. $k(x) = |x|$ [*Hint:* When $x \ge 0$, the rule of the function is $k(x) = x$ and when $x < 0$, the rule is $k(x) = -x$ (see the box on page 6). Consequently, the graph consists of two straight line segments.]

In Exercises 7–10, state the approximate intervals on which the function whose graph is shown is increasing and the approximate intervals on which it is decreasing.

7.

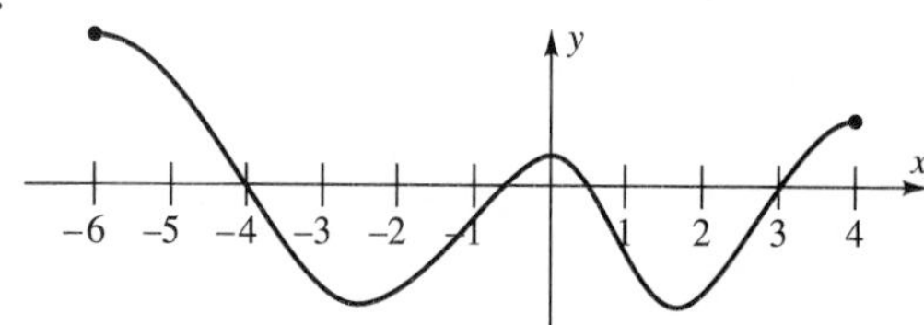

8.

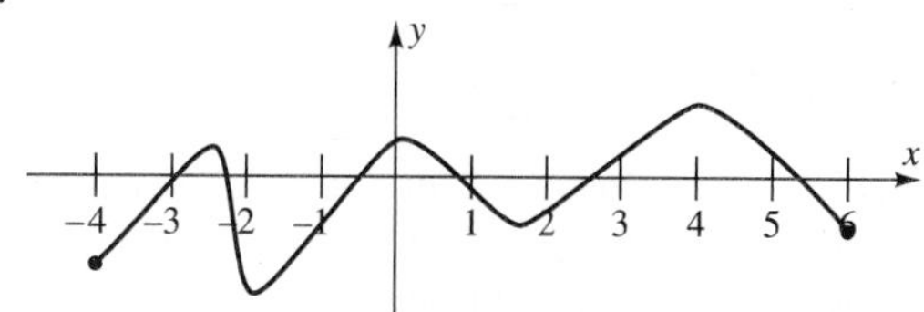

9.

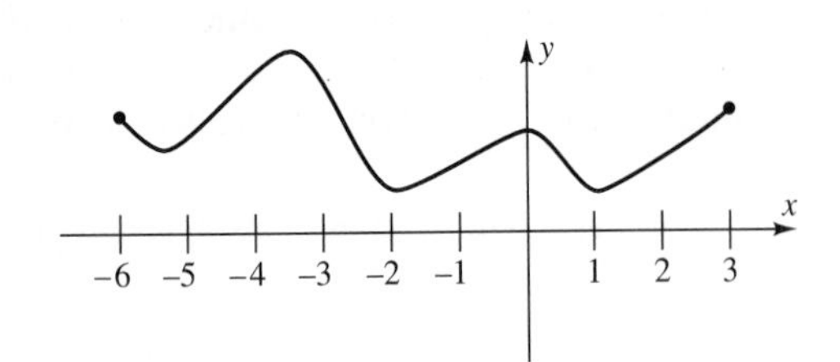

10.

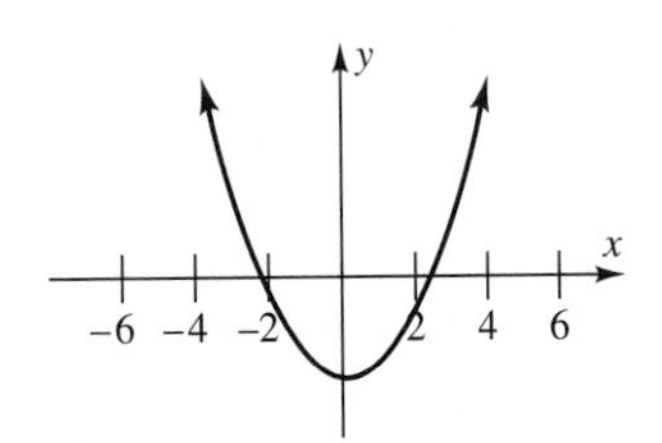

In Exercises 11–32, sketch the graph of the function.

11. $f(x) = x + 2$ **12.** $g(x) = \frac{3}{2}x + \frac{1}{2}$

13. $h(x) = 2 - x^2$ **14.** $f(x) = 1 - x^3$

15. $f(x) = \sqrt{-x}$ **16.** $f(x) = \sqrt{3 - x}$

17. $f(x) = x^3 - x$ **18.** $g(t) = -\sqrt{16 - t^2}$

19. $h(x) = \frac{x}{x^2 + 1}$ **20.** $k(x) = x^3 - 3x + 1$

Note: *The hint for Exercise 6 may be helpful in Exercises 21–24.*

21. $f(x) = |x| + 2$ **22.** $g(x) = |x| - 4$

23. $f(x) = |x - 5|$ **24.** $g(x) = |x + 3|$

25. $f(x) = \begin{cases} x^2 & \text{if } x \geq -1 \\ 2x + 3 & \text{if } x < -1 \end{cases}$

26. $g(x) = \begin{cases} |x| & \text{if } x < 1 \\ -3x + 4 & \text{if } x \geq 1 \end{cases}$

27. $k(u) = \begin{cases} -2u - 2 & \text{if } u < -3 \\ u - [u] & \text{if } -3 \leq u \leq 1 \\ 2u^2 & \text{if } u > 1 \end{cases}$

28. $f(x) = \begin{cases} x^2 & \text{if } x < -2 \\ x & \text{if } -2 \leq x < 4 \\ \sqrt{x} & \text{if } x \geq 4 \end{cases}$

29. $f(x) = [x]$ [*Hint:* What does the graph look like on the interval $[-2, -1)$? On $[-1, 0)$? On $[0, 1)$? On $[1, 2)$? The graph contains no vertical line segments, regardless of what a calculator may show.]

30. $f(x) = -[x]$, when $-3 \leq x \leq 3$ [See the hint for Exercise 29.]

31. $g(x) = [-x]$, when $-3 \leq x \leq 3$ [This is *not* the same as the function in Exercise 30.]

32. $h(x) = [x] + [-x]$, when $-3 \leq x \leq 3$.

33. At this writing first-class postage rates are 32¢ for the first ounce or fraction thereof, plus 23¢ for each additional ounce or fraction thereof. Assume that each first-class letter carries one 32¢ stamp and as many 23¢ stamps as are necessary. Then the *number* of stamps required for a first-class letter is a function of the weight of the letter in ounces. Call this function the *postage stamp function.*

(a) Describe the rule of the postage stamp function algebraically.

(b) Sketch the graph of the postage stamp function.

(c) Sketch the graph of the function whose rule is $f(x) = p(x) - [x]$, where p is the postage stamp function.

34. A plane flies from Austin, Texas, to Cleveland, Ohio, a distance of 1200 miles. Let f be the function whose rule is $f(t)$ = distance (in miles) from Austin at time t hours. Draw a plausible graph of f under the given circumstances. [There are many possible correct answers for each part.]

(a) The flight is nonstop and takes less than 4 hours.

(b) Bad weather forces the plane to land in Dallas (about 200 miles from Austin), remain overnight (for 8 hours) and continue the next day.

(c) The flight is nonstop, but due to heavy traffic the plane must fly in a holding pattern over Cincinnati (about 200 miles from Cleveland) for an hour before going on to Cleveland.

35. A bacteria population in a laboratory culture contains about a million bacteria at 8 A.M. The culture grows very rapidly until noon, when a bactericide is introduced and the bacteria population plunges. By 4 P.M. the bacteria have adapted to the bactericide and the culture slowly increases in population until 9 P.M. when the culture is accidentally destroyed by the clean-up crew. Let $g(t)$ denote the bacteria population at time t and draw a plausible graph of the function g. [Many correct answers are possible.]

In Exercises 36–42, determine the approximate intervals on which the function is increasing and the approximate intervals on which it is decreasing (as in Example 5).

36. $f(x) = x^3 + 2$ **37.** $g(x) = 3 - x^3$

38. $g(x) = 2x^3 - x^2 - 6x + 3$

39. $f(x) = -x^3 - 8x^2 + 8x + 5$

40. $f(x) = x^4 - .7x^3 - .6x^2 + 1$

41. $g(x) = .2x^4 - x^3 + x^2 - 2$

42. $g(x) = x^4 + x^3 - 4x^2 + x - 1$

In Exercises 43–45, use algebra to show that the function is increasing on the interval $(0, 10]$. *Some of them may also be increasing on other intervals. You may assume the usual facts about inequalities, including the following:*

if $0 \le c < d$, then $c^2 < d^2$

if $c < d$, then $c^3 < d^3$

43. $f(x) = x^2 + 3$ **44.** $g(x) = x^3 - 10{,}000$

45. $h(t) = t^2 + t + 5$

In Exercises 46–52, use the techniques of Examples 6 and 7 to graph the equation.

46. $9x^2 + 4y^2 = 36$ **47.** $4x^2 - 9y^2 = 36$

48. $9y^2 - x^2 = 9$ **49.** $9x^2 + 5y^2 = 45$

50. $x = 2y^2$ **51.** $x = y^2 - 2$

52. $x = 2(y + 3)^2 - 4$

In Exercises 53–58, use parametric graphing. Find a viewing window that shows a complete graph of the equation.

53. $x = y^3 + 5y^2 - 4y - 5$

54. $\sqrt[3]{y^2 - y + 1} - x + 2 = 0$

55. $xy^2 + xy + x = y^3 - 2y^2 + 4$ [*Hint:* First solve for x.]

56. $2y = xy^2 + 180x$ **57.** $x - \sqrt{y} + y^2 + 8 = 0$

58. $y^2 - x - \sqrt{y + 5} + 4 = 0$

In Exercises 59–64, find a viewing window that shows a complete graph of the curve determined by the parametric equations.

59. $x = 3t^2 - 5$ and $y = t^3$ $(-4 \le t \le 4)$

60. The Zorro curve: $x = .1t^3 - .2t^2 - 2t + 4$ and $y = 1 - t$ $(-5 \le t \le 6)$

61. $x = t^2 - 3t + 2$ and $y = 8 - t^3$ $(-4 \le t \le 4)$

62. $x = t^2 - 6t$ and $y = \sqrt{t + 7}$ $(-5 \le t \le 9)$

63. $x = 1 - t^2$ and $y = t^3 - t - 1$ $(-4 \le t \le 4)$

64. $x = t^2 - t - 1$ and $y = 1 - t - t^2$ $(-5 \le t \le 4)$

65. Graph the curve given by

$$x = (t^2 - 1)(t^2 - 4)(t + 5) + t + 3$$

$$y = (t^2 - 1)(t^2 - 4)(t^3 + 4) + t - 1$$

$$(-2.5 \le t \le 2.5)$$

How many times does this curve cross itself?

66. Use parametric equations to describe a curve that crosses itself more times than the curve in Exercise 65. [Many correct answers are possible.]

In Exercises 67–69, graph both curves on the same screen and determine, as best you can, how the two curves are related to the line $y = x$ *(whose parametric equations are* $x = t$ *and* $y = t$*).*

67. $x = t, y = t^3 + t + 1$ and $x = t^3 + t + 1, y = t$ (all real numbers t)

68. $x = t, y = 5\sqrt[3]{t^2 - 1}$ and $x = 5\sqrt[3]{t^2 - 1}, y = t$ (all real numbers t)

69. $x = t, y = t^2$ and $x = t^2, y = t$ (all real numbers t)

Thinkers

70. Show that a graph that passes the Vertical Line Test is necessarily the graph of a function. [*Hint:* See Example 10 in Section 2.1.]

In Exercises 71 and 72, sketch the graph of the equation.

71. $|x| + |y| = 1$ **72.** $|y| = x^2$

2.4 GRAPH READING

In order to use functions effectively, you must be able to translate statements from any one of the following "languages" to any other:

The English language;

Formula language (algebraic and functional notation); and

Graphical language (graphs).

In previous sections we have translated English into functional notation and functional notation into graphs. Now we consider *graph reading,* that is, translating graphical information into equivalent statements in English or functional notation.

EXAMPLE 1 What are the domain and range of the functions whose complete graphs are shown in Figure 2–17?

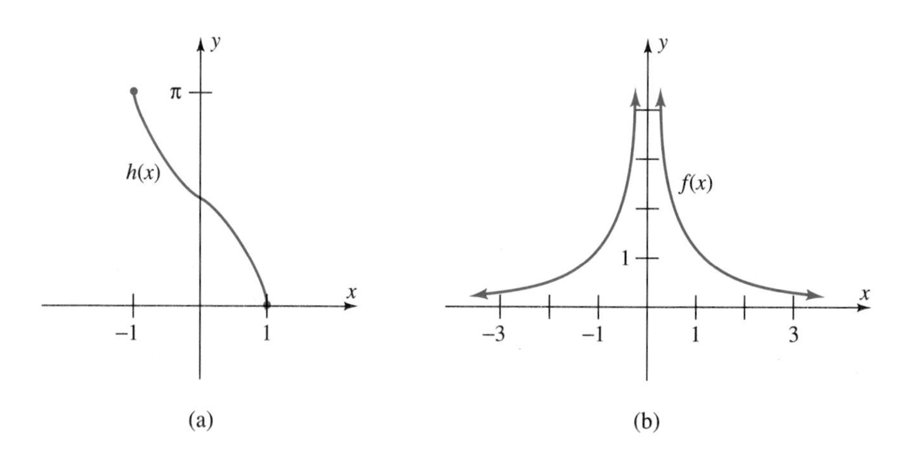

Figure 2-17

Solution

(a) The graph of h consists of all points $(x, h(x))$ with x in the domain of h. Thus the domain of h consists of all those real numbers that are first coordinates of points on the graph of h. Figure 2–17(a) shows that these are the numbers x with $-1 \leq x \leq 1$. In other words, the domain of h is the interval $[-1, 1]$. The range of h consists of all possible values of $h(x)$ when x is a number in the domain. Thus the range consists of all those real numbers that are second coordinates of points on the graph. In Figure 2–17(a), the second coordinates of points on the graph are the numbers y with $0 \leq y \leq \pi$, so the range of h is the interval $[0, \pi]$.

(b) The domain of f consists of the numbers that are first coordinates of points on the graph, namely, all real numbers *except* 0 (Figure 2–17(b)). The range of f consists of the numbers that are second coordinates of points on the graph. Figure 2–17(b) shows that the graph moves upward forever near $x = 0$ and moves closer and closer to the x-axis as x takes larger and larger values. Thus, every *positive* real number is the second coordinate of some point on the graph and the range of f consists of all positive real numbers. ■

EXAMPLE 2 A recording device in the weather bureau in city F produces the graph of the function that relates time of day to temperature, as shown in Figure 2–18. We shall adopt this functional notation: tem(t) denotes the temperature at time t (measured in hours after midnight). For example, since (4, 39) is on the graph* we know that tem(4) = 39, that is, the temperature at 4 A.M. was 39°. At what times during the day was the temperature below 50°?

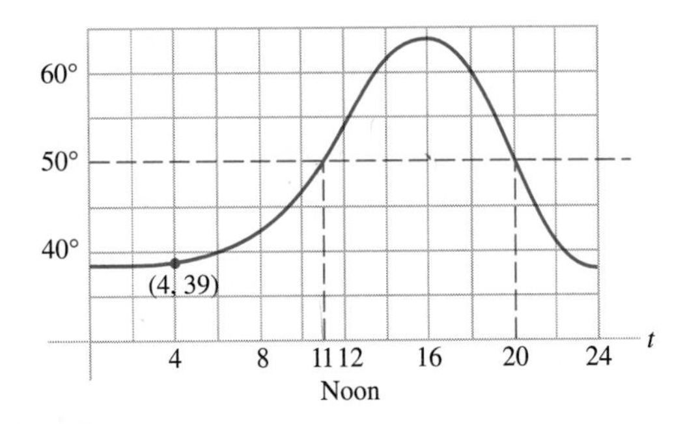

Figure 2-18

Solution To answer this question we translate it into functional and graphical terms. To say that the temperature is below 50° at time t means that the second coordinate of the point (t, tem(t)) is less than 50. The points with this property are the ones that lie *below* the horizontal line through 50, as shown in Figure 2–18. The first coordinates of all such points are the *times* when the temperature was below 50. The graph shows that

$$\text{tem}(t) < 50 \qquad \text{whenever} \qquad 0 \leq t < 11 \qquad \text{or} \qquad 20 < t \leq 24.$$

In other words, the temperature was below 50° from midnight to 11 A.M. and again from 8 P.M. ($t = 20$) to midnight. ■

EXAMPLE 3 In the time-temperature graph of Example 2, a slightly more complicated problem is to determine the time period *before* 4 P.M. during which the temperature was at least 60°. Remembering that 4 P.M. is 16 hours after midnight, we make these translations:

Statement	*Functional Notation*	*Graph*
The time is before 4 P.M.	$t < 16$	(t, tem(t)) lies to the left of the vertical line through $t = 16$.
The temperature is at least 60°.	tem(t) ≥ 60	(t, tem(t)) lies on or above the horizontal line through 60.

*Here and below our results are only as accurate as our measuring ability. But the basic idea should be clear.

Figure 2–19 below shows that the points $(t, \text{tem}(t))$ with $t < 16$ and $\text{tem}(t) \geq 60$ are those with first coordinates between 13 and 16. So the temperature was at least 60° from 1 P.M. ($t = 13$) to 4 P.M. ($t = 16$). ■

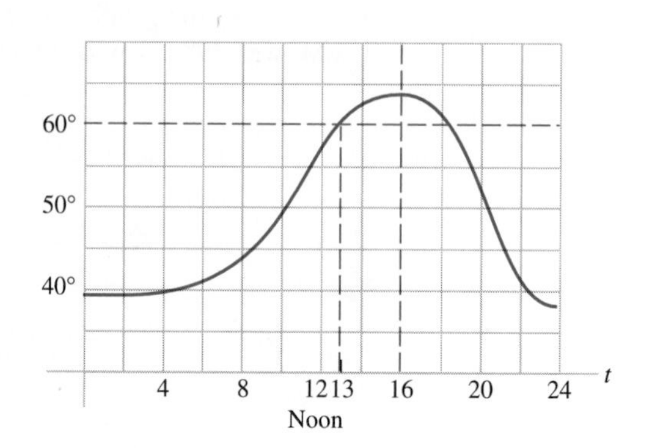

Figure 2-19

EXAMPLE 4 Suppose the temperature graph for city S is superimposed on the one for city F, as in Figure 2–20. Denote the temperature at time t in city S by $\text{tem}_S(t)$ and answer these questions:

(a) At what times was it warmer in city F than in city S?

(b) Was there any time when it was at least 10° warmer in city S than in city F?

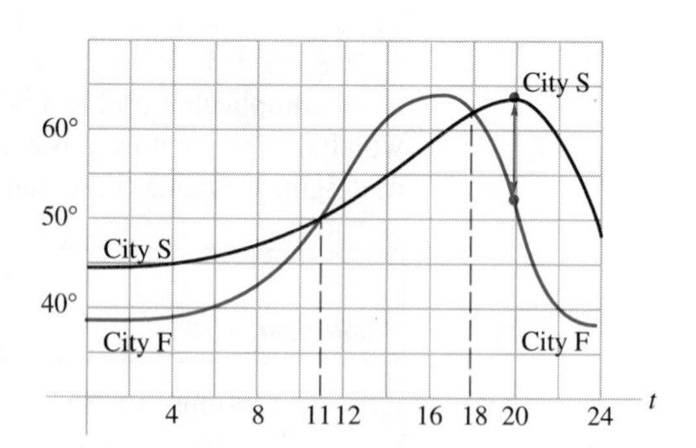

Figure 2-20

Once again, it's a matter of three-way translation:

Statement	*Functional Notation*	*Graph*
City F is warmer than City S at time t.	$\text{tem}(t) > \text{tem}_S(t)$	The point $(t, \text{tem}(t))$ lies directly above $(t, \text{tem}_S(t))$.
City S is at least 10° warmer than City F at time t.	$\text{tem}_S(t) \geq \text{tem}(t) + 10$	The point $(t, \text{tem}_S(t))$ is at least 10 units above $(t, \text{tem}(t))$.

Careful measurement in Figure 2–20 shows that

(a) $\text{tem}(t) > \text{tem}_S(t)$ for all t in the interval $(11, 18)$.

(b) $\text{tem}_S(t) \geq \text{tem}(t) + 10$ for many values of t, including $t = 20$.

In other words,

(a) It was warmer in city F between 11 A.M. and 6 P.M.

(b) It was at least 10° warmer in city S at 8 P.M. ($t = 20$) and at other times. ■

EXAMPLE 5 Use the graphs of the functions h and g in Figure 2–21 to work the following problems.

(a) Find all numbers x such that $h(x) < 0$.

(b) Find all numbers x in the interval $[-3, 3]$ such that $g(x) = 2$.

(c) Find the largest interval over which g is increasing and h is decreasing and $h(x) \geq g(x)$ for every number x in the interval.

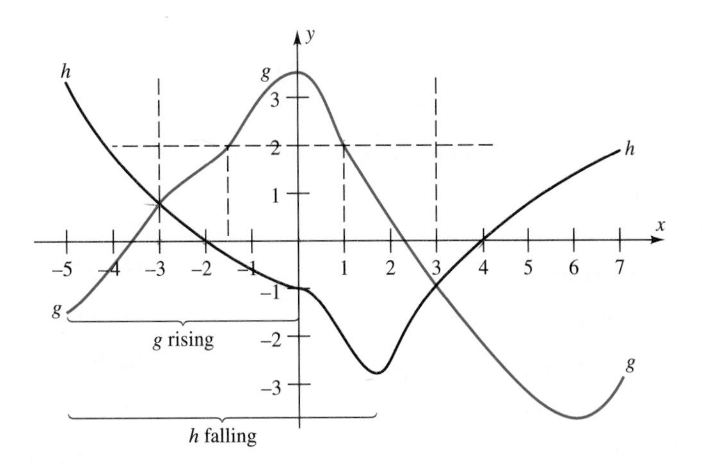

Figure 2-21

Solution

(a) The graph of h consists of all points $(x, h(x))$. The numbers x such that $h(x) < 0$ correspond to the points on the graph with negative second coordinates, that is, the points that lie below the x-axis. The graph of h shows that these are the points whose first coordinates satisfy $-2 < x < 4$. So the answer is all numbers in the interval $(-2, 2)$.

(b) The graph of g consists of the points $(x, g(x))$. The points on the graph that lie *between* the vertical lines through -3 and 3 and *on* the horizontal line through 2 have first coordinate x in the interval $[-3, 3]$ and second coordinate $g(x) = 2$. Figure 2–21 shows that the only such points are $(-1.5, 2)$ and $(1, 2)$. So the answer is $x = -1.5$ and $x = 1$.

(c) Figure 2–21 shows that $[-5, 0]$ is the only interval over which the graph of h is falling *and* the graph of g is rising. Clearly, $h(x) \geq g(x)$ exactly when the point $(x, h(x))$ lies above the point $(x, g(x))$. The only time this occurs in the interval $[-5, 0]$ is when $-5 \leq x \leq -3$. Therefore, the answer is the interval $[-5, -3]$. ■

EXERCISES 2.4

In Exercises 1–6, use your calculator to determine the domain and range of the function by reading its graph as best you can (the trace feature may be helpful). Careful approximate answers are acceptable.

1. $f(x) = 3x - 2$ **2.** $g(x) = x^2 - 4$

3. $h(x) = \sqrt{x^2 - 4}$ **4.** $k(x) = \sqrt{x^2 + 4}$

5. $f(x) = \dfrac{x^2 + 1}{x}$ **6.** $g(x) = \dfrac{4x^4 - 3x^2 - 6}{2x^4 - x^2 + x + 2}$

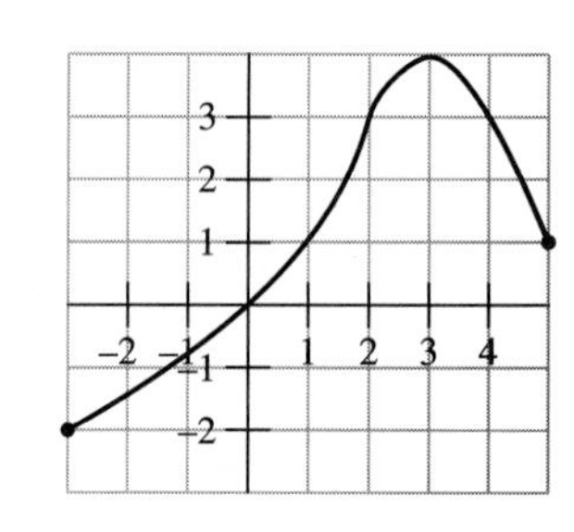

Figure 2-22

Exercises 7–20 deal with the function g whose entire graph is shown in Figure 2–22.

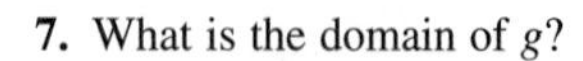

7. What is the domain of g?

8. What is the range of g?

9. If $t = 1.5$, then $g(2t) = ?$

10. If $t = 1.5$, then $2g(t) = ?$

11. If $y = 2$, then $g(y + 1.5) = ?$

12. If $y = 2$, then $g(y) + g(1.5) = ?$

13. If $y = 2$, then $g(y) + 1.5 = ?$

14. For what values of x is $g(x) < 0$?

15. If $v = 1.5$, then $g(3v - 1.5) = ?$

16. If $s = 2$, then $g(-s) = ?$

17. For what values of z is $g(z) = 1$?

18. For what values of z is $g(z) = -1$?

19. What is the largest interval over which g is increasing?

20. At what number t in the interval $[-1, 2]$ is $g(t)$ largest?

21. Draw the graph of a function f that satisfies the following four conditions:

(i) domain $f = [-2, 4]$

(ii) range $f = [-5, 6]$

(iii) $f(-1) = f(2)$

(iv) $f\left(\frac{1}{2}\right) = 0$

22. Draw the graph of a function different from your answer to Exercise 21 that also satisfies all the conditions of Exercise 21.

Exercises 23–32 deal with the function f whose entire graph is shown in Figure 2–23.

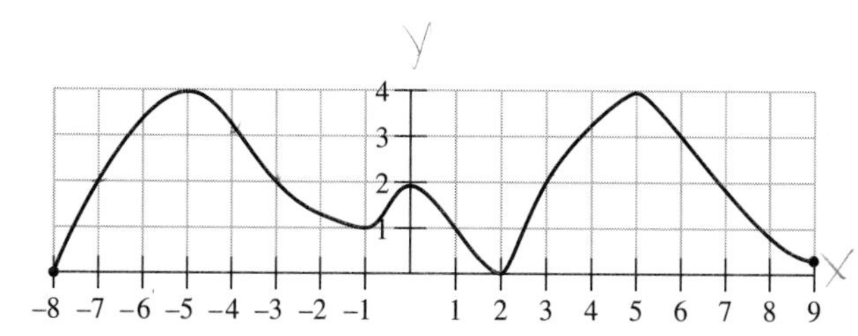

Figure 2-23

23. What is the domain of f?

24. What is the range of f?

25. Find all numbers x such that $f(x) = 2$.

26. Find all numbers x such that $f(x) > 2$.

27. Find at least three numbers x such that $f(x) = f(-x)$.

28. Find all numbers x such that $f(x) = f(7)$.

29. Find a number x such that $f(x + 1) = 0$.

30. Find two numbers x such that $f(x - 2) = 4$.

31. Find a number x such that $f(x + 1) = f(x - 2)$.

32. Find a number x such that $f(x) + 1 = f(x - 4)$.

Exercises 33–40 deal with the two functions f and g whose entire graphs are shown in Figure 2–24.

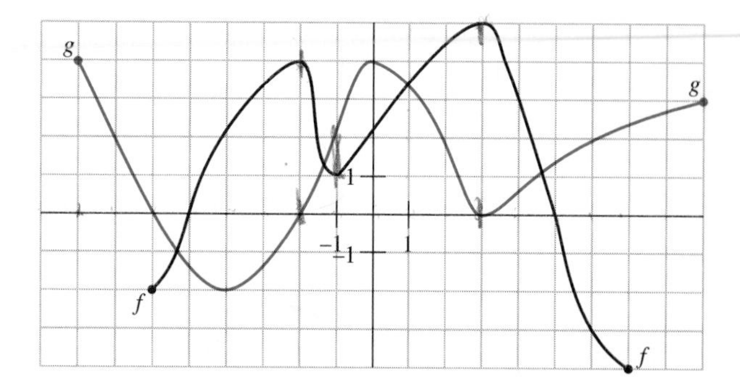

Figure 2-24

33. What is the domain of f? The domain of g?

34. What is the range of f? The range of g?

35. Find all numbers x in the interval $[-3, 1]$ such that $f(x) = 2$.

36. Find all numbers x in the interval $[-3, 3]$ such that $g(x) \geq 2$.

37. Find the number x for which $f(x) - g(x)$ is largest.

38. For how many values of x is it true that $f(x) = g(x)$?

39. Find all intervals over which both functions are defined, f is decreasing, and g is increasing.

40. Find all intervals over which g is decreasing.

In Exercises 41–45, find one or more functions (among those whose graphs appear in Figure 2–25 on the next page) for which the given statement is true.

41. $f(2) < f(1)$.

42. $f(x)$ is negative and increasing from $x = 1$ to $x = 2$.

43. $f(0) < 0$ but $f(2) > 0$.

44. $f(x)$ is negative and decreasing from $x = -4$ to $x = -3$.

45. $f(x) > f(-x)$ for some number x with $|x| \leq 5$.

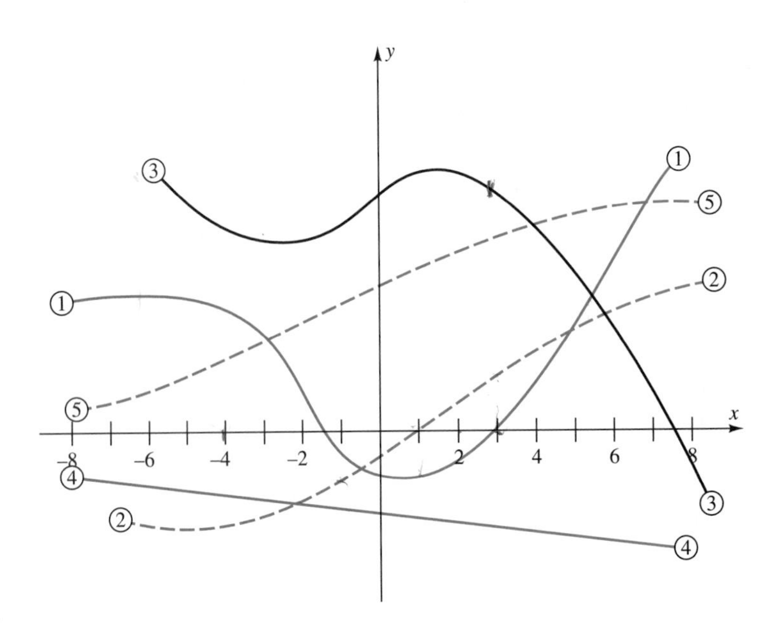

Figure 2-25

In Exercises 46–51, find a pair of functions (or several pairs) from among the five whose graphs appear in Figure 2–25 for which the given statement is true.

46. $f(1) - g(1) > 0$.

47. $f(x) < g(3)$ for $0 \le x \le 2$.

48. $f(x)g(x) < 0$ for $x > 4$.

49. $\dfrac{f(x)}{g(x)} < 1$ for some $x < 0$.

50. $f(x) = g(x)$ for some x with $|x| \ge 2$.

51. $f(x) \le g(-x)$ for some x with $|x| < 4$.

52. Sketch the graph of a function f that satisfies these five conditions:

(i) $f(-1) = 2$
(ii) $f(x) \ge 2$ when x is in the interval $(-1, \frac{1}{2})$
(iii) $f(x)$ starts decreasing when $x = 1$
(iv) $f(3) = 3 = f(0)$
(v) $f(x)$ starts increasing when $x = 5$

[*Note:* The function whose graph you sketch need not be given by an algebraic formula.]

Exercises 53–58 deal with this situation: The owners of the Melville & Pluth Hammer Factory have determined that both their weekly manufacturing expenses and their weekly sales income are functions of the number of hammers manufactured each week. Figure 2–26 shows the graphs of these two functions.

53. Use careful measurement on the graph and the fact that profit = income − expenses to determine the weekly profit if 5000 hammers are manufactured.

54. Do the same if 10,000, 14,000, 18,000, or 22,000 hammers are manufactured.

55. What is the smallest number of hammers that can be manufactured each week without losing money?

56. What is the largest number of hammers that can be manufactured without losing money?

57. The owners build a new lounge and swimming pool for their employees. This raises their expenses by approximately \$5000 per week. Draw the graph of the new "expense function."

58. Owing to competitive pressure, hammer prices cannot be increased and the income function remains the same. Answer Exercises 53–56 with the expense function of Exercise 57 in place of the old one.

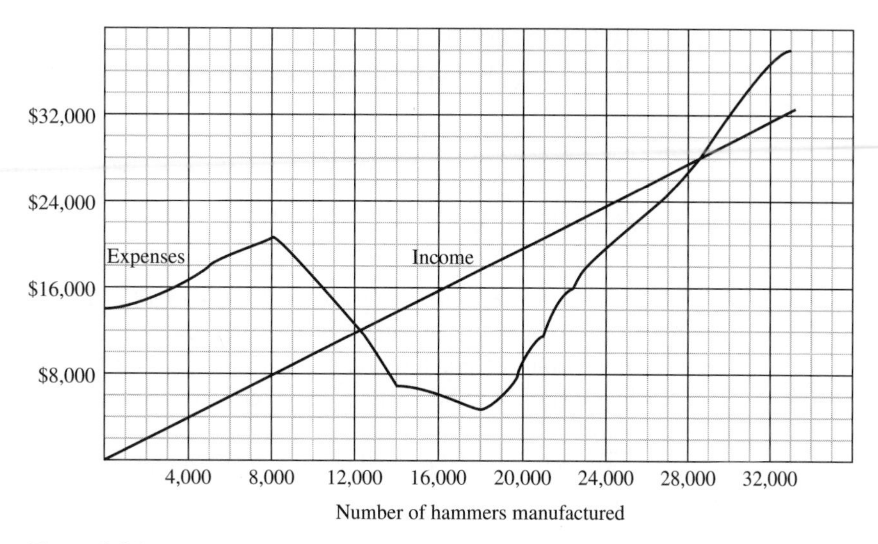

Figure 2-26

Exercises 59–67 deal with the weather bureau graph of the temperature as a function of time (measured in hours after midnight), as shown in Figure 2–27. We write tem(t) for the temperature at t hours after midnight.

59. Find tem(10). Find tem(3 + 12).

60. Is tem(6) bigger than, equal to, or less than tem(18)?

61. At which time is the temperature 50°?

62. Find a 4-hour period for which tem(h) > 40 for all h in this 4-hour period.

63. Find the difference in temperature at 10 and 16 hours. Identify this difference in the graph; that is, express difference in terms of points and their location on the graph.

64. Is it true that tem(6) = tem(8)? Explain in terms of the graph.

65. Is it true that tem(6 · 2) = 6 · tem(2)?

66. The temperature graph in Figure 2–27 was recorded in city F. City B is 500 miles to the south of city F, and its temperature is 7° higher all day long. Sketch into Figure 2–27 the temperature for city B during the same day.

67. Find an hour h at which the temperature in city B is the same as tem(12) in city F. (See Exercise 66.)

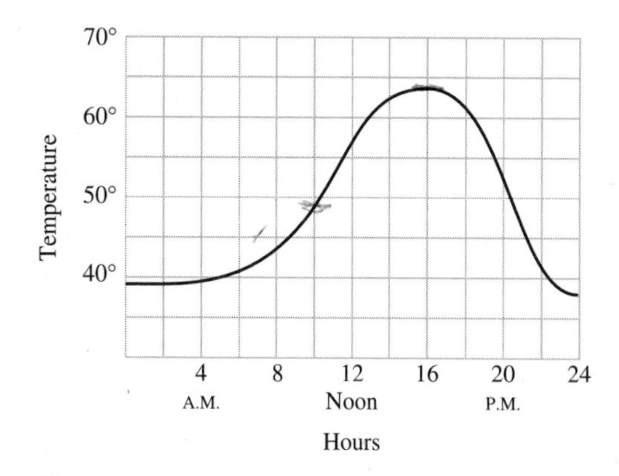

Figure 2-27

2.5 GRAPHS AND TRANSFORMATIONS

There are a number of algebraic operations that can be applied to the rule of a function. For instance, if $f(x) = x^2 - x + 2$, then a constant can be added or subtracted from the rule of f or the rule can be multiplied by a constant to obtain new functions, such as:

$$g(x) = f(x) + 5 = (x^2 - x + 2) + 5 = x^2 - x + 7$$

$$h(x) = -4f(x) = -4(x^2 - x + 2) = -4x^2 + 4x - 8.$$

Similarly, replacing the variable x by $x + 3$ produces the new function

$$k(x) = f(x + 3) = (x + 3)^2 - (x + 3) + 2 = x^2 + 6x + 9 - x - 3 + 2$$
$$= x^2 + 5x + 8$$

In this section we shall see that when the rule of a function is changed algebraically to produce a new function, then the graph of the new function can be obtained from the graph of the original function by a geometric transformation, such as a vertical or horizontal shift, a reflection in the x-axis, or by stretching or shrinking.

The same procedure will be used for each topic in this section:

First, you will be asked to assemble some evidence by doing a graphing exploration.

Next, general conclusions deduced from the evidence will be summarized in the colored boxes.

Last, there may be a discussion of how these conclusions can be proved in particular cases and possibly some additional examples.

Vertical Shifts

GRAPHING EXPLORATION Using the standard viewing window, graph these three functions on the same screen:

$$f(x) = x^2 \qquad g(x) = x^2 + 5 \qquad h(x) = x^2 - 7$$

and answer these questions:

Do the graphs of g and h look very similar to the graph of f in *shape*?

How do their vertical positions differ?

Where would you predict that the graph of $k(x) = x^2 - 9$ is located relative to the graph of $f(x) = x^2$ and what is its shape?

Confirm your prediction by graphing k on the same screen as f, g, and h.

The results of this Exploration should make the following statement plausible:

Vertical Shifts ►

If $c > 0$, then the graph of $g(x) = f(x) + c$ is the graph of f shifted upward c units.

If $c > 0$, then the graph of $h(x) = f(x) - c$ is the graph of f shifted downward c units.

You can see why the first statement is true by considering an example with $f(x) = x^2$, $c = 5$, and $g(x) = x^2 + 5$. Suppose the point (x, x^2) is on the graph of $f(x) = x^2$. On the graph of $g(x) = x^2 + 5$, the point with first coordinate x has second coordinate $x^2 + 5$. Since the points (x, x^2) and $(x, x^2 + 5)$ have the same first coordinate, they are on the same vertical line. Since their second coordinates differ by 5, the point $(x, x^2 + 5)$ on the graph of g lies 5 units directly above the point (x, x^2) on the graph of f. Since this is true for every x, the graph of g is just the graph of f shifted 5 units upward. Similar arguments apply to downward shifts.

EXAMPLE 1 A calculator was used to obtain a complete graph of $f(x) = .04x^3 - x - 3$ in Figure 2–28. The graph of

$$h(x) = f(x) - 4 = (.04x^3 - x - 3) - 4 = .04x^3 - x - 7$$

is the graph of f shifted 4 units downward, as shown in Figure 2–29.

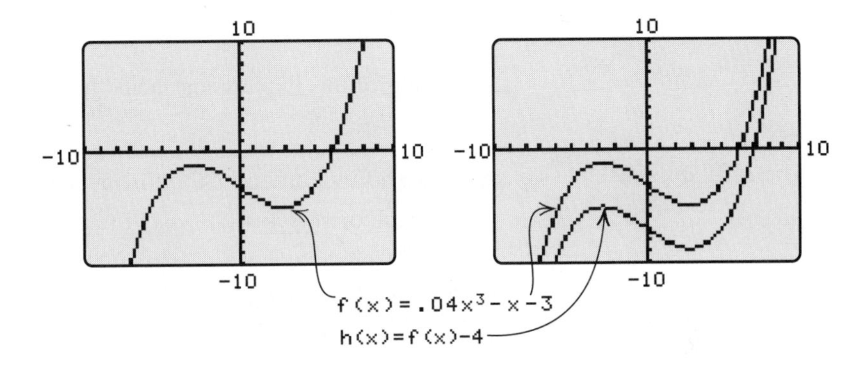

Figure 2-28 **Figure 2-29**

Although it may appear that the graph of h is closer to the graph of f at the outer edges of Figure 2–29 than in the center, this is an optical illusion:

the *vertical* distance between the graphs is always 4 units.

GRAPHING EXPLORATION Use the trace feature of your calculator to confirm this fact:*

Move the cursor to any point on the graph of f and note its coordinates.

Use the down arrow to drop the cursor to the graph of h and note the coordinates of the cursor in its new position.

The x-coordinates will be the same in both cases and the new y-coordinate will be 4 less than the original y-coordinate. ■

*On most calculators the trace cursor can be moved vertically from graph to graph by using the up and down arrows. If your trace feature works only on the last function graphed, however, you won't be able to do this.

Horizontal Shifts

GRAPHING EXPLORATION Using the standard viewing window, graph these three functions on the same screen:

$$f(x) = 2x^3 \qquad g(x) = 2(x+6)^3 \qquad h(x) = 2(x-8)^3$$

and answer these questions:

Do the graphs of g and h look very similar to the graph of f in *shape*?

How do their horizontal positions differ?

Where would you predict that the graph of $k(x) = 2(x+2)^3$ is located relative to the graph of $f(x) = 2x^3$ and what is its shape?

Confirm your prediction by graphing k on the same screen as f, g, and h.

The results of this Exploration should make the following statement plausible:

Horizontal Shifts ▶

Let f be a function and c a positive constant.

The graph of $g(x) = f(x+c)$ is the graph of f shifted horizontally c units to the left.

The graph of $h(x) = f(x-c)$ is the graph of f shifted horizontally c units to the right.

To understand why the first statement in the box is true, consider the function $f(x) = x^3 + 1$ and $c = 4$, so that $g(x) = (x+4)^3 + 1$. Note, for example, that

$$f(3) = 3^3 + 1 = 28 \qquad \text{and} \qquad g(-1) = (-1+4)^3 + 1 = 28,$$

so that $(3, 28)$ is on the graph of f and $(-1, 28)$ is on the graph of g. The points have the same second coordinate, so they are on the same horizontal line. The point $(-1, 28)$ on the graph of g is exactly 4 units to the *left* of the point $(3, 28)$ on the graph of f. The same thing happens for any value of x, say $x = a$:

$$f(a) = a^3 + 1 \qquad \text{and} \qquad g(a-4) = [(a-4)+4]^3 + 1 = a^3 + 1,$$

so that $(a, a^3 + 1)$ is on the graph of f and $(a-4, a^3+1)$ is on the graph of g. These two points are on the same horizontal line (why?) and the point on the graph of g is exactly 4 units to the *left* of the point on the graph of f (because its first coordinate is 4 units smaller). Thus, every point on the graph of g is 4 units horizontally to the left of a point on the graph of f, that is, the graph of g is the graph of f shifted 4 units to the left. Similar arguments work for the second statement in the box and horizontal shifts to the right.

EXAMPLE 2 In some cases, shifting the graph of a function f horizontally may produce a graph that overlaps the graph of f. For instance, a complete graph of $f(x) = x^2 - 7$ is shown in Figure 2–30. The graph of

$$g(x) = f(x + 5) = (x + 5)^2 - 7 = x^2 + 10x + 25 - 7 = x^2 + 10x + 18$$

is the graph of f shifted 5 units to the left and the graph of

$$h(x) = f(x - 4) = (x - 4)^2 - 7 = x^2 - 8x + 16 - 7 = x^2 - 8x + 9$$

is the graph of f shifted 4 units to the right, as shown in Figure 2–30. ■

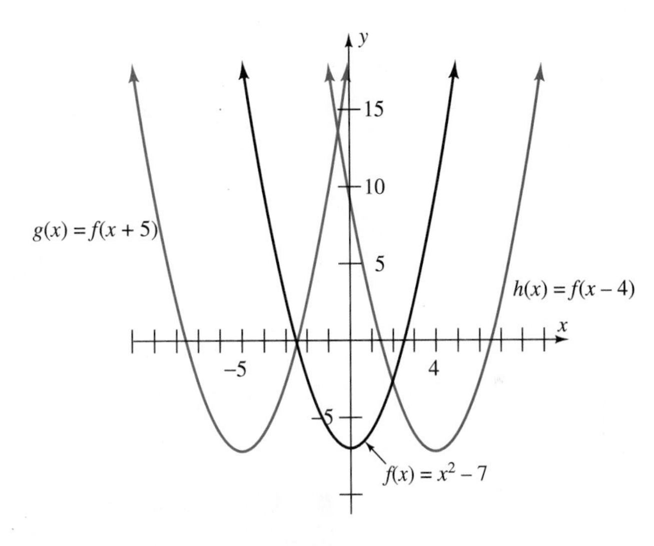

Figure 2-30

Expansions and Contractions

> **▶ TECHNOLOGY TIP**
> On TI calculators (except TI-81) you can graph both functions in the Exploration at the same time by keying in
>
> $y = \{1,3\}(x^2 - 4)$.

GRAPHING EXPLORATION In the viewing window with $-5 \leq x \leq 5$ and $-15 \leq y \leq 15$, graph these functions on the same screen:

$$f(x) = x^2 - 4 \qquad g(x) = 3f(x) = 3(x^2 - 4).$$

One way to understand the relationship between the two graphs is to imagine that the graph of f is nailed to the x-axis at its intercepts (± 2) and that you can vertically "stretch" the graph by pulling from the top and bottom away from the x-axis (with the nails holding the x-intercepts in place) so that it fits onto the graph of g. In this process (Figure 2-31), the point $(0, -4)$ on the graph of f is stretched down to the point $(0, -12)$ on the graph of g—that is, it is stretched away from the x-axis by a factor of 3. Similarly, the point $(3, 5)$ on the graph of f is stretched up by a factor of 3 to the point $(3, 15)$ on the graph of g, shown in Figure 2–31.

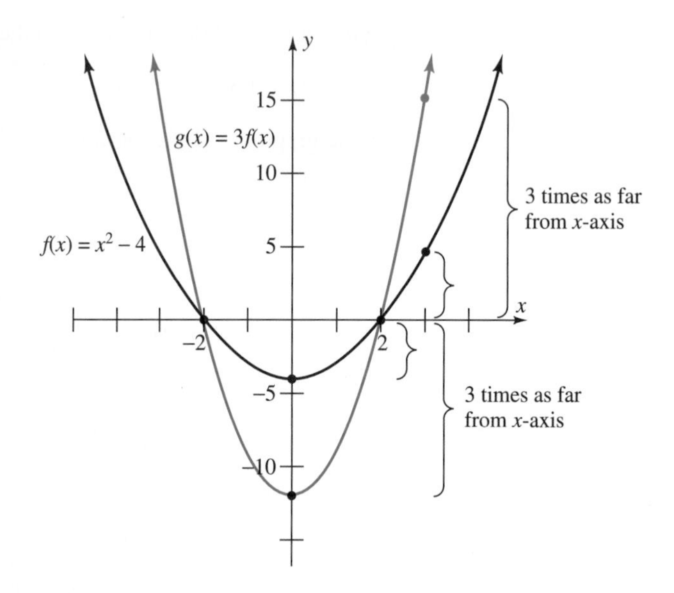

Figure 2-31

GRAPHING EXPLORATION In the viewing window with $-5 \le x \le 5$ and $-5 \le y \le 10$, graph these functions on the same screen:

$$f(x) = x^2 - 4 \qquad h(x) = \frac{1}{4}(x^2 - 4).$$

Your screen should suggest that the graph of h is the graph of f "shrunk" vertically toward the x-axis by a factor of 1/4.

Analogous facts are true in the general case:

Expansions and Contractions ▶

If $c > 1$, then the graph of $g(x) = cf(x)$ is the graph of f stretched vertically away from the x-axis by a factor of c.

If $0 < c < 1$, then the graph of $h(x) = cf(x)$ is the graph of f shrunk vertically toward the x-axis by a factor of c.

Reflections

GRAPHING EXPLORATION In the standard viewing window, graph these functions on the same screen:

$$f(x) = .04x^3 - x \qquad g(x) = -f(x) = -(.04x^3 - x).$$

If your trace cursor can be moved from graph to graph, verify that for every point on the graph of f there is a point on the graph of g with the same first coordinate that is on the opposite side of the x-axis, the same distance from the x-axis.

This Exploration shows that the graph of g is the mirror image (reflection) of the graph of f, with the x-axis being the (two-way) mirror. The same thing is true in the general case:

Reflections ▶

The graph of $g(x) = -f(x)$ is the graph of f reflected in the x-axis.

EXAMPLE 3 If $f(x) = x^2 - 3$, then the graph of $g(x) = -f(x) = -(x^2 - 3)$ is the reflection of the graph of f in the x-axis, as shown in Figure 2–32. ■

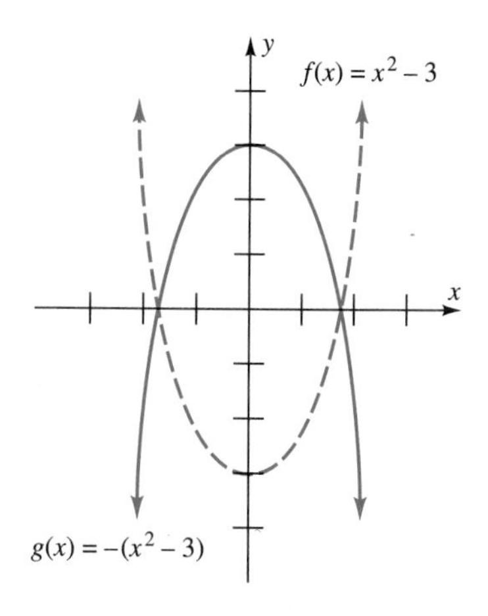

Figure 2-32

GRAPHING EXPLORATION In the standard viewing window graph these functions on the same screen:

$$f(x) = \sqrt{5x + 10} \quad \text{and} \quad h(x) = f(-x) = \sqrt{5(-x) + 10}.$$

Reflect carefully: How are the two graphs related to the y-axis?
Now graph these two functions on the same screen:

$$f(x) = x^2 + 3x - 3 \quad \text{and}$$

$$h(x) = f(-x) = (-x)^2 + 3(-x) - 3 = x^2 - 3x - 3.$$

Are the graphs of f and h related in the same way as the first pair?

This Exploration shows that the graph of h in each case is the mirror image (reflection) of the graph of f, with the y-axis as the mirror. The same thing is true in the general case.

Reflections ▶

The graph of $h(x) = f(-x)$ is the graph of f reflected in the y-axis.

To see why this statement is true, let a be a number. Then $(a, f(a))$ is on the graph of f. On the other hand, note that $h(-a) = f(-(-a)) = f(a)$ so that

$(-a, h(-a)) = (-a, f(a))$ is on the graph of h. The points $(a, f(a))$ and $(-a, f(a))$ are on the same horizontal line (because they have the same second coordinate) and lie on opposite sides of the y-axis at the same distance from the y-axis (because their first coordinates are negatives of each other). Thus every point on the graph of f has a "mirror image point" on the graph of h, with the y-axis being the mirror.

In addition to vertical and horizontal shifts, stretching, shrinking, and reflecting in the x- or y-axis, there are several other transformations that can be performed on the graph of a function. Each of these transformations also corresponds to an algebraic operation on the rule of the function. See Exercises 37–48 for details.

Combining Transformations

The transformations described above may be used in sequence to analyze the graphs of functions whose rules are algebraically complicated.

EXAMPLE 4 To understand the graph of $g(x) = 2(x - 3)^2 - 1$, note that the rule of g may be obtained from the rule of $f(x) = x^2$ in three steps:

$$f(x) = x^2 \xrightarrow{\text{Step 1}} (x-3)^2 \xrightarrow{\text{Step 2}} 2(x-3)^2 \xrightarrow{\text{Step 3}} 2(x-3)^2 - 1 = g(x)$$

Step 1 shifts the graph of f horizontally 3 units to the right; step 2 stretches the resulting graph away from the x-axis by a factor of 2; step 3 shifts this graph 1 unit downward, thus producing the graph of g in Figure 2–33. ■

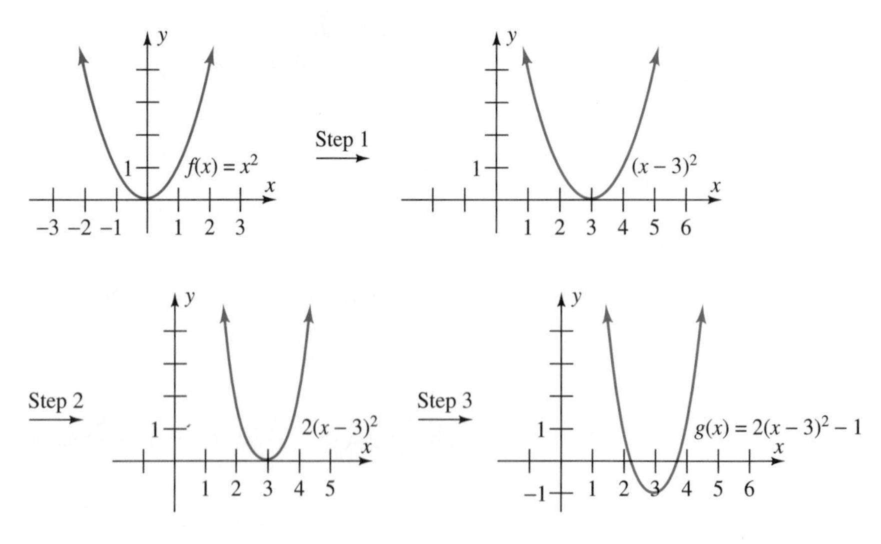

Figure 2-33

CALCULATOR INVESTIGATIONS 2.5

Investigations 1 and 2 show that, because of its small screen size, a calculator may not always display clearly what it should.

1. Graph $f(x) = x^2 - x - 6$ in the standard viewing window. Describe verbally what the graph of $h(x) = f(x - 1000)$ should look like (see the box on page 126). Now find an appropriate viewing window and graph h. Can you find a viewing window that clearly displays both the graph of f and the graph of h?

2. Graph $f(x) = x^2 - x - 6$ in the standard viewing window. Describe verbally what the graph of $g(x) = 1000f(x)$ should look like (see the box on page 128). Now find an appropriate viewing window and graph g. Can you find a viewing window that clearly displays both the graph of f and the graph of g?

3. Multiple Graphs on TI

(a) On TI calculators (except TI-81), use the viewing window with $-5 \le x \le 5$ and $-1 \le y \le 5$ and enter $y_1 = \{1, 2, .3\}x^2$ on the $y=$ list. When you touch GRAPH, what functions does the calculator graph?

(b) Which ones does it graph if you enter $\{-1, 2\}x^2$?

(c) Graph the following functions by entering just one function in the $y=$ list:

$$f(x) = x^2 + 4x - 1, \quad g(x) = \tfrac{1}{2}x^2 + 2x - \tfrac{1}{2}, \quad h(x) = -2x^2 - 8x + 2.$$

Check to see that you have done it correctly by entering f, g, and h separately in the $y=$ list and graphing them on the same screen.

EXERCISES 2.5

In Exercises 1–4, find a single viewing window that shows complete graphs of the functions f, g, h.

1. $f(x) = .25x^3 - 9x + 5;\quad g(x) = f(x) + 15;\quad h(x) = f(x) - 20$

2. $f(x) = \sqrt{x^2 - 9} - 5;\quad g(x) = 3f(x);\quad h(x) = .5f(x)$

3. $f(x) = |x^2 - 5|;\quad g(x) = f(x + 8);\quad h(x) = f(x - 6)$

4. $f(x) = .125x^3 - .25x^2 - 1.5x + 5;\quad g(x) = f(x) - 5;\quad h(x) = 5 - f(x)$

In Exercises 5 and 6, find complete graphs of the functions f and g in the same viewing window.

5. $f(x) = \dfrac{4 - 5x^2}{x^2 + 1};\quad g(x) = -f(x)$

6. $f(x) = x^4 - 4x^3 + 2x^2 + 3;\quad g(x) = f(-x)$

In Exercises 7–12, describe a sequence of transformations that will transform the graph of the function f into the graph of the function g.

7. $f(x) = x^2 + x;\quad g(x) = (x - 3)^2 + (x - 3) + 2$

8. $f(x) = x^2 + 5;\quad g(x) = (x + 2)^2 + 10$

9. $f(x) = \sqrt{x^3 + 5};\quad g(x) = -\dfrac{1}{2}\sqrt{x^3 + 5} - 6$

10. $f(x) = \sqrt{x^4 + x^2 + 1};\quad g(x) = 10 - \sqrt{4x^4 + 4x^2 + 4}$

11. $f(x) = \dfrac{3x}{x^2 + 10};\quad g(x) = \dfrac{-6x + 12}{(x - 2)^2 + 10}$

12. $f(x) = \dfrac{1}{2x^2 + 2};\quad g(x) = \dfrac{-1}{(x - 1)^2 + 1}$

In Exercises 13–18, write the rule of a function g whose graph can be obtained from the graph of the function f by performing the transformations in the order given.

13. $f(x) = x^2 + 2$; shift the graph horizontally 5 units to the left and then vertically upward 4 units.

14. $f(x) = x^2 - x + 1$; reflect the graph in the x-axis, then shift it vertically upward 3 units.

15. $f(x) = \sqrt{x}$; shift the graph horizontally 6 units to the right, stretch it away from the x-axis by a factor of 2, and shift it vertically downward 3 units.

16. $f(x) = \sqrt{-x}$; shift the graph horizontally 3 units to the left, then reflect it in the x-axis, and shrink it toward the x-axis by a factor of 1/2.

17. $f(x) = x^2 + 3x + 1$; stretch the graph away from the x-axis by a factor of 2, shift it horizontally 2 units to the right, and shift it vertically upward 2 units.

18. $f(x) = x^2 + 3x + 1$; shift the graph horizontally 2 units to the right, then shift it vertically upward 2 units, and stretch it away from the x-axis by a factor of 2. Compare the result with Exercise 17.

In Exercises 19–22, use the graph of the function f in Figure 2–34 to sketch the graph of the function g.

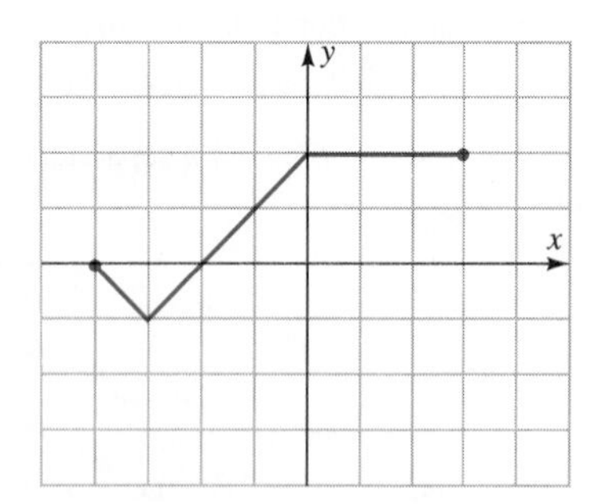

Figure 2-34

19. $g(x) = f(x) + 3$ **20.** $g(x) = f(x) - 1$

21. $g(x) = 3f(x)$ **22.** $g(x) = .25f(x)$

In Exercises 23–26, use the graph of the function f in Figure 2–35 to sketch the graph of the function h.

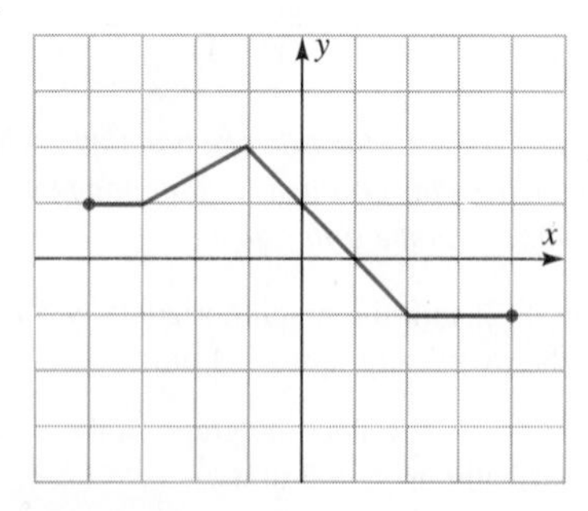

Figure 2-35

23. $h(x) = -f(x)$ **24.** $h(x) = -4f(x)$

25. $h(x) = f(-x)$ **26.** $h(x) = f(-x) + 2$

In Exercises 27–32, use the graph of the function f in Figure 2–36 to sketch the graph of the function g.

27. $g(x) = f(x + 3)$ **28.** $g(x) = f(x - 2)$

29. $g(x) = f(x - 2) + 3$ **30.** $g(x) = f(x + 1) - 3$

31. $g(x) = 2 - f(x)$ **32.** $g(x) = f(-x) + 2$

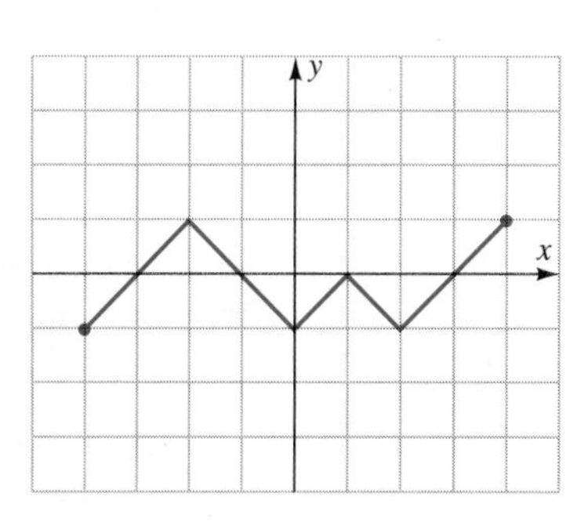

Figure 2-36

In Exercises 33–35, use the standard viewing window to graph the function f and the function $g(x) = |f(x)|$ on the same screen. Exercise 36 may be helpful for interpreting the results.

33. $f(x) = .5x^2 - 5$ **34.** $f(x) = x^3 - 4x^2 + x + 3$

35. $f(x) = x + 3$

36. (a) Let f be a function and let g be the function defined by $g(x) = |f(x)|$. Use the definition of absolute value (page 6) to explain why the following statement is true:

$$g(x) = \begin{cases} f(x) & \text{if } f(x) \geq 0 \\ -f(x) & \text{if } f(x) < 0 \end{cases}$$

(b) Use part (a) and your knowledge of transformations to explain why the graph of g consists of those parts of the graph of f that lie above the x-axis together with the reflection in the x-axis of those parts of the graph of f that lie below the x-axis.

In Exercises 37–39, assume $f(x) = (.2x)^6 - 4$. Use the standard viewing window to graph the functions f and g on the same screen.

37. $g(x) = f(2x)$ **38.** $g(x) = f(3x)$ **39.** $g(x) = f(4x)$

40. Based on the results of Exercises 37–39, describe the transformation that transforms the graph of a function $f(x)$ into the graph of the function $f(cx)$, where c is a constant with $c > 1$. [*Hint:* How are the two graphs related to the y-axis? Stretch your mind.]

In Exercises 41–43, assume $f(x) = x^2 - 3$. Use the standard viewing window to graph the functions f and g on the same screen.

41. $g(x) = f\left(\frac{1}{2}x\right)$ **42.** $g(x) = f\left(\frac{1}{3}x\right)$

43. $g(x) = f\left(\frac{1}{4}x\right)$

44. Based on the results of Exercises 41–43, describe the transformation that transforms the graph of a function $f(x)$ into the graph of the function $f(cx)$, where c is a constant with $0 < c < 1$. [*Hint:* How are the two graphs related to the y-axis?]

In Exercises 45–47, use the standard viewing window to graph the function f and the function $g(x) = f(|x|)$ on the same screen.

45. $f(x) = x - 4$ **46.** $f(x) = x^3 - 3$

47. $f(x) = .5(x - 4)^2 - 9$

48. Based on the results of Exercises 45–47, describe the relationship between the graph of a function $f(x)$ and the graph of the function $f(|x|)$.

2.5.A *Excursion* SYMMETRY

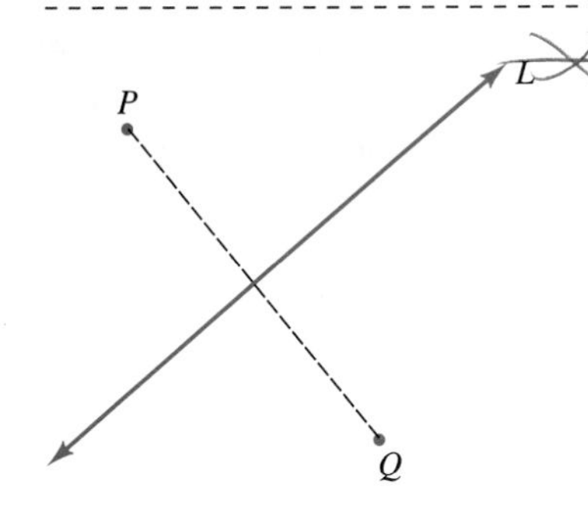

Figure 2-37

Two points P and Q are said to be **symmetric with respect to a line** L if L is the perpendicular bisector of the line segment PQ, as shown in Figure 2–37. You can think of P and Q as "mirror images" of each other, with the line L being the mirror.

A graph is **symmetric with respect to a line** L if for every point P on the graph, there is another point Q on the graph such that P and Q are symmetric with respect to L. In this case, the part of the graph on one side of L will be the mirror image of the part on the other side, with the line L being the mirror. There are two important types of line symmetry, the first of which is **symmetry with respect to the y-axis:**

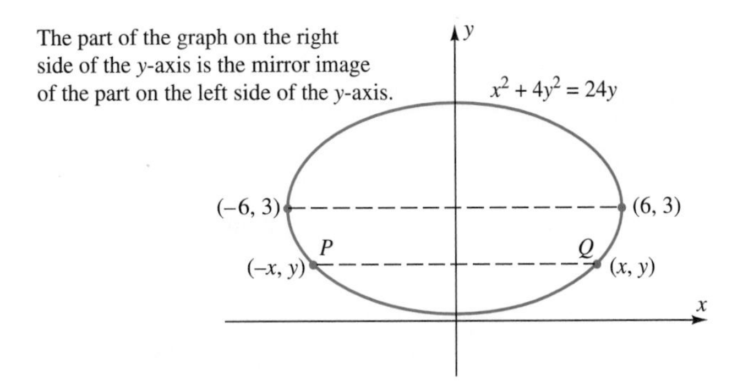

Figure 2-38

As shown in Figure 2–38, points P and Q are symmetric with respect to the y-axis exactly when

their second coordinates are the same (P and Q are on the same side of the x-axis and the same distance from it);

their first coordinates are negatives of each other (P and Q lie on opposite sides of the y-axis and the same distance from it).

Thus, a graph is symmetric with respect to the y-axis provided that

Whenever (x, y) is on the graph, then $(-x, y)$ is also on it.

In algebraic terms, this means that replacing x by $-x$ in the equation leads to the same number y. In other words, replacing x by $-x$ produces an equivalent equation.

EXAMPLE 1 Replacing x by $-x$ in the equation $y = x^4 - 5x^2 + 3$ produces $(-x)^4 - 5(-x)^2 + 3$, which is the same equation because $(-x)^2 = x^2$ and $(-x)^4 = x^4$. Therefore, the graph is symmetric with respect to the y-axis.

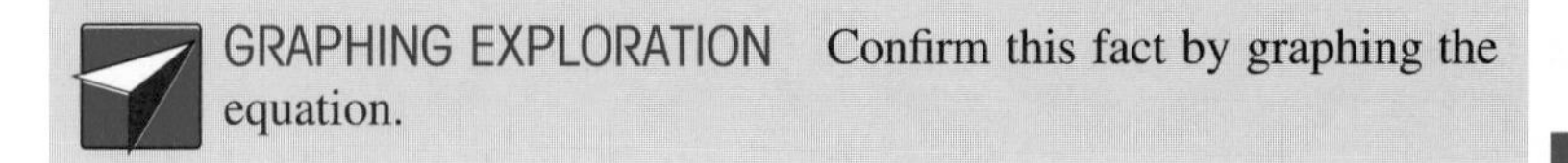

GRAPHING EXPLORATION Confirm this fact by graphing the equation. ■

The second important type of line symmetry is **symmetry with respect to the x-axis:**

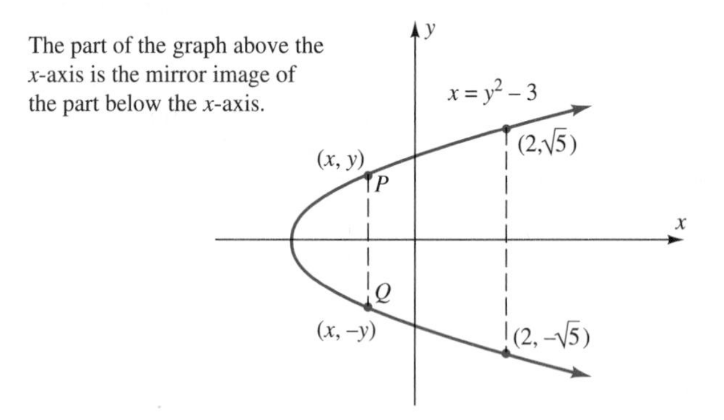

Figure 2-39

Using Figure 2–39 and argument analogous to the one preceding Example 1, we see that a graph is symmetric with respect to the x-axis provided that

Whenever (x, y) is on the graph, then $(x, -y)$ is also on it.

In algebraic terms, this means that replacing y by $-y$ in the equation leads to the same number x. In other words, replacing y by $-y$ produces an equivalent equation.

EXAMPLE 2 Replacing y by $-y$ in the equation $y^2 = 4x - 12$ produces $(-y)^2 = 4x - 12$, which is the same equation, so the graph is symmetric with respect to the x-axis.

> GRAPHING EXPLORATION Confirm this fact by graphing the equation. In order to do this, note that every point on the graph of $y^2 = 4x - 12$ is also on the graph of either $y = \sqrt{4x - 12}$ or $y = -\sqrt{4x - 12}$. Each of these latter equations defines a function; graph them both on the same screen as in Example 6 on page 111.

■

Two points P and Q are said to be **symmetric with respect to a point** K if K is the midpoint of the line segment PQ. A *graph* is **symmetric with respect to a point** K if for each point P on the graph, there is another point Q on the graph such that P and Q are symmetric with respect to K. The most useful type of point symmetry is **symmetry with respect to the origin:**

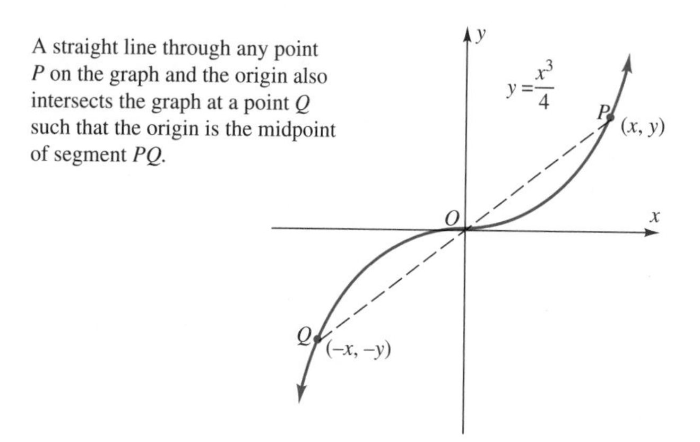

Figure 2-40

Using Figure 2–40, we can also describe symmetry with respect to the origin in terms of coordinates and equations (Exercise 34):

Whenever (x, y) is on the graph, then $(-x, -y)$ is also on it.

In algebraic terms, this means that replacing x by $-x$ and y by $-y$ in the equation produces an equivalent equation.

EXAMPLE 3 Replacing x by $-x$ and y by $-y$ in the equation $y = x^3/10 - x$ yields

$$-y = (-x)^3/10 - (-x), \qquad \text{that is,} \qquad -y = -x^3/10 + x.$$

This equation is equivalent to $y = x^3/10 - x$ since it can be obtained from it by multiplying by -1. Therefore, the graph of $y = x^3/10 - x$ is symmetric with respect to the origin.

GRAPHING EXPLORATION Confirm this fact by graphing the equation. ■

Here is a summary of the various tests for line and point symmetry:

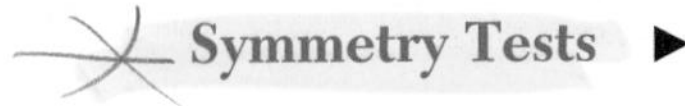

Symmetry Tests ▶

Symmetry with Respect to	*Coordinate Test for Symmetry*	*Algebraic Test for Symmetry*
y-axis	(x, y) on graph implies $(-x, y)$ on graph.	Replacing x by $-x$ produces an equivalent equation.
x-axis	(x, y) on graph implies $(x, -y)$ on graph.	Replacing y by $-y$ produces an equivalent equation.
origin	(x, y) on graph implies $(-x, -y)$ on graph.	Replacing x by $-x$ *and* y by $-y$ produces an equivalent equation.

Even and Odd Functions

For *functions,* the algebraic description of symmetry takes a different form. A function f whose graph is symmetric with respect to the y-axis is called an **even function.** To say that the graph of $y = f(x)$ is symmetric with respect to the y-axis means that replacing x by $-x$ produces the same y value. In other words, the function takes the same value at both x and $-x$. Therefore,

Even Functions ▶

A function f is even provided that

$$f(x) = f(-x) \text{ for every number } x \text{ in the domain of } f.$$

For example, $f(x) = x^4 + x^2$ is even because

$$f(-x) = (-x)^4 + (-x)^2 = x^4 + x^2 = f(x).$$

Thus, the graph of f is symmetric with respect to the y-axis, as you can easily verify with your calculator (do it!).

Except for zero functions ($f(x) = 0$ for every x in the domain), *the graph of a function is never symmetric with respect to the x-axis.* The reason is the Vertical Line Test: The graph of a function never contains two points with the same first coordinate. If both $(5, 3)$ and $(5, -3)$, for instance, were on the graph, this would say that $f(5) = 3$ and $f(5) = -3$, which is impossible when f is a function.

A function whose graph is symmetric with respect to the origin is called an **odd function.** If both (x, y) and $(-x, -y)$ are on the graph of such a function f, then we must have both

$$y = f(x) \qquad \text{and} \qquad -y = f(-x)$$

so that $f(-x) = -y = -f(x)$. Therefore,

Odd Functions ▶

A function f is odd provided that

$$f(-x) = -f(x) \text{ for every number } x \text{ in the domain of } f.$$

For example, $f(x) = x^3$ is an odd function because

$$f(-x) = (-x)^3 = -x^3 = -f(x).$$

Hence, the graph of f is symmetric with respect to the origin (verify this with your calculator).

EXERCISES 2.5.A

In Exercises 1–4, find the graph of the equation. If the graph is symmetric with respect to a line or a point, find the line or point.

1. $y = (x + 2)^2$ **2.** $x = (y - 3)^2 + 2$

3. $y = x^3 + 2$ **4.** $y = (x + 2)^3$

In Exercises 5–14, determine whether the given function is even, odd, or neither (that is, whether its graph is symmetric with respect to the y-axis, the origin, or neither).

5. $f(x) = 4x$ **6.** $k(t) = -5t$

7. $f(x) = x^2 - |x|$ **8.** $h(u) = |3u|$

9. $k(t) = t^4 - 6t^2 + 5$ **10.** $f(x) = x(x^4 - x^2) + 4$

11. $f(t) = \sqrt{t^2 - 5}$ **12.** $h(x) = \sqrt{7 - 2x^2}$

13. $f(x) = \dfrac{x^2 + 2}{x - 7}$ **14.** $g(x) = \dfrac{x^2 + 1}{x^2 - 1}$

In Exercises 15–18, determine algebraically whether or not the graph of the given equation is symmetric with respect to the x-axis.

15. $x^2 - 6x + y^2 + 8 = 0$ **16.** $x^2 + 8x + y^2 = -15$

17. $x^2 - 2x + y^2 + 2y = 2$ **18.** $x^2 - x + y^2 - y = 0$

In Exercises 19–24, determine whether the given graph is symmetric with respect to the y-axis, the x-axis, or the origin.

19.

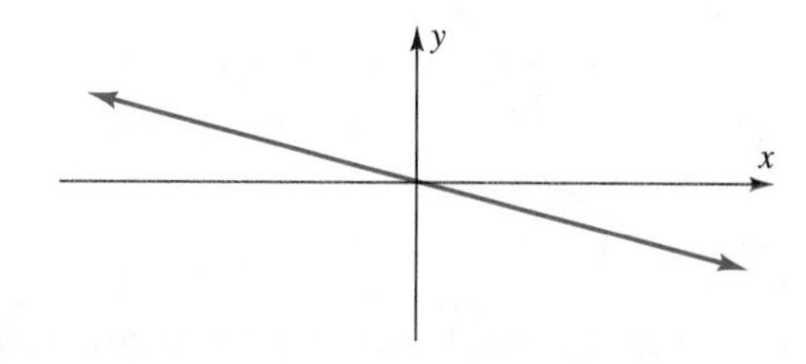

20.

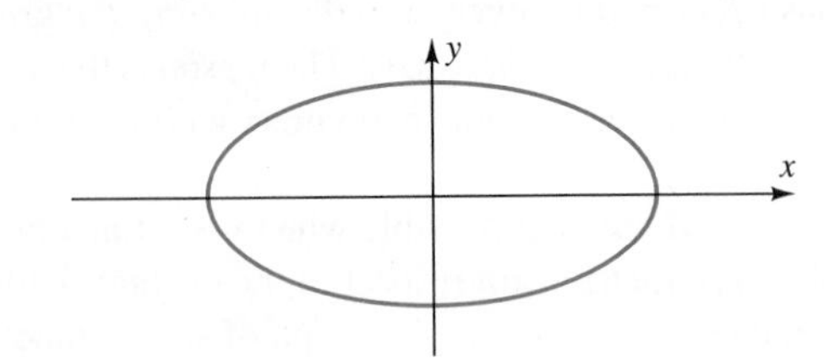

21.

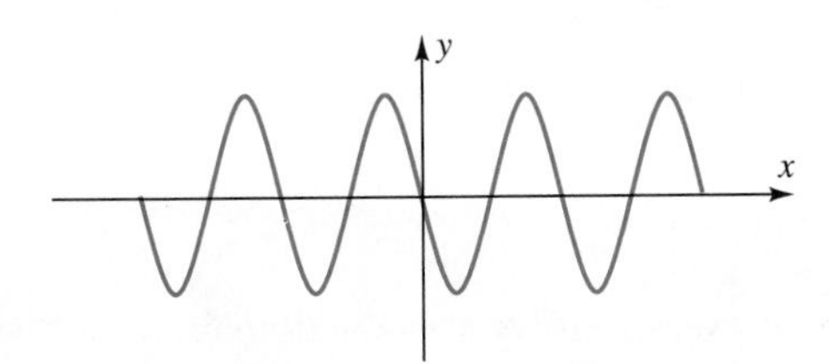

22.

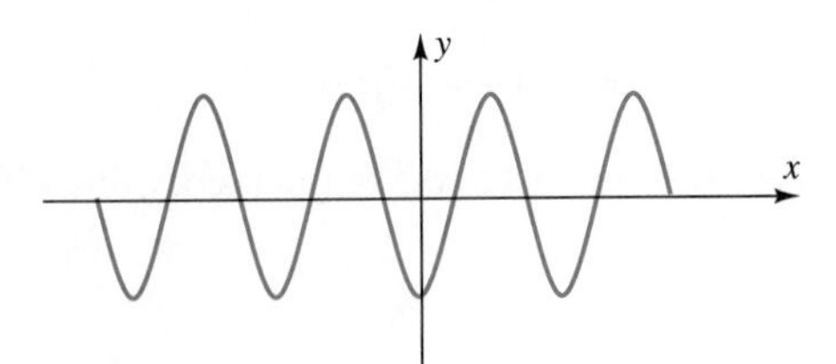

23.

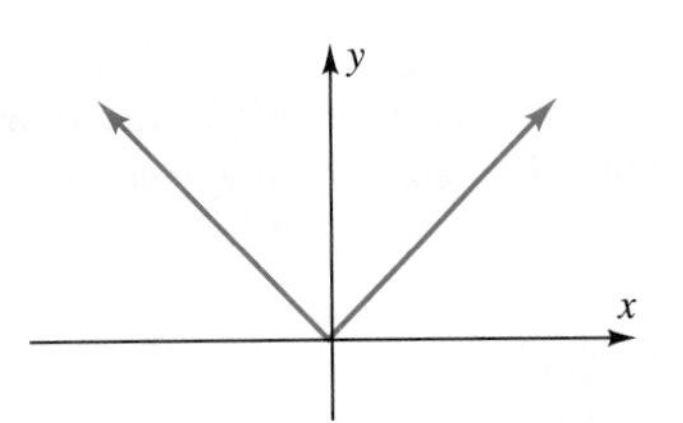

24.

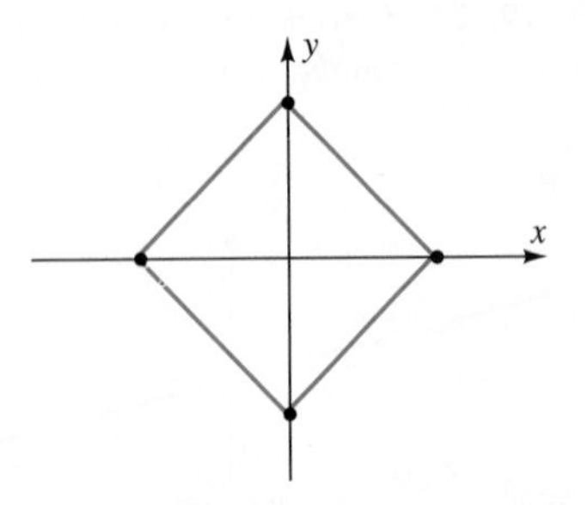

In Exercises 25–28, complete the graph of the given function, assuming that it satisfies the given symmetry condition.

25. even

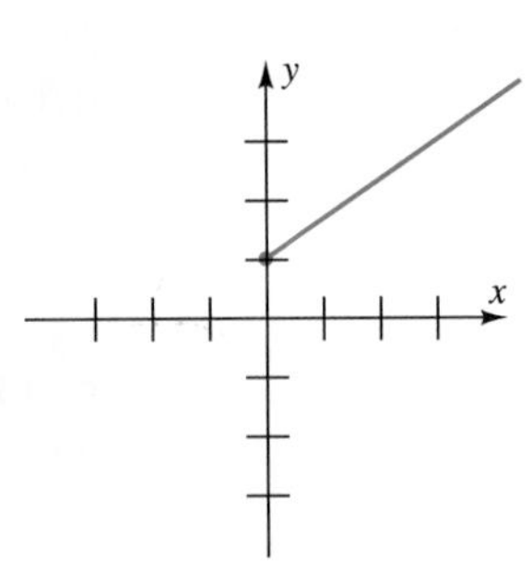

26. even

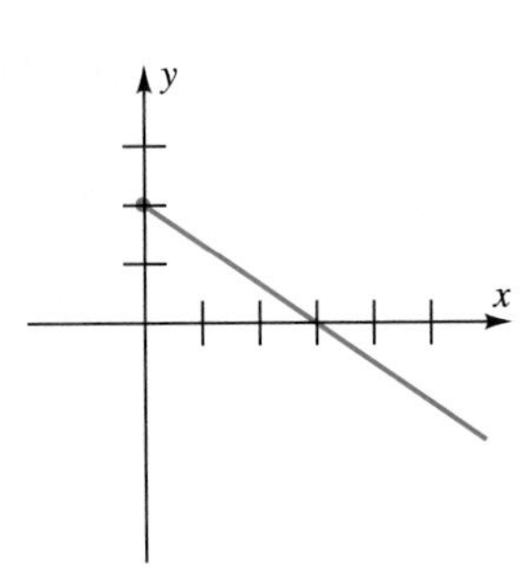

27. odd

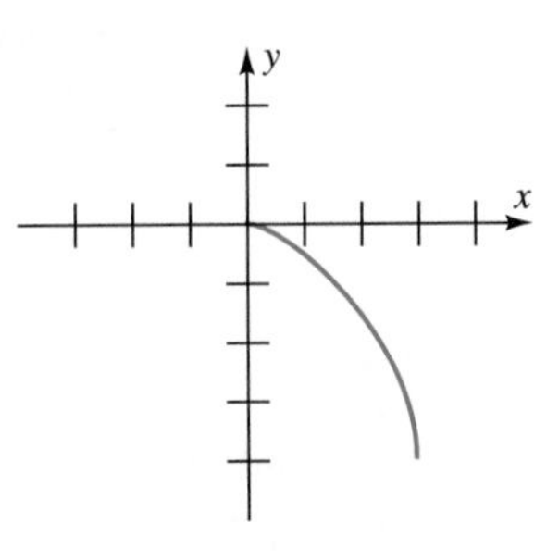

28. odd

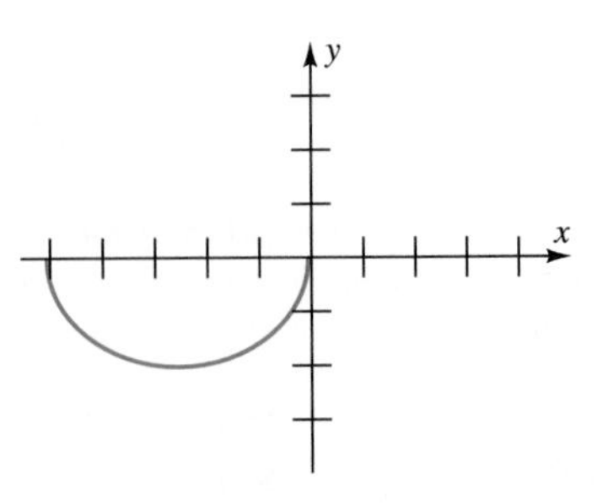

29. (a) Draw some coordinate axes and plot the points $(0, 1)$, $(1, -3)$, $(-5, 2)$, $(-3, 5)$, $(2, 3)$, and $(4, 1)$.

(b) Suppose the points in part (a) lie on the graph of an *even* function f. Plot the points $(0, f(0))$, $(-1, f(-1))$, $(5, f(5))$, $(3, f(3))$, $(-2, f(-2))$, and $(-4, f(-4))$.

30. Draw the graph of an *even* function that includes the points $(0, -3)$, $(-3, 0)$, $(2, 0)$, $(1, -4)$, $(2.5, -1)$, $(-4, 3)$, and $(-5, 3)$.*

31. (a) Plot the points $(0, 0)$; $(2, 3)$; $(3, 4)$; $(5, 0)$; $(7, -3)$; $(-1, -1)$; $(-4, -1)$; $(-6, 1)$.

(b) Suppose the points in part (a) lie on the graph of an *odd* function f. Plot the points $(-2, f(-2))$; $(-3, f(-3))$; $(-5, f(-5))$; $(-7, f(-7))$; $(1, f(1))$; $(4, f(4))$; $(6, f(6))$.

(c) Draw the graph of an odd function f that includes all the points plotted in parts (a) and (b).*

32. Draw the graph of an odd function that includes the points $(-3, 5)$, $(-1, 1)$, $(2, -6)$, $(4, -9)$, and $(5, -5)$.*

33. Show that any graph that has two of the three types of symmetry (x-axis, y-axis, origin) necessarily has the third type also.

34. Use the midpoint formula to show that $(0, 0)$ is the midpoint of the segment joining (x, y) and $(-x, -y)$. Conclude that the coordinate test for symmetry with respect to the origin (page 136) is correct.

2.6 OPERATIONS ON FUNCTIONS

We now examine ways in which two or more given functions can be used to create new functions. If f and g are functions, then their **sum** is the function h defined by the rule

$$h(x) = f(x) + g(x).$$

For example, if $f(x) = 3x^2 + x$ and $g(x) = 4x - 2$, then

$$h(x) = f(x) + g(x) = (3x^2 + x) + (4x - 2) = 3x^2 + 5x - 2.$$

Instead of using a different letter h for the sum function, we shall usually denote it by $f + g$. Thus, the sum $f + g$ is defined by the rule

$$(f + g)(x) = f(x) + g(x).$$

This rule is *not* just a formal manipulation of symbols. If x is a number, then so are $f(x)$ and $g(x)$. The plus sign in $f(x) + g(x)$ is addition of *numbers* and the result is a number. But the plus sign in $f + g$ is addition of *functions* and the result is a new function.

The **difference** $f - g$ and the **product** fg of functions f and g are the functions defined by the rules

$$(f - g)(x) = f(x) - g(x) \qquad \text{and} \qquad (fg)(x) = f(x)g(x).$$

The domain of the sum, difference, and product functions is the set of all real numbers that are in both the domain of f and the domain of g. The **quotient** f/g is the function defined by

$$\left(\frac{f}{g}\right)(x) = \frac{f(x)}{g(x)}.$$

Its domain is the set of all real numbers x in both the domain of f and the domain of g such that $g(x) \neq 0$.

> **▶ TECHNOLOGY TIP**
>
> If you have two functions entered in the equation memory as y_1 and y_2, you can graph their sum as follows. On most TI's and Casio 9800, enter $y_1 + y_2$ as y_3 in the equation memory and graph y_3. On TI-92 and HP-38, use $y_1(x) + y_2(x)$ as y_3 (using f in place of y on HP-38).
>
> Differences, products, and quotients of functions can be graphed by a similar procedure.

*There are many correct answers.

EXAMPLE 1 If $f(x) = \sqrt{3x}$ and $g(x) = x^2 - 1$, then

$$(f + g)(x) = \sqrt{3x} + x^2 - 1$$
$$(f - g)(x) = \sqrt{3x} - (x^2 - 1) = \sqrt{3x} - x^2 + 1$$
$$(fg)(x) = \sqrt{3x}(x^2 - 1) = \sqrt{3x}\cdot x^2 - \sqrt{3x}$$
$$\left(\frac{f}{g}\right)(x) = \frac{\sqrt{3x}}{x^2 - 1}.$$

The domain of $f + g$, $f - g$, and fg consists of all x in both the domain of g (all real numbers) and the domain of f (all nonnegative reals), that is, all $x \geq 0$. The domain of f/g consists of the nonnegative x for which $g(x) \neq 0$, that is, all nonnegative reals *except* $x = 1$. ■

If c is a real number and f is a function, then the product of f and the constant function $g(x) = c$ is usually denoted cf. For example, if the function $f(x) = x^3 - x + 2$, and $c = 5$, then $5f$ is the function given by

$$(5f)(x) = 5\cdot f(x) = 5(x^3 - x + 2) = 5x^3 - 5x + 10.$$

Composition of Functions

Another way of combining functions is illustrated by the function $h(x) = \sqrt{x^3}$. To compute $h(4)$, for example, you first find $4^3 = 64$ and then take the square root $\sqrt{64} = 8$. So the rule of h may be rephrased as:

First apply the function $f(x) = x^3$,
Then apply the function $g(t) = \sqrt{t}$ to the result.

The same idea can be expressed in functional notation like this:

$$x \xrightarrow{\textit{first apply } f} f(x) \xrightarrow{\textit{then apply } g \textit{ to the result}} g(f(x))$$
$$x \qquad\qquad x^3 \qquad\qquad \sqrt{x^3}$$

apply h

So the rule of h may be written as $h(x) = g(f(x))$, where $f(x) = x^3$ and $g(t) = \sqrt{t}$. We can think of h as being made up of two simpler functions f and g, or we can think of f and g being "composed" to create the function h. Both viewpoints are useful.

EXAMPLE 2 Suppose $f(x) = 4x^2 + 1$ and $g(t) = \dfrac{1}{t + 2}$. Define a new function h whose rule is "first apply f; then apply g to the result." In functional notation

$$x \xrightarrow{\text{first apply } f} f(x) \xrightarrow{\text{then apply } g \text{ to the result}} g(f(x))$$

So the rule of the function h is $h(x) = g(f(x))$. Evaluating $g(f(x))$ means that whenever t appears in the formula for $g(t)$, we must replace it by $f(x) = 4x^2 + 1$:

$$h(x) = g(f(x)) = \frac{1}{f(x) + 2} = \frac{1}{(4x^2 + 1) + 2} = \frac{1}{4x^2 + 3}. \quad ■$$

The function h in Example 2 is an illustration of the following definition.

Composite Functions ▶

Let f and g be functions. The *composite function* of f and g is the function which assigns to x the number $g(f(x))$. It is denoted $g \circ f$.

▶ TECHNOLOGY TIP

Evaluating composite functions is easy on TI-82/83, TI-92, and HP-38. If the functions are entered in the equation memory as $y_1 = g(x)$ and $y_2 = h(x)$ (with f in place of y on HP-38), then keying in $y_2(y_1(5))$ ENTER produces the number $h(g(5))$.

On other calculators (including other TI models) this syntax does *not* produce $h(g(5))$; it produces $h(x) \cdot g(x) \cdot 5$ for whatever number is stored in the x-memory.

The symbol "$g \circ f$" is read "g circle f" or "f followed by g." (Note the order carefully; the functions are applied *right* to *left*.) So the rule of the composite function is:

$$(g \circ f)(x) = g(f(x)).$$

EXAMPLE 3 If $f(x) = 2x + 5$ and $g(t) = 3t^2 + 2t + 4$, then

$$(f \circ g)(2) = f(g(2)) = f(3 \cdot 2^2 + 2 \cdot 2 + 4) = f(20) = 2 \cdot 20 + 5 = 45.$$

Similarly,

$$(g \circ f)(-1) = g(f(-1)) = g(2(-1) + 5) = g(3) = 3 \cdot 3^2 + 2 \cdot 3 + 4 = 37.$$

The value of a composite function can also be computed like this:

$$(g \circ f)(5) = g(f(5)) = 3(f(5)^2) + 2(f(5)) + 4$$
$$= 3(15^2) + 2(15) + 4 = 709. \quad ■$$

The domain of $g \circ f$ is determined by the usual convention:

Domains of Composite Functions ▶

The domain of the composite function $g \circ f$ is the set of all real numbers x such that x is in the domain of f and $f(x)$ is in the domain of g.

EXAMPLE 4 If $f(x) = \sqrt{x}$ and $g(t) = t^2 - 5$, then

$$(g \circ f)(x) = g(f(x)) = (f(x))^2 - 5 = (\sqrt{x})^2 - 5 = x - 5.$$

Although $x - 5$ is defined for every real number x, the domain of $g \circ f$ is *not* the set of all real numbers. The domain of g is the set of all real numbers, but the function $f(x) = \sqrt{x}$ is defined only when $x \geq 0$. So the domain of $g \circ f$ is the set of nonnegative real numbers, that is, the interval $[0, \infty)$. ■

EXAMPLE 5 If $h(x) = \sqrt{3x^2 + 1}$, then h may be considered as the composite $g \circ f$, where $f(x) = 3x^2 + 1$ and $g(u) = \sqrt{u}$ because

$$(g \circ f)(x) = g(f(x)) = g(3x^2 + 1) = \sqrt{3x^2 + 1} = h(x).$$

There are other ways to consider $h(x) = \sqrt{3x^2 + 1}$ as a composite function. For instance, h is also the composite $j \circ k$, where $j(x) = \sqrt{x + 1}$ and $k(x) = 3x^2$:

$$(j \circ k)(x) = j(k(x)) = j(3x^2) = \sqrt{3x^2 + 1} = h(x). \blacksquare$$

EXAMPLE 6 If $k(x) = (x^2 - 2x + \sqrt{x})^3$, then k is $g \circ f$, where $f(x) = x^2 - 2x + \sqrt{x}$ and $g(t) = t^3$ because

$$(g \circ f)(x) = g(f(x)) = g(3x^2 - 2x + \sqrt{x}) = (x^2 - 2x + \sqrt{x})^3 = k(x). \blacksquare$$

As you may have noticed, there are two possible ways to form a composite function from two given functions. If f and g are functions, we can consider either

$$(g \circ f)(x) = g(f(x)), \quad \text{[the composite of } f \text{ and } g]$$

$$(f \circ g)(x) = f(g(x)), \quad \text{[the composite of } g \text{ and } f]$$

The *order is important,* as we shall now see:

$g \circ f$ and $f \circ g$ usually are *not* the same function.

EXAMPLE 7 If $f(x) = x^2$ and $g(x) = x + 3$,* then

$$(g \circ f)(x) = g(f(x)) = g(x^2) = x^2 + 3$$

but

$$(f \circ g)(x) = f(g(x)) = f(x + 3) = (x + 3)^2 = x^2 + 6x + 9.$$

Obviously, $g \circ f \neq f \circ g$ since, for example, they have different values at $x = 0$. ■

WARNING Don't confuse the product function fg with the composite function $f \circ g$ (g followed by f). For instance, if $f(x) = 2x^2$ and $g(x) = x - 3$, then the product fg is given by:

$$(fg)(x) = f(x)g(x) = 2x^2(x - 3) = 2x^3 - 6x^2.$$

It is *not* the same as the composite $f \circ g$ because

$$(f \circ g)(x) = f(g(x)) = f(x - 3) = 2(x - 3)^2 = 2x^2 - 12x + 18.$$

*Now that you have the idea of composite functions, we'll use the same letter for the variable in both functions.

By using the operations above, a complicated function may be considered as being built up from simple parts.

EXAMPLE 8 The function $f(x) = \sqrt{\dfrac{3x^2 - 4x + 5}{x^3 + 1}}$ may be considered as the composite $f = g \circ h$, where

$$h(x) = \frac{3x^2 - 4x + 5}{x^3 + 1} \quad \text{and} \quad g(x) = \sqrt{x}$$

since

$$(g \circ h)(x) = g(h(x)) = g\left(\frac{3x^2 - 4x + 5}{x^3 + 1}\right) = \sqrt{\frac{3x^2 - 4x + 5}{x^3 + 1}} = f(x).$$

The function $h(x) = \dfrac{3x^2 - 4x + 5}{x^3 + 1}$ is the quotient $\dfrac{p}{q}$, where

$$p(x) = 3x^2 - 4x + 5 \quad \text{and} \quad q(x) = x^3 + 1.$$

The function $p(x) = 3x^2 - 4x + 5$ may be written $p = k - s + r$, where

$$k(x) = 3x^2, \qquad s(x) = 4x, \qquad r(x) = 5.$$

The function k, in turn, can be considered as the product $3I^2$, where I is the *identity function* [whose rule is $I(x) = x$]:

$$(3I^2)(x) = 3(I^2(x)) = 3(I(x)I(x)) = 3 \cdot x \cdot x = 3x^2 = k(x).$$

Similarly, $s(x) = (4I)(x) = 4I(x) = 4x$. The function $q(x) = x^3 + 1$ may be "decomposed" in the same way.

Thus the complicated function f is just the result of performing suitable operations on the identity function I and various constant functions. ■

Applications

Composition of functions arises in applications involving several functional relationships simultaneously. In such cases one quantity may have to be expressed as a function of another.

EXAMPLE 9 A circular puddle of liquid is evaporating and slowly shrinking in size. After t minutes, the radius r of the puddle measures $\dfrac{18}{2t + 3}$ inches; in other words, the radius is a function of time. The area A of the puddle is given by $A = \pi r^2$, that is, area is a function of the radius r. We can express the area as a function of time by substituting $r = \dfrac{18}{2t + 3}$ in the area equation:

$$A = \pi r^2 = \pi\left(\frac{18}{2t+3}\right)^2.$$

This amounts to forming the composite function $f \circ g$, where $f(r) = \pi r^2$ and $g(t) = \dfrac{18}{2t+3}$:

$$(f \circ g)(t) = f(g(t)) = f\left(\frac{18}{2t+3}\right) = \pi\left(\frac{18}{2t+3}\right)^2.$$

When area is expressed as a function of time, it is easy to compute the area of the puddle at any time. For instance, after 12 minutes the area of the puddle is

$$A = \pi\left(\frac{18}{2t+3}\right)^2 = \pi\left(\frac{18}{2 \cdot 12 + 3}\right)^2 = \frac{4\pi}{9} \approx 1.396 \text{ square inches.} \quad \blacksquare$$

EXAMPLE 10 At noon a car leaves Podunk on a straight road, heading south at 45 miles per hour, and a plane three miles above the ground passes over Podunk heading east at 350 miles per hour.

(a) Express the distance r traveled by the car and the distance s traveled by the plane as functions of time.

(b) Express the distance d between the plane and the car in terms of r and s.

(c) Express d as a function of time.

(d) How far apart were the plane and the car at 1:30 P.M.?

Solution

(a) Traveling at 45 miles per hour for t hours the car will go a distance of $45t$ miles. Hence the equation $r = 45t$ expresses the distance r as a function of the time t. Similarly, the equation $s = 350t$ expresses the distance s as a function of the time t.

(b) In order to express the distance d as a function of r and s, consider Figure 2–41.

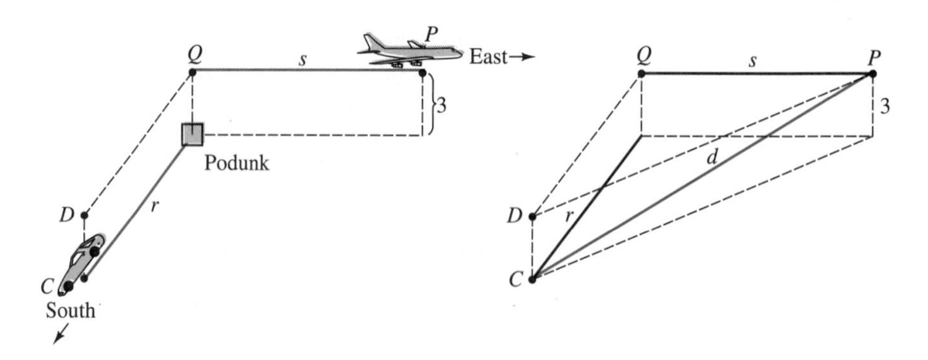

Figure 2-41

Right triangle PQD and the Pythagorean Theorem show that $(PD)^2 = r^2 + s^2$; hence $PD = \sqrt{r^2 + s^2}$. Applying the Pythagorean Theorem to right triangle PDC, we have

$$d^2 = 3^2 + (PD)^2$$
$$d^2 = 3^2 + \left(\sqrt{r^2 + s^2}\right)^2$$
$$d^2 = 9 + r^2 + s^2$$
$$d = \sqrt{9 + r^2 + s^2}$$

(c) The preceding equation expresses d in terms of r and s. By substituting $r = 45t$ and $s = 350t$ in this equation, we can express d as a function of the time t:

$$d = \sqrt{9 + r^2 + s^2}$$
$$d = \sqrt{9 + (45t)^2 + (350t)^2}$$
$$d = \sqrt{9 + 2025t^2 + 122{,}500t^2} = \sqrt{9 + 124{,}525t^2}.$$

(d) At 1:30 P.M. we have $t = 1.5$ (since noon is $t = 0$). At this time

$$d = \sqrt{9 + 124{,}525t^2} = \sqrt{9 + 124{,}525(1.5)^2} = \sqrt{280{,}190.25}$$
$$\approx 529.33 \text{ miles.} \quad \blacksquare$$

EXERCISES 2.6

In Exercises 1–4, find the indicated values, where $g(t) = t^2 - t$ and $f(x) = 1 + x$.

1. $g(f(0))$ **2.** $(f \circ g)(3)$

3. $g(f(2) + 3)$ **4.** $f(2g(1))$

In Exercises 5–8, find $(g \circ f)(3)$, $(f \circ g)(1)$, and $(f \circ f)(0)$.

5. $f(x) = 3x - 2, \quad g(x) = x^2$

6. $f(x) = |x + 2|, \quad g(x) = -x^2$

7. $f(x) = x, \quad g(x) = -3$

8. $f(x) = x^2 - 1, \quad g(x) = \sqrt{x}$

In Exercises 9–14, find the rule of the function $f \circ g$ and the rule of $g \circ f$.

9. $f(x) = x^2, \quad g(x) = x + 3$

10. $f(x) = -3x + 2, \quad g(x) = x^3$

11. $f(x) = 1/x, \quad g(x) = \sqrt{x}$

12. $f(x) = \dfrac{1}{2x + 1}, \quad g(x) = x^2 - 1$

13. $f(x) = \sqrt[3]{x}, \quad g(x) = x^2 + 1$

14. $f(x) = 2x^2 + 2x - 1, \quad g(x) = |x - 1| + 2$

In Exercises 15–18, verify that $f \circ g = I$ and $g \circ f = I$, where I is the identity function whose rule is $I(x) = x$ for every x.

15. $f(x) = 9x + 2, \quad g(x) = \dfrac{x - 2}{9}$

16. $f(x) = \sqrt[3]{x - 1}, \quad g(x) = x^3 + 1$

17. $f(x) = \sqrt[3]{x} + 2, \quad g(x) = (x - 2)^3$

18. $f(x) = 2x^3 - 5, \quad g(x) = \sqrt[3]{\dfrac{x + 5}{2}}$

In Exercises 19–22, find $(f + g)(x)$, $(f - g)(x)$, and $(g - f)(x)$.

19. $f(x) = -3x + 2, \quad g(x) = x^3$

20. $f(x) = x^2 + 2, \quad g(x) = -4x + 7$

21. $f(x) = 1/x, \quad g(x) = x^2 + 2x - 5$

22. $f(x) = \sqrt{x}, \quad g(x) = x^2 + 1$

In Exercises 23 and 24, find $(fg)(x)$, $(f/g)(x)$, and $(g/f)(x)$.

23. $f(x) = -3x + 2, \quad g(x) = x^3$

24. $f(x) = 4x^2 + x^4, \quad g(x) = \sqrt{x^2 + 4}$

Exercises 25 and 26 refer to the function f whose graph is shown in Figure 2–42.

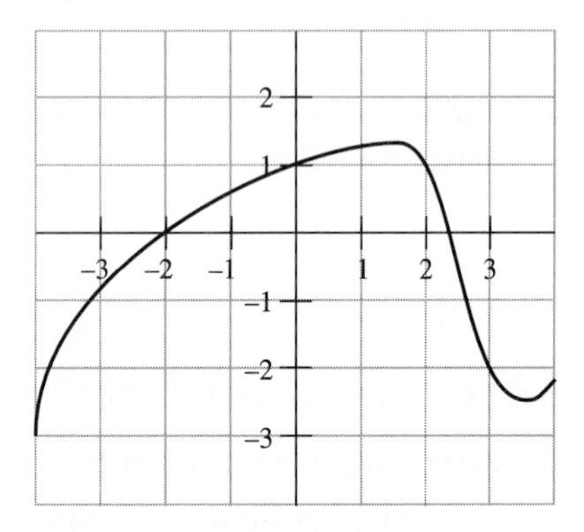

Figure 2-42

25. Let g be the composite function $f \circ f$ [that is, $g(x) = (f \circ f)(x) = f(f(x))$]. Use the graph of f to fill in the following table (approximate where necessary).

x	$f(x)$	$g(x) = f(f(x))$
−4		
−3		
−2	0	1
−1		
0		
1		
2		
3		
4		

26. Use the information obtained in Exercise 25 to sketch the graph of the function g.

In Exercises 27–30, fill the blanks in the given table. In each case the values of the functions f and g are given by these tables:

x	$f(x)$
1	3
2	5
3	1
4	2
5	3

t	$g(t)$
1	5
2	4
3	4
4	3
5	2

27.

x	$(g \circ f)(x)$
1	4
2	
3	5
4	
5	

28.

t	$(f \circ g)(t)$
1	
2	2
3	
4	
5	

29.

x	$(f \circ f)(x)$
1	
2	
3	3
4	
5	

30.

t	$(g \circ g)(t)$
1	
2	
3	
4	4
5	

In Exercises 31–36, write the given function as the composite of two functions, neither of which is the identity function, as in Examples 5 and 6. (There may be more than one way to do this.)

31. $f(x) = \sqrt[3]{x^2 + 2}$

32. $g(x) = \sqrt{x + 3} - \sqrt[3]{x + 3}$

33. $h(x) = (7x^3 - 10x + 17)^7$

34. $k(x) = \sqrt[3]{(7x - 3)^2}$

35. $f(x) = \dfrac{1}{3x^2 + 5x - 7}$

36. $g(t) = \dfrac{3}{\sqrt{t - 3}} + 7$

37. If $f(x) = x + 1$ and $g(t) = t^2$, then

$$(g \circ f)(x) = g(f(x)) = g(x + 1) = (x + 1)^2$$
$$= x^2 + 2x + 1$$

Find two other functions $h(x)$ and $k(t)$ such that $(k \circ h)(x) = x^2 + 2x + 1$.

38. If f is any function and I is the identity function, what are $f \circ I$ and $I \circ f$?

In Exercises 39–42, determine whether the functions $f \circ g$ and $g \circ f$ are defined. If a composite function is defined, find its domain.

39. $f(x) = x^3, \quad g(x) = \sqrt{x}$

40. $f(x) = x^2 + 1, \quad g(x) = \sqrt{x}$

41. $f(x) = \sqrt{x + 10}, \quad g(x) = 5x$

42. $f(x) = -x^2, \quad g(x) = \sqrt{x}$

43. **(a)** If $f(x) = 2x^3 + 5x - 1$, find $f(x^2)$.
(b) If $f(x) = 2x^3 + 5x - 1$, find $(f(x))^2$.
(c) Are the answers in parts (a) and (b) the same? What can you conclude about $f(x^2)$ and $(f(x))^2$?

44. Give an example of a function f such that $f\left(\dfrac{1}{x}\right) \neq \dfrac{1}{f(x)}$.

In Exercises 45 and 46, graph both $f \circ g$ and $g \circ f$ on the same screen. Use the graphs to determine whether $f \circ g$ is the same function as $g \circ f$.

45. $f(x) = x^5 - x^3 - x; \quad g(x) = x - 2$

46. $f(x) = x^3 + x; \quad g(x) = \sqrt[3]{x - 1}$

47. **(a)** What is the area of the puddle in Example 9 after one day? After a week? After a month?
(b) Does the puddle ever totally evaporate? Is this realistic? Under what circumstances might this area function be an accurate model of reality?

48. In a laboratory culture, the number $N(d)$ of bacteria (in thousands) at temperature d degrees Celsius is given by the function $N(d) = \dfrac{-90}{d + 1} + 20 \qquad (4 \le d \le 32)$.
The temperature $D(t)$ at time t hours is given by the function $D(t) = 2t + 4 \qquad (0 \le t \le 14)$.
(a) What does the composite function $N \circ D$ represent?
(b) How many bacteria are in the culture after 4 hours? After 10 hours?

49. A certain fungus grows in a circular shape. Its diameter after t weeks is $6 - \dfrac{50}{t^2 + 10}$ inches.
(a) Express the area covered by the fungus as a function of time.
(b) What is the area covered by the fungus when $t = 0$? What area does it cover at the end of 8 weeks?
(c) When is its area 25 square inches?

50. Tom left point P at 6 A.M. walking south at 4 miles per hour. Anne left point P at 8 A.M. walking west at 3.2 miles per hour.
(a) Express the distance between Tom and Anne as a function of the time t elapsed since 6 A.M.
(b) How far apart are Tom and Anne at noon?
(c) At what time are they 35 miles apart?

51. As a weather balloon is inflated its radius increases at the rate of 4 cm per second. Express the volume of the balloon as a function of time and determine the volume of the balloon after 4 seconds. [*Hint:* The volume of a sphere of radius r is $4\pi r^3/3$.]

52. Express the surface area of the weather balloon in Exercise 51 as a function of time. [*Hint:* The surface area of a sphere of radius r is $4\pi r^2$.]

53. Charlie, who is 6 feet tall, walks away from a streetlight that is 15 ft high at a rate of 5 feet per second, as shown in Figure 2–43 on the next page. Express the length s of Charlie's shadow as a function of time. [*Hint:* First use similar triangles to express s as a function of the distance d from the streetlight to Charlie.]

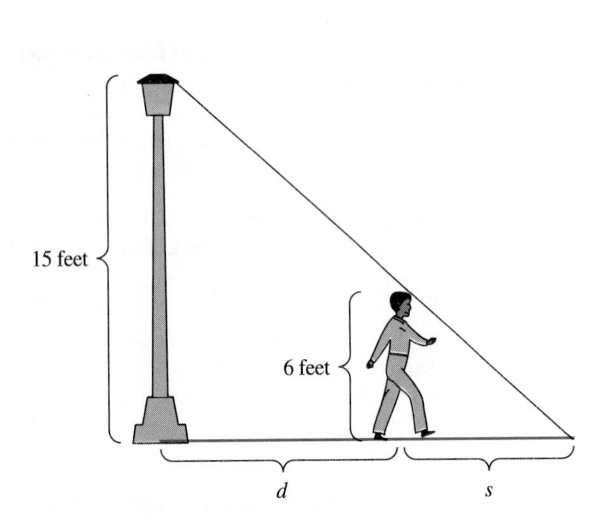

Figure 2-43

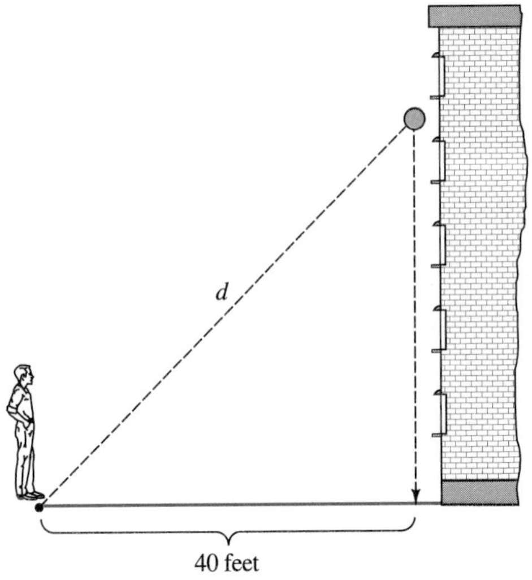

Figure 2-44

54. A water-filled balloon is dropped from a window 120 feet above the ground. Its height above the ground after t seconds is $120 - 16t^2$ feet. Howie is standing on the ground 40 feet from the point where the balloon will hit the ground, as shown in Figure 2–44.
(a) Express the distance d between Howie and the balloon as a function of time.
(b) When is the balloon exactly 90 ft from Howie?

Thinker

55. Find a function f (other than the identity function) such that $(f \circ f \circ f)(x) = x$ for every x in the domain of f. [Several correct answers are possible.]

2.7 INVERSE FUNCTIONS*

The function $h(x) = x^2$ can take the same value at two different numbers; for example, since $h(2) = 4$ and $h(-2) = 4$, the function takes the same value at both 2 and -2. On the other hand, the function $f(x) = x^3$ never takes the same value at two different numbers (because $a^3 = b^3$ is possible only when $a = b$). Functions that have this property are given a special name.

A function f is said to be **one-to-one** if it never takes the same value at two different numbers, that is, different inputs always produce different outputs:

$$\text{if } a \neq b, \text{ then } f(a) \neq f(b).$$

In graphical terms this means that two points on the graph, $(a, f(a))$ and $(b, f(b))$, that have different x-coordinates can't have the same y-coordinate. Hence, these points cannot lie on the same horizontal line, because all points on a horizontal

*This material is not needed until Chapter 5.

line have the same y-coordinate. Therefore, we have this geometric test to determine if a function is one-to-one.

The Horizontal Line Test ▶

If a function f is one-to-one, then it has this property:

No horizontal line intersects the graph of f more than once.

Conversely, if the graph of a function has this property, then the function is one-to-one.

EXAMPLE 1 Which of the following functions are one-to-one?

(a) $f(x) = 7x^5 + 3x^4 - 2x^3 + 2x + 1$

(b) $g(x) = x^3 - 3x - 1$

(c) $h(x) = 1 - .2x^3$

Solution Complete graphs of each function are shown in Figure 2–45.

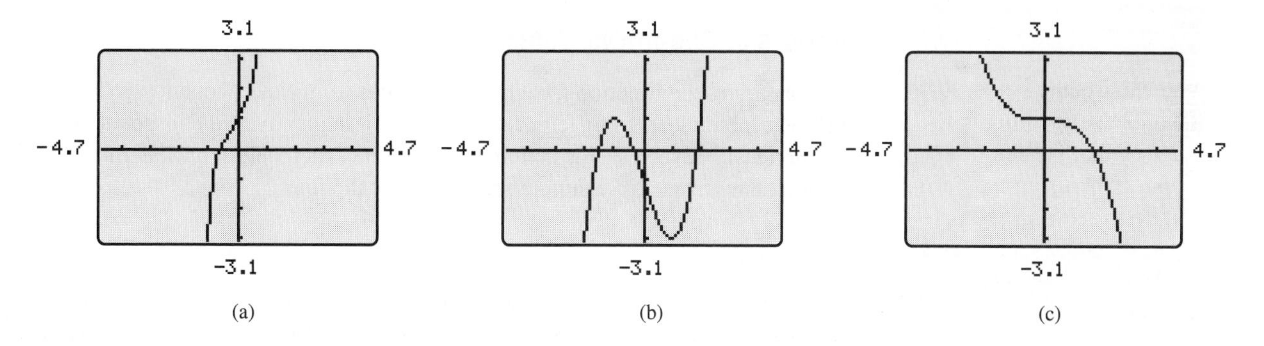

Figure 2-45

(a) The graph of f in Figure 2–45(a) passes the Horizontal Line Test since no horizontal line intersects the graph more than once. Hence, f is one-to-one.

(b) The graph of g in Figure 2–45(b) obviously fails the Horizontal Line Test because many horizontal lines (including the x-axis) intersect the graph more than once. Therefore, g is not one-to-one.

(c) The graph of h in Figure 2–45(c) appears to contain a horizontal line segment. So, h appears to fail the Horizontal Line Test because the horizontal line through (0, 1) seems to intersect the graph infinitely many times. But appearances are deceiving.

GRAPHING EXPLORATION Graph $h(x) = 1 - .2x^3$ and use the trace feature to move from left to right along the "horizontal" segment. Do the y-coordinates stay the same, or do they decrease?

The Exploration shows that the graph is actually falling from left to right, so that each horizontal line intersects it only once. (It appears to have a horizontal segment because the amount the graph falls there is less than the height of a pixel on the screen.) Therefore, h is a one-to-one function. ■

▶ TECHNOLOGY TIP

Although a horizontal segment may appear on a calculator screen when the graph is actually rising or falling, there is another possibility. The graph may have a tiny wiggle (less than the height of a pixel) and thus fail the Horizontal Line Test:

You can usually detect such a wiggle by zooming in to magnify that portion of the graph, or by using the trace feature to see if the y-coordinates increase, then decrease (or vice versa) along the "horizontal" segment.

The function f in Example 1 is an **increasing function** (its graph is always rising from left to right) and the function h is a **decreasing function** (its graph is always falling from left to right). Every increasing or decreasing function is necessarily one-to-one because its graph can never touch the same horizontal line twice (it would have to change from rising to falling, or vice versa, to do so).

Except for some piecewise-defined functions, such as the greatest integer function, and constant functions (such as $f(x) = 5$, whose graph is a horizontal line), the graphs of functions in this book do not contain horizontal line segments. However, a calculator screen may suggest otherwise as in Example 1(c).

Inverses of One-to-One Functions

With a one-to-one function f, each output comes from exactly one input (because different inputs lead to different outputs).* Consequently, we can define a new function that "reverses" the action of f by sending each output back to the unique input it came from. An example should clarify the idea.

EXAMPLE 2 The graph of the function $f(x) = 3x - 2$ is a straight line that certainly passes the Horizontal Line Test, so f is one-to-one. We can think of the rule of f as being given by the equation $y = 3x - 2$, where x is an input and y is the corresponding output. We want to define a function g that takes y back to x. We can do this by solving the equation $y = 3x - 2$ for x:

$$3x - 2 = y$$

(*)

Add 2 to both sides: $$3x = y + 2$$

Divide both sides by 3: $$x = \frac{y + 2}{3}$$

*This isn't true for functions that are not one-to-one. For example, with the function $h(x) = x^2$, the output 4 comes from two different inputs, namely, 2 and -2.

Then the function g, whose rule is $g(y) = \dfrac{y+2}{3}$, takes each output $f(x)$ back to the input it came from:

$$\text{If} \quad f(x) = 3x - 2, \quad \text{then} \quad g(f(x)) = g(3x-2) = \frac{(3x-2)+2}{3} = x.$$

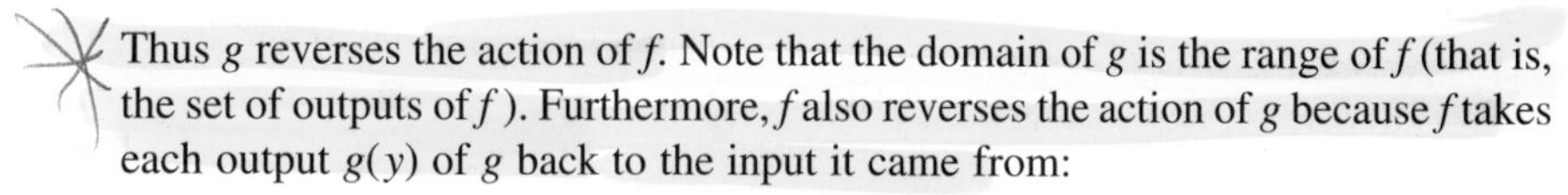

Thus g reverses the action of f. Note that the domain of g is the range of f (that is, the set of outputs of f). Furthermore, f also reverses the action of g because f takes each output $g(y)$ of g back to the input it came from:

$$\text{If} \quad g(y) = \frac{y+2}{3}, \quad \text{then} \quad f(g(y)) = f\left(\frac{y+2}{3}\right) = 3\left(\frac{y+2}{3}\right) - 2 = y. \quad ■$$

Example 2 serves as a model for the general case. Given a one-to-one function f, we write its rule as an equation that expresses y as a function of x. Then we rewrite the equation so that it expresses x as a function of y, as in statement $(*)$ above.† In effect, statement $(*)$ says

$$\frac{y+2}{3} = x \qquad \text{exactly when} \qquad 3x - 2 = y$$

which can be written in functional notation like this:

$$g(y) = x \qquad \text{exactly when} \qquad f(x) = y.$$

Essentially the same process works in the general case:

Inverse Functions ▶

Let f be a one-to-one function. Then f has an inverse function g, whose domain is the range of f and whose rule is determined by:

$$g(y) = x \qquad \textbf{exactly when} \qquad f(x) = y.$$

EXAMPLE 3 Use your calculator to verify that the function $f(x) = x^3 + 5$ passes the Horizontal Line Test and hence is one-to-one. Its inverse can be found by solving for x in the equation $y = x^3 + 5$:

Subtract 5 from both sides: $\quad x^3 = y - 5$

Take cube roots on both sides: $\quad x = \sqrt[3]{y-5}$

†This process won't work with functions that aren't one-to-one. For instance, the equation $y = x^2$ does express y as a function of x (each value of x leads to exactly one value of y), but it cannot be rewritten to express x as a function of y because substituting a number for y produces two values of x, namely, $x = \sqrt{y}$ and $x = -\sqrt{y}$.

Therefore, $g(y) = \sqrt[3]{y - 5}$ is the inverse function of f. The functions f and g have the same reversal properties that we saw in Example 2:

$$g(f(x)) = g(x^3 + 5) = \sqrt[3]{(x^3 + 5) - 5} = \sqrt[3]{x^3} = x \qquad \text{for every } x$$

$$f(g(y)) = f(\sqrt[3]{y - 5}) = (\sqrt[3]{y - 5})^3 + 5 = y - 5 + 5 = y \qquad \text{for every } y$$

Since the letter used for the variable of a function doesn't matter, the same variable is usually used for both f and its inverse function g and the rule of g is written as $g(x) = \sqrt[3]{x - 5}$. ■

EXAMPLE 4 The function $f(x) = \sqrt{x - 3}$ is one-to-one, as you can verify with your calculator. To find its inverse we solve the equation:

$$y = \sqrt{x - 3}$$

Square both sides: $\qquad y^2 = x - 3$

Add 3 to both sides: $\qquad x = y^2 + 3$

Although this last equation is defined for all real numbers y, the original equation $y = \sqrt{x - 3}$ has $y \geq 0$ (since square roots are nonnegative). In other words, the range of the function f (the possible values of y) consists of all nonnegative real numbers. Consequently, the domain of the inverse function g is the set of all nonnegative real numbers and its rule is:

$$g(y) = y^2 + 3 \qquad (y \geq 0).$$

Once again, it's customary to use the same variable to describe both f and its inverse function, so we write the rule of g as $g(x) = x^2 + 3$ $(x \geq 0)$. Note that f and g have the reversal properties: For every x in the domain of f,

$$g(f(x)) = g(\sqrt{x - 3}) = (\sqrt{x - 3})^2 + 3 = x - 3 + 3 = x$$

and for every x in the domain of g (that is, every $x \geq 0$),

$$f(g(x)) = f(x^2 + 3) = \sqrt{(x^2 + 3) - 3} = \sqrt{x^2} = x. \quad ■$$

As illustrated in Examples 2–4, a one-to-one function f and its inverse function g have these reversal properties:

$$g(f(x)) = x \text{ for every number } x \text{ in the domain of } f;$$

$$f(g(y)) = y \text{ for every number } y \text{ in the domain of } g.$$

Actually, somewhat more is true (as is proved in Exercise 53):

Inverse Theorem ▶

Suppose f and g are functions such that

$$\mathbf{g(f(x)) = x \text{ for every number } x \text{ in the domain of } f;}$$

$$\mathbf{f(g(x)) = x \text{ for every number } x \text{ in the domain of } g.}$$

Then f is one-to-one and its inverse is g.

EXAMPLE 5 If $f(x) = \dfrac{5}{2x - 4}$ and $g(x) = \dfrac{4x + 5}{2x}$, then for every x in the domain of f (that is, all $x \neq 2$),

$$g(f(x)) = g\left(\frac{5}{2x-4}\right) = \frac{4\left(\dfrac{5}{2x-4}\right) + 5}{2\left(\dfrac{5}{2x-4}\right)} = \frac{\dfrac{20 + 5(2x-4)}{2x-4}}{\dfrac{10}{2x-4}}$$

$$= \frac{20 + 5(2x-4)}{10} = \frac{20 + 10x - 20}{10} = \frac{10x}{10} = x$$

and for every x in the domain of g (all $x \neq 0$),

$$f(g(x)) = f\left(\frac{4x+5}{2x}\right) = \frac{5}{2\left(\dfrac{4x+5}{2x}\right) - 4} = \frac{5}{\dfrac{4x+5}{x} - 4}$$

$$= \frac{5}{\dfrac{4x + 5 - 4x}{x}} = \frac{5}{\dfrac{5}{x}} = x.$$

Therefore, f is a one-to-one function with inverse g. ■

Graphs of Inverse Functions

In theory, the algebraic method for finding the inverse of a one-to-one function illustrated in Examples 2–4 can always be used. In practice, however, you may not be able to solve the equation $y = f(x)$ for x; (for instance, try solving $y = .7x^5 + .3x^4 - .2x^3 + 2x + .5$ for x). Nevertheless, if you have the graph of f, it is possible to find the graph of its inverse function, even if you don't know the rule of the inverse. In order to do this, you must understand the relationship between the graph of f and the graph of its inverse function.

Suppose f is a one-to-one function and g is its inverse function. If (a, b) is on the graph of f, then by definition $f(a) = b$. Since the inverse function g takes each output of f back to its corresponding input, we know that $g(b) = a$. Hence, (b, a) is on the graph of g. A similar argument works in the other direction and leads to this conclusion:

(*) **(a, b) is on the graph of f exactly when (b, a) is on the graph of the inverse function g.**

This fact makes it very easy to graph an inverse function by using parametric graphing mode.

EXAMPLE 6 The function $f(x) = .7x^5 + .3x^4 - .2x^3 + 2x + .5$ can be graphed in parametric mode by letting $x = t$ and $y = f(t) = .7t^5 + .3t^4 - .2t^3 + 2t + .5$. Its complete graph in Figure 2–46 shows that f has an

> **▶ TECHNOLOGY TIP**
> The graph of the inverse function of $y_1(x)$ can be drawn automatically on most TI's. Use DRAW DRAWINV Y_1 on TI-82/83; GRAPH DRAW DRINV y_1 on TI-85; and DRAWINV $y_1(x)$ on TI-92.

inverse function g (why?), but it would be difficult to find its rule algebraically. According to statement (∗) above, the graph of g can be obtained by taking each point on the graph of f and reversing its coordinates. In other words, g can be graphed parametrically by letting

$$x = f(t) = .7t^5 + .3t^4 - .2t^3 + 2t + .5 \qquad \text{and} \qquad y = t.$$

Figure 2–47 shows the graphs of g and f on the same screen. ■

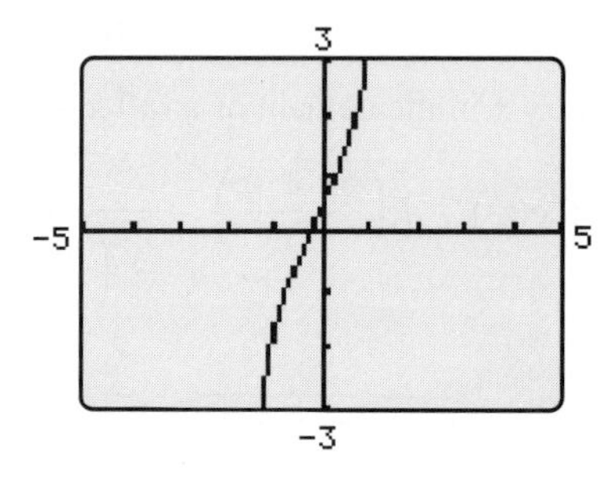

Figure 2-46

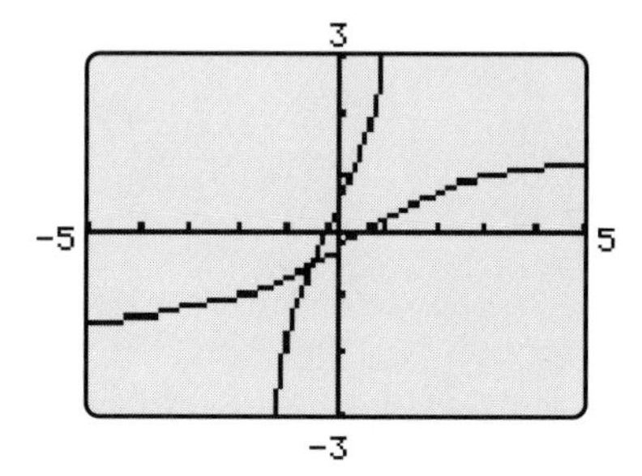

Figure 2-47

Statement (∗) on page 153 leads to a useful description of the graph of an inverse function. Recall that two points P and Q are symmetric with respect to a line L if L is the perpendicular bisector of the line segment PQ. Exercise 51 in this section or Exercise 65 in Section 3.1 shows that the points (a, b) and (b, a) are symmetric with respect to the line $y = x$. Consequently, the graphs of f and g are symmetric with respect to the line $y = x$, which means that the graph of the inverse function g is the mirror image of the graph of f, with the line $y = x$ being the mirror (see Excursion 2.5.A).

> GRAPHING EXPLORATION Illustrate this fact by graphing the line $y = x$, the function $f(x) = x^3 + 5$ of Example 3, and its inverse $g(x) = \sqrt[3]{x - 5}$ on the same screen (use a square viewing window so that the mirror effect won't be distorted).

In formal terms:

Graphs of Inverse Functions ▶

> **If g is the inverse function of f, then the graph of g is the reflection of the graph of f in the line $y = x$.**

Note In many texts the inverse function of a function f is denoted f^{-1}. In this notation, for instance, the inverse of the function $f(x) = x^3 + 5$ in Example 3 would be written as $f^{-1}(x) = \sqrt[3]{x - 5}$. Similarly, the reversal properties of inverse functions become

$$f^{-1}(f(x)) = x \text{ for every } x \text{ in the domain of } f\text{; and}$$
$$f(f^{-1}(x)) = x \text{ for every } x \text{ in the domain of } f^{-1}.$$

In this context, f^{-1} does *not* mean $1/f$ (see Exercise 47).

EXERCISES 2.7

In Exercises 1–8, use a calculator and the Horizontal Line Test to determine whether or not the function f is one-to-one.

1. $f(x) = x^4 - 4x^2 + 3$

2. $f(x) = x^4 - 4x + 3$

3. $f(x) = x^3 + x - 5$

4. $f(x) = \begin{cases} x - 3 & \text{for } x \le 3 \\ 2x - 6 & \text{for } x > 3 \end{cases}$

5. $f(x) = x^5 + 2x^4 - x^2 + 4x - 5$

6. $f(x) = x^3 - 4x^2 + x - 10$

7. $f(x) = .1x^3 - .1x^2 - .005x + 1$

8. $f(x) = .1x^3 + .005x + 1$

In Exercises 9–22, use algebra to find the inverse of the given one-to-one function, if it has one.

9. $f(x) = -x$

10. $f(x) = -x + 1$

11. $f(x) = 5x - 4$

12. $f(x) = -3x + 5$

13. $f(x) = 5 - 2x^3$

14. $f(x) = (x^5 + 1)^3$

15. $f(x) = \sqrt{4x - 7}$

16. $f(x) = 5 + \sqrt{3x - 2}$

17. $f(x) = 1/x$

18. $f(x) = 1/\sqrt{x}$

19. $f(x) = \dfrac{1}{2x + 1}$

20. $f(x) = \dfrac{x}{x + 1}$

21. $f(x) = \dfrac{x^3 - 1}{x^3 + 5}$

22. $f(x) = \sqrt[5]{\dfrac{3x - 1}{x - 2}}$

In Exercises 23–28, use the Inverse Theorem on page 152 to show that g is the inverse of f.

23. $f(x) = x + 1, \quad g(x) = x - 1$

24. $f(x) = 2x - 6, \quad g(x) = \dfrac{x}{2} + 3$

25. $f(x) = \dfrac{1}{x + 1}, \quad g(x) = \dfrac{1 - x}{x}$

26. $f(x) = \dfrac{-3}{2x + 5}, \quad g(x) = \dfrac{-3 - 5x}{2x}$

27. $f(x) = x^5, \quad g(x) = \sqrt[5]{x}$

28. $f(x) = x^3 - 1, \quad g(x) = \sqrt[3]{x + 1}$

29. Show that the inverse function of the function f whose rule is $f(x) = \dfrac{2x + 1}{3x - 2}$ is f itself.

30. List three different functions (other than the one in Exercise 29), each of which is its own inverse. [Many correct answers are possible.]

In Exercises 31 and 32, the graph of a function f is given. Sketch the graph of the inverse function of f. [Reflect carefully.]

31.

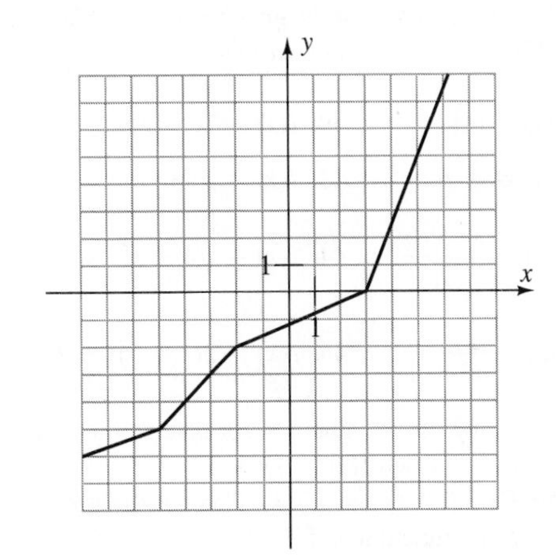

32.

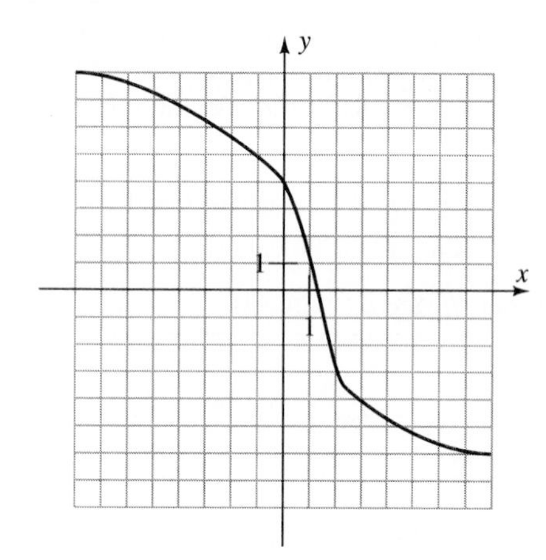

In Exercises 33–38, each given function has an inverse function. Sketch the graph of the inverse function.

33. $f(x) = \sqrt{x + 3}$ **34.** $f(x) = \sqrt{3x - 2}$

35. $f(x) = .3x^5 + 2$ **36.** $f(x) = \sqrt[3]{x + 3}$

37. $f(x) = \sqrt[5]{x^3 + x - 2}$

38. $f(x) = \begin{cases} x^2 - 1 & \text{for } x \le 0 \\ -.5x - 1 & \text{for } x > 0 \end{cases}$

In Exercises 39–46, none of the functions has an inverse. State at least one way of restricting the domain of the function (that is, find a function with the same rule and a smaller domain) so that the restricted function has an inverse. Then find the rule of the inverse function.

Example: $f(x) = x^2$ *has no inverse. But the function h with domain all* $x \ge 0$ *and rule* $h(x) = x^2$ *is increasing (its graph is the right half of the graph of f—see Figure 1–32 on page 37) and therefore has an inverse.*

39. $f(x) = |x|$ **40.** $f(x) = |x - 3|$

41. $f(x) = -x^2$ **42.** $f(x) = x^2 + 4$

43. $f(x) = \dfrac{x^2 + 6}{2}$ **44.** $f(x) = \sqrt{4 - x^2}$

45. $f(x) = \dfrac{1}{x^2 + 1}$ **46.** $f(x) = 3(x + 5)^2 + 2$

47. **(a)** Using the f^{-1} notation for inverse functions, find $f^{-1}(x)$ when $f(x) = 3x + 2$.

(b) Find $f^{-1}(1)$ and $1/f(1)$. Conclude that f^{-1} is not the same function as $1/f$.

48. Let C be the temperature in degrees Celsius. Then the temperature in degrees Fahrenheit is given by $f(C) = \frac{9}{5}C + 32$. Let g be the function that converts degrees Fahrenheit to degrees Celsius. Show that g is the inverse function of f and find the rule of g.

Thinkers

49. Let m and b be constants with $m \neq 0$. Show that the function $f(x) = mx + b$ has an inverse function g and find the rule of g.

50. Prove that the function $f(x) = 1 - .2x^3$ of Example 1(c) is one-to-one by showing that it satisfies the definition on page 148, namely, that

$$\text{If } a \neq b, \text{ then } f(a) \neq f(b).$$

[*Hint:* Use the rule of f to show that when $f(a) = f(b)$, then $a = b$. If this is the case, then it is impossible to have $f(a) = f(b)$ when $a \neq b$.]

51. Let P, with coordinates (a, b), be any point in the plane which is not on either of the lines $y = x$ or $y = -x$. Prove that P and the point Q with coordinates (b, a) are symmetric with respect to the line $y = x$ as follows. Let O be the origin and R the point where the line $y = x$ intersects the line segment PQ, as illustrated in Figure 2–48. Since R is on the line $y = x$, it has coordinates (c, c) for some number c.

(a) Use the distance formula to show that segments OP and OQ have the same length and that segments RP and RQ have the same length.

(b) Explain why triangles ORP and ORQ are congruent. {*Hint:* side-side-side.]

(c) Explain why angles ORP and ORQ are right angles. [*Hint:* Their sum is 180° (why?) and they are congruent.]

(d) Conclude that the line $y = x$ is the perpendicular bisector of segment PQ. Hence, P and Q are symmetric with respect to this line.

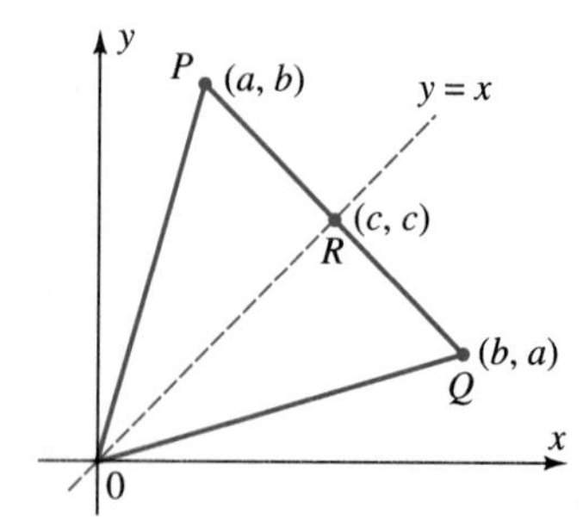

Figure 2-48

52. **(a)** Experiment with your calculator or use some of the preceding exercises to find four different increasing functions. For each function, sketch the graph of the function and the graphs of its inverse on the same set of axes.

(b) Based on the evidence in part (a), do you think the following statement true or false: The inverse function of every increasing function is also an increasing function.

(c) Do parts (a) and (b) with ''increasing'' replaced by ''decreasing.''

53. Prove the Inverse Theorem (page 152) as follows. By hypothesis, f and g have these properties:

(1) $g(f(x)) = x$ for every number x in the domain of f;

(2) $f(g(y)) = y$ for every number y in the domain of g.

(a) Prove that f is one-to-one by showing that

$$\text{if} \quad a \neq b, \qquad \text{then} \qquad f(a) \neq f(b).$$

[*Hint:* If $f(a) = f(b)$, apply g to both sides and use **(1)** to show that $a = b$. Consequently, if $a \neq b$, it is impossible to have $f(a) = f(b)$.]

(b) If $g(y) = x$, show that $f(x) = y$. [*Hint:* Use **(2)**.]

(c) If $f(x) = y$, show that $g(y) = x$. [*Hint:* Use **(1)**.]

Parts (b) and (c) prove that

$$g(y) = x \qquad \text{exactly when} \qquad f(x) = y.$$

Hence, g is the inverse function of f (see page 151).

CHAPTER 2 *Review*

Important Concepts ▶

Review Questions ▶

1. Let $[x]$ denote the greatest integer function and evaluate
   (a) $[-5/2] =$ ______. (b) $[1755] =$ ______.
   (c) $[18.7] + [-15.7] =$ ______. (d) $[-7] - [7] =$ ______.
2. If $f(x) = x + |x| + [x]$, then find $f(0)$, $f(-1)$, $f(1/2)$, and $f(-3/2)$.
3. Let f be the function given by the rule $f(x) = 7 - 2x$. Complete this table:

x	0	1	2	-4	t	k	$b - 1$	$1 - b$	$6 - 2u$
$f(x)$	7								

4. What is the domain of the function g given by
$$g(t) = \frac{\sqrt{t-2}}{t-3}?$$
5. In each case give a *specific* example of a function and numbers a, b to show that the given statement may be *false*.
   (a) $f(a + b) = f(a) + f(b)$ (b) $f(ab) = f(a)f(b)$
6. If $f(x) = |3 - x|\sqrt{x - 3} + 7$, then $f(7) - f(4) =$ ______.
7. What is the domain of the function given by
$$g(r) = \sqrt{r - 4} + \sqrt{r} - 2?$$
8. What is the domain of the function $f(x) = \sqrt{-x + 2}$?
9. If $h(x) = x^2 - 3x$, then $h(t + 2) =$ ______.
10. Which of the following statements about the greatest integer function $f(x) = [x]$ is true for *every* real number x?
    (a) $x - [x] = 0$ (b) $x - [x] \leq 0$
    (c) $[x] + [-x] \leq 0$ (d) $[-x] \geq [x]$
    (e) $3[x] = [3x]$
11. If $f(x) = 2x^3 + x + 1$, then $f(x/2) =$ ______.
12. If $g(x) = x^2 - 1$, then $g(x - 1) - g(x + 1) =$ ______.
13. Sketch the graph of the function f given by
$$f(x) = \begin{cases} x^2 & \text{if } x \leq 0 \\ x + 1 & \text{if } 0 < x < 4 \\ \sqrt{x} & \text{if } x \geq 4 \end{cases}$$
14. Without using your calculator, describe the general shape of the graph of $f(x) = x|x|$. [What does the graph look like when x is positive? When x is negative?]
15. Determine a viewing window that shows a complete graph of the function $f(x) = \dfrac{.1x^5 + .1x - .5}{x^2 + 1}$ and justify your answer.

In Questions 16–18, find a viewing window that shows a complete graph of the equation.

16. $x^2 + 3y^2 = 144$
17. $9y^2 - 4x^2 + 32x - 65 = 0$
18. $y^2 - x - 6y + 19 = 0$

In Questions 19 and 20, sketch the graph of the curve given by the parametric equations.

19. $x = t^2 - 4$ and $y = 2t + 1 \quad (-3 \le t \le 3)$

20. $x = t^3 + 3t^2 - 1$ and $y = t^2 + 1 \quad (-3 \le t \le 2)$

21. Sketch a graph that is symmetric with respect to both the x-axis and the y-axis. (There are many correct answers and your graph need not be the graph of an equation.)

22. Sketch the graph of a function that is symmetric with respect to the origin. (There are many correct answers and you don't have to state the rule of your function.)

In Questions 23–28, determine algebraically whether the graph of the given equation is symmetric with respect to the x-axis, and y-axis, or the origin.

23. $y = (x + 3)^4 - 2$ **24.** $y = x(x - 1)^2(x + 5)^2$

25. $y = 2x^3(x + 2)$ **26.** $x^2 = y^2 + 2$

27. $x^2 + y^4 + 5 = 0$ **28.** $5y = 7x^2 - 2x$

In Questions 29–31, determine whether the given function is even, odd, or neither.

29. $g(x) = 9 - x^2$ **30.** $f(x) = |x|x + 1$

31. $h(x) = 3x^5 - x(x^4 - x^2)$

32. **(a)** Draw some coordinate axes and plot the points $(-2, 1)$, $(-1, 3)$, $(0, 1)$, $(3, 2)$, $(4, 1)$.

(b) Suppose the points plotted in part (a) lie on the graph of an *even* function f. Plot these points: $(2, f(2))$, $(1, f(1))$, $(0, f(0))$, $(-3, f(-3))$, $(-4, f(-4))$.

33. Determine whether the circle with equation $x^2 + y^2 + 6y = -5$ is symmetric with respect to the x-axis, the y-axis, or the origin.

34. Sketch the graph of a function f that satisfies all of these conditions:

(i) domain of $f = [-3, 4]$ **(ii)** range of $f = [-2, 5]$

(iii) $f(-2) = 0$ **(iv)** $f(1) > 2$

[*Note:* There are many possible correct answers and the function whose graph you sketch need *not* have a simple algebraic rule.]

35. Sketch the graph of $g(x) = 5 + \dfrac{4}{x - 5}$.

Use the graph of the function f in Figure 2–49 to answer Questions 36–39.

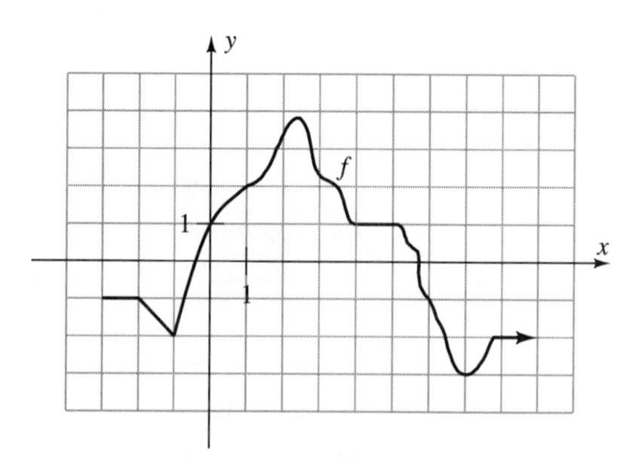

Figure 2-49

36. What is the domain of f? **37.** What is the range of f?

38. Find all numbers x such that $f(x) = 1$.

39. Find a number x such that $f(x + 1) < f(x)$. (Many correct answers are possible.)

Use the graph of the function f in Figure 2–50 to answer Questions 40–46.

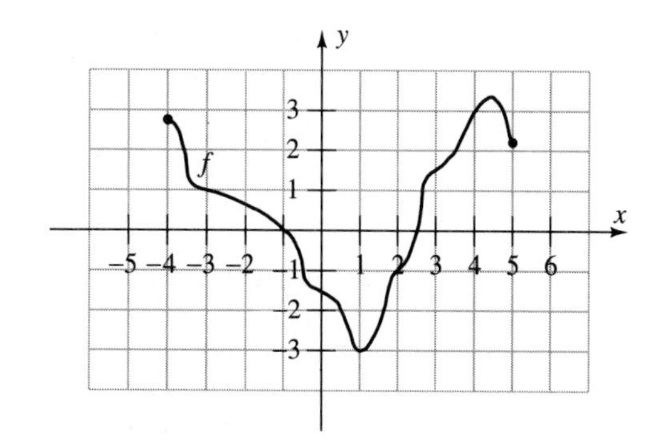

Figure 2-50

40. What is the domain of f? **41.** $f(-3) =$ ______.

42. $f(2 + 2) =$ ______. **43.** $f(-1) + f(1) =$ ______.

44. True or false: $2f(2) = f(4)$. **45.** True or false: $3f(2) = -f(4)$.

46. True or false: $f(x) = 3$ for exactly one number x.

Use the graphs of the functions f and g in Figure 2–51 to answer Questions 47–52.

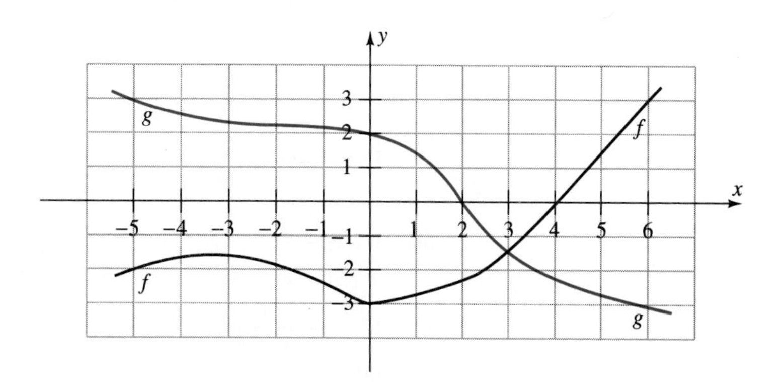

Figure 2-51

47. For which values of x is $f(x) = 0$?

48. True or false: If a and b are numbers such that $-5 \leq a < b \leq 6$, then $g(a) < g(b)$.

49. For which values of x is $g(x) \geq f(x)$? **50.** Find $f(0) - g(0)$.

51. For which values of x is $f(x + 1) < 0$?

52. What is the distance from the point $(-5, g(-5))$ to the point $(6, g(6))$?

In Questions 53–57, list the transformations, in the order they should be performed on the graph of $g(x) = x^2$, so as to produce a complete graph of the function f.

53. $f(x) = x^2 - 4$ **54.** $f(x) = (x - 2)^2$

55. $f(x) = .25x^2 + 2$ **56.** $f(x) = -(x + 4)^2 - 5$

57. $f(x) = -3(x - 7)^2 + 2$

58. The graph of a function f is shown in Figure 2–52. On the same coordinate plane, carefully draw the graphs of the functions g and h whose rules are:

$$g(x) = -f(x) \quad \text{and} \quad h(x) = 1 - f(x)$$

59. Figure 2–53 shows the graph of a function f. If g is the function given by $g(x) = f(x + 2)$, then which of these statements about the graph of g is true?
(a) It does not cross the x-axis. **(b)** It does not cross the y-axis.
(c) It crosses the y-axis at $y = 4$. **(d)** It crosses the y-axis at the origin.
(e) It crosses the x-axis at $x = -3$.

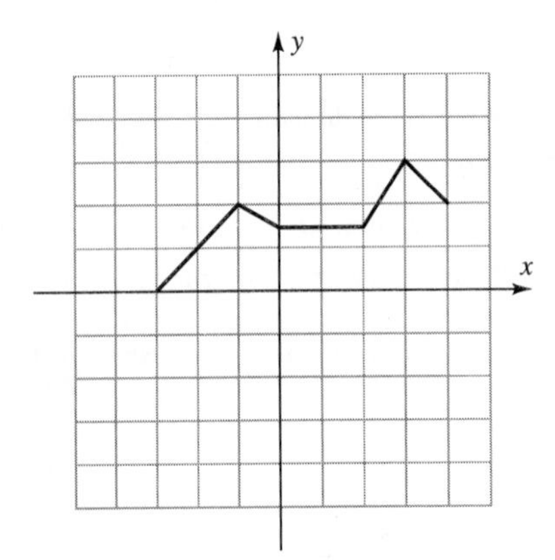

Figure 2-52

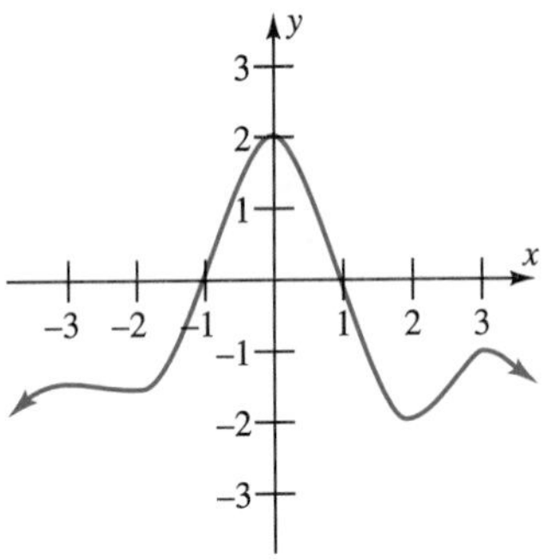

Figure 2-53

60. If $f(x) = 3x + 2$ and $g(x) = x^3 + 1$, find:
(a) $(f + g)(-1)$ **(b)** $(f - g)(2)$ **(c)** $(fg)(0)$

61. If $f(x) = \dfrac{1}{x - 1}$ and $g(x) = \sqrt{x^2 + 5}$, find:
(a) $(f/g)(2)$ **(b)** $(g/f)(x)$
(c) $(fg)(c + 1)$ $(c \neq 1)$

62. Find two functions f and g such that neither is the identity function and

$$(f \circ g)(x) = (2x + 1)^2$$

63. Use the graph of the function g in Figure 2–54 to fill in the table below, in which h is the composite function $g \circ g$.

x	-4	-3	-2	-1	0	1	2	3	4
$g(x)$					-1				
$h(x) = g(g(x))$									

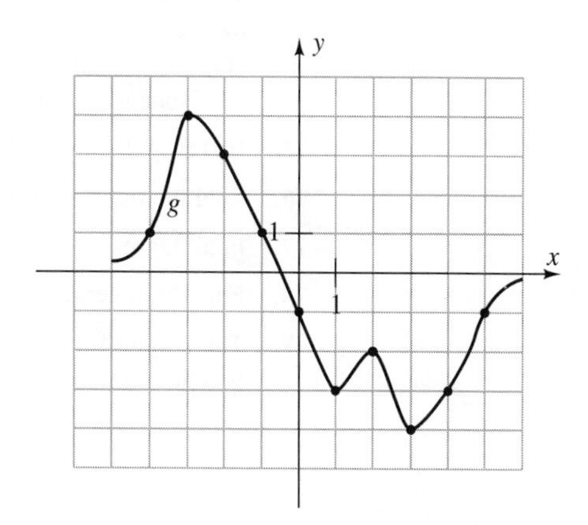

Figure 2-54

Questions 64–69 refer to the functions $f(x) = \dfrac{1}{x+1}$ *and* $g(t) = t^3 + 3$.

64. $(f \circ g)(1) =$ ______.

65. $(g \circ f)(2) =$ ______.

66. $g(f(-2)) =$ ______.

67. $(g \circ f)(x-1) =$ ______.

68. $g(2 + f(0)) =$ ______.

69. $f(g(1) - 1) =$ ______.

70. Let f and g be the functions given by

$$f(x) = 4x + x^4 \quad \text{and} \quad g(x) = \sqrt{x^2 + 1}$$

(a) $(f \circ g)(x) =$ ______. **(b)** $(g - f)(x) =$ ______.

71. If $f(x) = \dfrac{1}{x}$ and $g(x) = x^2 - 1$, then

$$(f \circ g)(x) = \_\_\_\_\_\_ \quad \text{and} \quad (g \circ f)(x) = \_\_\_\_\_\_.$$

72. Let $f(x) = x^2$. Give an example of a function g with domain all real numbers such that $g \circ f \neq f \circ g$.

73. If $f(x) = \dfrac{1}{1-x}$ and $g(x) = \sqrt{x}$, then find the domain of the composite function $f \circ g$.

74. These tables show the values of the functions f and g at certain numbers:

x	-1	0	1	2	3
$f(x)$	1	0	1	3	5

and

t	0	1	2	3	4
$g(t)$	-1	0	1	2	5

Which of the following statements are *true*?
(a) $(g - f)(1) = 1$ **(b)** $(f \circ g)(2) = (f - g)(0)$
(c) $f(1) + f(2) = f(3)$ **(d)** $(g \circ f)(2) = 1$
(e) None of the above is true.

75. If $f(x) = 2x^2$ and $g(x) = -1$, then sketch the graph of the composite function $f \circ g$.

76. Find the inverse of the function $f(x) = 2x + 1$.

77. Find the inverse of the function $f(x) = \sqrt{5 - x} + 7$.

78. Find the inverse of the function $f(x) = \sqrt[5]{x^3 + 1}$.

79. The graph of a function f is shown in Figure 2–55. Sketch the graph of the inverse function of f.

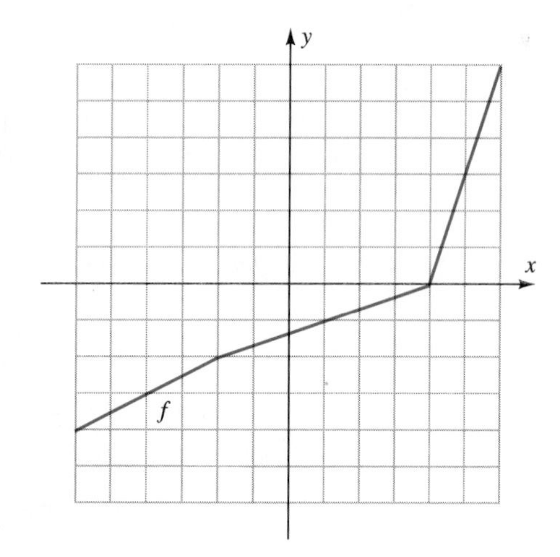

Figure 2-55

80. Which of the following functions have inverse functions (give reasons for your answers):
(a) $f(x) = x^3$ **(b)** $f(x) = 1 - x^2, \quad x \leq 0$
(c) $f(x) = |x|$

In Exercises 81–83, determine whether or not the given function has an inverse function [give reasons for your answer]. If it does, find the graph of the inverse function.

81. $f(x) = 1/x$

82. $f(x) = .02x^3 - .04x^2 + .6x - 4$

83. $f(x) = .2x^3 - 4x^2 + 6x - 15$

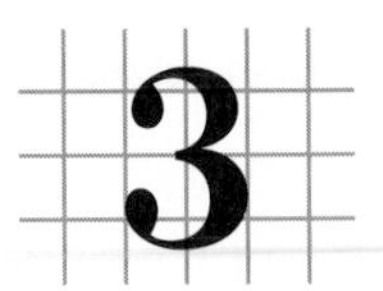

Linear and Quadratic Functions

Roadmap

Sections 1 and 3 are independent of each other and may be read in either order. Section 2 depends on Section 1.

In order to deal effectively with tangent lines and rates of change, which play a crucial role in calculus, you need to know the essential facts about functions whose graphs are straight lines. These "linear functions" and their connection with rates of change are studied in this chapter.

The rule of each linear function is given by a first-degree polynomial. Quadratic functions are those whose rules are given by second-degree polynomials; these functions also have some special properties, which are discussed here. Polynomial functions of higher degree will be considered in Chapter 4.

3.1 LINES AND LINEAR FUNCTIONS

A **linear function** is a function whose rule can be written in the form

$$f(x) = mx + b$$

for some constants m and b, or equivalently, a function whose rule is given by an equation of the form $y = mx + b$.

GRAPHING EXPLORATION Using the standard viewing window, graph the following linear functions on the same screen:

$$f(x) = 1.5x + 2, \quad f(x) = 5,^{*} \quad y = -2x,^{**} \quad y = .8x + 1.$$

What you see should explain why these functions are called "linear."

*This is the case $f(x) = mx + b$, with $m = 0$ and $b = 5$.

**This is the case $y = mx + b$, with $m = -2$ and $b = 0$.

The results of this Exploration make the following statements plausible:

Linear Function Theorem ▶

The graph of the linear function $f(x) = mx + b$ is a nonvertical straight line. Conversely, every nonvertical straight line is the graph of a linear function.

We shall assume this theorem for now; its proof is on page 174 and may be read at any time. For convenience we shall sometimes refer to "the line $y = mx + b$" or "the line with equation $y = mx + b$," meaning the line that is the graph of the linear function $f(x) = mx + b$.

A key geometric feature of any nonvertical straight line is how steeply it rises or falls as you move from left to right. There is an easy numerical way to measure this steepness, as we now see.

GRAPHING EXPLORATION Using the standard viewing window, graph the following linear functions on the same screen:

$$f(x) = .5x, \qquad g(x) = x, \qquad h(x) = 3x, \qquad k(x) = 7x$$

and answer the following questions:

Which graph rises least steeply (from left to right)? Which one rises most steeply?

How is the coefficient of x in each function rule (.5 for f, 1 for g, etc.) related to how steeply the graph rises?

Now graph the following linear functions on the same screen:

$$f(x) = 2, \quad g(x) = -x + 2, \quad h(x) = -2.5x + 2, \quad k(x) = -5x + 2$$

and answer these questions:

Which graph falls least steeply (from left to right)? Which one falls most steeply?

How is the coefficient of x in each function rule (0 for f, -1 for g, etc.) related to how steeply the graph falls?

The graph of the function $f(x) = mx + b$ is a straight line and the number m is called the **slope** of the line. Your answers to the questions in the preceding Exploration should indicate that the slope measures the steepness of the line, as summarized here:

Properties of Slope ▶

The graph of $y = mx + b$ is a nonvertical straight line with slope m. The slope measures how steeply the line rises or falls:

If $m > 0$, the line rises from left to right; the larger m is, the more steeply the line rises.

If $m = 0$, the line is horizontal.

If $m < 0$, the line falls from left to right; the larger $|m|$ is, the more steeply the line falls.

If the graph of an equation intersects the y-axis at the point $(0, c)$, then the number c is called a **y-intercept** of the graph. To find the y-intercepts without graphing, just set $x = 0$ in the equation and solve for y: These values are precisely the y-coordinates of points with x-coordinate 0 (that is, points on the graph that are also on the y-axis). For instance, the line $y = 3x + 4$ has y-intercept $y = 3 \cdot 0 + 4 = 4$, meaning that the line crosses the y-axis at $(0, 4)$. Similarly, the graph of $y = mx + b$ crosses the y-axis when $x = 0$ and $y = m \cdot 0 + b$. In other words,

Intercepts ▶

The graph of $y = mx + b$ is a nonvertical straight line with slope m and y-intercept b.

EXAMPLE 1 Show that the graph of $\sqrt{7}x = 4y + 10$ is a straight line and find its slope and y-intercept.

Solution Write the equation in the form $y = mx + b$ by rearranging terms:

$$4y + 10 = \sqrt{7}x$$

$$4y = \sqrt{7}x - 10$$

$$y = \frac{\sqrt{7}}{4}x - \frac{10}{4} = \frac{\sqrt{7}}{4}x - 2.5.$$

The equation is now in the form $y = mx + b$, with $m = \sqrt{7}/4$ and $b = -2.5$. Therefore, the graph is a straight line with slope $\sqrt{7}/4$ and y-intercept -2.5. ■

EXAMPLE 2 The equation $y = 3$ can be written as $y = 0x + 3$. So, its graph is a line with slope 0 and y-intercept 3, that is, a horizontal line through $(0, 3)$, as you can readily verify with your calculator. ■

The Equation of a Line

When the rule of a linear function is given, you can easily read off the slope and y-intercept of its graph, as in Examples 1 and 2. Conversely, when information about a straight line is given, it is often possible to determine its slope and

y-intercept, and hence to determine the equation of the line, that is, the equation whose graph is this line. For example, if (x_1, y_1) and (x_2, y_2) are distinct points on a nonvertical straight line, then the slope of this line can be found as follows. The line is the graph of an equation $y = mx + b$ for some constants m and b. Since (x_1, y_1) and (x_2, y_2) are points on the graph of $y = mx + b$, their coordinates satisfy the equation:

$$y_2 = mx_2 + b$$
$$y_1 = mx_1 + b.$$

Subtracting the second equation from the first and solving for m, we have:

$$y_2 - y_1 = mx_2 - mx_1$$
$$y_2 - y_1 = m(x_2 - x_1)$$
$$m = \frac{y_2 - y_1}{x_2 - x_1}.$$

We have proved the following useful result:

Computing the Slope ▶

The slope of a nonvertical straight line is the number

$$\frac{y_2 - y_1}{x_2 - x_1}$$

where (x_1, y_1) and (x_2, y_2) are any two distinct points on the line.

EXAMPLE 3 To find the slope of the line through $(0, -1)$ and $(4, 1)$ we apply the formula in the box above with $x_1 = 0$, $y_1 = -1$ and $x_2 = 4$, $y_2 = 1$:

$$\text{slope} = \frac{y_2 - y_1}{x_2 - x_1} = \frac{1 - (-1)}{4 - 0} = \frac{2}{4} = \frac{1}{2}.$$

The order of the points makes no difference; if you use $(4, 1)$ for (x_1, y_1) and $(0, -1)$ for (x_2, y_2) we obtain the same number:

$$\text{slope} = \frac{y_2 - y_1}{x_2 - x_1} = \frac{-1 - 1}{0 - 4} = \frac{-2}{-4} = \frac{1}{2}. \quad ■$$

If you know the slope m and y-intercept b of a straight line, then you immediately know that the line is the graph of the equation $y = mx + b$. We say that $y = mx + b$ is the **slope–intercept form** of the equation of the line. The equation of the line can also be found when you know its slope and any point on the line.

Point–Slope Form of the Equation of a Line ▶

The line L with slope m through the point (x_1, y_1) is the graph of the equation

$$y - y_1 = m(x - x_1).$$

Proof If b is the y-intercept of L, then we know that L is the graph of $y = mx + b$. Since (x_1, y_1) is on L, its coordinates satisfy this equation, that is,

$$y_1 = mx_1 + b, \qquad \text{or equivalently,} \qquad b = y_1 - mx_1.$$

Consequently, the equation can be rewritten in the desired form:

$$\begin{aligned} y &= mx + b \\ y &= mx + (y_1 - mx_1) \\ y - y_1 &= mx - mx_1 \\ y - y_1 &= m(x - x_1). \quad \square \end{aligned}$$

EXAMPLE 4 The line with slope 2 through the point $(1, -6)$ is the graph of the equation $y - y_1 = m(x - x_1)$ with $m = 2$ and $(x_1, y_1) = (1, -6)$, that is,

$$\begin{aligned} y - (-6) &= 2(x - 1) \qquad &&\textit{[Point–Slope Form]} \\ y + 6 &= 2x - 2 \\ y &= 2x - 8. &&\textit{[Slope–Intercept Form]} \ \blacksquare \end{aligned}$$

EXAMPLE 5 To find the equation of the line M through $(5, 3)$ and $(-3, 1)$ we must first find the slope of M:

$$\frac{1 - 3}{-3 - 5} = \frac{-2}{-8} = \frac{1}{4}.$$

Now we use the slope 1/4 and the point $(5, 3)$ to find the point–slope form of the equation:

$$\begin{aligned} y - y_1 &= m(x - x_1) \\ y - 3 &= \frac{1}{4}(x - 5) \end{aligned}$$

which simplifies to $y = \frac{1}{4}x + \frac{7}{4}$. If you use the point $(-3, 1)$ instead of $(5, 3)$, you get the same equation:

$$y - 1 = \frac{1}{4}(x - (-3)), \qquad \text{or equivalently,} \qquad y = \frac{1}{4}x + \frac{7}{4}. \ \blacksquare$$

Vertical Lines

A vertical straight line cannot be the graph of any function.* However, every vertical line is the graph of an equation. If L is a vertical line, then L crosses the x-axis at some point $(c, 0)$, as shown in Figure 3–1. The points on L are exactly those points with first coordinate c, that is, the points (x, y) with $x = c$. Therefore,

Vertical Lines ▶

The graph of the equation $x = c$ is the vertical line through $(c, 0)$.

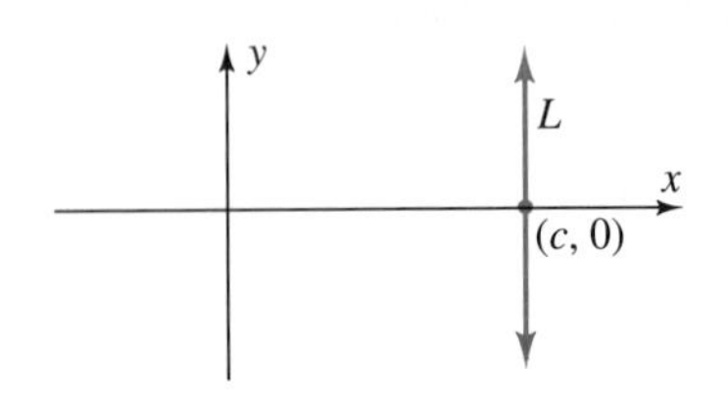

Figure 3-1

For example, the vertical line through (14, 79) also goes through (14, 0), so it is the graph of $x = 14$. Note that the formula for calculating slope (page 168) doesn't work for vertical lines. If you try to apply it using the points (14, 79) and (14, 0) on the line $x = 14$, the result is $\frac{79 - 0}{14 - 14} = \frac{79}{0}$, which is *not* a real number. Consequently,

The slope of a vertical line is not defined.

Parallel and Perpendicular Lines

GRAPHING EXPLORATION Using the viewing window with $-6 \le x \le 6$ and $-6 \le y \le 6$, graph the following linear functions on the same screen:

$$f(x) = 1.5x + 5, \quad y = 1.5x + 2, \quad g(x) = 1.5x, \quad y = 1.5x - 4.$$

All four of the lines in the Exploration have the same slope and illustrate this fact:

Slopes of Parallel Lines ▶

Two nonvertical straight lines are parallel exactly when they have the same slope.

This result (whose proof is in Exercises 60 and 61) should not be surprising: The slope of a line measures how steeply it rises or falls and parallel lines rise or fall equally steeply.

EXAMPLE 6 Find the equation of the line L through $(2, -1)$ that is parallel to the line M with equation $3x - 2y + 6 = 0$.

Solution First, find the slope of M by rewriting its equation in slope–intercept form:

*For the reason why, see the Vertical Line Test on page 105.

$$2y = 3x + 6, \qquad \text{or equivalently,} \qquad y = \frac{3}{2}x + 3.$$

Therefore, M has slope 3/2. The parallel line L must have the same slope, 3/2. Since (2, − 1) is on L, the equation of L is

$$y - (-1) = \frac{3}{2}(x - 2), \qquad \text{or equivalently,} \qquad y = \frac{3}{2}x - 4. \quad ■$$

Two lines that meet in a right angle (90° angle) are said to be **perpendicular.** As you might suspect, there is a close relationship between the slopes of two perpendicular lines.

Slopes of Perpendicular Lines ►

Two nonvertical lines are perpendicular exactly when the product of their slopes is − 1.

One proof of this fact is outlined in Exercise 62; another is given at the end of Excursion 8.2.A.

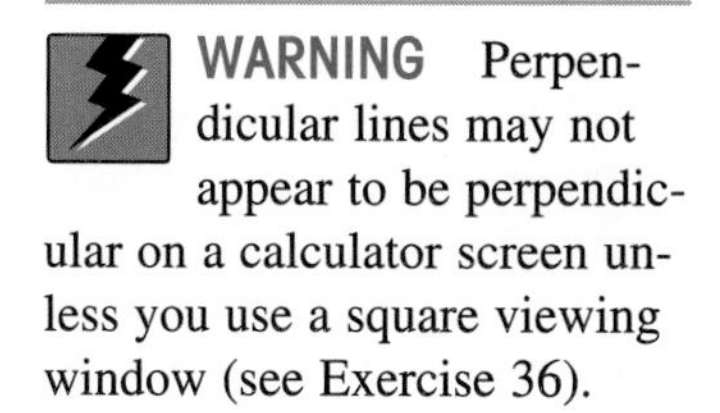
WARNING Perpendicular lines may not appear to be perpendicular on a calculator screen unless you use a square viewing window (see Exercise 36).

EXAMPLE 7 Let L be the line through (0, 2) and (1, 5). Let M be the line through (− 6, − 1) and (3, − 4). Then,

$$\text{slope } L = \frac{5 - 2}{1 - 0} = 3 \qquad \text{and} \qquad \text{slope } M = \frac{-4 - (-1)}{3 - (-6)} = \frac{-3}{9} = -\frac{1}{3}.$$

Since $3(-\frac{1}{3}) = -1$, the lines L and M are perpendicular. ■

EXAMPLE 8 Find the equation of the perpendicular bisector of the line segment with endpoints (− 5, − 4) and (7, 2).

Solution The perpendicular bisector M goes through the midpoint of the line segment from (− 5, − 4) and (7, 2). The Midpoint Formula (page 27) shows that this midpoint is

$$\left(\frac{x_1 + x_2}{2}, \frac{y_1 + y_2}{2}\right) = \left(\frac{-5 + 7}{2}, \frac{-4 + 2}{2}\right) = (1, -1).$$

The line L through (− 5, − 4) and (7, 2) has slope

$$\frac{y_2 - y_1}{x_2 - x_1} = \frac{2 - (-4)}{7 - (-5)} = \frac{6}{12} = \frac{1}{2}.$$

Since M is perpendicular to L,

$$\text{slope } M = \frac{-1}{\text{slope } L} = \frac{-1}{1/2} = -2.$$

Thus, M is the line through $(1, -1)$ with slope -2, and its equation is:

$$y - (-1) = -2(x - 1) \qquad [\textit{Point–Slope Form}]$$

$$y = -2x + 1. \qquad [\textit{Slope–Intercept Form}] \quad ■$$

Applications

EXAMPLE 9 An office buys a new computer for \$7000. Five years later its value is \$800. Assuming linear depreciation, what was its value after two years?

Solution Let $f(x)$ be the value of the computer in year x. Since f is a linear function, we know that $f(x) = mx + b$ for some constants m and b. Since the computer is worth \$7000 new (that is, $f(0) = 7000$), we have

$$7000 = f(0) = m0 + b = b,$$

so that $f(x) = mx + 7000$. Since $f(5) = 800$ (why?), we see that

$$800 = f(5) = m5 + 7000$$

$$-6200 = 5m$$

$$m = \frac{-6200}{5} = -1240.$$

Therefore, the depreciation function is $f(x) = -1240x + 7000$. The value of the computer after two years is $f(2) = -1240 \cdot 2 + 7000 = \$4520$. ■

EXAMPLE 10 A factory manufactures can openers. The fixed costs (building, fixtures, machinery, etc.) are \$26,000. The variable cost (material and labor) for producing each can opener is \$2.75. What is the total cost of making 1000 can openers? 20,000? 40,000? What is the average cost per can opener in each case?

Solution Let f be the cost function, that is, $f(x)$ is the total cost of manufacturing x can openers. The variable cost of making x can openers is $2.75x$. So the rule of the cost function is

$$f(x) = \text{variable costs} + \text{fixed costs} = 2.75x + 26{,}000.$$

Thus, the cost of making 1000 can openers is $f(1000) = 2.75(1000) + 26000 = \$28{,}750$. Similarly,

$$f(20{,}000) = 2.75(20{,}000) + 26000 = \$81{,}000$$

$$f(40{,}000) = 2.75(40{,}000) + 26000 = \$136{,}000.$$

Now the *average* cost per can opener is the cost divided by the number of can openers. So the average cost per can opener is as follows:

For 1000: $f(1000)/1000 = 28{,}750/1000 = \$28.75$

For 20,000: $f(20{,}000)/20{,}000 = 81{,}000/20{,}000 = \$4.05$

For 40,000: $f(40{,}000)/40{,}000 = 136{,}000/40{,}000 = \$3.40.$ ■

EXAMPLE 11 How many can openers must be manufactured in Example 10 in order for the average cost per can opener to fall to \$3?

Solution As we saw in Example 10, the average cost per can opener for x can openers is $f(x)/x$, where $f(x)$ is the total cost of manufacturing x can openers. So the average cost will be 3 when $f(x)/x = 3$. We can answer the question either algebraically or graphically.

Algebraic Solution:

$$\frac{f(x)}{x} = 3$$

$$\frac{2.75x + 26{,}000}{x} = 3$$

$$2.75x + 26{,}000 = 3x$$

$$-.25x = -26{,}000$$

$$x = \frac{-26{,}000}{-.25} = 104{,}000$$

▶ TECHNOLOGY TIP

The graphical intersection finder is in the CALC menu on TI-82/83, in the MATH submenu of the GRAPH menu on TI-85/92, in the plot FCN menu on HP-38, in the G-SOLVE menu on CASIO 9800, and in the JUMP menu on Sharp 9300. An intersection-finder program for TI-81 is in the Program Appendix.

Graphical Solution: The point where $f(x)/x = 3$ is the point where the graph of $y = f(x)/x$ intersects the horizontal line $y = 3$, as shown in Figure 3–2. The point is hard to read in this window, so you can either use zoom-in and trace to find it or use your calculator's "intersection finder" if it has one (see the Tip in the margin). The intersection finder on a TI-82 produces Figure 3–3, which once again shows that when 104,000 can openers are made, the average cost per can opener is \$3. ■

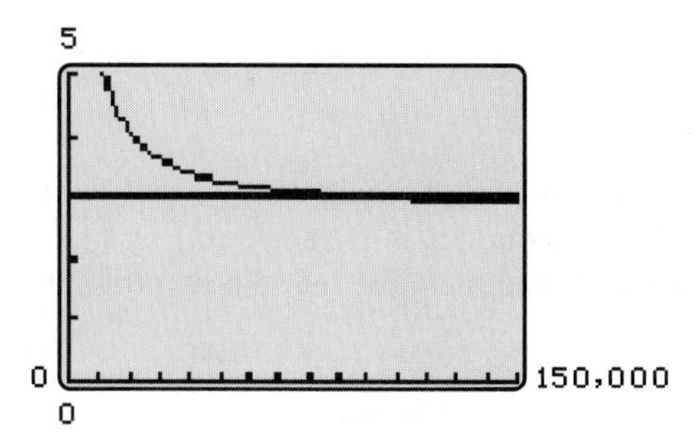

Figure 3-2

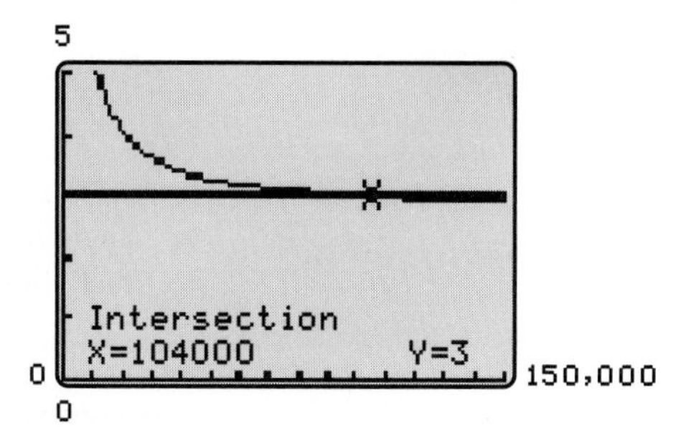

Figure 3-3

Proof of the Linear Function Theorem

We first prove that the graph of the linear function $f(x) = mx + b$ is a nonvertical straight line. Let $P = (x_1, y_1)$ and $Q = (x_2, y_2)$ be distinct points on the graph of f. Then,

$$(*) \qquad \begin{aligned} y_1 &= f(x_1) = mx_1 + b \\ y_2 &= f(x_2) = mx_2 + b. \end{aligned}$$

As on page 168, subtracting the second equation from the first and solving for m shows that

$$(**) \qquad m = \frac{y_2 - y_1}{x_2 - x_1}.$$

The points P and Q determine a straight line L, which is nonvertical because no two points on the graph of the function f can lie on the same vertical line (see page 105). Consequently, for each real number x there is exactly one point with first coordinate x on the line L. Similarly, there is exactly one point on the graph of f with first coordinate x because the domain of f consists of all real numbers. We shall show that the graph of f is actually the line L by showing that for every real number x, the point (x, y) on L is the point $(x, f(x))$ on the graph of f.

The points P, Q, and (x, y) all lie on the line L, as shown in Figure 3–4.†

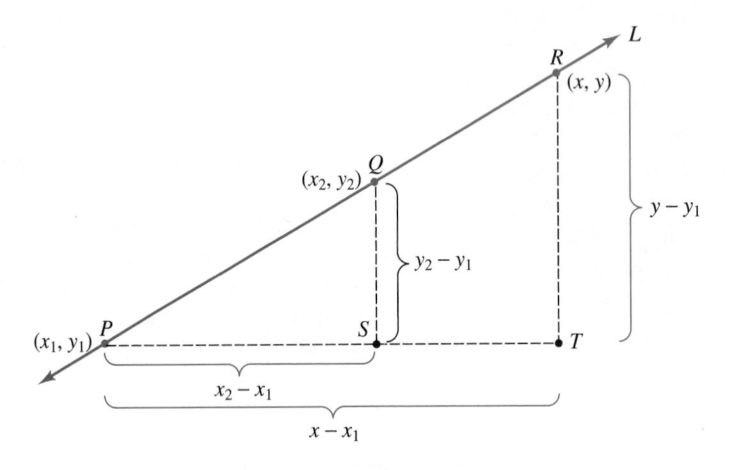

Figure 3-4

In triangles PTR and PSQ, the right angles S and T are equal and obviously angle P = angle P. Since the sum of the angles of any triangle is 180°, it follows that angle R = angle Q and therefore that the two triangles are similar.‡ Con-

†Figure 3–4 shows the case when $x_1 < x_2 < x$. The picture may differ in other possible cases, but only minor modifications are needed in the argument.

‡Similar triangles are discussed in the Geometry Review Appendix.

sequently, the ratios of the corresponding sides of the triangles are equal. In particular,

$$\frac{\text{length } RT}{\text{length } PT} = \frac{\text{length } QS}{\text{length } PS}.$$

These lengths can be expressed in terms of the coordinates of the points, as shown in Figure 3–4:

$$\frac{y - y_1}{x - x_1} = \frac{\text{length } RT}{\text{length } PT} = \frac{\text{length } QS}{\text{length } PS} = \frac{y_2 - y_1}{x_2 - x_1}.$$

Since $m = \dfrac{y_2 - y_1}{x_2 - x_1}$ by (**), we can rewrite the first and last terms of this equation:

$$\frac{y - y_1}{x - x_1} = m$$

$$y - y_1 = m(x - x_1)$$

$$y = mx - mx_1 + y_1.$$

However, $y_1 = mx_1 + b$ by (*), so that $b = -mx_1 + y_1$. Substituting this expression in the last equation above, we have

$$y = mx - mx_1 + y_1 = mx + b = f(x).$$

Therefore, the point (x, y) is the point $(x, f(x))$ and the graph of f is the line L.

The proof of the converse, namely, that every nonvertical straight line is the graph of a linear function, is outlined in Exercise 64.

EXERCISES 3.1

1. On one graph sketch five line segments, not all meeting at a single point, whose slopes are five different positive numbers. Do this in such a way that the left-hand line as the largest slope, the second line from the left the next largest slope, and so on.

2. On one graph, sketch five lines that meet at a single point and satisfy this condition: One line has slope 0, two lines have positive slope, and two lines have negative slope.

3. For which of the line segments in Figure 3–5 is the slope
(a) largest? **(b)** smallest?
(c) largest in absolute value? **(d)** closest to zero?

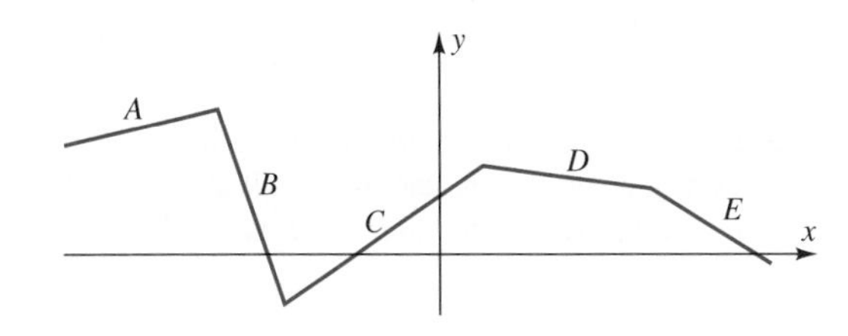

Figure 3-5

4. The doorsill of a campus building is 5 ft above ground level. To allow wheelchair access, the steps in front of the door are to be replaced by a straight ramp with constant slope 1/12, as shown in Figure 3–6. How long must the ramp be? [The answer is *not* 60 ft.]

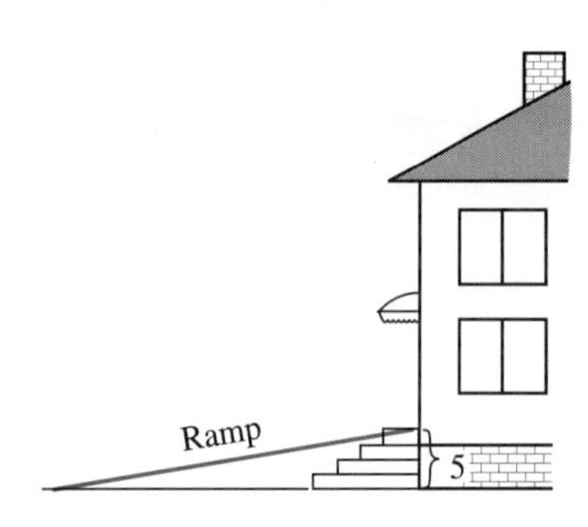

Figure 3-6

In Exercises 5–8, find the slope and y-intercept of the line whose equation is given.

5. $2x - y + 5 = 0$ **6.** $3x + 4y = 7$

7. $3(x - 2) + y = 7 - 6(y + 4)$

8. $2(y - 3) + (x - 6) = 4(x + 1) - 2$

In Exercises 9–12, find the slope of the line through the given points.

9. $(1, 2); (3, 7)$ **10.** $(-1, -2); (2, -1)$

11. $(1/4, 0); (3/4, 2)$ **12.** $(\sqrt{2}, -1); (2, -9)$

In Exercises 13–16, find a number t such that the line passing through the two given points has slope -2.

13. $(0, t); (9, 4)$ **14.** $(1, t); (-3, 5)$

15. $(t + 1, 5); (6, -3t + 7)$ **16.** $(t, t); (5, 9)$

17. Let L be a nonvertical straight line through the origin. L intersects the vertical line through $(1, 0)$ at a point P. Show that the second coordinate of P is the slope of L.

18. **(a)** Graph the linear function $y = 2x + 4$ in the standard viewing window.
(b) Graph the linear function $y = 6x - 3$ in the viewing window with $-10 \le x \le 10$ and $-30 \le y \le 30$.
(c) Which of the lines in parts (a) and (b) *appears* to rise more steeply? Which has the largest slope? Explain why this does not contradict the properties of slope in the box on page 167.

In Exercises 19–22, find the equation of the line with slope m that passes through the given point.

19. $m = 1;\ (3, 5)$ **20.** $m = 2;\ (-2, 1)$

21. $m = -1;\ (6, 2)$ **22.** $m = 0;\ (-4, -5)$

In Exercises 23–26, find the equation of the line through the given points.

23. $(0, -5)$ and $(-3, -2)$ **24.** $(4, 3)$ and $(2, -1)$

25. $(4/3, 2/3)$ and $(1/3, 3)$ **26.** $(6, 7)$ and $(6, 15)$

In Exercises 27–30, determine whether the line through P and Q is parallel or perpendicular to the line through R and S, or neither.

27. $P = (2, 5), Q = (-1, -1)$ and $R = (4, 2), S = (6, 1)$

28. $P = (0, 3/2), Q = (1, 1)$ and $R = (2, 7), S = (3, 9)$

29. $P = (-3, 1/3), Q = (1, -1)$ and $R = (2, 0), S = (4, -2/3)$

30. $P = (3, 3), Q = (-3, -1)$ and $R = (2, -2), S = (4, -5)$

In Exercises 31 and 32, determine whether the lines whose equations are given are parallel, perpendicular, or neither.

31. $2x + y - 2 = 0$ and $4x + 2y + 18 = 0$

32. $3x + y - 3 = 0$ and $6x + 2y + 17 = 0$

33. Use slopes to show that the points $(-5, -2), (-3, 1), (3, 0)$, and $(5, 3)$ are the vertices of a parallelogram.

34. Use slopes to show that the points $(-4, 6), (-1, 12)$, and $(-7, 0)$ all lie on the same straight line.

35. Use slopes to determine if $(9, 6), (-1, 2)$, and $(1, -3)$ are the vertices of a right triangle.

36. **(a)** Show that the lines $y = 2x + 4$ and $y = -.5x - 3$ are perpendicular.
(b) Graph the lines in part (a), using the standard viewing window. Do the lines look perpendicular?
(c) Find a viewing window in which the lines in part (a) appear to be perpendicular.

In Exercises 37–44, find an equation for the line satisfying the given conditions.

37. Through $(-2, 1)$ with slope 3.

38. y-intercept -7 and slope 1.

39. Through $(2, 3)$ and parallel to $3x - 2y = 5$.

40. Through $(1, -2)$ and perpendicular to $y = 2x - 3$.

41. x-intercept 5 and y-intercept -5.

42. Through $(-5, 2)$ and parallel to the line through $(1, 2)$ and $(4, 3)$.

43. Through $(-1, 3)$ and perpendicular to the line through $(0, 1)$ and $(2, 3)$.

44. y-intercept 3 and perpendicular to $2x - y + 6 = 0$.

45. Find a real number k such that $(3, -2)$ is on the line $kx - 2y + 7 = 0$.

46. Find a real number k such that the line $3x - ky + 2 = 0$ has y-intercept -3.

If P is a point on a circle with center C, then the tangent line to the circle at P is the straight line through P that is perpendicular to the radius CP. In Exercises 47–50, find the equation of the tangent line to the circle at the given point.

47. $x^2 + y^2 = 25$ at $(3, 4)$ [*Hint:* Here C is $(0, 0)$ and P is $(3, 4)$; what is the slope of radius CP?]

48. $x^2 + y^2 = 169$ at $(-5, 12)$

49. $(x - 1)^2 + (y - 3)^2 = 5$ at $(2, 5)$

50. $x^2 + y^2 + 6x - 8y + 15 = 0$ at $(-2, 1)$

51. Let A, B, C, D be nonzero real numbers. Show that the lines $Ax + By + C = 0$ and $Ax + By + D = 0$ are parallel.

52. Let L be a line that is neither vertical nor horizontal and which does not pass through the origin. Show that L is the graph of $\frac{x}{a} + \frac{y}{b} = 1$, where a is the x-intercept and b is the y-intercept of L.

53. The sales of a small software company are a linear function of time. Sales were \$120,000 in 1994 and \$180,000 in 1997.
(a) Find the rule of the sales function f, where $f(x)$ is the amount of sales in year x and $x = 0$ represents 1993.
(b) Assuming that this model remains valid, what will sales be in 1999?
(c) The owner plans to hire another staff member when sales reach \$250,000 per year. When will this occur?

54. The Celsius and Fahrenheit temperature scales are related as follows: 0°C is the same as 32°F (the temperature at which water freezes) and 100°C is 212°F (the temperature at which water boils).
(a) Express the temperature F in degrees Fahrenheit as a linear function of the temperature C in degrees Celsius.
(b) Use the function from part (a) to determine the Fahrenheit equivalents of the following Celsius temperatures: $-10°$, $10°$, and $25°$.
(c) Express C as a function of F.

55. The Whismo Hat Company has fixed costs of \$50,000 and variable costs of \$8.50 per hat.
(a) What is the cost function?
(b) What is the average cost per hat when 20,000 are manufactured? 50,000? 100,000?

56. In Exercise 55, how many hats must be manufactured in order to have an average cost per hat of \$15?

57. A publisher has fixed costs of \$180,000 for a certain book. The variable costs are \$25 per book. The book sells for \$40. Find the rule of the
(a) Cost function.
(b) Revenue function.
(c) Profit function.
(d) What is the publisher's break-even point (that is, how many books must be sold in order for revenue to equal cost)?

58. The profit function in thousands of dollars for x thousand units of a specialty item is $f(x) = .6x - 14.5$. The cost function is $g(x) = .8x + 14.5$.
(a) Find the revenue function.
(b) How many units must be sold for the company to break even?

59. A road is being built along the floor of a valley. At the end of the valley it must pass over the hill, as shown in Figure 3–7. Because of soil conditions and other factors, the road is to lie along a straight line of slope 1/25

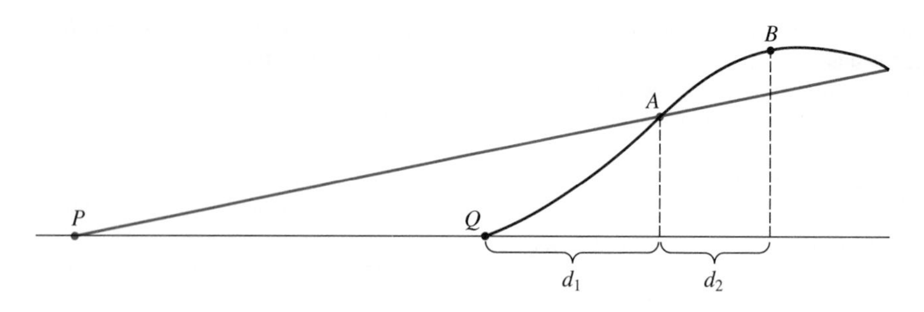

Figure 3-7

that passes through point A (as indicated by the colored line above). The top of the hill will be cut away, and the bottom of the hill will be filled in to road level. The point A lies 400 ft above the floor of the valley. The peak of the hill (point B) lies 500 ft above the valley floor. The horizontal distance d_1 is 600 ft; distance d_2 is 300 ft.

(a) Find the point where the road should meet the valley floor by determining the distance from P to Q. [*Hint:* Sketch the picture on a coordinate system, with the valley floor as the x-axis and the vertical line through A as the y-axis.]

(b) Find out how much of the hilltop must be cut off by determining the vertical distance from point B to the road.

60. Prove that nonvertical parallel lines L and M have the same slope, as follows. Suppose M lies above L and choose two points (x_1, y_1) and (x_2, y_2) on L.

(a) Let P be the point on M with first coordinate x_1. Let b denote the vertical distance from P to (x_1, y_1). Show that the second coordinate of P is $y_1 + b$.

(b) Let Q be the point on M with first coordinate x_2. Use the fact that L and M are parallel to show that the second coordinate of Q is $y_2 + b$.

(c) Compute the slope of L using (x_1, y_1) and (x_2, y_2). Compute the slope of M using the points P and Q. Verify that the two slopes are the same.

61. Show that two nonvertical lines with the same slope are parallel. [*Hint:* The equations of distinct lines with the same slope must be of the form $y = mx + b$ and $y = mx + c$ with $b \neq c$ (why?). If (x_1, y_1) were a point on both lines, its coordinates would satisfy both equations. Show that this leads to a contradiction and conclude that the lines have no point in common.]

62. This exercise provides a proof of the statement about slopes of perpendicular lines in the box on page 171. Let L be a line with slope k and M a line with slope m and assume *both* L and M pass through the origin.

(a) Show that L passes through $(1, k)$ and M passes through $(1, m)$.

(b) Compute the length of each side of the triangle with vertices $(0, 0)$, $(1, k)$, and $(1, m)$.

(c) Suppose L and M are perpendicular. Then the triangle of part (b) has a right angle at $(0, 0)$. Use part (b) and the Pythagorean Theorem to find an equation involving k, m, and various constants. Simplify this equation to show that $km = -1$.

(d) Suppose instead that $km = -1$ and prove that L and M are perpendicular. [*Hint:* You may assume that a triangle whose sides a, b, c satisfy $a^2 + b^2 = c^2$ is a right triangle with hypotenuse c. Use this fact and $km = -1$ to "reverse" the argument in part (c)].

(e) Finally, assume L and M are any two nonvertical lines (which don't necessarily go through the origin), with slope $L = k$ and slope $M = m$. Use the preceding material to prove that L is perpendicular to M exactly when $km = -1$. [*Hint:* Every line is parallel to a line through the origin and parallel lines have the same slope.]

63. Show that the diagonals of a square are perpendicular. [*Hint:* Place the square in the first quadrant of the plane, with one vertex at the origin and sides on the positive axes. Label the coordinates of the vertices appropriately.]

64. Show that every nonvertical straight line L is the graph of a linear function. [*Hint:* Choose distinct points (x_1, y_1) and (x_2, y_2) on L. Let $m = \dfrac{y_2 - y_1}{x_2 - x_1}$ and $b = y_1 - mx_1$. The graph of the linear function $f(x) = mx + b$ is a straight line, as proved on page 174. Show that (x_1, y_1) and (x_2, y_2) are on the graph of f. Since a straight line is completely determined by two points, this shows that L is the graph of f.]

65. Suppose a and b are real numbers with $a \neq b$. Use slopes to show that the line through (a, b) and (b, a) is perpendicular to the line $y = x$. This fact, together with Exercise 51(a) of Section 2.7, shows that $y = x$ is the perpendicular bisector of the line segment joining (a, b) and (b, a).

3.2 RATES OF CHANGE

In Sections 2.1 and 2.2 we saw that when a rock is dropped straight down from a high place, then the distance the rock travels (ignoring wind resistance) is given by the function

$$d(t) = 16t^2$$

with distance $d(t)$ measured in feet and time t in seconds. The following table shows the distance the rock has fallen at various times:

Time t	0	1	2	3	3.5	4	4.5	5
Distance $d(t)$	0	16	64	144	196	256	324	400

To find the distance the rock falls from time $t = 1$ to $t = 3$, we note that at the end of three seconds, the rock has fallen $d(3) = 144$ feet, whereas it had only fallen $d(1) = 16$ feet at the end of one second. So during this time interval the rock traveled

$$d(3) - d(1) = 144 - 16 = 128 \text{ feet.}$$

The distance traveled by the rock during other time intervals can be found similarly:

Time Interval	*Distance Traveled*
$t = 1$ to $t = 4$	$d(4) - d(1) = 256 - 16 = 240$
$t = 2$ to $t = 3.5$	$d(3.5) - d(2) = 196 - 64 = 132$
$t = 2$ to $t = 4.5$	$d(4.5) - d(2) = 324 - 64 = 260$

The same procedure works in general:

The distance traveled from time $t = a$ to time $t = b$ is $d(b) - d(a)$ feet.

In the preceding chart, the length of each time interval can be computed by taking the difference between the two times. For example, from $t = 1$ to $t = 4$ is a time interval of length $4 - 1 = 3$ seconds. Similarly, the interval from $t = 2$ to $t = 3.5$ is of length $3.5 - 2 = 1.5$ seconds and in general,

The time interval from $t = a$ to $t = b$ is an interval of $b - a$ seconds.

Since distance = average speed × time,

$$\text{average speed} = \frac{\text{distance traveled}}{\text{time interval}}.$$

Hence, the average speed over the time interval from $t = a$ to $t = b$ is

$$\text{average speed} = \frac{\text{distance traveled}}{\text{time interval}} = \frac{d(b) - d(a)}{b - a}.$$

For example, to find the average speed from $t = 1$ to $t = 4$, apply the preceding formula with $a = 1$ and $b = 4$:

$$\text{average speed} = \frac{d(4) - d(1)}{4 - 1} = \frac{256 - 16}{4 - 1} = \frac{240}{3} = 80 \text{ ft per sec.}$$

Similarly, the average speed from $t = 2$ to $t = 4.5$ is

$$\frac{d(4.5) - d(2)}{4.5 - 2} = \frac{324 - 64}{4.5 - 2} = \frac{260}{2.5} = 104 \text{ ft per sec.}$$

The units in which average speed is measured here (feet per second) indicate the number of units of distance traveled during each unit of time, that is, the *rate of change* of distance (feet) with respect to time (seconds). The preceding discussion can be summarized by saying that the average speed (rate of change of distance with respect to time) as time changes from $t = a$ to $t = b$ is given by

$$\text{average speed} = \text{average rate of change} = \frac{\text{change in distance}}{\text{change in time}} = \frac{d(b) - d(a)}{b - a}.$$

Although speed is the most familiar example, rates of change play a role in many other situations as well, as illustrated in Examples 1–3 below. Consequently, we define the average rate of change of any function as follows.

Average Rate of Change ▶

Let f be a function. The *average rate of change of $f(x)$* with respect to x as x changes from a to b is the number

$$\frac{\text{change in } f(x)}{\text{change in } x} = \frac{f(b) - f(a)}{b - a}.$$

EXAMPLE 1 A large heavy-duty balloon is being filled with water. Its approximate volume (in gallons) is given by

$$V(x) = \frac{x^3}{55}$$

where x is the radius of the balloon (in inches). The average rate of change of the volume of the balloon as the radius increases from 5 to 10 inches is

$$\frac{\text{change in volume}}{\text{change in radius}} = \frac{V(10) - V(5)}{10 - 5} \approx \frac{18.18 - 2.27}{10 - 5} = \frac{15.91}{5}$$

$$= 3.182 \text{ gallons per inch.} \quad \blacksquare$$

EXAMPLE 2 A small manufacturing company makes specialty office desks. The cost (in thousands of dollars) of producing x desks is given by the function

$$c(x) = .0009x^3 - .06x^2 + 1.6x + 5.$$

For example, you can readily verify that $c(10) = 15.9$ and $c(30) = 23.3$. Since $c(x)$ is measured in thousands, this means that the cost of making 10 desks is \$15,900 and the cost of making 30 is \$23,300.

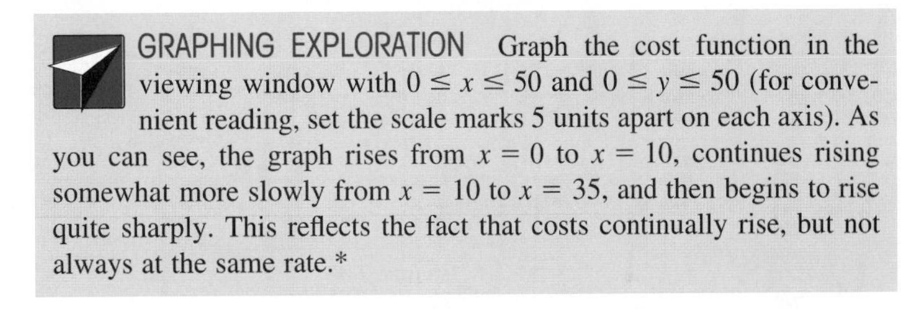

GRAPHING EXPLORATION Graph the cost function in the viewing window with $0 \le x \le 50$ and $0 \le y \le 50$ (for convenient reading, set the scale marks 5 units apart on each axis). As you can see, the graph rises from $x = 0$ to $x = 10$, continues rising somewhat more slowly from $x = 10$ to $x = 35$, and then begins to rise quite sharply. This reflects the fact that costs continually rise, but not always at the same rate.*

As production increases from 0 to 10 desks, the average rate of change of cost is

$$\frac{\text{change in cost}}{\text{change in production}} = \frac{c(10) - c(0)}{10 - 0} = \frac{15.9 - 5}{10} = \frac{10.9}{10} = 1.09.$$

This means that costs are rising at an average rate of 1.09 thousand dollars (that is, \$1090) per desk. As production goes from 10 to 30 desks, the average rate of change of cost is

$$\frac{c(30) - c(10)}{30 - 10} = \frac{23.3 - 15.9}{30 - 10} = \frac{7.4}{20} = .37$$

so that costs are rising at an average rate of only \$370 per desk. The rate increases as production goes from 30 to 50:

$$\frac{c(50) - c(30)}{50 - 30} = \frac{47.5 - 23.3}{50 - 30} = \frac{24.2}{20} = 1.21, \quad \text{that is,} \quad \$1210 \text{ per desk.} \quad \blacksquare$$

*Costs are high at the beginning because of initial setup costs (rent, equipment, etc.); in fact, it costs \$5000 even if no desks are manufactured ($c(0) = 5$). As more desks are produced, costs increase more slowly because of efficiencies of scale. Then they climb again (increasing production past a certain point might require a new building, or more machines, or a second shift of workers, etc.).

EXAMPLE 3 Figure 3–8 is the graph of the temperature function f during a particular day; $f(x)$ is the temperature at x hours after midnight. What is the average rate of change of the temperature **(a)** from 4 A.M. to noon? **(b)** from 3 P.M. to 8 P.M.?

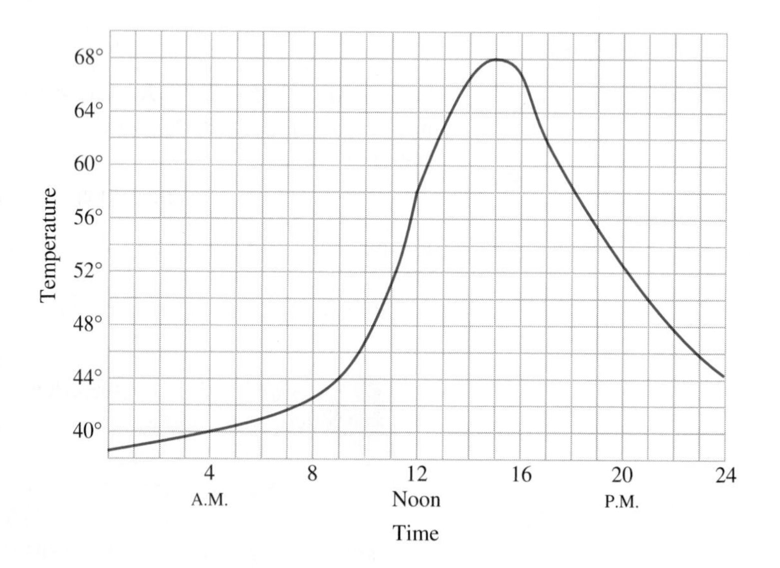

Figure 3-8

Solution

(a) The graph shows that the temperature at 4 A.M. is $f(4) = 40°$ and the temperature at noon is $f(12) = 58°$. The average rate of change of temperature is

$$\frac{\text{change in temperature}}{\text{change in time}} = \frac{f(12) - f(4)}{12 - 4} = \frac{58 - 40}{12 - 4} = \frac{18}{8}$$
$$= 2.25 \text{ degrees per hour.}$$

The rate of change is positive because the temperature is increasing at an average rate of 2.25° per hour.

(b) Now 3 P.M. corresponds to $x = 15$ and 8 P.M. to $x = 20$. The graph shows that $f(15) = 68°$ and $f(20) = 53°$. Hence the average rate of change of temperature is:

$$\frac{\text{change in temperature}}{\text{change in time}} = \frac{f(20) - f(15)}{20 - 15} = \frac{53 - 68}{20 - 15} = \frac{-15}{5}$$
$$= -3 \text{ degrees per hour.}$$

The rate of change is negative because the temperature is decreasing at an average rate of 3° per hour. ■

The Difference Quotient

Average rates of change are often computed for very small intervals. For instance, we might compute the rate from 4 to 4.01 or from 4 to 4.001. Since $4.01 = 4 + .01$ and $4.001 = 4 + .001$, we are doing essentially the same thing in both cases: computing the rate of change over the interval from 4 to $4 + h$ for some small nonzero quantity h. Furthermore, it's often possible to use a single calculation to determine the average rate for all possible values of h.

EXAMPLE 4 Consider the falling rock with which this section began. The distance the rock has traveled at time t is given by $d(t) = 16t^2$ and its average speed (rate of change) from $t = 4$ to $t = 4 + h$ is:

$$\begin{aligned}\text{average speed} &= \frac{d(4+h) - d(4)}{(4+h) - 4} = \frac{16(4+h)^2 - 16\cdot 4^2}{h}\\ &= \frac{16(16 + 8h + h^2) - 256}{h} = \frac{256 + 128h + 16h^2 - 256}{h}\\ &= \frac{128h + 16h^2}{h} = \frac{h(128 + 16h)}{h} = 128 + 16h.\end{aligned}$$

Thus, we can quickly compute the average speed over the interval from 4 to $4 + h$ seconds for any value of h by using the formula: average speed $= 128 + 16h$. For example, the average speed from 4 seconds to 4.001 seconds (here $h = .001$) is

$$128 + 16h = 128 + 16(.001) = 128 + .016 = 128.016 \text{ ft per sec.} \quad \blacksquare$$

Similar calculations can be done with any number in place of 4. In each such case, we are dealing with an interval from x to $x + h$ for some number x. As in Example 4, a single computation can often be used for all possible x and h.

EXAMPLE 5 The average speed of the falling rock of Example 4 from time x to time $x + h$ is:*

$$\begin{aligned}\text{average speed} &= \frac{d(x+h) - d(x)}{(x+h) - x} = \frac{16(x+h)^2 - 16x^2}{h}\\ &= \frac{16(x^2 + 2xh + h^2) - 16x^2}{h} = \frac{16x^2 + 32xh + 16h^2 - 16x^2}{h}\\ &= \frac{32xh + 16h^2}{h} = \frac{h(32x + 16h)}{h} = 32x + 16h.\end{aligned}$$

When $x = 4$, then this result states that the average speed from 4 to $4 + h$ is $32(4) + 16h = 128 + 16h$, which is exactly what we found in Example 4. To

*Note that this calculation is the same as in Example 4, except that 4 has been replaced by x.

find the average speed from 3 to 3.1 seconds, apply the formula average speed $= 32x + 16h$ with $x = 3$ and $h = .1$:

$$\text{average speed} = 32 \cdot 3 + 16(.1) = 96 + 1.6 = 97.6 \text{ ft per sec.} \quad ■$$

More generally, we can compute the average rate of change of any function f over the interval from x to $x + h$ just as we did in Example 5: Apply the definition of average rate of change in the box on page 180 with x in place of a and $x + h$ in place of b:

$$\textbf{average rate of change} = \frac{f(b) - f(a)}{b - a} = \frac{f(x+h) - f(x)}{(x+h) - x} = \frac{f(x+h) - f(x)}{h}.$$

When the average rate of change from x to $x + h$ is expressed in this way, it is given a special name:

The Difference Quotient ▶

If f is a function, then the *difference quotient of f* is the quantity

$$\frac{f(x+h) - f(x)}{h}.$$

The difference quotient is the average rate of change of f over the interval from x to $x + h$.

For instance, Example 5 shows that the difference quotient of the function $f(x) = 16x^2$ is $32x + h$.

EXAMPLE 6 Find the difference quotient of $V(x) = x^3/55$.

Solution Use the definition of the difference quotient and algebra:

$$\frac{V(x+h) - V(x)}{h} = \frac{\overbrace{\frac{(x+h)^3}{55}}^{V(x+h)} - \overbrace{\frac{x^3}{55}}^{V(x)}}{h} = \frac{\frac{1}{55}[(x+h)^3 - x^3]}{h}$$

$$= \frac{1}{55} \cdot \frac{(x+h)^3 - x^3}{h} = \frac{1}{55} \cdot \frac{x^3 + 3x^2h + 3xh^2 + h^3 - x^3}{h}$$

$$= \frac{1}{55} \cdot \frac{3x^2h + 3xh^2 + h^3}{h} = \frac{1}{55} \cdot \frac{h(3x^2 + 3xh + h^2)}{h}$$

$$= \frac{3x^2 + 3xh + h^2}{55}. \quad ■$$

Instantaneous Rate of Change

Rates of change are a major theme in calculus—not just the average rate of change discussed above, but also the *instantaneous rate of change* of a function (that is, its rate of change at a particular instant). Even without calculus, however, we can obtain quite accurate approximations of instantaneous rates of change by using average rates appropriately.

EXAMPLE 7 A rock is dropped from a high place. What is its speed exactly 3 seconds after it is dropped?

Solution The distance the rock has fallen at time t is given by the function $d(t) = 16t^2$. The exact speed at $t = 3$ can be approximated by finding the average speed over very small time intervals, say, 3 to 3.01 or even shorter intervals. Over a very short time span, such as a hundredth of a second, the rock cannot change speed very much so these average speeds should be a reasonable approximation of its speed at the instant $t = 3$. Example 5 shows that the average speed is given by the difference quotient $32x + 16h$. When $x = 3$, the difference quotient is $32 \cdot 3 + 16h = 96 + 16h$ and we have:

Change in Time *3 to 3 + h*	*h*	*Average Speed* *[Difference Quotient at x = 3]* *96 + 16h*
3 to 3.1	.1	$96 + 16(.1) = 97.6$ ft per sec
3 to 3.01	.01	$96 + 16(.01) = 96.16$ ft per sec
3 to 3.005	.005	$96 + 16(.005) = 96.08$ ft per sec
3 to 3.00001	.00001	$96 + 16(.00001) = 96.00016$ ft per sec

The table suggests that exact speed of the rock at the instant $t = 3$ seconds is very close to 96 ft per sec. ■

EXAMPLE 8 A balloon is being filled with water in such a way that when its radius is x inches, then its volume is $V(x) = x^3/55$ gallons. In Example 1 we saw that the average rate of change of the volume as the radius increases from 5 inches to 10 inches is 3.182 gallons per inch. What is the rate of change at the instant when the radius is 7 inches?

Solution The average rate of change when the radius goes from x to $x + h$ inches is given by the difference quotient of $V(x)$, which was found in Example 6:

$$\frac{V(x+h) - V(x)}{h} = \frac{3x^2 + 3xh + h^2}{55}.$$

Therefore, when $x = 7$ the difference quotient is $\frac{3 \cdot 7^2 + 3 \cdot 7 \cdot h + h^2}{55} = \frac{147 + 21h + h^2}{55}$ and we have these average rates of change over small intervals near 7:

Change in Radius 7 *to* $7 + h$	h	*Average Rate of Change of Volume [Difference Quotient at* $x = 7$*]* $\frac{147 + 21h + h^2}{55}$
7 to 7.01	.01	2.6765 gallons per inch
7 to 7.001	.001	2.6731 gallons per inch
7 to 7.0001	.0001	2.6728 gallons per inch
7 to 7.00001	.00001	2.6727 gallons per inch

The chart suggests that at the instant the radius is 7 inches the volume is changing at a rate of approximately 2.673 gallons per inch. ■

Secant Lines

If P and Q are points on the graph of a function f, then the straight line determined by P and Q is called a **secant line.** Figure 3–9 shows the secant line joining the points (4, 40) and (12, 58) on the graph of the temperature function f of Example 3.

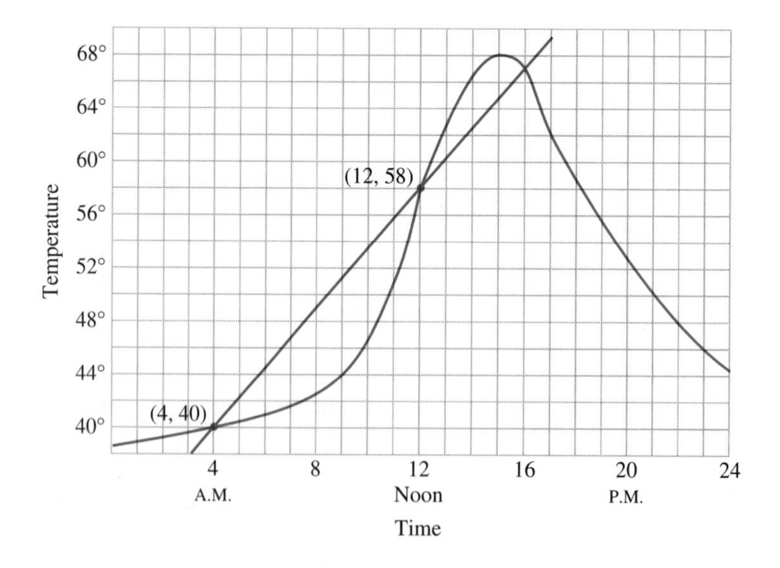

Figure 3-9

Using the points (4, 40) and (12, 58), we see that the slope of this secant line is $\frac{58-40}{12-4} = \frac{18}{8} = 2.25$. To say that (4, 40) and (12, 58) are on the graph of f means that $f(4) = 40$ and $f(12) = 58$. Thus,

$$\text{slope of secant line} = 2.25 = \frac{58-40}{12-4} = \frac{f(12)-f(4)}{12-4}$$
$$= \text{average rate of change as } x \text{ goes from 4 to 12.}$$

The same thing happens in the general case:

Secant Lines and Average Rates of Change ▶

If f is a function, then the average rate of change of $f(x)$ with respect to x as x changes from $x = a$ to $x = b$ is the slope of the secant line joining the points $(a, f(a))$ and $(b, f(b))$ on the graph of f.

EXAMPLE 9 Consider the temperature graph discussed above and repeated in Figure 3–10. We saw that the average rate of change of temperature from 4 A.M. to noon was 2.25 degrees per hour. Find two (approximate) time intervals, one beginning at noon and the other at 9 A.M., over which the average rate of change of temperature is also 2.25 degrees per hour.

Solution We saw that the average rate of change of the temperature from 4 A.M. to noon (namely, 2.25) is the slope of the secant line L joining the points (4, 40) and (12, 58). Figure 3–10 shows that L also intersects the temperature graph at the point P, whose first coordinate is approximately 16.

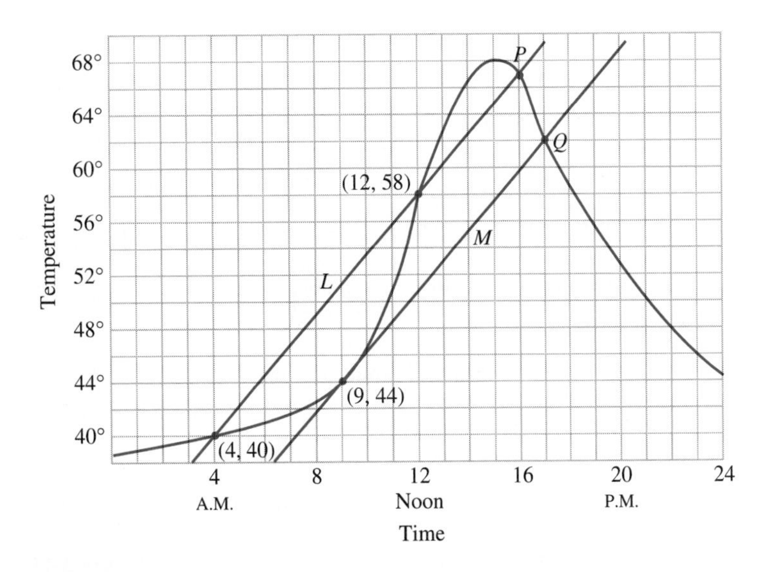

Figure 3-10

Thus, L is also the secant line joining (12, 58) and P. Hence the average rate of change from 12 to approximately 16, that is, from noon to approximately 4 P.M. is 2.25 (the slope of L).

In order to find a time interval beginning at 9 A.M. over which the average rate of change is 2.25 degrees, we must find a secant line M through (9, 44) whose slope is 2.25 (the slope of L). Since L and M have the same slope, M is the line parallel to L through (9, 44), as shown in Figure 3–10. M intersects the graph at the point Q, whose first coordinate is approximately 17. Therefore 2.25 (the slope of M) is the average rate of change from 10 to approximately 17, that is, from 10 A.M. to approximately 5 P.M. ■

EXERCISES 3.2

1. A car moves along a straight test track. The distance traveled by the car at various times is shown in this table:

Time (sec)	0	5	10	15	20	25	30
Distance (ft)	0	20	140	400	680	1400	1800

Find the average speed of the car over the interval from
(a) 0 to 10 sec **(b)** 10 to 20 sec
(c) 20 to 30 sec **(d)** 15 to 30 sec

2. The yearly profit of a small manufacturing firm is shown in the following tables.

Year	1986	1987	1988	1989
Profit	\$5000	\$6000	\$6500	\$6800

Year	1990	1991	1992	1993
Profit	\$7200	\$6700	\$6500	\$7000

What is the average rate of change of profits over the given time span?
(a) 1986–1990 **(b)** 1986–1993
(c) 1989–1992 **(d)** 1988–1992

3. Find the average rate of change of the volume of the balloon in Example 1 as the radius increases from
(a) 2 to 5 inches **(b)** 4 to 8 inches

4. Find the average rate of change of cost for the company in Example 2 when production increases from
(a) 5 to 25 desks **(b)** 0 to 40 desks

5. The graph in Figure 3–11 shows the annual sales of floral pattern ties (in thousands of ties) made by Neckwear, Inc., over a 48-month period. Sales are very low when the ties are first introduced, increase significantly, hold steady for a while, and then drop off as the ties go out of fashion. Find the average rate of change of sales (in ties per month) over the interval:
(a) 0 to 12 **(b)** 8 to 24 **(c)** 12 to 24
(d) 20 to 28 **(e)** 28 to 36 **(f)** 32 to 44
(g) 36 to 40 **(h)** 40 to 48

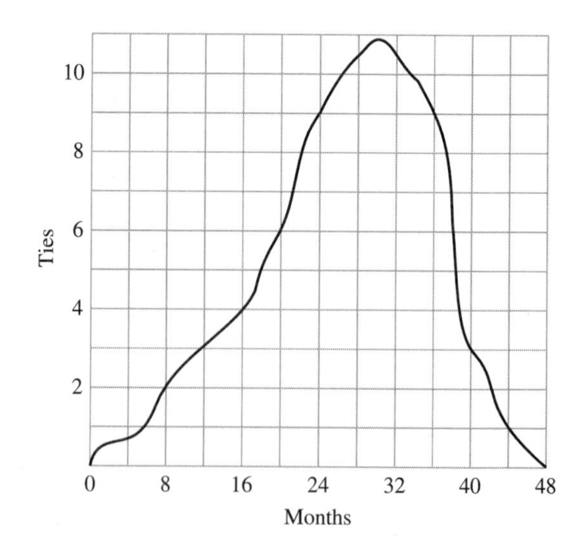

Figure 3-11

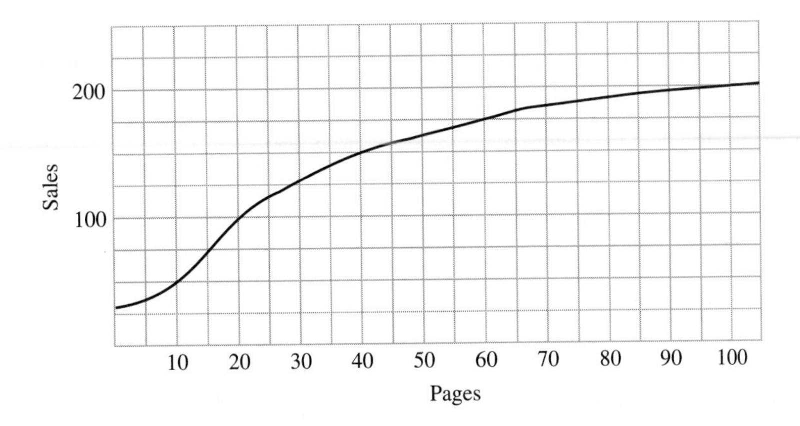

Figure 3-12

6. The XYZ Company has found that its sales are related to the amount of advertising it does in trade magazines. The graph in Figure 3–12 shows the sales (in thousands of dollars) as a function of the amount of advertising (in number of magazine ad pages). Find the average rate of change of sales when the number of ad pages increases from

(a) 10 to 20 **(b)** 20 to 60
(c) 60 to 100 **(d)** 0 to 100
(e) Is it worthwhile to buy more than 70 pages of ads, if the cost of a one-page ad is \$2000? If the cost is \$5000? If the cost is \$8000?

7. When blood flows through an artery (which can be thought of as a cylindrical tube) its velocity is greatest at the center of the artery. Because of friction along the walls of the tube, the blood's velocity decreases as the distance r from the center of the artery increases, finally becoming 0 at the wall of the artery. The velocity (in cm per sec) is given by the function $v = 18{,}500(.000065 - r^2)$, where r is measured in centimeters. Find the average rate of change of the velocity as the distance from the center changes from

(a) $r = .001$ to $r = .002$ **(b)** $r = .002$ to $r = .003$
(c) $r = 0$ to $r = .025$

8. A car is stopped at a traffic light and begins to move forward along a straight road when the light turns green. The distance (in feet) traveled by the car in t seconds is given by $s(t) = 2t^2$ $(0 \le t \le 30)$. What is the average speed of the car from

(a) $t = 0$ to $t = 5$? **(b)** $t = 5$ to $t = 10$?
(c) $t = 10$ to $t = 30$? **(d)** $t = 10$ to $t = 10.1$?

In Exercises 9–14, find the average rate of change of the function f over the given interval.

9. $f(x) = 2 - x^2$ from $x = 0$ to $x = 2$

10. $f(x) = .25x^4 - x^2 - 2x + 4$ from $x = -1$ to $x = 4$

11. $f(x) = x^3 - 3x^2 - 2x + 6$ from $x = -1$ to $x = 3$

12. $f(x) = -\sqrt{x^4 - x^3 + 2x^2 - x + 4}$ from $x = 0$ to $x = 3$

13. $f(x) = \sqrt{x^3 + 2x^2 - 6x + 5}$ from $x = 1$ to $x = 2$

14. $f(x) = \dfrac{x^2 - 3}{2x - 4}$ from $x = 3$ to $x = 6$

In Exercises 15–22, compute the difference quotient of the function.

15. $f(x) = x + 5$ **16.** $f(x) = 7x + 2$

17. $f(x) = x^2 + 3$ **18.** $f(x) = x^2 + 3x - 1$

19. $f(t) = 160{,}000 - 8000t + t^2$ **20.** $V(x) = x^3$

21. $A(r) = \pi r^2$ **22.** $V(p) = 5/p$

23. Water is draining from a large tank. After t minutes there are $160{,}000 - 8000t + t^2$ gallons of water in the tank.

(a) Use the results of Exercise 19 to find the average rate at which the water runs out in the interval from 10 to 10.1 minutes.
(b) Do the same for the interval from 10 to 10.01 minutes.
(c) Estimate the rate at which the water runs out after exactly 10 minutes.

24. Use the results of Exercise 20 to find the average rate of change of the volume of a cube whose side has length x as x changes from
(a) 4 to 4.1 **(b)** 4 to 4.01 **(c)** 4 to 4.001
(d) Estimate the rate of change of the volume at the instant when $x = 4$.

25. Use the results of Exercise 21 to find the average rate of change of the area of a circle of radius r as r changes from
(a) 3 to 3.5 **(b)** 3 to 3.2 **(c)** 3 to 3.1
(d) Estimate the rate of change at the instant when $r = 3$.
(e) How is your answer in part (d) related to the circumference of a circle of radius 3?

26. Under certain conditions, the volume V of a quantity of air is related to the pressure p (which is measured in kilopascals) by the equation $V = 5/p$. Use the results of Exercise 22 to estimate the rate at which the volume is changing at the instant when the pressure is 50 kilopascals.

27. Two cars race on a straight track, beginning from a dead stop. The distance (in ft) each car has covered at each time during the first 16 seconds is shown in Figure 3–13.

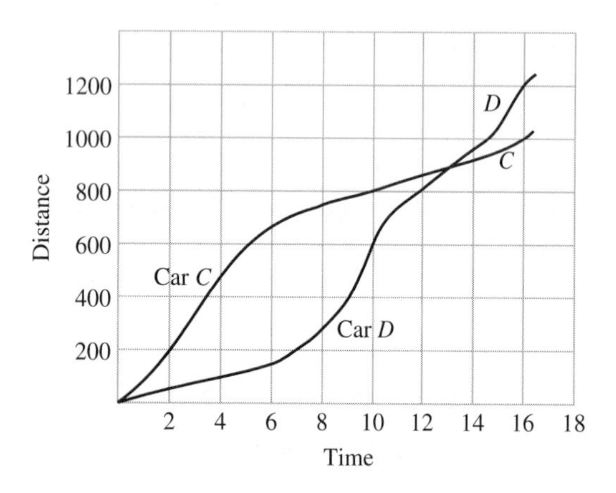

Figure 3-13

(a) What is the average speed of each car during this 16-second interval?
(b) Find an interval beginning at $t = 4$ during which the average speed of car D was approximately the same as the average speed of car C from $t = 2$ to $t = 10$.
(c) Use secant lines and slopes to justify the statement "car D traveled at a higher average speed than car C from $t = 4$ to $t = 10$."

28. Figure 3–14 shows the profits earned by a certain company during the last quarters of three consecutive years.

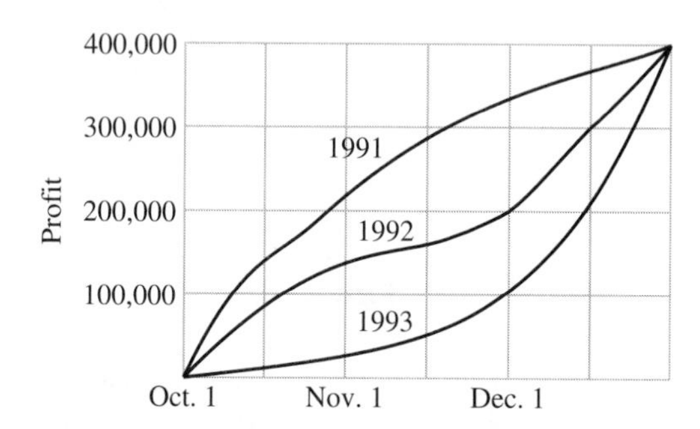

Figure 3-14

(a) Explain why the average rate of change of profits from October 1 to December 31 was the same in all three years.
(b) During what month in what year was the average rate of change of profits the greatest?

29. The graph in Figure 3–15 shows the chipmunk population in a certain wilderness area. The population increases as the chipmunks reproduce, but then decreases sharply as predators move into the area.
(a) During what approximate time period (beginning on day 0) is the average growth rate of the chipmunk population positive?
(b) During what approximate time period, beginning on day 0, is the average growth rate of the chipmunk population 0?
(c) What is the average growth rate of the chipmunk population from day 50 to day 100? What does this number mean?
(d) What is the average growth rate from day 45 to day 50? From day 50 to day 55? What is the approximate average growth rate from day 49 to day 51?

30. Lucy has a viral flu. How bad she feels depends primarily on how fast her temperature is rising at that time. Figure 3–16 shows her temperature during the first day of the flu.

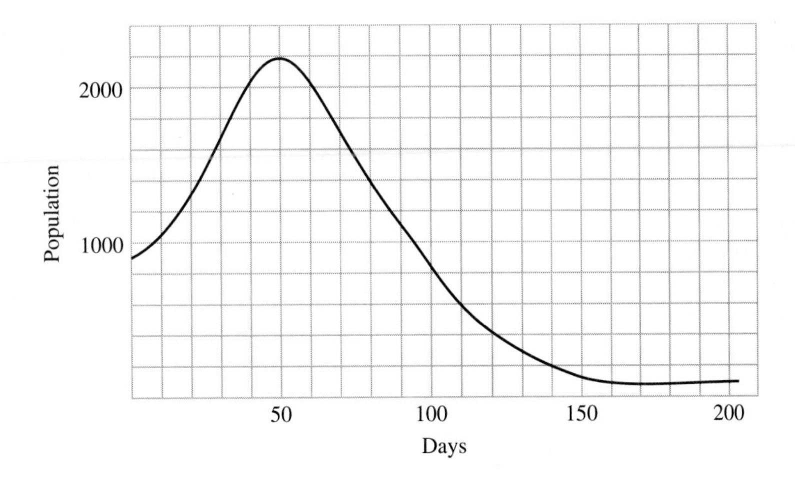

Figure 3-15

(a) At what average rate does her temperature rise during the entire day?
(b) During what two-hour period during the day does she feel worst?
(c) Find two time intervals, one in the morning and one in the afternoon, during which she feels about the same (that is, during which her temperature is rising at the same average rate).

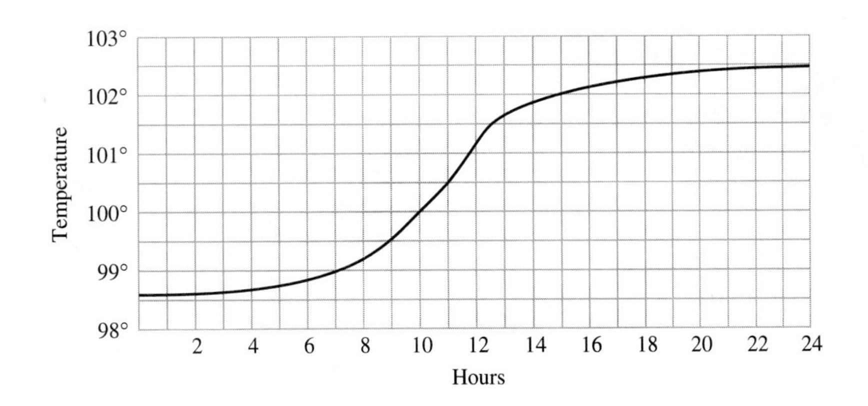

Figure 3-16

3.2.A *Excursion* THE SECANT METHOD

As we saw in Section 1.5, the solutions of the equation $f(x) = 0$ are the x-intercepts of the graph of $y = f(x)$. We shall now see how secant lines to the graph of $f(x)$ can be used to approximate its x-intercepts and thus provide another method of solving equations.

Suppose that f is a function with an x-intercept at $x = c$, as shown below. We shall approximate c as follows.

Geometric Description	Picture	Algebraic Description

1. Make two initial guesses, x_1 and x_2. These are the first two approximations of c.

2. Draw the secant line through $(x_1, f(x_1))$ and $(x_2, f(x_2))$; it intersects the x-axis at x_3. This is the third approximation of c.

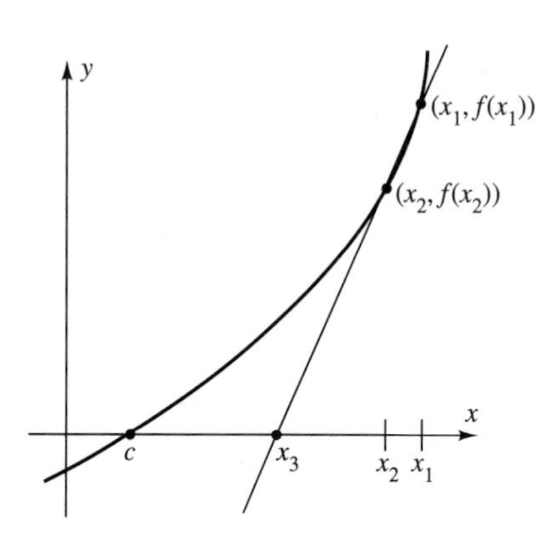

Slope of secant line through $(x_1, f(x_1))$ and $(x_2, f(x_2))$ is

$$M = \frac{f(x_2) - f(x_1)}{x_2 - x_1}.$$

Equation of secant line is

$$y - f(x_2) = M(x - x_2)$$
$$y = M(x - x_2) + f(x_2).$$

The x-intercept of this line (found by setting $y = 0$ and solving for x) is

$$x_3 = x_2 - \frac{f(x_2)}{M}.$$

3. Repeat the process, using x_2 and x_3: Draw the secant line through $(x_2, f(x_2))$ and $(x_3, f(x_3))$; it intersects the x-axis at x_4. This is the fourth approximation of c.

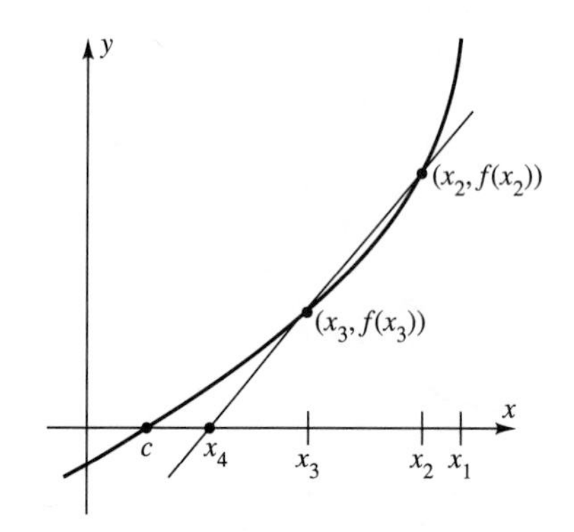

Slope of secant line through $((x_2, f(x_2))$ and $((x_3, f(x_3))$ is

$$M = \frac{f(x_3) - f(x_2)}{x_3 - x_2}.$$

Equation of secant line is

$$y - f(x_3) = M(x - x_3)$$
$$y = M(x - x_3) + f(x_3).$$

The x-intercept of this line is

$$x_4 = x_3 - \frac{f(x_3)}{M}.$$

4. Repeat the process, using x_3 and x_4, to obtain the fifth approximation x_5.

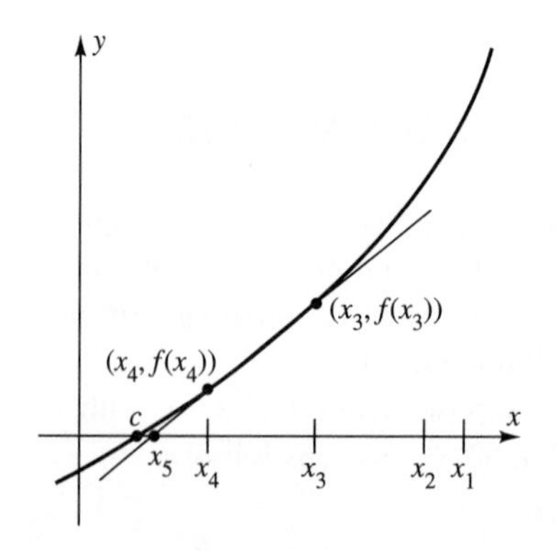

The x-intercept of this line is

$$x_5 = x_4 - \frac{f(x_4)}{M}$$

where

$$M = \frac{f(x_4) - f(x_3)}{x_4 - x_3}.$$

This process can be continued as long as you want, with each successive approximation being obtained from the two preceding ones, as summarized here:

The Secant Method ▶

Let f be a function and let x_1 and x_2 be initial guesses (approximations) for an x-intercept. Then further approximations $x_3, x_4, x_5, x_6, x_7, \ldots$ may be obtained from the formula

$$x_{n+1} = x_n - \frac{f(x_n)}{M}, \quad \text{where} \quad M = \frac{f(x_n) - f(x_{n-1})}{x_n - x_{n-1}}.$$

The calculation in step 2 above is the case when $n = 2$ and $n + 1 = 3$; the calculation in step 3 is the case when $n = 3$ and $n + 1 = 4$. In most cases the procedure works as outlined above, although the picture may be different.

How many steps are needed for the secant method to produce a good approximation? To answer this, consider what happens if two successive approximations are equal, say $x_k = x_{k+1}$. Then by the formula in the box above

$$x_k = x_{k+1} = x_k - \frac{f(x_k)}{M}$$

$$0 = -\frac{f(x_k)}{M}$$

$$0 = f(x_k).$$

Thus, when $x_k = x_{k+1}$, the x-intercept is x_k. Similarly, whenever two successive approximations are very close to one another, then these approximations are very close to the actual x-intercept. As a general rule you should follow this guideline:

Accuracy Guideline ▶

Whenever successive approximations x_k and x_{k+1} agree to eight decimal places, then x_{k+1} is a good approximation of the actual x-intercept.

EXAMPLE 1 Use the secant method to solve the equation $x^3 - 2x - 8 = 0$.

Solution You can easily verify that the graph of $f(x) = x^3 - 2x - 8$ has only one x-intercept, which is between 2 and 3. So we use $x_1 = 3$ and $x_2 = 2$ as our first two guesses. Repeated use of formula in the box above produces the following table.*

n	x_n	x_{n-1}	$x_{n+1} = x_n - \dfrac{f(x_n)}{M} = x_n - \dfrac{f(x_n)}{\dfrac{f(x_n) - f(x_{n-1})}{x_n - x_{n-1}}}$
1	3		
2	2	3	$x_3 = 2 - \dfrac{f(2)}{\dfrac{f(2) - f(3)}{2 - 3}} = 2.235294118$
3	2.2353	2	$x_4 = 2.2353 - \dfrac{f(2.2353)}{\dfrac{f(2.2353) - f(2)}{2.2353 - 2}} = 2.348823174$
4	2.3488	2.2353	$x_5 = 2.3488 - \dfrac{f(2.3488)}{\dfrac{f(2.3488) - f(2.2353)}{2.3488 - 2.2353}} = 2.329879208$
5	2.3299	2.3488	$x_6 = 2.3299 - \dfrac{f(2.3299)}{\dfrac{f(2.3299) - f(2.3488)}{2.3299 - 2.3488}} = 2.330738468$
6	2.33074	2.3299	$x_7 = 2.33074 - \dfrac{f(2.33074)}{\dfrac{f(2.33074) - f(2.3299)}{2.33074 - 2.3299}} = 2.330746089$
7	2.33075	2.33074	$x_8 = 2.33075 - \dfrac{f(2.33075)}{\dfrac{f(2.33075) - f(2.33074)}{2.33075 - 2.33074}} = 2.330746086$

▶ TECHNOLOGY TIP

Programs for implementing the secant method on most calculators are in the Program Appendix.

The approximations x_7 and x_8 agree to eight decimal places, which indicates that x_8 is a very good approximation of the x-intercept. So, the solution of the equation is $x \approx 2.330746086$. ■

The program mentioned in the Tip in the margin was used for the next example, which demonstrates the efficiency and accuracy of the secant method.

*Most entries in the table are rounded for convenient reading; however, the full decimal expansion of each x_n (as given in the last column) was used in all computations.

EXAMPLE 2 One of the solutions of $x^2 - 3 = 0$ is $\sqrt{3}$, whose decimal expansion to 11 places is 1.73205080757. Here is what the secant method produces for various initial guesses:

$x_1 = 0$ and $x_2 = 3$	$x_1 = 1$ and $x_2 = 2$	$x_1 = 1.7$ and $x_2 = 1.8$
$x_3 = 1$	$x_3 = 1.666666667$	$x_3 = 1.731428571$
$x_4 = 1.5$	$x_4 = 1.727272727$	$x_4 = 1.732038835$
$x_5 = 1.8$	$x_5 = 1.732142857$	$x_5 = 1.732050810$
$x_6 = 1.727272727$	$x_6 = 1.732050680$	$x_6 = 1.732050808$
$x_7 = 1.731958763$	$x_7 = 1.732050808$	
$x_8 = 1.732050935$		
$x_9 = 1.732050808$		

In each case a solution accurate to nine decimal places is readily obtained, with more rapid convergence when the initial guesses are closer to the actual solution. ■

Although the secant method is generally very efficient and accurate, it has some limitations:

1. If $f(x_n) = f(x_{n-1})$ at some stage, then the method fails because the slope $M = \dfrac{f(x_n) - f(x_{n-1})}{x_n - x_{n-1}}$ is 0 and $x_{n+1} = x_n - f(x_n)/M$ is not defined.
2. The approximations may not converge, that is, you may never obtain two successive approximations that agree to eight decimal places.
3. If an equation has several solutions, the secant method may not produce the desired one.

These problems can usually be overcome by making different initial guesses, although in a few cases a fair amount of trial and error may be necessary (see Exercises 7–9).

Finally, when the secant method (or any other solution method) is programmed for calculator or computer use, round-off errors when functional values are exceptionally large may produce bizarre results (see Exercise 10). So be sure to check your answers in the original equation.

EXERCISES 3.2.A

Note: *A secant method program for a calculator or computer is needed for most of these exercises.*

In Exercises 1–6, use the secant method to solve the equation.

1. $x^3 - 2x - 4 = 0$

2. $x^6 - 2x^4 + 3x^2 - 1 = 0$

3. $x^5 - 3x^2 + 2x - 7 = 0$

4. $x^4 - 6x^3 + 2x^2 - 4 = 0$

5. $x^3 - 10x^2 + 20x + 5 = 0$

6. $x^5 - 7x^3 + 2 = 0$

7. The solution of $\sqrt[3]{x} = 0$ is obviously $x = 0$, but you may have trouble verifying this by the secant method. Find at least three pairs of initial guesses that do not produce the solution and at least one pair of initial guesses that does. [The difficulty here is related to the fact that the graph of $y = \sqrt[3]{x}$ is almost vertical at $x = 0$.]

8. Use the secant method, with initial guesses of -1 and 1, to solve $x^6 - x - 1 = 0$. What happens? [*Hint:* Compute x_3 by hand; then compute the slope M of the secant line through $(x_2, f(x_2))$ and $(x_3, f(x_3))$].

9. (a) Graph $y = x^6 - x - 1$ and verify that it has one x-intercept between -1 and 0 and another between 1 and 2.

(b) If your initial guesses for the solutions of $x^6 - x - 1 = 0$ are -2 and 1, which solution do you think the secant method will produce?

(c) If your initial guesses for the solutions of $x^6 - x - 1 = 0$ are 0 and 2, which solution do you think the secant method will produce?

(d) Check your answers to parts (b) and (c) by using the secant method with the given initial guesses.

10. Use the secant method, with initial guesses of -5 and 5, to solve $x^6 - x - 1 = 0$. Why do you get the wrong answer? [*Hint:* Verify that $x_3 = 15{,}624$. Does your calculator tell you that $f(x_3) - f(x_2) = f(x_3)$? What's going on?]

3.3 QUADRATIC FUNCTIONS*

A **quadratic function** is a function whose rule can be written in the form

$$f(x) = ax^2 + bx + c$$

for some constants a, b, c, with $a \neq 0$. The graph of a quadratic function is called a **parabola.**

GRAPHING EXPLORATION Using the standard viewing window, graph the following quadratic functions on the same screen:

$f(x) = x^2,\quad f(x) = 3x^2 + 30x + 77,\quad f(x) = -x^2 + 4x,$
$f(x) = -.2x^2 + 1.5x - 5$

As the Exploration illustrates, all parabolas have the same basic "cup" shape, though the cup may be broad or narrow. The parabola opens upward when the coefficient of x^2 is positive and downward when this coefficient is negative.

If a parabola opens upward, its **vertex** is the lowest point on the graph. For instance, (0, 0) is the vertex of $f(x) = x^2$ since every other point on the graph has a positive y-coordinate. If a parabola opens downward, its **vertex** is the highest point on the graph. Every parabola is symmetric with respect to the vertical line through its vertex; this line is called the **axis** of the parabola.

The vertex of a parabola can always be located approximately by graphing it and using trace or a min/max finder on a calculator. However, there are algebraic techniques for finding the vertex precisely.

*This section may be omitted or postponed if desired. Section 2.5 (Graphs and Transformations) is a prerequisite.

EXAMPLE 1 The function $g(x) = 2(x - 3)^2 + 1$ is quadratic because its rule can be written in the required form:

$$g(x) = 2(x - 3)^2 + 1 = 2(x^2 - 6x + 9) + 1 = 2x^2 - 12x + 19.$$

Graphing the function in the standard viewing window (Figure 3–17) and using the trace feature shows that the vertex is approximately (3.0526, 1.0055).

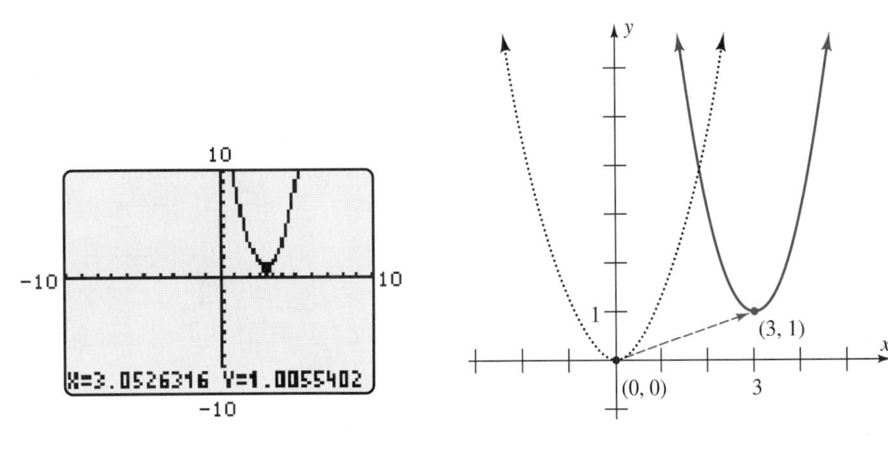

Figure 3-17

Figure 3-18

As we saw in Section 2.5, the graph of $g(x) = 2(x - 3)^2 + 1$ is the graph of $f(x) = x^2$ shifted horizontally 3 units to the right, stretched by a factor of 2, and shifted vertically 1 unit upward, as shown in Figure 3–18 above. When the vertex of f, namely (0, 0), is shifted 3 units right and 1 unit up, it moves to (3, 1). Thus, (3, 1) is the vertex of $g(x) = 2(x - 3)^2 + 1$. Note how the coordinates of the vertex of g are related to the constants in its rule:

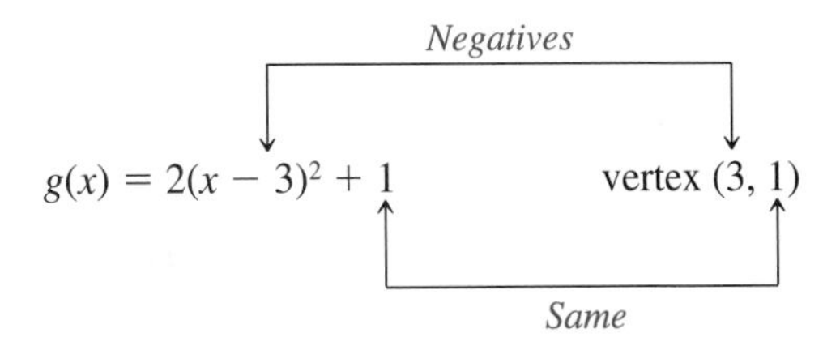

■

The vertex of the function g in Example 1 was easily determined because the rule of g had a special algebraic form. The vertex of the graph of any quadratic function can be determined in a similar fashion by first rewriting its rule.

EXAMPLE 2 To find the vertex of the graph of $g(x) = 3x^2 + 30x + 77$, we first rewrite its rule as $g(x) = 3(x^2 + 10x) + 77$. The next step is to complete the square in the expression in parentheses by adding 25 (the square of half the coefficient of x).* In order not to change the rule of the function, we must also *subtract* 25:

$$g(x) = 3(x^2 + 10x + 25 - 25) + 77.$$

Using the distributive law and factoring, we have:

$$g(x) = 3(x^2 + 10x + 25) - 3 \cdot 25 + 77$$

$$g(x) = 3(x + 5)^2 + 2.$$

As we saw in Section 2.5, the graph of $g(x) = 3(x + 5)^2 + 2$ is the graph of $f(x) = x^2$ shifted horizontally 5 units to the left, stretched by a factor of 3, then shifted vertically 2 units upward, as shown in Figure 3–19. In this process, the vertex (0, 0) of f moves to (− 5, 2), which is therefore the vertex of g. Once again, note how the coordinates of the vertex are related to the rule of the function:

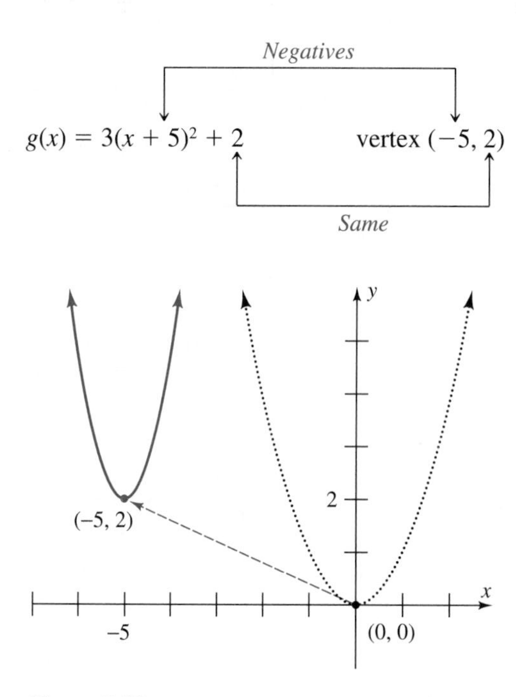

Figure 3-19

The technique in Example 2 works for any quadratic function $f(x) = ax^2 + bx + c$.** First, rewrite the rule as $f(x) = a\left(x^2 + \frac{b}{a}x\right) + c$. Then complete the square in the expression in parentheses by adding the square of half the coefficient of x, namely $b^2/4a^2$. In order not to change the rule of the function, subtract the same quantity:

*Completing the square is discussed on page 17.

**The following argument is exactly the one used in Example 2, with a in place of 3, b in place of 30, c in place of 77, and $b^2/4a^2$ in place of 25.

$$f(x) = a\left(x^2 + \frac{b}{a}x + \frac{b^2}{4a^2} - \frac{b^2}{4a^2}\right) + c.$$

Then use the distributive law and factor:

$$\begin{aligned} f(x) &= a\left(x^2 + \frac{b}{a}x + \frac{b^2}{4a^2}\right) - a\frac{b^2}{4a^2} + c \\ &= a\left(x + \frac{b}{2a}\right)^2 + \left(c - \frac{b^2}{4a}\right). \end{aligned}$$

As in the preceding examples, the graph of f is just the graph of x^2 shifted horizontally, stretched by a factor of a, and shifted vertically. As above, the vertex of this parabola can be read from the rule of the function:

Negatives

$$f(x) = a\left(x + \frac{b}{2a}\right)^2 + \left(c - \frac{b^2}{4a}\right) \qquad \text{vertex}\left(\frac{-b}{2a}, c - \frac{b^2}{4a}\right)$$

Same

Consequently, the preceding discussion can be summarized as follows:

Quadratic Functions ▶

The graph of the quadratic function $f(x) = ax^2 + bx + c$ is a parabola that opens upward if $a > 0$ and downward if $a < 0$. The vertex of this parabola has x-coordinate $-b/2a$.

It isn't necessary to memorize the y-coordinate of the vertex here because you can always compute $f(-b/2a)$ to find it.

EXAMPLE 3 The graph of $f(x) = -4x^2 + 12x - 8$ is a downward-opening parabola because the coefficient of x^2 is negative. Its vertex has x-coordinate

$$-\frac{b}{2a} = -\frac{12}{2(-4)} = \frac{-12}{-8} = \frac{3}{2}$$

and y-coordinate

$$f\left(\frac{3}{2}\right) = -4\left(\frac{3}{2}\right)^2 + 12\left(\frac{3}{2}\right) - 8 = 1. \quad \blacksquare$$

Applications

Suppose $(r, f(r))$ is the vertex of the graph of a quadratic function f. Depending on whether the parabola opens upward or downward, one of these cases must hold:

$(r, f(r))$ is the highest point on the graph

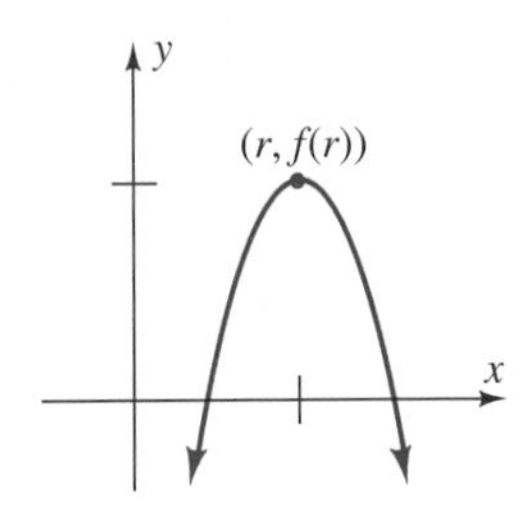

Every other point $(x, f(x))$ on the graph lies below $(r, f(r))$ and hence has smaller y-coordinate. Thus,

$$f(x) \leq f(r) \qquad \text{for every } x.$$

In this case $f(r)$ is said to be the **maximum value** of the function.

$(r, f(r))$ is the lowest point on the graph

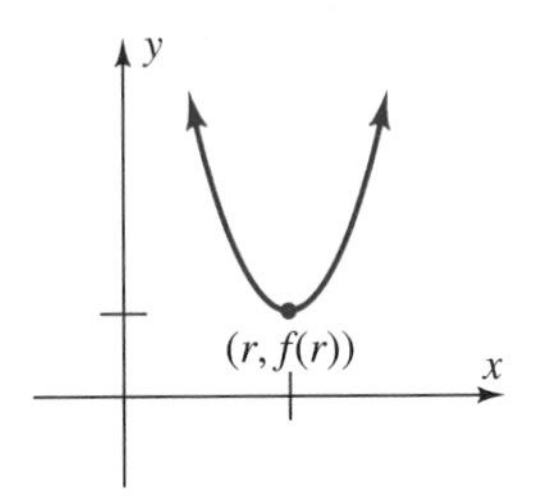

Every other point $(x, f(x))$ on the graph lies above $(r, f(r))$ and hence has larger y-coordinate. Thus,

$$f(x) \geq f(r) \qquad \text{for every } x.$$

In this case $f(r)$ is said to be the **minimum value** of the function.

The solution of many applied problems depends on finding the maximum or minimum value of a quadratic function.

EXAMPLE 4 Find the area and dimensions of the largest rectangular field that can be enclosed with 3000 feet of fence.

Solution Let x denote the length and y the width of the field, as shown in Figure 3–20.

Perimeter $= x + y + x + y = 2x + 2y$
Area $= xy$

x
y y
x

Figure 3-20

Since the perimeter is the length of the fence, $2x + 2y = 3000$. Hence, $2y = 3000 - 2x$ and $y = 1500 - x$. Consequently, the area is

$$A = xy = x(1500 - x) = 1500x - x^2 = -x^2 + 1500x.$$

The largest possible area is just the maximum value of the quadratic function $A(x) = -x^2 + 1500x$. This maximum occurs at the vertex of the graph of $A(x)$ (which is a downward-opening parabola because the coefficient of x^2 is negative). The vertex may be found by using the fact in the box on page 199 (with $a = -1$ and $b = 1500$):

$$\text{The } x\text{-coordinate of the vertex is } -\frac{1500}{2(-1)} = 750 \text{ feet.}$$

Hence, the y-coordinate of the vertex, the maximum value of $A(x)$, is

$$A(750) = -750^2 + 1500 \cdot 750 = 562{,}500 \text{ sq ft.}$$

It occurs when the length is $x = 750$. In this case the width is $y = 1500 - x = 1500 - 750 = 750$. ■

E X A M P L E 5 Find real numbers c and d whose difference is 5 and whose product is as small as possible.

Solution Since $c - d = 5$, we have $c = d + 5$. We want the product

$$cd = (5 + d)d = d^2 + 5d$$

to be a minimum. Since the graph of $f(d) = d^2 + 5d$ is an upward-opening parabola, the minimum value of $f(d)$ occurs at the vertex of the parabola. We can find the vertex by using the fact in the box on page 199 (with $a = 1$ and $b = 5$):

$$\text{The first coordinate of the vertex is } d = \frac{-5}{2 \cdot 1} = -\frac{5}{2}.$$

Hence, the smallest product occurs when $d = -5/2$ and $c = 5 + d = 5 - \frac{5}{2} = 5/2$. The smallest value of cd is $(\frac{5}{2})(-\frac{5}{2}) = -25/4$. ■

E X A M P L E 6 When the fare on an airport bus is \$3, there are, on the average, 20 passengers per trip. With each fare increase of 25¢ the average number of passengers decreases by 1. What fare should be charged to have the largest possible revenue per trip?

Solution For each *dollar* the fare is raised, there are 4 fewer passengers per trip. So if the fare is increased by x dollars to $(3 + x)$ dollars, then there will be $20 - 4x$ passengers per trip. Then,

$$\begin{aligned} \text{revenue} &= (\text{fare}) \cdot (\text{number of passengers}) \\ &= (3 + x)(20 - 4x) \\ &= -4x^2 + 8x + 60. \end{aligned}$$

We must find the value of x at which the quadratic function $r(x) = -4x^2 + 8x + 60$ takes its maximum value. The graph of $r(x)$ is a downward-opening parabola (why?), and the maximum value occurs at the vertex, whose x-coordinate is

$$x = \frac{-b}{2a} = \frac{-8}{2(-4)} = 1.$$

Therefore, the fare should be $3 + 1 = \$4$. ■

EXERCISES 3.3

In Exercises 1–16, without graphing, *determine the vertex of the given parabola and state whether it opens upward or downward.*

1. $f(x) = 3(x-5)^2 + 2$

2. $g(x) = -6(x-2)^2 - 5$

3. $y = -(x-1)^2 + 2$

4. $h(x) = -x^2 + 1$

5. $f(x) = x^2 - 6x + 3$

6. $g(x) = x^2 + 8x - 1$

7. $h(x) = x^2 + 3x + 6$

8. $f(x) = x^2 - 5x - 7$

9. $y = 2x^2 + 12x - 3$

10. $y = 3x^2 + 6x + 1$

11. $f(x) = -x^2 + 8x - 2$

12. $g(x) = -x^2 - 6x + 4$

13. $f(x) = -3x^2 + 4x + 5$

14. $g(x) = 2x^2 - x - 1$

15. $y = -x^2 + x$

16. $y = -2x^2 + 2x - 1$

17. The graph of the quadratic function g is obtained from the graph of $f(x) = x^2$ by vertically stretching it by a factor of 2 and then shifting vertically 5 units downward. What is the rule of the function g? What is the vertex of its graph?

18. The graph of the quadratic function g is obtained from the graph of $f(x) = x^2$ by shifting it horizontally 4 units to the left, then vertically stretching it by a factor of 3, and then shifting vertically 2 units upward. What is the rule of the function g? What is the vertex of its graph?

19. If the graph of the quadratic function h is shifted vertically 4 units downward, then shrunk by a factor of 1/2, and then shifted horizontally 3 units to the left, the resulting graph is the parabola $f(x) = x^2$. What is the rule of the function h? What is the vertex of its graph?

20. If the graph of the quadratic function h is shifted vertically 3 units upward, then reflected in the x-axis, and then shifted horizontally 5 units to the right, the resulting graph is the parabola $f(x) = x^2$. What is the rule of the function h? What is the vertex of its graph?

21. Find the rule of the quadratic function whose graph is the parabola with vertex at the origin that passes through (2, 12).

22. Find the rule of the quadratic function whose graph is the parabola with vertex (0, 1) that passes through (2, −7).

23. Find the number b such that the vertex of the parabola $y = x^2 + bx + c$ lies on the y-axis.

24. Find the number c such that the vertex of the parabola $y = x^2 + 8x + c$ lies on the x-axis.

25. If the vertex of the parabola $f(x) = x^2 + bx + c$ is at (2, 4), find b and c.

26. If the vertex of the parabola $f(x) = -x^2 + bx + 8$ has second coordinate 17 and is in the second quadrant, find b.

27. What is the minimum product of two numbers whose difference is 4? What are the numbers?

28. Find numbers c and d whose sum is −18 and whose product is as large as possible.

29. Find two numbers such that the sum of one and twice the other is 36 and the product of the two numbers is as large as possible.

30. Find two numbers such that the difference of one and three times the other is 48 and the product of the two numbers is as small as possible.

31. The sum of the height h and the base b of a triangle is 30. What height and base will produce a triangle of maximum area?

32. A trough is to be made by bending a long, flat piece of tin 10 inches wide into a rectangular shape. What depth should the trough be in order to have the maximum possible cross-sectional area?

33. A field bounded on one side by a river is to be fenced on three sides so as to form a rectangular enclosure. If the total length of fence to be used is 200 feet, what dimensions will yield an enclosure of the largest possible area?

34. A rectangular box (with top) has a square base. The sum of the lengths of its 12 edges is 8 feet. What dimensions should the box have in order that its surface area be as large as possible?

35. A salesperson finds that her sales average 40 cases per store when she visits 20 stores a week. Each time she visits an additional store per week, the average sales per store decrease by 1 case. How many stores should she visit each week if she wants to maximize her sales?

36. A miniature golf course averages 200 patrons per evening when it charges $2 per person. For each 5¢ increase in the admission price, the average attendance drops by 2 people. What admission price will produce the largest ticket revenue?

37. A potter can sell 120 bowls per week at $4 per bowl. For each 50¢ decrease in price 20 more bowls are sold. What price should be charged in order to maximize sales income?

38. A vendor can sell 200 souvenirs per day at a price of $2 each. Each 10¢ price increase decreases the number of sales by 25 per day. Souvenirs cost the vendor $1.50 each. What price should be charged in order to maximize the profit?

39. When a basketball team charges $4 per ticket, average attendance is 500 people. For each 20¢ decrease in ticket price, average attendance increases by 30 people. What should the ticket price be to insure maximum income?

40. A ballpark concessions manager finds that each salesperson sells an average of 40 boxes of popcorn per game when there are 20 salespeople working. When an additional salesperson is employed, each salesperson averages 1 less box per game. How many salespeople should be hired to insure maximum income?

In Exercises 41–44, use the formula for the height h of an object (that is traveling vertically subject only to gravity) at time t: $h = -16t^2 + v_0t + h_0$, *where* h_0 *is the initial height and* v_0 *the initial velocity.*

41. A ball is thrown upward from the top of a 96-foot-high tower with an initial velocity of 80 feet per second. When does the ball reach its maximum height and how high is it at that time?

42. A rocket is fired upward from ground level with an initial velocity of 1600 feet per second. When does it attain its maximum height and what is that height?

43. A ball is thrown upward from a height of 6 feet with an initial velocity of 32 feet per second. Find its maximum height.

44. A bullet is fired upward from ground level with an initial velocity of 1500 feet per second. How high does it go?

45. A projectile is fired at an angle of 45° upward. Exactly t sec after firing, its vertical height above the ground is $500t - 16t^2$. What is the greatest height the projectile reaches and at what times does this occur?

CHAPTER 3 *Review*

Important Concepts ▶

Important Facts and Formulas ▶

- The slope of the line through (x_1, y_1) and (x_2, y_2) (where $x_1 \neq x_2$) is

$$\frac{y_2 - y_1}{x_2 - x_1}.$$

- Nonvertical parallel lines have the same slope.
- Two lines (neither vertical) are perpendicular exactly when the product of their slopes is -1.
- The equation of the line through (x_1, y_1) with slope m is

$$y - y_1 = m(x - x_1).$$

- The equation of the line with slope m and y-intercept b is

$$y = mx + b.$$

- The average rate of change of a function f as x changes from a to b is the number

$$\frac{f(b) - f(a)}{b - a}.$$

- The difference quotient of the function f is the quantity

$$\frac{f(x + h) - f(x)}{h}.$$

- The average rate of change of a function f as x changes from a to b is the slope of the secant line joining the points $(a, f(a))$ and $(b, f(b))$.
- The graph of $f(x) = ax^2 + bx + c$ is a parabola whose vertex has x-coordinate $-b/2a$.

Review Questions ▶

1. **(a)** What is the y-intercept of the graph of the linear function $f(x) = x - \frac{x - 2}{5} + \frac{3}{5}$?
   (b) What is the slope of the graph of f?
2. Find the equation of the line passing through (1, 3) and (2, 5).
3. Find the equation of the line passing through $(2, -1)$ with slope 3.
4. Find a point on the graph of $y = 3x$ whose distance to the origin is 2.
5. Find the equation of the line that crosses the y-axis at $y = 1$ and is perpendicular to the line $2y - x = 5$.
6. **(a)** Find the y-intercept of the line $2x + 3y - 4 = 0$.
   (b) Find the equation of the line through (1, 3) that has the same y-intercept as the line in part (a).
7. Find the equation of the line through $(-4, 5)$ that is parallel to the line through $(1, 3)$ and $(-4, 2)$.
8. Sketch the graph of the line $3x + y - 1 = 0$.
9. As a balloon is launched from the ground, the wind is blowing it due east. The conditions are such that the balloon is ascending along a straight line with slope 1/5. After 1 hour the balloon is 5000 ft vertically above the ground. How far east has the balloon blown?
10. The point (u, v) lies on the line $y = 5x - 10$. What is the slope of the line passing through (u, v) and the point $(0, -10)$?

Questions 11–17, determine whether the statement is true or false.

11. The graph of $x = 5y + 6$ has y-intercept 6.
12. The average rate of change of the function $f(x) = 3x + 2$ as x changes from a to any other number is 3.
13. The lines $3x + 4y = 12$ and $4x + 3y = 12$ are perpendicular.

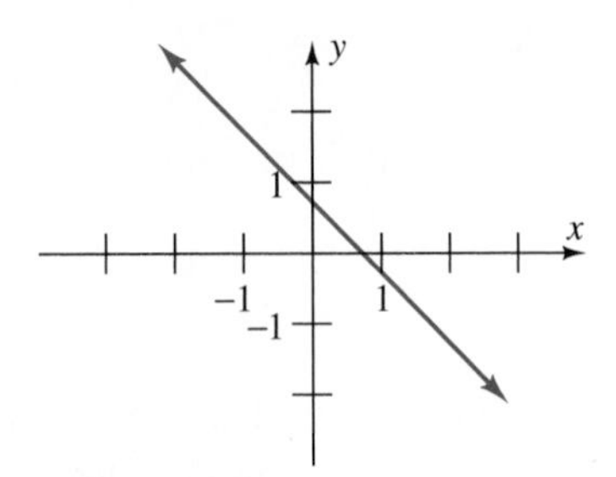

Figure 3-21

14. Slope is not defined for horizontal lines.

15. The line in Figure 3–21 has positive slope.

16. The line in Figure 3–21 does not pass through the third quadrant.

17. The y-intercept of the line in Figure 3–21 is negative.

18. Consider the *slopes* of the lines shown in Figure 3–22. Which slope has the largest *absolute value?*

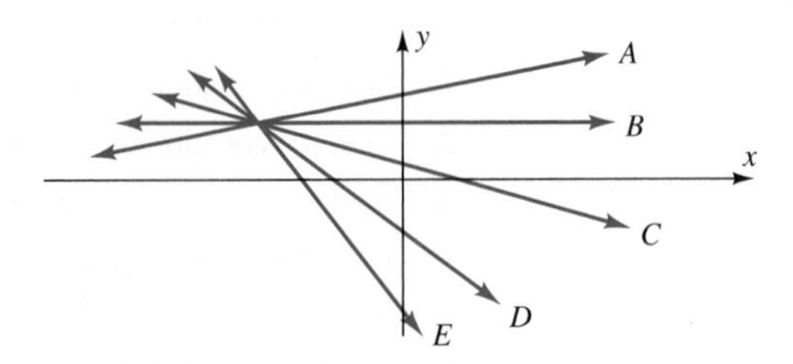

Figure 3-22

19. Which of the following lines rises most steeply from left to right?
(a) $y = -4x - 10$ **(b)** $y = 3x + 4$
(c) $20x + 2y - 20 = 0$ **(d)** $4x = y - 1$
(e) $4x = 1 - y$

20. Which of the following lines is *not* perpendicular to the line $y = x + 5$?
(a) $y = 4 - x$ **(b)** $y + x = -5$
(c) $4 - 2x - 2y = 0$ **(d)** $x = 1 - y$
(e) $y - x = \frac{1}{5}$

21. Which of the following lines does *not* pass through the third quadrant?
(a) $y = x$ **(b)** $y = 4x - 7$
(c) $y = -2x - 5$ **(d)** $y = 4x + 7$
(e) $y = -2x + 5$

22. Let a, b be fixed real numbers. Where do the lines $x = a$ and $y = b$ intersect?
(a) Only at (b, a).
(b) Only at (a, b).

(c) These lines are parallel, so they don't intersect.

(d) If $a = b$, then these are the same line, so they have infinitely many points of intersection.

(e) Since these equations are not of the form $y = mx + b$, the graphs are not lines.

23. What is the y-intercept of the line $2x - 3y + 5 = 0$?

24. For what values of k will the graphs of $2y + x + 3 = 0$ and $3y + kx + 2 = 0$ be perpendicular lines?

25. Find the average rate of change of the function $g(x) = \dfrac{x^3 - x + 1}{x + 2}$ as x changes from

(a) -1 to 1 **(b)** 0 to 2

26. Find the average rate of change of the function $f(x) = \sqrt{x^2 - x + 1}$ as x changes from

(a) -3 to 0 **(b)** -3 to 3.5 **(c)** -3 to 5

27. If $f(x) = 2x + 1$ and $g(x) = 3x - 2$, find the average rate of change of the composite function $f \circ g$ as x changes from 3 to 5.

28. If $f(x) = x^2 + 1$ and $g(x) = x - 2$, find the average rate of change of the composite function $f \circ g$ as x changes from -1 to 1.

In Questions 29–32, find the difference quotient of the function.

29. $f(x) = 3x + 4$ **30.** $g(x) = \sqrt{x}$

31. $g(x) = x^2 - 1$ **32.** $f(x) = x^2 + x$

33. The profit (in hundreds of dollars) from selling x tons of Wonderchem is given by $P(x) = .2x^2 + .5x - 1$. What is the average rate of change of profit when the number of tons of Wonderchem sold increases from

(a) 4 to 8 tons? **(b)** 4 to 5 tons? **(c)** 4 to 4.1 tons?

34. On the planet Mars, the distance traveled by a falling rock (ignoring atmospheric resistance) in t seconds is $6.1t^2$ ft. How far must a rock fall in order to have an average speed of 25 ft per sec over that time interval?

35. The graph in Figure 3–23 shows the population of fruit flies during a 50-day experiment in a controlled atmosphere.

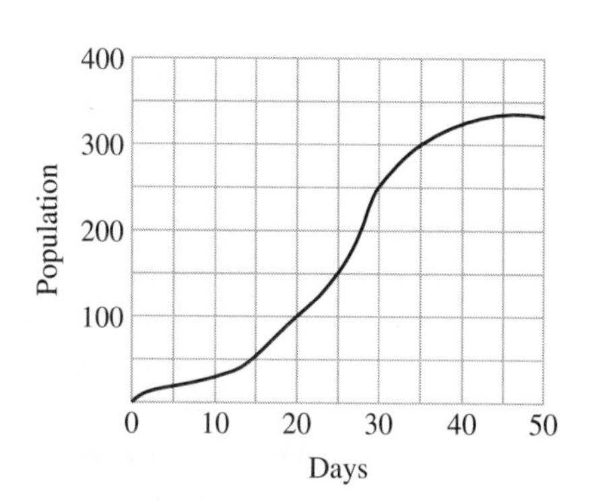

Figure 3-23

(a) During what 5-day period is the average rate of population growth the slowest?
(b) During what 10-day period is the average rate of population growth the fastest?
(c) Find an interval beginning at the 30th day during which the average rate of population growth is the same as the average rate from day 10 to day 20.

36. The graph of the function g in Figure 3–24 consists of straight line segments. Find an interval over which the average rate of change of g is
(a) 0 **(b)** -3 **(c)** .5
(d) Explain why the average rate of change of g is the same from -3 to -1 as it is from -2.5 to 0.

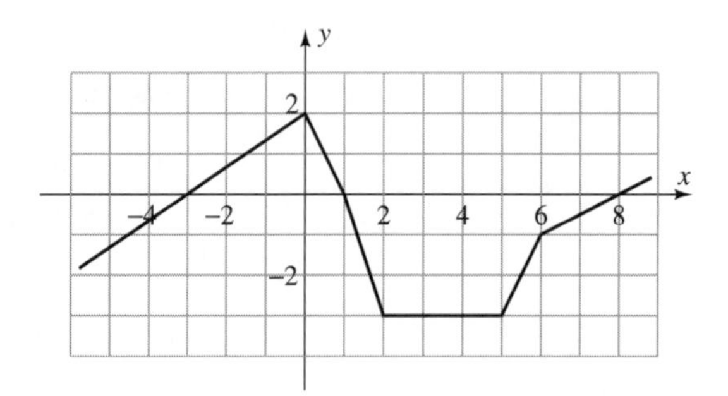

Figure 3-24

In Questions 37–40, find the vertex of the graph of the quadratic function.

37. $f(x) = (x - 2)^2 + 3$ **38.** $f(x) = 2(x + 1)^2 - 1$

39. $f(x) = x^2 - 8x + 12$ **40.** $f(x) = x^2 - 7x + 6$

41. Which of the following statements about the functions

$$f(x) = 3x^2 + 2 \quad \text{and} \quad g(x) = -3x^2 + 2$$

is *false*?
(a) The graphs of f and g are parabolas.
(b) The graphs of f and g have the same vertex.
(c) The graphs of f and g open in opposite directions.
(d) The graph of f is the graph of $y = 3x^2$ shifted 2 units to the right.

42. A model rocket is launched straight up from a platform at time $t = 0$ (where t is time measured in seconds). The altitude $h(t)$ of the rocket above the ground at given time (t) is given by $h(t) = 10 + 112t - 16t^2$ (where $h(t)$ is measured in feet).
(a) What is the altitude of the rocket the instant it is launched?
(b) What is the altitude of the rocket 2 seconds after launching?
(c) What is the maximum *altitude* attained by the rocket?
(d) At what *time* does the rocket return to the altitude at which it was launched?

43. A rectangular garden next to a building is to be fenced with 120 feet of fencing. The side against the building will not be fenced. What should the lengths of the other three sides be in order to assure the largest possible area?

44. A factory offers 100 calculators to a retailer at a price of \$20 each. The price per calculator on the entire order will be reduced 5¢ for each additional calculator over 100. What number of calculators will produce the largest possible sales revenue for the factory?

CHAPTER 4

Polynomial and Rational Functions

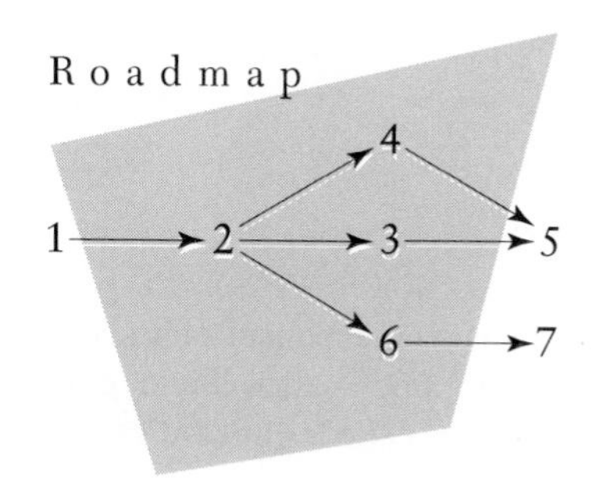

Polynomial functions arise naturally in many applications. Many complicated functions in applied mathematics can be approximated by polynomial functions or their quotients (rational functions).

4.1 POLYNOMIALS AND POLYNOMIAL FUNCTIONS

Informally, a **polynomial** is an algebraic expression such as

$$x^3 - 6x^2 + \tfrac{1}{2}x \qquad \text{or} \qquad x^{15} + x^{10} + 7 \qquad \text{or} \qquad x - \pi \qquad \text{or} \qquad 12.$$

We assume $x^0 = 1$ so that a real number, such as 12, may be thought of as $12x^0$. Consequently, a polynomial is a sum of terms of the form cx^k, where c is a constant and k is a *nonnegative* integer. Thus $x^{-2} + 6$ and $\sqrt{x^2 + 1}$ and $x^{1/4} + x - 3$ are not polynomials, but $\sqrt{5}$ and x^7 and $x + \frac{1}{2}$ are.

The formal definition is that a **polynomial in x** is an algebraic expression that can be written in the form

$$a_n x^n + a_{n-1}x^{n-1} + \cdots + a_3x^3 + a_2x^2 + a_1x + a_0$$

where n is a nonnegative integer, x is a variable, and each of $a_0, a_1, \ldots, a_n$ is a constant.* The numbers $a_0, a_1, \ldots, a_n$ are called the **coefficients** of the polynomial. The coefficient a_0 is called the **constant term.**

We omit any term with zero coefficient, we write x^0 as 1, and we don't write a coefficient or exponent if it is the number 1; for instance, we write $x^2 + 3$ instead of $1x^2 + 0x^1 + 3x^0$. A polynomial that consists of only a constant term,

*Any letter may be used as the variable in a polynomial, but we shall usually use x.

such as 12 or $-1/2$, is called a **constant polynomial.** The **zero polynomial** is the constant polynomial 0.

The *exponent* of the highest power of x appears with *nonzero* coefficient is the **degree** of the polynomial, and the nonzero coefficient of this highest power of x is the **leading coefficient.** For example,

Polynomial	*Degree*	*Leading Coefficient*	*Constant Term*
$6x^7 + 4x^3 + 5x^2 - 7x + 10$	7	6	10
$-x^4 + 2x^3 + \frac{1}{2}$	4	-1	$\frac{1}{2}$
x^3	3	1	0
12	0	12	12
$7x^5 - 3x^3 + x^2 + 4x$	5	7	0
$0x^4 + 5x^3 - 6x^2 + 2x - \frac{1}{4}$	3 (note well)	5	$-\frac{1}{4}$
$2x^6 + 3x^7 + x^8 - 2x - 4 + 3x^2$	8 (be careful)	1	-4

The degree of the zero polynomial is *not defined* since no exponent of x occurs with nonzero coefficient. First-degree polynomials are often called **linear polynomials.** Second- and third-degree polynomials are called **quadratics** and **cubics,** respectively.

A **polynomial function** is a function whose rule is given by a polynomial; for example,

$$f(x) = x^2 \qquad g(x) = -x^3 + 2x - 1 \qquad h(x) = 3x^5 + x^4 - 2x^3.$$

When dealing with polynomial functions, such as $f(x) = 7x^4 + 4x^2 - 6x + 5$, most mathematicians are pretty casual about their language. They may refer to "the polynomial $f(x)$" or "the function $7x^4 + 4x^2 + 6x + 5$."

Polynomial Arithmetic

You should be familiar with addition, subtraction, and multiplication of polynomials, which are presented in the Algebra Review Appendix. Long division of polynomials is quite similar to long division of numbers, as we now see.

EXAMPLE 1 To divide $2x^5 + 5x^4 - 4x^3 + 8x^2 + 1$ by $2x^2 - x + 1$, we first write:

$$2x^2 - x + 1 \overline{\big)\, 2x^5 + 5x^4 - 4x^3 + 8x^2 + 1}$$

We call $2x^2 - x + 1$ the **divisor** and $2x^5 + 5x^4 - 4x^3 + 8x^2 + 1$ the **dividend.** The **quotient** of the division will be written above the horizontal line. We begin by dividing the first term of the divisor ($2x^2$) into the first term of the dividend ($2x^5$) and putting the result $\left(\text{namely, } \dfrac{2x^5}{2x^2} = x^3\right)$ on the top line, as shown below. Then multiply x^3 times the entire divisor, put the result on the third line, and subtract:

$$\begin{array}{rl}
 & x^3 \\
2x^2 - x + 1 \,\big) & 2x^5 + 5x^4 - 4x^3 + 8x^2 + 1 \\
 & \underline{2x^5 - x^4 + x^3} \qquad \longleftarrow x^3 \cdot (2x^2 - x + 1) \\
 & \phantom{2x^5 +{}} 6x^4 - 5x^3 + 8x^2 + 1 \longleftarrow \textit{Subtraction*}
\end{array}$$

Next, divide the first term of the divisor $(2x^2)$ into $6x^4$ and put the result $\left(\frac{6x^4}{2x^2} = 3x^2\right)$ on the top line, as shown below. Then multiply $3x^2$ times the entire divisor, put the result on the fifth line, and subtract. Continuing this procedure, we obtain:

$$\begin{array}{rlll}
 & x^3 + 3x^2 - x + 2 & & \longleftarrow \textit{Quotient} \\
2x^2 - x + 1 \,\big) & 2x^5 + 5x^4 - 4x^3 + 8x^2 & + 1 & \\
 & 2x^5 - x^4 + x^3 & & \longleftarrow x^3 \cdot (2x^2 - x + 1) \\
 & \quad 6x^4 - 5x^3 + 8x^2 & + 1 & \longleftarrow \textit{Subtraction} \\
 & \quad 6x^4 - 3x^3 + 3x^2 & & \longleftarrow 3x^2 \cdot (2x^2 - x + 1) \\
 & \qquad -2x^3 + 5x^2 & + 1 & \longleftarrow \textit{Subtraction} \\
 & \qquad -2x^3 + x^2 - x & & \longleftarrow (-x)(2x^2 - x + 1) \\
 & \qquad\qquad 4x^2 + x & + 1 & \longleftarrow \textit{Subtraction} \\
 & \qquad\qquad 4x^2 - 2x & + 2 & \longleftarrow 2 \cdot (2x^2 - x + 1) \\
 & \qquad\qquad\qquad 3x & - 1 & \longleftarrow \textit{Subtraction}
\end{array}$$

The polynomial $3x - 1$ is called the **remainder.** The division process stops when the remainder is zero or has *smaller degree* than the divisor (here the divisor $2x^2 - x + 1$ has degree 2 and the remainder $3x - 1$ has degree 1). ■

When the divisor in polynomial division is a first-degree polynomial such as $x - 2$ or $x + 5$, there is a convenient shorthand method of doing the division called **synthetic division.** See Excursion 4.1.A for details.

Recall how you check a long division problem with numbers:

$$\begin{array}{rr}
 & 145 \\
31 \,\big) & 4509 \\
 & \underline{31} \\
 & 140 \\
 & \underline{124} \\
 & 169 \\
 & \underline{155} \\
 & 14
\end{array}
\qquad \textit{Check:} \qquad
\begin{array}{rl}
145 & \longleftarrow \textit{Quotient} \\
\underline{\times 31} & \longleftarrow \textit{Divisor} \\
145 & \\
\underline{435} & \\
4495 & \\
\underline{+14} & \longleftarrow \textit{Remainder} \\
4509 & \longleftarrow \textit{Dividend}
\end{array}$$

We can summarize this process in one line:

*If this subtraction is confusing, write it out horizontally and watch the signs carefully:

$$\begin{aligned}
&(2x^5 + 5x^4 - 4x^3 + 8x^2 + 1) - (2x^5 - x^4 + x^3) \\
&\qquad = 2x^5 + 5x^4 - 4x^3 + 8x^2 + 1 - 2x^5 + x^4 - x^3 \\
&\qquad = 6x^4 - 5x^3 + 8x^2 + 1
\end{aligned}$$

$$\text{Dividend} = \text{Divisor} \cdot \text{Quotient} + \text{Remainder}.$$

The same process can be used with polynomial division. In Example 1 above, you can easily verify that the divisor times the quotient is

$$(2x^2 - x + 1)(x^3 + 3x^2 - x + 2) = 2x^5 + 5x^4 - 4x^3 + 8x^2 - 3x + 2.$$

Adding the remainder $3x - 1$ to this result yields the original dividend:

$$(2x^5 + 5x^4 - 4x^3 + 8x^2 - 3x + 2) + (3x - 1)$$
$$= 2x^5 + 5x^4 - 4x^3 + 8x^2 + 1.$$

So just as with division of numbers we have:

$$\text{Dividend} = \text{Divisor} \cdot \text{Quotient} + \text{Remainder}.$$

Because this fact is so important it is given a special name and a formal statement:

The Division Algorithm ▶

If a polynomial $f(x)$ is divided by a nonzero polynomial $h(x)$, then there is a quotient polynomial $q(x)$ and a remainder polynomial $r(x)$ such that

$$f(x) = h(x)q(x) + r(x)$$

where either $r(x) = 0$ or $r(x)$ has degree less than the degree of the divisor $h(x)$.

The Division Algorithm can be used to determine if one polynomial is a factor of another polynomial.

▶ TECHNOLOGY TIP

The TI-92 does polynomial division (use PROPFRAC in the ALGEBRA menu). It displays the answer as the sum of a fraction and a polynomial:

$$\frac{\text{remainder}}{\text{divisor}} + \text{quotient}.$$

EXAMPLE 2 To determine if $2x^2 + 1$ is a factor of $6x^3 - 4x^2 + 3x - 2$, we divide:

$$\begin{array}{r|l} & 3x \quad - 2 \\ \hline 2x^2 + 1 & 6x^3 - 4x^2 + 3x - 2 \\ & 6x^3 \qquad\quad + 3x \\ \hline & \quad - 4x^2 \qquad\; - 2 \\ & \quad - 4x^2 \qquad\; - 2 \\ \hline & \qquad\qquad\qquad 0 \end{array}$$

Since the remainder is 0, the Division Algorithm tells us that:

$$\text{Dividend} = \text{Divisor} \cdot \text{Quotient} + \text{Remainder}$$

$$6x^3 - 4x^2 + 3x - 2 = (2x^2 + 1)(3x - 2) + 0$$
$$= (2x^2 + 1)(3x - 2).$$

Therefore, $2x^2 + 1$ is a factor of $6x^3 - 4x^2 + 3x - 2$ and the other factor is the quotient $3x - 2$. ■

The same argument works in the general case:

Remainders and Factors ▶

The remainder in polynomial division is 0 exactly when the divisor is a factor of the dividend. In this case the other factor is the quotient.

Remainders and Roots

When a polynomial $f(x)$ is divided by a first-degree polynomial, such as $x - 3$ or $x + 5$, the remainder is a constant (because constants are the only polynomials of degree less than 1, the degree of the divisor). For example, you can verify that when $f(x) = x^3 - 2x^2 - 4x + 5$ is divided by $x - 3$, the quotient is $x^2 + x - 1$ and the remainder is 2. Hence, by the Division Algorithm

$$\text{Dividend} = \text{Divisor} \cdot \text{Quotient} + \text{Remainder}$$

$$f(x) = x^3 - 2x^2 - 4x + 5 = (x - 3)(x^2 + x - 1) + 2$$

Observe that

$$f(3) = (3 - 3)(3^2 + 3 - 1) + 2 = 0 + 2 = 2.$$

Thus, value of $f(x)$ at 3 (namely, $f(3) = 2$) is the same as the remainder when $f(x)$ is divided by $x - 3$. A similar argument works in the general case and proves:

The Remainder Theorem ▶

If a polynomial $f(x)$ is divided by $x - c$, then the remainder is the number $f(c)$.

EXAMPLE 3 To find the remainder when $f(x) = x^{79} + 3x^{24} + 5$ is divided by $x - 1$, we apply the Remainder Theorem with $c = 1$. The remainder is

$$f(1) = 1^{79} + 3 \cdot 1^{24} + 5 = 1 + 3 + 5 = 9. \blacksquare$$

EXAMPLE 4 To find the remainder when $f(x) = 3x^4 - 8x^2 + 11x + 1$ is divided by $x + 2$, we must apply the Remainder Theorem *carefully*. The divisor in the theorem is $x - c$, not $x + c$. So we rewrite $x + 2$ as $x - (-2)$ and apply the theorem with $c = -2$. The remainder is

$$f(-2) = 3(-2)^4 - 8(-2)^2 + 11(-2) + 1 = 48 - 32 - 22 + 1 = -5. \blacksquare$$

If $f(x)$ is a polynomial, then a solution of the equation $f(x) = 0$ is called a **root** (or a **zero***) of the polynomial $f(x)$. Thus, a number c is a root of $f(x)$ if $f(c) = 0$. A root that is a real number is called a **real root.** As we saw in Section 1.5

*We use the term "root" because this is the term used in advanced algebra texts, but many mathematicians prefer the term "zero."

The real roots of a polynomial $f(x)$ (the solutions of $f(x) = 0$) are the x-intercepts of the graph of the polynomial function $f(x)$.

Techniques for finding roots of polynomials are discussed in Section 4.2. For now, we examine the connection between roots and factors.

EXAMPLE 5 The remainder when $f(x) = x^3 - 4x^2 + 2x + 3$ is divided by $x - 3$ is

$$f(3) = 3^3 - 4 \cdot 3^2 + 2 \cdot 3 + 3 = 27 - 36 + 6 + 3 = 0.$$

Since division by $x - 3$ leaves remainder 0, $x - 3$ is a *factor* of $f(x)$, as explained above. Division shows that the other factor is $x^2 - x - 1$, so that

$$f(x) = (x - 3)(x^2 + x + 1).$$

Furthermore, 3 is a *root* of $f(x)$ because $f(3) = 0$. ■

Example 5 is an illustration of:

The Factor Theorem ▶

The number c is a root of the polynomial $f(x)$ exactly when $x - c$ is a factor of $f(x)$.

Proof of the Factor Theorem Divide $f(x)$ by $x - c$. The Remainder Theorem shows that the remainder is $f(c)$. Hence, by the Division Algorithm,

$$f(x) = (x - c)q(x) + f(c).$$

If c is a root of $f(x)$, then $f(c) = 0$ and we have $f(x) = (x - c)q(x)$. Hence, $x - c$ is a factor. Conversely, if $x - c$ is a factor, then the remainder $f(c)$ must be 0, so c is a root. ❑

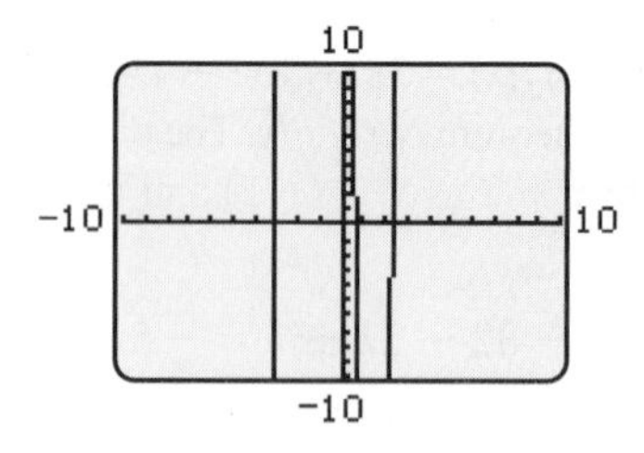

Figure 4-1

EXAMPLE 6 The Factor Theorem and a calculator can sometimes be used to factor polynomials. The graph of $f(x) = 15x^3 - x^2 - 114x + 72$ in the standard viewing window (Figure 4–1) is obviously not complete, but suggests that -3 is an x-intercept, and hence a root of $f(x)$. It is easy to verify that this is indeed the case:

$$f(-3) = 15(-3)^3 - (-3)^2 - 114(-3) + 72 = -405 - 9 + 342 + 72 = 0.$$

Since -3 is a root, $x - (-3) = x + 3$ is a factor of $f(x)$. Use synthetic or long division to verify that the other factor is $15x^2 - 46x + 24$. By factoring this quadratic, we obtain a complete factorization of $f(x)$:

$$f(x) = (x + 3)(15x^2 - 46x + 24) = (x + 3)(3x - 2)(5x - 12).$$ ■

EXAMPLE 7 Find three polynomials of different degrees that have 1, 2, 3, and -5 as roots.

Solution A polynomial that has 1, 2, 3 and -5 as roots must have $x-1$, $x-2$, $x-3$, and $x-(-5)=x+5$ as factors. Many polynomials satisfy these conditions, such as

$$g(x)=(x-1)(x-2)(x-3)(x+5)=x^4-x^3-19x^2+49x-30$$

$$h(x)=8(x-1)(x-2)(x-3)^2(x+5)$$

$$k(x)=2(x+4)^2(x-1)(x-2)(x-3)(x+5)(x^2+x+1).$$

Note that g has degree 4. When h is multiplied out, its leading term is $8x^5$, so h has degree 5. Similarly, k has degree 8 since its leading term is $2x^8$. ■

If a polynomial $f(x)$ has four roots, say a, b, c, d, then by the same argument used in Example 7, it must have

$$(x-a)(x-b)(x-c)(x-d)$$

as a factor. Since $(x-a)(x-b)(x-c)(x-d)$ has degree 4 (multiply it out—its leading term is x^4), $f(x)$ must have degree at least 4. In particular, this means that no polynomial of degree 3 can have four or more roots. A similar argument works in the general case.

Number of Roots ▶

A polynomial of degree n has at most n roots.

EXERCISES 4.1

In Exercises 1–8, determine whether the given algebraic expression is a polynomial. If it is, list its leading coefficient, constant term, and degree.

1. $1+x^3$ **2.** -7 **3.** $(x-1)(x^2+1)$

4. 7^x+2x+1 **5.** $(x+\sqrt{3})(x-\sqrt{3})$

6. $4x^2+3\sqrt{x}+5$ **7.** $\dfrac{7}{x^2}+\dfrac{5}{x}-15$

8. $(x-1)^k$ (where k is a fixed positive integer)

In Exercises 9–14, state the quotient and remainder when the first polynomial is divided by the second. Check your division by calculating (divisor) (quotient) + remainder.

9. $3x^4+2x^2-6x+1;\quad x+1$

10. $x^5-x^3+x-5;\quad x-2$

11. $x^5+2x^4-6x^3+x^2-5x+1;\quad x^3+1$

12. $3x^4-3x^3-11x^2+6x-1;\quad x^3+x^2-2$

13. $5x^4+5x^2+5;\quad x^2-x+1$

14. $x^5-1;\quad x-1$

In Exercises 15–18, determine whether the first polynomial is a factor of the second.

15. $x^2+3x-1;\quad x^3+2x^2-5x-6$

16. $x^2+9;\quad x^5+x^4-81x-81$

17. $x^2+3x-1;\quad x^4+3x^3-2x^2-3x+1$

18. $x^2-5x+7;\quad x^3-3x^2-3x+9$

In Exercises 19–22, determine which of the given numbers are roots of the given polynomial.

19. $2, 3, 0, -1;\quad g(x)=x^4+6x^3-x^2-30x$

20. $1, 1/2, 2, -1/2, 3;\quad f(x)=6x^2+x-1$

21. $2\sqrt{2}, \sqrt{2}, -\sqrt{2}, 1, -1;\quad h(x)=x^3+x^2-8x-8$

22. $\sqrt{3}, -\sqrt{3}, 1, -1;\quad k(x)=8x^3-12x^2-6x+9$

In Exercises 23–32, find the remainder when $f(x)$ is divided by $g(x)$, without *using division.*

23. $f(x)=x^{10}+x^8;\quad g(x)=x-1$

24. $f(x)=x^6-10;\quad g(x)=x-2$

25. $f(x) = 3x^4 - 6x^3 + 2x - 1; \quad g(x) = x + 1$

26. $f(x) = x^5 - 3x^2 + 2x - 1; \quad g(x) = x - 2$

27. $f(x) = x^3 - 2x^2 + 5x - 4; \quad g(x) = x + 2$

28. $f(x) = 10x^{75} - 8x^{65} + 6x^{45} + 4x^{32} - 2x^{15} + 5;$
$g(x) = x - 1$

29. $f(x) = 2x^5 - 3x^4 + x^3 - 2x^2 + x - 8;$
$g(x) = x - 10$

30. $f(x) = x^3 + 8x^2 - 29x + 44; \quad g(x) = x + 11$

31. $f(x) = 2x^5 - 3x^4 + 2x^3 - 8x - 8; \quad g(x) = x - 20$

32. $f(x) = x^5 - 10x^4 + 20x^3 - 5x - 95; \quad g(x) = x + 10$

In Exercises 33–38, use the Factor Theorem to determine whether or not $h(x)$ is a factor of $f(x)$.

33. $h(x) = x - 1; \quad f(x) = x^5 + 1$

34. $h(x) = x - 1/2; \quad f(x) = 2x^4 + x^3 + x - 3/4$

35. $h(x) = x + 2; \quad f(x) = x^3 - 3x^2 - 4x - 12$

36. $h(x) = x + 1; \quad f(x) = x^3 - 4x^2 + 3x + 8$

37. $h(x) = x - 1; \quad f(x) = 14x^{99} - 65x^{56} + 51$

38. $h(x) = x - 2; \quad f(x) = x^3 + x^2 - 4x + 4$

In Exercises 39–42, use the Factor Theorem and a calculator to factor the polynomial.

39. $f(x) = 6x^3 - 7x^2 - 89x + 140$

40. $g(x) = x^3 - 5x^2 - 5x - 6$

41. $h(x) = 4x^4 + 4x^3 - 35x^2 - 36x - 9$

42. $f(x) = x^5 - 5x^4 - 5x^3 + 25x^2 + 6x - 30$

In Exercises 43–46, each graph is of a polynomial function $f(x)$ of degree 5 whose leading coefficient is 1. The graph is not drawn to scale. Use the Factor Theorem to find the polynomial. [Hint: What are the roots of $f(x)$? What does the Factor Theorem tell you?]

43.

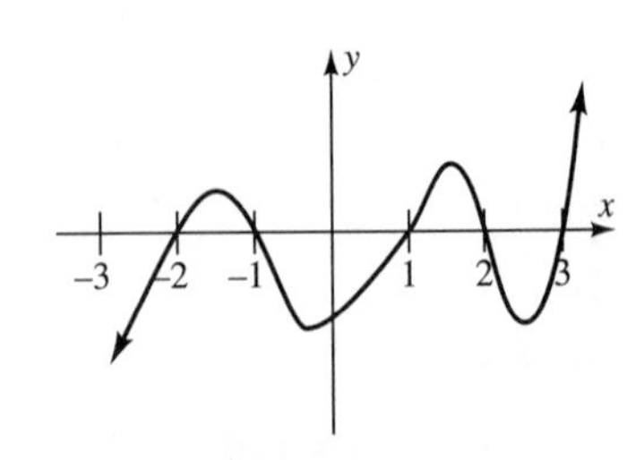

44.

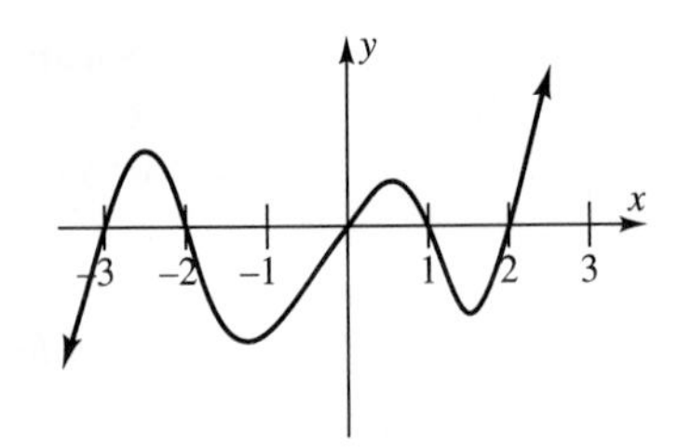

45.

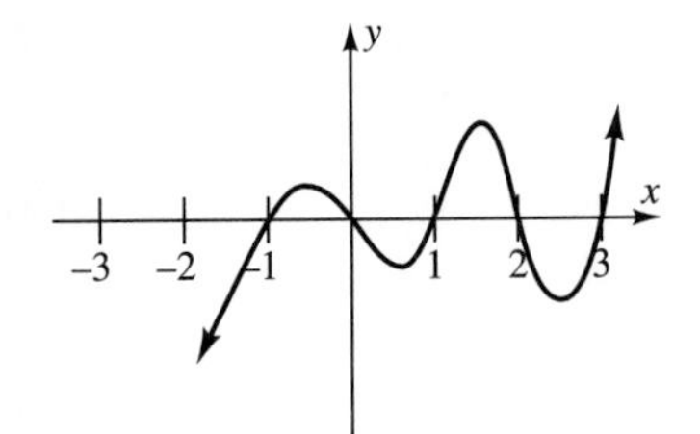

46.

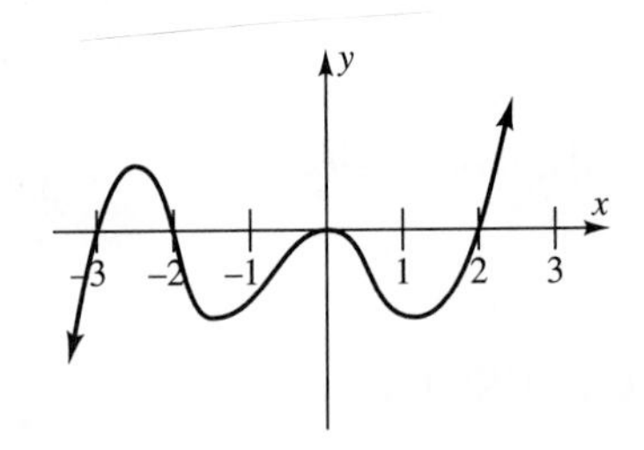

In Exercises 47–50, find a polynomial with the given degree n, the given roots, and no other roots.

47. $n = 3$; roots $1, 7, -4$

48. $n = 3$; roots $1, -1$

49. $n = 6$; roots $1, 2, \pi$

50. $n = 5$; root 2

51. Find a polynomial function f of degree 3 such that $f(10) = 17$ and the roots of $f(x)$ are 0, 5, and 8.

52. Find a polynomial function g of degree 4 such that the roots of g are $0, -1, 2, -3$ and $g(3) = 288$.

In Exercises 53–56, find a number k satisfying the given condition.

53. $x + 2$ is a factor of $x^3 + 3x^2 + kx - 2$.

54. $x - 3$ is a factor of $x^4 - 5x^3 - kx^2 + 18x + 18$.

55. $x - 1$ is a factor of $k^2x^4 - 2kx^2 + 1$.

56. $x + 2$ is a factor of $x^3 - kx^2 + 3x + 7k$.

57. Use the Factor Theorem to show that for every real number c, $x - c$ is *not* a factor of $x^4 + x^2 + 1$.

58. Let c be a real number and n a positive integer.
(a) Show that $x - c$ is a factor of $x^n - c^n$.
(b) If n is even, show that $x + c$ is a factor of $x^n - c^n$. [*Remember:* $x + c = x - (-c)$.]

59. (a) If c is a real number and n an odd positive integer, give an example to show that $x + c$ may not be a factor of $x^n - c^n$.
(b) If c and n are as in part (a), show that $x + c$ is a factor of $x^n + c^n$.

Thinkers

60. For what value of k is the difference quotient of $g(x) = kx^2 + 2x + 1$ equal to $7x + 2 - (3.5)h$?

61. For what value of k is the difference quotient of $f(x) = x^2 + kx$ equal to $2x + 5 + h$?

4.1.A *Excursion* SYNTHETIC DIVISION

Synthetic division is a fast method of doing polynomial division when the divisor is a first-degree polynomial of the form $x - c$ for some real number c. To see how it works, we first consider an example of ordinary long division:

$$
\begin{array}{rrrrrrl}
 & 3x^3 & +6x^2 & +4x & -3 & & \longleftarrow \textit{Quotient} \\
\textit{Divisor} \longrightarrow x-2 \,\big)\, & 3x^4 & & -8x^2 & -11x & +1 & \longleftarrow \textit{Dividend} \\
 & 3x^4 & -6x^3 & & & & \\
\hline
 & & 6x^3 & -8x^2 & & & \\
 & & 6x^3 & -12x^2 & & & \\
\hline
 & & & 4x^2 & -11x & & \\
 & & & 4x^2 & -8x & & \\
\hline
 & & & & -3x & +1 & \\
 & & & & -3x & +6 & \\
\hline
 & & & & & -5 & \longleftarrow \textit{Remainder}
\end{array}
$$

This calculation obviously involves a lot of repetitions. If we insert 0 coefficients for terms that don't appear above and keep the various coefficients in the proper columns, we can eliminate the repetitions and all the x's:

$$
\begin{array}{rrrrrrl}
 & 3 & 6 & 4 & -3 & & \longleftarrow \textit{Quotient} \\
\textit{Divisor} \longrightarrow 1-2 \,\big)\, & 3 & 0 & -8 & -11 & 1 & \longleftarrow \textit{Dividend} \\
 & & -6 & & & & \\
\hline
 & & 6 & & & & \\
 & & & -12 & & & \\
\hline
 & & & 4 & & & \\
 & & & & -8 & & \\
\hline
 & & & & -3 & & \\
 & & & & & +6 & \\
\hline
 & & & & & -5 & \longleftarrow \textit{Remainder}
\end{array}
$$

We can save space by moving the lower lines upward and writing 2 in the divisor position (since that's enough to remind us that the divisor is $x - 2$):

$$\begin{array}{r|rrrrrl} & 3 & 6 & 4 & -3 & & \longleftarrow \textit{Quotient} \\ \textit{Divisor} \longrightarrow 2 & 3 & 0 & -8 & -11 & 1 & \longleftarrow \textit{Dividend} \\ & & -6 & -12 & -8 & 6 & \\ \hline & & 6 & 4 & -3 & \boxed{-5} & \longleftarrow \textit{Remainder} \end{array}$$

Since the last line contains most of the quotient line, we can save more space and still preserve the essential information by inserting a 3 in the last line and omitting the top line:

$$\begin{array}{r|rrrrrl} \textit{Divisor} \longrightarrow 2 & 3 & 0 & -8 & -11 & 1 & \longleftarrow \textit{Dividend} \\ & & -6 & -12 & -8 & 6 & \\ \hline & 3 & 6 & 4 & -3 & \boxed{-5} & \longleftarrow \textit{Remainder} \end{array}$$

$\underbrace{3 \quad 6 \quad 4 \quad -3}_{\textit{Quotient}}$

Synthetic division is a quick method for obtaining the last row of this array. Here is a step-by-step explanation of the division of $3x^4 - 8x^2 - 11x + 1$ by $x - 2$:

STEP 1 In the first row list the 2 from the divisor and the coefficients of the dividend in order of decreasing powers of x (insert 0 coefficients for missing powers of x).

$$\begin{array}{r|rrrrr} 2 & 3 & 0 & -8 & -11 & 1 \end{array}$$

STEP 2 Bring down the first divided coefficient (namely, 3) to the third row.

$$\begin{array}{r|rrrrr} 2 & 3 & 0 & -8 & -11 & 1 \\ & & & & & \\ \hline & 3 & & & & \end{array}$$

STEP 3 Multiply $2 \cdot 3$ and insert the answer 6 in the second row, in the position shown here.

$$\begin{array}{r|rrrrr} 2 & 3 & 0 & -8 & -11 & 1 \\ & & 6 & & & \\ \hline & 3 & & & & \end{array}$$

STEP 4 Add $0 + 6$ and write the answer 6 in the third row.

$$\begin{array}{r|rrrrr} 2 & 3 & 0 & -8 & -11 & 1 \\ & & 6 & & & \\ \hline & 3 & 6 & & & \end{array}$$

STEP 5 Multiply $2 \cdot 6$ and insert the answer 12 in the second row.

$$\begin{array}{r|rrrrr} 2 & 3 & 0 & -8 & -11 & 1 \\ & & 6 & 12 & & \\ \hline & 3 & 6 & & & \end{array}$$

STEP 6 Add $-8 + 12$ and write the answer 4 in the third row.

$$\begin{array}{r|rrrrr} 2 & 3 & 0 & -8 & -11 & 1 \\ & & 6 & 12 & & \\ \hline & 3 & 6 & 4 & & \end{array}$$

STEP 7 Multiply $2 \cdot 4$ and insert the answer 8 in the second row.

$$\begin{array}{r|rrrrr} 2 & 3 & 0 & -8 & -11 & 1 \\ & & 6 & 12 & 8 & \\ \hline & 3 & 6 & 4 & & \end{array}$$

STEP 8 Add $-11 + 8$ and write the answer -3 in the third row.

$$\begin{array}{r|rrrrr} 2 & 3 & 0 & -8 & -11 & 1 \\ & & 6 & 12 & 8 & \\ \hline & 3 & 6 & 4 & -3 & \end{array}$$

STEP 9 Multiply $2\cdot(-3)$ and insert the answer -6 in the second row.

$$\begin{array}{r|rrrrr} 2 & 3 & 0 & -8 & -11 & 1 \\ & & 6 & 12 & 8 & -6 \\ \hline & 3 & 6 & 4 & -3 & \end{array}$$

STEP 10 Add $1 + (-6)$ and write the answer -5 in the third row.

$$\begin{array}{r|rrrrr} 2 & 3 & 0 & -8 & -11 & 1 \\ & & 6 & 12 & 8 & -6 \\ \hline & 3 & 6 & 4 & -3 & \boxed{-5} \end{array}$$

Except for the signs in the second row, this last array is the same as the array obtained from the long division process, and we can read off the quotient and remainder:

> **▶ TECHNOLOGY TIP**
> Synthetic division programs for TI calculators are in the Program Appendix.

The last number in the third row is the remainder.

The other numbers in the third row are the coefficients of the quotient (arranged in order of decreasing powers of x).

Since we are dividing the *fourth*-degree polynomial $3x^4 - 8x^2 - 11x + 1$ by the *first*-degree polynomial $x - 2$, the quotient must be a polynomial of degree *three* with coefficients $3, 6, 4, -3$, namely, $3x^3 + 6x^2 + 4x - 3$. The remainder is -5.

WARNING Synthetic division can be used *only* when the divisor is a first-degree polynomial of the form $x - c$. In the example above, $c = 2$. If you want to use synthetic division with a divisor such as $x + 3$, you must write it as $x - (-3)$, which is of the form $x - c$ with $c = -3$.

EXAMPLE 1 To divide $x^5 + 5x^4 + 6x^3 - x^2 + 4x + 29$ by $x + 3$, we write the divisor as $x - (-3)$ and proceed as above:

$$\begin{array}{r|rrrrrr} -3 & 1 & 5 & 6 & -1 & 4 & 29 \\ & & -3 & -6 & 0 & 3 & -21 \\ \hline & 1 & 2 & 0 & -1 & 7 & \boxed{8} \end{array}$$

The last row shows that the quotient is $x^4 + 2x^3 - x + 7$ and the remainder is 8. ■

EXAMPLE 2 Show that $x - 7$ is a factor of $8x^5 - 52x^4 + 2x^3 - 198x^2 - 86x + 14$ and find the other factor.

Solution $x - 7$ is a factor exactly when division by $x - 7$ leaves remainder 0, in which case the quotient is the other factor. Using synthetic division we have:

$$\begin{array}{r|rrrrrr} 7 & 8 & -52 & 2 & -198 & -86 & 14 \\ & & 56 & 28 & 210 & 84 & -14 \\ \hline & 8 & 4 & 30 & 12 & -2 & \boxed{0} \end{array}$$

Since the remainder is 0, the divisor $x - 7$ and the quotient $8x^4 + 4x^3 + 30x^2 + 12x - 2$ are factors:

$$8x^5 - 52x^4 + 2x^3 - 198x^2 - 86x + 14 = (x - 7)(8x^4 + 4x^3 + 30x^2 + 12x - 2).$$ ■

EXERCISES 4.1.A

In Exercises 1–8, use synthetic division to find the quotient and remainder.

1. $(3x^4 - 8x^3 + 9x + 5) \div (x - 2)$
2. $(4x^3 - 3x^2 + x + 7) \div (x - 2)$
3. $(2x^4 + 5x^3 - 2x - 8) \div (x + 3)$
4. $(3x^3 - 2x^2 - 8) \div (x + 5)$
5. $(5x^4 - 3x^2 - 4x + 6) \div (x - 7)$
6. $(3x^4 - 2x^3 + 7x - 4) \div (x - 3)$
7. $(x^4 - 6x^3 + 4x^2 + 2x - 7) \div (x - 2)$
8. $(x^6 - x^5 + x^4 - x^3 + x^2 - x + 1) \div (x + 3)$

In Exercises 9–12, use synthetic division to find the quotient and the remainder. In each divisor $x - c$, the number c is not an integer, but the same technique will work.

9. $(3x^4 - 2x^2 + 2) \div \left(x - \frac{1}{4}\right)$
10. $(2x^4 - 3x^2 + 1) \div \left(x - \frac{1}{2}\right)$
11. $(2x^4 - 5x^3 - x^2 + 3x + 2) \div \left(x + \frac{1}{2}\right)$
12. $\left(10x^5 - 3x^4 + 14x^3 + 13x^2 - \frac{4}{3}x + \frac{7}{3}\right) \div \left(x + \frac{1}{5}\right)$

In Exercises 13–16, use synthetic division to show that the first polynomial is a factor of the second and find the other factor.

13. $x + 4$; $3x^3 + 9x^2 - 11x + 4$
14. $x - 5$; $x^5 - 8x^4 + 17x^2 + 293x - 15$
15. $x - 1/2$; $2x^5 - 7x^4 + 15x^3 - 6x^2 - 10x + 5$
16. $x + 1/3$; $3x^6 + x^5 - 6x^4 + 7x^3 + 3x^2 - 15x - 5$

In Exercises 17 and 18, use a calculator and synthetic division to find the quotient and remainder.

17. $(x^3 - 5.27x^2 + 10.708x - 10.23) \div (x - 3.12)$
18. $(2.79x^4 + 4.8325x^3 - 6.73865x^2 + .9255x - 8.125) \div (x - 1.35)$

Thinkers

19. When $x^3 + cx + 4$ is divided by $x + 2$, the remainder is 4. Find c.
20. If $x - d$ is a factor of $2x^3 - dx^2 + (1 - d^2)x + 5$, what is d?

4.2 ROOTS OF POLYNOMIALS*

Finding the roots of polynomials is the same as solving polynomial equations. The root of a first-degree polynomial, such as $f(x) = 5x - 3$, is easily found by solving the first-degree equation $5x - 3 = 0$. Similarly, the roots of $f(x) =$

*In this section ''root'' means ''real root'' unless specified otherwise.

$ax^2 + bx + c$ are the solutions of the quadratic equation $ax^2 + bx + c = 0$. As we saw in Section 1.2 they can always be found by using the quadratic formula.

EXAMPLE 1 The roots of $f(x) = 3x^2 - 5x + 1$ are

$$x = \frac{-b \pm \sqrt{b^2 - 4ac}}{2a} = \frac{-(-5) \pm \sqrt{(-5)^2 - 4 \cdot 3 \cdot 1}}{2 \cdot 3} = \frac{5 \pm \sqrt{13}}{6}.$$ ■

Rational Roots

Although there are complicated analogues of the quadratic formula for finding the roots of third- and fourth-degree polynomials, there are no formulas that work for all polynomials of degree n when $n \geq 5$. However, when a polynomial has integer coefficients, you can find all of its **rational roots** (roots that are rational numbers) by using:

The Rational Root Test ▶

If a rational number r/s (in lowest terms) is a root of the polynomial

$$a_n x^n + \cdots + a_1 x + a_0,$$

where the coefficients $a_n, \ldots, a_1, a_0$ are integers with $a_n \neq 0$, $a_0 \neq 0$, then

r is a factor of the constant term a_0 and

s is a factor of the leading coefficient a_n.

The test states conditions that a rational root must satisfy.* Of course, there may be numbers satisfying these conditions that are not roots. By finding all the numbers that meet these conditions, we produce a list of the *possible* rational roots. It is then straightforward, but tedious, to evaluate the polynomial function at each number on the list to see if it actually is a root. As we shall see in the following examples, this testing process can be considerably shortened by using a graphing calculator and the fact that the real roots of a polynomial are the x-intercepts of its graph.

EXAMPLE 2 If $f(x) = 2x^5 - 3x^4 - 3x^3 + 2x^2 - 12x + 6$ has a rational root r/s, then by the Rational Root Test r must be a factor of the constant term 6. Therefore, r must be one of ± 1, ± 2, ± 3, or ± 6 (the only factors of 6). Similarly, s must be a factor of the leading coefficient 2, so s must be one of ± 1 or ± 2 (the only factors of 2). Consequently, the only *possibilities* for r/s are

$$\frac{\pm 1}{\pm 1}, \frac{\pm 2}{\pm 1}, \frac{\pm 3}{\pm 1}, \frac{\pm 6}{\pm 1}, \frac{\pm 1}{\pm 2}, \frac{\pm 2}{\pm 2}, \frac{\pm 3}{\pm 2}, \frac{\pm 6}{\pm 2}.$$

*Since the proof of the Rational Root Test sheds very little light on how the result is actually used to solve equations, it will be omitted.

Eliminating duplications from this list, we see that the only *possible* rational roots are

$$1,\ -1,\ 2,\ -2,\ 3,\ -3,\ 6,\ -6,\ \frac{1}{2},\ -\frac{1}{2},\ \frac{3}{2},\ -\frac{3}{2}.$$

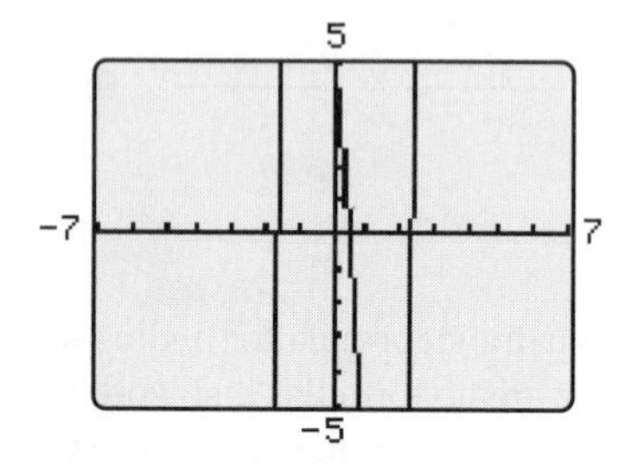

Figure 4-2

Now graph $f(x)$ in a viewing window that includes all of these numbers on the x-axis, say $-7 \le x \le 7$ and $-5 \le y \le 5$ (Figure 4–2). A complete graph isn't necessary since we are interested only in the x-intercepts. The only x-intercepts of the graph in this window are between -2 and -1, between 0 and 1, and between 2 and 3. Consequently, the only numbers on our list that could possibly be roots are $-3/2$ and $1/2$ and these are the only ones that need to be tested:

$$f(-3/2) = f(-1.5) = 2(-1.5)^5 - 3(-1.5)^4 - 3(-1.5)^3 + 2(-1.5)^2 - 12(-1.5) + 6 = 8.25$$

$$f(1/2) = f(.5) = 2(.5)^5 - 3(.5)^4 - 3(.5)^3 + 2(.5)^2 - 12(.5) + 6 = 0.$$

Therefore, 1/2 is the only rational root of $f(x)$. The two other real roots (x-intercepts) in Figure 4–2 must be irrational numbers. ■

EXAMPLE 3 If $g(x) = x^4 - 11x^2 - 2x + 12$ has a rational root r/s, then s must be a factor of the leading coefficient 1. Hence $s = \pm 1$, so that r/s is either r or $-r$. Since r must be a factor of the constant term 12, the only *possible* rational roots are $\pm 1, \pm 2, \pm 3, \pm 4, \pm 6, \pm 12$. Although the x-intercepts of the graph of $g(x) = x^4 - 11x^2 - 2x + 12$ in Figure 4–3 are not easily read, they are all between -4 and 4. Consequently, we change the viewing window and obtain Figure 4–4, where it is easy to see that the only numbers on our list that could possibly be rational roots are -3 and 1.

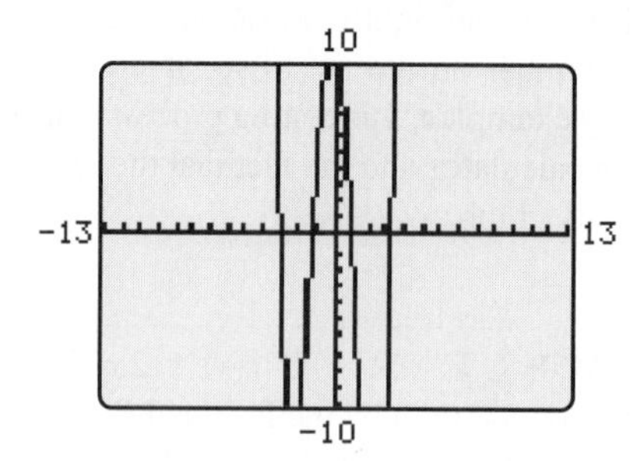

Figure 4-3

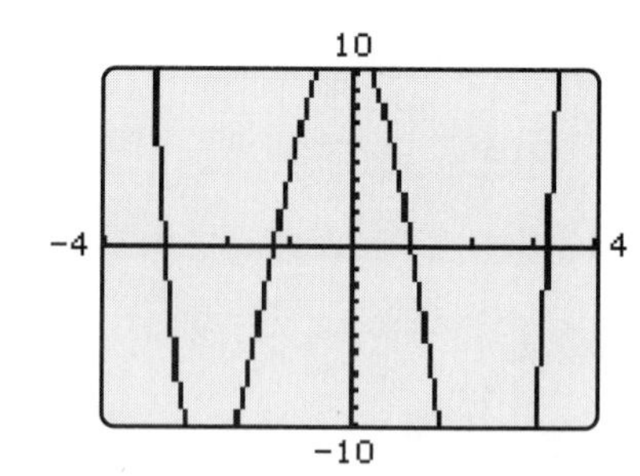

Figure 4-4

It is now easy to check that both -3 and 1 are roots of $g(x)$:

$$g(-3) = (-3)^4 - 11(-3)^2 - 2(-3) + 12 = 0$$

$$g(1) = 1^4 - 11 \cdot 1^2 - 2 \cdot 1 + 12 = 0.$$ ■

Roots and the Factor Theorem

Once some roots of a polynomial have been found, the Factor Theorem can be used to factor the polynomial, which may lead to additional roots.

EXAMPLE 4 In Example 3 we saw that 1 and -3 are roots of $g(x) = x^4 - 11x^2 - 2x + 12$. By the Factor Theorem, $x - 1$ and $x - (-3) = x + 3$ are factors of $g(x)$. Using synthetic or long division, we find the other factors and two more roots of $g(x)$:

$$x^4 - 11x^2 - 2x + 12 = 0$$

$$(x - 1)(x^3 + x^2 - 10x - 12) = 0$$

$$(x - 1)(x + 3)(x^2 - 2x - 4) = 0$$

$$x - 1 = 0 \quad \text{or} \quad x + 3 = 0 \quad \text{or} \quad x^2 - 2x - 4 = 0$$

$$x = 1 \qquad\qquad x = -3$$

Using the quadratic formula on the last equation, we have:

$$x = \frac{-(-2) \pm \sqrt{(-2)^2 - 4 \cdot 1 \cdot (-4)}}{2 \cdot 1}$$

$$= \frac{2 \pm \sqrt{20}}{2} = \frac{2 \pm 2\sqrt{5}}{2} = 1 \pm \sqrt{5}.$$

So, $g(x)$ has rational roots 1 and -3 and irrational roots $1 + \sqrt{5}$ and $1 - \sqrt{5}$. ■

When multiplied out, the polynomial $f(x) = (x - 5)^2(x - 7)^3$ has degree 5, but it is clear that it only has two roots, 5 and 7, each of which is a **repeated root.*** When using the Rational Root Test, you should check each rational root you find to see if it is a repeated root, as in the next example.

EXAMPLE 5 To find the roots of $f(x) = 3x^5 - 17x^4 + 31x^3 - 18x^2 + 4x - 8$, we first check for rational ones. The possible rational roots are

$$\pm 1, \ \pm 2, \ \pm 4, \ \pm 8, \ \pm\frac{1}{3}, \ \pm\frac{2}{3}, \ \pm\frac{4}{3}, \ \pm\frac{8}{3}.$$

GRAPHING EXPLORATION Graph $f(x)$ in the viewing window with $-9 \le x \le 9$ and $-6 \le y \le 6$ and verify that the only number on the list that could possibly be a root is 2. Then show that 2 actually is a root of $f(x)$ by evaluating $f(2)$.

*We say that 5 is a root of **multiplicity** 2 because the factor $x - 5$ appears twice in the factorization of $f(x)$. Similarly, 7 is a root of **multiplicity** 3.

By the Factor Theorem, $x - 2$ is a factor of $f(x)$. Synthetic or long division can be used to find the other factor:

$$f(x) = (x - 2)(3x^4 - 11x^3 + 9x^2 + 4).$$

The other roots of $f(x)$ are the roots of $g(x) = 3x^4 - 11x^3 + 9x^2 + 4$. Evaluating $g(x)$ at 2, we have $g(2) = 3 \cdot 2^4 - 11 \cdot 2^3 + 9 \cdot 2^2 + 4 = 0$. Therefore, 2 is also a root of $g(x)$ and $x - 2$ is a factor. Dividing $g(x)$ by $x - 2$ leads to this factorization:

$$f(x) = (x - 2)(3x^4 - 11x^3 + 9x^2 + 4) = (x - 2)(x - 2)(3x^3 - 5x^2 - x - 2).$$

Evaluating $h(x) = 3x^3 - 5x^2 - x - 2$ shows that 2 is a root of $h(x)$ and hence $x - 2$ is a factor. Dividing $h(x)$ by $x - 2$ produces

$$\begin{aligned} f(x) = (x - 2)^2(3x^3 - 5x^2 - x - 2) &= (x - 2)^2(x - 2)(3x^2 + x + 1) \\ &= (x - 2)^3(3x^2 + x + 1). \end{aligned}$$

The remaining roots of $f(x)$ are the roots of the factor $3x^2 + x + 1$. The quadratic formula shows that this polynomial has no real roots. Therefore, the repeated root 2 is the only real root of $f(x)$. ■

Note A calculator can be used to check the accuracy of factoring. For instance, if you have factored $f(x) = x^5 - 8x^4 + 16x^3 - 5x^2 + 4x - 20$ as $(x - 2)^2(x - 5)(x^2 + x + 1)$, then graph both of the equations

$$y = x^5 - 8x^4 + 16x^3 - 5x^2 + 4x - 20$$
$$y = (x - 2)^2(x - 5)(x^2 + x + 1)$$

on the same screen. If the two graphs coincide, then the factorization is correct. If the graphs differ, then the factorization is in error and should be recalculated.

Irrational Roots

The Rational Root Test enables us to find all the rational roots of a polynomial. Sometimes, we can also find the roots that are irrational numbers, as in Example 4. Even when all the roots of a polynomial cannot be found exactly, they can always be approximated by the graphical techniques of Section 1.5.

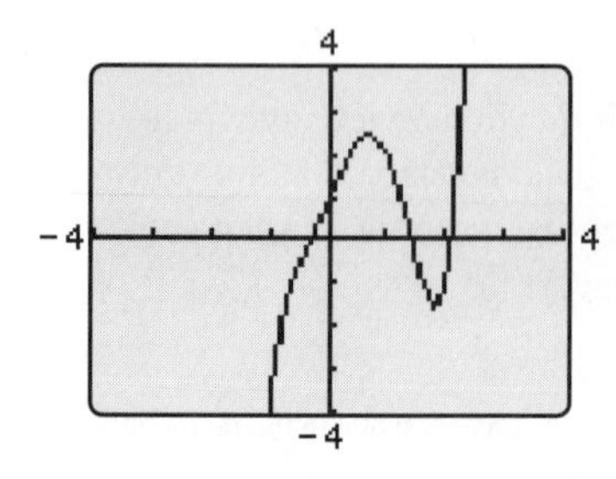

Figure 4-5

EXAMPLE 6 The only possible rational roots of $f(x) = x^5 - 2x^4 - x^3 + 3x + 1$ are ± 1 (why?) and it is easy to verify that neither is actually a root. If $f(x)$ has any real roots, they must be irrational. The graph of $f(x)$ in Figure 4–5 has three x-intercepts, one between -1 and 0, one between 1 and 2, and one between 2 and 3. Therefore, $f(x)$ has at least three real roots. Using zoom-in or a root finder we see that these roots are $-.336$, 1.427, and 2.101 (with a maximum error of .01). ■

How can we be sure that the three roots found in Example 6 are the only real roots of $f(x) = x^5 - 2x^4 - x^3 + 3x + 1$? One way might be a graph $f(x)$ in very large viewing windows. If no additional x-intercepts were found, we might conclude that there were no more real roots. However, there is an easy algebraic way to prove that all the real roots of $f(x)$ are between -1 and 3, and hence that the roots found in Example 6 are the only real roots. Before presenting this technique, we introduce some terminology.

Bounds

Since a polynomial of degree n has at most n real roots, it has a smallest and a largest real root (unless it has no real roots). Hence, there are **lower** and **upper bounds** for the real roots, that is, numbers r and s such that $r < s$ and every root lies between r and s.* Here is a technique for finding such bounds.

EXAMPLE 7 Prove that all the real roots of $f(x) = x^5 - 2x^4 - x^3 + 3x + 1$ lie between -1 and 3.

Solution We first prove that 3 is an upper bound for the roots of $f(x)$ as follows. Use synthetic division** to divide $f(x)$ by $x - 3$:

$$\begin{array}{r|rrrrrr} 3 & 1 & -2 & -1 & 0 & 3 & 1 \\ & & 3 & 3 & 6 & 18 & 63 \\ \hline & 1 & 1 & 2 & 6 & 21 & \boxed{64} \end{array}$$

Thus, the quotient is $x^4 + x^3 + 2x^2 + 6x + 21$ and the remainder is 64. Applying the Division Algorithm, we have:

$$f(x) = (x - 3)(x^4 + x^3 + 2x^2 + 6x + 21) + 64.$$

When $x > 3$, then the factor $x - 3$ is positive and the quotient $x^4 + x^3 + 2x^2 + 6x + 21$ is also positive (because all its coefficients are). The remainder 64 is also positive. Therefore, $f(x)$ is positive whenever $x > 3$. In particular, $f(x)$ is never zero when $x > 3$ and so there are no roots of $f(x)$ greater than 3. Hence 3 is an upper bound for the roots of $f(x)$.

To prove that -1 is a lower bound for the roots, divide $f(x)$ by $x - (-1) = x + 1$:

$$\begin{array}{r|rrrrrr} -1 & 1 & -2 & -1 & 0 & 3 & 1 \\ & & -1 & 3 & -2 & 2 & -5 \\ \hline & 1 & -3 & 2 & -2 & 5 & \boxed{-4} \end{array}$$

Read off the quotient and remainder and apply the Division Algorithm:

$$f(x) = (x + 1)(x^4 - 3x^3 + 2x^2 - 2x + 5) - 4.$$

When $x < -1$, then the factor $x + 1$ is negative. When x is negative, its odd powers are negative and its even powers are positive. Consequently, the quotient

*The bounds are not unique. Any number smaller than r is also a lower bound and any number larger than s is also an upper bound.

**If you haven't read Excursion 4.1.A, use long division to find the quotient and remainder.

$x^4 - 3x^3 + 2x^2 - 2x + 5$ is positive (because the odd powers of x are multiplied by negative coefficients). The product of the positive quotient with the negative factor $x + 1$ is negative. The remainder -4 is also negative. Hence, $f(x)$ is negative whenever $x < -1$. So there are no real roots less than -1, and -1 is a lower bound for the real roots of $f(x)$. ■

The technique in Example 7 works because the coefficients of the quotient and the remainder (the last row in the synthetic division process) have the right signs (all nonnegative for the upper bound and alternating for the lower bound). The same is true in the general case:

Bounds Test ▶

> **Let $f(x)$ be a polynomial with positive leading coefficient.**
>
> **If $d > 0$ and every number in the last row in the synthetic division of $f(x)$ by $x - d$ is nonnegative,* then d is an upper bound for the real roots of $f(x)$.**
>
> **If $c < 0$ and the numbers in the last row of the synthetic division of $f(x)$ by $x - c$ are alternately positive and negative (with 0 considered as either,** then c is a lower bound for the real roots of $f(x)$.**

EXAMPLE 8 Find all the real roots of $f(x) = 2x^5 - 3x^4 - 3x^3 + 2x^2 - 12x + 6$.

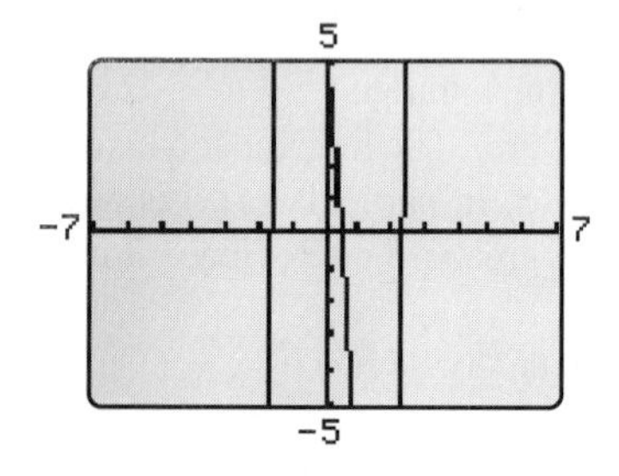

Figure 4-6

Solution First, apply the Rational Root Test to find the rational roots. This was done in Example 2, where we used the graph in Figure 4–6 and found that 1/2 is the only rational root. Verify that 1/2 is not a repeated root (Exercise 37). Figure 4–6 shows x-intercepts only between -2 and 3, which suggests that these numbers might be lower and upper bounds for the roots of $f(x)$. Synthetic division and the Bounds Test show that this is indeed the case:

-2	2	-3	-3	2	-12	6
		-4	14	-22	40	-56
	2	-7	11	-20	28	-50

Alternating Signs
-2 is a lower bound

3	2	-3	-3	2	-12	6
		6	9	18	60	144
	2	3	6	20	48	150

All Nonnegative
3 is an upper bound

Therefore, the three x-intercepts in Figure 4–6 are the only real roots of $f(x)$. Zoom-in or a root finder shows that the other two roots are -1.626 and 2.331. ■

*Equivalently, all the coefficients of the quotient and the remainder are nonnegative.

**Equivalently, the coefficients of the quotient are alternatively positive and negative, with the last one and the remainder having opposite signs.

Summary

The strategy used to find all the roots of a polynomial, as illustrated in the preceding examples, can be summarized as follows.

1. Use the Rational Root Test to list the possible rational roots of $f(x)$. Graph $f(x)$ to see which numbers on the list could possibly be roots and test each one of these to find the rational roots [Examples 2, 3, 5].
2. Use the Factor Theorem and division to write $f(x)$ as a product of linear factors (one for each rational root) and another factor $g(x)$ [Example 4].
3. The remaining roots of $f(x)$ are the roots of the factor $g(x)$. If $g(x)$ is quadratic, use the quadratic formula to find its roots [Examples 1, 4].
4. If $g(x)$ has degree 3 or more, check for repeated rational roots by evaluating $g(x)$ at each of the rational roots found in step 1.
   (a) If repeated roots are found, use the Factor Theorem and division to factor further and write $f(x)$ as a product of linear factors (one for each rational root, repeated as often as necessary) and a factor $h(x)$ (possibly a constant) which has no rational roots [Example 5].
   (b) If no repeated roots are found, proceed to step 5 (using $g(x)$ in place of $h(x)$ there) [Example 8].
5. The remaining roots of $f(x)$ are the roots of $h(x)$. Use the Bounds Test to find upper and lower bounds for its roots [Examples 7, 8] and then graphically approximate these roots [Examples 6, 8].

Shortcuts and variations in these steps are sometimes possible. For instance, some fourth-degree polynomials can be easily factored. Similarly, if the graph of a cubic shows three x-intercepts, then it has three different roots (the maximum possible) and there is no point in checking for repeated roots or bounds.

► TECHNOLOGY TIP

All the real roots of a polynomial can be found simultaneously by using POLY on TI-85, SOLVE on TI-92 (in the ALGEBRA menu), or POLYROOT on HP-38 (in the POLYNOMIAL submenu of the MATH menu). The roots may be found one at a time with other solvers (Casio 9800 can do quadratics and cubics in one step). In many cases, you may get decimal approximations rather than exact answers for some rational roots, square roots, etc.

EXERCISES 4.2

Directions: *When asked to find the roots of a polynomial, find exact roots whenever possible and approximate the other roots with a maximum error of .01.*

In Exercises 1–12, find all the rational roots of the polynomial.

1. $x^3 + 3x^2 - x - 3$

2. $x^3 - x^2 - 3x + 3$

3. $x^3 + 5x^2 - x - 5$

4. $3x^3 + 8x^2 - x - 20$

5. $f(x) = 2x^5 + 5x^4 - 11x^3 + 4x^2$ [*Hint:* The Rational Root Test can only be used on polynomials with nonzero constant terms. Factor $f(x)$ as a product of a power of x and a polynomial $g(x)$ with nonzero constant term. Then use the Rational Root Test on $g(x)$.]

6. $2x^6 - 3x^5 - 7x^4 - 6x^3$

7. $f(x) = \frac{1}{12}x^3 - \frac{1}{12}x^2 - \frac{2}{3}x + 1$ [*Hint:* The Rational Root Test can only be used on polynomials with integer coefficients. Note that $f(x)$ and $12f(x)$ have the same roots. (Why?)]

8. $\frac{2}{3}x^4 + \frac{1}{2}x^3 - \frac{5}{4}x^2 - x - \frac{1}{6}$

9. $\frac{1}{3}x^4 - x^3 - x^2 + \frac{13}{3}x - 2$

10. $\frac{1}{3}x^7 - \frac{1}{2}x^6 - \frac{1}{6}x^5 + \frac{1}{6}x^4$ **11.** $.1x^3 - 1.9x + 3$

12. $.05x^3 + .45x^2 - .4x + 1$

In Exercises 13–18, factor the polynomial as a product of linear factors and a factor g(x) such that g(x) is either a constant or a polynomial that has no rational roots.

13. $2x^3 - 4x^2 + x - 2$ **14.** $6x^3 - 5x^2 + 3x - 1$

15. $x^6 + 2x^5 + 3x^4 + 6x^3$

16. $x^5 - 2x^4 + 2x^3 - 3x + 2$

17. $x^5 - 4x^4 + 8x^3 - 14x^2 + 15x - 6$

18. $x^5 + 4x^3 + x^2 + 6x$

In Exercises 19–22, use the Bounds Test to find lower and upper bounds for the real roots of the polynomial.

19. $x^3 + 2x^2 - 7x + 20$ **20.** $x^3 - 15x^2 - 16x + 12$

21. $-x^5 - 5x^4 + 9x^3 + 18x^2 - 68x + 176$ [*Hint:* The Bounds Test applies only to polynomials with positive leading coefficient. The polynomial $f(x)$ has the same roots as $-f(x)$ (why?).]

22. $-.002x^3 - 5x^2 + 8x - 3$

In Exercises 23–36, find all real roots of the polynomial.

23. $2x^3 - 5x^2 + x + 2$ **24.** $t^4 - t^3 + 2t^2 - 4t - 8$

25. $6x^3 - 11x^2 + 6x - 1$ **26.** $z^3 + z^2 + 2z + 2$

27. $x^4 + x^3 - 19x^2 + 32x - 12$

28. $3x^5 + 2x^4 - 7x^3 + 2x^2$

29. $2x^5 - x^4 - 10x^3 + 5x^2 + 12x - 6$

30. $x^5 - x^3 + x$

31. $x^6 - 4x^5 - 5x^4 - 9x^2 + 36x + 45$

32. $x^5 + 3x^4 - 4x^3 - 11x^2 - 3x + 2$

33. $3x^4 + 2x^3 - 4x^2 + 4x - 1$

34. $x^5 + 8x^4 + 20x^3 + 9x^2 - 27x - 27$

35. $x^4 - 48x^3 - 101x^2 + 49x + 50$

36. $3x^7 + 8x^6 - 13x^5 - 36x^4 - 10x^3 + 21x^2 + 41x + 10$

37. Example 2 showed that 1/2 is the only rational root of $f(x) = 2x^5 - 3x^4 - 3x^3 + 2x^2 - 12x + 6$. Hence, $x - \frac{1}{2}$ is a factor of $f(x)$ (why?).
(a) Use long or synthetic division to find the other factor $g(x)$.
(b) Show that 1/2 is not a root of $g(x)$ and hence is not a repeated root of $f(x)$.

38. Graph $f(x) = .001x^3 - .199x^2 - .23x + 6$ in the standard viewing window.
(a) How many roots does $f(x)$ appear to have? Without changing the viewing window, explain why $f(x)$ must have an additional root. [*Hint:* Each root corresponds to a factor of $f(x)$. What does the rest of the factorization consist of?]
(b) Find all the roots of $f(x)$.

39. **(a)** Show that $\sqrt{2}$ is an irrational number. [*Hint:* $\sqrt{2}$ is a root of $x^2 - 2$. Does this polynomial have any rational roots?]
(b) Show that $\sqrt{3}$ is irrational.

40. During the first 150 hours of an experiment, the growth rate of a bacteria population at time t hours is $g(t) = -.0003t^3 + .04t^2 + .3t + .2$ bacteria per hour.
(a) What is the growth rate at 50 hours? At 100 hours?
(b) What is the growth rate at 145 hours? What does this mean?
(c) At what time is the growth rate 0?
(d) At what time is the growth rate -50 bacteria per hour?
(e) Approximately at what time does the highest growth rate occur?

41. An open-top reinforced box is to be made from a 12-by-36-inch piece of cardboard by cutting along the marked lines, discarding the shaded pieces, and folding, as shown in Figure 4–7. If the box must be less than 2.5 inches high, what size squares should be cut from the corners in order for the box to have a volume of 448 cubic inches?

42. A box with a lid is to be made from a 48-by-24-inch piece of cardboard by cutting and folding, as shown in

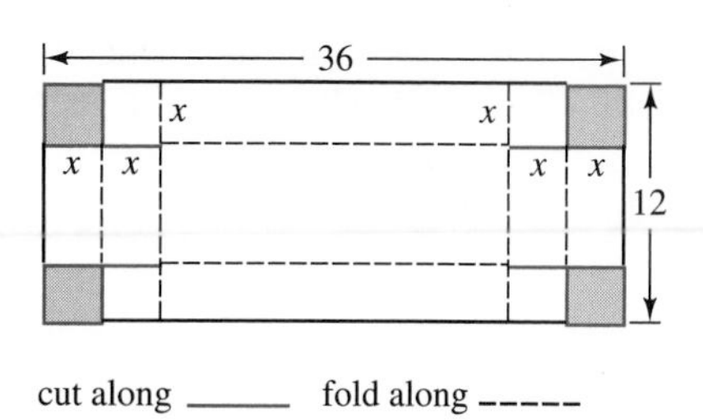

Figure 4-7

Figure 4–8. If the box must be at least 6 inches high, what size squares should be cut from the two corners in order for the box to have a volume of 1000 cubic inches?

43. In a sealed chamber where the temperature varies, the instantaneous rate of change of temperature with respect to time over an 11-day period is given by $F(t) = .0035t^4 - .4t^2 - .2t + 6$, where time is measured in days and temperature in degrees Fahrenheit (so that rate of change is in degrees per day).

(a) At what rate is the temperature changing at the beginning of the period ($t = 0$)? At the end of the period ($t = 11$)?

(b) When is the temperature increasing at a rate of 4 degrees per day?

(c) When is the temperature decreasing at a rate of 3 degrees per day?

(d) When is the temperature decreasing at the fastest rate?

44. (a) If c is a root of $f(x) = 5x^4 - 4x^3 + 3x^2 - 4x + 5$, show that $1/c$ is also a root.

(b) Do part (a) with $f(x)$ replaced by $g(x) = 2x^6 + 3x^5 + 4x^4 - 5x^3 + 4x^2 + 3x + 2$.

(c) Let $f(x) = a_{12}x^{12} + a_{11}x^{11} + \cdots + a_2x^2 + a_1x + a_0$. What conditions must the coefficients a_i satisfy in order that this statement be true: If c is a root of $f(x)$, then $1/c$ is also a root.

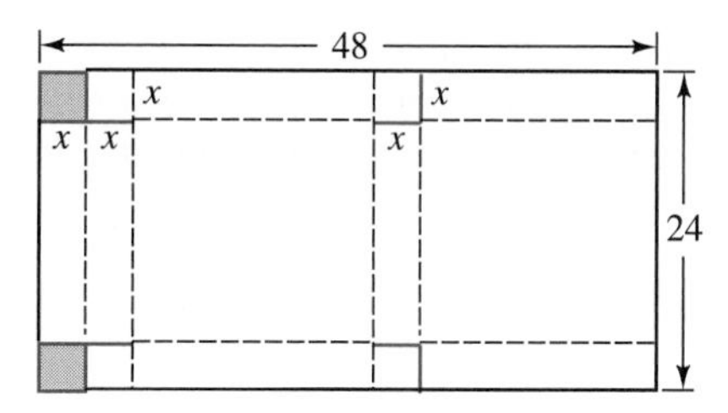

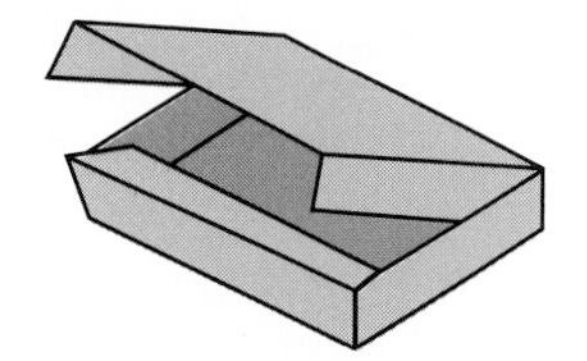

Figure 4-8

4.3 GRAPHS OF POLYNOMIAL FUNCTIONS

The graphs of first-degree polynomial functions are straight lines (Section 3.1) and the graphs of second-degree polynomial functions are parabolas (Section 3.3).* The emphasis here will be on higher degree polynomial functions.

When looking at polynomial graphs, particularly on a calculator screen, remember that every such graph is *continuous,* meaning that you can draw the

*It is not necessary to have read Sections 3.1 and 3.3 in order to understand the material here.

graph between any two points without lifting your pencil from the paper. In other words, the graph is an unbroken curve, with no jumps, gaps, or holes. In addition, polynomial graphs have no sharp corners. Thus none of the graphs in Figure 4–9 is the graph of a polynomial function.

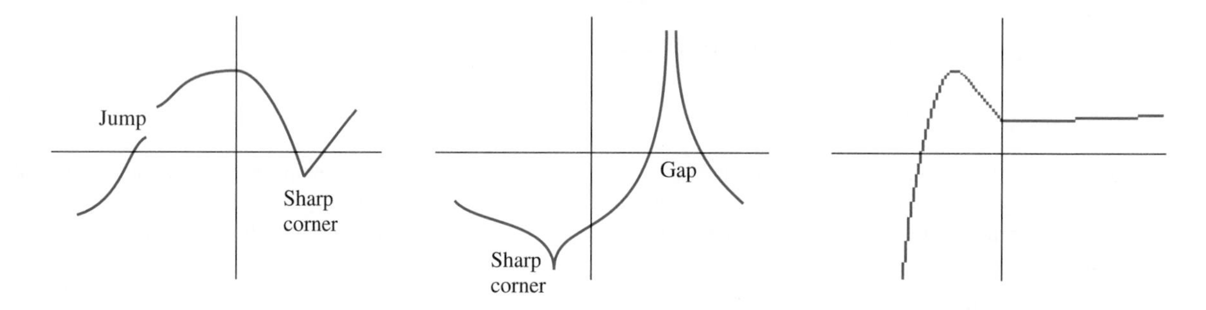

Figure 4-9

On a calculator screen, however, a polynomial graph may look like the graph on the right in Figure 4–9 rather than a continuous curve.

Monomial Graphs

The simplest polynomials are the monomials, those of the form $f(x) = ax^n$ (where a is a constant). Their graphs are of four types, as shown in the following chart. You should verify the accuracy of this summary by graphing the examples in the viewing window with $-5 \le x \le 5$ and $-30 \le y \le 30$.

Graph of $f(x) = ax^n$

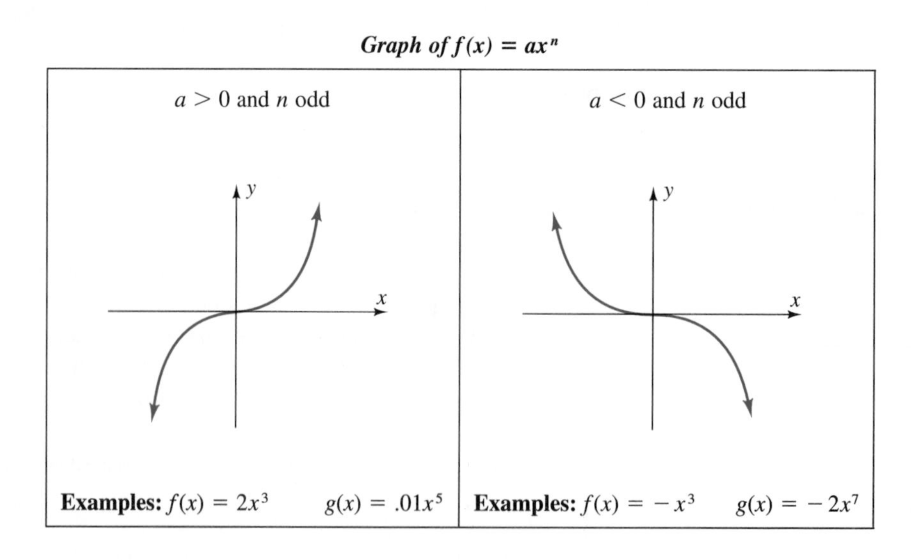

Graph of f(x) = axⁿ

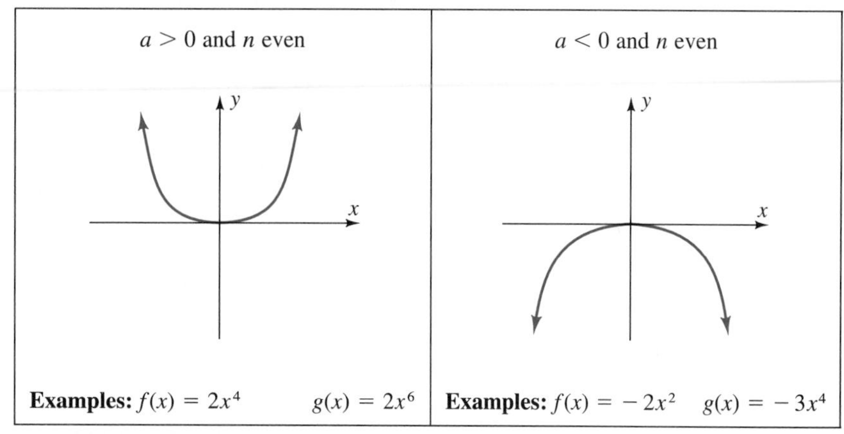

Properties of Polynomial Graphs

The graphs of more complicated polynomial functions, which can vary considerably, are largely determined by the following factors:

1. The shape of the graph when $|x|$ is large (that is, the shape of the graph far from the origin).
2. The x-intercepts of the graph.
3. The ''peaks and valleys'' of the graph.
4. The ''bending'' of the graph.

Understanding these characteristics, each of which is discussed below, will assist you in correctly interpreting the images on the calculator screen and determining when a polynomial graph is complete.

Shape of the Graph When $|x|$ is Large The shape of a polynomial graph at the far left and far right of the coordinate plane is easily determined by using our knowledge of graphs of functions of the form $f(x) = ax^n$.

EXAMPLE 1 Consider the function $f(x) = 2x^3 + x^2 - 6x$ and the function determined by its leading term $g(x) = 2x^3$.

GRAPHING EXPLORATION Using the standard viewing window, graph f and g on the same screen.

Do the graphs look different? Now graph f and g in the viewing window with $-20 \le x \le 20$ and $-10{,}000 \le y \le 10{,}000$. Do the graphs look almost the same?

Finally, graph f and g in the viewing window with $-100 \le x \le 100$ and $-1{,}000{,}000 \le y \le 1{,}000{,}000$. Do the graphs look virtually identical?

The reason the answer to the last question is "yes" can be understood from this table:

x	-100	-50	70	100
$-6x$	600	300	-420	-600
x^2	10,000	2,500	4,900	10,000
$g(x) = 2x^3$	$-2{,}000{,}000$	$-250{,}000$	686,000	2,000,000
$f(x) = 2x^3 + x^2 - 6x$	$-1{,}989{,}400$	$-247{,}200$	690,480	2,009,400

It shows that when $|x|$ is large, the terms x^2 and $-6x$ are insignificant compared with $2x^3$ and play a very minor role in determining the value of $f(x)$. Hence the values of $f(x)$ and $g(x)$ are relatively close. ■

Example 1 is typical of what happens in every case: When $|x|$ is very large, the highest power of x totally overwhelms all lower powers and plays the greatest role in determining the value of the function.

Behavior When $|x|$ is Large ▶

When $|x|$ is very large, the graph of a polynomial function closely resembles the graph of its highest degree term.

In particular, when the polynomial function has odd degree, one end of its graph shoots upward and the other end downward. When the polynomial function has even degree, both ends of its graph shoot upward or both ends shoot downward.

x-Intercepts As we saw in Section 4.1, the x-intercepts of the graph of a polynomial function are the real roots of the polynomial. Since a polynomial of degree n has at most n roots (see page 215), we have:

x-Intercepts ▶

The graph of a polynomial function of degree n meets the x-axis at most n times.

Peaks and Valleys Some polynomial graphs always rise from left to right and others always fall from left to right. The typical polynomial graph, however, rises in some places and falls in others, so that the graph has "peaks" and "valleys" (Figure 4-10):

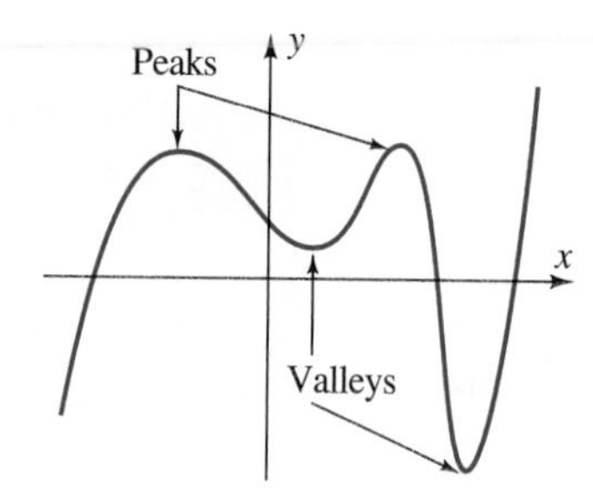

Figure 4-10

A peak is not necessarily the highest point on the graph, but it is the highest point in its locality. Similarly, a valley is the lowest point in the locality, but not necessarily the lowest point on the graph.

In more formal terminology, we say that a function f has a **local maximum** (or **relative maximum**) at $x = c$ and that $f(c)$ is a local maximum value of f provided that

$$f(x) \leq f(c) \qquad \text{for all } x \text{ in some open interval around } c.$$

What this means is that the graph of f has a peak at $(c, f(c))$, as shown in Figure 4–11.

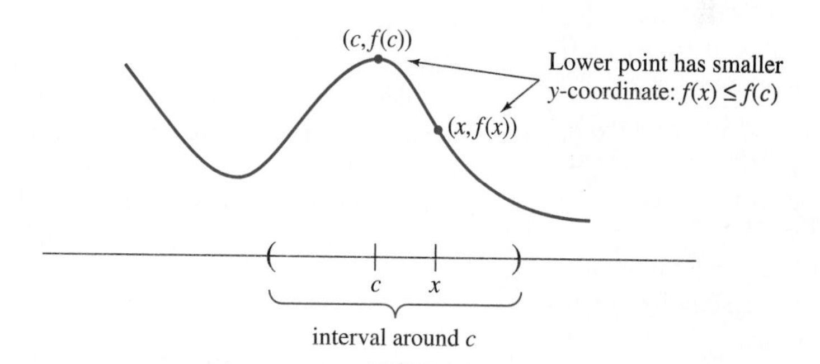

Figure 4-11

Similarly we say that a function f has a **local minimum** (or **relative minimum**) at $x = d$ and that $f(d)$ is a local minimum value of f provided that

$$f(x) \geq f(d) \qquad \text{for all } x \text{ in some open interval around } d.$$

In other words, the graph of f has a valley at $(d, f(d))$ (all nearby points have larger y-coordinates).

The term **local extremum** (plural, extrema) refers to either a local maximum (plural, maxima) or a local minimum (plural, minima). So at a local extremum, the graph has either a peak or a valley.

> GRAPHING EXPLORATION Graph $f(x) = x^3 + 2x^2 - 4x - 3$ in the standard viewing window. What is the total number of peaks and valleys on the graph? What is the degree of $f(x)$? Now graph $g(x) = x^4 - 3x^3 - 2x^2 + 4x + 5$ in the standard viewing window. What is the total number of peaks and valleys on the graph? What is the degree of $g(x)$? ■

The two polynomials you have just graphed are illustrations of the following fact, which is proved in calculus:

Local Extrema ▶

A polynomial function of degree n has at most $n - 1$ local extrema. In other words, the total number of peaks and valleys on the graph is at most $n - 1$.

Calculus is usually needed to determine the exact locations of local extrema, but a calculator can be used to obtain very accurate approximations.

> **▶ TECHNOLOGY TIP**
> The graphical min/max finder is in the CALC menu on TI-82/83, in the MATH submenu of the GRAPH menu on TI-85 and 92, in the G-SOLVE menu on Casio 9800, and in the JUMP menu on Sharp 9300. It is labeled EXTREMUM in the HP-38 PLOT FCN menu. A min/max finder program for TI-81 is in the Program Appendix.

EXAMPLE 2 Find the local extrema of $f(x) = .2x^3 + .1x^2 - 4x + 1$.

Solution The graph of $f(x)$ in the standard viewing window (Figure 4–12) shows one peak and one valley. These are the only local extrema of f because $f(x)$ has degree 3 and therefore has at most 2 local extrema. Approximations of the locations of the local maximum and local minimum can be found by using zoom-in or by using the graphical min/max finder on some calculators. The max/min finder on a TI-82 shows that f has a local minimum at $x \approx 2.4207$ and that the local minimum value is approximately $f(2.4207) \approx -5.2599$ (Figure 4–13).

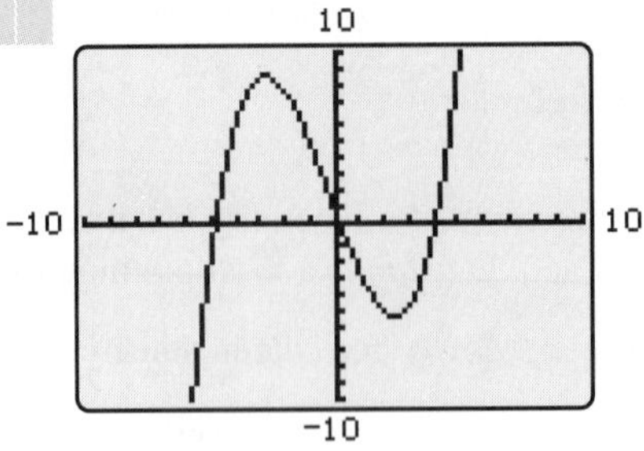

Figure 4-12

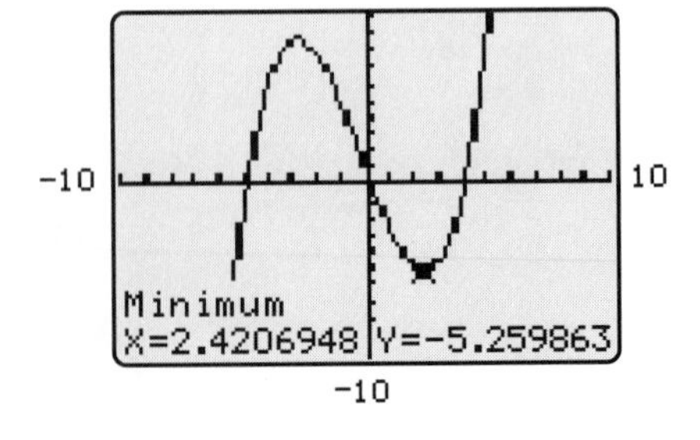

Figure 4-13

Similarly, f has a local maximum at $x \approx -2.7540$ and a local maximum value of approximately $f(-2.7540) \approx 8.5969$. ■

Bending A polynomial graph may bend upward or downward as indicated here by the vertical arrows (Figure 4–14):

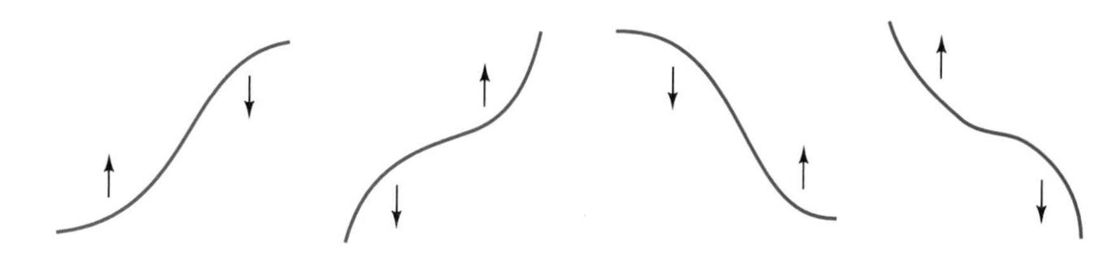

Figure 4-14

A point at which the graph changes from bending downward to bending upward (or vice versa) is called a **point of inflection.** The direction in which a graph bends may not always be clear on a calculator screen and calculus is usually required to determine the exact location of points of inflection. The number of inflection points and hence the amount of bending in the graph are governed by these facts, which are proved in calculus:

Points of Inflection ▶

The graph of a polynomial function of degree n has at most $n - 2$ points of inflection. The graph of a polynomial function of odd degree has at least one point of inflection.

▶ **TECHNOLOGY TIP**

Points of inflection can be found on TI-85/92 by using INFLC or INFLECTION in the GRAPH MATH menu.

Thus, the graph of a quadratic function (degree 2) has no points of inflection ($n - 2 = 2 - 2 = 0$) and the graph of a cubic has exactly one (since it has at least one and at most $3 - 2 = 1$).

Graphs of Cubic Polynomials

The information presented above can be used to determine in advance the possibilities for the general shape of a polynomial graph. For example, if

$$f(x) = ax^3 + bx^2 + cx + d$$

is a cubic polynomial, then its graph must resemble the graph of ax^3 when $|x|$ is large. Consequently, one end of the graph shoots upward and the other downward and the graph bends in the same way as the graph of ax^3. Since $f(x)$ has degree 3, it has at most two local extrema. However, it cannot have just one local extremum (if it had only a local minimum, for instance, both ends of the graph would shoot upward). Hence, the graph has either two local extremum or none. If it has two, one must be a local minimum and the other a local maximum (why?). Consequently, there are only four possibilities (Figure 4–15):

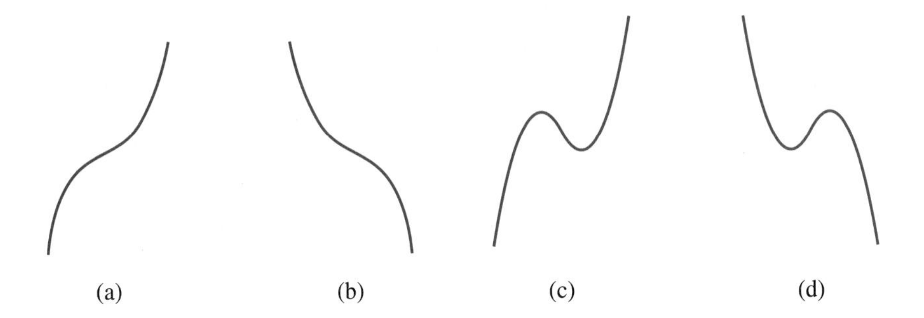

Figure 4-15

Of course, the heights and depths of the peaks and valleys can vary greatly, as can the distance between them, but the basic pattern is one of these four.

EXAMPLE 3 The shape of the graph of $f(x) = x^3 - 1.8x^2 + x + 2$ in the standard viewing window (Figure 4–16 below) is very similar to pattern (a) in Figure 4–15 above. However if you use the trace feature on the flat portion of the graph just to the right of the y-axis, you see that the y-coordinates increase slightly, then decrease slightly, then increase again. Zooming in on the portion of the graph between $x = 0$ and $x = 1$ (Figure 4–17), we see that the graph actually has a tiny peak and valley. Thus, the shape of the graph is pattern (c) in Figure 4–15. ■

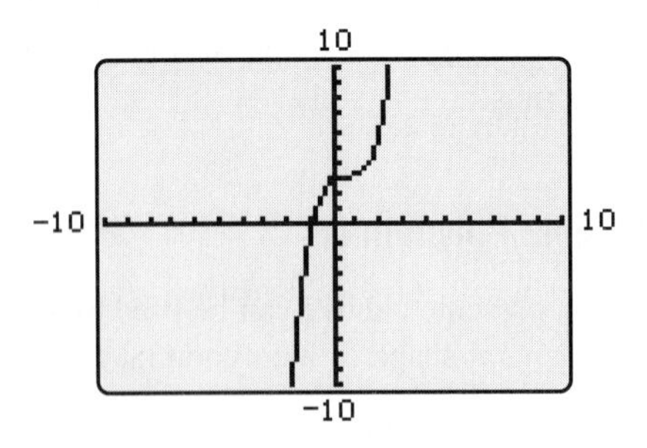

Figure 4-16

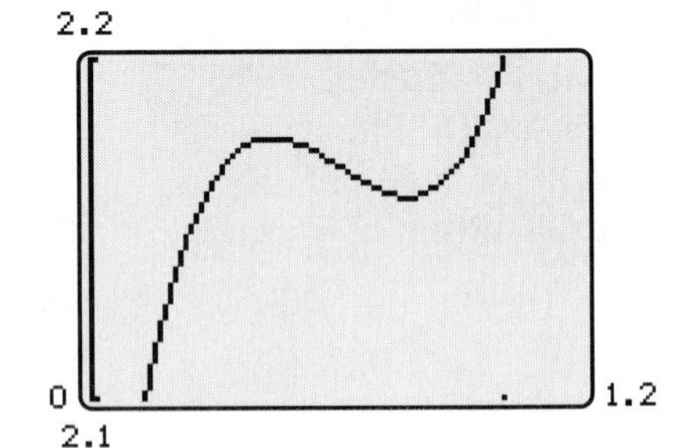

Figure 4-17

The graph in Figure 4–16 illustrates an important point: A calculator may erroneously show horizontal line segments as part of the graph of a polynomial function. Except for graphs of constant functions such as $f(x) = 5$, however, *no polynomial graph contains any horizontal line segments* (Exercise 49).

Graphs of Higher Degree Polynomial Functions

Although it is possible to analyze the general shapes of fourth-, fifth-, and higher-degree polynomial graphs, as was done for cubics, it isn't practical because there are so many possible shapes. So, you should deal with such graphs on a case-by-case basis, using the properties discussed above.

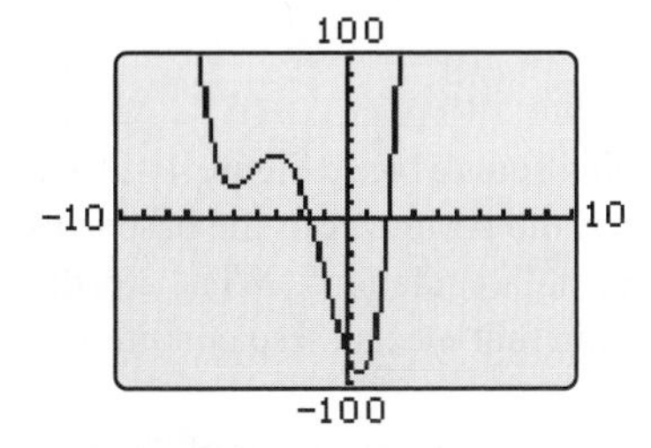

Figure 4-18

EXAMPLE 4 Find a complete graph of $f(x) = x^4 + 10x^3 + 21x^2 - 40x - 80$.

Solution Since $f(0) = -80$, the standard viewing window probably won't show a complete graph, so we try the window with $-10 \le x \le 10$ and $-100 \le y \le 100$ and obtain Figure 4–18. The three peaks and valleys shown here are the only ones because a fourth-degree polynomial graph has at most three local extrema. There cannot be more x-intercepts than the two shown here because if the graph turned toward the x-axis farther out, there would be an additional peak, which is impossible. Finally, the outer ends of the graph resemble the graph of x^4, the highest degree term (see the chart on page 231). Hence, Figure 4–18 includes all the important features of the graph and is therefore complete. ■

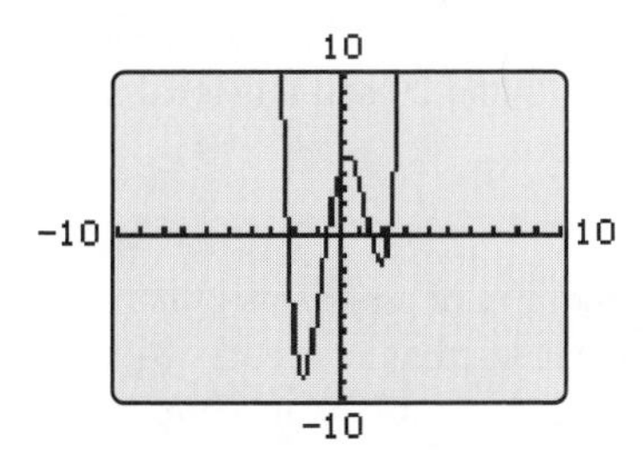

Figure 4-19

EXAMPLE 5 In the standard viewing window, the graph of $f(x) = .01x^5 + x^4 - x^3 - 6x^2 + 5x + 4$ looks like Figure 4–19. This cannot be a complete graph because, when $|x|$ is large, the graph of $f(x)$ must resemble the graph of $g(x) = .01x^5$, whose left end goes downward (see the chart on page 230). So the graph of $f(x)$ must turn downward and cross the x-axis somewhere to the left of the origin. Even without graphing, we can see that there must be one more peak (where the graph turns downward), making a total of four local extrema (the most a fifth-degree polynomial can have), and another x-intercept for a total of five. When these additional features are shown, we will have a complete graph.

> GRAPHING EXPLORATION Find a viewing window that includes the local maximum and x-intercept not shown in Figure 4–19. When you do, the scale will be such that the local extrema and x-intercepts shown in Figure 4–19 will no longer be visible.

Consequently, a complete graph of $f(x)$ requires several viewing windows in order to see all the important features. ■

The graphs in Examples 3 through 5 were known to be complete because in each case they included the maximum possible number of local extrema. In many cases, however, a graph may not have the largest possible number of peaks and valleys. For instance, the graph of the sixth-degree polynomial $f(x) = x^6$ has only one local extremum, a valley at (0, 0). In such cases, use whatever other information you may have and try several viewing windows to obtain the most likely complete graph.

Applications

The solution of many applied problems reduces to finding a local extremum of a polynomial function.

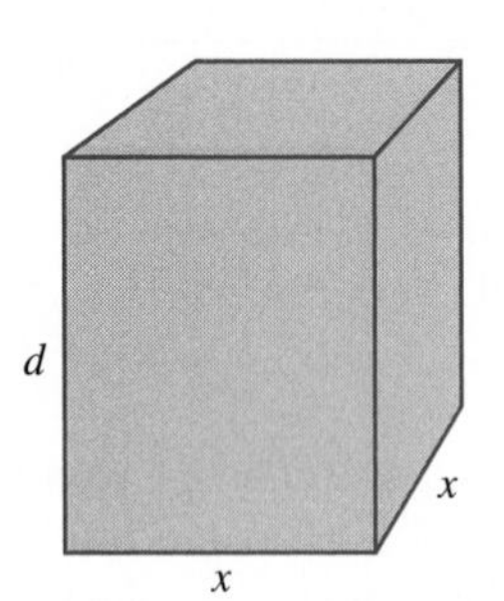

Figure 4-20

EXAMPLE 6 A rectangular box with a square base (Figure 4–20) is to be mailed. The sum of the height of the box and the perimeter of the base is to be 84 inches, the maximum allowable under postal regulations. What are the dimensions of the box with largest possible volume that meets these conditions?

Solution If the length of one side of the base is x, then the perimeter of the base (the sum of the lengths of its four sides) is $4x$. If the height of the box is d, then $4x + d = 84$, so that $d = 84 - 4x$ and hence the volume is

$$V = x \cdot x \cdot d = x \cdot x \cdot (84 - 4x) = 84x^2 - 4x^3.$$

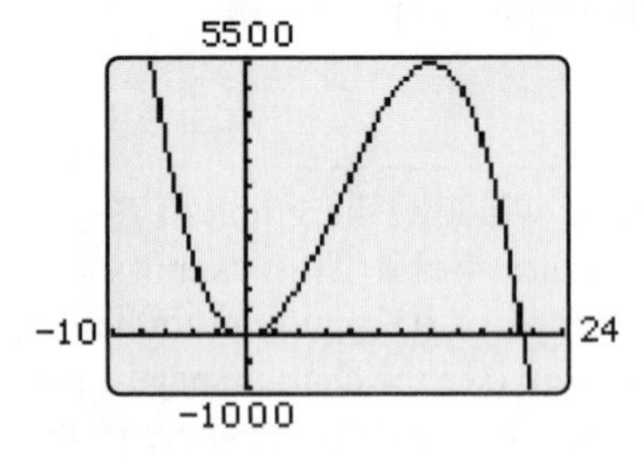

Figure 4-21

The graph of the polynomial function $V(x) = 84x^2 - 4x^3$ in Figure 4–21 is complete (why?). However, the only relevant part of the graph in this situation is the portion with x and $V(x)$ positive (because x is a length and $V(x)$ is a volume). The graph of $V(x)$ has a local maximum between 10 and 20 and this local maximum value is the largest possible volume for the box.

GRAPHING EXPLORATION Use zoom-in or a min/max finder to find the x-value at which the local maximum occurs, with a maximum error of .01. State the dimensions of the box in this case. ■

EXERCISES 4.3

In Exercises 1–6, decide whether the given graph could possibly *be the graph of a polynomial function.*

1.

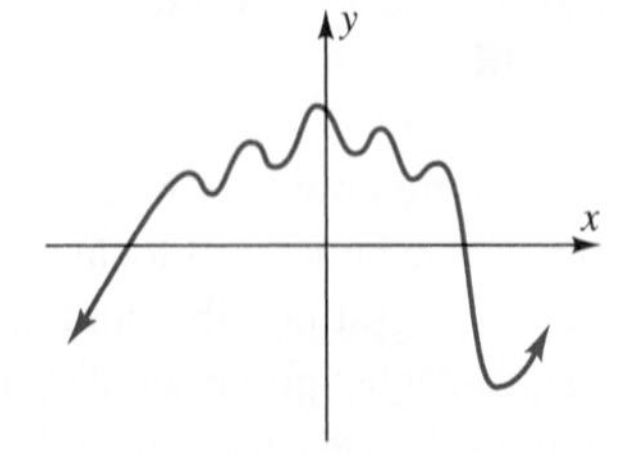

2.

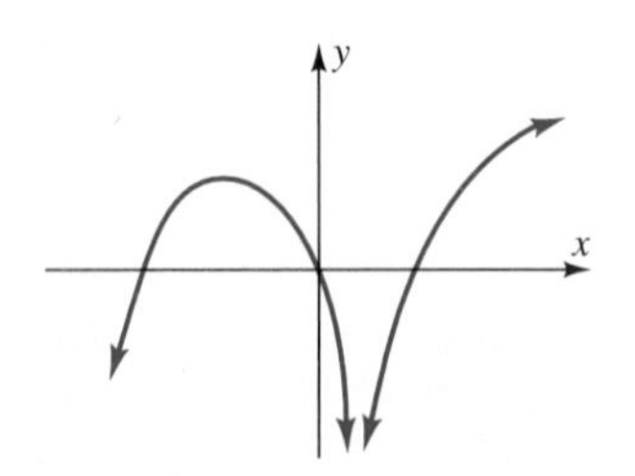

3.

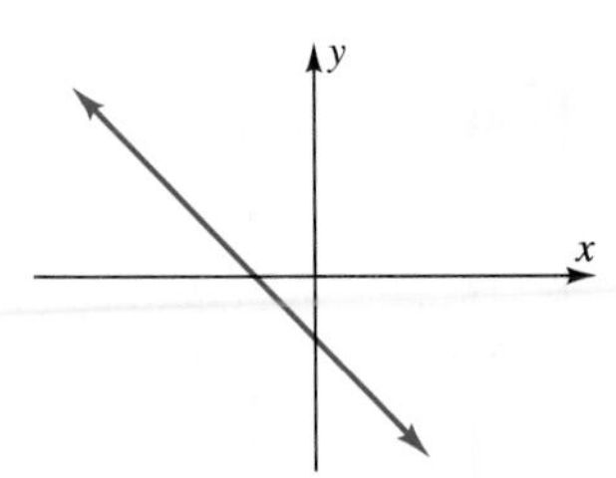

4.

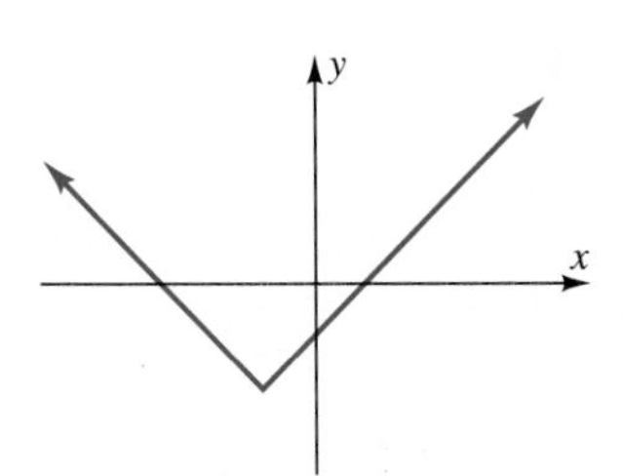

5.

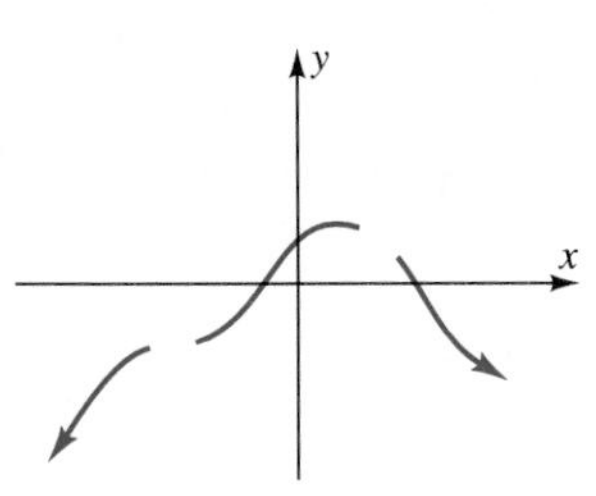

6.

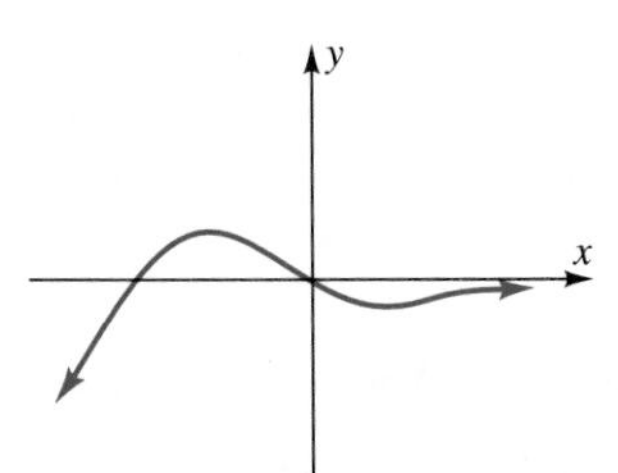

In Exercises 7–12, determine whether the given graph could possibly be the graph of a polynomial function of degree 3, of degree 4, or of degree 5.

7.

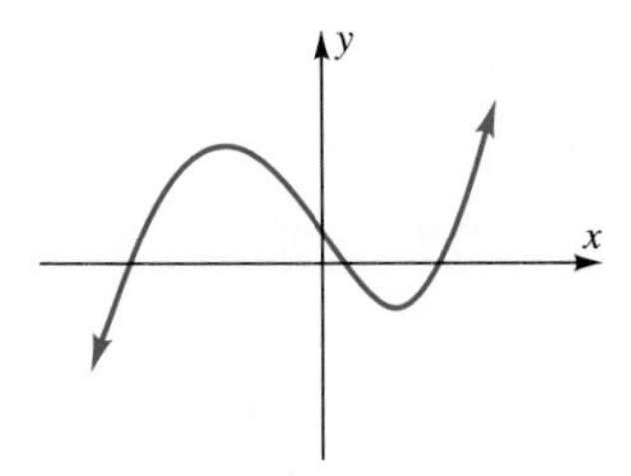

8.

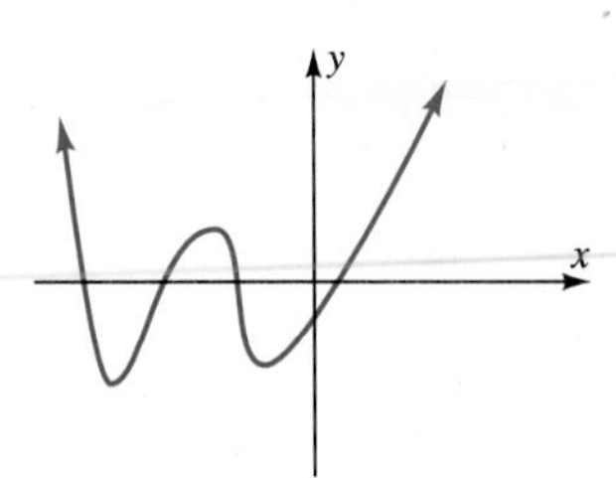

9.

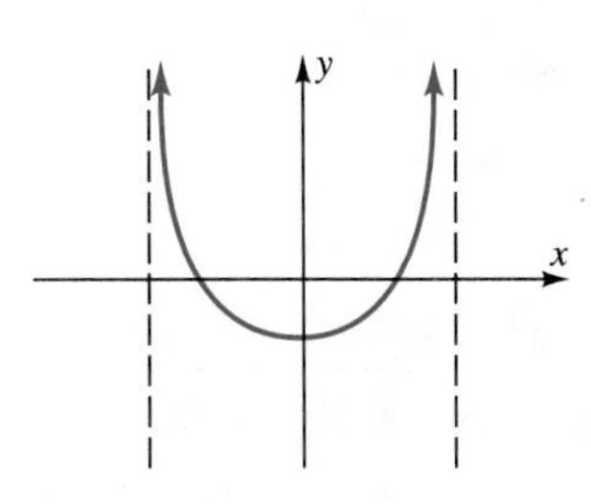

10.

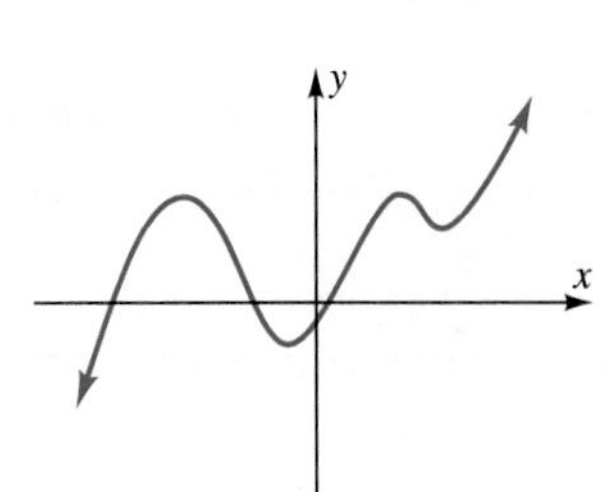

11.

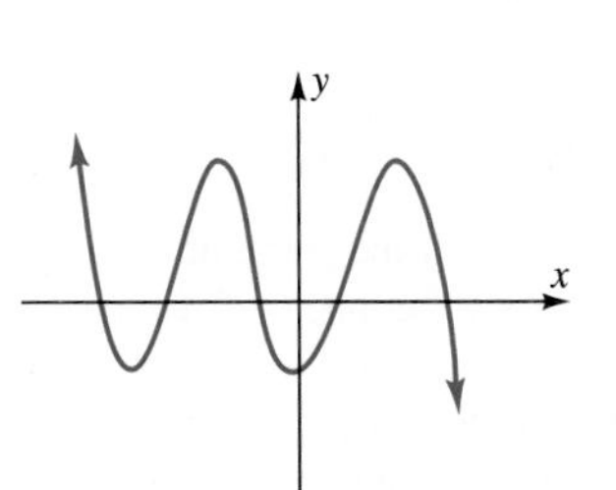

12.

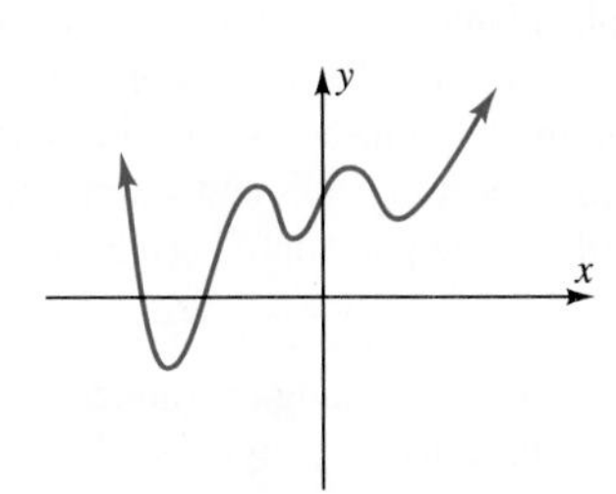

In Exercises 13–16, graph the function in the standard viewing window and explain why that graph cannot possibly be complete.

13. $f(x) = .01x^3 - .2x^2 - .4x + 7$

14. $g(x) = .01x^4 + .1x^3 - .8x^2 - .7x + 9$

15. $h(x) = .005x^4 - x^2 + 5$

16. $f(x) = .001x^5 - .01x^4 - .2x^3 + x^2 + x - 5$

In Exercises 17–22, find a single viewing window that shows a complete graph of the function.

17. $f(x) = x^3 + 8x^2 + 5x - 14$

18. $g(x) = x^3 - 3x^2 - 4x - 5$

19. $g(x) = -x^4 - 3x^3 + 24x^2 + 80x + 15$

20. $f(x) = x^4 - 10x^3 + 35x^2 - 50x + 24$

21. $f(x) = 2x^5 - 3.5x^4 - 10x^3 + 5x^2 + 12x + 6$

22. $g(x) = x^5 + 8x^4 + 20x^3 + 9x^2 - 27x - 7$

In Exercises 23–28, find a complete graph of the function and list the viewing window(s) that show this graph.

23. $f(x) = .1x^5 + 3x^4 - 4x^3 - 11x^2 + 3x + 2$

24. $g(x) = x^4 - 48x^3 - 101x^2 + 49x + 50$

25. $g(x) = .03x^3 - 1.5x^2 - 200x + 5$

26. $f(x) = .25x^6 + .25x^5 - 35x^4 - 7x^3 + 823x^2 + 25x - 2750$

27. $g(x) = 2x^3 - .33x^2 - .006x + 5$

28. $f(x) = .3x^5 + 2x^4 - 7x^3 + 2x^2$

29. **(a)** Explain why the graph of a fourth-degree polynomial has either one or three local extrema.
(b) Ignoring possible bends, list the four general shapes (with respect to peaks and valleys) that the graph of a fourth-degree polynomial can have (as was done for cubic polynomials in Figure 4–15).

30. **(a)** Explain why the graph of a fifth-degree polynomial has an even number of local extrema (possibly 0).
(b) Ignoring possible bends, list the general shapes (with respect to peaks and valleys) that the graph of a fifth-degree polynomial can have.

31. Figure 4–22 is a partial view of the graph of a cubic polynomial whose leading coefficient is negative. Which of the patterns in Figure 4–15 does this graph have?

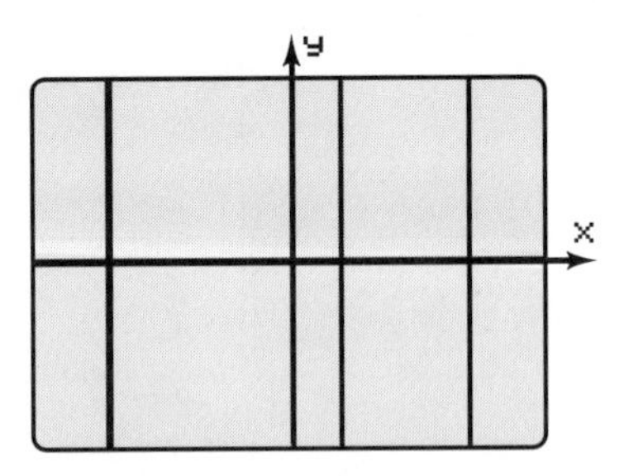

Figure 4-22

32. Figure 4–23 is a partial view of the graph of a fourth-degree polynomial. Sketch the general shape of the graph (as in Exercise 29) and state whether the leading coefficient is positive or negative.

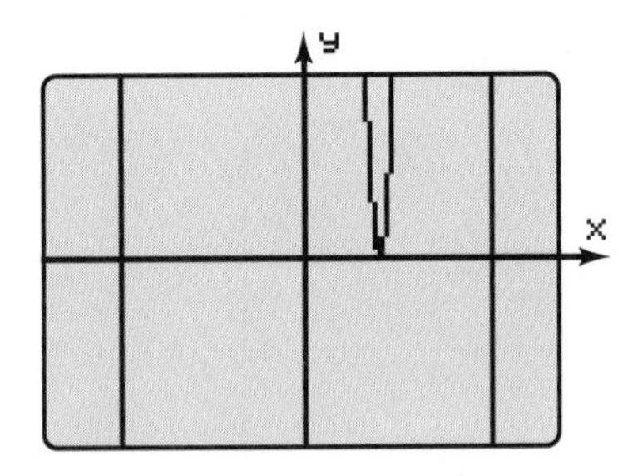

Figure 4-23

In Exercises 33–42, sketch a complete graph of the function. Label each x-intercept and the coordinates of each local extremum; find intercepts and coordinates exactly when possible and otherwise approximate them with a maximum error of .1.

33. $f(x) = x^3 - 3x^2 + 4$

34. $g(x) = 4x - 4x^3/3$

35. $h(x) = .25x^4 - 2x^3 + 4x^2$

36. $f(x) = .25x^4 - 2x^3/3$

37. $g(x) = 3x^3 - 18.5x^2 - 4.5x - 45$

38. $h(x) = 2x^3 + x^2 - 4x - 2$

39. $f(x) = x^5 - 3x^3 + x + 1$

40. $g(x) = .25x^4 - x^2 + .5$

41. $h(x) = 8x^4 + 22.8x^3 - 50.6x^2 - 94.8x + 138.6$

42. $f(x) = 32x^6 - 48x^4 + 18x^2 - 1$

43. Name tags can be sold for $29 per thousand. The cost of manufacturing x thousand tags is

$.001x^3 + .06x^2 - 1.5x$ dollars. Assuming that all tags manufactured are sold,

(a) What number of tags should be made to guarantee a maximum profit? What will that profit be?

(b) What is the largest number of tags that can be made without losing money?

44. When there are 22 apple trees per acre, the average yield has been found to be 500 apples per tree. For each additional tree planted per acre, the yield per tree decreases by 15 apples per tree. How many additional trees per acre should be planted to maximize the yield?

45. An auto parts manufacturer makes radiators that sell for \$350 each. The cost of producing x radiators is approximated by the function $C(x) = 600{,}000 - 25x + .01x^2$.

(a) What is the revenue function in this situation? What is the profit function?

(b) How many radiators must be sold in order to make any profit?

(c) What is the maximum possible number of radiators that can be sold without losing money?

(d) What number of radiators will produce the largest possible profit?

46. The top of a 12-ounce can of soda pop is three times thicker than the sides and bottom (so that the flip top opener will work properly) and the can has a volume of 355 cubic cm. What should the radius and height of the can be in order to use the least possible amount of metal? [Assume that the entire can is made from a single sheet of metal, with three layers being used for the top. Example 4 in Section 1.7 may be helpful.]

47. An open-top reinforced box is to be made from a 12-by-36-inch piece of cardboard as in Exercise 41 of Section 4.2. What size squares should be cut from the corners in order to have a box with maximum volume?

48. A box with a lid is to be made from a 48-by-24-inch piece of cardboard by cutting and folding, as in Exercise 42 of Section 4.2. What size squares should be cut from the two corners in order to have a box of maximum volume?

Thinkers

49. (a) Graph $g(x) = .01x^3 - .06x^2 + .12x + 3.92$ in the viewing window with $-3 \le x \le 3$ and $0 \le y \le 6$ and verify that the graph appears to coincide with the horizontal line $y = 4$ between $x = 1$ and $x = 3$. In other words, it appears that every x with $1 \le x \le 3$ is a solution of the equation

$$.01x^3 - .06x^2 + .12x + 3.92 = 4.$$

Explain why this is impossible. Conclude that the actual graph is not horizontal between $x = 1$ and $x = 3$.

(b) Use the trace feature to verify that the graph is actually rising from left to right between $x = 1$ and $x = 3$. Find a viewing window that shows this.

(c) Show that it is not possible for the graph of a polynomial $f(x)$ to contain a horizontal segment. [*Hint:* A horizontal line segment is part of the horizontal line $y = k$ for some constant k. Adapt the argument in part (a), which is the case $k = 4$.]

50. (a) Using the viewing window with $0 \le x \le 6$ and $-4 \le y \le 4$, graph each of the six polynomials $f(x) = (x^2 - 5x - 6)(x - 3)$, $f(x) = (x^2 - 5x - 6)(x - 3)^2$, $f(x) = (x^2 - 5x - 6)(x - 3)^3, \ldots,$ $f(x) = (x^2 - 5x - 6)(x - 3)^6$, each of which has an x-intercept at $x = 3$. For which values of n does the graph of $f(x) = (x^2 - 5x - 6)(x - 3)^n$ cross the x-axis at $x = 3$?

(b) Based on the evidence in part (a), would you say that the graph of $f(x) = (x^2 - 5x - 6)(x - 3)^7$ crosses the x-axis at $x = 3$? What about the graph of $f(x) = (x^2 - 5x - 6)(x - 3)^8$? Confirm your answers by graphing.

(c) Using the evidence in parts (a) and (b), complete the blanks in this conjecture: If $f(x) = g(x)(x - c)^n$ is a polynomial and c is not a root of $g(x)$, then the graph of f crosses the x-axis at $x = c$ when n is ______ and does not cross when n is ______.

51. For x-values in a particular interval, a nonpolynomial function $g(x)$ can often be approximated by a polynomial function, meaning that there is some polynomial $f(x)$ such that $g(x) \approx f(x)$ for every x in the interval.

(a) In the standard viewing window, graph both $g(x) = \sqrt{x}$ (which is not a polynomial function) and the polynomial function

$$f(x) = .26705x^3 - .78875x^2 + 1.3021x + .22033.$$

Are the graphs similar?

(b) Graph $f(x)$ and $g(x)$ in the viewing window with $0.26 \le x \le 1$ and $0 \le y \le 1$. Does it now appear that $f(x)$ is a good approximation of $g(x)$ over the interval $[.26, 1]$?

(c) For any particular value of x, the error in this approximation is the difference between $f(x)$ and $g(x)$. In other words, $h(x) = f(x) - g(x)$ measures the error in the approximation. Graph the function $h(x)$ in the viewing window with $.26 \le x \le 1$ and

$-.001 \le y \le .001$ and use the trace feature to determine the maximum error in the approximation.

52. **(a)** Graph $f(x) = x^3 - 4x$ in the viewing window with $-3 \le x \le 3$ and $-5 \le y \le 5$.
(b) Graph the difference quotient of $f(x)$ (with $h = .01$) on the same screen.
(c) Find the x-coordinates of the relative extrema of $f(x)$. How do these numbers compare with the x-intercepts of the difference quotient?

53. The graph of $f(x) = (x + 18)(x^2 - 20)(x - 2)^2(x - 10)$ has x-intercepts at each of its roots, that is, at $x = -18$, $\pm\sqrt{20} \approx \pm 4.472$, 2, and 10. It is also true that $f(x)$ has a relative minimum at $x = 2$.
(a) Draw the x-axis and mark the roots of $f(x)$. Then use the fact that $f(x)$ has degree 6 (why?) to sketch the general shape of the graph (as was done for cubics in Figure 4–15).
(b) Now graph $f(x)$ in the standard viewing window. Does the graph resemble your sketch? Does it even show all the x-intercepts between -10 and 10?
(c) Graph $f(x)$ in the viewing window with $-19 \le x \le 11$ and $-10 \le y \le 10$. Does this window include all the x-intercepts as it should?
(d) List viewing windows that give a complete graph of $f(x)$.

4.4 RATIONAL FUNCTIONS

A **rational function** is a function whose rule is the quotient of two polynomials, such as

$$f(x) = \frac{1}{x}, \qquad t(x) = \frac{4x - 3}{2x + 1}, \qquad k(x) = \frac{2x^3 + 5x + 2}{x^2 - 7x + 6}.$$

A polynomial function is defined for every real number, but the rational function $f(x) = g(x)/h(x)$ is defined only when its denominator is nonzero. Hence,

Domain ▶

The domain of the rational function $f(x) = \dfrac{g(x)}{h(x)}$ is the set of all real numbers that are *not* roots of the denominator $h(x)$.

For instance, the domain of $f(x) = \dfrac{x^2 + 3x + 1}{x^2 - x - 6}$ is the set of all real numbers except -2 and 3 (the roots of $x^2 - x - 6 = (x + 2)(x - 3)$).

In many cases calculators do a poor job of graphing rational functions. If you know how rational functions behave, however, and you analyze each function before graphing it, you will be able to interpret the screen images correctly. So, the emphasis here will be on algebraic analysis of rational functions.

Linear Rational Functions

The simplest rational functions are those whose denominator is a first-degree polynomial and whose numerator is either a constant or a first-degree polynomial, such as

$$f(x) = \frac{3x + 5}{2x - 4}, \qquad g(x) = \frac{-3}{2x + 5}, \qquad h(x) = \frac{1}{x}.$$

The key to understanding the behavior of these functions is this simple fact from arithmetic:

The Big–Little Principle ▶

> **If c is a number far from 0, then $1/c$ is a number close to 0. Conversely, if c is close to 0, then $1/c$ is far from 0. In less precise, but more suggestive terms:**
>
> $$\frac{1}{\text{big}} = \text{little} \quad \text{and} \quad \frac{1}{\text{little}} = \text{big}.$$

For example, 5000 is big (far from 0) and 1/5000 is little (close to 0). Similarly, $-1/1000$ is very close to 0, but $\dfrac{1}{-1/1000} = -1000$ is far from 0. To see the role played by the Big–Little Principle, we consider a typical example.

EXAMPLE 1 Without using a calculator, describe the graph of $f(x) = \dfrac{x+1}{2x-4}$ near $x = 2$ and far from $x = 2$. Then sketch the graph.

Solution The function is not defined when the denominator is zero, that is, when $x = 2$. When $x > 2$ and is very close to 2

the numerator $x + 1$ is very close to $2 + 1 = 3$;

the denominator $2x - 4$ is positive and very close to $2 \cdot 2 - 4 = 0$.

Therefore,

$$f(x) = \frac{x+1}{2x-4} \approx \frac{3}{\text{little}} = 3 \cdot \frac{1}{\text{little}} = 3 \cdot \text{big} = \text{BIG!}$$

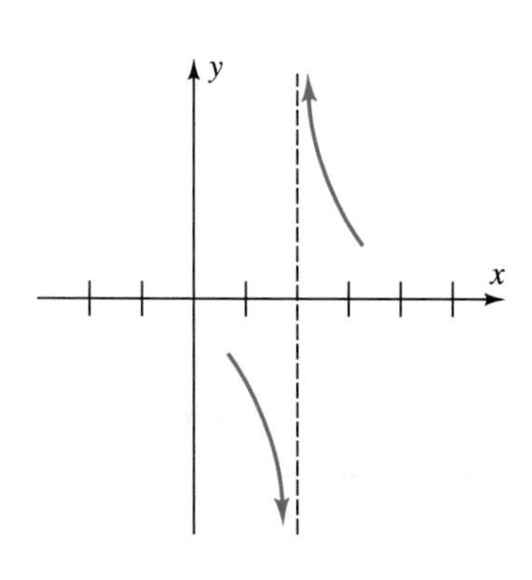

Figure 4-24

In graphical terms this means that the points with x-coordinates slightly larger than 2 have gigantic y-coordinates and the graph shoots upward just to the right of $x = 2$ (Figure 4–24).

Similarly, when $x < 2$ and very close to 2, the numerator $x + 1$ is very close to 3 and the denominator $2x - 4$ is negative and very close to 0, so that the quotient $f(x)$ is a negative number far from 0. Thus, the points on the graph with x-coordinate slightly smaller than 2 are very far below the x-axis and the graph shoots downward just to the left of $x = 2$ (Figure 4–24).

The dashed vertical line in Figure 4–24 is included for easier visualization; it is *not* part of the graph, but is called a **vertical asymptote** of the graph. The graph gets closer and closer to the vertical asymptote, but never crosses it because $f(x)$ is not defined when $x = 2$.

To see what the graph looks like far from $x = 2$, that is, when x is large in absolute value, we rewrite the rule of f like this:

$$f(x) = \frac{x+1}{2x-4} = \frac{\dfrac{x+1}{x}}{\dfrac{2x-4}{x}} = \frac{1+\dfrac{1}{x}}{2-\dfrac{4}{x}}$$

As x gets larger in absolute value (far from 0), both $1/x$ and $4/x$ get very close to 0 by the Big–Little Principle. Consequently, $f(x) = \dfrac{1+(1/x)}{2-(4/x)}$ gets very close to $\dfrac{1+0}{2-0} = \dfrac{1}{2}$. So when $|x|$ is large, the graph gets closer and closer to the horizontal line $y = 1/2$, but never touches it, as shown in Figure 4–25. The line $y = 1/2$ is called a **horizontal asymptote** of the graph.

The x-intercepts of the graph occur where the numerator is 0 and the denominator is nonzero (because then the fraction $f(x) = 0$). Setting $x + 1 = 0$ we see that the only x-intercept is at $x = -1$. The y-intercept occurs where $x = 0$, that is, at $f(0) = 1/2$. This information together with a few hand-plotted points produces the graph in Figure 4–25. ■

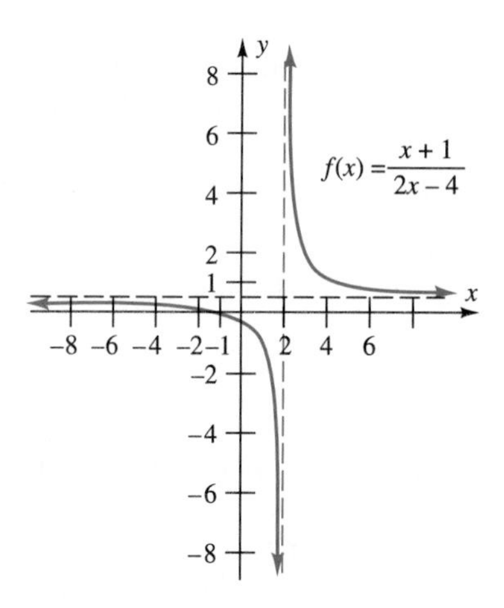

Figure 4-25

Depending on the viewing window, a calculator may not accurately represent the graph of a rational function, particularly near a vertical asymptote. For example, a TI-82 produced the following graphs of $f(x) = \dfrac{x+1}{2x-4}$, two of which do not look like Figure 4–25 as they should:

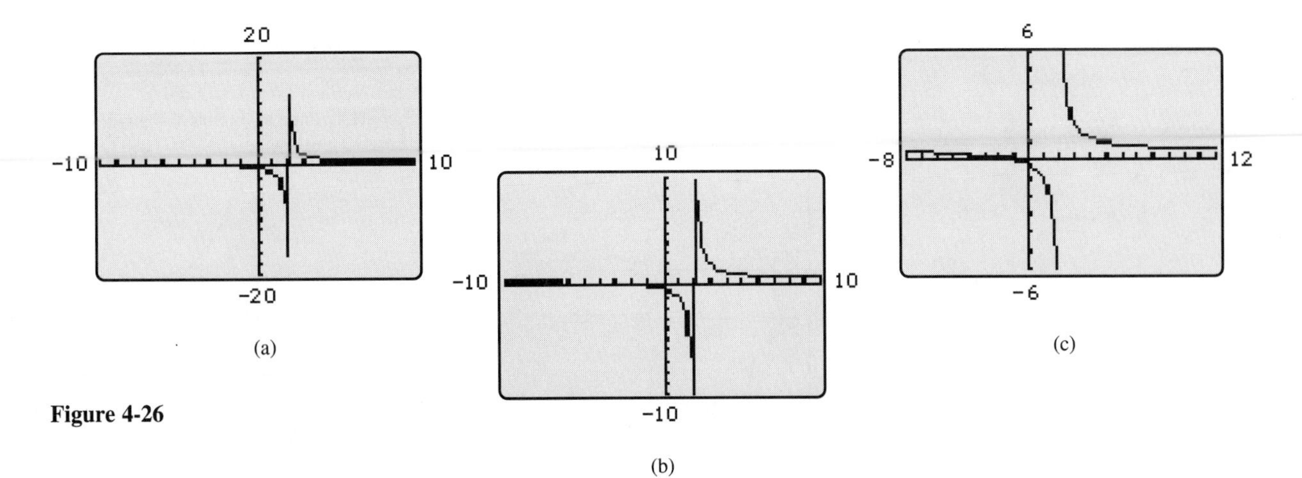

(a)

(b)

(c)

Figure 4-26

The vertical segments in graphs (a) and (b) are *not* representations of the vertical asymptote. They are a result of the calculator evaluating $f(x)$ just to the left of $x = 2$ and just to the right of $x = 2$, but not at $x = 2$, and then erroneously connecting these points with a near vertical segment that looks like an asymptote.

In the accurate graph (c) the calculator attempted to plot a point with $x = 2$ and when it found that $f(2)$ was not defined, skipped a pixel and did not join the points on either side of the skipped one.

> **TECHNOLOGY TIP**
> On some calculators you can avoid an erroneous vertical line in a graph by choosing a window that has the vertical asymptote at its center. In Figure 4–26(c), for instance, the vertical asymptote is at $x = 2$, which is at the center of the window with $-8 \le x \le 12$.
> Also, see the Tip on page 251.

GRAPHING EXPLORATION Find a viewing window on your calculator (other than the one in Figure 4–26) that displays the graph of $f(x)$ without any erroneous vertical line segments being shown. The Tip in the margin may be helpful.

The analysis in Example 1 works in the general case:

Linear Rational Functions ▶

The graph of $f(x) = \dfrac{ax + b}{cx + d}$ (with $c \neq 0$ and $ad \neq bc$) has two asymptotes:

The vertical asymptote occurs at the root of the denominator.

The horizontal asymptote is the line $y = a/c$.

Example 1 is the case where $a = 1$, $b = 1$, $c = 2$, and $d = -4$. Figure 4–27 on the next page shows some additional examples, in which the asymptotes are indicated by dashed lines that are not part of the graph.

$f(x) = \dfrac{-5x + 12}{2x - 4}$

Vertical asymptote $x = 2$

Horizontal asymptote $y = -\frac{5}{2}$

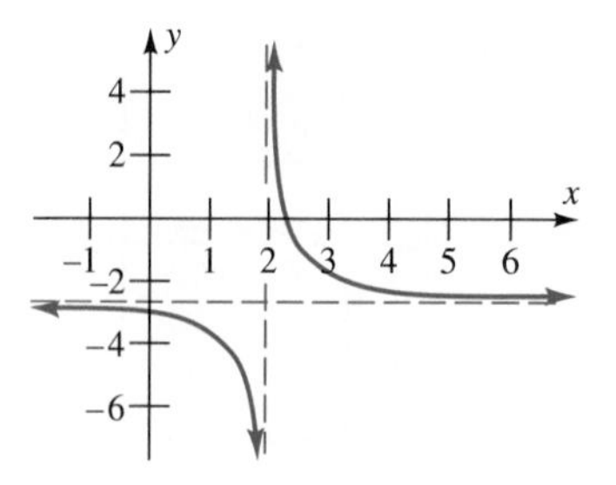

$k(x) = \dfrac{3x + 6}{x} = \dfrac{3x + 6}{1x + 0}$

Vertical asymptote $x = 0$

Horizontal asymptote $y = \frac{3}{1} = 3$

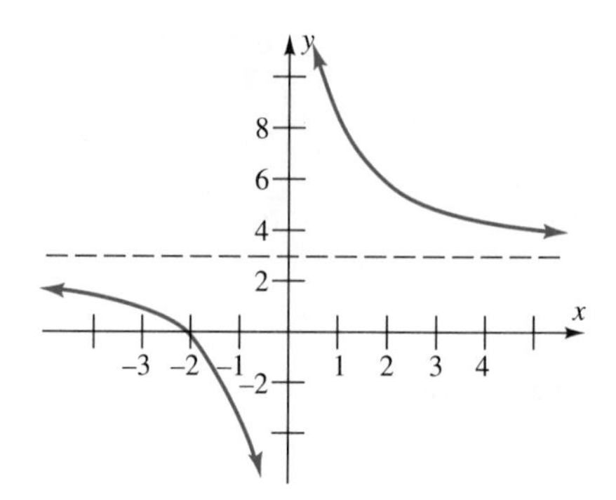

$k(x) = \dfrac{-3}{2x + 5} = \dfrac{0x - 3}{2x + 5}$

Vertical asymptote $x = -\frac{5}{2}$

Horizontal asymptote $y = \frac{0}{2} = 0$

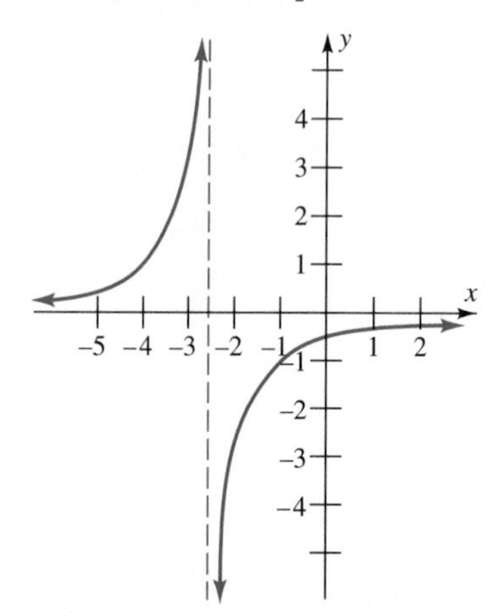

$f(x) = \dfrac{1}{x} = \dfrac{0x + 1}{1x + 0}$

Vertical asymptote $x = 0$

Horizontal asymptote $y = \frac{0}{1} = 0$

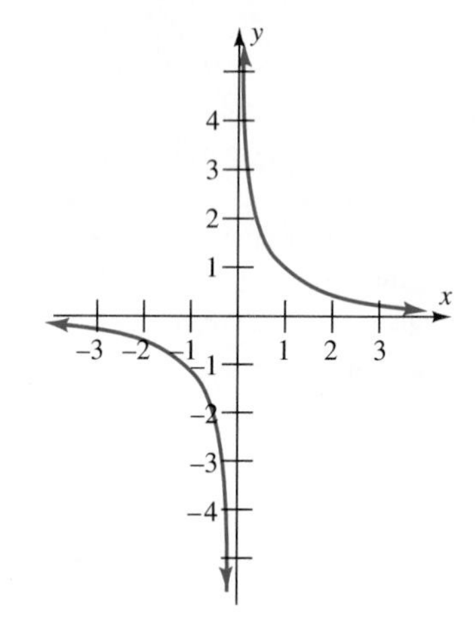

Figure 4-27

Properties of Rational Graphs

Here is a summary of the important characteristics of graphs of more complicated rational functions.

Continuity There will be breaks in the graph of a rational function wherever the function is not defined. Except for breaks at these undefined points, the graph is a continuous unbroken curve. In addition, the graph has no sharp corners.

Local Maxima and Minima The graph may have some local extrema (peaks and valleys) and calculus is needed to determine their exact location. There are no simple rules for the possible number of peaks and valleys as there were with polynomial functions.

Points of Inflection There may be points of inflection, where the graph changes from bending downward to bending upward (or vice versa). However, there is no easy algebraic way to determine the maximum number of inflection points.

Intercepts As with any function, the y-intercept of the graph of a rational function f occurs at $f(0)$, provided that f is defined at $x = 0$. The x-intercepts of the graph of any function f occur at each number c for which $f(c) = 0$. Now a fraction is 0 only when its numerator is 0 and its denominator nonzero (since division by 0 is not defined). Thus,

Intercepts ▶

The x-intercepts of the graph of the rational function $f(x) = \dfrac{g(x)}{h(x)}$ occur at the numbers that are roots of the numerator $g(x)$ but *not* of the denominator $h(x)$. If f has a y-intercept, it occurs at $f(0)$.

For example, the graph of $f(x) = \dfrac{x^2 - x - 2}{x - 5}$ has x-intercepts at $x = -1$ and $x = 2$ (which are the roots of $x^2 - x - 2 = (x + 1)(x - 2)$, but not of $x - 5$) and y-intercept at $y = 2/5$ (the value of f at $x = 0$).

Vertical Asymptotes In Example 1 we saw that the graph of $f(x) = \dfrac{x + 1}{2x - 4}$ had a vertical asymptote at $x = 2$. Note that $x = 2$ is a root of the denominator $2x - 4$, but not of the numerator $x + 1$. The same thing occurs in the general case:

Vertical Asymptotes ▶

The function $f(x) = \dfrac{g(x)}{h(x)}$ has a vertical asymptote at every number that is a root of the denominator $h(x)$, but *not* of the numerator $g(x)$.

Near a vertical asymptote, the graph of a rational function may look like the graph in Example 1, or like one of these graphs (Figure 4–28):

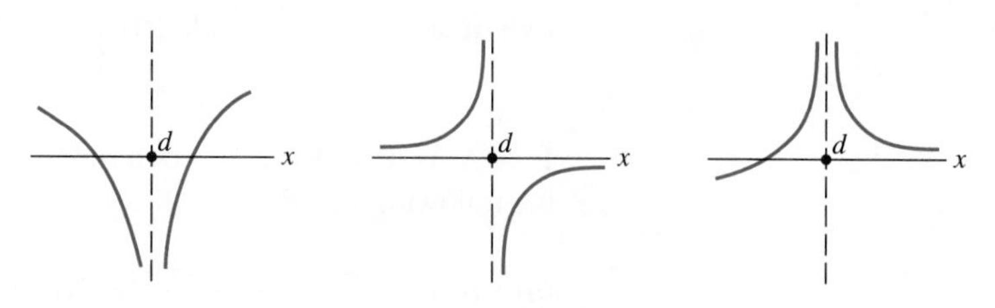

Figure 4-28 Vertical asymptotes at $x = d$

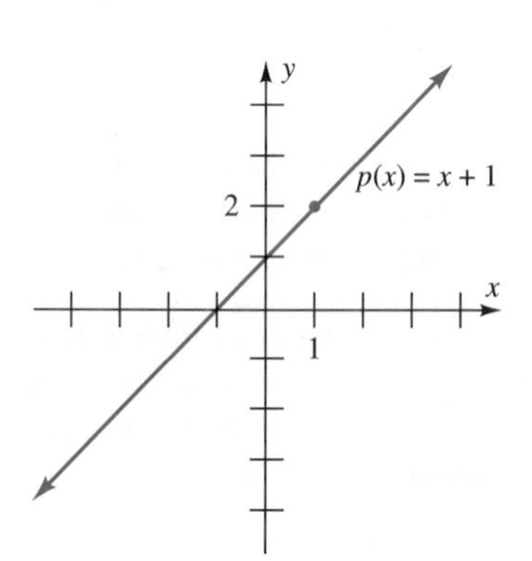

Figure 4-29

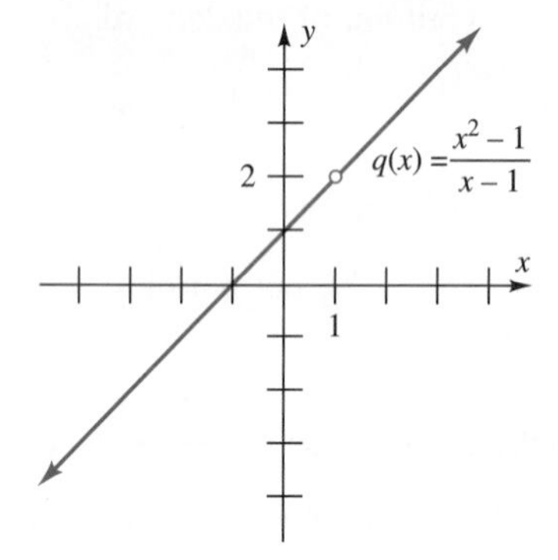

Figure 4-30

Holes You have often cancelled factors in an algebraic expression, as here:

$$\frac{x^2 - 1}{x - 1} = \frac{(x + 1)(x - 1)}{x - 1} = x + 1.$$

But the function with the rule $p(x) = x + 1$ is *not* the same as the function with the rule $q(x) = \frac{x^2 - 1}{x - 1}$ because p and q have different domains. When $x = 1$, then $p(1) = 1 + 1 = 2$. However, $q(1)$ is not defined:

$$q(1) = \frac{1^2 - 1}{1 - 1} = \frac{0}{0}.$$

For any number except 1, the two functions have the same value, and hence the same graph. Now the graph of $p(x) = x + 1$ is the straight line in Figure 4–29, which includes the point (1, 2). The graph of $q(x)$ is the same straight line, with the point (1, 2) omitted, so that the graph of q has a **hole** when $x = 1$. In hand-drawn graphs a hole is usually indicated by an open circle, as in Figure 4–30.

A calculator-drawn graph may not show holes where it should. If the calculator actually attempts to compute an undefined quantity, it indicates a hole in a graph by skipping a pixel; otherwise, it may erroneously show a continuous graph with no hole.

> GRAPHING EXPLORATION Find a viewing window in which the graph of $q(x) = \frac{x^2 - 1}{x - 1}$ has a hole when $x = 1$. Then, find a viewing window including the point (1, 2) in which no hole appears in the graph of q.

In general,

Holes ▶

> **If $f(x) = \frac{g(x)}{h(x)}$ is a rational function and d is a root of both $g(x)$ and $h(x)$, then the graph of f has either a hole* or a vertical asymptote at $x = d$.**

Behavior When |x| Is Large The shape of a rational graph at the far left and far right (that is, when $|x|$ is large) can usually be found by algebraic analysis.

EXAMPLE 2 Determine the shape of the graph when $|x|$ is large for the following functions.

(a) $f(x) = \frac{7x^4 - 6x^3 + 4}{2x^4 + x^2}$ **(b)** $g(x) = \frac{x^2 - 2}{x^3 - 3x^2 + x - 3}$

*The technical term for "hole" is "removable singularity" or "removable discontinuity."

Solution

(a) When $|x|$ is very large, a polynomial function behaves in essentially the same way as its highest degree term, as we saw on page 232. Consequently, we have this approximation

$$f(x) = \frac{7x^4 - 6x^3 + 4}{2x^4 + x^2} \approx \frac{7x^4}{2x^4} = \frac{7}{2} = 3.5.$$

Thus, when $|x|$ is large, the graph of $f(x)$ is very close to the horizontal line $y = 3.5$, which is a horizontal asymptote of the graph. This means that at the far left and far right, the graph of $f(x)$ is almost flat and very close to the line $y = 3.5$.

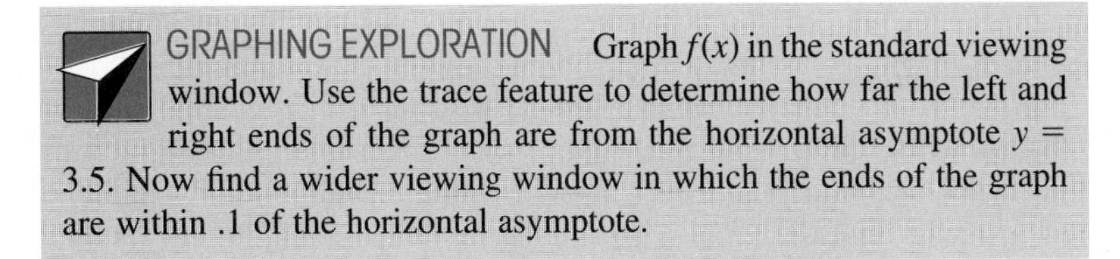

GRAPHING EXPLORATION Graph $f(x)$ in the standard viewing window. Use the trace feature to determine how far the left and right ends of the graph are from the horizontal asymptote $y = 3.5$. Now find a wider viewing window in which the ends of the graph are within .1 of the horizontal asymptote.

(b) When $|x|$ is large, the graph of g closely resembles the graph of $y = \frac{x^2}{x^3} = \frac{1}{x}$. By the Big–Little Principle, $1/x$ is very close to 0 when $|x|$ is large. So, the line $y = 0$ (that is, the x-axis) is the horizontal asymptote.

GRAPHING EXPLORATION Find a viewing window that provides the answer to this question: as you move to the left from the origin does the graph of $g(x)$ approach its horizontal asymptote from above or from below? Does the graph ever cross the horizontal asymptote? ■

Arguments similar to those in the preceding example, using the highest degree terms in the numerator and denominator, carry over to the general case and lead to this conclusion:

Horizontal Asymptotes ▶

Let $f(x) = \dfrac{ax^n + \cdots}{cx^k + \cdots}$ be a rational function whose numerator has degree n and whose denominator has degree k.

If $n = k$, then the line $y = a/c$ is a horizontal asymptote.

If $n < k$, then the x-axis (the line $y = 0$) is a horizontal asymptote.

The asymptotes of rational functions in which the denominator has smaller degree than the numerator are discussed in Excursion 4.4.A.

Graphs of Rational Functions

The facts presented above can be used in conjunction with a calculator to find accurate, complete graphs of rational functions whose numerators have degree less than or equal to the degree of their denominators. The basic procedure is as follows.

Graphing the Rational Function $f(x) = \dfrac{g(x)}{h(x)}$ When Degree $g(x) \leq$ Degree $h(x)$

1. **Analyze the function algebraically to determine its vertical asymptotes, holes, and intercepts.**
2. **Determine the horizontal asymptote of the graph when $|x|$ is large by using the facts in the box on page 249.**
3. **Use the preceding information to select an appropriate viewing window (or windows), to interpret the calculator's version of the graph (if necessary), and to sketch an accurate graph.**

EXAMPLE 3 If you ignore the preceding advice and simply graph $f(x) = \dfrac{x-1}{x^2 - x - 6}$ in the standard viewing window, you get garbage (Figure 4–31). So let's try analyzing the function. We begin by factoring:

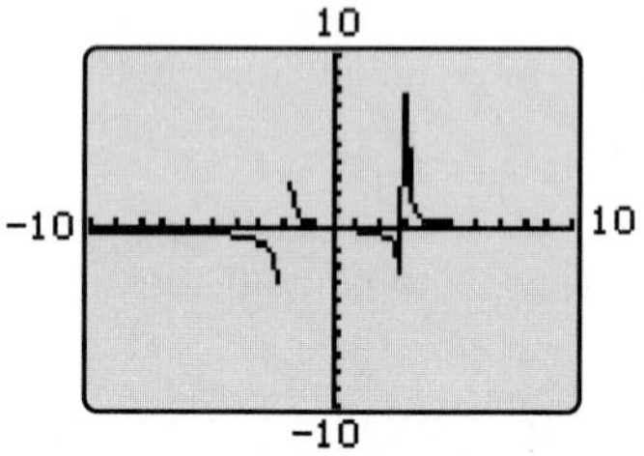

$$f(x) = \frac{x-1}{x^2 - x - 6} = \frac{x-1}{(x+2)(x-3)}.$$

Figure 4-31

The factored form allows us to read off the necessary information:

Vertical Asymptotes: $x = -2$ and $x = 3$ (roots of the denominator but not of the numerator).

Horizontal Asymptote: x-axis (because denominator has larger degree than the numerator).

Intercepts: y-intercept at $f(0) = \dfrac{0-1}{0^2 - 0 - 6} = \dfrac{1}{6}$; x-intercept at $x = 1$ (root of the numerator but not of the denominator).

Interpreting Figure 4–31 in the light of this information suggests that a complete graph of f looks something like Figure 4–32.

GRAPHING EXPLORATION Find a viewing window in which the graph of f looks similar to Figure 4–32. The Tip in the margin on the next page may be helpful. ■

▶ TECHNOLOGY TIP

When the horizontal distance between adjacent pixels in a viewing window is .1, then the calculator evaluates the function at . . . 2.8, 2.9, 3, 3.1. . . , etc. If there is a vertical asymptote at one of these numbers, the graph is accurately displayed near it.

Windows that have this property include the DECIMAL windows on most TI's (in the ZOOM menu) and on HP-38 (in the VIEWS menu); the INIT window on Casio; the DEFAULT window on Sharp 9300; and a window with $-4.7 \leq x \leq 4.8$ on TI-81. You may need to adjust the y-range on these windows to see the graph.

By doubling the Xmin and Xmax settings (for instance, from ± 4.7 to ± 9.4), you obtain a window in which the distance between adjacent pixels is .2 and the calculator evaluates at . . . 2.8, 3, 3.2, etc.

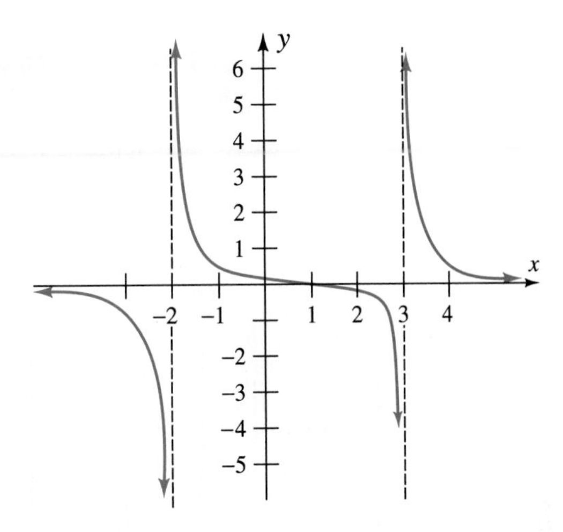

Figure 4-32

Note The graph of a rational function never touches a horizontal asymptote when x is large in absolute value. For smaller values of x, however, the graph may cross the asymptote, as in Example 3.

EXAMPLE 4 Find a complete graph of $g(x) = \dfrac{-2x^2 + 3x + 2}{x^2 - x - 12}$.

Solution Write the rule of $g(x)$ in factored form and determine the necessary information:

$$g(x) = \frac{-2x^2 + 3x + 2}{x^2 - x - 12} = \frac{(2x + 1)(2 - x)}{(x + 3)(x - 4)}.$$

Vertical Asymptotes: $x = -3$ and $x = 4$ (roots of the denominator but not the numerator).

Horizontal Asymptote: $y = -2/1 = 2$ (because the numerator and denominator have the same degree; see the box on page 249).

Intercepts: x-intercepts at $x = -1/2$ and $x = 2$ (roots of the numerator); y-intercept at $g(0) = 2/(-12) = -1/6$.

The graph of $g(x)$ in the standard viewing window (Figure 4–33 on the next page) contains erroneous vertical lines because the calculator incorrectly connected points on either side of the vertical asymptotes. By adjusting the window slightly, we obtain Figure 4–34, which accurately portrays the function's behavior at the vertical and horizontal asymptotes, but still obscures the local minimum near the origin.

> GRAPHING EXPLORATION Find a viewing window that shows the local minimum near the origin clearly. Then, find the location of the local minimum. ■

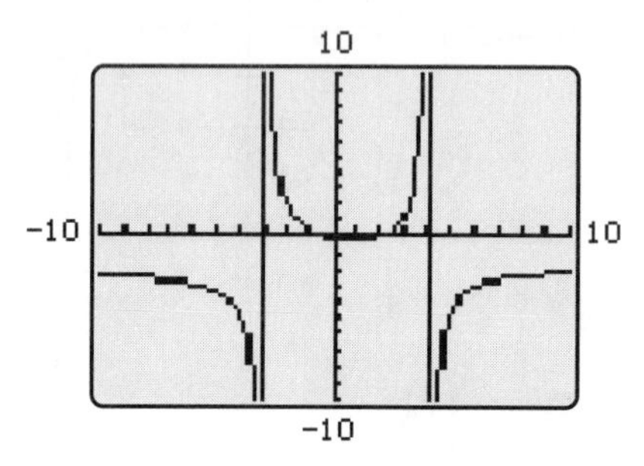

Figure 4-33

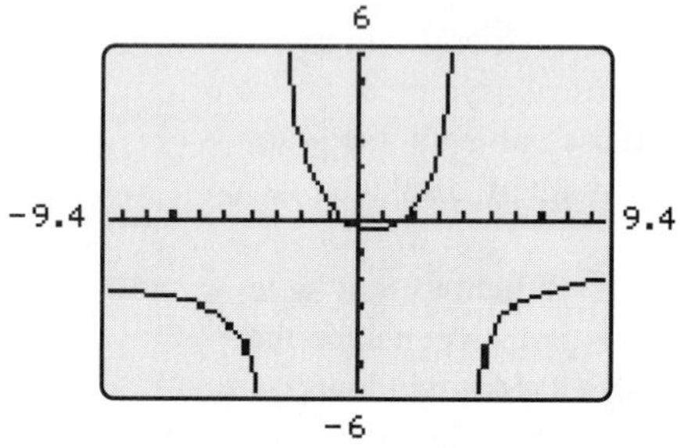

6
-9.4
9.4
-6

Figure 4-34

EXAMPLE 5 To graph $f(x) = \dfrac{2x^2}{x^2 + x - 2}$ we factor and then read off the necessary information:

$$f(x) = \frac{2x^2}{x^2 + x - 2} = \frac{2x^2}{(x + 2)(x - 1)}.$$

Vertical Asymptotes: $x = -2$ and $x = 1$ (roots of denominator).

Horizontal Asymptote: $y = 2/1 = 2$ (because numerator and denominator have the same degree; see the box on page 249).

Intercepts: x-intercept at $x = 0$ (root of numerator); y-intercept at $f(0) = 0$.

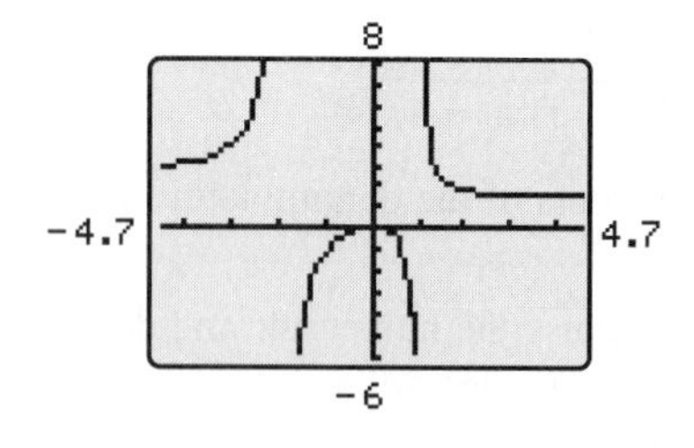

Figure 4-35

Using this information and selecting a viewing window that will accurately portray the graph near the vertical asymptotes, we obtain what seems to be a reasonably complete graph in Figure 4–35. The graph appears to be falling to the right of $x = 1$, but this is deceptive.

> GRAPHING EXPLORATION Graph f in this same viewing window and use the trace feature, beginning at approximately $x = 1.1$ and moving to the right. For what values of x is the graph above the horizontal asymptote $y = 2$? For what values of x is the graph below the horizontal asymptote?

This use of the trace feature indicates that there is some *hidden behavior* of the graph that is not visible in Figure 4–34.

GRAPHING EXPLORATION To see this hidden behavior, graph both f and the line $y = 2$ in the viewing window with $1 \leq x \leq 50$ and $1.7 \leq y \leq 2.1$.

This Exploration shows that the graph has a local minimum near $x = 4$ and then stays below the asymptote, moving closer and closer to it as x takes larger values. ■

EXAMPLE 6 Describe and sketch the graph of

$$f(x) = \frac{x^3 - x - 5}{x^3 + x^2 - 10x - 10}.$$

Solution

Vertical Asymptotes: To find the location of the vertical asymptotes we must determine the roots of the denominator $h(x) = x^3 + x^2 - 10x - 10$. The partial graph of h in Figure 4–36 shows that it has three real roots (x-intercepts), one of which appears to be at or close to $x = -1$. Verify that $h(-1) = 0$, so that $x = -1$ is a root of $h(x)$.

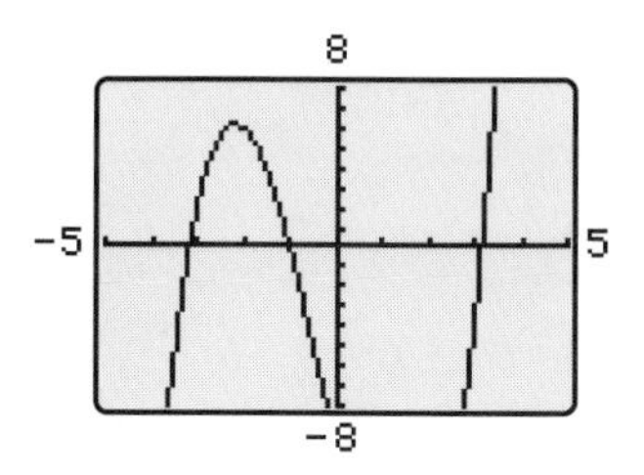

Figure 4-36

Consequently, $x - (-1) = x + 1$ is a factor of $h(x)$ by the Factor Theorem (page 214). Using synthetic or long division to divide $h(x)$ by $x + 1$, we find that

$$x^3 + x^2 - 10x - 10 = (x + 1)(x^2 - 10) = (x + 1)(x + \sqrt{10})(x - \sqrt{10}),$$

so that

$$f(x) = \frac{x^3 - x - 5}{x^3 + x^2 - 10x - 10} = \frac{x^3 - x - 5}{(x + 1)(x + \sqrt{10})(x - \sqrt{10})}.$$

You can easily verify that roots of the denominator, -1, $\sqrt{10}$ and $-\sqrt{10}$, are not roots of the numerator. Hence the graph of $f(x)$ has vertical asymptotes at $x = -\sqrt{10}$, at $x = -1$, and at $x = \sqrt{10}$.

Horizontal Asymptote: Since the numerator and denominator have the same degree, the horizontal asymptote is $y = 1/1 = 1$.

Intercepts: The y-intercept is at $f(0) = \frac{-5}{-10} = -\frac{1}{2}$. The x-intercepts occur at the roots of the numerator.

GRAPHING EXPLORATION Verify that the standard viewing window shows a complete graph of the numerator $g(x) = x^3 - x - 5$ (why?) and verify that the only root of $g(x)$ is $x \approx 1.9042$.

Therefore, the only x-intercept of $f(x)$ is at $x \approx 1.9042$.

Using the preceding information and a calculator, we obtain Figure 4–37.

GRAPHING EXPLORATION Find a viewing window in which the graph of $f(x)$ looks like Figure 4–37. You may not be able to avoid erroneous vertical line segments. ■

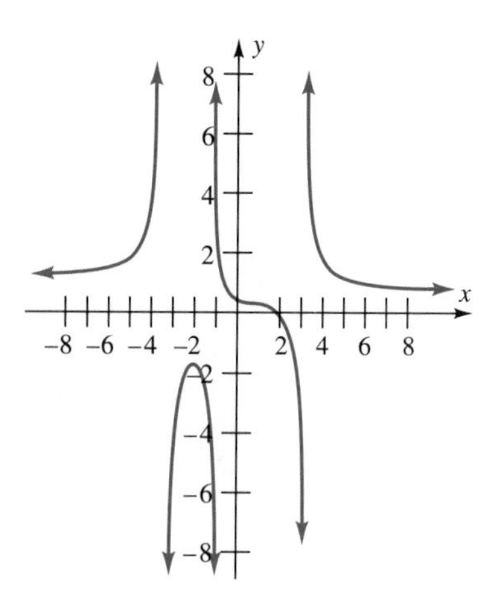

Figure 4-37

Applications

Several applications of rational functions were considered in Section 1.7. Here is another one.

EXAMPLE 7 A cardboard box with a square base and a volume of 1000 cubic inches is to be constructed (Figure 4–38). The box must be at least 2 inches in height.

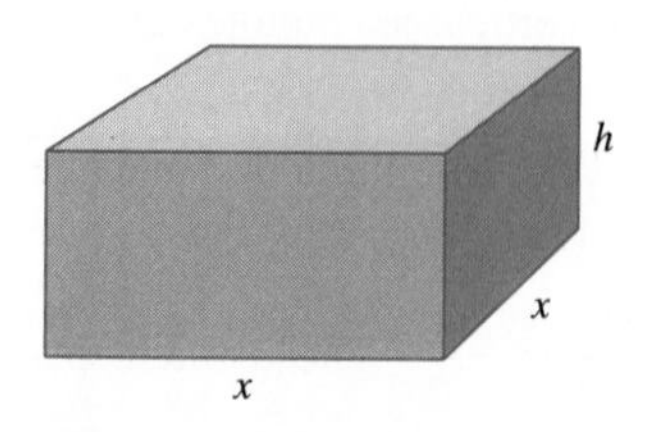

Figure 4-38

(a) What are the possible lengths for a side of the base if no more than 1100 square inches of cardboard can be used to construct the box?

(b) What is the least possible amount of cardboard that can be used?

(c) What are the dimensions of the box that uses the least possible amount of cardboard?

Solution The amount of cardboard needed to construct the box is given by the surface area of the box. Since the top and bottom each have area x^2 (why?) and each of the four sides has area xh, the surface area S is given by

$$S = x^2 + x^2 + xh + xh + xh + xh = 2x^2 + 4xh$$

Since the volume of the box is given by:

$$\text{length} \times \text{width} \times \text{height} = \text{volume}$$

we have

$$x \cdot x \cdot h = 1000 \qquad \text{or equivalently,} \qquad h = \frac{1000}{x^2}.$$

Substituting this into the surface area formula allows us to express the surface area as a function of x:

$$S(x) = 2x^2 + 4xh = 2x^2 + 4x\left(\frac{1000}{x^2}\right) = 2x^2 + \frac{4000}{x} = \frac{2x^3 + 4000}{x}.$$

Although the rational function $S(x)$ is defined for all nonzero real numbers, x is a length here and must be positive. Furthermore, $x^2 \le 500$ because if $x^2 > 500$, then $h = \dfrac{1000}{x^2}$ would be less than 2, contrary to specifications. Hence, the only values of x that make sense in this context are those with $0 < x \le \sqrt{500}$. Since $\sqrt{500} \approx 22.4$, we choose the viewing window in Figure 4-39.

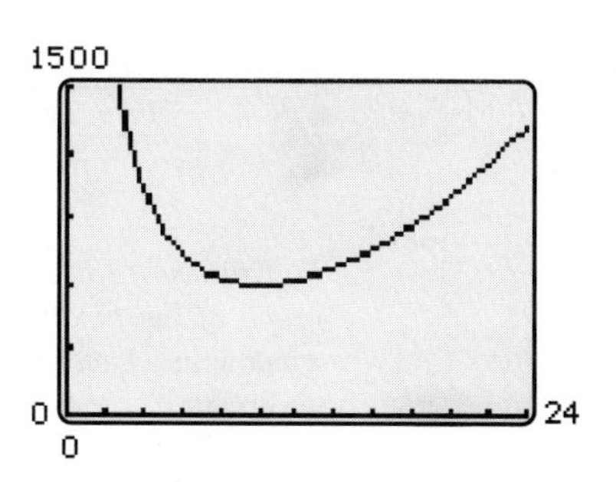

Figure 4-39

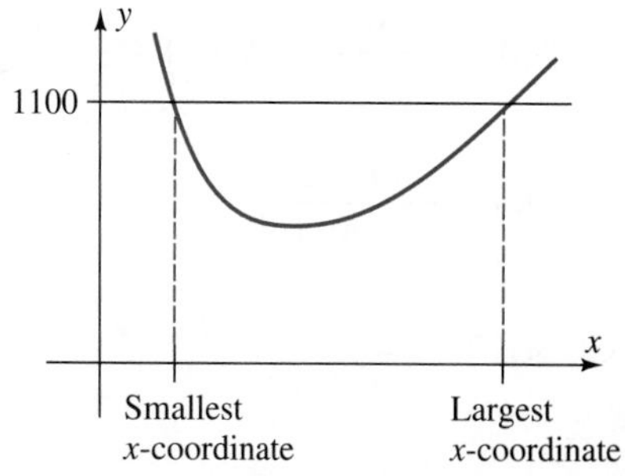

Figure 4-40

For each point (x, y) on the graph, x is a possible side length for the base of the box and y is the corresponding surface area.

(a) The points on the graph corresponding to the requirement that no more than 1100 square inches of cardboard be used are those whose y-coordinates are less than or equal to 1100. The x-coordinates of these points are the possible side lengths. The x-coordinates of the points where the graph of S meets the horizontal line $y = 1100$ are the smallest and largest possible values for x, as indicated schematically in Figure 4–40 above.

GRAPHING EXPLORATION Graph $S(x)$ and $y = 1100$ on the same screen. Use zoom-in or an intersection finder to show that the possible side lengths that use no more than 1100 square inches of cardboard are those with $3.73 \le x \le 21.36$.

(b) The least possible amount of cardboard corresponds to the point on the graph of $S(x)$ with the smallest y-coordinate.

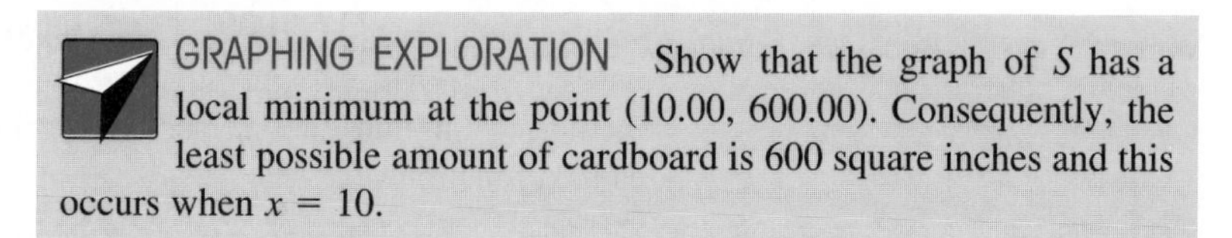

GRAPHING EXPLORATION Show that the graph of S has a local minimum at the point (10.00, 600.00). Consequently, the least possible amount of cardboard is 600 square inches and this occurs when $x = 10$.

(c) When $x = 10$, $h = 1000/10^2 = 10$. So the dimensions of the box using the least amount of cardboard are $10 \times 10 \times 10$. ■

EXERCISES 4.4

In Exercises 1–6, find the domain of the function. You may need to use some of the techniques of Section 4.2.

1. $f(x) = \dfrac{-3x}{2x+5}$ **2.** $g(x) = \dfrac{x^3+x+1}{2x^2-5x-3}$

3. $h(x) = \dfrac{6x-5}{x^2-6x+4}$

4. $g(x) = \dfrac{x^3-x^2-x-1}{x^5-36x}$

5. $f(x) = \dfrac{x^5-2x^3+7}{x^3-x^2-2x+2}$

6. $h(x) = \dfrac{x^5-5}{x^4+12x^3+60x^2+50x-125}$

In Exercises 7–12, use algebra and a calculator to determine the location of the vertical asymptotes and holes in the graph of the function.

7. $f(x) = \dfrac{x^2+4}{x^2-5x-6}$ **8.** $g(x) = \dfrac{x-5}{x^3+7x^2+2x}$

9. $f(x) = \dfrac{x}{x^3+2x^2+x}$ **10.** $g(x) = \dfrac{x}{x^3+5x}$

11. $f(x) = \dfrac{x^2-4x+4}{(x+2)(x-2)^3}$ **12.** $h(x) = \dfrac{x-3}{x^2-x-6}$

In Exercises 13–18, find the horizontal asymptote of the graph of the function when $|x|$ is large and find a viewing window in which the ends of the graph are within .1 of this asymptote.

13. $f(x) = \dfrac{3x-2}{x+3}$ **14.** $g(x) = \dfrac{3x^2+x}{2x^2-2x+4}$

15. $h(x) = \dfrac{5-x}{x-2}$ **16.** $f(x) = \dfrac{4x^2-5}{2x^3-3x^2+x}$

17. $g(x) = \dfrac{5x^3-8x^2+4}{2x^3+2x}$

18. $h(x) = \dfrac{8x^5-6x^3+2x-1}{.5x^5+x^4+3x^2+x}$

In Exercises 19–38, analyze the function algebraically: List its vertical asymptotes, holes, and horizontal asymptote. Then, sketch a complete graph of the function.

19. $f(x) = \dfrac{1}{x+5}$ **20.** $q(x) = \dfrac{-7}{x-6}$

21. $k(x) = \dfrac{-3}{2x+5}$ **22.** $g(x) = \dfrac{-4}{2-x}$

23. $f(x) = \dfrac{3x}{x-1}$ **24.** $p(x) = \dfrac{x-2}{x}$

25. $f(x) = \dfrac{2-x}{x-3}$ **26.** $g(x) = \dfrac{3x-2}{x+3}$

27. $f(x) = \dfrac{1}{x(x+1)^2}$

28. $g(x) = \dfrac{x}{2x^2 - 5x - 3}$

29. $f(x) = \dfrac{x-3}{x^2 + x - 2}$

30. $g(x) = \dfrac{x+2}{x^2 - 1}$

31. $h(x) = \dfrac{(x^2 + 6x + 5)(x+5)}{(x+5)^3(x-1)}$

32. $f(x) = \dfrac{x^2 - 1}{x^3 - 2x^2 + x}$

33. $f(x) = \dfrac{-4x^2 + 1}{x^2}$

34. $k(x) = \dfrac{x^2 + 1}{x^2 - 1}$

35. $q(x) = \dfrac{x^2 + 2x}{x^2 - 4x - 5}$

36. $F(x) = \dfrac{x^2 + x}{x^2 - 2x + 4}$

37. $p(x) = \dfrac{(x+3)(x-3)}{(x-5)(x+4)(x+3)}$

38. $p(x) = \dfrac{x^3 + 3x^2}{x^4 - 4x^2}$

In Exercises 39–48, find a viewing window, or windows, that shows a complete graph of the function (if possible, with no erroneous vertical line segments). Be alert for hidden behavior, such as that in Example 5.

39. $f(x) = \dfrac{x^3 + 4x^2 - 5x}{(x^2 - 4)(x^2 - 9)}$

40. $g(x) = \dfrac{x^2 + x - 6}{x^3 - 19x + 30}$

41. $h(x) = \dfrac{2x^2 - x - 6}{x^3 + x^2 - 6x}$

42. $f(x) = \dfrac{x^3 - x + 1}{x^4 - 2x^3 - 2x^2 + x - 1}$

43. $f(x) = \dfrac{2x^4 - 3x^2 + 1}{3x^4 - x^2 + x - 1}$

44. $g(x) = \dfrac{x^4 + 2x^3}{x^5 - 25x^3}$

45. $h(x) = \dfrac{3x^2 + x - 4}{2x^2 - 5x}$

46. $f(x) = \dfrac{2x^2 - 1}{3x^3 + 2x + 1}$

47. $g(x) = \dfrac{x - 4}{2x^3 - 5x^2 - 4x + 12}$

48. $h(x) = \dfrac{x^2 - 9}{x^3 + 2x^2 - 23x - 60}$

49. **(a)** Graph $f(x) = 1/x$ in the viewing window with $-6 \le x \le 6$ and $-6 \le y \le 6$.

(b) Without using a calculator, describe how the graph of $g(x) = 2/x$ can be obtained from the graph of $f(x)$. [*Hint:* $g(x) = 2f(x)$; see Section 2.5.]

(c) Without using a calculator, describe how the graphs of each of the following functions can be obtained from the graph of $f(x)$:

$$h(x) = \frac{1}{x} + 4, \qquad k(x) = \frac{1}{x-3}, \qquad t(x) = \frac{1}{x+2}.$$

[*Hint:* $h(x) = f(x) + 4$; $k(x) = f(x-3)$; and $t(x) = f(x+2)$.]

(d) Without using a calculator, describe how the graph of $p(x) = \dfrac{2}{x-3} + 4$ can be obtained from the graph of $f(x) = 1/x$.

(e) Show that the function $p(x)$ of part (d) is a rational function by rewriting its rule as the quotient of two first-degree polynomials.

(f) If r, s, t are constants, describe how the graph of $q(x) = \dfrac{r}{x+s} + t$ can be obtained from the graph of $f(x) = 1/x$.

(g) Show that the function $q(x)$ of part (f) is a rational function by rewriting its rule as the quotient of two first-degree polynomials.

50. (a) Write the rule of the rational function $g(x) = \frac{4x - 5}{2x + 3}$ in the form $g(x) = r + \frac{1}{sx + t}$ for some constants r, s, t. [*Hint:* Divide the numerator by the denominator and use the Division Algorithm.]

(b) Write the rule of the rational function $h(x) = \frac{3x + 1}{2x - 4}$ in the form $g(x) = r + \frac{1}{sx + t}$ for some constants r, s, t.

(c) Describe how the graphs of the functions $g(x)$ of part (a) and $h(x)$ of part (b) can be obtained from the graph of $f(x) = 1/x$. [*Hint:* Exercise 49.]

(d) If a, b, c, d are constants (with $ad - bc \neq 0$), describe how the graph of $k(x) = \frac{ax + b}{cx + d}$ can be obtained from the graph of $f(x) = 1/x$.

51. (a) Find the difference quotient of the function $f(x) = 1/x$ and express it as a single fraction in lowest terms. [*Hint:* Section 3.2.]

(b) Use the difference quotient in part (a) to determine the average rate of change of $f(x)$ as x changes from 2 to 2.1, from 2 to 2.01, and from 2 to 2.001. Estimate the instantaneous rate of change of $f(x)$ at $x = 2$.

(c) Use the difference quotient in part (a) to determine the average rate of change of $f(x)$ as x changes from 3 to 3.1, from 3 to 3.01, and from 3 to 3.001. Estimate the instantaneous rate of change of $f(x)$ at $x = 3$.

(d) How are the estimated instantaneous rates of change of $f(x)$ at $x = 2$ and $x = 3$ related to the values of the function $g(x) = -1/x^2$ at $x = 2$ and $x = 3$?

52. Do Exercise 51 for the functions $f(x) = 1/x^2$ and $g(x) = -2/x^3$.

53. (a) When $x \geq 0$, what rational function has the same graph as $f(x) = \frac{x - 1}{|x| - 2}$? [*Hint:* Use the definition of absolute value on page 6.]

(b) When $x < 0$, what rational function has the same graph as $f(x) = \frac{x - 1}{|x| - 2}$? [See the hint for part (a).]

(c) Use parts (a) and (b) to explain why the graph of $f(x) = \frac{x - 1}{|x| - 2}$ has two vertical asymptotes. What are they? Confirm your answer by graphing the function.

54. The graph of $f(x) = \frac{2x^3 - 2x^2 - x + 1}{3x^3 - 3x^2 + 2x - 1}$ has a vertical asymptote. Find a viewing window that demonstrates this fact.

55. It costs 2.5 cents per square inch to make the top and bottom of the box in Example 7. The sides cost 1.5 cents per square inch. What are the dimensions of the cheapest possible box?

56. A box with a square base and a volume of 1000 cubic inches is to be constructed. The material for the top and bottom of the box costs \$3 per 100 square inches and the material for the sides costs \$1.25 per 100 square inches.

(a) If x is the length of a side of the base, express the cost of constructing the box as a function of x.

(b) If the side of the base must be at least 6 inches long, for what values of x will the cost of the box be less than \$7.50?

57. A truck traveling at a constant speed on a reasonably straight, level road burns fuel at the rate of $g(x)$ gallons per mile, where x is the speed of the truck (in miles per hour) and $g(x)$ is given by $g(x) = \frac{800 + x^2}{200x}$.

(a) If fuel costs \$1.40 per gallon, find the rule of the cost function $c(x)$ that expresses the cost of fuel for a 500-mile trip as a function of the speed. [*Hint:* $500g(x)$ gallons of fuel are needed to go 500 miles (why?).]

(b) What driving speeds will make the cost of fuel for the trip less than \$250?

(c) What driving speed will minimize the cost of fuel for the trip?

58. Pure alcohol is being added to 50 gallons of a coolant mixture that is 40% alcohol.

(a) Find the rule of the concentration function $c(x)$ that expresses the percentage of alcohol in the resulting mixture as a function of the number x of gallons of pure alcohol that are added. [*Hint:* The final mixture contains $50 + x$ gallons (why?). So $c(x)$ is the amount of alcohol in the final mixture divided by the total amount $50 + x$. How much alcohol is in the original 50-gallon mixture? How much is in the final mixture?]

(b) How many gallons of pure alcohol should be added to produce a mixture that is at least 60% alcohol and no more than 80% alcohol?
(c) Determine algebraically the exact amount of pure alcohol that must be added to produce a mixture that is 70% alcohol.

59. A rectangular garden with an area of 250 square meters is to be located next to a building and fenced on three sides, with the building acting as a fence on the fourth side.
(a) If the side of the garden parallel to the building has length x meters, express the amount of fencing needed as a function of x.
(b) For what values of x will less than 60 meters of fencing be needed?
(c) What value of x will result in the least possible amount of fencing being used? What are the dimensions of the garden in this case?

60. A certain company has fixed costs of $40,000 and variable costs of $2.60 per unit.
(a) Let x be the number of units produced. Find the rule of the average cost function. [The average cost is the cost of the units divided by the number of units.]
(b) Graph the average cost function in a window with $0 \le x \le 100{,}000$ and $0 \le y \le 20$.
(c) Find the horizontal asymptote of the average cost function. Explain what the asymptote means in this situation [how low can the average cost possibly be?].

61. Radioactive waste is stored in a cylindrical tank, whose exterior has radius r and height h (Figure 4–41). The sides, top, and bottom of the tank are one foot thick and the tank has a volume of 150 cubic feet (including top, bottom, and walls).

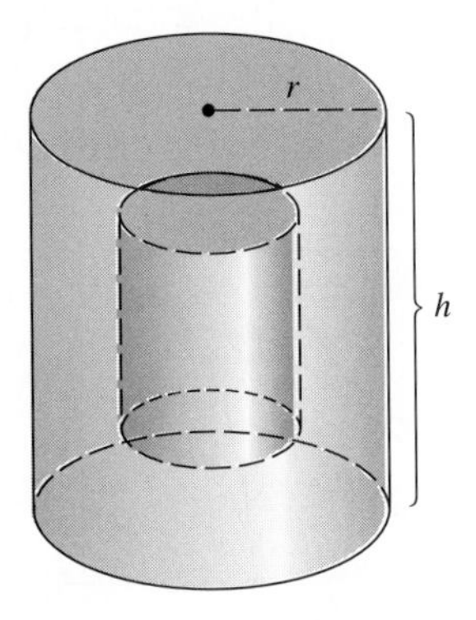

Figure 4-41

(a) Express the interior height h_1 (that is, the height of the storage area) as a function of h.
(b) Express the interior height as a function of r.
(c) Express the volume of the interior as a function of r.
(d) Explain why r must be greater than 1.
(e) What should the dimensions of the tank be in order for it to hold as much as possible?

62. The relationship between the fixed focal length F of a camera, the distance u from the object being photographed to the lens, and the distance v from the lens to the film (Figure 4–42) is given by $\frac{1}{F} = \frac{1}{u} + \frac{1}{v}$.

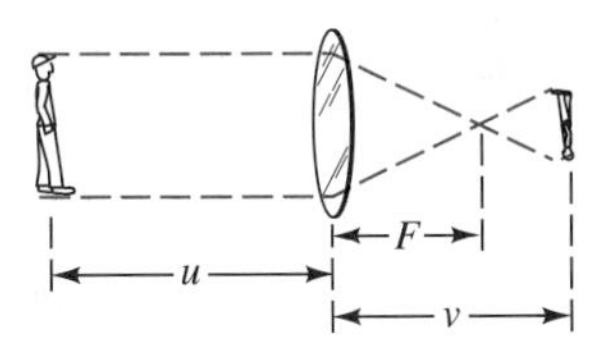

Figure 4-42

(a) If the focal length is 50 mm, express v as a function of u.
(b) What is the horizontal asymptote of the graph of the function in part (a)?
(c) Graph the function in part (a) when $50 \text{ mm} < u < 35{,}000 \text{ mm}$.
(d) When you focus the camera on an object, the distance between the lens and the film is changed. If the distance from the lens to the camera changes by less than .1 mm, the object will remain in focus. Explain why you have more latitude in focusing on distant objects than on very close ones.

63. The formula for the gravitational acceleration (in units of meters/sec^2) of an object relative to the earth is

$$g(r) = \frac{3.987 \times 10^{14}}{(6.378 \times 10^6 + r)^2}$$

where r is the distance in meters above the earth's surface.
(a) What is the gravitational acceleration at the earth's surface?
(b) Graph the function $g(r)$ for $r \ge 0$.
(c) Can you ever escape the pull of gravity? [Does the graph have any r-intercepts?]

4.4.A *Excursion* OTHER RATIONAL FUNCTIONS

We now examine the graphs of rational functions in which the degree of the denominator is smaller than the degree of the numerator. Such a graph has no horizontal asymptote. However, it does have some polynomial curve as an asymptote, which means that the graph will get very close to this curve when $|x|$ is very large.

EXAMPLE 1 To graph $f(x) = \dfrac{x^3 + 3x^2 + x + 1}{x^2 + 2x - 1}$, we begin by finding the vertical asymptotes and the x- and y-intercepts. The quadratic formula can be used to find the roots of the denominator:

$$x = \frac{-2 \pm \sqrt{2^2 - 4 \cdot 1(-1)}}{2 \cdot 1} = \frac{-2 \pm \sqrt{8}}{2} = \frac{-2 \pm 2\sqrt{2}}{2} = -1 \pm \sqrt{2}.$$

It is easy to verify that neither of these numbers is a root of the numerator, so the graph has vertical asymptotes at $x = -1 - \sqrt{2}$ and $x = -1 + \sqrt{2}$. The y-intercept is $f(0) = -1$. The x-intercepts are the roots of the numerator.

GRAPHING EXPLORATION Use a calculator to verify that $x^3 + 3x^2 + x + 1$ has exactly one real root, located between -3 and -2.

Therefore, the graph of $f(x)$ has one x-intercept. Using this information and the calculator graph in Figure 4–43 (which erroneously shows some vertical segments), we conclude that the graph looks approximately like Figure 4–44.

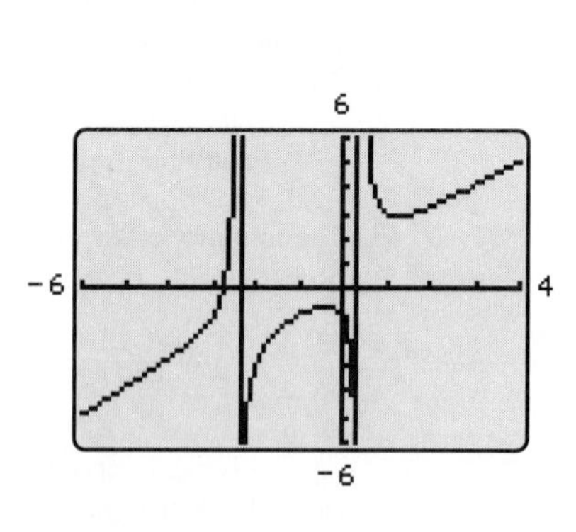

Figure 4-43

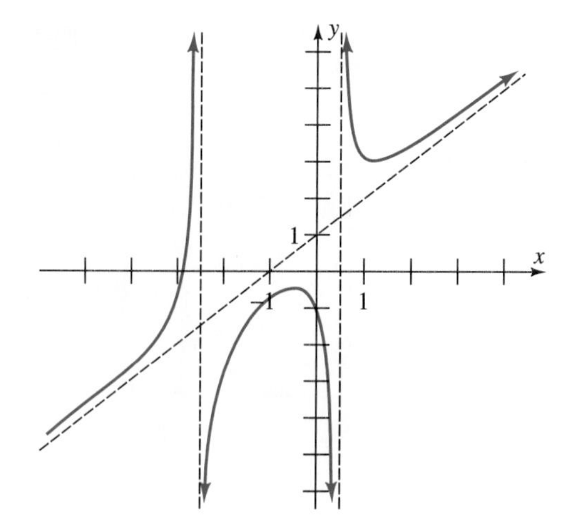

Figure 4-44

At the left and right ends, the graph moves away from the x-axis. To understand the behavior of the graph when $|x|$ is large, divide the numerator of $f(x)$ by its denominator:

$$
\begin{array}{r|l}
 & x + 1 \\
\hline
x^2 + 2x - 1 & x^3 + 3x^2 + x + 1 \\
 & x^3 + 2x^2 - x \\
\hline
 & \qquad x^2 + 2x + 1 \\
 & \qquad x^2 + 2x - 1 \\
\hline
 & \qquad\qquad\qquad 2
\end{array}
$$

By the Division Algorithm

$$x^3 + 3x^2 + x + 1 = (x^2 + 2x - 1)(x + 1) + 2$$

so that

$$
\begin{aligned}
f(x) &= \frac{x^3 + 3x^2 + x + 1}{x^2 + 2x - 1} = \frac{(x^2 + 2x - 1)(x + 1) + 2}{x^2 + 2x - 1} \\
&= (x + 1) + \frac{2}{x^2 + 2x - 1}.
\end{aligned}
$$

Now when x is very large in absolute value, so is $x^2 + 2x - 1$. Hence, $2/(x^2 + 2x - 1)$ is very close to 0 by the Big–Little Principle and $f(x)$ is very close to $(x + 1) + 0$. Therefore, as x gets larger in absolute value, the graph of $f(x)$ gets closer and closer to the line $y = x + 1$ (the dashed slanted line in Figure 4–44) and this line is an asymptote of the graph.* Note that $x + 1$ is just the quotient obtained in the long division above. ■

It is instructive to examine the graph in Example 1 further to see that the asymptote accurately indicates the behavior of the function when $|x|$ is large.

GRAPHING EXPLORATION Using the viewing window with $-20 \le x \le 20$ and $-20 \le y \le 20$, graph both $f(x)$ and $y = x + 1$ on the same screen.

Except near the vertical asymptotes of $f(x)$, the two graphs are virtually identical.

GRAPHING EXPLORATION Now change the range values, so that the viewing window has $-100 \le x \le 100$ and $-100 \le y \le 100$.

In this viewing window, the vertical asymptotes of $f(x)$ are no longer visible and the graph is indistinguishable from the graph of the asymptote $y = x + 1$.

*An asymptote that is a nonvertical and nonhorizontal straight line is called an **oblique asymptote.**

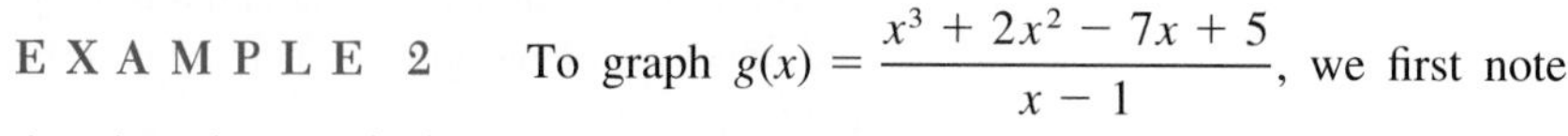

EXAMPLE 2 To graph $g(x) = \dfrac{x^3 + 2x^2 - 7x + 5}{x - 1}$, we first note that there is a vertical asymptote at $x = 1$ (root of the denominator, but not the numerator). The y-intercept is at $g(0) = -5$. By carefully choosing a viewing window that accurately portrays the behavior of $g(x)$ near its vertical asymptote, we obtain Figure 4–45.

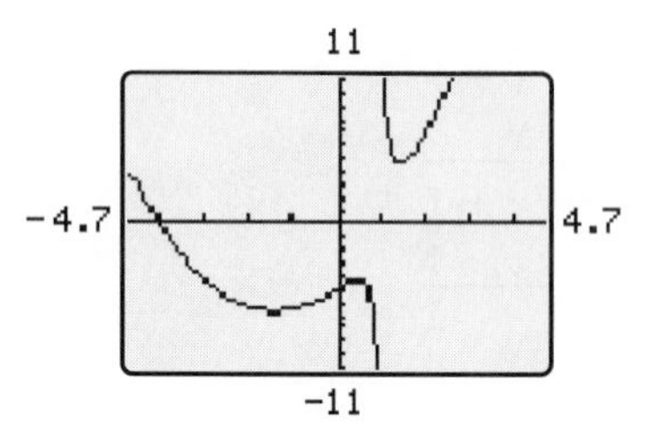

Figure 4-45

> GRAPHING EXPLORATION Verify that the x-intercept near $x = -4$ is the only one by showing graphically that the numerator of $g(x)$ has exactly one real root.

To confirm that Figure 4–45 is a complete graph, we find its asymptote when $|x|$ is large. Divide the denominator by the numerator and use the Division Algorithm to rewrite $g(x)$:

$$
\begin{array}{r|l}
 & x^2 + 3x - 4 \\
\hline
x - 1 & x^3 + 2x^2 - 7x + 5 \\
 & x^3 - x^2 \\
\hline
 & \quad 3x^2 - 7x + 5 \\
 & \quad 3x^2 - 3x \\
\hline
 & \qquad -4x + 5 \\
 & \qquad -4x + 4 \\
\hline
 & \qquad\qquad 1
\end{array}
$$

Hence,

$$
\begin{aligned}
g(x) &= \frac{x^3 + 2x^2 - 7x + 5}{x - 1} = \frac{(x - 1)(x^2 + 3x - 4) + 1}{x - 1} \\
&= (x^2 + 3x - 4) + \frac{1}{x - 1}.
\end{aligned}
$$

When $|x|$ is large, $1/(x - 1)$ is very close to 0 (why?), so that $y = x^2 + 3x - 4$ is the asymptote. Once again, the asymptote is given by the quotient of the division.

> GRAPHING EXPLORATION Graph $g(x)$ and $y = x^2 + 3x - 4$ on the same screen to show that the graph of $g(x)$ does get very close to the asymptote when $|x|$ is large. Then find a large enough viewing window so that the two graphs appear to be identical. ■

The procedures used in the preceding examples may be summarized as follows.

Graphing the Rational Function

$$f(x) = \frac{g(x)}{h(x)}$$

When Degree $g(x) >$ Degree $h(x)$

1. **Analyze the function algebraically to determine its vertical asymptotes, holes, and intercepts.**
2. **Divide the numerator $g(x)$ by the denominator $h(x)$. The quotient $q(x)$ is the nonvertical asymptote of the graph, which describes the behavior of the graph when $|x|$ is large.**
3. **Use the preceding information to select an appropriate viewing window (or windows), to interpret the calculator's version of the graph (if necessary), and to sketch an accurate graph.**

EXERCISES 4.4.A

In Exercises 1–4, find the nonvertical asymptote of the graph of the function when $|x|$ is large and find a viewing window in which the ends of the graph are within .1 of this asymptote.

1. $f(x) = \dfrac{x^3 - 1}{x^2 - 4}$

2. $g(x) = \dfrac{x^3 - 4x^2 + 6x + 5}{x - 2}$

3. $h(x) = \dfrac{x^3 + 3x^2 - 4x + 1}{x + 4}$

4. $f(x) = \dfrac{x^3 + 3x^2 - 4x + 1}{x^2 - x}$

In Exercises 5–12, analyze the function algebraically: List its vertical asymptotes and holes, and determine its nonvertical asymptote. Then sketch a complete graph of the function.

5. $f(x) = \dfrac{x^2 - x - 6}{x - 2}$

6. $k(x) = \dfrac{x^2 + x - 2}{x}$

7. $Q(x) = \dfrac{4x^2 + 4x - 3}{2x - 5}$

8. $K(x) = \dfrac{3x^2 - 12x + 15}{3x + 6}$

9. $f(x) = \dfrac{x^3 - 2}{x - 1}$

10. $p(x) = \dfrac{x^3 + 8}{x + 1}$

11. $q(x) = \dfrac{x^3 - 1}{x - 2}$

12. $f(x) = \dfrac{x^4 - 1}{x^2}$

In Exercises 13–18, find a viewing window (or windows) that shows a complete graph of the function (if possible, with no erroneous vertical line segments). Be alert for hidden behavior.

13. $f(x) = \dfrac{2x^2 + 5x + 2}{2x + 7}$

14. $g(x) = \dfrac{2x^3 + 1}{x^2 - 1}$

15. $h(x) = \dfrac{x^3 - 2x^2 + x - 2}{x^2 - 1}$

16. $f(x) = \dfrac{3x^3 - 11x - 1}{x^2 - 4}$

17. $g(x) = \dfrac{2x^4 + 7x^3 + 7x^2 + 2x}{x^3 - x + 50}$

18. $h(x) = \dfrac{2x^3 + 7x^2 - 4}{x^2 + 2x - 3}$

19. **(a)** Show that when $0 < x < 4$, the rational function

$$r(x) = \frac{4096x^3 + 34560x^2 + 19440x + 729}{18432x^2 + 34560x + 5832}$$

is a good approximation of the function $s(x) = \sqrt{x}$ by graphing both functions in the viewing window with $0 \le x \le 4$ and $0 \le y \le 2$.

(b) For what values of x is $r(x)$ within .01 of $s(x)$?

20. Find a rational function f that has these properties:

(i) The curve $y = x^3 - 8$ is an asymptote of the graph of f.

(ii) $f(2) = 1$.

(iii) The line $x = 1$ is a vertical asymptote of the graph of f.

4.5 POLYNOMIAL AND RATIONAL INEQUALITIES

Inequalities may be solved using algebraic or geometric methods, both of which are discussed here. Whenever possible we shall use algebra to obtain exact solutions. When algebraic methods are too difficult, approximate graphical solutions will be found. The basic tools for working with inequalities are the following principles.

Basic Principles for Solving Inequalities ▶

Performing any of the following operations on an inequality produces an equivalent inequality:*

1. **Add or subtract the same quantity on both sides of the inequality.**
2. **Multiply or divide both sides of the inequality by the same *positive* quantity.**
3. **Multiply or divide both sides of the inequality by the same *negative* quantity and *reverse the direction of the inequality.***

Note Principle 3 carefully. It says, for example, that if you multiply both sides of $-3 < 5$ by -2, the equivalent inequality is $6 > -10$ (direction of inequality is reversed).

Linear Inequalities

EXAMPLE 1 To solve $5x + 3 \leq 6 + 7x$ we use the Basic Principles to transform it into an inequality whose solutions are obvious:

Subtract 7x from both sides: $\quad -2x + 3 \leq 6$

Subtract 3 from both sides: $\quad -2x \leq 3$

Divide both sides by -2 and reverse the direction *of the inequality:* $\quad x \geq -3/2$

Therefore, the solutions are all real numbers greater than or equal to $-3/2$, that is, the interval $[-3/2, \infty)$, as shown in Figure 4–46. ■

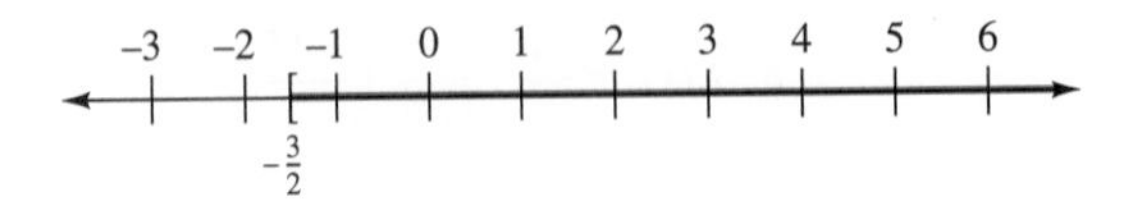

Figure 4-46

EXAMPLE 2 A solution of the inequality $2 \leq 3x + 5 < 2x + 11$ is any number that is a solution of *both* of these inequalities:

*Two inequalities are **equivalent** if they have the same solutions.

$$2 \le 3x + 5 \qquad \text{and} \qquad 3x + 5 < 2x + 11$$

Each of these inequalities can be solved by the methods used above. For the first one we have:

$$2 \le 3x + 5$$

Subtract 5 from both sides: $\quad -3 \le 3x$

Divide both sides by 3: $\quad -1 \le x$

The second inequality is solved similarly:

$$3x + 5 < 2x + 11$$

Subtract 5 from both sides: $\quad 3x < 2x + 6$

Subtract 2x from both sides: $\quad x < 6$

The solutions of the original inequality are the numbers x that satisfy *both* $-1 \le x$ *and* $x < 6$, that is, all x with $-1 \le x < 6$. Thus, the solutions are precisely the numbers in the interval $[-1, 6)$, as shown in Figure 4–47. ■

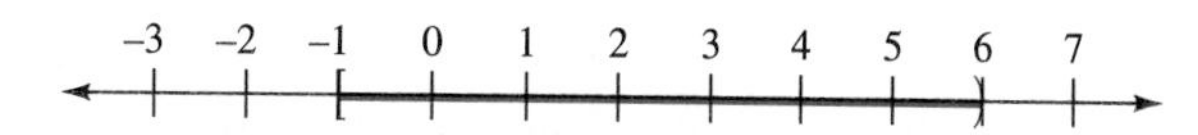

Figure 4-47

EXAMPLE 3 When solving the inequality $4 < 3 - 5x < 18$, in which the variable appears only in the middle part, you can proceed as follows:

$$4 < 3 - 5x < 18$$

Subtract 3 from each part: $\quad 1 < -5x < 15$

Divide each part by -5 *and* reverse the direction *of the inequalities:* $\quad -\frac{1}{5} > x > -3$

Reading this last inequality from right to left we see that $-3 < x < -1/5$, so that the solutions are precisely the numbers in the interval $(-3, -1/5)$. ■

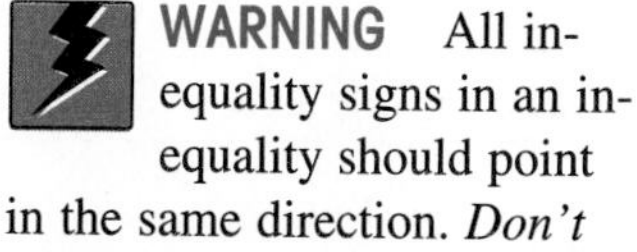

WARNING All inequality signs in an inequality should point in the same direction. *Don't* write things like $4 < x > 2$ or $-3 \ge x < 5$.

Polynomial Inequalities

Although the basic principles play a role in the solution of nonlinear inequalities, the key to solving such inequalities is this geometric fact:

> **The graph of $y = f(x)$ lies above the x-axis exactly when $f(x) > 0$ and below the x-axis exactly when $f(x) < 0$.**

Consequently, the solutions of $f(x) > 0$ are the numbers x for which the graph of f lies above the x-axis and the solutions $f(x) < 0$ are the numbers x for which the graph of f lies below the x-axis.

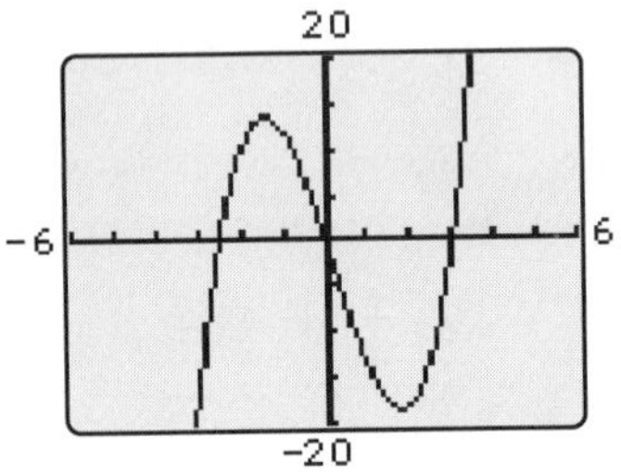

EXAMPLE 4 To solve $2x^3 - 15x < x^2$, replace it by the equivalent inequality

$$2x^3 - x^2 - 15x < 0$$

and consider the graph of the function $f(x) = 2x^3 - x^2 - 15x$ (Figure 4–48). Since $f(x)$ factors as

$$f(x) = 2x^3 - x^2 - 15x = x(2x^2 - x - 15) = x(2x + 5)(x - 3)$$

its roots (the x-intercepts of its graph) are $x = 0, x = -5/2$, and $x = 3$. The graph of $f(x) = 2x^3 - x^2 - 15x$ in Figure 4–48 is complete (why?) and lies below the x-axis when $x < -5/2$ or $0 < x < 3$. Therefore, the solutions of

$$2x^3 - x^2 - 15x < 0,$$

and hence of the original inequality, are all numbers x such that $x < -5/2$ or $0 < x < 3$. ■

Figure 4-48

EXAMPLE 5 Solve $2x^3 - x^2 - 15x \geq 0$.

Solution Figure 4–48 shows that the solutions of $2x^3 - x^2 - 15x > 0$ (that is, the numbers x for which the graph of $f(x) = 2x^3 - x^2 - 15x$ lies above the x-axis) are all x such that $-5/2 < x < 0$ or $x > 3$. The solutions of the equation $2x^3 - x^2 - 15x = 0$ are the roots of $f(x) = 2x^3 - x^2 - 15x$, namely, $0, -5/2$, and 3 as we saw in Example 4. Therefore, the solutions of the given inequality are all numbers x such that $-5/2 \leq x \leq 0$ or $x \geq 3$. ■

When the roots of a polynomial $f(x)$ cannot be determined exactly, zoom-in or a root finder can be used to approximate them and to find approximate solutions of the inequalities $f(x) > 0$ and $f(x) < 0$.

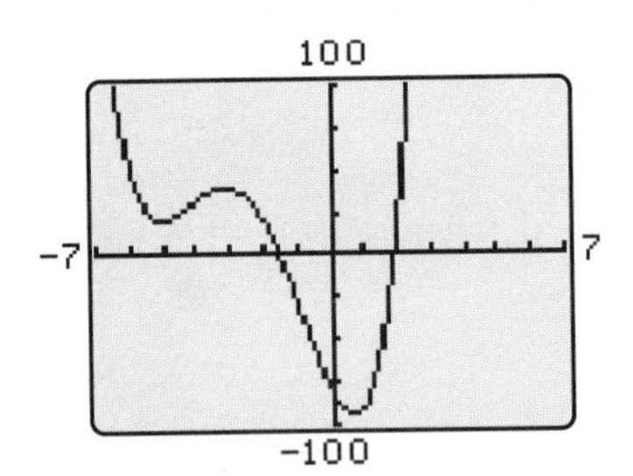

Figure 4-49

EXAMPLE 6 To solve $x^4 + 10x^3 + 21x^2 + 8 > 40x + 88$, we note that this inequality is equivalent to

$$x^4 + 10x^3 + 21x^2 - 40x - 80 > 0.$$

The graph $f(x) = x^4 + 10x^3 + 21x^2 - 40x - 80$ in Figure 4–49 is complete (why?) and shows that $f(x)$ has two roots, one between -2 and -1 and the other near 2.

GRAPHING EXPLORATION Use zoom-in or a root finder to show that the approximate roots of $f(x)$ are -1.53 and 1.89.

Therefore, the approximate solutions of the inequality (the numbers x for which the graph is above the x-axis) are all numbers x such that $x < -1.53$ or $x > 1.89$. ■

WARNING Do not attempt to write the solution in Example 6, namely, "$x < -1.53$ or $x > 1.89$" as a single inequality. If you do, the result will be a *nonsense statement* such as $-1.53 > x > 1.89$ (which says, among other things, that $-1.53 > 1.89$).

Quadratic and Factorable Inequalities

The preceding examples show that solving a polynomial inequality depends only on knowing the roots of a polynomial and the places where its graph is above or below the x-axis. In the case of quadratic inequalities or completely factored polynomial inequalities a calculator is not needed to determine this information.

EXAMPLE 7* The solutions of $2x^2 + 3x - 4 \leq 0$ are the numbers x at which the graph of $f(x) = 2x^2 + 3x - 4$ lies on or below the x-axis. The points where the graph meets the x-axis are the roots of $f(x) = 2x^2 + 3x - 4$, which can be found by means of the quadratic formula:

$$x = \frac{-3 \pm \sqrt{3^2 - 4 \cdot 2(-4)}}{2 \cdot 2} = \frac{-3 \pm \sqrt{41}}{4}.$$

From Section 3.3 we know that the graph of $f(x)$ is an upward opening parabola, so the graph must have the general shape shown in Figure 4-50:

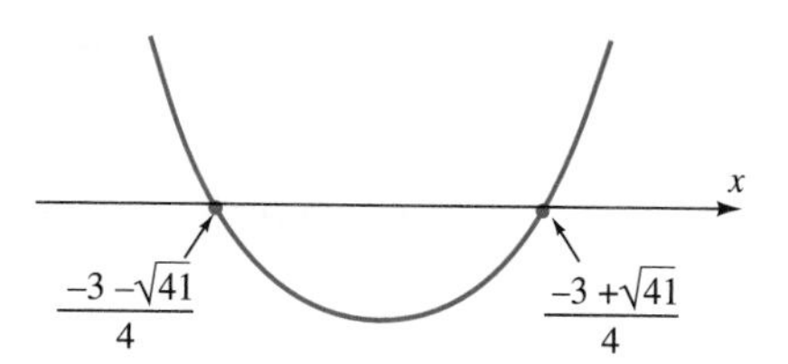

Figure 4-50

The graph lies below the x-axis between the two roots. Therefore, the solutions of the original inequality are all numbers x such that

$$\frac{-3 - \sqrt{41}}{4} \leq x \leq \frac{-3 + \sqrt{41}}{4}. \quad ■$$

EXAMPLE 8 Solve $(x + 15)(x - 2)^6(x - 10) \leq 0$.

Solution The roots of $f(x) = (x + 15)(x - 2)^6(x - 10)$ are easily read from the factored form: -15, 2, and 10. So we need only determine where the graph of

*If you haven't read Section 3.3 (Quadratic Functions), skip this example.

$f(x)$ is on or below the x-axis. To do this without a calculator, note that the three roots of $f(x)$ divide the x-axis into four intervals:

$$x < -15, \qquad -15 < x < 2, \qquad 2 < x < 10, \qquad x > 10.$$

For each of these intervals, we shall determine whether the graph is above or below the x-axis.

Consider, for example, the interval between the roots 2 and 10. The graph of $f(x)$ touches the x-axis at $x = 2$ and $x = 10$, but does not touch the axis at any point in between since the only other root (x-intercept) is -15. Furthermore, the graph of $f(x)$ is continuous, which means that between the points (2, 0) and (10, 0) it can be drawn without lifting the pencil from the paper. If you experiment with Figure 4–51, you will find that the only way to draw a graph from (2, 0) to (10, 0) without lifting your pencil and without touching the x-axis is to stay entirely on one side of the x-axis (try it!). In other words, between the two adjacent roots 2 and 10 the graph of $f(x)$ is either entirely above the x-axis or entirely below it.

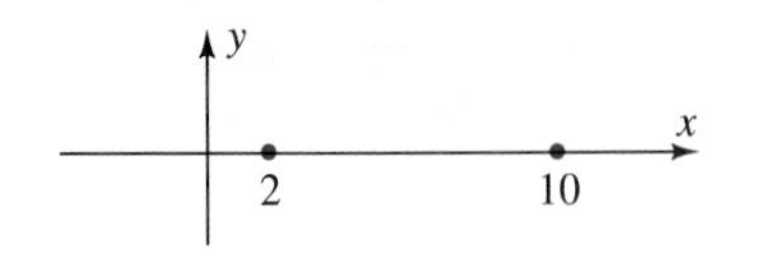

Figure 4-51

In order to determine which is the case, choose any number between 2 and 10, say $x = 4$, and evaluate $f(4)$:

$$f(4) = (4 + 15)(4 - 2)^6(4 - 10) = 19(2^6)(-6).$$

You don't even have to finish the computation to see that $f(4)$ is a negative number. Therefore the point $(4, f(4))$ on the graph of $f(x)$ lies below the x-axis. Since one point of the graph between 2 and 10 lies below the x-axis, the entire graph must be below the x-axis between 2 and 10.

The location of the graph on the other intervals can be determined similarly, by choosing a "test number" in each interval, as summarized in this chart:

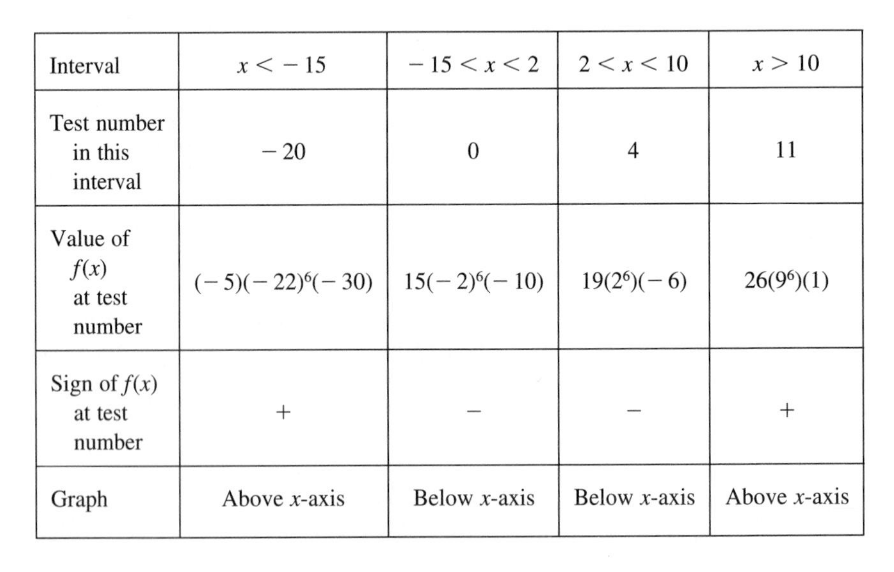

Interval	$x < -15$	$-15 < x < 2$	$2 < x < 10$	$x > 10$
Test number in this interval	-20	0	4	11
Value of $f(x)$ at test number	$(-5)(-22)^6(-30)$	$15(-2)^6(-10)$	$19(2^6)(-6)$	$26(9^6)(1)$
Sign of $f(x)$ at test number	+	−	−	+
Graph	Above x-axis	Below x-axis	Below x-axis	Above x-axis

The last line of the chart shows that the intervals where the graph is below the x-axis are $-15 < x < 2$ and $2 < x < 10$. Since the graph touches the x-axis at the roots -15, 2, and 10, the solutions of the original inequality (the numbers x

for which the graph is on or below the x-axis) are all numbers x such that $-15 \leq x \leq 10$. ■

The procedures used in Examples 4–8 may be summarized as follows.

Solving Polynomial Inequalities ▶

1. **Write the inequality in one of these forms:**

$$f(x) > 0, \qquad f(x) \geq 0, \qquad f(x) < 0, \qquad f(x) \leq 0.$$

2. **Determine the roots of $f(x)$, exactly if possible, approximately otherwise.**
3. **Use a calculator (as in Examples 4–6), your knowledge of quadratic functions (as in Example 7), or a sign chart (as in Example 8) to determine whether the graph of $f(x)$ is above or below the x-axis on each of the intervals determined by the roots.**
4. **Use the information in step 3 to find the solutions of the inequality.**

Rational Inequalities

Rational inequalities are solved in essentially the same way that polynomial inequalities are solved, with one difference. The solution of the polynomial inequality $f(x) < 0$, for example, depends on the x-intercepts (roots) of $f(x)$ since these are the only places where the graph of $f(x)$ can possibly move from one side of the x-axis to the other. The graph of a rational function may cross the x-axis at an x-intercept, but there is another possibility: The graph may be above the x-axis on one side of a vertical asymptote and below it on the other side (see, for instance, Examples 3–6 in Section 4.4).* Since the x-intercepts of the graph of the rational function $g(x)/h(x)$ are determined by the roots of its numerator $g(x)$ and the vertical asymptotes by the roots of its denominator $h(x)$, all of these roots must be considered in determining the solution of an inequality involving $g(x)/h(x)$.

EXAMPLE 9 Solve $\dfrac{x}{x-1} > -6$.

Solution There are three ways to solve this inequality.

Geometric: The fastest way to get an approximate solution is to replace the given inequality by an equivalent one

*These are the only ways that the graph of a rational function can change from one side of the x-axis to the other because a rational function is continuous everywhere that it is defined. On any interval on which the function is defined, its graph can be drawn without lifting pencil from paper; see the discussion in the second paragraph on page 268.

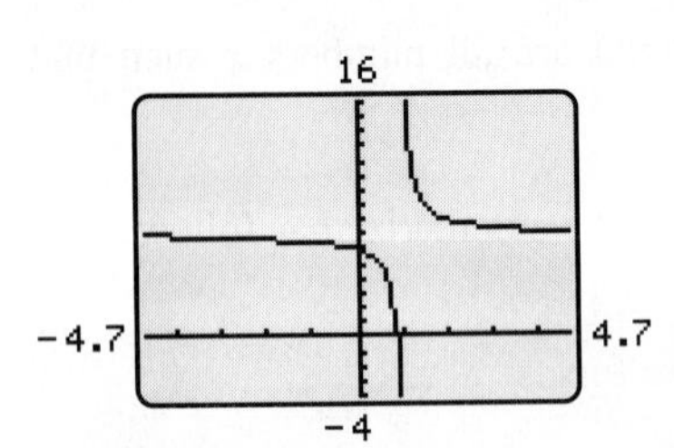

Figure 4-52

$$(*) \qquad \frac{x}{x-1} + 6 > 0$$

and graph the function $f(x) = \dfrac{x}{x-1} + 6$ as in Figure 4–52.

The graph is above the x-axis everywhere except between the x-intercept and the vertical asymptote $x = 1$. Using zoom-in or a root finder, we see that the x-intercept is approximately .857. Therefore, the approximate solutions of the original inequality are all numbers x such that $x < .857$ or $x > 1$.

Algebraic/Geometric: Proceed as above, but rewrite the rule of the function f as a single rational expression before graphing:

$$f(x) = \frac{x}{x-1} + 6 = \frac{x}{x-1} + \frac{6(x-1)}{x-1} = \frac{x + 6x - 6}{x-1} = \frac{7x-6}{x-1}.$$

When the rule of f is written in this form, it is easy to see that the x-intercept of the graph (the root of the numerator) is $x = 6/7$ (whose decimal approximation begins .857). Therefore, the exact solutions of the original inequality (the numbers x for which the graph in Figure 4–52 is above the x-axis) are all numbers x such that $x < 6/7$ or $x > 1$.

Algebraic: Write the rule of the function f as a single rational expression $f(x) = \dfrac{7x-6}{x-1}$. The roots of the numerator and denominator (6/7 and 1) divide the x-axis into three intervals. Use test numbers and a sign chart instead of graphing to determine the location of the graph on each interval:*

Interval	$x < 6/7$	$6/7 < x < 1$	$x > 1$
Test number in this interval	0	.9	2
Value of $f(x)$ at test number	$\dfrac{7 \cdot 0 - 6}{0 - 1}$	$\dfrac{7(.9) - 6}{.9 - 1}$	$\dfrac{7 \cdot 2 - 6}{2 - 1}$
Sign of $f(x)$ at test number	+	−	+
Graph	Above x-axis	Below x-axis	Above x-axis

The last line of the chart shows that the solutions of the original inequality (the numbers x for which the graph is above the x-axis) are all such that $x < 6/7$ or $x > 1$. ■

*The justification for this approach is essentially the same as that in Example 8: Because f is continuous everywhere that it is defined, the graph can change from one side of the x-axis to the other only at x-intercepts or vertical asymptotes, so testing one number in each interval is sufficient to determine the side on which the graph lies.

WARNING Don't treat rational inequalities as if they are equations, as in this *incorrect* ''solution'' of the preceding example:

$$\frac{x}{x-1} > -6$$
$$x > -6(x-1) \qquad \textit{[Both sides multiplied by } x-1\textit{]}$$
$$x > -6x + 6$$
$$7x > 6$$
$$x > \frac{6}{7}$$

According to this, the inequality has no negative solution and $x = 1$ is a solution, but as we saw in Example 9, *every* negative number is a solution and $x = 1$ is not.*

The algebraic technique of writing the left side of the inequality as a single rational expression is useful whenever the resulting numerator has low degree (so that its roots can be found exactly), but can usually be omitted when the roots of the numerator must be approximated.

EXAMPLE 10 To solve

$$\frac{x^4 - 2x^3 + 4x - 3}{x^4 - 16} \geq 2$$

we replace the inequality by an equivalent one

$$\frac{x^4 - 2x^3 + 4x - 3}{x^4 - 16} - 2 \geq 0$$

and graph the function f whose rule is given by the left side of this inequality.

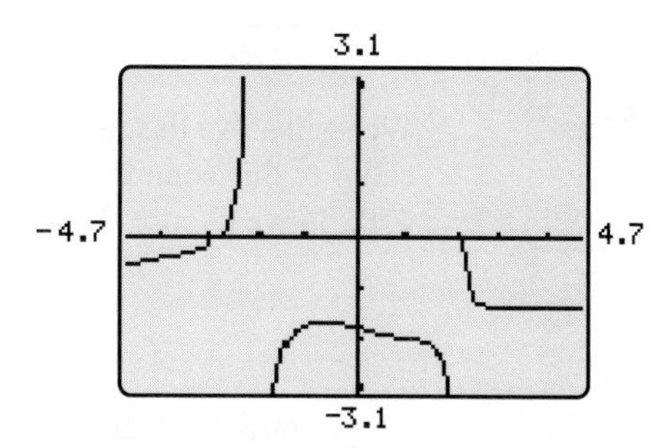

Figure 4-53

*The source of the error is multiplying by $x - 1$. This quantity is negative for some values of x and positive for others. To do this calculation correctly, you must consider two separate cases and reverse the direction of the inequality when $x - 1$ is negative.

Figure 4–53 suggests that the graph has vertical asymptotes at $x = -2$ and $x = 2$. This is indeed the case, since these are the real roots of the denominator:

$$x^4 - 16 = (x^2 + 4)(x^2 - 4) = (x^2 + 4)(x + 2)(x - 2).$$

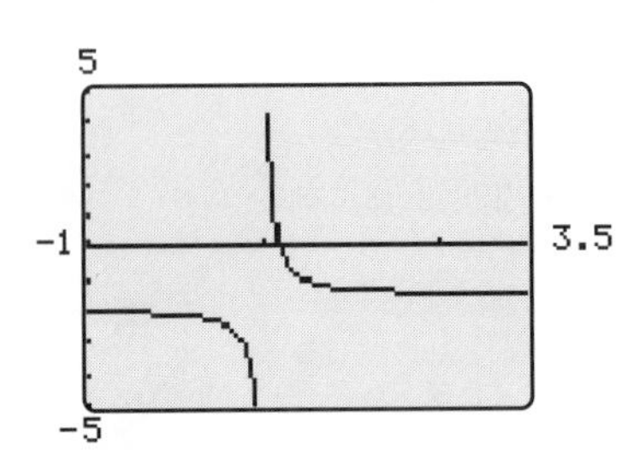

Figure 4-54

The graph has an x-intercept between -3 and -2 and by using zoom-in or a root finder we see that this root is approximately $x = -2.81$. Another viewing window is needed to see what's going on near $x = 2$; so, we use Figure 4–54.

We find that the root is approximately $x = 2.09$. The solutions of the original inequality (the numbers x for which the graph is on or above the x-axis) can now be read from Figures 4–53 and 4–54:

$$-2.81 \leq x < -2 \qquad \text{or} \qquad 2 < x \leq 2.09.$$

Note that the function is not defined at $x = \pm 2$, so that $<$ cannot be replaced by $\leq$. ■

Applications

EXAMPLE 11 A computer store has determined that the cost C of ordering and storing x laser printers is given by

$$C = 2x + \frac{300{,}000}{x}.$$

If the delivery truck can bring at most 450 printers per order, how many printers should be ordered at a time to keep the cost below \$1600?

Solution To find the values of x that make C less than 1600, we must solve the inequality

$$2x + \frac{300{,}000}{x} < 1600 \qquad \text{or equivalently,} \qquad 2x + \frac{300{,}000}{x} - 1600 < 0.$$

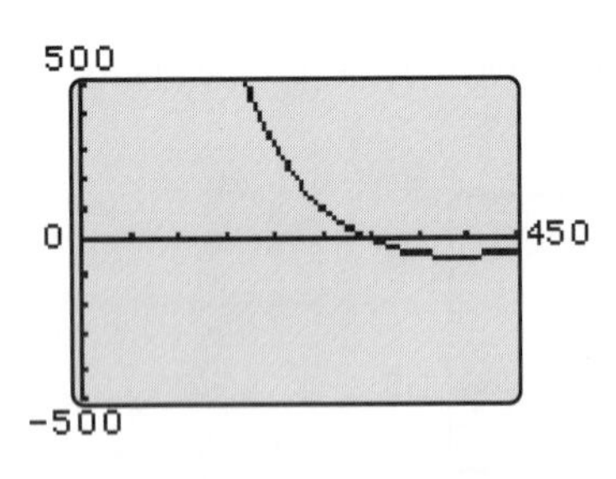

Figure 4-55

We shall solve this inequality graphically, although it can also be solved algebraically. In this context, the only solutions that make sense are those between 0 and 450. So we choose the viewing window in Figure 4–55 and graph

$$f(x) = 2x + \frac{300{,}000}{x} - 1600.$$

Figure 4–55 is consistent with the fact that $f(x)$ has a vertical asymptote at $x = 0$ and shows that the desired solutions (numbers where the graph is below the x-axis) are all numbers x between the root and 450. Zoom-in or a root finder shows that the root is $x \approx 300$. In fact this is the exact root since a simple computation shows that $f(300) = 0$. (Do it!) Therefore, to keep costs under \$1600, x printers should be ordered each time, with $300 < x \leq 450$. ■

EXERCISES 4.5

In Exercises 1–20, solve the inequality and express your answer in interval notation.

1. $2x + 4 \le 7$ **2.** $3x - 5 > -6$

3. $3 - 5x < 13$ **4.** $2 - 3x < 11$

5. $6x + 3 \le x - 5$ **6.** $5x + 3 \le 2x + 7$

7. $5 - 7x < 2x - 4$ **8.** $5 - 3x > 7x - 3$

9. $2 < 3x - 4 < 8$ **10.** $1 < 5x + 6 < 9$

11. $0 < 5 - 2x \le 11$ **12.** $-4 \le 7 - 3x < 0$

13. $2x + 7(3x - 2) < 2(x - 1)$

14. $x + 3(x - 5) \ge 3x + 2(x + 1)$

15. $\dfrac{x+1}{2} - 3x \le \dfrac{x+5}{3}$

16. $\dfrac{x-1}{4} + 2x \ge \dfrac{2x-1}{3} + 2$

17. $2x + 3 \le 5x + 6 < -3x + 7$

18. $4x - 2 < x + 8 < 9x + 1$

19. $3 - x < 2x + 1 \le 3x - 4$

20. $2x + 5 \le 4 - 3x < 1 - 4x$

In Exercises 21–24, a, b, c, and d are positive constants. Solve the inequality for x.

21. $ax - b < c$ **22.** $d - cx > a$

23. $0 < x - c < a$ **24.** $-d < x - c < d$

In Exercises 25–54, solve the inequality. Find exact solutions when possible and approximate ones otherwise.

25. $x^2 - 4x + 3 \le 0$ **26.** $x^2 - 7x + 10 \le 0$

27. $x^2 + 9x + 14 \ge 0$ **28.** $x^2 + 8x + 15 \ge 0$

29. $6 + x - x^2 \le 0$ **30.** $4 - 3x - x^2 \ge 0$

31. $x^3 - x \ge 0$ **32.** $x^3 + 2x^2 + x > 0$

33. $x^3 - 2x^2 - 3x < 0$ **34.** $x^4 - 14x^3 + 48x^2 \ge 0$

35. $x^4 - 5x^2 + 4 < 0$ **36.** $x^4 - 10x^2 + 9 \le 0$

37. $x^3 - 2x^2 - 5x + 7 \ge 2x + 1$

38. $x^4 - 6x^3 + 2x^2 < 5x - 2$

39. $2x^4 + 3x^3 < 2x^2 + 4x - 2$

40. $x^5 + 5x^4 > 4x^3 - 3x^2 + 2$

41. $\dfrac{3x+1}{2x-4} > 0$ **42.** $\dfrac{2x-1}{5x+3} \ge 0$

43. $\dfrac{x^2+x-2}{x^2-2x-3} < 0$ **44.** $\dfrac{2x^2+x-1}{x^2-4x+4} \ge 0$

45. $\dfrac{x-2}{x-1} < 1$ **46.** $\dfrac{-x+5}{2x+3} \ge 2$

47. $\dfrac{x-3}{x+3} \le 5$ **48.** $\dfrac{2x+1}{x-4} > 3$

49. $\dfrac{2}{x+3} \ge \dfrac{1}{x-1}$ **50.** $\dfrac{1}{x-1} < \dfrac{-1}{x+2}$

51. $\dfrac{x^3 - 3x^2 + 5x - 29}{x^2 - 7} > 3$

52. $\dfrac{x^4 - 3x^3 + 2x^2 + 2}{x - 2} > 15$

53. $\dfrac{2x^2 + 6x - 8}{2x^2 + 5x - 3} < 1$ [Be alert for hidden behavior.]

54. $\dfrac{1}{x^2 + x - 6} + \dfrac{x-2}{x+3} > \dfrac{x+3}{x-2}$

55. The temperature F in degrees Fahrenheit is related to the temperature C in degrees Celsius by the formula $F = \dfrac{9}{5}C + 32$. Find the temperature range in degrees Celsius corresponding to the Fahrenheit temperatures from 32° (where water freezes) through 212° (where water boils).

56. A business executive leases a car for \$300 per month. She decides to lease another brand for \$250 per month, but has to pay a penalty of \$1000 for breaking the first lease. How long must she keep the second car in order to come out ahead?

57. One freezer costs \$623.95 and uses 90 kilowatt hours (kwh) of electricity each month. A second freezer costs \$500 and uses 100 kwh of electricity each month. The expected life of each freezer is 12 years. What is the minimum electric rate (in *cents* per kwh) for which the 12-year total cost (purchase price + electricity costs) will be less for the first freezer?

58. A Gas Guzzler automobile has a 26-gallon gas tank and gets 12 miles per gallon. If the car travels more than

210 miles and runs out of gas, what are the possible amounts of gas that were in the tank at the beginning of the trip?

59. One salesperson is paid a salary of \$1000 per month plus a commission of 2% of her total sales. A second salesperson receives no salary, but is paid a commission of 10% of her total sales. What dollar amount of sales must the second salesperson have in order to earn more per month than the first?

60. A developer subdivided 60 acres of a 100-acre tract, leaving 20% of the 60 acres as a park. Zoning laws require that at least 25% of the total tract be set aside for parks. For financial reasons the developer wants to have no more than 30% of the tract as parks. How many one-quarter-acre lots can the developer sell in the remaining 40 acres and still meet the requirements for the whole tract?

61. If \$5000 is invested at 8%, how much more should be invested at 10% in order to guarantee a total annual interest income between \$800 and \$940?

62. How many gallons of a 12% salt solution should be added to 10 gallons of an 18% salt solution in order to produce a solution whose salt content is between 14 and 16%?

63. Find all pairs of numbers that satisfy these two conditions: Their sum is 20 and the sum of their squares is less than 362.

64. The length of a rectangle is 6 inches longer than its width. What are the possible widths if the area of the rectangle is at least 667 square inches?

65. It costs a craftsman \$5 in materials to make a medallion. He has found that if he sells the medallions for $50 - x$ dollars each, where x is the number of medallions produced each week, then he can sell all that he makes. His fixed costs are \$350 per week. If he wants to sell all he makes and show a profit each week, what are the possible numbers of medallions he should make?

66. A retailer sells file cabinets for $80 - x$ dollars each, where x is the number of cabinets she receives from the supplier each week. She pays \$10 for each file cabinet and has fixed costs of \$600 per week. How many file cabinets should she order from the supplier each week in order to guarantee that she makes a profit?

In Exercises 67–70, you will need the formula for the height h of an object above the ground at time t seconds: $h = -16t^2 + v_0t + h_0$; this formula was explained on page 70.

67. A toy rocket is fired straight up from ground level with an initial velocity of 80 feet per second. During what time interval will it be at least 64 feet above the ground?

68. A projectile is fired straight up from ground level with an initial velocity of 72 feet per second. During what time interval is it at least 37 feet above the ground?

69. A ball is dropped from the roof of a 120-foot-high building. During what time period will it be strictly between 56 and 39 feet above the ground?

70. A ball is thrown straight up from a 40-foot-high tower with an initial velocity of 56 feet per second.
(a) During what time interval is the ball at least 8 feet above the ground?
(b) During what time interval is the ball between 53 feet and 80 feet above the ground?

71. (a) Solve the inequalities $x^2 < x$ and $x^2 > x$.
(b) Use the results of part (a) to show that for any nonzero real number c with $|c| < 1$, it is always true that $c^2 < |c|$.
(c) Use the results of part (a) to show that for any nonzero real number c with $|c| > 1$, it is always true that $c^2 > c$.

72. (a) If $0 < a \leq b$, prove that $1/a \geq 1/b$.
(b) If $a \leq b < 0$, prove that $1/a \geq 1/b$.
(c) If $a < 0 < b$, how are $1/a$ and $1/b$ related?

4.5.A *Excursion* ABSOLUTE VALUE INEQUALITIES

Polynomial and rational inequalities involving absolute value can be solved graphically just as was done above: Rewrite the inequality in an equivalent form that has 0 on the right side of the inequality sign; then graph the function whose rule is given by the left side and determine where the graph is above or below the x-axis.

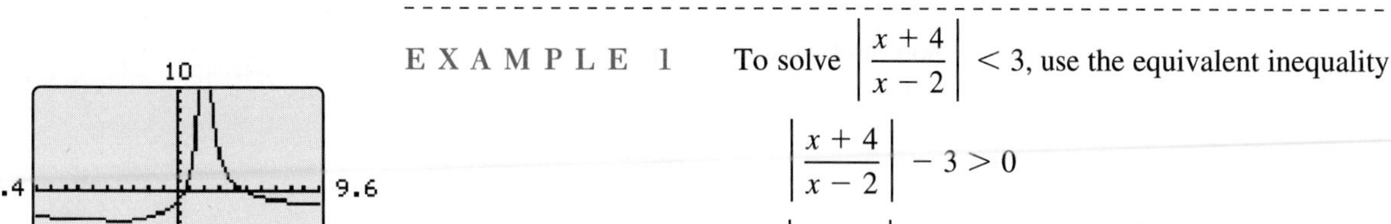

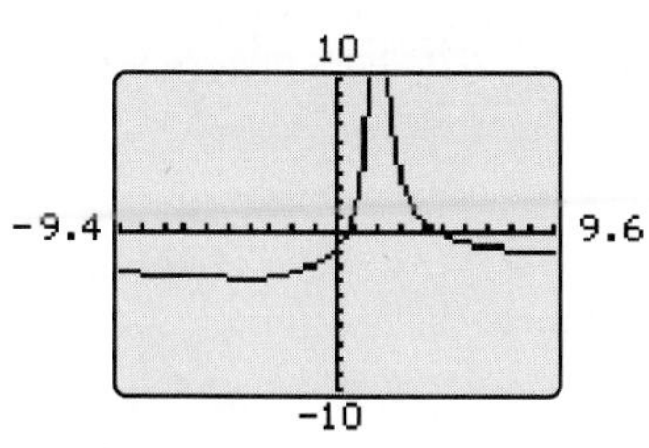

Figure 4-56

EXAMPLE 1 To solve $\left|\frac{x+4}{x-2}\right| < 3$, use the equivalent inequality

$$\left|\frac{x+4}{x-2}\right| - 3 > 0$$

and graph the function $f(x) = \left|\frac{x+4}{x-2}\right| - 3$ (Figure 4–56). The graph is above the x-axis between the two x-intercepts, which can be found algebraically or graphically.

GRAPHING EXPLORATION Verify that the x-intercepts are $x = 1/2$ and $x = 5$.

Since $f(x)$ is not defined at $x = 2$ (where the graph has a vertical asymptote), the solutions of the original inequality are all x such that $1/2 < x < 2$ or $2 < x < 5$. ■

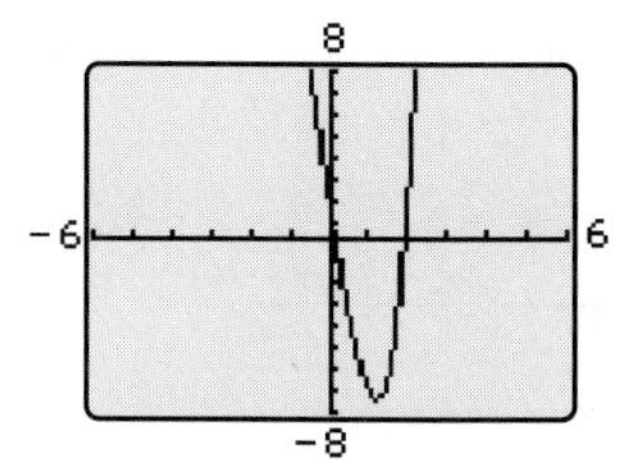

Figure 4-57

EXAMPLE 2 The solutions of $|x^4 + 2x^2 - x + 2| < 11x$ can be found by determining the numbers for which the graph of

$$f(x) = |x^4 + 2x^2 - x + 2| - 11x$$

lies below the x-axis (why?). Convince yourself that the graph of $f(x)$ in Figure 4–57 is complete.

Zoom-in or a root finder shows that the approximate x-intercepts are $x = .17$ and $x = 1.92$. Therefore, the approximate solutions of the original inequality (the numbers where the graph is below the x-axis) are all x such that $.17 < x < 1.92$. ■

Algebraic Methods

Most linear and quadratic inequalities involving absolute values can be solved exactly by algebraic means. In fact, this is often the easiest way to solve such inequalities. The key to the algebraic method is the fact that the absolute value of a number can be interpreted as distance on the number line. For example, the inequality $|r| \le 5$ states that the distance from r to 0 (namely, $|r|$) is 5 units or less. A glance at the number line in Figure 4–58 shows that these are the numbers r with $-5 \le r \le 5$:

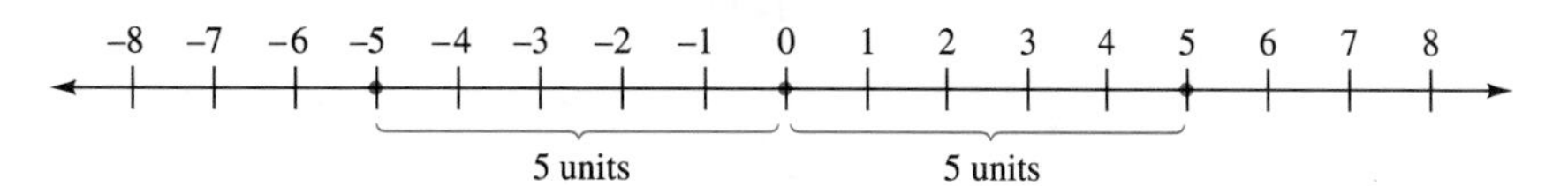

Figure 4-58

Similarly, the numbers r such that $|r| \geq 5$ are those whose distance to 0 is 5 or more units, that is, the numbers r with $r \leq -5$ or $r \geq 5$. This argument works with any positive number k in place of 5 and proves the following facts (which are also true with $<$ and $>$ in place of $\leq$ and $\geq$):

Absolute Value Inequalities ▶

Let k be a positive number and r any real number.

$$|r| \leq k \quad \text{is equivalent to} \quad -k \leq r \leq k.$$

$$|r| \geq k \quad \text{is equivalent to} \quad r \leq -k \quad \text{or} \quad r \geq k.$$

EXAMPLE 3 To solve $|3x - 7| \leq 11$, apply the first fact in the box, with $3x - 7$ in place of r and 11 in place of k, and obtain this equivalent inequality $-11 \leq 3x - 7 \leq 11$. Then,

Add 7 to each part: $\quad -4 \leq 3x \leq 18$

Divide each part by 3: $\quad -4/3 \leq x \leq 6.$

Therefore, the solutions of the original inequality are all numbers in the interval $[-4/3, 6]$, as shown in Figure 4–59. ■

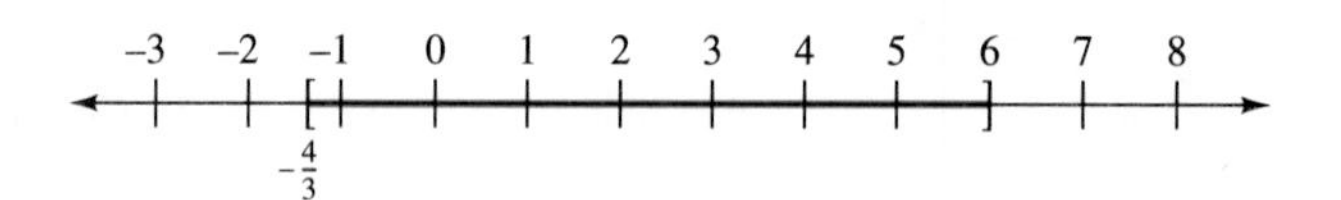

Figure 4–59

EXAMPLE 4 To solve $|5x + 2| > 3$, apply the second fact in the box, with $5x + 2$ in place of r, and 3 in place of k, and $>$ in place of $\geq$. This produces the equivalent statement:

$$5x + 2 < -3 \quad \text{or} \quad 5x + 2 > 3$$

$$5x < -5 \qquad\qquad 5x > 1$$

$$x < -1 \quad \text{or} \quad x > 1/5.$$

Therefore, the solutions of the original inequality are the numbers in *either* of the intervals $(-\infty, -1)$ or $(1/5, \infty)$. ■

EXAMPLE 5 If a and δ are real numbers with δ positive, then the inequality $|x - a| < \delta$ is equivalent to $-\delta < x - a < \delta$. Adding a to each part shows that $a - \delta < x < a + \delta$. ■

E X A M P L E 6 To solve $|x^2 - x - 4| \geq 2$, we use the fact in the box on the facing page to replace it by an equivalent inequality:

$$x^2 - x - 4 \leq -2 \qquad \text{or} \qquad x^2 - x - 4 \geq 2$$

which is the same as

$$x^2 - x - 2 \leq 0 \qquad \text{or} \qquad x^2 - x - 6 \geq 0.$$

The solutions are all numbers that are solutions of *either one* of the two inequalities.

To solve the first of these inequalities, note that the graph of $f(x) = x^2 - x - 2$ is an upward-opening parabola, which crosses the x-axis at the roots of $f(x)$. The roots are -1 and 2, as can be seen from the factorization: $x^2 - x - 2 = (x + 1)(x - 2)$. Therefore, the solutions of $x^2 - x - 2 \leq 0$ (the numbers for which the graph of $f(x)$ is on or below the x-axis) are all x with $-1 \leq x \leq 2$. The second inequality above, $x^2 - x - 6 \geq 0$, is solved similarly.

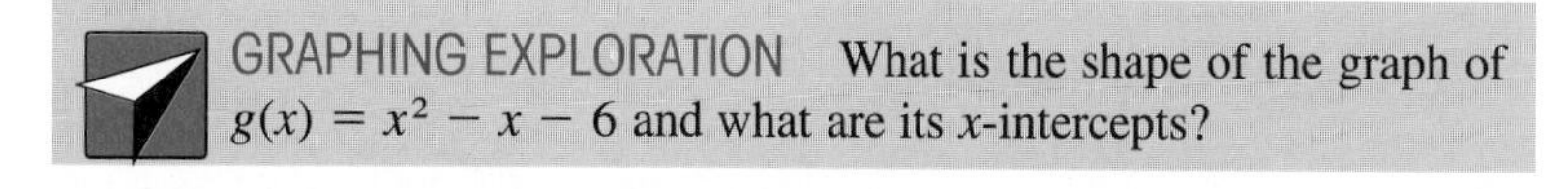

This Exploration shows that the solutions of the second inequality (the numbers for which the graph of $g(x)$ is on or above the x-axis) are all x with $x \leq -2$ or $x \geq 3$.

Consequently, the solutions of the original inequality are all numbers x such that $x \leq -2$ or $-1 \leq x \leq 2$ or $x \geq 3$, as shown in Figure 4–60. ■

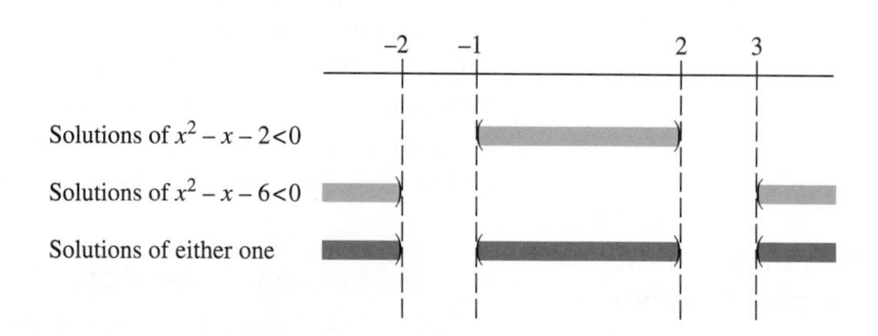

Figure 4–60

EXERCISES 4.5.A

In Exercises 1–32, solve the inequality. Find exact solutions when possible and approximate ones otherwise.

1. $|3x + 2| \leq 2$

2. $|5x - 1| < 3$

3. $|3 - 2x| < 2/3$

4. $|4 - 5x| \leq 4$

5. $|2x + 3| > 1$

6. $|3x - 1| \geq 2$

7. $|5x + 2| \geq \frac{3}{4}$

8. $|2 - 3x| > 4$

9. $\left|\frac{12}{5} + 2x\right| > \frac{1}{4}$

10. $\left|\frac{5}{6} + 3x\right| < \frac{7}{6}$

11. $\left|\frac{x - 1}{x + 2}\right| \leq 3$

12. $\left|\frac{x + 1}{3x + 5}\right| < 2$

13. $\left|\frac{2x-1}{x+5}\right| > 1$

14. $\left|\frac{x+1}{x+2}\right| \geq 2$

15. $\left|\frac{1-4x}{2+3x}\right| < 1$

16. $\left|\frac{3x+1}{1-2x}\right| \geq 2$

17. $|x^2 - 2| < 1$

18. $|x^2 - 4| \leq 3$

19. $|x^2 - 2| > 4$

20. $\left|\frac{1}{x^2-1}\right| \leq 2$

21. $|x^2 + x - 1| \geq 1$

22. $|x^2 + x - 4| \leq 2$

23. $|3x^2 - 8x + 2| < 2$

24. $|x^2 + 3x - 4| < 6$

25. $|x^5 - x^3 + 1| < 2$

26. $|4x - x^3| > 1$

27. $|x^4 - x^3 + x^2 - x + 1| > 4$

28. $|x^3 - 6x^2 + 4x - 5| < 3$

29. $\frac{x+2}{|x-3|} \leq 4$

30. $\frac{x^2-9}{|x^2-4|} < -2$

31. $\left|\frac{2x^2+2x-12}{x^3-x^2+x-2}\right| > 2$

32. $\left|\frac{x^2-x-2}{x^2+x-2}\right| > 3$

Thinkers

33. Let E be a fixed real number. Show that every solution of $|x - 3| < E/5$ is also a solution of $|(5x - 4) - 11| < E$.

34. Let a and b be fixed real numbers with $a < b$. Show that the solutions of

$$\left|x - \frac{a+b}{2}\right| < \frac{b-a}{2}$$

are all x with $a < x < b$.

4.6 COMPLEX NUMBERS

If you are restricted to nonnegative integers, you can't solve the equation $x + 5 = 0$. Enlarging the number system to include negative integers makes it possible to solve this equation ($x = -5$). Enlarging it again, to include rational numbers, makes it possible to solve equations like $3x = 7$, which have no integer solutions. Similarly, the equation $x^2 = 2$ has no solutions in the rational number system, but has $\sqrt{2}$ and $-\sqrt{2}$ as solutions in the real number system. So the idea of enlarging a number system in order to solve an equation that can't be solved in the present system is a natural one.

Equations such as $x^2 = -1$ and $x^2 = -4$ have no solutions in the real number system because $\sqrt{-1}$ and $\sqrt{-4}$ are not real numbers. In order to solve such equations (or equivalently, in order to find square roots of negative numbers) we must enlarge the number system again. We claim that there is a number system, called the **complex number system,** with these properties:

Properties of the Complex Number System ▶

1. **The complex number system contains all real numbers.**
2. **Addition, subtraction, multiplication, and division of complex numbers obey the same rules of arithmetic that hold in the real number system.**
3. **The complex number system contains a number (usually denoted by i) such that $i^2 = -1$.**
4. **Every complex number can be written in the *standard form* $a + bi$, where a and b are real numbers.***
5. **Two complex numbers $a + bi$ and $c + di$ are equal exactly when $a = c$ and $b = d$.**

*Hereafter whenever we write $a + bi$ or $c + di$, it is assumed that a, b, c, d are real numbers and $i^2 = -1$.

In view of our past experience with enlarging the number system, this claim *ought* to appear plausible. But the mathematicians who invented the complex numbers in the 17th century were very uneasy about a number i such that $i^2 = -1$ (that is, $i = \sqrt{-1}$). Consequently, they called numbers of the form bi (b any real number), such as $5i$ and $-\frac{1}{4}i$, **imaginary numbers.** The old familiar numbers (integers, rationals, irrationals) were called **real numbers.** Sums of real and imaginary numbers, numbers of the form $a + bi$, such as

$$5 + 2i, \qquad 7 - 4i, \qquad 18 + \frac{3}{2}i, \qquad \sqrt{3} - 12i$$

were called **complex numbers.***

Every real number is a complex number; for instance, $7 = 7 + 0i$. Similarly, every imaginary number bi is a complex number since $bi = 0 + bi$. Since the usual laws of arithmetic still hold, it's easy to add, subtract, and multiply complex numbers. As the following examples demonstrate, *all symbols can be treated as if they were real numbers, provided that i^2 is replaced by* -1. Unless directed otherwise, express your answers in the standard form $a + bi$.

EXAMPLE 1

(a) $(1 + i) + (3 - 7i) = 1 + i + 3 - 7i$
$= (1 + 3) + (i - 7i) = 4 - 6i.$

(b) $(4 + 3i) - (8 - 6i) = 4 + 3i - 8 - (-6i)$
$= (4 - 8) + (3i + 6i) = -4 + 9i.$

(c) $4i(2 + \frac{1}{2}i) = 4i \cdot 2 + 4i(\frac{1}{2}i) = 8i + 4 \cdot \frac{1}{2} \cdot i^2$
$= 8i + 2i^2 = 8i + 2(-1) = -2 + 8i.$

(d) $(2 + i)(3 - 4i) = 2 \cdot 3 + 2(-4i) + i \cdot 3 + i(-4i)$
$= 6 - 8i + 3i - 4i^2 = 6 - 8i + 3i - 4(-1)$
$= (6 + 4) + (-8i + 3i) = 10 - 5i.$ ■

The familiar multiplication patterns and exponent laws for integer exponents hold in the complex number system.

EXAMPLE 2

(a) $(3 + 2i)(3 - 2i) = 3^2 - (2i)^2$
$= 9 - 4i^2 = 9 - 4(-1) = 9 + 4 = 13.$

*This terminology is still used, even though there is nothing complicated, unreal, or imaginary about complex numbers—they are just as valid mathematically as are real numbers. See Exercise 80 for a formal construction of the complex numbers and proofs of the claims made above.

> ▶ **TECHNOLOGY TIP**
> TI-83/85/92, HP-38, Casio 9800, and Sharp 9300 can do complex number arithmetic directly. The complex number $a + bi$ is entered as it stands on TI-83/92, Casio, and Sharp, using a special i key; on TI-85 and HP-38 it is entered as (a, b).
>
> Other calculators can do complex arithmetic if you use matrices; see Calculator Investigation 1 for details.

(b) $(4 + i)^2 = 4^2 + 2 \cdot 4 \cdot i + i^2 = 16 + 8i + (-1) = 15 + 8i.$

(c) To find i^{54}, we first note that $i^4 = i^2 i^2 = (-1)(-1) = 1$ and that $54 = 52 + 2 = 4 \cdot 13 + 2$. Consequently,

$$i^{54} = i^{52+2} = i^{52} i^2 = i^{4 \cdot 13} i^2 = (i^4)^{13} i^2 = 1^{13}(-1) = -1. \quad ■$$

The **conjugate** of the complex number $a + bi$ is the number $a - bi$, and the conjugate of $a - bi$ is $a + bi$. For example, the conjugate of $3 + 4i$ is $3 - 4i$ and the conjugate of $-3i = 0 - 3i$ is $0 + 3i = 3i$. *Every real number is its own conjugate;* for instance, the conjugate of $17 = 17 + 0i$ is $17 - 0i = 17$.

For any complex number $a + bi$, we have

$$(a + bi)(a - bi) = a^2 - (bi)^2 = a^2 - b^2 i^2 = a^2 - b^2(-1) = a^2 + b^2.$$

Since a^2 and b^2 are nonnegative real numbers, so is $a^2 + b^2$. Therefore *the product of a complex number and its conjugate is a nonnegative real number.* This fact enables us to express quotients of complex numbers in standard form.

EXAMPLE 3 To express $\dfrac{3 + 4i}{1 + 2i}$ in the form $a + bi$, *multiply both numerator and denominator by the conjugate of the denominator,* namely, $1 - 2i$:

$$\frac{3 + 4i}{1 + 2i} = \frac{3 + 4i}{1 + 2i} \cdot \frac{1 - 2i}{1 - 2i} = \frac{(3 + 4i)(1 - 2i)}{(1 + 2i)(1 - 2i)}$$
$$= \frac{3 + 4i - 6i - 8i^2}{1^2 - (2i)^2} = \frac{3 + 4i - 6i - 8(-1)}{1 - 4i^2} = \frac{11 - 2i}{1 - 4(-1)}$$
$$= \frac{11 - 2i}{5} = \frac{11}{5} - \frac{2}{5} i.$$

This is the form $a + bi$ with $a = 11/5$ and $b = -2/5$. ■

EXAMPLE 4 To express $\dfrac{1}{1 - i}$ in standard form, note that the conjugate of the denominator is $1 + i$ and therefore:

$$\frac{1}{1 - i} = \frac{1 \cdot (1 + i)}{(1 - i)(1 + i)} = \frac{1 + i}{1^2 - i^2} = \frac{1 + i}{1 - (-1)} = \frac{1 + i}{2} = \frac{1}{2} + \frac{1}{2} i.$$

We can check this result by multiplying $\frac{1}{2} + \frac{1}{2} i$ by $1 - i$ to see if the product is 1 $\left(\text{which it should be if } \frac{1}{2} + \frac{1}{2} i = \frac{1}{1 - i}\right)$:

$$\left(\frac{1}{2} + \frac{1}{2} i\right)(1 - i) = \frac{1}{2} \cdot 1 - \frac{1}{2} i + \frac{1}{2} i \cdot 1 - \frac{1}{2} i^2 = \frac{1}{2} - \frac{1}{2}(-1) = 1. \quad ■$$

Since $i^2 = -1$, we define $\sqrt{-1}$ to be the complex number i. Similarly, since $(5i)^2 = 5^2i^2 = 25(-1) = -25$, we define $\sqrt{-25}$ to be $5i$. In general,

Square Roots of Negative Numbers ▶

For any positive real number b, $\sqrt{-b}$ is defined to be $\sqrt{b}\,i$.

This is because $(\sqrt{b}\,i)^2 = (\sqrt{b})^2 i^2 = b(-1) = -b$.

WARNING $\sqrt{b}\,i$ is *not* the same as $\sqrt{bi}$. To avoid confusion it may help to write $\sqrt{b}\,i$ as $i\sqrt{b}$.

EXAMPLE 5

(a) $\sqrt{-3} = \sqrt{3}i = i\sqrt{3}$.

(b) $\dfrac{1-\sqrt{-7}}{3} = \dfrac{1-\sqrt{7}i}{3} = \dfrac{1}{3} - \dfrac{\sqrt{7}}{3}i.$ ■

WARNING The property $\sqrt{cd} = \sqrt{c}\sqrt{d}$, which is valid for positive real numbers, *does not hold* when both c and d are negative. For example, according to the definition above

$$\sqrt{-20}\sqrt{-5} = \sqrt{20}i\cdot\sqrt{5}i = \sqrt{20}\sqrt{5}\cdot i^2 = \sqrt{20\cdot 5}\,(-1)$$
$$= \sqrt{100}\,(-1) = -10$$

But $\sqrt{(-20)(-5)} = \sqrt{100} = 10$, so that

$$\sqrt{(-20)(-5)} \neq \sqrt{-20}\sqrt{-5}.$$

To avoid difficulty, *always write square roots of negative numbers in terms of i before doing any simplification.*

EXAMPLE 6

$$(7-\sqrt{-4})(5+\sqrt{-9}) = (7-\sqrt{4}i)(5+\sqrt{9}i)$$
$$= (7-2i)(5+3i)$$
$$= 35 + 21i - 10i - 6i^2$$
$$= 35 + 11i - 6(-1) = 41 + 11i.$$ ■

▶ TECHNOLOGY TIP

Most calculators that do complex number arithmetic automatically return a complex number when asked for the square root of a negative number.

Since every negative real number has a square root in the complex number system, we can now find complex solutions for equations that have no real solutions. For example, the solutions of $x^2 = -25$ are $x = \pm\sqrt{-25} = \pm 5i$. In fact,

every quadratic equation with real coefficients has solutions in the complex number system.

EXAMPLE 7 To solve the equation $2x^2 + x + 3 = 0$, we apply the quadratic formula:

$$x = \frac{-1 \pm \sqrt{1^2 - 4 \cdot 2 \cdot 3}}{2 \cdot 2} = \frac{-1 \pm \sqrt{-23}}{4}.$$

Since $\sqrt{-23}$ is not a real number, this equation has no real number solutions. But $\sqrt{-23}$ *is* a complex number, namely, $\sqrt{-23} = \sqrt{23}i$. Thus the equation does have solutions in the complex number system:

$$x = \frac{-1 \pm \sqrt{-23}}{4} = \frac{-1 \pm \sqrt{23}i}{4} = -\frac{1}{4} \pm \frac{\sqrt{23}}{4}i.$$

Note that the two solutions, $-\frac{1}{4} + \frac{\sqrt{23}}{4}i$ and $-\frac{1}{4} - \frac{\sqrt{23}}{4}i$, are conjugates of each other. ■

EXAMPLE 8 To find *all* solutions of $x^3 = 1$, we rewrite the equation and use the Difference of Cubes pattern (see the Algebra Review Appendix) to factor:

$$x^3 = 1$$
$$x^3 - 1 = 0$$
$$(x - 1)(x^2 + x + 1) = 0$$
$$x - 1 = 0 \quad \text{or} \quad x^2 + x + 1 = 0.$$

► TECHNOLOGY TIP

The solutions of polynomial equations (both real and complex) can be found by using POLY on TI-85 or C-SOLVE on TI-92 (in the COMPLEX submenu of the ALGEBRA menu) or POLYROOT on HP-38 (in the POLYNOMIAL submenu of the MATH menu). The Casio 9800 solver finds all solutions of quadratics and cubics. Other equation solvers generally find only real solutions.

The solution of the first equation is $x = 1$. The solutions of the second can be obtained from the quadratic formula:

$$x = \frac{-1 \pm \sqrt{1^2 - 4 \cdot 1 \cdot 1}}{2 \cdot 1} = \frac{-1 \pm \sqrt{-3}}{2} = \frac{-1 \pm \sqrt{3}i}{2} = -\frac{1}{2} \pm \frac{\sqrt{3}}{2}i.$$

Therefore, the equation $x^3 = 1$ has one real solution ($x = 1$) and two nonreal complex solutions [$x = -1/2 + (\sqrt{3}/2)i$ and $x = -1/2 - (\sqrt{3}/2)i$]. Each of these solutions is said to be a **cube root of one** or a **cube root of unity.** Observe that the two nonreal complex cube roots of unity are conjugates of each other. ■

The preceding examples illustrate this useful fact (whose proof is discussed in Section 4.7):

Conjugate Solutions ►

If $a + bi$ is a solution of a polynomial equation with *real* coefficients, then its conjugate $a - bi$ is also a solution of this equation.

CALCULATOR INVESTIGATION 4.6

The following investigation of complex number arithmetic is for use with TI-81/82 and other calculators that do not do complex number arithmetic, but do have matrix capabilities. Before doing it, look up "matrix" or "matrices" in your instruction manual and learn how to enter and store 2 by 2 matrices and how to do addition, subtraction, and multiplication with them.

1. The complex number $a + bi$ is expressed in matrix notation as the matrix $\begin{pmatrix} a & b \\ -b & a \end{pmatrix}$. For example, $-3 + 6i$ is written as $\begin{pmatrix} -3 & 6 \\ -6 & -3 \end{pmatrix}$.

(a) Write $3 + 4i$, $1 + 2i$, and $1 - i$ in matrix form and enter them in your calculator as $[A]$, $[B]$, $[C]$.

(b) We know that $(3 + 4i) + (1 + 2i) = 4 + 6i$. Verify that $[A] + [B]$ is $\begin{pmatrix} 4 & 6 \\ -6 & 4 \end{pmatrix}$, which represents the complex number $4 + 6i$.

(c) Use matrix addition, subtraction, and multiplication to find the following. Interpret the answers as complex numbers: $[A] - [C]$, $[B] + [C]$, $[A][B]$, $[B][C]$.

(d) In Example 3 we saw that $\frac{3 + 4i}{1 + 2i} = \frac{11}{5} - \frac{2}{5}i = 2.2 - .4i$. Do this problem in matrix form by computing $[A] \cdot [B]^{-1}$ (use the x^{-1} key for the exponent).

(e) Do each of the following calculations and interpret the problem in terms of complex numbers: $[A] \cdot [C]^{-1}$, $[B][A]^{-1}$, $[B][C]^{-1}$.

EXERCISES 4.6

In Exercises 1–54, perform the indicated operation and write the result in the form $a + bi$.

1. $(2 + 3i) + (6 - i)$

2. $(-5 + 7i) + (14 + 3i)$

3. $(2 - 8i) - (4 + 2i)$

4. $(3 + 5i) - (3 - 7i)$

5. $\frac{5}{4} - \left(\frac{7}{4} + 2i\right)$

6. $(\sqrt{3} + i) + (\sqrt{5} - 2i)$

7. $\left(\frac{\sqrt{2}}{2} + i\right) - \left(\frac{\sqrt{3}}{2} - i\right)$

8. $\left(\frac{1}{2} + \frac{\sqrt{3}i}{2}\right) + \left(\frac{3}{4} - \frac{5\sqrt{3}i}{2}\right)$

9. $(2 + i)(3 + 5i)$

10. $(2 - i)(5 + 2i)$

11. $(-3 + 2i)(4 - i)$

12. $(4 + 3i)(4 - 3i)$

13. $(2 - 5i)^2$

14. $(1 + i)(2 - i)i$

15. $(\sqrt{3} + i)(\sqrt{3} - i)$

16. $\left(\frac{1}{2} - i\right)\left(\frac{1}{4} + 2i\right)$

17. i^{15}

18. i^{26}

19. i^{33}

20. $(-i)^{53}$

21. $(-i)^{107}$

22. $(-i)^{213}$

23. $\frac{1}{5 - 2i}$

24. $\frac{1}{i}$

25. $\frac{1}{3i}$

26. $\frac{i}{2 + i}$

27. $\frac{3}{4 + 5i}$

28. $\frac{2 + 3i}{i}$

29. $\frac{1}{i(4 + 5i)}$

30. $\frac{1}{(2 - i)(2 + i)}$

31. $\frac{2 + 3i}{i(4 + i)}$

32. $\frac{2}{(2 + 3i)(4 + i)}$

33. $\frac{2 + i}{1 - i} + \frac{1}{1 + 2i}$

34. $\dfrac{1}{2-i} + \dfrac{3+i}{2+3i}$ **35.** $\dfrac{i}{3+i} - \dfrac{3+i}{4+i}$

36. $6 + \dfrac{2i}{3+i}$ **37.** $\sqrt{-36}$ **38.** $\sqrt{-81}$

39. $\sqrt{-14}$ **40.** $\sqrt{-50}$ **41.** $-\sqrt{-16}$

42. $-\sqrt{-12}$ **43.** $\sqrt{-16} + \sqrt{-49}$

44. $\sqrt{-25} - \sqrt{-9}$ **45.** $\sqrt{-15} - \sqrt{-18}$

46. $\sqrt{-12}\sqrt{-3}$ **47.** $\sqrt{-16}/\sqrt{-36}$

48. $-\sqrt{-64}/\sqrt{-4}$ **49.** $(\sqrt{-25} + 2)(\sqrt{-49} - 3)$

50. $(5 - \sqrt{-3})(-1 + \sqrt{-9})$

51. $(2 + \sqrt{-5})(1 - \sqrt{-10})$

52. $\sqrt{-3}(3 - \sqrt{-27})$ **53.** $1/(1 + \sqrt{-2})$

54. $(1 + \sqrt{-4})/(3 - \sqrt{-9})$

In Exercises 55–58, find x and y. Remember that $a + bi = c + di$ exactly when $a = c$ and $b = d$.

55. $3x - 4i = 6 + 2yi$ **56.** $8 - 2yi = 4x + 12i$

57. $3 + 4xi = 2y - 3i$ **58.** $8 - xi = \dfrac{1}{2}y + 2i$

In Exercises 59–70, solve the equation and express each solution in the form $a + bi$.

59. $3x^2 - 2x + 5 = 0$ **60.** $5x^2 + 2x + 1 = 0$

61. $x^2 + x + 2 = 0$ **62.** $5x^2 - 6x + 2 = 0$

63. $2x^2 - x = -4$ **64.** $x^2 + 1 = 4x$

65. $2x^2 + 3 = 6x$ **66.** $3x^2 + 4 = -5x$

67. $x^3 - 8 = 0$ **68.** $x^3 + 125 = 0$

69. $x^4 - 1 = 0$ **70.** $x^4 - 81 = 0$

71. Simplify: $i + i^2 + i^3 + \cdots + i^{15}$

72. Simplify: $i - i^2 + i^3 - i^4 + i^5 - \cdots + i^{15}$

Thinkers

If $z = a + bi$ is a complex number, then its conjugate is usually denoted $\bar{z}$, that is, $\bar{z} = a - bi$. In Exercises 73–77, prove that for any complex numbers $z = a + bi$ and $w = c + di$:

73. $\overline{z + w} = \bar{z} + \bar{w}$ **74.** $\overline{zw} = \bar{z} \cdot \bar{w}$

75. $\overline{\left(\dfrac{z}{w}\right)} = \dfrac{\bar{z}}{\bar{w}}$ **76.** $\bar{\bar{z}} = z$

77. z is a real number exactly when $\bar{z} = z$.

78. The **real part** of the complex number $a + bi$ is defined to be the real number a. The **imaginary part** of $a + bi$ is defined to be the real number b *(not bi)*.

(a) Show that the real part of $z = a + bi$ is $\dfrac{z + \bar{z}}{2}$.

(b) Show that the imaginary part of $z = a + bi$ is $\dfrac{z + \bar{z}}{2i}$.

79. If $z = a + bi$ (with a, b real numbers, not both 0), express $1/z$ in standard form.

80. Construction of the Complex Numbers. We assume that the real number system is known. In order to construct a new number system with the desired properties, we must do the following:

(i) Define a set C (whose elements will be called complex numbers).

(ii) The set C must contain the real numbers or at least a copy of them.

(iii) Define addition and multiplication in the set C in such a way that the usual laws of arithmetic are valid.

(iv) Show that C has the other properties listed on page 278.

We begin by defining C to be the set of all ordered pairs of real numbers. Thus, $(1, 5)$, $(-6, 0)$, $(4/3, -17)$, and $(\sqrt{2}, 12/5)$ are some of the elements of the set C. More generally, a complex number (= element of C) is any pair (a, b) where a and b are real numbers. By definition, two complex numbers are *equal* exactly when they have the same first and the same second coordinate.

(a) *Addition in C* is defined by this rule:

$$(a, b) + (c, d) = (a + c, b + d)$$

For example,

$$(3, 2) + (5, 4) = (3 + 5, 2 + 4) = (8, 6).$$

Verify that this addition has the following properties. For any complex numbers (a, b), (c, d), (e, f) in C:

(i) $(a, b) + (c, d) = (c, d) + (a, b)$

(ii) $((a, b) + (c, d)) + (e, f) = (a, b) + ((c, d) + (e, f))$

(iii) $(a, b) + (0, 0) = (a, b)$

(iv) $(a, b) + (-a, -b) = (0, 0)$

(b) *Multiplication in C* is defined by this rule:

$$(a, b)(c, d) = (ac - bd, bc + ad)$$

For example,

$$(3, 2)(4, 5) = (3 \cdot 4 - 2 \cdot 5, 2 \cdot 4 + 3 \cdot 5)$$
$$= (12 - 10, 8 + 15) = (2, 23).$$

Verify that this multiplication has the following properties. For any complex numbers (a, b), (c, d), (e, f) in C:

(i) $(a, b)(c, d) = (c, d)(a, b)$
(ii) $((a, b)(c, d))(e, f) = (a, b)((c, d)(e, f))$
(iii) $(a, b)(1, 0) = (a, b)$
(iv) $(a, b)(0, 0) = (0, 0)$

(c) Verify that for any two elements of C with second coordinate zero:

(i) $(a, 0) + (c, 0) = (a + c, 0)$
(ii) $(a, 0)(c, 0) = (ac, 0)$

Identify $(t, 0)$ with the real number t. Statements (i) and (ii) show that when addition or multiplication in C is performed on two real numbers (that is, elements of C with second coordinate 0), the result is the usual sum or product of real numbers. Thus, C contains (a copy of) the real number system.

(d) *New Notation.* Since we are identifying the complex number $(a, 0)$ with the real number a, we shall hereafter denote $(a, 0)$ simply by the symbol a. Also, let i denote the complex number $(0, 1)$.

(i) Show that $i^2 = -1$ [that is, $(0, 1)(0, 1) = (-1, 0)$].
(ii) Show that for any complex number $(0, b)$, $(0, b) = bi$ [that is, $(0, b) = (b, 0)(0, 1)$].
(iii) Show that any complex number (a, b) can be written: $(a, b) = a + bi$ [that is, $(a, b) = (a, 0) + (b, 0)(0, 1)$].

In this new notation, every complex number is of the form $a + bi$ with a, b real and $i^2 = -1$, and our construction is finished.

4.7 THEORY OF EQUATIONS*

The complex numbers were constructed in order to obtain a solution for the equation $x^2 = -1$, that is, a root of the polynomial $x^2 + 1$. In Section 4.6 we saw that much more is true: *Every* quadratic polynomial with real coefficients has roots in the complex number system. A natural question now arises:

> Do we have to enlarge the complex number system (perhaps many times) to find roots for higher degree polynomials?

In this section we shall see that the somewhat surprising answer is no.

In order to give the full answer, we shall consider not just polynomials with real coefficients, but also those with complex coefficients, such as

$$x^3 - ix^2 + (4 - 3i)x + 1 \qquad \text{or} \qquad (-3 + 2i)x^6 - 3x + (5 - 4i).$$

The discussion of polynomial division in Section 4.1 can easily be extended to include polynomials with complex coefficients. In fact, *all of the results in Section 4.1 are valid for polynomials with complex coefficients.* For example, you can check that i is a root of $f(x) = x^2 + (i - 1)x + (2 + i)$ and that $x - i$ is a factor of $f(x)$:

$$f(x) = x^2 + (i - 1)x + (2 + i) = (x - i)(x - (1 - 2i)).$$

Since every real number is also a complex number, polynomials with real coefficients are just special cases of polynomials with complex coefficients. So in

*Section 4.6 is a prerequisite for this section.

the rest of this section, ''polynomial'' means ''polynomial with complex (possibly real) coefficients'' unless specified otherwise. We can now answer the question posed in the first paragraph.

The Fundamental Theorem of Algebra ▶

Every nonconstant polynomial has a root in the complex number system.

Although this is obviously a powerful result, neither the Fundamental Theorem nor its proof provides a practical method for *finding* a root of a given polynomial.* The proof of the Fundamental Theorem is beyond the scope of this book, but we shall explore some of the useful implications of the theorem, such as this one:

Factorization Over the Complex Numbers ▶

Let $f(x)$ be a polynomial of degree $n > 0$ with leading coefficients d. Then there are (not necessarily distinct) complex numbers $c_1, c_2, \ldots, c_n$ such that

$$f(x) = d(x - c_1)(x - c_2)(x - c_3) \cdots (x - c_n)$$

Furthermore, $c_1, c_2, \ldots, c_n$ are the only roots of $f(x)$.

Proof By the Fundamental Theorem, $f(x)$ has a complex root c_1. The Factor Theorem shows that $x - c_1$ must be a factor of $f(x)$, say,

$$f(x) = (x - c_1)g(x)$$

where $g(x)$ has degree $n - 1$.** If $g(x)$ is nonconstant, then it has a complex root c_2 by the Fundamental Theorem. Hence $x - c_2$ is a factor of $g(x)$, so that

$$f(x) = (x - c_1)(x - c_2)h(x)$$

for some $h(x)$ of degree $n - 2$ (1 less than the degree of $g(x)$). If $h(x)$ is nonconstant, then it has a complex root c_3 and the argument can be repeated. Continuing in this way, with the degree of the last factor going down by 1 at each step, we reach a factorization in which the last factor is a constant (degree 0 polynomial):

(*) $$f(x) = (x - c_1)(x - c_2)(x - c_3) \cdots (x - c_n)d.$$

If the right side were multiplied out, it would look like

$$dx^n + \text{lower degree terms.}$$

So the constant factor d is the leading coefficient of $f(x)$.

▶ **TECHNOLOGY TIP**

To find all the roots (both real and complex) of a polynomial, use POLY on TI-85 or C-SOLVE on TI-92 (in the COMPLEX submenu of the ALGEBRA menu) or POLYROOT on HP-38 (in the POLYNOMIAL submenu of the MATH menu). The Casio 9800 solver finds all roots of quadratics and cubics.

TI-92 can also factor a polynomial directly into linear factors.

*It may seem strange that you can prove that a root exists without actually exhibiting one. But such ''existence theorems'' are quite common. A rough analogy is the situation that occurs when someone is killed by a sniper's bullet. The police know that there *is* a killer, but *finding* the killer may be impossible.

**The degree of $g(x)$ is 1 less than the degree n of $f(x)$ because $f(x)$ is the product of $g(x)$ and $x - c_1$ (which has degree 1).

It is easy to see from the factored form (*) that the numbers $c_1, c_2, \ldots, c_n$ are roots of $f(x)$. If k is *any* root of $f(x)$, then

$$0 = f(k) = d(k - c_1)(k - c_2)(k - c_3) \cdots (k - c_n).$$

The product on the right is 0 only when one of the factors is 0. Since the leading coefficient d is nonzero, we must have

$$k - c_1 = 0 \quad \text{or} \quad k - c_2 = 0 \quad \text{or} \quad \cdots \quad k - c_n = 0$$
$$k = c_1 \quad \text{or} \quad k = c_2 \quad \text{or} \quad \cdots \quad k = c_n.$$

Therefore, k is one of the c's and $c_1, \ldots, c_n$ are the only roots of $f(x)$. This completes the proof. ❑

Since the n roots $c_1, \ldots, c_n$ of $f(x)$ may not all be distinct, we see that:

Number of Roots ▶

Every polynomial of degree $n > 0$ has at most n different roots in the complex number system.

Suppose c is a root of a polynomial $f(x)$ of degree n. If $f(x)$ is written in factored form (*) as above, then c appears one or more times on the list of roots $c_1, c_2, \ldots, c_n$. If c appears exactly k times, then c is said to be a **root of multiplicity** k. Equivalently, c is a root of multiplicity k exactly when

$$(x - c)^k \text{ is a factor of } f(x)\text{, but } (x - c)^{k+1} \text{ is not a factor,}$$

as can be seen from the factored form (*) above. Consequently, if every root is counted as many times as its multiplicity, then the statement in the preceding box implies that

A polynomial of degree n has exactly n roots.

EXAMPLE 1 Find a polynomial $f(x)$ of degree 5 such that 1, -2, and 5 are roots, 1 is a root of multiplicity 3, and $f(2) = -24$.

Solution Since 1 is a root of multiplicity 3, $(x - 1)^3$ must be a factor of $f(x)$. There are at least two other factors corresponding to the roots -2 and 5: $x - (-2) = x + 2$ and $x - 5$. The product of these factors $(x - 1)^3(x + 2)(x - 5)$ has degree 5, as does $f(x)$, so $f(x)$ must look like this:

$$f(x) = d(x - 1)^3(x + 2)(x - 5)$$

where d is the leading coefficient. Since $f(2) = -24$ we have:

$$d(2 - 1)^3(2 + 2)(2 - 5) = f(2) = -24$$

which reduces to $-12d = -24$. Therefore, $d = (-24)/(-12) = 2$ and

$$\begin{aligned} f(x) &= 2(x-1)^3(x+2)(x-5) \\ &= 2x^2 - 12x^4 + 4x^3 + 40x^2 - 54x + 20. \quad \blacksquare \end{aligned}$$

Polynomials with Real Coefficients

Recall that the **conjugate** of the complex number $a + bi$ is the number $a - bi$ (see page 280). We usually write a complex number as a single letter, say z, and indicate its conjugate by $\bar{z}$ (sometimes read "z bar"). For instance, if $z = 3 + 7i$, then $\bar{z} = 3 - 7i$. Conjugates play a role whenever a quadratic polynomial with real coefficients has complex roots.

EXAMPLE 2 The quadratic formula shows that $x^2 - 6x + 13$ has two complex roots:

$$\frac{-(-6) \pm \sqrt{(-6)^2 - 4 \cdot 1 \cdot 13}}{2 \cdot 1} = \frac{6 \pm \sqrt{-16}}{2} = \frac{6 \pm 4i}{2} = 3 \pm 2i.$$

The complex roots are $z = 3 + 2i$ and its conjugate $\bar{z} = 3 - 2i$. ■

The preceding example is a special case of a more general theorem, whose proof is outlined in Exercises 59 and 60:

Conjugate Roots Theorem ▶

Let $f(x)$ be a polynomial with *real* coefficients. If the complex number z is a root of $f(x)$, then its conjugate $\bar{z}$ is also a root of $f(x)$.

EXAMPLE 3 Find a polynomial with real coefficients whose roots include the numbers 2 and $3 + i$.

Solution Since $3 + i$ is a root, its conjugate $3 - i$ must also be a root. Consider the polynomial

$$f(x) = (x - 2)(x - (3 + i))(x - (3 - i)).$$

Obviously 2, $3 + i$, and $3 - i$ are roots of $f(x)$. Multiplying out this factored form shows that $f(x)$ *does* have real coefficients:

$$\begin{aligned} f(x) &= (x-2)(x^2 - (3-i)x - (3+i)x + (3+i)(3-i)) \\ &= (x-2)(x^2 - 3x + ix - 3x - ix + 9 - i^2) \\ &= (x-2)(x^2 - 6x + 10) \\ &= x^3 - 8x^2 + 22x - 20. \end{aligned}$$

The next-to-last line of this calculation also shows that $f(x)$ can be factored as a product of a linear and a quadratic polynomial, each with *real* coefficients. ■

The technique in Example 3 works because the polynomial

$$(x - (3 + i))(x - (3 - i))$$

turns out to have real coefficients. The proof of the following result shows why this must always be the case:

Factorization Over the Real Numbers ▶

Every nonconstant polynomial with real coefficients can be factored as a product of linear and quadratic polynomials with real coefficients in such a way that the quadratic factors, if any, have no real roots.*

Proof The box on page 286 shows that

$$f(x) = d(x - c_1)(x - c_2) \cdots (x - c_n)$$

where $c_1, \ldots, c_n$ are the roots of $f(x)$. If some c_i is a real number, then the factor $x - c_i$ is a linear polynomial with real coefficients.** If some c_j is a nonreal complex root, then its conjugate must also be a root. Thus some c_k is the conjugate of c_j, say, $c_j = a + bi$ (with a, b real) and $c_k = a - bi$.† In this case,

$$\begin{aligned}(x - c_j)(x - c_k) &= (x - (a + bi))(x - (a - bi)) \\ &= x^2 - (a - bi)x - (a + bi)x + (a + bi)(a - bi) \\ &= x^2 - ax + bix - ax - bix + a^2 - (bi)^2 \\ &= x^2 - 2ax + (a^2 + b^2).\end{aligned}$$

Therefore, the factor $(x - c_j)(x - c_k)$ of $f(x)$ is a quadratic with real coefficients (because a and b are real numbers). Its roots (c_j and c_k) are nonreal. By taking the real roots of $f(x)$ one at a time and the nonreal ones in conjugate pairs in this fashion, we obtain the desired factorization of $f(x)$. ❑

EXAMPLE 4 Given that $1 + i$ is a root of $f(x) = x^4 - 2x^3 - x^2 + 6x - 6$, factor $f(x)$ completely over the real numbers.

Solution Since $1 + i$ is a root of $f(x)$, so is its conjugate $1 - i$, and hence $f(x)$ has this quadratic factor:

$$(x - (1 + i))(x - (1 - i)) = x^2 - 2x + 2.$$

*A "real root" is a root that is a real number.

**In Example 3, for instance, 2 is a real root and $x - 2$ a linear factor.

†In Example 3, for instance, $c_j = 3 + i$ and $c_k = 3 - i$ are conjugate roots.

Dividing $f(x)$ by $x^2 - 2x + 2$ shows that the other factor is $x^2 - 3$, which factors as $(x + \sqrt{3})(x - \sqrt{3})$. Therefore

$$f(x) = (x + \sqrt{3})(x - \sqrt{3})(x^2 - 2x + 2). \quad \blacksquare$$

EXERCISES 4.7

In Exercises 1–6, find the remainder when $f(x)$ is divided by $g(x)$ without using synthetic or long division.

1. $f(x) = x^{10} + x^8; \quad g(x) = x - 1$
2. $f(x) = x^6 - 10; \quad g(x) = x - 2$
3. $f(x) = 3x^4 - 6x^3 + 2x - 1; \quad g(x) = x + 1$
4. $f(x) = x^5 - 3x^2 + 2x - 1; \quad g(x) = x - 2$
5. $f(x) = x^3 - 2x^2 + 5x - 4; \quad g(x) = x + 2$
6. $f(x) = 10x^{75} - 8x^{65} + 6x^{45} + 4x^{32} - 2x^{15} + 5;$ $g(x) = x - 1$

In Exercises 7–10, list the roots of the polynomial and state the multiplicity of each root.

7. $f(x) = x^{54}\left(x + \frac{4}{5}\right)$
8. $g(x) = 3\left(x + \frac{1}{6}\right)\left(x - \frac{1}{5}\right)\left(x + \frac{1}{4}\right)$
9. $h(x) = 2x^{15}(x - \pi)^{14}(x - (\pi + 1))^{13}$
10. $k(x) = (x - \sqrt{7})^7(x - \sqrt{5})^5(2x - 1)$

In Exercises 11–22, find all the roots of $f(x)$ in the complex number system; then write $f(x)$ as a product of linear factors.

11. $f(x) = x^2 - 2x + 5$
12. $f(x) = x^2 - 4x + 13$
13. $f(x) = 3x^2 + 2x + 7$
14. $f(x) = 3x^2 - 5x + 2$
15. $f(x) = x^3 - 27$ [*Hint:* Factor first.]
16. $f(x) = x^3 + 125$
17. $f(x) = x^3 + 8$
18. $f(x) = x^6 - 64$ [*Hint:* Let $u = x^3$ and factor $u^2 - 64$ first.]
19. $f(x) = x^4 - 1$
20. $f(x) = x^4 - x^2 - 6$
21. $f(x) = x^4 - 3x^2 - 10$
22. $f(x) = 2x^4 - 7x^2 - 4$

In Exercises 23–44, find a polynomial $f(x)$ with real *coefficients that satisfies the given conditions. Some of these problems have many correct answers.*

23. Degree 3; only roots are $1, 7, -4$.
24. Degree 3; only roots are 1 and -1.
25. Degree 6; only roots are $1, 2, \pi$.
26. Degree 5; only root is 2.
27. Degree 3; roots $-3, 0, 4$; $f(5) = 80$.
28. Degree 3; roots $-1, 1/2, 2$; $f(0) = 2$.
29. Roots include $2 + i$ and $2 - i$.
30. Roots include $1 + 3i$ and $1 - 3i$.
31. Roots include 2 and $2 + i$.
32. Roots include 3 and $4i - 1$.
33. Roots include $-3, 1 - i, 1 + 2i$.
34. Roots include $1, 2 + i, 3i - 1$.
35. Degree 2; roots $1 + 2i$ and $1 - 2i$.
36. Degree 4; roots $3i$ and $-3i$, each of multiplicity 2.
37. Degree 4; only roots are $4, 3 + i$, and $3 - i$.
38. Degree 5; roots 2 (of multiplicity 3), i, and $-i$.
39. Degree 6; roots 0 (of multiplicity 3) and $3, 1 + i, 1 - i$, each of multiplicity 1.
40. Degree 6; roots include i (of multiplicity 2) and 3.
41. Degree 2; roots include $1 + i$; $f(0) = 6$.
42. Degree 2; roots include $3 + i$; $f(2) = 3$.
43. Degree 3; roots include i and 1; $f(-1) = 8$.
44. Degree 3; roots include $2 + 3i$ and -2; $f(2) = -3$.

In Exercises 45–48, find a polynomial with complex coefficients that satisfies the given conditions.

45. Degree 2; roots i and $1 - 2i$.

46. Degree 2; roots $2i$ and $1 + i$.

47. Degree 3; roots 3, i, and $2 - i$.

48. Degree 4; roots $\sqrt{2}$, $-\sqrt{2}$, $1 + i$, and $1 - i$.

In Exercises 49–56, one root of the polynomial is given; find all the roots.

49. $x^3 - 2x^2 - 2x - 3$; root 3.

50. $x^3 + x^2 + x + 1$; root i.

51. $x^4 + 3x^3 + 3x^2 + 3x + 2$; root i.

52. $x^4 - x^3 - 5x^2 - x - 6$; root i.

53. $x^4 - 2x^3 + 5x^2 - 8x + 4$; root 1 of multiplicity 2.

54. $x^4 - 6x^3 + 29x^2 - 76x + 68$; root 2 of multiplicity 2.

55. $x^4 - 4x^3 + 6x^2 - 4x + 5$; root $2 - i$.

56. $x^4 - 5x^3 + 10x^2 - 20x + 24$; root $2i$.

57. Let $z = a + bi$ and $w = c + di$ be complex numbers (a, b, c, d are real numbers). Prove the given equality by computing each side and comparing the results:

(a) $\overline{z + w} = \bar{z} + \bar{w}$ (The left side says: First find $z + w$ and then take the conjugate. The right side says: First take the conjugates of z and w and then add.)

(b) $\overline{z \cdot w} = \bar{z} \cdot \bar{w}$

58. Let $g(x)$ and $h(x)$ be polynomials of degree n and assume that there are $n + 1$ numbers $c_1, c_2, \ldots, c_n, c_{n+1}$ such that

$$g(c_i) = h(c_i) \qquad \text{for every } i.$$

Prove that $g(x) = h(x)$. [*Hint:* Show that each c_i is a root of $f(x) = g(x) - h(x)$. If $f(x)$ is nonzero, what is its largest possible degree? To avoid a contradiction, conclude that $f(x) = 0$.]

59. Suppose $f(x) = ax^3 + bx^2 + cx + d$ has real coefficients and z is a complex root of $f(x)$.

(a) Use Exercise 57 and the fact that $\bar{r} = r$, when r is a real number, to show that

$$\begin{aligned}\overline{f(z)} &= \overline{az^3 + bz^2 + cz + d}\\ &= a\bar{z}^3 + b\bar{z}^2 + c\bar{z} + d = f(\bar{z}).\end{aligned}$$

(b) Conclude that $\bar{z}$ is also a root of $f(x)$. [*Note:* $f(\bar{z}) = \overline{f(z)} = \bar{0} = 0$.]

60. Let $f(x)$ be a polynomial with real coefficients and z a complex root of $f(x)$. Prove that the conjugate $\bar{z}$ is also a root of $f(x)$. [*Hint:* Exercise 59 is the case when $f(x)$ has degree 3; the proof in the general case is similar.]

61. Use the statement in the last box on page 289 to show that every polynomial with real coefficients and *odd* degree must have at least one real root.

62. Give an example of a polynomial $f(x)$ with complex, nonreal coefficients and a complex number z such that z is a root of $f(x)$, but its conjugate is not. Hence, the conclusion of the Conjugate Roots Theorem (page 288) may be *false* if $f(x)$ doesn't have real coefficients.

CHAPTER 4 *Review*

Important Concepts ▶

Section 4.1	Polynomial 209 Coefficient 209 Constant term 209 Zero polynomial 210 Degree of a polynomial 210 Leading coefficient 210 Polynomial function 210 Division of polynomials 210–211 Division Algorithm 212 Remainder Theorem 213 Root or zero of a polynomial 213 Real root 213 Roots and x-intercepts 214 Factor Theorem 214 Number of roots of a polynomial 215
Excursion 4.1.A	Synthetic division 217
Section 4.2	The Rational Root Test 221 Repeated root 223 Irrational roots 224 Bounds Test for roots 226
Section 4.3	Continuity 229 Monomial graphs 230–231 Behavior when $\lvert x \rvert$ is large 231 x-intercepts 232 Local extrema 232–234 Points of inflection 235
Section 4.4	Rational function 242 Domain 242 The Big–Little Principle 243 Linear rational functions 245 Intercepts 247 Vertical asymptotes 247 Holes 248 Horizontal asymptotes 249
Excursion 4.4.A	Asymptote when $\lvert x \rvert$ is large 260
Section 4.5	Basic principles for solving inequalities 264 Linear inequalities 264 Polynomial inequalities 265 Rational inequalities 269

Review Questions ▶

1. Which of the following are polynomials?

(a) $2^3 + x^2$ **(b)** $x + \frac{1}{x}$

(c) $x^3 - \frac{1}{\sqrt{2}}$ **(d)** $\sqrt[3]{x^4}$

(e) $\pi^3 - x$ **(f)** $\sqrt{2} + 2x^2$

(g) $\sqrt{x} + 2x^2$ **(h)** $|x|$

2. What is the remainder when $x^4 + 3x^3 + 1$ is divided by $x^2 + 1$?

3. What is the remainder when $x^{112} - 2x^8 + 9x^5 - 4x^4 + x - 5$ is divided by $x - 1$?

4. Is $x - 1$ a factor of $f(x) = 14x^{87} - 65x^{56} + 51$? Justify your answer.

5. Use synthetic division to show that $x - 2$ is a factor of $x^6 - 5x^5 + 8x^4 + x^3 - 17x^2 + 16x - 4$ and find the other factor.

6. List the roots of this polynomial and the multiplicity of each root:

$$f(x) = 5(x - 4)^3(x - 2)(x + 17)^3(x^2 - 4).$$

7. Find a polynomial f of degree 3 such that $f(-1) = 0$, $f(1) = 0$, and $f(0) = 5$.

8. Find the root(s) of $2\left(\frac{x}{5} + 7\right) - 3x - \frac{x + 2}{5} + 4$.

9. Find the roots of $3x^2 - 2x - 5$.

10. Factor the polynomial $x^3 - 8x^2 + 9x + 6$. [*Hint:* 2 is a root.]

11. Find all real roots of $x^6 - 4x^3 + 4$.

12. Find all real roots of $9x^3 - 6x^2 - 35x + 26$. [*Hint:* Try $x = -2$.]

13. Find all real roots of $3y^3(y^4 - y^2 - 5)$.

14. Find the rational roots of $x^4 - 2x^3 - 4x^2 + 1$.

15. Consider the polynomial $2x^3 - 8x^2 + 5x + 3$.

(a) List the only *possible* rational roots.

(b) Find one rational root.

(c) Find all the roots of the polynomial.

16. **(a)** Find all rational roots of $x^3 + 2x^2 - 2x - 2$.
(b) Find two consecutive integers such that an irrational root of $x^3 + 2x^2 - 2x - 2$ lies between them.

17. How many distinct real roots does $x^3 + 4x$ have?

18. How many distinct real roots does $x^3 - 6x^2 + 11x - 6$ have?

19. Find the roots of $x^4 - 11x^2 + 18$.

20. The polynomial $x^3 - 2x + 1$ has
(a) no real roots.
(b) only one real root.
(c) three rational roots.
(d) only one rational root.
(e) none of the above.

21. Show that 5 is an upper bound for the real roots of $x^4 - 4x^3 + 16x - 16$.

22. Show that -1 is a lower bound for the real roots of $x^4 - 4x^3 + 15$.

In Questions 23 and 24, find the real roots of the polynomial.

23. $x^6 - 2x^5 - x^4 + 3x^3 - x^2 - x + 1$

24. $x^5 - 3x^4 - 2x^3 - x^2 - 23x - 20$

In Questions 25 and 26, compute and simplify the difference quotient of the function.

25. $f(x) = x^2 + x$ **26.** $g(x) = x^3 - x + 1$

27. Draw the graph of a function that could not possibly be the graph of a polynomial function and explain why.

28. Draw a graph that could be the graph of a polynomial function of degree 5. You need not list a specific polynomial, nor do any computation.

29. Which of the statements below is *not* true about the polynomial function f whose graph is shown in Figure 4–61?

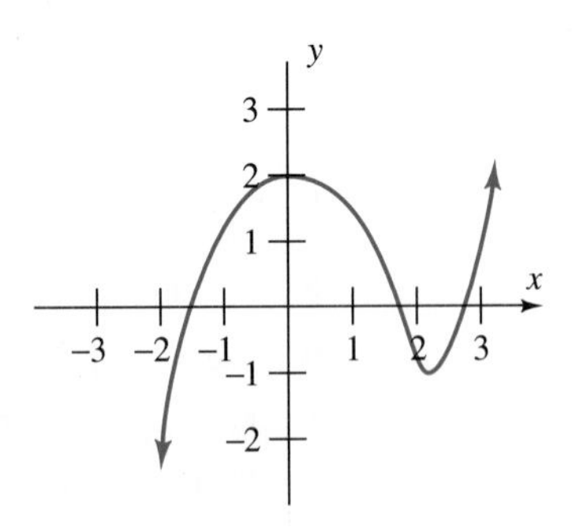

Figure 4-61

(a) f has three roots between -2 and 3.
(b) $f(x)$ could possibly be a fifth-degree polynomial.
(c) $(f \circ f)(0) > 0$.
(d) $f(2) - f(-1) < 3$.
(e) $f(x)$ is positive for all x in the interval $[-1, 0]$.

30. Which of the statements (i)–(v) below about the polynomial function f whose graph is shown in Figure 4–62 are *false*?

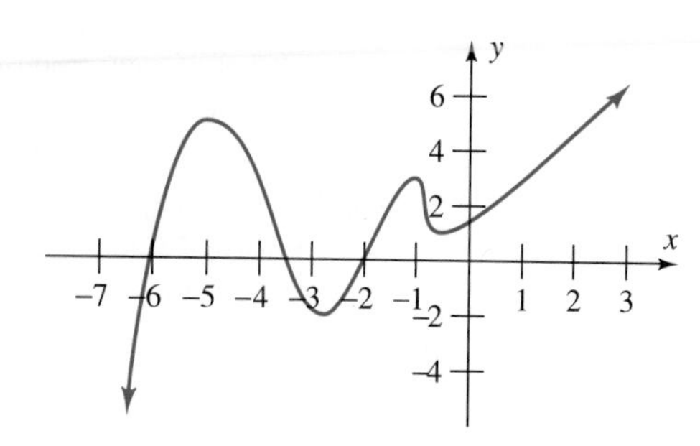

Figure 4-62

(i) f has 2 roots in the interval $[-6, -3)$.

(ii) $f(-3) - f(-6) < 0$.

(iii) $f(0) < f(1)$.

(iv) $f(2) - 2 = 0$.

(v) f has degree ≤ 4.

In Questions 31–34, find a viewing window (or windows) that shows a complete graph of the function. Be alert for hidden behavior.

31. $f(x) = .5x^3 - 4x^2 + x + 1$

32. $g(x) = .3x^5 - 4x^4 + x^3 - 4x^2 + 5x + 1$

33. $h(x) = 4x^3 - 100x^2 + 600x$

34. $f(x) = 32x^3 - 99x^2 + 100x + 2$

In Questions 35–42, sketch a complete graph of the function. Label the x-intercepts, all local extrema, holes, and asymptotes.

35. $f(x) = x^3 - 9x$

36. $g(x) = x^3 - 2x^2 + 3$

37. $h(x) = x^4 - x^3 - 4x^2 + 4x + 2$

38. $f(x) = x^4 - 3x - 2$

39. $g(x) = \dfrac{-2}{x + 4}$

40. $h(x) = \dfrac{3 - x}{x - 2}$

41. $k(x) = \dfrac{4x + 10}{3x - 9}$

42. $f(x) = \dfrac{x + 1}{x^2 - 1}$

In Questions 43 and 44, list all asymptotes of the graph of the function.

43. $f(x) = \dfrac{x^2 - 1}{x^3 - 2x^2 - 5x + 6}$

44. $g(x) = \dfrac{x^4 - 6x^3 + 2x^2 - 6x + 2}{x^2 - 3}$

In Questions 45–48, find a viewing window (or windows) that shows a complete graph of the function. Be alert for hidden behavior.

45. $f(x) = \dfrac{x - 3}{x^2 + x - 2}$

46. $g(x) = \dfrac{x^2 - x - 6}{x^3 - 3x^2 + 3x - 1}$

47. $h(x) = \dfrac{x^4 + 4}{x^4 - 99x^2 - 100}$

48. $k(x) = \dfrac{x^3 - 2x^2 - 4x + 8}{x - 10}$

49. It costs The Junkfood Company 50¢ to produce a bag of Munchies. There are fixed costs of $500 per day for building, equipment, etc. The Company has found that if the price of a bag of Munchies is set at $1.95 - \frac{x}{2000}$ dollars, where x is the number of bags produced per day, then all the bags that are produced will be sold. What number of bags can be produced each day if all are to be sold and the Company is to make a profit? What are the possible retail prices?

50. Highway engineers have found that a good model of the relationship between the density of automobile traffic and the speed at which it moves along a particular section of highway is given by the function

$$s = \frac{100}{1 + d^2} \qquad (0 \leq d \leq 3)$$

where the density of traffic d is measured in hundreds of cars per mile and the speed s at which traffic moves is measured in miles per hour. Then the traffic flow q is given by the product of the density and the speed, that is, $q = ds$, with q being measured in hundreds of cars per hour.

(a) Express traffic flow as a function of traffic density.
(b) For what densities will traffic flow be at least 3000 cars per hour?
(c) What traffic density will maximize traffic flow? What is the maximum flow?

51. Charlie lives 150 miles from the city. He drives 40 miles to the station and catches a train to the city. The average speed of the train is 25 mph faster than the average speed of the car.

(a) Express the total time for the journey as a function of the speed of the car. What speeds make sense in this context?
(b) How fast should Charlie drive if the entire journey is to take no more than two and a half hours?

52. The survival rate s of seedlings in the vicinity of a parent tree is given by

$$s = \frac{.5x}{1 + .4x^2}$$

where x is the distance from the seedling to the tree (in meters) and $0 < x \leq 10$.

(a) For what distances is the survival rate at least .21?
(b) What distance produces the maximum survival rate?

In Questions 53 and 54, find the average rate of change of the function between x and $x + h$.

53. $f(x) = \frac{x}{x + 1}$ **54.** $g(x) = \frac{1}{x^2 + 1}$

55. Which of these statements about the graph of $f(x) = \frac{(x - 1)(x + 3)}{(x^2 + 1)(x^2 - 1)}$ is *true*?

(a) The graph has two vertical asymptotes.
(b) The graph touches the x-axis at $x = 3$.
(c) The graph lies above the x-axis when $x < -1$.
(d) The graph has a hole at $x = 1$.
(e) The graph has no horizontal asymptotes.

56. Solve for x: $-3(x-4) \leq 5 + x$.

57. Solve for y: $\left|\dfrac{y+2}{3}\right| \geq 5$.

58. Solve for x: $-4 < 2x + 5 < 9$.

59. On which intervals is $\dfrac{2x-1}{3x+1} < 1$?

60. On which intervals is $\dfrac{2}{x+1} < x$?

61. Solve for x: $\left|\dfrac{1}{1-x^2}\right| \geq \dfrac{1}{2}$.

62. Solve for x: $x^2 + x > 12$.

63. Solve for x: $(x-1)^2(x^2-1)x \leq 0$.

64. If $0 < r \leq s - t$, then which of these statements is *false*?

(a) $s \geq r + t$ **(b)** $t - s \leq -r$

(c) $-r \geq s - t$ **(d)** $\dfrac{s-t}{r} > 0$

(e) $s - r \geq t$

65. If $\dfrac{x+3}{2x-3} > 1$, then which of these statements is *true*?

(a) $\dfrac{x-3}{2x+3} < -1$ **(b)** $\dfrac{2x-3}{x+3} < -1$

(c) $\dfrac{3-2x}{x+3} > 1$ **(d)** $2x + 3 < x - 3$

(e) none of these

66. Solve and express your answer in interval notation:

$$2x - 3 \leq 5x + 9 < -3x + 4.$$

In Questions 67–74, solve the inequality.

67. $|3x + 2| \geq 2$ **68.** $x^2 + x - 20 > 0$

69. $\dfrac{x-2}{x+4} \leq 3$

70. $(x+1)^2(x-3)^4(x+2)^3(x-7)^5 > 0$

71. $\dfrac{x^2+x-9}{x+3} < 1$ **72.** $\dfrac{x^2-x-6}{x-3} > 1$

73. $\dfrac{x^2-x-5}{x^2+2} > -2$ **74.** $\dfrac{x^4-3x^2+2x-3}{x^2-4} < -1$

In Questions 75–82, solve the equation in the complex number system.

75. $x^2 + 3x + 10 = 0$ **76.** $x^2 + 2x + 5 = 0$

77. $5x^2 + 2 = 3x$ **78.** $-3x^2 + 4x - 5 = 0$

79. $3x^4 + x^2 - 2 = 0$

80. $8x^4 + 10x^2 + 3 = 0$

81. $x^3 + 8 = 0$

82. $x^3 - 27 = 0$

83. One root of $x^4 - x^3 - x^2 - x - 2$ is i. Find all the roots.

84. One root of $x^4 + x^3 - 5x^2 + x - 6$ is i. Find all the roots.

85. Give an example of a fourth-degree polynomial with real coefficients whose roots include 0 and $1 + i$.

86. Find a fourth-degree polynomial f whose only roots are $2 + i$ and $2 - i$, such that $f(-1) = 50$.

Exponential and Logarithmic Functions

Roadmap

Each section of this chapter depends on the preceding one.

Exponential and logarithmic functions are essential for the mathematical description of a variety of phenomena in physical science, engineering, and economics. Some of these applications will be considered in this chapter. Except for carefully chosen numbers, the values of these functions cannot be computed by hand, so a calculator is essential. However, you won't be able to use your calculator efficiently or to interpret its answers unless you understand the properties of these functions. Furthermore, when calculations can readily be done by hand, you will be expected to do them without a calculator.

5.1 RADICALS AND RATIONAL EXPONENTS

Our goal is to define c^r, where c is a real number and r is a rational number, and to develop the properties of such rational exponents. The first step is to define the nth root of a real number. Let n be a positive integer. Depending on whether n is odd or even, the graph of the function $f(x) = x^n$ has one of the shapes shown in Figure 5–1.

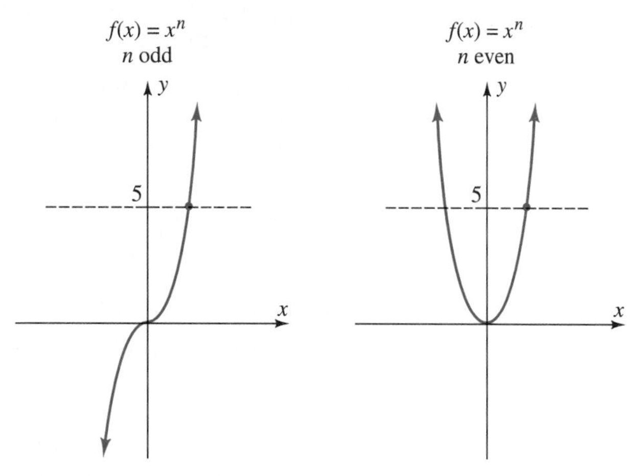

Figure 5-1

Figure 5–1 shows that the horizontal line $y = 5$ intersects the graph of $f(x) = x^n$ at exactly one point in the first quadrant. If (u, v) is this intersection point, then $u^n = v$ (because the point is on the graph of $f(x) = x^n$) and $v = 5$ (because the point is on the horizontal line $y = 5$). In other words, the x-coordinate of this intersection point is the unique positive solution of the equation $x^n = 5$. The same argument works with any nonnegative number c in place of 5 and leads to this conclusion:

> If c is a nonnegative real number, then the equation $x^n = c$ has a unique nonnegative solution.

Recall that the unique nonnegative solution of $x^2 = c$ is called the *square root* of c and the unique solution of $x^3 = c$ is called the *cube root* of c. Similar terminology is used in the general case.

***n*th Roots** ▶

Let c be a nonnegative real number and n a positive integer. Then, the *nth root of c* is the unique nonnegative solution of the equation $x^n = c$. The nth root of c is denoted by either of these symbols:

$$c^{1/n} \quad \text{or} \quad \sqrt[n]{c}.$$

The radical notation for roots is familiar: $\sqrt[3]{c}$ for the cube root of c and $\sqrt{c}$ (rather than $\sqrt[2]{c}$) for the square root of c. The exponent notation, however, will prove more useful, as we shall see below. The exponent notation is also easier to implement on most calculators.

EXAMPLE 1

(a) To find $\sqrt[11]{225} = 225^{1/11}$, key in [225] [∧] [(] [1] [÷] [11] [)] [ENTER].* The calculator then shows that $225^{1/11} \approx 1.636193919$. In other words, 1.636193919 is the (approximate) solution of $x^{11} = 225$.

(b) A calculator isn't needed to compute $\sqrt[4]{81} = 81^{1/4}$ because it's easy to see that the unique positive solution of $x^4 = 81$ is $x = 3$. Hence, $81^{1/4} = 3$. ■

Since the nth root of c is a solution of $x^n = c$ by definition, we see that

$$(\sqrt[n]{c})^n = c, \qquad \text{or equivalently,} \qquad (c^{1/n})^n = c.$$

The exponential form of this statement says that the familiar property of integer exponents, $(c^m)^n = c^{mn}$, remains true with $1/n$ in place of m, that is,

$$(c^{1/n})^n = c^{(1/n)n} = c^1 = c.$$

This fact suggests that it may be possible to define rational number exponents in such a way that the various exponent laws for integer exponents continue to hold.

*The [∧] key is labeled [x^y] or [y^x] or [a^b] on some calculators. The [ENTER] key is labeled [EXE] on Casio.

Consider, for example, how the expression $4^{3/2}$ might be defined. The exponent 3/2 can be written as either 3(1/2) or (1/2)3. If the property $(c^m)^n = c^{mn}$ is to hold, we might define $4^{3/2}$ as either $(4^3)^{1/2}$ or $(4^{1/2})^3$. The result is the same in both cases:

$$(4^3)^{1/2} = 64^{1/2} = \sqrt{64} = 8 \qquad \text{and} \qquad (4^{1/2})^3 = (\sqrt{4})^3 = 2^3 = 8.$$

It can be proved that for any nonnegative real number c and any integers t, k with k positive:

$$(c^t)^{1/k} = (c^{1/k})^t, \qquad \text{or in radical notation,} \qquad \sqrt[k]{c^t} = (\sqrt[k]{c})^t.$$

Consequently, we make this definition:

Rational Exponents ▶

Let c be a positive real number and let t/k be a rational number with positive denominator. Then,

$c^{t/k}$ is defined to be the number $(c^t)^{1/k} = (c^{1/k})^t$.

EXAMPLE 2 Every terminating decimal is a rational number, so expressions such as $13^{3.78}$ now have meaning: $13^{3.78} = 13^{378/100}$. Using the [∧] key on a calculator and entering the exponent in decimal form, we find that $13^{3.78} \approx 16{,}244.482$. ■

EXAMPLE 3 To express $5^{-2/3}$ in terms of radicals, use the definition:

$$5^{-2/3} = (5^{-2})^{1/3} = \sqrt[3]{5^{-2}} = \sqrt[3]{\frac{1}{5^2}} = \frac{\sqrt[3]{1}}{\sqrt[3]{5^2}} = \frac{1}{\sqrt[3]{25}}.$$

To compute a decimal approximation for $5^{-2/3}$, use the [∧] key and parentheses:

[5] [∧] [(] [(−)] [2] [÷] [3] [)] [ENTER] [.341995] (approximately). ■

Rational exponents were defined in a way that guaranteed that one of the familiar exponent laws would remain valid. In fact, all the exponent laws developed for integer exponents are valid for rational exponents, as summarized here and illustrated in Examples 4 through 7.

Exponent Laws ▶

Let c and d be nonnegative real numbers and let r and s be any rational numbers. Then,

1. $c^r c^s = c^{r+s}$
2. $\dfrac{c^r}{c^s} = c^{r-s} \quad (c \neq 0)$
3. $(c^r)^s = c^{rs}$
4. $(cd)^r = c^r d^r$
5. $\left(\dfrac{c}{d}\right)^r = \dfrac{c^r}{d^r} \quad (d \neq 0)$
6. $\dfrac{1}{c^{-r}} = c^r \quad (c \neq 0)$

EXAMPLE 4

$$(8r^{3/4}s^{-3})^{2/3} = 8^{2/3}(r^{3/4})^{2/3}(s^{-3})^{2/3} = \sqrt[3]{8^2}(r^{2/4})(s^{-2})$$
$$= \sqrt[3]{64}r^{1/2}s^{-2} = \frac{4r^{1/2}}{s^2}.$$

We can leave the answer in exponential form, or if it is more convenient, write it as $4\sqrt{r}/s^2$. ■

WARNING The exponent laws deal only with products and quotients. There are no analogous properties for sums. In particular, if both c and d are nonzero, then

$(c + d)^r$ is ***not*** equal to $c^r + d^r$.

EXAMPLE 5

$$x^{1/2}(x^{3/4} - x^{3/2}) = x^{1/2}x^{3/4} - x^{1/2}x^{3/2}$$
$$= x^{1/2+3/4} - x^{1/2+3/2} = x^{5/4} - x^2. \quad ■$$

EXAMPLE 6

$$(x^{5/2}y^4)(xy^{7/4})^{-2} = (x^{5/2}y^4)x^{-2}(y^{7/4})^{-2}$$
$$= x^{5/2}y^4x^{-2}y^{(7/4)(-2)} = x^{5/2}x^{-2}y^4y^{-7/2}$$
$$= x^{(5/2)-2}y^{4-(7/2)} = x^{1/2}y^{1/2}. \quad ■$$

EXAMPLE 7 Let k be a positive rational number and express $\sqrt[10]{c^{5k}}\sqrt{(c^{-k})^{1/2}}$ without radicals, using only positive exponents.

Solution

$$\sqrt[10]{c^{5k}}\sqrt{(c^{-k})^{1/2}} = (c^{5k})^{1/10}[(c^{-k})^{1/2}]^{1/2} = c^{k/2}c^{-k/4} = c^{k/2-k/4} = c^{k/4}. \quad ■$$

Negative Bases

The expression c^r can sometimes be defined when the base c is negative. If n is an odd positive integer ($n = 1, 3, 5$, etc.), then the equation $x^n = c$ has a unique solution even when c is negative.* Consequently, the nth root of c can be defined in this case. For example,

$$\sqrt[3]{-125} = (-125)^{1/3} = -5 \qquad \text{since} \qquad (-5)^3 = -125.$$

Other fractional exponents can be defined when the base is negative and the fraction is in lowest terms with an odd denominator. For example,

$$(-8)^{2/3} = \sqrt[3]{(-8)^2} = \sqrt[3]{64} = 4$$
$$(-24)^{3/5} = \sqrt[5]{(-24)^3} = \sqrt[5]{-13{,}824} \approx -6.7317.$$

*Figure 5–1 shows that when n is odd, the horizontal line $y = c$ intersects the graph of $f(x) = x^n$ exactly once, so the equation has a solution.

Rational powers of a negative number are not defined when the exponent has an even denominator. For example, $(-4)^{1/2} = \sqrt{-4}$ is meaningless in the real number system since the equation $x^2 = -4$ has no real solutions.

WARNING Most calculators will not raise a negative number to a fractional exponent unless the numerator of the fraction is 1. For instance, we know that

$$(-8)^{2/3} = ((-2)^3)^{2/3} = (-2)^2 = 4,$$

but entering $(-8)^{(2/3)}$ produces an error message on most calculators (the TI-92 is an exception). However, entering $(-8)^{2/3}$ by keying in either

$$((-8)^2)^{(1/3)} \quad \text{or} \quad ((-8)^{(1/3)})^2$$

produces the correct answer 4.*

GRAPHING EXPLORATION The preceding warning can be confirmed graphically by graphing $f(x) = x^{2/3}$ (which is defined for every real number x) in the standard viewing window. Does the screen show a graph to the left of the y-axis? Now rewrite the rule of f as $f(x) = (x^2)^{1/3}$ and graph again.

Scientific Notation

Any real number may be written as the product of a power of 10 and a number between 1 and 10. For example,

$$356 = 3.56 \times 100 = 3.56 \times 10^2$$

$$1{,}563{,}427 = 1.563427 \times 1{,}000{,}000 = 1.563427 \times 10^6$$

$$.072 = 7.2 \div 100 = 7.2 \times 1/100 = 7.2 \times 10^{-2}$$

$$.000862 = 8.62 \div 10{,}000 = 8.62 \times 10^{-4}$$

A number written in this form is said to be in **scientific notation.** Scientific notation is very useful for computations with very large or very small numbers.

EXAMPLE 8

$$\begin{aligned}(.00000002)(4{,}300{,}000{,}000) &= (2 \times 10^{-8})(4.3 \times 10^9) \\ &= 2(4.3)10^{-8+9} = (8.6)10^1 = 86. \quad \blacksquare\end{aligned}$$

*On HP-38, entering $((-8)^2)^{1/3}$ produces 3.9999 · · · , but entering any of the other forms results in a complex number.

EXAMPLE 9

$$\frac{(50{,}000{,}000)^3(.000002)^5}{(.000008)} = \frac{(5 \times 10^7)^3(2 \times 10^{-6})^5}{8 \times 10^{-6}}$$

$$= \frac{(5^3 \cdot 10^{21})(2^5 \cdot 10^{-30})}{8 \cdot 10^{-6}} = \frac{125 \cdot 32 \cdot 10^{21-30}}{8 \cdot 10^{-6}}$$

$$= 125 \cdot 4 \cdot 10^{21-30+6} = 500 \cdot 10^{-3} = \frac{500}{1000} = \frac{1}{2}. \quad \blacksquare$$

Calculators display numbers in scientific notation without showing the 10. For instance, the calculator display 7.235 E -12 indicates the number 7.235×10^{-12}. To enter a number in scientific notation into a calculator, for example 5.6×10^{73}, enter 5.6, then press the EE (or Exp or EEX) key and enter 73. Calculators automatically switch to scientific notation when a number is too large or too small to be displayed in the standard way.

WARNING Your calculator may not always obey the laws of arithmetic when dealing with very large or very small numbers. For instance, the associative law shows that

$$(1 + 10^{19}) - 10^{19} = 1 + (10^{19} - 10^{19}) = 1 + 0 = 1.$$

But the calculator may round off $1 + 10^{19}$ as 10^{19} (instead of the correct number 10,000,000,000,000,000,001). So the calculator computes $(1 + 10^{19}) - 10^{19}$ as $10^{19} - 10^{19} = 0$.

Irrational Exponents

An example will illustrate how a^t is defined when t is an irrational number.* To compute $10^{\sqrt{2}}$ we use the infinite decimal expansion $\sqrt{2} \approx 1.4142135623\cdots$ (see Excursion 1.1.A). Each of

$$1.4,\ 1.41,\ 1.414,\ 1.4142,\ 1.41421,\ \ldots$$

is a rational number approximation of $\sqrt{2}$, and each is a more accurate approximation than the preceding one. We know how to raise 10 to each of these rational numbers:

$$10^{1.4} \approx 25.1189 \qquad 10^{1.4142} \approx 25.9537$$

$$10^{1.41} \approx 25.7040 \qquad 10^{1.41421} \approx 25.9543$$

$$10^{1.414} \approx 25.9418 \qquad 10^{1.414213} \approx 25.9545$$

*This example is not a proof, but should make the idea plausible. Calculus is required for a rigorous proof.

It appears that as the exponent r gets closer and closer to $\sqrt{2}$, 10^r gets closer and closer to a real number whose decimal expansion begins $25.954\cdots$. We define $10^{\sqrt{2}}$ to be this number.

Similarly, for any $a > 0$.

a^t is a well-defined *positive* number for each real exponent t.

We shall also assume this fact:

The exponent laws (page 301) are valid for *all* real exponents.

EXERCISES 5.1

Note: *Unless directed otherwise, assume all letters represent positive real numbers.*

In Exercises 1–12, simplify the expression without using a calculator. For example,

$$\sqrt{8}\sqrt{12} = \sqrt{8\cdot 12} = \sqrt{96} = \sqrt{16\cdot 6} = \sqrt{16}\sqrt{6} = 4\sqrt{6}.$$

1. $\sqrt{.0081}$ **2.** $\sqrt{.000169}$ **3.** $\sqrt{(.08)^{12}}$

4. $\sqrt{(-11)^{28}}$ **5.** $\sqrt{6}\sqrt{12}$ **6.** $\sqrt{8}\sqrt{96}$

7. $\dfrac{\sqrt{10}}{\sqrt{8}\sqrt{5}}$ **8.** $\dfrac{\sqrt{6}}{\sqrt{14}\sqrt{63}}$

9. $(1+\sqrt{3})(2-\sqrt{3})$ **10.** $(3+\sqrt{2})(3-\sqrt{2})$

11. $(4-\sqrt{3})(5+2\sqrt{3})$ **12.** $(2\sqrt{5}-4)(3\sqrt{5}+2)$

In Exercises 13–18, write the given expression without using radicals.

13. $\sqrt[3]{a^2+b^2}$ **14.** $\sqrt[4]{a^3-b^3}$ **15.** $\sqrt[4]{\sqrt[4]{a^3}}$

16. $\sqrt{\sqrt[3]{a^3b^4}}$ **17.** $\sqrt[5]{t\sqrt{16t^5}}$ **18.** $\sqrt{x}(\sqrt[3]{x^2})(\sqrt[4]{x^3})$

In Exercises 19–38, simplify the given expression.

19. $\sqrt{16a^8b^{-2}}$ **20.** $\sqrt{24x^6y^{-4}}$ **21.** $\dfrac{\sqrt{c^2d^6}}{\sqrt{4c^3d^{-4}}}$

22. $\dfrac{\sqrt{a^{-10}b^{-12}}}{\sqrt{a^{14}d^{-4}}}$ **23.** $5\sqrt{20}-\sqrt{45}+2\sqrt{80}$

24. $\sqrt[3]{40}+2(\sqrt[3]{135})-5(\sqrt[3]{320})$ **25.** $\sqrt[4]{(4x+2y)^8}$

26. $(\sqrt[3]{a+b})(\sqrt[3]{-(a+b)^2})+\sqrt[3]{a+b}$

27. $\dfrac{2^{11/2}\cdot 2^{-7}\cdot 2^{-5}}{2^3\cdot 2^{1/2}\cdot 2^{-10}}$ **28.** $\dfrac{(3^2)^{-1/2}(9^4)^{-1}}{27^{-3}}$

29. $\sqrt{x^7}\cdot x^{5/2}\cdot x^{-3/2}$ **30.** $(x^{1/2}y^3)(x^0y^7)^{-2}$

31. $(c^{2/5}d^{-2/3})(c^6d^3)^{4/3}$ **32.** $\left(\dfrac{r^{2/3}}{s^{1/5}}\right)^{15/9}$

33. $\dfrac{(7a)^2(5b)^{3/2}}{(5a)^{3/2}(7b)^4}$ **34.** $\dfrac{(6a)^{1/2}\sqrt{ab}}{a^2b^{3/2}}$

35. $\dfrac{(2a)^{1/2}(3b)^{-2}(4a)^{3/5}}{(4a)^{-3/2}(3b)^2(2a)^{1/5}}$ **36.** $\dfrac{(a^{3/4}b)^2(ab^{1/4})^3}{(ab)^{1/2}(bc)^{-1/4}}$

37. $(a^{x^2})^{1/x}$ **38.** $\dfrac{(b^x)^{x-1}}{b^{-x}}$

In Exercises 39–42, write the given expression without radicals, using only positive exponents.

39. $(\sqrt[3]{xy^2})^{-3/5}$ **40.** $(\sqrt[4]{r^{14}s^{-21/5}})^{-3/7}$

41. $\dfrac{c}{(c^{5/6})^{42}(c^{51})^{-2/3}}$ **42.** $(c^{5/6}-c^{-5/6})^2$

In Exercises 43–48, compute and simplify.

43. $x^{1/2}(x^{2/3}-x^{4/3})$ **44.** $x^{1/2}(3x^{3/2}+2x^{-1/2})$

45. $(x^{1/2}+y^{1/2})(x^{1/2}-y^{1/2})$

46. $(x^{1/3}+y^{1/2})(2x^{1/3}-y^{3/2})$

47. $(x+y)^{1/2}[(x+y)^{1/2}-(x+y)]$

48. $(x^{1/3}+y^{1/3})(x^{2/3}-x^{1/3}y^{1/3}+y^{2/3})$

In Exercises 49–54, factor the given expression. For example,

$$x - x^{1/2} - 2 = (x^{1/2}-2)(x^{1/2}+1).$$

49. $x^{2/3}+x^{1/3}-6$ **50.** $x^{2/5}+11x^{1/5}+30$

51. $x+4x^{1/2}+3$ **52.** $x^{1/3}+7x^{1/6}+10$

53. $x^{4/5} - 81$ **54.** $x^{2/3} - 6x^{1/3} + 9$

In Exercises 55–60, express the number in scientific notation.

55. 79,327 **56.** 5,200,000 **57.** .002

58. .00000079 **59.** 5,963,000,000,000

60. .00000000000035

In Exercises 61–66, express the given number in decimal notation.

61. 7.4×10^5 **62.** 6.53×10^7

63. 3.8×10^{-12} **64.** 6.02×10^{-8}

65. 3.457×10^{10} **66.** 13.23×10^{13}

67. Write exponent laws 3, 4, and 5 in radical notation in the case when $r = 1/m$ and $s = 1/n$.

68. Write the warning in the margin on page 302 in radical notation in the case when $r = 1/n$.

69. **(a)** Graph $f(x) = x^5$ and explain why this function has an inverse function.
(b) Show algebraically that the inverse function is $g(x) = x^{1/5}$.

70. If n is an odd positive integer, show that $f(x) = x^n$ has an inverse function and find the rule of the inverse function. [*Hint*: Exercise 69 is the case when $n = 5$.]

71. Using the viewing window with $0 \le x \le 4$ and $0 \le y \le 2$, graph the following functions on the same screen:

$$f(x) = x^{1/2}, \qquad g(x) = x^{1/4}, \qquad h(x) = x^{1/6}.$$

In each of the following cases, arrange $x^{1/2}$, $x^{1/4}$, and $x^{1/6}$ in order of increasing size and justify your answer by using the graphs.
(a) $0 < x < 1$ **(b)** $x > 1$

72. Using the viewing window with $-3 \le x \le 3$ and $-1.5 \le y \le 1.5$, graph the following functions on the same screen:

$$f(x) = x^{1/3}, \qquad g(x) = x^{1/5}, \qquad h(x) = x^{1/7}.$$

In each of the following cases, arrange $x^{1/3}$, $x^{1/5}$, and $x^{1/7}$ in order of increasing size and justify your answer by using the graphs.
(a) $x < -1$ **(b)** $-1 < x < 0$ **(c)** $0 < x < 1$
(d) $x > 1$

73. Graph $f(x) = \sqrt{x}$ in the standard viewing window. Then, without doing any more graphing, describe the graphs of these functions:
(a) $g(x) = \sqrt{x + 3}$ [*Hint*: $g(x) = f(x + 3)$; see Section 2.5.]
(b) $h(x) = \sqrt{x} - 2$ **(c)** $k(x) = \sqrt{x + 3} - 2$

74. Do Exercise 73 with $\sqrt[3]{}$ in place of $\sqrt{}$.

5.1.A *Excursion* RADICAL EQUATIONS

Equations involving radicals and rational exponents can be solved either graphically or algebraically.

EXAMPLE 1 There are two ways to solve

$$\sqrt[5]{x^2 - 6x + 2} = x - 4$$

graphically. One is to rewrite the equation as

$$\sqrt[5]{x^2 - 6x + 2} - x + 4 = 0$$

and graph the function $h(x) = \sqrt[5]{x^2 - 6x + 2} - x + 4$ (Figure 5–2). The x-intercept of the graph is the solution of the equation.

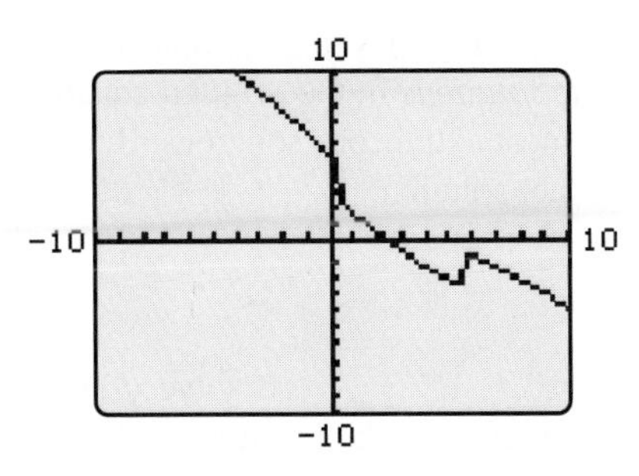

Figure 5-2

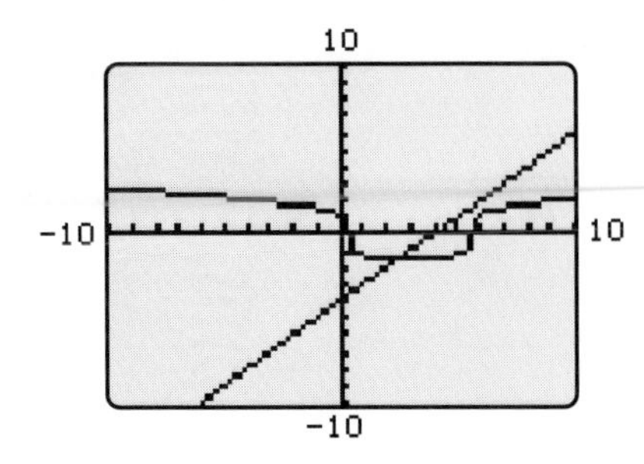

Figure 5-3

Alternatively, you can graph both $f(x) = \sqrt[5]{x^2 - 6x + 2}$ and $g(x) = x - 4$ on the same screen (Figure 5–3). A point on the graph of f has coordinates $(x, f(x))$ and a point on the graph of g has coordinates $(x, g(x))$. The x-coordinate of the point where the graphs intersect must satisfy $f(x) = g(x)$ because it lies on both graphs. Thus the x-coordinate of a point of intersection is a solution of the original equation.

GRAPHING EXPLORATION Use zoom-in or a root finder to find the x-intercept in Figure 5–2 or use zoom-in or a graphical intersection finder to find the intersection point in Figure 5–3, to verify that the solution of the equation is $x \approx 2.534$. ■

Although the (approximate) solutions of any radical equation can be found graphically, exact solutions can be found algebraically in many cases. The algebraic solution of radical equations depends on the following fact. If two numbers are equal, say $a = b$, then $a^r = b^r$ for every positive integer r. The same is true of algebraic expressions. For example,

$$\text{if} \quad x - 2 = 3, \quad \text{then} \quad (x - 2)^2 = 3^2 = 9.$$

Thus, every solution of $x - 2 = 3$ is also a solution of $(x - 2)^2 = 9$. But *be careful:* This only works in *one* direction. For instance, -1 is a solution of $(x - 2)^2 = 9$, but not of $x - 2 = 3$. Similarly, 1 is the only solution of $x = 1$, but $x^3 = 1^3$ (that is, $x^3 - 1 = 0$) has two additional complex solutions, as shown in Example 8 on page 282. Therefore,

Power Principle ▶

If both sides of an equation are raised to the same positive integer power, every solution of the original equation is also a solution of the new equation. But the new equation may have solutions that are *not* solutions of the original one.*

*Such solutions are called **extraneous solutions.**

Consequently, if you raise both sides of an equation to a power, you must *check your solutions* in the *original* equation. Graphing provides a quick way to eliminate most extraneous solutions. But only an algebraic computation can confirm an exact solution.

EXAMPLE 2 To solve $\sqrt[3]{2x^2 + 7x - 6} = 3^{2/3}$, we first cube both sides, and then solve the resulting equation:

$$(\sqrt[3]{2x^2 + 7x - 6})^3 = (3^{2/3})^3$$
$$2x^2 + 7x - 6 = 9$$
$$2x^2 + 7x - 15 = 0$$
$$(2x - 3)(x + 5) = 0$$
$$2x - 3 = 0 \quad \text{or} \quad x + 5 = 0$$
$$2x = 3 \qquad\qquad x = -5$$
$$x = \frac{3}{2}$$

Substituting 3/2 and -5 in the left side of the original equation shows that:

$$\sqrt[3]{2\left(\frac{3}{2}\right)^2 + 7\left(\frac{3}{2}\right) - 6} = \sqrt[3]{\frac{30}{2} - 6} = \sqrt[3]{9} = \sqrt[3]{3^2} = 3^{2/3}$$
$$\sqrt[3]{2(-5)^2 + 7(-5) - 6} = \sqrt[3]{9} = \sqrt[3]{3^2} = 3^{2/3}$$

Therefore, both 3/2 and -5 are solutions. ■

EXAMPLE 3 To solve $5 + \sqrt{3x - 11} = x$, we first rearrange terms to get the radical expression alone on one side:

$$\sqrt{3x - 11} = x - 5.$$

Then, square both sides and solve the resulting equation:

$$(\sqrt{3x - 11})^2 = (x - 5)^2$$
$$3x - 11 = x^2 - 10x + 25$$
$$0 = x^2 - 13x + 36$$
$$0 = (x - 4)(x - 9)$$
$$x - 4 = 0 \quad \text{or} \quad x - 9 = 0$$
$$x = 4 \quad \text{or} \quad x = 9$$

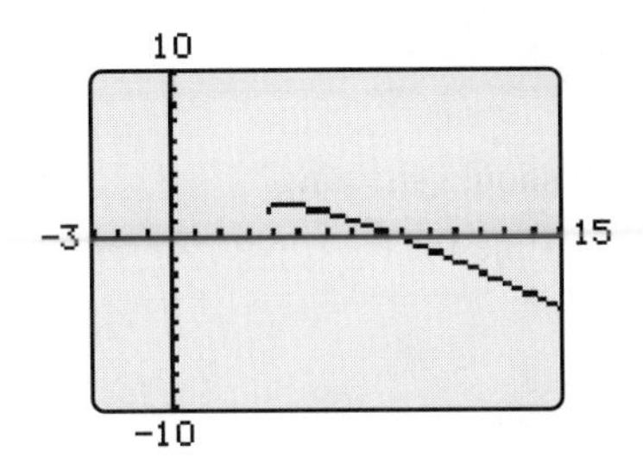

Figure 5-4

If these numbers are solutions of the original equation, they should be x-intercepts of the graph of $f(x) = \sqrt{3x - 11} - x + 5$ (why?). But the graph of f in Figure 5–4 doesn't have an x-intercept at $x = 4$, so 4 is not a solution of the original equation. The graph appears to show that $x = 9$ is a solution of the original equation, and direct calculation confirms this:

Left side: $5 + \sqrt{3 \cdot 9 - 11} = 5 + \sqrt{16} = 9$ *Right side:* 9

Hence, 9 is the only solution of the original equation. ■

EXAMPLE 4 To solve $\sqrt{2x - 3} - \sqrt{x + 7} = 2$, we first rearrange terms so that one side contains only a single radical term:

$$\sqrt{2x - 3} = \sqrt{x + 7} + 2.$$

Then, square both sides and simplify:

$$(\sqrt{2x - 3})^2 = (\sqrt{x + 7} + 2)^2$$
$$2x - 3 = (\sqrt{x + 7})^2 + 2 \cdot 2 \cdot \sqrt{x + 7} + 2^2$$
$$2x - 3 = x + 7 + 4\sqrt{x + 7} + 4$$
$$x - 14 = 4\sqrt{x + 7}.$$

Now, square both sides and solve the resulting equation:

$$(x - 14)^2 = (4\sqrt{x + 7})^2$$
$$x^2 - 28x + 196 = 4^2 \cdot (\sqrt{x + 7})^2$$
$$x^2 - 28x + 196 = 16(x + 7)$$
$$x^2 - 28x + 196 = 16x + 112$$
$$x^2 - 44x + 84 = 0$$
$$(x - 2)(x - 42) = 0$$
$$x - 2 = 0 \quad \text{or} \quad x - 42 = 0$$
$$x = 2 \qquad\qquad x = 42$$

Substituting 2 and 42 in the left side of the original equation shows that

$$\sqrt{2 \cdot 2 - 3} - \sqrt{2 + 7} = \sqrt{1} - \sqrt{9} = 1 - 3 = -2$$
$$\sqrt{2 \cdot 42 - 3} - \sqrt{42 + 7} = \sqrt{81} - \sqrt{49} = 9 - 7 = 2$$

Therefore, 42 is the only solution of the equation. ■

EXAMPLE 5 Stella, who is standing at point A on the bank of a 2.5-km-wide river wants to reach point B, 15 km downstream on the opposite bank. She plans to row to a point C on the opposite shore and then run to B, as shown in Figure 5–5. She can row at a rate of 4 km per hour and can run at 8 km per hour.

(a) If her trip is to take 3 hours, how far from B should she land?

(b) How far from B should she land to make the time for the trip as short as possible?

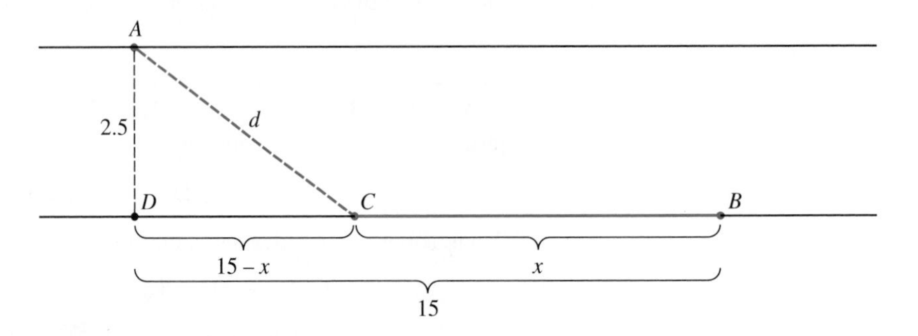

Figure 5-5

Solution Let x be the distance that Stella must run from C to B. Using the basic formula for distance, we have:

$$\text{rate} \times \text{time} = \text{distance}$$

$$\text{time} = \frac{\text{distance}}{\text{rate}} = \frac{x}{8}.$$

Similarly, the time required to row distance d is

$$\text{time} = \frac{\text{distance}}{\text{rate}} = \frac{d}{4}.$$

Since $15 - x$ is the distance from D to C, the Pythagorean Theorem applied to right triangle ADC shows that

$$d^2 = (15 - x)^2 + 2.5^2 \qquad \text{or equivalently} \qquad d = \sqrt{(15 - x)^2 + 6.25}.$$

Therefore, the total time for the trip is given by

$$T(x) = \text{rowing time} + \text{running time} = \frac{d}{4} + \frac{x}{8} = \frac{\sqrt{(x - 15)^2 + 6.25}}{4} + \frac{x}{8}.$$

(a) If the trip is to take 3 hours, then $T(x) = 3$ and we must solve the equation

$$\frac{\sqrt{(x - 15)^2 + 6.25}}{4} + \frac{x}{8} = 3.$$

GRAPHING EXPLORATION Using the viewing window with $0 \le x \le 15$ and $-2 \le y \le 2$, graph the function

$$f(x) = \frac{\sqrt{(x-15)^2 + 6.25}}{4} + \frac{x}{8} - 3$$

and use zoom-in or a root finder to find its x-intercept (the solution of the equation).

This exploration shows that Stella should land approximately 6.74 km from B in order to make the trip in 3 hours.

(b) To find the shortest possible time, we must find the value of x that makes $T(x) = \dfrac{\sqrt{(x-15)^2 + 6.25}}{4} + \dfrac{x}{8}$ as small as possible.

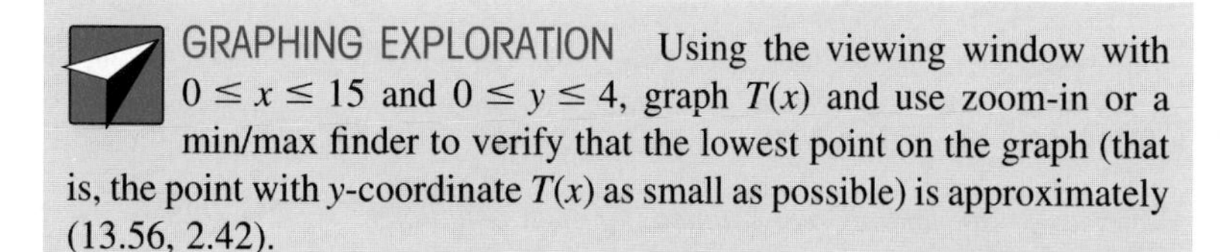

GRAPHING EXPLORATION Using the viewing window with $0 \le x \le 15$ and $0 \le y \le 4$, graph $T(x)$ and use zoom-in or a min/max finder to verify that the lowest point on the graph (that is, the point with y-coordinate $T(x)$ as small as possible) is approximately (13.56, 2.42).

Therefore, the shortest time for the trip will be 2.42 hours and will occur if Stella rows to a point 13.56 km from B. ■

Equations involving rational exponents can be solved graphically, provided that the function to be graphed is entered properly. Many such equations can also be solved algebraically by making an appropriate substitution.

EXAMPLE 6 Solve $x^{2/3} - 2x^{1/3} - 15 = 0$ algebraically and graphically.

Solution *Algebraic:* Let $u = x^{1/3}$, rewrite the equation and solve:

$$x^{2/3} - 2x^{1/3} - 15 = 0$$

$$(x^{1/3})^2 - 2x^{1/3} - 15 = 0$$

$$u^2 - 2u - 15 = 0$$

$$(u+3)(u-5) = 0$$

$$u + 3 = 0 \quad \text{or} \quad u - 5 = 0$$

$$u = -3 \qquad\qquad u = 5$$

$$x^{1/3} = -3 \qquad\qquad x^{1/3} = 5$$

Cubing both sides of these last equations shows that

$$(x^{1/3})^3 = (-3)^3 \qquad \text{or} \qquad (x^{1/3})^3 = 5^3$$
$$x = -27 \qquad\qquad x = 125.$$

Since we cubed both sides, we must check these numbers in the original equation. Verify that both *are* solutions.

Geometric: Graph the function $f(x) = x^{2/3} - 2x^{1/3} - 15$ and find the x-intercepts, namely $x = -27$ and $x = 125$. The only difficulty is the one mentioned in the Warning on page 303: Be careful to enter the function f in one of these forms:

$$f(x) = (x^2)^{1/3} - 2x^{1/3} - 15 \qquad \text{or} \qquad f(x) = (x^{1/3})^2 - 2x^{1/3} - 15.$$

Otherwise, the calculator may not produce a graph when x is negative. ■

EXERCISES 5.1.A

In Exercises 1–30, find all real solutions of each equation. Find exact solutions when possible and approximate ones otherwise.

1. $\sqrt{x+2} = 3$ **2.** $\sqrt{x-7} = 4$ **3.** $\sqrt{4x+9} = 5$

4. $\sqrt{3x-2} = 7$ **5.** $\sqrt[3]{5-11x} = 3$

6. $\sqrt[3]{6x-10} = 2$ **7.** $\sqrt[3]{x^2-1} = 2$

8. $(x+1)^{2/3} = 4$ **9.** $\sqrt{x^2-x-1} = 1$

10. $\sqrt{x^2-5x+4} = 2$ **11.** $\sqrt{x+7} = x-5$

12. $\sqrt{x+5} = x-1$ **13.** $\sqrt{3x^2+7x-2} = x+1$

14. $\sqrt{4x^2-10x+5} = x-3$

15. $\sqrt[3]{x^3+x^2-4x+5} = x+1$

16. $\sqrt[3]{x^3-6x^2+2x+3} = x-1$

17. $\sqrt[5]{9-x^2} = x^2+1$ **18.** $\sqrt[4]{x^3-x+1} = x^2-1$

19. $\sqrt[3]{x^5-x^3-x} = x+2$

20. $\sqrt{x^3+2x^2-1} = x^3+2x-1$

21. $\sqrt{x^2+3x-6} = x^4-3x^2+2$

22. $\sqrt[3]{x^4+x^2+1} = x^2-x-5$

23. $\sqrt{5x+6} = 3+\sqrt{x+3}$

24. $\sqrt{3y+1} - 1 = \sqrt{y+4}$

25. $\sqrt{2x-5} = 1+\sqrt{x-3}$ **26.** $\sqrt{x-3}+\sqrt{x+5} = 4$

27. $\sqrt{3x+5}+\sqrt{2x+3}+1 = 0$

28. $\sqrt{20-x} = \sqrt{9-x}+3$

29. $\sqrt{6x^2+x+7} - \sqrt{3x+2} = 2$

30. $\sqrt{x^3+x^2-3} = \sqrt{x^3-x+3} - 1$

In Exercises 31–34, assume that all letters represent positive numbers and solve each equation for the required letter.

31. $A = \sqrt{1+\dfrac{a^2}{b^2}}$ for b **32.** $T = 2\pi\sqrt{\dfrac{m}{g}}$ for g

33. $K = \sqrt{1-\dfrac{x^2}{u^2}}$ for u **34.** $R = \sqrt{d^2+k^2}$ for d

In Exercises 35–46, solve each equation algebraically.

35. $x - 4x^{1/2} + 4 = 0$ [*Hint:* Let $u = x^{1/2}$.] **36.** $x - x^{1/2} - 12 = 0$

37. $2x - \sqrt{x} - 6 = 0$ **38.** $3x - 11\sqrt{x} - 4 = 0$

39. $x^{2/3} + 3x^{1/3} + 2 = 0$ [*Hint:* Let $u = x^{1/3}$.]

40. $x^{2/3} - 4x^{1/3} + 3 = 0$ **41.** $\sqrt[3]{x^2} + 2\sqrt[3]{x} - 8 = 0$

42. $2\sqrt[3]{x^2} - \sqrt[3]{x} - 6 = 0$ **43.** $x^{1/2} - x^{1/4} - 2 = 0$ [*Hint:* Let $u = x^{1/4}$.]

44. $x^{1/3} + x^{1/6} - 2 = 0$

45. $x^{-2} - x^{-1} - 6 = 0$
[*Hint*: Let $u = x^{-1}$.]

46. $x^{-2} - 6x^{-1} + 5 = 0$

In Exercises 47–50, solve each equation graphically.

47. $x^{3/5} - 2x^{2/5} + x^{1/5} - 6 = 0$

48. $x^{5/3} + x^{4/3} - 3x^{2/3} + x = 5$

49. $x^{-3} + 2x^{-2} - 4x^{-1} + 5 = 0$

50. $x^{-2/3} - 3x^{-1/3} = 4$

51. A rope is to be stretched at uniform height from a tree to a 35-foot-long fence, which is 20 feet from the tree, and then to the side of a building at a point 30 feet from the fence (Figure 5–6).
(a) If 63 feet of rope is to be used, how far from the building wall should the rope meet the fence?
(b) How far from the building wall should the rope meet the fence if as little rope as possible is to be used?

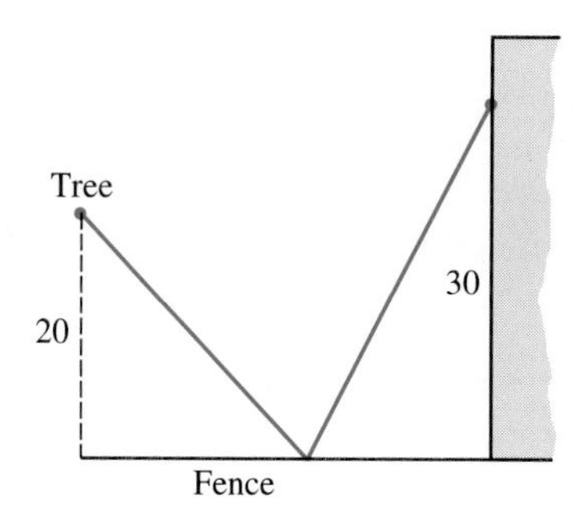

Figure 5-6

52. Anne is standing on a straight road and wants to reach her helicopter, which is located two miles down the road from her, a mile from the road in a field (Figure 5–7). She can run 5 miles per hour on the road and 3 miles per hour in the field. She plans to run down the road, then cut diagonally across the field to reach the helicopter.
(a) Where should she leave the road in order to reach the helicopter in exactly 42 minutes (.7 hour)?
(b) Where should she leave the road in order to reach the helicopter as soon as possible?

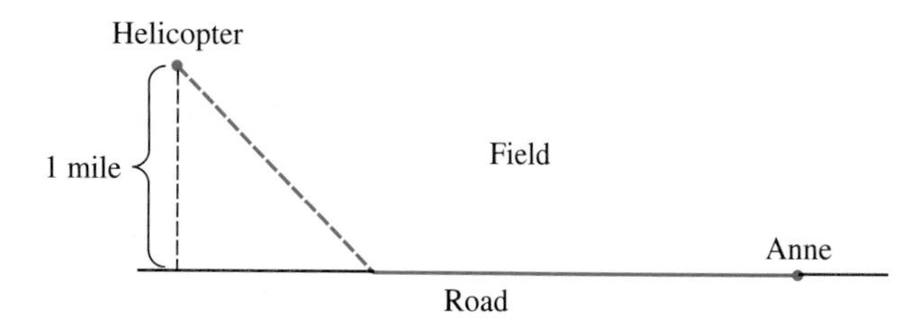

Figure 5-7

53. A spotlight is to be placed on a building wall to illuminate a bench that is 32 feet from the base of the wall. The intensity I of light at the bench is known to be x/d^3, where x is the height of the spotlight above the ground and d is the distance from the bench to the spotlight.
(a) Express I as a function of x. [It may help to draw a picture.]
(b) How high should the spotlight be in order to provide maximum illumination at the bench?

5.2 EXPONENTIAL FUNCTIONS

For each positive real number a there is a function (called the **exponential function with base a**) whose domain is all real numbers and whose rule is $f(x) = a^x$. For example,

$$f(x) = 10^x, \qquad g(x) = 2^x, \qquad h(x) = \left(\frac{1}{2}\right)^x, \qquad k(x) = \left(\frac{3}{2}\right)^x.$$

The shape of the graph of the exponential function $f(x) = a^x$ depends only on the size of a, as shown below.

Graph of $f(x) = a^x$

$a > 1$ 

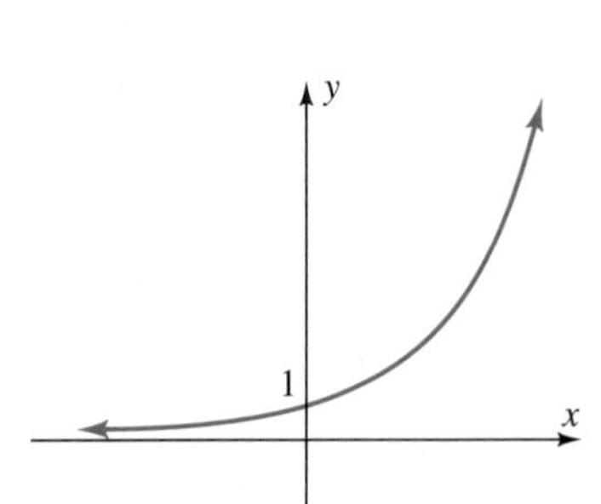

Graph is above x-axis.

y-intercept is 1.

$f(x)$ is increasing.

Negative x-axis is a horizontal asymptote.

$0 > a > 1$

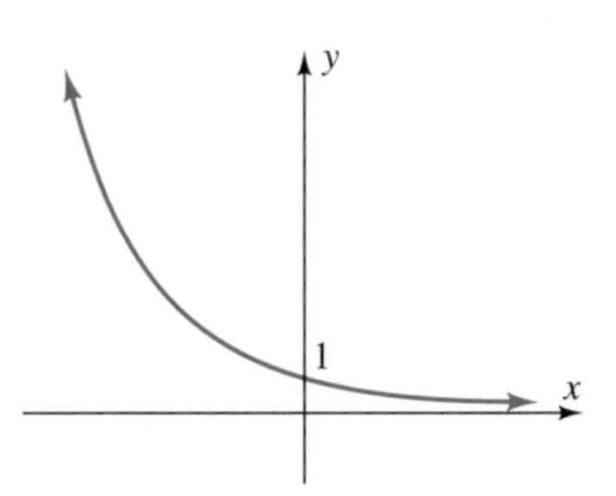

Graph is above x-axis.

y-intercept is 1.

$f(x)$ is decreasing.

Positive x-axis is a horizontal asymptote.

GRAPHING EXPLORATION

(a) Using the viewing window with $-3 \leq x \leq 7$ and $-2 \leq y \leq 18$, graph

$$f(x) = 1.3^x, \qquad g(x) = 2^x, \qquad h(x) = 10^x$$

on the same screen and observe their behavior to the *right* of the y-axis.

Which one rises least steeply? Which one most steeply?

How does the steepness of the graph of $f(x) = a^x$ seem to be related to the size of the base a?

Where would you expect the graph of $k(x) = 4^x$ to be?

Verify your conjecture by graphing f, g, h, and k on the same screen.

(b) To see what's going on to the *left* of the y-axis, graph the same four functions in the viewing window with $-4 \leq x \leq 2$ and $-.5 \leq y \leq 2$.

As you move to the left, how does size of the base a seem to be related to how quickly the graph of $f(x) = a^x$ falls toward the x-axis?

GRAPHING EXPLORATION Using the viewing window with $-4 \leq x \leq 4$ and $-1 \leq y \leq 4$, graph

$$f(x) = .2^x, \quad g(x) = .4^x, \quad h(x) = .6^x, \quad k(x) = .8^x$$

on the same screen. Note that the bases of these exponential functions are increasing in size: $0 < .2 < .4 < .6 < .8 < 1$.

How is the size relationship of the bases related to the graphs? Which graph falls least steeply? Which one falls most steeply?

The preceding Explorations show that the graph of $f(x) = a^x$ rises or falls less steeply when the base a is close to 1.

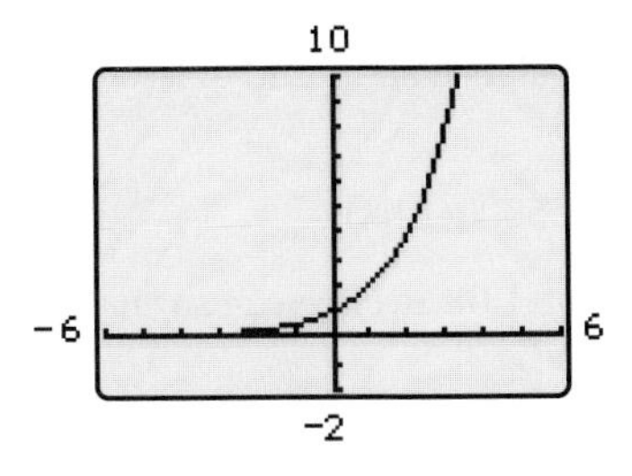

Figure 5-8

EXAMPLE 1 The graph of $f(x) = 2^x$ is shown in Figure 5–8. Without graphing, describe the shape of each of the following graphs:

(a) $g(x) = 2^{x+3}$ **(b)** $h(x) = 2^{x-3} - 4$ **(c)** $k(x) = 5 \cdot 2^x$

Solution

(a) Since $f(x) = 2^x$, we have $f(x + 3) = 2^{x+3} = g(x)$. So the graph of $g(x)$ is just the graph of $f(x) = 2^x$ shifted horizontally 3 units to the left. [See page 126.]

GRAPHING EXPLORATION Verify the preceding statement graphically by graphing $f(x)$ and $g(x)$ on the same screen.

(b) In this case $f(x - 3) - 4 = 2^{x-3} - 4 = h(x)$. So the graph of $h(x)$ is the graph of $f(x) = 2^x$ shifted horizontally 3 units to the right and vertically 4 units downward. [See pages 124 and 126.]

GRAPHING EXPLORATION Verify the preceding statement graphically by graphing $f(x)$ and $h(x)$ on the same screen.

(c) The graph of $k(x) = 5 \cdot 2^x$ is the graph of $f(x) = 2^x$ stretched away from the x-axis by a factor of 5 (so that the graph of $k(x)$ climbs much more steeply than the graph of $f(x)$). [See page 128.]

GRAPHING EXPLORATION Verify the preceding statement graphically by graphing $f(x)$ and $k(x)$ on the same screen. ■

Exponential Growth and Decay

Exponential functions can be used to model a number of real life situations in which a quantity increases or decreases by a fixed factor.

EXAMPLE 2 *(Compound Interest)* If you deposit \$5000 in a savings account that pays 8% interest, compounded annually, then after one year the account balance is

$$5000 + 8\% \text{ of } 5000 = 5000 + .08 \cdot 5000 = 5000(1 + .08) = 5000(1.08).$$

In other words, the account balance has changed by a factor of 1.08, from \$5000 to \$5000(1.08) = \$5400. If you leave the \$5400 in the account for another year, your new balance is

$$5400 + 8\% \text{ of } 5400 = 5400 + .08 \cdot 5400 = 5400(1 + .08) = 5400(1.08).$$

Since 5400 = 5000(1.08), we see that the account balance at the end of two years is

$$5400(1.08) = 5000(1.08)(1.08) = 5000(1.08)^2.$$

Similarly, your balance will change by a factor of 1.08 every year so that the balance in the account at the end of year x is given by

$$B(x) = 5000 \cdot (1.08)^x.$$

(a) How much money is in the account after nine years?

(b) When will the balance reach one million dollars?

Solution

(a) To find the account balance after nine years evaluate $B(x)$ at $x = 9$:

$$B(9) = 5000 \cdot (1.08)^9 = \$9995.02.$$

The balance has just about doubled in nine years.

(b) In the graph of $B(x)$ in Figure 5–9, the x-coordinate of each point represents the year, and the y-coordinate the balance at that time. So we must find the point on the graph with y-coordinate 1,000,000, that is, the intersection of the graph of $B(x)$ with the horizontal line $y = 1{,}000{,}000$. Zoom-in or a graphical intersection finder shows that the approximate coordinates of the point of intersection are (68.84, 1000000). Therefore, the balance will reach one million dollars in about 68.84 years. ■

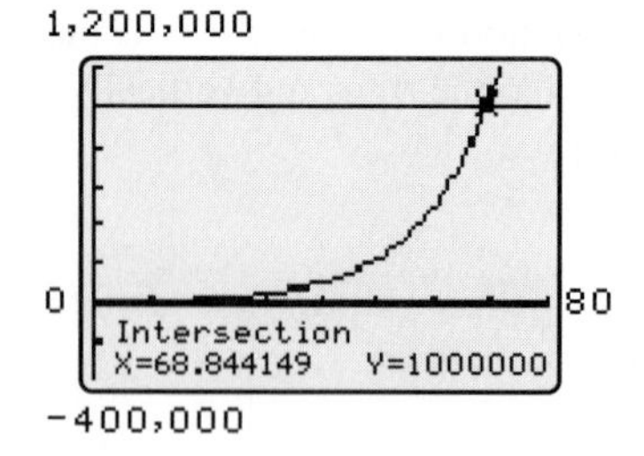

Figure 5-9

EXAMPLE 3 *(Population Growth)* The world population in 1950 was about 2.5 billion people and has been increasing at approximately 1.85% per year. So the population (in billions) in 1951 was

$$2.5 + 1.85\% \text{ of } 2.5 = 2.5 + .0185(2.5) = 2.5(1 + .0185) = 2.5(1.0185).$$

Similarly, in each successive year the population changed by a factor of 1.0185 so that the population in billions in year x is given by the function

$$S(x) = 2.5(1.0185)^x,$$

where $x = 0$ corresponds to 1950, $x = 1$ to 1951, etc. Assuming that this function continues to be accurate:

(a) Find the world population in 2000.

(b) In what year will the population be double what it is in 2000?

Solution

(a) The year 2000 corresponds to $x = 50$, so the population will be

$$S(50) = 2.5(1.0185)^{50} \approx 6.25 \text{ billion people.}$$

(b) Twice the population in 2000 is $2(6.25) = 12.5$ billion. We must find the number x such that $S(x) = 12.5$, that is, solve the equation

$$2.5(1.0185)^x = 12.5.$$

This can be done with an equation solver or by either of the following graphical methods:

(i) Write the equation as $2.5(1.0185)^x - 12.5 = 0$ and find the x-intercept of the graph of $f(x) = 2.5(1.0185)^x - 12.5$.

(ii) Proceed as in Example 1 and find the x-coordinate of the intersection point of the graph of $S(x) = 2.5(1.0185)^x$ and the line $y = 12.5$ (Figure 5–10).

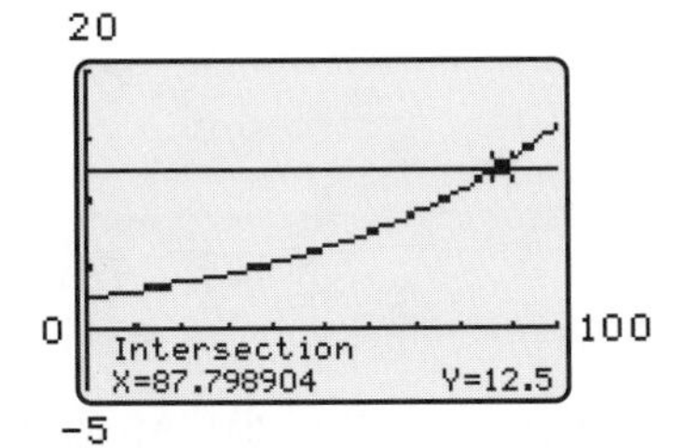

Figure 5-10

Both methods show that $x \approx 87.8$, or 88 when rounded to the nearest year. This corresponds to the year 2038. In other words, the world population will double in just 38 years. This is what is meant by the population explosion. ■

The preceding examples illustrate **exponential growth.** The general case is similar. Suppose some quantity P is increasing at a rate r per period ($r = .08$ in Example 1 and $r = .0185$ in Example 2 and in each case the period was a year). At the end of the period the amount will be $P + rP = P(1 + r)$, that is,

> The amount at the end of the period is the starting amount multiplied by $(1 + r)$.

At the beginning of the second period, the amount is $P(1 + r)$ and by the end of the second period it has been multiplied by a factor of $(1 + r)$, so the total is

$$[P(1 + r)](1 + r) = P(1 + r)^2.$$

The total continues to increase by a factor of $1 + r$ each time period, leading to this conclusion:

Exponential Growth ▶

Exponential growth can be described by a function of the form

$$f(x) = Pa^x,$$

where $f(x)$ is the quantity at time x, P is the initial quantity (the amount when $x = 0$), and $a > 1$ is the factor by which the quantity changes when x increases by 1. If the quantity is growing at rate r per time period, then $a = 1 + r$ and

$$f(x) = Pa^x = P(1 + r)^x.$$

EXAMPLE 4 At the beginning of an experiment a culture contains 1000 bacteria. Five hours later there are 7600 bacteria. Assuming that the bacteria grow exponentially, how many will there be after 24 hours?

Solution The bacteria population is given by $f(x) = Pa^x$, where P is the initial population, a is the change factor, and x is the time in hours. We are given that $P = 1000$, so that $f(x) = 1000a^x$. The next step is to determine a. Since there are 7600 bacteria when $x = 5$, we have

$$7600 = f(5) = 1000a^5$$

so that

$$\begin{aligned} 1000a^5 &= 7600 \\ a^5 &= 7.6 \\ a &= \sqrt[5]{7.6} \approx 1.5. \end{aligned}$$

Therefore, the population function is approximately $f(x) = 1000(1.5)^x$. After 24 hours the bacteria population will be

$$f(24) = 1000(1.5)^{24} \approx 16{,}834{,}112. \quad ■$$

In some situations a quantity decreases by a fixed factor as time goes on, as in the next example.

EXAMPLE 5 When tap water is filtered through a layer of charcoal and other purifying agents, 30% of the chemical impurities in the water are removed and 70% remain. If the water is filtered through a second purifying layer, then the amount of impurities remaining is 70% of 70%, that is, $(.7)(.7) = .7^2 = .49$ or 49%. A third layer results in $.7^3$ of the impurities remaining. Thus, the function

$$f(x) = .7^x$$

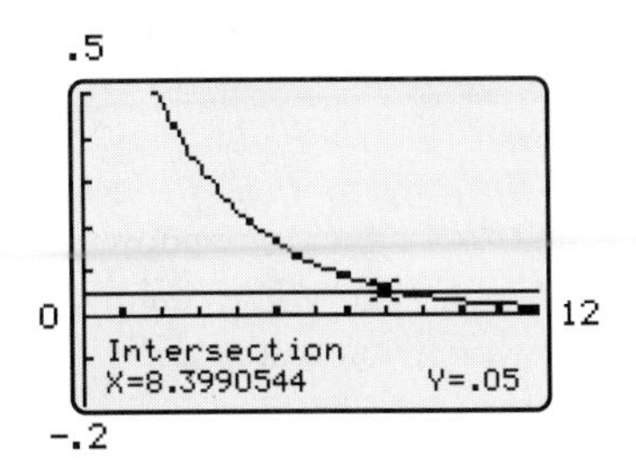

Figure 5-11

gives the percentage of impurities remaining in the water after it passes through x layers of purifying material. How many layers are needed to ensure that 95% of the impurities are removed from the water?

Solution If 95% of the impurities are removed, then 5% will remain. Hence, we must find x such that $f(x) = .05$, that is, we must solve the equation $.7^x = .05$. This can be done numerically or graphically. Figure 5–11 shows that the solution is $x \approx 8.4$, so 8.4 layers of material are needed.

GRAPHING EXPLORATION How many layers are needed to ensure that 99% of the impurities are removed? ■

Example 5 illustrates **exponential decay.** Note that the impurities were removed at a rate of $30\% = .3$ and that the amount of impurities remaining in the water were changing by a factor of $1 - .30 = .7$. The same thing is true in the general case:

Exponential Decay ▶

Exponential decay can be described by a function of the form

$$f(x) = Pa^x,$$

where $f(x)$ is the quantity at time x, P is the initial quantity (the amount when $x = 0$), and $0 < a < 1$; here a is the factor by which the quantity changes when x increases by 1. If the quantity is decaying at rate r per period, then $a = 1 - r$ and

$$f(x) = Pa^x = P(1 - r)^x.$$

Exponential functions are particularly useful for dealing with radioactive decay. The **half-life** of a radioactive element is the time it takes a given quantity to decay to one half of its original mass. The half-life depends only on the substance and does not depend on the size of the sample. Exercise 68 shows that the function

$$M(x) = c\left(\frac{1}{2}\right)^{x/h} = c(.5)^{x/h},$$

gives the mass at time x, where h is the half-life and c is the original mass of the element.

EXAMPLE 6 *(Radioactive Decay)* Plutonium (^{239}Pu) has a half-life of 24,360 years. So the amount remaining from 1 kilogram after x years is given by the function $M(x) = 1(.5)^{x/24360}$. The rule of M can be written as

$$M(x) = (.5)^{x/24360} = (.5^{1/24360})^x \approx (.99997)^x.$$

Since M is an exponential function with base smaller than but very close to 1, its graph falls *very slowly* from left to right.

GRAPHING EXPLORATION Verify that in a viewing window with $-500 \leq x \leq 500$, the graph of M looks like a horizontal straight line. Find a viewing window in which you can actually see the graph falling to the right of the y-axis.

The fact that the graph falls so slowly as x gets large means that even after an extremely long time, a substantial amount of plutonium will remain. Most of the original kilogram is still there after *ten thousand* years because

$$M(10{,}000) = .5^{10000/24360} \approx .7524 \text{ kg}.$$

This is the reason that nuclear waste disposal is such a serious problem. ■

The Number *e* and the Natural Exponential Function

You have often worked with the irrational number π, which is the ratio of the circumference of a circle to its diameter. The number π is built into reality. You can't have a circle without π any more than you can repeal the law of gravity.

There is another irrational number, denoted e, which also arises naturally in a variety of phenomena* and plays a central role in the mathematical description of the physical universe. The infinite decimal expansion of e begins

$$e = 2.718281828459045 \cdots.$$

The most important exponential function is $f(x) = e^x$, which is sometimes called the **natural exponential function**. One indication of the wide use of this function is the fact that your calculator has an e^x key. To display the first part of the decimal expansion of e, key in e^x 1 ENTER, so that the display shows e^1. Since $2 < e < 3$, the graph of $f(x) = e^x$ looks very much like the graphs of $g(x) = 2^x$ and $h(x) = 3^x$, as you can readily verify.

▶ **TECHNOLOGY TIP**

On most calculators you use the e^x key, but not the x^y or ∧ keys to enter the function $f(x) = e^x$. Check your instruction manual.

GRAPHING EXPLORATION Graph $f(x) = e^x$, $g(x) = 2^x$, and $h(x) = 3^x$ on the same screen in a window with $-5 \leq x \leq 5$. The Tip in the margin may be helpful.

EXAMPLE 7 *(Population Growth)* If the population of the United States continues to grow as it has recently, then the approximate population of the United States (in millions) in years t will be given by the function

$$P(t) = 227e^{.0093t},$$

*See Excursion 5.2.A for one example.

where 1980 corresponds to $t = 0$.

(a) Estimate the population in 2015.

(b) When will the population reach half a billion?

Solution

(a) The population in 2015 (that is, $t = 35$) will be approximately

$$P(35) = 227e^{.0093(35)} \approx 314.3 \text{ million people.}$$

(b) Half a billion is 500 million people. So we must find the value of t for which $P(t) = 500$, that is, we must solve the equation

$$227e^{.0093t} = 500.$$

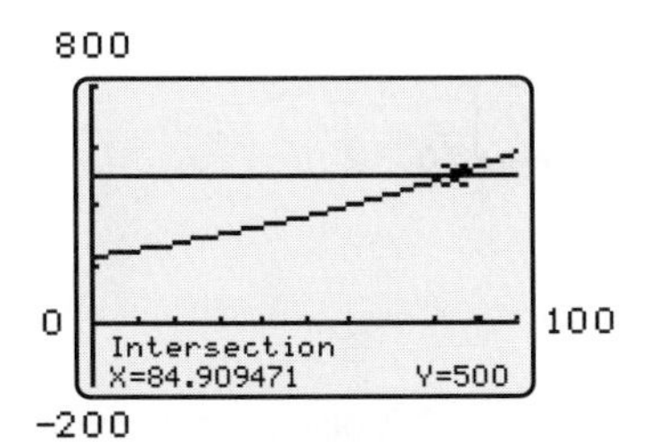

Figure 5-12

This can be done graphically by finding the intersection of the graph of $P(t)$ and the horizontal line $y = 500$, which occurs when $t \approx 84.9$ (Figure 5–12). Therefore, the population will reach half a billion late in the year 2064. ■

The following fact, which will be proved at the end of Section 5.3, is one reason that exponential functions involving e are important.

Exponential Growth and Decay ▶

Any exponential growth or decay function can be written in the form

$$f(x) = Pe^{kx},$$

for some constant k, where $f(x)$ is the amount at time x, P is the initial quantity, and k is positive for exponential growth and negative for exponential decay.

For instance, the rule of world population function of Example 3, $S(x) = 2.5(1.0185)^x$, can be rewritten by using the fact that $e^{.01833} \approx 1.0185$ so that

$$S(x) = 2.5(1.0185)^x \approx 2.5(e^{.01833})^x = 2.5e^{.01833x}.$$

Similarly, the exponential decay function $f(x) = .7^x$ can be written as $f(x) = e^{-.3567x}$ because $e^{-.3567} \approx .7$. For an explanation of what the constant k represents, see the end of Excursion 5.2.A.

EXAMPLE 8 *(Inhibited Population Growth)* There is an upper limit on the fish population in a certain lake due to the oxygen supply, available food, etc. The population of fish in this lake at time t months is given by the function

$$p(t) = \frac{20{,}000}{1 + 24e^{-t/4}} \qquad [t \geq 0].$$

What is the upper limit on the fish population?

Solution The graph of $p(t)$ in Figure 5–13 suggests that the horizontal line $y = 20{,}000$ is a horizontal asymptote of the graph.

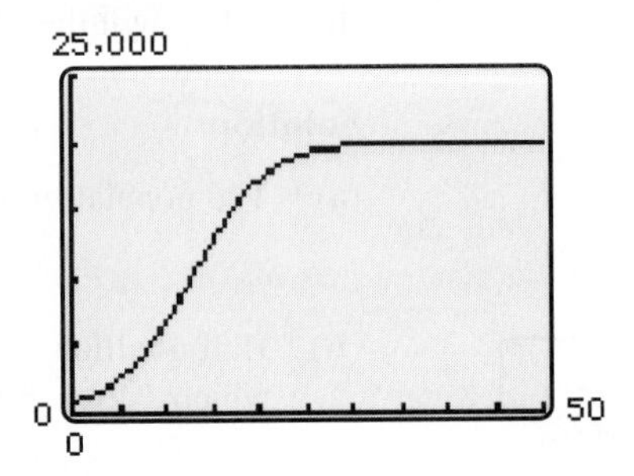

Figure 5-13

In other words, the fish population never goes above 20,000. You can confirm this algebraically by rewriting the rule of p in this form:

$$p(t) = \frac{20{,}000}{1 + 24e^{-t/4}} = \frac{20{,}000}{1 + \frac{24}{e^{t/4}}}.$$

When t is very large, so is $t/4$, which means that $e^{t/4}$ is huge. Hence, by the Big–Little Principle (page 243), $\frac{24}{e^{t/4}}$ is very close to 0 and $p(t)$ is very close to $\frac{20{,}000}{1+0} = 20{,}000$. Since $e^{t/4}$ is positive, the denominator of $p(t)$ is slightly bigger than 1, so that $p(t)$ is always less than 20,000. ■

WARNING The preceding examples are mathematical models of reality that sometimes ignore important factors in order to keep the functions involved relatively simple. Such simplicity may lead to poor predictions. For instance, the human population growth model in Example 3 does not take into account factors that may limit growth in the future (major wars, unusual diseases, changing ethical standards, etc.). Hence this model is unlikely to be accurate for long-term predictions over several centuries.

EXERCISES 5.2

In Exercises 1–6, sketch a complete graph of the function.

1. $f(x) = 4^{-x}$ **2.** $f(x) = (5/2)^{-x}$ **3.** $f(x) = 2^{3x}$

4. $g(x) = 3^{x/2}$ **5.** $f(x) = 2^{x^2}$ **6.** $g(x) = 2^{-x^2}$

In Exercises 7–12, list the transformations needed to transform the graph of $h(x) = 2^x$ into the graph of the given function. [Section 2.5 may be helpful.]

7. $f(x) = 2^x - 5$ **8.** $g(x) = -(2^x)$

9. $k(x) = 3(2^x)$ **10.** $g(x) = 2^{x-1}$

11. $f(x) = 2^{x+2} - 5$

12. $g(x) = -5(2^{x-1}) + 7$

In Exercises 13–16, use your knowledge of exponential graphs to solve the inequality. A calculator shouldn't be necessary.

13. $\left(\frac{1}{4}\right)^x < \left(\frac{1}{2}\right)^x$ **14.** $\left(\frac{1}{4}\right)^x > \left(\frac{3}{5}\right)^x$

15. $\left(\frac{1}{5}\right)^x \geq 5^x$ **16.** $7^x \geq \left(\frac{1}{7}\right)^x$

In Exercises 17–21, determine whether the function is even, odd, or neither (see Excursion 2.5.A).

17. $f(x) = 10^x$ **18.** $g(x) = 2^x - x$

19. $f(x) = \dfrac{e^x + e^{-x}}{2}$ **20.** $f(x) = \dfrac{e^x - e^{-x}}{2}$

21. $f(x) = e^{-x^2}$

22. Use the Big–Little Principle to explain why $e^x + e^{-x}$ is approximately equal to e^x when x is large.

In Exercises 23–26, find the average rate of change of the function.

23. $f(x) = x2^x$ as x goes from 1 to 3

24. $g(x) = 3^{x^2-x}$ as x goes from -1 to 1

25. $h(x) = 5^{-x^2}$ as x goes from -1 to 0

26. $f(x) = e^x - e^{-x}$ as x goes from -3 to -1

In Exercises 27–30, find the difference quotient of the function.

27. $f(x) = 10^x$ **28.** $g(x) = 5^{x^2}$

29. $f(x) = 2^x + 2^{-x}$ **30.** $f(x) = e^x - e^{-x}$

In Exercises 31–38, find a viewing window (or windows) that shows a complete graph of the function.

31. $k(x) = e^{-x}$ **32.** $f(x) = e^{-x^2}$

33. $f(x) = \dfrac{e^x + e^{-x}}{2}$ **34.** $h(x) = \dfrac{e^x - e^{-x}}{2}$

35. $g(x) = 2^x - x$ **36.** $k(x) = \dfrac{2}{e^x + e^{-x}}$

37. $f(x) = \dfrac{5}{1 + e^{-x}}$ **38.** $g(x) = \dfrac{10}{1 + 9e^{-x/2}}$

In Exercises 39–44, list all asymptotes of the graph of the function and the approximate coordinates of each local extremum.

39. $f(x) = x2^x$ **40.** $g(x) = x2^{-x}$ **41.** $h(x) = e^{x^2/2}$

42. $k(x) = 2^{x^2-6x+2}$ **43.** $f(x) = e^{-x^2}$

44. $g(x) = -xe^{x^2/20}$

45. The oxygen consumption of yearling salmon (in appropriate units), which changes with the speed of swimming, is given by $f(x) = 100 \cdot 3^{.6x}$, where x is the speed in feet per second. What is the oxygen consumption when the fish are still (that is, $x = 0$)? What is the consumption when they swim at 3 ft/sec? At 5 ft/sec?

46. If you deposit \$750 at 8.2% interest, compounded annually and paid from the day of deposit to the day of withdrawal, your balance at time t is given by $B(t) = 750(1.082)^t$. How much will you have after 2 years? After 3 years and 9 months?

47. Water and salt are continuously added to a tank in such a way that the number of kilograms of salt in the tank at time t minutes is $g(t) = 200 - 100e^{-t/20}$. How much salt is in the tank at the beginning ($t = 0$)? After 10 minutes? After 20 minutes? After 40 minutes?

48. In a room where the temperature is 70°, a certain hot object cools according to the function $T(x) = 70 + 100e^{-.04x}$, where $T(x)$ is the temperature of the object at time x minutes. What is the temperature after 40 minutes? After an hour and a half?

49. The population of a colony of fruit flies t days from now is given by the function $p(t) = 100 \cdot 3^{t/10}$.
(a) What will the population be in 15 days? In 25 days?
(b) When will the population reach 2500?

50. A certain type of bacteria grows according to the function $f(x) = 5000e^{.4055x}$, where the time x is measured in hours.
(a) What will the population be in 8 hours?
(b) When will the population reach one million?

51. The population of Mexico was 67.4 million in 1980 and has been increasing by approximately 2.6% each year.
(a) If $g(x)$ is the population of Mexico (in millions) in year x (with $x = 0$ being 1980), find the rule of the function f. [See Example 3.]
(b) Estimate the population of Mexico in the year 2000.

52. There are now 3.2 million people who play bridge and the number increases by 3.5% a year.
(a) Write the rule of a function that gives the number of bridge players in year x.
(b) How many people will be playing bridge in 15 years?
(c) When will there be ten million bridge players?

53. The number of dandelions in your lawn increases by 5% a week and there are 75 dandelions now.
(a) If $f(x)$ is the number of dandelions in week x, find the rule of the function f.
(b) How many dandelions will there be in 16 weeks?

54. An eccentric billionaire offers you a job for the month of September. She says that she will pay you 2¢ on the first day, 4¢ on the second day, 8¢ on the third day, and so on, doubling your pay on each successive day.
(a) Let $P(x)$ denote your salary in *dollars* on day x. Find the rule of the function P.
(b) Would you be better off financially if instead you were paid $10,000 per day? [*Hint*: Consider $P(30)$.]

55. Kerosene is passed through a pipe filled with clay in order to remove various pollutants. Each foot of pipe removes 25% of the pollutants.
(a) Write the rule of a function that gives the percentage of pollutants remaining in the kerosene after it has passed through x feet of pipe. [See Example 5.]
(b) How many feet of pipe are needed to ensure that 90% of the pollutants have been removed from the kerosene?

56. If inflation runs at a steady 3% per year, then the amount a dollar is worth decreases by 3% each year.
(a) Write the rule of a function that gives the value of a dollar in year x.
(b) How much will the dollar be worth in 5 years? In 10 years?
(c) How many years will it take before today's dollar is worth only a dime?

57. A weekly census of the tree-frog population in Frog Hollow State Park produces the following results:

Week	1	2	3	4	5	6
Population	18	54	162	486	1458	4374

(a) Find a function of the form $f(x) = Pa^x$ that describes the frog population at time x weeks.
(b) What is the growth factor in this situation (that is, by what number must this week's population be multiplied to obtain next week's population)?
(c) Each tree frog requires 10 sq ft of space and the park has an area of 6.2 square miles. Will the space required by the frog population exceed the size of the park in 12 weeks? In 14 weeks? [Remember: 1 sq mi $= 5280^2$ sq ft.]

58. The fruit fly population in a certain laboratory triples every day. Today there are 200 fruit flies.
(a) Make a table showing the number of fruit flies present for the first 4 days (today is day 0, tomorrow is day 1, etc.).
(b) Find a function of the form $f(x) = Pa^x$ that describes the fruit fly population at time x days.
(c) What is the growth factor here (that is, by what number must each day's population be multiplied to obtain the next day's population)?
(d) How many fruit flies will there be a week from now?

59. Take an ordinary piece of typing paper and fold it in half; then the folded sheet is twice as thick as the single sheet was. Fold it in half again, so that it is twice as thick as before. Keep folding it in half as long as you can. Soon the folded paper will be so thick and small that you will be unable to continue, but suppose you could keep folding the paper as long as you wanted. Assume the paper is .002 inches thick.
(a) Make a table showing the thickness of the folded paper for the first four folds (with fold 0 being the thickness of the original unfolded paper).
(b) Find a function of the form $f(x) = Pa^x$ that describes the thickness of the folded paper after x folds.
(c) How thick would the paper be after 20 folds?
(d) How many folds would it take to reach the moon (which is 243,000 miles from the earth)? [*Hint:* One mile is 5280 ft.]

60. Figure 5–14 is the graph of an exponential growth function $f(x) = Pa^x$.
(a) In this case, what is P? [*Hint:* What is $f(0)$?]
(b) Find the rule of the function f by finding a. [*Hint:* What is $f(2)$?]

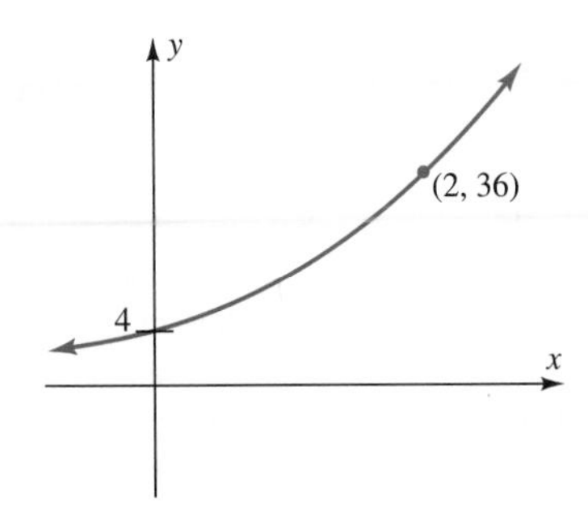

Figure 5-14

61. At the beginning of an experiment a culture contains 200 *E. coli* bacteria. An hour later there are 205 bacteria. Assuming that the *E. coli* bacteria grow exponentially, how many will there be after 10 hours? After 2 days? [See Example 4.]

62. If the population of India was 650 million a decade ago and is now 790 million people and continues to grow exponentially at the same rate, what will the population be in 5 years?

63. You have 5 grams of carbon-14, whose half-life is 5730 years.
(a) Write the rule of the function that gives the amount of carbon-14 remaining after x years. [See the discussion preceding Example 6.]
(b) How much carbon-14 will be left after 4000 years? After 8000 years?
(c) When will there be just one gram left?

64. (a) The half-life of radium is 1620 years. If you start with 100 mg of radium, what is the rule of the function that gives the amount remaining after t years?
(b) How much radium is left after 800 years? After 1600 years? After 3200 years?

65. The estimated number of units that will be sold by a certain company t months from now is given by $N(t) = 100{,}000e^{-.09t}$.
(a) What are current sales ($t = 0$)? What will sales be in 2 months? In 6 months?
(b) Will sales ever start to increase again? (What does the graph of $N(t)$ look like?)

66. (a) The function $g(t) = 1 - e^{-.0479t}$ gives the percentage of the population (expressed as a decimal) that has seen a new TV show t weeks after it goes on the air. What percentage of people have seen the show after 24 weeks?
(b) Approximately when will 90% of the people have seen it?

67. (a) The beaver population near a certain lake in year t is approximately $p(t) = \dfrac{2000}{1 + 199e^{-.5544t}}$. What is the population now ($t = 0$) and what will it be in 5 years?
(b) Approximately when will there be 1000 beavers?

Thinkers

68. This exercise provides a justification for the claim that the function $M(x) = c(.5)^{x/h}$ gives the mass after x years of a radioactive element with half-life h years. Suppose we have c grams of an element that has a half-life of 50 years. Then after 50 years we would have $c(\frac{1}{2})$ grams. After another 50 years we would have half of that, namely $c(\frac{1}{2})(\frac{1}{2}) = c(\frac{1}{2})^2$.
(a) How much remains after a third 50-year period? After a fourth 50-year period?
(b) How much remains after t 50-year periods?
(c) If x is the number of years, then $x/50$ is the number of 50-year periods. By replacing the number of periods t in part (b) by $x/50$, you obtain the amount remaining after x years. This gives the function $M(x)$ when $h = 50$. The same argument works in the general case (just replace 50 by h).

69. Find a function $f(x)$ with the property $f(r + s) = f(r)f(s)$ for all real numbers r and s. [*Hint*: Think exponential.]

70. Find a function $g(x)$ with the property $g(2x) = (g(x))^2$ for every real number x.

71. (a) Using the viewing window with $-4 \le x \le 4$ and $-1 \le y \le 8$, graph $f(x) = (½)^x$ and $g(x) = 2^x$ on the same screen. If you think of the y-axis as a mirror, how would you describe the relationship between the two graphs?
(b) Without graphing, explain how the graphs of $g(x) = 2^x$ and $k(x) = 2^{-x}$ are related.

72. Look back at Section 4.3, where the basic properties of graphs of polynomial functions were discussed. Then review the basic properties of the graph of $f(x) = a^x$ discussed in this section. Using these various properties, give an argument to show that for any fixed positive number a ($\neq 1$), it is *not* possible to find a polynomial function $g(x) = c_n x^n + \cdots + c_1 x + c_0$ such that $a^x = g(x)$ for *all* numbers x. In other words, *no exponential function is a polynomial function.* However, see Exercise 73.

73. *Approximating exponential functions by polynomials.* For each positive integer n, let f_n be the polynomial function whose rule is

$$f_n(x) = 1 + x + \frac{x^2}{2!} + \frac{x^3}{3!} + \frac{x^4}{4!} + \cdots + \frac{x^n}{n!}.$$

(a) Using the viewing window with $-4 \le x \le 4$ and $-5 \le y \le 55$ graph $g(x) = e^x$ and $f_4(x)$ on the same screen. Do the graphs appear to coincide?

(b) Replace the graph of $f_4(x)$ by that of $f_5(x)$, then by $f_6(x)$, $f_7(x)$, and so on until you find a polynomial $f_n(x)$ whose graph appears to coincide with the graph of $g(x) = e^x$ in this viewing window. Use the trace feature to move from graph to graph at the same value of x to see how accurate this approximation is.

(c) Change the viewing window so that $-6 \le x \le 6$ and $-10 \le y \le 400$. Is the polynomial you found in part (b) a good approximation for $g(x)$ in this viewing window? What polynomial is?

5.2.A *Excursion* COMPOUND INTEREST AND THE NUMBER *e*

When money earns compound interest, as in Example 2 on page 316, the exponential growth function can be described as follows.

Compound Interest Formula ▶

If P dollars is invested at interest rate r per time period (expressed as a decimal), then the amount A after t periods is

$$A = P(1 + r)^t.$$

Interest is often paid from day of deposit to day of withdrawal, regardless of the period used for compounding the interest. So the formula is used even when t is not an integer.

EXAMPLE 1 *(Compound Interest)* If \$7500 is invested at 12% interest compounded yearly, how much is in the account **(a)** after 5 years? **(b)** after 9 years and 3 months?

Solution

(a) Apply the compound interest formula with $P = 7500$, $r = .12$, and $t = 5$. Then $A = 7500(1.12)^5 = \$13{,}217.56$.

(b) Since 9 years and 3 months is 9.25 years, the amount in the account then is $A = 7500(1.12)^{9.25} = \$21{,}395.77$. ■

Banks often state an annual interest rate, but compound it more than once a year—for instance, 8% compounded quarterly. If the interest rate i is compounded n times a year, then the interest rate is i/n per period and there are n periods in 1 year.

EXAMPLE 2 *(Compound Interest)* If \$9000 is invested at 8% annual interest, compounded monthly, how much will the investment be worth after 6 years?

Solution Interest is compounded 12 times per year, so the time period is 1/12 of a year and the interest rate per period is $r = .08/12$. The number of periods in 6 years is $t = 6 \cdot 12 = 72$. Using these numbers in the compound interest formula shows that

$$A = 9000(1 + .08/12)^{72} \approx 9000(1.0067)^{72} = \$14{,}521.52.$$

This is more money than there would be if the interest were compounded annually ($r = .08$ and $t = 6$ in the formula):

$$A = 9000(1 + .08)^6 = 9000(1.08)^6 = \$14{,}281.87. \quad ■$$

EXAMPLE 3 *(Compound Interest)* If \$5000 is invested at 6.5% annual interest compounded monthly, how long will it take for the investment to double in value?

Solution The compound interest formula (with $P = 5000$ and $r = .065/12$) shows that the amount in the account at time t months is $5000(1 + .065/12)^t$. We must find the value of t such that

$$5000(1 + .065/12)^t = 10{,}000.$$

GRAPHING EXPLORATION Graph

$$f(t) = 5000(1 + .065/12)^t - 10{,}000$$

in a viewing window with $0 \le t \le 240$ (that's 20 years). Show that the t-intercept is approximately $t = 128.3$.

Therefore, it will take 128.3 months (approximately 10.7 years) for the investment to double. ■

EXAMPLE 4 *(Compound Interest)* What interest rate, compounded annually, is needed in order that a \$16,000 investment grow to \$50,000 in 18 years?

Solution In the compound interest formula, we have $A = 50{,}000$, $P = 16{,}000$, and $t = 18$ and must find r. The equation

$$16{,}000(1 + r)^{18} = 50{,}000$$

may be solved either graphically by finding the r-intercept of the graph of $f(r) = 16{,}000(1 + r)^{18} - 50{,}000$ or algebraically as follows:

Divide both sides by 16,000: $$(1 + r)^{18} = \frac{50{,}000}{16{,}000} = 3.125$$

Take 18th roots on both sides: $$\sqrt[18]{(1 + r)^{18}} = \sqrt[18]{3.125}$$

$$1 + r = \sqrt[18]{3.125}$$

$$r = \sqrt[18]{3.125} - 1 \approx .06535$$

So, the necessary interest rate is about 6.535%. ■

As a general rule, the more often your interest is compounded, the better off you are. But there is, alas, a limit.

EXAMPLE 5 *(The Number e)* You have $1 to invest for 1 year. The Exponential Bank offers to pay 100% annual interest, compounded n times per year and rounded to the nearest penny. You may pick any value you want for n. Can you choose n so large that your $1 will grow to some huge amount?

Solution Since the interest rate 100% (=1.00) is compounded n times per year, the interest rate per period is $r = 1/n$ and the number of periods in 1 year is n. According to the formula, the amount at the end of the year will be $A = \left(1 + \frac{1}{n}\right)^n$. Here's what happens for various values of n:*

Interest is Compounded	$n =$	$\left(1 + \frac{1}{n}\right)^n =$
Annually	1	$(1 + \frac{1}{1})^1 = 2$
Semiannually	2	$(1 + \frac{1}{2})^2 = 2.25$
Quarterly	4	$(1 + \frac{1}{4})^4 \approx 2.4414$
Monthly	12	$(1 + \frac{1}{12})^{12} \approx 2.6130$
Daily	365	$(1 + \frac{1}{365})^{365} \approx 2.71457$
Hourly	8760	$(1 + \frac{1}{8760})^{8760} \approx 2.718127$
Every minute	525,600	$(1 + \frac{1}{525{,}600})^{525{,}600} \approx 2.7182792$
Every second	31,536,000	$(1 + \frac{1}{31{,}536{,}000})^{31{,}536{,}000} \approx 2.7182818$

*The calculations in the table were done on a large computer, using double precision. The interest rate 100% was chosen for computational convenience. Essentially the same point can be made with a more realistic rate.

Since interest is rounded to the nearest penny, your dollar will grow no larger than \$2.72, no matter how big n is. ■

The last entry in the preceding table, 2.7182818, is the number e to seven decimal places. This is just one example of how e arises naturally in real-world situations. In calculus, it is proved that e is the *limit* of $\left(1 + \frac{1}{n}\right)^n$, meaning that as n gets larger and larger, $\left(1 + \frac{1}{n}\right)^n$ gets closer and closer to e.

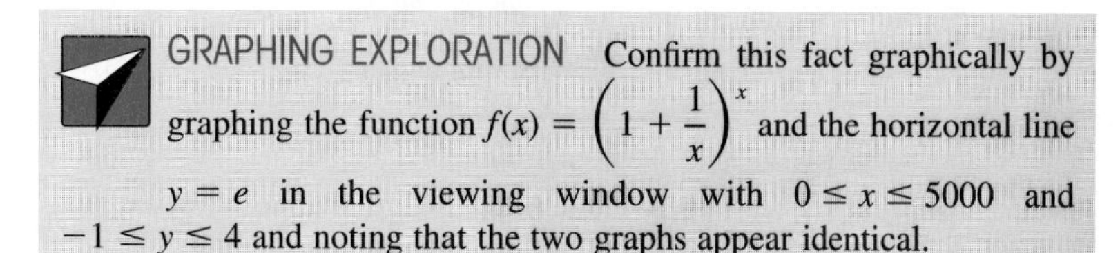

GRAPHING EXPLORATION Confirm this fact graphically by graphing the function $f(x) = \left(1 + \frac{1}{x}\right)^x$ and the horizontal line $y = e$ in the viewing window with $0 \le x \le 5000$ and $-1 \le y \le 4$ and noting that the two graphs appear identical.

When interest is compounded n times per year for larger and larger values of n, as in Example 5, we say that the interest is **continuously compounded.** In this terminology, Example 5 says that \$1 will grow to \$2.72 in 1 year at an interest rate of 100% compounded continuously.

Now consider a more realistic interest rate. Suppose P dollars is invested at 8% interest compounded continuously. If interest is compounded m times per year, then the interest rate per period is $.08/m$ and there are m periods in 1 year. Therefore at the end of 1 year, P dollars will have grown to

$$A = P\left(1 + \frac{.08}{m}\right)^m. \tag{1}$$

To see what happens to this amount as m gets larger, let n denote the number $\frac{m}{.08}$.

Algebraic manipulation shows that $n = \frac{m}{.08}$ is equivalent to

$$m = .08n \qquad \text{and} \qquad \frac{1}{n} = \frac{.08}{m}$$

and, therefore, equation (1) may be rewritten as:

$$A = P\left(1 + \frac{.08}{m}\right)^m = P\left(1 + \frac{1}{n}\right)^{.08n} = P\left[\left(1 + \frac{1}{n}\right)^n\right]^{.08}.$$

As m gets larger and larger, so does $.08m = n$. Example 5 shows that when n is very large,

$$\left(1 + \frac{1}{n}\right)^n \quad \text{is very close to the number } e.$$

Therefore, as m gets larger and larger,

$$A = P\left[\left(1+\frac{1}{n}\right)^n\right]^{.08} \quad \text{gets closer and closer to } Pe^{.08}.$$

Since dollar amounts are rounded, we can say that after 1 year P dollars grows to $Pe^{.08}$ dollars. In other words,

The amount at the end of a year is the beginning amount multiplied by $e^{.08}$.

Applying this fact repeatedly, we have:

Amount at end of year 1	$Pe^{.08}$
Amount at end of year 2	$(Pe^{.08})e^{.08} = P(e^{.08})^2 = Pe^{(.08)2}$
Amount at end of year 3	$(Pe^{(.08)2})e^{.08} = P(e^{.08})^2e^{.08} = P(e^{.08})^3 = Pe^{(.08)3}$

Continuing in this manner, we see that P dollars invested at 8% compounded continuously for 6 years will grow to

$$Pe^{(.08)6} \text{ dollars.}$$

When $P = 250$, for example, this says that \$250 invested at 8% compounded continuously for 6 years will grow to

$$250e^{(.08)6} = 250e^{.48} \approx \$404.02.$$

If we use interest rate r in place of 8% and t years in place of 6 years in the discussion above, then a virtually identical argument leads to this conclusion:

Continuous Compounding ▶

If P dollars is invested at interest rate r, compounded continuously, then the amount A after t years is

$$A = Pe^{rt}$$

EXAMPLE 6 *(Continuous Compounding)* If \$3800 is invested at 9.2% compounded continuously, then after seven and a half years the amount in the account will be

$$A = 3800e^{(.092)7.5} = 3800e^{.69} \approx \$7576.12.*$$

To find out how long it will take for the value of the investment to reach \$5000, you need only solve the equation

$$3800e^{.092t} = 5000, \qquad \text{or equivalently,} \qquad 3800e^{.092t} - 5000 = 0.$$

*This is the continuous compounding formula with $P = 3800$, $r = .092$, and $t = 7.5$.

GRAPHING EXPLORATION Solve the equation graphically and verify that it will take just a few days less than 3 years for the investment to be worth \$5000. ■

Continuous Growth and Decay

The discussion of continuous compounding of interest applies to all exponential growth and decay situations: Just replace P dollars by a quantity P and the continuous interest rate r by the rate k at which the quantity P is continuously increasing or decreasing ($k > 0$ for growth, $k < 0$ for decay). The resulting function $f(x) = Pe^{kx}$ (as in the box on page 321) describes the continuous exponential growth.

EXERCISES 5.2.A

1. If \$1000 is invested at 8%, find the value of the investment after 5 years if interest is compounded
(a) annually. **(b)** quarterly. **(c)** monthly.
(d) weekly.

2. If \$2500 is invested at 11.5%, what is the value of the investment after 10 years if interest is compounded
(a) annually? **(b)** monthly? **(c)** daily?

In Exercises 3–12, determine how much money will be in a savings account if the initial deposit was \$500 and the interest rate is:

3. 5% compounded annually for 8 years.

4. 5% compounded annually for 10 years.

5. 5% compounded quarterly for 10 years.

6. 9.3% compounded monthly for 9 years.

7. 8.9% compounded daily for 8.5 years.

8. 12.5% compounded weekly for 7 years and 7 months.

9. 7% compounded continuously for 4 years.

10. 8.5% compounded continuously for 10 years.

11. 8.45% compounded continuously for 6.2 years.

12. 9.25% compounded continuously for 11.6 years.

13. A typical credit card company charges 18% annual interest, compounded monthly, on the unpaid balance. If your current balance is \$520 and you don't make any payments for 6 months, how much will you owe (assuming they don't sue you in the meantime)?

14. How long will it take to double an investment of \$100 if the interest rate is 8% compounded annually? Answer the same question for \$500 investment.

15. How long will it take to double an investment of \$100 if the interest rate is 7% compounded continuously?

16. How long will it take to triple an investment of \$5000 if the interest rate is 8% compounded continuously?

17. If an investment of \$1000 grows to \$1407.10 in seven years with interest compounded annually, what is the interest rate?

18. If an investment of \$2000 grows to \$2700 in three and a half years, with an annual interest rate that is compounded quarterly, what is the annual interest rate?

19. If you put \$3000 in a savings account today, what interest rate (compounded annually) must you receive in order to have \$4000 after five years?

20. If interest is compounded continuously, what annual rate must you receive if your investment of \$1500 is to grow to \$2100 in six years?

5.3 COMMON AND NATURAL LOGARITHMIC FUNCTIONS*

Roadmap

We begin with the only logarithms that are in widespread use, common and natural logarithms. Natural logarithms are emphasized because of their central role in calculus. Those who prefer to begin with logarithms to an arbitrary base b should cover Excursion 5.3.A before reading this section.

From their invention in the 17th century until the development of computers and calculators, logarithms were the only effective tool for numerical computation in astronomy, chemistry, physics, and engineering. Although they are no longer needed for computation, logarithmic functions still play an important role in the sciences and engineering. In this section we examine the two most important types of logarithms, those to base 10 and those to base e. Logarithms to other bases are considered in Excursion 5.3.A.

Common Logarithms

The exponential function $f(x) = 10^x$, whose graph is shown in Figure 5–15, is an increasing function and hence is one-to-one (as explained on page 150). Therefore, f has an inverse function g whose graph is the reflection of the graph of f in the line $y = x$ (see page 154), as shown in Figure 5–16.**

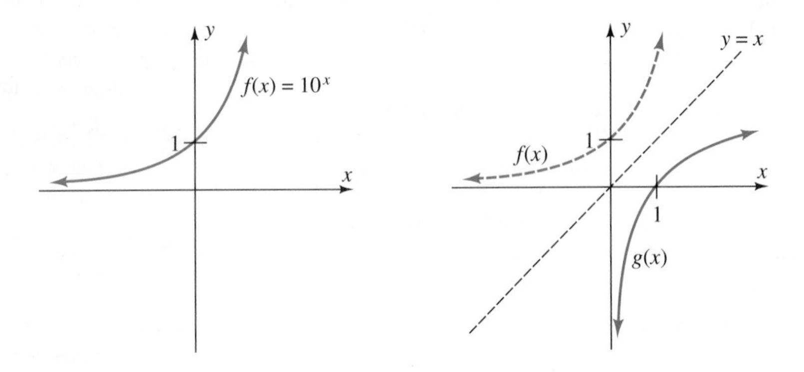

Figure 5-15

Figure 5-16

This inverse function g is called the **common logarithmic function.** The value $g(x)$ of this function at the number x is denoted **log x** and called the **common**

*Section 2.7 (Inverse Functions) is a prerequisite for this section.

**The graph of $f(x) = 10^x$ can be obtained in parametric mode by letting

$$x = t \quad \text{and} \quad y = 10^t \quad (t \text{ any real number}).$$

As explained on page 153, the graph of the inverse function g can then be obtained by letting

$$x = 10^t \quad \text{and} \quad y = t \quad (t \text{ any real number}).$$

logarithm of the number x. Every calculator has a LOG key for evaluating the function $g(x) = \log x$. For instance,

$$\log .6 = -.2218 \quad \text{and} \quad \log 327 = 2.5145.*$$

Although many properties of the logarithmic function $g(x) = \log x$ can be read from its graph, we also need an algebraic description of this function. On page 151 we saw that the relationship between a function f and its inverse function g is given by:

$$g(v) = u \quad \text{exactly when} \quad f(u) = v.$$

When $f(x) = 10^x$ and $g(x) = \log x$, this statement says

$$\mathbf{\log v = u} \quad \textbf{exactly when} \quad \mathbf{10^u = v.}$$

In other words,

log v is the exponent to which 10 must be raised to produce v.

EXAMPLE 1 Without using a calculator, find

(a) $\log 1000$ **(b)** $\log 1$ **(c)** $\log\sqrt{10}$

Solution

(a) To find log 1000, ask yourself "what power of 10 equals 1000?" The answer is 3 because $10^3 = 1000$. Therefore, $\log 1000 = 3$.

(b) What power must 10 be raised to in order to produce 1? Since $10^0 = 1$, we conclude that $\log 1 = 0$.

(c) $\text{Log}\sqrt{10} = 1/2$ because 1/2 is the exponent to which 10 must be raised to produce $\sqrt{10}$, that is, $10^{1/2} = \sqrt{10}$. ■

A calculator is necessary to find most logarithms, but even then you should proceed as in Example 1 to get a rough estimate. For instance, log 795 is the exponent to which 10 must be raised to produce 795. Since $10^2 = 100$ and $10^3 = 1000$, this exponent must be between 2 and 3, that is, $2 < \log 795 < 3$.

Since logarithms are a special kind of exponent, every statement about logarithms is equivalent to a statement about exponents. For instance,

Logarithmic Statement	**Equivalent Exponential Statement**
$\log v = u$	$10^u = v$
$\log 29 = 1.4624$	$10^{1.4624} = 29$
$\log 378 = 2.5775$	$10^{2.5775} = 378$

*Here and below all logarithms are rounded to four decimal places and an equal sign is used rather than the more correct "approximately equal." The word "common" will be omitted except when it is necessary to distinguish these logarithms from other types that are introduced below.

EXAMPLE 2 To solve the equation $\log x = 2$, note that the equation is equivalent to the exponential statement $10^2 = x$. So the solution is $x = 100$. ■

Natural Logarithms

The exponential function $f(x) = 10^x$ and the resulting common logarithms played an important role in mathematics and still do. With the advent of calculus, however, it became clear that the exponential function $f(x) = e^x$ was even more useful in science and engineering. Consequently, a new type of logarithm, based on the number e instead of 10, was developed.* This development is essentially a carbon copy of what was done above.

The definition of the logarithmic function $g(x) = \log x$ used only the fact that the exponential function $f(x) = 10^x$ was an increasing function that had an inverse function. The same idea can be applied to the exponential function $f(x) = e^x$ whose graph is shown in Figure 5–17. The function f is increasing and hence one-to-one, so f has an inverse function g whose graph is the reflection of the graph of f in the line $y = x$, as shown in Figure 5–18.

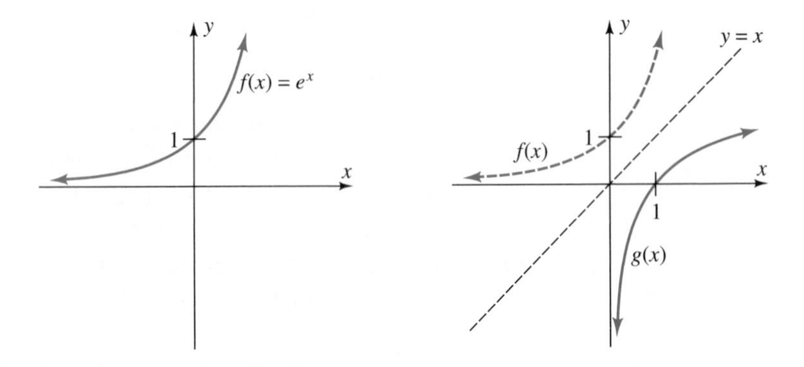

Figure 5-17 **Figure 5-18**

This inverse function g is called the **natural logarithmic function.** The value $g(x)$ of this function at a number x is denoted **ln *x*** and called the **natural logarithm** of the number x. When the relationship of inverse functions (page 151)

$$g(v) = u \qquad \text{exactly when} \qquad f(u) = v$$

is applied to the function $f(x) = e^x$ and its inverse $g(x) = \ln x$, it says

$$\ln v = u \qquad \textbf{exactly when} \qquad e^u = v.$$

In other words,

ln *v* is the exponent to which *e* must be raised to produce *v*.

*The number e was introduced on page 320 and discussed further in Excursion 5.2.A.

EXAMPLE 3 Every calculator has an LN key for evaluating the function $g(x) = \ln x$. For instance,

$$\ln .15 = -1.8971 \qquad \text{and} \qquad \ln 186 = 5.2257.$$

In a few cases you can determine $\ln x$ without a calculator. For example,

$$\ln 1 = 0$$

because $e^0 = 1$, that is, 0 is the exponent to which e must be raised to produce 1. Similarly, $\ln e = 1$ because 1 is the exponent to which e must be raised to produce e. ■

Once again, (natural) logarithms are a special kind of exponent and every statement about logarithms is equivalent to a statement about exponents:

Logarithmic Statement	Equivalent Exponential Statement
$\ln v = u$	$e^u = v$
$\ln 14 = 2.6391$	$e^{2.6391} = 14$
$\ln 158 = 5.0626$	$e^{5.0626} = 158$

Properties of Logarithms

Since common and natural logarithms have almost identical definitions (just replace 10 by e), it is not surprising that they share the same essential properties. You don't need a calculator to understand these properties. You need only translate logarithmic statements into equivalent exponential ones (or vice versa).

EXAMPLE 4 What is $\ln(-10)$? *Translation:* To what power must e be raised to produce -10? *Answer:* The graph of $f(x) = e^x$ in Figure 5–17 shows that every power of e is *positive*. So e^x can *never* be -10 or any negative number or zero and hence $\ln(-10)$ is not defined. Similarly, $\log(-10)$ is not defined because every power of 10 is positive. Therefore,

$\ln v$ and $\log v$ are defined only when $v > 0$. ■

EXAMPLE 5 What is log 1? *Translation:* To what power must 10 be raised to produce 1? *Answer:* 0 because $10^0 = 1$. So $\log 1 = 0$. Example 3 shows that the same thing is true for natural logarithms: $\ln 1 = 0$. ■

EXAMPLE 6 What is $\ln e^9$? *Translation:* To what power must e be raised to produce e^9? Obviously, the answer is 9. So $\ln e^9 = 9$ and in general

$\ln e^k = k$ for every real number k.

Similarly,

$$\log 10^k = k \qquad \text{for every real number } k$$

because k is the exponent to which 10 must be raised to produce 10^k. ■

EXAMPLE 7 What are $10^{\log 678}$ and $e^{\ln 678}$? By definition log 678 is the exponent to which 10 must be raised to produce 678, so if you raise 10 to this exponent the answer will be 678, that is, $10^{\log 678} = 678$. Similarly, ln 678 is the exponent to which e must be raised to produce 678 and hence $e^{\ln 678} = 678$. Similarly, for every $v > 0$

$$e^{\ln v} = v \qquad \text{and} \qquad 10^{\log v} = v. \quad ■$$

The facts presented in the preceding examples may be summarized as follows.

Properties of Logarithms ▶

	Natural Logarithms	Common Logarithms
1.	$\ln v$ is defined only when $v > 0$;	$\log v$ is defined only when $v > 0$.
2.	$\ln 1 = 0$ and $\ln e = 1$;	$\log 1 = 0$ and $\log 10 = 1$.
3.	$\ln e^k = k$ for every real number k;	$\log 10^k = k$ for every real number k.
4.	$e^{\ln v} = v$ for every $v > 0$;	$10^{\log v} = v$ for every $v > 0$.

Properties 3 and 4 (for natural logarithms) are restatements of the fact that the exponential function $f(x) = e^x$ and its inverse function $g(x) = \ln x$ have the usual "reversal properties" (see page 152):

$$(g \circ f)(x) = g(f(x)) = g(e^x) = \ln e^x = x \qquad \text{for all } x;$$

$$(f \circ g)(x) = f(g(x)) = f(\ln x) = e^{\ln x} = x \qquad \text{for all } x > 0.*$$

Analogous statements hold for common logarithms.

The properties of logarithms can be used to simplify expressions and solve equations.

EXAMPLE 8 Applying Property 3 with $k = 2x^2 + 7x + 9$ shows that $\ln e^{2x^2+7x+9} = 2x^2 + 7x + 9$. ■

*A calculator provides a visual demonstration of these facts. For instance, if you key in [LN] [e^x] [52] [ENTER] the calculator displays 52 (that is, $g(f(52)) = 52$). Similarly, keying in [e^x] [LN] [167] [ENTER] produces 167 (that is $f(g(167)) = 167$).

EXAMPLE 9 Solve the equation $\ln(x+1) = 2$.

Solution Since $\ln(x+1) = 2$, we have:

$$e^{\ln(x+1)} = e^2.$$

Applying Property 4 with $v = x + 1$ shows that

$$x + 1 = e^{\ln(x+1)} = e^2$$

$$x = e^2 - 1 \approx 6.3891. \blacksquare$$

Logarithm Laws

The first law of exponents states that $b^m b^n = b^{m+n}$, or in words,

The exponent of a product is the sum of the exponents of the factors.

Since logarithms are just particular kinds of exponents, this statement translates as:

The logarithm of a product is the sum of the logarithms of the factors.

Here is the same statement in symbolic language:

Product Law for Logarithms ▶

For all $v, w > 0$,

$$\ln(vw) = \ln v + \ln w \quad \text{and} \quad \log(vw) = \log v + \log w.$$

Proof According to Property 4 of logarithms (in the box on page 336),

$$e^{\ln v} = v \quad \text{and} \quad e^{\ln w} = w.$$

Therefore, by the first law of exponents (with $m = \ln v$ and $n = \ln w$):

$$vw = e^{\ln v}e^{\ln w} = e^{\ln v + \ln w}.$$

So raising e to the exponent $(\ln v + \ln w)$ produces vw. But the definition of logarithm says that $\ln vw$ is the exponent to which e must be raised to produce vw. Therefore, we must have $\ln vw = \ln v + \ln w$. A similar argument works for common logarithms (just replace e by 10 and "ln" by "log"). ❑

EXAMPLE 10 A calculator shows that $\ln 7 = 1.9459$ and $\ln 9 = 2.1972$. Therefore,

$$\ln 63 = \ln(7 \cdot 9) = \ln 7 + \ln 9 = 1.9459 + 1.1972 = 4.1341.$$

Similarly, you can readily verify that

$$\log 33 = \log(3 \cdot 11) = \log 3 + \log 11 = .4771 + 1.0414 = 1.5185. \blacksquare$$

WARNING A common error in applying the Product Law for Logarithms is to write the *false* statement

$$\ln 7 + \ln 9 = \ln(7+9) = \ln 16$$

instead of the correct statement

$$\ln 7 + \ln 9 = \ln(7 \cdot 9) = \ln 63.$$

GRAPHING EXPLORATION Illustrate the warning in the margin graphically by graphing both

$$f(x) = \ln x + \ln 9 \quad \text{and} \quad g(x) = \ln(x+9)$$

in the standard viewing window and verifying that the graphs are not the same. In particular, the functions have different values at $x = 7$.

The second law of exponents, namely, $b^m/b^n = b^{m-n}$, may be roughly stated in words as

The exponent of the quotient is the difference of exponents.

When the exponents are logarithms, this says

The logarithm of a quotient is the difference of the logarithms.

In other words,

Quotient Law for Logarithms ▶

For all $v, w > 0$,

$$\ln\left(\frac{v}{w}\right) = \ln v - \ln w \quad \text{and} \quad \log\left(\frac{v}{w}\right) = \log v - \log w.$$

The proof of the Quotient Law is very similar to the proof of the Product Law (see Exercise 92).

EXAMPLE 11 $\ln\left(\frac{17}{44}\right) = \ln 17 - \ln 44$ and $\log\left(\frac{297}{39}\right) = \log 297 - \log 39$. ■

EXAMPLE 12 For any $w > 0$,

$$\ln\left(\frac{1}{w}\right) = \ln 1 - \ln w = 0 - \ln w = -\ln w$$

and

$$\log\left(\frac{1}{w}\right) = \log 1 - \log w = 0 - \log w = -\log w.$$ ■

WARNING Do not confuse $\ln\left(\frac{v}{w}\right)$ with the quotient $\frac{\ln v}{\ln w}$. They are *different* numbers. For example,

$$\ln\left(\frac{36}{3}\right) = \ln(12) = 2.4849, \qquad \text{but} \qquad \frac{\ln 36}{\ln 3} = \frac{3.5835}{1.0986} = 3.2619.$$

GRAPHING EXPLORATION Illustrate the preceding warning graphically by graphing both $f(x) = \ln(x/3)$ and $g(x) = (\ln x)/(\ln 3)$ and verifying that the graphs are not the same at $x = 36$ (or anywhere else, for that matter).

The third law of exponents, namely $(b^m)^k = b^{mk}$, can also be translated into logarithmic language:

Power Law for Logarithms ▶

For all k and all $v > 0$,

$$\ln(v^k) = k(\ln v) \qquad \text{and} \qquad \log(v^k) = k(\log v).$$

Proof Since $v = 10^{\log v}$ (why?), the third law of exponents (with $b = 10$ and $m = \log v$) shows that

$$v^k = (10^{\log v})^k = 10^{(\log v)k} = 10^{k(\log v)}.$$

So raising 10 to the exponent $k(\log v)$ produces v^k. But the exponent to which 10 must be raised to produce v^k is, by definition, $\log(v^k)$. Therefore $\log(v^k) = k(\log v)$, and the proof is complete. A similar argument with e in place of 10 and "ln" in place of "log," works for natural logarithms. ❑

EXAMPLE 13 $\ln\sqrt{19} = \ln 19^{1/2} = \frac{1}{2}(\ln 19)$. ■

The logarithm laws can be used to simplify various expressions.

EXAMPLE 14

$$\begin{aligned}
\ln 3x + 4\cdot\ln x - \ln 3xy &= \ln 3x + \ln x^4 - \ln 3xy && [\textit{Power Law}]\\
&= \ln(3x\cdot x^4) - \ln 3xy && [\textit{Product Law}]\\
&= \ln\frac{3x^5}{3xy} && [\textit{Quotient Law}]\\
&= \ln\frac{x^4}{y} && [\textit{Cancel } 3x]
\end{aligned}$$
■

EXAMPLE 15 To simplify $\ln(\sqrt{x}/x) + \ln \sqrt[4]{ex^2}$, we begin by changing to exponential notation:

$$\ln\left(\frac{x^{1/2}}{x}\right) + \ln(ex^2)^{1/4} = \ln(x^{-1/2}) + \ln(ex^2)^{1/4}$$

$$= -\frac{1}{2}\cdot \ln x + \frac{1}{4}\cdot \ln ex^2 \qquad \text{[Power Law]}$$

$$= -\frac{1}{2}\cdot \ln x + \frac{1}{4}(\ln e + \ln x^2) \qquad \text{[Product Law]}$$

$$= -\frac{1}{2}\cdot \ln x + \frac{1}{4}(\ln e + 2\cdot \ln x) \qquad \text{[Power Law]}$$

$$= -\frac{1}{2}\cdot \ln x + \frac{1}{4}\cdot \ln e + \frac{1}{2}\cdot \ln x$$

$$= \frac{1}{4}\cdot \ln e = \frac{1}{4} \qquad [\ln e = 1]$$ ■

Applications

Because logarithmic growth is slow, measurements on a logarithmic scale (that is, on a scale determined by a logarithmic function) can sometimes be deceptive.

EXAMPLE 16 *(Earthquakes)* The magnitude $R(i)$ of an earthquake on the Richter scale is given by $R(i) = \log(i/i_0)$, where i is the amplitude of the ground motion of the earthquake and i_0 is the amplitude of the ground motion of the so-called zero earthquake.* A moderate earthquake might have 1000 times the ground motion of the zero earthquake (that is, $i = 1000i_0$). So its magnitude would be

$$\log(1000i_0/i_0) = \log 1000 = \log 10^3 = 3.$$

An earthquake with 10 times this ground motion (that is, $i = 10\cdot 1000i_0 = 10{,}000i_0$) would have a magnitude of

$$\log(10{,}000i_0/i_0) = \log 10{,}000 = \log 10^4 = 4.$$

So a *tenfold* increase in ground motion produces only a 1-point change on the Richter scale. In general,

Increasing the ground motion by a factor of 10^k increases the Richter magnitude by k units.**

*The zero earthquake has ground motion amplitude of less than 1 micron on a standard seismograph 100 kilometers from the epicenter.

***Proof:* If one quake has ground motion amplitude i and the other $10^k i$, then

$$R(10^k i) = \log 10^k i/i_0 = \log 10^k + \log(i/i_0)$$
$$= k + \log(i/i_0) = k + R(i).$$

For instance, the 1989 World Series earthquake in San Francisco measured 7.0 on the Richter scale, and the great earthquake of 1906 measured 8.3. The difference of 1.3 points means that the 1906 quake was $10^{1.3} \approx 20$ times more intense than the 1989 one in terms of ground motion. ■

Exponential Functions

The box at the end of Section 5.2 states that any exponential function $f(x) = Pa^x$ can be written in the form $f(x) = Pe^{kx}$. Here is a proof of that fact. Let $k = \ln a$. Then, $e^k = e^{\ln a} = a$ by the fourth property of natural logarithms (box on page 336). Thus,

$$Pa^x = P(e^k)^x = Pe^{kx}.$$

Furthermore, the graph of $g(x) = \ln x$ (see Figure 5–18) shows that when $0 < a < 1$ (exponential decay), then $k = \ln a$ is negative and when $a > 1$ (exponential growth), then $k = \ln a$ is positive.

EXERCISES 5.3

Unless stated otherwise, all letters represent positive numbers.

In Exercises 1–4, find the logarithm, without using a calculator.

1. $\log 10{,}000$ **2.** $\log .001$

3. $\log \dfrac{\sqrt{10}}{1000}$ **4.** $\log \sqrt[3]{.01}$

In Exercises 5–16, translate the given logarithmic statement into an equivalent exponential statement.

5. $\log 1000 = 3$ **6.** $\log .001 = -3$

7. $\log 750 = 2.88$ **8.** $\log (.8) = -.097$

9. $\ln 3 = 1.0986$ **10.** $\ln 10 = 2.3026$

11. $\ln 1000 = 6.9078$ **12.** $\ln(1/4) = -1.3863$

13. $\ln .01 = -4.6052$ **14.** $\ln s = r$

15. $\ln(x^2 + 2y) = z + w$ **16.** $\log(a + c) = d$

In Exercises 17–24, translate the given exponential statement into an equivalent logarithmic one.

17. $10^{-2} = .01$ **18.** $10^3 = 1000$

19. $10^{.4771} = 3$ **20.** $10^{7k} = r$

21. $e^{3.25} = 25.79$ **22.** $e^{-4} = .0183$

23. $e^{12/7} = 5.5527$ **24.** $e^k = t$

25. $e^{2/r} = w$ **26.** $e^{4uv} = m$

In Exercises 27–40, evaluate the given expression without using a calculator.

27. $\log 10^{\sqrt{43}}$ **28.** $\log 10^{\sqrt{x^2+y^2}}$ **29.** $\ln e^{15}$

30. $\ln e^{3.78}$ **31.** $\ln \sqrt{e}$ **32.** $\ln \sqrt[5]{e}$

33. $e^{\ln 931}$ **34.** $e^{\ln 34.17}$ **35.** $e^{\ln \sqrt{37}}$

36. $e^{\ln 7/5}$ **37.** $\ln e^{x+y}$ **38.** $\ln e^{x^2+2y}$

39. $e^{\ln x^2}$ **40.** $e^{\ln \sqrt{x+3}}$

In Exercises 41–46, write the given expression as a single logarithm, as in Example 14.

41. $\ln x^2 + 3 \ln y$

42. $\ln 2x + 2(\ln x) - \ln 3y$

43. $\log(x^2 - 9) - \log(x + 3)$

44. $\log 3x - 2(\log x - \log(2 + y))$

45. $2(\ln x) - 3(\ln x^2 + \ln x)$

46. $\ln(e/\sqrt{x}) - \ln \sqrt{ex}$

In Exercises 47–52, let $u = \ln x$ and $v = \ln y$. Write the given expression in terms of u and v. For example, $\ln x^3y = \ln x^3 + \ln y = 3(\ln x) + \ln y = 3u + v$.

47. $\ln(x^2y^5)$ **48.** (x^3/y^2) **49.** $\ln(\sqrt{x}\cdot y^2)$

50. $\ln(\sqrt{x}/y)$ **51.** $\ln(\sqrt[3]{x^2\sqrt{y}})$ **52.** $\ln(\sqrt{x^2y}/\sqrt[3]{y})$

In Exercises 53–56, find the domain of the given function (that is, the largest set of real numbers for which the rule produces well-defined real numbers).

53. $f(x) = \ln(x + 1)$ **54.** $g(x) = \ln(x + 2)$

55. $h(x) = \log(-x)$ **56.** $k(x) = \log(2 - x)$

57. **(a)** Graph $y = x$ and $y = e^{\ln x}$ in separate viewing windows [or use a split-screen, if your calculator has that feature]. For what values of x are the graphs identical?

(b) Use the properties of logarithms to explain your answer in part (a).

58. **(a)** Graph $y = x$ and $y = \ln(e^x)$ in separate viewing windows [or a split-screen, if your calculator has that feature]. For what values of x are the graphs identical?

(b) Use the properties of logarithms to explain your answer in part (a).

In Exercises 59–64, use graphical or algebraic means to determine whether the statement is true or false.

59. $\ln|x| = |\ln x|$? **60.** $\ln\left(\frac{1}{x}\right) = \frac{1}{\ln x}$?

61. $\log x^5 = 5(\log x)$? **62.** $e^{x \ln x} = x^x \quad (x > 0)$?

63. $\ln x^3 = (\ln x)^3$? **64.** $\log \sqrt{x} = \sqrt{\log x}$?

In Exercises 65–70, list the transformations that will change the graph of $g(x) = \ln x$ into the graph of the given function. [Section 2.5 may be helpful.]

65. $f(x) = 2\cdot \ln x$ **66.** $f(x) = \ln x - 7$

67. $h(x) = \ln(x - 4)$ **68.** $k(x) = \ln(x + 2)$

69. $h(x) = \ln(x + 3) - 4$ **70.** $k(x) = \ln(x - 2) + 2$In

Exercises 71–74, sketch the graph of the function.

71. $f(x) = \log(x - 3)$ **72.** $g(x) = 2 \ln x + 3$

73. $h(x) = -2 \log x$ **74.** $f(x) = \ln(-x) - 3$

In Exercises 75–80, find a viewing window (or windows) that shows a complete graph of the function.

75. $f(x) = \frac{x}{\ln x}$ **76.** $g(x) = \frac{\ln x}{x}$

77. $h(x) = \frac{\ln x^2}{x}$ **78.** $k(x) = e^{2/\ln x}$

79. $f(x) = 10 \log x - x$ **80.** $f(x) = \frac{\log x}{x}$

In Exercises 81–84, find the average rate of change of the function.

81. $f(x) = \ln(x - 2)$, as x goes from 3 to 5.

82. $g(x) = x - \ln x$, as x goes from .5 to 1.

83. $g(x) = \log(x^2 + x + 1)$, as x goes from -5 to -3.

84. $f(x) = x \log|x|$, as x goes from 1 to 4.

85. **(a)** What is the average rate of change of $f(x) = \ln x$, as x goes from 3 to $3 + h$?

(b) What is the value of h when the average rate of $f(x) = \ln x$, as x goes from 3 to $3 + h$, is .25?

86. **(a)** Find the average rate of change of $f(x) = \ln x^2$, as x goes from .5 to 2.

(b) Find the average rate of change of $g(x) = \ln(x - 3)^2$, as x goes from 3.5 to 5.

(c) What is the relationship between your answers in parts (a) and (b) and why is this so?

87. The concentration of hydrogen ions in a given solution is denoted $[H^+]$ and is measured in moles per liter. For example, $[H^+] = .00008$ for beer and $[H^+] = .0004$ for wine. Chemists define the pH of the solution be the number $pH = -\log[H^+]$. The solution is said to be an *acid* if $pH < 7$ and a *base* if $pH > 7$.

(a) Is beer an acid or a base? What about wine?

(b) If a solution has a pH of 2, what is its $[H^+]$?

(c) For hominy, $[H^+] = 5\cdot 10^{-8}$. Is hominy a base?

88. The doubling function $D(x) = \frac{\ln 2}{\ln(1 + x)}$ gives the years required to double your money when it is invested at interest rate x (expressed as a decimal), compounded annually.

(a) Find the time it takes to double your money at each of these interest rates: 4%, 6%, 8%, 12%, 18%, 24%, 36%.

(b) Round the answers in part (a) to the nearest year and compare them with these numbers: 72/4, 72/6, 72/8, 72/12, 72/18, 72/24, 72/36. Use this evidence to state a "rule of thumb" for determining approximate doubling time, without using the function D. This rule of thumb, which has long been used by bankers, is called the **rule of 72.**

89. Suppose $f(x) = A \ln x + B$, where A and B are constants. If $f(1) = 10$ and $f(e) = 1$, what are A and B?

90. If $f(x) = A \ln x + B$ and $f(e) = 5$ and $f(e^2) = 8$, what are A and B?

91. Show that $g(x) = \ln\left(\frac{x}{1-x}\right)$ is the inverse function of $f(x) = \frac{1}{1+e^{-x}}$. (See Section 2.7.)

92. Prove the Quotient Law for Logarithms: for $v, w > 0$, $\ln\left(\frac{v}{w}\right) = \ln v - \ln w$. (Use properties of exponents and the fact that $v = e^{\ln v}$ and $w = e^{\ln w}$.)

In Exercises 93–96, state the magnitude on the Richter scale of an earthquake that satisfies the given condition.

93. 100 times stronger than the zero quake.

94. $10^{4.7}$ times stronger than the zero quake.

95. 350 times stronger than the zero quake.

96. 2500 times stronger than the zero quake.

Exercises 97–100 deal with the energy intensity i of a sound, which is related to the loudness of the sound by the function $L(i) = 10 \cdot \log(i/i_0)$, *where* i_0 *is the minimum intensity detectable by the human ear and* $L(i)$ *is measured in decibels. Find the decibel measure of the sound.*

97. Ticking watch (intensity is 100 times i_0).

98. Soft music (intensity is 10,000 times i_0).

99. Loud conversation (intensity is 4 million times i_0).

100. Victoria Falls in Africa (intensity is 10 billion times i_0).

101. The height h above sea level (in meters) is related to air temperature t (in degrees Celsius), the atmospheric pressure p (in centimeters of mercury at height h), and the atmospheric pressure c at sea level by

$$h = (30t + 8000)\ln(c/p).$$

If the pressure at the top of Mount Rainier is 44 cm on a day when the sea level pressure is 75.126 cm and the temperature is 7°, what is the height of Mount Rainier?

102. Mount Everest is 8850 meters high. What is the atmospheric pressure at the top of the mountain on a day when the temperature is $-25°$ and the atmospheric pressure at sea level is 75 cm? [See Exercise 101.]

103. A class in elementary Sanskrit is tested at the end of the semester and weekly thereafter on the same material. The average score on the exam taken after t weeks is given by the "forgetting function"

$$g(t) = 77 - 10 \cdot \ln(t+1).$$

(a) What was the average score on the original exam?
(b) What was the average score after 2 weeks? After 5 weeks?
(c) When did the average score drop below 50?

104. Students in a precalculus class were given a final exam. Each month thereafter, they took an equivalent exam. The class average on the exam taken after t months is given by $F(t) = 82 - 8 \cdot \ln(t+1)$.
(a) What was the class average after 6 months?
(b) After a year?
(c) When did the class average drop below 55?

105. One person with a flu virus visited the campus. The number T of days it took for the virus to infect x people was given by:

$$T = -.93 \ln\left[\frac{7000 - x}{6999x}\right].$$

(a) How many days did it take for 6000 people to become infected?
(b) After two weeks, how many people were infected?

106. *Approximating logarithmic functions by polynomials.* For each positive integer n, let f_n be the polynomial function whose rule is

$$f_n(x) = x - \frac{x^2}{2} + \frac{x^3}{3} - \frac{x^4}{4} + \frac{x^5}{5} - \cdots \pm \frac{x^n}{n}$$

where the sign of the last term is + if n is odd and − if n is even. In the viewing window with $-1 \le x \le 1$ and $-4 \le y \le 1$, graph $g(x) = \ln(1+x)$ and $f_4(x)$ on the same screen. For what values of x does f_4 appear to be a good approximation of g?

107. Using the viewing window in Exercise 106, find a value of n for which the graph of the function f_n (as defined in Exercise 106) appears to coincide with the graph of $g(x) = \ln(1+x)$. Use the trace feature to move from graph to graph to see how good this approximation actually is.

108. The perceived loudness L of a sound of intensity I is given by $L = k \cdot \ln I$, where k is a certain constant. By how much must the intensity be increased to double the loudness? (That is, what must be done to I to produce $2L$?)

109. A bicycle store finds that the number N of bikes sold is related to the number d of dollars spent on advertising by $N = 51 + 100 \cdot \ln(d/100 + 2)$.

(a) How many bikes will be sold if nothing is spent on advertising? If \$1000 is spent? If \$10,000 is spent?

(b) If the average profit is \$25 per bike, is it worthwhile to spend \$1000 on advertising? What about \$10,000?

(c) What are the answers in part (b) if the average profit per bike is \$35?

110. The number N of days of training needed for a factory worker to produce x tools per day is given by

$$N = -25 \cdot \ln\left(1 - \frac{x}{60}\right).$$

(a) How many training days are needed for the worker to be able to produce 40 tools a day?

(b) It costs \$135 to train 1 worker for 1 day. If the profit on 1 tool is \$1.85, how many work days does it take before the factory breaks even on the training costs for a worker who can produce 40 tools a day?

5.3.A *Excursion* LOGARITHMIC FUNCTIONS TO OTHER BASES*

Common and natural logarithms were defined by considering the inverse functions of the exponential functions $f(x) = 10^x$ and $f(x) = e^x$. We now show that a similar procedure can be carried out with any positive number b in place of 10 and e.

Throughout this excursion, b is a fixed positive number with $b > 1$.**

The exponential function $f(x) = b^x$, whose graph is shown in Figure 5–19, is an increasing function and hence is one-to-one (as explained on page 150). Therefore, f has an inverse function g whose graph is the reflection of the graph of f in the line $y = x$ (see page 154), as shown in Figure 5–20.

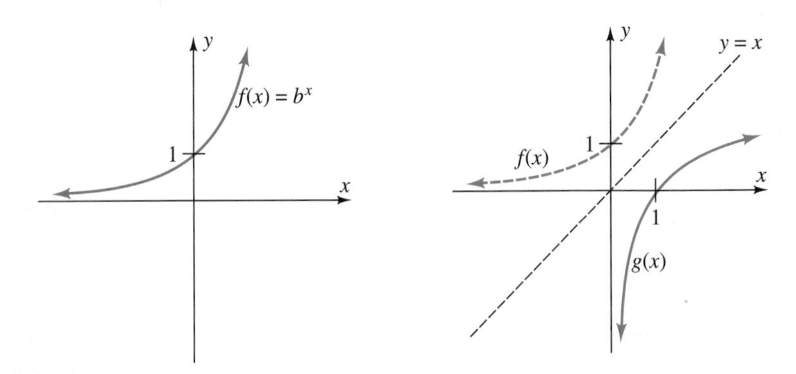

Figure 5-19

Figure 5-20

*This material is not needed in the sequel and may be read before Section 5.3 if desired. Section 2.7 (Inverse Functions) is a prerequisite for this section, which replicates the discussion of Section 5.3 in a more general context.

**The discussion is also valid when $0 < b < 1$, but in that case the graphs have a different shape.

This inverse function g is called the **logarithmic function to the base *b*.** The value $g(x)$ of this function at a number x is denoted $\log_b x$ and called the **logarithm to the base *b*** of the number x.

Although many properties of the logarithmic function $g(x) = \log_b x$ can be read from its graph, we also need an algebraic description of this function. On page 151 we saw that the relationship between a function f and its inverse function g is given by:

$$g(v) = u \qquad \text{exactly when} \qquad f(u) = v.$$

When $f(x) = b^x$ and $g(x) = \log_b x$, this statement says

$$\mathbf{\log_b v = u \qquad \text{exactly when} \qquad b^u = v.}$$

In other words, the number $\log_b v$ is the answer to the question

To what exponent must *b* be raised to produce *v*?

EXAMPLE 1 To find $\log_2 16$, ask yourself "what power of 2 equals 16?" Since $2^4 = 16$, we see that $\log_2 16 = 4$. Similarly, $\log_2(1/8) = -3$ because $2^{-3} = 1/8$. ■

EXAMPLE 2 Since logarithms are just exponents, every logarithmic statement can be translated into exponential language:

Logarithmic Statement	Equivalent Exponential Statement
$\log_3 81 = 4$	$3^4 = 81$
$\log_4 64 = 3$	$4^3 = 64$
$\log_{125} 5 = \frac{1}{3}$	$125^{1/3} = 5$*
$\log_8\left(\frac{1}{4}\right) = -\frac{2}{3}$	$8^{-2/3} = \frac{1}{4}$ (verify!) ■

EXAMPLE 3 The equation $\log_5 x = 3$ is equivalent to the exponential statement $5^3 = x$, so the solution is $x = 125$. ■

EXAMPLE 4 Logarithms to the base 10 are called **common logarithms.** It is customary to write $\log v$ instead of $\log_{10} v$. Then,

$$\log 100 = 2 \qquad \text{because} \qquad 10^2 = 100;$$

$$\log .001 = -3 \qquad \text{because} \qquad 10^{-3} = \frac{1}{10^3} = \frac{1}{1000} = .001.$$

*Because $125^{1/3} = \sqrt[3]{125} = 5$.

Scientific calculators have a LOG key for evaluating common logarithms. For instance,*

$$\log .4 = -0.3979, \qquad \log 45.3 = 1.6561, \qquad \log 685 = 2.8357. \quad ■$$

EXAMPLE 5 The most frequently used base for logarithms in modern applications is the number e ($\approx 2.71828 \cdots$). Logarithms to the base e are called **natural logarithms** and use a different notation: We write $\ln v$ instead of $\log_e v$. Scientific calculators also have an LN key for evaluating natural logarithms. For example,

$$\ln .5 = -0.6931, \qquad \ln 65 = 4.1744, \qquad \ln 158 = 5.0626. \quad ■$$

You *don't* need a calculator to understand the essential properties of logarithms. You need only translate logarithmic statements into exponential ones (or vice versa).

EXAMPLE 6 What is $\log(-25)$? *Translation:* To what power must 10 be raised to produce -25? The graph of $f(x) = 10^x$ lies entirely above the x-axis (see Figure 5–15 on page 332 or use your calculator), which means that *every* power of 10 is *positive.* So 10^x can *never* be -25, or any negative number, or zero. The same argument works for any base b:

$$\mathbf{\log_b v \text{ is defined only when } v > 0.} \quad ■$$

EXAMPLE 7 What is $\log_5 1$? *Translation:* To what power must 5 be raised to produce 1? The answer, of course, is $5^0 = 1$. So $\log_5 1 = 0$. Similarly, $\log_5 5 = 1$ because 1 is the answer to "what power of 5 equals 5?" In general,

$$\mathbf{\log_b 1 = 0 \qquad \text{and} \qquad \log_b b = 1.} \quad ■$$

EXAMPLE 8 What is $\log_2 2^9$? *Translation:* To what power must 2 be raised to produce 2^9? Obviously, the answer is 9. So $\log_2 2^9 = 9$ and, in general,

$$\mathbf{\log_b b^k = k \qquad \text{for every real number } k.}$$

This property holds even when k is a complicated expression. For instance, if x and y are positive, then

$$\log_6 6^{\sqrt{3x+y}} = \sqrt{3x + y} \qquad (\text{here } k = \sqrt{3x + y}). \quad ■$$

*Here and below, all logarithms are rounded to four decimal places. So strictly speaking, the equal sign should be replaced by an "approximately equal" sign ($\approx$).

EXAMPLE 9 What is $10^{\log 439}$? Well, log 439 is the power to which 10 must be raised to produce 439, that is, $10^{\log 439} = 439$. Similarly,

$$b^{\log_b v} = v \qquad \text{for every } v > 0. \quad ■$$

Here is a summary of the facts illustrated in the preceding examples.

Properties of Logarithms ▶

1. **$\log_b v$ is defined only when $v > 0$.**
2. **$\log_b 1 = 0$ and $\log_b b = 1$.**
3. **$\log_b(b^k) = k$ for every real number k.**
4. **$b^{\log_b v} = v$ for every $v > 0$.**

Property 1 is simply a restatement of the fact that the domain of the logarithmic function $g(x) = \log_b x$ is the set of all positive real numbers, as can be seen from the fact that its graph (Figure 5–20 on page 344) lies entirely to the right of the y-axis. Properties 3 and 4 are a restatement of the fact that the exponential function $f(x) = b^x$ and its inverse function $g(x) = \log_b x$ have the usual "reversal properties" (see page 152):

$$(g \circ f)(x) = g(f(x)) = g(b^x) = \log_b b^x = x \qquad \text{for all } x;$$

$$(f \circ g)(x) = f(g(x)) = f(\log_b x) = b^{\log_b x} = x \qquad \text{for all } x > 0.^*$$

Logarithm Laws

The first law of exponents states that $b^m b^n = b^{m+n}$, or in words,

The exponent of a product is the sum of the exponents of the factors.

Since logarithms are just particular kinds of exponents, this statement translates as:

The logarithm of a product is the sum of the logarithms of the factors.

The second and third laws of exponents, namely, $b^m/b^n = b^{m-n}$ and $(b^m)^k = b^{mk}$, can also be translated into logarithmic language:

Logarithm Laws ▶

Let b, v, w, k be real numbers, with b, v, w positive and $b \neq 1$.

Product Law: $\log_b(vw) = \log_b v + \log_b w$.

Quotient Law: $\log_b\left(\frac{v}{w}\right) = \log_b v - \log_b w$.

Power Law: $\log_b(v^k) = k(\log_b v)$.

*A calculator provides a visual demonstration of these facts for base 10 logarithms. For instance, if you key in LOG 10^x 52 ENTER the calculator displays 52 (that is, $g(f(52)) = 52$). Similarly, keying in 10^x LOG 167 ENTER produces 167 (that is, $f(g(167)) = 167$).

Proof According to Property 4 in the box at the top of page 347,

$$b^{\log_b v} = v \qquad \text{and} \qquad b^{\log_b w} = w.$$

Therefore, by the first law of exponents (with $m = \log_b v$ and $n = \log_b w$):

$$vw = b^{\log_b v} b^{\log_b w} = b^{\log_b v + \log_b w}.$$

So raising b to the exponent $(\log_b v + \log_b w)$ produces vw. But the definition of logarithm says that $\log_b vw$ is the exponent to which b must be raised to produce vw. Therefore we must have $\log_b vw = \log_b v + \log_b w$. This completes the proof of the Product Law.

Similarly, by the second law of exponents (with $m = \log_b v$ and $n = \log_b w$) we have:

$$\frac{v}{w} = \frac{b^{\log_b v}}{b^{\log_b w}} = b^{\log_b v - \log_b w}.$$

Since $\log_b (v/w)$ is the exponent to which b must be raised to produce v/w, we must have $\log_b (v/w) = \log_b v - \log_b w$. This proves the Quotient Law. The Power Law is proved in a similar fashion. ❑

EXAMPLE 10 Given that $\log_7 2 = .3562$, $\log_7 3 = .5646$, and $\log_7 5 = .8271$, find: **(a)** $\log_7 10$; **(b)** $\log_7 2.5$; **(c)** $\log_7 48$.

Solution

(a) By the Product Law,

$$\log_7 10 = \log_7 (2 \cdot 5) = \log_7 2 + \log_7 5 = .3562 + .8271 = 1.1833.$$

(b) By the Quotient Law,

$$\log_7 2.5 = \log_7 \left(\frac{5}{2}\right) = \log_7 5 - \log_7 2 = .8271 - .3562 = .4709.$$

(c) By the Product and Power Laws,

$$\begin{aligned}\log_7 48 = \log_7 (3 \cdot 16) &= \log_7 3 + \log_7 16 = \log_7 3 + \log_7 2^4 \\ &= \log_7 3 + 4 \cdot \log_7 2 = .5646 + 4(.3562) \\ &= 1.9894.\end{aligned}$$ ■

WARNING

1. A common error in using the Product Law is to write something like $\log 6 + \log 7 = \log(6 + 7) = \log 13$ instead of the correct statement $\log 6 + \log 7 = \log(6 \cdot 7) = \log 42$.

2. Do not confuse $\log_b\left(\frac{v}{w}\right)$ with the quotient $\frac{\log_b v}{\log_b w}$. They are *different* numbers. For example, when $b = 10$

$$\log\left(\frac{48}{4}\right) = \log 12 = 1.0792, \text{ but } \frac{\log 48}{\log 4} = \frac{1.6812}{0.6021} = 2.7922.$$

For graphic illustrations of the errors mentioned in the Warning, see Exercises 76 and 77.

Example 10 worked because we were *given* several logarithms to base 7. But there's no $\log_7$ key on the calculator, so how do you find logarithms to base 7, or to any base other than e or 10? *Answer:* Use the LN key on the calculator and the following formula:

Change of Base Formula ▶

For any positive number v,

$$\log_b v = \frac{\ln v}{\ln b} = \frac{1}{\ln b} \cdot \ln v.$$

Proof By Property 4 in the box on page 347, $b^{\log_b v} = v$. Take the natural logarithm of each side of this equation:

$$\ln(b^{\log_b v}) = \ln v.$$

Apply the Power Law for natural logarithms on the left side:

$$(\log_b v)(\ln b) = \ln v.$$

Dividing both sides by $\ln b$ finishes the proof:

$$\log_b v = \frac{\ln v}{\ln b}. \quad \square$$

EXAMPLE 11 To find $\log_7 3$, apply the change of base formula with $b = 7$:

$$\log_7 3 = \frac{\ln 3}{\ln 7} = \frac{1.0986}{1.9459} = .5646. \quad ■$$

EXERCISES 5.3.A

Note: Unless stated otherwise, all letters represent positive numbers and $b \neq 1$.

In Exercises 1–8, fill in the missing entries in each table.

1.

x	0	1	2	4
$f(x) = \log_4 x$				

2.

x	1/25	5	25	$\sqrt{5}$
$g(x) = \log_5 x$				

3.

x		1/6	1	216
$h(x) = \log_6 x$	-2			

4.

x	10/3	4	6	12
$k(x) = \log_3(x - 3)$				

5.

x	0	1/7	$\sqrt{7}$	49
$f(x) = 2\log_7 x$				

6.

x			100	1000
$g(x) = 3\log x$	6	3		

7.

x	-2.75	-1	1	29
$h(x) = 3\log_2(x + 3)$				

8.

x	$1/e$	1	e	e^2
$k(x) = 2\ln x$				

In Exercises 9–18, translate the given exponential statement into an equivalent logarithmic one.

9. $10^{-2} = .01$ **10.** $10^3 = 1000$

11. $\sqrt[3]{10} = 10^{1/3}$ **12.** $10^{.4771} \approx 3$

13. $10^{7k} = r$ **14.** $10^{(a+b)} = c$

15. $7^8 = 5{,}764{,}801$ **16.** $2^{-3} = 1/8$

17. $3^{-2} = 1/9$ **18.** $b^{14} = 3379$

In Exercises 19–28, translate the given logarithmic statement into an equivalent exponential one.

19. $\log 10{,}000 = 4$ **20.** $\log .001 = -3$

21. $\log 750 \approx 2.88$ **22.** $\log(.8) = -.097$

23. $\log_5 125 = 3$ **24.** $\log_8(1/4) = -2/3$

25. $\log_2(1/4) = -2$ **26.** $\log_2\sqrt{2} = 1/2$

27. $\log(x^2 + 2y) = z + w$ **28.** $\log(a + c) = d$

In Exercises 29–36, evaluate the given expression without using a calculator.

29. $\log 10^{\sqrt{43}}$ **30.** $\log_{17}(17^{17})$

31. $\log 10^{\sqrt{x^2+y^2}}$ **32.** $\log_{3.5}(3.5^{(x^2-1)})$

33. $\log_{16}4$ **34.** $\log_2 64$

35. $\log_{\sqrt{3}}(27)$ **36.** $\log_{\sqrt{3}}(1/9)$

In Exercises 37–40, a graph or a table of values for the function $f(x) = \log_b x$ is given. Find b.

37.

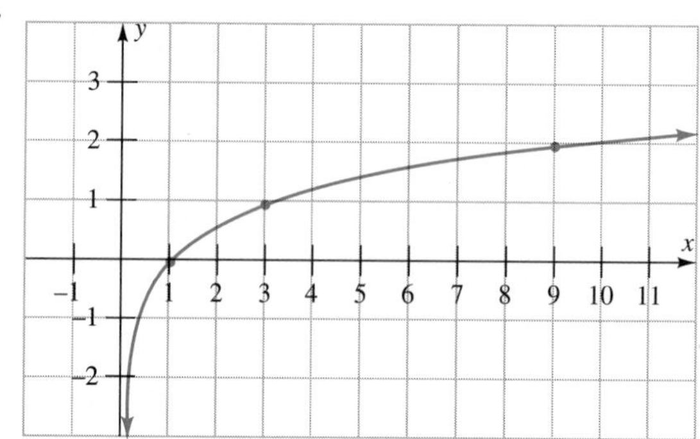

38.

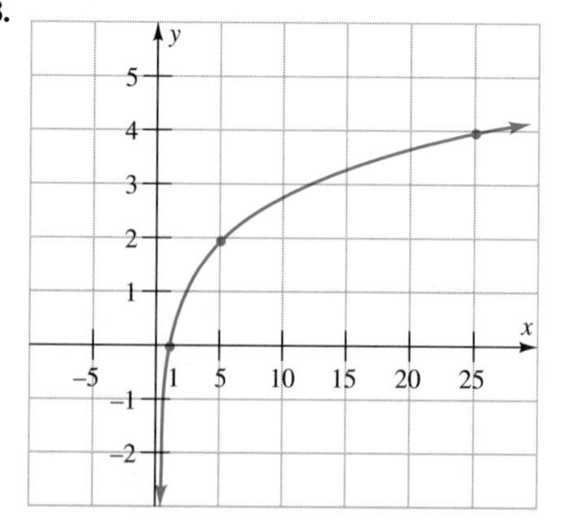

39.

x	.05	1	400	$2\sqrt{5}$
$f(x)$	-1	0	2	1/2

40.

x	1/25	1	5	125
$f(x)$	-4	0	2	6

In Exercises 41–46, find x.

41. $\log_3 243 = x$ **42.** $\log_{81} 27 = x$ **43.** $\log_{27} x = 1/3$

44. $\log_5 x = -4$ **45.** $\log_x 64 = 3$

46. $\log_x(1/9) = -2/3$

In Exercises 47–54, b is a positive number such that $\log_b 2 = .13$, $\log_b 3 = .2$, and $\log_b 5 = .3$. Find

47. $\log_b 10$ **48.** $\log_b 15$ **49.** $\log_b 4$

50. $\log_b 27$ **51.** $\log_b(5/3)$ **52.** $\log_b 48$

53. $\log_b 45$ **54.** $\log_b 36$

In Exercises 55–62, use a calculator and the change of base formula to find the logarithm.

55. $\log_2 10$ **56.** $\log_2 22$ **57.** $\log_7 5$

58. $\log_5 7$ **59.** $\log_{500} 1000$ **60.** $\log_{500} 250$

61. $\log_{12} 56$ **62.** $\log_{12} 725$

In Exercises 63–68, answer true *or* false *and give reasons for your answer.*

63. $\log_b(r/5) = \log_b r - \log_b 5$

64. $\dfrac{\log_b a}{\log_b c} = \log_b\left(\dfrac{a}{c}\right)$

65. $(\log_b r)/t = \log_b(r^{1/t})$

66. $\log_b(cd) = \log_b c + \log_b d$

67. $\log_5(5x) = 5(\log_5 x)$

68. $\log_b(ab)^t = t(\log_b a) + t$

69. Which is larger: 397^{398} or 398^{397}? [*Hint*: $\log 397 \approx 2.5988$ and $\log 398 \approx 2.5999$ and $f(x) = 10^x$ is an increasing function.]

70. If $\log_b 9.21 = 7.4$ and $\log_b 359.62 = 19.61$, then what is $\log_b 359.62/\log_b 9.21$?

In Exercises 71–74, assume that a and b are positive with $a \neq 1$ and $b \neq 1$.

71. Express $\log_b u$ in terms of logarithms to the base a.

72. Show that $\log_b a = 1/\log_a b$.

73. How are $\log_{10} u$ and $\log_{100} u$ related?

74. Show that $a^{\log b} = b^{\log a}$.

75. If $\log_b x = \dfrac{1}{2}\log_b v + 3$, show that $x = (b^3)\sqrt{v}$.

76. Graph the functions $f(x) = \log x + \log 7$ and $g(x) = \log(x + 7)$ on the same screen. For what values of x is it true that $f(x) = g(x)$? What do you conclude about the statement "$\log 6 + \log 7 = \log(6 + 7)$"?

77. Graph the functions $f(x) = \log(x/4)$ and $g(x) = (\log x)/(\log 4)$. Are they the same? What does this say about a statement such as "$\log\left(\dfrac{48}{4}\right) = \dfrac{\log 48}{\log 4}$"?

In Exercises 78–80, sketch a complete graph of the function, labeling any holes, asymptotes, or local extrema.

78. $f(x) = \log_5 x + 2$ **79.** $h(x) = x\log x^2$

80. $g(x) = \log_{20} x^2$

5.4 ALGEBRAIC SOLUTIONS OF EXPONENTIAL AND LOGARITHMIC EQUATIONS

Earlier in this chapter, many exponential and logarithmic equations were solved by graphical means. Most of them could also have been solved algebraically by the techniques presented in this section, which depend primarily on the properties of logarithms.

The easiest exponential equations to solve are those in which both sides are powers of the same base.

EXAMPLE 1 The equation $8^x = 2^{x+1}$ can be rewritten $(2^3)^x = 2^{x+1}$ or, equivalently, $2^{3x} = 2^{x+1}$. Since the powers of 2 are the same, the exponents must be equal, that is,

$$3x = x + 1$$

$$x = \frac{1}{2}. \quad \blacksquare$$

When different bases are involved in an exponential equation, the solution technique is to take the logarithm of each side and then apply the Power Law for Logarithms.*

EXAMPLE 2 To solve $5^x = 2$,

Take logarithms on each side: $\ln 5^x = \ln 2$

Use the Power Law: $x(\ln 5) = \ln 2$

Divide both sides by ln 5: $x = \dfrac{\ln 2}{\ln 5} \approx \dfrac{.6931}{1.6094} \approx .4307.$

Remember: $\dfrac{\ln 2}{\ln 5}$ is *not* $\ln \dfrac{2}{5}$ or $\ln 2 - \ln 5$. $\blacksquare$

EXAMPLE 3 To solve $2^{4x-1} = 3^{1-x}$,

Take logarithms of each side: $\ln 2^{4x-1} = \ln 3^{1-x}$

Use the Power Law: $(4x - 1)(\ln 2) = (1 - x)(\ln 3)$

Multiply out both sides: $4x(\ln 2) - \ln 2 = \ln 3 - x(\ln 3)$

Rearrange terms: $4x(\ln 2) + x(\ln 3) = \ln 2 + \ln 3$

Factor left side: $(4 \cdot \ln 2 + \ln 3)x = \ln 2 + \ln 3$

Divide both sides by $(4 \cdot \ln 2 + \ln 3)$: $x = \dfrac{\ln 2 + \ln 3}{4 \cdot \ln 2 + \ln 3} \approx .4628. \quad \blacksquare$

EXAMPLE 4 To solve $e^x - e^{-x} = 4$, we first multiply both sides by e^x. Since $e^x > 0$ for every x, we get an equivalent equation:

$$e^x e^x - e^{-x} e^x = 4e^x$$

$$(e^x)^2 - 4e^x - 1 = 0$$

*We shall use natural logarithms, but the same techniques are valid for logarithms to other bases (Exercise 34).

Let $u = e^x$ so that the equation becomes

$$u^2 - 4u - 1 = 0$$

The solutions are given by the quadratic formula:

$$u = \frac{-(-4) \pm \sqrt{(-4)^2 - 4 \cdot 1 \cdot (-1)}}{2 \cdot 1} = \frac{4 \pm \sqrt{20}}{2} = \frac{4 \pm 2\sqrt{5}}{2} = 2 \pm \sqrt{5}.$$

Since $u = e^x$, we have

$$e^x = 2 - \sqrt{5} \qquad \text{or} \qquad e^x = 2 + \sqrt{5}.$$

But e^x is always positive and $2 - \sqrt{5} < 0$, so the first equation has no solutions. The second can be solved as above:

$$\begin{aligned} \ln e^x &= \ln(2 + \sqrt{5}) \\ x(\ln e) &= \ln(2 + \sqrt{5}) \\ x(1) &= \ln(2 + \sqrt{5}) && [\ln e = 1] \\ x &= \ln(2 + \sqrt{5}) \approx 1.4436. \end{aligned}$$

■

Applications of Exponential Equations

EXAMPLE 5 *(Radiocarbon Dating)* When a living organism dies, its carbon-14 decays. The half-life of carbon-14 is 5730 years. As explained before Example 6 on page 319, the amount left at time t is given by $M(t) = c(.5)^{t/5730}$, where c is the mass of carbon-14 that was present initially.

The skeleton of a mastodon has lost 58% of its original carbon-14.* When did the mastodon die?

Solution Time is measured from the death of the mastodon. The present mass of carbon-14 is $.58c$ less than the original mass c. So the present value of $M(t)$ is $c - .58c = .42c$ and we have:

$$\begin{aligned} M(t) &= c(.5)^{t/5730} \\ .42c &= c(.5)^{t/5730} \\ .42 &= (.5)^{t/5730} \end{aligned}$$

The solution of this equation is the time elapsed from the mastodon's death to the present. It can be solved as above:

$$\begin{aligned} \ln .42 &= \ln(.5)^{t/5730} \\ \ln .42 &= \frac{t}{5730}(\ln .5) \end{aligned}$$

*Archeologists can determine how much carbon-14 has been lost by a technique that involves measuring the ratio of carbon-14 to carbon-12 in the skeleton.

$$5730(\ln .42) = t(\ln .5)$$

$$t = \frac{5730(\ln .42)}{\ln .5} \approx 7171.32$$

Therefore, the mastodon died approximately 7200 years ago. ■

EXAMPLE 6 *(Compound Interest)** \$3000 is to be invested at 8% per year, compounded quarterly. In how many years will the investment be worth \$10,680?

Solution The interest rate per quarter is .08/4 = .02. The compound interest formula (page 326) shows that the value of the investment after t quarters is $A = 3000(1 + .02)^t = 3000(1.02)^t$. So we must solve the equation

$$3000(1.02)^t = 10{,}680$$

$$1.02^t = \frac{10{,}680}{3000} = 3.56$$

$$\ln 1.02^t = \ln 3.56$$

$$t(\ln 1.02) = \ln 3.56$$

$$t = \frac{\ln 3.56}{\ln 1.02} \approx 64.12 \text{ quarters}$$

Therefore, it will take $\frac{64.12}{4} = 16.03$ years. ■

EXAMPLE 7 *(Population Growth)* A biologist knows that if there are no inhibiting or stimulating factors, a certain type of bacteria will continuously increase exponentially, with the population at time t given by a function of the form $S(t) = Pe^{kt}$, where P is the original population and k is the continuous growth rate.** The biologist has a culture that contains 1000 bacteria. Seven hours later there are 5000 bacteria.

(a) Find the rule of the population function S.

(b) When will the bacteria population reach one billion?

Solution

(a) The original population is $P = 1000$, so the population function is $S(t) = 1000e^{kt}$. To determine the growth rate k we use the fact that $S(7) = 5000$, that is,

$$1000e^{k7} = 5000.$$

*Skip this example if you haven't read Excursion 5.2.A.

**The biologist has read the end of Excursion 5.2.A, where continuous growth is discussed.

This equation can be solved for k as above; first, divide both sides by 1000:

$$\begin{aligned} e^{7k} &= 5 \\ \ln e^{7k} &= \ln 5 \\ 7k \cdot \ln e &= \ln 5 \\ k &= \frac{\ln 5}{7 \cdot \ln e} = \frac{\ln 5}{7 \cdot 1} \approx .2299. \qquad [\ln e = 1] \end{aligned}$$

Therefore, the population function is $S(t) = 1000e^{.2299t}$.

(b) To find the value of t for which $S(t)$ is one billion, we must solve:

$$\begin{aligned} 1000e^{.2299t} &= 1{,}000{,}000{,}000 \\ e^{.2299t} &= 1{,}000{,}000 \\ \ln e^{.2299t} &= \ln 1{,}000{,}000 \\ .2299t \cdot \ln e &= \ln 1{,}000{,}000 \\ t &= \frac{\ln 1{,}000{,}000}{.2299 \cdot \ln e} = \frac{\ln 1{,}000{,}000}{.2299} \approx 60.09 \text{ hours.} \quad ■ \end{aligned}$$

EXAMPLE 8 *(Inhibited Population Growth)* The population of fish in a lake at time t months is given by the function

$$p(t) = \frac{20{,}000}{1 + 24e^{-t/4}}.$$

How long will it take for the population to reach 15,000?

Solution We must solve this equation for t:

$$\begin{aligned} 15{,}000 &= \frac{20{,}000}{1 + 24e^{-t/4}} \\ 15{,}000(1 + 24e^{-t/4}) &= 20{,}000 \\ 1 + 24e^{-t/4} &= \frac{20{,}000}{15{,}000} = \frac{4}{3} \\ 24e^{-t/4} &= \frac{1}{3} \\ e^{-t/4} &= \frac{1}{3} \cdot \frac{1}{24} = \frac{1}{72} \\ \ln e^{-t/4} &= \ln\left(\frac{1}{72}\right) \end{aligned}$$

$$\left(-\frac{t}{4}\right)(\ln e) = \ln 1 - \ln 72$$

$$-\frac{t}{4} = -\ln 72 \qquad [\ln e = 1 \text{ and } \ln 1 = 0]$$

$$t = 4(\ln 72) \approx 17.1067.$$

So the population reaches 15,000 in a little over 17 months. ■

Logarithmic Equations

Whenever $\ln u = \ln v$, the properties of logarithms show that $u = e^{\ln u} = e^{\ln v} = v$. This fact is useful for solving certain logarithmic equations.

EXAMPLE 9 To solve $\ln(x - 3) + \ln(2x + 1) = 2 \cdot \ln x$, use the Product Law on the left side and the Power Law on the right side:

$$\ln[(x - 3)(2x + 1)] = \ln x^2$$

$$\ln(2x^2 - 5x - 3) = \ln x^2$$

Since the logarithms are equal, we must have:

$$2x^2 - 5x - 3 = x^2$$

$$x^2 - 5x - 3 = 0$$

$$x = \frac{-(-5) \pm \sqrt{(-5)^2 - 4 \cdot 1 \cdot (-3)}}{2 \cdot 1} = \frac{5 \pm \sqrt{37}}{2}$$

But $x - 3$ is negative when $x = (5 - \sqrt{37})/2$, so $\ln(x - 3)$ is not defined in that case. Therefore the only solution of the original equation is $x = (5 + \sqrt{37})/2 \approx 5.5414$. ■

A different technique is needed when some terms of the equation involve logarithms and others don't.

EXAMPLE 10 To solve $\ln(3x + 2) - \ln x = 2$, apply the Quotient Law on the left side:

$$\ln \frac{3x + 2}{x} = 2$$

Now use the fact that $\ln v = u$ means $e^u = v$ $\left(\text{with } v = \frac{3x + 2}{x} \text{ and } u = 2\right)$ to rewrite the equation in exponential form:

$$e^2 = \frac{3x + 2}{x}$$

$$e^2 x = 3x + 2$$

$$e^2x - 3x = 2$$
$$(e^2 - 3)x = 2$$
$$x = \frac{2}{e^2 - 3} \approx .4557. \quad ■$$

EXAMPLE 11 To solve $\log(x - 8) + \log(x - 9) = 2$, we apply the Product Law to obtain the equivalent equation

$$\log(x - 8)(x - 9) = 2$$
$$\log(x^2 - 17x + 72) = 2$$

Now rewrite this equation in exponential form by using the fact that $\log v = u$ means $10^u = v$):

$$10^2 = x^2 - 17x + 72$$
$$100 = x^2 - 17x + 72$$
$$x^2 - 17x - 28 = 0$$

The quadratic formula shows that

$$x = \frac{17 \pm \sqrt{17^2 - 4 \cdot 1(-28)}}{2} = \frac{17 \pm \sqrt{401}}{2} \approx \begin{cases} 18.5125 \\ -1.5125 \end{cases}$$

But $x - 8$ and $x - 9$ are negative when $x = -1.5125$, so their logarithms are not defined. Therefore $x \approx 18.5125$ is the only solution of the original equation. ■

EXERCISES 5.4

In Exercises 1–8, solve the equation without using logarithms.

1. $3^x = 81$ **2.** $3^x + 3 = 30$ **3.** $3^{x+1} = 9^{5x}$

4. $4^{5x} = 16^{2x-1}$ **5.** $3^{5x}9^{x^2} = 27$

6. $2^{x^2+5x} = 1/16$ **7.** $9^{x^2} = 3^{-5x-2}$

8. $4^{x^2-1} = 8^x$

In Exercises 9–30, solve the equation. Express your answer, in terms of natural logarithms (for instance, $x = (2 + \ln 5)/(\ln 3)$). Then use a calculator to find an approximation for the answer.

9. $3^x = 5$ **10.** $5^x = 4$ **11.** $2^x = 3^{x-1}$

12. $4^{x+2} = 2^{x-1}$ **13.** $3^{1-2x} = 5^{x+5}$

14. $4^{3x-1} = 3^{x-2}$ **15.** $2^{1-3x} = 3^{x+1}$

16. $3^{z+3} = 2^z$ **17.** $e^{2x} = 5$

18. $e^{-3x} = 2$ **19.** $6e^{-1.4x} = 21$

20. $3.4e^{-x/3} = 5.6$ **21.** $2.1e^{(x/2)\ln 3} = 5$

22. $7.8e^{(x/3)\ln 5} = 14$

23. $9^x - 4 \cdot 3^x + 3 = 0$ [*Hint*: Note that $9^x = (3^x)^2$; let $u = 3^x$.]

24. $4^x - 6 \cdot 2^x = -8$

25. $e^{2x} - 5e^x + 6 = 0$ [*Hint*: Let $u = e^x$.]

26. $2e^{2x} - 9e^x + 4 = 0$ **27.** $6e^{2x} - 16e^x = 6$

28. $8e^{2x} + 8e^x = 6$ **29.** $4^x + 6 \cdot 4^{-x} = 5$

30. $5^x + 3 = 10 \cdot 5^{-x}$

In Exercises 31–33, solve the equation for x.

31. $\dfrac{e^x - e^{-x}}{2} = t$ **32.** $\dfrac{e^x + e^{-x}}{e^x - e^{-x}} = t$

33. $\dfrac{e^x - e^{-x}}{e^x + e^{-x}} = t$

34. **(a)** Solve $7^x = 3$, using natural logarithms. Leave your answer in logarithmic form; don't approximate with a calculator.
(b) Solve $7^x = 3$, using common (base 10) logarithms. Leave your answer in logarithmic form.
(c) Use the change of base formula in Excursion 5.3.A to show that your answers in parts (a) and (b) are the same.

In Exercises 35–44, solve the equation as in Example 9.

35. $\ln(3x - 5) = \ln 11 + \ln 2$

36. $\log(4x - 1) = \log(x + 1) + \log 2$

37. $\log(3x - 1) + \log 2 = \log 4 + \log(x + 2)$

38. $\ln(x + 6) - \ln 10 = \ln(x - 1) - \ln 2$

39. $2 \ln x = \ln 36$

40. $2 \log x = 3 \log 4$

41. $\ln x + \ln(x + 1) = \ln 3 + \ln 4$

42. $\ln(6x - 1) + \ln x = \frac{1}{2}\ln 4$

43. $\ln x = \ln 3 - \ln(x + 5)$

44. $\ln(2x + 3) + \ln x = \ln e$

In Exercises 45–52, solve the equation.

45. $\ln(x + 9) - \ln x = 1$

46. $\ln(2x + 1) - 1 = \ln(x - 2)$

47. $\log x + \log(x - 3) = 1$ [*Remember*: $\log v = u$ means $10^u = v$.]

48. $\log(x - 1) + \log(x + 2) = 1$

49. $\log\sqrt{x^2 - 1} = 2$

50. $\log\sqrt[3]{x^2 + 21x} = 2/3$

51. $\ln(x^2 + 1) - \ln(x - 1) = 1 + \ln(x + 1)$

52. $\dfrac{\ln(x + 1)}{\ln(x - 1)} = 2$

Exercises 53–62 deal with the half-life function $M(x) = c(.5)^{x/h}$, which was discussed on page 319 and used in Example 5 of this section.

53. How old is a piece of ivory that has lost 36% of its carbon-14?

54. How old is a mummy that has lost 49% of its carbon-14?

55. A Native American mummy was found recently. If it has lost 26.4% of its carbon-14, approximately how long ago did the Native American die?

56. How old is a wooden statue that has only one third of its original carbon-14?

57. A quantity of uranium decays to two thirds of its original mass in .26 billion years. Find the half-life of uranium.

58. A certain radioactive substance loses one third of its original mass in 5 days. Find its half-life.

59. Krypton-85 loses 6.44% of its mass each year. What is its half-life?

60. Strontium-90 loses 2.5% of its mass each year. What is its half-life?

61. The half-life of a certain substance is 3.6 days. How long will it take for 20 grams to decay to 3 grams?

62. The half-life of cobalt-60 is 4.945 years. How long will it take for 25 grams to decay to 15 grams?

Exercises 63–68 deal with the compound interest formula $P = (1 + r)^t$, which was discussed in Excursion 5.2.A and used in Example 6 of this section.

63. At what annual rate of interest should \$1000 be invested so that it will double in 10 years, if interest is compounded quarterly?

64. How long does it take \$500 to triple if it is invested at 6% compounded: **(a)** annually, **(b)** quarterly, **(c)** daily?

65. **(a)** How long will it take to triple your money if you invest \$500 at a rate of 5% per year compounded annually?
(b) How long will it take at 5% compounded quarterly?

66. At what rate of interest (compounded annually) should you invest \$500 if you want to have \$1500 in 12 years?

67. How much money should be invested at 5% interest, compounded quarterly, so that 9 years later the investment will be worth \$5000? This amount is called the **present value** of \$5000 at 5% interest.

68. Find a formula that gives the time needed for an investment of P dollars to double, if the interest rate is r% compounded annually. [*Hint*: Solve the compound interest formula for t, when $A = 2P$.]

Exercises 69–76 deal with functions of the form $f(x) = Pe^{kx}$, where k is the continuous exponential growth rate (see Example 7).

69. The present concentration of carbon dioxide in the atmosphere is 364 parts per million (ppm) and is increasing exponentially at a continuous yearly rate of .4%

(that is, $k = .004$). How many years will it take for the concentration to reach 500 ppm?

70. The amount P of ozone in the atmosphere is currently decaying exponentially each year at a continuous rate of ¼% (that is, $k = -.0025$). How long will it take for half the ozone to disappear (that is, when will the amount be $P/2$)? [Your answer is the half-life of ozone.]

71. The population of Brazil increased from 122 million in 1980 to 158 million in 1992.
(a) At what continuous rate was the population growing during this period?
(b) Assuming that Brazil's population continues to increase at this rate, when will it reach 250 million?

72. A colony of 1000 weevils grows exponentially to 1750 in one week.
(a) At what continuous rate is the population growing?
(b) How many weeks does it take for the weevil population to reach 3000?

73. The probability P percent of having an accident while driving a car is related to the alcohol level of the driver's blood by the formula $P = e^{kt}$, where k is a constant. Accident statistics show that the probability of an accident is 25% when the blood alcohol level is $t = .15$.
(a) Find k. [Use $P = 25$, not .25.]
(b) At what blood alcohol level is the probability of having an accident 50%?

74. Under normal conditions, the atmospheric pressure (in millibars) at height h feet above sea level is given by $P(h) = 1015e^{-kh}$, where k is a positive constant.
(a) If the pressure at 18,000 feet is half the pressure at sea level, find k.
(b) Using the information from part (a), find the atmospheric pressure at 1000 feet, 5000 feet, and 15,000 feet.

75. One hour after an experiment begins, the number of bacteria in a culture is 100. An hour later there are 500.
(a) Find the number of bacteria at the beginning of the experiment and the number 3 hours later.
(b) How long does it take the number of bacteria at any given time to double?

76. If the population at time t is given by $S(t) = ce^{kt}$, find a formula that gives the time it takes for the population to double.

77. The spread of a flu virus in a community of 45,000 people is given by the function $f(t) = \dfrac{45,000}{1 + 224e^{-.899t}}$, where $f(t)$ is the number of people infected in week t.
(a) How many people had the flu at the outbreak of the epidemic? After 3 weeks?
(b) When will half the town be infected?

78. The beaver population near a certain lake in year t is approximately $p(t) = \dfrac{2000}{1 + 199e^{-.5544t}}$.
(a) When will the beaver population reach 1000?
(b) Will the population ever reach 2000? Why?

Thinkers

79. According to one theory of learning, the number of words per minute N that a person can type after t weeks of practice is given by $N = c(1 - e^{-kt})$, where c is an upper limit that N cannot exceed and k is a constant that must be determined experimentally for each person.
(a) If a person can type 50 wpm (words per minute) after 4 weeks of practice and 70 wpm after 8 weeks, find the values of k and c for this person. According to the theory, this person will never type faster than c wpm.
(b) Another person can type 50 wpm after 4 weeks of practice and 90 wpm after 8 weeks. How many weeks must this person practice to be able to type 125 wpm?

80. Wendy has been offered two jobs, each with the same starting salary of $24,000 and identical benefits. Assuming satisfactory performance, she will receive a $1200 raise each year at the Great Gizmo Company, whereas the Wonder Widget Company will give her a 4% raise each year.
(a) In what year (after the first year) would her salary be the same at either company? Until then, which company pays better? After that, which company pays better?
(b) Answer the questions in part (a) assuming that the annual raise at Great Gizmo is $1800.

CHAPTER 5 *Review*

Important Concepts ▶

Important Facts and Formulas ▶

- *Laws of Exponents:*

$$c^r c^s = c^{r+s} \qquad (cd)^r = c^r d^r$$

$$\frac{c^r}{c^s} = c^{r-s} \qquad \left(\frac{c}{d}\right)^r = \frac{c^r}{d^r}$$

$$(c^r)^s = c^{rs} \qquad c^{-r} = \frac{1}{c^r}$$

- $g(x) = \log x$ is the inverse function of $f(x) = 10^x$:

$$10^{\log v} = v \text{ for all } v > 0 \qquad \text{and} \qquad \log 10^u = u \text{ for all } u.$$

- $g(x) = \ln x$ is the inverse function of $f(x) = e^x$:

$$e^{\ln v} = v \text{ for all } v > 0 \qquad \text{and} \qquad \ln(e^u) = u \text{ for all } u.$$

- $h(x) = \log_b x$ is the inverse function of $k(x) = b^x$:

$$b^{\log_b v} = v \text{ for all } v > 0 \qquad \text{and} \qquad \log_b(b^u) = u \text{ for all } u.$$

- *Logarithm Laws:* For all $v, w > 0$ and any k:

$$\ln(vw) = \ln v + \ln w \qquad \log_b(vw) = \log_b v + \log_b w$$

$$\ln\left(\frac{v}{w}\right) = \ln v - \ln w \qquad \log_b\left(\frac{v}{w}\right) = \log_b v - \log_b w$$

$$\ln(v^k) = k(\ln v) \qquad \log_b(v^k) = k(\log_b v)$$

- *Exponential Growth Functions:*

$$f(x) = P(1 + r)^x \quad (r > 0),$$
$$f(x) = Pa^x \quad (a > 1),$$
$$f(x) = Pe^{kx} \quad (k > 0)$$

- *Exponential Decay Functions:*

$$f(x) = P(1 - r)^x \quad (0 < r < 1),$$
$$f(x) = Pa^x \quad (0 < a < 1),$$
$$f(x) = Pe^{kx} \quad (k < 0)$$

- *Compound Interest Formula:* $A = P(1 + r)^t$
- *Change of Base Formula:* $\log_b v = \dfrac{\ln v}{\ln b}$

Review Questions ▶

In Questions 1–6, simplify the expression.

1. $\sqrt{\sqrt[3]{c^{12}}}$
2. $(\sqrt[3]{4c^3d^2})^3(c\sqrt{d})^2$
3. $(a^{-2/3}b^{2/5})(a^3b^6)^{4/3}$
4. $\dfrac{(3c)^{3/5}(2d)^{-2}(4c)^{1/2}}{(4c)^{1/5}(2d)^4(2c)^{-3/2}}$
5. $(u^{1/4} - v^{1/4})(u^{1/4} + v^{1/4})$
6. $c^{3/2}(2c^{1/2} + 3c^{-3/2})$

In Exercises 7 and 8, simplify and write the expression without radicals or negative exponents:

7. $\dfrac{\sqrt[3]{6c^4d^{14}}}{\sqrt[3]{48c^{-2}d^2}}$
8. $\dfrac{(8u^5)^{1/4}2^{-1}u^{-3}}{2u^8}$

In Questions 9–14, find all real solutions of the equation.

9. $\sqrt{x - 1} = 2 - x$
10. $\sqrt[3]{1 - t^2} = -2$
11. $\sqrt{x + 1} + \sqrt{x - 1} = 1$
12. $x^{1/4} - 5x^{1/2} + 6 = 0$
13. $\sqrt[3]{x^4 - 2x^3 + 6x - 7} = x + 3$
14. $x^{4/3} + x - 2x^{2/3} + x^{1/3} = 4$

In Questions 15 and 16, find a viewing window (or windows) that shows a complete graph of the function.

15. $f(x) = 2^{x^3-x-2}$ **16.** $g(x) = \ln\left(\frac{x}{x-2}\right)$

In Questions 17–20, sketch a complete graph of the function. Indicate all asymptotes clearly.

17. $g(x) = 2^x - 1$ **18.** $f(x) = 2^{x-1}$

19. $h(x) = \ln(x + 4) - 2$ **20.** $k(x) = \ln\left(\frac{x+4}{x}\right)$

In Questions 21–26, translate the given exponential statement into an equivalent logarithmic one.

21. $e^{6.628} = 756$ **22.** $e^{5.8972} = 364$

23. $e^{r^2-1} = u + v$ **24.** $e^{a-b} = c$

25. $10^{2.8785} = 756$ **26.** $10^{c+d} = t$

In Questions 27–32, translate the given logarithmic statement into an equivalent exponential one.

27. $\ln 1234 = 7.118$ **28.** $\ln(ax + b) = y$

29. $\ln(rs) = t$ **30.** $\log 1234 = 3.0913$

31. $\log_5(cd - k) = u$ **32.** $\log_d(uv) = w$

In Questions 33–36, evaluate the given expression without using a calculator.

33. $\ln e^3$ **34.** $\ln \sqrt[3]{e}$

35. $e^{\ln 3/4}$ **36.** $e^{\ln(x+2y)}$

37. Simplify: $3 \ln \sqrt{x} + (1/2)\ln x$

38. Simplify: $\ln(e^{4e})^{-1} + 4e$

39. Write as a single logarithm: $\ln 3x - 3 \ln x + \ln 3y$.

40. Write as a single logarithm: $4 \ln x - 2(\ln x^3 + 4 \ln x)$.

41. $\log(-.01) = ?$ **42.** $\log_{20}400 = ?$

43. Assume $\ln 5 = 1.6$ and $\ln 11 = 2.4$. Find $\frac{\ln 11}{\ln 5}$.

In Questions 44–48, assume $\log_b 2 = .30$ and $\log_b 3 = .48$ and find the logarithm.

44. $\log_b 6$ **45.** $\log_b(3/2)$

46. $\log_b 27$ **47.** $\log_b(b/2)$

48. $\frac{\log_b 3}{\log_b 4}$

49. Which of the following statements are *true?*
(a) $\ln 10 = (\ln 2)(\ln 5)$ **(b)** $\ln(e/6) = \ln e + \ln 6$
(c) $\ln(1/7) + \ln 7 = 0$ **(d)** $\ln(-e) = -1$
(e) none of them

50. Which of the following statements are *false?*
(a) $10(\log 5) = \log 50$ **(b)** $\log 100 + 3 = \log 10^5$
(c) $\log 1 = \ln 1$ **(d)** $\log 6/\log 3 = \log 2$
(e) all of them

Use the graphs in Figure 5–21 for Questions 51 and 52.

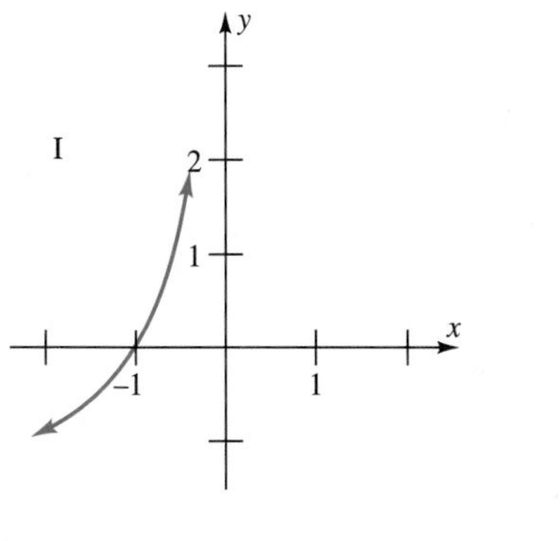

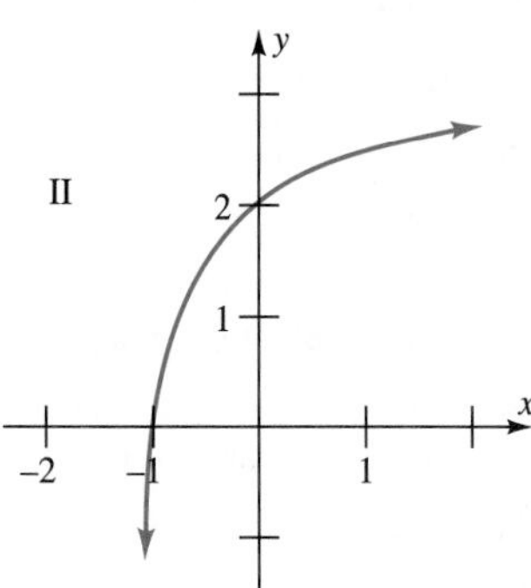

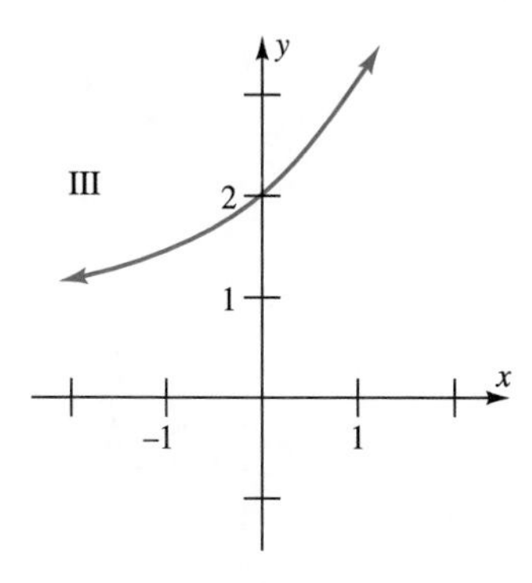

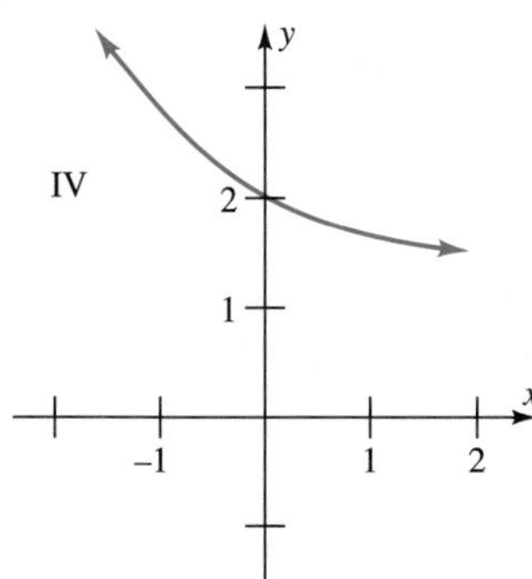

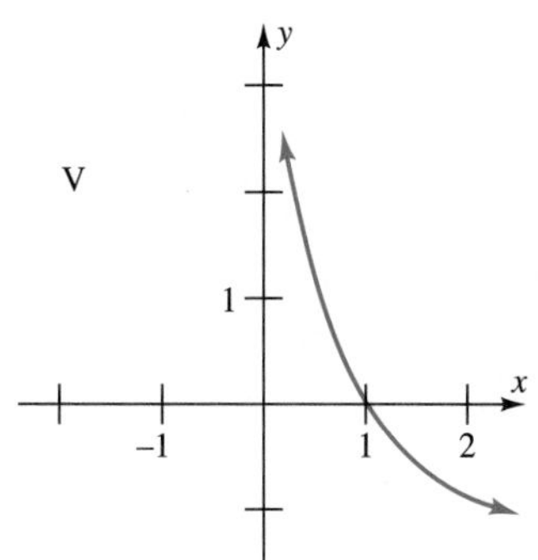

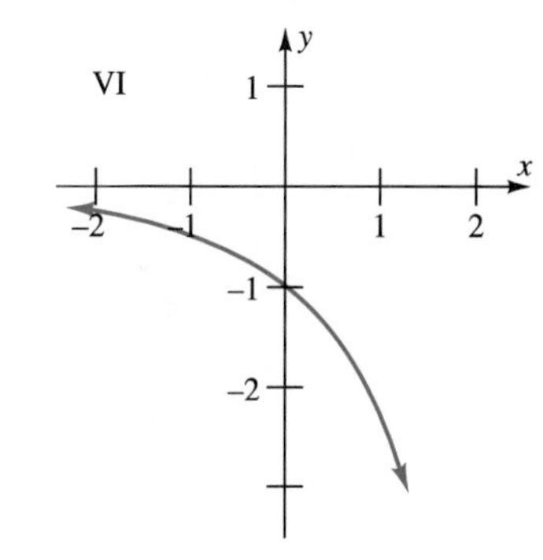

Figure 5-21

51. If $b > 1$, then the graph of $f(x) = -\log_b x$ could possibly be:
(a) I **(b)** IV
(c) V **(d)** VI
(e) none of these

52. If $0 < b < 1$, then the graph of $g(x) = b^x + 1$ could possibly be:
(a) II **(b)** III
(c) IV **(d)** VI
(e) none of these

53. If $\log_3 9^{x^2} = 4$, what is x?

54. What is the domain of the function $f(x) = \ln\left(\frac{x}{x-1}\right)$?

55. Solve: $8^x = 4^{x^2-3}$

56. Solve: $e^{3x} = 4$

57. Solve: $2 \cdot 4^x - 5 = -4$

58. Solve: $725e^{-4x} = 1500$

59. Solve for x: $u = c + d \ln x$

60. Solve: $2^x = 3^{x+3}$

61. Solve: $\ln x + \ln(3x - 5) = \ln 2$

62. Solve: $\ln(x + 8) - \ln x = 1$

63. Solve: $\log(x^2 - 1) = 2 + \log(x + 1)$

64. Solve: $\log_4 \sqrt{x^2 + 1} = 1$

65. The half-life of polonium (^{210}Po) is 140 days. If you start with 10 mg, how much will be left at the end of a year?

66. An insect colony grows exponentially from 200 to 2000 in 3 months' time. How long will it take for the insect population to reach 50,000?

67. Hydrogen-3 decays at a rate of 5.59% per year. Find its half-life.

68. The half-life of radium-88 is 1590 years. How long will it take for 10 grams to decay to 1 gram?

69. How much money should be invested at 8% per year, compounded quarterly, in order to have $1000 in 10 years?

70. At what annual interest rate should you invest your money if you want to double it in 6 years?

71. One earthquake measures 4.6 on the Richter scale. A second earthquake is 1000 times more intense than the first. What does it measure on the Richter scale?

6

Trigonometric Functions

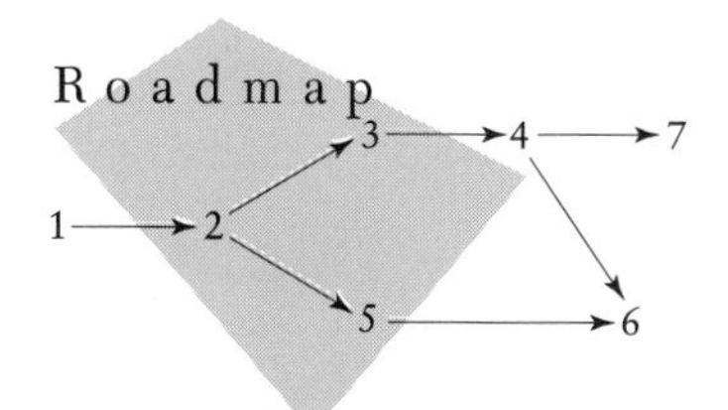

The ancient Greeks developed trigonometry for measuring angles and sides of triangles in order to solve problems in astronomy, navigation, and surveying.* But with the invention of calculus in the 17th century and the subsequent explosion of knowledge in the physical sciences, a different viewpoint toward trigonometry arose.

Whereas the ancients dealt only with *angles,* the classical trigonometric concepts of sine and cosine are now considered as *functions* with domain the set of all *real numbers.* The advantage of this switch in viewpoint is that almost any phenomena involving rotation or vibration can be described in terms of trigonometric functions, including light rays, sound waves, electron orbitals, planetary orbits, radio transmission, vibrating strings, pendulums, and many more.

The presentation of trigonometry here reflects this modern viewpoint. Nevertheless, angles still play an important role in defining the trigonometric functions, so the chapter begins with them.

6.1 ANGLES AND THEIR MEASUREMENT

In geometry, an angle is a static figure consisting of two half-lines (rays) that begin at the same point.** But in trigonometry an **angle** is thought of as being formed dynamically by *rotating* a half-line around its endpoint (the **vertex**). Its first position is the **initial side** and its final position is the **terminal side** of the angle. A small arrow is often used to indicate the direction of rotation and the number of revolutions, as in Figure 6–1:

*In fact ''trigonometry'' means ''triangle measurement.''

**The basic facts about such angles are discussed in the Geometry Review Appendix.

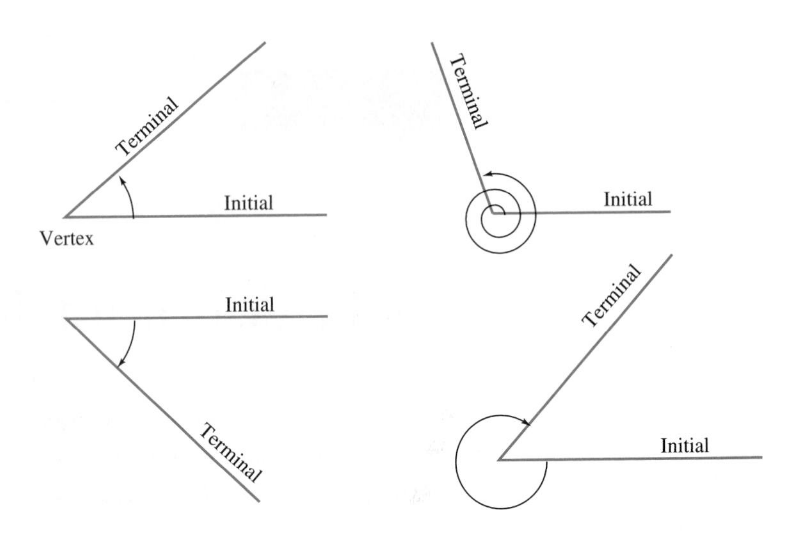

Figure 6-1

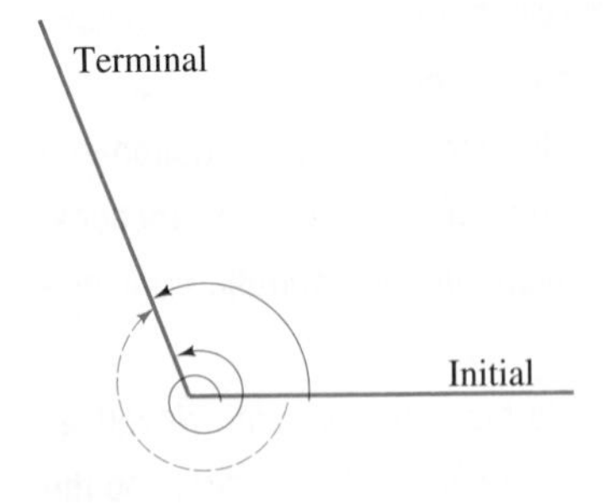

Figure 6-2

Figure 6–2 shows that different angles (that is, angles obtained by different rotations) may have the same initial and terminal side.* Such angles are said to be **coterminal.**

An angle in the coordinate plane is said to be in **standard position** if its vertex is at the origin and its initial side on the positive x-axis, as in Figure 6–3. When measuring angles in standard position, we use positive numbers for angles obtained by counterclockwise rotation **(positive angles)** and negative numbers for ones obtained by clockwise rotation **(negative angles).**

The classical unit for angle measurement is the **degree** (in symbols, °), which is the size of an angle obtained by 1/360 of a complete revolution.** Figure 6–3 shows the degree measure of several angles in standard position. Note that a 360° angle corresponds to a full revolution (the initial and terminal sides coincide), as shown on the left in Figure 6–3.

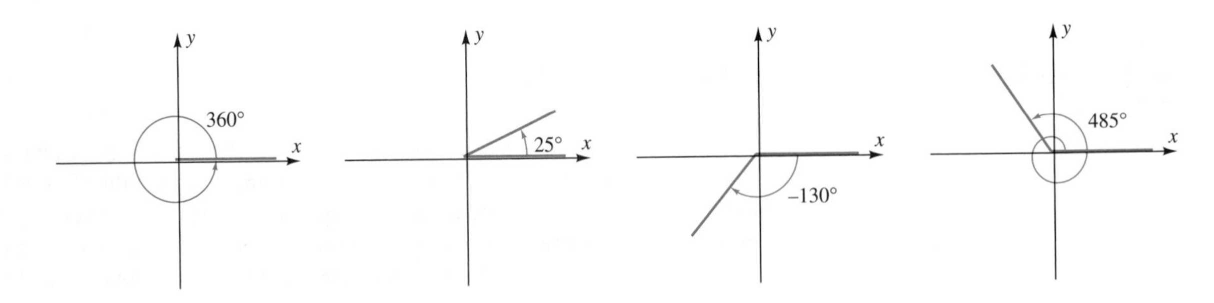

Figure 6-3

*They are *not* the same angle, however. For instance, both a 1/2 turn and $1\frac{1}{2}$ turns put a circular faucet handle in the same position, but the water flow is quite different.

**Degree measure is discussed in the Geometry Review Appendix.

Decimal notation is used in angle measurement whenever necessary—for instance, an angle of 159.483 degrees.* You should be familiar with all of the positive angles in standard position shown in Figure 6–4.

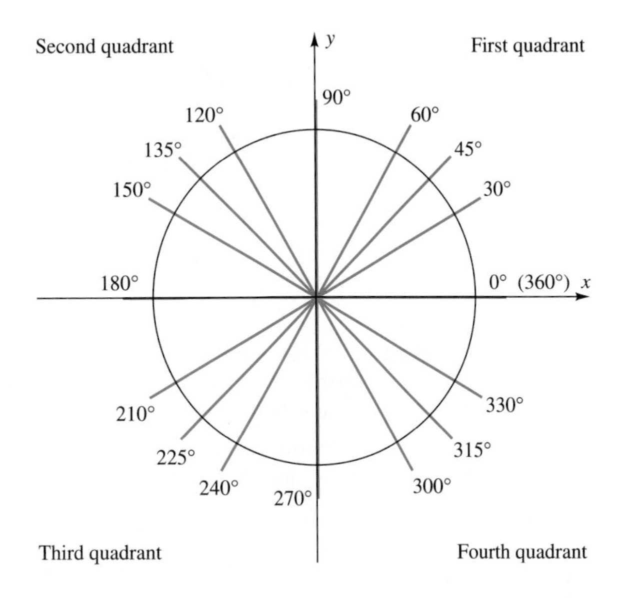

Figure 6-4

Radian Measure

A different unit of angle measure (called a **radian**) is commonly used in scientific and mathematical applications.** The **radian measure** of a positive angle in standard position is defined to be

> **The distance (measured in the counterclockwise direction along the unit circle†) from the point (1, 0) to the point where the terminal side of the angle intersects the unit circle.**

The radian measure of a negative angle in standard position is found in the same way, except that you move clockwise along the unit circle. For example, Figure 6–5 on the next page shows angles of radian measure 3.75 and -2:

*Until modern times, small angles were measured in other units. A **minute** is 1/60 of a degree and a **second** is 1/60 of a minute (= 1/3600 of a degree). For example, $215°56'34''$ denotes an angle of 215 degrees, 56 minutes, 34 seconds ($= 215 + 56/60 + 34/3600$ degrees $\approx 215.943°$). Many calculators have a key for converting from decimal to degree–minute–second notation.

**The reason is that radian measure greatly simplifies many formulas in calculus and physics. Measuring angles in radians instead of degrees is analogous to measuring length in meters rather than feet.

†The circle with center (0, 0) and radius 1, whose equation is $x^2 + y^2 = 1$; see page 29.

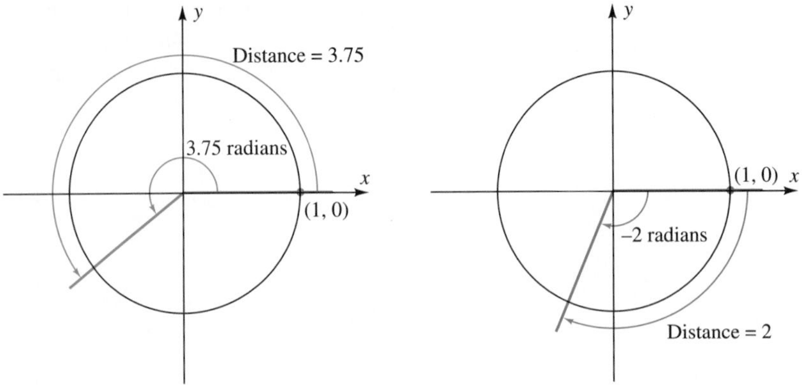

Figure 6-5

The circumference of the unit circle is 2π. So the radian measure of an angle formed when the terminal side makes a full-circle revolution is 2π radians, as shown in Figure 6–6. Hence, the radian measure of an angle of 360° (a full revolution) is 2π and the radian measure of an angle of 180° (half a revolution) is π radians. Therefore,

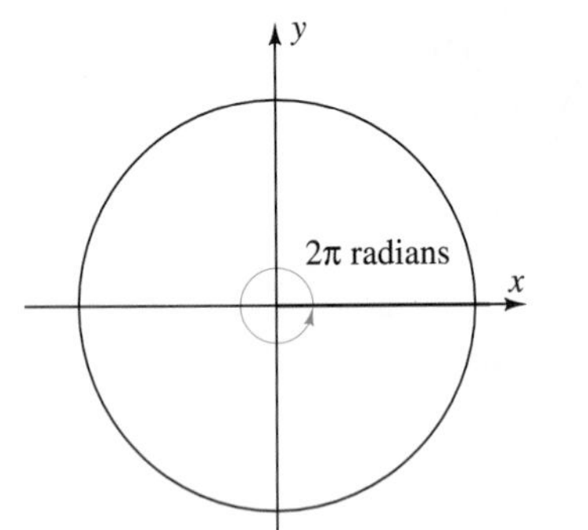

Figure 6-6

$$(*) \qquad \boldsymbol{\pi} \textbf{ radians} = \mathbf{180°.}$$

Dividing both sides of (∗) by π shows that

$$1 \text{ radian} = \frac{180}{\pi} \text{ degrees} \approx 57.3°,$$

and dividing both sides of (∗) by 180 shows that

$$1° = \frac{\pi}{180} \text{ radians} \approx .0175 \text{ radians}.$$

Consequently, we have these rules:

Conversion Rules for Degrees and Radians ▶

To convert radians to degrees, multiply by $180/\pi$.
To convert degrees to radians, multiply by $\pi/180$.

EXAMPLE 1

(a) To find the degree measure of an angle of 2.4 radians, multiply by $\frac{180}{\pi}$:

$$(2.4)\left(\frac{180}{\pi}\right) = \frac{432}{\pi} \approx 137.51°.$$

(b) An angle of $-.3$ radians has degree measure

$$(-.3)\left(\frac{180}{\pi}\right) = \frac{-54}{\pi} \approx -17.19°. \quad ■$$

The preceding formulas can be used with a calculator to convert angles from radians to degrees, or vice versa. Such conversions can also be done directly on a calculator, without referring to the formula. Since procedures vary, depending on the calculator, check your instruction manual. Naturally, a calculator uses a decimal approximation for π and displays its final results in decimal form. In many cases, however, it is more convenient *not* to use a decimal approximation for π when expressing angles in radian measure.

EXAMPLE 2 The radian measure of an angle of 45° is

$$45 \cdot \frac{\pi}{180} = \frac{\pi}{4} \text{ radians.}$$

It's customary to use $\pi/4$ rather than the decimal approximation .7854. Similarly, the radian measure of an angle of 210° is usually expressed as

$$210 \cdot \frac{\pi}{180} = \frac{7\pi}{6} \text{ radians.} \quad ■$$

You should become as comfortable with radian measure as you are with degree measure. The first step is to memorize the radian measure of these familiar angles:

Degrees	30°	45°	60°	90°	120°	135°	150°	180°
Radians	$\pi/6$	$\pi/4$	$\pi/3$	$\pi/2$	$2\pi/3$	$3\pi/4$	$5\pi/6$	π

Next you must learn to "think radian" rather than mentally translating from radians to degrees and back.

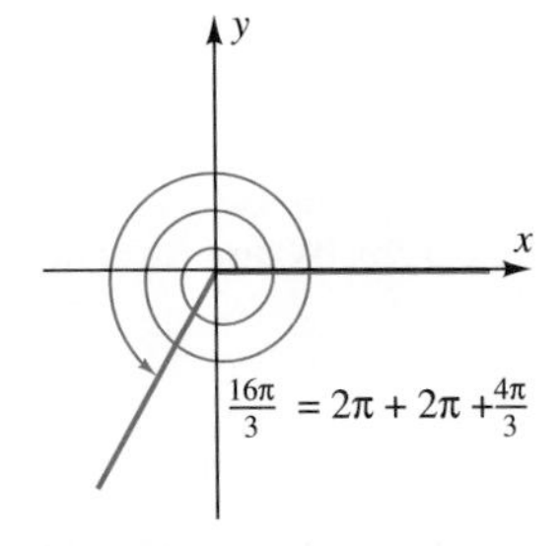

Figure 6-7

EXAMPLE 3 To construct an angle of $16\pi/3$ radians in standard position, note that

$$\frac{16\pi}{3} = \frac{12\pi}{3} + \frac{4\pi}{3} = 4\pi + \frac{4\pi}{3} = 2\pi + 2\pi + \frac{4\pi}{3}.$$

So, the terminal side must be rotated counterclockwise through 2 complete revolutions (each full-circle revolution is 2π radians) and then rotated an additional 2/3 of a revolution (since $4\pi/3$ is 2/3 of a complete revolution of 2π radians), as shown in Figure 6–7. ■

E X A M P L E 4 Since $-5\pi/4 = -\pi - \pi/4$, an angle of $-5\pi/4$ radians in standard position is obtained by rotating the terminal side *clockwise* for half a revolution (π radians) plus an additional 1/8 of a revolution (since $\pi/4$ is 1/8 of a full-circle revolution of 2π radians), as shown in Figure 6–8. ■

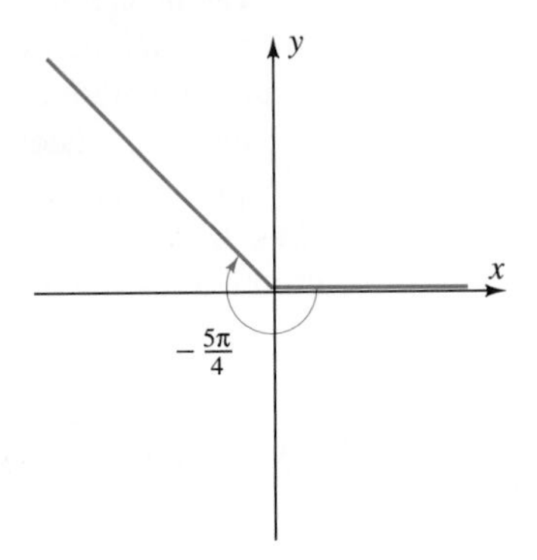

Figure 6-8

Consider an angle of t radians in standard position (Figure 6–9). Since 2π radians corresponds to a full revolution of the terminal side, this angle has the same terminal side as an angle of $t + 2\pi$ radians or $t - 2\pi$ radians or $t + 4\pi$ radians:

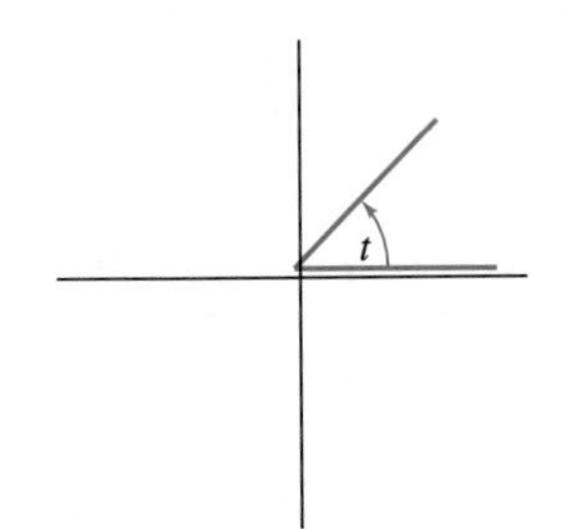

$t + 2\pi$

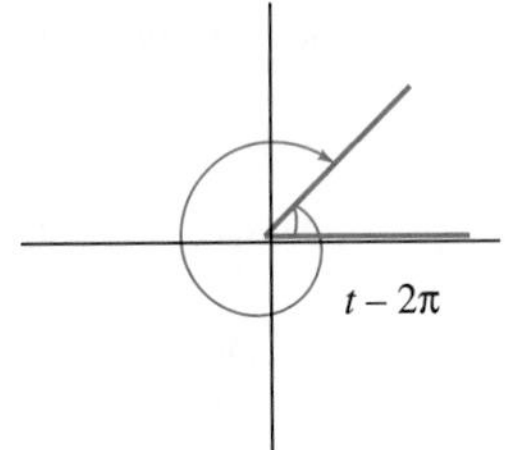

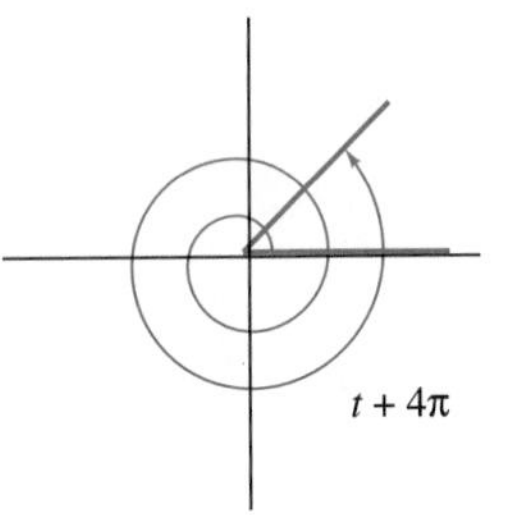

Figure 6-9

The same thing is true in general:

Coterminal Angles ▶

Increasing or decreasing the radian measure of an angle by an integer multiple of 2π results in a coterminal angle.

E X A M P L E 5 Find angles in standard position that are coterminal with an angle of: **(a)** $23\pi/5$ radians **(b)** $-\pi/12$ radians.

Solution

(a) We can subtract 2π to obtain a coterminal angle whose measure is $23\pi/5 - 2\pi = 23\pi/5 - 10\pi/5 = 13\pi/5$ radians, or we can subtract 4π to obtain a coterminal angle of measure $23\pi/5 - 4\pi = 3\pi/5$ radians. Subtracting 6π produces a coterminal angle of $23\pi/5 - 6\pi = -7\pi/5$ radians.

(b) An angle of $-\pi/12$ radians is coterminal with an angle of $-\pi/12 + 2\pi = 23\pi/12$ radians and with an angle of $-\pi/12 - 2\pi = -25\pi/12$ radians. ■

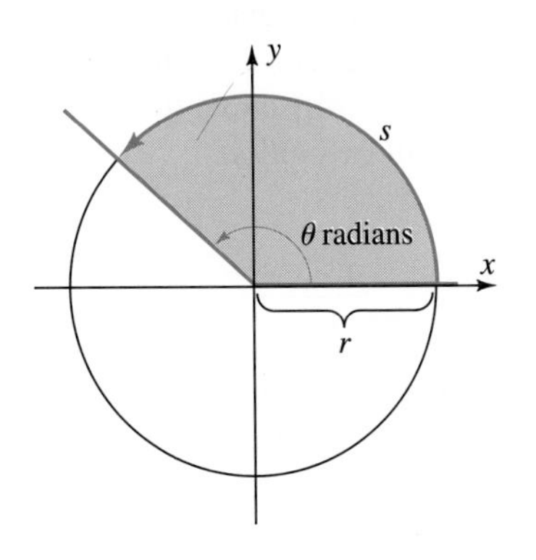

Figure 6-10

Arc Length and Angular Speed*

Consider a circle of radius r, with center at the origin, and an angle of θ radians in standard position, as shown in Figure 6–10. As you can see, the sides of the angle of θ radians determine an arc length s along the circle. We say that the **central angle** of θ radians **subtends an arc** of length s on the circle. It can be shown that the ratio of the arc length s to the circumference of the entire circle (namely, $2\pi r$) is the same as the ratio of the angle of θ radians to the full-circle angle of 2π radians; that is, $\frac{s}{2\pi r} = \frac{\theta}{2\pi}$. Solving this equation for s, we obtain this fact:

Arc Length ▶

A central angle of θ radians in a circle of radius r subtends an arc of length

$$s = \theta r,$$

that is, the length s of the arc is the product of the radian measure of the angle and the radius of the circle.

EXAMPLE 6 The second hand on a large clock is 6 inches long. How far does the tip of the second hand move in 15 seconds?

Solution The second hand makes a full revolution every 60 seconds, that is, it moves through an angle of 2π radians. During a 15 second interval it will make $\frac{15}{60} = \frac{1}{4}$ of a revolution, moving through an angle of $\pi/2$ radians (Figure 6–11). If we think of the second hand as the radius of a circle, then during a 15-second interval its tip travels along the arc subtended by an angle of $\pi/2$ radians. Therefore, the distance (arc length) traveled by the tip of the second hand is

$$s = \theta r = \frac{\pi}{2} \cdot 6 = 3\pi \approx 9.425 \text{ inches.} \quad ■$$

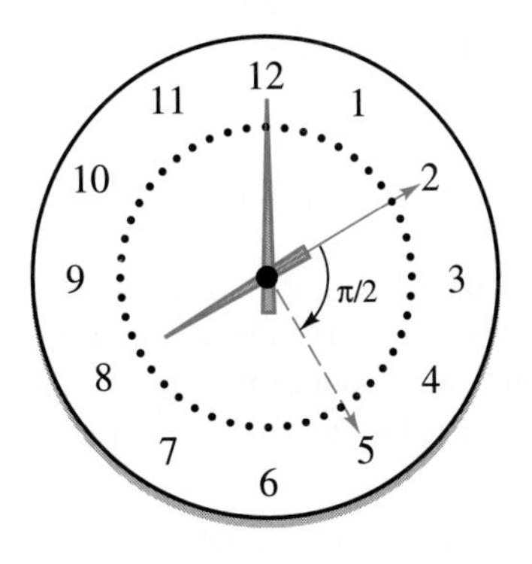

Figure 6-11

*The remainder of this section is not needed in the sequel and may be omitted if desired.

EXAMPLE 7 Suppose the circle in Figure 6–10 has radius 6 and the arc length s is 14.64. We can find the measure of the central angle that subtends the arc s by solving $s = \theta r$ for θ:

$$\theta = \frac{s}{r} = \frac{14.64}{6} = 2.44 \text{ radians.} \quad ■$$

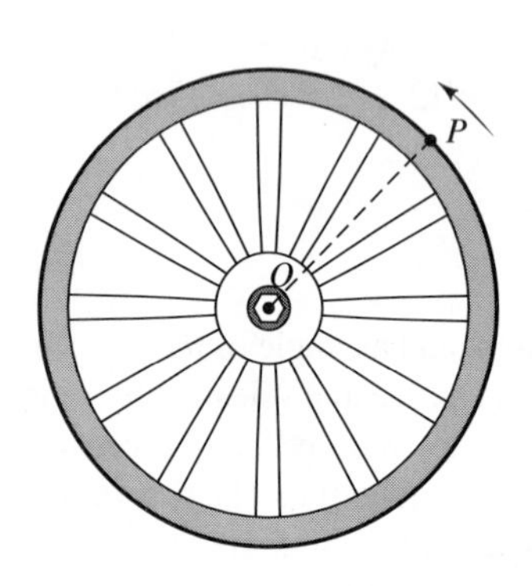

Figure 6-12

Suppose a wheel is rotating at a constant rate around its center O. If P is a point on the circumference of the wheel, as in Figure 6–12, consider the angle θ through which radius OP moves as the wheel rotates. It will move through an angle of 2π radians each time the wheel makes one full revolution.

Recall that the speed of a moving object is

$$\frac{\text{distance traveled}}{\text{time elapsed}}, \quad \text{or more briefly,} \quad \frac{\text{distance}}{\text{time}}.$$

Thus, if an object moves s feet in t seconds, its speed is s/t ft per second. Similarly, the **angular speed** of the wheel is the radian measure of the angle through which OP travels divided by the time, that is,

$$\textbf{angular speed} = \frac{\textbf{angle}}{\textbf{time}} = \frac{\theta}{t}.$$

EXAMPLE 8 A merry-go-round makes 8 revolutions per minute.

(a) What is the angular speed of the merry-go-round in radians per minute?

(b) How fast is a horse 12 feet from the center traveling?

(c) How fast is a horse 4 feet from the center traveling?

Solution

(a) Each revolution of the merry-go-round corresponds to a central angle of 2π radians, so it travels through an angle of $8 \cdot 2\pi = 16\pi$ radians in one minute. Therefore, its

$$\text{angular speed} = \frac{\text{angle}}{\text{time}} = \frac{16\pi}{1} = 16\pi \text{ radians per minute.}$$

(b) The horse 12 feet from the center travels along a circle of radius 12. As we saw in part (a), the angle through which the horse travels in one minute is 16π radians. By the arc length formula, the distance traveled by the horse is $r\theta = 12(16\pi) = 192\pi$ ft. Consequently, its linear speed is 192π ft/min (approximately 6.9 miles per hour).

(c) Proceeding as in part (b), with 4 in place of 12, we see that the linear speed of the horse is

$$\frac{\text{distance}}{\text{time}} = \frac{4(16\pi)}{1} = 64\pi \text{ ft/min}, \quad \text{or equivalently,} \quad 2.28 \text{ mph.} \quad ■$$

In Example 8(a), note that the angular speed is $8 \cdot 2\pi$, that is, the number of revolutions per minute times 2π. In Example 8(b), the linear speed of the horse is $12(16\pi)$, that is, the radius times the angular speed. The same relationships are valid for the angular speed of any rotating disc and the linear speed of a point on its circumference:

$$\textbf{angle of rotation} = \textbf{(number of revolutions)} \cdot (2\pi)$$

$$\textbf{angular speed} = \textbf{(revolutions per time period)} \cdot (2\pi)$$

$$\textbf{linear speed} = \textbf{(radius)} \cdot \textbf{(angular speed).}$$

EXERCISES 6.1

In Exercises 1–5, find the radian measure of the angle in standard position formed by rotating the terminal side the given amount.

1. 1/9 of a circle **2.** 1/24 of a circle

3. 1/18 of a circle **4.** 1/72 of a circle

5. 1/36 of a circle

6. State the radian measure of *every* standard position angle in Figure 6–13.*

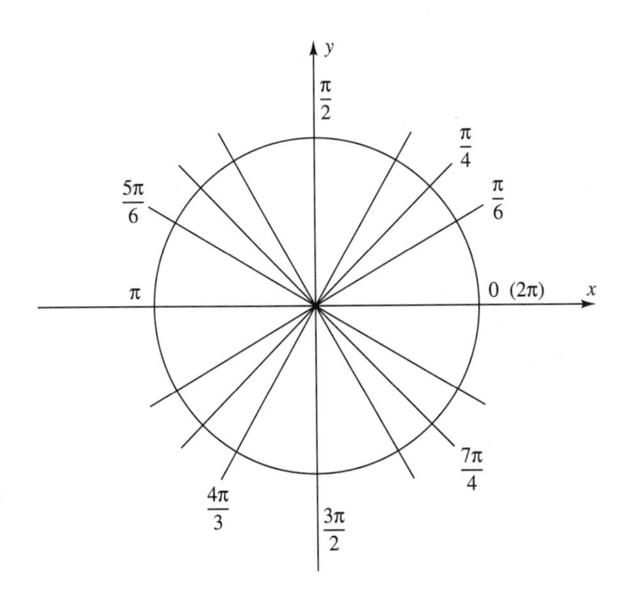

Figure 6-13

In Exercises 7–18, convert the given degree measure to radians.

7. 6° **8.** − 10° **9.** − 12° **10.** 36°

11. 75° **12.** − 105° **13.** 135° **14.** − 165°

15. − 225° **16.** 252° **17.** 930° **18.** − 585°

In Exercises 19–30, convert the given radian measure to degrees.

19. $\pi/5$ **20.** $-\pi/6$ **21.** $-\pi/10$

22. $2\pi/5$ **23.** $3\pi/4$ **24.** $-5\pi/3$

25. $\pi/45$ **26.** $-\pi/60$ **27.** $-5\pi/12$

28. $7\pi/15$ **29.** $27\pi/5$ **30.** $-41\pi/6$

In Exercises 31–34, find the radian measure of four angles in standard position that are coterminal with the angle in standard position whose measure is given.

31. $\pi/4$ **32.** $7\pi/5$ **33.** $-\pi/6$ **34.** $-9\pi/7$

In Exercises 35–42, find the radian measure of an angle in standard position that has measure between 0 and 2π and is coterminal with the angle in standard position whose measure is given.

35. $-\pi/3$ **36.** $-3\pi/4$ **37.** $19\pi/4$ **38.** $16\pi/3$

39. $-7\pi/5$ **40.** $45\pi/8$ **41.** 7 **42.** 18.5

*This is the same diagram that appears in Figure 6–4 on page 367, showing positive angles in standard position.

In Exercises 43–48, determine the positive radian measure of the angle that the second hand of a clock traces out in the given time.

43. 40 seconds **44.** 50 seconds **45.** 35 seconds

46. 2 minutes and 15 seconds

47. 3 minutes and 25 seconds

48. 1 minute and 55 seconds

49. The second hand on a clock is 6 cm long. How far does its tip travel in 40 seconds? (See Exercise 43.)

50. The second hand on a clock is 5 cm long. How far does its tip travel in 2 minutes and 15 seconds? (See Exercise 46.)

51. If the radius of the circle in Figure 6–14 is 20 cm and the length of arc s is 85 cm, what is the radian measure of the angle θ?

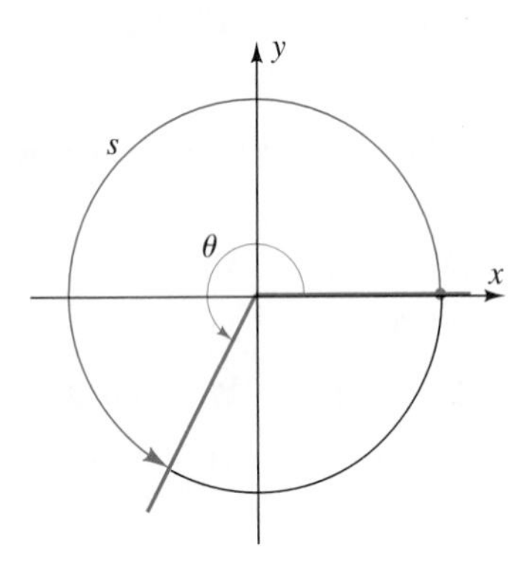

Figure 6-14

52. Find the radian measure of the angle θ in Figure 6–14 if the *diameter* of the circle is 150 cm and s has length 360 cm.

In Exercises 53–56, assume that a wheel on a car has radius 36 cm. Find the angle (in radians) that the wheel turns while the car travels the given distance.

53. 2 meters (= 200 cm) **54.** 5 meters

55. 720 meters

56. 1 kilometer (= 1000 meters)

In Exercises 57–60, find the length of the circular arc subtended by the central angle whose radian measure is given. Assume the circle has diameter 10.

57. 1 radian **58.** 2 radians **59.** 1.75 radians

60. 2.2 radians

In Exercises 61–64, the latitudes of a pair of cities are given. Assume that one city is directly south of the other and that the earth is a perfect sphere of radius 4000 miles. Find the distance between the two cities. [The ***latitude*** *of a point P on the earth is the degree measure of the angle θ between the point and the plane of the equator (with the center of the earth being the vertex), as shown in Figure 6–15. Remember that angles are measured in radians in the arc length formula.]*

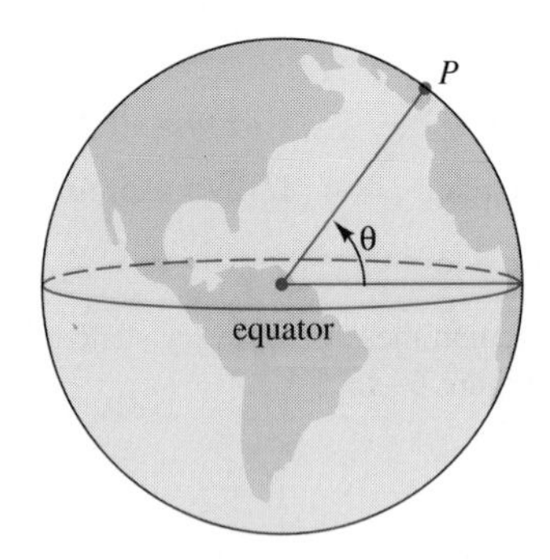

Figure 6-15

61. The North Pole (latitude 90° north) and Springfield, Illinois (latitude 40° north).

62. San Antonio, Texas (latitude 29.5° north) and Mexico City, Mexico (latitude 20° north).

63. Cleveland, Ohio (latitude 41.5° north) and Tampa, Florida (latitude 28° north).

64. Copenhagen, Denmark (latitude 54.3° north) and Rome, Italy (latitude 42° north).

In Exercises 65–72, a wheel is rotating around its axle. Find the angle (in radians) through which the wheel turns in the given time when it rotates at the given number of revolutions per minute (rpm).

65. 3.5 minutes, 1 rpm **66.** t minutes ($t > 0$), 1 rpm

67. 1 minute, 2 rpm **68.** 3.5 minutes, 2 rpm

69. 4.25 minutes, 5 rpm

70. t minutes $(t > 0)$, 5 rpm

71. 1 minute, k rpm $(k > 0)$

72. t minutes $(t > 0)$, k rpm $(k > 0)$

73. One end of a rope is attached to a winch (circular drum) of radius 2 ft and the other to a steel beam on the ground. When the winch is rotated, the rope wraps around the drum and pulls the object upward (Figure 6–16). Through what angle must the winch be rotated in order to raise the beam 6 ft above the ground?

Figure 6-16

74. A circular saw blade has an angular speed of 15,000 radians per minute.
- **(a)** How many revolutions per minute does the saw make?
- **(b)** How long will it take the saw to make 6000 revolutions?

75. A circular gear rotates at the rate of 200 revolutions per minute (rpm).
- **(a)** What is the angular speed of the gear in radians per minute?
- **(b)** What is the linear speed of a point on the gear 2 inches from the center in inches per minute and in feet per minute?

76. A wheel in a large machine is 2.8 feet in diameter and rotates at 1200 rpm.
- **(a)** What is the angular speed of the wheel?
- **(b)** How fast is a point on the circumference of the wheel traveling in feet per minute? In miles per hour?

77. A riding lawn mower has wheels that are 15 inches in diameter. If the wheels are making 2.5 revolutions per second
- **(a)** What is the angular speed of a wheel?
- **(b)** How fast is the lawn mower traveling in miles per hour?

78. A bicycle has wheels that are 26 inches in diameter. If the bike is traveling at 14 mph, what is the angular speed of each wheel?

79. A merry-go-round horse is traveling at 10 ft per second and the merry-go-round is making 6 revolutions per minute. How far is the horse from the center of the merry-go-round?

80. The pedal sprocket of a bicycle has radius 4.5 inches and the rear wheel sprocket has radius 1.5 inches (Figure 6–17). If the rear wheel has a radius of 13.5 inches and the cyclist is pedaling at the rate of 80 revolutions per minute, how fast is the bicycle traveling in ft per minute? In miles per hour?

Figure 6-17

6.2 THE SINE, COSINE, AND TANGENT FUNCTIONS

The sine and cosine functions are the basis of trigonometry. Both functions have the set of all real numbers as domain. Their rules are given by the following three-step geometric process:*

*This definition is geared to the eventual use of these functions in calculus and may seem quite different from the one you learned previously (especially if the earlier definition involved triangles). If so, just concentrate on this definition and don't worry about relating it to any definitions you remember from the past. The other ways of defining trigonometric functions will be discussed in Section 6.5.

1. Given a real number t, construct an angle of t radians in standard position.

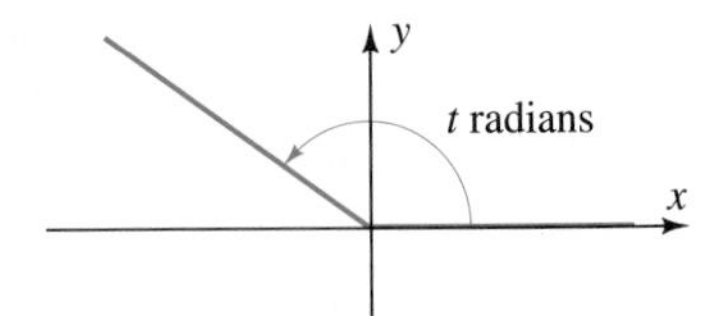

2. Find the coordinates of the point P where the terminal side of this angle meets the unit circle $x^2 + y^2 = 1$, say $P = (a, b)$.

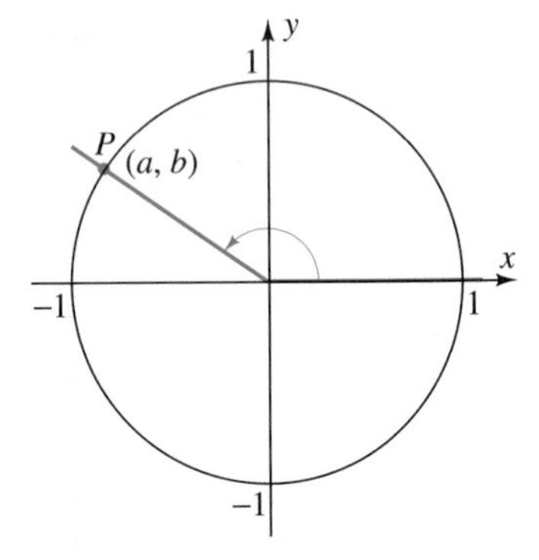

3. The value of the **cosine function** at t (denoted $\cos t$) is the x-coordinate of P:

$$\cos t = a.$$

The value of the **sine function** at t (denoted $\sin t$) is the y-coordinate of P:

$$\sin t = b.$$

In other words,

Sine and Cosine ▶

If P is the point where the terminal side of an angle of t radians in standard position meets the unit circle, then

$$P \text{ has coordinates } (\cos t, \sin t).$$

GRAPHING EXPLORATION With your calculator in radian mode and parametric graphing mode, set the range values as follows:

$$0 \leq t \leq 2\pi, \qquad -1.8 \leq x \leq 1.8, \qquad -1.2 \leq y \leq 1.2.^*$$

Then, graph the curve given by the parametric equations

$$x = \cos t \qquad \text{and} \qquad y = \sin t.$$

The graph is the unit circle. Use the trace to move around the circle. At each point, the screen will display three numbers: the values of t, x, and y. For each t, the cursor is on the point where the terminal side of an angle of t radians meets the unit circle, so the corresponding x is the number $\cos t$ and the corresponding y is the number $\sin t$.

*These settings give a square viewing window on calculators with a screen measuring approximately 95 by 63 pixels, and hence the unit circle will look like a circle. For wider screens, adjust the x range settings to obtain a square window.

The preceding exploration provides one way of evaluating sin t and cos t for selected values of t. In most cases, however, it is more convenient to use the SIN and COS keys on your calculator (in radian mode). For instance, a calculator shows that

$$\sin 2 \approx .91, \qquad \cos 10 = -.84, \qquad \sin(-7.3) = -.85,$$

where the results have been rounded to two places for convenience.

Although a calculator is usually necessary to evaluate the sine and cosine functions, it is equally important to understand how the definition can be used to evaluate these functions at certain numbers.

EXAMPLE 1 To evaluate the sine and cosine functions at $t = \pi$, construct an angle of π radians, as in Figure 6–18. It intersects the unit circle at $P = (-1, 0)$. Hence,

$$\cos \pi = x\text{-coordinate of } P = -1$$

$$\sin \pi = y\text{-coordinate of } P = 0$$

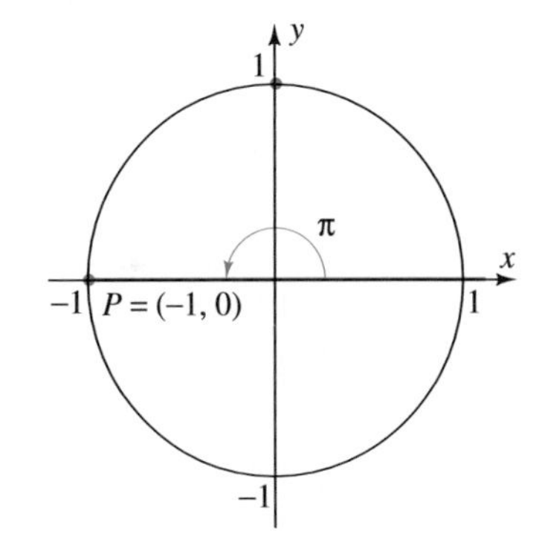

Figure 6-18 ■

EXAMPLE 2 An angle of 0 radians in standard position has its terminal side on the positive x-axis and meets the unit circle at $P = (1, 0)$. Therefore,

$$\cos 0 = 1 \ (x\text{-coordinate of } P) \qquad \text{and} \qquad \sin 0 = 0 \ (y\text{-coordinate of } P).$$ ■

The third basic trigonometric function is the **tangent function,** which is defined in terms of the sine and cosine functions. Its value at the number t (denoted tan t) is given by:

$$\tan t = \frac{\sin t}{\cos t}.$$

Every calculator has a TAN key for evaluating the tangent function directly. Nevertheless, you should be able to use the definition to do so whenever feasible.

EXAMPLE 3 From Example 1, we see that

$$\tan \pi = \frac{\sin \pi}{\cos \pi} = \frac{0}{-1} = 0. \quad \blacksquare$$

EXAMPLE 4 Evaluate the three trigonometric functions at $t = \pi/2$.

Solution Construct an angle of $\pi/2$ radians in standard position (Figure 6–19). It intersects the unit circle at $P = (0, 1)$. Hence,

$$\cos \frac{\pi}{2} = x\text{-coordinate of } P = 0$$

$$\sin \frac{\pi}{2} = y\text{-coordinate of } P = 1$$

$$\tan \frac{\pi}{2} = \frac{\sin (\pi/2)}{\cos (\pi/2)} = \frac{1}{0} \textit{ undefined}$$

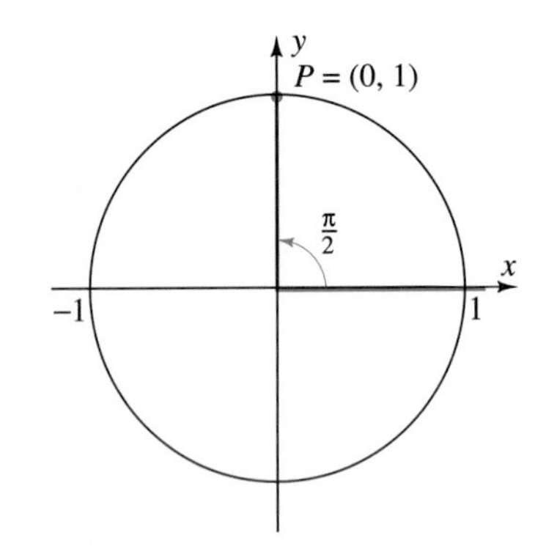

Figure 6-19 $\blacksquare$

Example 4 shows that the domain of the tangent function must exclude all numbers t for which $\cos t = 0$. Now $\cos t = 0$ whenever the point P in the definition has coordinates $(0, 1)$, as in Figure 6–19, or $(0, -1)$. That happens when t is $\pm\pi/2$, $\pm 3\pi/2$, $\pm 5\pi/2$, . . . and so on. Therefore, the domain of the tangent function consists of all real numbers *except* odd integer multiples of $\pi/2$.

Algebra with Trigonometric Functions

Since calculators make the numerical computations easy, we shall concentrate on the key properties of trigonometric functions and skills needed to use these functions effectively. We begin with two notational conventions:

1. **Parentheses are omitted whenever no confusion can result.** For example,

$$-(\sin t) \quad \text{is written} \quad -\sin t;$$
$$4(\tan t) \quad \text{is written} \quad 4\tan t.$$

2. **When dealing with powers of trigonometric functions, exponents are written between the function symbol and the variable.*** For example,

$$(\cos t)^3 \quad \text{is written} \quad \cos^3 t;$$
$$(\sin t)^4(\tan 7t)^2 \quad \text{is written} \quad \sin^4 t \tan^2 7t.$$

But be careful: $\cos t^3$ means $\cos(t^3)$, *not* $(\cos t)^3$ and *not* $\cos^3 t$ (see Exercise 56).

> **TECHNOLOGY TIP**
> Calculators do not use convention 2. If you try to enter $\cos^2(5)$ just as it reads, for example, you will get an error message. To obtain this number you must enter $(\cos 5)^2$.

Except for these conventions, the algebra of trigonometric functions is just like the algebra of any other functions. They may be added, multiplied, composed, etc.

EXAMPLE 5 If $f(t) = \sin^2 t + \tan t$ and $g(t) = \tan^3 t + 5$, then the product function fg is given by the rule

$$\begin{aligned}(fg)(t) = f(t)g(t) &= (\sin^2 t + \tan t)(\tan^3 t + 5)\\ &= \sin^2 t \tan^3 t + 5\sin^2 t + \tan^4 t + 5\tan t. \quad \blacksquare\end{aligned}$$

EXAMPLE 6 Factor $2\cos^2 t - 5\cos t - 3$.

Solution You can do this directly, but it may be easier to understand if you make a substitution. Let $u = \cos t$; then,

$$\begin{aligned}2\cos^2 t - 5\cos t - 3 &= 2(\cos t)^2 - 5\cos t - 3\\ &= 2u^2 - 5u - 3\\ &= (2u + 1)(u - 3)\\ &= (2\cos t + 1)(\cos t - 3). \quad \blacksquare\end{aligned}$$

*This convention is *not* used when the exponent is -1. The symbols $\sin^{-1} t$, $\cos^{-1} t$ and $\tan^{-1} t$ have an entirely different mathematical meaning, which is explained in Section 8.5.

EXAMPLE 7 If $f(t) = \cos^2 t - 9$ and $g(t) = \cos t + 3$, then the quotient function f/g is given by the rule

$$\left(\frac{f}{g}\right)(t) = \frac{f(t)}{g(t)} = \frac{\cos^2 t - 9}{\cos t + 3} = \frac{(\cos t + 3)(\cos t - 3)}{\cos t + 3} = \cos t - 3. \blacksquare$$

WARNING You are dealing with *functional notation* here, so the symbol $\sin t$ is a *single entity,* as are $\cos t$ and $\tan t$. Don't try some nonsensical "canceling" operation, such as

$$\frac{\sin t}{\cos t} = \frac{\sin}{\cos} \quad \text{or} \quad \frac{\cos t^2}{\cos t} = \frac{\cos t}{\cos} = t.$$

EXAMPLE 8 If $f(t) = \sin t$ and $g(t) = t^2 + 3$, then the composite function $g \circ f$ is given by the rule.

$$(g \circ f)(t) = g(f(t)) = g(\sin t) = \sin^2 t + 3.$$

The composite function $f \circ g$ is given by the rule

$$(f \circ g)(t) = f(g(t)) = f(t^2 + 3) = \sin(t^2 + 3).$$

The parentheses are absolutely necessary here because $\sin(t^2 + 3)$ is *not* the same function as $\sin t^2 + 3$. For instance, a calculator in radian mode shows that for $t = 5$,

$$\sin(5^2 + 3) = \sin(25 + 3) = \sin 28 \approx .2709$$

whereas

$$\sin 5^2 + 3 = \sin 25 + 3 \approx (-.1324) + 3 = 2.8676. \blacksquare$$

Signs

If f is a function, it is often useful to know which values of t make $f(t)$ positive and which make $f(t)$ negative. To determine this information for the trigonometric functions, we need only use the definition: $\cos t$ is the x-coordinate and $\sin t$ is the y-coordinate of a point on the terminal side of an angle of t radians in standard position. The quadrant in which the point $(\cos t, \sin t)$ lies determines which of its coordinates are positive and which are negative.

For instance, if $\pi/2 < t < \pi$, then the terminal side of an angle of t radians is in the second quadrant. A point (x, y) in the second quadrant has x negative and

y positive. Hence $\cos t$ is negative and $\sin t$ is positive and $\tan t = \sin t/\cos t$ is negative. Similar considerations lead to this summary:

Signs of the Trigonometric Functions ▶

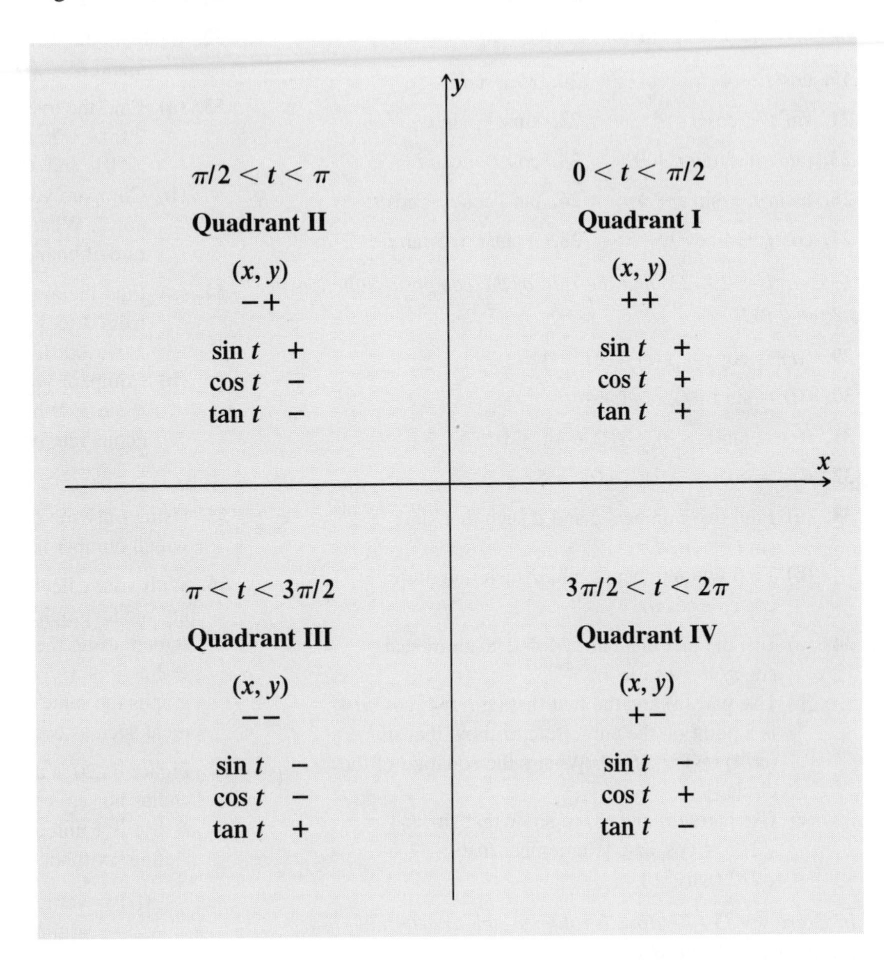

EXERCISES 6.2

Note: *Unless stated otherwise, all angles are in standard position.*

In Exercises 1–10, use the definition (not a calculator) to find the function value.

1. $\sin 3\pi/2$ **2.** $\sin(-\pi)$ **3.** $\cos 3\pi/2$

4. $\cos(-\pi/2)$ **5.** $\tan 4\pi$ **6.** $\tan(-\pi)$

7. $\cos(-3\pi/2)$ **8.** $\sin 9\pi/2$ **9.** $\cos(-11\pi/2)$

10. $\tan(-13\pi)$

In Exercises 11–14, assume that the terminal side of an angle of t radians passes through the given point on the unit circle. Find sin t, cos t, tan t.

11. $(-2/\sqrt{5}, 1/\sqrt{5})$ **12.** $(1/\sqrt{10}, -3/\sqrt{10})$

13. $(-3/5, -4/5)$ **14.** $(.6, -.8)$

In Exercises 15–18, find the rule of the product function fg.

15. $f(t) = 3 \sin t;\quad g(t) = \sin t + 2 \cos t$

16. $f(t) = 5 \tan t;\quad g(t) = \tan^3 t - 1$

17. $f(t) = 3\sin^2 t$; $g(t) = \sin t + \tan t$

18. $f(t) = \sin 2t + \cos^4 t$; $g(t) = \cos 2t + \cos^2 t$

In Exercises 19–28, factor the given expression.

19. $\cos^2 t - 4$ **20.** $25 - \tan^2 t$

21. $\sin^2 t - \cos^2 t$ **22.** $\sin^3 t - \sin t$

23. $\tan^2 t + 6\tan t + 9$ **24.** $\cos^2 t - \cos t - 2$

25. $6\sin^2 t - \sin t - 1$ **26.** $\tan t \cos t + \cos^2 t$

27. $\cos^4 t + 4\cos^2 t - 5$ **28.** $3\tan^2 t + 5\tan t - 2$

In Exercises 29–32, find the rule of the composite functions $f \circ g$ and $g \circ f$.

29. $f(t) = \cos t$; $g(t) = 2t + 4$

30. $f(t) = \sin t + 2$; $g(t) = t^2$

31. $f(t) = \tan(t + 3)$; $g(t) = t^2 - 1$

32. $f(t) = \cos^2(t - 2)$; $g(t) = 5t + 2$

33. **(a)** Find two numbers c and d such that $\sin(c + d) \neq \sin c + \sin d$.
(b) Find two numbers c and d such that $\cos(c + d) \neq \cos c + \cos d$.

34. **(a)** Use the fact that $\tan \pi/4 = 1$ to show that $\sin \pi/4 = \cos \pi/4$.
(b) Use part (a) and the fact that $(\sin \pi/4, \cos \pi/4)$ is a point on the unit circle to show that $\sin^2(\pi/4) = 1/2$. [*Hint:* What's the equation of the circle?]
(c) Use parts (a) and (b) to show that $\sin \pi/4 = \sqrt{2}/2 = \cos \pi/4$. [Remember that $\sqrt{1/2} = \sqrt{2}/2$ (why?).]

In Exercises 35–40, draw a rough sketch to determine if the given number is positive.

35. sin 1 [*Hint:* The terminal side of an angle of 1 radian lies in the first quadrant (why?), so any point on it will have a positive y-coordinate.]

36. cos 2 **37.** tan 3 **38.** (cos 2)(sin 2)

39. tan 1.5

40. cos 3 + sin 3

41. Find all numbers t such that $0 \leq t < 2\pi$ and $\sin t < 1$.

42. Find all numbers t such that $0 \leq t < 2\pi$ and $\cos t > 0$.

In Exercises 43–48, find all the solutions of the equation.

43. $\sin t = 1$ **44.** $\cos t = -1$ **45.** $\tan t = 0$

46. $\sin t = -1$ **47.** $|\sin t| = 1$ **48.** $|\cos t| = 1$

In Exercises 49–51, find the difference quotient of the function.

49. $f(t) = \sin t$ **50.** $g(t) = \cos t$ **51.** $k(t) = \tan t$

52. Evaluate the difference quotient of $f(t) = \sin^2 t$ when $t = 4$ and $h = .01$.

53. **(a)** Find the average rate of change of $f(t) = \sin t$ from 2 to $2 + h$, for each of these values of h: .01, .001, .0001, and .00001. [See page 98.]
(b) Compare your answers in part (a) with the number cos 2. What would you guess that the instantaneous rate of change of $f(t) = \sin t$ is at $t = 2$?

54. **(a)** Find the average rate of change of $g(t) = \cos t$ from 3 to $3 + h$, for each of these values of h: .01, .001, .0001, and .00001.
(b) Compare your answers in part (a) with the number $-\sin 3$. What would you guess that the instantaneous rate of change of $g(t) = \cos t$ is at $t = 3$?

Thinkers

55. Using only the definition and no calculator, determine which number is larger: sin(cos 0) or cos(sin 0).

56. With your calculator in radian mode and function graphing mode, graph the following functions on the same screen, using the viewing window with $0 \leq x \leq 2\pi$ and $-3 \leq y \leq 3$: $f(x) = \cos x^3$ and $g(x) = (\cos x)^3$. Are the graphs the same? What do you conclude about a statement such as $\cos x^3 = (\cos x)^3$?

57. Figure 6–20 is a diagram of a merry-go-round that includes horses A through F. The distance from the center P to A is 1 unit and the distance from P to D is 5 units. Define six functions as follows:

$A(t)$ = vertical distance from horse A to the x-axis at time t minutes;

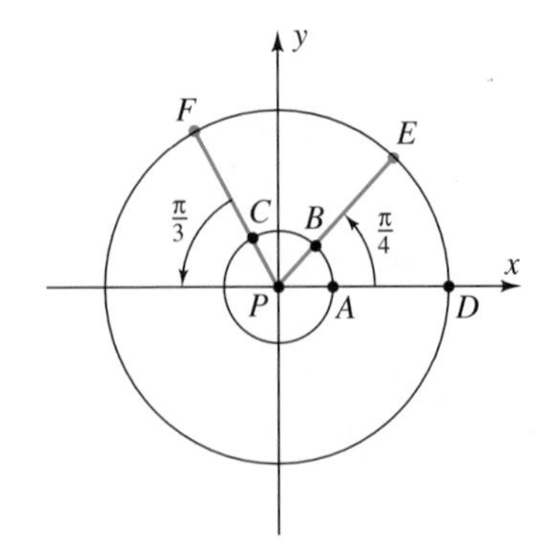

Figure 6-20

and similarly for $B(t)$, $C(t)$, $D(t)$, $E(t)$, $F(t)$. The merry-go-round rotates counterclockwise at a rate of 1 revolution per minute and the horses are in the positions shown in Figure 6–20 at the starting time $t = 0$. As the merry-go-round rotates, the horses move around the circles shown in the diagram.

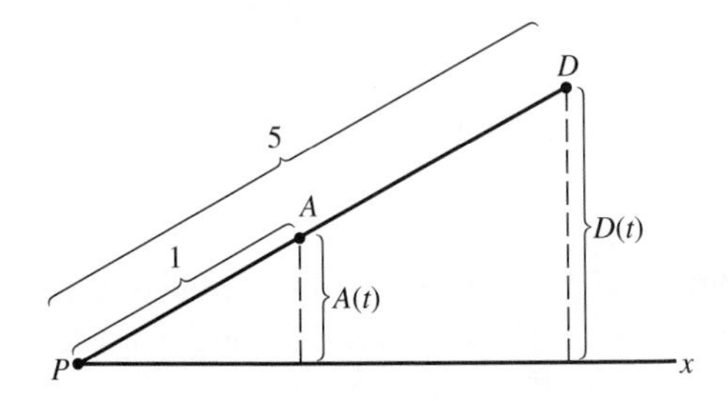

Figure 6-21

(a) Show that $B(t) = A(t + 1/8)$ for every t.

(b) In a similar manner, express $C(t)$ in terms of the function $A(t)$.

(c) Express $E(t)$ and $F(t)$ in terms of the function $D(t)$.

(d) Explain why Figure 6–21 is valid and use it and similar triangles to express $D(t)$ in terms of $A(t)$.

(e) In a similar manner, express $E(t)$ and $F(t)$ in terms of $A(t)$.

(f) Show that $A(t) = \sin(2\pi t)$ for every t. [*Hint:* Exercises 65–72 in Section 6.1 may be helpful.]

(g) Use parts (a), (b), and (f) to express $B(t)$ and $C(t)$ in terms of the sine function.

(h) Use parts (d), (e), and (f) to express $D(t)$, $E(t)$, and $F(t)$ in terms of the sine function.

6.3 BASIC GRAPHS

Although a graphing calculator will quickly sketch the graphs of the sine, cosine, and tangent functions, it will not give you much insight into why these graphs have the shapes they do and why these shapes are important. So the emphasis here is on the connection between the definition of these functions and their graphs. We begin with a property of sine and cosine that is the key to understanding their graphs.

Periodicity of Sine and Cosine

Let t be any real number and construct two angles in standard position of measure t and $t + 2\pi$ radians, respectively, as shown in Figure 6–22. As we saw in Section 6.1, both of these angles have the same terminal side. Therefore, the point P where the terminal side meets the unit circle is the *same* in both cases:

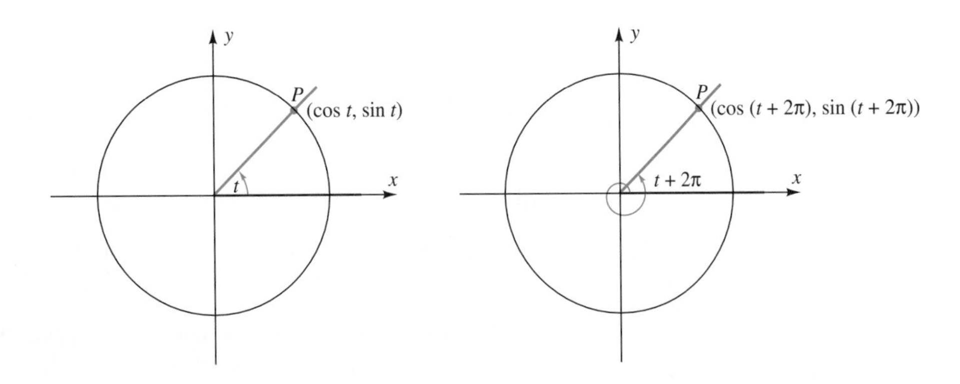

Figure 6-22

Since the second coordinate of P is the value of the sine function in each case, we see that $\sin t = \sin(t + 2\pi)$. Furthermore, since an angle of t radians in standard position has the same terminal side as angles of radian measure $t \pm 2\pi$, $t \pm 4\pi$, $t \pm 6\pi$, and so forth, essentially the same argument shows that for every number t

$$\sin t = \sin(t \pm 2\pi) = \sin(t \pm 4\pi) = \sin(t \pm 6\pi) = \cdots.$$

Similarly, the first coordinate of P is the value of the cosine function in each case so that

$$\cos t = \cos(t \pm 2\pi) = \cos(t \pm 4\pi) = \cos(t \pm 6\pi) = \cdots.$$

CALCULATOR EXPLORATION Illustrate the preceding discussion by computing each of the following numbers: $\sin 3$, $\sin(3 + 2\pi)$, $\sin(3 - 4\pi)$. Do the same for $\cos 4$, $\cos(4 + 2\pi)$, and $\cos(4 + 6\pi)$.

There is a special name for functions that repeat their values at regular intervals. A function f is said to be **periodic** if there is a positive constant k such that $f(t) = f(t + k)$ for every number t in the domain of f. There may be more than one constant k with this property; the smallest one is called the **period** of the function f. We have just seen that sine and cosine are periodic with $k = 2\pi$. Exercises 48 and 49 show that 2π is the smallest such positive constant k. Therefore,

Period of Sine and Cosine ▶

The sine and cosine functions are periodic with period 2π: For every real number t,

$$\sin t = \sin(t \pm 2\pi) \quad \text{and} \quad \cos t = \cos(t \pm 2\pi)$$

EXAMPLE 1 Given that $\sin \frac{\pi}{6} = \frac{1}{2}$, find $\sin \frac{13\pi}{6}$.

Solution The periodicity identity shows that

$$\sin \frac{13\pi}{6} = \sin\left(\frac{\pi}{6} + \frac{12\pi}{6}\right) = \sin\left(\frac{\pi}{6} + 2\pi\right) = \sin \frac{\pi}{6} = \frac{1}{2}. \quad ■$$

Periodicity means that sine and cosine repeat their values at regular intervals, and hence their graphs repeat the same pattern at regular intervals, as we now see.

Graph of the Sine Function

If P is the point where the unit circle meets the terminal side of an angle of t radians, then the y-coordinate of P is the number $\sin t$. We can get a rough sketch of the graph of $f(t) = \sin t$ by watching the y-coordinate of P:

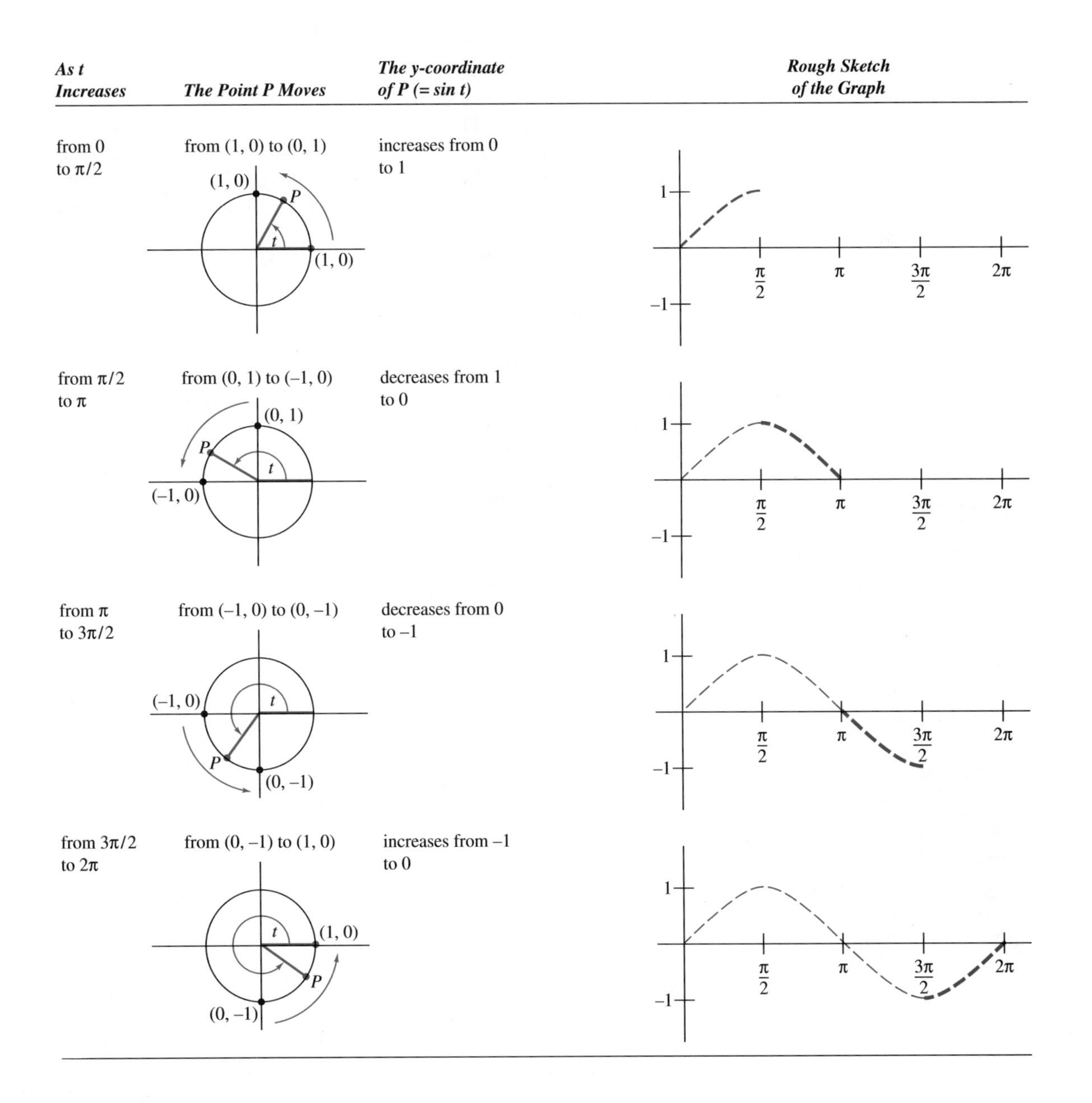

GRAPHING EXPLORATION Your calculator can provide a dynamic simulation of this process. Put it in parametric graphing mode and set the range values as follows:

$$0 \leq t \leq 6.28 \qquad -1 \leq x \leq 6.28 \qquad -2.5 \leq y \leq 2.5.$$

On the same screen, graph the two functions given by

$$x_1 = \cos t, \quad y_1 = \sin t \qquad \text{and} \qquad x_2 = t, \quad y_2 = \sin t.$$

Using the trace feature, move the cursor along the first graph (the unit circle). Stop at a point on the circle, note the value of t and the y-coordinate of the point. Then switch the trace to the second graph (the sine function) by using the up or down cursor arrows. The value of t remains the same. What are the x- and y-coordinates of the new point? How does the y-coordinate of the new point compare with the y-coordinate of the original point on the unit circle?

To complete the graph of the sine function, note that as t goes from 2π to 4π, the point P on the unit circle *retraces* the path it took from 0 to 2π, so *the same wave shape will repeat* on the graph. The same thing happens when t goes from 4π to 6π, or from -2π to 0, and so on. This repetition of the same pattern is simply the graphical expression of the fact that the sine function has period 2π: For any number t, the points

$$(t, \sin t) \qquad \text{and} \qquad (t + 2\pi, \sin(t + 2\pi))$$

on the graph have the same second coordinate.

A graphing calculator or some point plotting with an ordinary calculator now produces the graph of $f(t) = \sin t$ (Figure 6–23):

▶ TECHNOLOGY TIP

Graphing calculators have built-in viewing windows for trigonometric functions. Typically, they show the portion of the x-axis between -2π and 2π, with tic marks at intervals of $\pi/2$.

On TI, choose TRIG, ZTRIG, or ZOOMTRIG in the ZOOM menu. Choose TRIG in the HP-38 VIEWS menu or the Casio 9800 RANGE menu. On Sharp 9300, press RANGE MENU and choose TRIG.

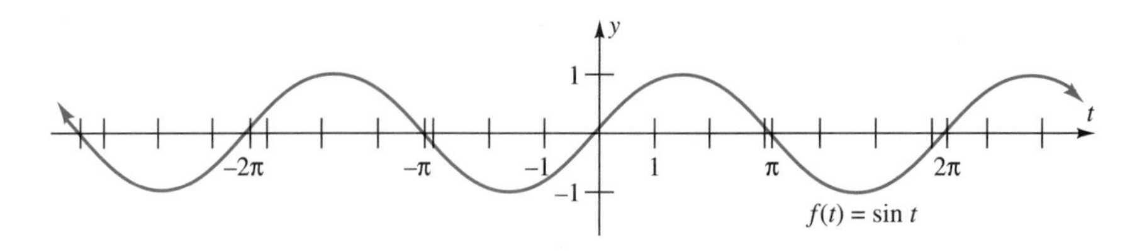

Figure 6-23

Note Throughout this chapter, we use t as the variable for trigonometric functions, to avoid any confusion with the x's and y's that are part of the definition of these functions. For calculator graphing in "function mode," however, you must use x as the variable: $f(x) = \sin x$, $g(x) = \cos x$, etc.

The graph of the sine function and the techniques of Section 2.5 can be used to graph other trigonometric functions.

EXAMPLE 2 The graph of $h(t) = 3 \sin t$ is the graph of $f(t) = \sin t$ stretched away from the horizontal axis by a factor of 3, as shown in Figure 6–24. ■

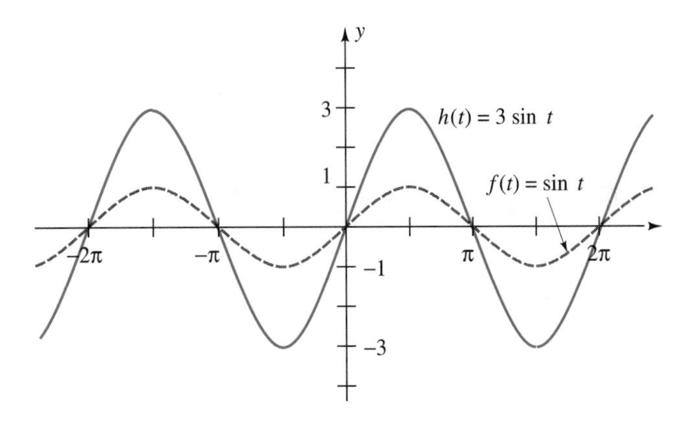

Figure 6-24

EXAMPLE 3 The graph of $k(t) = -\frac{1}{2} \sin t$ is the graph of $f(t) = \sin t$ shrunk by a factor of 1/2 toward the horizontal axis and then reflected in the horizontal axis, as shown in Figure 6–25. ■

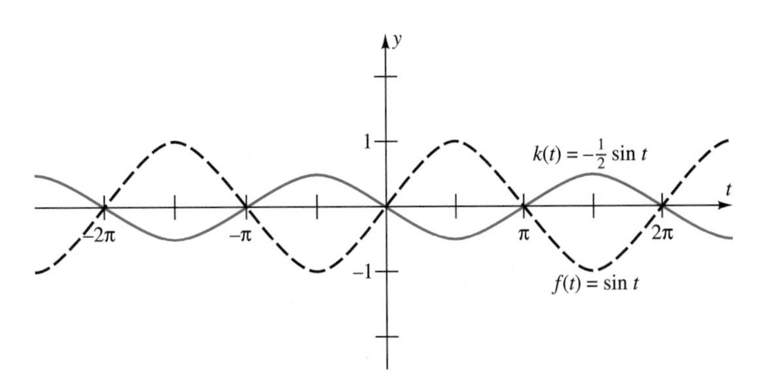

Figure 6-25

Graph of the Cosine Function

To obtain the graph of $g(t) = \cos t$, we follow the same procedure, except that we now watch the x-coordinate of P (which is $\cos t$).

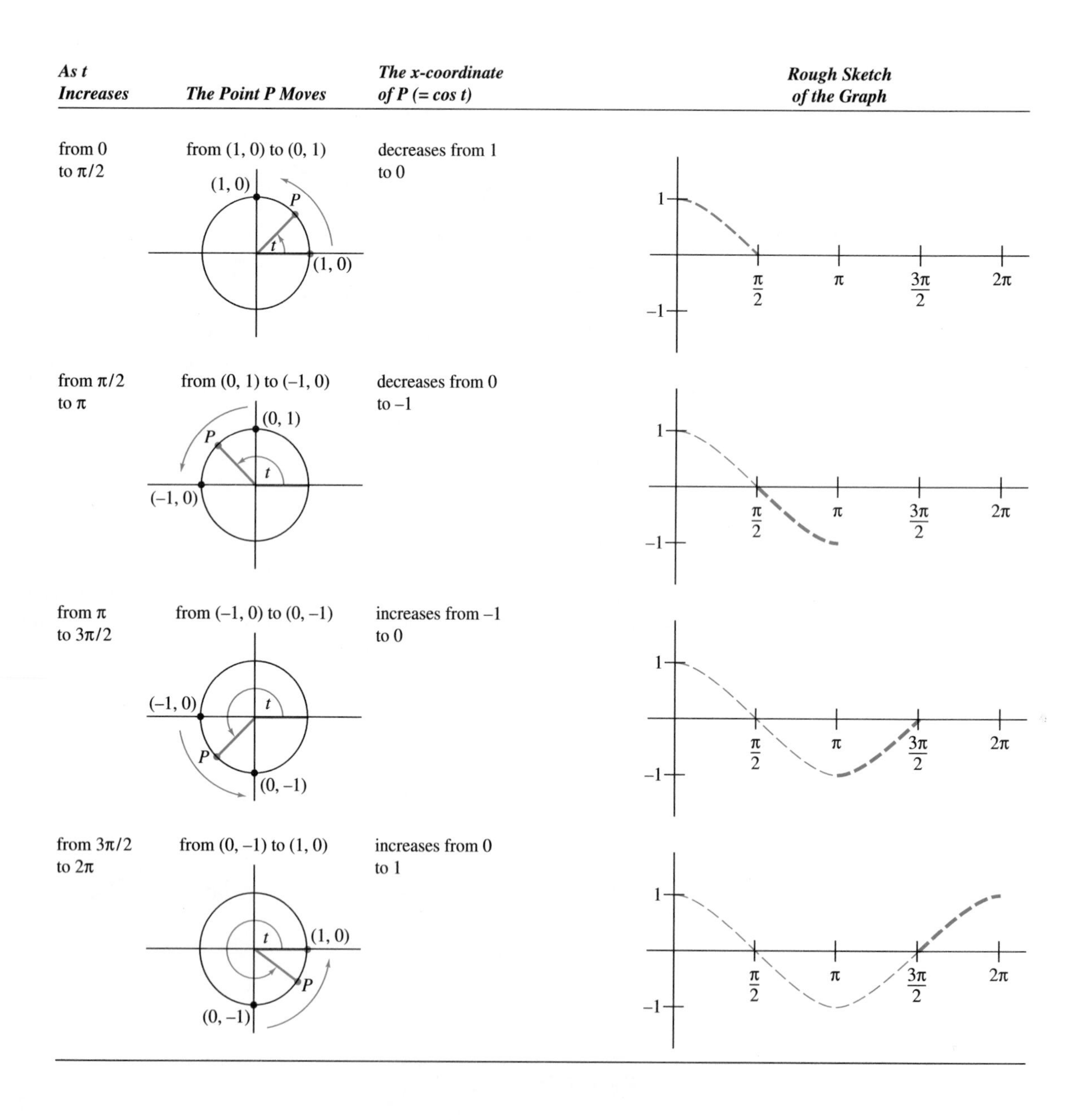

As t Increases	*The Point P Moves*	*The x-coordinate of P (= cos t)*	*Rough Sketch of the Graph*
from 0 to $\pi/2$	from (1, 0) to (0, 1)	decreases from 1 to 0	
from $\pi/2$ to π	from (0, 1) to (−1, 0)	decreases from 0 to −1	
from π to $3\pi/2$	from (−1, 0) to (0, −1)	increases from −1 to 0	
from $3\pi/2$ to 2π	from (0, −1) to (1, 0)	increases from 0 to 1	

As t takes larger values, P begins to retrace its path around the unit circle, so the graph of $g(t) = \cos t$ repeats the same wave pattern, and similarly for negative values of t. So, the graph looks like this (Figure 6–26):

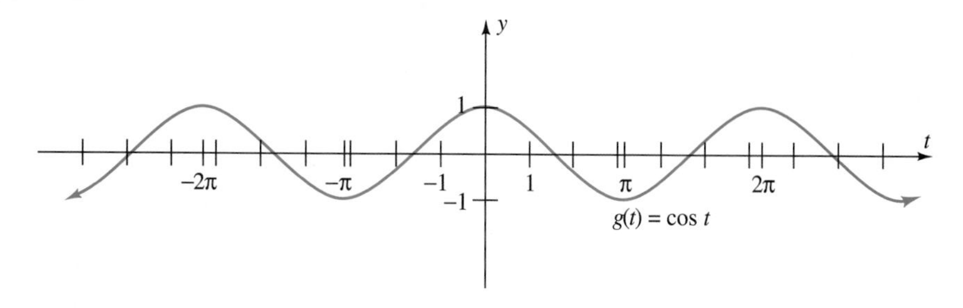

Figure 6-26

For a dynamic simulation of the cosine graphing process described above, see Exercise 47.

The graphs of the cosine function in Figure 6–26 and the sine function in Figure 6–23 lie between -1 and 1 on the y-axis. This is a graphical illustration of the following fact:

Range of Sine and Cosine ▶

$$-1 \leq \sin t \leq 1 \quad \text{and} \quad -1 \leq \cos t \leq 1$$
for every real number t

The techniques of Section 2.5 can be used to graph variations of the cosine function.

EXAMPLE 4 The graph of $h(t) = 4 \cos t$ is the graph of $g(t) = \cos t$ stretched away from the horizontal axis by a factor of 4, as shown in Figure 6–27. ■

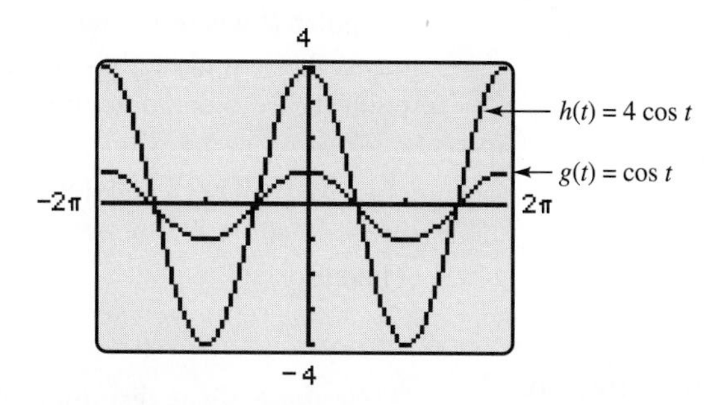

Figure 6-27

EXAMPLE 5 The graph of $k(t) = -2 \cos t + 3$ is the graph of $g(t) = \cos t$ stretched away from the horizontal axis by a factor of 2, reflected in the horizontal axis, and shifted vertically 3 units upward as shown in Figure 6–28. ■

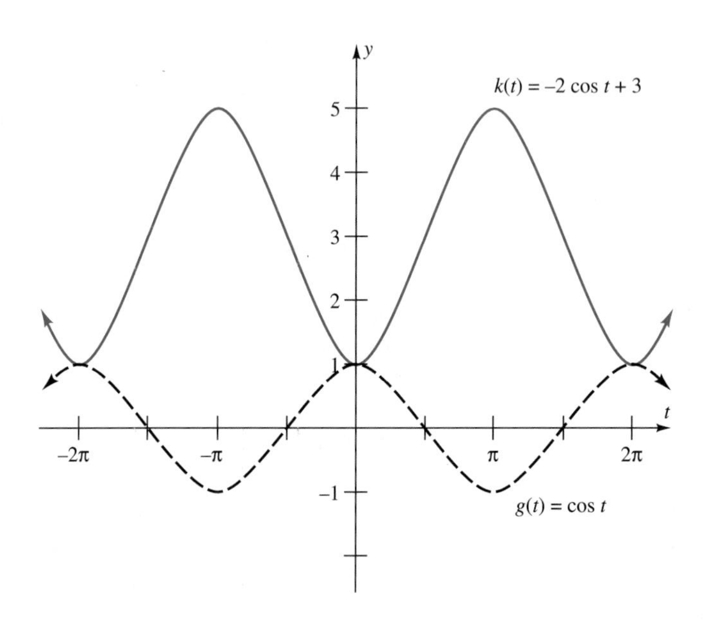

Figure 6-28

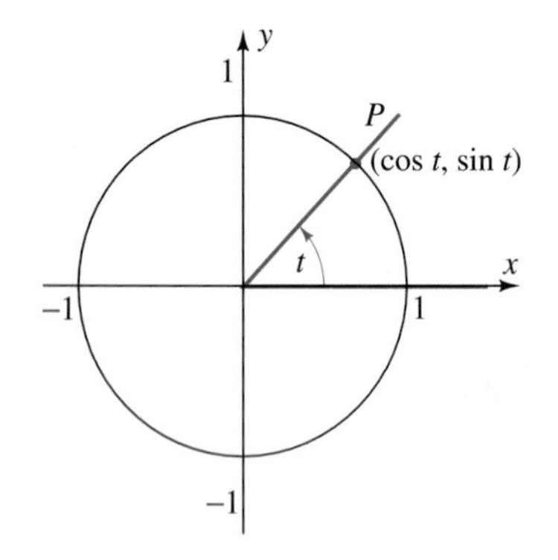

Figure 6-29

Graph of the Tangent Function

To determine the shape of the graph of $h(t) = \tan t$, we use an interesting connection between the tangent function and straight lines. As shown in Figure 6–29, the point P where the terminal side of an angle of t radians in standard position meets the unit circle has coordinates $(\cos t, \sin t)$. We can use this point and the point $(0, 0)$ to compute the *slope* of the terminal side:

$$\text{slope} = \frac{\sin t - 0}{\cos t - 0} = \frac{\sin t}{\cos t} = \tan t$$

Therefore,

Slope and Tangent ▶

The slope of the terminal side of an angle of t radians in standard position is the number $\tan t$.

The graph of $h(x) = \tan t$ can now be sketched by watching the slope of the terminal side of an angle of t radians, as t takes different values. Recall that the more steeply a line rises from left to right, the larger its slope. Similarly, lines that fall from left to right have negative slopes that increase in absolute value as the line falls more steeply.

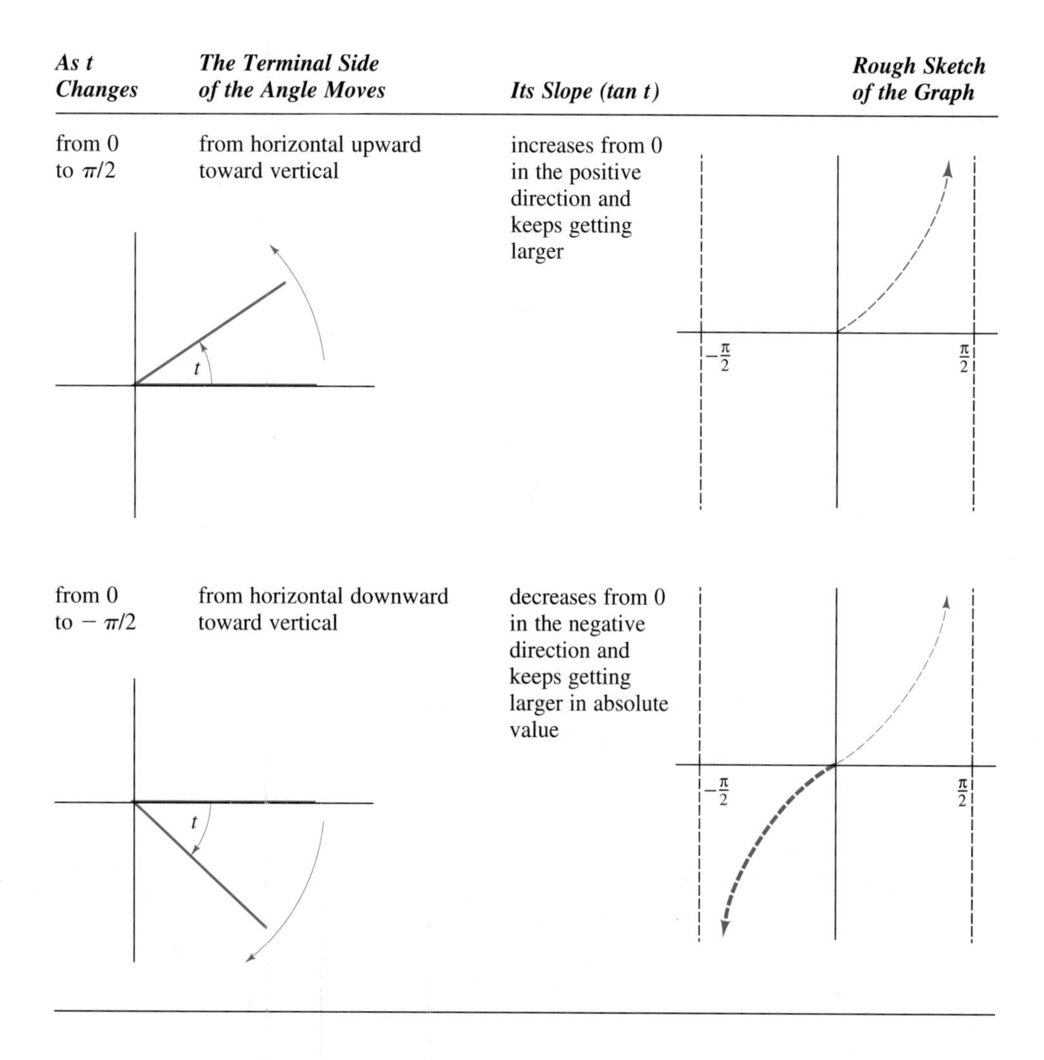

As t Changes	*The Terminal Side of the Angle Moves*	*Its Slope (tan t)*	*Rough Sketch of the Graph*
from 0 to $\pi/2$	from horizontal upward toward vertical	increases from 0 in the positive direction and keeps getting larger	
from 0 to $-\pi/2$	from horizontal downward toward vertical	decreases from 0 in the negative direction and keeps getting larger in absolute value	

When $t = \pm\pi/2$, the terminal side of the angle is vertical and hence its slope is not defined. This corresponds to the fact that the tangent function is not defined when $t = \pm\pi/2$. The vertical lines through $\pm\pi/2$ are vertical asymptotes of the graph: It gets closer and closer to these lines, but never touches them.

When t is slightly larger than $\pi/2$, the terminal side falls from left to right and has negative slope (draw a picture). As t goes from $\pi/2$ to $3\pi/2$, the terminal side goes from almost vertical with negative slope to horizontal to almost vertical with positive slope, exactly as it does between $-\pi/2$ and $\pi/2$. So, the graph repeats the same pattern. The same thing happens between $3\pi/2$ and $5\pi/2$, between $-3\pi/2$ and $-\pi/2$, etc. Therefore, the entire graph looks like this (Figure 6–30):

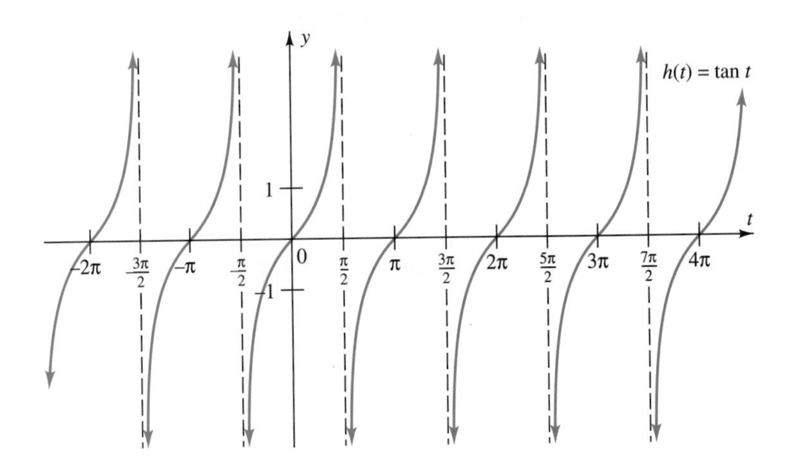

Figure 6-30

Because calculators sometimes do not graph accurately across vertical asymptotes, the graph may look slightly different on a calculator screen (with vertical line segments where the asymptotes should be).

The graph of the tangent function repeats the same pattern at intervals of length π. This means that the tangent function repeats its values at intervals of π.

Period of Tangent ▶

The tangent function is periodic with period π: For every real number t in its domain

$$\tan(t \pm \pi) = \tan t.$$

An algebraic proof of this fact will be given in Section 6.4.

EXAMPLE 6 The graph of $k(t) = 2\tan t + 4$ is the graph of $h(t) = \tan t$ stretched away from the horizontal axis by a factor of 2 and then shifted 4 units upward, as shown in Figure 6–31. ■

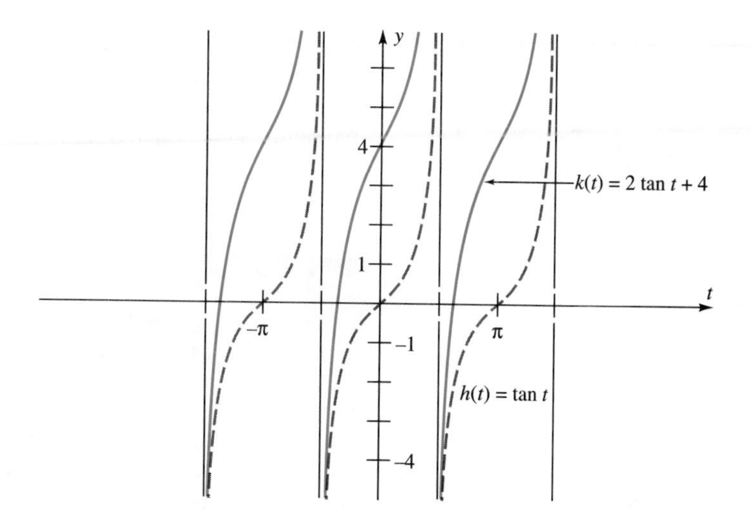

Figure 6-31

Graphs and Identities

As we shall see in the next section and in Chapter 8 there are numerous trigonometric identities.* In the meantime, however, you can use your calculator to determine that certain equations are not identities and to identify equations that possibly might be identities.

EXAMPLE 7 Which of the following equations could possibly be an identity?

(a) $\cos\left(\frac{\pi}{2} + t\right) = \sin t$ **(b)** $\cos\left(\frac{\pi}{2} - t\right) = \sin t$

4

$g(t) = \sin t$

$f(t) = \cos\left(\frac{\pi}{2} + t\right)$

−4

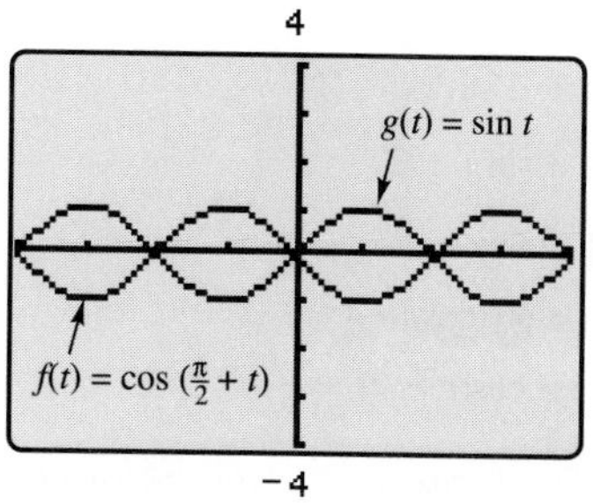

Figure 6-32

Solution

(a) Consider the functions $f(t) = \cos\left(\frac{\pi}{2} + t\right)$ and $g(t) = \sin t$, whose rules are given by the two sides of the equation $\cos\left(\frac{\pi}{2} + t\right) = \sin t$. If this equation is an identity, then $f(t) = g(t)$ for every real number t, and hence f and g have the same graph. But the graphs of f and g on the interval $[-2\pi, 2\pi]$ (Figure 6–32) are obviously different. Therefore, this equation is *not* an identity.

*An **identity** is an equation that is true for all values of the variable for which both sides of the equation are defined. For example $(x - 3)^2 = x^2 - 6x + 9$ for every real number x.

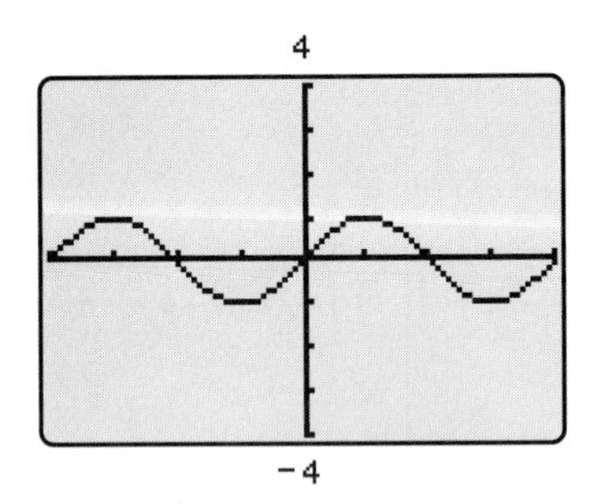

Figure 6-33

(b) We can test this equation in the same manner. The graph of the left side, that is, the graph of $h(t) = \cos\left(\frac{\pi}{2} - t\right)$, in Figure 6–33 appears to be the same as the graph of $g(t) = \sin t$ on the interval $[-2\pi, 2\pi]$ (Figure 6–32).

GRAPHING EXPLORATION Graph $h(t) = \cos\left(\frac{\pi}{2} - t\right)$ and $g(t) = \sin t$ on the same screen and use the trace feature to confirm that the graphs appear to be identical.

The fact that the graphs appear identical means that the two functions have the same value at every number t that the calculator computed in making the graphs (at least 95 numbers). This evidence strongly suggests that the equation $\cos\left(\frac{\pi}{2} - t\right) = \sin t$ is an identity, but does not prove it. All we can say at this point is that the equation could possibly be an identity. ■

Equation (b) in Example 7 is actually an identity, as we shall see in Section 8.2, but it's possible for two functions to have the same value for infinitely many numbers, but not for every number. Similarly, depending on the viewing window, two graphs that are actually quite different may appear identical. See Exercises 43 and 44 for some examples.

EXERCISES 6.3

In Exercises 1–6, use the graphs of the sine and cosine functions to find all the solutions of the equation.

1. $\sin t = 0$ **2.** $\cos t = 0$ **3.** $\sin t = 1$

4. $\sin t = -1$ **5.** $\cos t = -1$ **6.** $\cos t = 1$

In Exercises 7–10, find tan t, where the terminal side of an angle of t radians lies on the given line.

7. $y = 11x$ **8.** $y = 1.5x$ **9.** $y = 1.4x$

10. $y = .32x$

In Exercises 11–22, list the transformations needed to change the graph of f(t) into the graph of g(t). [See Section 2.5.]

11. $f(t) = \sin t$; $g(t) = \sin t + 3$

12. $f(t) = \cos t$; $g(t) = \cos t - 2$

13. $f(t) = \cos t$; $g(t) = -\cos t$

14. $f(x) = \sin t$; $g(t) = -3 \sin t$

15. $f(t) = \tan t$; $g(t) = \tan t + 5$

16. $f(t) = \tan t$; $g(t) = -\tan t$

17. $f(t) = \cos t$; $g(t) = 3 \cos t$

18. $f(t) = \sin t$; $g(t) = -2 \sin t$

19. $f(t) = \sin t$; $g(t) = 3 \sin t + 2$

20. $f(t) = \cos t$; $g(t) = 5 \cos t + 3$

21. $f(t) = \sin t$; $g(t) = \sin(t - 2)$

22. $f(t) = \cos t$; $g(t) = 3 \cos(t + 2) - 3$

In Exercises 23–30, use the graphs of the trigonometric functions to determine the number *of solutions of the equation between 0 and 2π.*

23. $\sin t = 3/5$ [*Hint:* How many points on the graph of $f(t) = \sin t$ between $t = 0$ and $t = 2\pi$ have second coordinate 3/5?]

24. $\cos t = -1/4$ **25.** $\tan t = 4$ **26.** $\cos t = 2/3$

27. $\sin t = -1/2$

28. $\sin t = k$, where k is a constant such that $-1 < k < 1$.

29. $\cos t = k$, where k is a constant such that $-1 < k < 1$.

30. $\tan t = k$, where k is any constant.

In Exercises 31–42, use graphs to determine whether the equation could possibly be an identity or definitely is not an identity.

31. $\sin(-t) = -\sin t$

32. $\cos(-t) = \cos t$

33. $\sin^2 t + \cos^2 t = 1$

34. $\sin(t + \pi) = -\sin t$

35. $\sin t = \cos(t - \pi/2)$

36. $\sin^2 t - \tan^2 t = -(\sin^2 t)(\tan^2 t)$

37. $\dfrac{\sin t}{1 + \cos t} = \tan t$

38. $\dfrac{\cos t}{1 - \sin t} = \dfrac{1}{\cos t} + \tan t$

39. $\cos\left(\dfrac{\pi}{2} + t\right) = -\sin t$

40. $\sin\left(\dfrac{\pi}{2} + t\right) = -\cos t$

41. $(1 + \tan t)^2 = \dfrac{1}{\cos t}$

42. $(\cos^2 t - 1)(\tan^2 t + 1) = -\tan^2 t$

Thinkers

Exercises 43–46 explore various ways in which a calculator can produce inaccurate or misleading graphs of trigonometric functions.

43. Choose a viewing window in which $-3 \le y \le 3$ and $0 \le x \le k$, with k chosen according to this table:

	TI-82/83 Sharp 9300 Casio 9800	*TI-81*	*TI-85*
Width of Screen	95 pixels	96 pixels	127 pixels
k	188π	190π	252π

	HP-38	*TI-92*
Width of Screen	131 pixels	239 pixels
k	260π	476π

(a) Graph $y = \cos x$ and the constant function $y = 1$ on the same screen. Do the graphs look identical? Are the functions the same?

(b) Use the trace feature to move the cursor along the graph of $y = \cos x$, starting at $x = 0$. For what values of x did the calculator plot points? [*Hint:* $2\pi \approx 6.28$.] Use this information to explain why the two graphs look identical.

44. Using the viewing window in Exercise 43, graph $y = \tan x + 2$ and $y = 2$ on the same screen. Explain why the graphs look identical even though the functions are not the same.

45. The graph of $g(x) = \cos x$ is a series of repeated waves (see Figure 6–26). A full wave (from the peak, down to the trough, and up to the peak again) starts at $x = 0$ and finishes at $x = 2\pi$.

(a) How many full waves will the graph make between $x = 0$ and $x = 502.65$ ($\approx 80 \cdot 2\pi$)?

(b) Graph $g(t) = \cos t$ in a viewing window with $0 \le t \le 502.65$. How many full waves are shown on the graph? Is your answer the same as in part (a)? What's going on?

46. Find a viewing window in which the graphs of $y = \cos x$ and $y = .54$ appear identical. [*Hint:* See the chart in Exercise 43 and note that $\cos 1 \approx .54$.]

47. With your calculator in parametric graphing mode and these range values

$$0 \le t \le 6.28 \qquad -1 \le x \le 6.28 \qquad -2.5 \le y \le 2.5,$$

graph the following two functions on the same screen:

$$x_1 = \cos t,\ y_1 = \sin t \quad \text{and} \quad x_2 = t,\ y_2 = \cos t.$$

Using the trace feature, move the cursor along the first graph (the unit circle). Stop at a point on the circle, note the value of t and the x-coordinate of the point. Then switch the trace to the second graph (the cosine function) by using the up or down cursor arrows. The value of t remains the same. How does the y-coordinate of the new point compare with the x-coordinate of the original point on the unit circle? Explain what's going on.

48. Here is proof that the sine function has period 2π. We saw in the text that $\sin(t + 2\pi) = \sin t$ for every t. We must show that there is no positive number smaller than 2π with this property. Do this as follows:

(a) Find a number t such that $\sin(t + \pi) \neq \sin t$.

(b) Find all numbers k such that $0 < k < 2\pi$ and $\sin k = 0$. [*Hint:* Draw a picture and use the definition of the sine function.]

(c) Suppose k is a number such that $\sin(t + k) = \sin t$ for every number t. Show that $\sin k = 0$. [*Hint:* Consider $t = 0$.]

(d) Use parts (a)–(c) to show that there is no positive number k less than 2π with the property that $\sin(t + k) = \sin t$ for *every* number t. Therefore, $k = 2\pi$ is the smallest such number and the sine function has period 2π.

49. Here is a proof that the cosine function has period 2π. We saw in the text that $\cos(t + 2\pi) = \cos t$ for every t. We must show that there is no positive number smaller than 2π with this property. Do this as follows:

(a) Find all numbers k such that $0 < k < 2\pi$ and $\cos k = 1$. [*Hint:* Draw a picture and use the definition of the cosine function.]

(b) Suppose k is a number such that $\cos(t + k) = \cos t$ for every number t. Show that $\cos k = 1$. [*Hint:* Consider $t = 0$.]

(c) Use parts (a) and (b) to show that there is no positive number k less than 2π with the property that $\cos(t + k) = \cos t$ for *every* number t. Therefore, $k = 2\pi$ is the smallest such number and the cosine function has period 2π.

50. Judging from their graphs, which of the functions $f(t) = \sin t$, $g(t) = \cos t$, $h(t) = \tan t$ appear to be even functions? Which appear to be odd functions? Which are neither even nor odd? [*Hint:* See Excursion 2.5.A.]

6.4 BASIC IDENTITIES

The trigonometric functions have numerous interrelationships that often are the key to the mathematical description of various phenomena. These interrelationships are usually expressed as identities.* The fundamental trigonometric identities are developed and examined in this section from both a geometric and an algebraic point of view.

GRAPHING EXPLORATION In the viewing window with $-2\pi \leq x \leq 2\pi$ and $-1 \leq y \leq 2$, graph $f(x) = \sin^2 x$ and $g(x) = \cos^2 x$ on the same screen.

(a) Use trace to move along the cosine graph. Stop at a point and note its y-coordinate. Now use the up or down arrow to move vertically to the sine graph. The x-coordinate remains the same, but note the y-coordinate of this point. What is the sum of these two y-coordinates?

(b) Use trace to move to another point on the cosine curve and repeat part (a). What is the sum of the y-coordinates this time? What do you think the result will be if you move to another point?

(c) Based on this information what do you think the graph of $y = \sin^2 x + \cos^2 x$ will look like? Confirm your conjecture by graphing.

*Recall that an identity is an equation that is true for all values of the variable for which both sides of the equation are defined.

The preceding Exploration suggests the following result:

The Pythagorean Identity ▶

$$\sin^2 t + \cos^2 t = 1 \qquad \text{for every real number } t.$$

The Pythagorean identity, which is also written as

$$\sin^2 t = 1 - \cos^2 t \qquad \text{or} \qquad \cos^2 t = 1 - \sin^2 t,$$

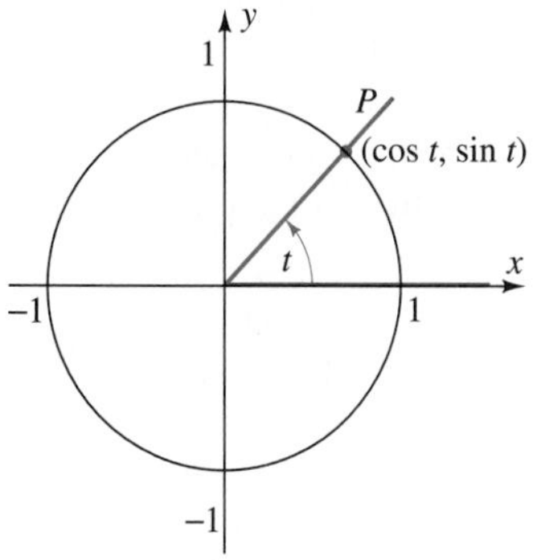

can be proved algebraically by recalling the definition of sine and cosine. For each real number t, $\cos t$ is the x-coordinate of P and $\sin t$ the y-coordinate of P, where P is the intersection of the unit circle and the terminal side of an angle of t radians in standard position, as in Figure 6–34. Consequently, the coordinates $(\cos t, \sin t)$ of P must satisfy the equation of the unit circle: $x^2 + y^2 = 1$, that is, $\cos^2 t + \sin^2 t = 1$.

Figure 6-34

EXAMPLE 1 If $\pi/2 < t < \pi$ and $\sin t = 2/3$, find $\cos t$ and $\tan t$.

Solution By the Pythagorean identity

$$\cos^2 t = 1 - \sin^2 t = 1 - \left(\frac{2}{3}\right)^2 = 1 - \frac{4}{9} = \frac{5}{9}.$$

So there are two possibilities: $\cos t = \sqrt{5/9} = \sqrt{5}/3$ or $\cos t = -\sqrt{5/9} = -\sqrt{5}/3$. Since $\pi/2 < t < \pi$, $\cos t$ is negative (see the chart on page 381). Therefore, $\cos t = -\sqrt{5}/3$ and

$$\tan t = \frac{\sin t}{\cos t} = \frac{2/3}{-\sqrt{5}/3} = \frac{-2\sqrt{5}}{5}. \quad \blacksquare$$

EXAMPLE 2 The Pythagorean identity is valid for *any* number t. For instance, if $t = 3k + 7$, then $\sin^2(3k + 7) + \cos^2(3k + 7) = 1$. ■

EXAMPLE 3 To simplify the expression $\tan^2 t \cos^2 t + \cos^2 t$, we use the definition of tangent and the Pythagorean identity:

$$\begin{aligned}\tan^2 t \cos^2 t + \cos^2 t &= \frac{\sin^2 t}{\cos^2 t}\cos^2 t + \cos^2 t \\ &= \sin^2 t + \cos^2 t = 1. \quad \blacksquare\end{aligned}$$

Negative Angle Identities

GRAPHING EXPLORATION

(a) In a viewing window with $-2\pi \le x \le 2\pi$, graph $y = \sin x$ and $y = \sin(-x)$ on the same screen. Use trace to move along $y = \sin x$. Stop at a point and note its y-coordinate. Use the up or down arrow to move vertically to the graph of $y = \sin(-x)$. The x-coordinate remains the same, but the y-coordinate is different. How are the two y-coordinates related? Is one the negative of the other? Repeat the procedure for other points. Are the results the same?

(b) In a viewing window with $-2\pi \le x \le 2\pi$, graph $y = \cos x$ and $y = \cos(-x)$ on the same screen. How do the graphs compare?

The preceding Exploration suggests the truth of the following statement:

Negative Angle Identities for Sine and Cosine ▶

For every real number t,

$$\sin(-t) = -\sin t \qquad \text{and} \qquad \cos(-t) = \cos t.$$

These identities can be proved by using the definition of sine and cosine and a geometric argument. Consider two angles in standard position, one of t radians and the other of $-t$ radians, as in Figure 6–35. By the definition of sine and cosine the point P, where the terminal side of the angle of t radians meets the unit circle, has coordinates $(\cos t, \sin t)$. Similarly, the point Q, where the terminal side of the angle of $-t$ radians meets the unit circle, has coordinates $(\cos(-t), \sin(-t))$:

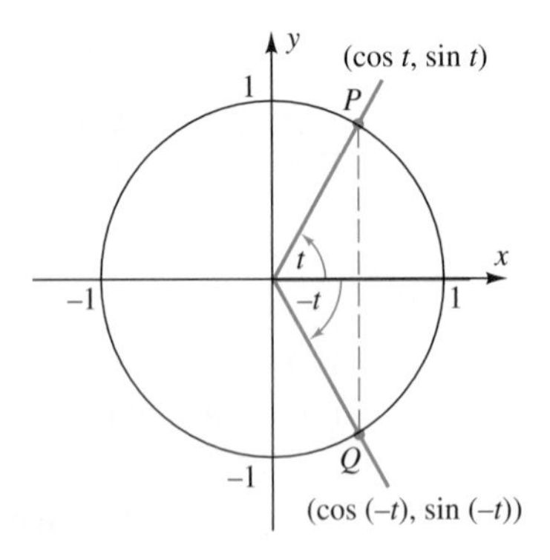

Figure 6-35

As the picture suggests, P and Q lie on the same vertical line. Therefore, they have the same first coordinate, that is, $\cos(-t) = \cos t$. Furthermore, as indicated in the picture, P and Q lie at equal distances from the x-axis.* So the y-coordinate of Q must be the negative of the y-coordinate of P, that is, $\sin(-t) = -\sin t$.

EXAMPLE 4 In Section 6.5 we shall see that $\sin \pi/6 = 1/2$ and that $\cos \pi/6 = \sqrt{3}/2$. Therefore,

$$\sin\left(-\frac{\pi}{6}\right) = -\sin\frac{\pi}{6} = -\frac{1}{2} \quad \text{and} \quad \cos\left(-\frac{\pi}{6}\right) = \cos\frac{\pi}{6} = \frac{\sqrt{3}}{2}. \quad \blacksquare$$

EXAMPLE 5 To simplify $(1 + \sin t)(1 + \sin(-t))$, we use the negative angle identity and the Pythagorean identity:

$$\begin{aligned}(1 + \sin t)(1 + \sin(-t)) &= (1 + \sin t)(1 - \sin t)\\ &= 1 - \sin^2 t\\ &= \cos^2 t. \quad \blacksquare\end{aligned}$$

The negative angle identity for the tangent function is now easy to prove: $\tan(-t) = \dfrac{\sin(-t)}{\cos(-t)} = \dfrac{-\sin t}{\cos t} = -\tan t$. Hence,

Negative Angle Identity for Tangent ▶

$\tan(-t) = -\tan t$ for every real number t.

Other Identities

GRAPHING EXPLORATION

(a) In a viewing window with $-2\pi \le x \le 2\pi$, graph $y = \sin x$ and $y = \sin(x + \pi)$ on the same screen. Use trace to see how the y-coordinate of a point on one graph is related to the y-coordinate of the point on the other graph with the same x-coordinate. Are they negatives of each other?

(b) Do part (a) for $y = \cos x$ and $y = \cos(x + \pi)$. Are the results the same?

*These facts can be proved by considering basic facts about congruent triangles and right angles.

The preceding Exploration suggests another pair of identities:

Diameter Identities ▶

$$\sin(t + \pi) = -\sin t \quad \text{and} \quad \cos(t + \pi) = -\cos t$$
for every real number *t*.

Here is a proof of these statements (which also explains the title "diameter identities"). Let t be any real number and construct two angles in standard position, one of measure t radians and the other of measure $t + \pi$ radians, as shown in Figure 6–36. Since P and Q are on the unit circle, the definition of the sine and cosine functions shows that

P has coordinates $(\cos t, \sin t)$; and

Q has coordinates $(\cos(t + \pi), \sin(t + \pi))$.

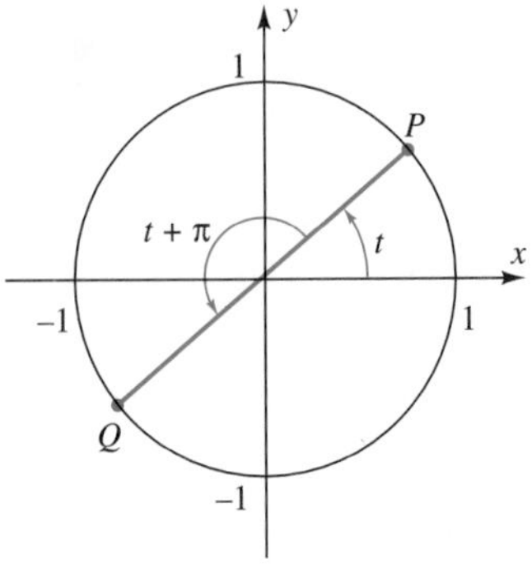

Figure 6-36

According to the Midpoint Formula (page 27), the coordinates of the midpoint of PQ are

$$\left(\frac{\cos t + \cos(t + \pi)}{2}, \frac{\sin t + \sin(t + \pi)}{2}\right).$$

But PQ is a diameter of the circle, so its midpoint is the center (0, 0). Hence,

$$\frac{\cos t + \cos(t + \pi)}{2} = 0 \qquad \frac{\sin t + \sin(t + \pi)}{2} = 0$$

$$\cos t + \cos(t + \pi) = 0 \qquad \sin t + \sin(t + \pi) = 0$$

$$\cos(t + \pi) = -\cos t \qquad \sin(t + \pi) = -\sin t$$

This completes the proof.

EXAMPLE 6 Given that $\sin \pi/6 = 1/2$ and $\cos \pi/6 = \sqrt{3}/2$, find

$$\sin\frac{7\pi}{6} \quad \text{and} \quad \cos\frac{7\pi}{6}.$$

Solution Since $\frac{7\pi}{6} = \frac{\pi}{6} + \frac{6\pi}{6} = \frac{\pi}{6} + \pi$ the diameter identities show that

$$\sin\frac{7\pi}{6} = \sin\left(\frac{\pi}{6} + \pi\right) = -\sin\frac{\pi}{6} = -\frac{1}{2}$$

and

$$\cos\frac{7\pi}{6} = \cos\left(\frac{\pi}{6} + \pi\right) = -\cos\frac{\pi}{6} = -\frac{\sqrt{3}}{2} \quad ■$$

The diameter identities enable us to show that the tangent function has period π (as stated in the box on page 392):

$$\tan(t + \pi) = \frac{\sin(t + \pi)}{\cos(t + \pi)} = \frac{-\sin t}{-\cos t} = \frac{\sin t}{\cos t} = \tan t.$$

They also lead to a number of additional identities.

EXAMPLE 7 If v is any real number, then

$$\sin(\pi - v) = \sin(-v + \pi)$$
$$= -\sin(-v) \qquad \text{[Diameter identity with } t = -v\text{]}$$
$$= -(-\sin v) = \sin v. \qquad \text{[Negative angle identity]}$$

This proves the identity: $\sin(\pi - v) = \sin v$. ■

EXERCISES 6.4

In Exercises 1–6, determine if it is possible for a number t to satisfy the given conditions. [Hint: *Think Pythagorean.*]

1. $\sin t = 5/13$ and $\cos t = 12/13$

2. $\sin t = -2$ and $\cos t = 1$

3. $\sin t = -1$ and $\cos t = 1$

4. $\sin t = 1/\sqrt{2}$ and $\cos t = -1/\sqrt{2}$

5. $\sin t = 1$ and $\tan t = 1$

6. $\cos t = 8/17$ and $\tan t = 15/8$

In Exercises 7–10, use the Pythagorean identity to find sin t.

7. $\cos t = -.5$ and $\pi < t < 3\pi/2$

8. $\cos t = -3/\sqrt{10}$ and $\pi/2 < t < \pi$

9. $\cos t = 1/2$ and $0 < t < \pi/2$

10. $\cos t = 2/\sqrt{5}$ and $3\pi/2 < t < \pi$

In Exercises 11–20, assume that $\sin t = 3/5$ and $0 < t < \pi/2$. Use identities in the text to find the number.

11. $\sin(-t)$ **12.** $\sin(t + 10\pi)$ **13.** $\sin(2\pi - t)$

14. $\cos t$ **15.** $\tan t$ **16.** $\cos(-t)$

17. $\tan(2\pi - t)$ **18.** $\tan(t + \pi)$

19. $\sin(t + \pi)$ **20.** $\cos(t + \pi)$

In Exercises 21–30, assume that $\cos t = -2/5$ and $\pi < t < 3\pi/2$. Use identities to find the number.

21. $\sin t$ **22.** $\tan t$ **23.** $\cos(2\pi - t)$

24. $\cos(-t)$ **25.** $\sin(4\pi + t)$ **26.** $\tan(4\pi - t)$

27. $\tan(t + \pi)$ **28.** $\sin(t + \pi)$ **29.** $\cos(t + \pi)$

30. $\sin(\pi - t)$

In Exercises 31–46, assume that

$$\sin \pi/4 = \sqrt{2}/2, \qquad \sin \pi/3 = \sqrt{3}/2,$$
$$\sin \pi/8 = \frac{\sqrt{2 - \sqrt{2}}}{2}$$

Use identities to find the functional value.

31. $\cos \pi/4$ **32.** $\cos \pi/3$ **33.** $\cos \pi/8$

34. $\tan \pi/4$ **35.** $\tan \pi/3$ **36.** $\tan \pi/8$

37. $\sin 9\pi/4$ **38.** $\sin 5\pi/3$ **39.** $\sin 17\pi/8$

40. $\sin(-4\pi/3)$ **41.** $\sin 5\pi/4$ **42.** $\cos 5\pi/4$

43. $\sin 9\pi/8$ **44.** $\tan(-\pi/4)$ **45.** $\cos 5\pi/3$

46. $\tan(-15\pi/8)$

In Exercises 47–50, use periodicity and the known values of sine and cosine at 0, $\pi/2$, π to find the number.

47. $\cos 9\pi/2$ **48.** $\sin(-11\pi)$ **49.** $\cos 45\pi/2$

50. $\cos(-17\pi/2)$

In Exercises 51–62, use algebra and identities in the text to simplify the expression. Assume all denominators are nonzero.

51. $(\sin t + \cos t)(\sin t - \cos t)$

52. $(\sin t - \cos t)^2$ **53.** $\tan t \cos t$ **54.** $(\sin t)/(\tan t)$

55. $\sqrt{\sin^3 t \cos t}\sqrt{\cos t}$

56. $(\tan t + 2)(\tan t - 3) - (6 - \tan t) + 2\tan t$

57. $\left(\dfrac{4\cos^2 t}{\sin^2 t}\right)\left(\dfrac{\sin t}{4\cos t}\right)^2$

58. $\dfrac{5\cos t}{\sin^2 t} \cdot \dfrac{\sin^2 t - \sin t \cos t}{\sin^2 t - \cos^2 t}$

59. $\dfrac{\cos^2 t + 4\cos t + 4}{\cos t + 2}$ **60.** $\dfrac{\sin^2 t - 2\sin t + 1}{\sin t - 1}$

61. $\dfrac{1}{\cos t} - \sin t \tan t$ **62.** $\dfrac{1 - \tan^2 t}{1 + \tan^2 t} + 2\sin^2 t$

In Exercises 63–67, show that the given function is periodic with period less than 2π. [Hint: *Find a positive number k with $k < 2\pi$, such that $f(t + k) = f(t)$ for every t in the domain of f.*]

63. $f(t) = \sin 2t$ **64.** $f(t) = \cos 3t$

65. $f(t) = \sin^2 t$ **66.** $f(t) = \cos(3\pi t/2)$

67. $f(t) = \tan 2t$

68. Fill the blanks with "even" or "odd" so that the resulting statement is true. Then prove the statement by using an appropriate identity. [Excursion 2.5.A may be helpful.]
(**a**) $f(t) = \sin t$ is an ______ function.
(**b**) $g(t) = \cos t$ is an ______ function.
(**c**) $h(t) = \tan t$ is an ______ function.

In Exercises 69–72, use algebra and the identities in the text to show that the given equation is an identity, as in Example 7. There are many correct ways to do these problems; each hint suggests one possibility.

69. $\sin(t - \pi) = -\sin t$ [*Hint:* $t - \pi = -(\pi - t)$; use a negative angle identity and Example 7.]

70. $\cos(t - \pi) = -\cos t$ [*Hint:* $t - \pi = (t - 2\pi) + \pi$; use a diameter identity and periodicity.]

71. $\cos(\pi - t) = -\cos t$ [*Hint:* Adapt Example 7.]

72. $\tan(\pi - t) = -\tan t$ [*Hint:* Example 7 and Exercise 71.]

In Exercises 73–76, determine graphically whether or not the equation could possibly be an identity. If it could, prove that it is.

73. $\sin(2\pi - t) = -\sin t$ **74.** $\cos(2\pi - t) = \cos t$

75. $\cos(-\pi - t) = \cos t$ **76.** $\sin(-\pi - t) = -\sin t$

6.5 ALTERNATE DESCRIPTIONS AND SPECIAL VALUES

The trigonometric functions were defined in terms of the unit circle. We now consider other methods of describing and evaluating these functions that do not use the unit circle. Among other things, these alternate descriptions will enable us to find *exact* values of sin t, cos t, and tan t when t is an integer multiple of $\pi/4$ or $\pi/6$.

Point-in-the-Plane Description

Consider an angle of t radians in standard position and choose any point Q (except the origin) on the terminal side of this angle. Then the values of the trigonometric functions can be described in terms of the coordinates of the point Q:

Point-in-the-Plane Description ▶

Let t be a real number. Let (x, y) be any point (except the origin) on the terminal side of an angle of t radians in standard position. Then,

$$\sin t = \frac{y}{r} \qquad \cos t = \frac{x}{r} \qquad \tan t = \frac{y}{x}$$

where $r = \sqrt{x^2 + y^2}$ is the distance from (x, y) to the origin.

Proof Let Q be the point on the terminal side of the standard position angle of t radians and let P be the point where the terminal side meets the unit circle, as in Figure 6–37. The definition of sine and cosine shows that P has coordinates $(\cos t, \sin t)$. The distance formula shows that the segment OQ has length $\sqrt{x^2 + y^2}$, which we denote by r.

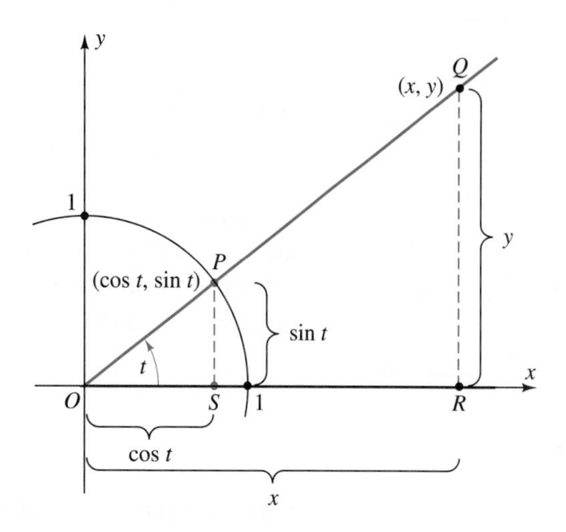

Figure 6-37

Both triangles QOR and POS are right triangles containing an angle of t radians. Therefore, these triangles are *similar.** Consequently,

$$\frac{\text{length } OP}{\text{length } OQ} = \frac{\text{length } PS}{\text{length } QR} \quad \text{and} \quad \frac{\text{length } OP}{\text{length } OQ} = \frac{\text{length } OS}{\text{length } OR}.$$

Figure 6–37 shows what each of these lengths is. Hence,

$$\frac{1}{r} = \frac{\sin t}{y} \quad \text{and} \quad \frac{1}{r} = \frac{\cos t}{x}$$

$$r \sin t = y \quad \text{and} \quad r \cos t = x$$

$$\sin t = \frac{y}{r} \qquad \cos t = \frac{x}{r}$$

Similar arguments work when the terminal side is not in the first quadrant. In every case, $\tan t = \dfrac{\sin t}{\cos t} = \dfrac{y/r}{x/r} = \dfrac{y}{x}$. This completes the proof of the statements in the box on page 402. ❑

*See the Geometry Review Appendix for the basic facts about similar triangles.

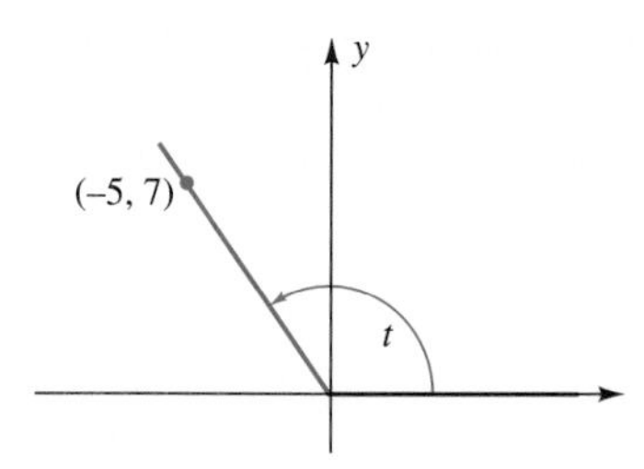

Figure 6-38

EXAMPLE 1 If $(-5, 7)$ is on the terminal side of an angle of t radians in standard position (Figure 6–38), then apply the facts in the box with $(x, y) = (-5, 7)$ and $r = \sqrt{x^2 + y^2} = \sqrt{(-5)^2 + 7^2} = \sqrt{74}$:

$$\sin t = \frac{y}{r} = \frac{7}{\sqrt{74}} \qquad \cos t = \frac{x}{r} = \frac{-5}{\sqrt{74}} \qquad \tan t = \frac{y}{x} = \frac{7}{-5} = -\frac{7}{5}. \quad ■$$

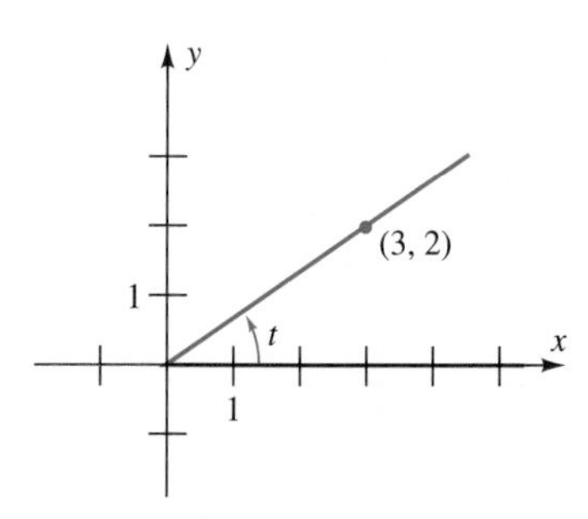

Figure 6-39

EXAMPLE 2 Find $\sin t$ and $\cos t$ when the terminal side of an angle of t radians in standard position is in the first quadrant and $\tan t = 2/3$.

Solution The terminal side passes through (0, 0) and has slope 2/3 ($\tan t$). But the slope of the line from (0, 0) to (3, 2) is 2/3, so (3, 2) must be on the terminal side of the angle, as shown in Figure 6–39. Apply the facts in the box with $(x, y) = (3, 2)$ and $r = \sqrt{x^2 + y^2} = \sqrt{3^2 + 2^2} = \sqrt{13}$:

$$\sin t = \frac{y}{r} = \frac{2}{\sqrt{13}} \quad \text{and} \quad \cos t = \frac{x}{r} = \frac{3}{\sqrt{13}}. \quad ■$$

Right Triangle Description

When t is a number strictly between 0 and $\pi/2$, we can use right triangles to evaluate the trigonometric functions at t. Take any right triangle with an angle of t radians. Move the triangle (flipping it over if necessary) so that the angle of t radians is in standard position, with the hypotenuse of the triangle as the terminal side, as shown in Figure 6–40.

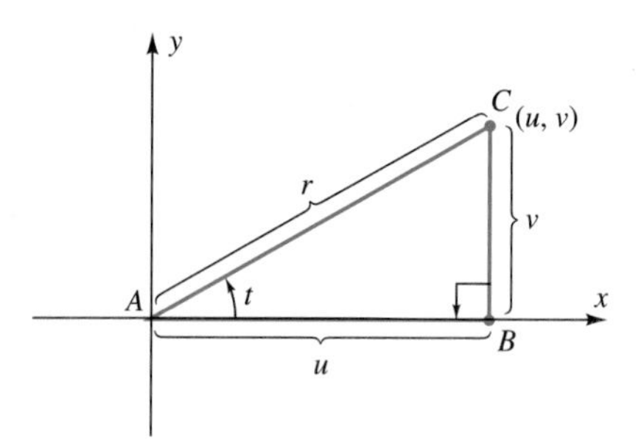

Figure 6-40

Denote the length of side AB (the one *adjacent* to the angle of t radians) by u and the length of side BC (the one *opposite* the angle of t radians) by v. Then the coordinates of C are (u, v). Let r be the length of the *hypotenuse* AC (the distance from (u, v) to the origin). Then the point-in-the-plane description shows that

$$\sin t = \frac{v}{r} = \frac{\text{length of opposite side}}{\text{length of hypotenuse}} \qquad \cos t = \frac{u}{r} = \frac{\text{length of adjacent side}}{\text{length of hypotenuse}}$$

$$\tan t = \frac{v}{u} = \frac{\text{length of opposite side}}{\text{length of adjacent side}}$$

These facts can be succinctly summarized as follows:

The Right Triangle Description ▶

Let t be a number strictly between 0 and $\pi/2$. Consider any right triangle with an angle of t radians. Then,

$$\sin t = \frac{\text{opposite}}{\text{hypotenuse}} \qquad \cos t = \frac{\text{adjacent}}{\text{hypotenuse}} \qquad \tan t = \frac{\text{opposite}}{\text{adjacent}}.$$

This is the description of the trigonometric functions that is usually presented first when trigonometry is studied from the viewpoint of angles and triangles, rather than from the viewpoint of functions. It has the advantage of being independent of both the unit circle and the coordinate system in the plane.

EXAMPLE 3 Consider a right triangle whose other two angles measure $\pi/6$ radians (30°) and $\pi/3$ radians (60°) and whose hypotenuse has length 2. As explained in Example 3 of the Geometry Review, the side opposite the angle of $\pi/6$ radians must have length 1 (half the hypotenuse) and the side adjacent to this angle must have length $\sqrt{3}$, as shown in Figure 6–41.

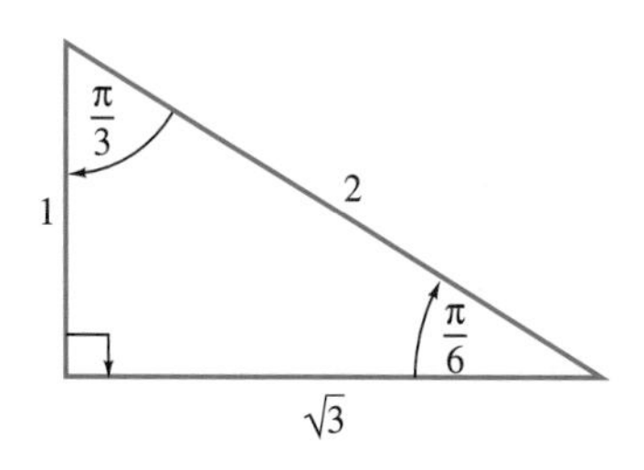

Figure 6-41

Now apply the right triangle description:

$$\sin \frac{\pi}{6} = \frac{\text{opposite}}{\text{hypotenuse}} = \frac{1}{2} \qquad \cos \frac{\pi}{6} = \frac{\text{adjacent}}{\text{hypotenuse}} = \frac{\sqrt{3}}{2}$$

$$\tan \frac{\pi}{6} = \frac{\text{opposite}}{\text{adjacent}} = \frac{1}{\sqrt{3}} = \frac{\sqrt{3}}{3}$$

We can also use the right triangle description with the angle of $\pi/3$ radians. For that angle the opposite side has length $\sqrt{3}$ and the adjacent side has length 1. Therefore,

$$\sin \frac{\pi}{3} = \frac{\text{opposite}}{\text{hypotenuse}} = \frac{\sqrt{3}}{2} \qquad \cos \frac{\pi}{3} = \frac{\text{adjacent}}{\text{hypotenuse}} = \frac{1}{2}$$

$$\tan \frac{\pi}{3} = \frac{\text{opposite}}{\text{adjacent}} = \frac{\sqrt{3}}{1} = \sqrt{3} \quad ■$$

EXAMPLE 4 Consider a right triangle with two angles of $\pi/4$ radians (45°) and hypotenuse 1. As shown in Example 2 of the Geometry Review, each side of the triangle has length $\sqrt{2}/2$ (Figure 6–42).

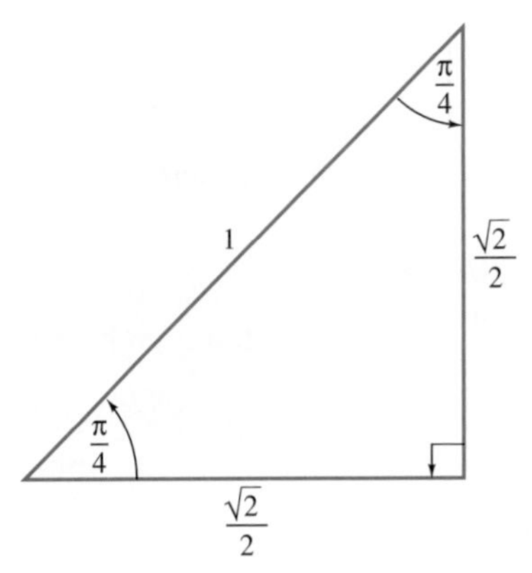

Figure 6-42

Therefore,

$$\sin\frac{\pi}{4} = \frac{\text{opposite}}{\text{hypotenuse}} = \frac{\sqrt{2}/2}{1} = \frac{\sqrt{2}}{2} \qquad \cos\frac{\pi}{4} = \frac{\text{adjacent}}{\text{hypotenuse}} = \frac{\sqrt{2}/2}{1} = \frac{\sqrt{2}}{2}$$

$$\tan\frac{\pi}{4} = \frac{\text{opposite}}{\text{adjacent}} = \frac{\sqrt{2}/2}{\sqrt{2}/2} = 1 \quad ■$$

Note You should memorize the values of the trigonometric functions found in Examples 3 and 4, that is, the values when $t = \pi/3$, $\pi/4$, or $\pi/6$.

EXAMPLE 5 The right triangle shown in Figure 6–43 has hypotenuse 7 and an angle of .6 radians. To find the lengths of the sides, note that

$$\sin .6 = \frac{\text{opposite}}{\text{hypotenuse}} = \frac{c}{7} \quad \text{and} \quad \cos .6 = \frac{\text{adjacent}}{\text{hypotenuse}} = \frac{d}{7}.$$

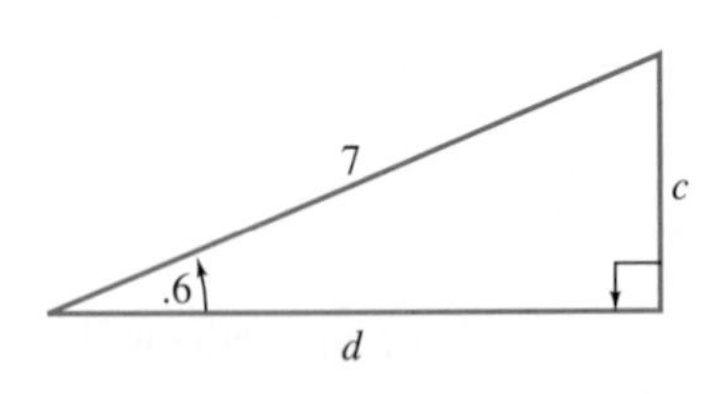

Figure 6-43

Now solve these equations for c and d and use a calculator (in radian mode) to evaluate sine and cosine:

$$c = 7\sin .6 \qquad d = 7\cos .6$$

$$c \approx 7(.5646) \approx 3.95 \qquad d \approx 7(.8253) \approx 5.78 \quad ■$$

EXAMPLE 6 The right triangle in Figure 6–44 has an angle of t radians. To find t, note that

$$\cos t = \frac{\text{adjacent}}{\text{hypotenuse}} = \frac{4}{5} = .8.$$

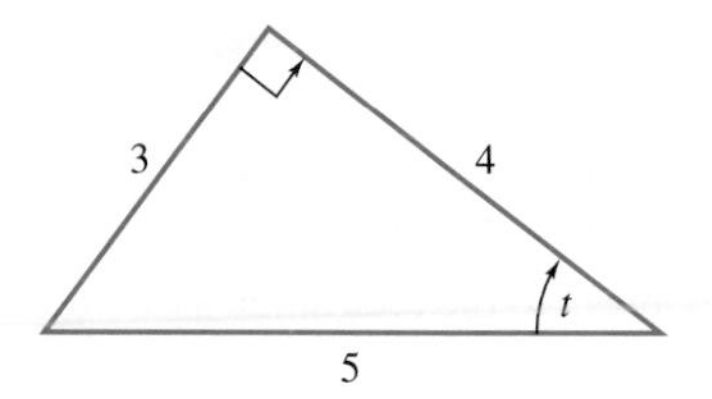

Figure 6-44

In order to find a number t whose cosine is .8, we could solve the equation $\cos t = .8$. It is easier, however, to use the COS^{-1} key on your calculator.* When you key in COS^{-1} .8 ENTER, the calculator produces a number between 0 and $\pi/2$ whose cosine is .8, namely, $t \approx .6435$ radians. (You can think of this as the electronic equivalent of searching through a table of cosine values until you find a number whose cosine is .8, although the calculator actually computes it directly by using infinite series.) ■

Note For now we will use the COS^{-1} key, and the analogous keys SIN^{-1} and TAN^{-1}, as they were used in the last example—as a way to find the number t when $\sin t$ or $\cos t$ or $\tan t$ is known.* In this chapter t will always be the radian measure of an angle in a right triangle. The other important uses of these keys will be discussed in Section 8.5 when we deal with the inverse functions of sine, cosine, and tangent.

Special Values

The following examples show how to find the exact values of $\sin t$, $\cos t$, and $\tan t$, when t is any integer multiple of $\pi/4$ or $\pi/6$, such as $t = -\dfrac{5\pi}{4} = -5 \cdot \dfrac{\pi}{4}$ and $t = \dfrac{11\pi}{3} = 22 \cdot \dfrac{\pi}{6}$.

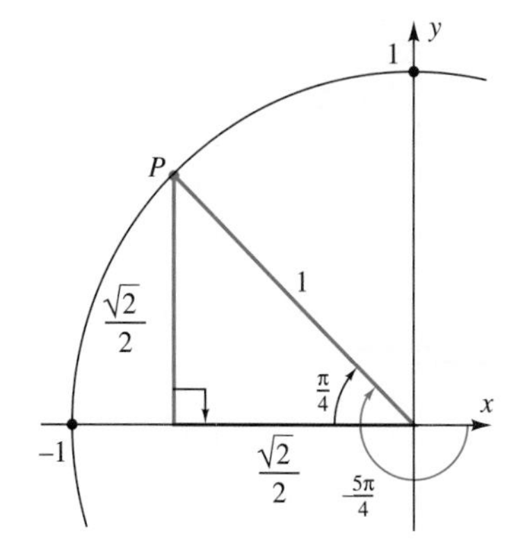

Figure 6-45

EXAMPLE 7 To evaluate the trigonometric functions at $-5\pi/4$, construct an angle of $-5\pi/4$ radians in standard position and let P be the point where the terminal side intersects the unit circle. Draw a vertical line from P to the x-axis, forming a right triangle with hypotenuse 1 and two angles of $\pi/4$ radians, as shown in Figure 6–45.

Each side of this triangle has length $\sqrt{2}/2$, as explained in Example 2 of the Geometry Review. Therefore, the coordinates of P are $(-\sqrt{2}/2, \sqrt{2}/2)$, so that

$$\sin \frac{-5\pi}{4} = y\text{-coordinate of } P = \frac{\sqrt{2}}{2}$$

$$\cos \frac{-5\pi}{4} = x\text{-coordinate of } P = -\frac{\sqrt{2}}{2}$$

$$\tan \frac{-5\pi}{4} = \frac{\sin t}{\cos t} = \frac{\sqrt{2}/2}{-\sqrt{2}/2} = -1. \quad ■$$

EXAMPLE 8 To evaluate the trigonometric functions at $11\pi/3$ (which is $22 \cdot \pi/6$), construct an angle of $11\pi/3$ radians in standard position and

*On HP calculators, the COS^{-1}, SIN^{-1}, and TAN^{-1} keys are labeled ACOS, ASIN, and ATAN respectively.

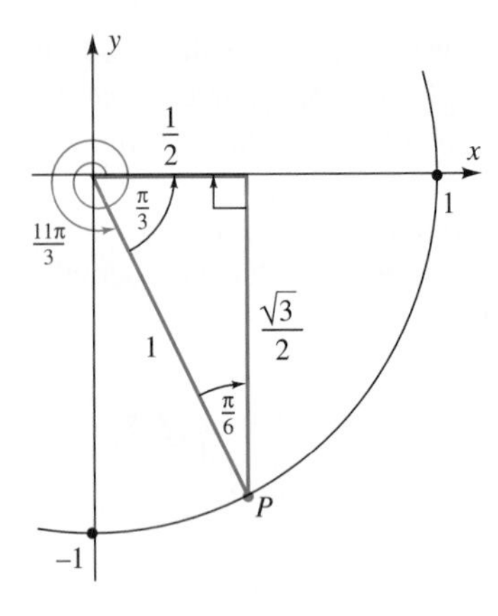

Figure 6-46

draw a vertical line from the x-axis to the point P where the terminal side of the angle meets the unit circle, as shown in Figure 6–46.

The right triangle formed in this way has hypotenuse 1 and angles of $\pi/6$ and $\pi/3$ radians. The sides of the triangle must have lengths 1/2 and $\sqrt{3}/2$ as shown in Figure 6–46 (see Example 4 of the Geometry Review). So the coordinates of P are $(1/2, -\sqrt{3}/2)$ and

$$\sin t = y\text{-coordinate of } P = -\sqrt{3}/2$$

$$\cos t = x\text{-coordinate of } P = 1/2$$

$$\tan t = \frac{\sin t}{\cos t} = \frac{-\sqrt{3}/2}{1/2} = -\sqrt{3}.$$ ■

Applications

The right triangle description of trigonometric functions has a number of practical applications.

EXAMPLE 9 A flagpole casts a 60-ft-long shadow and the angle from the tip of the shadow to the top of the pole measures .6 radians (Figure 6–47). Find the length of the flagpole.

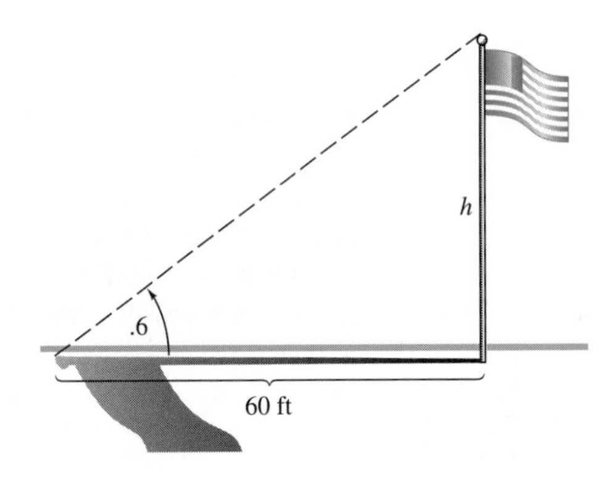

Figure 6-47

Solution Using the right triangle whose legs are the flagpole and its shadow, we see that

$$\tan .6 = \frac{h}{60}$$

$$h = 60 \tan .6 \approx 41.05 \text{ ft.}$$

So the length of the flagpole is about 41.05 ft. ■

Right triangle trigonometry will be considered further in Section 7.1.

EXERCISES 6.5

In Exercises 1–6, find sin t, cos t, tan t when the terminal side of an angle of t radians in standard position passes through the given point.

1. $(2, 7)$ **2.** $(-3, 2)$ **3.** $(-5, -6)$

4. $(4, -3)$ **5.** $(\sqrt{3}, -10)$ **6.** $(-\pi, 2)$

In Exercises 7–12, find $\sin \theta$, $\cos \theta$, $\tan \theta$, where θ is the radian measure of the indicated angle of the right triangle.

7.

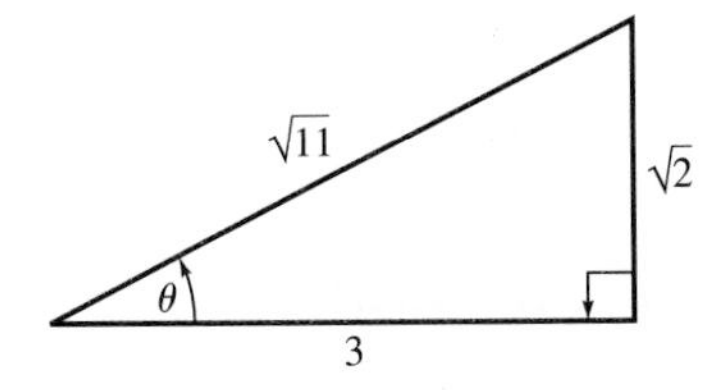

8.

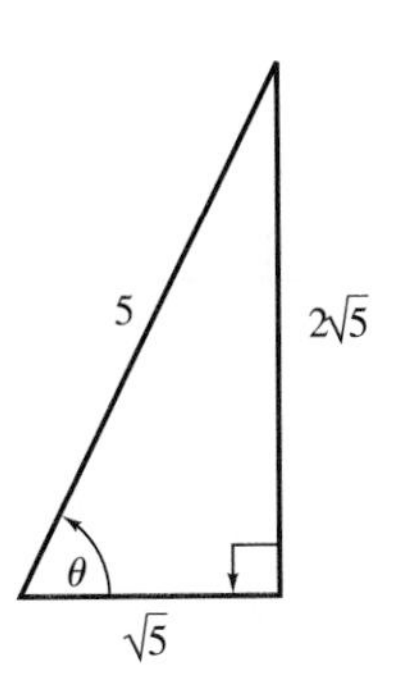

9.

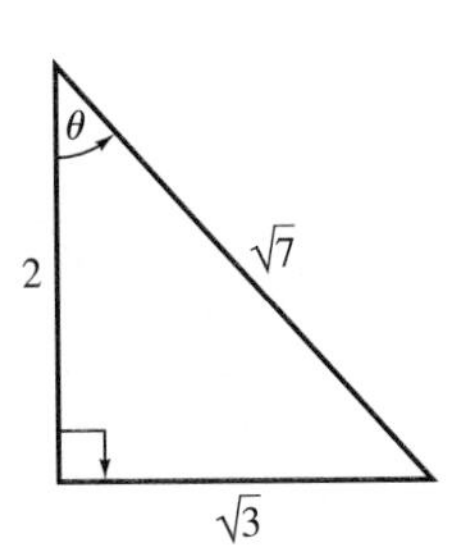

10.

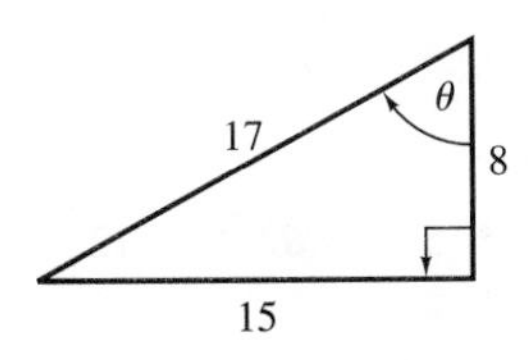

11.

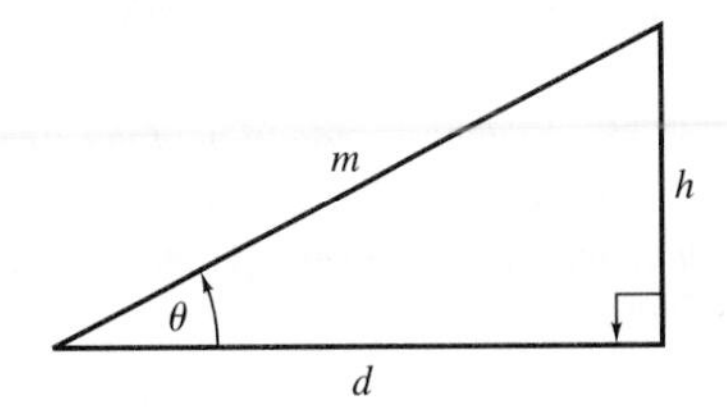

12.

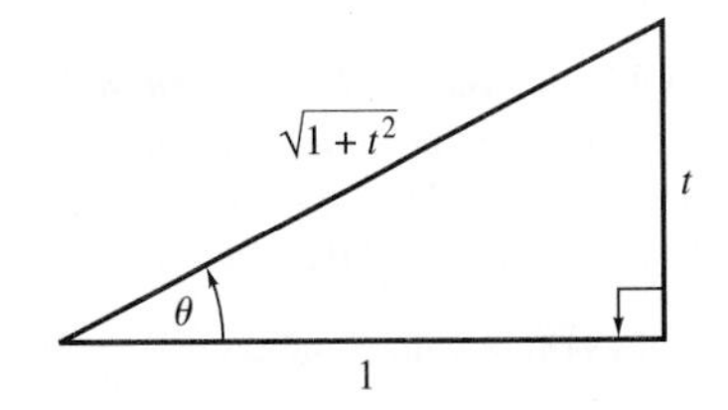

In Exercises 13–27, find the sine, cosine, and tangent of the number, without using a calculator.

13. $5\pi/6$ **14.** $7\pi/6$ **15.** $7\pi/3$ **16.** $17\pi/3$

17. $11\pi/4$ **18.** $5\pi/4$ **19.** $-3\pi/2$ **20.** 3π

21. $-23\pi/6$ **22.** $11\pi/6$ **23.** $-19\pi/3$

24. $-10\pi/3$ **25.** $-15\pi/4$ **26.** $-25\pi/4$

27. $-17\pi/2$

28. Fill the blanks in the table below. Write each entry as a fraction with denominator 2 and with a radical in the numerator. For example, $\sin \pi/2 = 1 = \sqrt{4}/2$. Some students find the resulting pattern an easy way to remember these functional values.

t	0	$\pi/6$	$\pi/4$	$\pi/3$	$\pi/2$
$\sin t$					
$\cos t$					

In Exercises 29–34, write the expression as a single real number. Do not use decimal approximations.

29. $\sin \pi/3 \cos \pi + \sin \pi \cos \pi/3$

30. $\sin \pi/6 \cos \pi/2 - \cos \pi/6 \sin \pi/2$

31. $\cos \pi/2 \cos \pi/4 - \sin \pi/2 \sin \pi/4$

32. $\cos 2\pi/3 \cos \pi + \sin 2\pi/3 \sin \pi$

33. $\sin 3\pi/4 \cos 5\pi/6 - \cos 3\pi/4 \sin 5\pi/6$

34. $\sin(-7\pi/3) \cos 5\pi/4 + \cos(-7\pi/3) \sin 5\pi/4$

In Exercises 35–40, find the equation of the straight line containing the terminal side of the angle in standard position whose radian measure is given. [Hint: *How are slope and tangent related?*]

35. $\pi/6$ **36.** $\pi/4$ **37.** $-\pi/3$ **38.** $-3\pi/4$

39. $7\pi/3$ **40.** $-19\pi/4$

In Exercises 41–46, assume that the terminal side of an angle of t radians in standard position lies in the given quadrant on the given straight line. Find sin t, cos t, tan t. [Hint: *Find a point on the terminal side of the angle.*]

41. Quadrant IV; line with equation $y = -3x$.

42. Quadrant III; line with equation $2y - 4x = 0$.

43. Quadrant IV; line through $(-3, 5)$ and $(-9, 15)$.

44. Quadrant III; line through the origin parallel to $7x - 2y = -6$.

45. Quadrant II; line through the origin parallel to $2y + x = 6$.

46. Quadrant I; line through the origin perpendicular to $3y + x = 6$.

In Exercises 47–54, find the required side of the right triangle in Figure 6–48. Use a calculator (in radian mode) and round your answers to two decimal places.

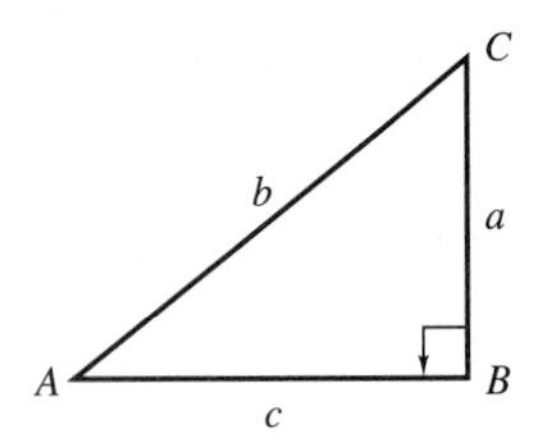

Figure 6-48

47. $b = 10$ and angle A measures .65 radians; find a. [*Hint:* $\sin .65 = a/10$.]

48. $b = 10$ and angle A measures .65 radians; find c. [*Hint:* $\cos .65 = c/10$.]

49. $c = 12$ and angle C measures .7 radians; find b.

50. $c = 4$ and angle A measures .5 radians; find a.

51. $c = 15$ and angle A measures .85 radians; find b.

52. $c = 10$ and angle A measures .7 radians; find a.

53. $a = 20$ and angle A measures .8 radians; find c.

54. $b = 16$ and angle C measures .55 radians; find c.

In Exercises 55–58, find the radian measure t of the indicated angle of the right triangle.

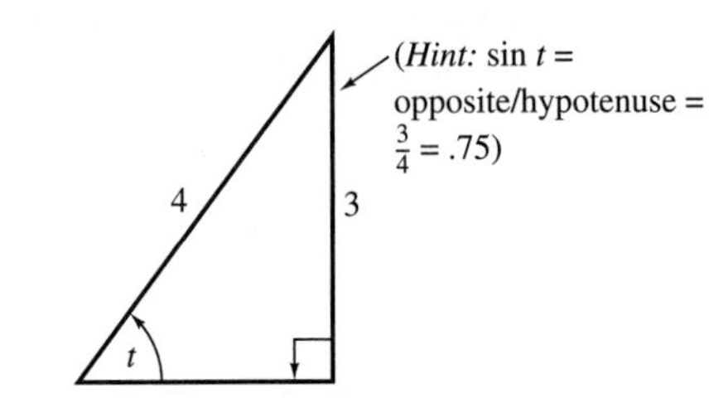

56.

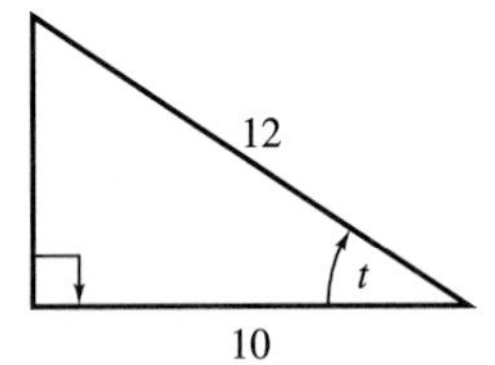

57.

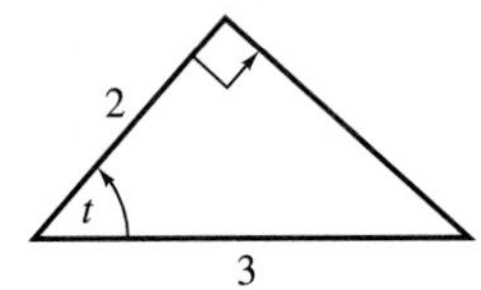

58.

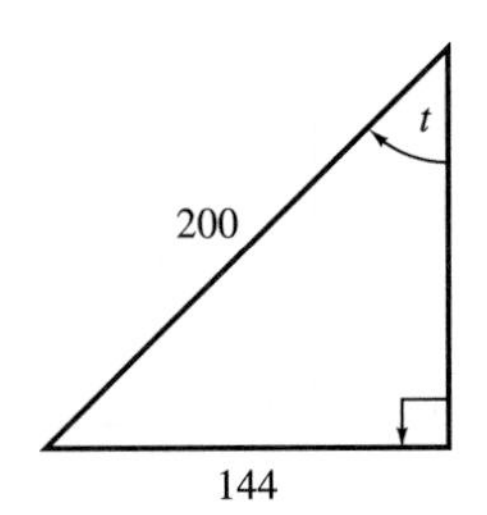

In Exercises 59–62, use Figure 6–48 and find the radian measure of angle A under the given conditions.

59. $a = 7.6$; $c = 10$ **60.** $a = 5.7$; $c = 5$

61. $a = 3.4$; $c = 5$ **62.** $a = 12.36$; $c = 12$

In Exercises 63–68, find the length h of the side of the right triangle, without using a calculator.

63.

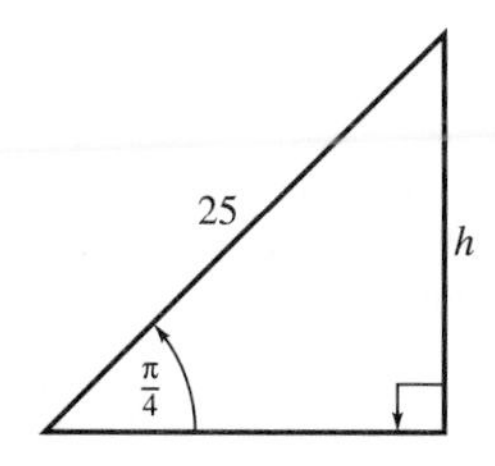

64.

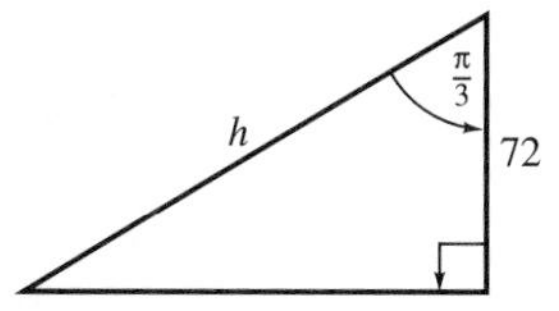

65.

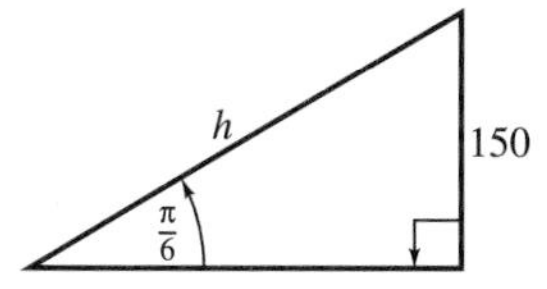

66.

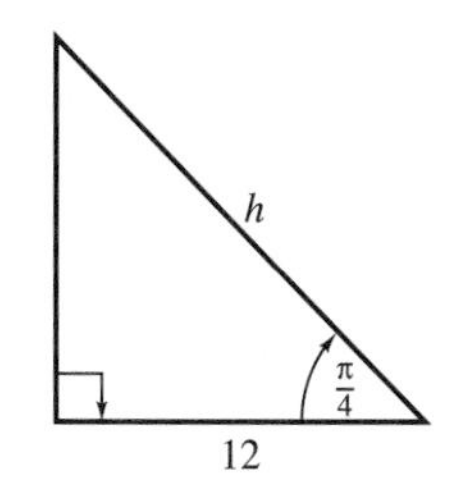

67.

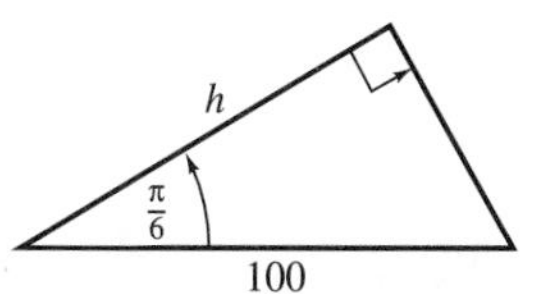

68.

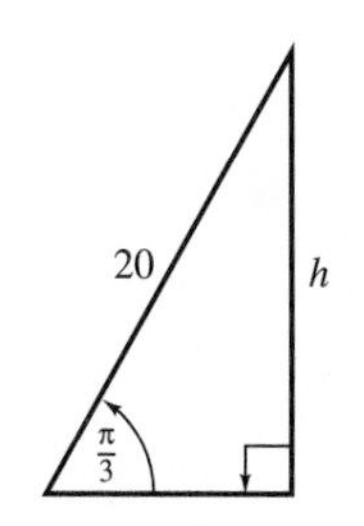

In Exercises 69–72, use Figure 6–48 and find the required side without using a calculator.

69. $a = 4$ and angle A measures $\pi/3$ radians; find c.

70. $c = 5$ and angle A measures $\pi/3$ radians; find a.

71. $c = 10$ and angle A measures $\pi/6$ radians; find a.

72. $a = 12$ and angle A measures $\pi/6$ radians; find c.

In Exercises 73–76, draw a picture of the situation, as in Example 9, then use an appropriate trigonometric function to answer the question.

73. A 20-ft-long ladder leans on a wall of a building. The foot of the ladder makes an angle of .9 radians with the ground. How far above the ground does the top of the ladder touch the wall?

74. A guy wire stretches from the top of an antenna tower to a point on level ground 18 ft from the base of the tower. The angle between the wire and the ground is 1.1 radians. How high is the tower?

75. A 150-ft-long ramp connects a ground-level parking lot with the entrance to a building. If the entrance is 8 ft above the ground, what angle does the ramp make with the ground?

76. A bullet is fired from the top of a 400-ft-high building and hits a point on level ground a quarter of a mile from the building. Assuming that the bullet travels in a straight line, what angle does its path make with the ground?

Thinkers

77. Imagine a 16-foot-long drawbridge on a medieval castle (Figure 6–49).

Figure 6-49

The royal army is engaged in ignominious retreat. The king would like to raise the end of the drawbridge 8 feet off the ground so that Sir Rodney can jump onto the drawbridge and scramble into the castle, while the enemy's cavalry are held at bay. Through how much of an angle must the drawbridge be raised in order for the end of it to be 8 feet off the ground?

78. Through what angle must the drawbridge in Exercise 77 be raised in order that its end be directly above the center of the moat?

79. The cross section of a tunnel is a semicircle with radius 10 meters. The interior walls of the tunnel form a rectangle, as shown in Figure 6–50.

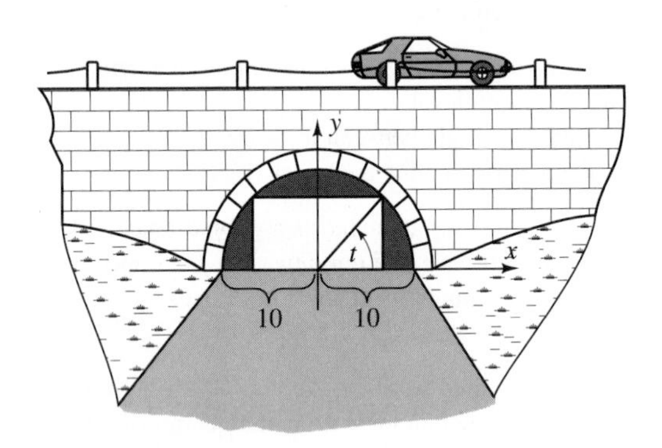

Figure 6-50

(a) Express the area of the rectangular cross section of the tunnel opening as a function of angle t.

(b) For what value of t is the cross-sectional area of the tunnel opening as large as possible? What are the dimensions of the tunnel opening in this case?

80. A gutter is to be made from a strip of metal 24 inches wide by bending up the sides to form a trapezoid, as shown in Figure 6–51.

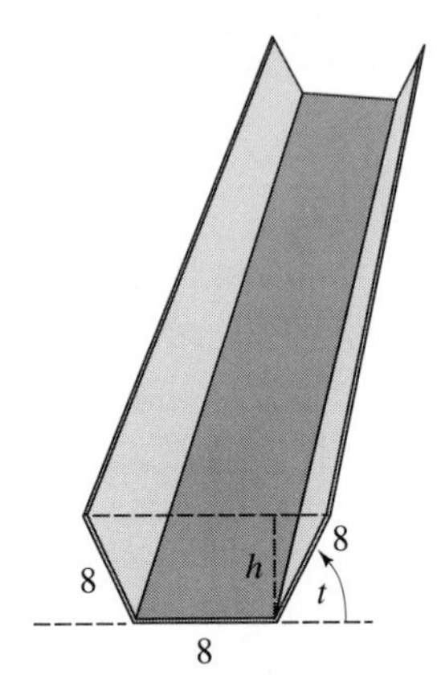

Figure 6-51

(a) Express the area of the cross section of the gutter as a function of the angle t. [*Hint:* The area of a trapezoid with bases b and b' and height h is $h(b + b')/2$.]

(b) For what value of t will this area be as large as possible?

6.6 OTHER TRIGONOMETRIC FUNCTIONS

The three remaining trigonometric functions are defined in terms of sine and cosine, as follows:

Definition of Cotangent, Secant, and Cosecant Functions ►

Name of Function	Value of Function at t Is Denoted	Rule of Function
cotangent	$\cot t$	$\cot t = \dfrac{\cos t}{\sin t}$
secant	$\sec t$	$\sec t = \dfrac{1}{\cos t}$
cosecant	$\csc t$	$\csc t = \dfrac{1}{\sin t}$

The domain of each function consists of all real numbers for which the denominator is nonzero. Since $\sin t = 0$ only when t is an integer multiple of π (see the graph on page 386), the domain of both the cotangent function and the cosecant function consists of all real numbers *except* $0, \pm\pi, \pm 2\pi, \pm 3\pi$, and so on. The domain of the secant function is the same as the domain of the tangent function: all real numbers except $\pm\pi/2, \pm 3\pi/2, \pm 5\pi/2$, and so on.

These functions are usually evaluated with a calculator, even though most calculators do not have keys for CSC, SEC, or COT.* To evaluate the secant and cosecant functions, use the SIN or COS keys, together with the x^{-1} key, to find the appropriate quotient. For instance, a calculator shows that

$$\sec 7 = \frac{1}{\cos 7} \approx 1.3264 \quad \text{and} \quad \csc 18.5 = \frac{1}{\sin 18.5} \approx -2.9199.$$

The cotangent function can be evaluated by using the SIN and COS keys, or by using a fact that will be proved on page 415, namely, that $\cot t = 1/\tan t$. For instance,

$$\cot(-5) = \frac{\cos(-5)}{\sin(-5)} \approx .2958 \quad \text{or} \quad \cot(-5) = \frac{1}{\tan(-5)} \approx .2958.$$

WARNING The calculator keys labeled SIN⁻¹, COS⁻¹, and TAN⁻¹ do *not* denote the functions $1/\sin t$, $1/\cos t$, and $1/\tan t$. For instance, if you key in

COS⁻¹ 7 ENTER

you will get an error message, not the number sec 7 and if you key in

TAN⁻¹ −5 ENTER

you will obtain -1.3734, which is *not* $\cot(-5)$.

Alternate Descriptions

The various descriptions of the sine, cosine, and tangent have analogues for the three new functions:

Point-in-the-Plane Description ▶

Let t be a real number and (x, y) any point (except the origin) on the terminal side of an angle of t radians in standard position. Let $r = \sqrt{x^2 + y^2}$. Then,

$$\cot t = \frac{x}{y} \qquad \sec t = \frac{r}{x} \qquad \csc t = \frac{r}{y}$$

for each number t in the domain of the given function.

*These keys are available on HP-38 (in the TRIG submenu of the MATH menu).

These statements are proved by using the similar descriptions of sine and cosine. For instance,

$$\cot t = \frac{\cos t}{\sin t} = \frac{x/r}{y/r} = \frac{x}{y}.$$

The proofs of the other statements are similar.

EXAMPLE 1 Evaluate all six trigonometric functions at $t = 3\pi/4$.

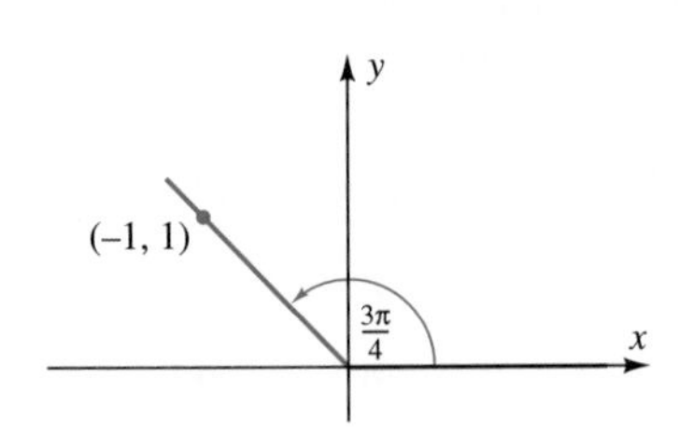

Figure 6-52

Solution The terminal side of an angle of $3\pi/4$ radians in standard position lies on the line $y = -x$, as shown in Figure 6–52.

We shall use the point $(-1, 1)$ on this line to compute the function values. In this case $r = \sqrt{x^2 + y^2} = \sqrt{(-1)^2 + 1^2} = \sqrt{2}$. Therefore,

$$\sin \frac{3\pi}{4} = \frac{y}{r} = \frac{1}{\sqrt{2}} = \frac{\sqrt{2}}{2} \qquad \cos \frac{3\pi}{4} = \frac{x}{r} = \frac{-1}{\sqrt{2}} = \frac{-\sqrt{2}}{2}$$

$$\tan \frac{3\pi}{4} = \frac{y}{x} = \frac{1}{-1} = -1 \qquad \csc \frac{3\pi}{4} = \frac{r}{y} = \frac{\sqrt{2}}{1} = \sqrt{2}$$

$$\sec \frac{3\pi}{4} = \frac{r}{x} = \frac{\sqrt{2}}{-1} = -\sqrt{2} \qquad \cot \frac{3\pi}{4} = \frac{x}{y} = \frac{-1}{1} = -1. \qquad \blacksquare$$

Right angles can be used to find $\cot t$, $\sec t$, and $\csc t$ when $0 < t < \pi/2$.

Right Triangle Description ▶

Let t be any real number strictly between 0 and $\pi/2$. Consider any right triangle containing an angle of t radians:

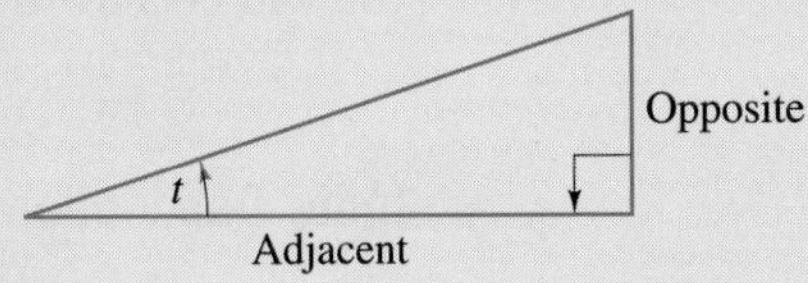

Then,

$$\cot t = \frac{\text{adjacent}}{\text{opposite}} \qquad \sec t = \frac{\text{hypotenuse}}{\text{adjacent}} \qquad \csc t = \frac{\text{hypotenuse}}{\text{opposite}}.$$

All three of these statements are consequences of the right triangle description of sine and cosine. For instance,

$$\sec t = \frac{1}{\cos t} = \frac{1}{\dfrac{\text{adjacent}}{\text{hypotenuse}}} = \frac{\text{hypotenuse}}{\text{adjacent}}.$$

EXAMPLE 2 To evaluate these new trigonometric functions at $\pi/3$, we use the right triangle with hypotenuse 2 in Figure 6–53*:

$$\cot\frac{\pi}{3} = \frac{\text{adjacent}}{\text{opposite}} = \frac{1}{\sqrt{3}} = \frac{\sqrt{3}}{3}$$

$$\sec\frac{\pi}{3} = \frac{\text{hypotenuse}}{\text{adjacent}} = \frac{2}{1} = 2$$

$$\csc\frac{\pi}{3} = \frac{\text{hypotenuse}}{\text{opposite}} = \frac{2}{\sqrt{3}} = \frac{2\sqrt{3}}{3}$$

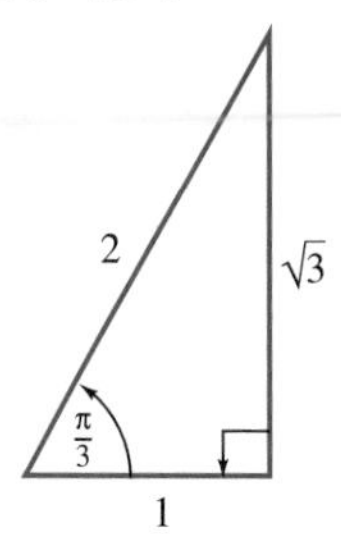

Figure 6-53

■

Basic Identities

We begin by noting the relationship between the cotangent and tangent functions.

Reciprocal Identities ▶

The cotangent and tangent functions are reciprocals; that is,

$$\cot t = \frac{1}{\tan t} \qquad \tan t = \frac{1}{\cot t}$$

for every number t in the domain of both functions.

The proof of these facts comes directly from the definitions; for instance,

$$\cot t = \frac{\cos t}{\sin t} = \frac{1}{\frac{\sin t}{\cos t}} = \frac{1}{\tan t}.$$

Period of Secant, Cosecant, Cotangent ▶

The secant and cosecant functions are periodic with period 2π and the cotangent function is periodic with period π. In symbols,

$$\sec(t + 2\pi) = \sec t \qquad \csc(t + 2\pi) = \csc t$$

$$\cot(t + \pi) = \cot t$$

for every number t in the domain of the given function.

*See Example 3 of the Geometry Review for the reasons why the sides have lengths 1 and $\sqrt{3}$

The proof of these statements uses the fact that each of these functions is the reciprocal of a function whose period is known. For instance,

$$\csc(t + 2\pi) = \frac{1}{\sin(t + 2\pi)} = \frac{1}{\sin t} = \csc t$$

$$\cot(t + \pi) = \frac{1}{\tan(t + \pi)} = \frac{1}{\tan t} = \cot t.$$

The other details are left as an exercise.

Pythagorean Identities ▶

$$\mathbf{1 + \tan^2 t = \sec^2 t \quad \text{and} \quad 1 + \cot^2 t = \csc^2 t}$$

for every number t in the domain of both functions.

The proof of these identities uses the definitions of the functions and the Pythagorean identity $\sin^2 t + \cos^2 t = 1$:

$$1 + \tan^2 t = 1 + \frac{\sin^2 t}{\cos^2 t} = \frac{\cos^2 t + \sin^2 t}{\cos^2 t} = \frac{1}{\cos^2 t} = \left(\frac{1}{\cos t}\right)^2 = \sec^2 t.$$

The second identity is proved similarly.

EXAMPLE 3 Simplify the expression $\dfrac{30\cos^3 t \sin t}{6\sin^2 t \cos t}$, assuming $\sin t \neq 0$, $\cos t \neq 0$:

$$\frac{30\cos^3 t \sin t}{6\sin^2 t \cos t} = \frac{5\cos^3 t \sin t}{\cos t \sin^2 t} = \frac{5\cos^2 t}{\sin t} = 5\frac{\cos t}{\sin t}\cos t = 5\cot t \cos t. \quad ■$$

EXAMPLE 4 Assume $\cos t \neq 0$ and simplify $\cos^2 t + \cos^2 t \tan^2 t$:

$$\cos^2 t + \cos^2 t \tan^2 t = \cos^2 t(1 + \tan^2 t) = \cos^2 t \sec^2 t = \cos^2 t \cdot \frac{1}{\cos^2 t} = 1. \quad ■$$

EXAMPLE 5 If $\tan t = 3/4$ and $\sin t < 0$, find $\cot t$, $\cos t$, $\sin t$, $\sec t$, $\csc t$.

Solution First we have $\cot t = 1/\tan t = 1/\frac{3}{4} = 4/3$. Next we use the Pythagorean identity to obtain:

$$\sec^2 t = 1 + \tan^2 t = 1 + \left(\frac{3}{4}\right)^2 = 1 + \frac{9}{16} = \frac{25}{16}$$

$$\sec t = \pm\sqrt{\frac{25}{16}} = \pm\frac{5}{4}$$

$$\frac{1}{\cos t} = \pm\frac{5}{4} \qquad \text{or equivalently} \qquad \cos t = \pm\frac{4}{5}.$$

Since $\sin t$ is given as negative and $\tan t = \sin t/\cos t$ is positive, $\cos t$ must be negative. Hence, $\cos t = -4/5$. Consequently,

$$\frac{3}{4} = \tan t = \frac{\sin t}{\cos t} = \frac{\sin t}{(-4/5)}$$

so that

$$\sin t = \left(-\frac{4}{5}\right)\left(\frac{3}{4}\right) = -\frac{3}{5}.$$

Therefore,

$$\sec t = \frac{1}{\cos t} = \frac{1}{(-4/5)} = -\frac{5}{4} \qquad \text{and} \qquad \csc t = \frac{1}{\sin t} = \frac{1}{(-3/5)} = -\frac{5}{3}. \quad \blacksquare$$

Graphs

The general shape of the graph of $g(t) = \sec t$ can be determined by using the graph of the cosine function and the fact that $\sec t = 1/\cos t$. First of all, $\sec t$ is not defined whenever $\cos t = 0$, that is, when $t = \pm\pi/2, \pm 3\pi/2, \pm 5\pi/2$, etc. When $\cos t$ is a number near 1 or -1, then so is $\sec t = 1/\cos t$ and their graphs are close together. When $\cos t$ is near 0 (so that its graph is close to the x-axis), then $\sec t = 1/\cos t$ is very large in absolute value* (so that its graph is far from the x-axis), as shown in Figure 6–54. The graph of $g(t) = \sec t$ has vertical asymptotes at those points where the function is not defined, that is, $x = \pi/2$, $x = -\pi/2$, $x = 3\pi/2$, $x = -3\pi/2$, etc.

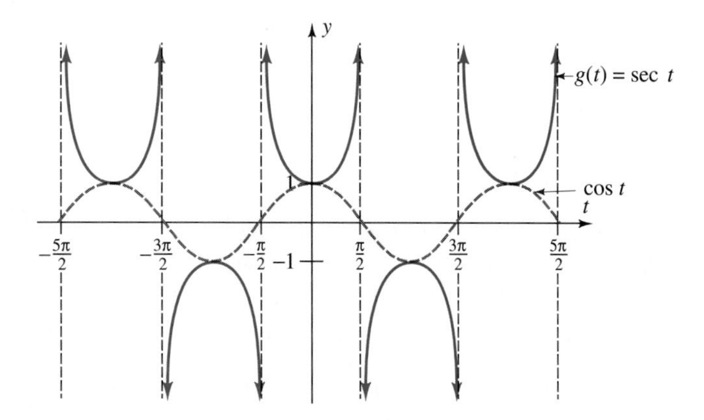

Figure 6-54

*See the Big–Little Principle on page 243.

The graphs of $h(t) = \csc t = 1/\sin t$ and $f(t) = \cos t = 1/\tan t$ can be obtained in a similar fashion (Figure 6–55).

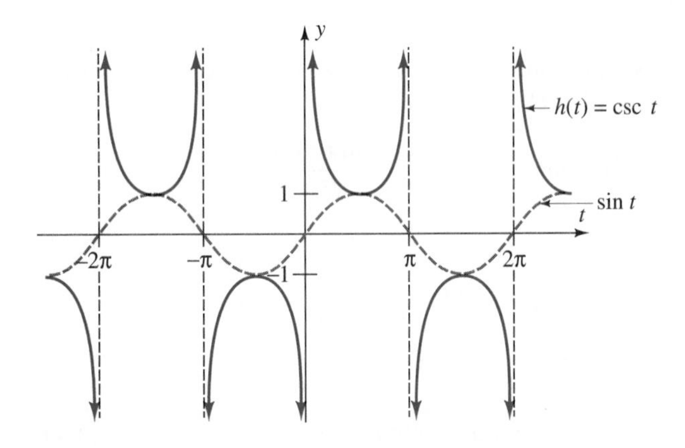

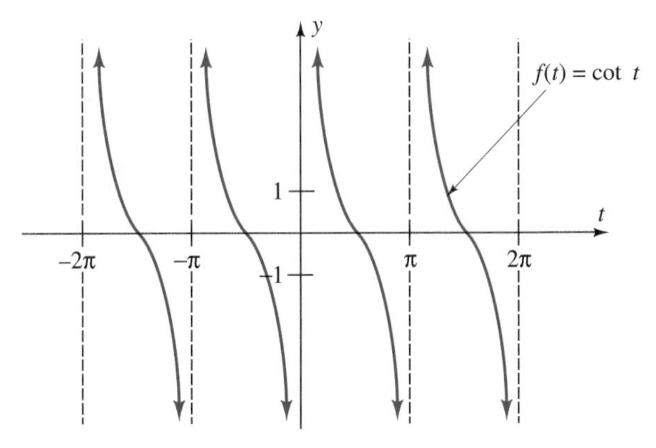

Figure 6-55

EXERCISES 6.6

In Exercises 1–6, determine the quadrant containing the terminal side of an angle of t radians in standard position under the given conditions.

1. $\cos t > 0$ and $\sin t < 0$
2. $\sin t < 0$ and $\tan t > 0$
3. $\sec t < 0$ and $\cot t < 0$
4. $\csc t < 0$ and $\sec t > 0$
5. $\sec t > 0$ and $\cot t < 0$
6. $\sin t > 0$ and $\sec t < 0$

In Exercises 7–16, evaluate all six trigonometric functions at t, where the given point lies on the terminal side of an angle of t radians in standard position.

7. $(3, 4)$ **8.** $(0, 6)$ **9.** $(-5, 12)$

10. $(-2, -3)$ **11.** $(-1/5, 1)$ **12.** $(4/5, -3/5)$

13. $(\sqrt{2}, \sqrt{3})$ **14.** $(-2\sqrt{3}, \sqrt{3})$ **15.** $(1 + \sqrt{2}, 3)$

16. $(1 + \sqrt{3}, 1 - \sqrt{3})$

In Exercises 17–20, evaluate all six trigonometric functions at the number t, which is the radian measure of angle A of the right triangle shown in Figure 6–56, under the given conditions.

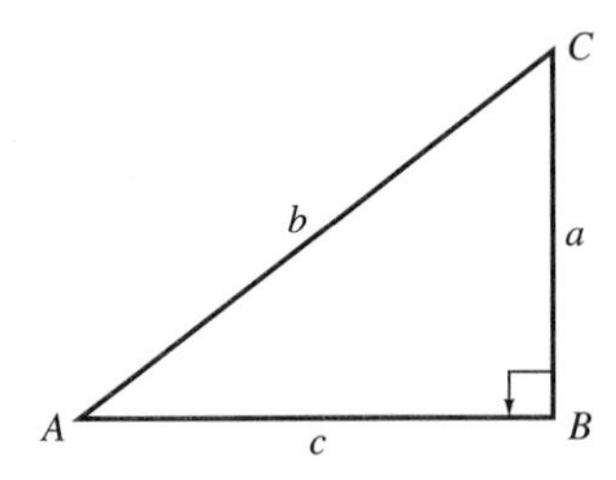

Figure 6-56

17. $a = 2$; $b = 2\sqrt{5}$; $c = 4$ **18.** $a = 5$; $b = 13$; $c = 12$

19. $a = 5$; $c = 3$ **20.** $a = 2$; $b = 10$

21. Find the average rate of change of $f(t) = \cot t$ from $t = 1$ to $t = 3$.

22. Find the average rate of change of $g(t) = \csc t$ from $t = 2$ to $t = 3$.

23. **(a)** Find the average rate of change of $f(t) = \tan t$ from $t = 2$ to $t = 2 + h$, for each of these values of h: .01, .001, .0001, and .00001.

(b) Compare your answers in part (a) with the number $(\sec 2)^2$. What would you guess that the instantaneous rate of change of $f(t) = \tan t$ is at $t = 2$?

In Exercises 24–30, perform the indicated operations, then simplify your answers by using appropriate definitions and identities.

24. $\tan t(\cos t - \csc t)$

25. $\cos t \sin t(\csc t + \sec t)$

26. $(1 + \cot t)^2$

27. $(1 - \sec t)^2$

28. $(\sin t - \csc t)^2$

29. $(\cot t - \tan t)(\cot^2 t + 1 + \tan^2 t)$

30. $(\sin t + \csc t)(\sin^2 t + \csc^2 t - 1)$

In Exercises 31–33, evaluate all six trigonometric functions at the given number without using a calculator.

31. $\frac{4\pi}{3}$ **32.** $-\frac{7\pi}{6}$ **33.** $\frac{7\pi}{4}$

34. Fill in the missing entries of the table below:

t	0	$\frac{\pi}{6}$	$\frac{\pi}{4}$	$\frac{\pi}{3}$	$\frac{\pi}{2}$	$\frac{2\pi}{3}$	$\frac{3\pi}{4}$	$\frac{5\pi}{6}$	π	$\frac{3\pi}{2}$
$\sin t$										
$\cos t$										
$\tan t$					—					—
$\cot t$	—								—	
$\sec t$					—					—
$\csc t$	—								—	

In Exercises 35–40, factor and simplify the given expression.

35. $\sec t \csc t - \csc^2 t$

36. $\tan^2 t - \cot^2 t$

37. $\tan^4 t - \sec^4 t$

38. $4\sec^2 t + 8\sec t + 4$

39. $\cos^3 t - \sec^3 t$

40. $\csc^4 t + 4\csc^2 t - 5$

In Exercises 41–46, simplify the given expression. Assume all denominators are nonzero.

41. $\dfrac{\cos^2 t \sin t}{\sin^2 t \cos t}$

42. $\dfrac{\sec^2 t + 2\sec t + 1}{\sec t}$

43. $\dfrac{4\tan t \sec t + 2\sec t}{6\sin t \sec t + 2\sec t}$

44. $\dfrac{\sec^2 t \csc t}{\csc^2 t \sec t}$

45. $(2 + \sqrt{\tan t})(2 - \sqrt{\tan t})$

46. $\dfrac{6\tan t \sin t - 3\sin t}{9\sin^2 t + 3\sin t}$

In Exercises 47–52, the radian measure of angle A and the length of one side of the right triangle in Figure 6–56 are given. Find the other side.

47. .5 radians; $a = 3$; $c = ?$ [*Hint:* $\cot .5 = c/3$.]

48. 1.1 radians; $a = 7$; $b = ?$

49. .3 radians; $b = 5$; $a = ?$

50. .7 radians; $b = 10$; $c = ?$

51. .9 radians; $c = 2$; $a = ?$

52. 1.3 radians; $c = 4$; $b = ?$

In Exercises 53–58, find the radian measure of angle A of the right triangle shown in Figure 6–56 under the given conditions.

53. $a = 5.88$; $c = 3$

54. $a = 4.8$; $b = 5$

55. $b = 7$; $c = 6.14$

56. $a = 3.125$; $c = 2$

57. $b = 7.4$; $c = 4$ **58.** $a = 6$; $b = 12.54$

In Exercises 59–62, use graphs to determine whether the equation could possibly be an identity or is definitely not an identity.

59. $\tan t = \cot\left(\frac{\pi}{2} - t\right)$ **60.** $\frac{\cos t}{\cos(t - \pi/2)} = \cot t$

61. $\frac{\sin t}{1 - \cos t} = \cot t$ **62.** $\frac{\sec t + \csc t}{1 + \tan t} = \csc t$

In Exercises 63–66, prove the given identity.

63. $1 + \cot^2 t = \csc^2 t$ [*Hint:* Look at the proof of the similar identity on page 416.]

64. $\cot(-t) = -\cot t$ [*Hint:* Express the left side in terms of sine and cosine; then use the negative angle identities and express the result in terms of cotangent.]

65. $\sec(-t) = \sec t$ [Adapt the hint for Exercise 64.]

66. $\csc(-t) = -\csc t$

In Exercises 67–71, prove the given identity. [Hint: *Express the left side in terms of sine and cosine, then use the diameter identities or use Example 7 and Exercises 71–72 of Section 6.4 and express the result in terms of secant, cosecant, or cotangent.*]

67. $\sec(\pi - t) = -\sec t$ **68.** $\sec(t + \pi) = -\sec t$

69. $\cot(\pi - t) = -\cot t$ **70.** $\csc(t + \pi) = -\csc t$

71. $\csc(\pi - t) = \csc t$

In Exercises 72–76, find the values of all six trigonometric functions at t if the given conditions are true.

72. $\cos t = -1/2$ and $\sin t > 0$
[*Hint:* $\sin^2 t + \cos^2 t = 1$.]

73. $\cos t = 0$ and $\sin t = 1$

74. $\sin t = -2/3$ and $\sec t > 0$

75. $\sec t = -13/5$ and $\tan t < 0$

76. $\csc t = 8$ and $\cos t < 0$

77. Show graphically that the equation $\sec t = t$ has infinitely many solutions, but none between $-\pi/2$ and $\pi/2$.

Thinker

78. In the diagram of the unit circle in Figure 6–57, find six line segments whose respective lengths are $\sin t$, $\cos t$, $\tan t$, $\cot t$, $\sec t$, $\csc t$. [*Hint:* $\sin t$ = length *CA*. Why? Note that *OC* has length 1 and various right triangles in the picture are similar.]

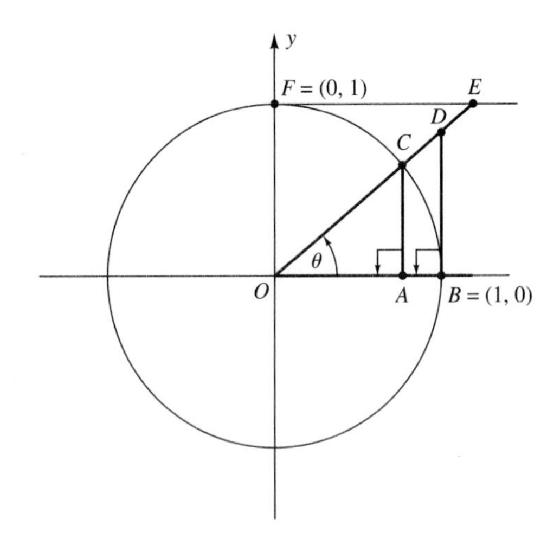

Figure 6-57

6.7 PERIODIC GRAPHS AND SIMPLE HARMONIC MOTION

Functions whose rules are of the form

$$f(t) = A\sin(bt + c) \qquad \text{or} \qquad g(t) = A\cos(bt + c)$$

with A, b, c constants, are used to describe many periodic physical phenomena. Figure 6–58 shows the graphs of three such functions:

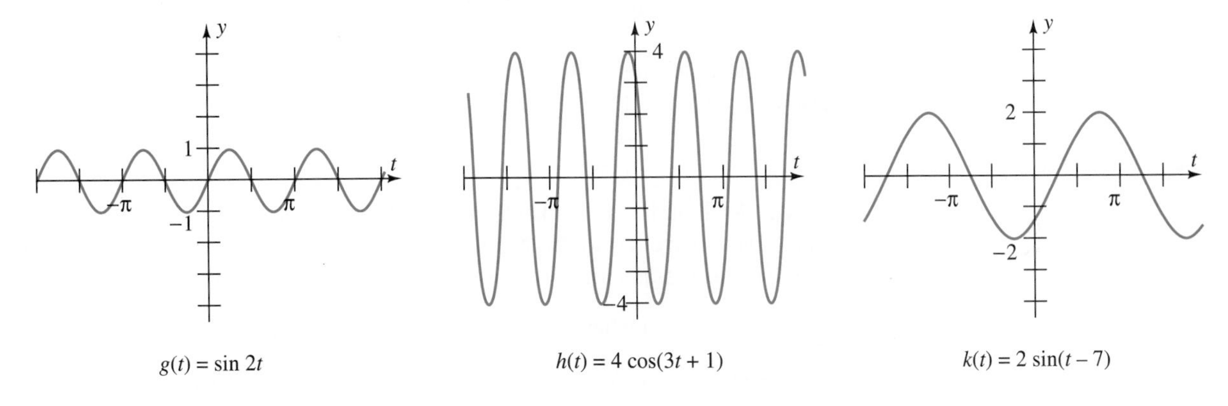

Figure 6-58

Each graph consists of uniform, repeating waves. This means that each function is periodic, with its period being the length of the interval over which the graph makes one complete wave and then begins to repeat. In addition to having different periods, these graphs also differ in the heights of their waves and in the place where the first wave begins. We now analyze each of these factors.

Period

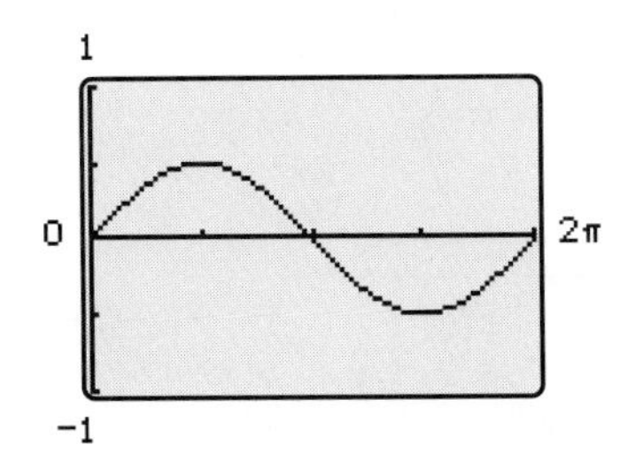

Figure 6-59

The graph of $f(t) = \sin t$ makes one complete wave between 0 and 2π: It begins on the horizontal axis, rises to height 1, falls to -1, and returns to the axis (see Figure 6–59). This corresponds to the fact that the sine function has period 2π. The graph of $g(t) = \sin 2t$ in Figure 6–58 above makes two complete waves between $t = 0$ and $t = 2\pi$. So its period (the length of one wave) is $2\pi/2 = \pi$.

> GRAPHING EXPLORATION Graph each of the following functions, one at a time, in a viewing window with $0 \le t \le 2\pi$ and determine the number of complete waves in each graph and the period of each function (the length of one wave):
>
> $$f(t) = \sin 3t, \qquad g(t) = \cos 4t, \qquad h(t) = \sin 5t.$$
>
> Keep in mind that a complete wave of the cosine function starts at height 1, falls to -1, then rises to height 1 again.

This exploration suggests

Period ▶

If $b > 0$, then the graph of either

$$f(t) = \sin bt \qquad \text{or} \qquad g(t) = \cos bt$$

makes b complete waves between 0 and 2π. Hence, each function has period $2\pi/b$.

Although we arrived at this statement by generalizing from several graphs, it can also be explained algebraically.

EXAMPLE 1 The graph of $g(t) = \cos t$ makes one complete wave as t takes values from 0 to 2π. Similarly, the graph of $k(t) = \cos 3t$ will complete one wave as the quantity $3t$ takes values from 0 to 2π. However,

$$3t = 0 \text{ when } t = 0 \qquad \text{and} \qquad 3t = 2\pi \text{ when } t = 2\pi/3.$$

So the graph of $k(t) = \cos 3t$ makes one complete wave between $t = 0$ and $t = 2\pi/3$, as shown in Figure 6–60, and hence k has period $2\pi/3$. Similarly, the graph makes a complete wave from $t = 2\pi/3$ to $t = 4\pi/3$ and another one from $t = 4\pi/3$ to $t = 2\pi$, as shown in Figure 6–60. ■

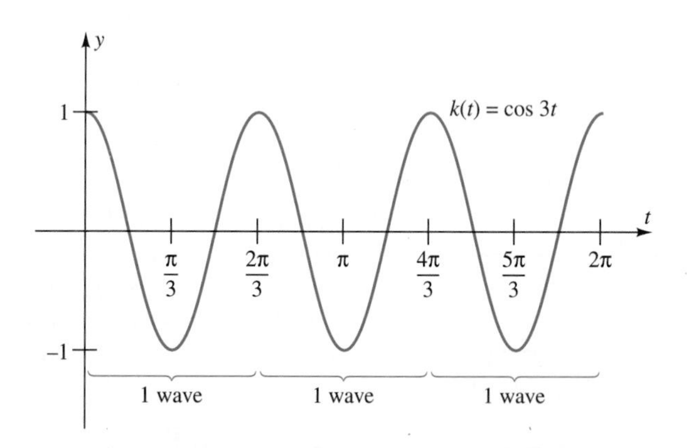

Figure 6-60

EXAMPLE 2 According to the box above, the function $f(t) = \sin \frac{1}{2}t$ has period $\dfrac{2\pi}{1/2} = 4\pi$. Its graph makes *half* a wave from $t = 0$ to $t = 2\pi$ (just as $\sin t$ does from $t = 0$ to $t = \pi$) and the other half of the wave from $t = 2\pi$ to $t = 4\pi$, as shown in Figure 6–61. ■

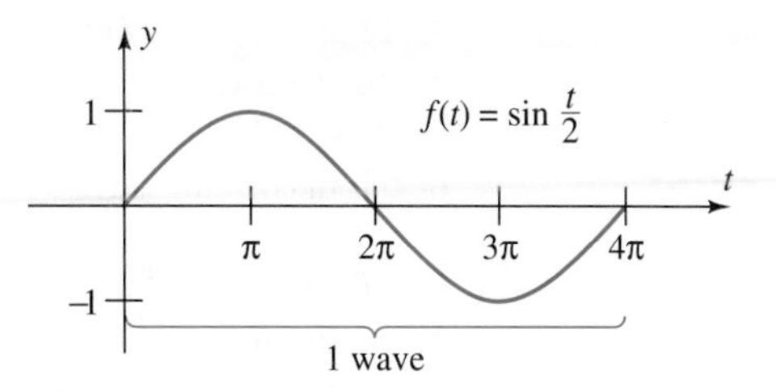

Figure 6-61

WARNING A calculator may produce a wildly inaccurate graph of sin bt or cos bt. For instance, the graph of $f(t) =$ sin $50t$ has 50 complete waves of the same height between 0 and 2π, but that's not what your calculator screen will show (try it!). Similarly, the graph of $g(t) = \cos 100t$ has 100 complete waves between 0 and 2π. How many does your calculator screen show? See Exercises 55–58 for an explanation.

Amplitude

As we saw in Section 2.5, multiplying the rule of a function by a positive constant has the effect of stretching its graph away from or shrinking it towards the horizontal axis.

EXAMPLE 3 The function $g(t) = 7 \cos 3t$ is just the function $k(t) = \cos 3t$ multiplied by 7. Consequently, the graph of g is just the graph of k (which was obtained in Example 1) stretched away from the horizontal axis by a factor of 7:

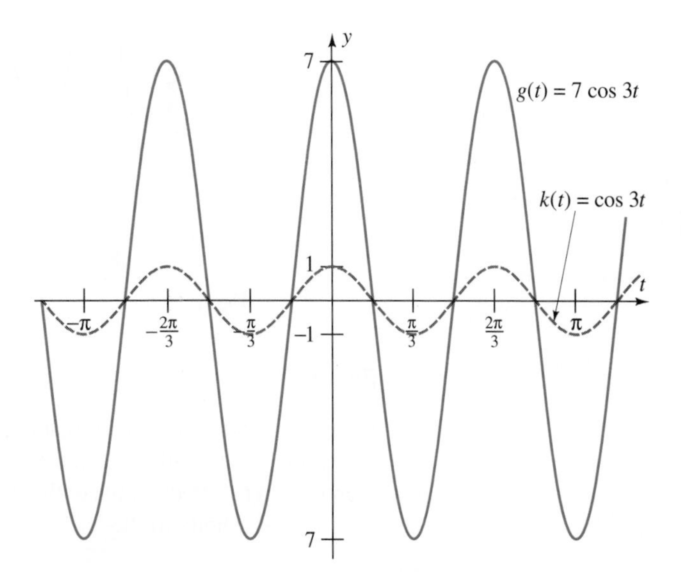

Figure 6-62

As Figure 6–62 shows, stretching the graph affects only the height of the waves, not the period of the function: Both graphs have period $2\pi/3$ and each full wave has length $2\pi/3$. The "stretching factor" 7 is maximum height of the wave, that is, the maximum vertical distance from the graph to the horizontal axis. ■

The waves of the graph of $g(t) = 7 \cos 3t$ in Figure 6–62 rise 7 units above the t-axis and drop 7 units below the axis. More generally, the waves of the graph of $f(t) = A \sin bt$ or $g(t) = A \cos bt$ move a distance of $|A|$ units above and below the t-axis and we say that these functions have **amplitude** $|A|$. In summary,

Amplitude and Period ▶

If $A \neq 0$ and $b > 0$, then each of the functions

$$f(t) = A \sin bt \quad \text{or} \quad g(t) = A \cos bt$$

has amplitude $|A|$ and period $2\pi/b$.

EXAMPLE 4 The function $f(t) = -2 \sin 4t$ has amplitude $|-2| = 2$ and period $2\pi/4 = \pi/2$. So the graph consists of waves of length $\pi/2$ that rise and fall between -2 and 2. But be careful: The waves in the graph of $2 \sin 4t$ (like the waves of $\sin t$) begin at height 0, rise, and then fall. But the graph of $f(t) = -2 \sin 4t$ is the graph of $2 \sin 4t$ reflected in the horizontal axis (see page 129). So its waves start at height 0, move *downward,* and then rise, as shown in Figure 6–63. ■

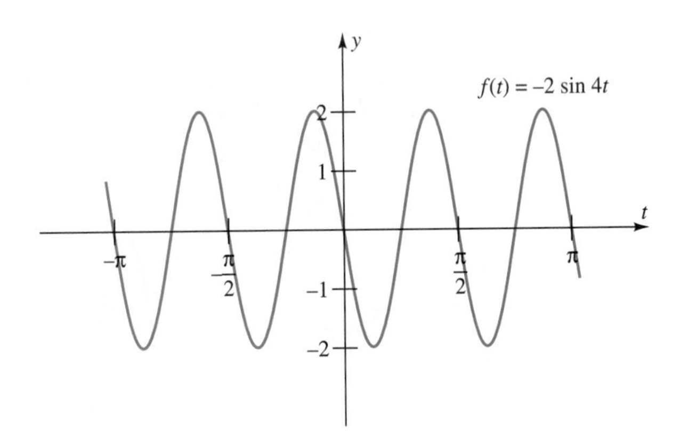

Figure 6-63

Phase Shift

In Section 2.5 we saw that replacing t by $t \pm 3$ in the rule of a function $f(t)$ shifts the graph horizontally (see page 126). Similarly, the graph of $\sin(t - 3)$ is the graph of $\sin t$ shifted 3 units to the right and the graph of $\sin(t + 3)$ is the graph of $\sin t$ shifted 3 units to the left.

EXAMPLE 5 **(a)** Find a sine function whose graph looks like Figure 6–64. **(b)** Find a cosine function whose graph looks like Figure 6–64.

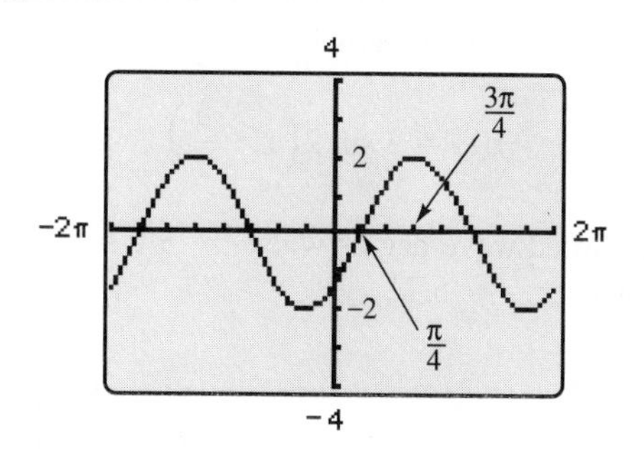

Figure 6-64

Solution

(a) Since each wave has height 2, Figure 6–64 looks like the graph of $2 \sin t$ shifted $\pi/4$ units to the right (so that a sine wave starts at $t = \pi/4$). Since the graph of $2\sin(t - \pi/4)$ is the graph of $2 \sin t$ shifted $\pi/4$ units to the right (see page 126), we conclude that Figure 6–64 closely resembles the graph of $f(t) = 2\sin(t - \pi/4)$.

(b) Figure 6–64 also looks like the graph of $2 \cos t$ shifted $3\pi/4$ units to the right (so that a cosine wave starts at $t = 3\pi/4$). Hence Figure 6–64 could also be the graph of $g(t) = 2\cos(t - 3\pi/4)$. ■

EXAMPLE 6 **(a)** Find the amplitude and the period of $f(t) = 3\sin(2t + 5)$. **(b)** Do the same for the function $f(t) = A \sin(bt + c)$, where A, b, c are constants.

Solution The analysis of $f(t) = 3\sin(2t + 5)$ is in the left-hand column below and the analysis of the general case $f(t) = A\sin(bt + c)$ is in the right-hand column. Observe that exactly the same procedure is used in both cases: just change 3 to A, 2 to b, and 5 to c.

(a) Rewrite the rule of $f(t) = 3\sin(2t + 5)$ as

$$f(t) = 3\sin(2t + 5) = 3\sin\left(2\left(t + \frac{5}{2}\right)\right).$$

Thus, the rule of f can be obtained from the rule of the function $k(t) = 3 \sin 2t$ by replacing t with $t + \frac{5}{2}$. Therefore, the graph of f is just the graph of k shifted horizontally 5/2 units to the left, as shown in Figure 6–65 on the next page.

(b) Rewrite the rule of $f(t) = A\sin(bt + c)$ as

$$f(t) = A\sin(bt + c) = A\sin\left(b\left(t + \frac{c}{b}\right)\right).$$

Thus, the rule of f can be obtained from the rule of the function $k(t) = A \sin bt$ by replacing t with $t + \frac{c}{b}$. Therefore, the graph of f is just the graph of k shifted horizontally by c/b units.

Hence, $f(t) = 3\sin(2t + 5)$ has the same amplitude as $k(t) = 3\sin 2t$, namely 3, and the same period, namely $2\pi/2 = \pi$.

On the graph of $k(t) = 3\sin 2t$ a wave begins when $t = 0$. On the graph of

$$f(t) = 3\sin 2\left(t + \frac{5}{2}\right)$$

the shifted wave begins when $t + 5/2 = 0$, that is, when $t = -5/2$.

Hence, $f(t) = A\sin(bt + c)$ has the same amplitude as $k(t) = A\sin bt$, namely $|A|$, and the same period, namely $2\pi/b$.

On the graph of $k(t) = A\sin bt$, a wave begins when $t = 0$. On the graph of

$$f(t) = A\sin b\left(t + \frac{c}{b}\right)$$

the shifted wave begins when $t + c/b = 0$, that is, when $t = -c/b$. ■

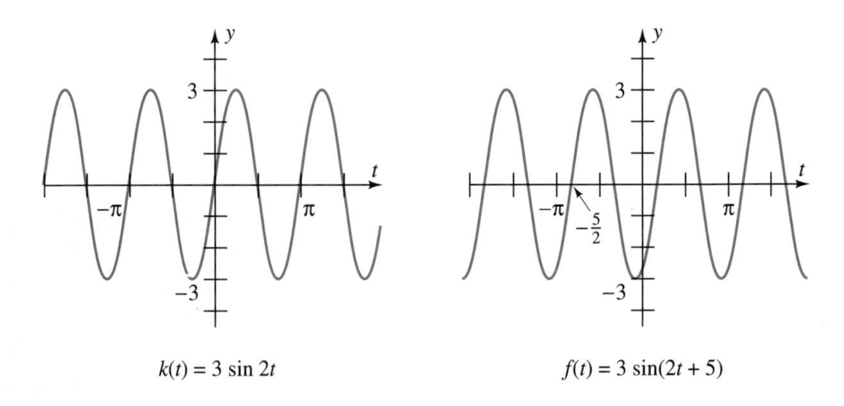

Figure 6-65

We say that the function $f(t) = A\sin(bt + c)$ has **phase shift** $-c/b$. A similar analysis applies to the function $g(t) = \cos(bt + c)$ and leads to this conclusion:

Amplitude, Period, and Phase Shift ▶

If $A \neq 0$ and $b > 0$, then each of the functions

$$f(t) = A\sin(bt + c) \quad \text{and} \quad g(t) = A\cos(bt + c)$$

has

amplitude $|A|$, period $2\pi/b$, phase shift $-c/b$.

A wave of the graph begins at $t = -c/b$.

EXAMPLE 7 The rule of the function $g(t) = 2\cos(3t - 4)$ can be rewritten as

$$g(t) = 2\cos(3t + (-4)).$$

This is the case described in the box above where $A = 2$, $b = 3$, and $c = -4$. Therefore, the function g has

$$\text{amplitude } |A| = |2| = 2, \quad \text{period } \frac{2\pi}{b} = \frac{2\pi}{3}, \quad \text{phase shift } -\frac{c}{b} = -\frac{-4}{3} = \frac{4}{3}.$$

Hence, the graph of g consists of waves of length $2\pi/3$, that run vertically between 2 and -2. A wave begins at $t = 4/3$.

GRAPHING EXPLORATION Verify the accuracy of this analysis by graphing $y = 2\cos(3t - 4)$ in the viewing window with $-2\pi \le t \le 2\pi$ and $-3 \le y \le 3$. Keeping in mind that a wave of the cosine graph begins at the maximum height above the horizontal axis, use the trace feature to show that a wave begins at $t = 4/3$. ■

Many other types of trigonometric graphs, including those consisting of waves of varying height and length, are considered in Excursion 6.7.A.

Applications

The sine and cosine functions, or variations on them, can be used to describe many different phenomena.

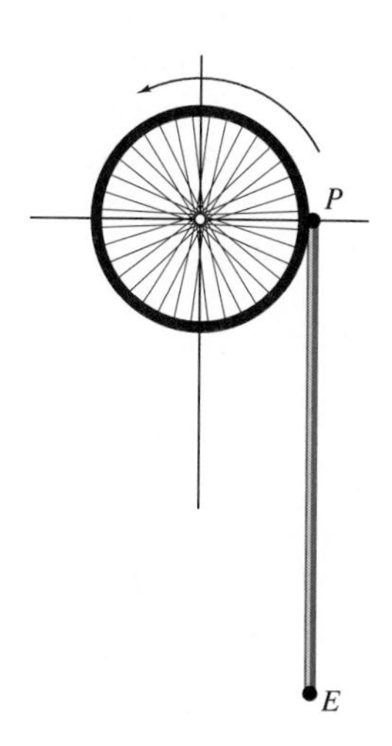

Figure 6-66

EXAMPLE 8 A wheel of radius 2 cm is rotating counterclockwise at 3 radians per second. A free-hanging rod 10 cm long is connected to the edge of the wheel at point P and remains vertical as the wheel rotates (Figure 6–66). Assuming that the center of the wheel is at the origin and that P is at $(2, 0)$ at time $t = 0$, find a function that describes the y-coordinate of the tip E of the rod at time t.

Solution The wheel is rotating at 3 radians per second, so after t seconds the point P has moved through an angle of $3t$ radians and is 2 units from the origin. Using the point-in-the-plane description, we see that the coordinates (x, y) of P satisfy

$$\frac{x}{2} = \cos 3t \qquad \frac{y}{2} = \sin 3t$$

$$x = 2\cos 3t \qquad y = 2\sin 3t.$$

Since E lies 10 cm directly below P, its y-coordinate is 10 less than the y-coordinate of P. Hence, the function giving the y-coordinate of E at time t is

$$f(t) = y - 10 = 2\sin 3t - 10. \quad ■$$

EXAMPLE 9 Suppose that a weight hanging from a spring is set in motion by an upward push, as shown in Figure 6–67.

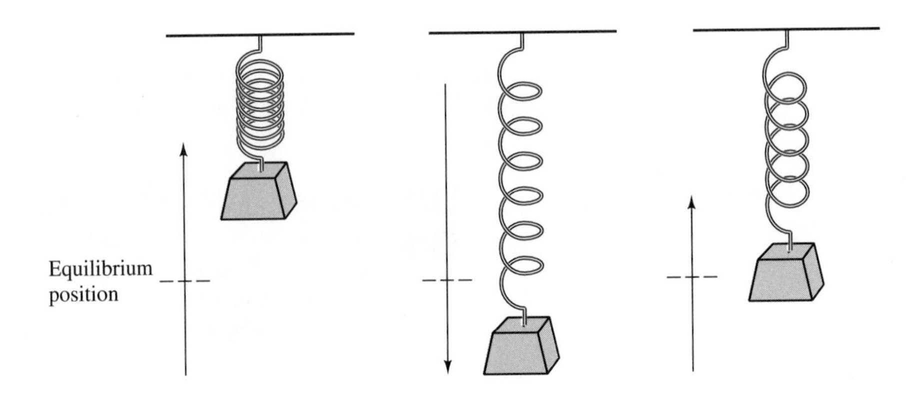

Figure 6-67

For now we shall consider only an idealized situation in which the spring has perfect elasticity, and friction, air resistance, and so forth are ignored. Suppose that the maximum distance the weight moves in either direction from its equilibrium point is 8 cm. Suppose that the length of time it takes for the weight to move upward from equilibrium to its maximum height, then downward to its maximum depth below equilibrium and upward to equilibrium again, is 5 sec.

Let $h(t)$ denote the height (or depth) of the weight above (or below) its equilibrium position at time t. Measure height in positive numbers and depth in negative numbers. Clearly, the height $h(t)$ is a function of the time t. The value of $h(t)$ is 0 when $t = 0$. As t takes increasing values from 0 to 5, $h(t)$ increases from 0 to 8, then decreases to 0, becomes negative and decreases to -8, then increases to 0 again. Consequently, the graph of h has some kind of wave shape, like those shown in Figure 6–68.

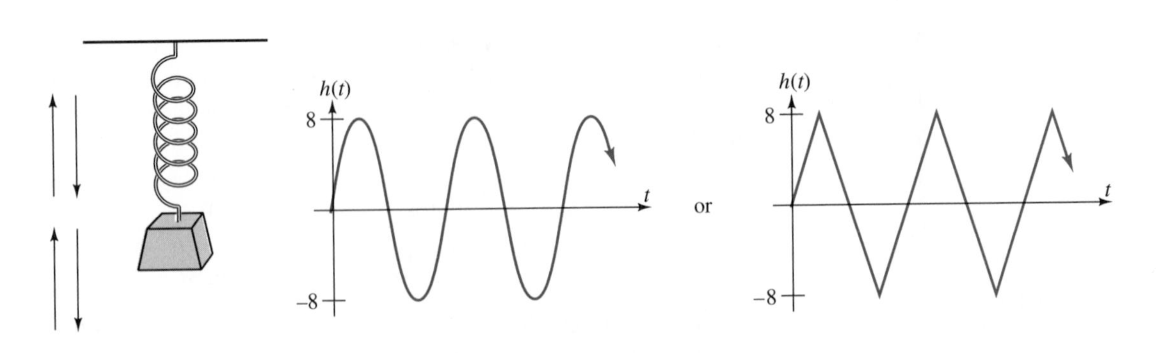

Figure 6-68

In each graph, time is measured on the horizontal axis and height on the vertical axis. The up-and-down motion of the *spring* is fixed in one place, but the up-and-down motion of the graph moves to the right as t increases. On the basis of the given information, it is not clear which graph accurately describes the situation. It may even be that some other wave-shaped graph is the best choice. But careful physical experimentation shows that the smooth curve on the left in Figure 6–68 is a reasonably accurate model of this process.

The graph of the height function h has the same general shape as the various sine graphs presented earlier in this section. A deeper study of the physics involved in the motion of the spring, as well as calculus and differential equations, shows that the rule of the function h actually has the form $h(t) = A\sin(bt + c)$ for some constants A, b, c. Since the amplitude of the function h is 8, its period is 5, and it has 0 phase shift, the constants A, b, c must satisfy

$$A = 8, \qquad \frac{2\pi}{b} = 5, \qquad -\frac{c}{b} = 0$$

or equivalently,

$$A = 8, \qquad b = \frac{2\pi}{5}, \qquad c = 0.$$

Therefore, the motion of the moving spring can be described by the function

$$h(t) = A\sin(bt + c) = 8\sin\left(\frac{2\pi}{5}t + 0\right) = 8\sin\frac{2\pi t}{5}. \quad \blacksquare$$

Motion that can be described by a function of the form $f(t) = A\sin(bt + c)$ or $f(t) = A\cos(bt + c)$ is called **simple harmonic motion.** Many kinds of physical motion are simple harmonic motions. Other periodic phenomena, such as sound waves, are more complicated to describe. Their graphs consist of waves of varying amplitude. Such graphs are discussed in Excursion 6.7.A.

EXERCISES 6.7

In Exercises 1–7, state the amplitude, period, and phase shift of the function.

1. $g(t) = 3\sin(2t - \pi)$
2. $h(t) = -4\cos(3t - \pi/6)$
3. $q(t) = -7\sin(7t + 1/7)$
4. $g(t) = 97\cos(14t + 5)$
5. $f(t) = \cos 2\pi t$
6. $k(t) = \cos(2\pi t/3)$
7. $p(t) = 6\cos(3\pi t + 1)$

8. **(a)** What is the period of $f(t) = \sin 2\pi t$?
   (b) For what values of t (with $0 \le t \le 2\pi$) is $f(t) = 0$?
   (c) For what values of t (with $0 \le t \le 2\pi$) is $f(t) = 1$? or $f(t) = -1$?

In Exercises 9–14, give the rule of a periodic function with the given numbers as amplitude, period, and phase shift (in this order).

9. $3, \pi/4, \pi/5$
10. $2, 3, 0$
11. $2/3, 1, 0$
12. $4/5, 2, 3$
13. $7, 5/3, -\pi/2$
14. $19, 4, -5$

In Exercises 15–18, state the rule of a function of the form $f(t) = A \sin bt$ or $g(t) = A \cos bt$ whose graph appears to be identical to the given graph.

15.

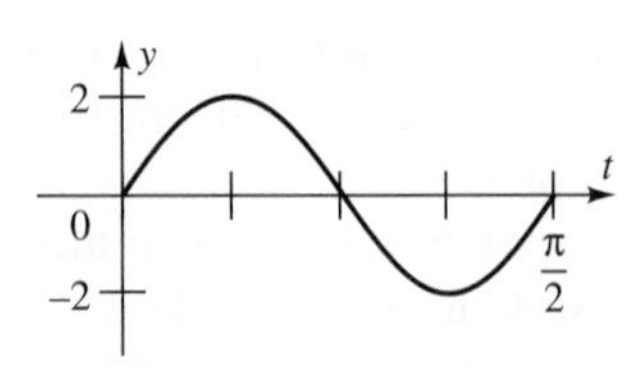

16.

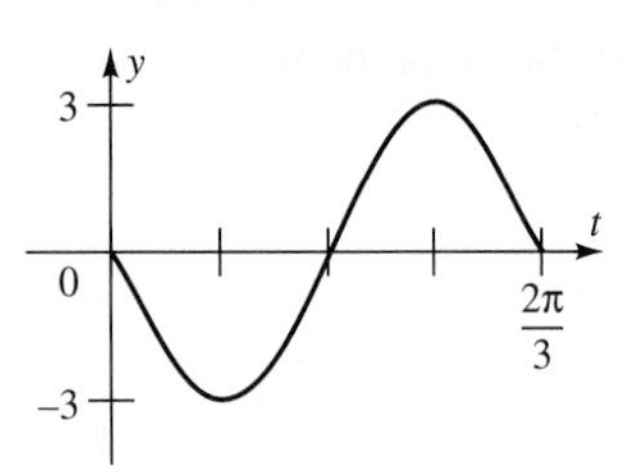

17.

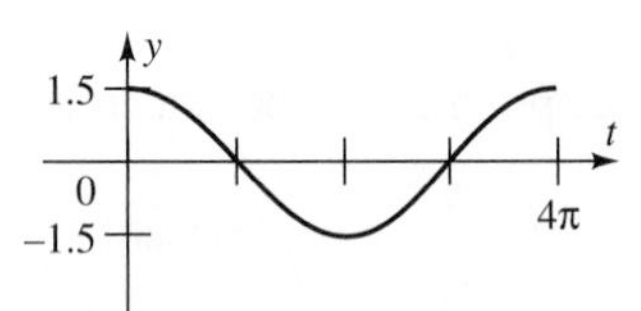

18.

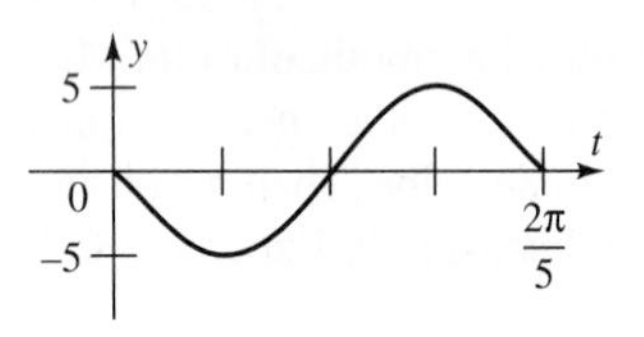

In Exercises 19–24,
(a) State the rule of a function of the form $f(t) = A \sin(bt + c)$ whose graph appears to be identical with the given graph.
(b) State the rule of a function of the form $g(t) = A \cos(bt + c)$ whose graph appears to be identical with the given graph.

19.

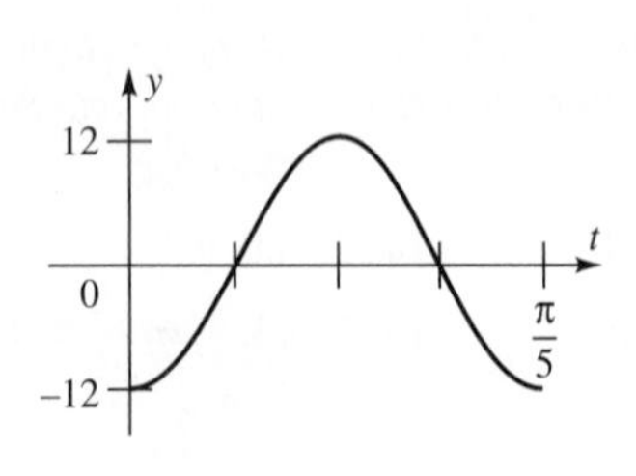

20.

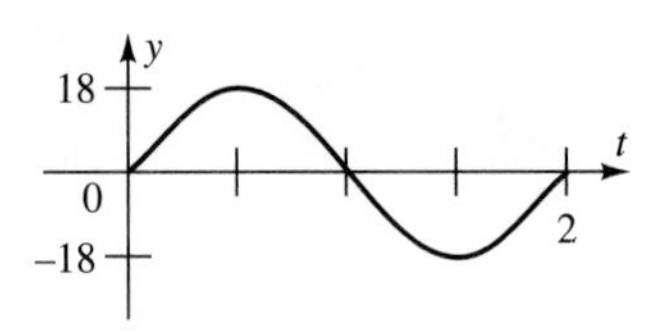

21.

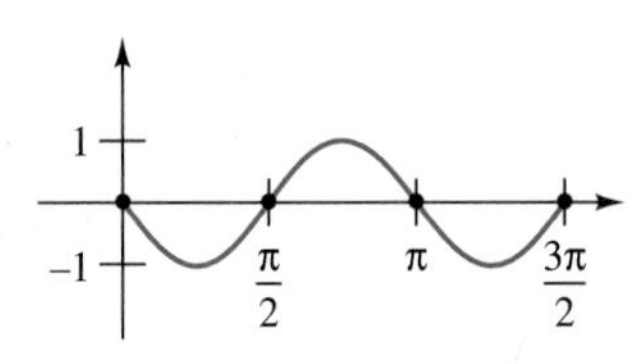

22.

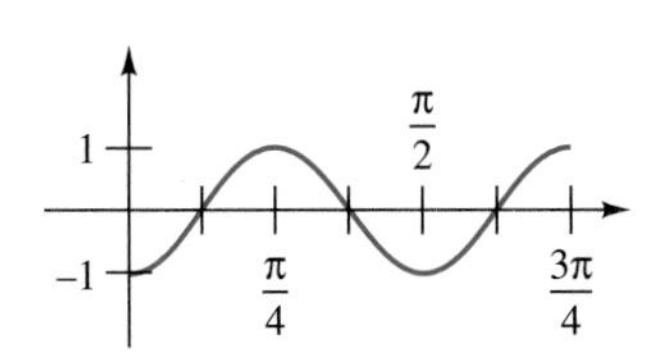

23.

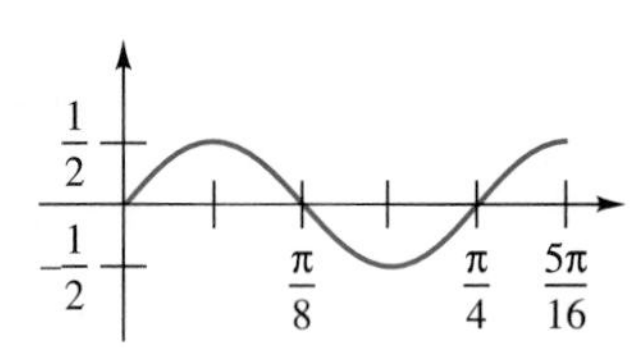

24.

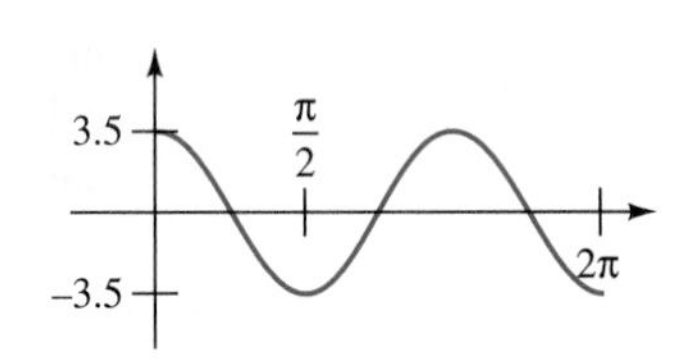

In Exercises 25–30, sketch a complete graph of the function.

25. $k(t) = -3 \sin t$

26. $y(t) = -2 \cos 3t$

27. $p(t) = -\dfrac{1}{2} \sin 2t$

28. $q(t) = \dfrac{2}{3} \cos \dfrac{3}{2} t$

29. $h(t) = 3 \sin(2t + \pi/2)$

30. $p(t) = 3 \cos(3t - \pi)$

In Exercises 31–34, graph the function over the interval $[0, 2\pi)$ and determine the location of all local maxima and minima. [This can be done either graphically or algebraically.]

31. $f(t) = \dfrac{1}{2} \sin\left(t - \dfrac{\pi}{3}\right)$

32. $g(t) = 2 \sin(2t/3 - \pi/9)$

33. $f(t) = -2 \sin(3t - \pi)$

34. $h(t) = \frac{1}{2} \cos\left(\frac{\pi}{2} t - \frac{\pi}{8}\right) + 1$

In Exercises 35–38, graph $f(t)$ in a viewing window with $-2\pi \leq t \leq 2\pi$. Use the trace feature to determine constants A, b, c such that the graph of $f(t)$ appears to coincide with the graph of $g(t) = A \sin(bt + c)$.

35. $f(t) = 2 \sin t + 5 \cos t$

36. $f(t) = -3 \sin t + 2 \cos t$

37. $f(t) = 3 \sin(4t + 2) + 2 \cos(4t - 1)$

38. $f(t) = 2 \sin(3t - 5) - 3 \cos(3t + 2)$

In Exercises 39 and 40, explain why there could not possibly be constants A, b, c such that the graph of $g(t) = A \sin(bt + c)$ coincides with the graph of $f(t)$.

39. $f(t) = \sin 2t + \cos 3t$

40. $f(t) = 2 \sin(3t - 1) + 3 \cos(4t + 1)$

41. The current generated by an AM radio transmitter is given by a function of the form $f(t) = A \sin 2000\pi mt$, where $550 \leq m \leq 1600$ is the location on the broadcast dial and t is measured in seconds. For example, a station at 980 on the AM dial has a function of the form $f(t) = A \sin 2000\pi(980)t = A \sin 1{,}960{,}000\pi t$. Sound information is added to this signal by varying (modulating) A, that is, by changing the amplitude of the waves being transmitted. (AM means amplitude modulation.) For a station at 980 on the dial, what is the period of the function f? What is the frequency (number of complete waves per second)?

42. Find the function f (as in Exercise 41), its period, and its frequency for a station at 1440 on the dial.

43. The original Ferris wheel, built by George Ferris for the Columbian Exposition of 1893, was much larger and slower than its modern counterparts: It had a diameter of 250 ft and contained 36 cars, each of which held 40 people; it made one revolution every 10 minutes. Imagine that the Ferris wheel revolves counterclockwise in the x-y plane with its center at the origin. Car D in Figure 6–69 had coordinates (125, 0) at time $t = 0$. Find the rule of a function that gives the y-coordinate of car D at time t.

44. Do Exercise 43 if the wheel turns at 2 radians per minute and car D is at $(0, -125)$ at time $t = 0$.

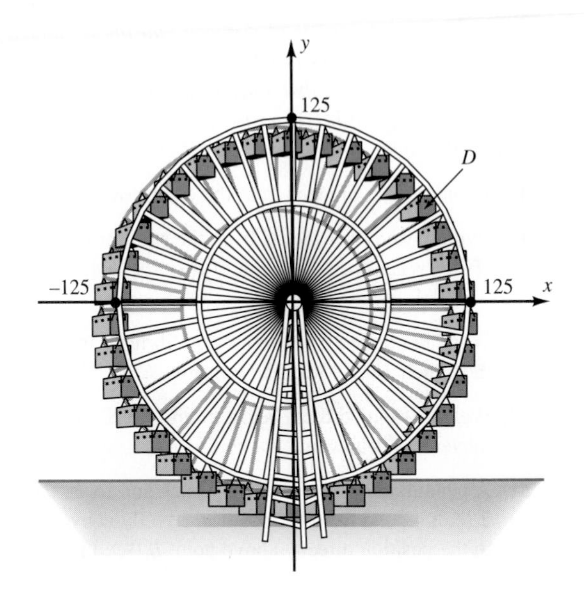

Figure 6-69

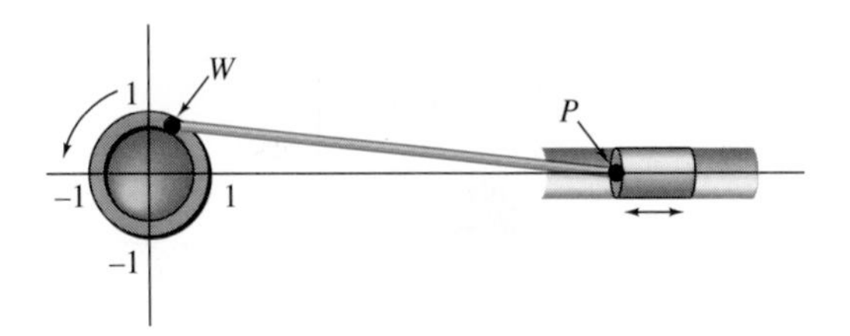

Figure 6-70

45. A circular wheel of radius 1 ft rotates counterclockwise. A 4-ft rod has one end attached to the edge of this wheel and the other end to the base of a piston (Figure 6–70). It transfers the rotary motion of the wheel into a back-and-forth linear motion of the piston. If the wheel is rotating at 10 revolutions per second, point W is at (1, 0) at time $t = 0$, and point P is always on the x-axis, find the rule of a function that gives the x-coordinate of P at time t.

46. Do Exercise 45 if the wheel has a radius of 2 ft, rotates at 50 revolutions per second, and is at (2, 0) when $t = 0$.

In Exercises 47–50, suppose there is a weight hanging from a spring (under the same idealized conditions described in Example 9). The weight is given a push to start it moving. At any time t, let h(t) be the height (or depth) of the weight above (or below) its equilibrium point. Assume that the maximum distance the weight moves in either direction from the equilibrium point is 6 cm and that it moves through a complete cycle every 4 sec. Express h(t) in terms of the sine or cosine function under the stated conditions.

47. Initial push is *upward* from the equilibrium.

48. Initial push is *downward* from the equilibrium point. [*Hint:* What does the graph of $A \sin bt$ look like when $A < 0$?]

49. Weight is pulled 6 cm above equilibrium, and the initial movement (at $t = 0$) is downward. [*Hint:* Think cosine.]

50. Weight is pulled 6 cm below equilibrium, and the initial movement is upward.

51. A pendulum swings uniformly back and forth, taking 2 sec to move from the position directly above point A to the position directly above point B (see Figure 6–71).

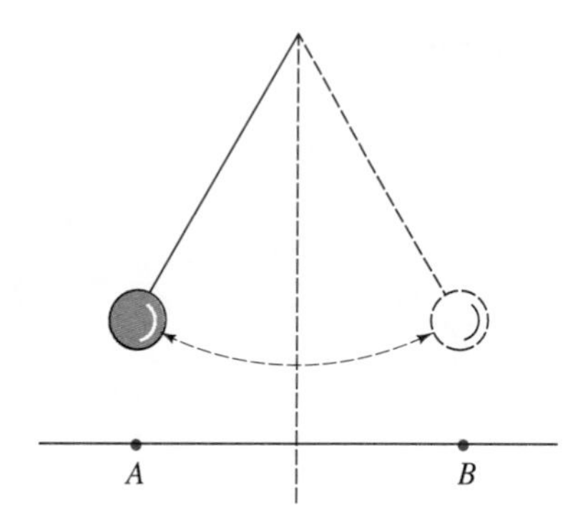

Figure 6-71

The distance from A to B is 20 cm. Let $d(t)$ be the horizontal distance from the pendulum to the (dotted) center line at time t seconds (with distances to the right of the line measured by positive numbers and distances to the left by negative ones). Assume that the pendulum is on the center line at time $t = 0$ and moving to the right. Assume the motion of the pendulum is simple harmonic motion. Find the rule of the function $d(t)$.

52. Figure 6–72 is a diagram of a merry-go-round that is turning counterclockwise at a constant rate, making 2 revolutions in 1 min. On the merry-go-round are horses A, B, C, and D at 4 meters from the center and horses E, F, and G at 8 meters from the center. There is a function $a(t)$, which gives the distance the horse A is from the y-axis (this is the x-coordinate of the position A is in) as a function of time t (measured in minutes). Similarly, $b(t)$ gives the x-coordinate for B as a function of time, and so on. Assume the diagram shows the situation at time $t = 0$.

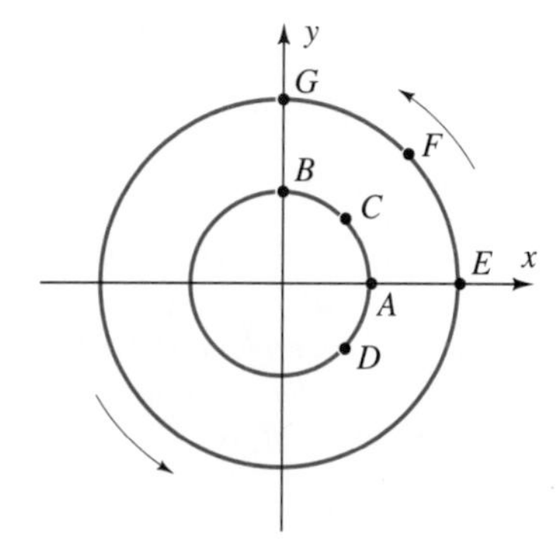

Figure 6-72

(a) Which of the following functions does $a(t)$ equal: $4 \cos t$, $4 \cos \pi t$, $4 \cos 2t$, $4 \cos 2\pi t$, $4 \cos(\frac{1}{2}t)$, $4 \cos((\pi/2)t)$, $4 \cos 4\pi t$? Explain.

(b) Describe the functions $b(t)$, $c(t)$, $d(t)$, and so on using the cosine function:

$b(t) =$ ______, $c(t) =$ ______, $d(t) =$ ______,

$e(t) =$ ______, $f(t) =$ ______, $g(t) =$ ______.

(c) Suppose the x-coordinate of a horse S is given by the function $4 \cos(4\pi t - (5\pi/6))$ and the x-coordinate of another horse T is given by $8 \cos(4\pi t + (\pi/3))$. Where are these horses located in relation to the rest of the horses? Mark the positions of T and S at $t = 0$ into Figure 6–72.

53. A grandfather clock has a pendulum length of k meters and its swing is given (as in Exercise 51) by the function $f(t) = .25 \sin(\omega t)$, where $\omega = \sqrt{\dfrac{9.8}{k}}$.

(a) Find k such that the period of the pendulum is 2 seconds.

(b) The temperature in the summer months causes the pendulum to increase its length by .01%. How much time will the clock lose in June, July, and

August? [*Hint:* These three months have a total of 92 days (7,948,800 seconds). If k is increased by .01%, what is $f(2)$?]

Thinkers

54. Based on the results of Exercises 35–38, under what conditions on the constants a, k, h, d, r, s does it appear that the graph of $f(t) = a\sin(kt + h) + d\cos(rt + s)$ coincides with the graph of the function $g(t) = A\sin(bt + c)$?

Exercises 55–58 explore various ways in which a calculator can produce inaccurate or misleading graphs of trigonometric functions.

55. **(a)** If you were going to draw a rough picture of a full wave of the sine function by plotting some points and connecting them with straight line segments, approximately how many points would you have to plot?

(b) If you were drawing a rough sketch of the graph of $f(t) = \sin 100t$ when $0 \le t \le 2\pi$, according to the method in part (a), approximately how many points would have to be plotted?

(c) How wide (in pixels) is your calculator screen? Your answer to this question is the maximum number of points that your calculator plots when graphing any function.

(d) Use parts (a)–(c) to explain why your calculator cannot possibly produce an accurate graph of $f(t) = \sin 100t$ in any viewing window with $0 \le t \le 2\pi$.

56. **(a)** Using a viewing window with $0 \le t \le 2\pi$, use the trace feature to move the cursor along the horizontal axis. [On some calculators it may be necessary to graph $y = 0$ in order to do this.] What is the distance between one pixel and the next (to the nearest hundredth)?

(b) What is the period of $f(t) = \sin 300t$? Since the period is the length of one full wave of the graph, approximately how many waves should there be between two adjacent pixels? What does this say about the possibility of your calculator's producing an accurate graph of this function between 0 and 2π?

57. **(a)** In Figure 6–73, use a pencil to mark the point on the graph that is directly above or below each hash mark on the horizontal axis (a few of them are already marked).

(b) Imagine that adjacent hash marks on the axis in Figure 6–73 represent adjacent pixels on a calculator screen. Then the graph produced by the calculator will consist of the points marked in part (a), except that these points will appear adjacent to each other rather than spread apart. Sketch the graph that will appear on the calculator screen. Then see Exercise 58.

58. Find a constant $k \ge 30$ such that the graph of $f(t) = \sin kt$ in a viewing window with $0 \le t \le 2\pi$ consists of between 5 and 8 complete waves of the same height. How many waves should the graph have in this viewing window? See Exercise 57 for an explanation.

59. *Approximating trigonometric functions by polynomials.* For each odd positive integer n, let f_n be the function whose rule is

$$f_n(t) = t - \frac{t^3}{3!} + \frac{t^5}{5!} - \frac{t^7}{7!} + \cdots - \frac{t^n}{n!}.$$

(Since the signs alternate, the sign of the last term might be + instead of −, depending on what n is.)

(a) Graph $f_7(t)$ and $g(t) = \sin t$ on the same screen in a viewing window with $-2\pi \le t \le 2\pi$. For what values of t does f appear to be a good approximation of g?

(b) What is the smallest value of n for which the graphs of f_n and g appear to coincide in this window? In this case, determine how accurate the approximation is by finding $f_n(2)$ and $g(2)$.

60. For each even positive integer n, let f_n be the function whose rule is

$$f_n(t) = 1 - \frac{t^2}{2!} + \frac{t^4}{4!} - \frac{t^6}{6!} + \frac{t^8}{8!} - \cdots + \frac{t^n}{n!}.$$

(The sign of the last term may be − instead of +, depending on what n is.)

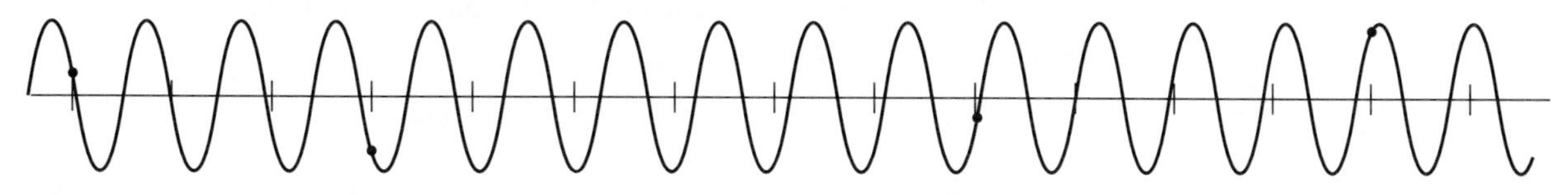

Figure 6-73

(a) In a viewing window with $-2\pi \le t \le 2\pi$, graph f_6, f_{10}, and f_{12}.

(b) Find a value of n for which the graph of f_n appears to coincide (in this window) with the graph of a well-known trigonometric function. What is the function?

61. Find a rational function whose graph appears to coincide with the graph of $h(t) = \tan t$ when $-2\pi \le t \le 2\pi$. [*Hint:* Exercises 59 and 60.]

62. Find a periodic function whose graph consists of "square waves." [*Hint:* Consider the sum $\sin \pi t + \frac{1}{3}\sin 3\pi t + \frac{1}{5}\sin 5\pi t + \frac{1}{7}\sin 7\pi t + \cdots$.]

6.7.A *Excursion* OTHER TRIGONOMETRIC GRAPHS

A graphing calculator enables you to explore with ease a wide variety of trigonometric functions. Some of the possibilities are considered here.

GRAPHING EXPLORATION Graph $g(t) = \cos t$ and $f(t) = \sin(t + \pi/2)$ on the same screen. Is there any apparent difference between the two graphs?

This exploration suggests that the equation $\cos t = \sin(t + \pi/2)$ is an identity and hence that the graph of the cosine function can be obtained by horizontally shifting the graph of the sine function. This is indeed the case, as will be proved in Section 8.2. Consequently, every graph in Section 6.7 is actually the graph of a function of the form $f(t) = A \sin(bt + c)$. In fact, considerably more is true.

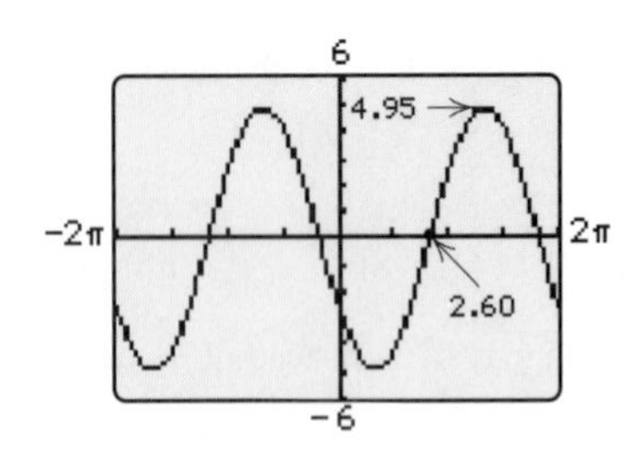

Figure 6-74

EXAMPLE 1 The function $g(t) = -2\sin(t + 7) + 3\cos(t + 2)$ has period 2π because this is the period of both $\sin(t + 7)$ and $\cos(t + 2)$. Its graph in Figure 6–74 consists of repeating waves of uniform height, as do graphs of functions of the form $f(t) = A \sin(bt + c)$.

By using the trace feature and zoom-in, we see that the maximum height of a wave is approximately 4.95 and that a wave similar to a sine wave begins at approximately $t = 2.60$, as indicated in Figure 6–74. Thus, the graph looks very much like a sine wave with amplitude 4.95 and phase shift 2.60. As we saw in Section 6.7, the function

$$f(t) = 4.95 \sin(t - 2.60)$$

has amplitude of 4.95, period $2\pi/1 = 2\pi$, and phase shift $-(-2.60)/1 = 2.60$.

GRAPHING EXPLORATION Graph

$$g(t) = -2\sin(t + 7) + 3\cos(t + 2)$$

and

$$f(t) = 4.95\sin(t - 2.60)$$

on the same screen. Do the graphs look identical? ■

Example 1 is an illustration (but not a proof) of the following fact.

Sinusoidal Graphs ▶

If b, D, E, r, s are constants, then the graph of the function

$$g(t) = D\sin(bt + r) + E\cos(bt + s)$$

is a sine curve: there exist constants A and c such that

$$D\sin(bt + r) + E\cos(bt + s) = A\sin(bt + c).$$

EXAMPLE 2 Estimate the constants A, b, c such that

$$A\sin(bt + c) = 4\sin(3t + 2) + 2\cos(3t - 4).$$

Solution The function $g(t) = 4\sin(3t + 2) + 2\cos(3t - 4)$ has period $2\pi/3$ because this is the period of both $\sin(3t + 2)$ and $\cos(3t - 4)$. The function $f(t) = A\sin(bt + c)$ has period $2\pi/b$. So we must have

$$\frac{2\pi}{b} = \frac{2\pi}{3}, \qquad \text{or equivalently,} \qquad b = 3.$$

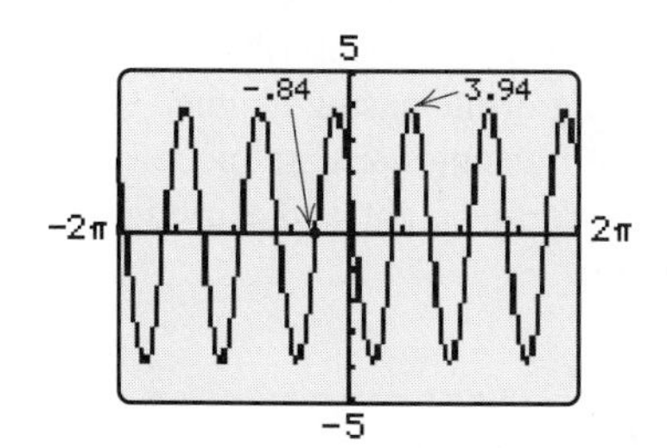

Figure 6-75

Using the trace feature and zoom-in on the graph of $g(t) = 4\sin(3t + 2) + 2\cos(3t - 4)$ in Figure 6–75, we see that the maximum height (amplitude) of a wave is approximately 3.94 and that a sine wave begins at approximately $t = -.84$. Therefore, the graph has (approximate) amplitude 3.94 and phase shift $-.84$. Since $b = 3$ and $f(t) = A\sin(bt + c)$ has amplitude $|A|$ and phase shift $-c/b = -c/3$, we have $A \approx 3.94$ and

$$-\frac{c}{3} = -.84, \qquad \text{or equivalently,} \qquad c \approx 3(.84) = 2.52.$$ ■

In the preceding examples, the variable t had the same coefficient in both the sine and cosine term of the function's rule. When this is not the case, the graph will consist of waves of varying size and shape, as you can readily illustrate.

GRAPHING EXPLORATION Graph each of the following functions separately in the viewing window with $-2\pi \leq t \leq 2\pi$ and $-6 \leq y \leq 6$:

$$f(t) = \sin 3t + \cos 2t, \qquad g(t) = -2\sin(3t + 5) + 4\cos(t + 2),$$
$$h(t) = 2\sin 2t - 3\cos 3t.$$

EXAMPLE 3 Find a complete graph of $f(t) = 4\sin 100\pi t + 2\cos 40\pi t$.

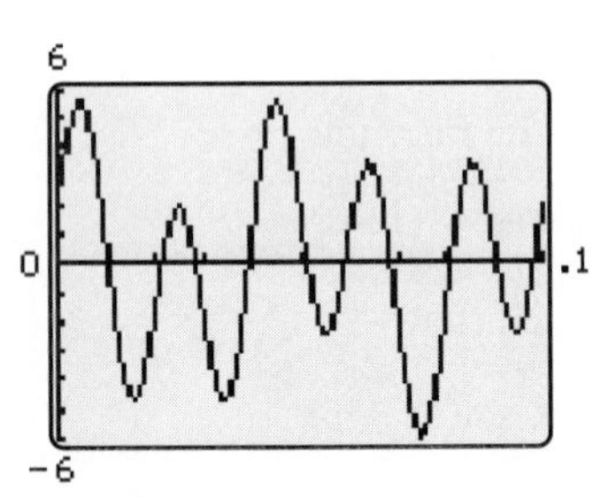

Figure 6-76

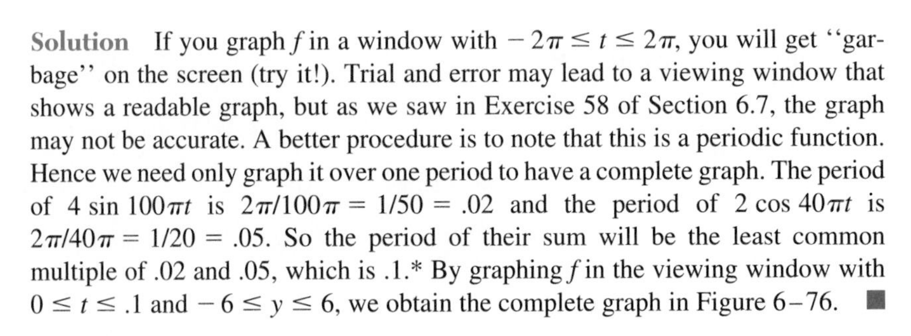

Solution If you graph f in a window with $-2\pi \leq t \leq 2\pi$, you will get "garbage" on the screen (try it!). Trial and error may lead to a viewing window that shows a readable graph, but as we saw in Exercise 58 of Section 6.7, the graph may not be accurate. A better procedure is to note that this is a periodic function. Hence we need only graph it over one period to have a complete graph. The period of $4\sin 100\pi t$ is $2\pi/100\pi = 1/50 = .02$ and the period of $2\cos 40\pi t$ is $2\pi/40\pi = 1/20 = .05$. So the period of their sum will be the least common multiple of .02 and .05, which is .1.* By graphing f in the viewing window with $0 \leq t \leq .1$ and $-6 \leq y \leq 6$, we obtain the complete graph in Figure 6–76. ■

Damped and Compressed Trigonometric Graphs

Suppose a weight hanging from a spring is set in motion by an upward push. No spring is perfectly elastic and friction acts to slow the motion of the weight as time goes on.** Consequently, the graph showing the height of the spring (above or below its equilibrium point) at time t will consist of waves that get smaller and smaller as t gets larger. Many other physical situations can be described by functions whose graphs consist of waves of diminishing or increasing heights. Other situations (for instance, sound waves in FM radio transmission) are modeled by functions whose graphs consist of waves of uniform height and varying frequency. Here are some examples of such functions.

EXAMPLE 4 Graph $f(t) = t\cos t$ in the viewing window with $-35 \leq t \leq 35$ and $-35 \leq y \leq 35$. You will see that the graph consists of waves that are quite small near the origin and get larger as you move away from the origin to the left or right. Some algebraic analysis may help to explain just what's going on. We know that

$$-1 \leq \cos t \leq 1 \qquad \text{for every } t.$$

*The multiples of .02 are .02, .04, .06, .08, .10, . . . and the multiples of .05 are .05, .10, ... Hence, the smallest common multiple is .10.

**These factors were ignored in Example 9 of Section 6.7.

If we multiply each term of this inequality by t and remember the rules for changing the direction of inequalities when multiplying by negatives, we see that

$$-t \leq t \cos t \leq t \qquad \text{when } t \geq 0$$

and

$$-t \geq t \cos t \geq t \qquad \text{when } t < 0.$$

In graphical terms this means that the graph of $f(x) = t \cos t$ lies between the straight lines $y = t$ and $y = -t$, with the waves growing larger or smaller to fill this space. The graph touches the lines $y = \pm t$ exactly when $t \cos t = \pm t$, that is, when $\cos t = \pm 1$. This occurs when $t = 0, \pm\pi, \pm 2\pi, \pm 3\pi, \ldots$.

GRAPHING EXPLORATION Illustrate this analysis by graphing $f(t)$, $y = t$, and $y = -t$ on the same screen. ■

EXAMPLE 5 No single viewing window gives a completely readable graph of $g(t) = .5^t \sin t$ (try some). To the left of the y-axis, the graph gets quite large, but to the right, it almost coincides with the horizontal axis. To get a better mental picture, note that $.5^t > 0$ for every t. Multiplying each term of the known inequality $-1 \leq \sin t \leq 1$ by $.5^t$ we see that

$$-.5^t \leq .5^t \sin t \leq .5^t \qquad \text{for every } t.$$

Hence, the graph of g lies between the graphs of the exponential functions $y = -.5^t$ and $y = .5^t$, which are shown in Figure 6–77. The graph of g will consist of sine waves rising and falling between these exponential graphs, as indicated in the sketch in Figure 6–78 (which is not to scale).

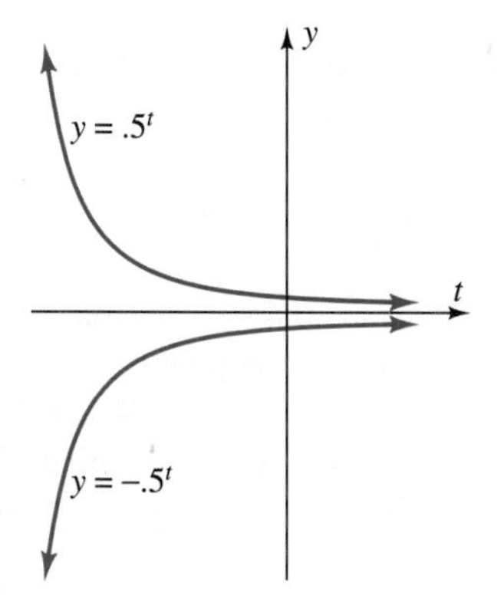

Figure 6-77

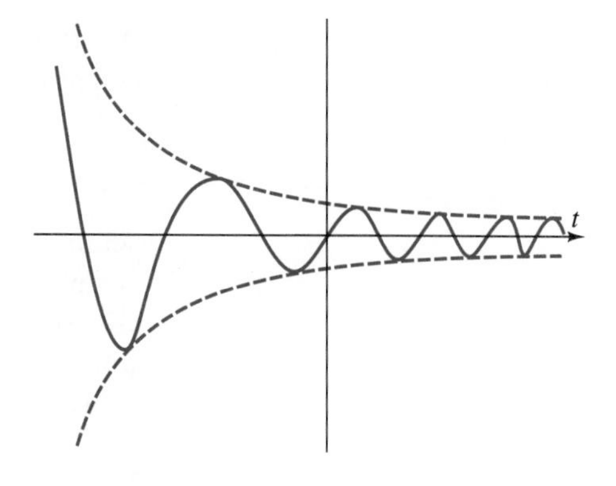

Figure 6-78

The best you can do with a calculator is to look at various viewing windows in which a portion of the graph is readable.

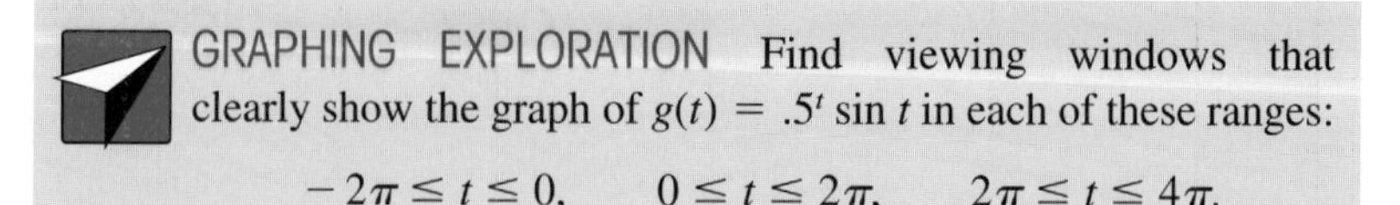

GRAPHING EXPLORATION Find viewing windows that clearly show the graph of $g(t) = .5^t \sin t$ in each of these ranges:

$$-2\pi \leq t \leq 0, \qquad 0 \leq t \leq 2\pi, \qquad 2\pi \leq t \leq 4\pi.$$

■

EXAMPLE 6 If you graph $f(t) = \sin(\pi/t)$ in a wide viewing window such as Figure 6–79, it is clear that the horizontal axis is an asymptote of the graph.* Near the origin, however, the graph is not very readable, even in a very narrow viewing window like Figure 6–80.

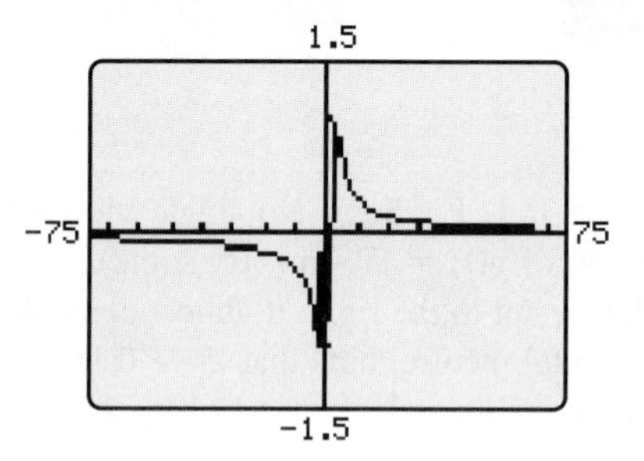

Figure 6-79

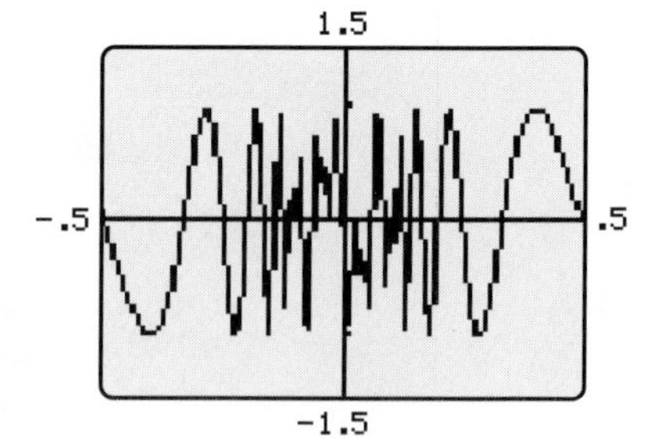

Figure 6-80

To understand the behavior of f near the origin, consider what happens as you move left from $t = 1/2$ to $t = 0$:

$$\text{As } t \text{ goes from } \frac{1}{2} \text{ to } \frac{1}{4}, \text{ then } \frac{\pi}{t} \text{ goes from } \frac{\pi}{1/2} = 2\pi \quad \text{to} \quad \frac{\pi}{1/4} = 4\pi.$$

As π/t takes all values from 2π to 4π, the graph of $f(t) = \sin(\pi/t)$ makes one complete sine wave. Similarly,

$$\text{As } t \text{ goes from } \frac{1}{4} \text{ to } \frac{1}{6}, \text{ then } \frac{\pi}{t} \text{ goes from } \frac{\pi}{1/4} = 4\pi \quad \text{to} \quad \frac{\pi}{1/6} = 6\pi.$$

As π/t takes all values from 4π to 6π, the graph of $f(t) = \sin(\pi/t)$ makes another complete sine wave. The same pattern continues, so that the graph of f makes a complete wave from $t = 1/2$ to $t = 1/4$, another from $t = 1/4$ to $t = 1/6$, another from $t = 1/6$ to $t = 1/8$, and so on. A similar phenomenon occurs as t takes values between $-1/2$ and 0. Consequently, the graph of f near 0 oscillates infinitely

*This can also be demonstrated algebraically: When t is very large in absolute value, then π/t is very close to 0 by the Big–Little Principle and hence $\sin(\pi/t)$ is very close to 0 as well.

often between -1 and 1, with the waves becoming more and more compressed as t gets closer to 0, as indicated in Figure 6–81. Since the function is not defined at $t = 0$, the left and right halves of the graph are not connected. ■

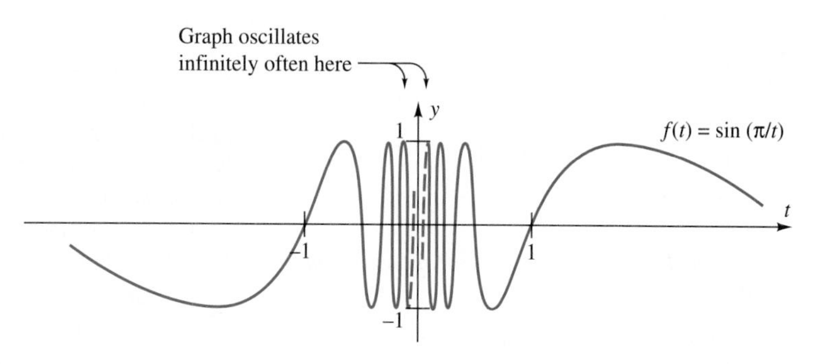

Figure 6-81

EXAMPLE 7 Describe the graph of $g(t) = \cos e^t$.

Solution When t is negative, then e^t is very close to 0 (why?) and hence $\cos e^t$ is very close to 1. Therefore the horizontal line $y = 1$ is an asymptote of the half of the graph to the left of the origin. As t takes increasing positive values, the corresponding values of e^t increase at a much faster rate (remember exponential growth). For instance,

> As t goes from 0 to 2π, e^t goes from $e^0 = 1$ to $e^{2\pi} \approx 535.5 \approx 107\pi = 85(2\pi)$.

Consequently, $\cos e^t$ runs through 85 periods, that is, the graph of g makes 85 full waves between 0 and 2π. As t gets larger, the graph of g makes waves at a faster and faster rate.

GRAPHING EXPLORATION To see how compressed the waves become, graph $g(t)$ in three viewing windows, with

$$0 \le t \le 3.5, \qquad 4.5 \le t \le 5, \qquad 6 \le t \le 6.2$$

and note how the number of waves increases in each succeeding window, even though the widths of the windows are getting smaller.

■

EXERCISES 6.7.A

In Exercises 1–6, estimate constants A, b, c such that $f(t) = A\sin(bt + c)$.

1. $f(t) = \sin t + 2\cos t$
2. $f(t) = 3\sin t + 2\cos t$
3. $f(t) = 2\sin 4t - 5\cos 4t$
4. $f(t) = 3\sin(2t - 1) + 4\cos(2t + 3)$
5. $f(t) = -5\sin(3t + 2) + 2\cos(3t - 1)$
6. $f(t) = .3\sin(2t + 4) - .4\cos(2t - 3)$

In Exercises 7–16, find a viewing window that shows a complete graph of the function.

7. $g(t) = (5\sin 2t)(\cos 5t)$
8. $h(t) = e^{\sin t}$
9. $f(t) = t/2 + \cos 2t$
10. $g(t) = \sin\left(\frac{t}{3} - 2\right) + 2\cos\left(\frac{t}{4} - 2\right)$
11. $h(t) = \sin 300t + \cos 500t$
12. $f(t) = 3\sin(200t + 1) - 2\cos(300t + 2)$
13. $g(t) = -5\sin(250\pi t + 5) + 2\cos(400\pi t - 7)$
14. $h(t) = 4\sin(600\pi t + 3) - 6\cos(500\pi t - 3)$
15. $f(t) = 4\sin .2\pi t - 5\cos .4\pi t$
16. $g(t) = 6\sin .05\pi t + 2\cos .04\pi t$

In Exercises 17–24, describe the graph of the function verbally (including such features as asymptotes, undefined points, amplitude and number of waves between 0 and 2π, etc.) as in Examples 5–7. Find viewing windows that illustrate the main features of the graph.

17. $g(t) = \sin e^t$
18. $h(t) = \dfrac{\cos 2t}{1 + t^2}$
19. $f(t) = \sqrt{|t|}\cos t$
20. $g(t) = e^{-t^2/8}\sin 2\pi t$
21. $h(t) = \dfrac{1}{t}\sin t$
22. $f(t) = t\sin\dfrac{1}{t}$
23. $g(t) = \ln|\cos t|$
24. $h(t) = \ln|\sin t + 1|$

CHAPTER 6 *Review*

Important Concepts ▶

Section 6.1	Angle 365 Vertex 365 Initial side 365 Terminal side 365 Coterminal angles 366 Standard position 366 Degree measure 366 Radian measure 367 Arc length 371 Angular speed 372
Section 6.2	Sine function 376 Cosine function 376 Tangent function 377 Algebra with trigonometric functions 378 Signs of trigonometric functions 381
Section 6.3	Periodicity 383 Graphs of sine, cosine and tangent functions 385, 388, 391
Section 6.4	Pythagorean identity 397 Negative angle identities 398, 399 Diameter identities 400
Section 6.5	Point-in-the-plane description 402 Right triangle description 405 Special values 407
Section 6.6	Cotangent function 412 Secant function 412 Cosecant function 412 Alternate descriptions 413 Identities 415 Graphs of cotangent, secant, and cosecant functions 417, 418
Section 6.7	Period 422 Amplitude 424 Phase shift 426 Simple harmonic motion 429
Excursion 6.7.A	Sinusoidal graphs 435 Damped and compressed graphs 436

Important Facts and Formulas ▶

- *Conversion Rules:* To convert radians to degrees, multiply by $180/\pi$. To convert degrees to radians, multiply by $\pi/180$.
- *Definition of Trigonometric Functions:* If P is the point where the terminal side of an angle of t radians in standard position meets the unit circle, then

$$\sin t = y\text{-coordinate of } P$$

$$\cos t = x\text{-coordinate of } P$$

$$\tan t = \frac{\sin t}{\cos t}$$

- *Point-in-the-Plane Description:* If (x, y) is any point other than the origin on the terminal side of an angle of t radians in standard position and $r = \sqrt{x^2 + y^2}$, then

$$\sin t = \frac{y}{r} \qquad \cos t = \frac{x}{r}$$

$$\tan t = \frac{y}{x} \qquad \cot t = \frac{x}{y}$$

$$\sec t = \frac{r}{x} \qquad \csc t = \frac{r}{y}$$

- *Right Triangle Description:* In a right triangle containing an angle of t radians $(0 < t < \pi/2)$,

$$\sin t = \frac{\text{opposite}}{\text{hypotenuse}} \qquad \cos t = \frac{\text{adjacent}}{\text{hypotenuse}}$$

$$\tan t = \frac{\text{opposite}}{\text{adjacent}} \qquad \cot t = \frac{\text{adjacent}}{\text{opposite}}$$

$$\sec t = \frac{\text{hypotenuse}}{\text{adjacent}} \qquad \csc t = \frac{\text{hypotenuse}}{\text{opposite}}$$

- *Basic Identities:*

$$\tan t = \frac{\sin t}{\cos t} \qquad \cot t = \frac{\cos t}{\sin t} \qquad \sec t = \frac{1}{\cos t}$$

$$\tan t = \frac{1}{\cot t} \qquad \cot t = \frac{1}{\tan t} \qquad \csc t = \frac{1}{\sin t}$$

$$\sin^2 t + \cos^2 t = 1 \qquad \sin(-t) = -\sin t$$

$$1 + \tan^2 t = \sec^2 t \qquad \cos(-t) = \cos t$$

$$1 + \cot^2 t = \csc^2 t \qquad \tan(-t) = -\tan t$$

$$\sin(t \pm 2\pi) = \sin t \qquad \csc(t \pm 2\pi) = \csc t$$
$$\cos(t \pm 2\pi) = \cos t \qquad \sec(t \pm 2\pi) = \sec t$$
$$\tan(t \pm \pi) = \tan t \qquad \cot(t \pm \pi) = \cot t$$
$$\sin(t + \pi) = -\sin t \qquad \cos(t + \pi) = -\cos t$$

- If $A \neq 0$ and $b > 0$, then each of $f(t) = A \sin(bt + c)$ and $g(t) = A \cos(bt + c)$ has

  amplitude $|A|$, period $2\pi/b$, and phase shift $-c/b$.

Review Questions ▶

1. Find a number t between 0 and 2π such that an angle of t radians in standard position is coterminal with an angle of $-23\pi/3$ radians in standard position.
2. Through how many radians does the second hand of a clock move in 2 minutes and 40 seconds?
3. $\frac{9\pi}{5}$ radians = ______ degrees.
4. 36 degrees = ______ radians.
5. $220° =$ ______ radians.
6. $\frac{17\pi}{12}$ radians = ______ degrees.
7. $-\frac{11\pi}{4}$ radians = ______ degrees.
8. $-135° =$ ______ radians.
9. If an angle of v radians has its terminal side in the second quadrant and $\sin v = \sqrt{8/9}$, then find $\cos v$.
10. $\cos \frac{47\pi}{2} = ?$
11. $\sin(-13\pi) = ?$
12. Simplify: $\frac{\tan(t + \pi)}{\sin(t + 2\pi)}$

Use Figure 6–82 in Questions 13–18.

13. $\cos \frac{\pi}{5} = ?$
14. $\sin\left(\frac{7\pi}{6}\right) = ?$
15. $\cos\left(\frac{-5\pi}{6}\right) = ?$
16. $\sin\left(\frac{16\pi}{6}\right) = ?$
17. $\sin\left(\frac{-4\pi}{3}\right) = ?$

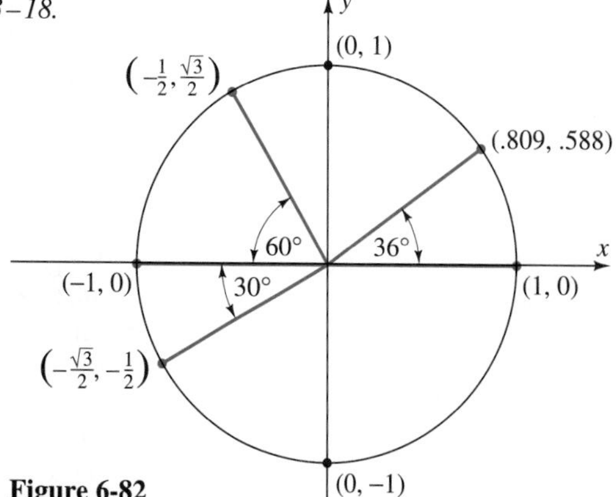

Figure 6-82

18. $\left[3 \sin\left(\frac{\pi}{5^{500}}\right)\right]^2 + \left[3 \cos\left(\frac{\pi}{5^{500}}\right)\right]^2 = ?$

19. Fill in the blanks (approximations not allowed):

t	0	$\frac{\pi}{6}$	$\frac{\pi}{4}$	$\frac{\pi}{3}$	$\frac{\pi}{2}$
sin t					
cos t					

20. Express as a single real number:

$$\cos\frac{3\pi}{4}\sin\frac{5\pi}{6} - \sin\frac{3\pi}{4}\cos\frac{5\pi}{6}$$

21. $\left(\sin\frac{\pi}{6} + 1\right)^2 = ?$

22. $\sin(\pi/2) + \sin 0 + \cos 0 = ?$

23. If $f(x) = \log_{10}x$ and $g(t) = -\cos t$, then $(f \circ g)(\pi) = ?$

24. Cos t is negative when the terminal side of an angle of t radians in standard position lies in which quadrants?

25. If $\sin t = 1/\sqrt{3}$ and the terminal side of an angle of t radians in standard position lies in the second quadrant, then $\cos t = ?$

26. Which of the following could possibly be a true statement about a real number t?
(a) $\sin t = -2$ and $\cos t = 1$
(b) $\sin t = 1/2$ and $\cos t = \sqrt{2}/2$
(c) $\sin t = -1$ and $\cos t = 1$
(d) $\sin t = \pi/2$ and $\cos t = 1 - (\pi/2)$
(e) $\sin t = 3/5$ and $\cos t = 4/5$

27. If $\sin t = -4/5$ and the terminal side of an angle of t radians in standard position lies in the third quadrant, then $\cos t =$ ______.

28. If $\sin(-101\pi/2) = -1$, then $\sin(-105\pi/2) = ?$

29. If $\pi/2 < t < \pi$ and $\sin t = 5/13$, then $\cos t = ?$

30. $\cos\left(-\frac{\pi}{6}\right) = ?$ **31.** $\cos\left(\frac{2\pi}{3}\right) = ?$

32. Suppose θ is a real number. Consider the right triangle with sides as shown in Figure 6–83. Then:
(a) $x = 1$
(b) $x = 2$
(c) $x = 4$
(d) $x = 2(\cos\theta + \sin\theta)$
(e) none of the above

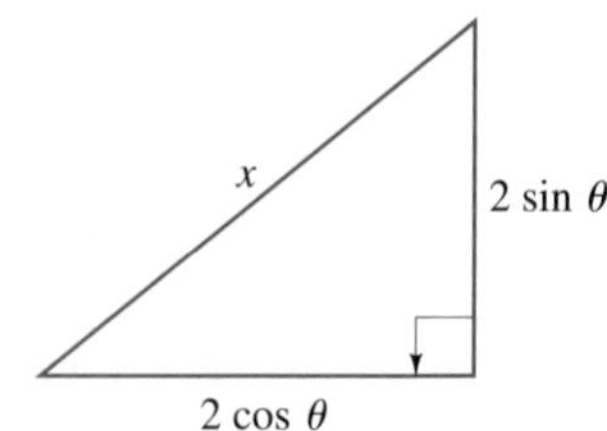

Figure 6-83

33. $\sin\left(\frac{\pi}{3}\right) = ?$ **34.** $\sin\left(-\frac{11\pi}{6}\right) = ?$

35. Which of the following is *not* true about the graph of $f(t) = \sin t$?
(a) It has no sharp corners.
(b) It crosses the horizontal axis more than once.
(c) It rises higher and higher as t gets larger.
(d) It is periodic.
(e) It has no vertical asymptotes.

36. Which of the functions below has the graph in Figure 6–84 between $-\pi$ and π?

(a) $f(x) = \begin{cases} \sin x, & \text{if } x \geq 0 \\ \cos x, & \text{if } x < 0 \end{cases}$

(b) $g(x) = \cos x - 1$

(c) $h(x) = \begin{cases} \sin x, & \text{if } x \geq 0 \\ \sin(-x), & \text{if } x < 0 \end{cases}$

(d) $k(x) = |\cos x|$

(e) $p(x) = \sqrt{1 - \sin^2 x}$

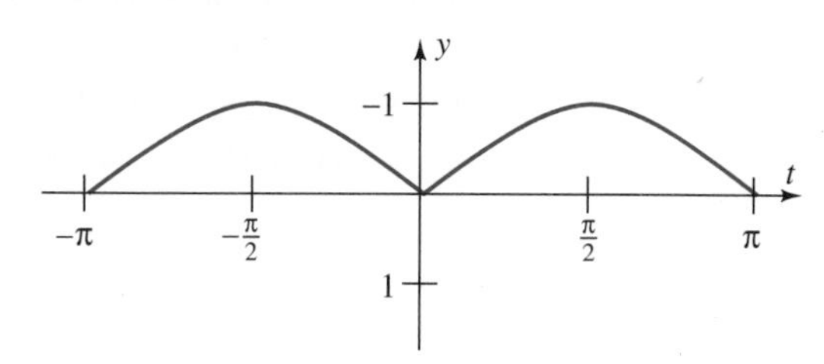

Figure 6-84

37. $\tan(5\pi/3) = ?$

The point $\left(-3/\sqrt{50}, 7/\sqrt{50}\right)$ lies on the terminal side of an angle of t radians (in standard position). Find:

38. $\sin t$ **39.** $\cos t$ **40.** $\tan t$

41. Find the equation of the straight line containing the terminal side of an angle of $5\pi/3$ radians (in standard position).

42. A tree casts a shadow of 60 ft when the sun's rays make an angle of .7 radians with the ground. How tall is the tree?

43. In the triangle in Figure 6–85 angle C is a right angle and angle B measures .6 radians.
(a) Angle A measures ______ radians.
(b) Side a has length ______.
(c) Side b has length ______.

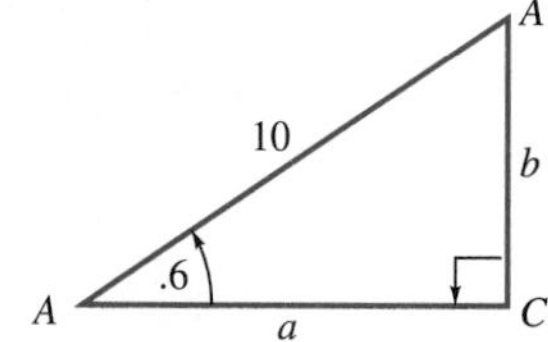

Figure 6-85

44. Find the length of side h in the triangle in Figure 6–86 where angle A measures .7 radians, and the distance from C to A is 25.

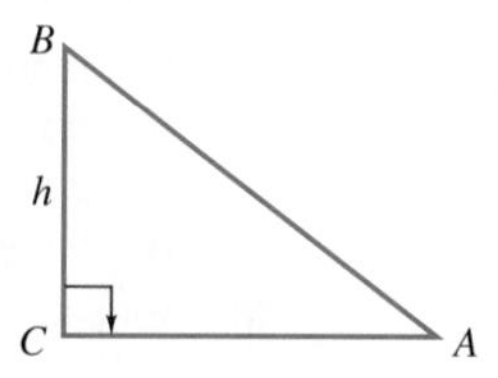

Figure 6-86

45. A wire is stretched from the top of a vertical tower to a point on the ground 80 ft from the base of the tower. The wire makes an angle of $\pi/3$ radians with the ground. How high is the tower?

Suppose that an angle of w radians has its terminal side in the fourth quadrant and $\cos w = 2/\sqrt{13}$. *Find:*

46. $\sin w$ **47.** $\tan w$ **48.** $\csc w$ **49.** $\cos(w + \pi)$

50. Fill in the blanks (approximations not allowed):

t	$\sin t$	$\tan t$	$\sec t$
$\pi/4$			
$2\pi/3$			
$5\pi/6$			

51. Sketch the graphs of $f(t) = \sin t$ and $h(t) = \csc t$ on the same set of coordinate axes $(-2\pi \le t \le 2\pi)$.

52. Let θ be the angle shown in Figure 6–87. Which of the following statements is true?
(a) $\sin \theta = \sqrt{2}/2$
(b) $\cos \theta = \sqrt{2}/2$
(c) $\tan \theta = 1$
(d) $\cos \theta = \sqrt{2}$
(e) $\tan \theta = -1$

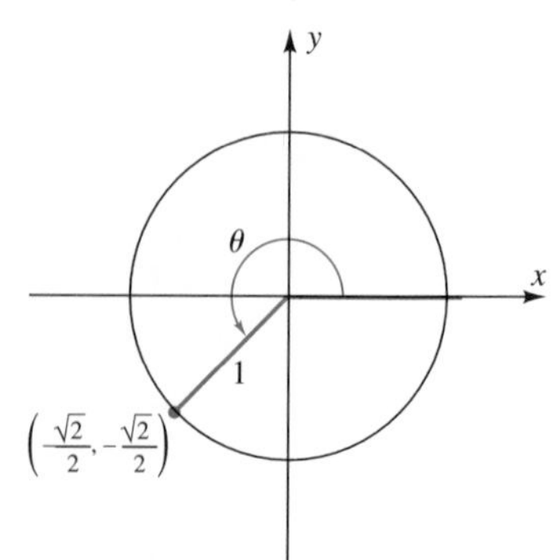

Figure 6-87

53. Let θ be as indicated in Figure 6–88. Which of the statements **(i)–(iii)** are *true*?

(i) $\cos\theta = -\frac{1}{3}$

(ii) $\tan\theta = \frac{2\sqrt{2}}{9}$

(iii) $\sin\theta = -\frac{2\sqrt{2}}{3}$

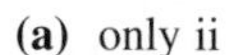

(a) only ii
(b) only ii and iii
(c) all of them
(d) only i and iii
(e) none of them

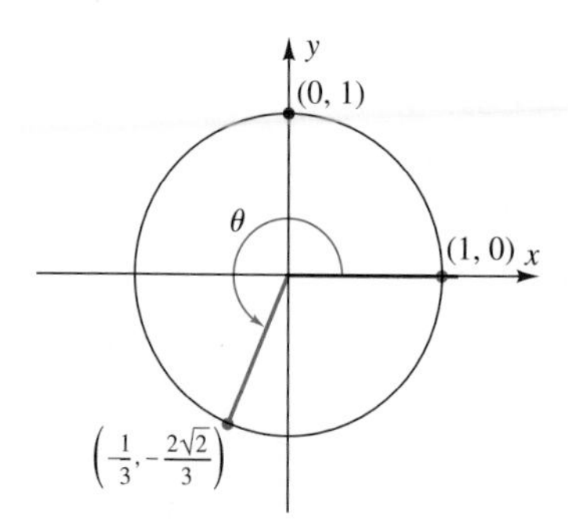

Figure 6-88

54. Between (and including) 0 and 2π, the function $h(t) = \tan t$ has:
(a) 3 roots and is undefined at 2 places.
(b) 2 roots and is undefined at 3 places.
(c) 2 roots and is undefined at 2 places.
(d) 3 roots and is defined everywhere.
(e) no roots and is undefined at 3 places.

55. If the terminal side of an angle of θ radians in standard position passes through the point $(-2, 3)$, then $\tan\theta =$ ______.

56. Which of the statements **(i)–(iii)** are true?

(i) $\sin(-x) = -\sin x$
(ii) $\cos(-x) = -\cos x$
(iii) $\tan(-x) = -\tan x$

(a) (i) and (ii) only
(b) (ii) only
(c) (i) and (iii) only
(d) all of them
(e) none of them

57. If $\sec x = 1$ and $-\pi/2 < x < \pi/2$, then $x =$?

58. Which of the following statements about the angle θ shown in Figure 6–89 is true?
(a) $\sin\theta = 3/4$
(b) $\cos\theta = 5/4$
(c) $\tan\theta = 3/5$
(d) $\sin\theta = 4/5$
(e) $\sin\theta = 4/3$

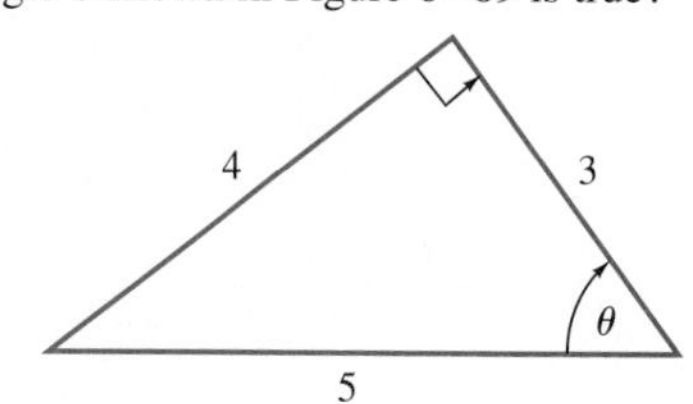

Figure 6-89

59. Which of the functions below has the graph in Figure 6–90?

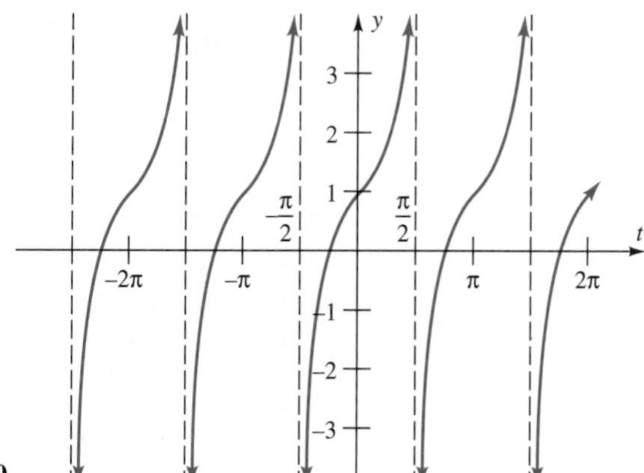

Figure 6-90

(a) $f(t) = \tan t$

(b) $g(t) = \tan\left(t + \dfrac{\pi}{2}\right)$

(c) $h(t) = 1 + \tan t$

(d) $k(t) = 3 \tan t$

(e) $p(t) = -\tan t$

60. If $\tan t = 4/3$ and $0 < t < \pi$, what is $\cos t$?

61. Which of the following is true about $\sec t$?
(a) $\sec(0) = 0$
(b) $\sec t = 1/\sin t$
(c) Its graph has no asymptotes.
(d) It is a periodic function.
(e) It is never negative.

62. If $\cot t = 0$ and $0 < t \leq \pi$, then $t =$ ______.

63. What is $\cot\left(\dfrac{2\pi}{3}\right)$?

64. Use the right triangle in Figure 6–91 to find $\sec t$.
(a) $\sec t = 7/4$
(b) $\sec t = 4/\sqrt{65}$
(c) $\sec t = 7/\sqrt{65}$
(d) $\sec t = \sqrt{65}/7$
(e) $\sec t = \sqrt{65}/4$

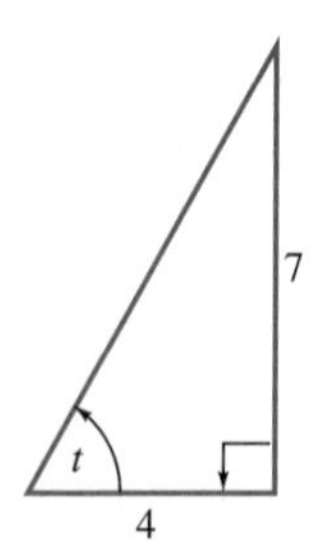

Figure 6-91

65. Let $f(t) = \frac{3}{2} \sin 5t$.
(a) What is the largest possible value of $f(x)$?
(b) Find the smallest positive number t such that $f(t) = 0$.

66. Sketch the graph of $g(t) = -2 \cos t$.

67. Sketch the graph of $f(t) = -\frac{1}{2}\sin 2t \quad (-2\pi \leq t \leq 2\pi)$.

68. Sketch the graph of $f(t) = \sin 4t \quad (0 \leq t \leq 2\pi)$.

In Questions 69–72, determine graphically whether the given equation could possibly be an identity.

69. $\cos t = \sin\left(t - \frac{\pi}{2}\right)$

70. $\tan \frac{t}{2} = \frac{\sin t}{1 + \cos t}$

71. $\frac{\sin t - \sin 3t}{\cos t + \cos 3t} = -\tan t$

72. $\cos 2t = \frac{1}{1 - 2\sin^2 t}$

73. What is the period of the function $g(t) = \sin 4\pi t$?

74. What are the amplitude, period, and phase shift of the function $h(t) = 13\cos(14t + 15)$?

75. State the rule of a periodic function whose graph from $t = 0$ to $t = 2\pi$ closely resembles the one in Figure 6–92.

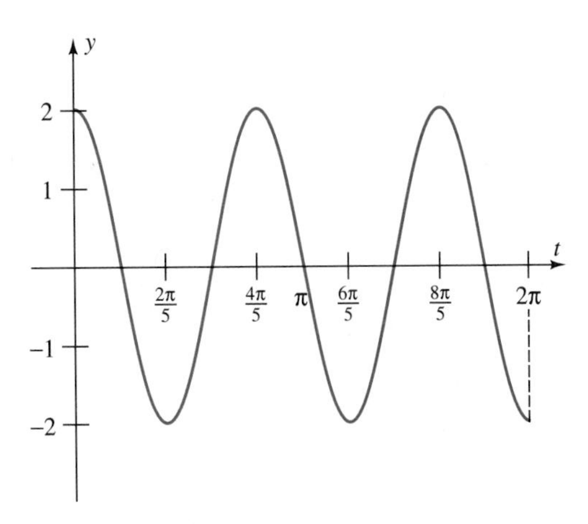

Figure 6-92

76. State the rule of a periodic function with amplitude 3, period π, and phase shift $\pi/3$.

77. State the rule of a periodic function with amplitude 8, period 5, and phase shift 14.

78. If $g(t) = 20\sin(200t)$, for how many values of t, with $0 \leq t \leq 2\pi$, is it true that $g(t) = 1$?

In Exercises 79 and 80, estimate constants A, b, c such that $f(t) = A\sin(bt + c)$.

79. $f(t) = 6\sin(4t + 7) - 5\cos(4t + 8)$

80. $f(t) = -5\sin(5t - 3) + 2\cos(5t + 2)$

In Exercises 81 and 82, find a viewing window that shows a complete graph of the function.

81. $f(t) = 3\sin(300t + 5) - 2\cos(500t + 8)$

82. $g(t) = -5\sin(400\pi t + 1) + 2\cos(150\pi t - 6)$

Triangle Trigonometry

Roadmap

Chapters 7, 8, and 9 are independent of each other and may be read in any order. In this chapter, the interdependence of sections is:

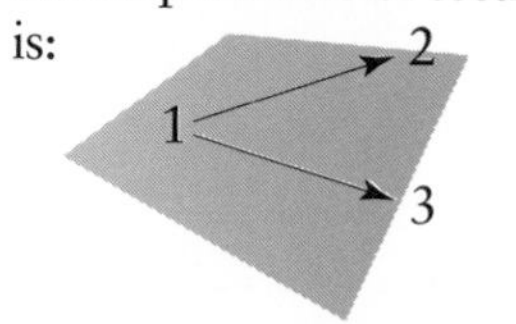

Trigonometry was first used by the ancients to solve practical problems in astronomy, navigation, and surveying that involved triangles. Trigonometric functions, as presented in Chapter 6, came much later. The early mathematicians took a somewhat different viewpoint than we have used up to now, but their approach is often the best way to deal with problems involving triangles.

7.1 RIGHT TRIANGLE TRIGONOMETRY

The process of evaluating the sine function may be summarized like this:

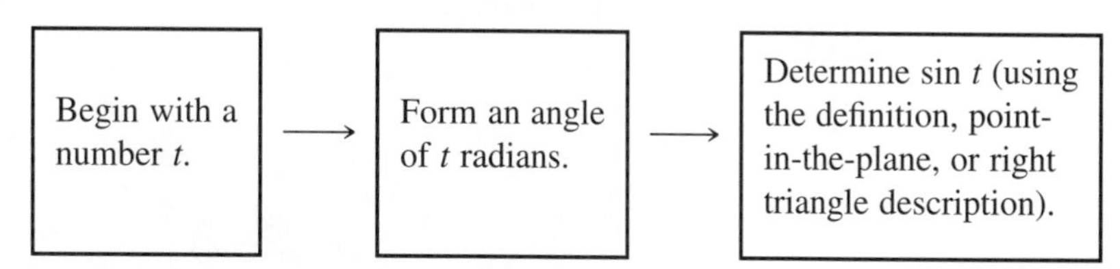

The classical approach begins at the second step, with *angles* rather than numbers. Although the ancients wouldn't have used these terms, their approach amounts to defining the sine function as a function whose domain consists of all *angles* instead of all real numbers. Analogous remarks apply to the other trigonometric functions.

In this chapter (and hereafter, whenever convenient) we shall take this classical approach. Instead of starting with numbers and *then* passing to angles, we shall just begin with the angles. From there on everything is essentially the same. The *values* of the various trigonometric functions are still *numbers* and are obtained as before. For example, the point-in-the-plane method yields this:

Point-in-the-Plane Description ▶

Let θ be an angle in standard position and let (x, y) be any point (except the origin) on the terminal side of θ. Let $r = \sqrt{x^2 + y^2}$. Then, the values of the six trigonometric functions of the angle θ are given by

$$\sin\theta = \frac{y}{r} \qquad \cos\theta = \frac{x}{r} \qquad \tan\theta = \frac{y}{x}$$

$$\csc\theta = \frac{r}{y} \qquad \sec\theta = \frac{r}{x} \qquad \cot\theta = \frac{x}{y}$$

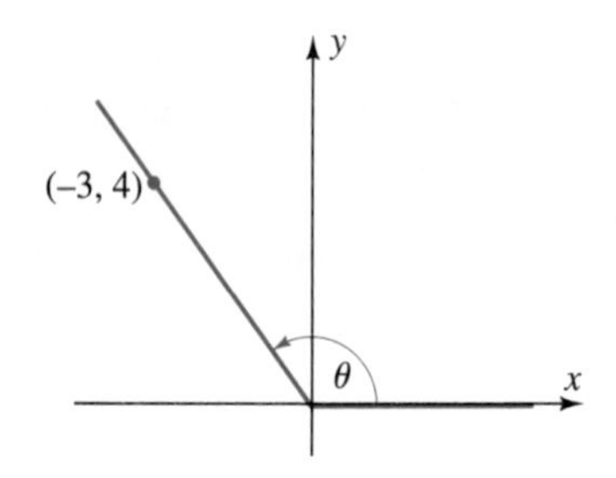

Figure 7-1

EXAMPLE 1 Evaluate the six trigonometric functions at the angle θ shown in Figure 7–1.

Solution We use $(-3, 4)$ as the point (x, y), so that $r = \sqrt{x^2 + y^2} = \sqrt{9 + 16} = \sqrt{25} = 5$. Thus,

$$\sin\theta = \frac{y}{r} = \frac{4}{5} \qquad \cos\theta = \frac{x}{r} = \frac{-3}{5} \qquad \tan\theta = \frac{y}{x} = \frac{4}{-3}$$

$$\csc\theta = \frac{r}{y} = \frac{5}{4} \qquad \sec\theta = \frac{r}{x} = \frac{5}{-3} \qquad \cot\theta = \frac{x}{y} = \frac{-3}{4}$$ ■

When θ is an angle between 0 and 90°, we can use right triangles:

Right Triangle Description ▶

Consider a right triangle containing an angle θ:

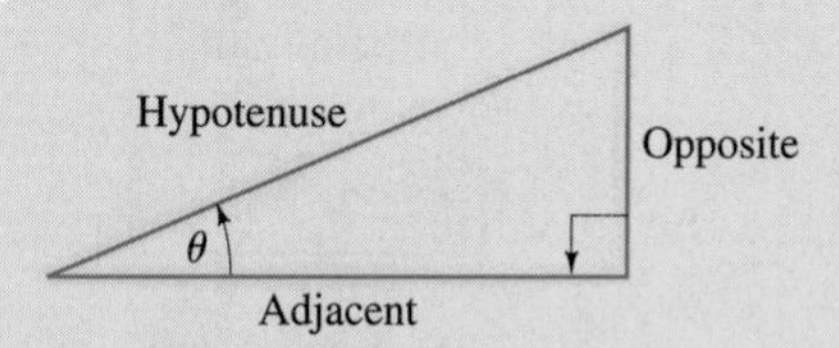

The values of the six trigonometric functions of the angle θ are given by:

$$\sin\theta = \frac{\text{opposite}}{\text{hypotenuse}} \qquad \cos\theta = \frac{\text{adjacent}}{\text{hypotenuse}} \qquad \tan\theta = \frac{\text{opposite}}{\text{adjacent}}$$

$$\csc\theta = \frac{\text{hypotenuse}}{\text{opposite}} \qquad \sec\theta = \frac{\text{hypotenuse}}{\text{adjacent}} \qquad \cot\theta = \frac{\text{adjacent}}{\text{opposite}}$$

From our work in Section 6.5 we know that both this method and the previous one result in the same functional values for a given angle θ.

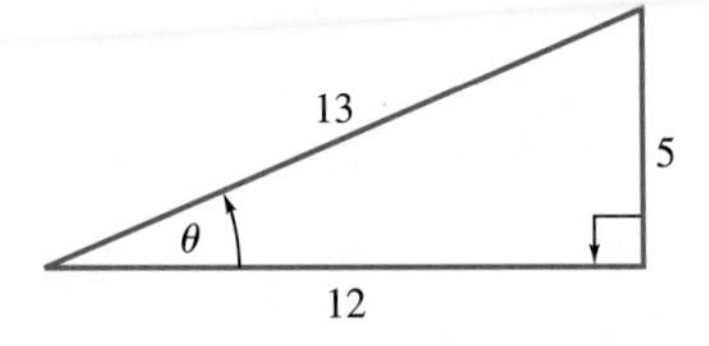

Figure 7-2

EXAMPLE 2 Evaluate the six trigonometric functions at the angle θ shown in Figure 7–2.

Solution

$$\sin\theta = \frac{\text{opposite}}{\text{hypotenuse}} = \frac{5}{13} \qquad \csc\theta = \frac{\text{hypotenuse}}{\text{opposite}} = \frac{13}{5}$$

$$\cos\theta = \frac{\text{adjacent}}{\text{hypotenuse}} = \frac{12}{13} \qquad \sec\theta = \frac{\text{hypotenuse}}{\text{adjacent}} = \frac{13}{12}$$

$$\tan\theta = \frac{\text{opposite}}{\text{adjacent}} = \frac{5}{12} \qquad \cot\theta = \frac{\text{adjacent}}{\text{opposite}} = \frac{12}{5}$$ ■

Degrees and Radians

Angles can be measured in either degrees or radians. If radian measure is used (as was the case in Chapter 6), then everything is the same as before. For example, sin 30 denotes the sine of an angle of 30 radians.

But when angles are measured in degrees (as will be done in the rest of this chapter), new notation is needed. In order to denote the value of the sine function at an angle of 30 *degrees,* we write

$$\sin 30^\circ \qquad \text{[note the degree symbol]}$$

The degree symbol here is absolutely essential to avoid error.

EXAMPLE 3 Since an angle of 30 degrees is the same as one of $\pi/6$ radians, $\sin 30^\circ$ is the same number as $\sin \pi/6$. Hence,

$$\sin 30^\circ = \sin \pi/6 = 1/2.$$

This is *different* from sin 30 (the sine of an angle of 30 *radians*); a calculator in radian mode shows that $\sin 30 = -.988$. ■

Angles of 0, $\pi/6$, $\pi/4$, $\pi/3$, and $\pi/2$ radians have degree measures 0°, 30°, 45°, 60°, and 90°, respectively. Using our knowledge of special values we see, for example, that $\sin 45^\circ = \sin \pi/4 = \sqrt{2}/2$. Similarly, we have the following table (in which a dash indicates that the function is not defined):

Trigonometric Functions of Special Angles ►

θ	$\sin\theta$	$\cos\theta$	$\tan\theta$	$\cot\theta$	$\sec\theta$	$\csc\theta$
0°	0	1	0	—	1	—
30°	$\frac{1}{2}$	$\frac{\sqrt{3}}{2}$	$\frac{\sqrt{3}}{3}$	$\sqrt{3}$	$\frac{2\sqrt{3}}{3}$	2
45°	$\frac{\sqrt{2}}{2}$	$\frac{\sqrt{2}}{2}$	1	1	$\sqrt{2}$	$\sqrt{2}$
60°	$\frac{\sqrt{3}}{2}$	$\frac{1}{2}$	$\sqrt{3}$	$\frac{\sqrt{3}}{3}$	2	$\frac{2\sqrt{3}}{3}$
90°	1	0	—	0	—	1

Except for these special values (and integer multiples of them), it's necessary to use a calculator in *degree mode* to evaluate trigonometric functions of angles measured in degrees. For instance,

$$\sin 346^\circ \approx -.2419 \qquad \cos 27^\circ \approx .891 \qquad \tan 268^\circ \approx 28.6363.$$

The various identities proved in earlier sections are valid for angles measured in degrees, provided that π radians is replaced by 180°. For any angle θ measured in degrees for which the functions are defined, the following identities hold.

Identities for Angles Measured in Degrees ►

Periodicity Identities

$$\sin(\theta + 360^\circ) = \sin\theta \qquad \csc(\theta + 360^\circ) = \csc\theta$$

$$\cos(\theta + 360^\circ) = \cos\theta \qquad \sec(\theta + 360^\circ) = \sec\theta$$

$$\tan(\theta + 180^\circ) = \tan\theta \qquad \cot(\theta + 180^\circ) = \cot\theta$$

Pythagorean Identities

$$\sin^2\theta + \cos^2\theta = 1 \qquad 1 + \tan^2\theta + \sec^2\theta \qquad 1 + \cot^2\theta = \csc^2\theta$$

Negative Angle Identities

$$\sin(-\theta) = -\sin\theta \qquad \cos(-\theta) = \cos\theta \qquad \tan(-\theta) = -\tan\theta$$

Diameter Identities

$$\sin(\theta + 180^\circ) = -\sin\theta \qquad \cos(\theta + 180^\circ) = -\cos\theta$$

Solving Right Triangles

Many applications of trigonometry involve "**solving a triangle.**" This means finding the lengths of all three sides and the measures of all three angles when only some of these quantities are given. Solving right triangles depends on this fact:

> **The right triangle description of a trigonometric function (such as sin θ = opposite/hypotenuse) relates three quantities: the angle θ and two sides of the right triangle.**

When two of these three quantities are known, then the third can always be found.

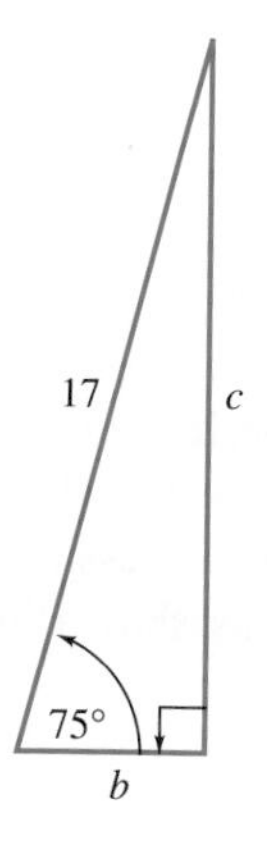

Figure 7-3

EXAMPLE 4 Find the lengths of sides b and c in the right triangle shown in Figure 7–3.

Solution Since the side c is opposite the 75° angle and the hypotenuse is 17, we have:

$$\sin 75^\circ = \frac{\text{opposite}}{\text{hypotenuse}} = \frac{c}{17}.$$

Solving the equation $\sin 75^\circ = c/17$ for c and using a calculator in degree mode, we have:

$$c = 17 \sin 75^\circ \approx 17(.9659) \approx 16.42.$$

Side b can now be found by the Pythagorean Theorem or by using the cosine function:

$$\cos 75^\circ = \frac{\text{adjacent}}{\text{hypotenuse}} = \frac{b}{17}$$

or equivalently, $b = 17 \cos 75^\circ \approx 4.40$. ■

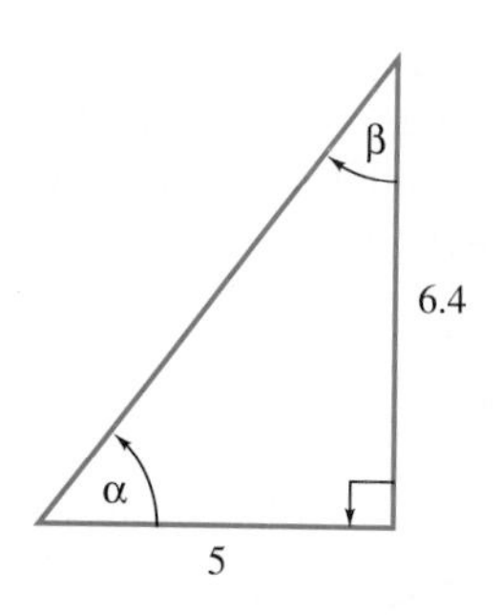

Figure 7-4

EXAMPLE 5 Find the measure of angles α and β in the right triangle in Figure 7–4.

Solution We have $\tan \alpha = \dfrac{\text{opposite}}{\text{adjacent}} = \dfrac{6.4}{5} = 1.28$. To find an angle between 0° and 90° whose tangent is 1.28, key in TAN⁻¹ 1.28 ENTER. The calculator displays the desired angle: $\alpha \approx 52^\circ$. Since the sum of the angles of a triangle is 180°, we have $180^\circ = 90^\circ + \alpha + \beta = 90^\circ + 52^\circ + \beta$. Hence, $\beta = 180^\circ - 90^\circ - 52^\circ = 38^\circ$. ■

EXAMPLE 6 A straight road leads from an ocean beach into the nearby hills. The road has a constant upward grade of 3°. After taking this road from the beach for 1 mile, how high above sea level are you?

Solution Figure 7–5 shows the situation [*Remember:* 1 mile = 5280 feet]:

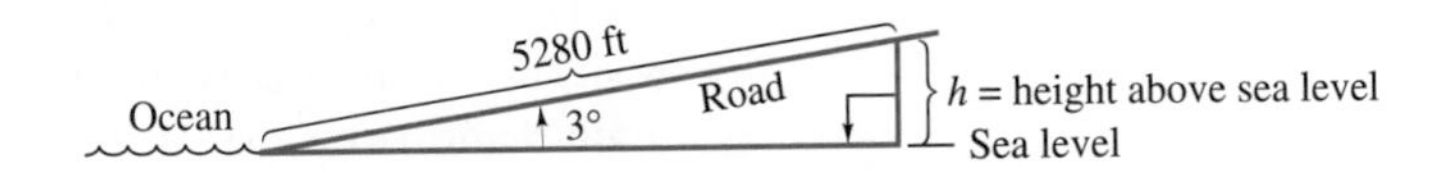

Figure 7-5

We know the 3° angle and the hypotenuse of this triangle and must find the side opposite the 3° angle. So we use the sine function:

$$\sin 3^\circ = \frac{\text{opposite}}{\text{hypotenuse}} = \frac{h}{5280}, \qquad \text{or equivalently,} \qquad h = 5280(\sin 3^\circ).$$

A calculator or table shows that $h = 5280(\sin 3^\circ) \approx 276.33$ feet. ■

In many practical applications, one uses the angle between the horizontal and some other line (for instance, the line of sight from an observer to a distant object). This angle is called the **angle of elevation** or the **angle of depression,** depending on whether the line is above or below the horizontal, as shown in Figure 7–6.

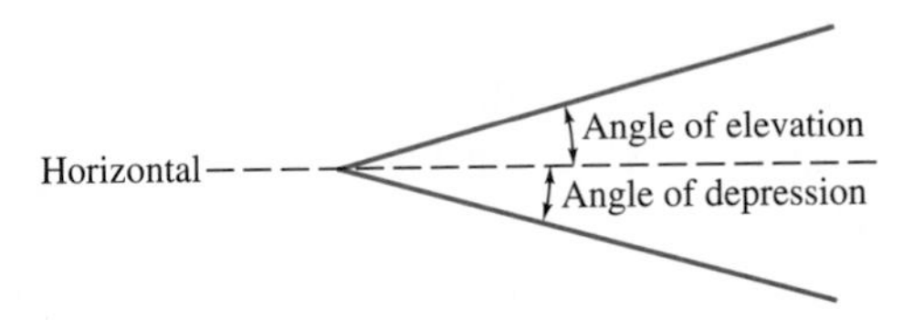

Figure 7-6

EXAMPLE 7 A wire is to be stretched from the top of a 10-meter-high building to a point on the ground. From the top of the building the angle of depression to the ground point is 22°. How long must the wire be?

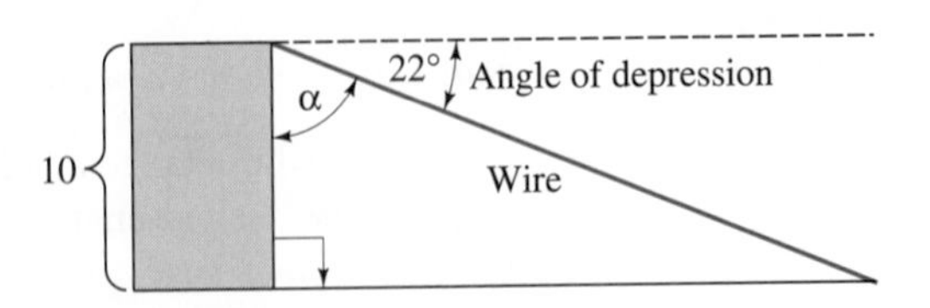

Figure 7-7

Solution Figure 7–7 shows that the sum of the angle of depression and the angle α is 90°. Hence α measures $90° - 22° = 68°$. We know the length of the side of the triangle adjacent to the angle α and must find the hypotenuse w (the length of the wire). Using the cosine function we see that

$$\cos 68° = \frac{\text{adjacent}}{\text{hypotenuse}} = \frac{10}{w}.$$

Solving the equation $\cos 68° = 10/w$ for w and using a calculator yields:

$$w = \frac{10}{\cos 68°} \approx 26.7 \text{ meters.} \quad ■$$

EXAMPLE 8 A person standing on the edge of one bank of a canal observes a lamp post on the edge of the other bank of the canal. The person's eye level is 152 cm above the ground (approximately 5 ft). The angle of elevation from eye level to the top of the lamp post is 12°, and the angle of depression from eye level to the bottom of the lamp post is 7°, as shown in Figure 7–8. How wide is the canal? How high is the lamp post?

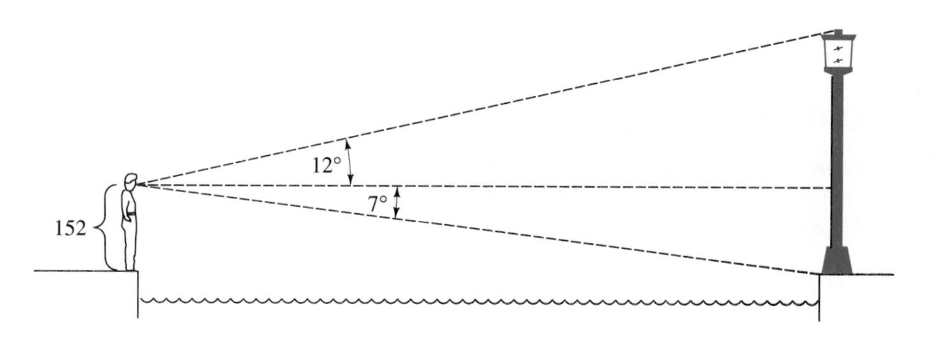

Figure 7-8

Solution Abstracting the essential information, we obtain the diagram in Figure 7–9.

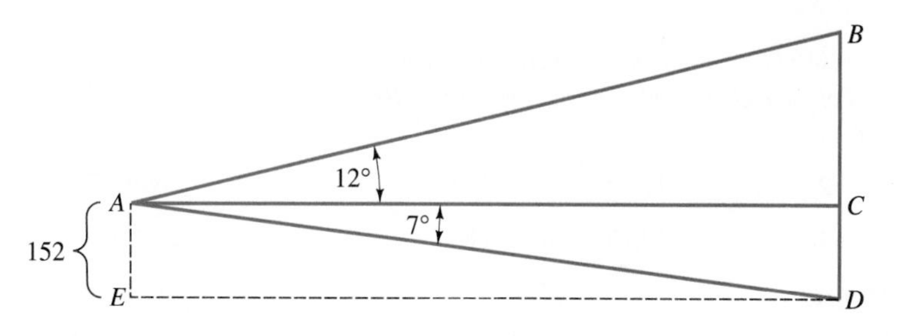

Figure 7-9

We must find the height of the lamp post BD and the width of the canal AC (or ED). The eye level height AE of the observer is 152 cm. Since AC and ED are parallel, CD also has length 152 cm. In right triangle ACD we know the angle of $7°$ and the side CD opposite it. We must find the adjacent side AC. The tangent function is needed:

$$\tan 7° = \frac{\text{opposite}}{\text{adjacent}} = \frac{152}{AC}, \qquad \text{or equivalently,} \qquad AC = \frac{152}{\tan 7°}$$

$$AC = \frac{152}{\tan 7°} \approx 1237.94 \text{ cm.}$$

So, the canal is approximately 12.3794 meters* wide (about 40.6 ft). Now using right triangle ACB we see that

$$\tan 12° = \frac{\text{opposite}}{\text{adjacent}} = \frac{BC}{AC} \approx \frac{BC}{1237.94}$$

or equivalently,

$$BC \approx 1237.94(\tan 12°) \approx 263.13 \text{ cm.}$$

Therefore, the height of the lamp post BD is $BC + CD \approx 263.13 + 152 = 415.13$ cm. ■

EXERCISES 7.1

Directions: *When solving triangles here, all decimal approximations should be rounded off to one decimal place at the end of the computation.*

In Exercises 1–10, evaluate all six trigonometric functions at the angle (in standard position) whose terminal side contains the given point.

1. $(2, 3)$ **2.** $(4, -2)$ **3.** $(-5, 6)$

4. $(\sqrt{2}, \sqrt{3})$ **5.** $(-3, -\sqrt{2})$ **6.** $(-5, 3)$

7. $(-\sqrt{5}, -\sqrt{7})$ **8.** $(7/2, -9/2)$ **9.** $(1 + \sqrt{2}, 1)$

10. $(1 + \sqrt{3}, 1 - \sqrt{3})$

In Exercise 11–16, evaluate all six trigonometric functions of angle A of the right triangle shown in Figure 7–10, under the given conditions.

11. $a = 2, c = 3$ **12.** $a = \sqrt{11}, b = 6$

13. $a = 4, b = 5$ **14.** $b = 10, c = 8$

15. $a = \sqrt{7}, c = 3$ **16.** $b = \sqrt{17}, c = \sqrt{13}$

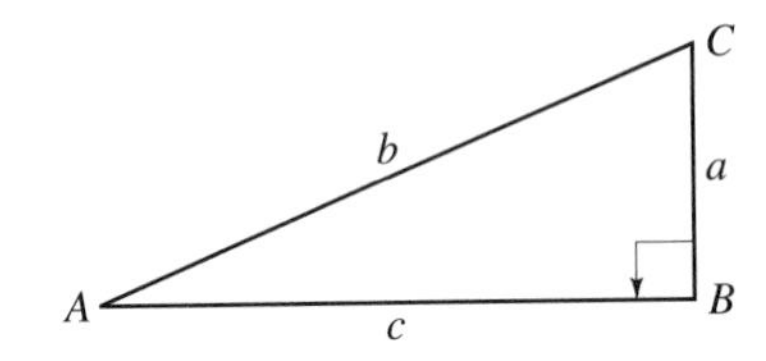

Figure 7-10

In Exercises 17–24, find the exact functional value without using a calculator.

17. $\sin 120°$ **18.** $\cos 240°$ **19.** $\tan 300°$

20. $\cot 135°$ **21.** $\sec 225°$ **22.** $\csc 315°$

23. $\sin 150°$ **24.** $\cos 210°$

In Exercises 25–30, find side c of the right triangle shown in Figure 7–10 under the given conditions.

25. $\cos A = 12/13$ and $b = 39$

26. $\sin C = 3/4$ and $b = 12$

*Remember, 100 cm = 1 meter.

27. $\tan A = 5/12$ and $a = 15$

28. $\sec A = 2$ and $b = 8$

29. $\cot A = 6$ and $a = 1.4$

30. $\csc C = 1.5$ and $b = 4.5$

In Exercises 31–38, solve the right triangle shown in Figure 7–10 under the given conditions.

31. $b = 10$ and $\measuredangle C = 50°$

32. $c = 12$ and $\measuredangle C = 37°$

33. $a = 6$ and $\measuredangle A = 14°$

34. $a = 8$ and $\measuredangle A = 40°$

35. $c = 5$ and $\measuredangle A = 65°$

36. $c = 4$ and $\measuredangle C = 28°$

37. $b = 3.5$ and $\measuredangle A = 72°$

38. $a = 4.2$ and $\measuredangle C = 33°$

In Exercises 39–46, find angles A and C of the right triangle shown in Figure 7–10 under the given conditions.

39. $a = 4$ and $c = 6$

40. $b = 14$ and $c = 5$

41. $a = 7$ and $b = 10$

42. $a = 5$ and $c = 3$

43. $b = 18$ and $c = 12$

44. $a = 4$ and $b = 9$

45. $a = 2.5$ and $c = 1.4$

46. $b = 3.7$ and $c = 2.2$

In Exercises 47–52, solve the right triangle shown in Figure 7–10 under the given conditions.

47. $a = 5, b = 10$

48. $a = 3, c = 4$

49. $b = 7.5, c = 2.5$

50. $a = 4, \measuredangle A = 22°$

51. $c = 10, \measuredangle A = 35°$

52. $b = 6.5, \measuredangle C = 57°$

53. For maximum safety, the distance from the base of a ladder to the building wall should be one fourth of the length of the ladder. If a ladder is in this position, what angle does it make with the ground?

54. **(a)** A 24-ft-long ladder leans on the side of a house, making an angle of 72° with the ground. How far up the side of the house does the ladder reach?
(b) Is the base of the ladder close enough to the house for safety (see Exercise 53)?

55. A plane takes off at an angle of 5° (Figure 7–11). After traveling one mile along this flight path, how high (in ft) is the plane above the ground?

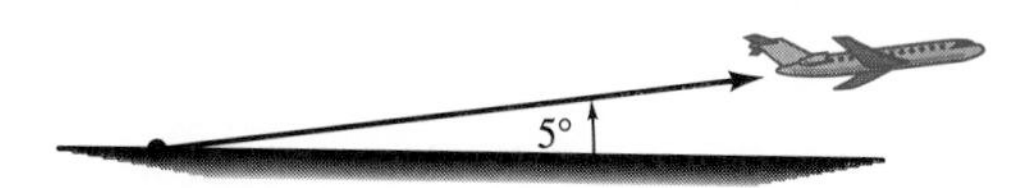

Figure 7-11

56. A plane takes off at an angle of 6° traveling at the rate of 200 ft/sec. If it continues on this flight path at the same speed, how many minutes will it take to reach an altitude of 8000 ft?

57. It is claimed that the Ohio Turnpike never has an uphill grade of more than 3°. How long must a straight uphill segment of the road be in order to allow a vertical rise of 450 ft?

58. Ruth is flying a kite. Her hand is 3 ft above ground level and is holding the end of a 300-ft-long kite string, which makes an angle of 57° with the horizontal. How high is the kite above the ground?

59. If you stand upright on a mountain side that makes a 62° angle with the horizontal and stretch your arm straight out at shoulder height, you may or may not be able to touch the mountain (Figure 7–12). Can a person with an arm reach of 27 inches, whose shoulder is 5 ft above the ground, touch the mountain?

Figure 7-12

60. A swimming pool is 3 ft deep in the shallow end. The bottom of the pool has a steady downward drop of

12°. If the pool is 50 ft long, how deep is it at the deep end?

61. Batman is on the edge of a 200-ft-deep chasm and wants to jump to the other side. A tree on the edge of the chasm is directly across from him. He walks 20 ft to his right and notes that the angle to the tree is 54° (Figure 7–13). His jet belt enables him to jump a maximum of 24 ft. How wide is the chasm and is it safe for Batman to jump?

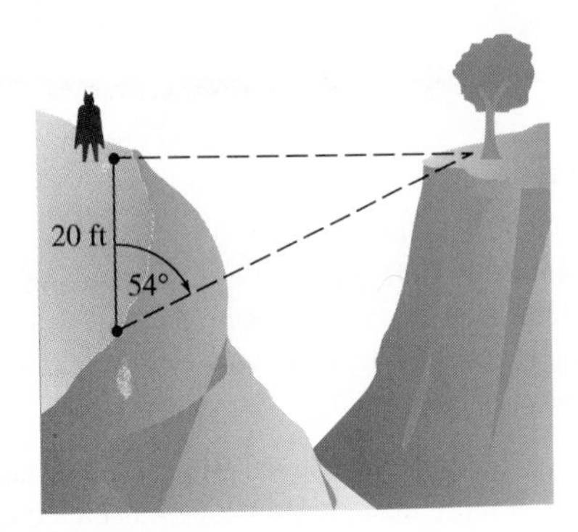

Figure 7-13

62. A wire from the top of a TV tower to the ground makes an angle of 49.5° with the ground and touches ground 225 ft from the base of the tower. How high is the tower?

63. A woman 5.5 ft tall stands 10 ft from a streetlight and casts a 4-ft-long shadow (Figure 7–14). How tall is the streetlight? What is angle θ?

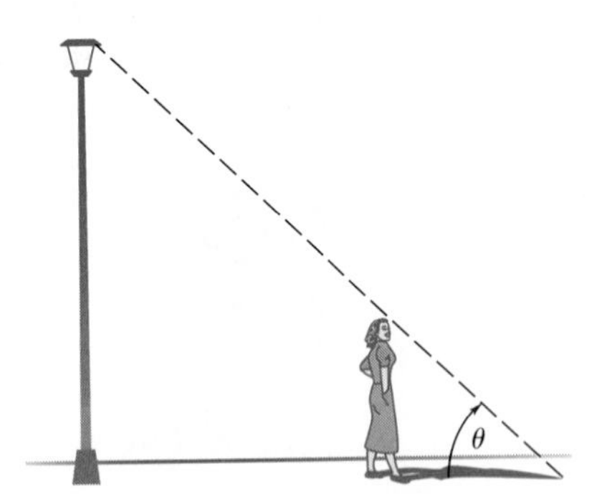

Figure 7-14

64. A plane flies a straight course. On the ground directly below the flight path, observers two miles apart spot the plane at the same time. The plane's angle of elevation is 46° from one observation point and 71° from the other (Figure 7–15). How high is the plane?

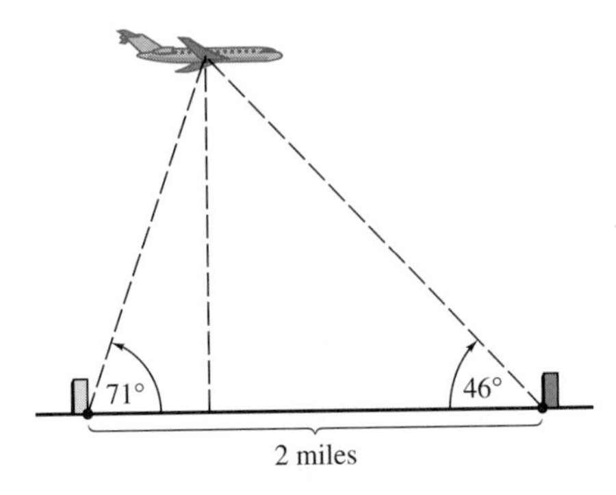

Figure 7-15

65. A buoy in the ocean is observed from the top of a 40-meter-high radar tower on shore. The angle of depression from the top of the tower to the base of the buoy is 6.5°. How far is the buoy from the base of the radar tower?

66. A plane passes directly over your head at an altitude of 500 ft. Two seconds later you observe that its angle of elevation is 42°. How far did the plane travel during those 2 sec?

67. A man stands 12 ft from a statue. The angle of elevation from eye level to the top of the statue is 30°, and the angle of depression to the base of the statue is 15°. How tall is the statue?

68. Two boats lie on a straight line with the base of a lighthouse. From the top of the lighthouse (21 meters above water level) it is observed that the angle of depression

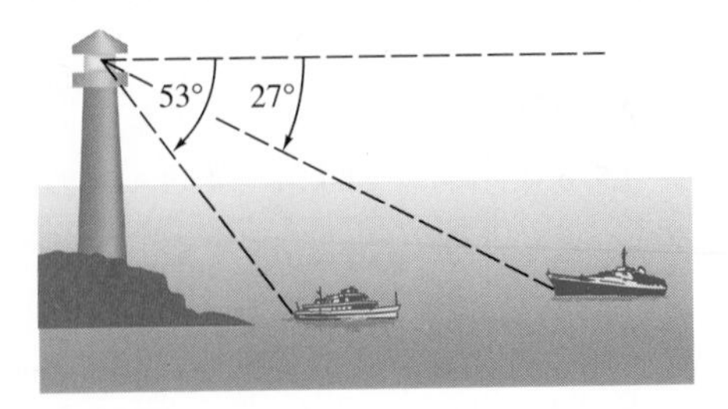

Figure 7-16

of the nearest boat is 53° and the angle of depression of the farthest boat is 27° (Figure 7–16). How far apart are the prows of the boats?

69. A rocket shoots straight up from the launchpad. Five seconds after lift-off an observer 2 miles away notes that the rocket's angle of elevation is 3.5°. Four seconds later the angle of elevation is 41°. How far did the rocket rise during those 4 sec?

70. From a 35-meter-high window, the angle of depression to the top of a nearby streetlight is 55°. The angle of depression to the base of the streetlight is 57.8°. How high is the streetlight?

71. A closed 60-ft-long drawbridge is 24 ft above water level. When the open the bridge makes an angle of 33° with the horizontal (Figure 7–17),

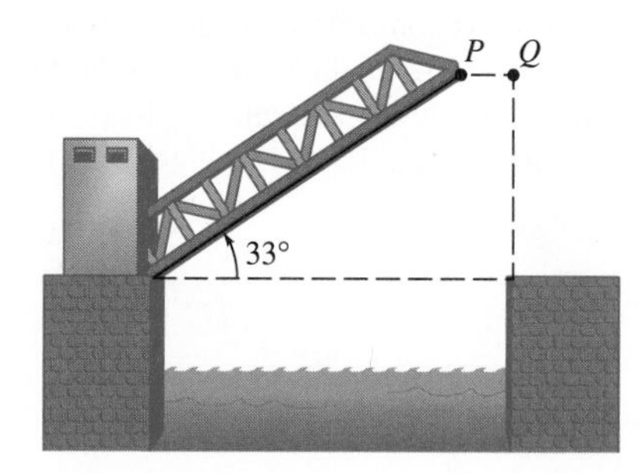

Figure 7-17

(a) How high is the tip P of the open bridge above the water?

(b) When the bridge is open, what is the distance from P to Q?

72. One plane flies straight east at an altitude of 31,000 ft. A second plane is flying west at an altitude of 14,000 ft on a course that lies directly below that of the first plane and directly above the straight road from Thomasville to Johnsburg. As the first plane passes over Thomasville, the second is passing over Johnsburg. At that instant both planes spot a beacon next to the road between Thomasville and Johnsburg. The angle of depression from the first plane to the beacon is 61° and the angle of depression from the second plane to the beacon is 34°. How far is Thomasville from Johnsburg?

73. A schematic diagram of a pedestrian overpass is shown in Figure 7–18. If you walk on the overpass from one end to the other, how far have you walked?

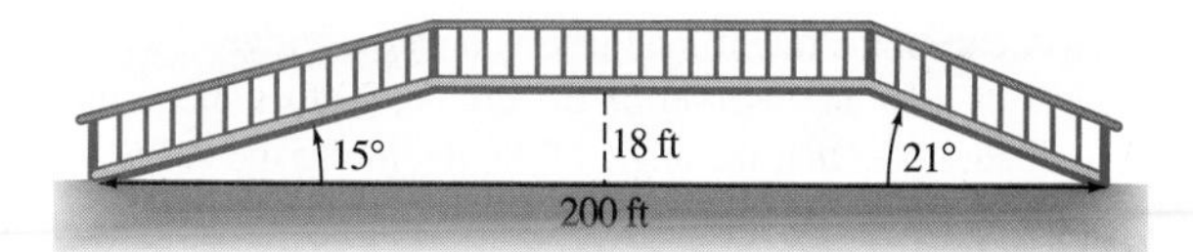

Figure 7-18

74. A 5-inch-high plastic beverage glass has a 2.5 inch diameter base. Its sides slope outward at a 4° angle as shown in Figure 7–19. What is the diameter of the top of the glass?

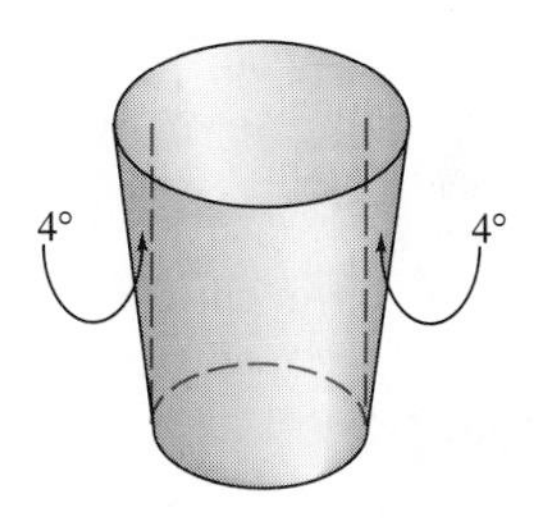

Figure 7-19

75. In aerial navigation, directions are given in degrees clockwise from north. Thus east is 90°, south is 180°, and so on, as shown in Figure 7–20. A plane travels from an airport for 200 mi in the direction 300°. How far west of the airport is the plane then?

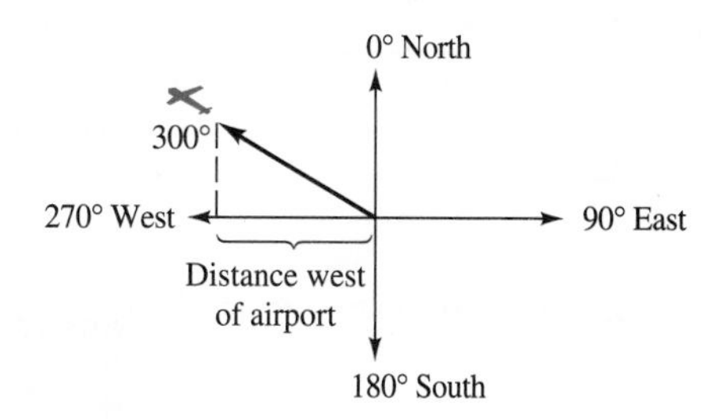

Figure 7-20

76. A plane travels at a constant 300 mph in the direction 65° (see Exercise 75).

(a) How far east of its starting point is the plane after half an hour?

(b) How far north of its starting point is the plane after 2 hr and 24 min?

77. A car on a straight road passes under a bridge. Two seconds later an observer on the bridge, 20 feet above the road, notes that the angle of depression to the car is 7.4°. How fast (in miles per hour) is the car traveling? [*Note:* 60 mph is equivalent to 88 ft/sec.]

Thinkers

78. A spy plane on a practice run over the midwest takes a picture that shows Cleveland, Ohio, on the eastern horizon and St. Louis, Missouri, 520 miles away on the western horizon (Figure 7–21, which is not to scale). Assuming that the radius of the earth is 3950 miles, how high was the plane when the picture was taken? [*Hint:* The sight lines from the plane to the horizons are tangent to the earth and a tangent line to a circle is perpendicular to the radius at that point. The arc of the earth between St. Louis and Cleveland is 520 miles long. Use this fact and the arc length formula to find angle θ (your answer will be in radians). Note that $\alpha = \theta/2$ (why?).]

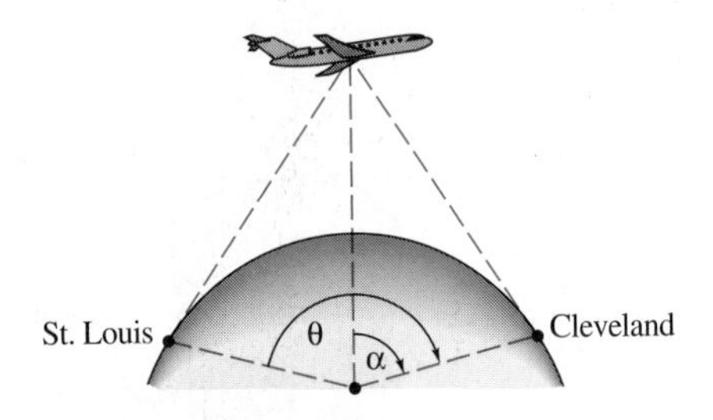

Figure 7-21

79. A 50-ft-high flagpole stands on top of a building. From a point on the ground the angle of elevation of the top of the pole is 43° and the angle of elevation of the bottom of the pole is 40°. How high is the building?

80. Two points on level ground are 500 meters apart. The angles of elevation from these points to the top of a nearby hill are 52° and 67°, respectively. The two points and the ground level point directly below the top of the hill lie on a straight line. How high is the hill?

7.2 THE LAW OF COSINES

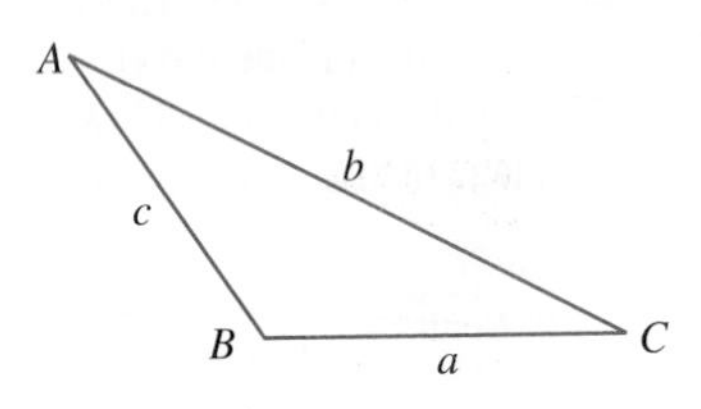

Figure 7-22

We now consider the solution of *oblique* triangles (ones that don't contain a right angle). We shall use **standard notation** for triangles: Each vertex is labeled with a capital letter, and the length of the side opposite that vertex is denoted by the same letter in lower case, as shown in Figure 7–22. The letter A will also be used to label the *angle* at vertex A, and similarly for B and C. So we shall make statements such as $A = 37°$ or $\cos B = .326$.

The first fact needed to solve oblique triangles* is the Law of Cosines, whose proof is given at the end of this section.

Law of Cosines ▶

In any triangle ABC, with sides of lengths a, b, c, as in Figure 7–22,

$$a^2 = b^2 + c^2 - 2bc \cos A$$

$$b^2 = a^2 + c^2 - 2ac \cos B$$

$$c^2 = a^2 + b^2 - 2ab \cos C$$

*Recall that "solving a triangle" means finding the lengths of all the sides and the measures of all the angles when only some of these quantities are given.

You need only memorize one of these equations since each of them provides essentially the same information: a description of one side of a triangle in terms of the angle opposite it and the other two sides.

Note When C is a right angle, then c is the hypotenuse and

$$\cos C = \cos 90° = 0,$$

so that the third equation in the Law of Cosines becomes the Pythagorean Theorem:

$$c^2 = a^2 + b^2$$

Solving the first equation in the Law of Cosines for $\cos A$, we obtain

Law of Cosines— Alternate Form ▶

$$\cos A = \frac{b^2 + c^2 - a^2}{2bc}.$$

The other two equations can be similarly rewritten. In this form, the Law of Cosines provides a description of each angle of a triangle in terms of the three sides. Consequently, the Law of Cosines can be used to solve triangles in these cases:

(i) Two sides and the angle between them are known (SAS).

(ii) Three sides are known (SSS).

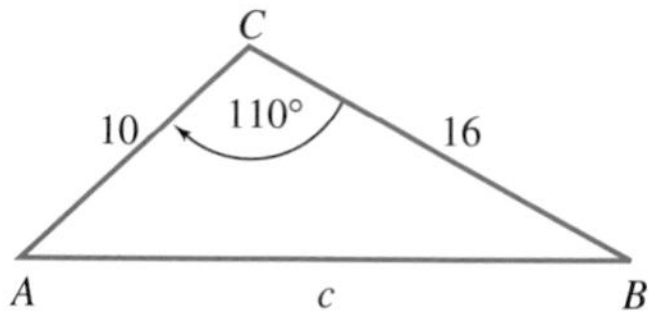

EXAMPLE 1 *(SAS)* Solve triangle ABC in Figure 7–23.

C, 10, 110°, 16, A, c, B

Figure 7-23

Solution We have $a = 16$, $b = 10$, and $C = 110°$. The right side of the third equation in the Law of Cosines involves only these known quantities. Hence,*

$$\begin{aligned} c^2 &= a^2 + b^2 - 2ab\cos C \\ &= 16^2 + 10^2 - 2\cdot 16\cdot 10\cos 110° \\ &\approx 256 + 100 - 320(-.342) \approx 465.4. \end{aligned}$$

Therefore, $c \approx \sqrt{465.4} \approx 21.6$. Now use the alternate form of the Law of Cosines:

$$\cos A = \frac{b^2 + c^2 - a^2}{2bc} \approx \frac{10^2 + (21.6)^2 - 16^2}{2\cdot 10\cdot 21.6} \approx .7172.$$

Keying in COS^{-1} .7172 ENTER in a calculator in degree mode produces an angle with cosine .7172: $A \approx 44.2°$. Hence, $B \approx 180° - (44.2° + 110°) = 25.8°$. ■

*Throughout this chapter all decimals are printed in rounded-off form for reading convenience, but no rounding is done in the actual computation until the final answer is obtained.

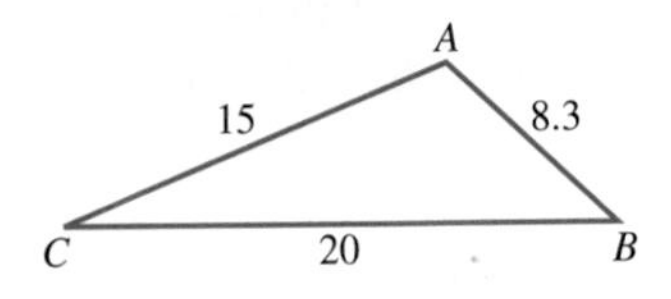

Figure 7-24

EXAMPLE 2 *(SSS)* Find the angles of triangle ABC in Figure 7–24.

Solution In this case, $a = 20$, $b = 15$, and $c = 8.3$. By the alternate form of the Law of Cosines:

$$\cos A = \frac{b^2 + c^2 - a^2}{2bc} = \frac{15^2 + 8.3^2 - 20^2}{2 \cdot 15 \cdot 8.3} = \frac{-106.11}{249} \approx -.4261.$$

The $\boxed{\text{COS}^{-1}}$ key shows that $A \approx 115.2°$. Similarly, the alternate form of the Law of Cosines yields:

$$\cos B = \frac{a^2 + c^2 - b^2}{2ac} = \frac{20^2 + 8.3^2 - 15^2}{2(20)(8.3)} = \frac{243.89}{332} \approx .7346$$

$$B \approx 42.7°.$$

Therefore, $C \approx 180° - (115.2° + 42.7°) = 180° - 157.9° = 22.1°$. ■

EXAMPLE 3 Two trains leave a station on different tracks. The tracks make an angle of 125° with the station as vertex. The first train travels at an average speed of 100 km/hr and the second an average of 65 km/hr. How far apart are the trains after 2 hours?

Solution The first train A traveling at 100 km/hr for 2 hours goes a distance of $100 \times 2 = 200$ km. The second train B travels a distance of $65 \times 2 = 130$ km. So, we have the situation shown in Figure 7–25.

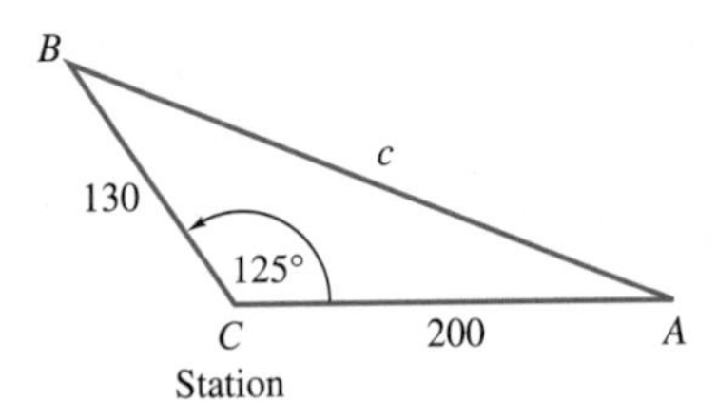

Figure 7-25

By the Law of Cosines

$$\begin{aligned} c^2 &= a^2 + b^2 - 2ab \cos C \\ &= 130^2 + 200^2 - 2 \cdot 130 \cdot 200 \cos 125° \\ &= 56{,}900 - 52{,}000 \cos 125° \approx 86{,}725.97 \end{aligned}$$

$$c \approx \sqrt{86{,}725.97} = 294.5 \text{ km}.$$

The trains are 294.5 km apart after 2 hours. ■

EXAMPLE 4 A 100-foot-tall antenna tower is to be placed on a hillside that makes an angle of 12° with the horizontal. It is to be anchored by two cables from the top of the tower to points 85 feet uphill and 95 feet downhill from the base. How much cable is needed?

Solution The situation is shown in Figure 7–26, where AB represents the tower and AC and AD the cables.

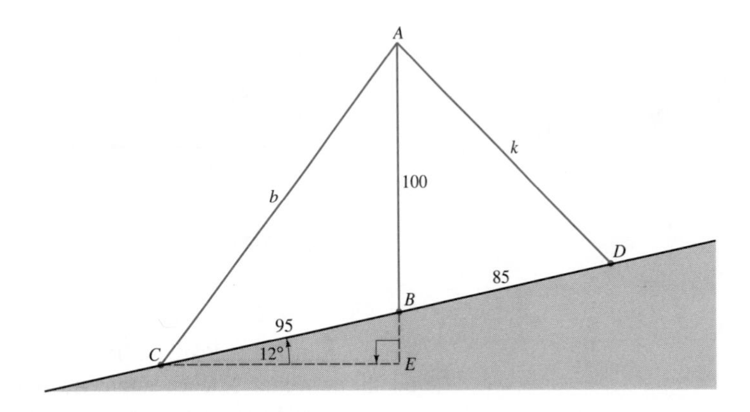

Figure 7-26

In triangle BEC, angle E is a right angle and by hypothesis, angle C measures 12°. Since the sum of the angles of a triangle is 180°, we must have

$$\measuredangle CBE = 180° - (90° + 12°) = 78°.$$

As shown in the figure, the sum of angles CBE and CBA is a straight angle (180°). Hence,

$$\measuredangle CBA = 180° - 78° = 102°.$$

Apply the Law of Cosines to triangle ABC:

$$\begin{aligned} b^2 &= 95^2 + 100^2 - 2 \cdot 95 \cdot 100 \cos 102° \\ &= 9025 + 10{,}000 - 19{,}000 \cos 102° \\ &\approx 22{,}975.32. \end{aligned}$$

Therefore, the length of the downhill cable is $b \approx \sqrt{22{,}975.32} \approx 151.58$ feet.

To find the length of the uphill cable, note that the sum of angles CBA and DBA is a straight angle, so that

$$\measuredangle DBA = 180° - \measuredangle CBA = 180° - 102° = 78°.$$

Applying the Law of Cosines to triangle DBA, we have:

$$\begin{aligned} k^2 &= 85^2 + 100^2 - 2 \cdot 85 \cdot 100 \cos 78° \\ &= 7225 + 10{,}000 - 17{,}000 \cos 78° \approx 13{,}690.50. \end{aligned}$$

Hence, the length of the uphill cable is $k \approx \sqrt{13{,}690.50} \approx 117.01$ feet. ■

Proof of the Law of Cosines

Given triangle ABC, position it on a coordinate plane so that angle A is in standard position with initial side c and terminal side b. Depending on the size of angle A, there are two possibilities, as shown in Figure 7–27.

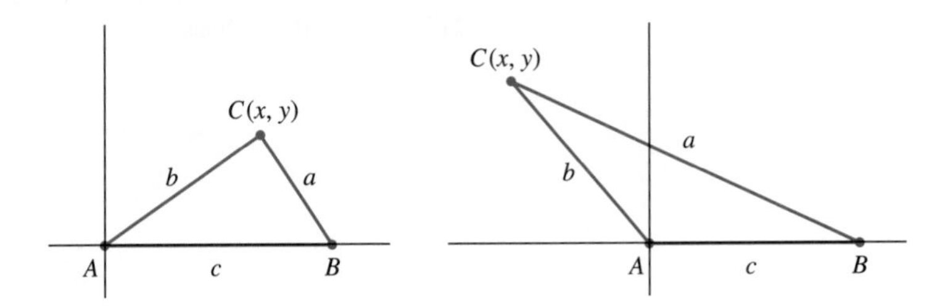

Figure 7-27

The coordinates of B are $(c, 0)$. Let (x, y) be the coordinates of C. Now C is a point on the terminal side of angle A, and the distance from C to the origin A is obviously b. Therefore, according to the point-in-the-plane description of sine and cosine, we have

$$\frac{x}{b} = \cos A, \qquad \text{or equivalently,} \qquad x = b \cos A$$

$$\frac{y}{b} = \sin A, \qquad \text{or equivalently,} \qquad y = b \sin A$$

Using the distance formula on the coordinates of B and C, we have

$$\begin{aligned} a &= \text{distance from } C \text{ to } B \\ &= \sqrt{(x-c)^2 + (y-0)^2} = \sqrt{(b \cos A - c)^2 + (b \sin A - 0)^2}. \end{aligned}$$

Squaring both sides of this last equation and simplifying, using the Pythagorean identity, yields:

$$\begin{aligned} a^2 &= (b \cos A - c)^2 + (b \sin A)^2 \\ a^2 &= b^2 \cos^2 A - 2bc \cos A + c^2 + b^2 \sin^2 A \\ a^2 &= b^2(\sin^2 A + \cos^2 A) + c^2 - 2bc \cos A \\ a^2 &= b^2 + c^2 - 2bc \cos A. \end{aligned}$$

This proves the first equation in the Law of Cosines. Similar arguments beginning with angles B or C in standard position prove the other two equations.

EXERCISES 7.2

Directions: *Standard notation for triangle ABC is used throughout. Use a calculator and round off your answers to one decimal place at the end of the computation.*

In Exercises 1–16, solve the triangle ABC under the given conditions.

1. $A = 20°, b = 10, c = 7$

2. $B = 40°, a = 12, c = 20$

3. $C = 118°, a = 6, b = 10$

4. $C = 52.5°, a = 6.5, b = 9$

5. $A = 140°, b = 12, c = 14$

6. $B = 25.4°, a = 6.8, c = 10.5$

7. $C = 78.6°, a = 12.1, b = 20.3$

8. $A = 118.2°, b = 16.5, c = 10.7$

9. $a = 7, b = 3, c = 5$

10. $a = 8, b = 5, c = 10$

11. $a = 16, b = 20, c = 32$

12. $a = 5.3, b = 7.2, c = 10$

13. $a = 7.2, b = 6.5, c = 11$

14. $a = 6.8, b = 12.4, c = 15.1$

15. $a = 12, b = 16.5, c = 21.3$

16. $a = 5.7, b = 20.4, c = 16.8$

17. Find the angles of the triangle whose vertices are $(0, 0)$, $(5, -2)$, $(1, -4)$.

18. Find the angles of the triangle whose vertices are $(-3, 4)$, $(6, 1)$, $(2, -1)$.

19. Two trains leave a station on different tracks. The tracks make a 112° angle with the station as vertex. The first train travels at an average speed of 90 km/hr and the second at an average speed of 55 km/hr. How far apart are the trains after 2 hr and 45 min?

20. One plane flies west from Cleveland at 350 mph. A second plane leaves Cleveland at the same time and flies southeast at 200 mph. How far apart are the planes after 1 hr and 36 min?

21. The pitcher's mound on a standard baseball diamond is 60.5 ft from home plate (Figure 7–28). How far is the pitcher's mound from first base?

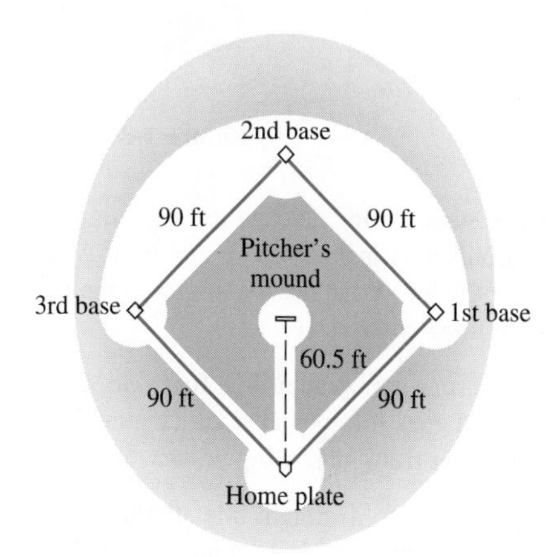

Figure 7-28

22. If the straight line distance from home plate over second base to the center field wall in a baseball stadium is 400 ft, how far is it from first base to the same point in center field. [Adapt Figure 7–28.]

23. A stake is located 10.8 ft from the end of a closed gate that is 8 ft long (Figure 7–29). The gate swings open, and its end hits the stake. Through what angle did the gate swing?

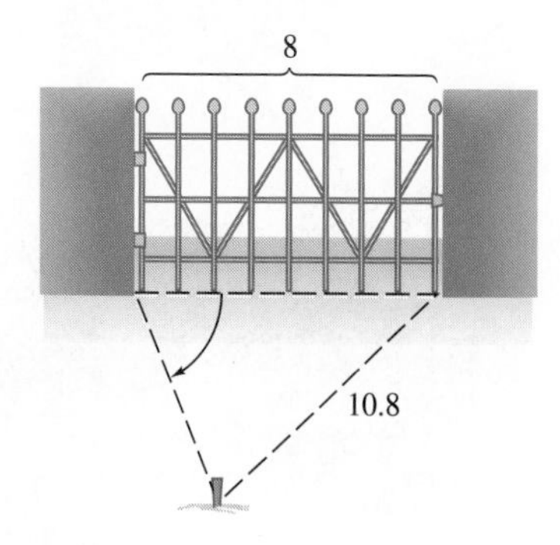

Figure 7-29

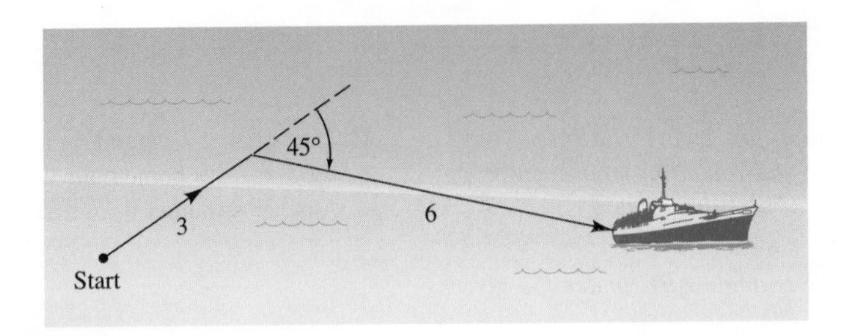

Figure 7-30

24. The distance from Chicago to St. Louis is 440 km, from St. Louis to Atlanta 795 km, and from Atlanta to Chicago 950 km. What are the angles in the triangle with these three cities as vertices?

25. A boat runs in a straight line for 3 km, then makes a 45° turn and goes for another 6 km (Figure 7–30). How far is the boat from its starting point?

26. A plane flies in a straight line at 400 mph for 1 hr and 12 min. It makes a 15° turn and flies at 375 mph for 2 hr and 27 min. How far is it from its starting point?

27. The side of a hill makes an angle of 12° with the horizontal. A wire is to be run from the top of a 175-ft tower on the top of the hill to a stake located 120 ft down the hillside from the base of the tower. How long a wire is needed?

28. Two ships leave port, one traveling in a straight course at 22 mph and the other traveling a straight course at 31 mph. Their courses diverge by 38°. How far apart are they after three hours?

29. An engineer wants to measure the width *CD* of a sink hole. So he places a stake at *B* and determines the measurements shown in Figure 7–31. How wide is the sink hole?

30. A straight tunnel is to be dug through a hill. Two people stand on opposite sides of the hill where the tunnel entrances are to be located. Both can see a stake located 530 m from the first person and 755 m from the second. The angle determined by the two people and the stake (vertex) is 77°. How long must the tunnel be?

31. One diagonal of a parallelogram is 6 cm long, and the other is 13 cm long. They form an angle of 42° with each other. How long are the sides of the parallelogram? [*Hint:* The diagonals of a parallelogram bisect each other.]

32. A parallelogram has diagonals of lengths 12 and 15 in. that intersect at an angle of 63.7°. How long are the sides of the parallelogram?

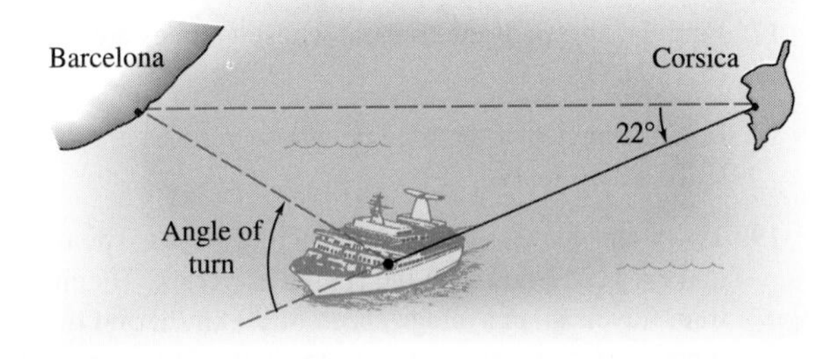

Figure 7-32

33. A ship is traveling at 18 mph from Corsica to Barcelona, a distance of 350 miles. To avoid bad weather, the ship leaves Corsica on a route 22° south of the direct route (Figure 7–32). After 7 hours the bad weather has been bypassed. Through what angle should the ship now turn to head directly to Barcelona?

34. A plane leaves South Bend for Buffalo, 400 miles away, intending to fly a straight course in the direction 70° (aerial navigation is explained in Exercise 75 of Section 7.1). After flying 180 miles, the pilot realizes that an error has been made and that he has actually been flying in the direction 55°.

(a) At that time, how far is the plane from Buffalo?

(b) In what direction should the plane now go to reach Buffalo?

35. Assume that the earth is a sphere of radius 3980 miles. A satellite travels in a circular orbit around the earth, 900 miles above the equator, making one full orbit every 6 hr. If it passes directly over a tracking station at 2 P.M., what is the distance from the satellite to the tracking station at 2:05 P.M.?

36. Two planes at the same altitude approach an airport. One plane is 16 miles from the control tower and the other is 22 miles from the tower. The angle determined by the planes and the tower, with the tower as vertex, is 11°. How far apart are the planes?

37. Assuming that the circles in Figure 7–33 are mutually tangent, find the lengths of the sides and the measures of the angles in triangle ABC.

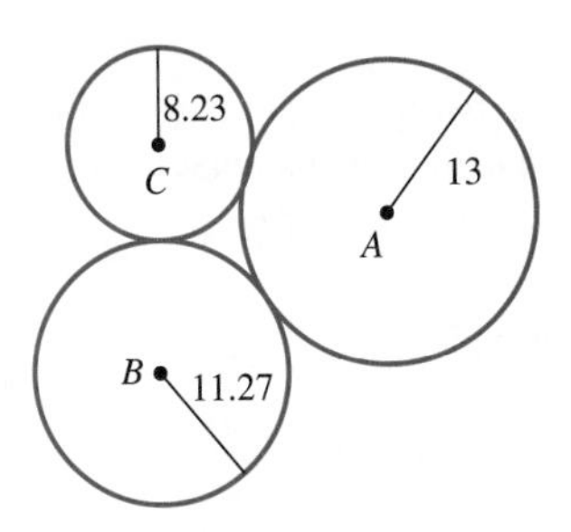

Figure 7-33

38. Assuming that the circles in Figure 7–34 are mutually tangent, find the lengths of the sides and the measures of the angles in triangle ABC.

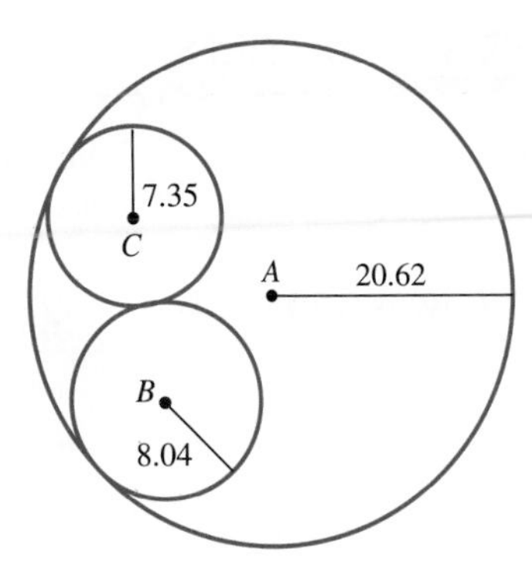

Figure 7-34

Thinkers

39. A rope is attached at points A and B and taut around a pulley whose center is at C, as shown in Figure 7–35 (in which AC has length 8 and BC length 7). The rope lies on the pulley from D to E and the radius of the pulley is 1 m. How long is the rope?

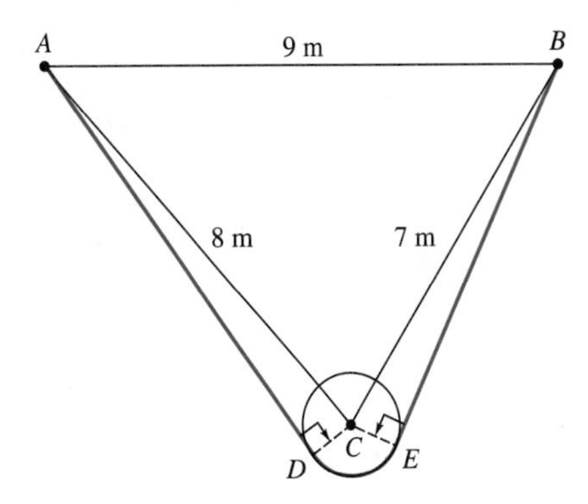

Figure 7-35

40. Use the Law of Cosines to prove that the sum of the squares of the lengths of the two diagonals of a parallelogram equals the sum of the squares of the lengths of the four sides.

7.3 THE LAW OF SINES

In order to solve oblique triangles in cases where the Law of Cosines cannot be used, we need this fact:

Law of Sines ▶

In any triangle *ABC* (in standard notation)*

$$\frac{a}{\sin A} = \frac{b}{\sin B} = \frac{c}{\sin C}.$$

Proof Position triangle ABC on a coordinate plane so that angle C is in standard position, with initial side b and terminal side a, as shown in Figure 7–36.

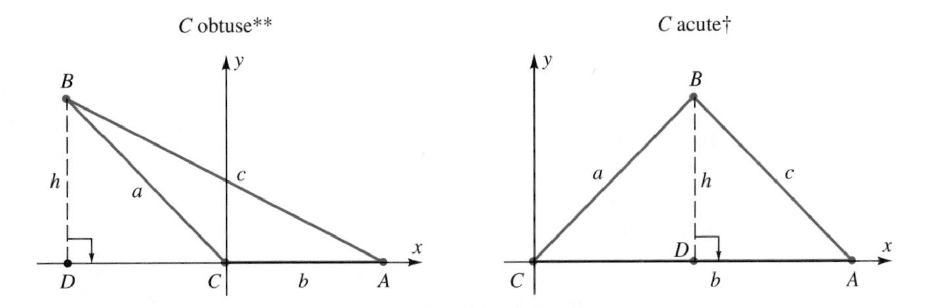

Figure 7-36

In each case we can compute $\sin C$ by using the point B on the terminal side of angle C. The second coordinate of B is h and the distance from B to the origin is a. Therefore, by the point-in-the-plane description of sine,

$$\sin C = \frac{h}{a}, \qquad \text{or equivalently,} \qquad h = a \sin C.$$

In each case, right triangle ADB shows that

$$\sin A = \frac{\text{opposite}}{\text{hypotenuse}} = \frac{h}{c}, \qquad \text{or equivalently,} \qquad h = c \sin A.$$

Combining this with the fact that $h = a \sin C$ we have

$$a \sin C = c \sin A.$$

*An equality of the form $u = v = w$ is shorthand for the statement $u = v$ and $v = w$ and $w = u$.

An **obtuse angle is an angle θ such that $90° < \theta < 180°$.

†If angle A is obtuse, then angle B is necessarily acute. Reposition the triangle so that a is the initial side, b is the terminal side, and vertex B is on the x-axis. Then reverse the roles of both A, B, and a, b in the proof.

Since angles in a triangle are nonzero, $\sin A \neq 0$ and $\sin C \neq 0$. Dividing both sides of the last equation by $\sin A \sin C$ yields

$$\frac{a}{\sin A} = \frac{c}{\sin C}.$$

This proves one equation in the Law of Sines. Similar arguments beginning with angles A or B in standard position prove the other equations. ❑

The Law of Sines can be used to solve triangles in these cases:

(i) Two angles and one side are known (AAS).

(ii) Two sides and the angle opposite one of them are known (SSA).

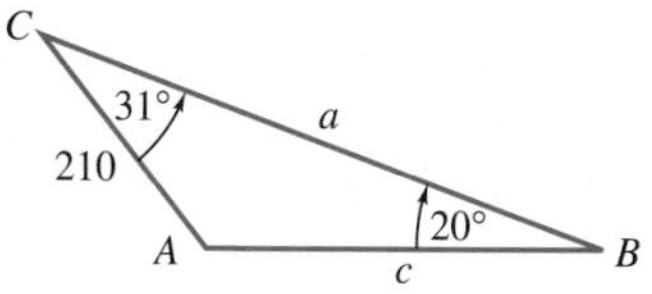

Figure 7-37

EXAMPLE 1 *(AAS)* If $B = 20°$, $C = 31°$, and $b = 210$ in Figure 7–37, find the other angles and sides.

Solution Since the sum of the angles of a triangle is 180°,

$$A = 180° - (20° + 31°) = 180° - 51° = 129°.$$

In order to find side a, we observe that we know three of the four quantities in one of the equations given by the Law of Sines:

$$\frac{a}{\sin A} = \frac{b}{\sin B}$$

$$\frac{a}{\sin 129°} = \frac{210}{\sin 20°}.$$

Solving this equation for a, and using a calculator, we obtain:

$$a = \frac{210(\sin 129°)}{\sin 20°} \approx 477.2.$$

Side c is found similarly. Beginning with an equation from the Law of Sines involving c and three known quantities, we have:

$$\frac{c}{\sin C} = \frac{b}{\sin B}$$

$$\frac{c}{\sin 31°} = \frac{210}{\sin 20°}$$

$$c = \frac{210 \sin 31°}{\sin 20°} \approx 316.2. \quad ■$$

In the AAS case, there is exactly one triangle satisfying the given data.* But when two sides of a triangle and the angle opposite one of them are known (SSA), there may be one, two, or no triangles that satisfy the given data (the **ambiguous case**). To see why this can happen, suppose sides a and b and angle A are given. Place angle A in standard position with terminal side b. If angle A is acute, then there are four possibilities for side a:

(i) Side a is too short to reach the third side: *no solution.*

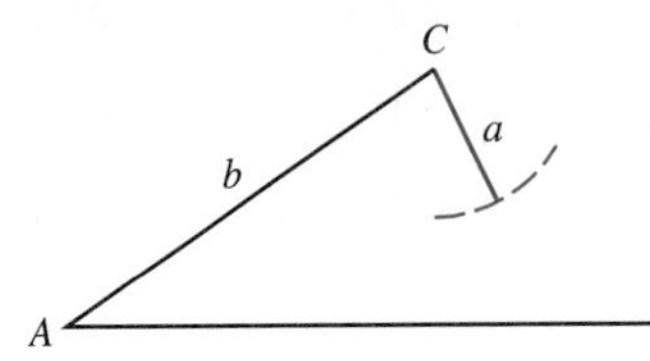

(ii) Side a just reaches the third side and is perpendicular to it: *one solution.*

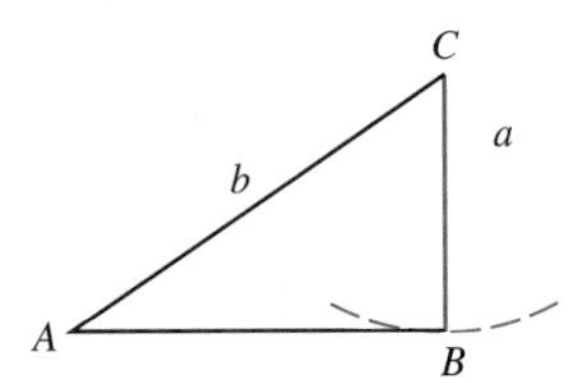

(iii) An arc of radius a meets the third side at 2 points to the right of A: *two solutions.*

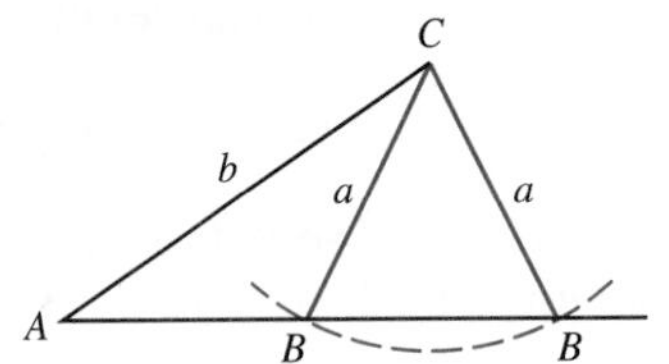

(iv) $a \geq b$, so that an arc of radius a meets the third side at just one point to the right of A: *one solution.*

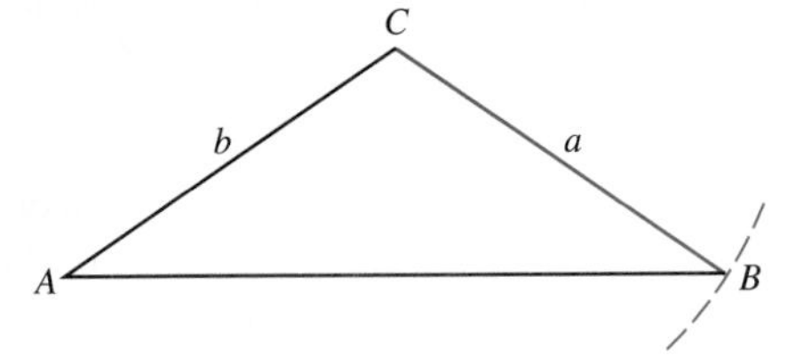

Figure 7-38

When angle A is obtuse, then there are only two possibilities:

(i) $a \leq b$, so that side a is too short to reach the third side: *no solution.*

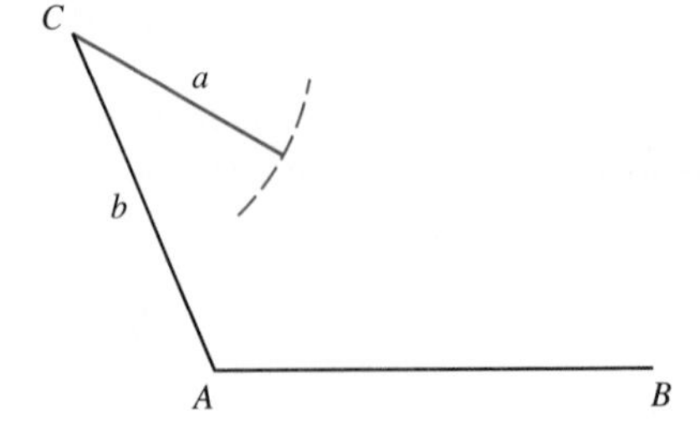

(ii) $a > b$ so that an arc of radius a meets the third side at just one point to the right of A: *one solution.*

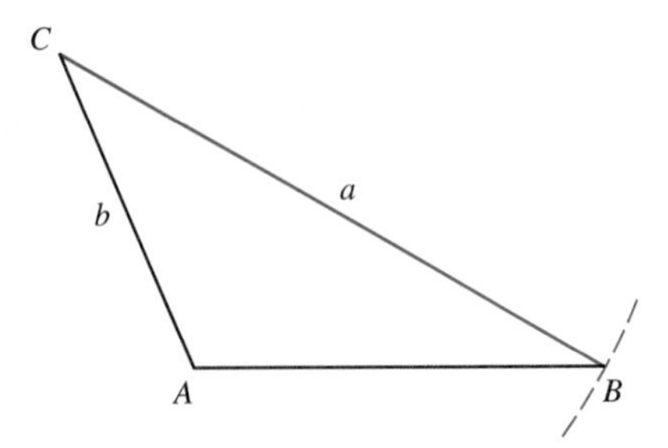

Figure 7-39

*Once you know two angles, you know all three (their sum must be 180°). Hence, you know two angles and the included side. Any two triangles satisfying these conditions will be congruent by the ASA Theorem of plane geometry.

In order to deal with the case of two solutions, we need this identity:

Supplementary Angle Identity ▶

> **If θ is an acute angle, then**
> $$\sin \theta = \sin(180^\circ - \theta).$$

Proof* Place the angle $180^\circ - \theta$ in standard position and choose a point D on its terminal side. Let r be the distance from D to the origin. The situation looks like this (Figure 7–40):

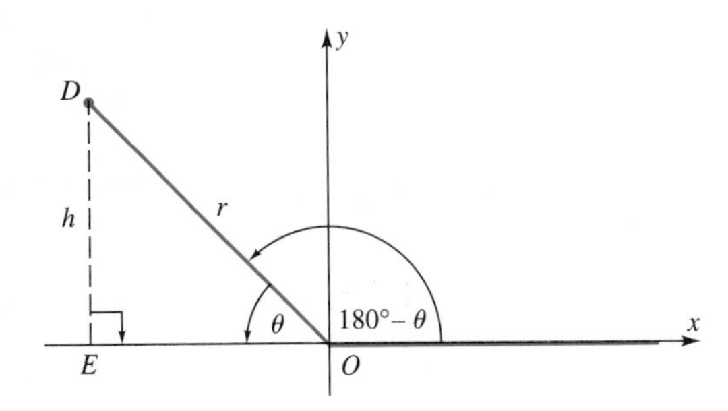

Figure 7-40

Since h is the second coordinate of D, we have $\sin(180^\circ - \theta) = h/r$. Right triangle OED shows that

$$\sin \theta = \frac{\text{opposite}}{\text{hypotenuse}} = \frac{h}{r} = \sin(180^\circ - \theta). \quad \square$$

EXAMPLE 2 *(SSA)* Given triangle ABC with $a = 6$, $b = 7$, and $A = 65^\circ$, find angle B.

Solution We use an equation from the Law of Sines involving the known quantities:

$$\frac{b}{\sin B} = \frac{a}{\sin A}$$

$$\frac{7}{\sin B} = \frac{6}{\sin 65^\circ}$$

$$\sin B = \frac{7 \sin 65^\circ}{6} \approx 1.06.$$

There is no angle B whose sine is greater than 1. Therefore, there is no triangle satisfying the given data (situation (i) in Figure 7–38). ■

*You may skip this proof if you've read Example 2 in Section 8.2, where the identity [in the form $\sin y = \sin(\pi - y)$] was proved for every angle.

EXAMPLE 3 An airplane, A, takes off from carrier B and flies in a straight line for 12 km. At that instant, an observer on destroyer C, located 5 km from the carrier, notes that the angle determined by the carrier, the destroyer (vertex), and the plane is 37°. How far is the plane from the destroyer?

Solution The given data provide Figure 7–41:

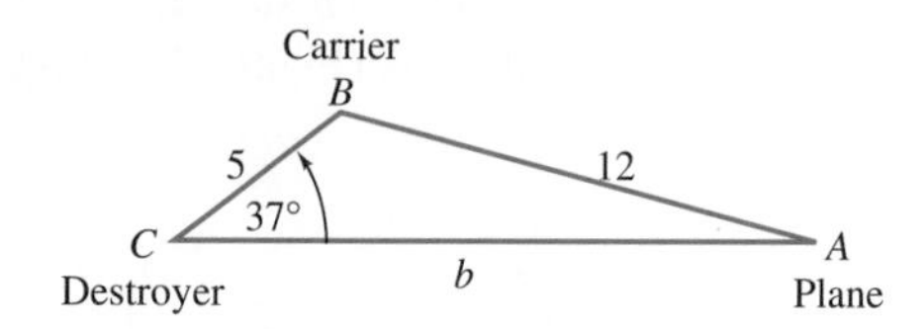

Figure 7-41

We must find side b. To do this, we first use the Law of Sines to find angle A:

$$\frac{a}{\sin A} = \frac{c}{\sin C}$$

$$\frac{5}{\sin A} = \frac{12}{\sin 37^\circ}$$

$$\sin A = \frac{5 \sin 37^\circ}{12} \approx .2508.$$

The SIN^{-1} key on a calculator in degree mode shows that 14.5° is an angle whose sine is .2508. The supplementary angle identity shows that $180^\circ - 14.5^\circ = 165.5^\circ$ is also an angle with sine .2508. But if $A = 165.5^\circ$ and $C = 37^\circ$, the sum of angles A, B, C would be greater than 180°. Since this is impossible, $A = 14.5^\circ$ is the only solution here (situation (iv) in Figure 7–38). Therefore,

$$B = 180^\circ - (37^\circ + 14.5^\circ) = 180^\circ - 51.5^\circ = 128.5^\circ.$$

Using the Law of Sines again, we have

$$\frac{b}{\sin B} = \frac{c}{\sin C}$$

$$\frac{b}{\sin 128.5^\circ} = \frac{12}{\sin 37^\circ}$$

$$b = \frac{12 \sin 128.5^\circ}{\sin 37^\circ} \approx 15.6.$$

Thus, the plane is approximately 15.6 km from the destroyer. ■

EXAMPLE 4 *(SSA)* Solve triangle ABC when $a = 7.5$, $b = 12$, and $A = 35°$.

Solution The Law of Sines shows that

$$\frac{b}{\sin B} = \frac{a}{\sin A}$$

$$\frac{12}{\sin B} = \frac{7.5}{\sin 35°}$$

$$\sin B = \frac{12 \sin 35°}{7.5} \approx .9177.$$

The SIN^{-1} key shows that 66.6° is a solution of $\sin B = .9177$. Therefore, $180° - 66.6° = 113.4°$ is also a solution of $\sin B = .9177$ by the supplementary angle identity. In each case the sum of angles A and B is less than 180°, so there are two triangles ABC satisfying the given data (situation (iii) in Figure 7–38, and shown here in Figure 7–42):

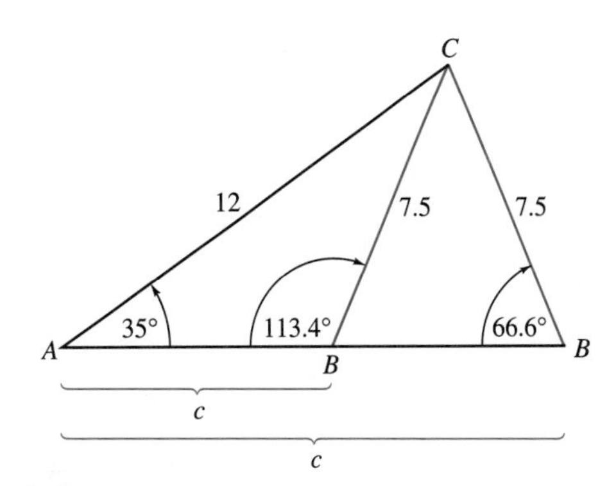

Figure 7-42

Case 1. If $B = 66.6°$, then $C = 180° - (35° + 66.6°) = 78.4°$. By the Law of Sines,

$$\frac{c}{\sin C} = \frac{a}{\sin A}$$

$$\frac{c}{\sin 78.4°} = \frac{7.5}{\sin 35°}$$

$$c = \frac{7.5 \sin 78.4°}{\sin 35°} \approx \frac{7.5(.9796)}{.5736} \approx 12.8.$$

Case 2. If $B = 113.4°$, then $C = 180° - (35° + 113.4°) = 31.6°$. Consequently,

$$\frac{c}{\sin C} = \frac{a}{\sin A}$$

$$\frac{c}{\sin 31.6°} = \frac{7.5}{\sin 35°}$$

$$c = \frac{7.5 \sin 31.6°}{\sin 35°} \approx \frac{7.5(.5240)}{.5736} \approx 6.9. \quad ■$$

EXAMPLE 5 A plane flying in a straight line passes directly over point A on the ground and later directly over point B, which is 3 miles from A. A few minutes after the plane passes over B, the angle of elevation from A to the plane is 43° and the angle of elevation from B to the plane is 67°. How high is the plane at that moment?

Solution If C represents the plane, then the situation is represented in Figure 7–43. We must find the length of h.

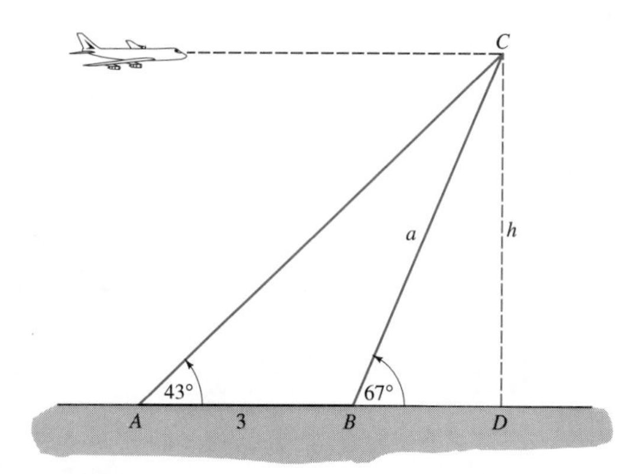

Figure 7-43

Note that angle ABC measures $180° - 67° = 113°$ and hence

$$\measuredangle BCA = 180° - (43° + 113°) = 24°.$$

Use the Law of Sines to find side a of triangle ABC:

$$\frac{a}{\sin 43°} = \frac{3}{\sin 24°}$$

$$a = \frac{3 \sin 43°}{\sin 24°} \approx 5.03.$$

Now in the right triangle CBD, we have

$$\sin 67^\circ = \frac{\text{opposite}}{\text{hypotenuse}} = \frac{h}{a} \approx \frac{h}{5.03}.$$

Therefore, $h \approx 5.03 \sin 67^\circ \approx 4.63$ miles. ■

The Area of a Triangle

The proof of the Law of Sines leads to this fact:

Area of a Triangle ▶

The area of a triangle containing an angle C with sides of lengths a and b is

$$\frac{1}{2}ab \sin C.$$

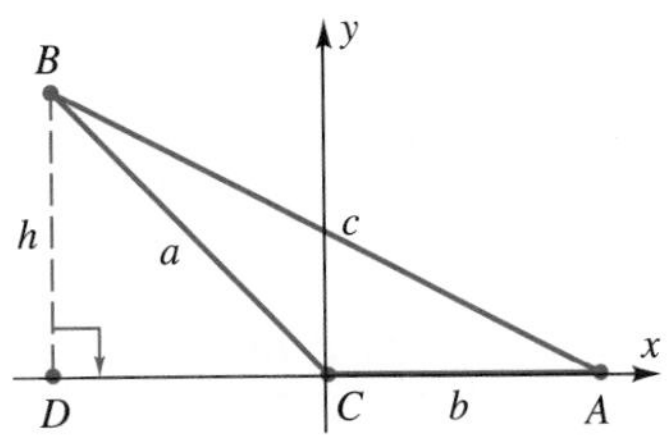

Figure 7-44

Proof Place the vertex of angle of C at the origin, with side b on the positive x-axis (Figure 7–44).* Then b is the base and h is the altitude of the triangle so that

$$\text{area of triangle } ABC = \frac{1}{2} \times \text{base} \times \text{altitude} = \frac{1}{2} \cdot b \cdot h.$$

The proof of the Law of Sines on page 470 shows that $h = a \sin C$. Therefore,

$$\text{area of triangle } ABC = \frac{1}{2} \cdot b \cdot h = \frac{1}{2} \cdot b \cdot a \sin C = \frac{1}{2} ab \sin C. \quad \square$$

EXAMPLE 6 The area of the triangle shown in Figure 7–45 is

$$\frac{1}{2} \cdot 8 \cdot 13 \sin 130^\circ \approx 39.83 \text{ sq cm.} \quad ■$$

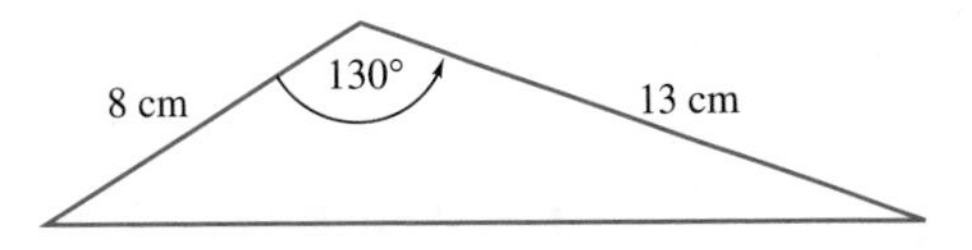

Figure 7-45

Here is a useful formula for the area of a triangle in terms of its sides:

*Figure 7–44 is the case when C is obtuse; the argument when C is acute is similar.

Heron's Formula ▶

The area of a triangle with sides a, b, c is

$$\sqrt{s(s-a)(s-b)(s-c)}$$

where $s = \frac{1}{2}(a+b+c)$.

Proof The preceding area formula and the Pythagorean identity

$$\sin^2 C = 1 - \cos^2 C = (1+\cos C)(1-\cos C)$$

show that the area of triangle ABC (standard notation) is

$$\frac{1}{2}ab\sin C = \sqrt{\left(\frac{1}{2}ab\sin C\right)^2} = \sqrt{\frac{1}{4}a^2b^2\sin^2 C}$$

$$= \sqrt{\frac{1}{4}a^2b^2(1-\cos^2 C)}$$

$$= \sqrt{\frac{1}{2}ab(1+\cos C)\frac{1}{2}ab(1-\cos C)}.$$

Exercise 62 uses the Law of Cosines to show that

$$\frac{1}{2}ab(1+\cos C) = \frac{(a+b)^2 - c^2}{4} = \frac{(a+b)+c}{2}\cdot\frac{(a+b)-c}{2}$$

$$= s(s-c)$$

and

$$\frac{1}{2}ab(1-\cos C) = \frac{c^2 - (a-b)^2}{4} = \frac{c-(a-b)}{2}\cdot\frac{c+(a-b)}{2}$$

$$= (s-a)(s-b).$$

Combining these facts completes the proof:

$$\text{Area} = \frac{1}{2}ab\sin C = \sqrt{\frac{1}{2}ab(1+\cos C)\frac{1}{2}ab(1-\cos C)}$$

$$= \sqrt{s(s-a)(s-b)(s-c)}. \quad \square$$

EXAMPLE 7 To find the area of the triangle whose sides have lengths 7, 9, and 12, apply Heron's Formula with $a = 7$, $b = 9$, $c = 12$, and

$$s = \frac{1}{2}(a+b+c) = \frac{1}{2}(7+9+12) = 14.$$

The area is

$$\sqrt{s(s-a)(s-b)(s-c)} = \sqrt{14(14-7)(14-9)(14-12)}$$

$$= \sqrt{980} \approx 31.3 \text{ square units.} \quad \blacksquare$$

EXERCISES 7.3

Directions: *Standard notation for triangle ABC is used throughout. Use a calculator and round off your answers to one decimal place at the end of the computation.*

In Exercises 1–8, solve triangle ABC under the given conditions.

1. $A = 48°, B = 22°, a = 5$

2. $B = 33°, C = 46°, b = 4$

3. $A = 116°, C = 50°, a = 8$

4. $A = 105°, B = 27°, b = 10$

5. $B = 44°, C = 48°, b = 12$

6. $A = 67°, C = 28°, a = 9$

7. $A = 102.3°, B = 36.2°, a = 16$

8. $B = 97.5°, C = 42.5°, b = 7$

In Exercises 9–16, find the area of triangle ABC under the given conditions.

9. $a = 4, b = 8, C = 27°$

10. $b = 10, c = 14, A = 36°$

11. $c = 7, a = 10, B = 68°$

12. $a = 9, b = 13, C = 75°$

13. $a = 11, b = 15, c = 18$

14. $a = 4, b = 12, c = 14$

15. $a = 7, b = 9, c = 11$

16. $a = 17, b = 27, c = 40$

In Exercises 17–36, solve the triangle. The Law of Cosines may be needed in Exercises 27–36.

17. $b = 15, c = 25, B = 47°$

18. $b = 30, c = 50, C = 60°$

19. $a = 12, b = 5, B = 20°$

20. $b = 12.5, c = 20.1, B = 37.3°$

21. $a = 5, c = 12, A = 102°$

22. $a = 9, b = 14, B = 95°$

23. $b = 11, c = 10, C = 56°$

24. $a = 12.4, c = 6.2, A = 72°$

25. $A = 41°, B = 67°, a = 10.5$

26. $a = 30, b = 40, A = 30°$

27. $b = 4, c = 10, A = 75°$

28. $a = 50, c = 80, C = 45°$

29. $a = 6, b = 12, c = 16$

30. $B = 20.67°, C = 34°, b = 185$

31. $a = 16.5, b = 18.2, C = 47°$

32. $a = 21, c = 15.8, B = 71°$

33. $b = 17.2, c = 12.4, B = 62.5°$

34. $b = 24.1, c = 10.5, C = 26.3°$

35. $a = 10.1, b = 18.2, A = 50.7°$

36. $b = 14.6, c = 7.8, B = 40.4°$

In Exercises 37 and 38, find the area of the triangle with the given vertices.

37. $(0, 0), (2, -5), (-3, 1)$ **38.** $(-4, 2), (5, 7), (3, 0)$

In Exercises 39 and 40, find the area of the polygonal region. [Hint: *divide the region into triangles.*]

39.

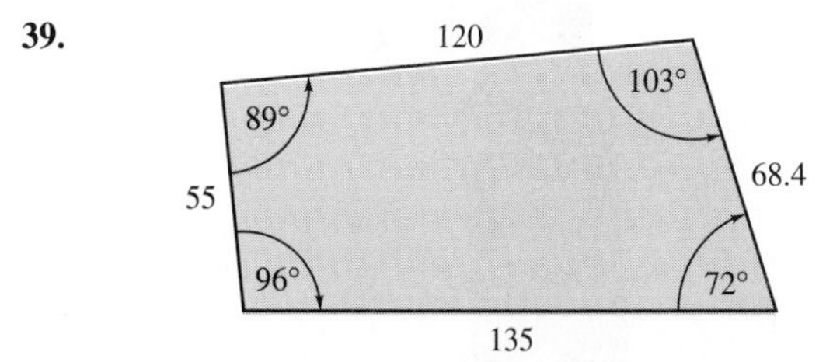

40.

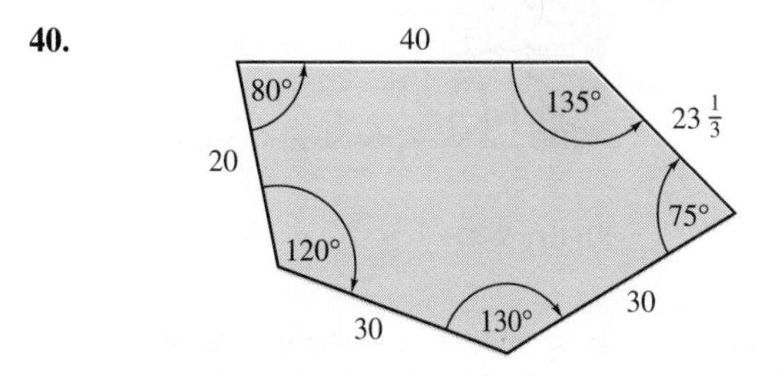

41. A surveyor marks points A and B 200 m apart on one bank of a river. She sights a point C on the opposite bank and determines the angles shown in Figure 7–46. What is the distance from A to C?

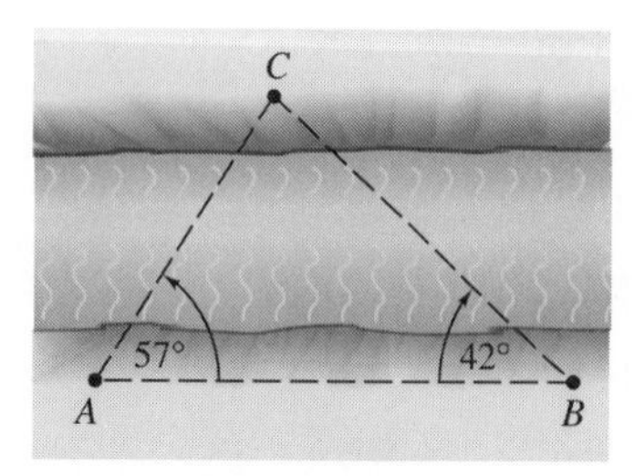

Figure 7-46

42. A forest fire is spotted from two fire towers. The triangle determined by the two towers and the fire has angles of 28° and 37° at the tower vertices. If the towers are 3000 m apart, which one is closest to the fire?

43. A visitor to the Leaning Tower of Pisa observed that the tower's shadow was 40 m long and that the angle of elevation from the tip of the shadow to the top of the tower was 57° (Figure 7–47). The tower is now 54 m tall (measured from the ground to the top along the center line of the tower). Approximate the angle α that the center line of the tower makes with the vertical.

Figure 7-47

44. A pole tilts at an angle 9° from the vertical, away from the sun, and casts a shadow 24 ft long. The angle of elevation from the end of the pole's shadow to the top of the pole is 53°. How long is the pole?

45. A side view of a bus shelter is shown in Figure 7–48. The brace d makes an angle of 37.25° with the back and an angle of 34.85° with the top of the shelter. How long is this brace?

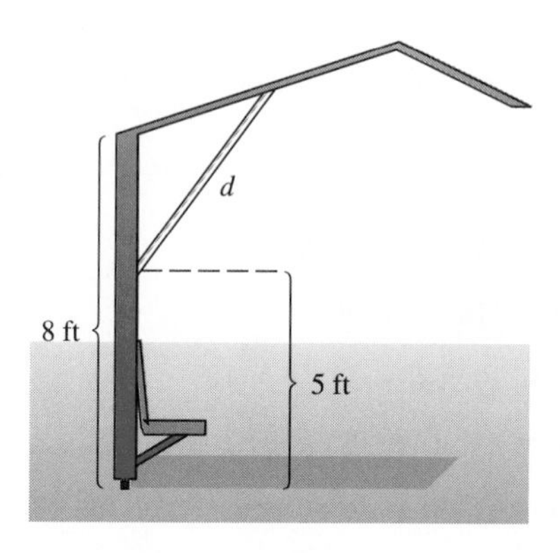

Figure 7-48

46. A straight path makes an angle of 6° with the horizontal. A statue at the higher end of the path casts a 6.5-m-long shadow straight down the path. The angle of elevation from the end of the shadow to the top of the statue is 32°. How tall is the statue?

47. A vertical statue 6.3 m high stands on the top of a hill. At a point on the side of the hill 35 m from the statue's base, the angle between the hillside and a line from the top of the statue is 10°. What angle does the side of the hill make with the horizontal?

48. A fence post is located 36 ft from one corner of a building and 40 ft from the adjacent corner. Fences are put up between the post and the building corners to form a triangular garden area. The 40-ft fence makes a 58° angle with the building. What is the area of the garden?

49. Two straight roads meet at an angle of 40° in Harville, one leading to Eastview and the other to Wellston (Figure 7–49). Eastview is 18 km from Harville and 20 km

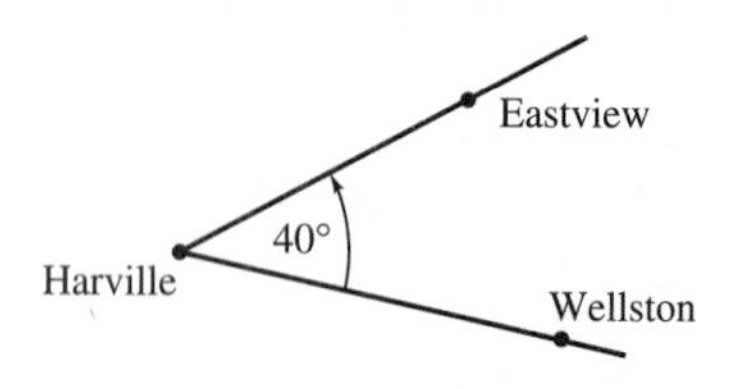

Figure 7-49

from Wellston. What is the distance from Harville to Wellston?

50. Each of two observers 400 ft apart measures the angle of elevation to the top of a tree that sits on the straight line between them. These angles are 51° and 65°, respectively. How tall is the tree? How far is the base of its trunk from each observer?

51. From the top of the 800-ft-tall Cartalk Tower, Tom sees a plane; the angle of elevation is 67°. At the same instant, Ray, who is on the ground, one mile from the building notes that his angle of elevation to the plane is 81° and that his angle of elevation to the top of Cartalk Tower is 8.6° (Figure 7–50). Assuming that Tom and Ray and the airplane are in a plane perpendicular to the ground, how high is the airplane?

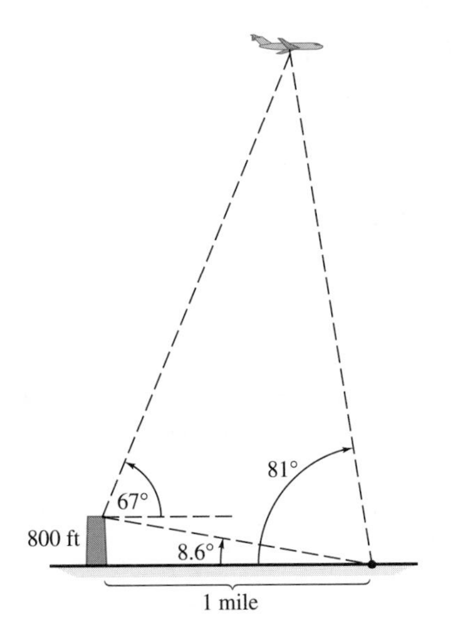

Figure 7-50

52. A plane flies in a direction of 105° from airport A. After a time, it turns and proceeds in a direction of 267°. Finally, it lands at airport B, 120 miles directly south of airport A. How far has the plane traveled? [*Note:* Aerial navigation directions are explained in Exercise 75 of Section 7.1.]

53. Charlie is afraid of water; he can't swim and refuses to get in a boat. However, he must measure the width of a river for his geography class. He has a long tape measure, but no way to measure angles. While pondering what to do, he paces along the side of the river using the five paths joining points A, B, C, D (Figure 7–51). If he can't determine the width of the river, he will flunk the course.

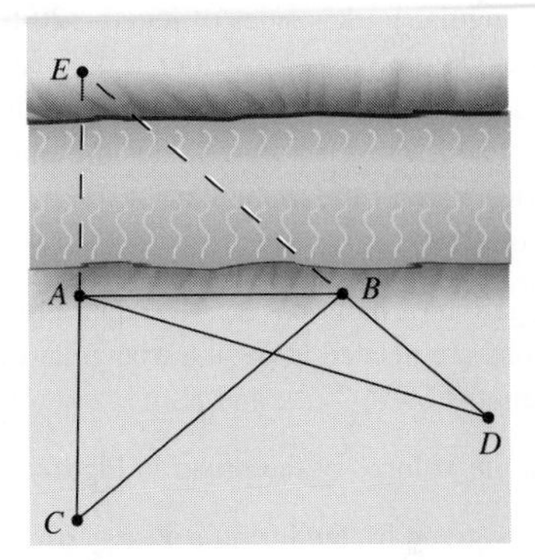

Figure 7-51

(a) Save Charlie from disaster by explaining how he can determine the width AE simply by measuring the lengths AB, AC, AD, BC, and BD and using trigonometry.

(b) Charlie determines that $AB = 75$ ft, $AC = 25$ ft, $AD = 90$ ft, $BC = 80$ ft, and $BD = 22$ ft. How wide is the river between A and E?

54. A plane flies in a direction of 85° from Chicago. It then turns and flies in the direction of 200° for 150 miles. It is then 195 miles from its starting point. How far did the plane fly in the direction of 85°? (See the note in Exercise 52.)

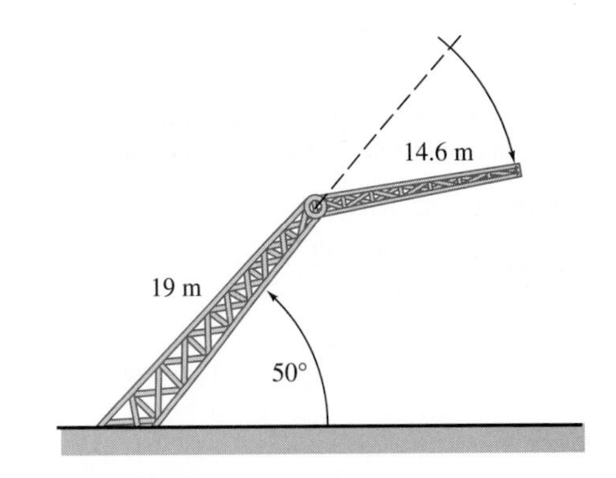

Figure 7-52

55. A hinged crane makes an angle of 50° with the ground (Figure 7–52). A malfunction causes the lock on the hinge to fail and the top part of the crane swings down. How far from the base of the crane does the top hit the ground?

56. A triangular lot has sides of length 120 ft and 160 ft. The angle between these sides is 42°. Adjacent to this lot is a rectangular lot whose longest side has length 200 ft and whose shortest side is the same length as the shortest side of the triangular lot. What is the total area of both lots?

57. If a gallon of paint covers 400 square feet, how many gallons are needed to paint a triangular deck with sides of lengths 65 ft, 72 ft, and 88 ft?

58. Find the volume of the prism in Figure 7–53. The volume is given by the formula $V = \frac{1}{3}Bh$, where B is the area of the base and h is the height.

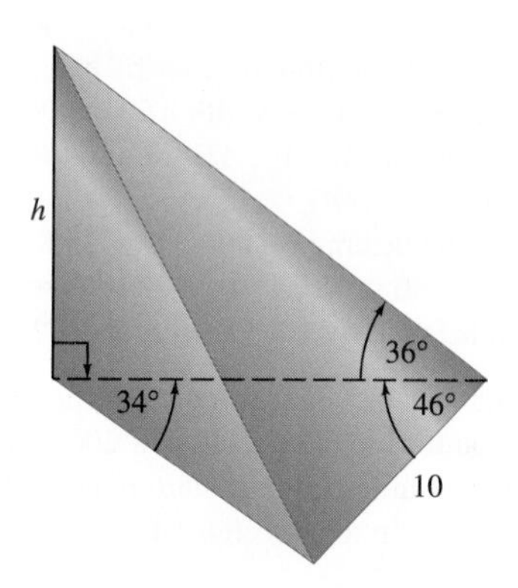

Figure 7-53

59. A rigid plastic triangle ABC rests on three vertical rods, as shown in Figure 7–54. What is its area?

60. Prove that the area of triangle ABC (standard notation) is given by $\dfrac{a^2 \sin B \sin C}{2 \sin A}$.

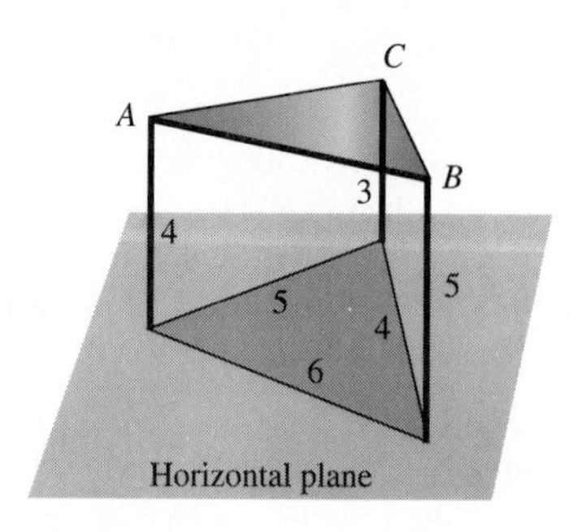

Figure 7-54

61. What is the area of a triangle whose sides have lengths 12, 20, and 36? (If your answer turns out strangely, try drawing a picture.)

62. Complete the proof of Heron's Formula as follows. Let $s = \frac{1}{2}(a + b + c)$.

(a) Show that

$$\frac{1}{2}ab(1 + \cos C) = \frac{(a + b)^2 - c^2}{4}$$
$$= \frac{(a + b) + c}{2} \cdot \frac{(a + b) - c}{2}$$
$$= s(s - c).$$

[*Hint:* Use the Law of Cosines to express $\cos C$ in terms of a, b, c; then simplify.]

(b) Show that

$$\frac{1}{2}ab(1 - \cos C) = \frac{c^2 - (a - b)^2}{4}$$
$$= \frac{c - (a - b)}{2} \cdot \frac{c + (a - b)}{2}$$
$$= (s - a)(s - b).$$

CHAPTER 7 *Review*

Important Concepts

Important Facts and Formulas

- *Law of Cosines:* $a^2 = b^2 + c^2 - 2bc \cos A$
- *Law of Cosines–Alternate Form:* $\cos A = \dfrac{b^2 + c^2 - a^2}{2bc}$
- *Law of Sines:* $\dfrac{a}{\sin A} = \dfrac{b}{\sin B} = \dfrac{c}{\sin C}$
- $\sin D = \sin(180° - D)$
- Area of triangle $ABC = \dfrac{ab \sin C}{2}$
- *Heron's Formula:*

$$\text{Area of triangle } ABC = \sqrt{s(s-a)(s-b)(s-c)}$$

where $s = \dfrac{1}{2}(a + b + c)$.

Review Questions

Note: *Standard notation is used for triangles.*

In Questions 1–4, angle B is a right angle. Solve triangle ABC.

1. $a = 12, c = 13$ **2.** $A = 40°, b = 10$

3. $C = 35°, a = 12$ **4.** $A = 56°, a = 11$

5. From a point on level ground 145 ft from the base of a tower, the angle of elevation to the top of the tower is 57.3°. How high is the tower?

6. A pilot in a plane at an altitude of 22,000 ft observes that the angle of depression to a nearby airport is 26°. How many *miles* is the airport from a point on the ground directly below the plane?

7. A road rises 140 ft per horizontal mile. What angle does the road make with the horizontal?

8. A lighthouse keeper 100 ft above the water sees a boat sailing in a straight line directly toward her. As she watches, the angle of depression to the boat changes from 25° to 40°. How far has the boat traveled during this time?

In Questions 9–12, use the Law of Cosines to solve triangle ABC.

9. $a = 12, b = 10, c = 15$

10. $a = 7.5, b = 3.2, c = 6.4$

11. $a = 10, c = 14, B = 130°$

12. $a = 7, b = 8.6, C = 72.4°$

13. Two trains depart simultaneously from the same station. The angle between the two tracks on which they leave is 120°. One train travels at an average speed of 45 mph and the other at 70 mph. How far apart are the trains after 3 hr?

14. A 40-ft-high flagpole sits on the side of a hill. The hillside makes a 17° angle with the horizontal. How long is a wire that runs from the top of the pole to a point 72 ft downhill from the base of the pole?

In Questions 15–20, use the Law of Sines to solve triangle ABC.

15. $B = 124°, C = 31°, c = 3.5$

16. $A = 96°, B = 44°, b = 12$

17. $a = 75, c = 84, C = 62°$

18. $a = 5, c = 2.5, C = 30°$

19. $a = 3.5, b = 4, A = 60°$

20. $a = 3.8, c = 2.8, C = 41°$

21. Find the area of triangle ABC if $b = 24$, $c = 15$, and $A = 55°$.

22. Find the area of triangle ABC if $a = 10$, $c = 14$, and $B = 75°$.

23. A boat travels for 8 km in a straight line from the dock. It is then sighted from a lighthouse which is 6.5 km from the dock. The angle determined by the dock, the lighthouse (vertex), and the boat is 25°. How far is the boat from the lighthouse?

24. A pole tilts 12° from the vertical, away from the sun, and casts a 34-ft-long shadow on level ground. The angle of elevation from the end of the shadow to the top of the pole is 64°. How long is the pole?

In Questions 25–28, solve triangle ABC.

25. $A = 48°, B = 57°, b = 47$

26. $A = 67°, c = 125, a = 100$

27. $a = 5, c = 8, B = 76°$

28. $a = 90, b = 70, c = 40$

29. Two surveyors, Joe and Alice, are 240 m apart on a river bank. Each sights a flagpole on the opposite bank. The angle from the pole to Joe (vertex) to Alice is 63°. The angle from the pole to Alice (vertex) to Joe is 54°. How far are Joe and Alice from the pole?

30. A surveyor stakes out points A and B on opposite sides of a building. Point C on the side of the building is 300 ft from A and 440 ft from B. Angle ACB measures 38°. What is the distance from A to B?

31. A woman on the top of a 488-ft-high building spots a small plane. As she views the plane, its angle of elevation is 62°. At the same instant a man at the ground level entrance to the building sees the plane and notes that its angle of elevation is 65°.

(a) How far is the woman from the plane?
(b) How far is the man from the plane?
(c) How high is the plane?

32. A straight road slopes at an angle of 10° with the horizontal. When the angle of elevation of the sun (from horizontal) is 62.5°, a telephone pole at the side of the road casts a 15-ft shadow downhill, parallel to the road. How high is the telephone pole?

33. Find the angle ABC in Figure 7–55.

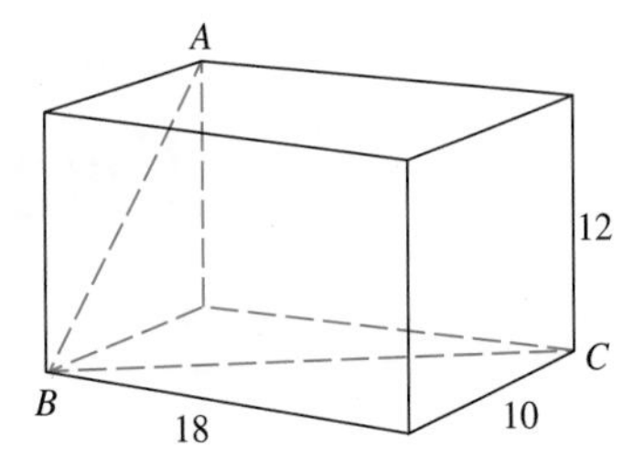

Figure 7-55

34. Use the Law of Sines to prove Engelsohn's equations:
For any triangle ABC (standard notation),

$$\frac{a+b}{c} = \frac{\sin A + \sin B}{\sin C}$$

and

$$\frac{a-b}{c} = \frac{\sin A - \sin B}{\sin C}.$$

In Questions 35–38, find the area of triangle ABC under the given conditions.

35. There is an angle of 30°, the sides of which have lengths 5 and 8.

36. There is an angle of 40°, the sides of which have lengths 3 and 12.

37. The sides have lengths 7, 11, and 14.

38. The sides have lengths 4, 8, and 10.

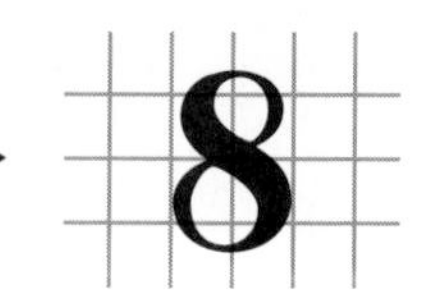

Trigonometric Identities and Equations

Roadmap

Chapters 8 and 9 are independent of each other and may be read in either order. The interdependence of the sections of this chapter is shown below. Sections 1, 4, and 5 are independent of one another and may be read in any order.

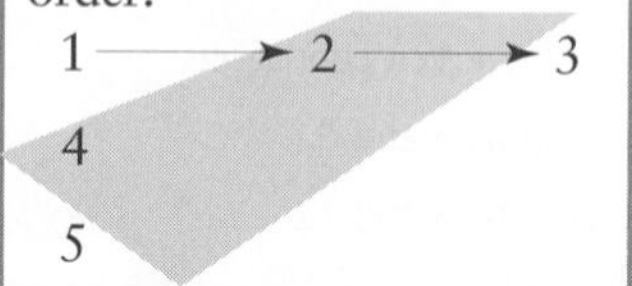

Until now the variable t has been used for trigonometric functions, to avoid confusion with the x's and y's that appear in their definitions. Now that you are comfortable with these functions, we shall usually use the letter x (or occasionally y) for the variable. Unless stated otherwise, all trigonometric functions in this chapter are considered as functions of real numbers, rather than functions of angles in degree measure.

Two kinds of trigonometric equations are considered here. *Identities* (Sections 8.1–8.3) are equations that are valid for all values of the variable for which the equation is defined, such as

$$\sin^2 x + \cos^2 x = 1 \qquad \text{and} \qquad \cot x = \frac{1}{\tan x}.$$

Conditional equations (Section 8.4) are valid only for certain values of the variable, such as $\sin x = 0$ and $\cos x = \frac{1}{2}$. Inverse trigonometric functions are discussed in Section 8.5.

8.1 BASIC IDENTITIES AND PROOFS

Trigonometric identities can be used for simplifying expressions, rewriting the rule of a trigonometric function, performing numerical calculations, and in other ways. There are no hard and fast rules for dealing with identities, but some suggestions are given below. The phrases "prove the identity" and "verify the identity" mean "prove that the given equation is an identity."

Graphical Testing

When presented with a trigonometric equation that *might* be an identity, you should first determine graphically whether or not this is possible. For instance, in Example 7 on page 393 we tested the equation $\cos\left(\frac{\pi}{2} + t\right) = \sin t$ by graphing the functions $f(t) = \cos\left(\frac{\pi}{2} + t\right)$ and $g(t) = \sin t$ on the same screen. Since the graphs were different, we concluded that the equation was not an identity.

Any equation can be tested in the same way, by simultaneously graphing the two functions whose rules are given by the left and right sides of the equation. If the graphs are different, the equation is not an identity. If the graphs appear to be the same, then it is *possible* that the equation is an identity. However,

> *The fact that the graphs appear identical does not prove that the equation is an identity, as the following exploration demonstrates.*

GRAPHING EXPLORATION In the viewing window with $-\pi \le x \le \pi$ and $-2 \le y \le 2$, graph both sides of the equation

$$\cos x = 1 - x^2/2 + x^4/24 - x^6/720 + x^8/40320.$$

Do the graphs appear identical? Now change the viewing window so that $-2\pi \le x \le 2\pi$. Is the equation an identity?

EXAMPLE 1 Is either of the following equations an identity?

(a) $2\sin^2 x - \cos x = 2\cos^2 x + \sin x$

(b) $\dfrac{1 + \sin x - \sin^2 x}{\cos x} = \cos x + \tan x$

Solution

(a) Test the equation graphically to see if it might be an identity.

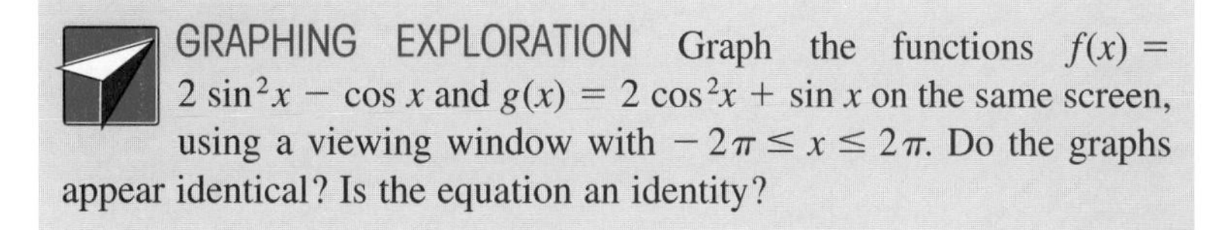

GRAPHING EXPLORATION Graph the functions $f(x) = 2\sin^2 x - \cos x$ and $g(x) = 2\cos^2 x + \sin x$ on the same screen, using a viewing window with $-2\pi \le x \le 2\pi$. Do the graphs appear identical? Is the equation an identity?

(b) Test the second equation graphically.

GRAPHING EXPLORATION Graph the functions $f(x) = \dfrac{1 + \sin x - \sin^2 x}{\cos x}$ and $g(x) = \cos x + \tan x$ on the same screen, using a viewing window with $-2\pi \le x \le 2\pi$. Do the graphs appear identical?

The Exploration suggests that the equation *may* be an identity, but the proof that it actually is an identity must be done algebraically. ■

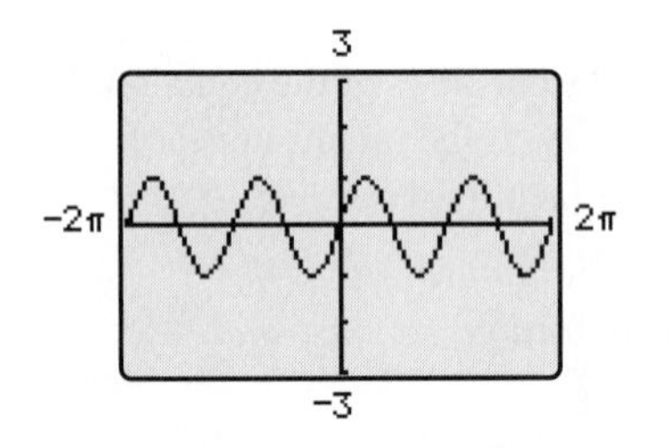

EXAMPLE 2 Find an equation involving $2 \sin x \cos x$ that could possibly be an identity.

Solution First graph $y = 2 \sin x \cos x$ and examine the graph (Figure 8–1). Does it look familiar? At first glance it looks like the graph of $\sin x$, but there's an important difference. The function graphed in Figure 8–1 has period π (a complete wave runs from $x = 0$ to $x = \pi$). As we saw in Section 6.7, the graph of $y = \sin 2x$ has waves like the sine graph and period π (see the box on page 422).

GRAPHING EXPLORATION Graph $y = 2 \sin x \cos x$ and $y = \sin 2x$ on the same screen. Do the graphs appear identical? What identity does this suggest? ■

Proving Identities

In the proofs presented below, we shall assume the elementary identities that were proved in Chapter 6 and are summarized here.

Basic Trigonometric Identities ▶

Reciprocal Identities

$$\sec x = \frac{1}{\cos x} \qquad \csc x = \frac{1}{\sin x}$$

$$\tan x = \frac{\sin x}{\cos x} \qquad \cot x = \frac{\cos x}{\sin x}$$

$$\cot x = \frac{1}{\tan x} \qquad \tan x = \frac{1}{\cot x}$$

Periodicity Identities

$$\sin(x + 2\pi) = \sin x \qquad \cos(x + 2\pi) = \cos x$$

$$\csc(x + 2\pi) = \csc x \qquad \sec(x + 2\pi) = \sec x$$

$$\tan(x + \pi) = \tan x \qquad \cot(x + \pi) = \cot x$$

Pythagorean Identities

$$\sin^2 x + \cos^2 x = 1 \qquad 1 + \tan^2 x = \sec^2 x \qquad 1 + \cot^2 x = \csc^2 x$$

Negative Angle Identities

$$\sin(-x) = -\sin x \qquad \cos(-x) = \cos x \qquad \tan(-x) = -\tan x$$

There are no cut and dried rules for simplifying trigonometric expressions or proving identities, but there are some common strategies that are often helpful. Four of these strategies are illustrated in the following examples. There are often a variety of ways to proceed and it will take some practice before you can easily decide which strategies are likely to be the most efficient in a particular case.

Strategy 1 ▶ **Express everything in terms of sine and cosine.**

EXAMPLE 3 Simplify $(\csc x + \cot x)(1 - \cos x)$.

Solution Using Strategy 1, we have

$$(\csc x + \cot x)(1 - \cos x)$$
$$= \left(\frac{1}{\sin x} + \frac{\cos x}{\sin x}\right)(1 - \cos x) \quad [\textit{Reciprocal identity}]$$
$$= \frac{(1 + \cos x)}{\sin x}(1 - \cos x)$$
$$= \frac{(1 + \cos x)(1 - \cos x)}{\sin x}$$
$$= \frac{1 - \cos^2 x}{\sin x} = \frac{\sin^2 x}{\sin x} \quad [\textit{Pythagorean identity}]$$
$$= \sin x. \quad \blacksquare$$

Strategy 2 ▶ **Use algebra and identities to transform the expression on one side of the equal sign into the expression on the other side.***

Example 3 shows how this strategy together with Strategy 1 could be used to prove the identity $(\csc x + \cot x)(1 - \cos x) = \sin x$. Here is another example.

EXAMPLE 4 In Example 1 we verified graphically that the equation

$$\frac{1 + \sin x - \sin^2 x}{\cos x} = \cos x + \tan x$$

might be an identity. Prove that it is.

*That is, start with expression A on one side and use identities and algebra to produce a sequence of equalities $A = B$, $B = C$, $C = D$, $D = E$, where E is the other side of the identity-to-be-proved; conclude that $A = E$.

Solution We use Strategy 2, beginning with the left side of the equation:

$$\frac{1 + \sin x - \sin^2 x}{\cos x} = \frac{(1 - \sin^2 x) + \sin x}{\cos x}$$

$$= \frac{\cos^2 x + \sin x}{\cos x} \qquad \text{[Pythagorean identity]}$$

$$= \frac{\cos^2 x}{\cos x} + \frac{\sin x}{\cos x}$$

$$= \cos x + \frac{\sin x}{\cos x}$$

$$= \cos x + \tan x. \quad \blacksquare$$

The strategies presented above and those to be considered below are ''plans of attack.'' By themselves they are not much help unless you also have some techniques for carrying out these plans. In the examples above we used the techniques of basic algebra and the use of known identities to change trigonometric expressions to equivalent ones. Here is another technique that is often useful when dealing with fractions:

Rewrite a fraction in equivalent form by multiplying its numerator and denominator by the same quantity.

EXAMPLE 5 Prove that $\dfrac{\sin x}{1 + \cos x} = \dfrac{1 - \cos x}{\sin x}$.

Solution We shall use Strategy 2, beginning with the left side, whose denominator is $1 + \cos x$. Multiply its numerator and denominator by $1 - \cos x$:*

$$\frac{\sin x}{1 + \cos x} = \frac{\sin x}{1 + \cos x} \cdot \frac{1 - \cos x}{1 - \cos x} = \frac{\sin x(1 - \cos x)}{(1 + \cos x)(1 - \cos x)}$$

$$= \frac{\sin x(1 - \cos x)}{1 - \cos^2 x}$$

$$= \frac{\sin x(1 - \cos x)}{\sin^2 x} \qquad \text{[Pythagorean identity]}$$

$$= \frac{1 - \cos x}{\sin x}.$$

Alternate Solution The numerators of the given equation look similar to the Pythagorean identity—with the squares missing. So we begin with the left side and introduce some squares by multiplying it by $\dfrac{\sin x}{\sin x} = 1$:

*This is analogous to the process used to rationalize the denominator of a fraction by multiplying its numerator and denominator by the conjugate of the denominator, as in this example:

$$\frac{1}{3 + \sqrt{2}} = \frac{1}{3 + \sqrt{2}} \cdot \frac{3 - \sqrt{2}}{3 - \sqrt{2}} = \frac{3 - \sqrt{2}}{3^2 - (\sqrt{2})^2} = \frac{3 - \sqrt{2}}{7}.$$

$$\frac{\sin x}{1+\cos x} = 1\cdot\frac{\sin x}{1+\cos x} = \frac{\sin x}{\sin x}\cdot\frac{\sin x}{1+\cos x} = \frac{\sin^2 x}{\sin x(1+\cos x)}$$

$$= \frac{1-\cos^2 x}{\sin x(1+\cos x)} \qquad \textit{[Pythagorean identity]}$$

$$= \frac{(1-\cos x)(1+\cos x)}{\sin x(1+\cos x)} \qquad \textit{[Factor numerator]}$$

$$= \frac{1-\cos x}{\sin x}. \quad ■$$

Strategy 3 ▶

Deal separately with each side of the equation $A = B$. First use identities and algebra to transform A into some expression C (so that $A = C$). Then use (possibly different) identities and algebra to transform B into the *same* expression C (so that $B = C$). Conclude that $A = B$.

EXAMPLE 6 Prove that $\csc x - \cot x = \dfrac{\sin x}{1+\cos x}$.

Solution We use Strategy 3, together with Strategy 1, beginning with the left side:

(*) $$\csc x - \cot x = \frac{1}{\sin x} - \frac{\cos x}{\sin x} = \frac{1-\cos x}{\sin x}.$$

Example 5 shows that the right side of the identity-to-be-proved can also be transformed into this same expression:

(**) $$\frac{\sin x}{1+\cos x} = \frac{1-\cos x}{\sin x}.$$

Combining the equalities (*) and (**) proves the identity:

$$\csc x - \cot x = \frac{1-\cos x}{\sin x} = \frac{\sin x}{1+\cos x}. \quad ■$$

Strategy 4 ▶

Replace a given equation involving fractions by an equivalent one that does not involve fractions, by using this fact: When $B \neq 0$ and $D \neq 0$,

$$\frac{A}{B} = \frac{C}{D} \qquad \text{exactly when } AD = BC.*$$

*Multiplying the first equation by BD produces the second; conversely, dividing the second by BD produces the first.

Many students misunderstand Strategy 4: It does *not* say that you begin with an equation involving fractions and cross multiply to eliminate the fractions. If you did that, you'd be assuming what has to be proved. What the strategy says is that to prove an identity involving fractions, you need only prove a different identity that does not involve fractions. In other words, if you prove that $AD = BC$ whenever $B \neq 0$ and $D \neq 0$, then you can conclude that $A/B = C/D$. Note that you do not *assume* that $AD = BC$; you use Strategy 2 or 3 or some other means to *prove* this statement.

EXAMPLE 7 Verify the identity $\dfrac{\sec x}{\tan x} = \dfrac{\tan x}{\sec x - \cos x}$.

Solution The given equation is defined only when $\tan x \neq 0$ and $\sec x - \cos x \neq 0$. According to Strategy 4 (with $A = \sec x$, $B = \tan x$, $C = \tan x$, and $D = \sec x - \cos x$), we can prove that this equation is an identity by proving instead that the following equation is an identity:

$$(***) \qquad \sec x(\sec x - \cos x) = \tan^2 x.$$

We shall use Strategy 2 to prove that (***) is an identity. Beginning with the left side of (***) and using Strategy 1 we have:

$$\begin{aligned}
\sec x(\sec x - \cos x) &= \sec^2 x - \sec x \cos x \\
&= \sec^2 x - \frac{1}{\cos x}\cos x \\
&= \sec^2 x - 1 \\
&= \tan^2 x \qquad \textit{[Pythagorean identity]}
\end{aligned}$$

Therefore, the original (equivalent) identity is valid. ■

EXAMPLE 8 To prove that $\dfrac{\cot x - 1}{\cot x + 1} = \dfrac{1 - \tan x}{1 + \tan x}$, we use Strategy 4 and prove the equivalent identity

$$(\cot x - 1)(1 + \tan x) = (\cot x + 1)(1 - \tan x).$$

Strategy 3 will be used to prove this identity. Multiplying out the left side shows that:

$$\begin{aligned}
(\cot x - 1)(1 + \tan x) &= \cot x - 1 + \cot x \tan x - \tan x \\
&= \cot x - 1 + \frac{1}{\tan x}\tan x - \tan x \\
&= \cot x - 1 + 1 - \tan x \\
&= \cot x - \tan x.
\end{aligned}$$

Similarly, on the right side:

$$\begin{aligned}(\cot x + 1)(1 - \tan x) &= \cot x + 1 - \cot x \tan x - \tan x \\ &= \cot x + 1 - 1 - \tan x \\ &= \cot x - \tan x.\end{aligned}$$

Since the left and right sides are equal to the same expression, the identity is proved. ■

It takes a good deal of practice, as well as *much* trial and error, to become proficient in proving identities. The more practice you have, the easier it will get. Since there are many correct methods, your proofs may be quite different from those of your instructor or the text answers.

If you don't see what to do immediately, try something and see where it leads: Multiply out or factor or multiply numerator and denominator by the same nonzero quantity. Even if this doesn't lead anywhere, it may give you some ideas on other things to try. When you do obtain a proof, check to see if it can be done more efficiently. In your final proof, don't include the side trips that may have given you some ideas but aren't themselves part of the proof.

EXERCISES 8.1

In Exercises 1–4, test the equation graphically to determine whether it might be an identity. You need not prove those equations that seem to be identities.

1. $\dfrac{\sec x - \cos x}{\sec x} = \sin^2 x$

2. $\tan x + \cot x = (\sin x)(\cos x)$

3. $\dfrac{1 - \cos(2x)}{2} = \sin^2 x$

4. $\dfrac{\tan x + \cot x}{\csc x} = \sec x$

In Exercises 5–8, insert one of A–F on the right of the equal sign so that the resulting equation appears to be an identity when you test it graphically. You need not prove the identity.

A. $\cos x$ **B.** $\sec x$ **C.** $\sin^2 x$

D. $\sec^2 x$ **E.** $\sin x - \cos x$ **F.** $\dfrac{1}{\sin x \cos x}$

5. $\csc x \tan x =$ ______

6. $\dfrac{\sin x}{\tan x} =$ ______

7. $\dfrac{\sin^4 x - \cos^4 x}{\sin x + \cos x} =$ ______

8. $\tan^2(-x) - \dfrac{\sin(-x)}{\sin x} =$ ______

In Exercises 9–18, prove the identity.

9. $\tan x \cos x = \sin x$

10. $\cot x \sin x = \cos x$

11. $\cos x \sec x = 1$

12. $\sin x \csc x = 1$

13. $\tan x \csc x = \sec x$

14. $\sec x \cot x = \csc x$

15. $\dfrac{\tan x}{\sec x} = \sin x$

16. $\dfrac{\cot x}{\csc x} = \cos x$

17. $(1 + \cos x)(1 - \cos x) = \sin^2 x$

18. $(\csc x - 1)(\csc x + 1) = \cot^2 x$

In Exercises 19–50, state whether or not the equation is an identity. If it is an identity, prove it.

19. $\sin x = \sqrt{1 - \cos^2 x}$

20. $\cot x = \dfrac{\csc x}{\sec x}$

21. $\dfrac{\sin(-x)}{\cos(-x)} = -\tan x$

22. $\tan x = \sqrt{\sec^2 x - 1}$

23. $\cot(-x) = -\cot x$ **24.** $\sec(-x) = \sec x$

25. $1 + \sec^2 x = \tan^2 x$

26. $\sec^4 x - \tan^4 x = 1 + 2\tan^2 x$

27. $\sec^2 x - \csc^2 x = \tan^2 x - \cot^2 x$

28. $\sec^2 x + \csc^2 x = \sec^2 x \csc^2 x$

29. $\sin^2 x(\cot x + 1)^2 = \cos^2 x(\tan x + 1)^2$

30. $\cos^2 x(\sec x + 1)^2 = (1 + \cos x)^2$

31. $\sin^2 x - \tan^2 x = -\sin^2 x \tan^2 x$

32. $\cot^2 x - 1 = \csc^2 x$

33. $(\cos^2 x - 1)(\tan^2 x + 1) = -\tan^2 x$

34. $(1 - \cos^2 x)\csc x = \sin x$

35. $\tan x = \dfrac{\sec x}{\csc x}$ **36.** $\dfrac{\cos(-x)}{\sin(-x)} = -\cot x$

37. $\cos^4 x - \sin^4 x = \cos^2 x - \sin^2 x$

38. $\cot^2 x - \cos^2 x = \cos^2 x \cot^2 x$

39. $(\sin x + \cos x)^2 = \sin^2 x + \cos^2 x$

40. $(1 + \tan x)^2 = \sec^2 x$

41. $\dfrac{\sec x}{\csc x} + \dfrac{\sin x}{\cos x} = 2\tan x$

42. $\dfrac{1 + \cos x}{\sin x} + \dfrac{\sin x}{1 + \cos x} = 2\csc x$

43. $\dfrac{\sec x + \csc x}{1 + \tan x} = \csc x$

44. $\dfrac{\cot x - 1}{1 - \tan x} = \dfrac{\csc x}{\sec x}$

45. $\dfrac{1}{\csc x - \sin x} = \sec x \tan x$

46. $\dfrac{1 + \csc x}{\csc x} = \dfrac{\cos^2 x}{1 - \sin x}$

47. $\dfrac{\sin x - \cos x}{\tan x} = \dfrac{\tan x}{\sin x + \cos x}$

48. $\dfrac{\cot x}{\csc x - 1} = \dfrac{\csc x + 1}{\cot x}$

49. $\dfrac{\cot x}{1 - \tan x} + \dfrac{\tan x}{1 - \cot x} = 1 + \sec x \csc x$

50. $\dfrac{\sin x - \cos x}{\sin x + \cos x} = \cot x$

In Exercises 51–54, half of an identity is given. Graph this half in a viewing window with $-2\pi \le x \le 2\pi$ and make a conjecture as to what the right side of the identity is. Then prove your conjecture.

51. $1 - \dfrac{\sin^2 x}{1 + \cos x} = ?$ [*Hint:* What familiar function has a graph that looks like this?]

52. $\dfrac{1 + \cos x - \cos^2 x}{\sin x} - \cot x = ?$

53. $(\sin x + \cos x)(\sec x + \csc x) - \cot x - 2 = ?$

54. $\cos^3 x(1 - \tan^4 x + \sec^4 x) = ?$

In Exercises 55–68, prove the identity.

55. $\dfrac{1 - \sin x}{\sec x} = \dfrac{\cos^3 x}{1 + \sin x}$

56. $\dfrac{\sin x}{1 - \cot x} + \dfrac{\cos x}{1 - \tan x} = \cos x + \sin x$

57. $\dfrac{\cos x}{1 - \sin x} = \sec x + \tan x$

58. $\dfrac{1 + \sec x}{\tan x + \sin x} = \csc x$

59. $\dfrac{\cos x \cot x}{\cot x - \cos x} = \dfrac{\cot x + \cos x}{\cos x \cot x}$

60. $\dfrac{\cos^3 x - \sin^3 x}{\cos x - \sin x} = 1 + \sin x \cos x$

61. $\log_{10}(\cot x) = -\log_{10}(\tan x)$

62. $\log_{10}(\sec x) = -\log_{10}(\cos x)$

63. $\log_{10}(\csc x + \cot x) = -\log_{10}(\csc x - \cot x)$

64. $\log_{10}(\sec x + \tan x) = -\log_{10}(\sec x - \tan x)$

65. $\tan x - \tan y = -\tan x \tan y(\cot x - \cot y)$

66. $\dfrac{\tan x - \tan y}{\cot x - \cot y} = -\tan x \tan y$

67. $\dfrac{\cos x - \sin y}{\cos y - \sin x} = \dfrac{\cos y + \sin x}{\cos x + \sin y}$

68. $\dfrac{\tan x + \tan y}{\cot x + \cot y} = \dfrac{\tan x \tan y - 1}{1 - \cot x \cot y}$

8.2 ADDITION AND SUBTRACTION IDENTITIES

A common student ERROR is to write

$$\sin\left(x+\frac{\pi}{6}\right) = \sin x + \sin\frac{\pi}{6} = \sin x + \frac{1}{2}.$$

GRAPHING EXPLORATION Verify graphically that the equation above is NOT an identity by graphing $y = \sin(x + \pi/6)$ and $y = \sin x + 1/2$ on the same screen.

The exploration shows that $\sin(x + y) = \sin x + \sin y$ is NOT an identity (because it's false when $y = \pi/6$). So, is there an identity involving $\sin(x + y)$ and $\sin x$ and $\sin y$?

GRAPHING EXPLORATION Graph $y = \sin\left(x+\frac{\pi}{6}\right)$ and $y = \frac{\sqrt{3}}{2}\sin x + \frac{1}{2}\cos x$ on the same screen. Do the graphs appear identical?

The Exploration suggests that

$$y = \sin\left(x+\frac{\pi}{6}\right) = y = \frac{\sqrt{3}}{2}\sin x + \frac{1}{2}\cos x$$

is an identity. Furthermore, note that the coefficients on the right side can be expressed in terms of $\pi/6$: $\frac{\sqrt{3}}{2} = \cos\frac{\pi}{6}$ and $\frac{1}{2} = \sin\frac{\pi}{6}$. In other words, the following equation appears to be an identity

$$\sin\left(x+\frac{\pi}{6}\right) = \sin x\left(\cos\frac{\pi}{6}\right) + \cos x\left(\sin\frac{\pi}{6}\right).$$

Is there something special about $\pi/6$ or would we get the same result with another number?

GRAPHING EXPLORATION Graph $y = \sin(x + 5)$ and $y = (\sin x)(\cos 5) + (\cos x)(\sin 5)$ on the same screen. Do the graphs appear identical. What identity does this suggest? Repeat the process with some other number in place of 5. Are the results the same?

The equations examined in the discussion and exploration above are examples of the first identity listed below (they are the cases when $y = \pi/6$ and when $y = 5$).

Addition and Subtraction Identities ▶

$$\sin(x + y) = \sin x \cos y + \cos x \sin y$$
$$\sin(x - y) = \sin x \cos y - \cos x \sin y$$
$$\cos(x + y) = \cos x \cos y - \sin x \sin y$$
$$\cos(x - y) = \cos x \cos y + \sin x \sin y$$

GRAPHING EXPLORATION Confirm the last identity listed above graphically in the case when $y = 3$ by graphing $\cos(x - 3)$ and $(\cos x)(\cos 3) + (\sin x)(\sin 3)$ on the same screen.

The addition and subtraction identities are probably the most important of all the trigonometric identities. Before reading their proofs at the end of this section, you should become familiar with the examples and special cases below.

EXAMPLE 1 Use the addition identities to find the *exact* value of $\sin(5\pi/12)$ and $\cos(5\pi/12)$.

Solution Since $5\pi/12 = 2\pi/12 + 3\pi/12 = \pi/6 + \pi/4$, we apply the addition identities with $x = \pi/6$ and $y = \pi/4$:

$$\sin\frac{5\pi}{12} = \sin\left(\frac{\pi}{6} + \frac{\pi}{4}\right) = \sin\frac{\pi}{6}\cos\frac{\pi}{4} + \cos\frac{\pi}{6}\sin\frac{\pi}{4}$$
$$= \frac{1}{2}\cdot\frac{\sqrt{2}}{2} + \frac{\sqrt{3}}{2}\cdot\frac{\sqrt{2}}{2} = \frac{\sqrt{2}(\sqrt{3}+1)}{4}$$

$$\cos\frac{5\pi}{12} = \cos\left(\frac{\pi}{6} + \frac{\pi}{4}\right) = \cos\frac{\pi}{6}\cos\frac{\pi}{4} - \sin\frac{\pi}{6}\sin\frac{\pi}{4}$$
$$= \frac{\sqrt{3}}{2}\cdot\frac{\sqrt{2}}{2} - \frac{1}{2}\cdot\frac{\sqrt{2}}{2} = \frac{\sqrt{2}(\sqrt{3}-1)}{4}.$$ ■

EXAMPLE 2 To find $\sin(\pi - y)$, apply the subtraction identity with $x = \pi$:

$$\sin(\pi - y) = \sin\pi\cos y - \cos\pi\sin y$$
$$= (0)(\cos y) - (-1)(\sin y)$$
$$= \sin y.$$ ■

EXAMPLE 3 In calculus, it will be necessary to show that for the function $f(x) = \sin x$ and any number $h \neq 0$,

$$\frac{f(x+h) - f(x)}{h} = \sin x\left(\frac{\cos h - 1}{h}\right) + \cos x\left(\frac{\sin h}{h}\right).$$

This can be done by using the addition identity for $\sin(x + y)$ with $y = h$:

$$\begin{aligned}\frac{f(x+h) - f(x)}{h} &= \frac{\sin(x+h) - \sin x}{h} \\ &= \frac{\sin x \cos h + \cos x \sin h - \sin x}{h} \\ &= \frac{\sin x(\cos h - 1) + \cos x \sin h}{h} \\ &= \sin x\left(\frac{\cos h - 1}{h}\right) + \cos x\left(\frac{\sin h}{h}\right).\end{aligned}$$ ■

EXAMPLE 4 Prove that

$$\cos x \cos y = \frac{1}{2}[\cos(x+y) + \cos(x-y)].$$

Solution We begin with the more complicated right side and use the addition and subtraction identities for cosine to transform it into the left side:

$$\begin{aligned}\frac{1}{2}[\cos(x+y) + \cos(x-y)] &= \frac{1}{2}[(\cos x \cos y - \sin x \sin y) \\ &\qquad + (\cos x \cos y + \sin x \sin y)] \\ &= \frac{1}{2}[\cos x \cos y + \cos x \cos y] \\ &= \frac{1}{2}[2 \cos x \cos y] = \cos x \cos y.\end{aligned}$$ ■

The addition and subtraction identities for sine and cosine can be used to obtain

Addition and Subtraction Identities for Tangent ►

$$\tan(x+y) = \frac{\tan x + \tan y}{1 - \tan x \tan y}$$

$$\tan(x-y) = \frac{\tan x - \tan y}{1 + \tan x \tan y}$$

A proof of these identities is outlined in Exercise 36.

EXAMPLE 5 Suppose x and y are numbers such that $0 < x < \pi/2$ and $\pi < y < 3\pi/2$. If $\sin x = 3/4$ and $\cos y = -1/3$, find the exact values of $\sin(x + y)$ and $\tan(x + y)$ and determine in which of the following intervals $x + y$ lies: $(0, \pi/2)$, $(\pi/2, \pi)$, $(\pi, 3\pi/2)$, or $(3\pi/2, 2\pi)$.

Solution Using the Pythagorean identity and the fact that $\cos x$ and $\tan x$ are positive when $0 < x < \pi/2$, we have

$$\cos x = \sqrt{1 - \sin^2 x} = \sqrt{1 - \left(\frac{3}{4}\right)^2} = \sqrt{1 - \frac{9}{16}}\ \sqrt{\frac{7}{16}} = \frac{\sqrt{7}}{4}$$

$$\tan x = \frac{\sin x}{\cos x} = \frac{3/4}{\sqrt{7}/4} = \frac{3}{4}\cdot\frac{4}{\sqrt{7}} = \frac{3}{\sqrt{7}} = \frac{3\sqrt{7}}{7}.$$

Since y lies between π and $3\pi/2$, its sine is negative; hence,

$$\sin y = -\sqrt{1 - \cos^2 y} = -\sqrt{1 - \left(-\frac{1}{3}\right)^2} = -\sqrt{\frac{8}{9}} = -\frac{\sqrt{8}}{3} = -\frac{2\sqrt{2}}{3}$$

$$\tan y = \frac{\sin y}{\cos y} = \frac{-2\sqrt{2}/3}{-1/3} = \frac{-2\sqrt{2}}{3}\cdot\frac{3}{-1} = 2\sqrt{2}.$$

The addition identities for sine and tangent now show that

$$\begin{aligned}\sin(x + y) &= \sin x \cos y + \cos x \sin y \\ &= \frac{3}{4}\cdot\frac{-1}{3} + \frac{\sqrt{7}}{4}\cdot\frac{-2\sqrt{2}}{3} = \frac{-1}{4} - \frac{2\sqrt{14}}{12} = \frac{-3 - 2\sqrt{14}}{12}\end{aligned}$$

$$\begin{aligned}\tan(x + y) &= \frac{\tan x + \tan y}{1 - \tan x \tan y} \\ &= \frac{\frac{3\sqrt{7}}{7} + 2\sqrt{2}}{1 - \left(\frac{3\sqrt{7}}{7}\right)(2\sqrt{2})} = \frac{\frac{3\sqrt{7} + 14\sqrt{2}}{7}}{\frac{7 - 6\sqrt{14}}{7}} = \frac{3\sqrt{7} + 14\sqrt{2}}{7 - 6\sqrt{14}}.\end{aligned}$$

So both the sine and tangent of $x + y$ are negative numbers. Therefore, $x + y$ must be in the interval $(3\pi/2, 2\pi)$ since the sign chart on page 381 shows that this is the only one of the four intervals in which both sine and tangent are negative. ■

Cofunction Identities

Other special cases of the addition and subtraction identities are the cofunction identities:

Cofunction Identities ▶

$$\sin x = \cos\left(\frac{\pi}{2} - x\right) \qquad \cos x = \sin\left(\frac{\pi}{2} - x\right)$$

$$\tan x = \cot\left(\frac{\pi}{2} - x\right) \qquad \cot x = \tan\left(\frac{\pi}{2} - x\right)$$

$$\sec x = \csc\left(\frac{\pi}{2} - x\right) \qquad \csc x = \sec\left(\frac{\pi}{2} - x\right)$$

The first cofunction identity is proved by using the identity for $\cos(x - y)$ with $\pi/2$ in place of x and x in place of y:

$$\cos\left(\frac{\pi}{2} - x\right) = \cos\frac{\pi}{2}\cos x + \sin\frac{\pi}{2}\sin x = (0)(\cos x) + (1)(\sin x) = \sin x.$$

Since the first cofunction identity is valid for *every* number x, it is also valid with the number $\pi/2 - x$ in place of x:

$$\sin\left(\frac{\pi}{2} - x\right) = \cos\left(\frac{\pi}{2} - \left(\frac{\pi}{2} - x\right)\right) = \cos x.$$

Thus, we have proved the second cofunction identity. The others now follow from these two. For instance,

$$\tan\left(\frac{\pi}{2} - x\right) = \frac{\sin((\pi/2) - x)}{\cos((\pi/2) - x)} = \frac{\cos x}{\sin x} = \cot x.$$

EXAMPLE 6 Verify that $\dfrac{\cos(x - \pi/2)}{\cos x} = \tan x.$

Solution Beginning on the left side, we see that the term $\cos(x - (\pi/2))$ looks almost, but not quite, like the term $\cos((\pi/2) - x)$ in the cofunction identity. But note that $-(x - (\pi/2)) = (\pi/2) - x$. Therefore,

$$\frac{\cos\left(x - \frac{\pi}{2}\right)}{\cos x} = \frac{\cos\left(-\left(x - \frac{\pi}{2}\right)\right)}{\cos x}$$ [*Negative angle identity with* $x - \frac{\pi}{2}$ *in place of* x]

$$= \frac{\cos\left(\frac{\pi}{2} - x\right)}{\cos x}$$

$$= \frac{\sin x}{\cos x}$$ [*Cofunction identity*]

$$= \tan x.$$ [*Reciprocal identity*] ■

Proof of the Addition and Subtraction Identities

We first prove that

$$\cos(x - y) = \cos x \cos y + \sin x \sin y.$$

If $x = y$, then this is true by the Pythagorean identity:

$$\cos(x - x) = \cos 0 = 1 = \cos^2 x + \sin^2 x = \cos x \cos x + \sin x \sin x.$$

Next we prove the identity in the case when $x > y$. Let P be the point where the terminal side of an angle of x radians in standard position meets the unit circle and let Q be the point where the terminal side of an angle of y radians in standard position meets the unit circle, as shown in Figure 8–2. According to the definitions of sine and cosine, P has coordinates $(\cos x, \sin x)$ and Q has coordinates $(\cos y, \sin y)$.

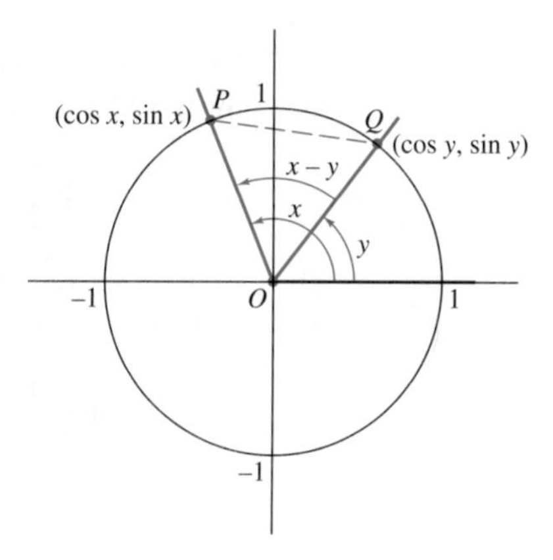

Figure 8-2

The angle QOP formed by the two terminal sides has radian measure $x - y$. Rotate this angle clockwise until side OQ lies on the horizontal axis, as shown in Figure 8–3. Angle QOP is now in standard position, and its terminal side meets the unit circle at P. Since angle QOP has radian measure $x - y$, the definitions of sine and cosine show that the point P, in this new location, has coordinates $(\cos(x - y), \sin(x - y))$. Q now has coordinates $(1, 0)$.

Using the coordinates of P and Q *before* the angle was rotated and the distance formula, we have:

distance from P to Q

$$\begin{aligned}
&= \sqrt{(\cos x - \cos y)^2 + (\sin x - \sin y)^2} \\
&= \sqrt{\cos^2 x - 2 \cos x \cos y + \cos^2 y + \sin^2 x - 2 \sin x \sin y + \sin^2 y} \\
&= \sqrt{(\cos^2 x + \sin^2 x) + (\cos^2 y + \sin^2 y) - 2 \cos x \cos y - 2 \sin x \sin y} \\
&= \sqrt{1 + 1 - 2 \cos x \cos y - 2 \sin x \sin y} \\
&= \sqrt{2 - 2 \cos x \cos y - 2 \sin x \sin y}.
\end{aligned}$$

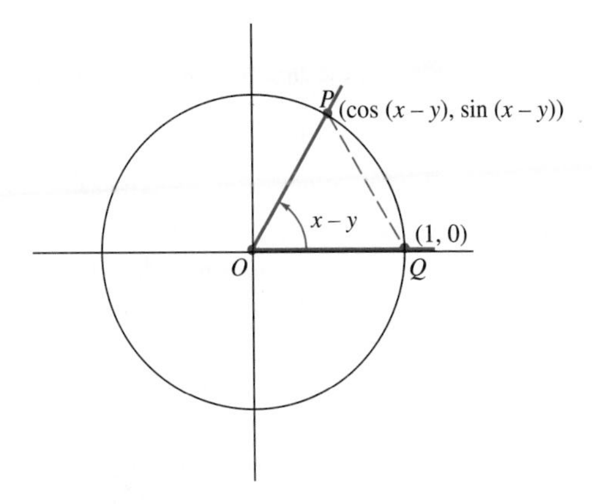

Figure 8-3

But using the coordinates of P and Q *after* the angle is rotated shows that

$$\begin{aligned}\text{distance from } P \text{ to } Q &= \sqrt{(\cos(x-y)-1)^2+(\sin(x-y)-0)^2}\\ &= \sqrt{\cos^2(x-y)-2\cos(x-y)+1+\sin^2(x-y)}\\ &= \sqrt{\cos^2(x-y)+\sin^2(x-y)-2\cos(x-y)+1}\\ &= \sqrt{1-2\cos(x-y)+1} \qquad [\textit{Pythagorean identity}]\\ &= \sqrt{2-2\cos(x-y)}.\end{aligned}$$

The two expressions for the distance from P to Q must be equal. Hence,

$$\sqrt{2-2\cos(x-y)} = \sqrt{2-2\cos x\cos y - 2\sin x\sin y}.$$

Squaring both sides of this equation and simplifying the result yields:

$$\begin{aligned}2-2\cos(x-y) &= 2-2\cos x\cos y-2\sin x\sin y\\ -2\cos(x-y) &= -2(\cos x\cos y+\sin x\sin y)\\ \cos(x-y) &= \cos x\cos y+\sin x\sin y.\end{aligned}$$

This completes the proof of the last addition identity when $x > y$. If $y > x$, then the proof just given is valid with the roles of x and y interchanged; it shows that

$$\begin{aligned}\cos(y-x) &= \cos y\cos x+\sin y\sin x\\ &= \cos x\cos y+\sin x\sin y.\end{aligned}$$

The negative angle identity with $x - y$ in place of x shows that

$$\cos(x-y) = \cos(-(x-y)) = \cos(y-x).$$

Combining this fact with the previous one shows that

$$\cos(x-y) = \cos x\cos y+\sin x\sin y$$

in this case also. Therefore, the last addition identity is proved.

The identity for $\cos(x + y)$ now follows from one for $\cos(x - y)$ by using the negative angle identities for sine and cosine:

$$\begin{aligned}\cos(x + y) &= \cos(x - (-y)) = \cos x \cos(-y) + \sin x \sin(-y)\\ &= \cos x \cos y + (\sin x)(-\sin y) = \cos x \cos y - \sin x \sin y.\end{aligned}$$

The proof of the first two cofunction identities on page 499 depended only on the addition identity for $\cos(x - y)$. Since that has been proved, we can validly use the first two cofunction identities in the remainder of the proof. In particular:

$$\sin(x - y) = \cos\left(\frac{\pi}{2} - (x - y)\right) = \cos\left(\left(\frac{\pi}{2} - x\right) + y\right).$$

Applying the proven identity for $\cos(x + y)$ with $(\pi/2) - x$ in place of x and the two cofunction identities now yields

$$\begin{aligned}\sin(x - y) &= \cos\left(\left(\frac{\pi}{2} - x\right) + y\right)\\ &= \cos\left(\frac{\pi}{2} - x\right)\cos y - \sin\left(\frac{\pi}{2} - x\right)\sin y\\ &= \sin x \cos y - \cos x \sin y.\end{aligned}$$

This proves the second of the addition and subtraction identities. The first is obtained from the second in the same way the third was obtained from the last.

EXERCISES 8.2

In Exercises 1–12, find the exact value.

1. $\sin \frac{\pi}{12}$ **2.** $\cos \frac{\pi}{12}$ **3.** $\tan \frac{\pi}{12}$

4. $\sin \frac{5\pi}{12}$ **5.** $\cot \frac{5\pi}{12}$ **6.** $\cos \frac{7\pi}{12}$

7. $\tan \frac{7\pi}{12}$ **8.** $\cos \frac{11\pi}{12}$ **9.** $\cot \frac{11\pi}{12}$

10. $\sin 75°$ [*Hint*: $75° = 45° + 30°$.]

11. $\sin 105°$ **12.** $\cos 165°$

In Exercises 13–18, rewrite the given expression in terms of sin x and cos x.

13. $\sin\left(\frac{\pi}{2} + x\right)$ **14.** $\cos\left(x + \frac{\pi}{2}\right)$

15. $\cos\left(x - \frac{3\pi}{2}\right)$ **16.** $\csc\left(x + \frac{\pi}{2}\right)$

17. $\sec(x - \pi)$ **18.** $\cot(x + \pi)$

In Exercises 19–24, simplify the given expression.

19. $\sin 3 \cos 5 - \cos 3 \sin 5$

20. $\sin 37° \sin 53° - \cos 37° \cos 53°$

21. $\cos(x + y) \cos y + \sin(x + y) \sin y$

22. $\sin(x - y) \cos y + \cos(x - y) \sin y$

23. $\cos(x + y) - \cos(x - y)$

24. $\sin(x + y) - \sin(x - y)$

25. If $\sin x = \frac{1}{3}$ and $0 < x < \frac{\pi}{2}$, then $\sin\left(\frac{\pi}{4} + x\right) = ?$

26. If $\cos x = -\frac{1}{4}$ and $\frac{\pi}{2} < x < \pi$, then $\cos\left(\frac{\pi}{6} - x\right) = ?$

27. If $\cos x = -\frac{1}{5}$ and $\pi < x < \frac{3\pi}{2}$, then $\sin\left(\frac{\pi}{3} - x\right) = ?$

28. If $\sin x = -\frac{3}{4}$ and $\frac{3\pi}{2} < x < 2\pi$, then $\cos\left(\frac{\pi}{4} + x\right) = ?$

In Exercises 29–34, assume $\sin x = .8$ and $\sin y = \sqrt{.75} \approx .866$, and that x, y lie between 0 and $\pi/2$. Evaluate the given expression.

29. $\cos(x + y)$ **30.** $\sin(x + y)$ **31.** $\cos(x - y)$

32. $\sin(x - y)$ **33.** $\tan(x + y)$ **34.** $\tan(x - y)$

35. If $f(x) = \cos x$ and h is a fixed nonzero number, prove that

$$\frac{f(x+h) - f(x)}{h} = \cos x\left(\frac{\cos h - 1}{h}\right) - \sin x\left(\frac{\sin h}{h}\right).$$

36. Prove the addition and subtraction identities for the tangent function (page 497). [*Hint:* $\tan(x + y) = \frac{\sin(x+y)}{\cos(x+y)}$. Use the addition identities on the numerator and denominator; then divide both numerator and denominator by $\cos x \cos y$ and simplify.]

37. If x is in the first and y in the second quadrant, $\sin x = 24/25$, and $\sin y = 4/5$, find the exact value of $\sin(x + y)$ and $\tan(x + y)$ and the quadrant in which $x + y$ lies.

38. If x and y are in the second quadrant, $\sin x = 1/3$, and $\cos y = -3/4$, find the exact value of $\sin(x + y)$, $\cos(x + y)$, $\tan(x + y)$, and find the quadrant in which $x + y$ lies.

39. If x is in the first and y in the second quadrant, $\sin x = 4/5$, and $\cos y = -12/13$, find the exact value of $\cos(x + y)$ and $\tan(x + y)$ and the quadrant in which $x + y$ lies.

40. If x is in the fourth and y in the first quadrant, $\cos x = 1/3$, and $\cos y = 2/3$, find the exact value of $\sin(x - y)$ and $\tan(x - y)$ and the quadrant in which $x - y$ lies.

41. Express $\sin(u + v + w)$ in terms of sines and cosines of u, v, and w. [*Hint*: First apply the addition identity with $x = u + v$ and $y = w$.]

42. Express $\cos(x + y + z)$ in terms of sines and cosines of x, y, and z.

43. If $x + y = \pi/2$, show that $\sin^2 x + \sin^2 y = 1$.

44. Prove that $\cot(x + y) = \frac{\cot x \cot y - 1}{\cot x + \cot y}$.

In Exercises 45–56, prove the identity.

45. $\sin(x - \pi) = -\sin x$ **46.** $\cos(x - \pi) = -\cos x$

47. $\cos(\pi - x) = -\cos x$ **48.** $\tan(\pi - x) = -\tan x$

49. $\sin(x + \pi) = -\sin x$ **50.** $\cos(x + \pi) = -\cos x$

51. $\tan(x + \pi) = \tan x$

52. $\sin x \cos y = \frac{1}{2}(\sin(x + y) + \sin(x - y))$

53. $\sin x \sin y = \frac{1}{2}(\cos(x - y) - \cos(x + y))$

54. $\cos x \sin y = \frac{1}{2}(\sin(x + y) - \sin(x - y))$

55. $\cos(x + y)\cos(x - y) = \cos^2 x \cos^2 y - \sin^2 x \sin^2 y$

56. $\sin(x + y)\sin(x - y) = \sin^2 x \cos^2 y - \cos^2 x \sin^2 y$

In Exercises 57–66, determine graphically whether the equation could possibly be an identity (by choosing a numerical value for y and graphing both sides). If it could, prove that it is.

57. $\frac{\cos(x - y)}{\sin x \cos y} = \cot x + \tan y$

58. $\frac{\cos(x + y)}{\sin x \cos y} = \cot x - \tan y$

59. $\sin(x - y) = \sin x - \sin y$

60. $\cos(x + y) = \cos x + \cos y$

61. $\frac{\sin(x + y)}{\sin(x - y)} = \frac{\tan x + \tan y}{\tan x - \tan y}$

62. $\frac{\sin(x + y)}{\sin(x - y)} = \frac{\cot y + \cot x}{\cot y - \cot x}$

63. $\frac{\cos(x + y)}{\cos(x - y)} = \frac{\cot x - \tan y}{\cot x + \tan y}$

64. $\frac{\cos(x - y)}{\cos(x + y)} = \frac{\cot y + \tan x}{\cot y - \tan x}$

65. $\tan(x + y) = \tan x + \tan y$

66. $\cot(x - y) = \cot x - \cot y$

8.2.A *Excursion* LINES AND ANGLES

Several interesting concepts dealing with lines are defined in terms of trigonometry. They lead to some useful facts whose proofs are based on the addition and subtraction identities for sine, cosine, and tangent.

If L is a nonhorizontal straight line, the **angle of inclination** of L is the angle θ formed by the part of L above the x-axis and the x-axis in the positive direction, as shown in Figure 8–4.

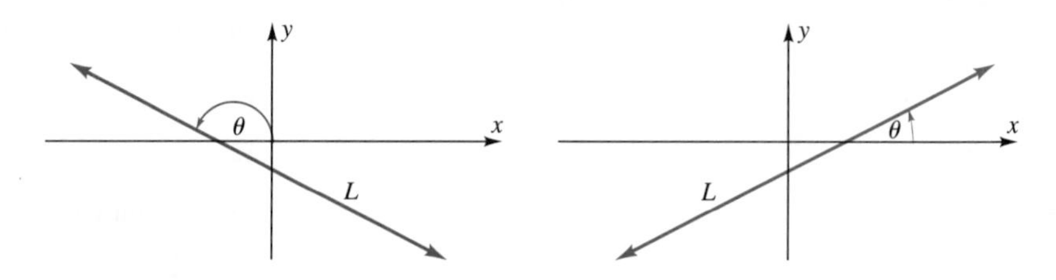

Figure 8-4

The angle of inclination of a horizontal line is defined to be $\theta = 0$. Thus, the radian measure of the angle of inclination of any line satisfies $0 \leq \theta < \pi$. Furthermore,

Angle of Inclination and Slope ▶

> **If L is a nonvertical line with angle of inclination θ, then**
>
> $$\tan \theta = \text{slope of } L.$$

Proof If L is horizontal, then L has slope 0 and angle of inclination $\theta = 0$. Hence, $\tan \theta = \tan 0 = 0 = \text{slope } L$. If L is not horizontal, then it intersects the x-axis at some point $(x_1, 0)$, as shown for two possible cases in Figure 8–5. Let (x_2, y_2) be any point on L above the x-axis.

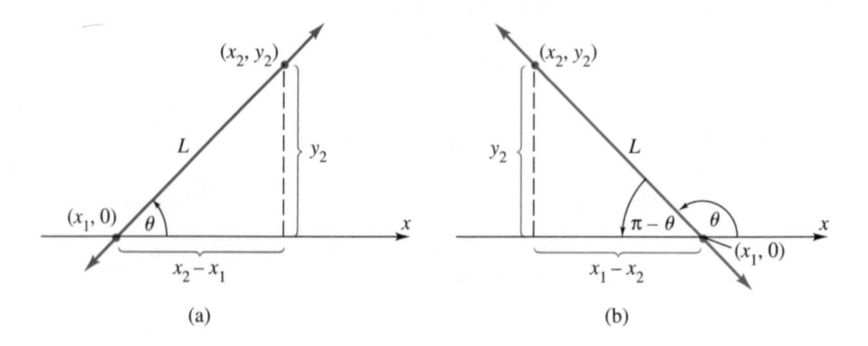

Figure 8-5

If θ is an acute angle, then the right triangle in Figure 8–5(a) shows that

$$\text{slope of } L = \frac{y_2 - 0}{x_2 - x_1} = \frac{y_2}{x_2 - x_1} = \frac{\text{opposite}}{\text{adjacent}} = \tan\theta.$$

If θ is an obtuse angle, then the right triangle in Figure 8–5(b) shows that

$$(*) \quad \text{slope of } L = \frac{0 - y_2}{x_1 - x_2} = -\frac{y_2}{x_1 - x_2} = -\frac{\text{opposite}}{\text{adjacent}} = -\tan(\pi - \theta).$$

Using the fact that the tangent function has period π and the negative angle identity for tangent, we have

$$-\tan(\pi - \theta) = -\tan(-\theta) = -(-\tan\theta) = \tan\theta.$$

Combining this fact with (*) shows that slope of $L = \tan\theta$ in this case also. ❑

EXAMPLE 1 The angle of inclination of a line of slope 5/3 satisfies $\tan\theta = 5/3$. The TAN^{-1} key on a calculator shows that $\theta \approx 1.03$ radians (equivalently, 59.04°). ■

A line with negative slope moves downward from left to right, and hence its angle of inclination lies between $\pi/2$ and π radians (as illustrated in Figure 8–5(b)).

EXAMPLE 2 If a line L has slope -2, then its angle of inclination is a solution of $\tan\theta = -2$ that lies between $\pi/2$ and π. A calculator shows that an approximate solution is -1.10715 and we know that $\tan t = \tan(t + \pi)$ for every t (see page 392). Hence, $\tan(-1.10715 + \pi) \approx -2$. Since $-1.10715 + \pi \approx 2.03444$ is between $\pi/2$ and π, the angle of inclination of L is $\theta \approx 2.03444$ radians (equivalently, 116.565°). ■

Angles Between Two Lines

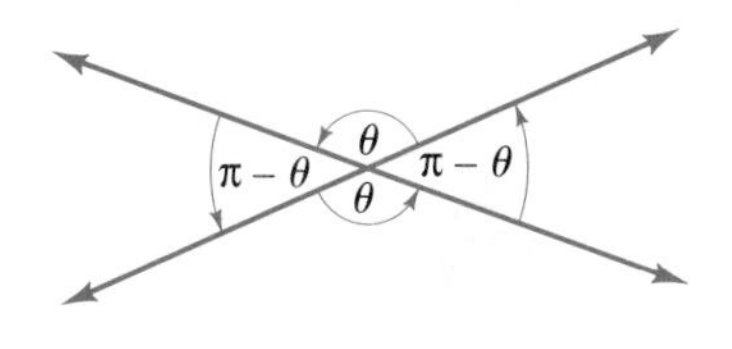

Figure 8-6

If two lines intersect, then they determine four angles with vertex at the point of intersection, as shown in Figure 8–6. If one of these angles measures θ radians, then each of the two angles adjacent to it measures $\pi - \theta$ radians (why?). The fourth angle also measures θ radians by the vertical angle theorem from plane geometry.

The angles between intersecting lines can be determined from the angles of inclination of the lines. Suppose L and M have angles of inclination α and β, respectively, such that $\beta \geq \alpha$. Basic facts about parallel lines, as illustrated in

Figure 8–7, show that $\beta - \alpha$ is one angle between L and M and $\pi - (\beta - \alpha)$ is the other one.*

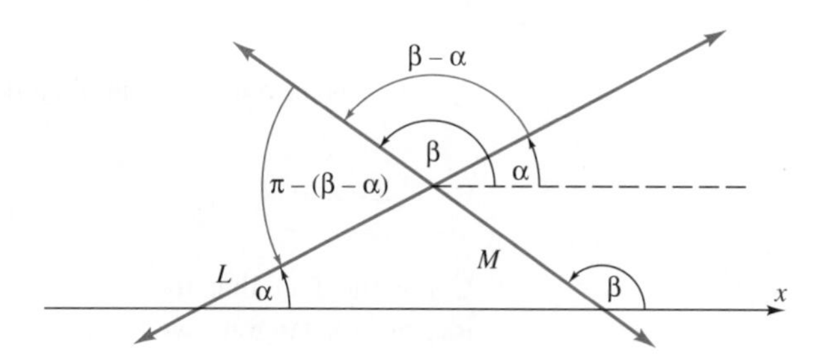

Figure 8-7

The angle between two lines can also be found from their slopes by using this fact:

Angle Between Two Lines ▶

If two nonvertical, nonperpendicular lines have slopes m and k, then one angle θ between them satisfies

$$\tan \theta = \left| \frac{m - k}{1 + mk} \right|.$$

Proof Suppose L has slope k and angle of inclination α and that M has slope m and angle of inclination β. By the definition of absolute value

$$\left| \frac{m - k}{1 + mk} \right| = \frac{m - k}{1 + mk} \quad \text{or} \quad -\frac{m - k}{1 + mk}$$

whichever is positive. We shall show that one angle between L and M has tangent $\frac{m - k}{1 + mk}$ and that the other one has tangent $-\frac{m - k}{1 + mk}$. Thus, one of them necessarily has tangent $\left| \frac{m - k}{1 + mk} \right|$. If $\beta \geq \alpha$, then $\beta - \alpha$ is one angle between L and M. By the subtraction identity for tangent,

$$\tan(\beta - \alpha) = \frac{\tan \beta - \tan \alpha}{1 + \tan \beta \tan \alpha} = \frac{m - k}{1 + mk}.$$

The other angle between L and M is $\pi - (\beta - \alpha)$ and by periodicity, the negative angle identity, and the addition identity, we have:

*Figure 8–7 illustrates the case when α is acute and β is obtuse. Analogous figures in the other possible cases lead to the same conclusion.

$$\begin{aligned}\tan(\pi - (\beta - \alpha)) &= \tan(-(\beta - \alpha)) \\ &= -\tan(\beta - \alpha) \\ &= -\frac{\tan \beta - \tan \alpha}{1 + \tan \beta \tan \alpha} = -\frac{m - k}{1 + mk}.\end{aligned}$$

This completes the proof when $\beta \geq \alpha$. The proof in the case $\alpha \geq \beta$ is similar. ❑

EXAMPLE 3 If L and M are lines of slopes 8 and -3, respectively, then the angle between them satisfies

$$\tan \theta = \left| \frac{8 - (-3)}{1 + 8(-3)} \right| = \left| \frac{11}{-23} \right| = \frac{11}{23}.$$

A calculator shows that $\theta \approx .446$ radians, or 25.56°. ■

The Slope Theorem for Perpendicular Lines

We can now prove the following fact, which was first presented in Section 3.1:

> **Let L be a line with slope k and M a line with slope m. Then L and M are perpendicular exactly when $km = -1$.**

First, suppose L and M are perpendicular. We must show that $km = -1$. If α and β (with $\beta \geq \alpha$) are the angles of inclination of L and M, then $\beta - \alpha$ is the angle between L and M, so that $\beta - \alpha = \pi/2$, or equivalently, $\beta = \alpha + \pi/2$. Therefore, by the addition identities for sine and cosine,

$$\begin{aligned}m = \tan \beta = \tan\left(\alpha + \frac{\pi}{2}\right) &= \frac{\sin[\alpha + (\pi/2)]}{\cos[\alpha + (\pi/2)]} \\ &= \frac{\sin \alpha \cos(\pi/2) + \cos \alpha \sin(\pi/2)}{\cos \alpha \cos(\pi/2) - \sin \alpha \sin(\pi/2)} \\ &= \frac{\sin \alpha\,(0) + \cos \alpha\,(1)}{\cos \alpha\,(0) - \sin \alpha\,(1)} \\ &= -\frac{\cos \alpha}{\sin \alpha} = -\cot \alpha = \frac{-1}{\tan \alpha} = \frac{-1}{k}.\end{aligned}$$

Thus, $m = -1/k$ and hence $mk = -1$.

Now suppose that $mk = -1$. We must show that L and M are perpendicular. If L and M are *not* perpendicular, then neither of the angles between them is $\pi/2$. In this case, if θ is either of the angles between L and M, then $\tan \theta$ is a well-defined real number. But we know that one of these angles must satisfy

$$\tan \theta = \left| \frac{m - k}{1 + mk} \right| = \left| \frac{m - k}{1 + (-1)} \right| = \frac{|m - k|}{0}$$

which is *not* defined. This contradiction shows that L and M must be perpendicular.

EXERCISES 8.2.A

In Exercises 1–6, find the angles of inclination of the straight line through the given points.

1. $(-1, 2), (3, 5)$
2. $(0, 4), (5, -1)$
3. $(1, 4), (6, 0)$
4. $(4, 2), (-3, -2)$
5. $(3, -7), (3, 5)$
6. $(0, 0), (-4, -5)$

In Exercises 7–12, find one of the angles between the straight lines L and M.

7. L has slope $3/2$ and M has slope -1.
8. L has slope 1 and M has slope 3.
9. L has slope -1 and M has slope 0.
10. L has slope -2 and M has slope -3.
11. $(3, 2)$ and $(5, 6)$ are on L; $(0, 3)$ and $(4, 0)$ are on M.
12. $(-1, 2)$ and $(3, -3)$ are on L; $(3, -3)$ and $(6, 1)$ are on M.

8.3 OTHER IDENTITIES

We now present a variety of identities that are special cases of the addition and subtraction identities of the last section, beginning with

Double-Angle Identities ▶

$$\sin 2x = 2 \sin x \cos x$$

$$\cos 2x = \cos^2 x - \sin^2 x$$

$$\tan 2x = \frac{2 \tan x}{1 - \tan^2 x}$$

To prove these identities, just let $x = y$ in the addition identities:

$$\sin 2x = \sin(x + x) = \sin x \cos x + \cos x \sin x = 2 \sin x \cos x$$

$$\cos 2x = \cos(x + x) = \cos x \cos x - \sin x \sin x = \cos^2 x - \sin^2 x$$

$$\tan 2x = \tan(x + x) = \frac{\tan x + \tan x}{1 - \tan x \tan x} = \frac{2 \tan x}{1 - \tan^2 x}.$$

EXAMPLE 1 If $\pi < x < 3\pi/2$ and $\cos x = -8/17$, find $\sin 2x$ and $\cos 2x$, and show that $5\pi/2 < 2x < 3\pi$.

Solution In order to use the double-angle identities, we first must determine $\sin x$. Now $\sin x$ can be found by using the Pythagorean identity:

$$\sin^2 x = 1 - \cos^2 x = 1 - \left(-\frac{8}{17}\right)^2 = 1 - \frac{64}{289} = \frac{225}{289}.$$

Since $\pi < x < 3\pi/2$, we know $\sin x$ is negative. Therefore,

$$\sin x = -\sqrt{\frac{225}{289}} = -\frac{15}{17}.$$

We now substitute these values in the double-angle identities:

$$\sin 2x = 2 \sin x \cos x = 2\left(-\frac{15}{17}\right)\left(-\frac{8}{17}\right) = \frac{240}{289} \approx .83$$

$$\cos 2x = \cos^2 x - \sin^2 x = \left(-\frac{8}{17}\right)^2 - \left(-\frac{15}{17}\right)^2$$

$$= \frac{64}{289} - \frac{225}{289} = -\frac{161}{289} \approx -.56.$$

Since $\pi < x < 3\pi/2$, we know that $2\pi < 2x < 3\pi$. The calculations above show that at $2x$ sine is positive and cosine is negative. This can occur only if $2x$ lies between $5\pi/2$ and 3π. ■

EXAMPLE 2 Express the rule of the function $f(x) = \sin 3x$ in terms of $\sin x$ and constants.

Solution We first use the addition identity for $\sin(x + y)$ with $y = 2x$:

$$f(x) = \sin 3x = \sin(x + 2x) = \sin x \cos 2x + \cos x \sin 2x.$$

Next apply the double-angle identities for $\cos 2x$ and $\sin 2x$:

$$\begin{aligned} f(x) = \sin 3x &= \sin x \cos 2x + \cos x \sin 2x \\ &= \sin x(\cos^2 x - \sin^2 x) + \cos x(2 \sin x \cos x) \\ &= \sin x \cos^2 x - \sin^3 x + 2 \sin x \cos^2 x \\ &= 3 \sin x \cos^2 x - \sin^3 x. \end{aligned}$$

Finally, use the Pythagorean identity:

$$\begin{aligned} f(x) = \sin 3x = 3 \sin x \cos^2 x - \sin^3 x &= 3 \sin x(1 - \sin^2 x) - \sin^3 x \\ = 3 \sin x - 3 \sin^3 x - \sin^3 x &= 3 \sin x - 4 \sin^3 x. \end{aligned}$$ ■

The double-angle identity for $\cos 2x$ can be rewritten in several useful ways. For instance, we can use the Pythagorean identity in the form $\cos^2 x = 1 - \sin^2 x$ to obtain:

$$\cos 2x = \cos^2 x - \sin^2 x = (1 - \sin^2 x) - \sin^2 x = 1 - 2 \sin^2 x.$$

Similarly, using the Pythagorean identity in the form $\sin^2 x = 1 - \cos^2 x$, we have:

$$\cos 2x = \cos^2 x - \sin^2 x = \cos^2 x - (1 - \cos^2 x) = 2 \cos^2 x - 1.$$

In summary:

More Double-Angle Identities ▶

$$\cos 2x = 1 - 2\sin^2 x$$
$$\cos 2x = 2\cos^2 x - 1$$

EXAMPLE 3 Prove that $\dfrac{1 - \cos 2x}{\sin 2x} = \tan x$.

Solution The first identity in the box above and the double-angle identity for sine show that

$$\frac{1 - \cos 2x}{\sin 2x} = \frac{1 - (1 - 2\sin^2 x)}{2\sin x \cos x} = \frac{2\sin^2 x}{2\sin x \cos x} = \frac{\sin x}{\cos x} = \tan x. \quad ■$$

If we solve the first equation in the box above for $\sin^2 x$ and the second one for $\cos^2 x$, we obtain a useful alternate form for these identities:

Power-Reducing Identities ▶

$$\sin^2 x = \frac{1 - \cos 2x}{2} \quad \text{and} \quad \cos^2 x = \frac{1 + \cos 2x}{2}$$

EXAMPLE 4 Express the rule of the function $f(x) = \sin^4 x$ in terms of constants and first powers of the cosine function.

Solution We begin by applying the power-reducing identity

$$f(x) = \sin^4 x = \sin^2 x \sin^2 x = \frac{1 - \cos 2x}{2} \cdot \frac{1 - \cos 2x}{2}$$
$$= \frac{1 - 2\cos 2x + \cos^2 2x}{4}.$$

Next we apply the power-reducing identity for cosine to $\cos^2 2x$. Note that this means using $2x$ in place of x in the identity:

$$\cos^2 2x = \frac{1 + \cos 2(2x)}{2} = \frac{1 + \cos 4x}{2}.$$

Finally, we substitute this last result in the expression for $\sin^4 x$ above:

$$f(x) = \sin^4 x = \frac{1 - 2\cos 2x + \cos^2 2x}{4} = \frac{1 - 2\cos 2x + \dfrac{1 + \cos 4x}{2}}{4}$$

$$= \frac{1}{4} - \frac{1}{2}\cos 2x + \frac{1}{8}(1 + \cos 4x)$$

$$= \frac{3}{8} - \frac{1}{2}\cos 2x + \frac{1}{8}\cos 4x. \quad \blacksquare$$

Half-Angle Identities

If we use the power-reducing identity with $x/2$ in place of x, we obtain

$$\sin^2\left(\frac{x}{2}\right) = \frac{1 - \cos 2\left(\dfrac{x}{2}\right)}{2} = \frac{1 - \cos x}{2}.$$

Consequently, we must have:

$$\sin\left(\frac{x}{2}\right) = \pm\sqrt{\frac{1 - \cos x}{2}}.$$

This proves the first of the half-angle identities.

Half-Angle Identities ▶

$$\sin\frac{x}{2} = \pm\sqrt{\frac{1 - \cos x}{2}} \qquad \cos\frac{x}{2} = \pm\sqrt{\frac{1 + \cos x}{2}}$$

$$\tan\frac{x}{2} = \pm\sqrt{\frac{1 - \cos x}{1 + \cos x}}$$

The half-angle identity for cosine is derived from a power-reducing identity, as was the half-angle identity for sine. The half-angle identity for tangent then follows immediately since $\tan(x/2) = \sin(x/2)/\cos(x/2)$. In all cases, *the sign in front of the radical depends on the quadrant in which $x/2$ lies.*

EXAMPLE 5 Find the exact value of

(a) $\cos 5\pi/8$ **(b)** $\sin \pi/12$.

Solution

(a) Since $5\pi/8 = \frac{1}{2}(5\pi/4)$, we use the half-angle identity with $x = 5\pi/4$ and the fact that $\cos(5\pi/4) = -\sqrt{2}/2$. The sign chart on page 381 shows that $\cos(5\pi/8)$ is negative because $5\pi/8$ is in the second quadrant. So we use the negative sign in front of the radical:

$$\cos\frac{5\pi}{8} = \cos\frac{5\pi/4}{2} = -\sqrt{\frac{1+\cos(5\pi/4)}{2}}$$
$$= -\sqrt{\frac{1+(-\sqrt{2}/2)}{2}} = -\sqrt{\frac{(2-\sqrt{2})/2}{2}}$$
$$= -\sqrt{\frac{2-\sqrt{2}}{4}}$$
$$= \frac{-\sqrt{2-\sqrt{2}}}{2}.$$

(b) Since $\pi/12 = \frac{1}{2}(\pi/6)$ and $\pi/12$ is in the first quadrant, where sine is positive, we have:

$$\sin\frac{\pi}{12} = \sin\frac{\pi/6}{2} = \sqrt{\frac{1-\cos(\pi/6)}{2}}$$
$$= \sqrt{\frac{1-\sqrt{3}/2}{2}} = \sqrt{\frac{(2-\sqrt{3})/2}{2}} = \sqrt{\frac{2-\sqrt{3}}{4}}$$
$$= \frac{\sqrt{2-\sqrt{3}}}{2}. \quad \blacksquare$$

The problem of determining signs in the half-angle formulas can be eliminated with tangent by using these:

Half-Angle Identities for Tangent ▶

$$\tan\frac{x}{2} = \frac{1-\cos x}{\sin x} \quad \text{and} \quad \tan\frac{x}{2} = \frac{\sin x}{1+\cos x}$$

The proof of the first of these identities follows from the identity $\tan x = \dfrac{1-\cos 2x}{\sin 2x}$, which was proved in Example 3; simply replace x by $x/2$:

$$\tan\left(\frac{x}{2}\right) = \frac{1-\cos 2(x/2)}{\sin 2(x/2)} = \frac{1-\cos x}{\sin x}.$$

The second identity in the box is proved in Exercise 71.

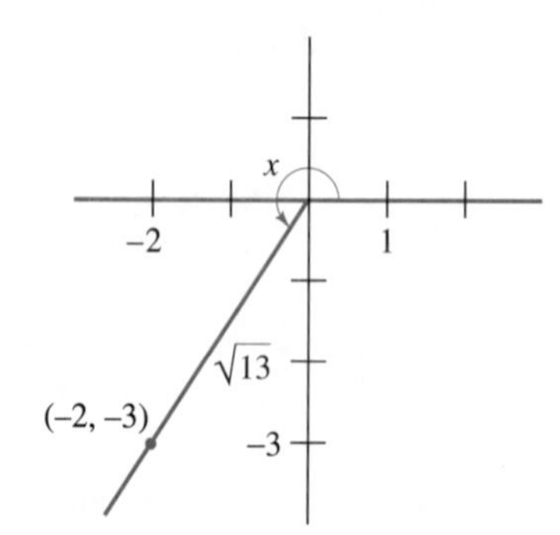

Figure 8-8

EXAMPLE 6 If $\tan x = \dfrac{3}{2}$ and $\pi < x < \dfrac{3\pi}{2}$, find $\tan\dfrac{x}{2}$.

Solution The terminal side of an angle of x radians in standard position lies in the third quadrant, as shown in Figure 8–8. The tangent of the angle in standard position whose terminal side passes through the point $(-2, -3)$ is $\dfrac{-3}{-2} = \dfrac{3}{2}$.

Since there is only one angle in the third quadrant with tangent 3/2, the point $(-2, -3)$ must lie on the terminal side of the angle of x radians.

Since the distance from $(-2, -3)$ to the origin is

$$\sqrt{(-2-0)^2 + (-3-0)^2} = \sqrt{13},$$

we have

$$\sin x = \frac{-3}{\sqrt{13}} \quad \text{and} \quad \cos x = \frac{-2}{\sqrt{13}}.$$

Therefore, by the first of the half-angle identities for tangent

$$\tan\frac{x}{2} = \frac{1-\cos x}{\sin x} = \frac{1-\left(\frac{-2}{\sqrt{13}}\right)}{\frac{-3}{\sqrt{13}}} = \frac{\frac{\sqrt{13}+2}{\sqrt{13}}}{\frac{-3}{\sqrt{13}}} = -\frac{\sqrt{13}+2}{3}. \quad ■$$

Product and Factoring Identities

Using the addition and subtraction identities to compute $\sin(x+y) + \sin(x-y)$, we see that

$$\begin{aligned}\sin(x+y) + \sin(x-y) &= \sin x \cos y + \cos x \sin y + \sin x \cos y - \cos x \sin y \\ &= 2 \sin x \cos y.\end{aligned}$$

Dividing both sides of this last equation by 2 produces the first of the following identities:

Product Identities ►

$$\sin x \cos y = \frac{1}{2}(\sin(x+y) + \sin(x-y))$$

$$\sin x \sin y = \frac{1}{2}(\cos(x-y) - \cos(x+y))$$

$$\cos x \cos y = \frac{1}{2}(\cos(x+y) + \cos(x-y))$$

$$\cos x \sin y = \frac{1}{2}(\sin(x+y) - \sin(x-y))$$

The proofs of the second and fourth product identities are similar to the proof of the first. The third product identity was proved in Example 4 of Section 8.2.

If we use the first product identity with $\frac{1}{2}(x + y)$ in place of x and $\frac{1}{2}(x - y)$ in place of y, we obtain:

$$\sin\left(\frac{1}{2}(x+y)\right)\cos\left(\frac{1}{2}(x-y)\right) = \frac{1}{2}\left[\sin\left(\frac{1}{2}(x+y) + \frac{1}{2}(x-y)\right) + \sin\left(\frac{1}{2}(x+y) - \frac{1}{2}(x-y)\right)\right]$$

$$= \frac{1}{2}[\sin x + \sin y].$$

Multiplying both sides of the last equation by 2 produces the first of the following identities:

Factoring Identities ►

$$\sin x + \sin y = 2\sin\left(\frac{x+y}{2}\right)\cos\left(\frac{x-y}{2}\right)$$

$$\sin x - \sin y = 2\cos\left(\frac{x+y}{2}\right)\sin\left(\frac{x-y}{2}\right)$$

$$\cos x + \cos y = 2\cos\left(\frac{x+y}{2}\right)\cos\left(\frac{x-y}{2}\right)$$

$$\cos x - \cos y = -2\sin\left(\frac{x+y}{2}\right)\sin\left(\frac{x-y}{2}\right)$$

The last three factoring identities are proved in the same way as the first.

EXAMPLE 7 The factoring identities can be used to prove the identity

$$\frac{\sin t + \sin 3t}{\cos t + \cos 3t} = \tan 2t.$$

Using the first factoring identity with $x = t$ and $y = 3t$ yields:

$$\sin t + \sin 3t = 2\sin\left(\frac{t+3t}{2}\right)\cos\left(\frac{t-3t}{2}\right) = 2\sin 2t\cos(-t).$$

Similarly,

$$\cos t + \cos 3t = 2\cos\left(\frac{t+3t}{2}\right)\cos\left(\frac{t-3t}{2}\right) = 2\cos 2t\cos(-t)$$

so that

$$\frac{\sin t + \sin 3t}{\cos t + \cos 3t} = \frac{2\sin 2t\cos(-t)}{2\cos 2t\cos(-t)} = \frac{\sin 2t}{\cos 2t} = \tan 2t. \quad \blacksquare$$

EXERCISES 8.3

In Exercises 1–12, use the half-angle identities to evaluate the given expression.

1. $\cos \frac{\pi}{8}$ **2.** $\tan \frac{\pi}{8}$ **3.** $\sin \frac{3\pi}{8}$ **4.** $\cos \frac{3\pi}{8}$

5. $\tan \frac{\pi}{12}$ **6.** $\sin \frac{5\pi}{8}$ **7.** $\cos \frac{\pi}{12}$ **8.** $\tan \frac{5\pi}{8}$

9. $\sin \frac{7\pi}{8}$ **10.** $\cos \frac{7\pi}{8}$ **11.** $\tan \frac{7\pi}{8}$ **12.** $\cot \frac{\pi}{8}$

In Exercises 13–18, write each expression as a sum or difference.

13. $\sin 4x \cos 6x$ **14.** $\sin 5x \sin 7x$.

15. $\cos 2x \cos 4x$ **16.** $\sin 3x \cos 5x$

17. $\sin 17x \sin(-3x)$ **18.** $\cos 13x \cos(-5x)$

In Exercises 19–22, write each expression as a product.

19. $\sin 3x + \sin 5x$ **20.** $\cos 2x + \cos 6x$

21. $\sin 9x - \sin 5x$ **22.** $\cos 5x - \cos 7x$

In Exercises 23–30, find sin 2x, cos 2x, and tan 2x under the given conditions.

23. $\sin x = \frac{5}{13} \quad \left(0 < x < \frac{\pi}{2}\right)$

24. $\sin x = -\frac{4}{5} \quad \left(\pi < x < \frac{3\pi}{2}\right)$

25. $\cos x = -\frac{3}{5} \quad \left(\pi < x < \frac{3\pi}{2}\right)$

26. $\cos x = -\frac{1}{3} \quad \left(\frac{\pi}{2} < x < \pi\right)$

27. $\tan x = \frac{3}{4} \quad \left(\pi < x < \frac{3\pi}{2}\right)$

28. $\tan x = -\frac{3}{2} \quad \left(\frac{\pi}{2} < x < \pi\right)$

29. $\csc x = 4 \quad \left(0 < x < \frac{\pi}{2}\right)$

30. $\sec x = -5 \quad \left(\pi < x < \frac{3\pi}{2}\right)$

In Exercises 31–36, find $\sin \frac{x}{2}$, $\cos \frac{x}{2}$, and $\tan \frac{x}{2}$ under the given conditions.

31. $\cos x = .4 \quad \left(0 < x < \frac{\pi}{2}\right)$

32. $\sin x = .6 \quad \left(\frac{\pi}{2} < x < \pi\right)$

33. $\sin x = -\frac{3}{5} \quad \left(\frac{3\pi}{2} < x < 2\pi\right)$

34. $\cos x = .8 \quad \left(\frac{3\pi}{2} < x < 2\pi\right)$

35. $\tan x = \frac{1}{2} \quad \left(\pi < x < \frac{3\pi}{2}\right)$

36. $\cot x = 1 \quad \left(-\pi < x < -\frac{\pi}{2}\right)$

In Exercises 37–42, assume $\sin x = .6$ and $0 < x < \pi/2$ and evaluate the given expression.

37. $\sin 2x$ **38.** $\cos 4x$ **39.** $\cos 2x$ **40.** $\sin 4x$

41. $\sin \frac{x}{2}$ **42.** $\cos \frac{x}{2}$

43. Express $\cos 3x$ in terms of $\cos x$.

44. **(a)** Express the rule of the function $f(x) = \cos^3 x$ in terms of constants and first powers of the cosine function as in Example 4.
(b) Do the same for $f(x) = \cos^4 x$.

In Exercises 45–50, simplify the given expression.

45. $\frac{\sin 2x}{2 \sin x}$ **46.** $1 - 2\sin^2\left(\frac{x}{2}\right)$

47. $2 \cos 2y \sin 2y$ (Think!)

48. $\cos^2\left(\frac{x}{2}\right) - \sin^2\left(\frac{x}{2}\right)$

49. $(\sin x + \cos x)^2 - \sin 2x$

50. $2 \sin x \cos^3 x - 2 \sin^3 x \cos x$

In Exercises 51–64, determine graphically whether the equation could possibly be an identity. If it could, prove that it is.

51. $\sin 16x = 2\sin 8x\cos 8x$

52. $\cos 8x = \cos^2 4x - \sin^2 4x$

53. $\cos^4 x - \sin^4 x = \cos 2x$

54. $\sec 2x = \dfrac{1}{1 - 2\sin^2 x}$

55. $\cos 4x = 2\cos 2x - 1$

56. $\sin^2 x = \cos^2 x - 2\sin x$

57. $\dfrac{1 + \cos 2x}{\sin 2x} = \cot x$

58. $\sin 2x = \dfrac{2\cot x}{\csc^2 x}$

59. $\sin 3x = (\sin x)(3 - 4\sin^2 x)$

60. $\sin 4x = (4\cos x\sin x)(1 - 2\sin^2 x)$

61. $\cos 2x = \dfrac{2\tan x}{\sec^2 x}$

62. $\cos 3x = (\cos x)(3 - 4\cos^2 x)$

63. $\csc^2\left(\dfrac{x}{2}\right) = \dfrac{2}{1 - \cos x}$

64. $\sec^2\left(\dfrac{x}{2}\right) = \dfrac{2}{1 + \cos x}$

In Exercises 65–70, prove the identity.

65. $\dfrac{\sin x - \sin 3x}{\cos x + \cos 3x} = -\tan x$

66. $\dfrac{\sin x - \sin 3x}{\cos x - \cos 3x} = -\cot 2x$

67. $\dfrac{\sin 4x + \sin 6x}{\cos 4x - \cos 6x} = \cot x$

68. $\dfrac{\cos 8x + \cos 4x}{\cos 8x - \cos 4x} = -\cot 6x\cot 2x$

69. $\dfrac{\sin x + \sin y}{\cos x - \cos y} = -\cot\left(\dfrac{x - y}{2}\right)$

70. $\dfrac{\sin x - \sin y}{\cos x + \cos y} = \tan\left(\dfrac{x - y}{2}\right)$

71. **(a)** Prove that $\dfrac{1 - \cos x}{\sin x} = \dfrac{\sin x}{1 + \cos x}$.

(b) Use part (a) and the half-angle identity proved in the text to prove that $\tan\dfrac{x}{2} = \dfrac{\sin x}{1 + \cos x}$.

8.4 TRIGONOMETRIC EQUATIONS

Any equation involving trigonometric functions can be solved graphically and many can be solved algebraically. We begin with the graphical method, which is the same as that used to solve other equations. Unlike the equations solved previously, however, trigonometric equations typically have an infinite number of solutions. In many cases, these solutions can be systematically determined by using periodicity, as we now see.

EXAMPLE 1 Solve: $3\sin^2 x - \cos x - 2 = 0$.

Solution Both sine and cosine have period 2π, so the period of $f(x) = 3\sin^2 x - \cos x - 2$ is at most 2π. The graph of f, which is shown in two viewing windows in Figure 8–9, does not repeat its pattern over any interval of less than 2π, so we conclude that f has period 2π.

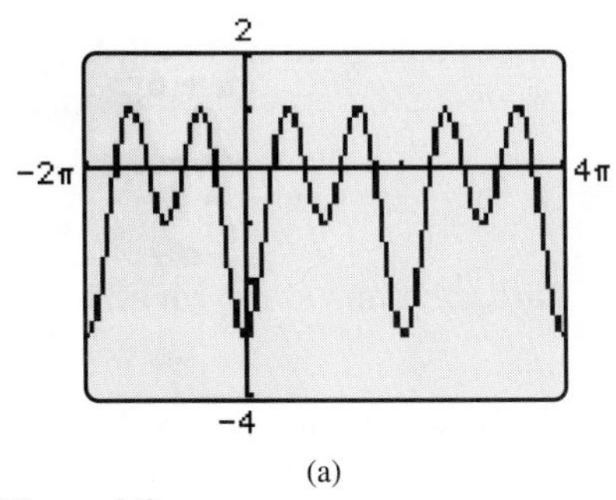

(a)

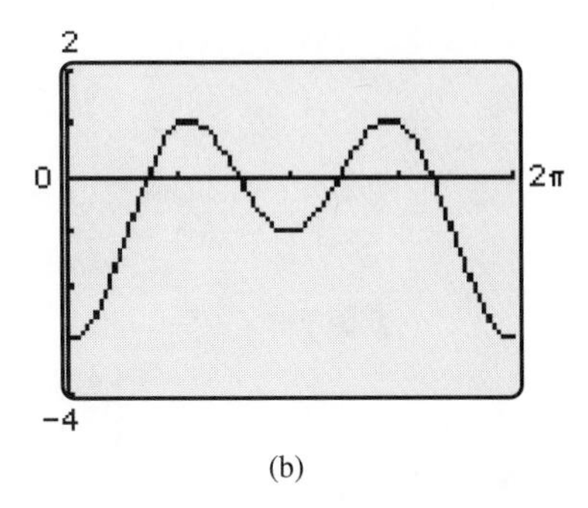

(b)

Figure 8-9

The function f makes one complete period in the interval $[0, 2\pi)$, as shown in Figure 8–9(b). The equation has four solutions between 0 and 2π, namely, the four x-intercepts of the graph in that interval, as explained in Section 1.5. Zoom-in or a graphical root finder shows that these four solutions are

$$x \approx 1.12, \qquad x \approx 2.45, \qquad x \approx 3.84, \qquad x \approx 5.16.$$

Since the graph repeats its pattern to the left and right, the other x-intercepts (solutions) will differ from these four by multiples of 2π. For instance, in addition to the x-intercept at 1.12, there will be x-intercepts (solutions) at

$$1.12 \pm 2\pi, \qquad 1.12 \pm 4\pi, \qquad 1.12 \pm 6\pi, \quad \text{etc.},$$

the first two of which can be seen in Figure 8–9(a). These solutions are customarily written like this:

$$x = 1.12 + 2k\pi \qquad (k = 0, \pm 1, \pm 2, \pm 3, \ldots).$$

A similar analysis applies to the other solutions between 0 and 2π. Hence, all solutions of the equation are given by:

$$x \approx 1.12 + 2k\pi, \quad x \approx 2.45 + 2k\pi, \quad x \approx 3.84 + 2k\pi, \quad x \approx 5.16 + 2k\pi,$$
$$\text{where } k = 0, \pm 1, \pm 2, \pm 3, \ldots . \quad ■$$

Although the graph in Figure 8–9(a) of Example 1 shows three periods of the function f, only one complete period, the interval $(0, 2\pi)$, was actually used to solve the equation. A similar procedure can be used to solve any trigonometric equation graphically:

Graphical Method for Solving Trigonometric Equations ▶

1. **Write the equation in the form $f(x) = 0$.**
2. **Determine the period p of $f(x)$.**
3. **Graph $f(x)$ over an interval of length p.**
4. **Use zoom-in or a graphical root finder to determine the x-intercepts of the graph in this interval.**
5. **For each x-intercept u, all of the numbers**

$$u + kp \qquad (k = 0, \pm 1, \pm 2, \pm 3, \ldots)$$

are solutions of the equation.

In Example 1, for instance, p was 2π. However, many trigonometric functions have periods other than 2π.

EXAMPLE 2 To solve the equation $\tan x = 3 \sin 2x$, we first rewrite it as

$$\tan x - 3 \sin 2x = 0.$$

Both $\tan x$ and $\sin 2x$ have period π (see pages 392 and 422). Hence the function given by the left side of the equation, $f(x) = \tan x - 3 \sin 2x$, also has period π. The graph of f on the interval $[0, \pi)$ (Figure 8–10) shows erroneous vertical line segments instead of the vertical asymptote at $x = \pi/2$, where tangent is not defined, as well as x-intercepts at the endpoints of the interval. Consequently, we use the more easily read graph f in Figure 8–11, which uses the interval $(-\pi/2, \pi/2)$.

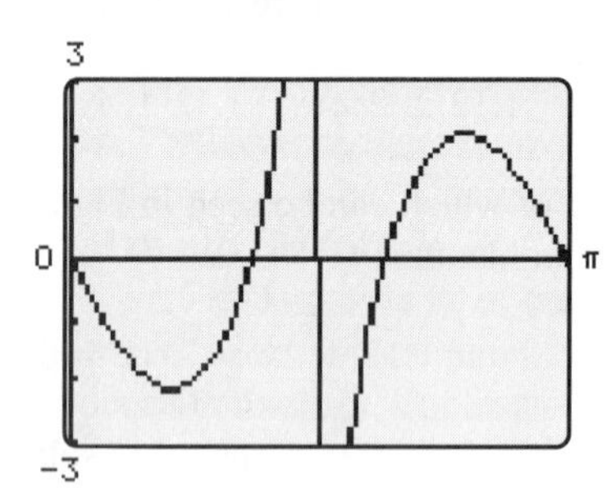

Figure 8-10

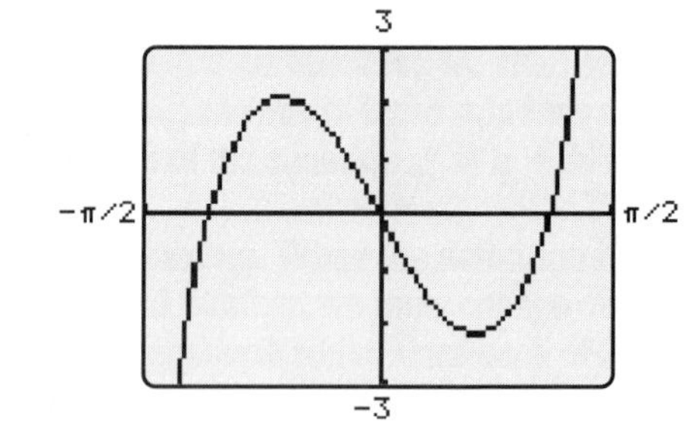

Figure 8-11

Even without the graph, we can verify that there is an x-intercept at the origin because $f(0) = \tan 0 - 3 \sin (2 \cdot 0) = 0$. Using zoom-in or a root finder on the other two x-intercepts in Figure 8–11 shows that they are $x \approx -1.15$ and $x \approx 1.15$. All other solutions differ from these by multiples of π (the period of f) and are given by:

$$x \approx -1.15 + k\pi, \qquad x \approx 0 + k\pi, \qquad x \approx 1.15 + k\pi$$

$$(k = 0, \pm 1, \pm 2, \pm 3, \ldots). \quad ■$$

Basic Equations

We shall call equations, such as

$$\sin x = .39, \qquad \cos x = .2, \qquad \tan x = -3,$$

basic equations. The following examples show how they can be quickly solved algebraically.

EXAMPLE 3 Solve $\tan x = 2$.

Solution The graph of $f(x) = \tan x$ in Figure 8–12 shows that there are infinitely many points with second coordinate 2 (each point where the graph intersects the horizontal line through 2). The first coordinate of every such point is a number whose tangent is 2, that is, a solution of $\tan x = 2$.

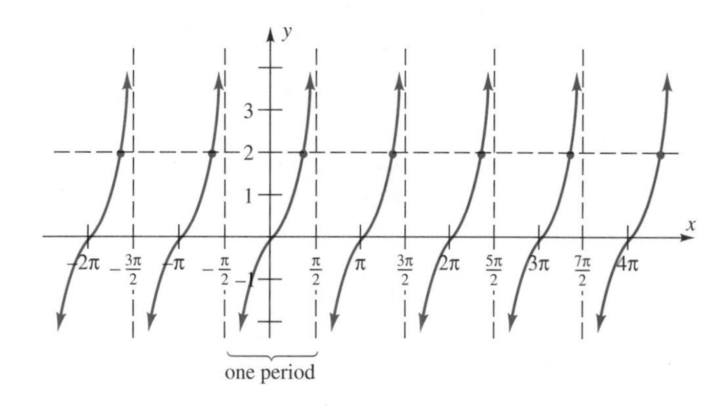

Figure 8-12

There is exactly one solution between $-\pi/2$ and $\pi/2$. Although this solution can be found graphically, it is faster to key in TAN⁻¹ 2 ENTER .* The calculator then displays the number between $-\pi/2$ and $\pi/2$ whose tangent is 2, namely, 1.1071. In other words, $x = 1.1071$ is a solution of $\tan x = 2$.** Since the interval $(-\pi/2, \pi/2)$ is one full period of the tangent function and $x = 1.1071$ is the only solution of $\tan x = -2$ in that interval, all solutions are given by

$$x = 1.1071 + k\pi \qquad (k = 0, \pm 1, \pm 2, \pm 3, \ldots). \quad \blacksquare$$

With only a slight modification, the procedure used in Example 3 also works for basic sine and cosine equations.

*Unless stated otherwise, radian mode is used throughout this section.

**For convenient reading all solutions in the text are rounded to four decimal places, but the full decimal expansion given by the calculator is used in all computations. We write = rather than ≈ even though these calculator solutions are approximations of the actual solutions.

EXAMPLE 4 Solve $\sin x = -.75$.

Solution There are infinitely many points on the graph of $f(x) = \sin x$ that have second coordinate $-.75$ (Figure 8–13). The first coordinate of every such point is a solution of the equation $\sin x = -.75$ (why?).

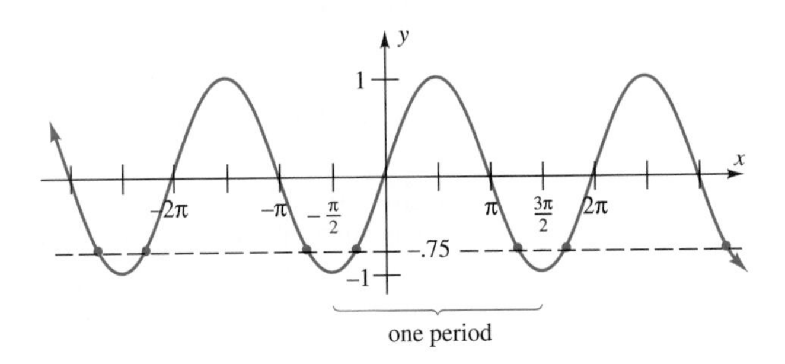

Figure 8-13

There is one solution between $-\pi/2$ and $\pi/2$, which can be found by keying in SIN^{-1} −.75 ENTER. Verify that this solution is $x = -.8481$. Note that there is a second solution in the period of the sine function from $-\pi/2$ to $3\pi/2$. It can be found by using the identity that was proved in Example 2 of Section 8.2:

$$\sin(\pi - y) = \sin y.$$

Applying this identity with $y = -.8481$ shows that

$$\sin(\pi - (-.8481)) = \sin(-.8481) = -.75.$$

Hence, $x = \pi - (-.8481) = 3.9897$ is also a solution of $\sin x = -.75$. Since the interval $[-\pi/2, 3\pi/2)$ is one full period of the sine function, all solutions of the equation $\sin x = -.75$ are given by

$$x = -.8481 + 2k\pi \qquad \text{and} \qquad x = 3.9897 + 2k\pi$$
$$(k = 0, \pm 1, \pm 2, \pm 3, \ldots).$$ ■

EXAMPLE 5 Solve $\cos x = .2$.

Solution A solution between 0 and π can be found by keying in COS^{-1} .2 ENTER, namely, $x = 1.3694$.

> ***Note*** The reason why the COS^{-1} key produces a solution in the interval from 0 to π, whereas the SIN^{-1} and TAN^{-1} keys produce solutions in the interval from $-\pi/2$ to $\pi/2$, is explained in Section 8.5.

GRAPHING EXPLORATION Verify that the equation $\cos x = .2$ has another solution between $-\pi$ and π by showing that the graph of $g(x) = \cos x$ intersects the horizontal line $y = .2$ at two points in that interval. Note that these points appear to be at equal distances from the y-axis, which suggests that their first coordinates are negatives of each other. ■

The second solution in the interval $[-\pi, \pi)$ is indeed the negative of the first one, because by the negative angle identity, $\cos(-x) = \cos x$, we have

$$\cos(-1.3694) = \cos 1.3694 = .2.$$

Thus, $x = 1.3694$ and $x = -1.3694$ are the solutions of $\cos x = .2$ in the interval $[-\pi, \pi)$. Since this interval is one full period of the cosine function, all solutions of the equation are given by

$$x = 1.3694 + 2k\pi \qquad \text{and} \qquad x = -1.3694 + 2k\pi$$
$$(k = 0, \pm 1, \pm 2, \pm 3, \ldots). \quad ■$$

The procedures in Examples 3–5 work in the general case and lead to the following conclusion.*

Solution Algorithm for Basic Trigonometric Equations ▶

If c is any real number, then the equation $\tan x = c$ has one solution between $-\pi/2$ and $\pi/2$, which can be found by using the TAN^{-1} key. All solutions can be found by adding or subtracting integer multiples of π to this one.

If $|c| < 1$, then the equation $\sin x = c$ has a solution between $-\pi/2$ and $\pi/2$, which can be found by using the SIN^{-1} key. A second solution is $\pi - u$, where u is the first solution. All solutions can be found by adding or subtracting integer multiples of 2π to these two.

If $|c| < 1$, then the equation $\cos x = c$ has a solution between 0 and π, which can be found by using the COS^{-1} key. A second solution is the negative of this one. All solutions can be found by adding or subtracting integer multiples of 2π to these two.

Note The solution algorithm for basic equations can be used to find solutions in degrees by replacing π by 180° and using a calculator in degree mode.

E X A M P L E 6 Find all solutions of $\sec x = 8$ in the interval $[0, 2\pi)$.

Solution Note that $\sec x = 8$ exactly when

$$\frac{1}{\cos x} = 8, \qquad \text{or equivalently,} \qquad \cos x = \frac{1}{8} = .125.$$

Since $\cos^{-1}(.125) = 1.4455$, the solutions of $\cos x = .125$, and hence of $\sec x = 8$ are

$$x = 1.4455 + 2k\pi \qquad \text{and} \qquad x = -1.4455 + 2k\pi$$
$$(k = 0, \pm 1, \pm 2, \pm 3, \ldots).$$

Of these solutions, the two between 0 and 2π are

$$x = 1.4455 \qquad \text{and} \qquad x = -1.4455 + 2\pi = 4.8377. \quad ■$$

*The equations $\sin x = \pm 1$ and $\cos x = \pm 1$ are special cases; see Exercises 13 and 14.

EXAMPLE 7 Solve $\sin u = \sqrt{2}/2$ exactly, without using a calculator.

Solution Our knowledge of special values (Section 6.5) shows that $u = \pi/4$ is one solution. As above, the identity $\sin(\pi - x) = \sin x$ shows that $u = \pi - \pi/4 = 3\pi/4$ is a second solution between $-\pi/2$ and $3\pi/2$. Therefore, the exact solutions of $\sin x = \sqrt{2}/2$ are

$$u = \frac{\pi}{4} + 2k\pi \quad \text{and} \quad u = \frac{3\pi}{4} + 2k\pi$$

$$(k = 0, \pm1, \pm2, \pm3, \ldots).$$ ■

EXAMPLE 8 Solve exactly: $\sin 2x = \sqrt{2}/2$.

Solution First, let $u = 2x$ and solve the basic equation $\sin u = \sqrt{2}/2$. As we saw in Example 7, the solutions are

$$u = \frac{\pi}{4} + 2k\pi \quad \text{and} \quad u = \frac{3\pi}{4} + 2k\pi \quad (k = 0, \pm1, \pm2, \pm3, \ldots).$$

Since $u = 2x$, each of these solutions leads to a solution of the original equation:

$$2x = u = \frac{\pi}{4} + 2k\pi, \quad \text{or equivalently,} \quad x = \frac{1}{2}\left(\frac{\pi}{4} + 2k\pi\right) = \frac{\pi}{8} + k\pi.$$

Similarly,

$$2x = u = \frac{3\pi}{4} + 2k\pi, \quad \text{or equivalently,} \quad x = \frac{1}{2}\left(\frac{3\pi}{4} + 2k\pi\right) = \frac{3\pi}{8} + k\pi.$$

Therefore, all solutions of $\sin 2x = \sqrt{2}/2$ are given by

$$x = \frac{\pi}{8} + k\pi \quad \text{and} \quad x = \frac{3\pi}{8} + k\pi \quad (k = 0, \pm1, \pm2, \pm3, \ldots).$$

The fact that the solutions are obtained by adding multiples of π rather than 2π is a reflection of the fact that the period of $\sin 2x$ is π. ■

Algebraic Solution of Other Trigonometric Equations

Many trigonometric equations can be solved algebraically by using factoring, the quadratic formula, and identities to reduce the problem to an equivalent one that involves only basic equations.

EXAMPLE 9 To solve $-10\cos^2 x - 3\sin x + 9 = 0$, we first use the Pythagorean identity to rewrite the equation in terms of the sine function:

$$-10\cos^2 x - 3\sin x + 9 = 0$$
$$-10(1 - \sin^2 x) - 3\sin x + 9 = 0$$
$$-10 + 10\sin^2 x - 3\sin x + 9 = 0$$
$$10\sin^2 x - 3\sin x - 1 = 0.$$

Now factor the left side:*

$$(2\sin x - 1)(5\sin x + 1) = 0$$
$$2\sin x - 1 = 0 \quad \text{or} \quad 5\sin x + 1 = 0$$
$$2\sin x = 1 \quad \text{or} \quad 5\sin x = -1$$
$$\sin x = 1/2 \quad \text{or} \quad \sin x = -1/5 = -.2$$

Each of these basic equations is readily solved. We note that $\sin(\pi/6) = 1/2$, so that $x = \pi/6$ and $x = \pi - \pi/6 = 5\pi/6$ are solutions of the first one. Since $\sin^{-1}(-.2) = -.2014$, both $x = -.2014$ and $x = \pi - (-.2014) = 3.3430$ are solutions of the second equation. Therefore, all solutions of the original equation are given by:

$$x = \frac{\pi}{6} + 2k\pi, \qquad x = \frac{5\pi}{6} + 2k\pi, \qquad x = -.2014 + 2k\pi,$$
$$x = 3.3430 + 2k\pi$$

where $k = 0, \pm 1, \pm 2, \pm 3, \ldots$. ■

EXAMPLE 10 To solve $\sec^2 x + 5\tan x = -2$, we use the Pythagorean identity $\sec^2 x = 1 + \tan^2 x$ to obtain an equivalent equation:

$$\sec^2 x + 5\tan x = -2$$
$$\sec^2 x + 5\tan x + 2 = 0$$
$$(1 + \tan^2 x) + 5\tan x + 2 = 0$$
$$\tan^2 x + 5\tan x + 3 = 0.$$

If we let $u = \tan x$, this last equation becomes $u^2 + 5u + 3 = 0$. Since the left side does not readily factor, we use the quadratic formula to solve the equation:

$$u = \frac{-5 \pm \sqrt{5^2 - 4 \cdot 1 \cdot 3}}{2} = \frac{-5 \pm \sqrt{13}}{2}.$$

*The factorization may be easier to see if you first substitute v for $\sin x$, so that $10\sin^2 x - 3\sin x - 1$ becomes $10v^2 - 3v - 1 = (2v - 1)(5v + 1)$.

Since $u = \tan x$, the original equation is equivalent to

$$\tan x = \frac{-5 + \sqrt{13}}{2} \approx -.6972 \qquad \text{or} \qquad \tan x = \frac{-5 - \sqrt{13}}{2} \approx -4.3028.$$

Solving these basic equations as above, we find that $x = -.6089$ is a solution of the first and $x = -1.3424$ is a solution of the second. Hence, the solutions of the original equation are

$$x = -.6089 + k\pi \qquad \text{and} \qquad x = -1.3424 + k\pi$$
$$(k = 0, \pm 1, \pm 2, \pm 3, \ldots). \quad ■$$

EXAMPLE 11 To solve $5 \cos x + 3 \cos 2x = 3$, we use the double-angle identity: $\cos 2x = 2 \cos^2 x - 1$ as follows.

$$5 \cos x + 3 \cos 2x = 3$$

Use double-angle identity: $5 \cos x + 3(2 \cos^2 x - 1) = 3$

Multiply out left side: $5 \cos x + 6 \cos^2 x - 3 = 3$

Rearrange terms: $6 \cos^2 x + 5 \cos x - 6 = 0$

Factor left side: $(2 \cos x + 3)(3 \cos x - 2) = 0$

$$2 \cos x + 3 = 0 \qquad \text{or} \qquad 3 \cos x - 2 = 0$$
$$2 \cos x = -3 \qquad \text{or} \qquad 3 \cos x = 2$$
$$\cos x = -\frac{3}{2} \qquad \text{or} \qquad \cos x = \frac{2}{3}$$

The equation $\cos x = -3/2$ has no solutions because $\cos x$ always lies between -1 and 1. A calculator shows that the solutions of $\cos x = 2/3$ are

$$x = .8411 + 2k\pi \qquad \text{and} \qquad x = -.8411 + 2k\pi$$
$$(k = 0, \pm 1, \pm 2, \pm 3, \ldots). \quad ■$$

EXAMPLE 12 Find all solutions of

$$\sin 6t + \sin 2t = \cos 2t.$$

Solution Although the double-angle and power-reducing identities don't seem promising here, observe that the left side has the form $\sin x + \sin y$. This suggests that we try the first factoring identity

$$\sin x + \sin y = 2 \sin\left(\frac{x+y}{2}\right) \cos\left(\frac{x-y}{2}\right)$$

with $x = 6t$ and $y = 2t$:

$$\sin 6t + \sin 2t = \cos 2t$$

$$2 \sin\left(\frac{6t + 2t}{2}\right) \cos\left(\frac{6t - 2t}{2}\right) = \cos 2t$$

$$2 \sin 4t \cos 2t = \cos 2t$$

$$2 \sin 4t \cos 2t - \cos 2t = 0$$

$$(2 \sin 4t - 1) \cos 2t = 0$$

$$2 \sin 4t - 1 = 0 \quad \text{or} \quad \cos 2t = 0$$

$$\sin 4t = \frac{1}{2}$$

Each of the preceding basic equations can be solved using our knowledge of special angles. The solutions of $\sin 4t = 1/2$ are

$$4t = \frac{\pi}{6} + 2k\pi \quad \text{and} \quad 4t = \frac{5\pi}{6} + 2k\pi \quad (k = 0, \pm 1, \pm 2, \ldots)$$

$$t = \frac{\pi}{24} + \frac{k\pi}{2} \quad \text{and} \quad t = \frac{5\pi}{24} + \frac{k\pi}{2} \quad (k = 0, \pm 1, \pm 2, \ldots).$$

Similarly, the solutions of $\cos 2t = 0$ are

$$2t = \frac{\pi}{2} + 2k\pi \quad \text{and} \quad 2t = \frac{3\pi}{2} + 2k\pi \quad (k = 0, \pm 1, \pm 2, \ldots)$$

$$t = \frac{\pi}{4} + k\pi \quad \text{and} \quad t = \frac{3\pi}{4} + k\pi \quad (k = 0, \pm 1, \pm 2, \ldots).$$ ■

EXERCISES 8.4

Directions: *Find exact solutions where possible (as in Examples 7, 8, and 12) and approximate ones otherwise. When the solution algorithms for basic equations are used, round your final answers (not any intermediate results) to four decimal places. When solutions are found graphically, they should be accurate to two decimal places.*

In Exercises 1–12, solve the equation graphically.

1. $4 \sin 2x - 3 \cos 2x = 2$

2. $5 \sin 3x + 6 \cos 3x = 1$

3. $3 \sin^3 2x = 2 \cos x$

4. $\sin^2 2x - 3 \cos 2x + 2 = 0$

5. $\tan x + 5 \sin x = 1$

6. $2 \cos^2 x + \sin x + 1 = 0$

7. $\cos^3 x - 3 \cos x + 1 = 0$

8. $\tan x = 3 \cos x$

9. $\cos^4 x - 3 \cos^3 x + \cos x = 1$

10. $\sec x + \tan x = 3$

11. $\sin^3 x + 2 \sin^2 x - 3 \cos x + 2 = 0$

12. $\csc^2 x + \sec x = 1$

13. Use the graph of the sine function to show the following:
(a) The solutions of $\sin x = 1$ are

$$x = \frac{\pi}{2}, \frac{5\pi}{2}, \frac{9\pi}{2}, \ldots \quad \text{and}$$
$$x = \frac{-3\pi}{2}, \frac{-7\pi}{2}, \frac{-11\pi}{2}, \ldots .$$

(b) The solutions of $\sin x = -1$ are

$$x = \frac{3\pi}{2}, \frac{7\pi}{2}, \frac{11\pi}{2}, \ldots \quad \text{and}$$
$$x = \frac{-\pi}{2}, \frac{-5\pi}{2}, \frac{-9\pi}{2}, \ldots .$$

14. Use the graph of the cosine function to show the following:
(a) The solutions of $\cos x = 1$ are

$$x = 0, \pm 2\pi, \pm 4\pi, \pm 6\pi, \ldots .$$

(b) The solutions of $\cos x = -1$ are

$$\pm\pi, \pm 3\pi, \pm 5\pi, \ldots .$$

In Exercises 15–22, use your knowledge of special angles to find the exact solutions of the equation.

15. $\sin x = \sqrt{3}/2$ **16.** $2 \cos x = \sqrt{2}$

17. $\tan x = -\sqrt{3}$ **18.** $\tan x = 1$

19. $2 \cos x = -\sqrt{3}$ **20.** $\sin x = 0$

21. $2 \sin x + 1 = 0$ **22.** $\csc x = \sqrt{2}$

In Exercises 23–26, approximate all solutions in $[0, 2\pi)$ of the given equation.

23. $\sin x = .119$ **24.** $\cos x = .958$

25. $\tan x = 5$ **26.** $\tan x = 17.65$

In Exercises 27–36, find all angles θ with $0° \leq \theta < 360°$ that are solutions of the given equation. [Hint: *Put your calculator in degree mode and replace π by 180° in the solution algorithms for basic equations.*]

27. $\tan \theta = 7.95$ **28.** $\tan \theta = 69.4$

29. $\cos \theta = -.42$ **30.** $\cot \theta = -2.4$

31. $2 \sin^2\theta + 3 \sin \theta + 1 = 0$

32. $4 \cos^2\theta + 4 \cos \theta - 3 = 0$

33. $\tan^2\theta - 3 = 0$ **34.** $2 \sin^2\theta = 1$

35. $4 \cos^2\theta + 4 \cos \theta + 1 = 0$

36. $\sin^2\theta - 3 \sin \theta = 10$

At the instant you hear a sonic boom from an airplane overhead, your angle of elevation α to the plane is given by the equation $\sin \alpha = 1/m$, where m is the Mach number for the speed of the plane (Mach 1 is the speed of sound, Mach 2.5 is 2.5 times the speed of sound, etc.). In Exercises 37–40, find the angle of elevation (in degrees) for the given Mach number. Remember that an angle of elevation must be between 0° and 90°.

37. $m = 1.1$ **38.** $m = 1.6$ **39.** $m = 2$

40. $m = 2.4$

When a light beam passes from one medium to another (for instance, from air to water), it changes both its speed and direction. According to Snell's Law of Refraction,

$$\frac{\sin \theta_1}{\sin \theta_2} = \frac{v_1}{v_2},$$

where v_1 is the speed of light in the first medium, v_2 its speed in the second medium, θ_1 the angle of incidence, and θ_2 the angle of refraction, as shown in Figure 8–14. The number v_1/v_2 is called the index of refraction. *Use this information to do Exercises 41–44.*

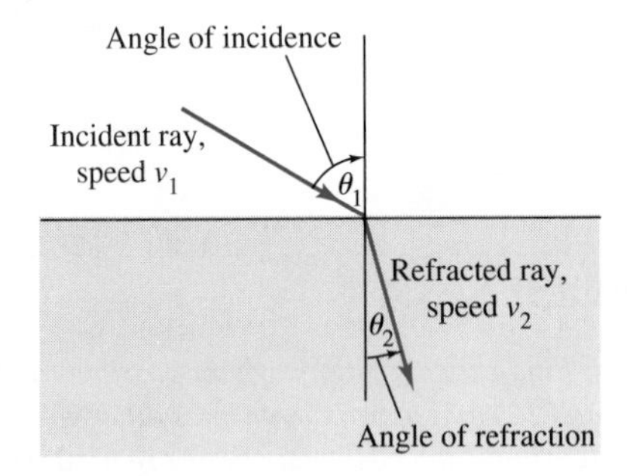

Figure 8-14

41. The index of refraction of light passing from air to water is 1.33. If the angle of incidence is 38°, find the angle of refraction.

42. The index of refraction of light passing from air to ordinary glass is 1.52. If the angle of incidence is 17°, find the angle of refraction.

43. The index of refraction of light passing from air to dense glass is 1.66. If the angle of incidence is 24°, find the angle of refraction.

44. The index of refraction of light passing from air to quartz is 1.46. If the angle of incidence is 50°, find the angle of refraction.

In Exercises 45–64, find all the solutions of the equation.

45. $\sin x = -.465$ **46.** $\sin x = -.682$

47. $\cos x = -.564$ **48.** $\cos x = -.371$

49. $\tan x = -.237$ **50.** $\tan x = -12.45$

51. $\cot x = 2.3$ [*Remember:* $\cot x = 1/\tan x$.]

52. $\cot x = -3.5$ **53.** $\sec x = -2.65$

54. $\csc x = 5.27$ **55.** $\sin 2x = -\sqrt{3}/2$

56. $\cos 2x = \sqrt{2}/2$ **57.** $2 \cos \frac{x}{2} = \sqrt{2}$

58. $2 \sin \frac{x}{3} = 1$ **59.** $\tan 3x = -\sqrt{3}$

60. $5 \sin 2x = 2$ **61.** $5 \cos 3x = -3$

62. $2 \tan 4x = 16$ **63.** $4 \tan \frac{x}{2} = 8$

64. $5 \sin \frac{x}{4} = 4$

In Exercises 65–90, use factoring, the quadratic formula, or identities to solve the equation. Find all solutions in the interval $[0, 2\pi)$.

65. $3 \sin^2 x - 8 \sin x - 3 = 0$

66. $5 \cos^2 x + 6 \cos x = 8$

67. $2 \tan^2 x + 5 \tan x + 3 = 0$

68. $3 \sin^2 x + 2 \sin x = 5$

69. $\cot x \cos x = \cos x$ (Be careful; see Exercise 99.)

70. $\tan x \cos x = \cos x$

71. $\cos x \csc x = 2 \cos x$

72. $\tan x \sec x + 3 \tan x = 0$

73. $4 \sin x \tan x - 3 \tan x + 20 \sin x - 15 = 0$
[*Hint*: One factor is $\tan x + 5$.]

74. $25 \sin x \cos x - 5 \sin x + 20 \cos x = 4$

75. $\sin^2 x + 2 \sin x - 2 = 0$

76. $\cos^2 x + 5 \cos x = 1$

77. $\tan^2 x + 1 = 3 \tan x$

78. $4 \cos^2 x - 2 \cos x = 1$

79. $2 \tan^2 x - 1 = 3 \tan x$

80. $6 \sin^2 x + 4 \sin x = 1$

81. $\sin^2 x + 3 \cos^2 x = 0$

82. $\sec^2 x - 2 \tan^2 x = 0$

83. $\sin 2x + \cos x = 0$

84. $\cos 2x - \sin x = 1$

85. $9 - 12 \sin x = 4 \cos^2 x$

86. $\sec^2 x + \tan x = 3$

87. $\cos^2 x - \sin^2 x + \sin x = 0$

88. $2 \tan^2 x + \tan x = 5 - \sec^2 x$

89. $\sin \frac{x}{2} = 1 - \cos x$

90. $4 \sin^2\left(\frac{x}{2}\right) + \cos^2 x = 2$

91. The number of hours of daylight in Detroit on day t of a non–leap year (with $t = 0$ being January 1) is given by the function $d(t) = 3 \sin\left(\frac{2\pi}{365}(t - 80)\right) + 12$.

(a) On what days of the year is there exactly 11 hours of daylight?

(b) What day has the maximum amount of daylight?

92. A weight hanging from a spring is set into motion (see Figure 6–67 on page 428), moving up and down. Its distance (in cm) above or below the equilibrium point at time t seconds is given by $d = 5(\sin 6t - 4 \cos 6t)$. At what times during the first two seconds is the weight at the equilibrium position ($d = 0$)?

In Exercises 93–96, use the following fact: When a projectile (such as a ball or a bullet) leaves its starting point at angle of elevation θ with velocity v, the horizontal distance d it travels is given by the equation $d = \frac{v^2}{32} \sin 2\theta$, where d is measured

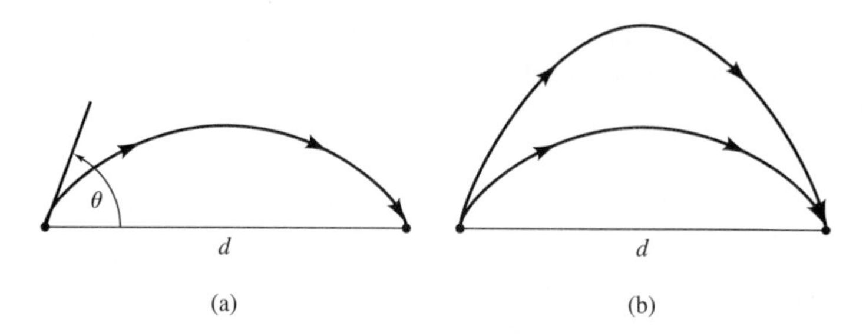

Figure 8-15

in feet and v in feet per second (Figure 8–15(a)). Note that the horizontal distance traveled may be the same for two different angles of elevation (Figure 8–15(b)), so that some of these exercises may have more than one correct answer.

93. If muzzle velocity of a rifle is 300 ft/sec, at what angle of elevation (in radians) should it be aimed in order for the bullet to hit a target 2500 ft away?

94. Is it possible for the rifle in Exercise 93 to hit a target that is 3000 ft away? [At what angle of elevation would it have to be aimed?]

95. A fly ball leaves the bat at a velocity of 98 miles per hour and is caught by an outfielder 288 feet away. At what angle of elevation (in degrees) did the ball leave the bat?

96. An outfielder throws the ball at a speed of 75 miles per hour to the catcher who is 200 ft away. At what angle of elevation was the ball thrown?

Thinkers

97. Under what conditions (on the constant) does a basic equation involving the sine or cosine function have *no* solutions?

98. Do Exercise 97 for the secant and cosecant functions.

99. What is wrong with this so-called solution?

$$\sin x \tan x = \sin x$$
$$\tan x = 1$$
$$x = \frac{\pi}{4} \quad \text{or} \quad \frac{5\pi}{4}.$$

[*Hint:* Solve the original equation by moving all terms to one side and factoring. Compare your answers with the ones above.]

100. Let n be a fixed positive integer. Describe *all* solutions of the equation $\sin nx = 1/2$. [*Hint:* See Exercises 55–64.]

8.5 INVERSE TRIGONOMETRIC FUNCTIONS

You should review the concept of inverse functions (Section 2.7) before reading this section. As explained there, a function cannot have an inverse function unless its graph has this property: No horizontal line intersects the graph more than once. The graphs of the sine, cosine, and tangent functions certainly don't have this property, as you can readily verify with your calculator. However, functions that are very closely related to these functions (same rules, but smaller domains) *do* have inverse functions.

We begin with the *restricted* sine function whose rule is $f(x) = \sin x$, but whose domain is restricted to the interval $[-\pi/2, \pi/2]$.* Its graph in Figure 8–16

*Other ways of restricting the domain are possible. Those presented here for sine, cosine, and tangent are the ones universally agreed on by mathematicians.

shows that for each number v between -1 and 1, there is exactly one number u between $-\pi/2$ and $\pi/2$ such that $\sin u = v$.

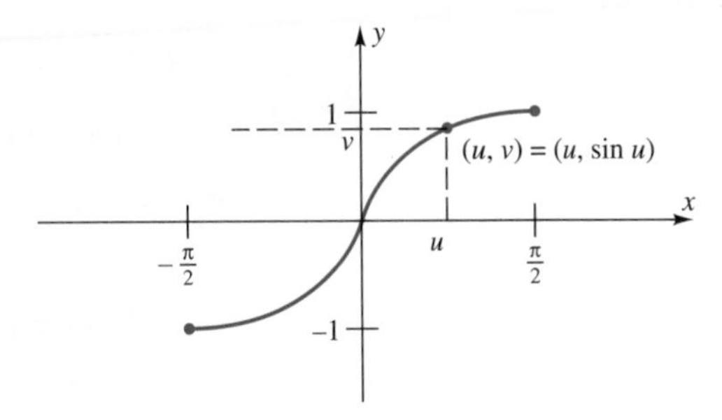

Figure 8-16

Since the graph of the restricted sine function passes the horizontal line test, we know that it has an inverse function. This inverse function is called the **inverse sine** (or **arcsine**) **function** and is denoted by $g(x) = \sin^{-1} x$ or $g(x) = \arcsin x$. The domain of the inverse sine function is the interval $[-1, 1]$ and its rule is

Inverse Sine Function ▶

For each v with $-1 \le v \le 1$,

$$\sin^{-1} v = \text{the unique number } u \text{ between } -\pi/2 \text{ and } \pi/2 \text{ whose sine is } v;$$

that is,

$$\sin^{-1} v = u \qquad \text{exactly when} \qquad \sin u = v.$$

EXAMPLE 1

(a) $\text{Sin}^{-1}(1/2)$ is the one number between $-\pi/2$ and $\pi/2$ whose sine is 1/2. From our study of special values, we know that $\sin \pi/6 = 1/2$, *and* $\pi/6$ is between $-\pi/2$ and $\pi/2$. Hence, $\sin^{-1}(1/2) = \pi/6$.

(b) $\text{Sin}^{-1}(-\sqrt{2}/2) = -\pi/4$ because $\sin(-\pi/4) = -\sqrt{2}/2$ and $-\pi/4$ is between $-\pi/2$ and $\pi/2$. ■

EXAMPLE 2 Except for special values you should use the $\boxed{\text{SIN}^{-1}}$ key (labeled ASIN on some calculators) in *radian mode* to evaluate the inverse sine function. For instance,

$$\sin^{-1}(-.67) = -.7342 \qquad \text{and} \qquad \sin^{-1}(.42) = .4334. \quad ■$$

EXAMPLE 3 If you key in SIN^{-1} 2 ENTER you will get an error message, because 2 is not in the domain of the inverse sine function.* ■

WARNING The notation $\sin^{-1}x$ is *not* exponential notation. It does *not* mean either $(\sin x)^{-1}$ or $\frac{1}{\sin x}$. For instance, Example 1 shows that $\sin^{-1}(1/2) = \pi/6 \approx .5236$, but

$$\left(\sin\frac{1}{2}\right)^{-1} = \frac{1}{\sin\frac{1}{2}} \approx \frac{1}{.4794} \approx 2.0858.$$

Suppose $-1 \leq v \leq 1$ and $\sin^{-1}v = u$. Then by the definition of the inverse sine function we know that $-\pi/2 \leq u \leq \pi/2$ and $\sin u = v$. Therefore,

$$\sin^{-1}(\sin u) = \sin^{-1}(v) = u \qquad \text{and} \qquad \sin(\sin^{-1}v) = \sin u = v.$$

This shows that the restricted sine function and the inverse sine function have the usual ''reversal properties'' of inverse functions. In summary,

Properties of Inverse Sine ▶

$$\sin^{-1}(\sin u) = u \quad \text{if} \quad -\frac{\pi}{2} \leq u \leq \frac{\pi}{2}$$

$$\sin(\sin^{-1}v) = v \quad \text{if} \quad -1 \leq v \leq 1$$

EXAMPLE 4 Since $\sin \pi/6 = 1/2$, we see that

$$\sin^{-1}(\sin \pi/6) = \sin^{-1}\left(\frac{1}{2}\right) = \pi/6$$

because $\pi/6$ is between $-\pi/2$ and $\pi/2$. On the other hand, $\sin 5\pi/6$ is also $1/2$, so an expression such as $\sin^{-1}(\sin 5\pi/6)$ *is* well defined. But

$$\sin^{-1}\left(\sin\frac{5\pi}{6}\right) = \sin^{-1}\left(\frac{1}{2}\right) = \frac{\pi}{6}, \qquad \textit{not} \quad \frac{5\pi}{6}.$$

The identity in the box is valid only when u is between $-\pi/2$ and $\pi/2$. ■

EXAMPLE 5 A calculator can visually demonstrate these identities. For instance, for any number v between -1 and 1 key in $\sin(\sin^{-1}v)$. The display will show the number you started with (except for minor round-off errors). ■

*TI-85 and HP-38 display the complex number $(1.5707\cdots, -1.3169\cdots)$ for $\sin^{-1}(2)$. For our purposes this is equivalent to an error message since we only deal with functions whose values are real numbers.

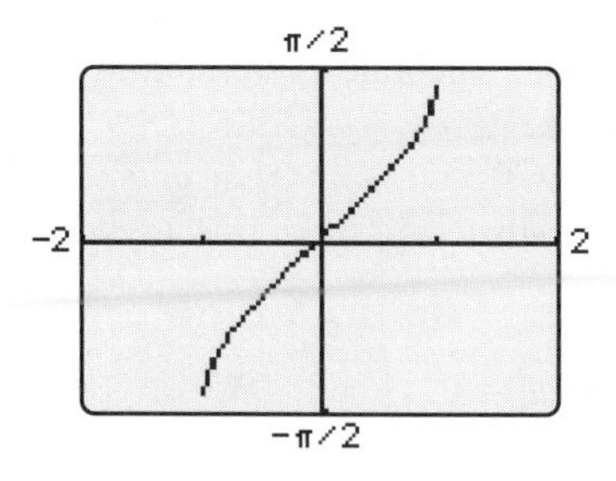

Figure 8-17

The graph of the inverse sine function is readily obtained from a calculator (Figure 8–17). Since the inverse sine function is the inverse of the restricted sine function, its graph is the reflection of the graph of the restricted sine function (Figure 8–16) in the line $y = x$ (as explained on page 154).

The Inverse Cosine Function

The rule of the *restricted* cosine function is $f(x) = \cos x$, and its domain is the interval $[0, \pi]$. Its graph in Figure 8–18 shows that for each v between -1 and 1 there is exactly one number u between 0 and π such that $\cos u = v$.

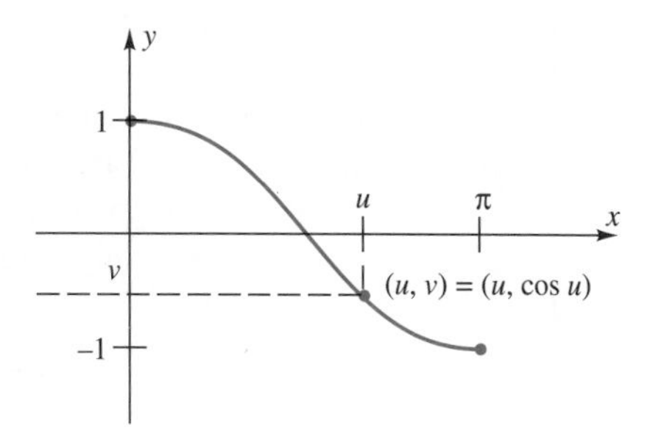

Figure 8-18

Since the graph of the restricted cosine function passes the horizontal line test, we know that it has an inverse function. This inverse function is called the **inverse cosine** (or **arccosine**) **function** and is denoted by $\boldsymbol{h(x) = \cos^{-1}x}$ or $\boldsymbol{h(x) = \arccos x}$. The domain of the inverse cosine function is the interval $[-1, 1]$ and its rule is

Inverse Cosine Function ▶

For each v with $-1 \le v \le 1$,

$$\cos^{-1}v = \text{the unique number } u \text{ between 0 and } \pi \text{ whose cosine is } v;$$

that is,

$$\cos^{-1}v = u \quad \text{exactly when} \quad \cos u = v.$$

The inverse cosine function has these properties:

$$\cos^{-1}(\cos u) = u \quad \text{if} \quad 0 \le u \le \pi;$$
$$\cos(\cos^{-1}v) = v \quad \text{if} \quad -1 \le v \le 1.$$

EXAMPLE 6

(a) $\text{Cos}^{-1}(1/2) = \pi/3$ since $\pi/3$ is the unique number between 0 and π whose cosine is 1/2.

(b) $\text{Cos}^{-1}(0) = \pi/2$ because $\cos \pi/2 = 0$ and $0 \le \pi/2 \le \pi$.

(c) The $\boxed{\text{COS}^{-1}}$ key on a calculator in *radian mode* shows that $\cos^{-1}(-.63) = 2.2523$. ■

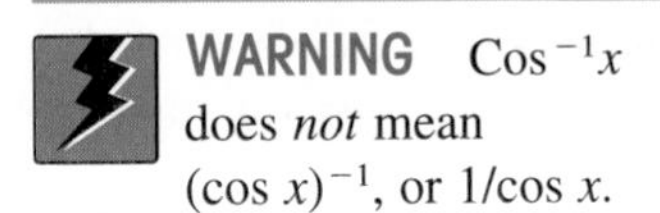

WARNING $\text{Cos}^{-1}x$ does *not* mean $(\cos x)^{-1}$, or $1/\cos x$.

EXAMPLE 7 Write $\sin(\cos^{-1}v)$ as an algebraic expression in v.

Solution $\text{Cos}^{-1}v = u$, where $\cos u = v$ and $0 \le u \le \pi$. Hence, $\sin u$ is nonnegative, and by the Pythagorean identity, $\sin u = \sqrt{\sin^2 u} = \sqrt{1 - \cos^2 u}$. Also, $\cos^2 u = v^2$. Therefore,

$$\sin(\cos^{-1}v) = \sin u = \sqrt{1 - \cos^2 u} = \sqrt{1 - v^2}. \quad ■$$

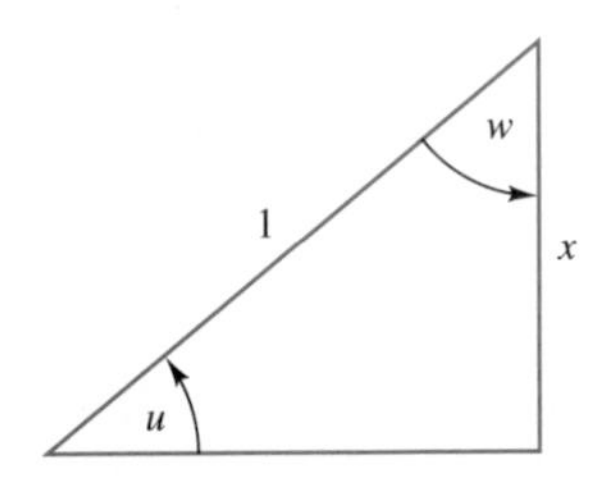

Figure 8-19

EXAMPLE 8 Prove the identity $\sin^{-1}x + \cos^{-1}x = \pi/2$.

Solution The identity is defined only when $-1 \le x \le 1$ and is true for $x = 0$ since $\sin^{-1}0 = 0$ and $\cos^{-1}0 = \pi/2$. Assume that $0 < x \le 1$ and consider a right triangle with hypotenuse 1, side x, and acute angles that measure u radians and w radians, respectively (Figure 8–19). We have

$$\sin u = \frac{\text{opposite}}{\text{hypotenuse}} = \frac{x}{1} = x \qquad \text{and} \qquad \cos w = \frac{\text{adjacent}}{\text{hypotenuse}} = \frac{x}{1} = x.$$

Therefore,

$$\sin^{-1}x = u \qquad \text{and} \qquad \cos^{-1}x = w.$$

But $u + w = \pi/2$; hence,

$$\sin^{-1}x + \cos^{-1}x = u + w = \pi/2.$$

The proof when $-1 \le x < 0$ is given in Exercise 62. ■

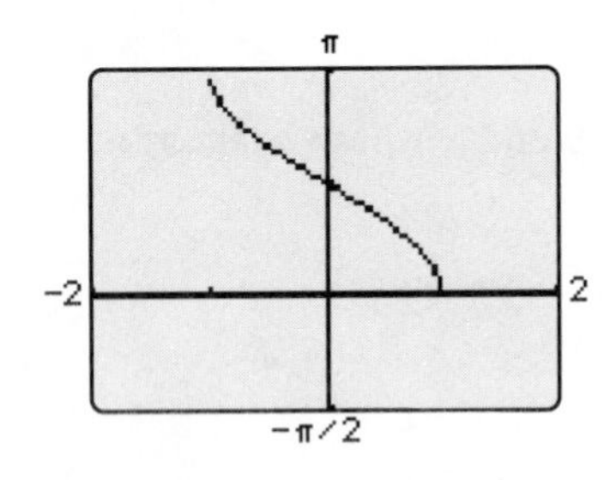

Figure 8-20

The graph of the inverse cosine function, which is the reflection of the graph of the restricted cosine function (Figure 8–18) in the line $y = x$, is shown in Figure 8–20.

The Inverse Tangent Function

The rule of the *restricted* tangent function is $f(x) = \tan x$, and its domain is the open interval $(-\pi/2, \pi/2)$. Its graph in Figure 8–21 shows that for every real number v, there is exactly one number u between $-\pi/2$ and $\pi/2$ such that $\tan u = v$.

Since the graph of the restricted tangent function passes the horizontal line test, we know that it has an inverse function. This inverse function is called the **inverse tangent** (or **arctangent**) **function** and is denoted by $\boldsymbol{g(x) = \tan^{-1}x}$ or $\boldsymbol{g(x) = \arctan x}$. The domain of the inverse tangent function is the set of all real numbers and its rule is given in the box on the next page.

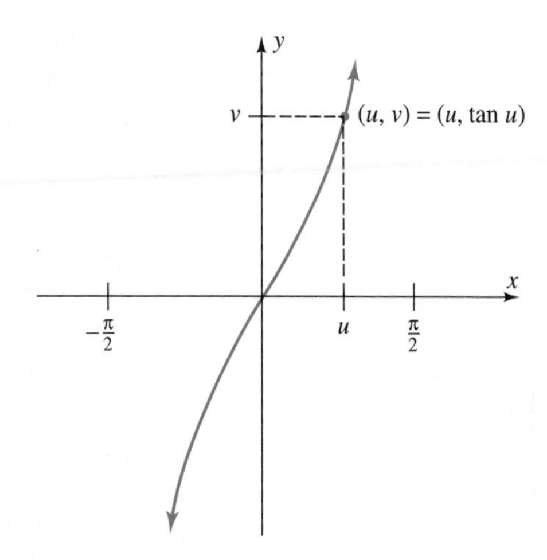

Figure 8-21

Inverse Tangent Function ►

For each real number v,

$\tan^{-1} v$ = the unique number u between $-\pi/2$ and $\pi/2$ whose tangent is v;

that is,

$$\tan^{-1} v = u \qquad \text{exactly when} \qquad \tan u = v.$$

The inverse tangent function has these properties:

$$\tan^{-1}(\tan u) = u \qquad \text{if} \qquad -\frac{\pi}{2} \leq u \leq \frac{\pi}{2};$$
$$\tan(\tan^{-1} v) = v \qquad \text{for every number } v.$$

WARNING $\text{Tan}^{-1} x$ does *not* mean $(\tan x)^{-1}$, or $1/\tan x$.

EXAMPLE 9 $\text{Tan}^{-1} 1 = \pi/4$ because $\pi/4$ is the unique number between $-\pi/2$ and $\pi/2$ such that $\tan \pi/4 = 1$. A calculator in *radian mode* shows that $\tan^{-1}(136) = 1.5634$. ■

EXAMPLE 10 Find the exact value of $\cos(\tan^{-1}(\sqrt{5}/2))$.

Solution Let $\tan^{-1}(\sqrt{5}/2) = u$; then $\tan u = \sqrt{5}/2$ and $-\pi/2 < u < \pi/2$. Since $\tan u$ is positive, u must be between 0 and $\pi/2$. Construct a right triangle containing an angle of u radians whose tangent is $\sqrt{5}/2$ (Figure 8–22):

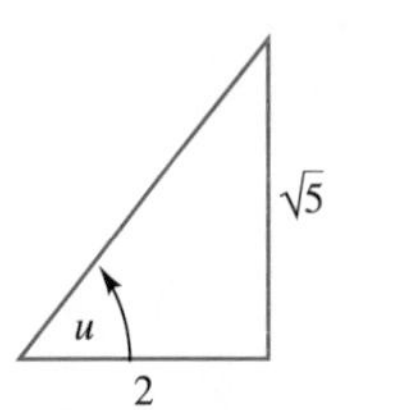

$$\tan u = \frac{\text{opposite}}{\text{adjacent}} = \frac{\sqrt{5}}{2}.$$

Figure 8-22

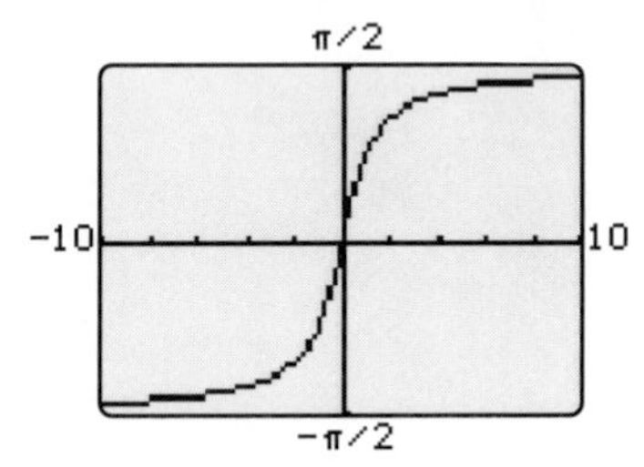

Figure 8-23

The hypotenuse has length $\sqrt{2^2 + (\sqrt{5})^2} = \sqrt{4+5} = 3$. Therefore,

$$\cos\left(\tan^{-1}\left(\frac{\sqrt{5}}{2}\right)\right) = \cos u = \frac{\text{adjacent}}{\text{hypotenuse}} = \frac{2}{3}. \quad \blacksquare$$

The graph of the inverse tangent function (Figure 8–23) is the reflection in the line $y = x$ of the graph of the restricted tangent function.

EXERCISES 8.5

In Exercises 1–14, find the exact functional value without using a calculator.

1. $\sin^{-1} 1$ **2.** $\cos^{-1} 0$ **3.** $\tan^{-1}(-1)$

4. $\sin^{-1}(-1)$ **5.** $\cos^{-1} 1$ **6.** $\tan^{-1} 1$

7. $\tan^{-1}(\sqrt{3}/3)$ **8.** $\cos^{-1}(\sqrt{3}/2)$

9. $\sin^{-1}(-\sqrt{2}/2)$ **10.** $\sin^{-1}(\sqrt{3}/2)$

11. $\tan^{-1}(-\sqrt{3})$ **12.** $\cos^{-1}(\sqrt{2}/2)$

13. $\cos^{-1}\left(-\frac{1}{2}\right)$ **14.** $\sin^{-1}\left(-\frac{1}{2}\right)$

In Exercises 15–24, use a calculator in radian mode to approximate the functional value.

15. $\sin^{-1} .35$ **16.** $\cos^{-1} .76$

17. $\tan^{-1}(-3.256)$ **18.** $\sin^{-1}(-.795)$

19. $\sin^{-1}(\sin 7)$ [The answer is *not* 7.]

20. $\cos^{-1}(\cos 3.5)$ **21.** $\tan^{-1}(\tan(-4))$

22. $\sin^{-1}(\sin(-2))$ **23.** $\cos^{-1}(\cos(-8.5))$

24. $\tan^{-1}(\tan 12.4)$

25. Given that $u = \sin^{-1}(-\sqrt{3}/2)$, find the exact value of $\cos u$ and $\tan u$.

26. Given that $u = \tan^{-1}(4/3)$, find the exact value of $\sin u$ and $\sec u$.

In Exercises 27–42, find the exact functional value without using a calculator.

27. $\sin^{-1}(\cos 0)$ **28.** $\cos^{-1}(\sin \pi/6)$

29. $\cos^{-1}(\sin 4\pi/3)$ **30.** $\tan^{-1}(\cos \pi)$

31. $\sin^{-1}(\cos 7\pi/6)$ **32.** $\cos^{-1}(\tan 7\pi/4)$

33. $\sin^{-1}(\sin 2\pi/3)$ (See Exercise 19.)

34. $\cos^{-1}(\cos 5\pi/4)$ **35.** $\cos^{-1}(\cos(-\pi/6))$

36. $\tan^{-1}(\tan(-4\pi/3))$

37. $\sin(\cos^{-1}(3/5))$ (See Example 10.)

38. $\tan(\sin^{-1}(3/5))$ **39.** $\cos(\tan^{-1}(-3/4))$

40. $\cos(\sin^{-1}(\sqrt{3}/5))$ **41.** $\tan(\sin^{-1}(5/13))$

42. $\sin(\cos^{-1}(3/\sqrt{13}))$

In Exercises 43–46, write the expression as an algebraic expression in v, as in Example 7.

43. $\cos(\sin^{-1}v)$ **44.** $\cot(\cos^{-1}v)$

45. $\tan(\sin^{-1}v)$ **46.** $\sin(2\sin^{-1}v)$

In Exercises 47–50, graph the function.

47. $f(x) = \cos^{-1}(x + 1)$ **48.** $g(x) = \tan^{-1}x + \pi$

49. $h(x) = \sin^{-1}(\sin x)$ **50.** $k(x) = \sin(\sin^{-1}x)$

51. In an alternating current circuit, the voltage is given by the formula

$$V = V_{max} \cdot \sin(2\pi ft + \phi),$$

where V_{max} is the maximum voltage, f is the frequency (in cycles per second), t is the time in seconds, and ϕ is the phase angle.

(a) If the phase angle is 0, solve the voltage equation for t.

(b) If $\phi = 0$, $V_{max} = 20$, $V = 8.5$, and $f = 120$, find the smallest positive value of t.

52. A model plane 40 ft above the ground is flying away from an observer (Figure 8–24).

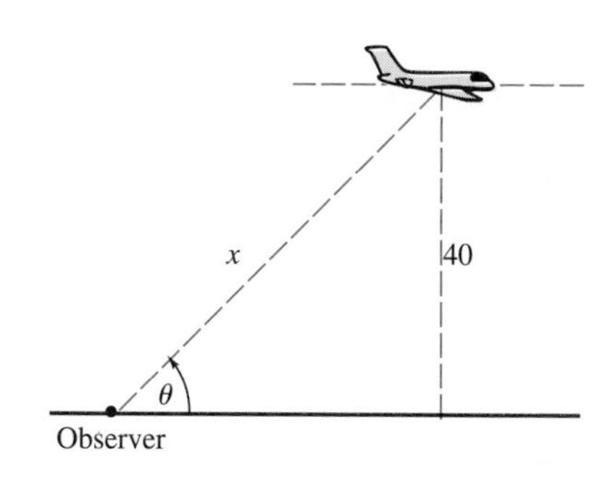

Figure 8-24

(a) Express the angle of elevation θ of the plane as a function of the distance x from the observer to the plane.

(b) What is θ when the plane is 250 ft from the observer?

53. A 15-ft-wide highway sign is placed 10 ft from a road, perpendicular to the road (Figure 8–25). A spotlight at the edge of the road is aimed at the sign.

(a) Express θ as a function of the distance x from point A to the spotlight.

(b) How far from point A should the spotlight be placed so that the angle θ is as large as possible?

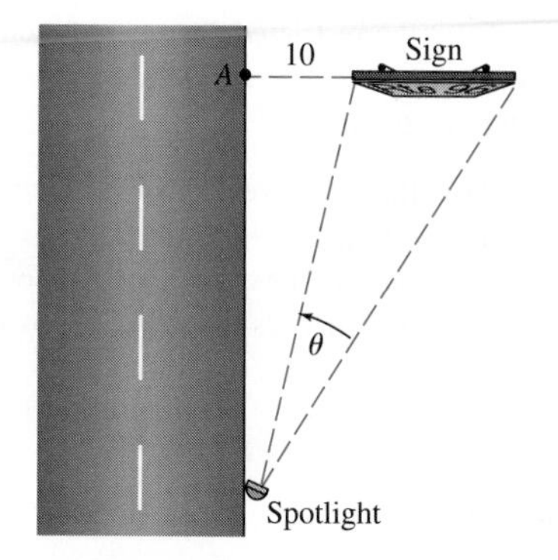

Figure 8-25

54. A camera on a 5-ft-high tripod is placed in front of a 6-ft-high picture that is mounted 3 ft above the floor (Figure 8–26).

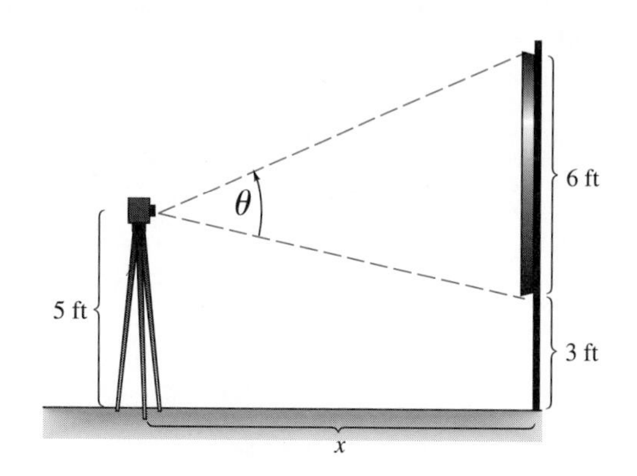

Figure 8-26

(a) Express angle θ as a function of the distance x from the camera to the wall.

(b) The photographer wants to use a particular lens, for which $\theta = 36°$ ($\pi/5$ radians). How far should she place the camera from the wall to be sure the entire picture will show in the photograph?

55. Show that the restricted secant function, whose domain consists of all numbers x such that $0 \leq x \leq \pi$ and $x \neq \pi/2$, has an inverse function. Sketch its graph.

56. Show that the restricted cosecant function, whose domain consists of all numbers x such that $-\pi/2 \le x \le \pi/2$ and $x \ne 0$, has an inverse function. Sketch its graph.

57. Show that the restricted cotangent function, whose domain is the interval $(0, \pi)$, has an inverse function. Sketch its graph.

58. **(a)** Show that the inverse cosine function actually has the two properties listed in the box on page 531.
(b) Show that the inverse tangent function actually has the two properties listed in the box on page 533.

In Exercises 59–67, prove the identity.

59. $\sin^{-1}(-x) = -\sin^{-1}x$ [*Hint:* Let $u = \sin^{-1}(-x)$ and show that $\sin^{-1}x = -u$.]

60. $\tan^{-1}(-x) = -\tan^{-1}x$

61. $\cos^{-1}(-x) = \pi - \cos^{-1}x$ [*Hint:* Let $u = \cos^{-1}(-x)$ and show that $0 \le \pi - u \le \pi$; use the identity $\cos(\pi - u) = -\cos u$.]

62. $\sin^{-1}x + \cos^{-1}x = \pi/2$ [*Hint:* The case when $0 \le x \le 1$ was demonstrated in Example 8. If $-1 < x < 0$, then $0 < -x \le 1$; use Exercises 59 and 61.]

63. $\tan^{-1}(\cot x) = \pi/2 - x \qquad (0 < x < \pi)$

64. $\sin^{-1}(\cos x) = \pi/2 - x \qquad (0 \le x \le \pi)$

65. $\sin^{-1}x = \tan^{-1}\left(\dfrac{x}{\sqrt{1 - x^2}}\right) \qquad (-1 < x < 1)$

[*Hint:* Let $u = \sin^{-1}x$ and show that $\tan u = x/\sqrt{1 - x^2}$. Since $\sin u = x$, $\cos u = \pm\sqrt{1 - x^2}$. Show that in this case, $\cos u = \sqrt{1 - x^2}$.]

66. $\cos^{-1}x = \dfrac{\pi}{2} - \tan^{-1}\left(\dfrac{x}{\sqrt{1 - x^2}}\right) \qquad (-1 < x < 1)$

[*Hint:* See Exercises 62 and 65.]

67. $\tan^{-1}x + \tan^{-1}\left(\dfrac{1}{x}\right) = \dfrac{\pi}{2}$

68. Using the viewing window with $-2\pi \le x \le 2\pi$ and $-4 \le y \le 4$ graph the functions $f(x) = \cos(\cos^{-1}x)$ and $g(x) = \cos^{-1}(\cos x)$. How do you explain the shapes of the two graphs?

69. Is it true that $\tan^{-1}x = \dfrac{\sin^{-1}x}{\cos^{-1}x}$? Justify your answer.

CHAPTER 8 *Review*

Important Concepts ▶

Important Facts and Formulas ▶

- All identities in the Chapter 6 Review
- *Addition and Subtraction Identities:*

$$\sin(x + y) = \sin x \cos y + \cos x \sin y$$

$$\sin(x - y) = \sin x \cos y - \cos x \sin y$$

$$\cos(x + y) = \cos x \cos y - \sin x \sin y$$

$$\cos(x - y) = \cos x \cos y + \sin x \sin y$$

$$\tan(x + y) = \frac{\tan x + \tan y}{1 - \tan x \tan y}$$

$$\tan(x - y) = \frac{\tan x - \tan y}{1 + \tan x \tan y}$$

- *Cofunction Identities:*

$$\sin x = \cos\left(\frac{\pi}{2} - x\right) \qquad \cos x = \sin\left(\frac{\pi}{2} - x\right)$$

$$\tan x = \cot\left(\frac{\pi}{2} - x\right) \qquad \cot x = \tan\left(\frac{\pi}{2} - x\right)$$

$$\sec x = \csc\left(\frac{\pi}{2} - x\right) \qquad \csc x = \sec\left(\frac{\pi}{2} - x\right)$$

- *Double-Angle Identities:*

$$\sin 2x = 2 \sin x \cos x$$

$$\cos 2x = \cos^2 x - \sin^2 x$$

$$\cos 2x = 2\cos^2 x - 1 \qquad \cos 2x = 1 - 2\sin^2 x$$

$$\tan 2x = \frac{2 \tan x}{1 - \tan^2 x}$$

- *Half-Angle Identities:*

$$\sin \frac{x}{2} = \pm\sqrt{\frac{1 - \cos x}{2}}$$

$$\cos \frac{x}{2} = \pm\sqrt{\frac{1 + \cos x}{2}}$$

$$\tan \frac{x}{2} = \frac{1 - \cos x}{\sin x} \qquad \tan \frac{x}{2} = \frac{\sin x}{1 + \cos x}$$

- $\sin^{-1} v = u$ exactly when $\sin u = v$ $\left(-\frac{\pi}{2} \leq u \leq \frac{\pi}{2}, -1 \leq v \leq 1\right)$
- $\cos^{-1} v = u$ exactly when $\cos u = v$ $(0 \leq u \leq \pi, -1 \leq v \leq 1)$
- $\tan^{-1} v = u$ exactly when $\tan u = v$ $\left(-\frac{\pi}{2} < u < \frac{\pi}{2}, \text{any } v\right)$

Review Questions ▶

In Questions 1–4, simplify the given expression.

1. $\dfrac{\sin^2 t + (\tan^2 t + 2\tan t - 4) + \cos^2 t}{3\tan^2 t - 3\tan t}$

2. $\dfrac{\sec^2 t \csc t}{\csc^2 t \sec t}$

3. $\dfrac{\tan^2 x - \sin^2 x}{\sec^2 x}$

4. $\dfrac{(\sin x + \cos x)(\sin x - \cos x) + 1}{\sin^2 x}$

In Questions 5–12, determine graphically whether the equation could possibly be an identity. If it could, prove that it is.

5. $\sin^4 t - \cos^4 t = 2\sin^2 t - 1$

6. $1 + 2\cos^2 t + \cos^4 t = \sin^4 t$

7. $\dfrac{\sin t}{1 - \cos t} = \dfrac{1 + \cos t}{\sin t}$

8. $\dfrac{\sin^2 t}{\cos^2 t} + 1 = \dfrac{1}{\cos^2 t}$

9. $\dfrac{\cos^2(\pi + t)}{\sin^2(\pi + t)} - 1 = \dfrac{1}{\sin^2 t}$

10. $\tan x + \cot x = \sec x \csc x$

11. $(\sin x + \cos x)^2 - \sin 2x = 1$

12. $\dfrac{1 - \cos 2x}{\tan x} = \sin 2x$

In Questions 13–22, prove the given identity.

13. $\dfrac{\tan x - \sin x}{2\tan x} = \sin^2\left(\dfrac{x}{2}\right)$

14. $2\cos x - 2\cos^3 x = \sin x \sin 2x$

15. $\cos(x + y)\cos(x - y) = \cos^2 x - \sin^2 y$

16. $\dfrac{\cos(x - y)}{\cos x \cos y} = 1 + \tan x \tan y$

17. $\dfrac{\sec x + 1}{\tan x} = \dfrac{\tan x}{\sec x - 1}$

18. $\dfrac{\cos^4 x - \sin^4 x}{1 - \tan^4 x} = \cos^4 x$

19. $\dfrac{1 + \tan^2 x}{\tan^2 x} = \csc^2 x$

20. $\sec x - \cos x = \sin x \tan x$

21. $\tan^2 x - \sec^2 x = \cot^2 x - \csc^2 x$

22. $\sin 2x = \dfrac{1}{\tan x + \cot 2x}$

23. If $\tan x = 5/12$ and $\sin x > 0$, find $\sin 2x$.

24. If $\cos x = 15/17$ and $0 < x < \pi/2$, find $\sin(x/2)$.

25. If $\tan x = 4/3$ and $\pi < x < 3\pi/2$, and $\cot y = -5/12$ with $3\pi/2 < y < 2\pi$, find $\sin(x - y)$.

26. If $\sin x = -12/13$ with $\pi < x < 3\pi/2$, and $\sec y = 13/12$ with $3\pi/2 < y < 2\pi$, find $\cos(x + y)$.

27. If $\sin x = 1/4$ and $0 < x < \pi/2$, then $\sin(\pi/3 + x) = ?$

28. If $\sin x = -2/5$ and $3\pi/2 < x < 2\pi$, then $\cos(\pi/4 + x) = ?$

29. If $\sin x = 0$, is it true that $\sin 2x = 0$? Justify your answer.

30. If $\cos x = 0$, is it true that $\cos 2x = 0$? Justify your answer.

31. Show that $\sqrt{2 + \sqrt{3}} = \dfrac{\sqrt{2} + \sqrt{6}}{2}$ by computing $\cos(\pi/12)$ in two ways, using the half-angle identity and the subtraction identity for cosine.

32. True or false: $2\sin x = \sin 2x$. Justify your answer.

33. $\sin(5\pi/12) = ?$

34. Express $\sec(x - \pi)$ in terms of $\sin x$ and $\cos x$.

35. $\sqrt{\dfrac{1 - \cos^2 x}{1 - \sin^2 x}} =$ ______ .

(a) $|\tan x|$ **(b)** $|\cot x|$

(c) $\sqrt{\dfrac{1 - \sin^2 x}{1 - \cos^2 x}}$ **(d)** $\sec x$

(e) undefined

36. $\dfrac{1}{(\csc x)(\sec^2 x)} =$ ______.

(a) $\dfrac{1}{(\sin x)(\cos^2 x)}$ **(b)** $\sin x - \sin^3 x$

(c) $\dfrac{1}{(\sin x)(1 + \tan^2 x)}$ **(d)** $\sin x - \dfrac{1}{1 + \tan^2 x}$

(e) $1 + \tan^3 x$

37. If $\sin x = .6$ and $0 < x < \pi/2$, find $\sin 2x$.

38. If $\sin x = .6$ and $0 < x < \pi/2$, find $\sin(x/2)$.

39. Find the angle of inclination of the straight line through the points $(2, 6)$ and $(-2, 2)$.

40. Find one of the angles between the line L through the points $(-3, 2)$ and $(5, 1)$ and the line M, which has slope 2.

In Questions 41–44, solve the equation graphically.

41. $5 \tan x = 2 \sin 2x$

42. $\sin^3 x + \cos^2 x - \tan x = 2$

43. $\sin x + \sec^2 x = 3$

44. $\cos^2 x - \csc^2 x + \tan(x - \pi/2) + 5 = 0$

In Questions 45–60, solve the equation by any means. Find exact solutions when possible and approximate ones otherwise.

45. $2 \sin x = 1$

46. $\cos x = \sqrt{3}/2$

47. $\tan x = -1$

48. $\sin 3x = -\sqrt{3}/2$

49. $\sin x = .7$

50. $\cos x = -.8$

51. $\tan x = 13$

52. $\cot x = .4$

53. $2 \sin^2 x + 5 \sin x = 3$

54. $4 \cos^2 x - 2 = 0$

55. $2 \sin^2 x - 3 \sin x = 2$

56. $\cos 2x = \cos x$
[*Hint:* First use an identity.]

57. $\sec^2 x + 3 \tan^2 x = 13$

58. $\sec^2 x = 4 \tan x - 2$

59. $2 \sin^2 x + \sin x - 2 = 0$

60. $\cos^2 x - 3 \cos x - 2 = 0$

61. Find all angles θ with $0° \leq \theta \leq 360°$ such that $\sin \theta = -.7133$.

62. Find all angles θ with $0° \leq \theta \leq 360°$ such that $\tan \theta = 3.7321$.

63. A cannon has a muzzle velocity of 600 ft per second. At what angle of elevation should it be fired in order to hit a target 3500 ft away? [*Hint:* Use the projectile equation preceding Exercise 93 of Section 8.4.]

64. A weight hanging from a spring is set into motion (see Figure 6–67 on page 428), moving up and down. Its distance (in cm) above or below the equilibrium point at time t seconds is given by $d = 5 \sin 3t - 3 \cos 3t$. At what times during the first two seconds is the weight at the equilibrium position ($d = 0$)?

65. $\cos^{-1}(\sqrt{2}/2) = ?$

66. $\sin^{-1}(\sqrt{3}/2) = ?$

67. $\tan^{-1}\sqrt{3} = ?$

68. $\sin^{-1}(\cos 11\pi/6) = ?$

69. $\cos^{-1}(\sin 5\pi/3) = ?$

70. $\tan^{-1}(\cos 7\pi/2) = ?$

71. $\sin^{-1}(\sin .75) = ?$

72. $\cos^{-1}(\cos 2) = ?$

73. $\sin^{-1}(\sin 8\pi/3) = ?$

74. $\cos^{-1}(\cos 13\pi/4) = ?$

75. Sketch the graph of $f(x) = \tan^{-1}x - \pi$.

76. Sketch the graph of $g(x) = \sin^{-1}(x - 2)$.

77. Find the exact value of $\sin(\cos^{-1}(1/4))$.

78. Find the exact value of $\sin(\tan^{-1}(1/2) - \cos^{-1}(4/5))$.

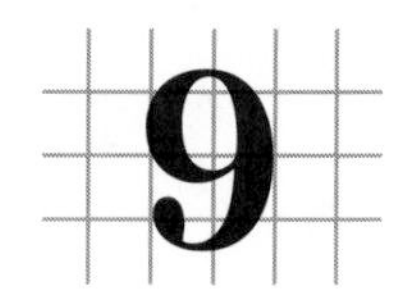

Applications of Trigonometry

Roadmap

The chapter is divided into two independent parts, each consisting of two sections. The parts may be read in either order:

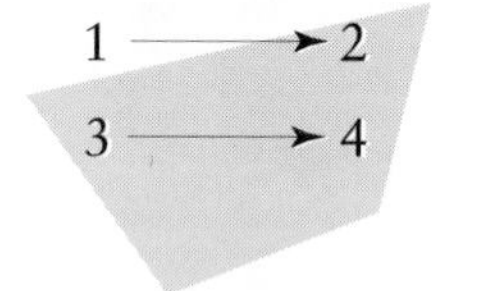

Trigonometry has a variety of useful applications in geometry, algebra, and the physical sciences, several of which are discussed in this chapter.

9.1 THE COMPLEX PLANE AND POLAR FORM FOR COMPLEX NUMBERS*

The real number system is represented geometrically by the number line. The complex number system can be represented geometrically by the coordinate plane:

> **The complex number $a + bi$ corresponds to the point (a, b) in the plane.**

For example, the point (2, 3) in Figure 9–1 on the next page is labeled by $2 + 3i$; and similarly for the other points shown:

*Section 4.6 is a prerequisite for this section.

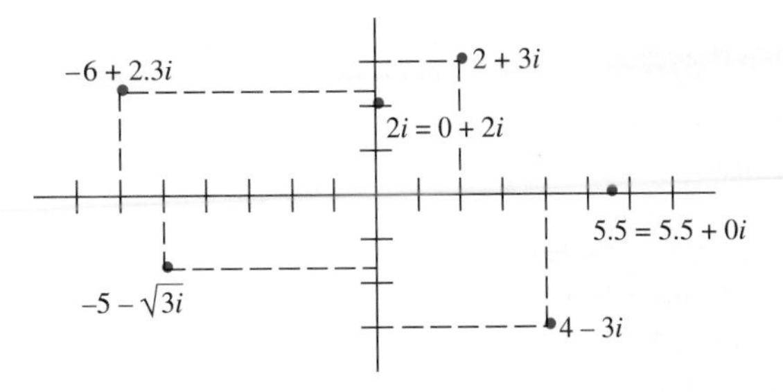

Figure 9-1

When the coordinate plane is labeled by complex numbers in this way, it is called the **complex plane.** Each real number $a = a + 0i$ corresponds to the point $(a, 0)$ on the horizontal axis; so this axis is called the **real axis.** The vertical axis is called the **imaginary axis** because every imaginary number $bi = 0 + bi$ corresponds to the point $(0, b)$ on the vertical axis.

The absolute value of a real number c is the distance from c to 0 on the number line (see page 8). So we define the **absolute value** (or **modulus**) of the complex number $a + bi$ to be the distance from $a + bi$ to the origin in the complex plane. Hence,

$$|a + bi| = \text{distance from } (a, b) \text{ to } (0, 0) = \sqrt{(a - 0)^2 + (b - 0)^2}.$$

Therefore,

Absolute Value ▶

The *absolute value* (or *modulus*) of the complex number $a + bi$ is

$$|a + bi| = \sqrt{a^2 + b^2}.$$

EXAMPLE 1

(a) $|3 + 2i| = \sqrt{3^2 + 2^2} = \sqrt{13}.$

(b) $|4 - 5i| = \sqrt{4^2 + (-5)^2} = \sqrt{41}.$ ■

Absolute values and trigonometry now lead to a useful way of representing complex numbers. Let $a + bi$ be a nonzero complex number and denote $|a + bi|$ by r. Then r is the length of the line segment joining (a, b) and $(0, 0)$ in the plane.

> **▶ TECHNOLOGY TIP**
> On calculators that handle complex numbers, you can use the ABS key to find the absolute value of a complex number. It is on the HP-38 keyboard, in the CMPLX menu of TI-85 and Casio 9800, in the CMPLX submenu of the MATH menu on TI-82/92, and in the Sharp 9300 MATH menu.

> **▶ TECHNOLOGY TIP**
> On calculators that handle complex numbers, you can use the ARG or ANGLE key to find the argument θ. It is in the CMPLX menu of TI-85 and Casio 9800, and in the CMPLX submenu of the MATH menu of TI-82/92, HP-38, and Sharp 9300.

Let θ be the angle in standard position with this line segment as terminal side (Figure 9–2).

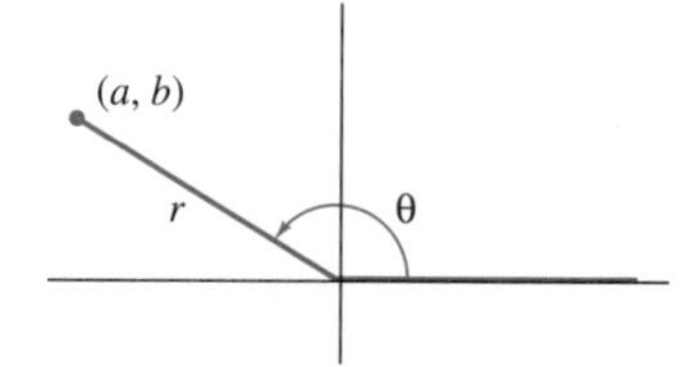

Figure 9-2

According to the point-in-the-plane description of sine and cosine

$$\cos\theta = \frac{a}{r} \qquad \text{and} \qquad \sin\theta = \frac{b}{r}$$

so that

$$a = r\cos\theta \qquad \text{and} \qquad b = r\sin\theta.$$

Consequently,

$$a + bi = r\cos\theta + (r\sin\theta)i = r(\cos\theta + i\sin\theta).^*$$

When a complex number $a + bi$ is written in this way, it is said to be in **polar form** or **trigonometric form.** The angle θ is called the **argument** and is usually expressed in radian measure. The number $r = |a + bi|$ is sometimes called the **modulus** (plural, moduli). The number 0 can also be written in polar notation by letting $r = 0$ and θ be any angle. Thus,

Polar Form ▶

> **Every complex number $a + bi$ can be written in polar form**
> $$r(\cos\theta + i\sin\theta)$$
> **where $r = |a + bi| = \sqrt{a^2 + b^2}$, $a = r\cos\theta$, and $b = r\sin\theta$.**

When a complex number is written in polar form, the argument θ is not uniquely determined since θ, $\theta \pm 2\pi$, $\theta \pm 4\pi$, etc., all satisfy the conditions in the box above.

EXAMPLE 2 Express $-\sqrt{3} + i$ in polar form.

Solution Here $a = -\sqrt{3}$ and $b = 1$, so that

$$r = \sqrt{a^2 + b^2} = \sqrt{(-\sqrt{3})^2 + 1^2} = \sqrt{3 + 1} = 2.$$

*It is customary to place i in front of $\sin\theta$ rather than after it. Some books abbreviate $r(\cos\theta + i\sin\theta)$ as $r \operatorname{cis} \theta$.

> **▶ TECHNOLOGY TIP**
> TI-85/92 and Sharp 9300 can automatically convert complex numbers from rectangular to polar form and vice versa. On TI-85 and Sharp, the polar form $r \cos \theta + i \sin \theta$ is entered and displayed as $r \angle \theta$. TI-82/92 use different notation. Check your instruction manual for details.

The angle θ must satisfy

$$\cos \theta = \frac{a}{r} = \frac{-\sqrt{3}}{2} \quad \text{and} \quad \sin \theta = \frac{b}{r} = \frac{1}{2}.$$

Since $-\sqrt{3} + i$ lies in the second quadrant (Figure 9–3), θ must be a second-quadrant angle. Our knowledge of special angles and Figure 9–3 show that $\theta = 5\pi/6$ satisfies these conditions. Hence,

$$-\sqrt{3} + i = 2\left(\cos \frac{5\pi}{6} + i \sin \frac{5\pi}{6}\right). \quad ■$$

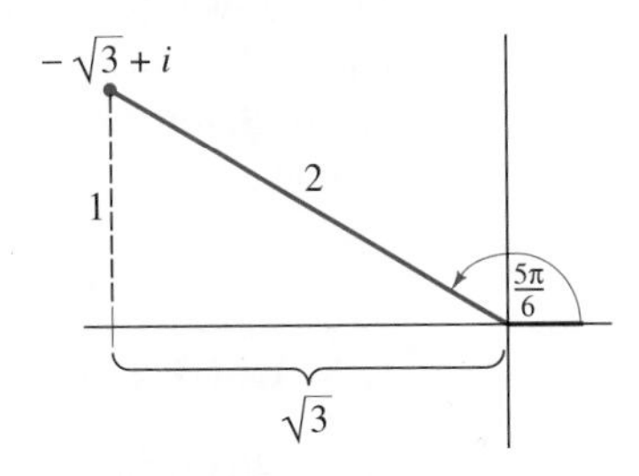

Figure 9-3

E X A M P L E 3 * Express $-2 + 5i$ in polar form.

Solution Since $a = -2$ and $b = 5$, $r = \sqrt{(-2)^2 + 5^2} = \sqrt{29}$. The angle θ must satisfy

$$\cos \theta = \frac{a}{r} = \frac{-2}{\sqrt{29}} \quad \text{and} \quad \sin \theta = \frac{b}{r} = \frac{5}{\sqrt{29}}$$

so that

$$\tan \theta = \frac{\sin \theta}{\cos \theta} = \frac{5/\sqrt{29}}{-2/\sqrt{29}} = -\frac{5}{2} = -2.5.$$

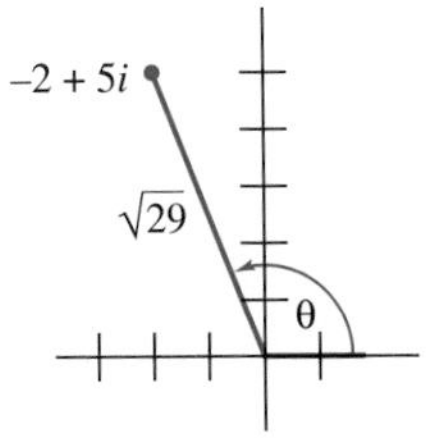

Since $-2 + 5i$ lies in the second quadrant (Figure 9–4), θ lies between $\pi/2$ and π. As we saw in Section 8.4, the only solution of the equation $\tan \theta = -2.5$ that lies between $\pi/2$ and π is $\theta \approx -1.1903 + \pi = 1.9513$. Therefore,

$$-2 + 5i \approx \sqrt{29}(\cos 1.9513 + i \sin 1.9513). \quad ■$$

$-2 + 5i$

$\sqrt{29}$

θ

Figure 9-4

Multiplication and division of complex numbers in polar form are done by the following rules, which are proved at the end of the section:

*Omit this example if you haven't read Section 8.4.

Polar Multiplication and Division Rules ▶

If $z_1 = r_1(\cos\theta_1 + i\sin\theta_1)$ and $z_2 = r_2(\cos\theta_2 + i\sin\theta_2)$ are any two complex numbers, then

$$z_1z_2 = r_1r_2(\cos(\theta_1 + \theta_2) + i\sin(\theta_1 + \theta_2))$$

and

$$\frac{z_1}{z_2} = \frac{r_1}{r_2}(\cos(\theta_1 - \theta_2) + i\sin(\theta_1 - \theta_2)) \qquad (z_2 \neq 0).$$

In other words, to multiply two numbers in polar form, just *multiply the moduli and add the arguments.* To divide, just *divide the moduli and subtract the arguments.* Before proving the statements in the box, we will illustrate them with some examples.

EXAMPLE 4 Suppose

$$z_1 = 2(\cos(5\pi/6) + i\sin(5\pi/6)) \qquad \text{and} \qquad z_2 = 3(\cos(7\pi/4) + i\sin(7\pi/4)).$$

Then r_1 is the number 2 and $\theta_1 = 5\pi/6$; similarly, $r_2 = 3$ and $\theta_2 = 7\pi/4$ and we have:

$$\begin{aligned} z_1z_2 &= r_1r_2(\cos(\theta_1 + \theta_2) + i\sin(\theta_1 + \theta_2)) \\ &= 2\cdot 3\left(\cos\left(\frac{5\pi}{6} + \frac{7\pi}{4}\right) + i\sin\left(\frac{5\pi}{6} + \frac{7\pi}{4}\right)\right) \\ &= 6\left(\cos\left(\frac{10\pi}{12} + \frac{21\pi}{12}\right) + i\sin\left(\frac{10\pi}{12} + \frac{21\pi}{12}\right)\right) \\ &= 6\left(\cos\frac{31\pi}{12} + i\sin\frac{31\pi}{12}\right). \quad \blacksquare \end{aligned}$$

▶ **TECHNOLOGY TIP**

TI-85 and Sharp 9300 do complex arithmetic in polar form when numbers are entered in the form $r \angle \theta$.

EXAMPLE 5 Suppose

$$z_1 = 10(\cos(\pi/3) + i\sin(\pi/3)) \qquad \text{and} \qquad z_2 = 2(\cos(\pi/4) + i\sin(\pi/4)).$$

Then:

$$\begin{aligned} \frac{z_1}{z_2} &= \frac{10\left(\cos\frac{\pi}{3} + i\sin\frac{\pi}{3}\right)}{2\left(\cos\frac{\pi}{4} + i\sin\frac{\pi}{4}\right)} = \frac{10}{2}\left(\cos\left(\frac{\pi}{3} - \frac{\pi}{4}\right) + i\sin\left(\frac{\pi}{3} - \frac{\pi}{4}\right)\right) \\ &= 5\left(\cos\frac{\pi}{12} + i\sin\frac{\pi}{12}\right). \quad \blacksquare \end{aligned}$$

Proof of the Polar Multiplication Rule

If $z_1 = r_1(\cos\theta_1 + i\sin\theta_1)$ and $z_2 = r_2(\cos\theta_2 + i\sin\theta_2)$, then

$$\begin{aligned} z_1z_2 &= r_1(\cos\theta_1 + i\sin\theta_1)r_2(\cos\theta_2 + i\sin\theta_2) \\ &= r_1r_2(\cos\theta_1 + i\sin\theta_1)(\cos\theta_2 + i\sin\theta_2) \\ &= r_1r_2(\cos\theta_1\cos\theta_2 + i\sin\theta_1\cos\theta_2 + i\cos\theta_1\sin\theta_2 + i^2\sin\theta_1\sin\theta_2) \\ &= r_1r_2((\cos\theta_1\cos\theta_2 - \sin\theta_1\sin\theta_2) + i(\sin\theta_1\cos\theta_2 + \cos\theta_1\sin\theta_2)). \end{aligned}$$

But the addition identities for sine and cosine (page 496) show that

$$\cos\theta_1\cos\theta_2 - \sin\theta_1\sin\theta_2 = \cos(\theta_1 + \theta_2)$$

and

$$\sin\theta_1\cos\theta_2 + \cos\theta_1\sin\theta_2 = \sin(\theta_1 + \theta_2).$$

Therefore,

$$\begin{aligned} z_1z_2 &= r_1r_2(\cos\theta_1\cos\theta_2 - \sin\theta_1\sin\theta_2) + i(\sin\theta_1\cos\theta_2 + \cos\theta_1\sin\theta_2)) \\ &= r_1r_2(\cos(\theta_1 + \theta_2) + i\sin(\theta_1 + \theta_2)). \end{aligned}$$

This completes the proof of the multiplication rule. The division rule is proved similarly (Exercise 51).

EXERCISES 9.1

In Exercises 1–8, plot the point in the complex plane corresponding to the number.

1. $3 + 2i$ **2.** $-7 + 6i$ **3.** $-\frac{8}{3} - \frac{5}{3}i$

4. $\sqrt{2} - 7i$ **5.** $(1 + i)(1 - i)$

6. $(2 + i)(1 - 2i)$ **7.** $2i\left(3 - \frac{5}{2}i\right)$

8. $\frac{4i}{3}(-6 - 3i)$

In Exercises 9–14, find the absolute value.

9. $|5 - 12i|$ **10.** $|2i|$ **11.** $|1 + \sqrt{2}i|$

12. $|2 - 3i|$ **13.** $|-12i|$ **14.** $|i^7|$

15. Give an example of complex numbers z and w such that $|z + w| \neq |z| + |w|$.

16. If $z = 3 - 4i$, find $|z|^2$ and $z\bar{z}$, where $\bar{z}$ is the conjugate of z (see page 280).

In Exercises 17–24, sketch the graph of the equation in the complex plane (z denotes a complex number of the form $a + bi$).

17. $|z| = 4$ [*Hint:* The graph consists of all points that lie 4 units from the origin.]

18. $|z| = 1$

19. $|z - 1| = 10$ [*Hint:* 1 corresponds to (1, 0) in the complex plane. What does the equation say about the distance from z to 1?]

20. $|z + 3| = 1$ **21.** $|z - 2i| = 4$

22. $|z - 3i + 2| = 9$
[*Hint:* Rewrite it as $|z - (-2 + 3i)| = 9$.]

23. $\text{Re}(z) = 2$ [The **real part** of the complex number $z = a + bi$ is defined to be the number a and is denoted $\text{Re}(z)$.]

24. $\text{Im}(z) = -5/2$ [The **imaginary part** of $z = a + bi$ is defined to be the number b *(not bi)* and is denoted $\text{Im}(z)$.]

In Exercises 25–32, express the number in polar form.

25. $3 + 4i$ **26.** $-4 + 3i$ **27.** $5 - 12i$

28. $-\sqrt{7} - 3i$ **29.** $1 + 2i$ **30.** $3 - 5i$

31. $-\frac{5}{2} + \frac{7}{2}i$ **32.** $\sqrt{5} + \sqrt{11}i$

In Exercises 33–38, perform the indicated multiplication or division; express your answer in both rectangular form $a + bi$ and polar form $r(\cos\theta + i\sin\theta)$.

33. $3\left(\cos\frac{\pi}{12} + i\sin\frac{\pi}{12}\right)\cdot 2\left(\cos\frac{7\pi}{12} + i\sin\frac{7\pi}{12}\right)$

34. $3\left(\cos\frac{\pi}{8} + i\sin\frac{\pi}{8}\right)\cdot 12\left(\cos\frac{3\pi}{8} + i\sin\frac{3\pi}{8}\right)$

35. $12\left(\cos\frac{11\pi}{12} + i\sin\frac{11\pi}{12}\right)\cdot\frac{7}{2}\left(\cos\frac{\pi}{4} + i\sin\frac{\pi}{4}\right)$

36. $\dfrac{8\left(\cos\frac{5\pi}{18} + i\sin\frac{5\pi}{18}\right)}{4\left(\cos\frac{\pi}{9} + i\sin\frac{\pi}{9}\right)}$

37. $\dfrac{6\left(\cos\frac{7\pi}{20} + i\sin\frac{7\pi}{20}\right)}{4\left(\cos\frac{\pi}{10} + i\sin\frac{\pi}{10}\right)}$

38. $\dfrac{\sqrt{54}\left(\cos\frac{9\pi}{4} + i\sin\frac{9\pi}{4}\right)}{\sqrt{6}\left(\cos\frac{7\pi}{12} + i\sin\frac{7\pi}{12}\right)}$

In Exercises 39–46, convert to polar form and then multiply or divide. Express your answer in polar form.

39. $(1 + i)(1 + \sqrt{3}i)$ **40.** $(1 - i)(3 - 3i)$

41. $\dfrac{1 + i}{1 - i}$ **42.** $\dfrac{2 - 2i}{-1 - i}$

43. $3i(2\sqrt{3} + 2i)$ **44.** $\dfrac{-4i}{\sqrt{3} + i}$

45. $i(i + 1)(-\sqrt{3} + i)$

46. $(1 - i)(2\sqrt{3} - 2i)(-4 - 4\sqrt{3}i)$

47. Explain what is meant by saying that multiplying a complex number $z = r(\cos\theta + i\sin\theta)$ by i amounts to rotating z 90° counterclockwise around the origin. [*Hint:* Express i and iz in polar form; what are their relative positions in the complex plane?]

48. Describe what happens geometrically when you multiply a complex number by 2.

Thinkers

49. The sum of two distinct complex numbers, $a + bi$ and $c + di$, can be found geometrically by means of the so-called **parallelogram rule:** Plot the points $a + bi$ and $c + di$ in the complex plane and form the parallelogram, three of whose vertices are 0, $a + bi$, and $c + di$; for example, see Figure 9–5. Then the fourth vertex of the parallelogram is the point whose coordinate is the sum

$$(a + bi) + (c + di) = (a + c) + (b + d)i.$$

Complete the following *proof* of the parallelogram rule when $a \neq 0$ and $c \neq 0$.

(a) Find the *slope* of the line K from 0 to $a + bi$. [*Hint:* K contains the points $(0, 0)$ and (a, b).]

(b) Find the *slope* of the line N from 0 to $c + di$.

(c) Find the *equation* of the line L, through $a + bi$ and parallel to line N of part (b). [*Hint:* The point (a, b) is on L; find the slope of L by using part (b) and facts about the slopes of parallel lines.]

(d) Find the *equation* of the line M, through $c + di$ and parallel to line K of part (a).

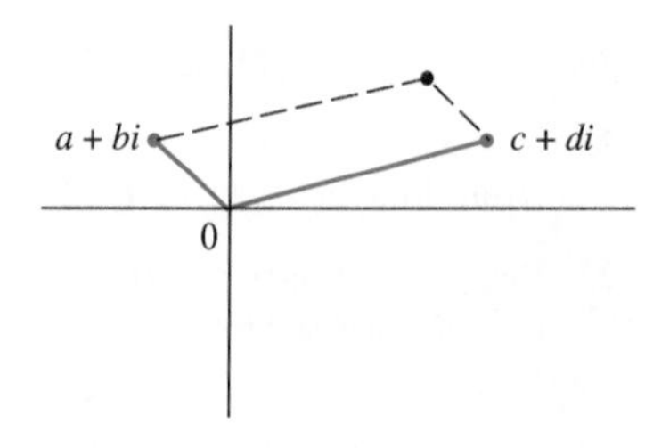

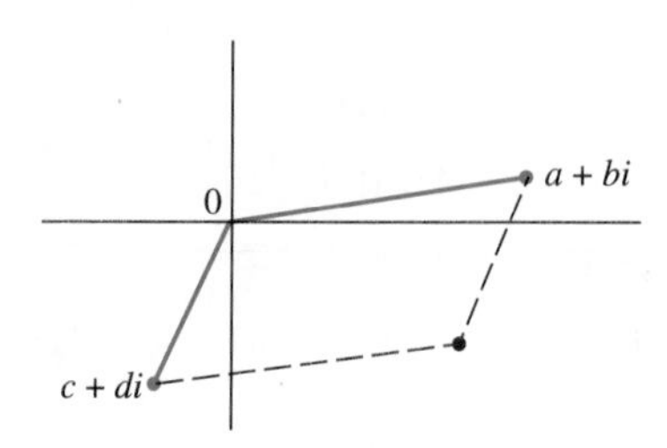

Figure 9-5

(e) Label the lines K, L, M, N in Figure 9–5.

(f) Show by using substitution that the point $(a + c, b + d)$ satisfies both the equation of line L and the equation of line M. Therefore, $(a + c, b + d)$ lies on both L and M. Since the only point on both L and M is the fourth vertex of the parallelogram (see Figure 9–5), this vertex must be $(a + c, b + d)$. Hence, this vertex has coordinate

$$(a + c) + (b + d)i = (a + bi) + (c + di).$$

50. Let $z = a + bi$ be a complex number and denote its conjugate $a - bi$ by $\bar{z}$. Prove that $|z|^2 = z\bar{z}$.

51. *Proof of the polar division rule.* Let $z_1 = r_1(\cos \theta_1 + i \sin \theta_1)$ and $z_2 = r_2(\cos \theta_2 + i \sin \theta_2)$. Then,

$$\begin{aligned}\frac{z_1}{z_2} &= \frac{r_1(\cos \theta_1 + i \sin \theta_1)}{r_2(\cos \theta_2 + i \sin \theta_2)} \\ &= \frac{r_1(\cos \theta_1 + i \sin \theta_1)}{r_2(\cos \theta_2 + i \sin \theta_2)} \cdot \frac{\cos \theta_2 - i \sin \theta_2}{\cos \theta_2 - i \sin \theta_2}.\end{aligned}$$

(a) Multiply out the denominator on the right side and use the Pythagorean identity to show that it is just the number r_2.

(b) Multiply out the numerator on the right side above; use the subtraction identities for sine and cosine (page 496) to show that it is

$$r_1(\cos(\theta_1 - \theta_2) + i \sin(\theta_1 - \theta_2)).$$

Therefore,

$$\frac{z_1}{z_2} = \left(\frac{r_1}{r_2}\right)(\cos(\theta_1 - \theta_2) + i \sin(\theta_1 - \theta_2)).$$

52. (a) If $s(\cos \beta + i \sin \beta) = r(\cos \theta + i \sin \theta)$, explain why we must have $s = r$. [*Hint:* Think distance.]

(b) If $r(\cos \beta + i \sin \beta) = r(\cos \theta + i \sin \theta)$, explain why $\cos \beta = \cos \theta$ and $\sin \beta = \sin \theta$. [*Hint:* See property 5 of the complex numbers on page 278.]

(c) If $\cos \beta = \cos \theta$ and $\sin \beta = \sin \theta$, show that angles β and θ in standard position have the same terminal side. [*Hint:* $(\cos \beta, \sin \beta)$ and $(\cos \theta, \sin \theta)$ are points on the unit circle.]

(d) Use parts (a)–(c) to prove this **equality rule for polar form:**

$$s(\cos \beta + i \sin \beta) = r(\cos \theta + i \sin \theta)$$

exactly when $s = r$ and $\beta = \theta + 2k\pi$ for some integer k. [*Hint:* Angles with the same terminal side must differ by an integer multiple of 2π.]

9.2 DEMOIVRE'S THEOREM AND *n*th ROOTS OF COMPLEX NUMBERS

Polar form provides a convenient way to calculate both powers and roots of complex numbers. If $z = r(\cos \theta + i \sin \theta)$, then the multiplication formula on page 546 shows that

$$\begin{aligned}z^2 = z \cdot z &= r \cdot r(\cos(\theta + \theta) + i \sin(\theta + \theta)) \\ &= r^2(\cos 2\theta + i \sin 2\theta).\end{aligned}$$

Similarly,

$$\begin{aligned}z^3 = z^2 \cdot z &= r^2 \cdot r(\cos(2\theta + \theta) + i \sin(2\theta + \theta)) \\ &= r^3(\cos 3\theta + i \sin 3\theta).\end{aligned}$$

Repeated application of the multiplication formula proves:

DeMoivre's Theorem ▶

For any complex number $z = r(\cos \theta + i \sin \theta)$ and any positive integer n,

$$z^n = r^n(\cos n\theta + i \sin n\theta).$$

EXAMPLE 1 To compute $(-\sqrt{3}+i)^5$, we express $-\sqrt{3}+i$ in polar form (as in Example 2 on page 544):

$$-\sqrt{3}+i = 2\left(\cos\frac{5\pi}{6} + i\sin\frac{5\pi}{6}\right).$$

By DeMoivre's Theorem,

$$(-\sqrt{3}+i)^5 = 2^5\left(\cos\left(5\cdot\frac{5\pi}{6}\right) + i\sin\left(5\cdot\frac{5\pi}{6}\right)\right) = 32\left(\cos\frac{25\pi}{6} + i\sin\frac{25\pi}{6}\right).$$

Since $25\pi/6 = (\pi/6) + (24\pi/6) = (\pi/6) + 4\pi$, we have

$$(-\sqrt{3}+i)^5 = 32\left(\cos\frac{25\pi}{6} + i\sin\frac{25\pi}{6}\right) = 32\left(\cos\frac{\pi}{6} + i\sin\frac{\pi}{6}\right)$$

$$= 32\left(\frac{\sqrt{3}}{2} + \frac{1}{2}i\right) = 16\sqrt{3} + 16i. \quad ■$$

EXAMPLE 2 To find $(1+i)^{10}$, first verify that the polar form of $1+i$ is $1+i = \sqrt{2}(\cos(\pi/4) + i\sin(\pi/4))$. Therefore, by DeMoivre's Theorem,

$$(1+i)^{10} = (\sqrt{2})^{10}\left(\cos\frac{10\pi}{4} + i\sin\frac{10\pi}{4}\right)$$

$$= (2^{1/2})^{10}\left(\cos\frac{5\pi}{2} + i\sin\frac{5\pi}{2}\right) = 2^5(0 + i\cdot 1) = 32i. \quad ■$$

If $a+bi$ is a complex number and n a positive integer, the equation $z^n = a+bi$ may have n different solutions in the complex numbers, as we shall see below. Furthermore, there is no obvious way to designate one of these solutions as *the* nth root of $a+bi$.* Consequently, *any* solution of the equation $z^n = a+bi$ is called an ***n*th root** of $a+bi$.

Every real number is, of course, a complex number. When the definition of an nth root of a complex number is applied to a real number, we must change our previous terminology. For instance, 16 now has *four* fourth roots, since each of 2, -2, $2i$, and $-2i$ is a solution of $z^4 = 16$, whereas we previously said that 2 was *the* fourth root of 16. In the context of complex numbers, this change will not cause any confusion.

Although nth roots are no longer unique, the radical symbol will be used only for nonnegative real numbers and will have the same meaning as before: If r is a nonnegative real number, then $\sqrt[n]{r}$ denotes the unique nonnegative real number whose nth power is r.

*You can't just choose the positive one, as we did in the real numbers, since "positive" and "negative" aren't meaningful terms in the complex numbers. For instance, should $3-2i$ be called positive or negative?

All the nth roots of a complex number $a + bi$ can easily be found if $a + bi$ is written in polar form, as illustrated in the next example.

EXAMPLE 3 Find the fourth roots of $-8 + 8\sqrt{3}i$.

Solution To solve $z^4 = -8 + 8\sqrt{3}i$, first verify that the polar form of $-8 + 8\sqrt{3}i$ is $16\left(\cos\frac{2\pi}{3} + i\sin\frac{2\pi}{3}\right)$. We must find numbers s and β such that

$$[s(\cos\beta + i\sin\beta)]^4 = 16\left(\cos\frac{2\pi}{3} + i\sin\frac{2\pi}{3}\right).$$

By DeMoivre's Theorem we must have:

$$s^4(\cos 4\beta + i\sin 4\beta) = 16\left(\cos\frac{2\pi}{3} + i\sin\frac{2\pi}{3}\right).$$

The equality rules for complex numbers (Exercise 52 in Section 9.1) show that this can happen only when

$$s^4 = 16 \qquad \text{and} \qquad 4\beta = \frac{2\pi}{3} + 2k\pi \qquad (k \text{ an integer})$$

$$s = \sqrt[4]{16} = 2 \qquad\qquad \beta = \frac{2\pi/3 + 2k\pi}{4}$$

Substituting these values in $s(\cos\beta + i\sin\beta)$ shows that the solutions of $z^4 = 16\left(\cos\frac{2\pi}{3} + i\sin\frac{2\pi}{3}\right)$ are

$$z = 2\left(\cos\frac{2\pi/3 + 2k\pi}{4} + i\sin\frac{2\pi/3 + 2k\pi}{4}\right) \qquad (k = 0, \pm 1, \pm 2, \pm 3, \ldots)$$

which can be simplified as

$$z = 2\left[\cos\left(\frac{\pi}{6} + \frac{k\pi}{2}\right) + i\sin\left(\frac{\pi}{6} + \frac{k\pi}{2}\right)\right] \qquad (k = 0, \pm 1, \pm 2, \pm 3, \ldots).$$

Letting $k = 0, 1, 2, 3$ produces four distinct solutions:

$$k = 0\colon z = 2\left(\cos\frac{\pi}{6} + i\sin\frac{\pi}{6}\right) = \sqrt{3} + i.$$

$$k = 1\colon z = 2\left(\cos\left(\frac{\pi}{6} + \frac{\pi}{2}\right) + i\sin\left(\frac{\pi}{6} + \frac{\pi}{2}\right)\right) = 2\left(\cos\frac{2\pi}{3} + i\sin\frac{2\pi}{3}\right)$$
$$= -1 + \sqrt{3}i.$$

$$k = 2: z = 2\left(\cos\left(\frac{\pi}{6} + \pi\right) + i \sin\left(\frac{\pi}{6} + \pi\right)\right) = 2\left(\cos\frac{7\pi}{6} + i \sin\frac{7\pi}{6}\right)$$
$$= -\sqrt{3} - i.$$

$$k = 3: z = 2\left(\cos\left(\frac{\pi}{6} + \frac{3\pi}{2}\right) + i \sin\left(\frac{\pi}{6} + \frac{3\pi}{2}\right)\right) = 2\left(\cos\frac{5\pi}{3} + i \sin\frac{5\pi}{3}\right)$$
$$= 1 - \sqrt{3}i.$$

Any other value of k produces an angle β with the same terminal side as one of the four angles used above, and hence leads to the same solution. For instance, when $k = 4$, then $\beta = \frac{\pi}{6} + \frac{4\pi}{2} = \frac{\pi}{6} + 2\pi$ and β has the same terminal side as $\pi/6$. Therefore, we have found *all* the solutions—the four fourth roots of $-8 + 8\sqrt{3}i$.* ■

The general equation $z^n = r(\cos\theta + i \sin\theta)$ can be solved by exactly the same method used in the preceding example—just substitute n for 4, r for 16, and θ for $2\pi/3$, as follows. A solution is a number $s(\cos\beta + i \sin\beta)$ such that

$$[s(\cos\beta + i \sin\beta)]^n = r(\cos\theta + i \sin\theta)$$
$$s^n(\cos n\beta + i \sin n\beta) = r(\cos\theta + i \sin\theta).$$

Therefore,

$$s^n = r \qquad \text{and} \qquad n\beta = \theta + 2k\pi \qquad (k \text{ any integer})$$
$$s = \sqrt[n]{r} \qquad\qquad \beta = \frac{\theta + 2k\pi}{n}$$

Taking $k = 0, 1, 2, \ldots, n - 1$ produces n distinct angles β. Any other value of k leads to an angle β with the same terminal side as one of these. Hence,

Formula for *n*th Roots ▶

For each positive integer n, a nonzero complex number

$$r(\cos\theta + i \sin\theta)$$

has exactly n distinct nth roots. They are given by:

$$\sqrt[n]{r}\left(\cos\left(\frac{\theta + 2k\pi}{n}\right) + i \sin\left(\frac{\theta + 2k\pi}{n}\right)\right)$$

where $k = 0, 1, 2, 3, \ldots, n - 1$.

EXAMPLE 4 To find the fifth roots of $4 + 4i$, first write it in polar form as $4\sqrt{2}\left(\cos\frac{\pi}{4} + i \sin\frac{\pi}{4}\right)$. Now apply the root formula with $n = 5$, $r =$

*Alternatively, page 286 shows that a fourth-degree equation, such as $z^4 = -8 + 8\sqrt{3}i$, has at most four distinct solutions.

> **TECHNOLOGY TIP**
>
> To solve polynomial equations with complex coefficients, such as $z^n = 4 + 4i$, use POLY on the TI-85 keyboard, C-SOLVE in the COMPLEX submenu of the TI-92 ALGEBRA menu, or POLYROOT in the POLYNOMIAL submenu of the HP-38 MATH menu.

$4\sqrt{2}$, $\theta = \pi/4$, and $k = 0, 1, 2, 3, 4$. Note that

$$\sqrt[5]{r} = \sqrt[5]{4\sqrt{2}} = (4\sqrt{2})^{1/5} = (2^2 2^{1/2})^{1/5} = (2^{5/2})^{1/5} = 2^{5/10} = 2^{1/2} = \sqrt{2}.$$

Therefore, the fifth roots are

$$\sqrt{2}\left(\cos\left(\frac{\pi/4 + 2k\pi}{5}\right) + i \sin\left(\frac{\pi/4 + 2k\pi}{5}\right)\right) \qquad k = 0, 1, 2, 3, 4.$$

That is,

$$k = 0\text{: } \sqrt{2}\left(\cos\left(\frac{\pi/4 + 0}{5}\right) + i \sin\left(\frac{\pi/4 + 0}{5}\right)\right) = \sqrt{2}\left(\cos\frac{\pi}{20} + i \sin\frac{\pi}{20}\right).$$

$$k = 1\text{: } \sqrt{2}\left(\cos\left(\frac{\pi/4 + 2\pi}{5}\right) + i \sin\left(\frac{\pi/4 + 2\pi}{5}\right)\right) = \sqrt{2}\left(\cos\frac{9\pi}{20} + i \sin\frac{9\pi}{20}\right).$$

$$k = 2\text{: } \sqrt{2}\left(\cos\left(\frac{\pi/4 + 4\pi}{5}\right) + i \sin\left(\frac{\pi/4 + 4\pi}{5}\right)\right) = \sqrt{2}\left(\cos\frac{17\pi}{20} + i \sin\frac{17\pi}{20}\right).$$

$$k = 3\text{: } \sqrt{2}\left(\cos\left(\frac{\pi/4 + 6\pi}{5}\right) + i \sin\left(\frac{\pi/4 + 6\pi}{5}\right)\right) = \sqrt{2}\left(\cos\frac{25\pi}{20} + i \sin\frac{25\pi}{20}\right).$$

$$k = 4\text{: } \sqrt{2}\left(\cos\left(\frac{\pi/4 + 8\pi}{5}\right) + i \sin\left(\frac{\pi/4 + 8\pi}{5}\right)\right) = \sqrt{2}\left(\cos\frac{33\pi}{20} + i \sin\frac{33\pi}{20}\right).$$ ■

Roots of Unity

The n distinct nth roots of 1 (the solutions of $z^n = 1$) are called the ***n*th roots of unity.** Since $\cos 0 = 1$ and $\sin 0 = 0$, the polar form of the number 1 is $\cos 0 + i \sin 0$. Applying the root formula with $r = 1$ and $\theta = 0$ shows that

Roots of Unity ▶

> **For each positive integer n, there are n distinct nth roots of unity:**
>
> $$\cos\frac{2k\pi}{n} + i \sin\frac{2k\pi}{n} \qquad (k = 0, 1, 2, \ldots, n - 1).$$

EXAMPLE 5 To find the cube roots of unity, apply the formula with $n = 3$ and $k = 0, 1, 2$:

$$k = 0\text{:} \qquad \cos 0 + i \sin 0 = 1$$

$$k = 1\text{:} \qquad \cos\frac{2\pi}{3} + i \sin\frac{2\pi}{3} = -\frac{1}{2} + \frac{\sqrt{3}}{2}i$$

$$k = 2\text{:} \qquad \cos\frac{4\pi}{3} + i \sin\frac{4\pi}{3} = -\frac{1}{2} - \frac{\sqrt{3}}{2}i.$$ ■

Denote by ω the first complex cube root of unity obtained in Example 5:

$$\omega = \cos \frac{2\pi}{3} + i \sin \frac{2\pi}{3}.$$

If we use DeMoivre's Theorem to find ω^2 and ω^3, we see that these numbers are the other two cube roots of unity found in Example 5:

$$\omega^2 = \left(\cos \frac{2\pi}{3} + i \sin \frac{2\pi}{3}\right)^2 = \cos \frac{4\pi}{3} + i \sin \frac{4\pi}{3}$$

$$\omega^3 = \left(\cos \frac{2\pi}{3} + i \sin \frac{2\pi}{3}\right)^3 = \cos \frac{6\pi}{3} + i \sin \frac{6\pi}{3} = \cos 2\pi + i \sin 2\pi$$
$$= 1 + 0 \cdot i = 1.$$

In other words, all the cube roots of unity are powers of ω. The same thing is true in the general case.

Roots of Unity ▶

Let n be a positive integer with $n > 1$. Then the number

$$z = \cos \frac{2\pi}{n} + i \sin \frac{2\pi}{n}$$

is an nth root of unity and all the nth roots of unity are

$$z, z^2, z^3, z^4, \ldots, z^{n-1}, z^n = 1.$$

The nth roots of unity have an interesting geometric interpretation. Every nth root of unity has absolute value 1 by the Pythagorean identity:

$$\left|\cos \frac{2k\pi}{n} + i \sin \frac{2k\pi}{n}\right| = \left(\cos \frac{2k\pi}{n}\right)^2 + \left(\sin \frac{2k\pi}{n}\right)^2$$
$$= \cos^2\left(\frac{2k\pi}{n}\right) + \sin^2\left(\frac{2k\pi}{n}\right) = 1.$$

Therefore, in the complex plane, every nth root of unity is exactly 1 unit from the origin. In other words, the nth roots of unity all lie on the unit circle.

EXAMPLE 6 The fifth roots of unity are

$$\cos \frac{2k\pi}{5} + i \sin \frac{2k\pi}{5} \qquad (k = 0, 1, 2, 3, 4)$$

that is,

$$\cos 0 + i \sin 0 = 1, \qquad \cos \frac{2\pi}{5} + i \sin \frac{2\pi}{5}, \qquad \cos \frac{4\pi}{5} + i \sin \frac{4\pi}{5},$$

$$\cos \frac{6\pi}{5} + i \sin \frac{6\pi}{5}, \qquad \cos \frac{8\pi}{5} + i \sin \frac{8\pi}{5}.$$

These five roots can be plotted in the complex plane by starting at $1 = 1 + 0i$ and moving counterclockwise around the unit circle, moving through an angle of $2\pi/5$ at each step, as shown in Figure 9–6. If you connect these five roots, they form the vertices of a regular pentagon (Figure 9–7). ■

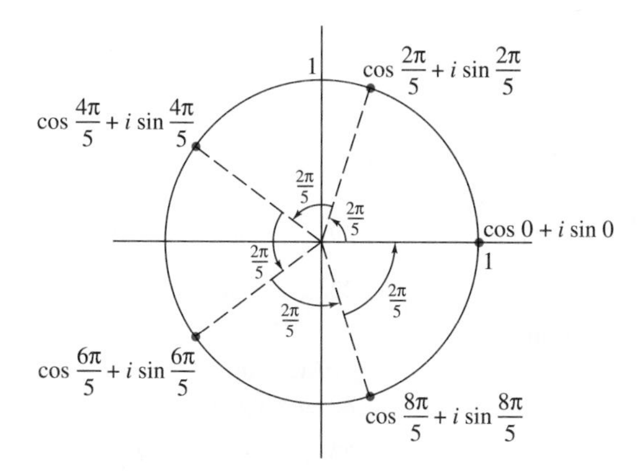

Figure 9-6

Figure 9-7

GRAPHING EXPLORATION With your calculator in parametric graphing mode, set these range values:

$$0 \le t \le 2\pi, \qquad t\text{-step} \approx .067,$$
$$-1.5 \le x \le 1.5, \qquad -1 \le y \le 1$$

and graph the unit circle, whose parametric equations are

$$x = \cos t \qquad \text{and} \qquad y = \sin t.\text{*}$$

Reset the t-step to be $2\pi/5$ and graph again. Your screen now looks exactly like the solid lines in Figure 9–7 because the calculator plotted only the five points corresponding to $t = 0, 2\pi/5, 4\pi/5, 6\pi/5, 8\pi/5$** and connected them with the shortest possible segments. Use the trace feature to move along the graph. The cursor will jump from vertex to vertex, that is, it will move from one fifth root of unity to the next.

EXAMPLE 7 Find the tenth roots of unity graphically.

Solution Graph the unit circle as in the preceding Exploration, but use $2\pi/10$ as the t-step. The result (Figure 9–8) is a regular decagon whose vertices are the tenth roots of unity. By using the trace feature you can approximate each of them.

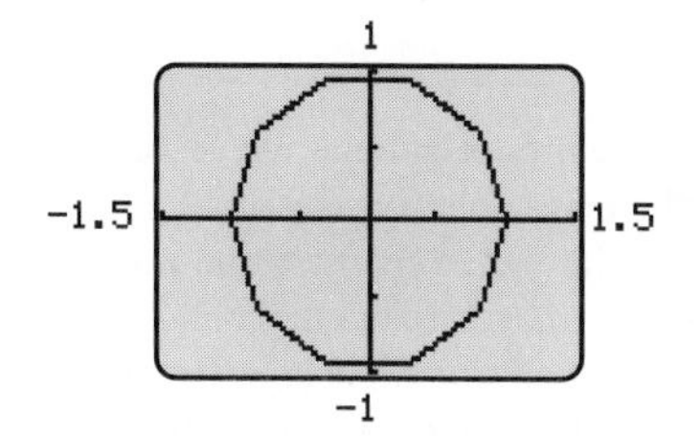

Figure 9-8

*On wide screen calculators (such as TI-92, TI-85, and HP-38), use $-2 \le x \le 2$ or $-1.7 \le x \le 1.7$ so that the unit circle looks like a circle.

**The point corresponding to $t = 10\pi/5 = 2\pi$ is the same as the one corresponding to $t = 0$.

GRAPHING EXPLORATION Verify that the two tenth roots of unity in the first quadrant are (approximately) $.8090 + .5878i$ and $.3090 + .9511i$. ■

EXERCISES 9.2

In Exercises 1–10, calculate the given product and express your answer in the form $a + bi$.

1. $\left(\cos\frac{\pi}{12} + i\sin\frac{\pi}{12}\right)^6$

2. $\left(\cos\frac{\pi}{5} + i\sin\frac{\pi}{5}\right)^{20}$

3. $\left(3\left(\cos\frac{7\pi}{30} + i\sin\frac{7\pi}{30}\right)\right)^5$

4. $\left(\sqrt[3]{4}\left(\cos\frac{7\pi}{36} + i\sin\frac{7\pi}{36}\right)\right)^{12}$

5. $(1 - i)^{12}$ [*Hint:* Use polar form and DeMoivre's Theorem.]

6. $(2 + 2i)^8$

7. $\left(\frac{\sqrt{3}}{2} + \frac{1}{2}i\right)^{10}$

8. $\left(-\frac{1}{2} + \frac{\sqrt{3}}{2}i\right)^{20}$

9. $\left(\frac{-1}{\sqrt{2}} + \frac{i}{\sqrt{2}}\right)^{14}$

10. $(-1 + \sqrt{3}i)^8$

In Exercises 11 and 12, find the indicated roots of unity and express your answers in the form $a + bi$.

11. Fourth roots of unity

12. Sixth roots of unity

In Exercises 13–22, find the nth roots of the given number in polar form.

13. $64\left(\cos\frac{\pi}{5} + i\sin\frac{\pi}{5}\right)$; $n = 3$

14. $8\left(\cos\frac{\pi}{10} + i\sin\frac{\pi}{10}\right)$; $n = 3$

15. $81\left(\cos\frac{\pi}{12} + i\sin\frac{\pi}{12}\right)$; $n = 4$

16. $16\left(\cos\frac{\pi}{7} + i\sin\frac{\pi}{7}\right)$; $n = 5$

17. -1; $n = 5$

18. 1; $n = 7$

19. i; $n = 5$

20. $-i$; $n = 6$

21. $1 + i$; $n = 2$

22. $1 - \sqrt{3}i$; $n = 3$

In Exercises 23–30, solve the given equation in the complex number system

23. $x^6 = -1$

24. $x^6 + 64 = 0$

25. $x^3 = i$

26. $x^4 = i$

27. $x^3 + 27i = 0$

28. $x^6 + 729 = 0$

29. $x^4 = -1 + \sqrt{3}i$

30. $x^4 = -8 - 8\sqrt{3}i$

In Exercises 31–35, represent the roots of unity graphically. Then use the trace feature to obtain approximations of the form $a + bi$ for each root (round to four places).

31. Seventh roots of unity

32. Fifth roots of unity

33. Eighth roots of unity

34. Twelfth roots of unity

35. Ninth roots of unity

36. Solve the equation $x^3 + x^2 + x + 1 = 0$. [*Hint:* First find the quotient when $x^4 - 1$ is divided by $x - 1$ and then consider solutions of $x^4 - 1 = 0$.]

37. Solve $x^5 + x^4 + x^3 + x^2 + x + 1 = 0$. [*Hint:* Consider $x^6 - 1$ and $x - 1$ and see Exercise 36.]

38. What do you think are the solutions of $x^{n-1} + x^{n-2} + \cdots + x^3 + x^2 + x + 1 = 0$? (See Exercises 36 and 37.)

Thinkers

39. In the complex plane, identify each point with its complex number label. The unit circle consists of all numbers (points) z such that $|z| = 1$. Suppose v and w are two points (numbers) that move around the unit circle in such a way that $v = w^{12}$ at all times. When w has made one complete trip around the circle, how many trips has v made? [*Hint:* Think polar and DeMoivre.]

40. Suppose u is an nth root of unity. Show that $1/u$ is also an nth root of unity. [*Hint:* Use the definition, *not* polar form.]

41. Let $u_1, u_2, \ldots, u_n$ be the distinct nth roots of unity and suppose v is a nonzero solution of the equation $z^n = r(\cos\theta + i\sin\theta)$. Show that $vu_1, vu_2, \ldots, vu_n$ are n distinct solutions of the equation. [*Remember:* Each u_i is a solution of $x^n = 1$.]

42. Use the formula for nth roots and the identities

$$\cos(x + \pi) = -\cos x \qquad \sin(x + \pi) = -\sin x$$

to show that the nonzero complex number $r(\cos\theta + i\sin\theta)$ has two square roots and that these square roots are negatives of each other.

9.3 VECTORS IN THE PLANE

Once a unit of measure has been agreed upon, quantities such as area, length, time, and temperature can be described by a single number. Other quantities, such as an east wind of 10 mph, require two numbers to describe them because they involve both *magnitude* and *direction.* Such quantities are called **vectors** and are represented geometrically by a directed line segment or arrow, as in Figure 9–9.

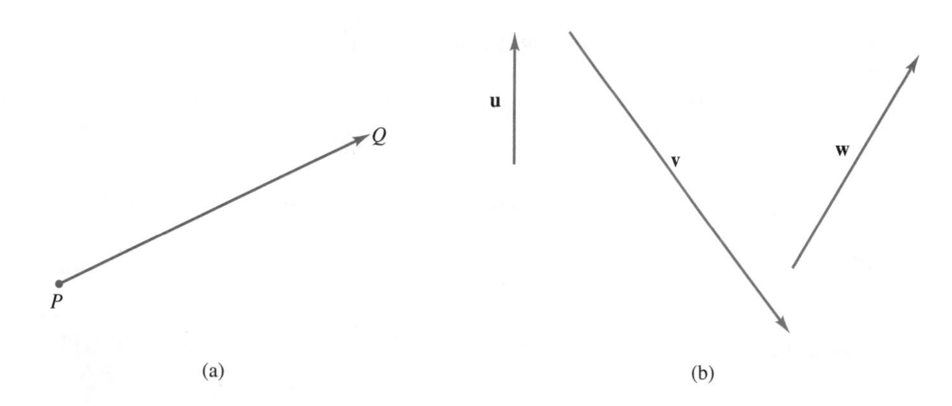

Figure 9-9

When a vector extends from a point P to a point Q, as in Figure 9–9(a), P is called the **initial point** of the vector and Q the **terminal point,** and the vector is written $\overrightarrow{PQ}$. Its **length** is denoted by $\|\overrightarrow{PQ}\|$. When the endpoints are not specified, as in Figure 9–9(b), vectors are denoted by boldface letters such as **u**, **v**, and **w**. The length of a vector **u** is denoted by $\|\mathbf{u}\|$ and is called the **magnitude** of **u**.

If **u** and **v** are vectors with the same magnitude and direction, we say that **u** and **v** are **equivalent** and write **u** = **v**. Some examples are shown in Figure 9–10.

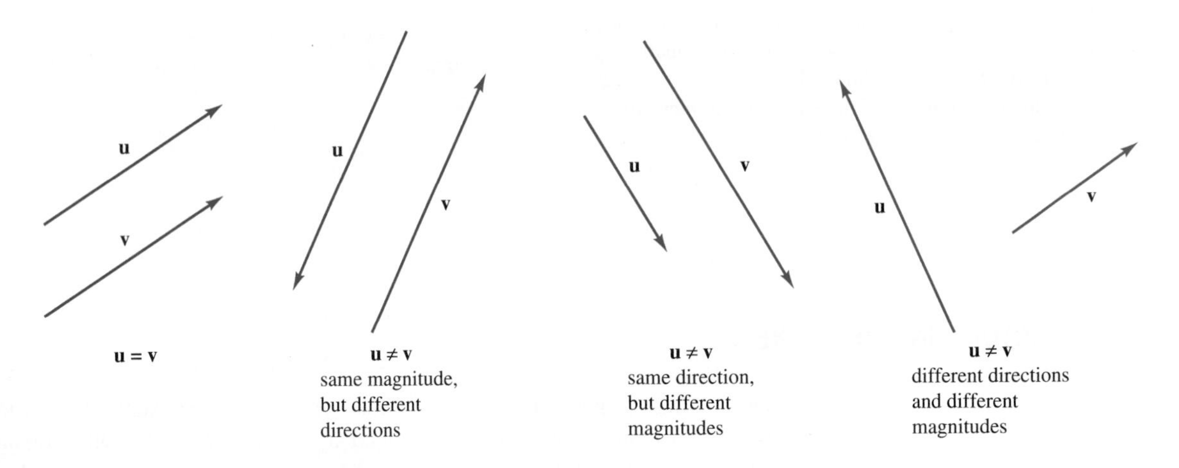

Figure 9-10

EXAMPLE 1 Let $P = (1, 2)$, $Q = (5, 4)$, $O = (0, 0)$, and $R = (4, 2)$, as in Figure 9–11. Show that $\overrightarrow{PQ} = \overrightarrow{OR}$.

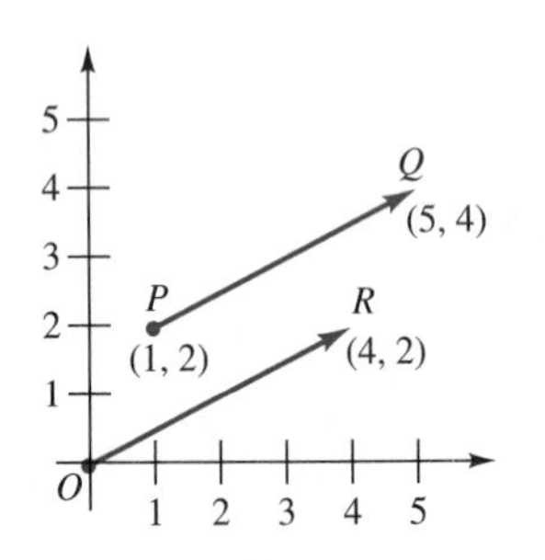

Figure 9-11

Solution The distance formula shows that $\overrightarrow{PQ}$ and $\overrightarrow{OR}$ have the *same length:*

$$\|\overrightarrow{PQ}\| = \sqrt{(5-1)^2 + (4-2)^2} = \sqrt{4^2 + 2^2} = \sqrt{20}.$$
$$\|\overrightarrow{OR}\| = \sqrt{(4-0)^2 + (2-0)^2} = \sqrt{4^2 + 2^2} = \sqrt{20}.$$

Furthermore, the lines through PQ and OR have the same slope:

$$\text{slope } PQ = \frac{4-2}{5-1} = \frac{2}{4} = \frac{1}{2} \qquad \text{slope } OR = \frac{2-0}{4-0} = \frac{2}{4} = \frac{1}{2}.$$

Since $\overrightarrow{PQ}$ and $\overrightarrow{OR}$ both point to the upper right on lines of the same slope, $\overrightarrow{PQ}$ and $\overrightarrow{OR}$ have the *same direction.* Therefore, $\overrightarrow{PQ} = \overrightarrow{OR}$. ■

According to the definition of equivalence, a vector may be moved from one location to another, provided that its magnitude and direction are not changed. Consequently, we have

Equivalent Vectors ▶

Every vector $\overrightarrow{PQ}$ is equivalent to a vector $\overrightarrow{OR}$ with initial point at the origin: If $P = (x_1, y_1)$ and $Q = (x_2, y_2)$, then

$$\overrightarrow{PQ} = \overrightarrow{OR}, \qquad \text{where } R = (x_2 - x_1, y_2 - y_1).$$

Proof The proof is similar to the one used in Example 1. It follows from the fact that $\overrightarrow{PQ}$ and $\overrightarrow{OR}$ have the same length:

$$\begin{aligned} \|\overrightarrow{OR}\| &= \sqrt{((x_2 - x_1) - 0)^2 + ((y_2 - y_1) - 0)^2} \\ &= \sqrt{(x_2 - x_1)^2 + (y_2 - y_1)^2} = \|\overrightarrow{PQ}\|; \end{aligned}$$

and that either the line segments PQ and OR are both vertical or they have the same slope:

$$\text{slope } OR = \frac{(y_2 - y_1) - 0}{(x_2 - x_1) - 0} = \frac{y_2 - y_1}{x_2 - x_1} = \text{slope } PQ$$

as shown in Figure 9–12. ❑

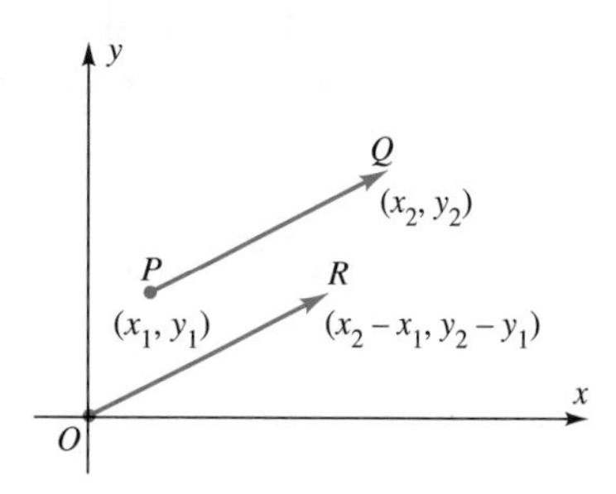

Figure 9-12

> **▶ TECHNOLOGY TIP**
> Vectors can be entered in component form on TI-85/92 and HP-38. Check your instruction manual.

The magnitude and direction of a vector with the origin as initial point are completely determined by the coordinates of its terminal point. Consequently, we denote the vector with initial point (0, 0) and terminal point (a, b) by $\langle a, b\rangle$. The numbers a and b are called the **components** of the vector $\langle a, b\rangle$.

Since the length of the vector $\langle a, b\rangle$ is the distance from (0, 0) to (a, b), the distance formula shows that:

Magnitude ▶

> **The *magnitude* (or *norm*) of the vector v = $\langle a, b\rangle$ is**
>
> $$|\mathbf{v}| = \sqrt{a^2 + b^2}.$$

EXAMPLE 2 Find the components and the magnitude of the vector with initial point $P = (-2, 6)$ and terminal point $Q = (4, -3)$.

Solution According to the fact in the box on the opposite page (with $x_1 = -2$, $y_1 = 6$, $x_2 = 4$, $y_2 = -3$):

$$\overrightarrow{PQ} = \overrightarrow{OR}, \qquad \text{where } R = (4 - (-2), -3 - 6) = (6, -9)$$

that is,

$$\overrightarrow{PQ} = \overrightarrow{OR} = \langle 6, -9 \rangle.$$

Therefore,

$$\|\overrightarrow{PQ}\| = \|\overrightarrow{OR}\| = \sqrt{6^2 + (-9)^2} = \sqrt{36 + 81} = \sqrt{117}. \quad ■$$

Vector Arithmetic

When dealing with vectors, it is customary to refer to ordinary real numbers as **scalars. Scalar multiplication** is an operation in which a scalar k is ''multiplied'' by a vector $\mathbf{v}$ to produce another *vector* denoted $k\mathbf{v}$. Here is the formal definition:

Scalar Multiplication ▶

If k is a real number and $\mathbf{v} = \langle a, b \rangle$ is a vector, then $k\mathbf{v}$ is the vector $\langle ka, kb \rangle$. The vector $k\mathbf{v}$ is called a *scalar multiple* of $\mathbf{v}$.

EXAMPLE 3 If $\mathbf{v} = \langle 3, 1 \rangle$, then

$$3\mathbf{v} = 3\langle 3, 1 \rangle = \langle 3 \cdot 3, 3 \cdot 1 \rangle = \langle 9, 3 \rangle$$

$$-2\mathbf{v} = -2\langle 3, 1 \rangle = \langle -2 \cdot 3, -2 \cdot 1 \rangle = \langle -6, -2 \rangle$$

as shown in Figure 9–13:

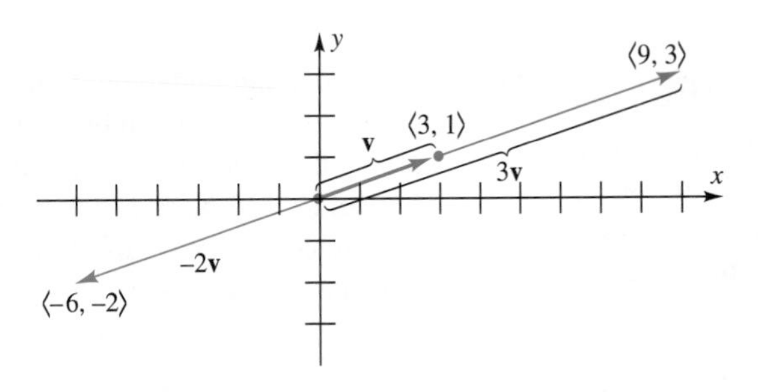

Figure 9-13

Figure 9–13 shows that $3\mathbf{v}$ has the same direction as $\mathbf{v}$, while $-2\mathbf{v}$ has the opposite direction. Also note that

$$\|\mathbf{v}\| = \|\langle 3, 1 \rangle\| = \sqrt{3^2 + 1^2} = \sqrt{10}$$

$$\|-2\mathbf{v}\| = \|\langle -6, -2 \rangle\| = \sqrt{(-6)^2 + (-2)^2} = \sqrt{40} = 2\sqrt{10}.$$

Therefore,

$$\|-2\mathbf{v}\| = 2\sqrt{10} = 2\|\mathbf{v}\| = |-2| \cdot \|\mathbf{v}\|.$$

Similarly, you can verify that $\|3\mathbf{v}\| = |3| \cdot \|\mathbf{v}\| = 3\|\mathbf{v}\|$. ■

Example 3 is an illustration of the following facts.

Geometric Interpretation of Scalar Multiplication ▶

> **The *magnitude* of the vector $k\mathbf{v}$ is $|k|$ times the length of v, that is,**
>
> $$\|k\mathbf{v}\| = |k| \cdot \|\mathbf{v}\|.$$
>
> **The *direction* of $k\mathbf{v}$ is the same as that of v when k is positive and opposite that of v when k is negative.**

See Exercise 75 for a proof of this statement.

Vector addition is an operation in which two vectors **u** and **v** are added to produce a new vector denoted **u** + **v**. Formally,

Vector Addition ▶

> **If $\mathbf{u} = \langle a, b \rangle$ and $\mathbf{v} = \langle c, d \rangle$, then $\mathbf{u} + \mathbf{v} = \langle a + c, b + d \rangle$.**

EXAMPLE 4 If $\mathbf{u} = \langle -5, 2 \rangle$ and $\mathbf{v} = \langle 3, 1 \rangle$, then

$$\mathbf{u} + \mathbf{v} = \langle -5, 2 \rangle + \langle 3, 1 \rangle = \langle -5 + 3, 2 + 1 \rangle = \langle -2, 3 \rangle$$

as shown in Figure 9–14. ■

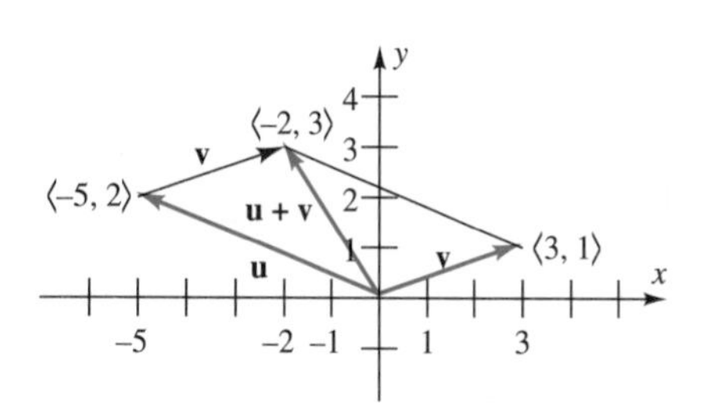

Figure 9-14

Example 4 is an illustration of these facts:

Geometric Interpretations of Vector Addition ▶

> 1. **If u and v are vectors with the same initial point P, then u + v is the vector $\overrightarrow{PQ}$, where $\overrightarrow{PQ}$ is the diagonal of the parallelogram with adjacent sides u and v.**
> 2. **If the vector v is moved (without changing its magnitude or direction) so that its initial point lies on the endpoint of the vector u, then u + v is the vector with the same initial point P as u and the same terminal point Q as v.**

See Exercise 76 for a proof of these statements.

The **negative** of a vector $\mathbf{v} = \langle c, d\rangle$ is defined to be the vector $(-1)\mathbf{v} = (-1)\langle c, d\rangle = \langle -c, -d\rangle$ and is denoted $-\mathbf{v}$. **Vector subtraction** is then defined as follows.

Vector Subtraction ▶

If $\mathbf{u} = \langle a, b\rangle$ and $\mathbf{v} = \langle c, d\rangle$, then $\mathbf{u} - \mathbf{v}$ is the vector

$$\mathbf{u} + (-\mathbf{v}) = \langle a, b\rangle + \langle -c, -d\rangle = \langle a - c, b - d\rangle.$$

A geometric interpretation of vector subtraction is given in Exercise 77.

EXAMPLE 5 If $\mathbf{u} = \langle 2, 5\rangle$ and $\mathbf{v} = \langle 6, 1\rangle$, then

$$\mathbf{u} - \mathbf{v} = \langle 2, 5\rangle - \langle 6, 1\rangle = \langle 2 - 6, 5 - 1\rangle = \langle -4, 4\rangle$$

as shown in Figure 9–15. ■

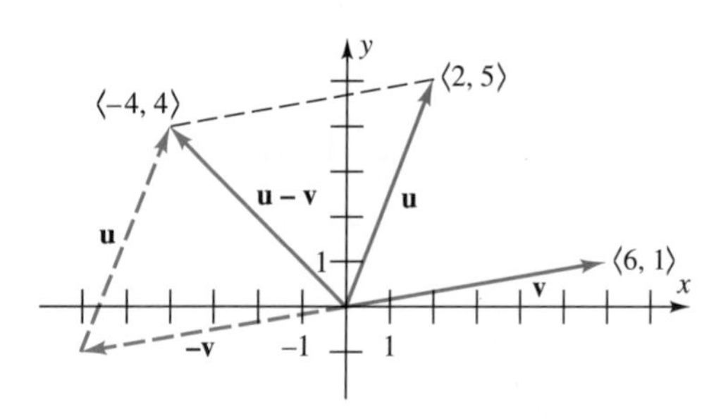

Figure 9-15

The vector $\langle 0, 0\rangle$ is called the **zero vector** and is denoted **0**.

EXAMPLE 6 If $\mathbf{u} = \langle -1, 6\rangle$, $\mathbf{v} = \langle 2/3, -4\rangle$, and $\mathbf{w} = \langle 2, 5/2\rangle$, then

$$2\mathbf{u} + 3\mathbf{v} = 2\langle -1, 6\rangle + 3\left\langle \frac{2}{3}, -4\right\rangle = \langle -2, 12\rangle + \langle 2, -12\rangle = \langle 0, 0\rangle = \mathbf{0}$$

and

$$4\mathbf{w} - 2\mathbf{u} = 4\left\langle 2, \frac{5}{2}\right\rangle - 2\langle -1, 6\rangle = \langle 8, 10\rangle - \langle -2, 12\rangle = \langle 8 - (-2), 10 - 12\rangle = \langle 10, -2\rangle. \quad ■$$

▶ **TECHNOLOGY TIP**

Vector arithmetic and other vector operations can be done on TI-85/92 and HP-38. Check your instruction manual.

Operations on vectors share many of the same properties as arithmetical operations on numbers.

Properties of Vector Addition and Scalar Multiplication ▶

For any vectors u, v, and w and any scalars r and s,

1. $\mathbf{u} + (\mathbf{v} + \mathbf{w}) = (\mathbf{u} + \mathbf{v}) + \mathbf{w}$
2. $\mathbf{u} + \mathbf{v} = \mathbf{v} + \mathbf{u}$
3. $\mathbf{v} + \mathbf{0} = \mathbf{v} = \mathbf{0} + \mathbf{v}$
4. $\mathbf{v} + (-\mathbf{v}) = \mathbf{0}$
5. $r(\mathbf{u} + \mathbf{v}) = r\mathbf{u} + r\mathbf{v}$
6. $(r + s)\mathbf{v} = r\mathbf{v} + s\mathbf{v}$
7. $(rs)\mathbf{v} = r(s\mathbf{v}) = s(r\mathbf{v})$
8. $1\mathbf{v} = \mathbf{v}$
9. $0\mathbf{v} = \mathbf{0}$ and $r\mathbf{0} = \mathbf{0}$

Proof If $\mathbf{u} = \langle a, b\rangle$ and $\mathbf{v} = \langle c, d\rangle$, then because addition of real numbers is commutative, we have:

$$\begin{aligned}\mathbf{u} + \mathbf{v} = \langle a, b\rangle + \langle c, d\rangle &= \langle a + c, b + d\rangle \\ &= \langle c + a, d + b\rangle = \langle c, d\rangle + \langle a, b\rangle = \mathbf{v} + \mathbf{u}.\end{aligned}$$

The other properties are proved similarly; see Exercises 53–58. ❑

Unit Vectors

A vector with length 1 is called a **unit vector.** For instance, $\langle 3/5, 4/5\rangle$ is a unit vector since

$$\left\|\left\langle \frac{3}{5}, \frac{4}{5}\right\rangle\right\| = \sqrt{\left(\frac{3}{5}\right)^2 + \left(\frac{4}{5}\right)^2} = \sqrt{\frac{9}{25} + \frac{16}{25}} = \sqrt{\frac{25}{25}} = 1.$$

> **▶ TECHNOLOGY TIP**
> A unit vector in the same direction as **v** can be found automatically by using UNITV in the MATH submenu of the TI-85 VECTOR menu or in the VECTOR OPS submenu of the MATRIX submenu of the TI-92 MATH menu.

EXAMPLE 7 Find a unit vector that has the same direction as the vector $\mathbf{v} = \langle 5, 12\rangle$.

Solution The length of **v** is

$$\|\mathbf{v}\| = \|\langle 5, 12\rangle\| = \sqrt{5^2 + 12^2} = \sqrt{169} = 13.$$

The vector $\mathbf{u} = \dfrac{1}{13}\mathbf{v} = \left\langle \dfrac{5}{13}, \dfrac{12}{13}\right\rangle$ has the same direction as **v** (since it is a scalar multiple by a positive number), and **u** is a unit vector because

$$\|\mathbf{u}\| = \left\|\frac{1}{13}\mathbf{v}\right\| = \left|\frac{1}{13}\right| \cdot \|\mathbf{v}\| = \frac{1}{13} \cdot 13 = 1. \quad ■$$

The procedure used in Example 7 (multiplying a vector by the reciprocal of its length) works in the general case:

Unit Vectors ▶

If v is a nonzero vector, then $\frac{1}{\|\mathbf{v}\|}\mathbf{v}$ is a unit vector with the same direction as v.

You can easily verify that the vectors $\mathbf{i} = \langle 1, 0\rangle$ and $\mathbf{j} = \langle 0, 1\rangle$ are unit vectors. The vectors **i** and **j** play a special role because they lead to a useful alternate notation for vectors. For example, if $\mathbf{u} = \langle 5, -7\rangle$, then

$$\mathbf{u} = \langle 5, 0\rangle + \langle 0, -7\rangle = 5\langle 1, 0\rangle - 7\langle 0, 1\rangle = 5\mathbf{i} - 7\mathbf{j}.$$

Similarly, if $\mathbf{v} = \langle a, b\rangle$ is any vector, then

$$\mathbf{v} = \langle a, b\rangle = \langle a, 0\rangle + \langle 0, b\rangle = a\langle 1, 0\rangle + b\langle 0, 1\rangle = a\mathbf{i} + b\mathbf{j}.$$

The vector **v** is said to be a **linear combination** of **i** and **j**. When vectors are written as linear combinations of **i** and **j**, then the properties in the box on page 563 can be used to write the rules for vector addition and scalar multiplication in this form:

$$(a\mathbf{i} + b\mathbf{j}) + (c\mathbf{i} + d\mathbf{j}) = (a + c)\mathbf{i} + (b + d)\mathbf{j}$$

and

$$c(a\mathbf{i} + b\mathbf{j}) = ca\mathbf{i} + cb\mathbf{j}.$$

EXAMPLE 8 If $\mathbf{u} = 2\mathbf{i} - 6\mathbf{j}$ and $\mathbf{v} = -5\mathbf{i} + 2\mathbf{j}$, then

$$\begin{aligned} 3\mathbf{u} - 2\mathbf{v} = 3(2\mathbf{i} - 6\mathbf{j}) - 2(-5\mathbf{i} + 2\mathbf{j}) &= 6\mathbf{i} - 18\mathbf{j} + 10\mathbf{i} - 4\mathbf{j} \\ &= 16\mathbf{i} - 22\mathbf{j}. \end{aligned}$$ ■

Direction Angles

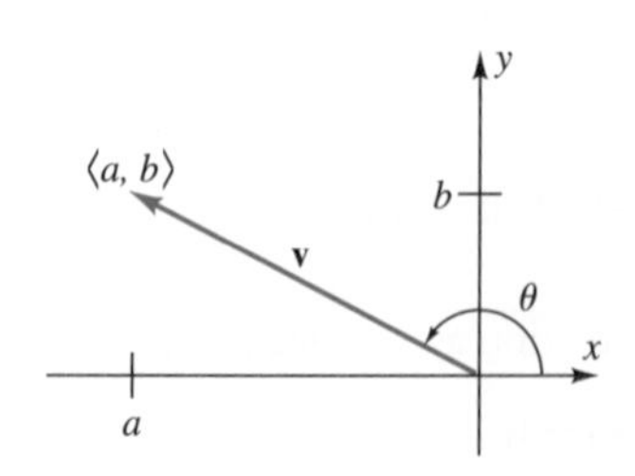

Figure 9-16

If $\mathbf{v} = \langle a, b\rangle = a\mathbf{i} + b\mathbf{j}$ is a vector, then the direction of **v** is completely determined by the standard position angle θ between 0° and 360°, whose terminal side is **v**, as shown in Figure 9–16. The angle θ is called the **direction angle** of the vector **v**. According to the point-in-the-plane description of the trigonometric functions,

$$\cos\theta = \frac{a}{\|\mathbf{v}\|} \quad \text{and} \quad \sin\theta = \frac{b}{\|\mathbf{v}\|}.$$

Rewriting each of these equations shows that

Components and the Direction Angle ▶

If v = ⟨*a*, *b*⟩ = *a*i + *b*j, then

$$a = ||\mathbf{v}|| \cos \theta \quad \text{and} \quad b = ||\mathbf{v}|| \sin \theta$$

where *θ* is the direction angle of v.

EXAMPLE 9 Find the component form of the vector that represents the velocity of an airplane at the instant its wheels leave the ground, if the plane is going 60 mph and the body of the plane makes a 7° angle with the horizontal.

Solution The velocity vector $\mathbf{v} = a\mathbf{i} + b\mathbf{j}$ has magnitude 60 and direction angle $\theta = 7°$, as shown in Figure 9–17. Hence,

$$\begin{aligned} \mathbf{v} &= (||\mathbf{v}|| \cos \theta)\mathbf{i} + (||\mathbf{v}|| \sin \theta)\mathbf{j} \\ &= (60 \cos 7°)\mathbf{i} + (60 \sin 7°)\mathbf{j} \\ &\approx (60 \cdot .9925)\mathbf{i} + (60 \cdot .1219)\mathbf{j} \\ &\approx 59.55\mathbf{i} + 7.31\mathbf{j} = \langle 59.55, 7.31 \rangle. \end{aligned}$$ ■

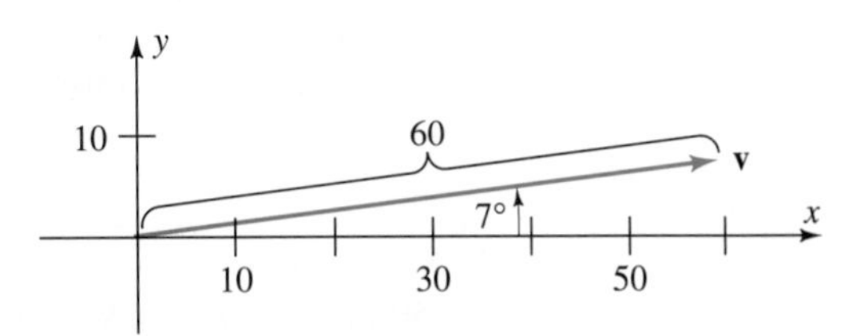

Figure 9-17

If $\mathbf{v} = a\mathbf{i} + b\mathbf{j}$ is a nonzero vector with direction angle θ, then

$$\tan \theta = \frac{\sin \theta}{\cos \theta} = \frac{b/||\mathbf{v}||}{a/||\mathbf{v}||} = \frac{b}{a}.$$

This fact provides a convenient way to find the direction angle of a vector.

EXAMPLE 10 Find the direction angle of

(a) $\mathbf{u} = 5\mathbf{i} + 13\mathbf{j}$ **(b)** $\mathbf{v} = -10\mathbf{i} + 7\mathbf{j}$.

Solution

(a) The direction angle θ of **u** satisfies $\tan \theta = b/a = 13/5 = 2.6$. Using the $\boxed{\text{TAN}^{-1}}$ key on a calculator we find that $\theta \approx 68.96°$, as shown in Figure 9–18 (a) on the next page.

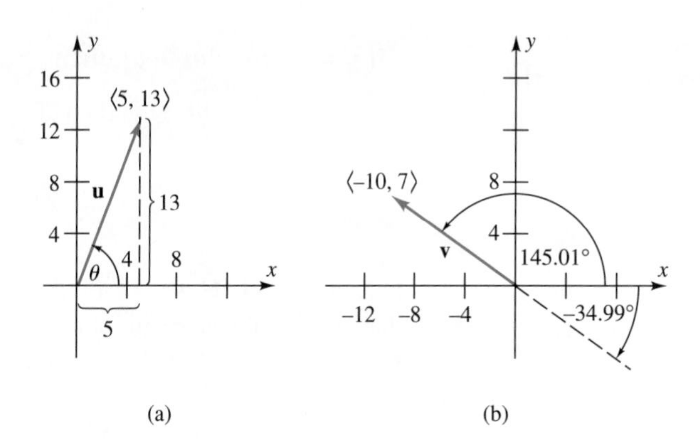

Figure 9-18

(b) The direction angle of **v** satisfies $\tan\theta = -7/10 = -.7$. Since **v** lies in the second quadrant, θ must be between 90° and 180°. A calculator shows that $-34.99°$ is an angle with tangent (approximately) $-.7$. Since tangent has period π (= 180°), we know that $\tan t = \tan(t + 180°)$ for every t. Therefore, $\theta = -34.99° + 180° = 145.01°$ is the angle between 90° and 180° such that $\tan\theta \approx -.7$. See Figure 9–18 (b). ■

EXAMPLE 11 An object at the origin is acted upon by two forces. A 150-pound force makes an angle of 20° with the positive x-axis, and the other force of 100 pounds makes an angle of 70°, as shown in Figure 9–19. Find the direction and magnitude of the resultant force.

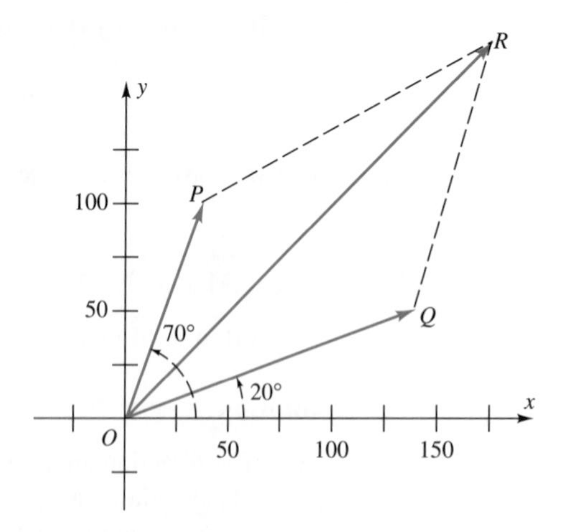

Figure 9-19

Solution The forces acting on the object are

$$\overrightarrow{OP} = (100 \cos 70°)\mathbf{i} + (100 \sin 70°)\mathbf{j}$$

$$\overrightarrow{OQ} = (150 \cos 20°)\mathbf{i} + (150 \sin 20°)\mathbf{j}.$$

The resultant force $\overrightarrow{OR}$ is the sum of $\overrightarrow{OP}$ and $\overrightarrow{OQ}$. Hence,

$$\begin{aligned}\overrightarrow{OR} &= (100 \cos 70° + 150 \cos 20°)\mathbf{i} + (100 \sin 70° + 150 \sin 20°)\mathbf{j} \\ &\approx 175.16\mathbf{i} + 145.27\mathbf{j}.\end{aligned}$$

Therefore, the magnitude of the resultant force is

$$\|\overrightarrow{OR}\| \approx \sqrt{(175.16)^2 + (145.27)^2} \approx 227.56.$$

The direction angle θ of the resultant force satisfies $\tan \theta \approx 145.27/175.16 \approx .8294$. A calculator shows that $\theta \approx 39.67°$. ■

Applications

EXAMPLE 12 A 200-pound box lies on a ramp that makes an angle of 24° with the horizontal. A rope is tied to the box from a post at the top of the ramp to keep it in position. Ignoring friction, how much force is being exerted on the rope by the box?

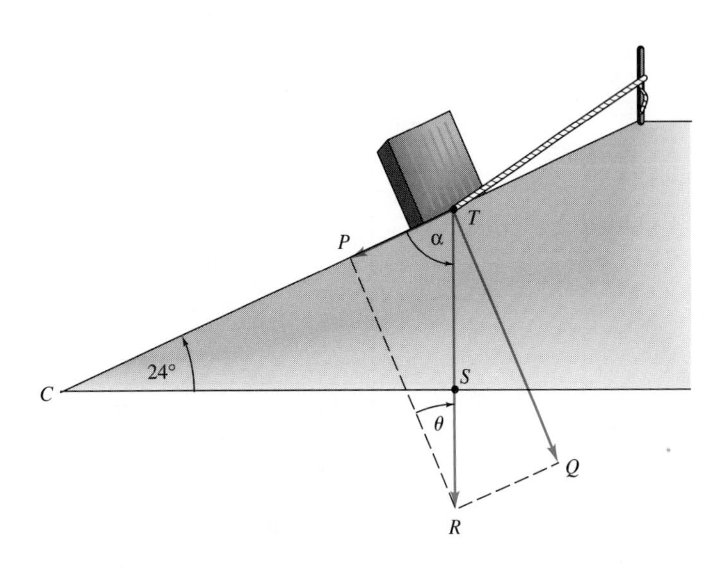

Figure 9-20

Solution Because of gravity the box exerts a 200-pound weight straight down (vector $\overrightarrow{TR}$). As Figure 9–20 shows, $\overrightarrow{TR}$ is the sum of $\overrightarrow{TP}$ and $\overrightarrow{TQ}$. The force on the rope is exerted by $\overrightarrow{TP}$, the vector of the force pulling the box down the ramp; so we must find $\|\overrightarrow{TP}\|$. In right triangle TSC, $\alpha + 24° = 90°$ and in right triangle TPR, $\alpha + \theta = 90°$. Hence,

$$\alpha + \theta = \alpha + 24°, \qquad \text{so that} \qquad \theta = 24°.$$

Therefore,

$$\frac{\|\overrightarrow{TP}\|}{\|\overrightarrow{TR}\|} = \sin\theta$$

$$\frac{\|\overrightarrow{TP}\|}{200} = \sin 24°$$

$$\|\overrightarrow{TP}\| = 200 \sin 24° \approx 81.35.$$

So, the force on the rope is 81.35 lb. ■

In aerial navigation, directions are given in terms of the angle measured in degrees clockwise from true north. Thus north is 0°, east is 90°, and so on.

EXAMPLE 13 An airplane is traveling in the direction 50° with an air speed of 300 mph, and there is a 35-mph wind from the direction 120°, as represented by the vectors **p** and **w** in Figure 9–21. Find the *course* and *ground speed* of the plane (that is, its direction and speed relative to the ground).

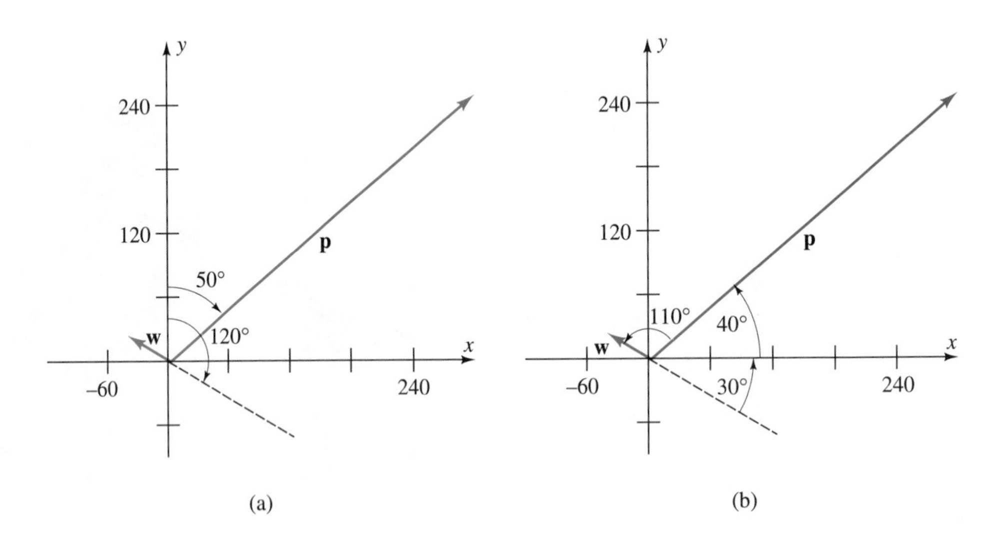

Figure 9-21

Solution The course of the plane is the direction of the vector **p** + **w**, and its ground speed is the magnitude of **p** + **w**. Figure 9–21(b) shows that the direction angle of **p** (the angle it makes with the positive x-axis) is 40° and that the direction angle of $\mathbf{w}$ is 150°. Therefore,

$$\begin{aligned}\mathbf{p}+\mathbf{w} &= [(300\cos 40°)\mathbf{i} + (300\sin 40°)\mathbf{j}] \\ &\qquad + [(35\cos 150°)\mathbf{i} + (35\sin 150°)\mathbf{j}] \\ &= (300\cos 40° + 35\cos 150°)\mathbf{i} + (300\sin 40° + 35\sin 150°)\mathbf{j} \\ &\approx 199.50\mathbf{i} + 210.34\mathbf{j}\end{aligned}$$

The direction angle θ of $\mathbf{p}+\mathbf{w}$ satisfies $\tan\theta = 210.34/199.50 \approx 1.0543$, and a calculator shows that $\theta \approx 46.5°$. This is the angle $\mathbf{p}+\mathbf{w}$ makes with the positive x-axis; hence, the course of the plane (the angle between true north and $\mathbf{p}+\mathbf{w}$) is $90° - 46.5° = 43.5°$. The ground speed of the plane is

$$\|\mathbf{p}+\mathbf{w}\| \approx \sqrt{(199.5)^2 + (210.34)^2} \approx 289.9 \text{ mph.} \quad \blacksquare$$

EXERCISES 9.3

In Exercises 1–4, find the magnitude of the vector $\overrightarrow{PQ}$.

1. $P = (2, 3)$, $Q = (5, 9)$

2. $P = (-3, 5)$, $Q = (7, -11)$

3. $P = (-7, 0)$, $Q = (-4, -5)$

4. $P = (30, 12)$, $Q = (25, 5)$

In Exercises 5–10, find a vector with the origin as initial point that is equivalent to the vector $\overrightarrow{PQ}$.

5. $P = (1, 5)$, $Q = (7, 11)$ **6.** $P = (2, 7)$, $Q = (-2, 9)$

7. $P = (-4, -8)$, $Q = (-10, 2)$

8. $P = (-5, 6)$, $Q = (-7, -9)$

9. $P = \left(\frac{4}{5}, -2\right)$, $Q = \left(\frac{17}{5}, -\frac{12}{5}\right)$

10. $P = (\sqrt{2}, 4)$, $Q = (\sqrt{3}, -1)$

In Exercises 11–20, find $\mathbf{u}+\mathbf{v}$, $\mathbf{u}-\mathbf{v}$, and $3\mathbf{u}-2\mathbf{v}$.

11. $\mathbf{u} = \langle -2, 4\rangle$, $\mathbf{v} = \langle 6, 1\rangle$

12. $\mathbf{u} = \langle 4, 0\rangle$, $\mathbf{v} = \langle 1, -3\rangle$

13. $\mathbf{u} = \langle 3, 3\sqrt{2}\rangle$, $\mathbf{v} = \langle 4\sqrt{2}, 1\rangle$

14. $\mathbf{u} = \left\langle \frac{2}{3}, 4\right\rangle$, $\mathbf{v} = \left\langle -7, \frac{19}{3}\right\rangle$

15. $\mathbf{u} = 2\langle -2, 5\rangle$, $\mathbf{v} = \frac{1}{4}\langle -7, 12\rangle$

16. $\mathbf{u} = \mathbf{i} - \mathbf{j}$, $\mathbf{v} = 2\mathbf{i} + \mathbf{j}$ **17.** $\mathbf{u} = 8\mathbf{i}$, $\mathbf{v} = 2(3\mathbf{i} - 2\mathbf{j})$

18. $\mathbf{u} = -4(-\mathbf{i} + \mathbf{j})$, $\mathbf{v} = -3\mathbf{i}$

19. $\mathbf{u} = -\left(2\mathbf{i} + \frac{3}{2}\mathbf{j}\right)$, $\mathbf{v} = \frac{3}{4}\mathbf{i}$

20. $\mathbf{u} = \sqrt{2}\mathbf{j}$, $\mathbf{v} = \sqrt{3}\mathbf{i}$

In Exercises 21–26, find the components of the given vector, where $\mathbf{u} = \mathbf{i} - 2\mathbf{j}$, $\mathbf{v} = 3\mathbf{i} + \mathbf{j}$, $\mathbf{w} = -4\mathbf{i} + \mathbf{j}$.

21. $\mathbf{u} + 2\mathbf{w}$ **22.** $\frac{1}{2}(3\mathbf{v} + \mathbf{w})$ **23.** $\frac{1}{2}\mathbf{w}$

24. $-2\mathbf{u} + 3\mathbf{v}$ **25.** $\frac{1}{4}(8\mathbf{u} + 4\mathbf{v} - \mathbf{w})$

26. $3(\mathbf{u} - 2\mathbf{v}) - 6\mathbf{w}$

In Exercises 27–34, find the component form of the vector $\mathbf{v}$ whose magnitude and direction angle θ are given.

27. $\|\mathbf{v}\| = 4$, $\theta = 0°$ **28.** $\|\mathbf{v}\| = 5$, $\theta = 30°$

29. $\|\mathbf{v}\| = 10$, $\theta = 225°$ **30.** $\|\mathbf{v}\| = 20$, $\theta = 120°$

31. $\|\mathbf{v}\| = 6$, $\theta = 40°$ **32.** $\|\mathbf{v}\| = 8$, $\theta = 160°$

33. $\|\mathbf{v}\| = 1/2$, $\theta = 250°$ **34.** $\|\mathbf{v}\| = 3$, $\theta = 310°$

In Exercises 35–42, find the magnitude and direction angle of the vector $\mathbf{v}$.

35. $\mathbf{v} = \langle 4, 4\rangle$ **36.** $\mathbf{v} = \langle 5, 5\sqrt{3}\rangle$

37. $\mathbf{v} = \langle -8, 0\rangle$ **38.** $\mathbf{v} = \langle 4, 5\rangle$

39. $\mathbf{v} = 6\mathbf{j}$ **40.** $\mathbf{v} = 4\mathbf{i} - 8\mathbf{j}$

41. $-2\mathbf{i} + 8\mathbf{j}$ **42.** $\mathbf{v} = -15\mathbf{i} - 10\mathbf{j}$

In Exercises 43–46, find a unit vector that has the same direction as $\mathbf{v}$.

43. $\langle 4, -5\rangle$ **44.** $-7\mathbf{i} + 8\mathbf{j}$

45. $5\mathbf{i} + 10\mathbf{j}$ **46.** $-3\mathbf{i} - 9\mathbf{j}$

In Exercises 47–50, an object at the origin is acted upon by two forces, $\mathbf{u}$ *and* $\mathbf{v}$*, with direction angle* $\theta_{\mathbf{u}}$ *and* $\theta_{\mathbf{v}}$*, respectively. Find the direction and magnitude of the resultant force.*

47. $\mathbf{u} = 30$ lb, $\theta_{\mathbf{u}} = 0°$; $\mathbf{v} = 90$ lb, $\theta_{\mathbf{v}} = 60°$

48. $\mathbf{u} = 6$ lb, $\theta_{\mathbf{u}} = 45°$; $\mathbf{v} = 6$ lb, $\theta_{\mathbf{v}} = 120°$

49. $\mathbf{u} = 12$ kg, $\theta_{\mathbf{u}} = 130°$; $\mathbf{v} = 20$ kg, $\theta_{\mathbf{v}} = 250°$

50. $\mathbf{u} = 30$ kg, $\theta_{\mathbf{u}} = 300°$; $\mathbf{v} = 80$ kg, $\theta_{\mathbf{v}} = 40°$

If forces $\mathbf{u}_1, \mathbf{u}_2, \ldots, \mathbf{u}_k$ *act on an object at the origin, the* resultant force *is the sum* $\mathbf{u}_1 + \mathbf{u}_2 + \cdots + \mathbf{u}_k$. *The forces are said to be in* equilibrium *if their resultant force is* $\mathbf{0}$. *In Exercises 51 and 52, find the resultant force and find an additional force* $\mathbf{v}$*, which, if added to the system, produces equilibrium.*

51. $\mathbf{u}_1 = \langle 2, 5\rangle, \mathbf{u}_2 = \langle -6, 1\rangle, \mathbf{u}_3 = \langle -4, -8\rangle$

52. $\mathbf{u}_1 = \langle 3, 7\rangle, \mathbf{u}_2 = \langle 8, -2\rangle, \mathbf{u}_3 = \langle -9, 0\rangle, \mathbf{u}_4 = \langle -5, 4\rangle$

In Exercises 53–58, let $\mathbf{u} = \langle a, b\rangle$, $\mathbf{v} = \langle c, d\rangle$, *and* $\mathbf{w} = \langle e, f\rangle$, *and let r and s be scalars. Prove that the stated property holds by calculating the vector on each side of the equal sign.*

53. $\mathbf{v} + \mathbf{0} = \mathbf{v} = \mathbf{0} + \mathbf{v}$ **54.** $\mathbf{v} + (-\mathbf{v}) = \mathbf{0}$

55. $r(\mathbf{u} + \mathbf{v}) = r\mathbf{u} + r\mathbf{v}$ **56.** $(r + s)\mathbf{v} = r\mathbf{v} + s\mathbf{v}$

57. $(rs)\mathbf{v} = r(s\mathbf{v}) = s(r\mathbf{v})$ **58.** $1\mathbf{v} = \mathbf{v}$ and $0\mathbf{v} = \mathbf{0}$

59. Two ropes are tied to a wagon. A child pulls one with a force of 20 pounds, while another child pulls the other with a force of 30 pounds (Figure 9–22). If the angle between the two ropes is 28°, how much force must be exerted by a third child, standing behind the wagon, to keep the wagon from moving? [*Hint:* Assume the wagon is at the origin and one rope runs along the positive x-axis. Proceed as in Example 11 to find the resultant force on the wagon from the ropes. The third child must use the same amount in the opposite direction.]

60. Two circus elephants, Bessie and Maybelle, are dragging a large container, as shown in Figure 9–23. If Bessie pulls with a force of 2200 pounds and Maybelle with a force of 1500 pounds and the container moves along the dashed line, what is angle θ?

Exercises 61–64 deal with an object on an inclined plane. The situation is similar to that in Figure 9–20 of Example 12, where $\|\overrightarrow{TP}\|$ *is the component of the weight of the object parallel to the plane and* $\|\overrightarrow{TQ}\|$ *is the component of the weight perpendicular to the plane.*

61. An object weighing 50 pounds lies on an inclined plane that makes a 40° angle with the horizontal. Find the components of the weight parallel and perpendicular to the plane. [*Hint:* Solve an appropriate triangle.]

62. Do Exercise 61 when the object weighs 200 pounds and the inclined plane makes a 20° angle with the horizontal.

63. If an object on an inclined plane weighs 150 pounds and the component of the weight perpendicular to the plane

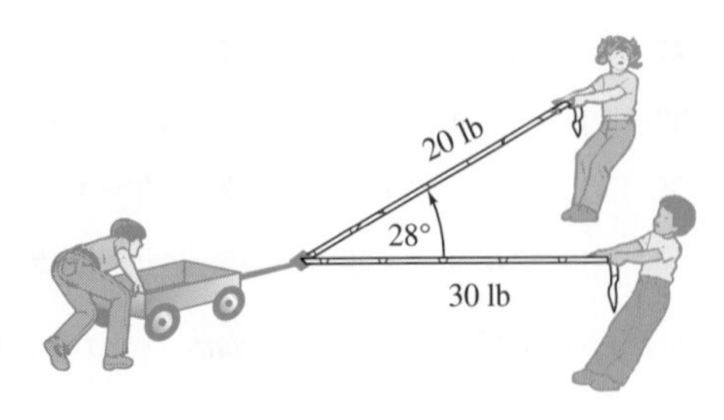

Figure 9-22

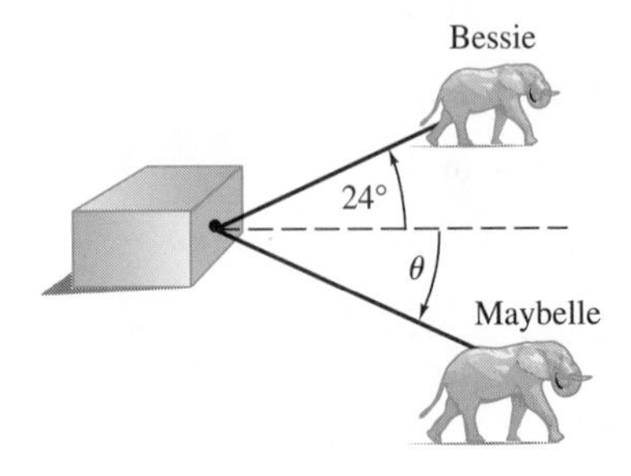

Figure 9-23

is 60 pounds, what angle does the plane make with the horizontal?

64. A force of 500 pounds is needed to pull a cart up a ramp that makes a 15° angle with the ground. Assume no friction is involved and find the weight of the cart. [*Hint:* Draw a picture similar to Figure 9–20, the 500-pound force is parallel to the ramp.]

In Exercises 65–70, find the course and ground speed of the plane under the given conditions. (See Example 13.)

65. Air speed 250 mph in the direction 60°; wind speed 40 mph from the direction 330°.

66. Air speed 400 mph in the direction 150°; wind speed 30 mph from the direction 60°.

67. Air speed 300 mph in the direction 300°; wind speed 50 mph in (*not* from) the direction 30°.

68. Air speed 500 mph in the direction 180°; wind speed 70 mph in the direction 40°.

69. The course and ground speed of a plane are 70° and 400 mph, respectively. There is a 60-mph wind blowing south. Find the (approximate) direction and air speed of the plane.

70. A plane is flying in the direction 200° with an air speed of 500 mph. Its course and ground speed are 210° and 450 mph, respectively. What is the direction and speed of the wind?

71. A river flows from east to west. A swimmer on the south bank wants to swim to a point on the opposite shore directly north of her starting point. She can swim at 2.8 mph and there is a 1-mph current in the river. In what direction should she head so as to travel directly north (that is, what angle should her path make with the south bank of the river)?

72. A river flows from west to east. A swimmer on the north bank swims at 3.1 mph along a straight course that makes a 75° angle with the north bank of the river and reaches the south bank at a point directly south of his starting point. How fast is the current in the river?

73. A 400-pound weight is suspended by two cables (Figure 9–24). What is the force (tension) on each cable? [*Hint:* Imagine that the weight is at the origin and that the dashed line is the x-axis. Then cable $\mathbf{v}$ is represented by the vector $(c \cos 65°)\mathbf{i} + (c \sin 65°)\mathbf{j}$, which has magnitude c (why?). Represent cable $\mathbf{u}$ similarly, denoting its magnitude by d. Use the fact that $\mathbf{u} + \mathbf{v} = 0\mathbf{i} + 400\mathbf{j}$ (why?) to set up a system of two equations in the unknowns c and d.]

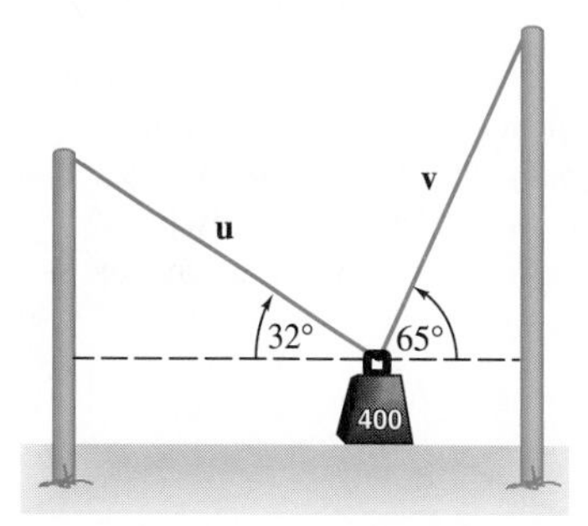

Figure 9-24

74. A 175-pound high wire artist stands balanced on a tightrope, which sags slightly at the point where he is standing. The rope in front of him makes a 6° angle with the horizontal and the rope behind him makes a 4° angle with the horizontal. Find the force on each end of the rope. [*Hint:* Use a picture and procedure similar to that in Exercise 73.]

75. Let $\mathbf{v}$ be the vector with initial point (x_1, y_1) and terminal point (x_2, y_2) and let k be any real number.
(a) Find the component form of $\mathbf{v}$ and $k\mathbf{v}$.
(b) Calculate $||\mathbf{v}||$ and $||k\mathbf{v}||$
(c) Use the fact that $\sqrt{k^2} = |k|$ to verify that $||k\mathbf{v}|| = |k| \cdot ||\mathbf{v}||$.
(d) Show that $\tan \theta = \tan \beta$, where θ is the direction angle of $\mathbf{v}$ and β is the direction angle of $k\mathbf{v}$. Use the fact that $\tan t = \tan (t + 180°)$ to conclude that $\mathbf{v}$ and $k\mathbf{v}$ have either the same or opposite directions.
(e) Use the fact that (c, d) and $(-c, -d)$ lie on the same straight line on opposite sides of the origin (Exercise 67 in Section 1.3) to verify that $\mathbf{v}$ and $k\mathbf{v}$ have the same direction if $k > 0$ and opposite directions if $k < 0$.

76. Let $\mathbf{u} = \langle a, b \rangle$, $\mathbf{v} = \langle c, d \rangle$. Verify the accuracy of the two geometric interpretations of vector addition given on page 561 as follows.
(a) Show that the distance from (a, b) to $(a + c, b + d)$ is the same as $||\mathbf{v}||$.

(b) Show that the distance from (c, d) to $(a + c, b + d)$ is the same as $\|\mathbf{u}\|$.

(c) Show that the line through (a, b) and $(a + c, b + d)$ is parallel to $\mathbf{v}$ by showing they have the same slope.

(d) Show that the line through (c, d) and $(a + c, b + d)$ is parallel to $\mathbf{u}$.

77. Let $\mathbf{u} = \langle a, b\rangle$ and $\mathbf{v} = \langle c, d\rangle$. Show that $\mathbf{u} - \mathbf{v}$ is equivalent to the vector $\mathbf{w}$ with initial point (c, d) and terminal point (a, b) as follows. (See Figure 9–25).

(a) Show that $\|\mathbf{u} - \mathbf{v}\| = \|\mathbf{w}\|$.

(b) Show that $\mathbf{u} - \mathbf{v}$ and $\mathbf{w}$ have the same direction.

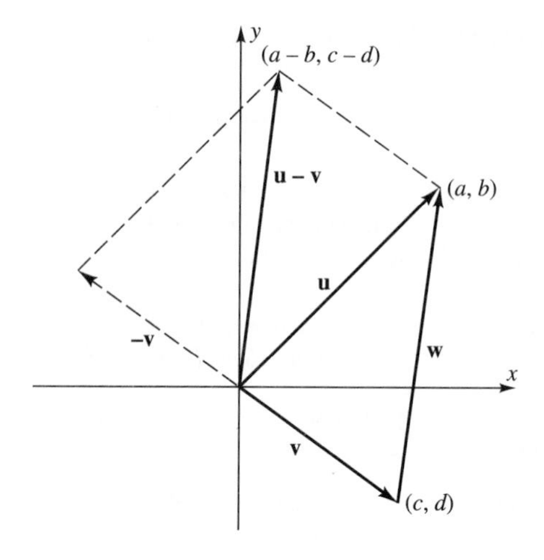

Figure 9-25

9.4 THE DOT PRODUCT

We now define a product for vectors and consider some of its many applications. When two vectors are added, their sum is another vector, but the situation is different with products. The dot product of two *vectors* is the *real number* defined as follows.

Dot Product ▶

The *dot product* of vectors $\mathbf{u} = \langle a, b\rangle = a\mathbf{i} + b\mathbf{j}$ and $\mathbf{v} = \langle c, d\rangle = c\mathbf{i} + d\mathbf{j}$ is denoted $\mathbf{u}\cdot\mathbf{v}$ and is defined to be the real number $ac + bd$. Thus,

$$\mathbf{u}\cdot\mathbf{v} = ac + bd.$$

▶ TECHNOLOGY TIP

Dot products can be computed automatically by using DOT or DOTP in the MATH submenu of the TI-85 VECTOR menu, or in the VECTOR OPS submenu of the MATRIX submenu of the TI-92 MATH menu, or in the MATRIX submenu of the HP-38 MATH menu.

EXAMPLE 1

(a) If $\mathbf{u} = \langle 5, 3\rangle$ and $\mathbf{v} = \langle -2, 6\rangle$, then

$$\mathbf{u}\cdot\mathbf{v} = 5(-2) + 3\cdot 6 = 8.$$

(b) If $\mathbf{u} = 4\mathbf{i} - 2\mathbf{j}$ and $\mathbf{v} = 3\mathbf{i} - \mathbf{j}$, then

$$\mathbf{u}\cdot\mathbf{v} = 4\cdot 3 + (-2)(-1) = 14.$$

(c) $\langle 2, -4\rangle\cdot\langle 6, 3\rangle = 2\cdot 6 + (-4)3 = 0.$ ■

The dot product has a number of useful properties:

Properties of the Dot Product ▶

If u, v, w are vectors and k is a real number, then

1. $\mathbf{u} \cdot \mathbf{u} = ||\mathbf{u}||^2$.
2. $\mathbf{u} \cdot \mathbf{v} = \mathbf{v} \cdot \mathbf{u}$.
3. $\mathbf{u} \cdot (\mathbf{v} + \mathbf{w}) = \mathbf{u} \cdot \mathbf{v} + \mathbf{u} \cdot \mathbf{w}$.
4. $k\mathbf{u} \cdot \mathbf{v} = k(\mathbf{u} \cdot \mathbf{v}) = \mathbf{u} \cdot k\mathbf{v}$.
5. $\mathbf{0} \cdot \mathbf{u} = 0$.

Proof

1. If $\mathbf{u} = \langle a, b \rangle$, then

$$||\mathbf{u}|| = \sqrt{a^2 + b^2}.$$

Hence,

$$\mathbf{u} \cdot \mathbf{u} = \langle a, b \rangle \cdot \langle a, b \rangle = a \cdot a + b \cdot b = a^2 + b^2 = \left(\sqrt{a^2 + b^2}\right)^2 = ||\mathbf{u}||^2.$$

2. If $\mathbf{u} = \langle a, b \rangle$ and $\mathbf{v} = \langle c, d \rangle$, then

$$\mathbf{u} \cdot \mathbf{v} = \langle a, b \rangle \cdot \langle c, d \rangle = ac + bd = ca + db = \langle c, d \rangle \cdot \langle a, b \rangle = \mathbf{v} \cdot \mathbf{u}.$$

The last three statements are proved similarly (Exercises 37–39). ❑

Angles

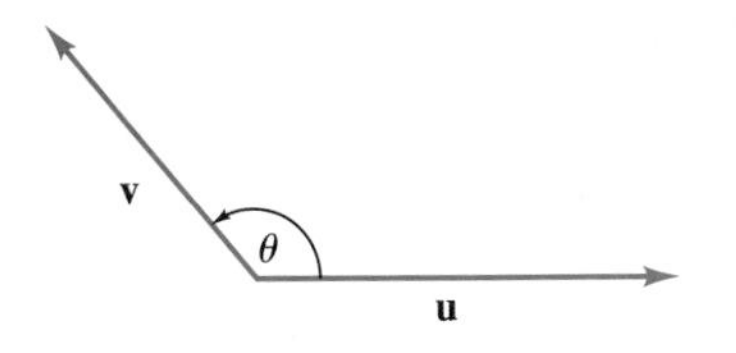

Figure 9-26

If $\mathbf{u} = \langle a, b \rangle$ and $\mathbf{v} = \langle c, d \rangle$ are nonzero vectors, then the **angle between u and v** is the smallest angle θ formed by these two line segments, as shown in Figure 9–26. We ignore clockwise or counterclockwise rotation and consider the angle between **v** and **u** to be the same as the angle between **u** and **v**. Thus, the radian measure of θ ranges from 0 to π.

Nonzero vectors **u** and **v** are said to be **parallel** if the angle between them is either 0 or π radians (that is, **u** and **v** lie on the same straight line through the origin and have either the same or opposite directions). The zero vector **0** is considered to be parallel to every vector.

Any scalar multiple of **u** is parallel to **u** since it lies on the same straight line as **u** (see Example 3 in Section 9.3). Conversely, if **v** is parallel to **u**, it is easy to show that **v** must be a scalar multiple of **u** (Exercise 40). Hence,

Parallel Vectors ▶

Vectors u and v are parallel exactly when

$$\mathbf{v} = k\mathbf{u} \text{ for some real number } k.$$

EXAMPLE 2 The vectors $\langle 2, 3 \rangle$ and $\langle 8, 12 \rangle$ are parallel because $\langle 8, 12 \rangle = 4\langle 2, 3 \rangle$. ■

The angle between nonzero vectors **u** and **v** is closely related to their dot product:

Angle Theorem ▶

> **If θ is the angle between the nonzero vectors u and v, then**
>
> $$\mathbf{u}\cdot\mathbf{v} = \|\mathbf{u}\|\,\|\mathbf{v}\|\cos\theta,$$
>
> **or equivalently,**
>
> $$\cos\theta = \frac{\mathbf{u}\cdot\mathbf{v}}{\|\mathbf{u}\|\,\|\mathbf{v}\|}.$$

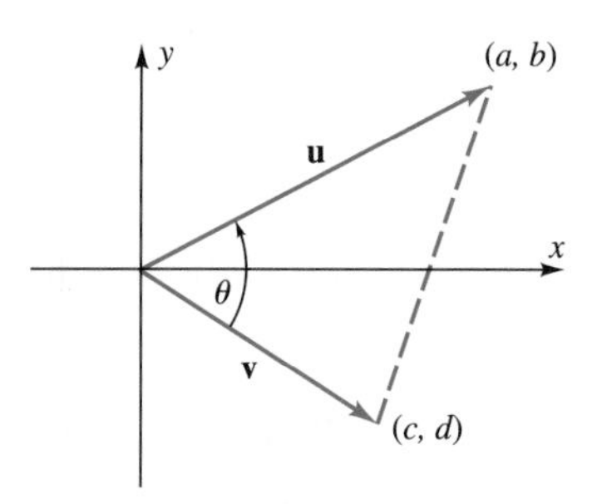

Figure 9-27

Proof If $\mathbf{u} = \langle a, b\rangle$, $\mathbf{v} = \langle c, d\rangle$, and the angle θ is not 0 or π, then **u** and **v** form two sides of a triangle, as shown in Figure 9–27.

The lengths of two sides of the triangle are $\|\mathbf{u}\| = \sqrt{a^2 + b^2}$ and $\|\mathbf{v}\| = \sqrt{c^2 + d^2}$. The distance formula shows that the length of the third side (opposite angle θ) is $\sqrt{(a - c)^2 + (b - d)^2}$. Therefore, by the Law of Cosines,

$$\left(\sqrt{(a-c)^2 + (b-d)^2}\right)^2 = \|\mathbf{u}\|^2 + \|\mathbf{v}\|^2 - 2\|\mathbf{u}\|\,\|\mathbf{v}\|\cos\theta$$

$$(a-c)^2 + (b-d)^2 = (a^2 + b^2) + (c^2 + d^2) - 2\|\mathbf{u}\|\,\|\mathbf{v}\|\cos\theta$$

$$a^2 - 2ac + c^2 + b^2 - 2bd + d^2 = (a^2 + c^2) + (b^2 + d^2) - 2\|\mathbf{u}\|\,\|\mathbf{v}\|\cos\theta$$

$$-2ac - 2bd = -2\|\mathbf{u}\|\,\|\mathbf{v}\|\cos\theta.$$

Dividing both sides by -2 shows that

$$ac + bd = \|\mathbf{u}\|\,\|\mathbf{v}\|\cos\theta.$$

Since the left side of this equation is precisely $\mathbf{u}\cdot\mathbf{v}$, the proof is complete in this case. The proof when θ is 0 or π is left to the reader (Exercise 41). ❑

EXAMPLE 3 Find the angle θ between the vectors $\langle -3, 1\rangle$ and $\langle 5, 2\rangle$ shown in Figure 9–28.

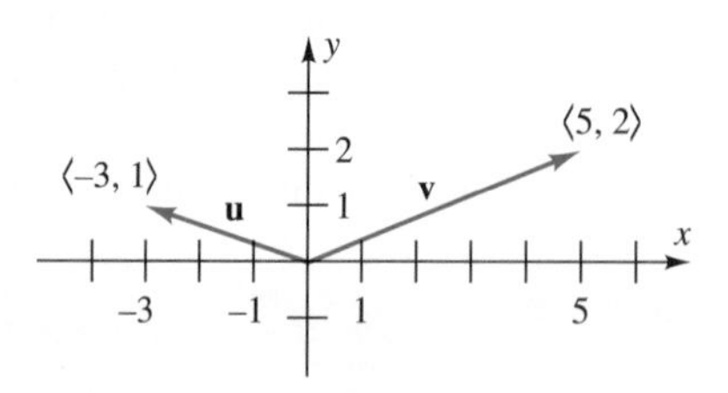

Figure 9-28

Solution Apply the formula in the box above with $\mathbf{u} = \langle -3, 1\rangle$ and $\mathbf{v} = \langle 5, 2\rangle$:

$$\cos\theta = \frac{\mathbf{u}\cdot\mathbf{v}}{\|\mathbf{u}\|\,\|\mathbf{v}\|} = \frac{(-3)5 + 1\cdot 2}{\sqrt{(-3)^2 + 1^2}\,\sqrt{5^2 + 2^2}} = \frac{-13}{\sqrt{10}\,\sqrt{29}} = \frac{-13}{\sqrt{290}}.$$

Using the COS^{-1} key, we see that

$$\theta \approx 2.4393 \text{ radians } (\approx 139.76°). \quad ■$$

The Angle Theorem has several useful consequences. For instance, by taking absolute values on both sides of $\mathbf{u} \cdot \mathbf{v} = ||\mathbf{u}|| \, ||\mathbf{v}|| \cos \theta$, and using the fact that $|\,||\mathbf{u}|| \, ||\mathbf{v}||\,| = ||\mathbf{u}|| \, ||\mathbf{v}||$ (because $||\mathbf{u}|| \, ||\mathbf{v}|| \geq 0$), we see that

$$|\mathbf{u} \cdot \mathbf{v}| = |\,||\mathbf{u}|| \, ||\mathbf{v}|| \cos \theta| = |\,||\mathbf{u}|| \, ||\mathbf{v}||\,| \, |\cos \theta| = ||\mathbf{u}|| \, ||\mathbf{v}|| \, |\cos \theta|.$$

But for any angle θ, $|\cos \theta| \leq 1$ so that

$$|\mathbf{u} \cdot \mathbf{v}| = ||\mathbf{u}|| \, ||\mathbf{v}|| \, |\cos \theta| \leq ||\mathbf{u}|| \, ||\mathbf{v}||.$$

This proves the Schwarz inequality:

Schwarz Inequality ▶

For any vectors u and v,

$$|\mathbf{u} \cdot \mathbf{v}| \leq ||\mathbf{u}|| \, ||\mathbf{v}||.$$

Vectors **u** and **v** are said to be **orthogonal** (or **perpendicular**) if the angle between them is $\pi/2$ radians (90°), or if at least one of them is **0**. Here is the key fact about orthogonal vectors:

Orthogonal Vectors ▶

The vectors u and v are orthogonal exactly when $\mathbf{u} \cdot \mathbf{v} = 0$.

Proof If **u** or **v** is **0**, then $\mathbf{u} \cdot \mathbf{v} = 0$, and if **u** and **v** are nonzero orthogonal vectors, then by the Angle Theorem

$$\mathbf{u} \cdot \mathbf{v} = ||\mathbf{u}|| \, ||\mathbf{v}|| \cos \theta = ||\mathbf{u}|| \, ||\mathbf{v}|| \cos (\pi/2) = ||\mathbf{u}|| \, ||\mathbf{v}|| \, (0) = 0.$$

Conversely, if **u** and **v** are vectors such that $\mathbf{u} \cdot \mathbf{v} = 0$, then Exercise 42 shows that **u** and **v** are orthogonal. ❑

EXAMPLE 4

(a) The vectors $\mathbf{u} = \langle 2, -6 \rangle$ and $\mathbf{v} = \langle 9, 3 \rangle$ are orthogonal because

$$\mathbf{u} \cdot \mathbf{v} = \langle 2, -6 \rangle \cdot \langle 9, 3 \rangle = 2 \cdot 9 + (-6)3 = 18 - 18 = 0.$$

(b) The vectors $\frac{1}{2}\mathbf{i} + 5\mathbf{j}$ and $10\mathbf{i} - \mathbf{j}$ are orthogonal since

$$\left(\frac{1}{2}\mathbf{i} + 5\mathbf{j}\right) \cdot (10\mathbf{i} - \mathbf{j}) = \frac{1}{2}(10) + 5(-1) = 5 - 5 = 0. \quad ■$$

Projections and Components

If **u** and **v** are nonzero vectors and θ is the angle between them, construct the perpendicular line segment from the terminal point P of **u** to the straight line on which **v** lies. This perpendicular segment intersects the line at a point Q, as shown in Figure 9–29.

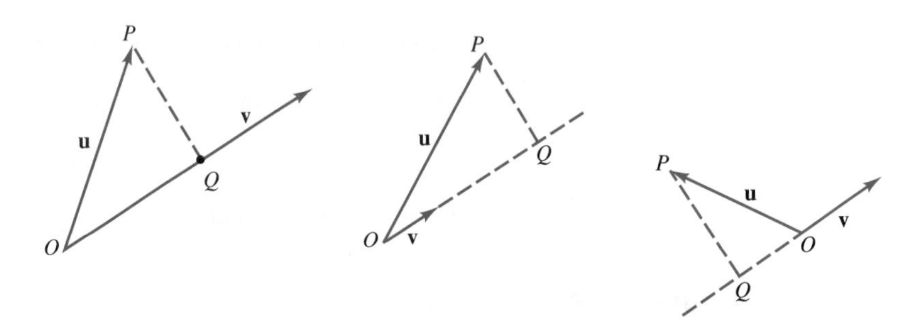

Figure 9-29

The vector $\overrightarrow{OQ}$ is called the **projection of u on v** and is denoted $\text{proj}_{\mathbf{v}}\mathbf{u}$. Here is a useful description of $\text{proj}_{\mathbf{v}}\mathbf{u}$.

Projection of u on v ▶

If u and v are nonzero vectors, then the projection of u on v is the vector

$$\text{proj}_{\mathbf{v}}\mathbf{u} = \left(\frac{\mathbf{u}\cdot\mathbf{v}}{\|\mathbf{v}\|^2}\right)\mathbf{v}.$$

Proof Since $\text{proj}_{\mathbf{v}}\mathbf{u}$ and **v** lie on the same straight line, they are parallel, and hence $\text{proj}_{\mathbf{v}}\mathbf{u} = k\mathbf{v}$ for some real number k. Let **w** be the vector with initial point at the origin and the same length and direction as $\overrightarrow{QP}$ as in the two cases shown in Figure 9–30:

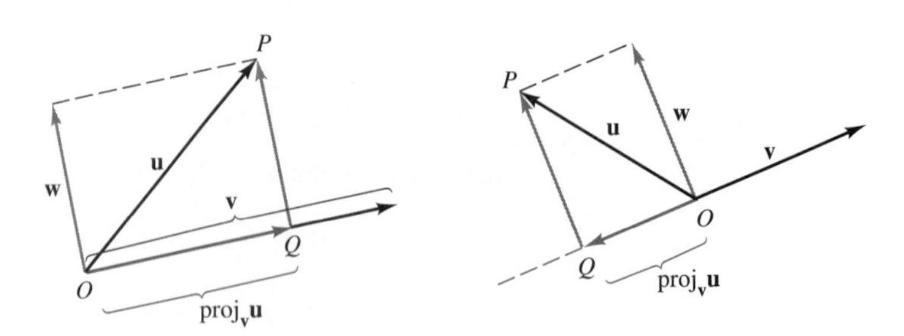

Figure 9-30

Note that $\mathbf{w}$ is parallel to $\overrightarrow{QP}$ and hence is orthogonal to $\mathbf{v}$. As shown in Figure 9–30, $\mathbf{u} = \text{proj}_{\mathbf{v}}\mathbf{u} + \mathbf{w} = k\mathbf{v} + \mathbf{w}$. Consequently, by the properties of the dot product

$$\begin{aligned}\mathbf{u}\cdot\mathbf{v} &= (k\mathbf{v} + \mathbf{w})\cdot\mathbf{v} = (k\mathbf{v})\cdot\mathbf{v} + \mathbf{w}\cdot\mathbf{v}\\ &= k(\mathbf{v}\cdot\mathbf{v}) + \mathbf{w}\cdot\mathbf{v} = k\|\mathbf{v}\|^2 + \mathbf{w}\cdot\mathbf{v}.\end{aligned}$$

But $\mathbf{w}\cdot\mathbf{v} = 0$ because $\mathbf{w}$ and $\mathbf{v}$ are orthogonal. Hence,

$$\mathbf{u}\cdot\mathbf{v} = k\|\mathbf{v}\|^2, \qquad \text{or equivalently,} \qquad k = \frac{\mathbf{u}\cdot\mathbf{v}}{\|\mathbf{v}\|^2}.$$

Therefore,

$$\text{proj}_{\mathbf{v}}\mathbf{u} = k\mathbf{v} = \left(\frac{\mathbf{u}\cdot\mathbf{v}}{\|\mathbf{v}\|^2}\right)\mathbf{v}$$

and the proof is complete. ❑

EXAMPLE 5 If $\mathbf{u} = 8\mathbf{i} + 3\mathbf{j}$ and $\mathbf{v} = 4\mathbf{i} - 2\mathbf{j}$, then

$$\mathbf{u}\cdot\mathbf{v} = 8\cdot 4 + 3(-2) = 26 \qquad \text{and} \qquad \|\mathbf{v}\|^2 = \mathbf{v}\cdot\mathbf{v} = 4^2 + (-2)^2 = 20.$$

Therefore,

$$\text{proj}_{\mathbf{v}}\mathbf{u} = \left(\frac{\mathbf{u}\cdot\mathbf{v}}{\|\mathbf{v}\|^2}\right)\mathbf{v} = \frac{26}{20}(4\mathbf{i} - 2\mathbf{j}) = \frac{26}{5}\mathbf{i} - \frac{13}{5}\mathbf{j},$$

as shown in Figure 9–31. We can also find the projection of $\mathbf{v}$ on $\mathbf{u}$ by noting that $\|\mathbf{u}\|^2 = \mathbf{u}\cdot\mathbf{u} = 8^2 + 3^2 = 73$, and hence

$$\text{proj}_{\mathbf{u}}\mathbf{v} = \left(\frac{\mathbf{v}\cdot\mathbf{u}}{\|\mathbf{u}\|^2}\right)\mathbf{u} = \frac{26}{73}(8\mathbf{i} + 3\mathbf{j}) = \frac{208}{73}\mathbf{i} + \frac{78}{73}\mathbf{j}.$$ ■

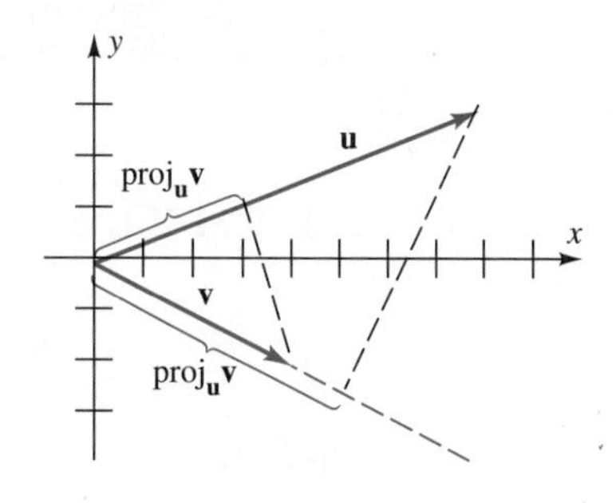

Figure 9-31

Recall that $\frac{1}{\|\mathbf{v}\|}\mathbf{v}$ is a unit vector in the direction of **v** (see page 564). We can express $\text{proj}_{\mathbf{v}}\mathbf{u}$ as a scalar multiple of this unit vector as follows:

$$\text{proj}_{\mathbf{v}}\mathbf{u} = \left(\frac{\mathbf{u}\cdot\mathbf{v}}{\|\mathbf{v}\|^2}\right)\mathbf{v} = \left(\frac{\mathbf{u}\cdot\mathbf{v}}{\|\mathbf{v}\|}\right)\left(\frac{1}{\|\mathbf{v}\|}\mathbf{v}\right).$$

The scalar $\frac{\mathbf{u}\cdot\mathbf{v}}{\|\mathbf{v}\|}$ is called the **component of u along v** and is denoted $\text{comp}_{\mathbf{v}}\mathbf{u}$. Thus,

$$\text{proj}_{\mathbf{v}}\mathbf{u} = \left(\frac{\mathbf{u}\cdot\mathbf{v}}{\|\mathbf{v}\|}\right)\left(\frac{1}{\|\mathbf{v}\|}\mathbf{v}\right) = \text{comp}_{\mathbf{v}}\mathbf{u}\left(\frac{1}{\|\mathbf{v}\|}\mathbf{v}\right).$$

Since $\frac{1}{\|\mathbf{v}\|}\mathbf{v}$ is a unit vector, the length of $\text{proj}_{\mathbf{v}}\mathbf{u}$ is

$$\|\text{proj}_{\mathbf{v}}\mathbf{u}\| = \left\|\text{comp}_{\mathbf{v}}\mathbf{u}\left(\frac{1}{\|\mathbf{v}\|}\mathbf{v}\right)\right\| = |\text{comp}_{\mathbf{v}}\mathbf{u}|\left\|\frac{1}{\|\mathbf{v}\|}\mathbf{v}\right\| = |\text{comp}_{\mathbf{v}}\mathbf{u}|.$$

Furthermore, since $\mathbf{u}\cdot\mathbf{v} = \|\mathbf{u}\|\,\|\mathbf{v}\|\cos\theta$, where θ is the angle between **u** and **v**, we have

$$\text{comp}_{\mathbf{v}}\mathbf{u} = \frac{\mathbf{u}\cdot\mathbf{v}}{\|\mathbf{v}\|} = \frac{\|\mathbf{u}\|\,\|\mathbf{v}\|\cos\theta}{\|\mathbf{v}\|}.$$

Cancelling $\|\mathbf{v}\|$ on the right side produces this result:

Projections and Components ▶

If u and v are nonzero vectors and θ is the angle between them, then

$$\text{comp}_{\mathbf{v}}\mathbf{u} = \frac{\mathbf{u}\cdot\mathbf{v}}{\|\mathbf{v}\|} = \|\mathbf{u}\|\cos\theta$$

and

$$\|\text{proj}_{\mathbf{v}}\mathbf{u}\| = |\text{comp}_{\mathbf{v}}\mathbf{u}|.$$

EXAMPLE 6 If $\mathbf{u} = 2\mathbf{i} + 3\mathbf{j}$ and $\mathbf{v} = -5\mathbf{i} + 2\mathbf{j}$, then

$$\text{comp}_{\mathbf{v}}\mathbf{u} = \frac{\mathbf{u}\cdot\mathbf{v}}{\|\mathbf{v}\|} = \frac{2(-5) + 3\cdot 2}{\sqrt{(-5)^2 + 2^2}} = \frac{-4}{\sqrt{29}}.$$

Similarly,

$$\text{comp}_{\mathbf{u}}\mathbf{v} = \frac{\mathbf{v}\cdot\mathbf{u}}{\|\mathbf{u}\|} = \frac{-4}{\sqrt{2^2 + 3^2}} = \frac{-4}{\sqrt{13}}.$$ ■

Applications

Vectors and the dot product can be used to solve a variety of physical problems.

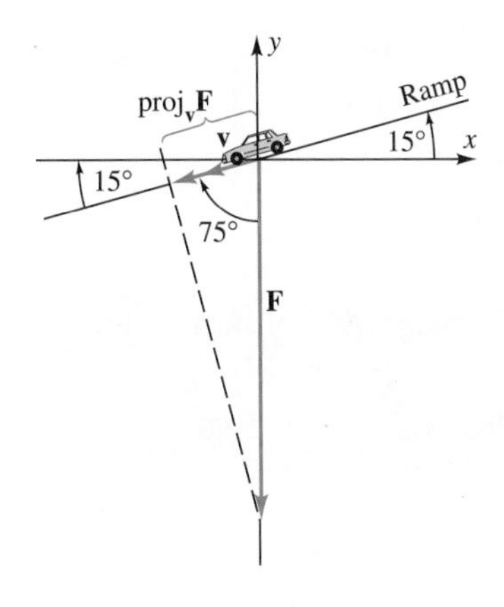

Figure 9-32

EXAMPLE 7 A 4000-pound automobile is on an inclined ramp that makes a 15° angle with the horizontal. Find the force required to keep it from rolling down the ramp, assuming that the only force that must be overcome is that due to gravity.

Solution The situation is shown in Figure 9–32, where the coordinate system is chosen so that the car is at the origin, the vector **F** representing the downward force of gravity is on the y-axis, and **v** is a unit vector from the origin down the ramp. Since the car weighs 4000 pounds, $\mathbf{F} = -4000\mathbf{j}$. Figure 9–32 shows that the angle between **v** and **F** is 75°. The vector $\text{proj}_{\mathbf{v}}\mathbf{F}$ is the force pulling the car down the ramp, so a force of the same magnitude in the opposite direction is needed to keep the car motionless. As we saw in the box above,

$$\|\text{proj}_{\mathbf{v}}\mathbf{F}\| = |\text{comp}_{\mathbf{v}}\mathbf{F}| = \big|\|\mathbf{F}\| \cos 75°\big|$$
$$\approx 4000(.25882) \approx 1035.3.$$

Therefore, a force of 1035.3 pounds is required to hold the car in place. ■

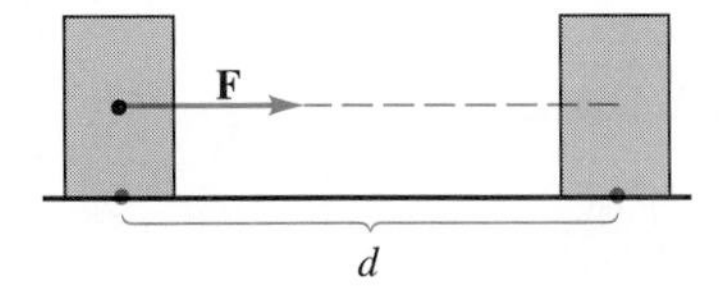

Figure 9-33

If a constant force **F** is applied to an object, pushing or pulling it a distance d in the direction of the force as shown in Figure 9–33, the amount of **work** done by the force is defined to be the product

$$W = (\text{magnitude of force})(\text{distance}) = \|\mathbf{F}\| \cdot d.$$

If the magnitude of **F** is measured in pounds and d in feet, then the units for W are foot-pounds. For example, if you push a car for 35 feet along a level driveway by exerting a constant force of 110 pounds, the amount of work done is $110 \cdot 35 = 3850$ foot-pounds.

When a force **F** moves an object in the direction of a vector **d** rather than in the direction of **F**, as shown in Figure 9–34, then the motion of the object can be considered as the result of the vector $\text{proj}_{\mathbf{d}}\mathbf{F}$, which is a force in the same direction as **d**.

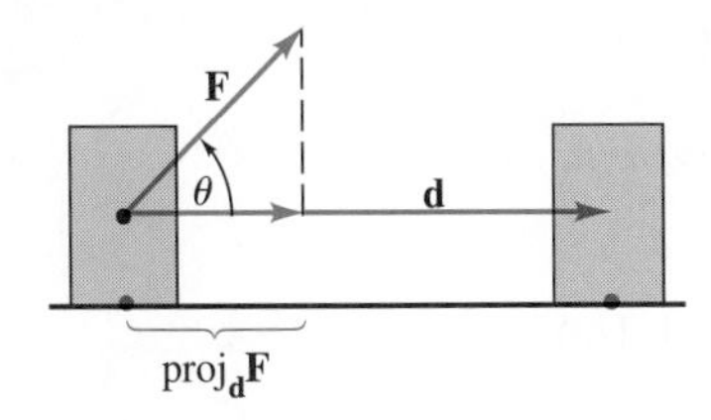

Figure 9-34

Therefore, the amount of work done by **F** is the same as the amount of work done by $\text{proj}_{\mathbf{d}}\mathbf{F}$, namely,

$$W = (\text{magnitude of proj}_{\mathbf{d}}\mathbf{F})(\text{length of } \mathbf{d}) = \|\text{proj}_{\mathbf{d}}\mathbf{F}\| \cdot \|\mathbf{d}\|.$$

The box on page 578 and the Angle Theorem (page 574) show that

$$\begin{aligned} W = \|\text{proj}_{\mathbf{d}}\mathbf{F}\| \cdot \|\mathbf{d}\| &= |\text{comp}_{\mathbf{d}}\mathbf{F}| \cdot \|\mathbf{d}\| \\ &= \|\mathbf{F}\|(\cos\theta)\|\mathbf{d}\| \\ &= \mathbf{F} \cdot \mathbf{d}.^* \end{aligned}$$

Consequently, we have these descriptions of work:

Work ▶

> **The work W done by a constant force F as its point of application moves along the vector d is**
>
> $$W = |\text{comp}_{\mathbf{d}}\mathbf{F}| \cdot \|\mathbf{d}\|, \quad \text{or equivalently,} \quad W = \mathbf{F} \cdot \mathbf{d}.$$

EXAMPLE 8 How much work is done by a child who pulls a sled 100 feet over level ground by exerting a constant 20-pound force on a rope that makes a 45° angle with the ground?

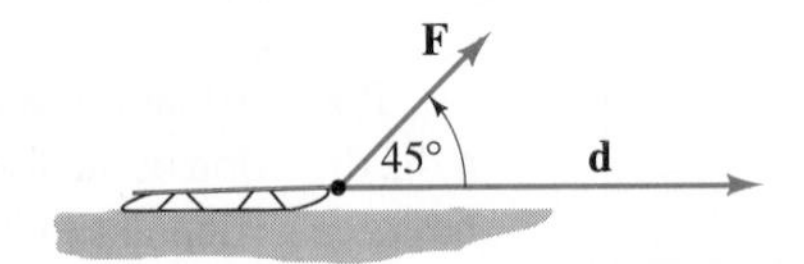

Figure 9-35

Solution The situation is shown in Figure 9–35, where the force **F** on the rope has magnitude 20 and the sled moves along vector **d** of length 100. The work done is

$$\begin{aligned} W = \mathbf{F} \cdot \mathbf{d} &= \|\mathbf{F}\|\,\|\mathbf{d}\| \cdot \cos\theta = 20 \cdot 100 \cdot \frac{\sqrt{2}}{2} \\ &= 100\sqrt{2} \approx 1414.2 \text{ foot-pounds.} \end{aligned}$$ ■

EXERCISES 9.4

In Exercises 1–6, find $\mathbf{u} \cdot \mathbf{v}$, $\mathbf{u} \cdot \mathbf{u}$, and $\mathbf{v} \cdot \mathbf{v}$.

1. $\mathbf{u} = \langle 3, 4 \rangle$, $\mathbf{v} = \langle -5, 2 \rangle$
2. $\mathbf{u} = \langle -1, 6 \rangle$, $\mathbf{v} = \langle -4, 1/3 \rangle$
3. $\mathbf{u} = 2\mathbf{i} + \mathbf{j}$, $\mathbf{v} = 3\mathbf{i}$
4. $\mathbf{u} = \mathbf{i} - \mathbf{j}$, $\mathbf{v} = 5\mathbf{j}$
5. $\mathbf{u} = 3\mathbf{i} + 2\mathbf{j}$, $\mathbf{v} = 2\mathbf{i} + 3\mathbf{j}$
6. $\mathbf{u} = 4\mathbf{i} - \mathbf{j}$, $\mathbf{v} = -\mathbf{i} + 2\mathbf{j}$

*This formula reduces to the previous one when **F** and **d** have the same direction because in that case $\cos\theta = \cos 0 = 1$, so that $W = \|\mathbf{F}\| \cdot \|\mathbf{d}\|$ = (magnitude of force)(distance moved).

In Exercises 7–12, find the dot product when $\mathbf{u} = \langle 2, 5\rangle$, $\mathbf{v} = \langle -4, 3\rangle$, $\mathbf{w} = \langle 2, -1\rangle$.

7. $\mathbf{u}\cdot(\mathbf{v}+\mathbf{w})$ **8.** $\mathbf{u}\cdot(\mathbf{v}-\mathbf{w})$

9. $(\mathbf{u}+\mathbf{v})\cdot(\mathbf{v}+\mathbf{w})$ **10.** $(\mathbf{u}+\mathbf{v})\cdot(\mathbf{u}-\mathbf{v})$

11. $(3\mathbf{u}+\mathbf{v})\cdot(2\mathbf{w})$ **12.** $(\mathbf{u}+4\mathbf{v})\cdot(2\mathbf{u}+\mathbf{w})$

In Exercises 13–18, find the angle between the two vectors.

13. $\langle 4, -3\rangle, \langle 1, 2\rangle$ **14.** $\langle 2, 4\rangle, \langle 0, -5\rangle$

15. $2\mathbf{i}-3\mathbf{j},\ -\mathbf{i}$ **16.** $2\mathbf{j},\ 4\mathbf{i}+\mathbf{j}$

17. $\sqrt{2}\mathbf{i}+\sqrt{2}\mathbf{j},\ \mathbf{i}-\mathbf{j}$ **18.** $3\mathbf{i}-5\mathbf{j},\ -2\mathbf{i}+3\mathbf{j}$

In Exercises 19–24, determine whether the given vectors are parallel, orthogonal, or neither.

19. $\langle 2, 6\rangle, \langle 3, -1\rangle$ **20.** $\langle -5, 3\rangle, \langle 2, 6\rangle$

21. $\langle 9, -6\rangle, \langle -6, 4\rangle$ **22.** $-\mathbf{i}+2\mathbf{j},\ 2\mathbf{i}-4\mathbf{j}$

23. $2\mathbf{i}-2\mathbf{j},\ 5\mathbf{i}+8\mathbf{j}$ **24.** $6\mathbf{i}-4\mathbf{j},\ 2\mathbf{i}+3\mathbf{j}$

In Exercises 25–28, find a real number k such that the two vectors are orthogonal.

25. $2\mathbf{i}+3\mathbf{j},\ 3\mathbf{i}-k\mathbf{j}$ **26.** $-3\mathbf{i}+\mathbf{j},\ 2k\mathbf{i}-4\mathbf{j}$

27. $\mathbf{i}-\mathbf{j},\ k\mathbf{i}+\sqrt{2}\mathbf{j}$ **28.** $-4\mathbf{i}+5\mathbf{j},\ 2\mathbf{i}+2k\mathbf{j}$

In Exercises 29–32, find $\text{proj}_{\mathbf{u}}\mathbf{v}$ *and* $\text{proj}_{\mathbf{v}}\mathbf{u}$.

29. $\mathbf{u} = 3\mathbf{i}-5\mathbf{j},\ \mathbf{v} = 6\mathbf{i}+2\mathbf{j}$

30. $\mathbf{u} = 2\mathbf{i}-3\mathbf{j},\ \mathbf{v} = \mathbf{i}+2\mathbf{j}$

31. $\mathbf{u} = \mathbf{i}+\mathbf{j},\ \mathbf{v} = \mathbf{i}-\mathbf{j}$

32. $\mathbf{u} = 5\mathbf{i}+\mathbf{j},\ \mathbf{v} = -2\mathbf{i}+3\mathbf{j}$

In Exercises 33–36, find $\text{comp}_{\mathbf{v}}\mathbf{u}$.

33. $\mathbf{u} = 10\mathbf{i}+4\mathbf{j},\ \mathbf{v} = 3\mathbf{i}-2\mathbf{j}$

34. $\mathbf{u} = \mathbf{i}-2\mathbf{j},\ \mathbf{v} = 3\mathbf{i}+\mathbf{j}$

35. $\mathbf{u} = 3\mathbf{i}+2\mathbf{j},\ \mathbf{v} = -\mathbf{i}+3\mathbf{j}$

36. $\mathbf{u} = \mathbf{i}+\mathbf{j},\ \mathbf{v} = -3\mathbf{i}-2\mathbf{j}$

In Exercises 37–39, let $\mathbf{u} = \langle a, b\rangle$, $\mathbf{v} = \langle c, d\rangle$, *and* $\mathbf{w} = \langle r, s\rangle$. *Verify that the given property of dot products is valid by calculating the quantities on each side of the equal sign.*

37. $\mathbf{u}\cdot(\mathbf{v}+\mathbf{w}) = \mathbf{u}\cdot\mathbf{v}+\mathbf{u}\cdot\mathbf{w}$

38. $k\mathbf{u}\cdot\mathbf{v} = k(\mathbf{u}\cdot\mathbf{v}) = \mathbf{u}\cdot k\mathbf{v}$

39. $\mathbf{0}\cdot\mathbf{u} = 0$

40. Suppose $\mathbf{u} = \langle a, b\rangle$ and $\mathbf{v} = \langle c, d\rangle$ are nonzero parallel vectors.

(a) If $c \neq 0$, show that $\mathbf{u}$ and $\mathbf{v}$ lie on the same nonvertical straight line through the origin.

(b) If $c \neq 0$, show that $\mathbf{v} = \frac{a}{c}\mathbf{u}$ (that is, $\mathbf{v}$ is a scalar multiple of $\mathbf{u}$). [*Hint:* The equation of the line on which $\mathbf{u}$ and $\mathbf{v}$ lie is $y = mx$ for some constant m (why?), which implies that $b = ma$ and $d = mc$.]

(c) If $c = 0$, show that $\mathbf{v}$ is a scalar multiple of $\mathbf{u}$. [*Hint:* If $c = 0$, then $a = 0$ (why?) and hence $b \neq 0$ (otherwise $\mathbf{u} = \mathbf{0}$).]

41. Prove the Angle Theorem in the case when θ is 0 or π.

42. If $\mathbf{u}$ and $\mathbf{v}$ are nonzero vectors such that $\mathbf{u}\cdot\mathbf{v} = 0$, show that $\mathbf{u}$ and $\mathbf{v}$ are orthogonal. [*Hint:* If θ is the angle between $\mathbf{u}$ and $\mathbf{v}$, what is $\cos\theta$ and what does this say about θ?]

43. Show that (1, 2), (3, 4), (5, 2) are the vertices of a right triangle by considering the sides of the triangle as vectors.

44. Find a number x such that the angle between the vectors $\langle 1, 1\rangle$ and $\langle x, 1\rangle$ is $\pi/4$ radians.

45. Find nonzero vectors $\mathbf{u}$, $\mathbf{v}$, and $\mathbf{w}$ such that $\mathbf{u}\cdot\mathbf{v} = \mathbf{u}\cdot\mathbf{w}$ and $\mathbf{v} \neq \mathbf{w}$ and neither $\mathbf{v}$ nor $\mathbf{w}$ is orthogonal to $\mathbf{u}$.

46. If $\mathbf{u}$ and $\mathbf{v}$ are nonzero vectors, show that the vectors $\|\mathbf{u}\|\mathbf{v}+\|\mathbf{v}\|\mathbf{u}$ and $\|\mathbf{u}\|\mathbf{v}-\|\mathbf{v}\|\mathbf{u}$ are orthogonal.

47. A 600-pound trailer is on an inclined ramp that makes a 30° angle with the horizontal. Find the force required to keep it from rolling down the ramp, assuming that the only force that must be overcome is that due to gravity.

48. In Example 7, find the vector that represents the force necessary to keep the car motionless.

In Exercises 49–52, find the work done by a constant force $\mathbf{F}$ *as the point of application of* $\mathbf{F}$ *moves along the vector* $\overrightarrow{PQ}$.

49. $\mathbf{F} = 2\mathbf{i}+5\mathbf{j},\ P = (0, 0),\ Q = (4, 1)$

50. $\mathbf{F} = \mathbf{i}-2\mathbf{j},\ P = (0, 0),\ Q = (-5, 2)$

51. $\mathbf{F} = 2\mathbf{i}+3\mathbf{j},\ P = (2, 3),\ Q = (5, 9)$ [*Hint:* Find the component form of $\overrightarrow{PQ}$.]

52. $\mathbf{F} = 5\mathbf{i}-\mathbf{j},\ P = (-1, 2),\ Q = (4, -3)$

53. A lawn mower handle makes an angle of 60° with the ground. A woman pushes on the handle with a force of 30 pounds. How much work is done in moving the lawn mower a distance of 75 feet on level ground?

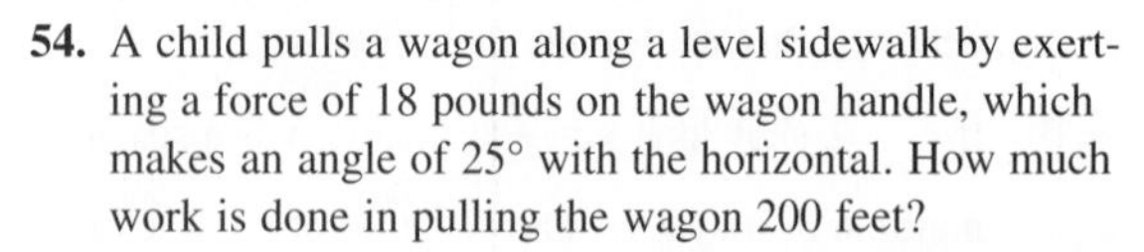

54. A child pulls a wagon along a level sidewalk by exerting a force of 18 pounds on the wagon handle, which makes an angle of 25° with the horizontal. How much work is done in pulling the wagon 200 feet?

55. A 40-pound cart is pushed 100 ft up a ramp that makes a 20° angle with the horizontal (Figure 9–36). How much work is done against gravity? [*Hint:* The amount of work done against gravity is the negative of the amount of work done *by* gravity. Coordinatize the situation so that the cart is at the origin. Then the cart moves along vector $\mathbf{d} = (100 \cos 20°)\mathbf{i} + (100 \sin 20°)\mathbf{j}$ and the downward force of gravity is $\mathbf{F} = 0\mathbf{i} - 40\mathbf{j}$.]

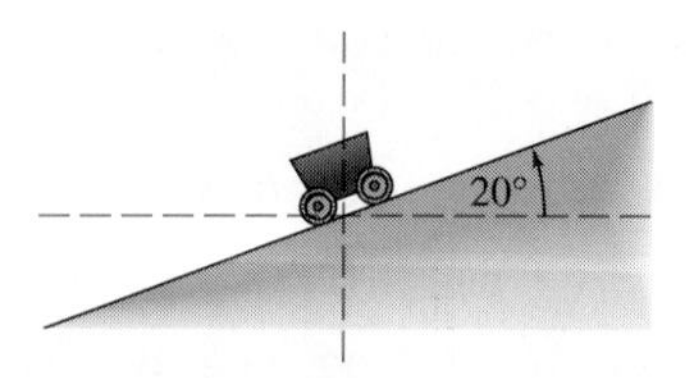

Figure 9-36

56. Suppose the child in Exercise 54 is pulling the wagon up a hill that makes an angle of 20° with the horizontal and all other facts remain the same. How much work is done in pulling the wagon 150 ft?

CHAPTER 9 *Review*

Important Concepts ▶

Section 9.1	Complex plane 542–543
	Real axis 543
	Imaginary axis 543
	Absolute value 543
	Polar form 544
	Argument 544
	Modulus 544
	Multiplication and division in polar form 546
Section 9.2	DeMoivre's Theorem 549
	Formula for *n*th roots 552
	Roots of unity 553
Section 9.3	Vector 557
	Magnitude 559
	Components 559
	Scalar multiplication 560
	Vector addition and subtraction 561, 562
	Unit vector 563
	Linear combination of **i** and **j** 564
	Direction angle of a vector 564
Section 9.4	Dot product 572
	Angle between vectors 573
	Parallel vectors 573
	Schwarz inequality 575
	Orthogonal vectors 576
	Projection of **u** on **v** 576
	Component of **u** along **v** 578
	Work 579

Important Facts and Formulas ▶

- $|a + bi| = \sqrt{a^2 + b^2}$
- $a + bi = r(\cos\theta + i\sin\theta)$, where

$$r = \sqrt{a^2 + b^2}, \qquad a = r\cos\theta, \qquad b = r\sin\theta$$

- $r_1(\cos\theta_1 + i\sin\theta_1)\cdot r_2(\cos\theta_2 + i\sin\theta_2) = r_1r_2(\cos(\theta_1 + \theta_2) + i\sin(\theta_1 + \theta_2))$
- $\dfrac{r_1(\cos\theta_1 + i\sin\theta_1)}{r_2(\cos\theta_2 + i\sin\theta_2)} = \dfrac{r_1}{r_2}(\cos(\theta_1 - \theta_2) + i\sin(\theta_1 - \theta_2))$

- *DeMoivre's Theorem:*

$$[r(\cos\theta + i\sin\theta)]^n = r^n(\cos(n\theta) + i\sin(n\theta))$$

- The distinct nth roots of $r(\cos\theta + i\sin\theta)$ are

$$\sqrt[n]{r}\left(\cos\left(\frac{\theta + 2k\pi}{n}\right) + \sin\left(\frac{\theta + 2k\pi}{n}\right)\right) \quad (k = 0, 1, 2, \ldots, n-1)$$

- The distinct nth roots of unity are

$$\cos\frac{2k\pi}{n} + i\sin\frac{2k\pi}{n} \qquad (k = 0, 1, 2, \ldots, n-1)$$

- If $P = (x_1, y_1)$ and $Q = (x_2, y_2)$, then $\overrightarrow{PQ} = \langle x_2 - x_1, y_2 - y_1\rangle$.
- $||\langle a, b\rangle|| = \sqrt{a^2 + b^2}$
- If $\mathbf{u} = \langle a, b\rangle$ and k is a scalar, then $k\mathbf{u} = \langle ka, kb\rangle$.
- If $\mathbf{u} = \langle a, b\rangle$ and $\mathbf{v} = \langle c, d\rangle$, then

$$\mathbf{u} + \mathbf{v} = \langle a + c, b + d\rangle \qquad \text{and} \qquad \mathbf{u} - \mathbf{v} = \langle a - c, b - d\rangle$$

- *Properties of Vector Addition and Scalar Multiplication:* For any vectors $\mathbf{u}$, $\mathbf{v}$, and $\mathbf{w}$ and any scalars r and s,
  1. $\mathbf{u} + (\mathbf{v} + \mathbf{w}) = (\mathbf{u} + \mathbf{v}) + \mathbf{w}$
  2. $\mathbf{u} + \mathbf{v} = \mathbf{v} + \mathbf{u}$
  3. $\mathbf{v} + \mathbf{0} = \mathbf{v} = \mathbf{0} + \mathbf{v}$
  4. $\mathbf{v} + (-\mathbf{v}) = \mathbf{0}$
  5. $r(\mathbf{u} + \mathbf{v}) = r\mathbf{u} + r\mathbf{v}$
  6. $(r + s)\mathbf{v} = r\mathbf{v} + s\mathbf{v}$
  7. $(rs)\mathbf{v} = r(s\mathbf{v}) = s(r\mathbf{v})$
  8. $1\mathbf{v} = \mathbf{v}$
  9. $0\mathbf{v} = \mathbf{0} = r\mathbf{0}$

- If $\mathbf{v} = \langle a, b\rangle = a\mathbf{i} + b\mathbf{j}$, then

$$a = ||\mathbf{v}||\cos\theta \qquad \text{and} \qquad b = ||\mathbf{v}||\sin\theta$$

  where θ is the direction angle of $\mathbf{v}$.
- If $\mathbf{v} = \langle a, b\rangle = a\mathbf{i} + b\mathbf{j}$ and $\mathbf{v} = \langle c, d\rangle = c\mathbf{i} + d\mathbf{j}$, then

$$\mathbf{u}\cdot\mathbf{v} = ac + bd.$$

- If θ is the angle between nonzero vectors $\mathbf{u}$ and $\mathbf{v}$, then

$$\mathbf{u}\cdot\mathbf{v} = ||\mathbf{u}||\,||\mathbf{v}||\cos\theta$$

- *Schwarz Inequality:* $|\mathbf{u}\cdot\mathbf{v}| \leq ||\mathbf{u}||\,||\mathbf{v}||$.

- Vectors **u** and **v** are orthogonal exactly when $\mathbf{u}\cdot\mathbf{v} = 0$.
- $\text{proj}_{\mathbf{v}}\mathbf{u} = \left(\dfrac{\mathbf{u}\cdot\mathbf{v}}{\|\mathbf{v}\|^2}\right)\mathbf{v}$
- $\text{comp}_{\mathbf{v}}\mathbf{u} = \dfrac{\mathbf{u}\cdot\mathbf{v}}{\|\mathbf{v}\|} = \|\mathbf{u}\|\cos\theta$, where θ is the angle between **u** and **v**.

Review Questions ▶

1. Simplify: $|i(4 + 2i)| + |3 - i|$.
2. Simplify: $|3 + 2i| - |1 - 2i|$.
3. Graph the equation $|z| = 2$ in the complex plane.
4. Graph the equation $|z - 3| = 1$ in the complex plane.
5. Express in polar form: $1 + \sqrt{3}i$.
6. Express in polar form: $4 - 5i$.

In Questions 7–11, express the given number in the form a + bi.

7. $2\left(\cos\dfrac{\pi}{12} + i\sin\dfrac{\pi}{12}\right)\cdot 4\left(\cos\dfrac{\pi}{6} + i\sin\dfrac{\pi}{6}\right)$
8. $3\left(\cos\dfrac{\pi}{8} + i\sin\dfrac{\pi}{8}\right)\cdot 2\left(\cos\dfrac{3\pi}{8} + i\sin\dfrac{3\pi}{8}\right)$
9. $\dfrac{12\left(\cos\dfrac{7\pi}{12} + i\sin\dfrac{7\pi}{12}\right)}{3\left(\cos\dfrac{5\pi}{12} + i\sin\dfrac{5\pi}{12}\right)}$
10. $\left(\cos\dfrac{\pi}{12} + i\sin\dfrac{\pi}{12}\right)^{18}$
11. $\left(\sqrt[3]{3}\left(\cos\dfrac{5\pi}{36} + i\sin\dfrac{5\pi}{36}\right)\right)^{12}$

In Questions 12–16, solve the given equation in the complex number system and express your answers in polar form.

12. $x^3 = i$
13. $x^6 = 1$
14. $x^8 = -\sqrt{3} - 3i$
15. $x^4 = i$
16. $x^3 = 1 - i$

In Questions 17–20, let $\mathbf{u} = \langle 3, -2\rangle$ *and* $\mathbf{v} = \langle 8, 1\rangle$. *Find*

17. $\mathbf{u} + \mathbf{v}$
18. $\|-3\mathbf{v}\|$
19. $\|2\mathbf{v} - 4\mathbf{u}\|$
20. $3\mathbf{u} - \dfrac{1}{2}\mathbf{v}$

In Questions 21–24, let $\mathbf{u} = -2\mathbf{i} + \mathbf{j}$ *and* $\mathbf{v} = 3\mathbf{i} - 4\mathbf{j}$. *Find*

21. $4\mathbf{u} - \mathbf{v}$
22. $\mathbf{u} + 2\mathbf{v}$
23. $\|\mathbf{u} + \mathbf{v}\|$
24. $\|\mathbf{u}\| + \|\mathbf{v}\|$
25. Find the components of the vector **v** such that $\|\mathbf{v}\| = 5$ and the direction angle of **v** is 45°.

26. Find the magnitude and direction angle of $3\mathbf{i} + 4\mathbf{j}$.

27. Find a unit vector whose direction is *opposite* the direction of $3\mathbf{i} - 6\mathbf{j}$.

28. An object at the origin is acted upon by a 10-pound force with direction angle 90° and a 20-pound force with direction angle 30°. Find the magnitude and direction of the resultant force.

29. A plane flies in the direction 120°, with an air speed of 300 mph. The wind is blowing from north to south at 40 mph. Find the course and ground speed of the plane.

30. An object weighing 40 pounds lies on an inclined plane that makes a 30° angle with the horizontal. Find the components of the weight parallel and perpendicular to the plane.

In Questions 31–34, $\mathbf{u} = \langle 3, -4\rangle$, $\mathbf{v} = \langle -2, 5\rangle$, *and* $\mathbf{w} = \langle 0, 3\rangle$. *Find*

31. $\mathbf{u}\cdot\mathbf{v}$ **32.** $\mathbf{u}\cdot\mathbf{u} - \mathbf{v}\cdot\mathbf{v}$ **33.** $(\mathbf{u} + \mathbf{v})\cdot\mathbf{w}$ **34.** $(\mathbf{u} + \mathbf{w})\cdot(\mathbf{w} - 3\mathbf{v})$

35. What is the angle between the vectors $5\mathbf{i} - 2\mathbf{j}$ and $3\mathbf{i} + \mathbf{j}$?

36. Is $3\mathbf{i} - 2\mathbf{j}$ orthogonal to $4\mathbf{i} + 6\mathbf{j}$?

In Questions 37 and 38, $\mathbf{u} = 4\mathbf{i} - 3\mathbf{j}$ *and* $\mathbf{v} = 2\mathbf{i} + \mathbf{j}$. *Find*

37. $\text{proj}_{\mathbf{v}}\mathbf{u}$ **38.** $\text{comp}_{\mathbf{u}}\mathbf{v}$

39. If $\mathbf{u}$ and $\mathbf{v}$ have the same magnitude, show that $\mathbf{u} + \mathbf{v}$ and $\mathbf{u} - \mathbf{v}$ are orthogonal.

40. If $\mathbf{u}$ and $\mathbf{v}$ are nonzero vectors, show that the vector $\mathbf{u} - k\mathbf{v}$ is orthogonal to $\mathbf{v}$, where $k = \dfrac{\mathbf{u}\cdot\mathbf{v}}{\|\mathbf{v}\|^2}$.

41. A 3500-pound automobile is on an inclined ramp that makes a 30° angle with the horizontal. Find the force required to keep it from rolling down the ramp, assuming that the only force that must be overcome is that due to gravity.

42. A sled is pulled along level ground by a rope that makes a 50° angle with the horizontal. If a force of 40 pounds is used to pull the sled, how much work is done in pulling it 100 feet?

CHAPTER 10

Roadmap

Chapters 10–12 are independent of each other and may be read in any order.
Sections 2.1–2.3 and 2.5 are prerequisites for this chapter. Except for the discussion of standard equations for conics in Sections 2 and 3, Chapter 6 (Trigonometry) is also a prerequisite.
The interdependence of sections in this chapter is:

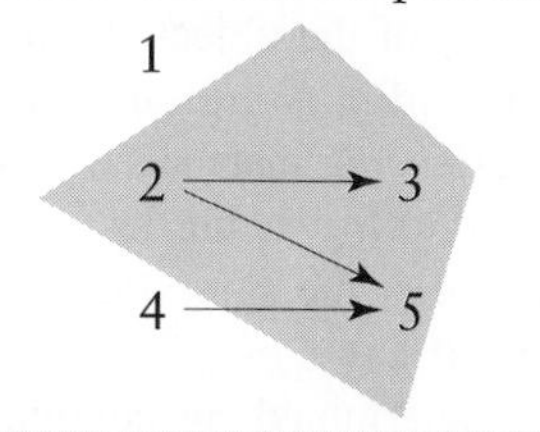

Analytic Geometry

The discussion of analytic geometry that was begun in Section 1.3 is continued here. A variety of interrelated topics are presented: a more thorough treatment of parametric graphing, an examination of conic sections (which have played a significant role in mathematics since ancient times), and an alternative method of coordinatizing the plane and graphing functions.

10.1 PLANE CURVES AND PARAMETRIC EQUATIONS

There are many curves in the plane that cannot be represented as the graph of a function $y = f(x)$. Parametric graphing enables us to represent such curves in terms of functions and also provides a formal definition of a curve in the plane.

Consider, for example, an object moving in the plane during a particular time interval. In order to describe both the path of the object and its location at any particular time, three variables are needed: the time t and the coordinates (x, y) of the object at time t. For instance, the coordinates x and y might be given by:

$$x = 4 \cos t + 5 \cos(3t) \qquad \text{and} \qquad y = \sin(3t) + t.$$

From $t = 0$ to $t = 12.5$, the object traces out the curve shown in Figure 10–1.

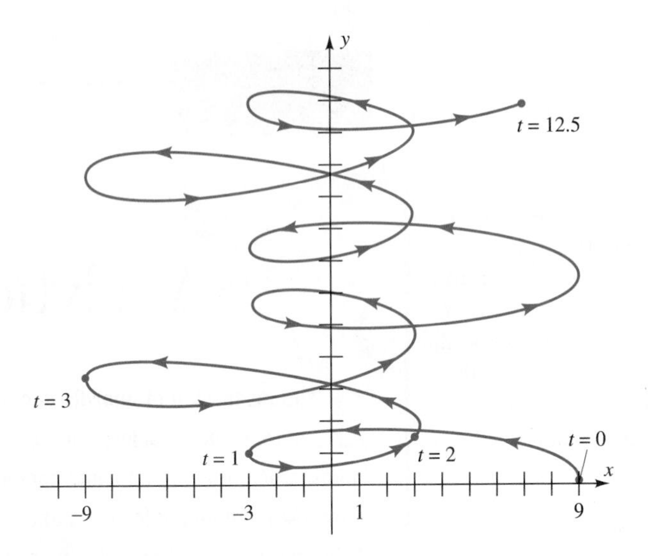

Figure 10-1

The points marked on the graph show the location of the object at various times. It begins at (9, 0) at time $t = 0$ and moves upward in an irregular spiral. The plotted points show that the speed of the object varies. For instance, it travels approximately 12 units in the time interval from $t = 0$ to $t = 1$, but less than 6 units from $t = 1$ to $t = 2$. Note that the object may be at the same location at different times (the points where the graph crosses itself).

In the preceding example, both x and y were determined by continuous functions of t, with t taking values in the interval [0, 12.5].* The example suggests the following definition.

Definition of Plane Curve ▶

Let f and g be continuous functions of t on an interval I. The set of all points (x, y) where

$$x = f(t) \qquad \text{and} \qquad y = g(t)$$

is called a *plane curve.* The variable t is called a *parameter* and the equations defining x and y are *parametric equations.*

In this general definition of "curve," the variable t need not represent time. As the examples below illustrate, different pairs of parametric equations may produce the same curve. Each such pair of parametric equations is called a **parameterization** of the curve.

*Intuitively, "continuous" means that the graph of the function that determines x, namely, $f(t) = 4\cos t + 5\cos 3t$, is a connected curve with no gaps or holes, and similarly for the function that determines y. Continuous functions are defined more precisely in Chapter 13, which is available as an optional supplement to this book.

▶ TECHNOLOGY TIP

The MODE key is labeled LIB on HP-38 and SET UP on Sharp and Casio 9800.

Graphing a curve given by parametric equations can be done by hand: evaluate x and y at many values of t, plot the corresponding points, and make an educated guess as to the shape of the curve. However, it's easier and more accurate to have your calculator do this. Use the "mode" key to change the graphing mode to "parametric" and enter the parametric equations for x and y. Set the viewing window by entering appropriate minimum and maximum values for x and y. You must also set minimum and maximum values for t and, on some calculators, specify the "t step" (or "t pitch"), which determines how much t changes before the next point is plotted.* If the t step is too large, relatively few points will be plotted and the graph may be quite rough. If the t step is too small, many points will be plotted and graphing may take a long time. As a general rule, a t step between .05 and .15 usually produces a satisfactory graph in a reasonable amount of time.

GRAPHING EXPLORATION Enter the parametric equations of the graph in Figure 10–1. Set the range values so that

$$-10 \le x \le 10, \qquad -5 \le y \le 15, \qquad 0 \le t \le 12.5.$$

Set t step = .1, unless your calculator sets it automatically. Graph the curve and use the trace feature to move along the curve to see just where the object is located at each time t between 0 and 12.5. ■

EXAMPLE 1 A calculator produced the graph in Figure 10–2 of the curve given by

$$x = t^3 - 10t^2 + 29t - 10 \quad \text{and} \quad y = t^2 - 7t + 14 \qquad (0 \le t \le 6.5).$$

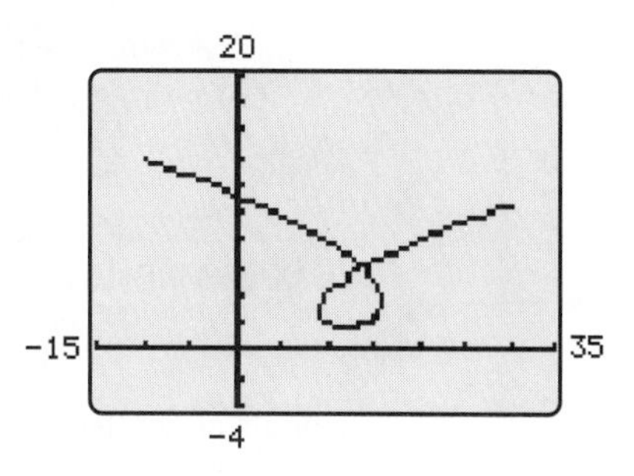

Figure 10-2

*A few calculators automatically determine the t step when the range of t values is entered.

GRAPHING EXPLORATION Set the same viewing window as in Figure 10–2 and graph the curve three separate times, using these ranges for t:

$$0 \leq t \leq 5; \qquad 1 \leq t \leq 6; \qquad -1 \leq t \leq 7.$$

What effect does changing the range of t have on the curve that is graphed? ■

EXAMPLE 2 The graph of the curve given by

$$x = -2t \qquad \text{and} \qquad y = 4t^2 - 4 \qquad (-1 \leq t \leq 2)$$

is shown in Figure 10-3.

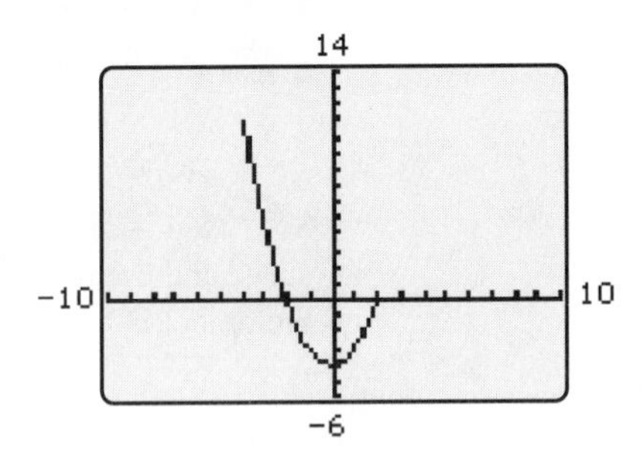

Figure 10-3

GRAPHING EXPLORATION Graph this curve on your calculator, using the same viewing window and range of t values as in Figure 10–3. In what direction is the curve traced out (that is, what is the first point graphed (corresponding to $t = -1$) and what is the last point graphed (corresponding to $t = 2$))? ■

EXAMPLE 3 The curve given by

$$x = t + 5 \cos t \qquad \text{and} \qquad y = 1 - 3 \sin t$$

spirals along the x-axis, as you will see for yourself.

GRAPHING EXPLORATION Graph this curve in the viewing window with $-25 \leq x \leq 25$, $-5 \leq y \leq 5$, and $-20 \leq t \leq 20$. ■

Some curves given by parametric equations can also be expressed as (part of) the graph of an equation in x and y. The process for doing this, called **eliminating the parameter,** is as follows:

Solve one of the parametric equations for t and substitute this result in the other parametric equation.

EXAMPLE 4 Consider the curve of Example 2, which was given by

$$x = -2t \quad \text{and} \quad y = 4t^2 - 4 \quad (-1 \le t \le 2).$$

Solving $x = -2t$ for t shows that $t = -x/2$. Substituting this in the second equation, we have

$$y = 4t^2 - 4 = 4\left(-\frac{x}{2}\right)^2 - 4 = 4\left(\frac{x^2}{4}\right) - 4 = x^2 - 4.$$

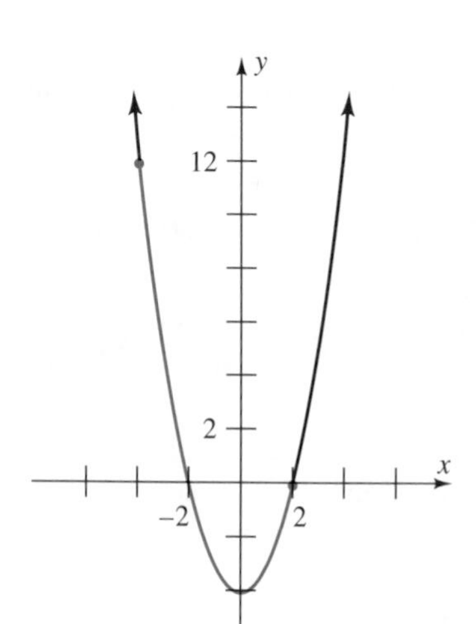

Figure 10-4

Therefore, every point on the curve is also on the graph of $y = x^2 - 4$. From Sections 2.5 or 3.3 we know that the graph of $y = x^2 - 4$ is the parabola in Figure 10–4. However, the curve given by the parametric equations is *not* the entire parabola, but only the part shown in color, which joins the points (2, 0) and (−4, 12) that correspond to the minimum and maximum values of t, namely, $t = -1$ and $t = 2$. ■

Having seen how to graph a curve given by parametric equations, we now consider the reverse problem: finding a parametric representation for the graph of an equation in x and y. This is easy when the equation defines y as a function of x, such as

$$y = x^3 + 5x^2 - 3x + 4.$$

A parametric description of this function can be obtained by changing the variable:

$$x = t \quad \text{and} \quad y = t^3 + 5t^2 - 3t + 4.$$

The same thing is true for equations that define x as a function of y, such as $x = y^2 + 3$, which can be parameterized by letting

$$x = t^2 + 3 \quad \text{and} \quad y = t.$$

In other cases, different techniques may be needed.

EXAMPLE 5 The graph of $(x - 4)^2 + (y - 1)^2 = 9$ is a circle with center (4, 1) and radius 3 (see page 28), so this equation does not define x as a function of y or y as a function of x. However, the graph of the equation is given by this parameterization:

$$(*) \qquad x = 3\cos t + 4 \quad \text{and} \quad y = 3\sin t + 1 \quad (0 \le t \le 2\pi)$$

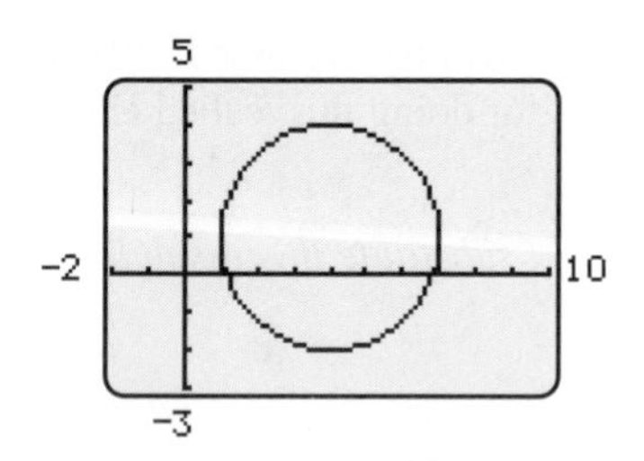

Figure 10-5

because by the Pythagorean identity

$$\begin{aligned}(x-4)^2+(y-1)^2 &= (3\cos t+4-4)^2+(3\sin t+1-1)^2\\ &= (3\cos t)^2+(3\sin t)^2 = 9\cos^2 t+9\sin^2 t\\ &= 9(\cos^2 t+\sin^2 t) = 9\cdot 1 = 9.\end{aligned}$$

With this parameterization the circle is traced out in a counterclockwise direction from the point (7, 1), as shown in Figure 10–5. Another parameterization is given by

$$x = 3\cos 2t + 4 \qquad \text{and} \qquad y = -3\sin 2t + 1 \qquad (0 \le t \le \pi).$$

GRAPHING EXPLORATION Verify that this parameterization traces out the circle in a clockwise direction, twice as fast as the parameterization given by (*), since t runs from 0 to π, rather than 2π. ■

The procedure in Example 5 works in the general case, as is proved in Exercise 28:

Parametric Equations of a Circle ▶

The circle with center (c, d) and radius r is given by the parametric equations:

$$x = r\cos t + c \qquad \text{and} \qquad y = r\sin t + d \qquad (0 \le t \le 2\pi).$$

EXAMPLE 6 Find three parameterizations of the straight line through $(1, -3)$ with slope -2.

Solution The point–slope form of the equation of this line is

$$y-(-3) = -2(x-1), \qquad \text{or equivalently,} \qquad y = -2x-1.$$

Its graph is shown in Figure 10–6. Since this equation defines y as a function of x, one parameterization is

$$x = t \qquad \text{and} \qquad y = -2t-1 \qquad (t \text{ any real number}).$$

A second parameterization is given by letting $x = t+1$; then

$$y = -2x-1 = -2(t+1)-1 = -2t-3 \qquad (t \text{ any real number}).$$

A third parameterization can be obtained by letting

$$x = \tan t \qquad \text{and} \qquad y = -2x-1 = -2\tan t - 1 \qquad (-\pi/2 < t < \pi/2).$$

When t runs from $-\pi/2$ to $\pi/2$, then $x = \tan t$ takes all possible real number values, and hence so does y. ■

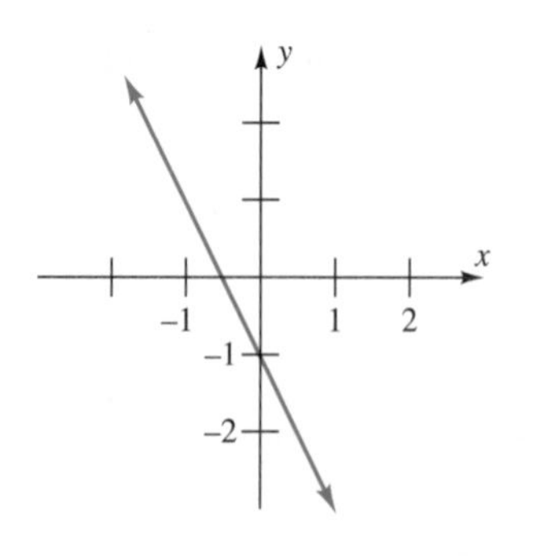

Figure 10-6

WARNING Some substitutions in an equation $y = f(x)$ do *not* lead to a parameterization of the entire graph. For instance, in Example 6, letting $x = t^2$ and substituting in the equation $y = -2x - 1$ leads to:

$$x = t^2 \qquad \text{and} \qquad y = -2t^2 - 1 \qquad \text{(any real number } t\text{)}.$$

Thus, x is always nonnegative and y is always negative. So the parameterization produces only the half of the line $y = -2x - 1$ to the right of the y-axis in Figure 10–6.

Applications

In the following applications we ignore air resistance and assume some facts about gravity that are proved in physics.

EXAMPLE 7 A golfer hits the ball with an initial velocity of 140 ft/sec so that its path as it leaves the ground makes an angle of 31° with the horizontal.

(a) When does the ball hit the ground?

(b) How far from its starting point does it land?

(c) What is the maximum height of the ball during its flight?

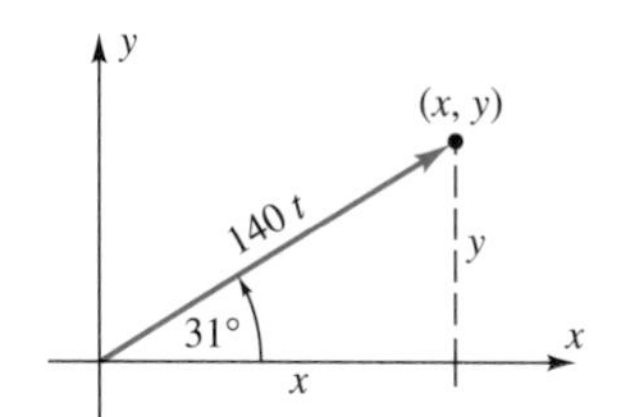

Figure 10-7

Solution Imagine that the golf ball starts at the origin and travels in the direction of the positive x-axis. If there were no gravity, the distance traveled by the ball in t seconds would be $140t$ ft. As shown in Figure 10–7, the coordinates (x, y) of the ball would satisfy

$$\frac{x}{140t} = \cos 31^\circ \qquad \frac{y}{140t} = \sin 31^\circ$$

$$x = (140 \cos 31^\circ)t \qquad y = (140 \sin 31^\circ)t.$$

However, there *is* gravity and at time t it exerts a force of $16t^2$ downward (that is, in the negative direction on the y-coordinate). Consequently, the coordinates of the golf ball at time t are

$$x = (140 \cos 31^\circ)t \qquad \text{and} \qquad y = (140 \sin 31^\circ)t - 16t^2.$$

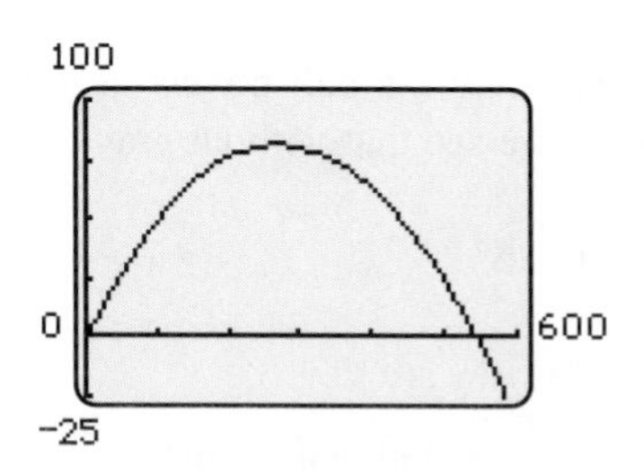

Figure 10-8

The path given by these parametric equations is graphed in Figure 10–8.*

(a) The ball is on the ground when $y = 0$, that is, at the x-intercepts of the graph. They can be found geometrically by using trace and zoom-in (the graphical

*Only the part of the graph on or above the x-axis represents the ball's path, since the ball does not go underground after it lands.

root finder does not operate in parametric mode), but this is very time consuming. To find the intercepts algebraically we need only set $y = 0$ and solve for t:

$$(140 \sin 31°)t - 16t^2 = 0$$
$$t(140 \sin 31° - 16t) = 0$$
$$t = 0 \quad \text{or} \quad 140 \sin 31° - 16t = 0$$
$$t = \frac{140 \sin 31°}{16} \approx 4.5066.$$

Thus, the ball hits the ground after approximately 4.5066 seconds.

(b) The horizontal distance traveled by the ball is given by the x-coordinate of the intercept. The x-coordinate when $t \approx 4.5066$ is

$$x = (140 \cos 31°)(4.5066) \approx 540.81 \text{ ft.}$$

(c) The graph in Figure 10–8 looks like a parabola and it is, as you can verify by eliminating the parameter t (Exercise 40). The y-coordinate of the vertex is the maximum height of the ball. It can be found geometrically by using trace and zoom-in (the min/max finder doesn't work in parametric mode) or algebraically as follows. The vertex occurs halfway between its two x-intercepts ($x = 0$ and $x \approx 540.81$), that is, when $x \approx 270.405$. Hence,

$$(140 \cos 31°)t = x = 270.405$$

so that

$$t = \frac{270.45}{140 \cos 31°} \approx 2.2533.$$

Therefore, the y-coordinate of the vertex (the maximum height of the ball) is

$$y = (140 \sin 31°)(2.2533) - 16(2.2533)^2 \approx 81.237 \text{ ft.} \quad ■$$

EXAMPLE 8 A batter hits a ball that is 3 ft above the ground and leaves the bat with an initial velocity of 138 ft/sec, making an angle of 26° with the horizontal and heading toward a 25-ft-high fence that is 400 ft away. Will the ball go over the fence?

Solution Imagine that home plate is at the origin and the ball travels in the direction of the positive x-axis. Using Figure 10–9 we see that without gravity

$$\frac{x}{138t} = \cos 26° \qquad \frac{y-3}{138t} = \sin 26°$$
$$x = (138 \cos 26°)t \qquad y = (138 \sin 26°)t + 3.$$

Allowing for the effect of gravity on the y-coordinate, we find that the ball's path is given by the parametric equations

$$x = (138 \cos 26°)t \quad \text{and} \quad y = (138 \sin 26°)t - 16t^2 + 3.$$

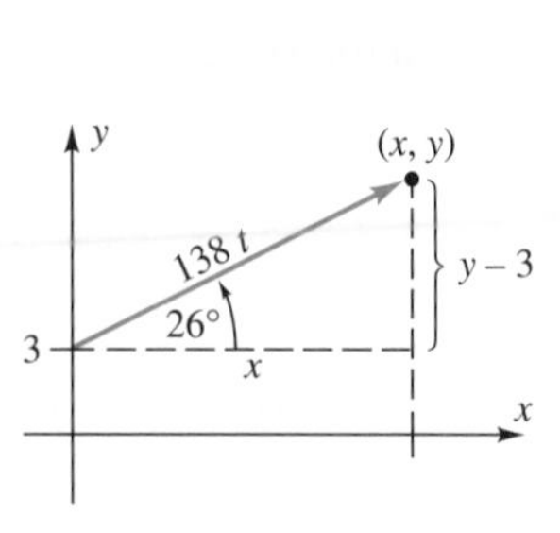

Figure 10-9

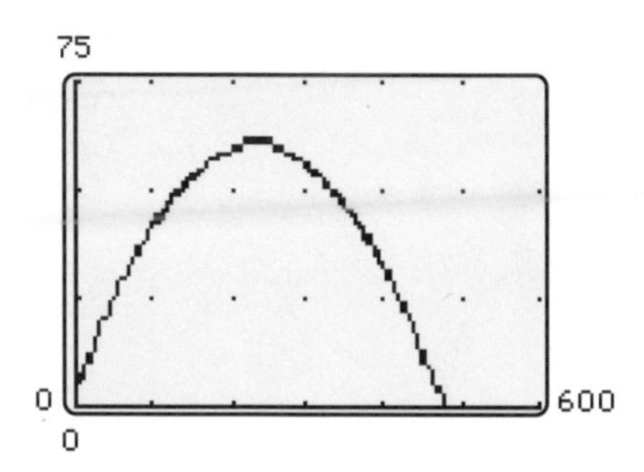

Figure 10-10

The graph of the ball's path in Figure 10–10 was made with the grid-on feature and vertical tic marks 25 units apart. It shows that the y-coordinate of the ball is greater than 25 when its x-coordinate is 400. So, the ball goes over the fence. ■

The procedure used in Example 8 applies to the general case. Replacing 3 by k, 26° by θ, and 138 by v leads to this conclusion:

Projectile Motion ▶

When a projectile is fired from the position $(0, k)$ on the positive y-axis at an angle θ with the horizontal, in the direction of the positive x-axis, with initial velocity v ft/sec, with negligible air resistance, then its position at time t seconds is given by the parametric equations:

$$x = (v \cos \theta)t \qquad \text{and} \qquad y = (v \sin \theta)t - 16t^2 + k.$$

▶ TECHNOLOGY TIP

In parametric graphing zoom-in can be very time consuming. It's often more effective to limit the t range to the values near the points you are interested in and set the t step very small. The picture may be hard to read, but trace can be used to determine coordinates.

GRAPHING EXPLORATION Will the ball in Example 8 go over the fence if its initial velocity is 135 ft/sec? Use degree mode and the viewing window of Figure 10–10 (with $0 \le t \le 4$ and t step = .1) to graph the ball's path. You may need to use trace if the graph is hard to read. If the answer still isn't clear, try changing the t step to .02.

Our final example is a curve that has several interesting applications.

EXAMPLE 9 Choose a point P on a circle of radius 3 and find a parametric description of the curve that is traced out by P as the circle rolls along the x-axis, as shown in Figure 10–11.

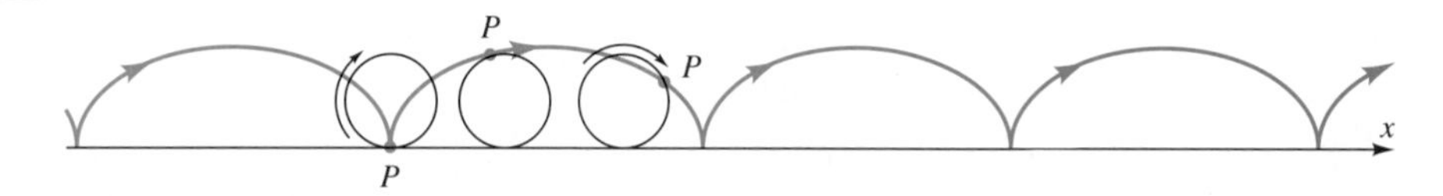

Figure 10-11

Solution This curve is called a **cycloid.** Begin with P at the origin and the center C of the circle at $(0, 3)$. As the circle rolls along the x-axis, the line segment CP moves from vertical through an angle of t radians, as shown in Figure 10–12.

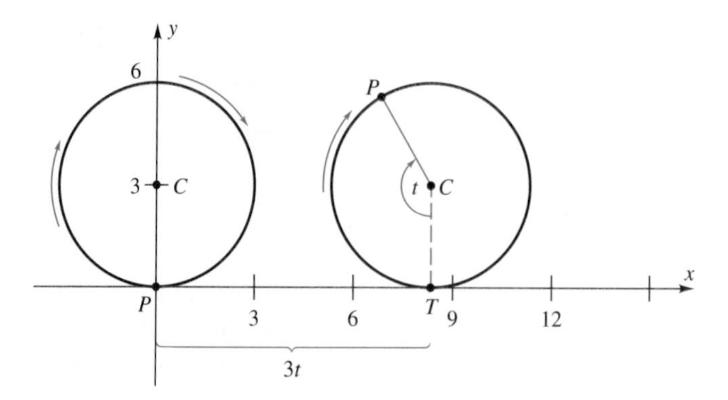

Figure 10-12

The distance from point T to the origin is the length of arc of the circle from T to P. As shown on page 371, this arc has length $3t$. Therefore the center C has coordinates $(3t, 3)$. When $0 < t < \pi/2$, the situation looks like Figure 10–13.

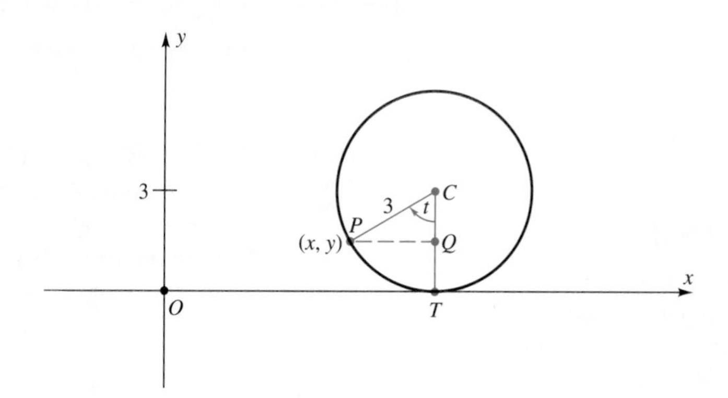

Figure 10-13

Right triangle PQC shows that

$$\sin t = \frac{PQ}{3}, \qquad \text{or equivalently,} \qquad PQ = 3 \sin t$$

and

$$\cos t = \frac{CQ}{3}, \qquad \text{or equivalently,} \qquad CQ = 3 \cos t.$$

In Figure 10–13, P has coordinates (x, y) and we have

$$x = OT - PQ = 3t - 3\sin t = 3(t - \sin t)$$

$$y = CT - CQ = 3 - 3\cos t = 3(1 - \cos t).$$

A similar analysis for other values of t (Exercises 47–49) shows that these equations are valid for every t. Therefore, the parametric equations of this cycloid are

$$x = 3(t - \sin t) \qquad \text{and} \qquad y = 3(1 - \cos t) \qquad (t \text{ any real number}).$$ ■

If a cycloid is traced out by a circle of radius r, then the argument given in Example 9, with r in place of 3, shows that the parametric equations of the cycloid are

$$x = r(t - \sin t) \qquad \text{and} \qquad y = r(1 - \cos t) \qquad (t \text{ any real number}).$$

Cycloids have a number of interesting applications. For example, among all the possible paths joining points P and Q in Figure 10–14, an arch of an inverted cycloid (shown in color) is the curve along which a particle (subject only to gravity) will slide from P to Q in the shortest possible time. This fact was first proved by J. Bernoulli in 1696.

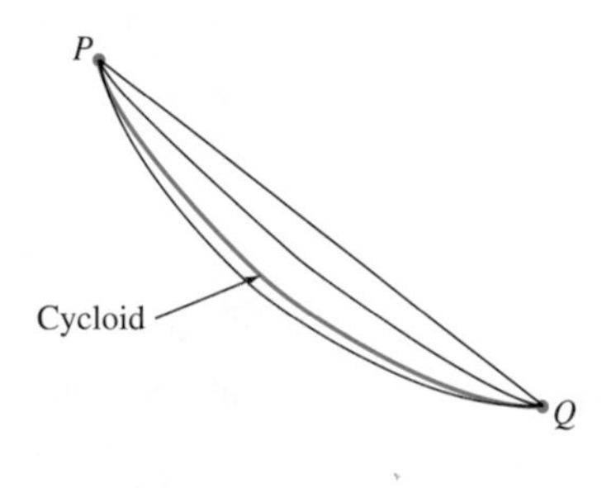

Figure 10-14

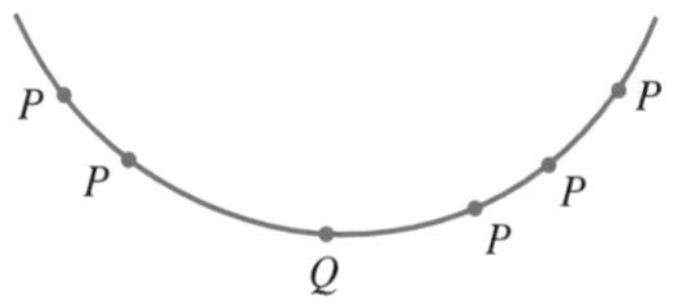

Figure 10-15

The Dutch physicist Huygens (who invented the pendulum clock) proved that a particle takes the same time to slide to the bottom point Q of an inverted cycloid arch (as in Figure 10–15) from *any* point P on the curve.

EXERCISES 10.1

In Exercises 1–14, find a viewing window that shows a complete graph of the curve.

1. $x = t^2 - 4, \quad y = t/2, \quad -2 \le t \le 3$

2. $x = 3t^2, \quad y = 2 + 5t, \quad 0 \le t \le 2$

3. $x = 2t, \quad y = t^2 - 1, \quad -1 \le t \le 2$

4. $x = t - 1, \quad y = \dfrac{t+1}{t-1}, \quad t \ge 1$

5. $x = 4\sin 2t + 9, \quad y = 6\cos t - 8, \quad 0 \le t \le 2\pi$

6. $x = t^3 - 3t - 8, \quad y = 3t^2 - 15, \quad -4 \le t \le 4$

7. $x = 6 \cos t + 12 \cos^2 t, \quad y = 8 \sin t + 8 \sin t \cos t,$
$0 \le t \le 2\pi$

8. $x = 12 \cos t, \quad y = 12 \sin 2t, \quad 0 \le t \le 2\pi$

9. $x = 6 \cos t + 5 \cos 3t, \quad y = 6 \sin t - 5 \sin 3t,$
$0 \le t \le 2\pi$

10. $x = 3t^2 + 10, \quad y = 4t^3,$ any real number t

11. $x = 12 \cos 3t \cos t + 6, \quad y = 12 \cos 3t \sin t - 7,$
$0 \le t \le 2\pi$

12. $x = 2 \cos 3t - 6, \quad y = 2 \cos 3t \sin t + 7, \quad 0 \le t \le 2\pi$

13. $x = t \sin t, \quad y = t \cos t, \quad 0 \le t \le 8\pi$

14. $x = 9 \sin t, \quad y = 9t \cos t, \quad 0 \le t \le 20$

In Exercises 15–24, the given curve is part of the graph of an equation in x and y. Find the equation by eliminating the parameter.

15. $x = t - 3, \quad y = 2t + 1, \quad t \ge 0$

16. $x = t + 5, \quad y = \sqrt{t}, \quad t \ge 0$

17. $x = -2 + t^2, \quad y = 1 + 2t^2,$ any real number t

18. $x = t^2 + 1, \quad y = t^2 - 1,$ any real number t

19. $x = e^t, \quad y = t,$ any real number t

20. $x = 2e^t, \quad y = 1 - e^t, \quad t \ge 0$

21. $x = 3 \cos t, \quad y = 3 \sin t, \quad 0 \le t \le 2\pi$

22. $x = 4 \sin 2t, \quad y = 2 \cos 2t, \quad 0 \le t \le 2\pi$

23. $x = 3 \cos t, \quad y = 4 \sin t, \quad 0 \le t \le 2\pi$

24. $x = 2 \sin t - 3, \quad y = 2 \cos t + 1, \quad 0 \le t \le \pi$

In Exercises 25 and 26, sketch the graphs of the given curves and compare them. Do they differ and if so, how?

25. **(a)** $x = -4 + 6t, \quad y = 7 - 12t, \quad 0 \le t \le 1$
(b) $x = 2 - 6t, \quad y = -5 + 12t, \quad 0 \le t \le 1$

26. **(a)** $x = t, \quad y = t^2$ **(b)** $x = \sqrt{t}, \quad y = t$
(c) $x = e^t, \quad y = e^{2t}$

27. By eliminating the parameter, show that the curve with parametric equations

$$x = a + (c - a)t, \qquad y = b + (d - b)t,$$

any real number t

is a straight line.

28. By proceeding as in Example 5, show that the curve with parametric equations

$$x = r \cos t + c, \qquad y = r \sin t + d, \qquad 0 \le t \le 2\pi$$

is a circle with center (c, d) and radius r.

In Exercises 29–35, find a parameterization of the given curve. Confirm your answer by graphing.

29. line segment from $(-6, 12)$ to $(12, -10)$ [*Hint:* Exercise 27.]

30. line segment from $(14, -5)$ to $(5, -14)$

31. line segment from $(18, 4)$ to $(-16, 14)$

32. circle with center $(7, -4)$ and radius 6

33. circle with center $(9, 12)$ and radius 5

34. $x^2 + y^2 - 14x + 8y + 29 = 0$ [*Hint:* Example 6 in Section 1.3.]

35. $x^2 + y^2 - 4x - 6y + 9 = 0$

In Exercises 36–39, locate all local maxima and minima (other than endpoints) of the curve.

36. $x = 4t^3 - \cos t - 5, \quad y = 3t^2 - 8, \quad -2 \le t \le 2$

37. $x = 4t - 6, \quad y = 3t^2 + 2, \quad -10 \le t \le 10$

38. $x = t^3 + \sin t + 4, \quad y = \cos t, \quad -1.5 \le t \le 2$

39. $x = 4t^3 - t + 4, \quad y = -3t^2 + 5, \quad -2 \le t \le 2$

40. Show that the ball's path in Example 7 is a parabola by eliminating the parameter in the parametric equations

$$x = (140 \cos 31°)t \quad \text{and} \quad y = (140 \sin 31°)t - 16t^2.$$

[*Hint*: Solve the first equation for x and substitute the result in the second equation.]

In Exercises 41–46, use a calculator in degree mode and assume that air resistance is negligible.

41. A skeet is fired from the ground with an initial velocity of 100 ft/sec at an angle of 28°.
(a) Graph the skeet's path
(b) How long is the skeet in the air?
(c) How high does it go?

42. A ball is thrown from a height of 5 ft above the ground with an initial velocity of 60 ft/sec at an angle of 50° with the horizontal.

(a) Graph the ball's path.
(b) When and where does the ball hit ground?

43. A medieval bowman shoots an arrow which leaves the bow 4 ft above the ground with an initial velocity of 88 ft/sec at an angle of 48° with the horizontal.
(a) Graph the arrow's path.
(b) Will the arrow go over the 40-ft-high castle wall that is 200 ft from the archer?

44. A golfer at a driving range stands on a platform 2 ft above the ground and hits the ball with an initial velocity of 120 ft/sec at an angle of 39° with the horizontal. There is a 32-ft-high fence 400 ft away. Will the ball fall short, hit the fence, or go over it?

45. A golf ball is hit off the tee at an angle of 30° and lands 300 ft away. What was its initial velocity? [*Hint:* The ball lands when $x = 300$ and $y = 0$. Use this fact and the parametric equations for the ball's path to find two equations in the variables t and v. Solve for v.]

46. A football kicked from the ground has an initial velocity of 75 ft/sec.
(a) Set up the parametric equations that describe the ball's path. Experiment graphically with different angles to find the smallest angle (within one degree) needed so that the ball travels at least 150 ft.
(b) Use algebra and trigonometry to find the angle needed for the ball to travel exactly 150 ft. [*Hint:* The ball lands when $x = 150$ and $y = 0$. Use this fact and the parametric equations for the ball's path to find two equations in the variables t and θ. Solve the "x equation" for t and substitute this result in the other one; then solve for θ. The double angle identity may be helpful for putting this equation in a form that is easy to solve.]

In Exercises 47–49, complete the derivation of the parametric equations of the cycloid in Example 9.

47. (a) If $\pi/2 < t < \pi$, verify that angle θ in Figure 10–16 has measure $t - \pi/2$ and that

$$x = OT - CQ = 3t - 3\cos\left(t - \frac{\pi}{2}\right)$$

$$y = CT + PQ = 3 + 3\sin\left(t - \frac{\pi}{2}\right).$$

(b) Use the addition and subtraction identities for sine and cosine to show that in this case

$$x = 3(t - \sin t) \quad \text{and} \quad y = 3(1 - \cos t).$$

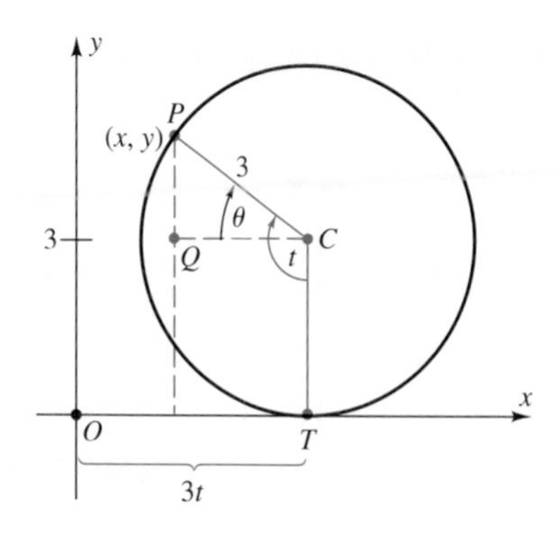

Figure 10-16

48. (a) If $\pi < t < 3\pi/2$, verify that angle θ in Figure 10–17 has measure $3\pi/2 - t$ and that

$$x = OT + CQ = 3t + 3\cos\left(\frac{3\pi}{2} - t\right)$$

$$y = CT + PQ = 3 + 3\sin\left(\frac{3\pi}{2} - t\right).$$

(b) Use the addition and subtraction identities for sine and cosine to show that in this case

$$x = 3(t - \sin t) \quad \text{and} \quad y = 3(1 - \cos t).$$

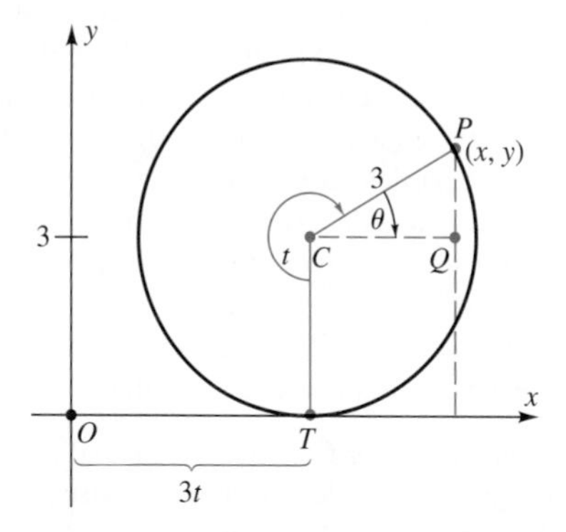

Figure 10-17

49. (a) If $3\pi/2 < t < 2\pi$, verify that angle θ in Figure 10–18 on the next page has measure $t - 3\pi/2$ and that

$$x = OT + CQ = 3t + 3\cos\left(t - \frac{3\pi}{2}\right)$$

$$y = CT - PQ = 3 - 3\sin\left(t - \frac{3\pi}{2}\right).$$

(b) Use the addition and subtraction identities for sine and cosine to show that in this case

$$x = 3(t - \sin t) \qquad \text{and} \qquad y = 3(1 - \cos t)$$

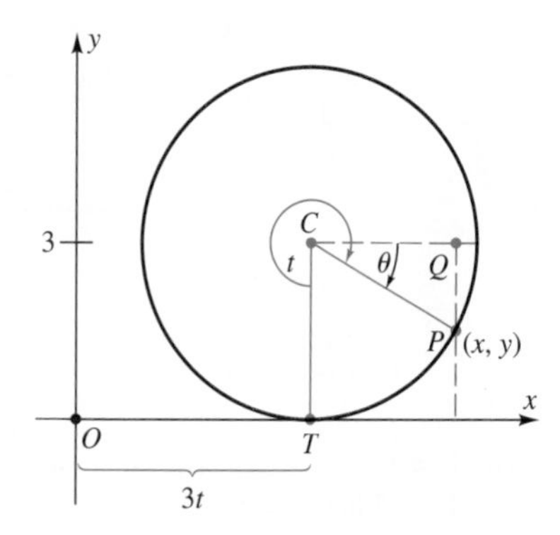

Figure 10-18

Thinkers

50. A particle moves on the horizontal line $y = 3$. Its x-coordinate at time t seconds is given by $x = 2t^3 - 13t^2 + 23t - 8$. This exercise explores the motion of the particle.

(a) Graph the path of the particle in the viewing window with $-10 \le x \le 10$, $-2 \le y \le 4$, $0 \le t \le 4.3$, and t step $= .05$. Note that the calculator seems to pause before completing the graph.

(b) Use trace (starting with $t = 0$) and watch the path of the particle as you press the right arrow key at regular intervals. How many times does it change direction? When does it appear to be moving the fastest?

(c) At what times t does the particle change direction? What are its x-coordinates at these times?

51. Set your calculator for radian mode and for simultaneous graphing mode [check your instruction manual for how to do this]. Particles A, B, and C are moving in the plane, with their positions at time t seconds given by:

A: $x = 8 \cos t$ and $y = 5 \sin t$

B: $x = 3t$ and $y = 5t$

C: $x = 3t$ and $y = 4t$.

(a) Graph the paths of A and B in the window with $0 \le x \le 12$, $0 \le y \le 6$, and $0 \le t \le 2$. The paths intersect, but do they actually collide? That is, are they at the same point at the same time? [For slow motion, choose a very small t step, such as .01.]

(b) Set t step $= .05$ and use trace to estimate the time at which A and B are closest to each other.

(c) Graph the paths of A and C and determine geometrically (as in part (b)) whether they collide. Approximately when are they closest?

(d) Confirm your answers in part (c) as follows. Explain why the distance between particles A and C at time t is given by

$$d = \sqrt{(8 \cos t - 3t)^2 + (5 \sin t - 4t)^2}.$$

A and C will collide if $d = 0$ at some time. Using function graphing mode, graph this distance function when $0 \le t \le 2$ and zoom-in if necessary, show that d is always positive. Find the value of t for which d is smallest.

52. Let P be a point at distance k from the center of a circle of radius r. As the circle rolls along the x-axis, P traces out a curve called a **trochoid.** [When $k \le d$, it may help to think of the circle as a bicycle wheel and P as a point on one of the spokes.]

(a) Assume that P is on the y-axis as close as possible to the x-axis when $t = 0$ and show that the parametric equations of the trochoid are

$$x = rt - k \sin t \qquad \text{and} \qquad y = r - k \cos t.$$

Note that when $k = r$, these are the equations of a cycloid.

(b) Sketch the graph of the trochoid with $r = 3$ and $k = 2$.

(c) Sketch the graph of the trochoid with $r = 3$ and $k = 4$.

53. A circle of radius b rolls along the inside of a larger circle of radius a. The curve traced out by a fixed point P on the smaller circle is called a **hypocycloid.**

(a) Assume that the larger circle has center at the origin and that the smaller circle starts with P located at $(a, 0)$. Use Figure 10–19 to show that the parametric equations of the hypocycloid are

$$x = (a - b)\cos t + b \cos\left(\frac{a - b}{b} t\right)$$

$$y = (a - b)\sin t - b \sin\left(\frac{a - b}{b} t\right).$$

(b) Sketch the graph of the hypocycloid with $a = 5$, $b = 1$, and $0 \le t \le 2\pi$.

(c) Sketch the graph of the hypocycloid with $a = 5$, $b = 2$, and $0 \leq t \leq 4\pi$.

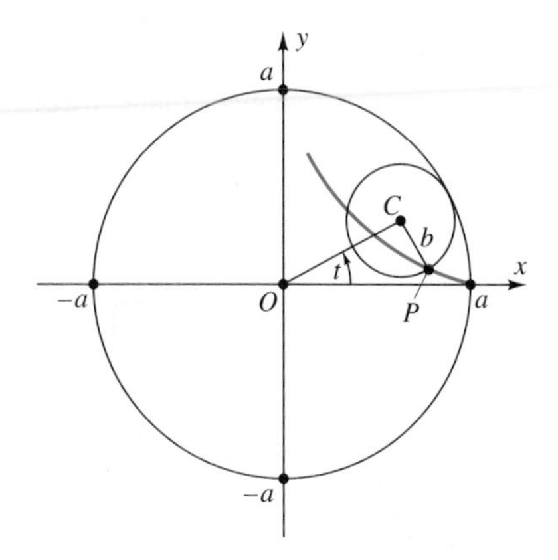

Figure 10-19

54. Jill is on a ferris wheel of radius 20 ft whose bottom just grazes the ground. The wheel is rotating counterclockwise at the rate of .7 radians per second. Jack is standing 100 ft from the bottom of the ferris wheel. When Jill is at point P, he throws a ball toward the wheel (Figure 10–20). The ball leaves Jack's hand 5 ft above the ground with an initial velocity of 62 ft/sec at an angle of 1.2 radians with the horizontal. Follow the steps below to answer the question: Will Jill be able to catch the ball?

(a) Imagine that Jack is at the origin and the bottom of the ferris wheel is at (100, 0). Then the ball leaves his hand from the point (0, 5) and the wheel is a circle with center (100, 20) and radius 20 (Figure 10–21). Therefore, the circle is the graph of the equation

$$(x - 100)^2 + (y - 20)^2 = 20^2.$$

Show that Jill's movement around the wheel is given by the parametric equations

$$x = 20\cos(.7t) + 100 \quad \text{and}$$

$$y = 20\sin(.7t) + 20$$

by verifying that these equations give a parameterization of the circle (wheel) [as in Example 5].

(b) Find the parametric equations that describe the position of the ball at time t.

(c) Set your calculator for parametric mode, radian mode, and simultaneous graphing mode [check your instruction manual for how to do this]. Using a square viewing window with $0 \leq x \leq 130$, $0 \leq t \leq 9$, and t step $= .1$ to graph both sets of parametric equations (Jill's motion and the ball's) simultaneously. [For slow motion, make the t step smaller.] Assuming that Jill can reach 2 ft in any direction, can she catch the ball? If not, use trace to estimate the time at which Jill is closest to the ball.

(d) Experiment by changing the angle and/or initial velocity of the ball to find values that will allow Jill to catch the ball.

Figure 10-20

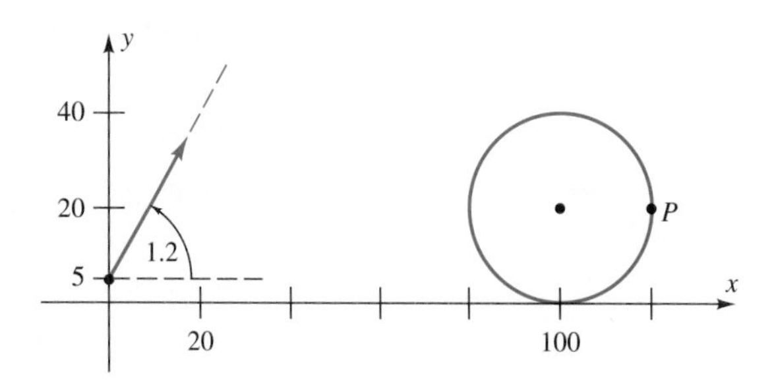

Figure 10-21

10.2 Conic Sections

When a right circular cone is cut by a plane, the intersection is a curve called a **conic section,** as shown in Figure 10–22.* Conic sections were studied by the ancient Greeks and are still of interest. For instance, planets travel in elliptical orbits, parabolic mirrors are used in telescopes, and certain atomic particles follow hyperbolic paths.

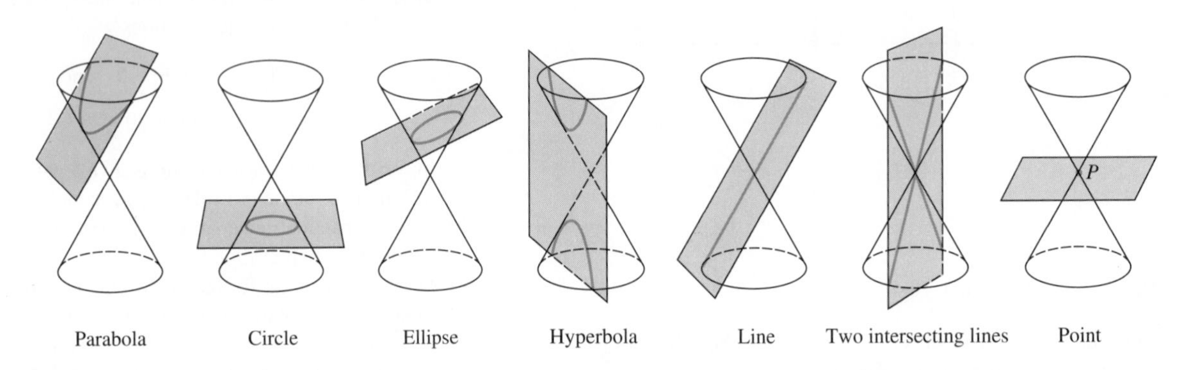

Figure 10-22

Although the Greeks studied conic sections from a purely geometric point of view, the modern approach is to describe them in terms of the coordinate plane and distance, or as the graphs of certain types of equations. This was done for circles in Section 1.3 and will be done here for ellipses, hyperbolas, and parabolas.

In each case the conic is defined in terms of points and distances and its equation determined. The standard form of the equation of a conic includes the key information necessary for a rough sketch of its graph, just as the standard form of the equation of a circle tells you its center and radius. Techniques for graphing conic sections with a calculator are discussed at the end of the section.

Ellipses

Definition Let P and Q be points in the plane and r a number greater than the distance from P to Q. The **ellipse** with **foci**** P and Q is the set of all points X such that

$$(\text{distance from } X \text{ to } P) + (\text{distance from } X \text{ to } Q) = r.$$

To draw this ellipse, take a piece of string of length r and pin its ends on P and Q. Put your pencil point against the string and move it, keeping the string taut. You will trace out the ellipse, as shown in Figure 10–23.†

*A point, a line, or two interesecting lines are sometimes called **degenerate conic sections.**

**''Foci'' is the plucal of ''focus.''

†If $P = Q$, you will trace out a circle of radius $r/2$. So a circle is just a special case of an ellipse.

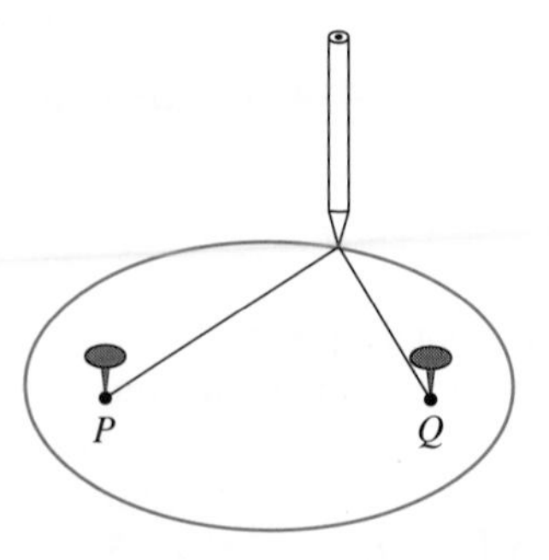

Figure 10-23

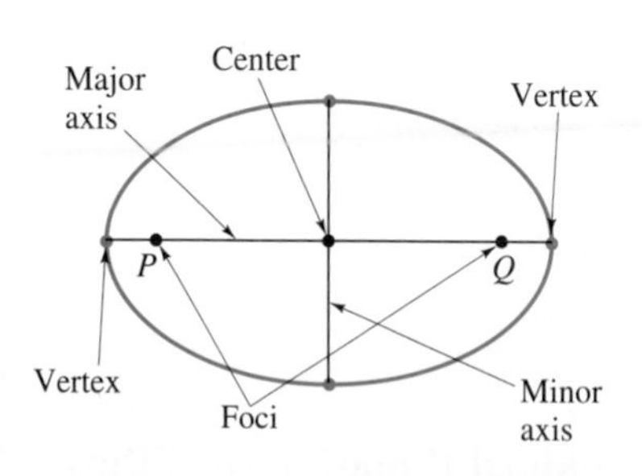

Figure 10-24

The midpoint of the line segment from P to Q is the **center** of the ellipse. The points where the straight line through the foci intersects the ellipse are its **vertices.** The **major axis** of the ellipse is the line segment joining the vertices; its **minor axis** is the line segment through the center, perpendicular to the major axis, as shown in Figure 10–24.

Equation Suppose that foci are $P = (-c, 0)$ and $Q = (c, 0)$ for some $c > 0$. Let $a = r/2$, so that $r = 2a$. Then (x, y) is on the ellipse exactly when

$$[\text{distance from } (x, y) \text{ to } P] + [\text{distance from } (x, y) \text{ to } Q] = r$$

$$\sqrt{(x + c)^2 + (y - 0)^2} + \sqrt{(x - c)^2 + (y - 0)^2} = 2a$$

$$\sqrt{(x + c)^2 + y^2} = 2a - \sqrt{(x - c)^2 + y^2}$$

Squaring both sides and simplifying (Exercise 54) we obtain

$$a\sqrt{(x - c)^2 + y^2} = a^2 - cx.$$

Again squaring both sides and simplifying, we have

$$(a^2 - c^2)x^2 + a^2y^2 = a^2(a^2 - c^2).$$

To simplify the form of this equation, let $b = \sqrt{a^2 - c^2}$* so that $b^2 = a^2 - c^2$ and the equation becomes

$$b^2x^2 + a^2y^2 = a^2b^2.$$

Dividing both sides by a^2b^2 shows that the coordinates of every point on the ellipse satisfy the equation

$$\frac{x^2}{a^2} + \frac{y^2}{b^2} = 1.$$

Conversely, it can be shown that every point whose coordinates satisfy this equation is on the ellipse. When the equation is in this form the x- and y-intercepts of the graph are easily found. For instance, to find the x-intercepts, we set $y = 0$ and solve:

*The distance between the foci is $2c$. Since $r = 2a$ and $r > 2c$ by definition, we have $2a > 2c$ and hence $a > c$. Therefore, $a^2 - c^2$ is a positive number and has a real square root.

$$\frac{x^2}{a^2} + \frac{0^2}{b^2} = 1$$
$$x^2 = a^2$$
$$x = \pm a.$$

Similarly, the y-intercepts are $\pm b$.

A similar argument applies when the foci are on the y-axis and leads to this conclusion:

Standard Equations of Ellipses Centered at the Origin ▼

Let a and b be real numbers with $a > b > 0$. Then the graph of each of the following equations is an ellipse centered at the origin:

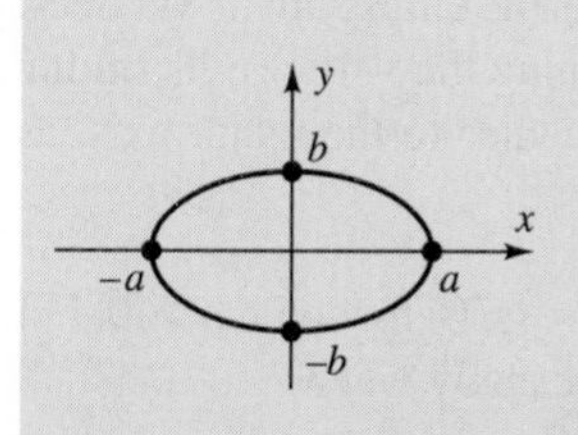

$$\frac{x^2}{a^2} + \frac{y^2}{b^2} = 1 \quad \begin{cases} \textbf{x-intercepts: } \pm a \quad \textbf{y-intercepts: } \pm b \\ \textbf{major axis on the x-axis, with vertices } (a, 0) \textbf{ and } (-a, 0) \\ \textbf{foci: } (c, 0) \textbf{ and } (-c, 0)\textbf{, where } c = \sqrt{a^2 - b^2}. \end{cases}$$

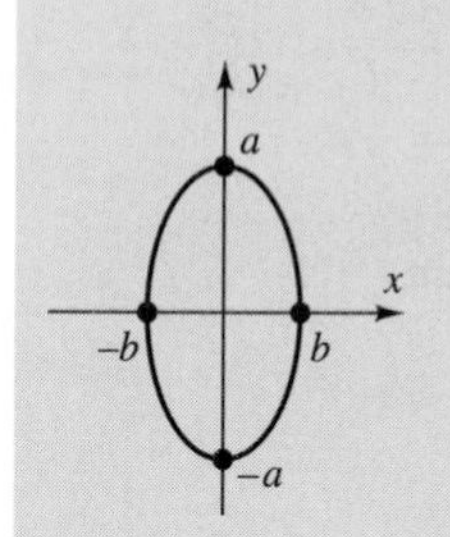

$$\frac{x^2}{b^2} + \frac{y^2}{a^2} = 1 \quad \begin{cases} \textbf{x-intercepts: } \pm b \quad \textbf{y-intercepts: } \pm a \\ \textbf{major axis on the y-axis, with vertices } (0, a) \textbf{ and } (0, -a) \\ \textbf{foci: } (0, c) \textbf{ and } (0, -c)\textbf{, where } c = \sqrt{a^2 - b^2}. \end{cases}$$

In the box above $a > b$, but don't let all the letters confuse you: When the equation is in standard form, the denominator of the x term tells you the x-intercepts, the denominator of the y term tells you the y-intercepts, and the major axis is the longer one, as illustrated in the following examples.

EXAMPLE 1 To identify the graph of $4x^2 + 9y^2 = 36$, we put the equation in standard form by dividing both sides by 36:

$$\frac{4x^2}{36} + \frac{9y^2}{36} = \frac{36}{36}$$
$$\frac{x^2}{9} + \frac{y^2}{4} = 1$$
$$\frac{x^2}{3^2} + \frac{y^2}{2^2} = 1.$$

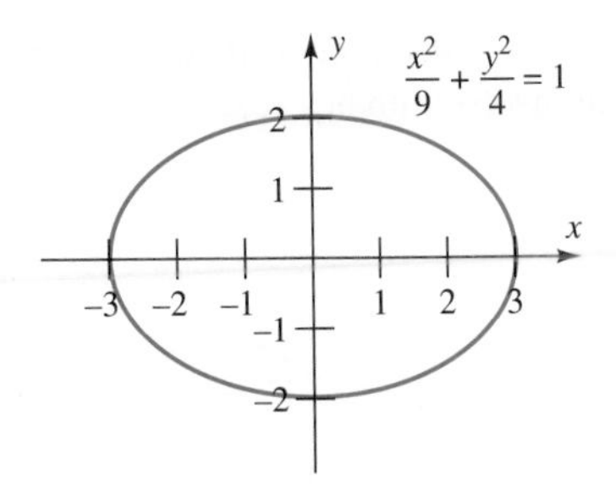

Figure 10-25

The graph now has the form of the first equation in the box above, with $a = 3$ and $b = 2$. So its graph is an ellipse with x-intercepts ± 3 and y-intercepts ± 2, as shown in Figure 10–25. Its major axis and foci lie on the x-axis, as do its vertices $(\pm 3, 0)$. ■

EXAMPLE 2 Find the equation of the ellipse with vertices $(0, \pm 6)$ and foci $(0, \pm 2\sqrt{6})$ and sketch its graph.

Solution Since the foci are $(0, 2\sqrt{6})$ and $(0, -2\sqrt{6})$, the center of the ellipse is $(0, 0)$ and its major axis lies on the y-axis. Hence, its equation is of the form

$$\frac{x^2}{b^2} + \frac{y^2}{a^2} = 1.$$

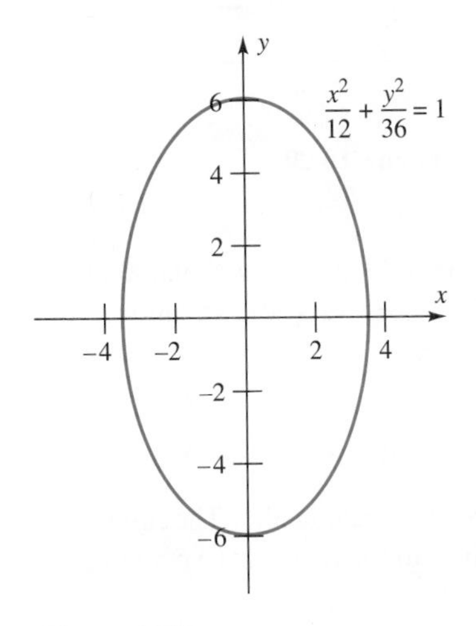

Figure 10-26

From the box on the opposite page, we see that $a = 6$ and $c = 2\sqrt{6}$. Since $c = \sqrt{a^2 - b^2}$, we have $c^2 = a^2 - b^2$, so that

$$b^2 = a^2 - c^2 = 6^2 - (2\sqrt{6})^2 = 36 - 4\cdot 6 = 12.$$

Hence, $b = \sqrt{12}$ and the equation of the ellipse is

$$\frac{x^2}{(\sqrt{12})^2} + \frac{y^2}{6^2} = 1, \qquad \text{or equivalently,} \qquad \frac{x^2}{12} + \frac{y^2}{36} = 1.$$

The graph has x-intercepts $\pm\sqrt{12} \approx 3.46$ and y-intercepts ± 6, as sketched in Figure 10–26. ■

Applications Elliptical surfaces have interesting reflective properties. If a sound or light ray passes through one focus and reflects off an ellipse, the ray will pass through the other focus, as shown in Figure 10–27. Exactly this situation occurs under the elliptical dome of the United States Capitol. A person who stands at one focus and whispers can be clearly heard by anyone at the other focus. Before this fact was widely known, when Congress used to sit under the dome, several political secrets were inadvertently revealed by congressmen to members of the other party.

The planets and many comets have elliptical orbits, with the sun as one focus. The moon travels in an elliptical orbit with the earth as one focus. Satellites are usually put into elliptical orbits around the earth.

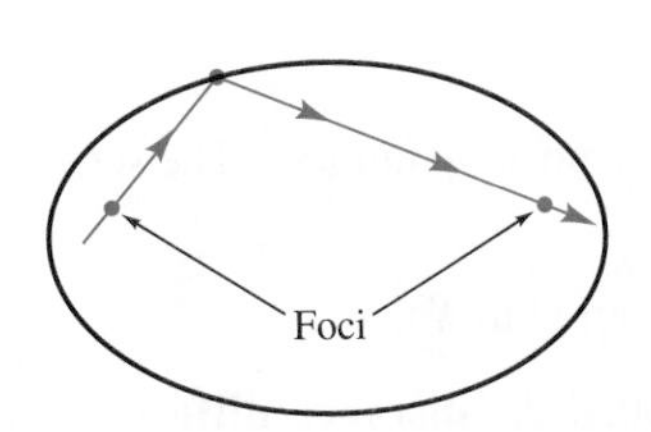

Figure 10-27

EXAMPLE 3 The earth's orbit around the sun is an ellipse that is almost a circle. The sun is one focus and the major and minor axes have lengths 186,000,000 miles and 185,974,062 miles, respectively. What are the minimum and maximum distances from the earth to the sun?

Solution The orbit is shown in Figure 10–28. If we use a coordinate system with the major axis on the x-axis and the sun having coordinates $(c, 0)$, then we obtain Figure 10–29.

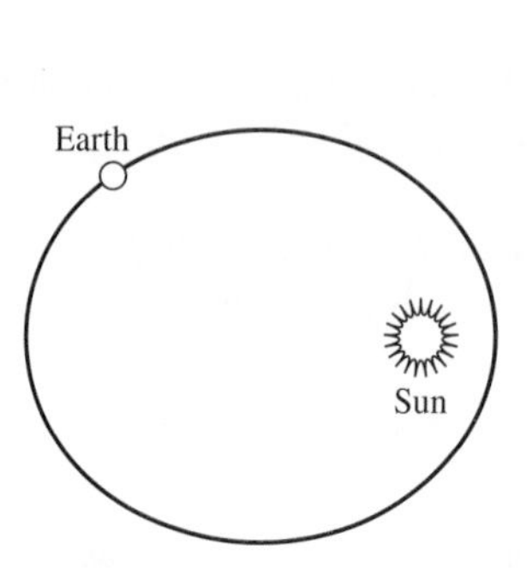

Figure 10-28

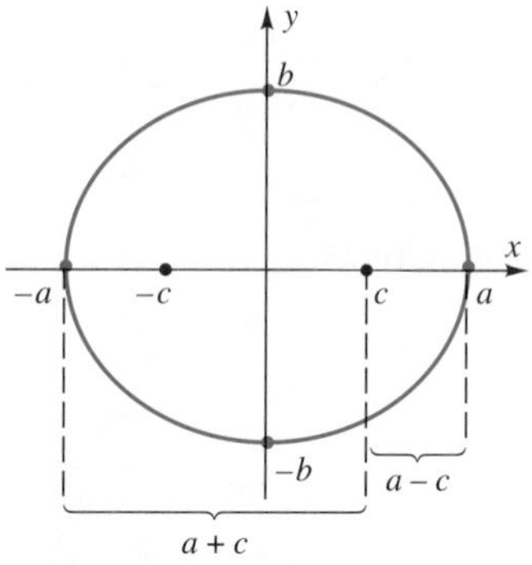

Figure 10-29

The length of the major axis is $2a = 186{,}000{,}000$, so that $a = 93{,}000{,}000$. Similarly, $2b = 185{,}974{,}062$, so that $b = 92{,}987{,}031$. As shown above, the equation of the orbit is $\frac{x^2}{a^2} + \frac{y^2}{b^2} = 1$, where

$$c = \sqrt{a^2 - b^2} = \sqrt{(93{,}000{,}000)^2 - (92{,}987{,}031)^2} \approx 1{,}553{,}083.$$

Figure 10–29 suggests a fact that can also be proven algebraically: The minimum and maximum distances from a point on the ellipse to the focus $(c, 0)$ occur at the endpoints of the major axis:

$$\text{minimum distance} = a - c \approx 93{,}000{,}000 - 1{,}553{,}083 = 91{,}446{,}917 \text{ miles}$$

$$\text{maximum distance} = a + c \approx 93{,}000{,}000 + 1{,}553{,}083 = 94{,}553{,}083 \text{ miles.} \quad ■$$

Hyperbolas

Definition Let P and Q be points in the plane and r a positive number. The set of all points X such that

$$|(\text{distance from } P \text{ to } X) - (\text{distance from } Q \text{ to } X)| = r$$

is the **hyperbola** with **foci** P and Q; r will be called the **distance difference.** Every hyperbola has the general shape shown by the colored lines in Figure 10–30. The dotted straight lines are the **asymptotes** of the hyperbola; it gets closer and closer to the asymptotes, but never touches them. The asymptotes intersect at the midpoint of the line segment from P to Q; this point is called the **center** of the hyperbola. The **vertices** of the hyperbola are the points where it intersects the line segment from P to Q. The line through P and Q is called the **focal axis.**

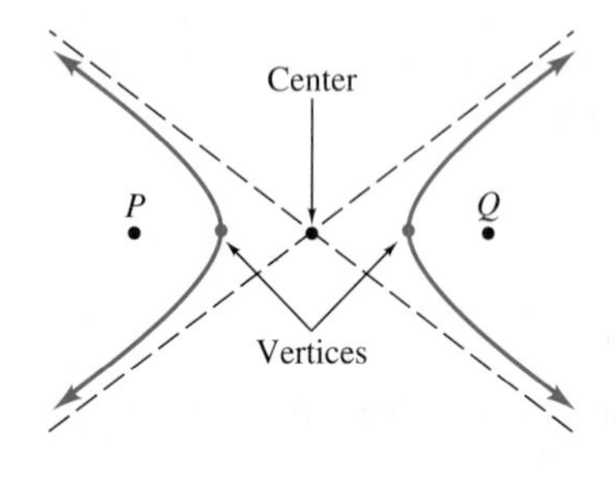

Figure 10-30

Equation Another complicated exercise in the use of the distance formula, which will be omitted here, leads to the following algebraic description:

Standard Equations of Hyperbolas Centered at the Origin ▼

Let a and b be positive real numbers. Then the graph of each of the following equations is a hyperbola centered at the origin:

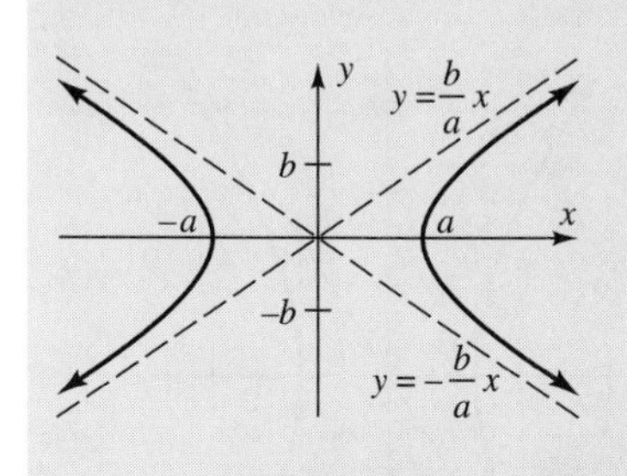

$$\frac{x^2}{a^2} - \frac{y^2}{b^2} = 1$$

- **x-intercepts: $\pm a$ $\quad$ y-intercepts: none**
- **focal axis on the x-axis, with vertices $(a, 0)$ and $(-a, 0)$**
- **foci: $(c, 0)$ and $(-c, 0)$, where $c = \sqrt{a^2 + b^2}$.**
- **asymptotes: $y = \frac{b}{a}x$ and $y = -\frac{b}{a}x$**

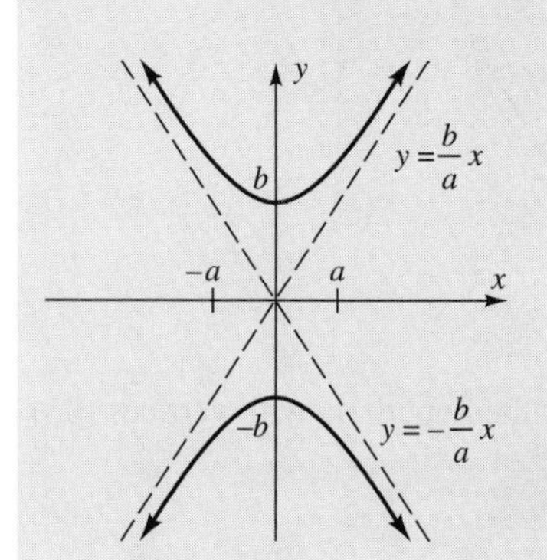

$$\frac{y^2}{a^2} - \frac{x^2}{b^2} = 1$$

- **x-intercepts: none $\quad$ y-intercepts: $\pm a$**
- **focal axis on the y-axis, with vertices $(0, a)$ and $(0, -a)$**
- **foci: $(0, c)$ and $(0, -c)$, where $c = \sqrt{a^2 + b^2}$.**
- **asymptotes: $y = \frac{a}{b}x$ and $y = -\frac{a}{b}x$**

Once again don't worry about all the letters in the box. When the equation is in standard form with the x term positive and y term negative, the hyperbola intersects the x-axis and opens from side to side. When the x term is negative and the y term positive, the hyperbola intersects the y-axis and opens up and down.

EXAMPLE 4 To graph $9x^2 - 4y^2 = 36$ we first rewrite the equation:

$$\frac{9x^2}{36} - \frac{4y^2}{36} = \frac{36}{36}$$

$$\frac{x^2}{4} - \frac{y^2}{9} = 1$$

$$\frac{x^2}{2^2} - \frac{y^2}{3^2} = 1.$$

Applying the fact in the box with $a = 2$ and $b = 3$ shows that the graph is a hyperbola with vertices (2, 0) and (− 2, 0) and asymptotes $y = \frac{3}{2}x$ and $y = -\frac{3}{2}x$. We first plot the vertices and sketch the rectangle determined by the vertical lines $x = \pm 2$ and the horizontal lines $y = \pm 3$. The asymptotes go through the origin and the corners of this rectangle, as shown on the left in Figure 10–31. It is then easy to sketch the hyperbola. ■

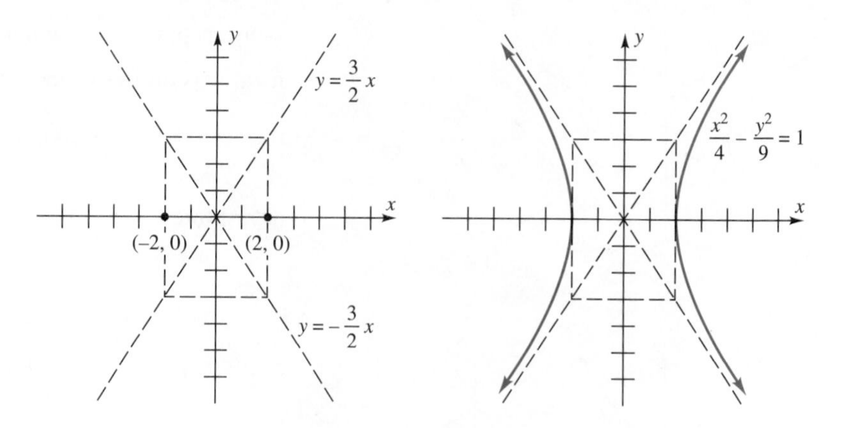

Figure 10-31

EXAMPLE 5 Find the equation of the hyperbola with vertices (0, 1) and (0, − 1) that passes through the point $(3, \sqrt{2})$.

Solution The vertices are on the y-axis and the equation is of the form

$$\frac{y^2}{a^2} - \frac{x^2}{b^2} = 1$$

with $a = 1$. Since $(3, \sqrt{2})$ is on the graph, we have

$$\frac{(\sqrt{2})^2}{1^2} - \frac{3^2}{b^2} = 1$$

$$2 - \frac{9}{b^2} = 1$$

$$b^2 = 9.$$

Therefore, $b = 3$ and the equation is

$$\frac{y^2}{1^2} - \frac{x^2}{3^2} = 1, \qquad \text{or equivalently,} \qquad y^2 - \frac{x^2}{9} = 1.$$

The asymptotes of the hyperbola are the lines $y = \pm\frac{1}{3}x$. ■

Applications The reflective properties of hyperbolas are used in the design of camera and telescope lenses. If a light ray passes through one focus of a hyperbola and reflects off the hyperbola at a point P, then the reflected ray moves along the straight line determined by P and the other focus, as shown in Figure 10–32. Other applications of hyperbolas are discussed in Section 10.3.

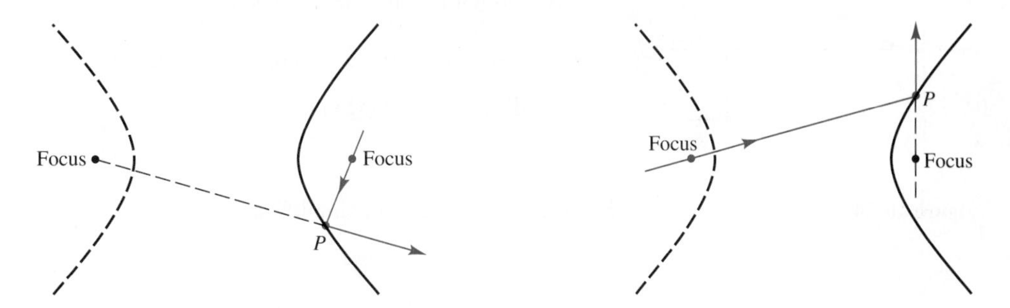

Figure 10-32

Parabolas

Definition Parabolas appeared in Section 3.3 as the graphs of quadratic functions. Parabolas of this kind are a special case of the following more general definition. Let L be a line in the plane and P a point not on L. If X is any point not on L, the distance from X to L is defined to be the length of the perpendicular line segment from X to L. The **parabola** with **focus** P and **directrix** L is the set of all point X such that

$$\text{distance from } X \text{ to } P = \text{distance from } X \text{ to } L$$

as shown in Figure 10–33.

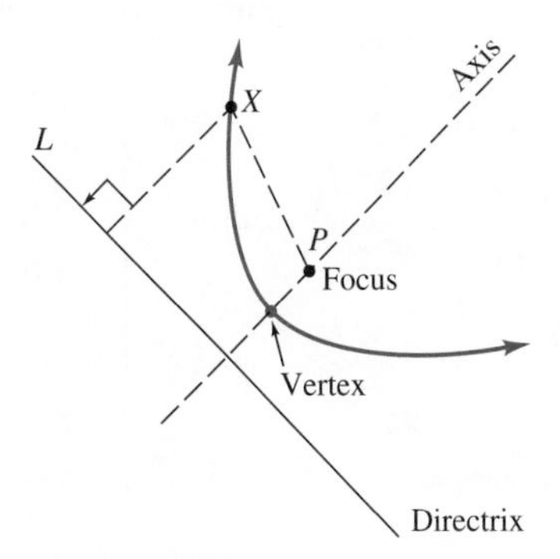

Figure 10-33

The line through P perpendicular to L is called the **axis.** The intersection of the axis with the parabola (the midpoint of the segment of the axis from P to L) is the **vertex** of the parabola, as illustrated in Figure 10–33. The parabola is symmetric with respect to its axis.

Figure 10-34

Equation Suppose that the focus is on the y-axis at the point $(0, p)$, where p is a nonzero constant and that the directrix is the horizontal line $y = -p$. If (x, y) is any point on the parabola, then the distance from (x, y) to the horizontal line $y = -p$ is the length of the vertical segment from (x, y) to $(x, -p)$, as shown in Figure 10–34.

By the definition of the parabola,

$$\text{distance from } (x, y) \text{ to } (0, p) = \text{distance from } (x, y) \text{ to } y = -p$$

$$\text{distance from } (x, y) \text{ to } (0, p) = \text{distance from } (x, y) \text{ to } (x, -p)$$

$$\sqrt{(x-0)^2 + (y-p)^2} = \sqrt{(x-x)^2 + (y-(-p))^2}.$$

Squaring both sides and simplifying, we have

$$(x-0)^2 + (y-p)^2 = (x-x)^2 + (y+p)^2$$

$$x^2 + y^2 - 2py + p^2 = 0^2 + y^2 + 2py + p^2$$

$$x^2 = 4py.$$

Conversely, it can be shown that every point whose coordinates satisfy this equation is on the parabola.

A similar argument works for the parabola with focus $(p, 0)$ on the x-axis and directrix the vertical line $x = -p$, and leads to this conclusion:

Standard Equations of Parabolas with Vertex at the Origin ▼

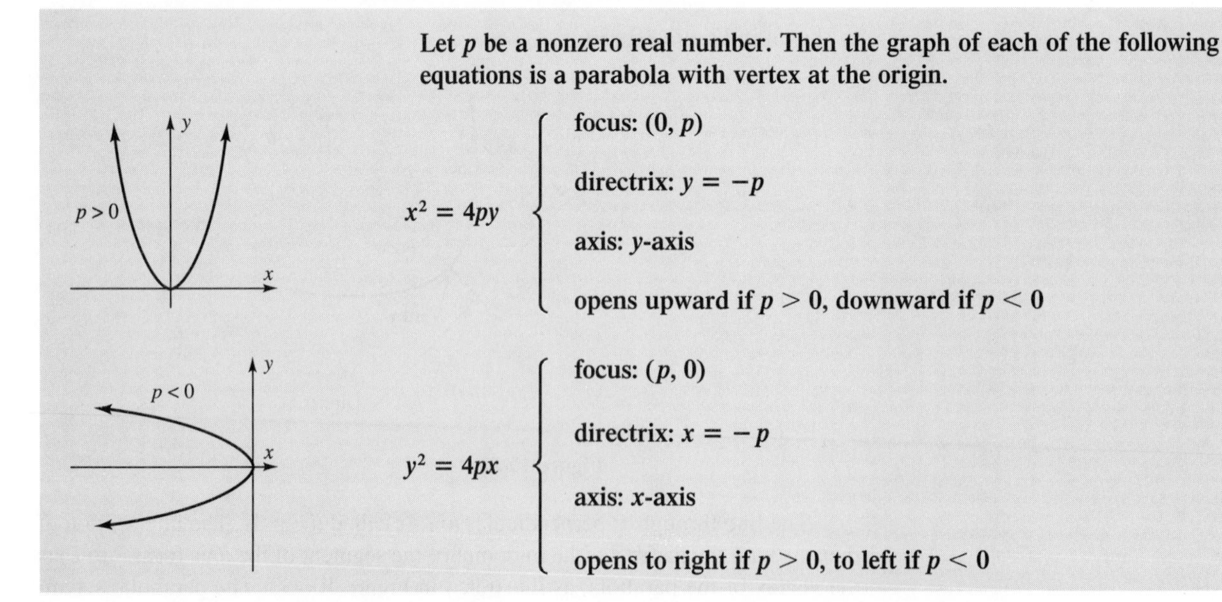

Let p be a nonzero real number. Then the graph of each of the following equations is a parabola with vertex at the origin.

$x^2 = 4py$
- **focus: $(0, p)$**
- **directrix: $y = -p$**
- **axis: y-axis**
- **opens upward if $p > 0$, downward if $p < 0$**

$y^2 = 4px$
- **focus: $(p, 0)$**
- **directrix: $x = -p$**
- **axis: x-axis**
- **opens to right if $p > 0$, to left if $p < 0$**

EXAMPLE 6 Show that the graph of $y = -x^2/8$ is a parabola and find its focus and directrix.

Solution The equation $y = -x^2/8$ can be rewritten as $x^2 = -8y$. This equation is of the form $x^2 = 4py$, with $4p = -8$, so that $p = -2$. Hence, the graph is a downward opening parabola with focus $(0, p) = (0, -2)$ and directrix $y = -p = -(-2) = 2$, as shown in Figure 10–35. ■

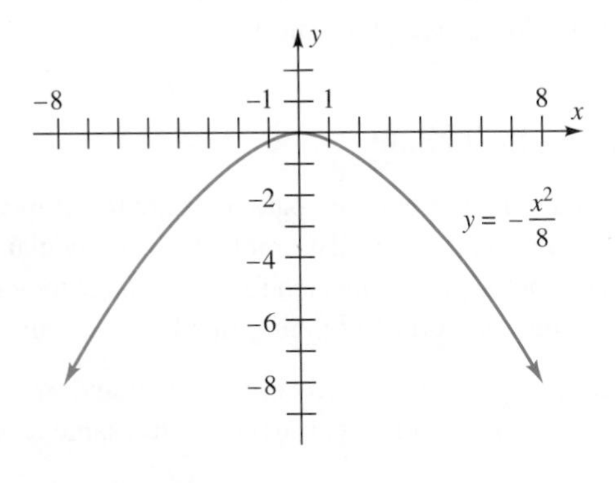

Figure 10-35

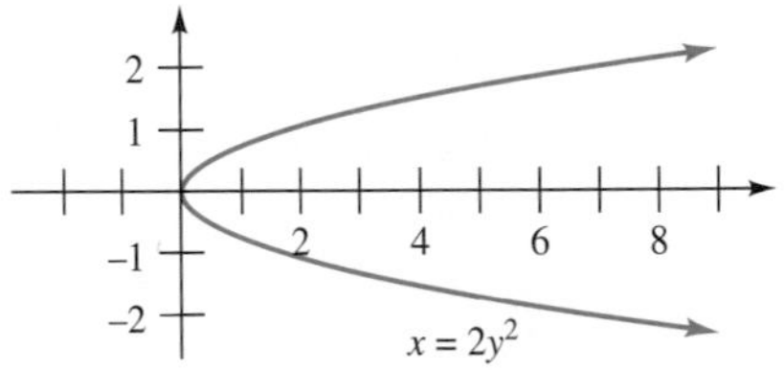

Figure 10-36

EXAMPLE 7 Find the focus, directrix, and equation of the parabola that passes through the point (8, 2), has vertex (0, 0), and focus on the x-axis.

Solution The equation is of the form $y^2 = 4px$. Since (8, 2) is on the graph, we have $2^2 = 4p \cdot 8$, so that $p = 1/8$. Therefore, the focus is (1/8, 0) and the directrix is the vertical line $x = -1/8$. The equation is $y^2 = 4(\frac{1}{8})x = \frac{1}{2}x$, or equivalently, $x = 2y^2$. Its graph is sketched in Figure 10–36. ■

Applications Certain laws of physics show that sound waves or light rays from a source at the focus of a parabola will reflect off the parabola in rays parallel to the axis of the parabola, as shown in Figure 10–37. This is the reason that parabolic reflectors are used in automobile headlights and search lights.

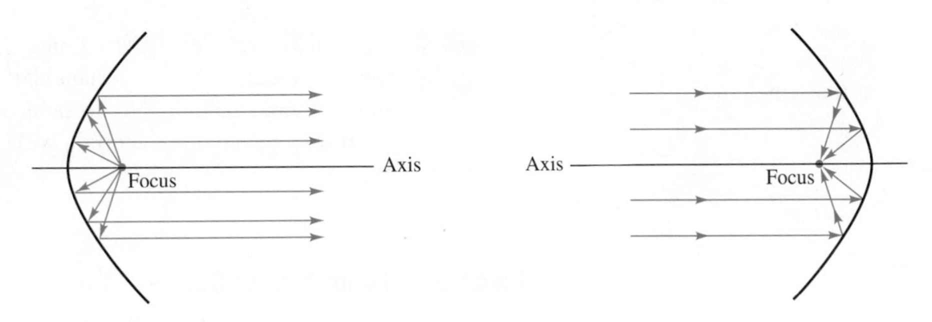

Figure 10-37

Figure 10-38

Conversely, a light ray coming toward a parabola will be reflected into the focus, as shown in Figure 10–38. This fact is used in the design of radar antennas, satellite dishes, and field microphones used at outdoor sporting events to pick up conversation on the field.

Projectiles follow a parabolic curve, a fact that is used in the design of water slides in which the rider slides down a sharp incline, then up and over a hill, before plunging downward into a pool. At the peak of the parabola-shaped hill, the rider shoots up along a parabolic arc several inches above the slide, experiencing a sensation of weightlessness.

Graphing Conic Sections

Upward and downward opening parabolas are the graphs of functions of the form $y = f(x)$, so they can be easily graphed on a calculator. Other conic sections, however, are not the graphs of such functions since they fail the Vertical Line Test (page 105). One method of graphing this kind of conic section is to

> Write its equation in the form $y^2 = f(x)$ and graph both of the functions $y = \sqrt{f(x)}$ and $y = -\sqrt{f(x)}$ on the same screen.

EXAMPLE 8 The graph of $4x^2 + 25y^2 = 100$ is an ellipse (divide both sides by 100 to put the equation in standard form). Solving this equation for y shows that

$$25y^2 = 100 - 4x^2$$

$$y^2 = \frac{100 - 4x^2}{25}$$

$$y = \sqrt{\frac{100 - 4x^2}{25}} \quad \text{or} \quad y = -\sqrt{\frac{100 - 4x^2}{25}}$$

Each of these last equations defines y as a function of x and can be entered in a calculator.

▶ TECHNOLOGY TIP

On TI calculators (except TI-81) you can graph both functions in the Exploration at the same time by keying in $y = \{1, -1\}\sqrt{\frac{100 - 4x^2}{25}}$.

GRAPHING EXPLORATION Using the standard viewing window in "function" (or "rectangular") graphing mode, graph both of these functions on the same screen and verify that the graph is an ellipse centered at the origin with x-intercepts ± 5 and y-intercepts ± 2. ■

Parametric Equations for Conics

Conic sections can be conveniently graphed in parametric graphing mode. As was the case with circles in Section 10.1, parameterizations of the other conic sections are a consequence of various Pythagorean identities in trigonometry. For example, the hyperbola with equation

$$\frac{x^2}{a^2} - \frac{y^2}{b^2} = 1$$

can be obtained from this parameterization:

$$x = a \sec t \qquad \text{and} \qquad y = b \tan t \qquad (0 \le t \le 2\pi)$$

because

$$\frac{x^2}{a^2} - \frac{y^2}{b^2} = \frac{(a \sec t)^2}{a^2} - \frac{(b \tan t)^2}{b^2}$$
$$= \frac{a^2\, sec^2 t}{a^2} - \frac{b^2 \tan^2 t}{b^2} = \sec^2 t - \tan^2 t = 1.$$

Similar arguments (Exercises 56 and 57) lead to the following table. The parameterizations listed in it are not the only possible ones (Exercise 58).

Parametric Equations for Conic Sections ▶

Conic Section	*Parameterization*
Ellipse: $\frac{x^2}{a^2} + \frac{y^2}{b^2} = 1$	$x = a \cos t, \quad y = b \sin t$ $(0 \le t \le 2\pi)$
Hyperbola: $\frac{x^2}{a^2} - \frac{y^2}{b^2} = 1$	$x = a \sec t = \frac{a}{\cos t}, \quad y = b \tan t$ $(0 \le t \le 2\pi)$
$\frac{y^2}{a^2} - \frac{x^2}{b^2} = 1$	$x = b \tan t, \quad y = a \sec t = \frac{a}{\cos t}$ $(0 \le t \le 2\pi)$
Parabola: $y^2 = 4px$	$x = t^2/4p, \quad y = t$ (any real number t)

EXAMPLE 9 Find parametric equations for each of these conic sections:

(a) $y^2 = 6x$ **(b)** $\frac{y^2}{9} - \frac{x^2}{16} = 1.$

Solution

(a) As in the last line of the box above, this is the equation of a parabola, with $4p = 6$. Let $y = t$ and $x = t^2/6$, where t is any real number.

> GRAPHING EXPLORATION With your calculator in parametric graphing mode, graph this parabola.

(b) This equation has the form of the second hyperbola equation in the box on the previous page, with $a = 3$ and $b = 4$. Let $x = 4 \tan t$ and $y = 3/\cos t$.

> GRAPHING EXPLORATION Set $0 \le t \le 2\pi$ and graph the hyperbola. How is it traced out? Now change the t range so that $-\pi/2 \le t \le 3\pi/2$. Now how is the graph traced out? ■

EXAMPLE 10 In order to use parametric equations to graph $4x^2 + 25y^2 = 100$, put the equation in standard form by dividing both sides by 100:

$$\frac{x^2}{25} + \frac{y^2}{4} = 1, \qquad \text{or equivalently,} \qquad \frac{x^2}{5^2} + \frac{y^2}{2^2} = 1.$$

Now parameterize by letting $x = 5 \cos t$ and $y = 2 \sin t$ with $0 \le t \le 2\pi$.

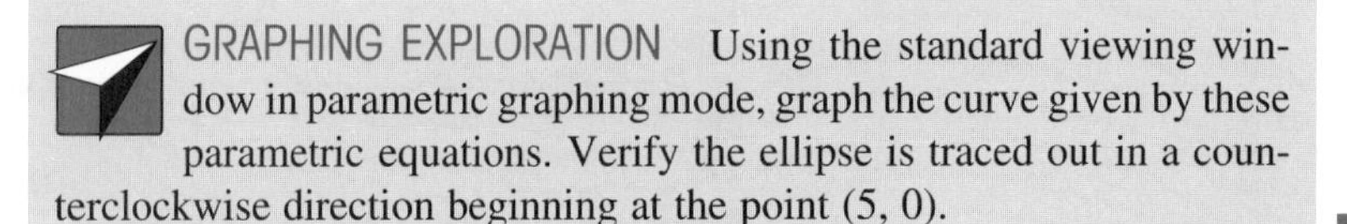

> GRAPHING EXPLORATION Using the standard viewing window in parametric graphing mode, graph the curve given by these parametric equations. Verify the ellipse is traced out in a counterclockwise direction beginning at the point (5, 0). ■

EXERCISES 10.2

In Exercises 1–14, find the equation of the conic section that satisfies the given conditions.

1. Ellipse with center (0, 0); foci on x-axis; x-intercepts ± 7; y-intercepts ± 2.
2. Ellipse with center (0, 0); foci on y-axis; x-intercepts ± 1; y-intercepts ± 8.
3. Ellipse with center (0, 0); foci on x-axis; major axis of length 12; minor axis of length 8.
4. Ellipse with center (0, 0); foci on y-axis; major axis of length 20; minor axis of length 18.
5. Hyperbola with center (0, 0); x-intercepts ± 3; asymptote $y = 2x$.
6. Hyperbola with center (0, 0); y-intercepts ± 12; asymptote $y = 3x/2$.
7. Hyperbola with center (0, 0); vertex (2, 0); passing through $(4, \sqrt{3})$.
8. Hyperbola with center (0, 0); vertex $(0, \sqrt{12})$; passing through $(2\sqrt{3}, 6)$.
9. Parabola with vertex (0, 0); axis $x = 0$; passing through (2, 12).
10. Parabola with vertex (0, 0); axis $y = 0$; passing through (2, 12).
11. Parabola with vertex (0, 0) and focus (5, 0).

12. Parabola with vertex (0, 0) and focus (0, 3.5).

13. Ellipse with center (0, 0); endpoints of major and minor axes: (− 3, 0), (3, 0), (0, − 7), (0, 7).

14. Ellipse with center (0, 0); vertices (8, 0) and (− 8, 0); minor axis of length 8.

In Exercises 15–20, determine which of the following equations could possibly have the given graph.

$$y = x^2/4, \quad x^2 = -8y, \quad 6x = y^2, \quad y^2 = -4x,$$
$$2x^2 + y^2 = 12, \quad x^2 + 6y^2 = 18,$$
$$6y^2 - x^2 = 6, \quad 2x^2 - y^2 = 8$$

15.

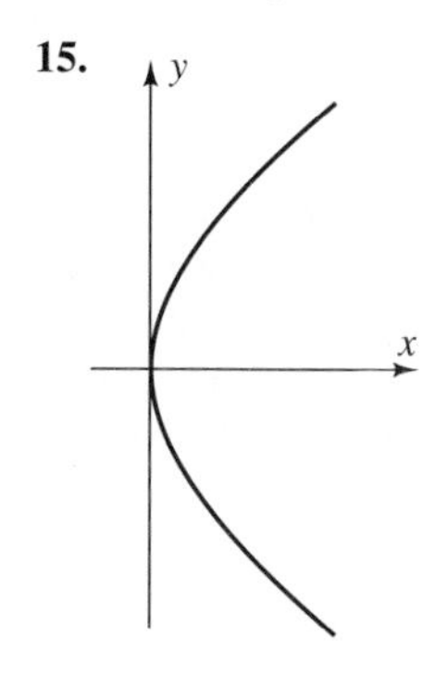

16.

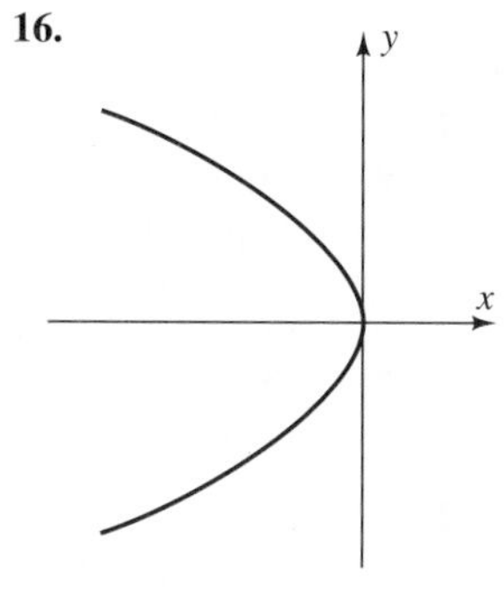

17.

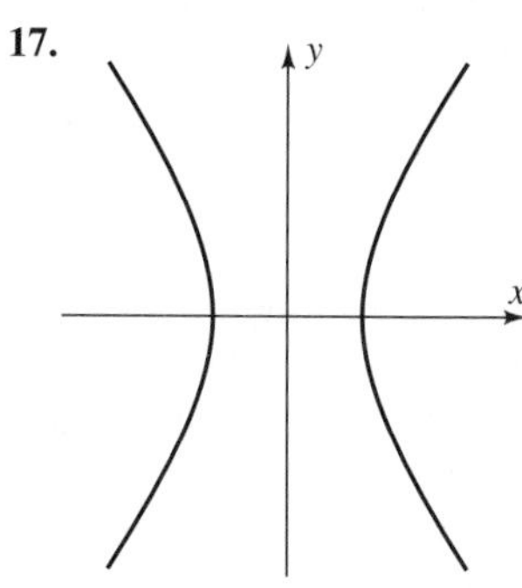

18.

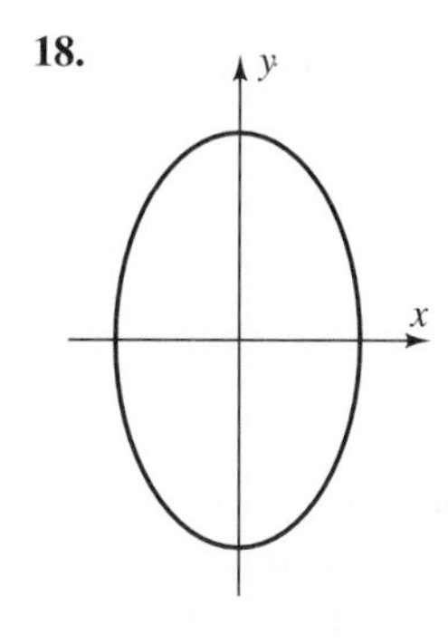

19.

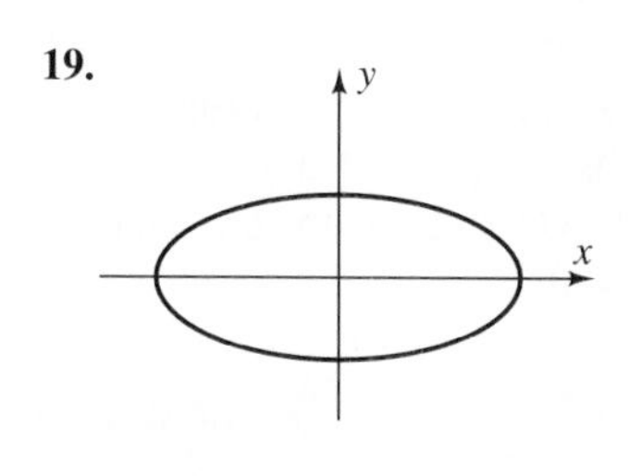

20.

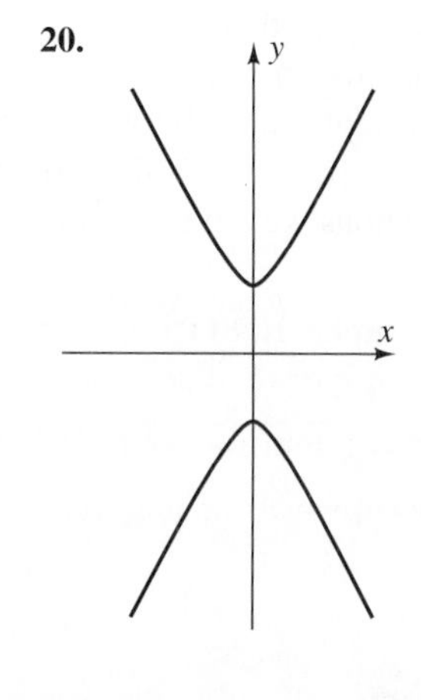

In Exercises 21–32, identify the conic section whose equation is given and find a complete graph of the equation, using the method of Example 8.

21. $\frac{x^2}{25} + \frac{y^2}{4} = 1$ **22.** $\frac{x^2}{6} + \frac{y^2}{16} = 1$

23. $4x^2 + 3y^2 = 12$ **24.** $9x^2 + 4y^2 = 72$

25. $\frac{x^2}{6} - \frac{y^2}{16} = 1$ **26.** $\frac{x^2}{4} - y^2 = 1$

27. $4x^2 - y^2 = 16$ **28.** $3y^2 - 5x^2 = 15$

29. $x = -6y^2$ **30.** $.5y^2 = 2x$

31. $18y^2 - 8x^2 - 2 = 0$ **32.** $x^2 - 2y^2 = -1$

In Exercises 33–42, find parametric equations for the curve whose equation is given and use these parametric equations to find a complete graph of the curve.

33. $\frac{x^2}{10} - 1 = \frac{-y^2}{36}$ **34.** $\frac{y^2}{49} + \frac{x^2}{81} = 1$

35. $4x^2 + 4y^2 = 1$ **36.** $x^2 + 4y^2 = 1$

37. $\frac{x^2}{10} - \frac{y^2}{36} = 1$ **38.** $\frac{y^2}{9} - \frac{x^2}{16} = 1$

39. $x^2 - 4y^2 = 1$ **40.** $2x^2 - y^2 = 4$

41. $8x = 2y^2$ **42.** $4y = x^2$

Calculus can be used to show that the area of the ellipse with equation $\frac{x^2}{a^2} + \frac{y^2}{b^2} = 1$ *is* πab. *Use this fact to find the area of each ellipse in Exercises 43–48.*

43. $\frac{x^2}{16} + \frac{y^2}{4} = 1$ **44.** $\frac{x^2}{9} + \frac{y^2}{5} = 1$

45. $3x^2 + 4y^2 = 12$ **46.** $7x^2 + 5y^2 = 35$

47. $6x^2 + 2y^2 = 14$ **48.** $5x^2 + y^2 = 5$

In Exercises 49–52, find the focus and directrix of the parabola.

49. $y = 3x^2$ **50.** $x = .5y^2$

51. $y = .25x^2$ **52.** $x = -6y^2$

53. Consider the ellipse whose equation is $\frac{x^2}{a^2} + \frac{y^2}{b^2} = 1$. Show that if $a = b$, then the graph is actually a circle.

54. Complete the derivation of the equation of the ellipse on page 603 as follows.

(a) By squaring both sides, show that the equation

$$\sqrt{(x+c)^2+y^2} = 2a - \sqrt{(x-c)^2+y^2}$$

may be simplified as

$$a\sqrt{(x-c)^2+y^2} = a^2 - cx.$$

(b) Show that the last equation in part (a) may be further simplified as

$$(a^2-c^2)x^2 + a^2y^2 = a^2(a^2-c^2).$$

55. Sketch the graph of $\frac{y^2}{4} - \frac{x^2}{b^2} = 1$ for $b = 2$, $b = 4$, $b = 8$, $b = 12$, and $b = 20$. What happens to the hyperbola as b takes larger and larger values? Could the graph ever degenerate into a pair of horizontal lines?

56. Show that the curve defined by the parametric equations $x = a \cos t$ and $y = b \sin t$ $(0 \le t \le 2\pi)$ is the ellipse whose equation is $\frac{x^2}{a^2} + \frac{y^2}{b^2} = 1$.

57. Let p be a nonzero constant. Show that the curve defined by the parametric equations $x = t^2/4p$ and $y = t$ (t any real number) is the parabola whose equation is $y^2 = 4px$.

58. Find parameterizations for the ellipse and hyperbola dif ferent than those given in the box on page 613. [*Hint*: Example 5 in Section 10.1.]

59. The orbit of the moon around the earth is an ellipse with the earth as one focus. If the length of the major axis of the orbit is 477,736 miles and the length of the minor axis is 477,078 miles, find the minimum and maximum distances from the earth to the moon.

60. Halley's comet has an elliptical orbit with the sun as one focus and a major axis that is 1,636,484,848 miles long. The closest the comet comes to the sun is 54,004,000 miles. What is the maximum distance from the comet to the sun?

Thinkers

61. The punch bowl and a table holding the punch cups are placed 50 ft apart at a yard party. A portable fence is then set up so that any guest inside the fence can walk straight to the table, then to the punch bowl, and then return to his or her starting point without traveling more than 150 ft. Describe the longest possible such fence that encloses the largest possible area.

62. An arched footbridge over a 100-ft-wide river is shaped like half an ellipse. The maximum height of the bridge over the river is 20 ft. Find the height of the bridge over a point in the river, exactly 25 ft from the center of the river.

10.3 Translations and Rotations of Conics

Now that you are familiar with conic sections centered at the origin, we expand the discussion to include both conics centered at other points in the plane and ones with axes that may not be parallel to the coordinate axes. We begin with a basic fact about graphing that plays a key role in the discussion.

In Section 2.5 we saw that replacing the variable x by $x - 5$ in the rule of the function $y = f(x)$ shifts the graph of the function horizontally 5 units to the right, whereas replacing x by $x + 5$ (that is, $x - (-5)$) shifts the graph horizontally 5 units to the left (see the box on page 126). Similarly, if the rule of a function is given by $y = f(x)$, then replacing y by $y - 4$ shifts the graph 4 units vertically upward (because $y - 4 = f(x)$ is equivalent to $y = f(x) + 4$; see the box on page 124). For arbitrary equations we have a similar result:

Vertical and Horizontal Shifts ▶

Let h and k be constants. Replacing x by $x - h$ and y by $y - k$ in an equation shifts the graph of the equation

$|h|$ units horizontally (right for positive h, left for negative h) and

$|k|$ units vertically (upward for positive k, downward for negative k).

EXAMPLE 1 Identify and sketch the graph of

$$\frac{(x-5)^2}{9} = \frac{(y+4)^2}{36} = 1.$$

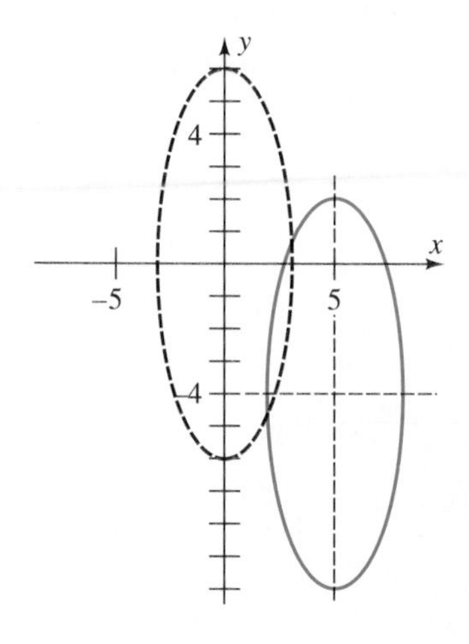

Figure 10-39

Solution This equation can be obtained from the equation $\frac{x^2}{9} + \frac{y^2}{36} = 1$ (whose graph is known to be an ellipse) as follows:

$$\text{replace } x \text{ by } x - 5 \qquad \text{and} \qquad \text{replace } y \text{ by } y - (-4) = y + 4.$$

This is the situation described in the box above with $h = 5$ and $k = -4$. Therefore, the graph is the ellipse $\frac{x^2}{9} + \frac{y^2}{36} = 1$ shifted horizontally 5 units to the right and vertically 4 units downward, as shown in Figure 10–39. The center of the ellipse is at $(5, -4)$. Its major axis (the longer one) lies on the vertical line $x = 5$, as do its foci. The minor axis is on the horizontal line $y = -4$. ■

EXAMPLE 2 To identify the graph of

$$4x^2 + 9y^2 - 32x - 90y + 253 = 0,$$

we first rewrite the equation:

$$(4x^2 - 32x) + (9y^2 - 90y) = -253$$
$$4(x^2 - 8x) + 9(y^2 - 10y) = -253.$$

Now complete the square in $x^2 - 8x$ and $y^2 - 10y$:

$$4(x^2 - 8x + 16) + 9(y^2 - 10y + 25) = -253 + ? + ?$$

Be careful here: On the left side we haven't just added 16 and 25. When the left side is multiplied out we have actually added in $4 \cdot 16 = 64$ and $9 \cdot 25 = 225$. Therefore, to leave the original equation unchanged, we must add these numbers on the right:

$$4(x^2 - 8x + 16) + 9(y^2 - 10y + 25) = -253 + 64 + 225$$
$$4(x-4)^2 + 9(y-5)^2 = 36$$
$$\frac{4(x-4)^2}{36} + \frac{9(y-5)^2}{36} = \frac{36}{36}$$
$$\frac{(x-4)^2}{9} + \frac{(y-5)^2}{4} = 1.$$

The graph of this equation is the ellipse $\frac{x^2}{9} + \frac{y^2}{4} = 1$ shifted horizontally 4 units to the right and vertically 5 units upward. Its center is at $(4, 5)$. Its major axis lies on the horizontal line $y = 5$ and its minor axis on the vertical line $x = 4$, as shown in Figure 10–40. ■

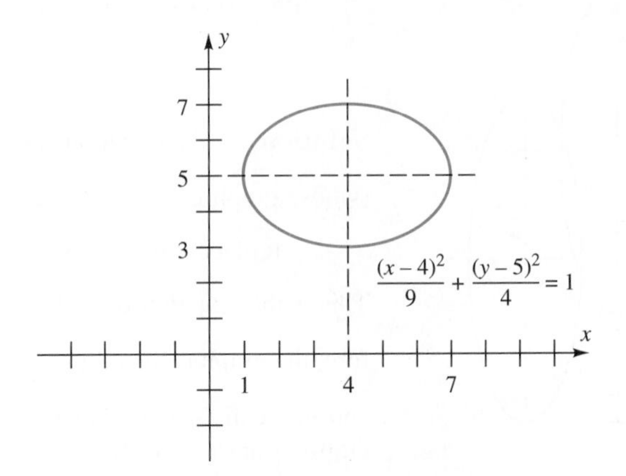

Figure 10-40

EXAMPLE 3 To identify the graph of $x = 2y^2 + 12y + 14$, we rewrite the equation and complete the square in y, being careful to add the appropriate amounts to both sides of the equation:

$$2y^2 + 12y = x - 14$$

$$2(y^2 + 6y) = x - 14$$

$$2(y^2 + 6y + 9) = x - 14 + 2 \cdot 9$$

$$2(y + 3)^2 = x + 4$$

$$(y - (-3))^2 = \frac{1}{2}(x - (-4)).$$

Thus, the graph is the graph of the parabola $y^2 = \frac{1}{2}x$ shifted 4 units horizontally to the left and 3 units vertically downward, as shown in Figure 10–41. The parabola $y^2 = \frac{1}{2}x$ has its vertex at (0, 0) and the x-axis as its axis. When the graph is shifted, the parabola will have its vertex at $(-4, -3)$ and the horizontal line $y = -3$ as its axis. ■

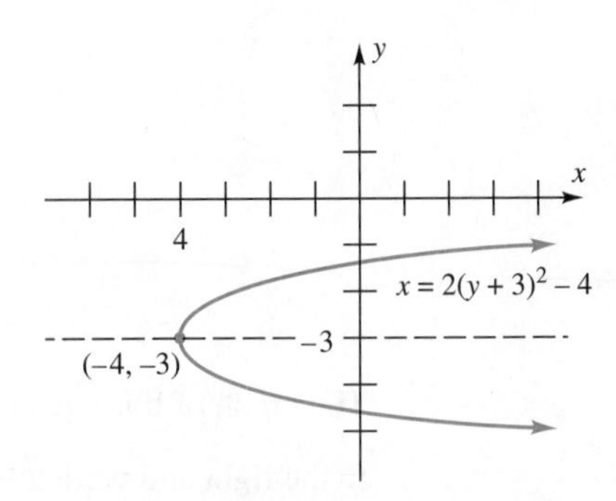

Figure 10-41

By translating the information about conic sections centered at the origin in the boxes on pages 604, 607, and 610, we obtain this summary of the standard equations of conic sections whose axes are parallel to the coordinate axes:

Standard Equations of Conic Sections ▶

Let (h, k) be any point in the plane. If a and b are real numbers with $a > b > 0$, then the graph of each of the following equations is an ellipse with center (h, k).

$$\frac{(x-h)^2}{a^2} + \frac{(y-k)^2}{b^2} = 1 \quad \begin{cases} \text{major axis on the horizontal line } y = k \\ \text{minor axis on the vertical line } x = h \\ \text{foci: } (h-c, k) \text{ and } (h+c, k), \text{ where} \\ \quad c = \sqrt{a^2 - b^2} \end{cases}$$

$$\frac{(x-h)^2}{b^2} + \frac{(y-k)^2}{a^2} = 1 \quad \begin{cases} \text{major axis on the vertical line } x = h \\ \text{minor axis on the horizontal line } y = k \\ \text{foci: } (h, k-c) \text{ and } (h, k+c), \text{ where} \\ \quad c = \sqrt{a^2 - b^2} \end{cases}$$

If a and b are positive real numbers, then the graph of each of the following equations is a hyperbola with center (h, k).

$$\frac{(x-h)^2}{a^2} - \frac{(y-k)^2}{b^2} = 1 \quad \begin{cases} \text{focal axis on the horizontal line } y = k \\ \text{foci: } (h-c, k) \text{ and } (h+c, k), \text{ where} \\ \quad c = \sqrt{a^2 + b^2} \\ \text{vertices: } (h-a, k) \text{ and } (h+a, k) \\ \text{asymptotes: } y = \pm\frac{b}{a}(x-h) + k \end{cases}$$

$$\frac{(y-k)^2}{a^2} - \frac{(x-h)^2}{b^2} = 1 \quad \begin{cases} \text{focal axis on the vertical line } x = h \\ \text{foci: } (h, k-c) \text{ and } (h, k+c), \text{ where} \\ \quad c = \sqrt{a^2 + b^2} \\ \text{vertices: } (h, k-a) \text{ and } (h, k+a) \\ \text{asymptotes: } y = \pm\frac{a}{b}(x-h) + k \end{cases}$$

If p is a nonzero real number, then the graph of each of the following equations is a parabola with vertex (h, k).

$$(x-h)^2 = 4p(y-k) \quad \begin{cases} \text{focus: } (h, k+p) \\ \text{directrix: the horizontal line } y = k - p \\ \text{axis: the vertical line } x = h \\ \text{opens upward if } p > 0, \text{ downward if } p < 0 \end{cases}$$

$$(y-k)^2 = 4p(x-h) \quad \begin{cases} \text{focus: } (h+p, k) \\ \text{directrix: the vertical line } x = h - p \\ \text{axis: the horizontal line } y = k \\ \text{opens to right if } p > 0, \text{ to left if } p < 0 \end{cases}$$

Graphing Techniques

When the equation of a conic section is in standard form, the techniques of Section 10.2 can be used to obtain its graph.

EXAMPLE 4 Graph the equation $\dfrac{(y+1)^2}{2} - \dfrac{(x-3)^2}{4} = 1.$

Solution The graph is a hyperbola centered at $(3, -1)$ (why?), which can be found in several ways. Solve the equation for y:

Method 1:

$$\frac{(y+1)^2}{2} = 1 + \frac{(x-3)^2}{4}$$

$$(y+1)^2 = 2\left(1 + \frac{(x-3)^2}{4}\right) = 2 + \frac{(x-3)^2}{2}$$

$$y + 1 = \pm\sqrt{2 + \frac{(x-3)^2}{2}}$$

$$y = \sqrt{2 + \frac{(x-3)^2}{2}} - 1 \qquad \text{or} \qquad y = -\sqrt{2 + \frac{(x-3)^2}{2}} - 1$$

Now graph the last two equations on the same screen. The graph of the first is the top half and the graph of the second the bottom half of the graph of the original equation.

GRAPHING EXPLORATION Find a viewing window that shows a complete graph of the original equation.

Method 2:

Use parametric equations. On page 613 we saw that the hyperbola $\dfrac{y^2}{a^2} - \dfrac{x^2}{b^2} = 1$ can be parameterized by letting $x = b \tan t$ and $y = a/\cos t$ $(0 \le t \le 2\pi)$. In this case we have $x - 3$ in place of x and $y + 1$ in place of y, with $a = \sqrt{2}$ and $b = 2$. So, we use this parameterization:

$$x - 3 = 2 \tan t \qquad \text{and} \qquad y + 1 = \frac{\sqrt{2}}{\cos t}$$

$$x = 2 \tan t + 3 \qquad\qquad y = \frac{\sqrt{2}}{\cos t} - 1 \qquad (0 \le t \le 2\pi).$$

GRAPHING EXPLORATION Use this parameterization to graph the equation. ■

When a second-degree equation is not in standard form, the fastest way to graph it is to use the first method in Example 4, modified as in the next example.

EXAMPLE 5 To graph the equation $x^2 + 8y^2 + 6x + 9y + 4 = 0$ without first putting it in standard form, rewrite it like this:

$$8y^2 + 9y + (x^2 + 6x + 4) = 0.$$

This is a quadratic equation of the form $ay^2 + by + c = 0$, with

$$a = 8, \qquad b = 9, \qquad c = x^2 + 6x + 4$$

and hence can be solved by using the quadratic formula:

$$y = \frac{-b \pm \sqrt{b^2 - 4ac}}{2a}$$

$$y = \frac{-9 \pm \sqrt{9^2 - 4 \cdot 8 \cdot (x^2 + 6x + 4)}}{2 \cdot 8}$$

$$= \frac{-9 \pm \sqrt{81 - 32(x^2 + 6x + 4)}}{16}.$$

▶ TECHNOLOGY TIP

TI-82/83/85 users can save keystrokes by entering the first function in the Exploration as y_1 and then using the RCL key to copy the text of y_1 to y_2. TI-92 users can do the same with CUT and PASTE on the F1 TOOLBAR menu. Then only one sign needs to be changed to make y_2 into the second function of the Exploration.

GRAPHING EXPLORATION Find a complete graph of the equation by graphing both of the following functions on the same screen and identify the conic:

$$y = \frac{-9 + \sqrt{81 - 32(x^2 + 6x + 4)}}{16}$$

$$y = \frac{-9 - \sqrt{81 - 32(x^2 + 6x + 4)}}{16}.$$

■

Rotations and Second-Degree Equations

A second-degree equation in x and y is one that can be written in the form

$$(*) \qquad Ax^2 + Bxy + Cy^2 + Dx + Ey + F = 0$$

for some constants A, B, C, D, E, F, with at least one of A, B, C nonzero. Every conic section is the graph of a second-degree equation. For instance, the ellipse equation $\frac{x^2}{4} + \frac{(y-3)^2}{6} = 1$ can be rewritten as

$$12\left(\frac{x^2}{4}\right) + 12\left(\frac{(y-3)^2}{6}\right) = 12, \quad \text{or equivalently,} \quad 3x^2 + 2(y-3)^2 = 12$$

which simplifies to $3x^2 + 2y^2 - 12y + 6 = 0$; this is equation (*) with $A = 3$, $B = 0, C = 2, D = 0, E = -12$, and $F = 6$. Conversely, it can be shown that the graph of every second-degree equation is a conic section (possibly degenerate). When the equation has an xy term, the conic may be rotated from standard position, so that its axis or axes are not parallel to the coordinate axes.

EXAMPLE 6 To graph the equation

$$3x^2 + 6xy + y^2 + x - 2y + 7 = 0,$$

we first rewrite it as:

$$y^2 + 6xy - 2y + 3x^2 + x + 7 = 0$$
$$y^2 + (6x - 2)y + (3x^2 + x + 7) = 0.$$

This equation has the form $ay^2 + by + c = 0$, with $a = 1, b = 6x - 2$, and $c = 3x^2 + x + 7$. It can be solved with the quadratic formula:

$$y = \frac{-b \pm \sqrt{b^2 - 4ac}}{2a} = \frac{-(6x-2) \pm \sqrt{(6x-2)^2 - 4 \cdot 1 \cdot (3x^2 + x + 7)}}{2 \cdot 1}.$$

The top half of the graph is obtained by graphing

$$y = \frac{-6x + 2 + \sqrt{(6x-2)^2 - 4(3x^2 + x + 7)}}{2}$$

and the bottom half by graphing

$$y = \frac{-6x + 2 - \sqrt{(6x-2)^2 - 4(3x^2 + x + 7)}}{2}.$$

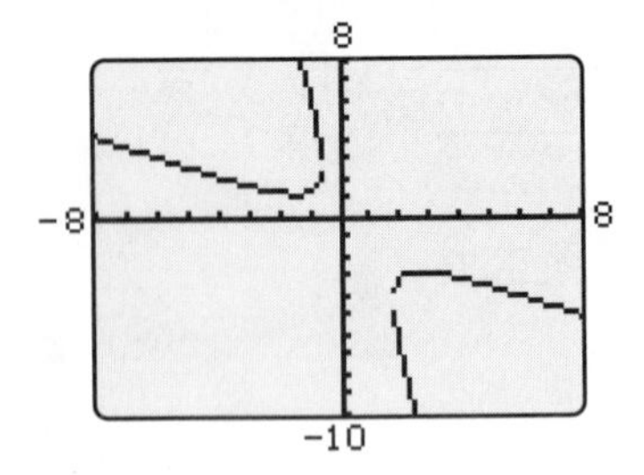

Figure 10-42

The graph is a hyperbola whose focal axis tilts upward to the left, as shown in Figure 10–42. ■

The following fact, which will not be proved here, makes it easy to identify the graphs of second-degree equations before graphing them:

Graphs of Second-Degree Equations ▶

> **If the equation**
>
> $$Ax^2 + Bxy + Cy^2 + Dx + Ey + F = 0 \qquad (A, B, C \text{ not all } 0)$$
>
> **has a graph, then that graph is**
>
> **a circle or an ellipse (or a point), if $B^2 - 4AC < 0$;**
>
> **a parabola (or a line or two parallel lines), if $B^2 - 4AC = 0$;**
>
> **a hyperbola (or two intersecting lines), if $B^2 - 4AC > 0$.**

The expression $B^2 - 4AC$ is called the **discriminant** of the equation.

EXAMPLE 7 To identify the graph of

$$2x^2 - 4xy + 3y^2 + 5x + 6y - 8 = 0$$

we compute the discriminant, with $A = 2$, $B = -4$, and $C = 3$:

$$B^2 - 4AC = (-4)^2 - 4 \cdot 2 \cdot 3 = 16 - 24 = -8.$$

Hence, the graph is an ellipse (possibly a circle or a single point). The graph can be found as above by using the quadratic formula to solve

$$3y^2 + (-4x + 6)y + (2x^2 + 5x - 8) = 0$$

and graphing both solutions on the same screen:

$$y = \frac{-b \pm \sqrt{b^2 - 4ac}}{2a}$$

$$= \frac{-(-4x + 6) \pm \sqrt{(-4x + 6)^2 - 4 \cdot 3 \cdot (2x^2 + 5x - 8)}}{2 \cdot 3}.$$

GRAPHING EXPLORATION Find a viewing window that shows a complete graph of the equation. In what direction does the major axis run? ■

EXAMPLE 8 The discriminant of

$$3x^2 + 6xy + 3y^2 + 13x + 9y + 53 = 0$$

is $B^2 - 4AC = 6^2 - 4 \cdot 3 \cdot 3 = 0$. Hence, the graph is a parabola (or a line or parallel lines in the degenerate case).

GRAPHING EXPLORATION Find a viewing window that shows a complete graph of the equation. ■

Applications

The long-range navigation system (LORAN) uses hyperbolas to enable a ship to determine its exact location by radio, as illustrated in the following example.

EXAMPLE 9 Three LORAN radio transmitters Q, P, and R are located 200 miles apart along a straight line and simultaneously transmit signals at regular intervals. These signals travel at a speed of 980 feet per microsecond. A ship S receives a signal from P and 305 microseconds later a signal from R. It also receives a signal from Q 528 microseconds after the one from P. The ship's location is then determined as follows.

Take the line through the LORAN stations as the x-axis, with the origin located midway between Q and P, so that the situation looks like Figure 10–43.

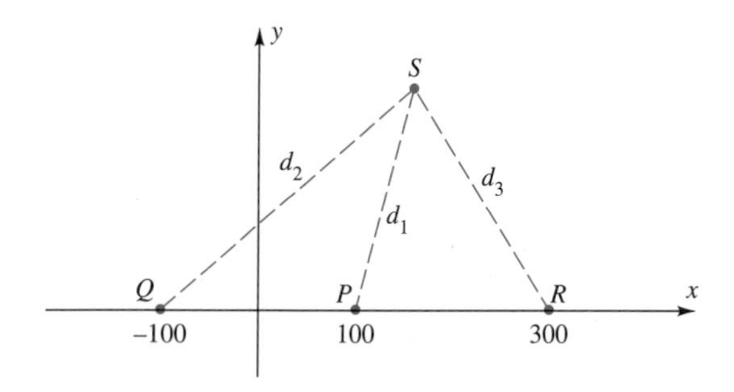

Figure 10-43

If the signal takes t microseconds to go from P to S, then

$$d_1 = 980t \qquad \text{and} \qquad d_2 = 980(t + 528)$$

so that

$$|d_1 - d_2| = |980t - 980(t + 528)| = 980 \cdot 528 = 517{,}440 \text{ feet.}$$

Since one mile is 5280 feet, this means that

$$|d_1 - d_2| = 517{,}440/5{,}280 \text{ miles} = 98 \text{ miles.}$$

In other words,

$$|(\text{distance from } P \text{ to } S) - (\text{distance from } Q \text{ to } S)| = |d_1 - d_2| = 98.$$

This is precisely the situation described in the definition of "hyperbola" on page 606: S is on the hyperbola with foci $P = (100, 0)$, $Q = (-100, 0)$ and distance difference $r = 98$. This hyperbola has an equation of the form

$$\frac{x^2}{a^2} - \frac{y^2}{b^2} = 1,$$

where $(\pm a, 0)$ are the vertices, $(\pm c, 0) = (\pm 100, 0)$ are the foci and $c^2 = a^2 + b^2$. Figure 10–44 and the fact that the vertex $(a, 0)$ is on the hyperbola show that

$$|(\text{distance from } P \text{ to } (a, 0)) - (\text{distance } Q \text{ to } (a, 0))| = r = 98$$

$$|(100 - a) - (100 + a)| = 98$$

$$|-2a| = 98$$

$$|a| = 49.$$

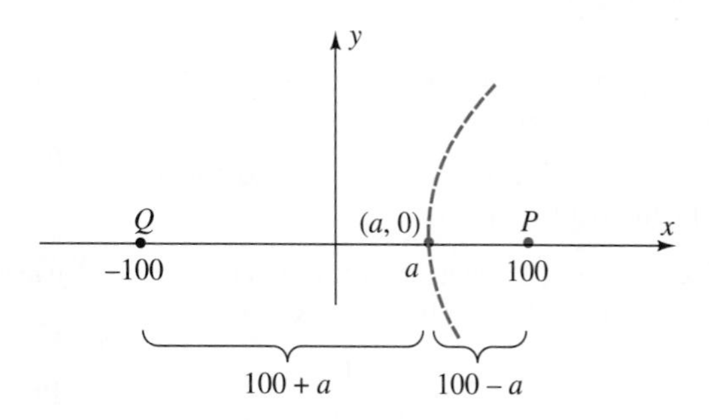

Figure 10-44

Consequently, $a^2 = 49^2 = 2401$ and hence $b^2 = c^2 - a^2 = 100^2 - 49^2 = 7599$. Thus the ship lies on the hyperbola

$$(*) \qquad \frac{x^2}{2401} - \frac{y^2}{7599} = 1.$$

A similar argument using P and R as foci shows that the ship also lies on the hyperbola with foci $P = (100, 0)$ and $R = (300, 0)$ and center $(200, 0)$, whose distance difference r is

$$|d_1 - d_3| = 980 \cdot 305 = 298{,}900 \text{ feet} \approx 56.61 \text{ miles}.$$

As above you can verify that $a = 56.61/2 = 28.305$ and hence $a^2 = 28.305^2 = 801.17$. This hyperbola has center $(200, 0)$ and its foci are $(200 - c, k) = (100, 0)$ and $(200 + c, k) = (300, 0)$, which implies that $c = 100$. Hence, $b^2 = c^2 - a^2 = 100^2 - 801.17 = 9198.83$ and the ship also lies on the hyperbola

$$(**) \qquad \frac{(x - 200)^2}{801.17} - \frac{y^2}{9198.83} = 1.$$

Since the ship lies on both hyperbolas, its coordinates are solutions of both the equations $(*)$ and $(**)$. They can be found algebraically by solving each of the

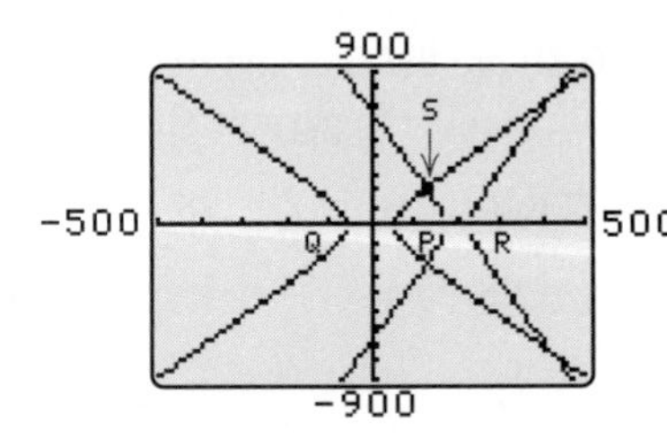

Figure 10-45

equations for y^2, setting the results equal, and solving for x. They can be found geometrically by graphing both hyperbolas and finding the intersection point. As shown in Figure 10–45, there are actually four points of intersection. However, the two below the x-axis represent points on land in our situation. Furthermore, since the signal from P was received first, the ship is closest to P. So it is located at the point S in Figure 10–45. Zoom-in or a graphical intersection finder shows that this point is approximately (130.48, 215.14), where the coordinates are in miles from the origin. ■

EXERCISES 10.3

In Exercises 1–16, find the equation of the conic sections satisfying the given conditions.

1. Ellipse with center (2, 3); endpoints of major and minor axes: (2, − 1), (0, 3), (2, 7), (4, 3).

2. Ellipse with center (− 5, 2); endpoints of major and minor axes: (0, 2), (− 5, 17), (− 10, 2), (− 5, − 13).

3. Ellipse with center (7, − 4); foci on the line $x = 7$; major axis of length 12; minor axis of length 5.

4. Ellipse with center (− 3, − 9); foci on the line $y = -9$; major axis of length 15; minor axis of length 7.

5. Hyperbola with center (− 2, 3); vertex (− 2, 1); passing through $(-2 + 3\sqrt{10}, 11)$.

6. Hyperbola with center (− 5, 1); vertex (− 3, 1); passing through $(-1, 1 - 4\sqrt{3})$.

7. Hyperbola with center (4, 2); vertex (7, 2); asymptote $3y = 4x - 10$.

8. Hyperbola with center (− 3, − 5); vertex (− 3, 0); asymptote $6y = 5x - 15$.

9. Parabola with vertex (1, 0); axis $x = 1$; (2, 13) on graph.

10. Parabola with vertex (− 3, 0); axis $y = 0$; (− 1, 1) on graph.

11. Parabola with vertex (2, 1); axis $y = 1$; (5, 0) on graph.

12. Parabola with vertex (1, − 3); axis $y = -3$; (− 1, − 4) on graph.

13. Ellipse with center (3, − 2); passing through (3, − 6) and (9, − 2).

14. Ellipse with center (2, 5); passing through (2, 4) and (− 3, 5).

15. Parabola with vertex (− 3, − 2) and focus (− 47/16, − 2).

16. Parabola with vertex (− 5, − 5) and focus (− 5, − 99/20).

In Exercises 17–22, assume that the graph of the equation is a nondegenerate conic section. Without graphing, determine whether the graph is a circle, ellipse, hyperbola, or parabola.

17. $x^2 - 2xy + 3y^2 - 1 = 0$ **18.** $xy - 1 = 0$

19. $x^2 + 2xy + y^2 + 2\sqrt{2}x - 2\sqrt{2}y = 0$

20. $2x^2 - 4xy + 5y^2 - 6 = 0$

21. $17x^2 - 48xy + 31y^2 + 50 = 0$

22. $2x^2 - 4xy - 2y^2 + 3x + 5y - 10 = 0$

In Exercises 23–34, find parametric equations for the conic section whose equation is given and use these parametric equations to find a complete graph.

23. $\dfrac{(x-1)^2}{4} + \dfrac{(y-5)^2}{9} = 1$

24. $\dfrac{(x-2)^2}{16} + \dfrac{(y+3)^2}{12} = 1$

25. $\dfrac{(x+1)^2}{16} + \dfrac{(y-4)^2}{8} = 1$

26. $\dfrac{(x+5)^2}{4} + \dfrac{(y+2)^2}{12} = 1$

27. $y = 4(x-1)^2 + 2$ **28.** $y = 3(x-2)^2 - 3$

29. $x = 2(y-2)^2$ **30.** $x = -3(y-1)^2 - 2$

31. $\dfrac{(y+3)^2}{25} - \dfrac{(x+1)^2}{16} = 1$

32. $\dfrac{(y+1)^2}{9} - \dfrac{(x-1)^2}{25} = 1$

33. $\dfrac{(x+3)^2}{1} - \dfrac{(y-2)^2}{4} = 1$

34. $\dfrac{(y+5)^2}{9} - \dfrac{(x-2)^2}{1} = 1$

In Exercises 35–52, use the discriminant to identify the conic section whose equation is given and find a viewing window that shows a complete graph.

35. $9x^2 + 4y^2 + 54x - 8y + 49 = 0$

36. $4x^2 + 5y^2 - 8x + 30y + 29 = 0$

37. $4y^2 - x^2 + 6x - 24y + 11 = 0$

38. $x^2 - 16y^2 = 0$ 39. $3y^2 - x - 2y + 1 = 0$

40. $x^2 - 6x + y + 5 = 0$

41. $41x^2 - 24xy + 34y^2 - 25 = 0$

42. $x^2 + 2\sqrt{3}xy + 3y^2 + 8\sqrt{3}x - 8y + 32 = 0$

43. $17x^2 - 48xy + 31y^2 + 49 = 0$

44. $52x^2 - 72xy + 73y^2 = 200$

45. $9x^2 + 24xy + 16y^2 + 90x - 130y = 0$

46. $x^2 + 10xy + y^2 + 1 = 0$

47. $23x^2 + 26\sqrt{3}xy - 3y^2 - 16x + 16\sqrt{3}y + 128 = 0$

48. $x^2 + 2xy + y^2 + 12\sqrt{2}x - 12\sqrt{2}y = 0$

49. $17x^2 - 12xy + 8y^2 - 80 = 0$

50. $11x^2 - 24xy + 4y^2 + 30x + 40y - 45 = 0$

51. $3x^2 + 2\sqrt{3}xy + y^2 + 4x - 4\sqrt{3}y - 16 = 0$

52. $3x^2 + 2\sqrt{2}xy + 2y^2 - 12 = 0$

In Exercises 53 and 54, find the equations of two distinct ellipses satisfying the given conditions.

53. Center at $(-5, 3)$; major axis of length 14; minor axis of length 8.

54. Center at $(2, -6)$; major axis of length 15; minor axis of length 6.

55. Show that the asymptotes of the hyperbola $\dfrac{x^2}{a^2} - \dfrac{y^2}{a^2} = 1$ are perpendicular to each other.

56. Find a number k such that $(-2, 1)$ is on the graph of $3x^2 + ky^2 = 4$. Then graph the equation.

57. Find the number b such that the vertex of the parabola $y = x^2 + bx + c$ lies on the y-axis.

58. Find the number d such that the parabola $(y+1)^2 = dx + 4$ passes through $(-6, 3)$.

59. Find the points of intersection of the parabola $4y^2 + 4y = 5x - 12$ and the line $x = 9$.

60. Find the points of intersection of the parabola $4x^2 - 8x = 2y + 5$ and the line $y = 15$.

61. Two listening stations 1 mile apart record an explosion. One microphone receives the sound 2 seconds after the other does. Use the line through the microphones as the x-axis, with the origin midway between the microphones, and the fact that sound travels at 1100 feet per second to find the equation of the hyperbola on which the explosion is located. Can you determine the exact location of the explosion?

62. Two transmission stations P and Q are located 200 miles apart on a straight shore line. A ship 50 miles from shore is moving parallel to the shore line. A signal from Q reaches the ship 400 microseconds after a signal from P. If the signals travel at 980 feet per microsecond, find the location of the ship (in terms of miles) in the coordinate system with x-axis through P and Q, and origin midway between them.

10.4 POLAR COORDINATES

In the past we have used a rectangular coordinate system in the plane, based on two perpendicular coordinate axes. Now we introduce another coordinate system for the plane, based on angles.

Choose a point O in the plane (called the **origin** or **pole**) and a half-line extending from O (called the **polar axis**). Choose a unit of length. A point P in the plane is given the coordinates (r, θ), where r is the length of segment OP and θ is the angle with the polar axis as initial side, vertex O, and terminal side OP, as shown in Figure 10–46.

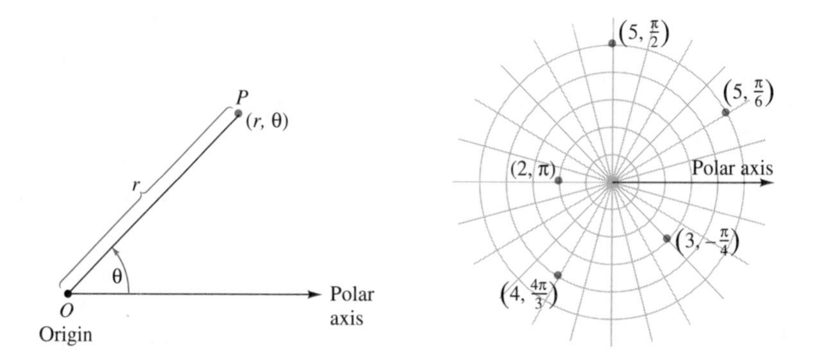

Figure 10-46 **Figure 10-47**

We shall usually measure the angle θ in radians; it may be either positive or negative, depending on whether it is generated by a clockwise or counterclockwise rotation. Some typical points are shown in Figure 10–47, which also illustrates the ‘‘circular grid’’ that a polar coordinate system imposes on the plane.

The polar coordinates of a point P are *not* unique. For instance, angles of radian measure $\pi/3$, $7\pi/3$, and $-5\pi/3$ all have the same terminal side,* so $(2, \pi/3)$, $(2, 7\pi/3)$, and $(2, -5\pi/3)$ represent the same point (Figure 10–48 below). Furthermore, we shall consider the coordinates of the origin to be $(0, \theta)$, where θ is *any* angle.

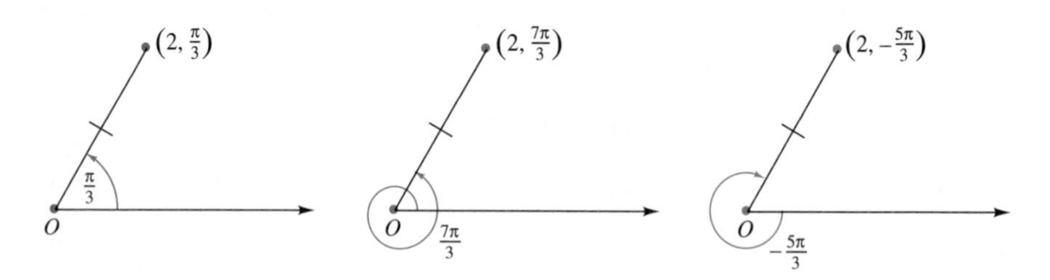

Figure 10-48

Negative values for the first coordinate will be allowed according to this convention: For each positive r, the point $(-r, \theta)$ lies on the straight line contain-

*Because $\frac{7\pi}{3} = \frac{\pi}{3} + 2\pi$ and $\frac{-5\pi}{3} = \frac{\pi}{3} - 2\pi$.

ing the terminal side of θ, at distance r from the origin, on the *opposite* side of the origin from the point (r, θ), as shown in Figure 10–49.

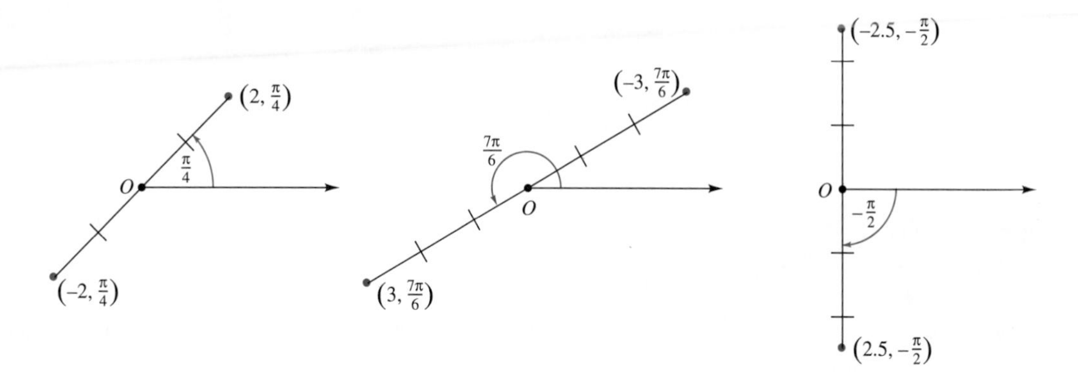

Figure 10-49

It is sometimes convenient to use both a rectangular and a polar coordinate system in the plane, with the polar axis coinciding with the positive x-axis. Then the y-axis is the polar line $\theta = \pi/2$. Suppose P has rectangular coordinates (x, y) and polar coordinates (r, θ), with $r > 0$, as in Figure 10–50.

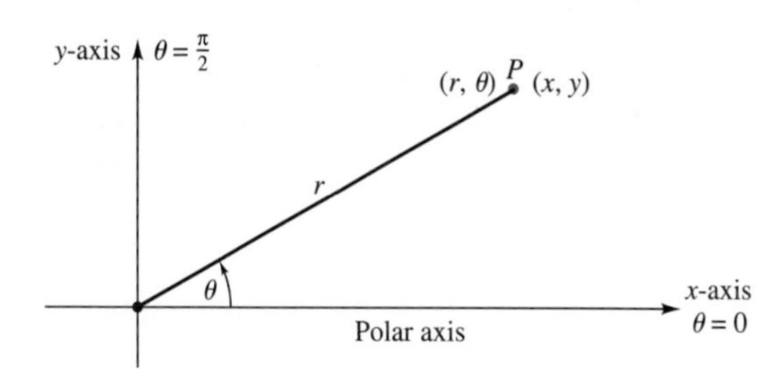

Figure 10-50

Since r is the distance from (x, y) to $(0, 0)$, the distance formula shows that $r = \sqrt{x^2 + y^2}$, and hence $r^2 = x^2 + y^2$. The point-in-the-plane description of the trigonometric functions shows that

$$\cos\theta = \frac{x}{r} \qquad \sin\theta = \frac{y}{r} \qquad \tan\theta = \frac{y}{x}.$$

Solving the first two equations for x and y, we obtain the relationship between polar and rectangular coordinates:*

*Although the discussion above dealt only with the case when $r > 0$, the conclusions in the box on the next page are also true when $r < 0$ (Exercise 50).

Coordinate Conversion Formulas ▶

If a point has rectangular coordinates (x, y) and polar coordinates (r, θ), then

$$x = r\cos\theta \quad \text{and} \quad y = r\sin\theta$$

$$r^2 = x^2 + y^2 \quad \text{and} \quad \tan\theta = \frac{y}{x}.$$

▶ TECHNOLOGY TIP

Conversions between polar and rectangular coordinates can be made automatically on most calculators. Use the MATH menu on TI-81, the ANGLE menu on TI-82/83, the VECTOR OPS menu on TI-85, the ANGLE submenu of MATH menu on TI-92, the MATH COR menu on Casio 9800, and the MATH CONV menu on Sharp 9300. Conversion programs for HP-38 are in the Program Appendix.

EXAMPLE 1 Convert each of the following points in polar coordinates to rectangular coordinates: **(a)** $(2, \pi/6)$ **(b)** $(3, 4)$.

Solution

(a) Apply the first set of equations in the box with $r = 2$ and $\theta = \pi/6$:

$$x = 2\cos\frac{\pi}{6} = 2\cdot\frac{\sqrt{3}}{2} = \sqrt{3} \quad \text{and} \quad y = 2\sin\frac{\pi}{6} = 2\cdot\frac{1}{2} = 1.$$

So the rectangular coordinates are $(\sqrt{3}, 1)$.

(b) The point with polar coordinates $(3, 4)$ has $r = 3$ and $\theta = 4$ radians. Therefore, its rectangular coordinates are

$$(3\cos 4, 3\sin 4) \approx (-1.9609, -2.2704). \quad ■$$

EXAMPLE 2 Convert each of the following points in rectangular coordinates to polar coordinates:
(a) $(2, -2)$ **(b)** $(3, 5)$ **(c)** $(-2, 4)$.

Solution

(a) The second set of equations in the box, with $x = 2$, $y = -2$, shows that

$$r = \sqrt{2^2 + (-2)^2} = \sqrt{8} = 2\sqrt{2} \quad \text{and} \quad \tan\theta = -2/2 = -1.$$

We must find an angle θ whose terminal side passes through $(2, -2)$ and whose tangent is -1. Two of the many possibilities are $\theta = -\pi/4$ and $\theta = 7\pi/4$. So $(2\sqrt{2}, -\pi/4)$ is one pair of polar coordinates and $(2\sqrt{2}, 7\pi/4)$ is another.

(b) Applying the second set of equations in the box, with $x = 3$, $y = 5$, we have

$$r = \sqrt{3^2 + 5^2} = \sqrt{34} \quad \text{and} \quad \tan\theta = 5/3.$$

The $\boxed{TAN^{-1}}$ key on a calculator shows that $\theta \approx 1.0304$ radians is an angle between 0 and $\pi/2$ with tangent 5/3. Since $(3, 5)$ is in the first quadrant, one pair of (approximate) polar coordinates is $(\sqrt{34}, 1.0304)$.

(c) In this case $r = \sqrt{(-2)^2 + 4^2} = \sqrt{20} = 2\sqrt{5}$ and $\tan\theta = 4/(-2) = -2$. The $\boxed{TAN^{-1}}$ key shows that $\theta \approx -1.1071$ is an angle between $-\pi/2$ and 0 with tangent -2. However, we want an angle between $\pi/2$ and π

with tangent -2 because $(-2, 4)$ is in the second quadrant. The tangent function has period π; hence,

$$-2 = \tan(-1.1071) = \tan(-1.1071 + \pi),$$

with $-1.1071 + \pi \approx 2.0344$ an angle between $\pi/2$ and π. Therefore, one pair of polar coordinates is $(2\sqrt{5}, 2.0344)$. ■

Polar Graphs

The graphs of a few polar coordinate equations can be easily determined from the appropriate definitions.

EXAMPLE 3 The graph of the equation $r = 3$* consists of all points (r, θ) with first coordinate 3, that is, all points whose distance from the origin is 3. So the graph is a circle with center O and radius 3 (Figure 10–51). ■

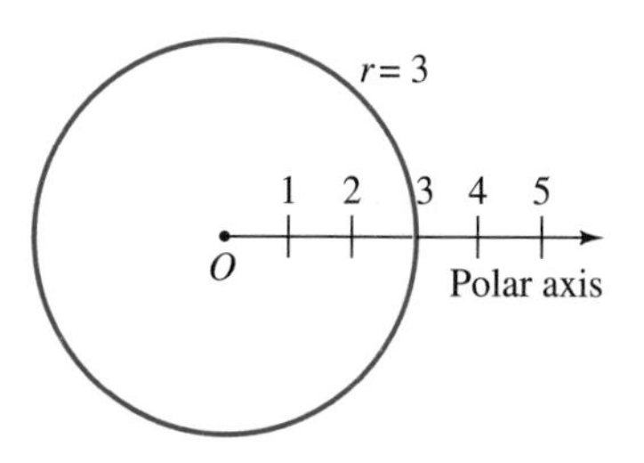

Figure 10-51

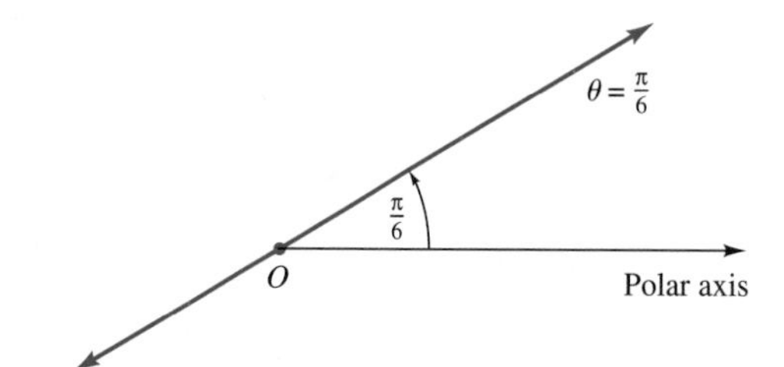

Figure 10-52

EXAMPLE 4 The graph of $\theta = \pi/6$ consists of all points $(r, \pi/6)$. If $r \geq 0$, then $(r, \pi/6)$ lies on the terminal side of an angle of $\pi/6$ radians, whose initial side is the polar axis. A point $(r, \pi/6)$, with $r < 0$, lies on the extension of this terminal side across the origin. So, the graph is the straight line in Figure 10–52 above. ■

Some polar graphs can be sketched by hand by using basic facts about trigonometric functions.

*Every equation here is understood to involve two variables, but one may have coefficient 0; in this case, $r = 3 + 0 \cdot \theta$. An analogous situation occurs in rectangular coordinates with equations such as $y = 3$ or $x = -2$.

EXAMPLE 5 To graph $r = 1 + \sin \theta$, remember the behavior of $\sin \theta$ between 0 and 2π:

As θ increases from 0 to $\pi/2$, $\sin \theta$ increases from 0 to 1. So $r = 1 + \sin \theta$ increases from 1 to 2.

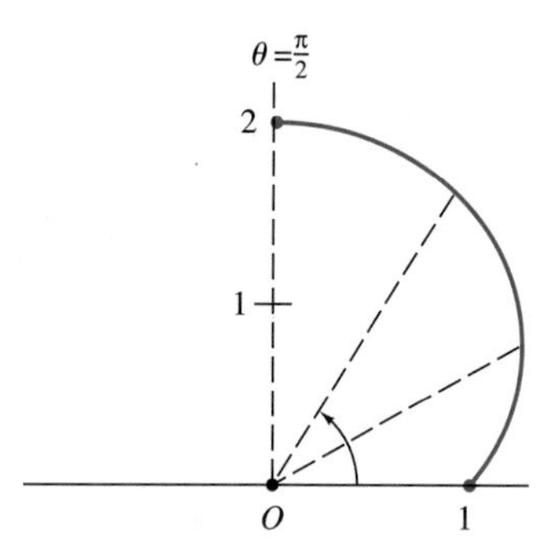

As θ increases from $\pi/2$ to π, $\sin \theta$ decreases from 1 to 0. So $r = 1 + \sin \theta$ decreases from 2 to 1.

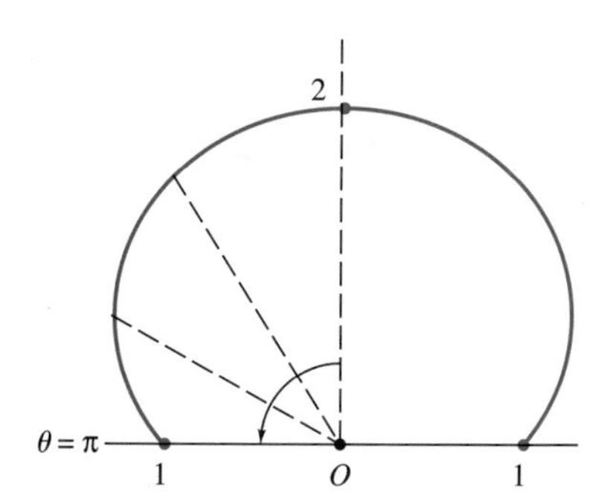

As θ increases from π to $3\pi/2$, $\sin \theta$ decreases from 0 to -1. So $r = 1 + \sin \theta$ decreases from 1 to 0.

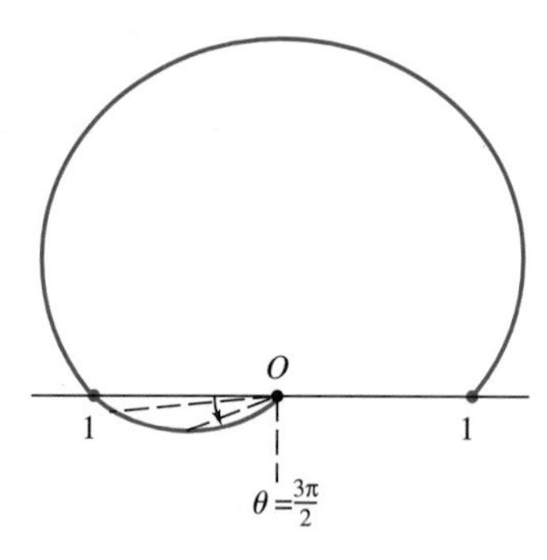

As θ increases from $3\pi/2$ to 2π, $\sin \theta$ increases from -1 to 0. So $r = 1 + \sin \theta$ increases from 0 to 1.

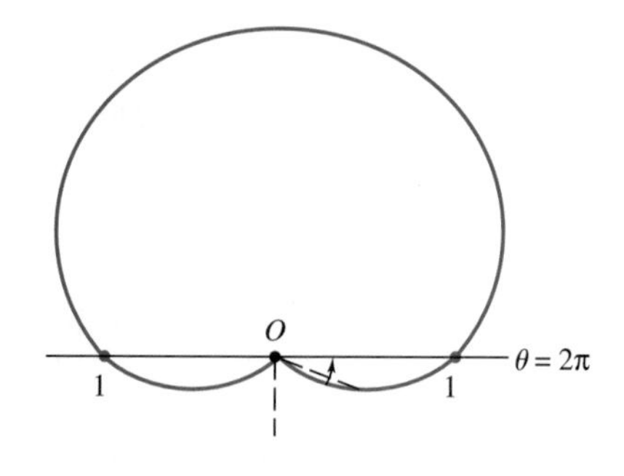

Figure 10-53

As θ takes values larger than 2π, $\sin \theta$ repeats the same pattern, and hence so does $r = 1 + \sin \theta$. The same is true for negative values of θ. The full graph (called a **cardioid**) is at the lower right above. ■

The easiest way to graph polar equation $r = f(\theta)$ is to use a calculator in polar graphing mode. A second way is to use parametric graphing mode, with the coordinate conversion formulas as a parameterization:

$$x = r\cos\theta = f(\theta)\cos\theta$$
$$y = r\sin\theta = f(\theta)\sin\theta.$$

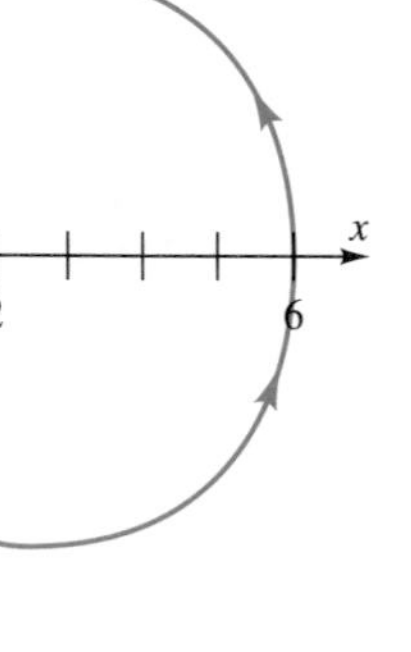

Figure 10-54

EXAMPLE 6 Graph $r = 2 + 4\cos\theta$.

Solution *Polar Method:* Put your calculator in polar graphing mode and enter $r = 2 + 4\cos\theta$ in the function memory. Set the viewing window by entering minimum and maximum values for x, y, and θ. Since sine has period 2π, a complete graph can be obtained by taking $0 \le \theta \le 2\pi$. You must also set the θ step (or θ pitch), which determines how many values of θ the calculator uses to plot the graph. With an appropriate θ step, the graph should look like Figure 10–54.

> GRAPHING EXPLORATION Graph $r = 2 + 4\cos\theta$, using the viewing window with $-4 \le x \le 8$, $-4 \le y \le 4$, $0 \le \theta \le 6.3$, and θ step $= 1$. If the graph does not resemble Figure 10–54, try a smaller θ step until you find one that produces a graph like Figure 10–54.
>
> To understand what the θ step does, set your calculator to graph in "dot" mode rather than "connected" mode, so you can see the points it actually plots. Now graph again, beginning with θ step $= 1$ and then using smaller values.

Parametric Method: Put your calculator in parametric graphing mode. The parametric equations for $r = 2 + 4\cos\theta$ are as follows (using t as the variable instead of θ with $0 \le t \le 2\pi$):

$$x = r\cos t = (2 + 4\cos t)\cos t = 2\cos t + 4\cos^2 t$$
$$x = r\sin t = (2 + 4\cos t)\sin t = 2\sin t + 4\sin t\cos t.$$

They also produce the graph in Figure 10–54. ■

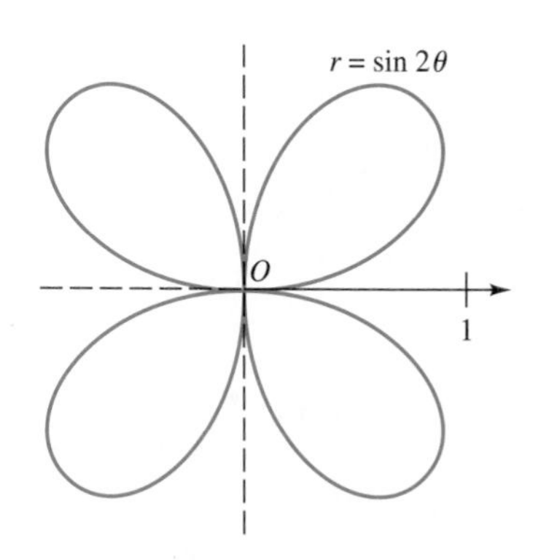

Figure 10-55

EXAMPLE 7 The graph of $r = \sin 2\theta$ in Figure 10–55 can be obtained either by graphing directly in polar mode or by using parametric mode and the equations:

$$x = r\cos t = \sin 2t\cos t \quad \text{and} \quad y = r\sin t = \sin 2t\sin t \quad (0 \le t \le 2\pi).$$ ■

Here is a summary of commonly encountered polar graphs (in each case, a and b are constants):

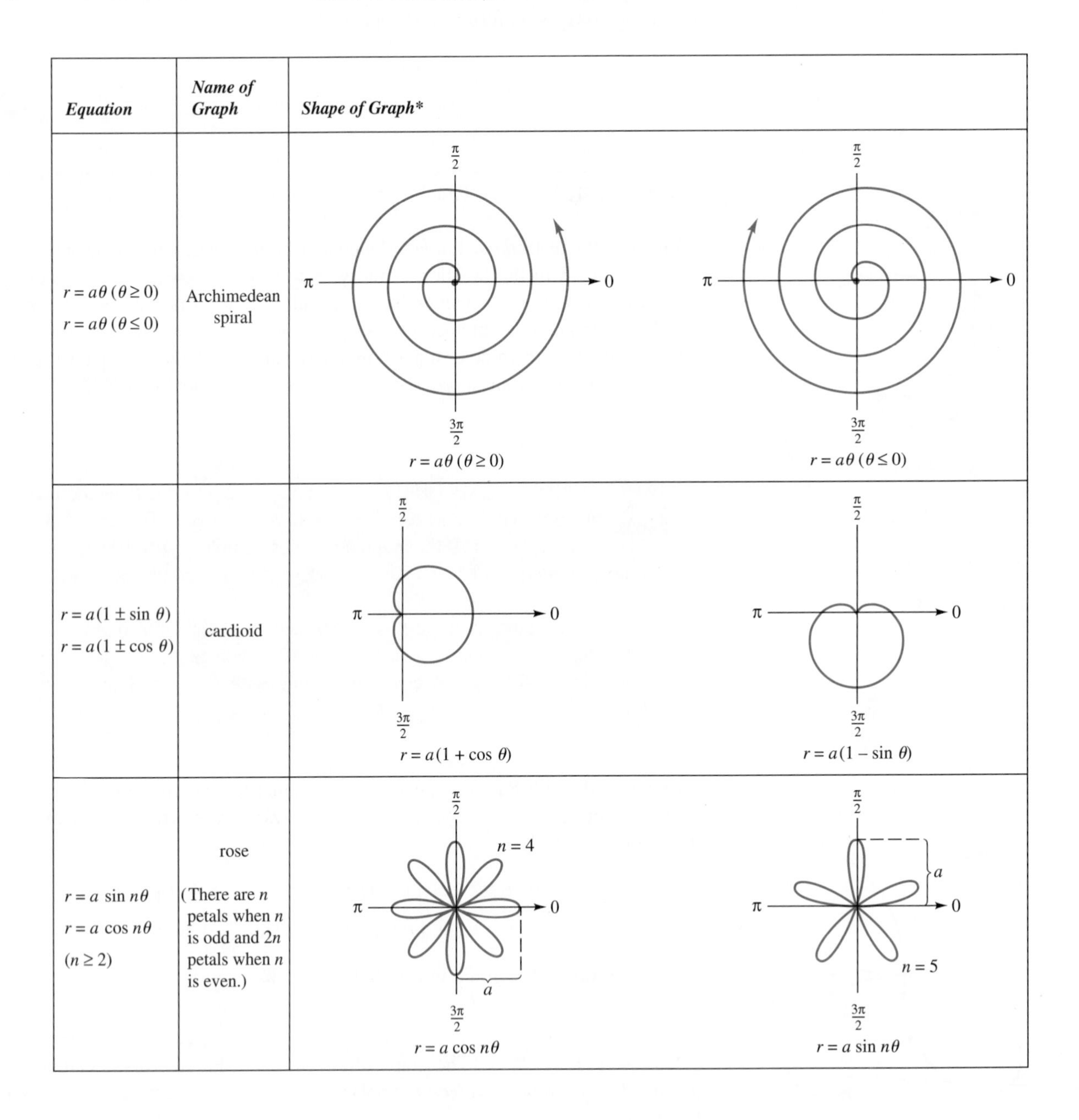

Equation	*Name of Graph*	*Shape of Graph**
$r = a\theta\ (\theta \geq 0)$ $r = a\theta\ (\theta \leq 0)$	Archimedean spiral	$r = a\theta\ (\theta \geq 0)$; $r = a\theta\ (\theta \leq 0)$
$r = a(1 \pm \sin\theta)$ $r = a(1 \pm \cos\theta)$	cardioid	$r = a(1 + \cos\theta)$; $r = a(1 - \sin\theta)$
$r = a \sin n\theta$ $r = a \cos n\theta$ $(n \geq 2)$	rose (There are n petals when n is odd and $2n$ petals when n is even.)	$r = a \cos n\theta$ ($n = 4$); $r = a \sin n\theta$ ($n = 5$)

*Depending on the plus or minus sign and whether sine or cosine is involved, the basic shape of a specific graph may differ from those shown by a rotation, reversal, or horizontal or vertical shift.

Equation	*Name of Graph*	*Shape of Graph**
$r = a \sin \theta$ $r = a \cos \theta$	circle	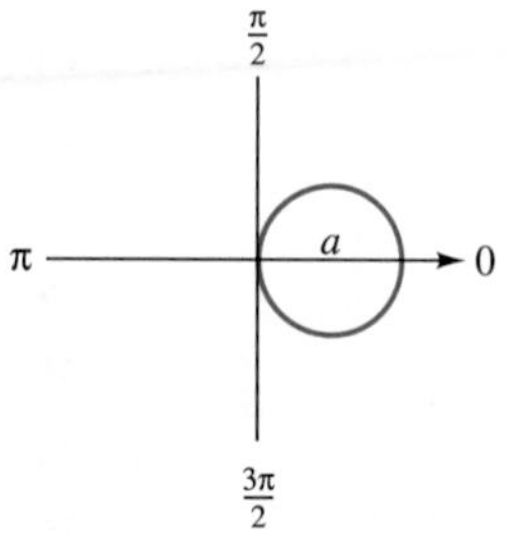 $\frac{\pi}{2}$, π, 0, $\frac{3\pi}{2}$, a $r = a \cos \theta$ 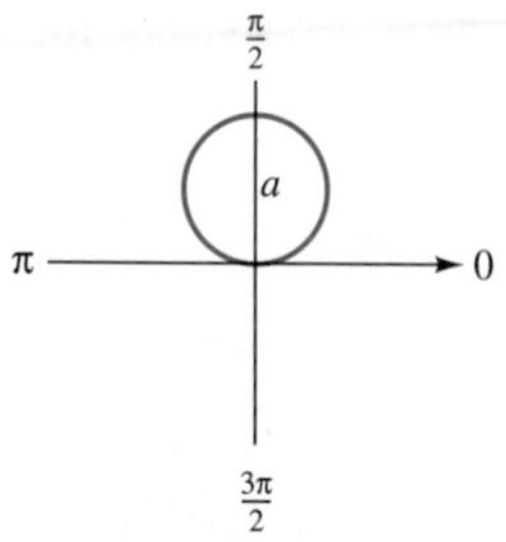 — $r = a \sin \theta$
$r^2 = \pm a^2 \sin 2\theta$ $r^2 = \pm a^2 \cos 2\theta$	lemniscate	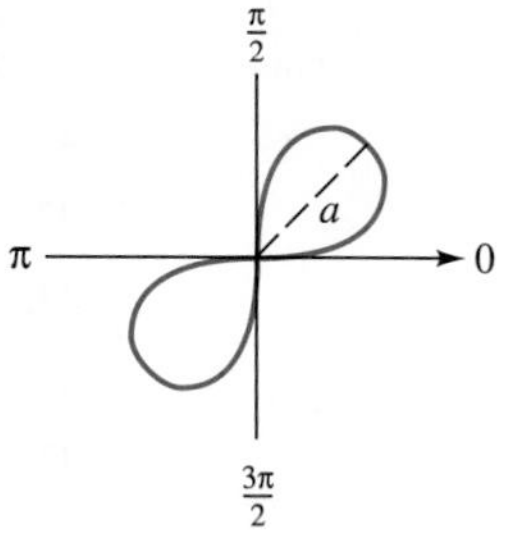 $\frac{\pi}{2}$, π, 0, $\frac{3\pi}{2}$, a $r^2 = a^2 \sin 2\theta$ 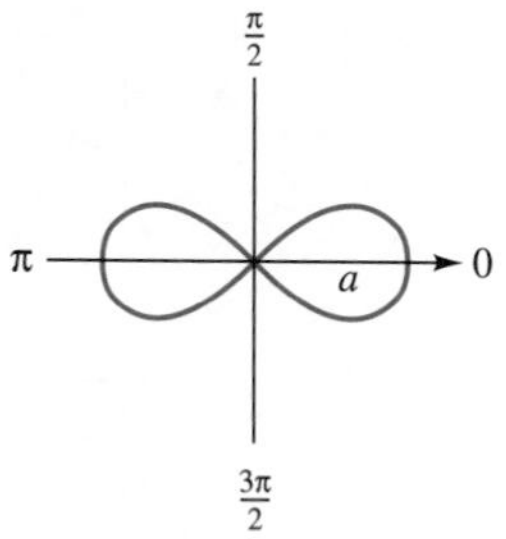 — $r^2 = a^2 \cos 2\theta$
$r = a \pm b \sin \theta$ $r = a \pm b \cos \theta$ $(a, b > 0;\ a \neq b)$	limaçon	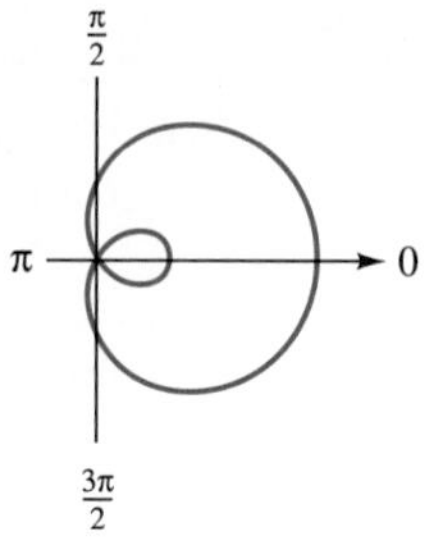 $\frac{\pi}{2}$, π, 0, $\frac{3\pi}{2}$ $a < b$, $r = a + b \cos \theta$ — $b < a < 2b$, $r = a + b \sin \theta$ — $a \geq 2b$, $r = a - b \sin \theta$

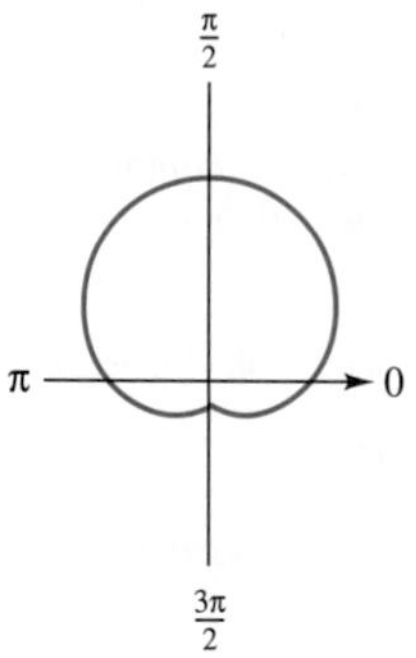

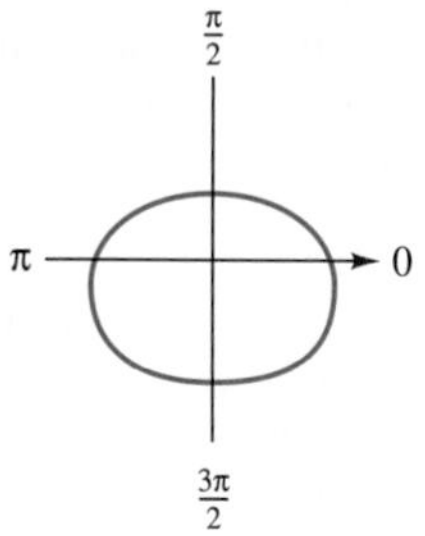

EXERCISES 10.4

1. What are the polar coordinates of the points P, Q, R, S, T, U, V in Figure 10–56?

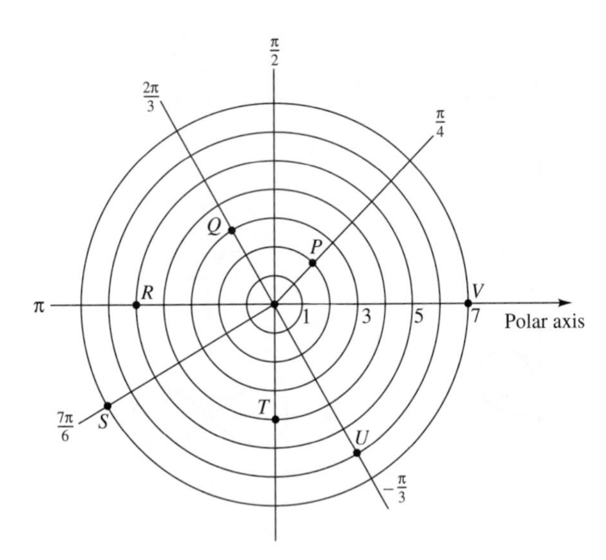

Figure 10-56

In Exercises 2–6, list four other pairs of polar coordinates for the given point, each with a different combination of signs (that is, $r > 0$, $\theta > 0$; $r > 0$, $\theta < 0$; $r < 0$, $\theta > 0$; $r < 0$, $\theta < 0$).

2. $(3, \pi/3)$
3. $(-5, \pi)$
4. $(2, -2\pi/3)$
5. $(-1, -\pi/6)$
6. $(\sqrt{3}, 3\pi/4)$

In Exercises 7–10, convert the polar coordinates to rectangular coordinates.

7. $(3, \pi/3)$
8. $(-2, \pi/4)$
9. $(-1, 5\pi/6)$
10. $(2, 0)$

In Exercises 11–16, convert the rectangular coordinates to polar coordinates.

11. $(3\sqrt{3}, -3)$
12. $(2\sqrt{3}, -2)$
13. $(2, 4)$
14. $(3, -2)$
15. $(-5, 2.5)$
16. $(-6.2, -3)$

In Exercises 17–22, sketch the graph of the equation without using a calculator.

17. $r = 4$
18. $r = -1$
19. $\theta = -\pi/3$
20. $\theta = 5\pi/6$
21. $\theta = 1$
22. $\theta = -4$

In Exercises 23–46, sketch the graph of the equation.

23. $r = \theta \quad (\theta \leq 0)$
24. $r = 3\theta \quad (\theta \geq 0)$
25. $r = 1 - \sin\theta$
26. $r = 3 - 3\cos\theta$
27. $r = -2\cos\theta$
28. $r = -6\sin\theta$
29. $r = \cos 2\theta$
30. $r = \cos 3\theta$
31. $r = \sin 3\theta$
32. $r = \sin 4\theta$
33. $r^2 = 4\cos 2\theta$
34. $r^2 = \sin 2\theta$
35. $r = 2 + 4\cos\theta$
36. $r = 1 + 2\cos\theta$
37. $r = \sin\theta + \cos\theta$
38. $r = 4\cos\theta + 4\sin\theta$
39. $r = \sin(\theta/2)$
40. $r = 4\tan\theta$
41. $r = \sin\theta\tan\theta$ (cissoid)
42. $r = 4 + 2\sec\theta$ (conchoid)
43. $r = e^{\theta}$ (logarithmic spiral)
44. $r^2 = 1/\theta$
45. $r = 1/\theta \quad (\theta > 0)$
46. $r^2 = \theta$

47. **(a)** Find a complete graph of $r = 1 - 2\sin 3\theta$.
(b) Predict what the graph of $r = 1 - 2\sin 4\theta$ will look like. Then check your prediction with a calculator.
(c) Predict what the graph of $r = 1 - 2\sin 5\theta$ will look like. Then check your prediction with a calculator.

48. **(a)** Find a complete graph of $r = 1 - 3\sin 2\theta$.
(b) Predict what the graph of $r = 1 - 3\sin 3\theta$ will look like. Then check your prediction with a calculator.
(c) Predict what the graph of $r = 1 - 3\sin 4\theta$ will look like. Then check your prediction with a calculator.

49. If a, b are constants such that $ab \neq 0$, show that the graph of $r = a\sin\theta + b\cos\theta$ is a circle. [*Hint*: Multiply both sides by r and convert to rectangular coordinates.]

50. Prove that the coordinate conversion formulas are valid when $r < 0$. [*Hint*: If P has coordinates (x, y) and (r, θ), with $r < 0$, verify that the point Q with rectangular coordinates $(-x, -y)$ has polar coordinates $(-r, \theta)$. Since $r < 0$, $-r$ is positive and the conversion formulas proved in the text apply to Q. For instance, $-x = -r\cos\theta$, which implies that $x = r\cos\theta$.]

51. *Distance Formula for Polar Coordinates:*
Prove that the distance from (r, θ) to (s, β) is

$\sqrt{r^2 + s^2 - 2rs\cos(\theta - \beta)}$. [*Hint*: If $r > 0$, $s > 0$, and $\theta > \beta$, then the triangle with vertices (r, θ), (s, β), $(0, 0)$ has an angle of $\theta - \beta$, whose sides have lengths r and s. Use the Law of Cosines.]

52. Explain why the following symmetry tests for the graphs of polar equations are valid.
(a) If replacing θ by $-\theta$ produces an equivalent equation, then the graph is symmetric with respect to the line $\theta = 0$ (the x-axis).
(b) If replacing θ by $\pi - \theta$ produces an equivalent equation, then the graph is symmetric with respect to the line $\theta = \pi/2$ (the y-axis).
(c) If replacing r by $-r$ produces an equivalent equation, then the graph is symmetric with respect to the origin.

10.5 POLAR EQUATIONS OF CONICS

In a rectangular coordinate system each type of conic section has a different definition. By using polar coordinates it is possible to give a unified treatment of conics and their equations. Before doing this we must first introduce a concept that will play a key role in the development.

Recall that both ellipses and hyperbolas are defined in terms of two foci and both have two vertices that lie on the line through the foci (see pages 602 and 606). The **eccentricity** of an ellipse or a hyperbola is denoted e and is defined to be the ratio

$$e = \frac{\text{distance between the foci}}{\text{distance between the vertices}}.$$

For conics centered at the origin, with foci on the x-axis, the situation is as follows:

Ellipse

$$\frac{x^2}{a^2} + \frac{y^2}{b^2} = 1 \qquad (a > b)$$

foci: $(\pm c, 0)$ vertices: $(\pm a, 0)$

$$c = \sqrt{a^2 - b^2}$$

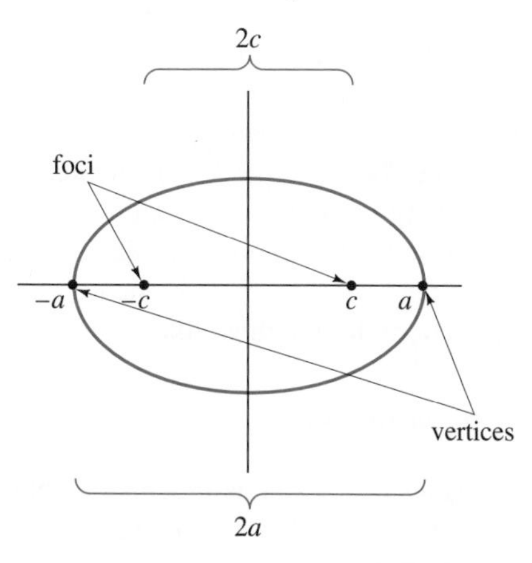

$$e = \frac{2c}{2a} = \frac{c}{a} = \frac{\sqrt{a^2 - b^2}}{a}$$

Hyperbola

$$\frac{x^2}{a^2} - \frac{y^2}{b^2} = 1$$

foci: $(\pm c, 0)$ vertices: $(\pm a, 0)$

$$c = \sqrt{a^2 + b^2}$$

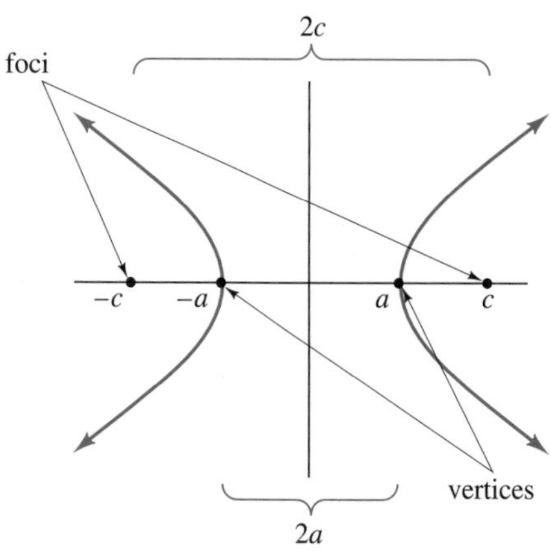

$$e = \frac{2c}{2a} = \frac{c}{a} = \frac{\sqrt{a^2 + b^2}}{a}$$

A similar analysis shows that the formulas for e are also valid for conics whose equations are of the form

$$\frac{x^2}{b^2} + \frac{y^2}{a^2} = 1 \quad (a > b) \qquad \text{or} \qquad \frac{y^2}{a^2} - \frac{x^2}{b^2} = 1.$$

These formulas can be used to compute the eccentricity of any ellipse or hyperbola whose equation can be put in standard form.

EXAMPLE 1 Find the eccentricity of the conic with equation

(a) $\dfrac{y^2}{4} - \dfrac{x^2}{21} = 1$ **(b)** $4x^2 + 9y^2 - 32x - 90y + 253 = 0.$

Solution

(a) In this case $a^2 = 4$ (so that $a = 2$) and $b^2 = 21$. Hence, the eccentricity is

$$e = \frac{\sqrt{a^2 + b^2}}{a} = \frac{\sqrt{4 + 21}}{2} = \frac{\sqrt{25}}{2} = \frac{5}{2} = 2.5.$$

(b) In Example 2 of Section 10.3 we saw that the equation can be put into this standard form: $\dfrac{(x-4)^2}{9} + \dfrac{(y-5)^2}{4} = 1$. Hence, its graph is just the ellipse $\dfrac{x^2}{9} + \dfrac{y^2}{4} = 1$ shifted vertically and horizontally. Since the shifting does not change the distances between foci or vertices, both ellipses have the same eccentricity, which can be computed using $a^2 = 9$ and $b^2 = 4$:

$$e = \frac{\sqrt{a^2 - b^2}}{a} = \frac{\sqrt{9-4}}{3} = \frac{\sqrt{5}}{3} \approx .745. \quad ■$$

Example 1 and the preceding pictures illustrate the following fact. For ellipses the distance between the foci (numerator of e) is less than the distance between the vertices (denominator), so $e < 1$. For hyperbolas, however, $e > 1$ because the distance between the foci is greater than that between the vertices.

The eccentricity of an ellipse measures its ''roundness.'' An ellipse whose eccentricity is close to 0 is almost circular (Exercise 19). The eccentricity of a hyperbola measures how ''flat'' its branches are. The branches of a hyperbola with large eccentricity look almost like parallel lines (Exercise 20).

Conics and Polar Equations

The polar analogues of the standard equations of ellipses, parabolas, and hyperbolas are given in the chart below. The proof of these statements is given at the end of the section. In the chart e and d are constants, with $e > 0$. Remember that in a rectangular coordinate system whose positive x-axis coincides with the polar axis, a point with polar coordinates (r, θ) is on the x-axis when $\theta = 0$ or π and on the y-axis when $\theta = \pi/2$ or $3\pi/2$.

Polar Equations for Conic Sections ▶

Equation		*Graph*
$r = \dfrac{ed}{1 + e\cos\theta}$ or $r = \dfrac{ed}{1 - e\cos\theta}$	$0 < e < 1$	*Ellipse* with eccentricity e One of the foci: (0, 0) Vertices at $\theta = 0$ and $\theta = \pi$
	$e = 1$	*Parabola* with focus (0, 0) Vertex at $\theta = 0$ or $\theta = \pi$; (r is not defined for the other value of θ)
	$e > 1$	*Hyperbola* with eccentricity e One of the foci: (0, 0) Vertices at $\theta = 0$ and $\theta = \pi$
$r = \dfrac{ed}{1 + e\sin\theta}$ or $r = \dfrac{ed}{1 - e\sin\theta}$	$0 < e < 1$	*Ellipse* with eccentricity e One of the foci: (0, 0) Vertices at $\theta = \pi/2$ and $\theta = 3\pi/2$
	$e = 1$	*Parabola* with focus (0, 0) Vertex at $\theta = \pi/2$ or $\theta = 3\pi/2$; (r is not defined for the other value of θ)
	$e > 1$	*Hyperbola* with eccentricity e One of the foci: (0, 0) Vertices at $\theta = \pi/2$ and $\theta = 3\pi/2$

EXAMPLE 2 Find a complete graph of $r = \dfrac{3e}{1 + e\cos\theta}$ when **(a)** $e = .7$ **(b)** $e = 1$ **(c)** $e = 2$.

Solution From the first equation in the chart above (with $d = 3$) we know that the graphs are an ellipse, parabola, and hyperbola, respectively, as shown in Figure 10–57.

(a) $e = .7$

$$r = \frac{2.1}{1 + .7\cos\theta}$$

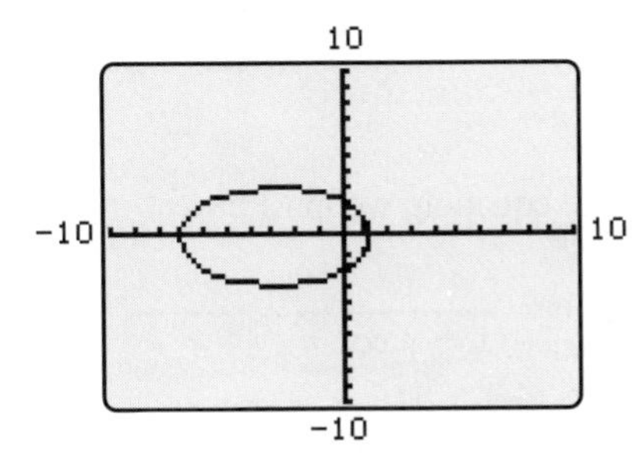

(b) $e = 1$

$$r = \frac{3}{1 + \cos\theta}$$

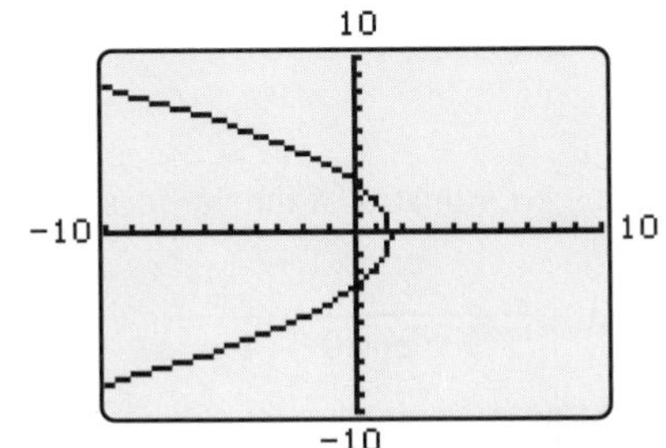

(c) $e = 2$

$$r = \frac{6}{1 + 2\cos\theta}$$ ■

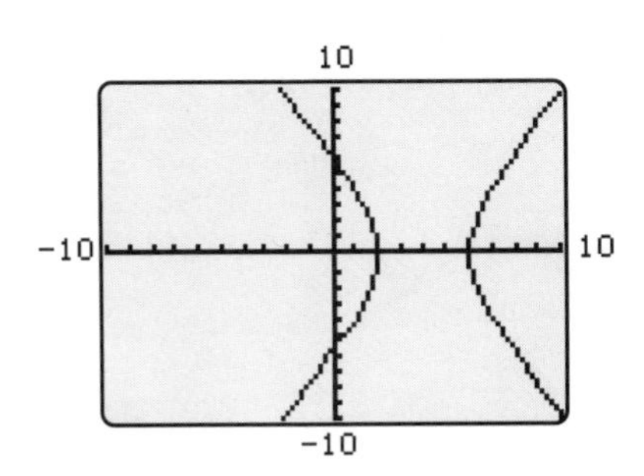

Figure 10-57

EXAMPLE 3 Identify the conic section that is the graph of

$$r = \frac{20}{4 - 10 \sin \theta}$$

and find its vertices and eccentricity.

Solution First, rewrite the equation in one of the forms listed in the box on the previous page:

$$r = \frac{20}{4 - 10 \sin \theta} = \frac{20}{4\left(1 - \frac{10}{4} \sin \theta\right)} = \frac{5}{1 - 2.5 \sin \theta}.$$

This is one such form, with $e = 2.5$ and $ed = 5$ (so that $d = 2$). Consequently, the graph is a hyperbola with eccentricity $e = 2.5$ whose vertices are at

$$\theta = \frac{\pi}{2}, \qquad r = \frac{20}{4 - 10 \sin \frac{\pi}{2}} = \frac{20}{4 - 10 \cdot 1} = -\frac{20}{6} = -\frac{10}{3}$$

and

$$\theta = \frac{3\pi}{2}, \qquad r = \frac{20}{4 - 10 \sin \frac{3\pi}{2}} = \frac{20}{4 - 10(-1)} = \frac{20}{14} = \frac{10}{7}.$$

GRAPHING EXPLORATION Find a viewing window that shows a complete graph of this hyperbola. ■

EXAMPLE 4 Find a polar equation of the ellipse with $(0, 0)$ as a focus and vertices $(3, 0)$ and $(6, \pi)$.

Solution Because of the location of the vertices, the polar equation is of the form $r = ed/(1 \pm e \cos \theta)$. We first consider the equation

$$r = \frac{ed}{1 + e \cos \theta}.$$

Since the coordinates of the vertices satisfy the equation, we must have:

$$3 = \frac{ed}{1 + e \cos 0} = \frac{ed}{1 + e} \qquad \text{and} \qquad 6 = \frac{ed}{1 + e \cos \pi} = \frac{ed}{1 - e}$$

which imply that

$$3(1 + e) = ed \qquad \text{and} \qquad 6(1 - e) = ed.$$

Therefore,

$$3(1 + e) = 6(1 - e)$$
$$3 + 3e = 6 - 6e$$
$$9e = 3$$
$$e = 1/3.$$

Substituting $e = 1/3$ in either of the original equations shows that $d = 12$. So an equation of the ellipse is

$$r = \frac{ed}{1 + e\cos\theta} = \frac{\frac{1}{3}\cdot 12}{1 + \frac{1}{3}\cos\theta} = \frac{12}{3 + \cos\theta}.$$

If we had started instead with the equation $r = ed/(1 - e\cos\theta)$ and solved for e as above, we would have obtained $e = -1/3$, which is impossible since $e > 0$.

Alternate Solution Verify that the vertex $(3, 0)$ also has polar coordinates $(-3, \pi)$. Similarly, $(6, \pi)$ also has polar coordinates $(-6, 0)$. If you begin with the equation $r = ed/(1 - e\cos\theta)$ and the vertices $(-3, \pi)$ and $(-6, 0)$ and proceed as before to find e and d, you obtain the equation

$$r = \frac{-12}{3 - \cos\theta}. \quad \blacksquare$$

Alternate Definition of Conics

The theorem stated below is sometimes used as a definition of the conic sections because it provides a unified approach instead of the variety of descriptions given in Section 10.2. Its proof also provides a proof of the statements in the box at the beginning of this section.

The basic idea is to describe every conic in terms of a straight line L(the **directrix**) and a point P not on L (the **focus**), in much the same way that parabolas were defined in Section 10.2. The number e in the theorem turns out to be the eccentricity of the conic.

Conic Section Theorem ▶

Let L be a straight line, P a point not on L, and e a positive constant. The set of all points X in the plane such that

$$\frac{\text{distance from } X \text{ to } P}{\text{distance from } X \text{ to } L} = e$$

is a conic section with P as one of the foci.* The conic is an ellipse if $0 < e < 1$, a parabola if $e = 1$, and a hyperbola if $e > 1$.**

*The distance from X to L is measured along the line through X that is perpendicular to L.

**When $e = 1$, the given condition is equivalent to

$$\text{distance from } X \text{ to } P = \text{distance from } X \text{ to } L$$

which is the definition of a parabola given in Section 10.2.

Proof Coordinatize the plane so that the pole is the point P, the polar axis is horizontal, and the directrix L is a vertical line to the left of the pole, as in Figure 10–58. Let d be the distance from P to L and (r, θ) the polar coordinates of X.

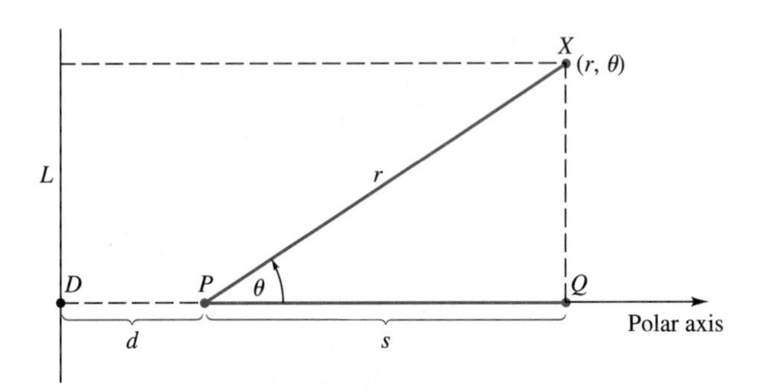

Figure 10-58

If X satisfies the condition

$$\frac{\text{distance from } X \text{ to } P}{\text{distance from } X \text{ to } L} = e$$

then

$$(*) \qquad \text{distance from } X \text{ to } P = e[\text{distance from } X \text{ to } L].$$

Figure 10–58 shows that r is the distance from P to X and that the distance from X to L is the same as that from D to Q, namely, $d + s$. Furthermore, $\cos\theta = s/r$ so that $s = r\cos\theta$. Consequently, equation $(*)$ can be written in polar coordinates as follows:

$$\begin{aligned}
\text{distance from } X \text{ to } P &= e[\text{distance from } X \text{ to } L] \\
r &= e[d + s] \\
r &= e[d + r\cos\theta] \\
r - er\cos\theta &= ed \\
r(1 - e\cos\theta) &= ed \\
r &= \frac{ed}{1 - e\cos\theta}.
\end{aligned}$$

To show that this is actually the equation of a conic, we translate it into rectangular coordinates using the conversion formulas from Section 10.4:

$$r^2 = x^2 + y^2 \quad \text{and} \quad \cos\theta = \frac{x}{r} = \frac{x}{\pm\sqrt{x^2 + y^2}}.$$

Then the polar coordinate equation becomes

$$\pm\sqrt{x^2 + y^2} = \frac{ed}{1 - e\left(\dfrac{x}{\pm\sqrt{x^2 + y^2}}\right)}$$

$$\pm\sqrt{x^2 + y^2}\left(1 - \frac{ex}{\pm\sqrt{x^2 + y^2}}\right) = ed$$

$$\pm\sqrt{x^2 + y^2} - ex = ed$$

$$\pm\sqrt{x^2 + y^2} = ed + ex.$$

Squaring both sides and rearranging terms, we have

$$x^2 + y^2 = e^2d^2 + 2de^2x + e^2x^2$$

$$(**) \qquad (1 - e^2)x^2 - 2de^2x + y^2 = e^2d^2.$$

Now we consider the two possibilities, $e = 1$ and $e \neq 1$.

Case 1. If $e = 1$, then equation (**) becomes

$$-2dx + y^2 = d^2$$

$$y^2 = 2dx + d^2$$

$$(y - 0)^2 = 2d\left(x + \frac{d}{2}\right)$$

$$(y - 0)^2 = 4\left(\frac{d}{2}\right)\left(x - \left(-\frac{d}{2}\right)\right).$$

The box on page 619 (with $k = 0$, $p = d/2$, $h = -d/2$) shows that this is the standard equation of a parabola with

$$\text{vertex } \left(-\frac{d}{2}, 0\right), \qquad \text{focus } (0, 0), \qquad \text{directrix } x = -d.$$

Case 2. If $e \neq 1$, then we can divide both sides of equation (**) by the nonzero number $1 - e^2$:

$$\left(x^2 - \frac{2de^2}{1 - e^2}x\right) + \frac{y^2}{1 - e^2} = \frac{e^2d^2}{1 - e^2}.$$

Next we complete the square on the expression in parentheses by adding the square of half the coefficient of x to both sides of the equation and simplify the result:

$$\left(x^2 - \frac{2de^2}{1-e^2}x + \left(\frac{de^2}{1-e^2}\right)^2\right) + \frac{y^2}{1-e^2} = \frac{e^2d^2}{1-e^2} + \left(\frac{de^2}{1-e^2}\right)^2$$

$$\left(x - \frac{de^2}{1-e^2}\right)^2 + \frac{y^2}{1-e^2} = \frac{(1-e^2)e^2d^2 + (de^2)^2}{(1-e^2)^2}$$

$$\left(x - \frac{de^2}{1-e^2}\right)^2 + \frac{y^2}{1-e^2} = \frac{e^2d^2}{(1-e^2)^2}.$$

Dividing both sides of the last equation by $e^2d^2/(1-e^2)^2$ produces the equation

$$(***) \qquad \frac{\left(x - \dfrac{de^2}{1-e^2}\right)^2}{\dfrac{e^2d^2}{(1-e^2)^2}} + \frac{y^2}{\dfrac{e^2d^2}{1-e^2}} = 1.$$

Now we consider the two possibilities, $e < 1$ and $e > 1$.

Case 2A. If $e < 1$, then $1 - e^2 > 0$ and the constants in the denominators on the left side of equation (***) are positive. Therefore, equation (***) is of the form

$$\frac{(x-h)^2}{a^2} + \frac{(y-k)^2}{b^2} = 1$$

with $h = de^2/(1-e^2)$, $k = 0$, and a and b positive numbers such that

$$a^2 = \frac{e^2d^2}{(1-e^2)^2} \qquad \text{and} \qquad b^2 = \frac{e^2d^2}{1-e^2}.$$

In this case $a > b$ by Exercise 47. According to the box on page 619 this is the standard equation of an ellipse with center $(h, 0)$ and foci $(h - c, 0)$ and $(h + c, 0)$, where $c^2 = a^2 - b^2$. Its eccentricity is the number

$$\frac{c}{a} = \sqrt{\frac{c^2}{a^2}} = \sqrt{\frac{a^2-b^2}{a^2}} = \sqrt{1 - b^2 \cdot \frac{1}{a^2}}$$

$$= \sqrt{1 - \frac{e^2d^2}{1-e^2} \cdot \frac{(1-e^2)^2}{e^2d^2}} = \sqrt{e^2} = e.$$

Case 2B. If $e > 1$, then $1 - e^2 < 0$. Therefore,

$$e^2 - 1 = -(1 - e^2) > 0$$

so that equation (***) may be written as

$$\frac{\left(x - \dfrac{de^2}{1-e^2}\right)^2}{\dfrac{e^2d^2}{(1-e^2)^2}} - \frac{y^2}{\dfrac{e^2d^2}{e^2-1}} = 1.$$

This is an equation of the form

$$\frac{(x-h)^2}{a^2} - \frac{(y-k)^2}{b^2} = 1$$

with a and b positive. The box on page 619 shows that it is the standard equation of a hyperbola with foci $(h-c, 0)$ and $(h+c, 0)$, where $c^2 = a^2 + b^2$. Exercise 48 shows that its eccentricity is e.

The preceding argument depends on coordinatizing the plane in a certain way and taking d to be the distance from the pole to L. Similar arguments, in which d is the distance from the pole to L and the plane is coordinatized so that L is to the right of the pole or parallel to the polar axis, lead to the other polar equations shown in Figure 10–59.

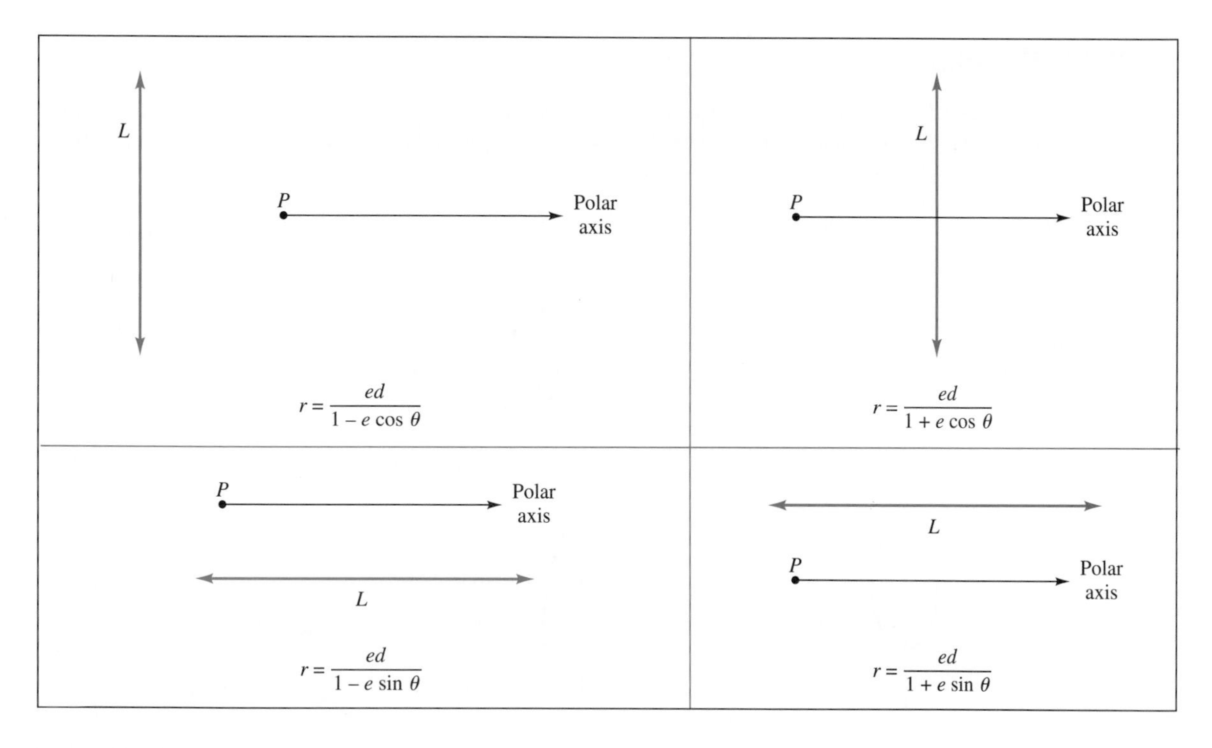

Figure 10-59

Analogous arguments when $d < 0$ (using $-d$ as the distance from the pole to L) then complete the proof. ❑

EXAMPLE 5 Find the polar equation of the hyperbola with focus at the pole, directrix $r = -4 \csc\theta$, and eccentricity 3.

Solution The equation of the directrix can be written as

$$r = \frac{-4}{\sin\theta}, \qquad \text{or equivalently,} \qquad r\sin\theta = -4.$$

With the conversion formulas for a rectangular coordinate system whose positive x-axis coincides with the polar axis, this equation becomes $y = -4$. So the directrix is a line parallel to the polar axis and 4 units below it. Using Figure 10–59, we see that $d = 4$ and the equation is

$$r = \frac{ed}{1 - e\sin\theta} = \frac{3\cdot 4}{1 - 3\sin\theta} = \frac{12}{1 - 3\sin\theta}. \quad ■$$

EXERCISES 10.5

In Exercises 1–6, which of the graphs (a)–(f) at the bottom of the page could possibly be the graph of the equation?

1. $r = \dfrac{3}{1 - \cos\theta}$ **2.** $r = \dfrac{6}{2 + \cos\theta}$

3. $r = \dfrac{6}{2 - 4\sin\theta}$ **4.** $r = \dfrac{15}{1 + 4\cos\theta}$

5. $r = \dfrac{6}{3 - 2\sin\theta}$ **6.** $r = \dfrac{6}{\frac{3}{2} + \frac{3}{2}\sin\theta}$

In Exercises 7–12, identify the conic section whose equation is given; if it is an ellipse or hyperbola, state its eccentricity.

7. $r = \dfrac{12}{3 + 4\sin\theta}$ **8.** $r = \dfrac{-10}{2 + 3\cos\theta}$

9. $r = \dfrac{8}{3 + 3\sin\theta}$ **10.** $r = \dfrac{20}{5 - 10\sin\theta}$

11. $r = \dfrac{2}{6 - 4\cos\theta}$ **12.** $r = \dfrac{-6}{5 + 2\cos\theta}$

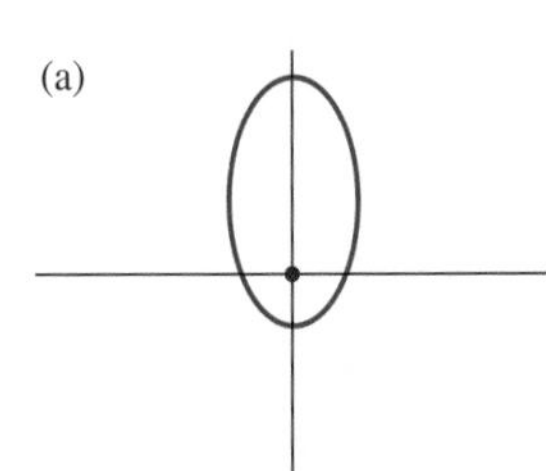

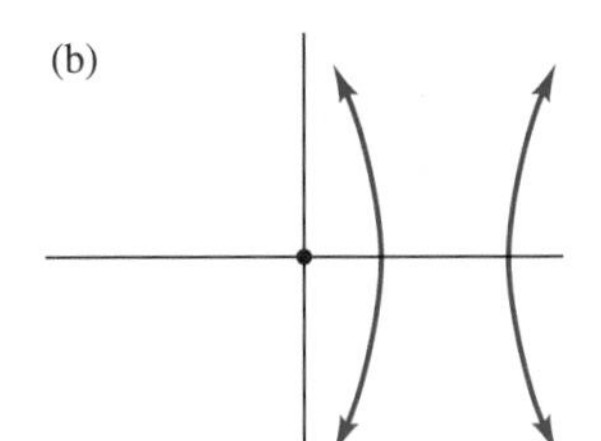

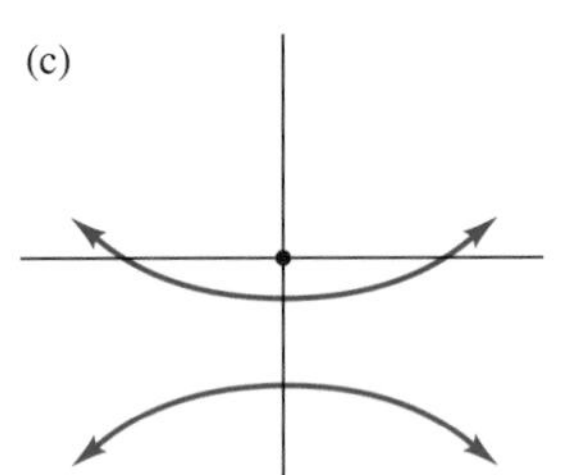

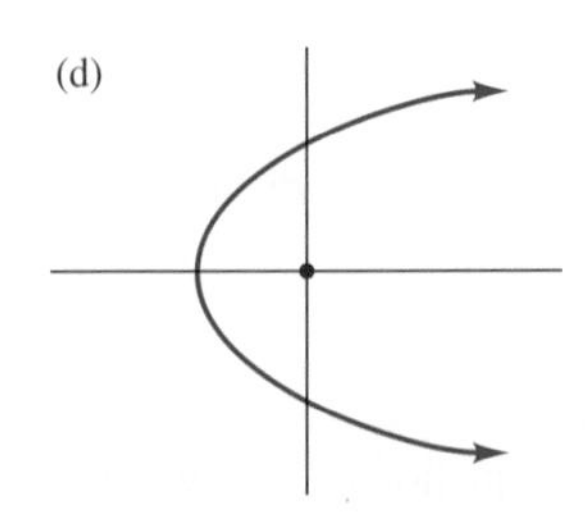

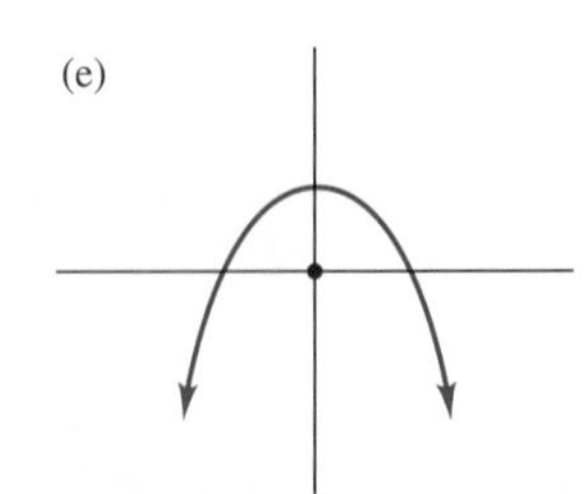

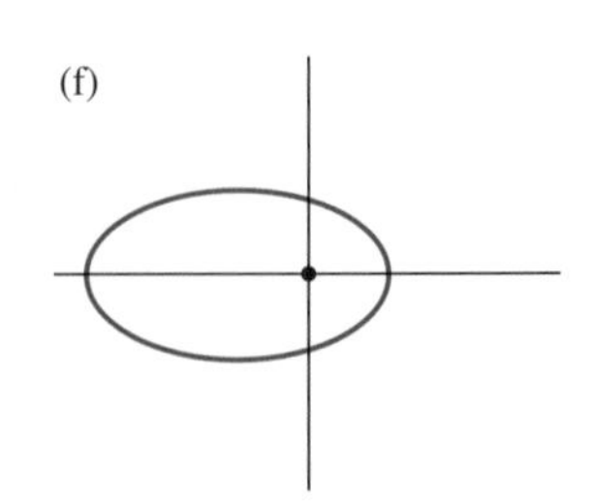

In Exercises 13–18, find the eccentricity of the conic whose equation is given.

13. $\dfrac{x^2}{100} + \dfrac{y^2}{99} = 1$ **14.** $\dfrac{(x-4)^2}{18} + \dfrac{(y+5)^2}{25} = 1$

15. $\dfrac{(x-6)^2}{10} - \dfrac{y^2}{40} = 1$

16. $4x^2 + 9y^2 - 24x + 36y + 36 = 0$

17. $16x^2 - 9y^2 - 32x + 36y + 124 = 0$

18. $4x^2 - 5y^2 - 16x - 50y + 71 = 0$

19. **(a)** Using a square viewing window (so that circles look like circles), graph these ellipses (on the same screen if possible):

$$\frac{x^2}{16} + \frac{y^2}{1} = 1 \qquad \frac{x^2}{16} + \frac{y^2}{6} = 1 \qquad \frac{x^2}{16} + \frac{y^2}{14} = 1$$

(b) Compute the eccentricity of each ellipse in part (a).

(c) Based on parts (a) and (b), how is the shape of an ellipse related to its eccentricity?

20. **(a)** Graph these hyperbolas (on the same screen if possible):

$$\frac{y^2}{4} - \frac{x^2}{1} = 1 \qquad \frac{y^2}{4} - \frac{x^2}{12} = 1 \qquad \frac{y^2}{4} - \frac{x^2}{96} = 1$$

(b) Compute the eccentricity of each hyperbola in part (a).

(c) Based on parts (a) and (b), how is the shape of a hyperbola related to its eccentricity?

In Exercises 21–32, sketch the graph of the equation and label the vertices.

21. $r = \dfrac{8}{1 - \cos\theta}$ **22.** $r = \dfrac{5}{3 + 2\sin\theta}$

23. $r = \dfrac{4}{2 - 4\cos\theta}$ **24.** $r = \dfrac{5}{1 + \cos\theta}$

25. $r = \dfrac{10}{4 - 3\sin\theta}$ **26.** $r = \dfrac{12}{3 + 4\sin\theta}$

27. $r = \dfrac{15}{3 - 2\cos\theta}$ **28.** $r = \dfrac{32}{3 + 5\sin\theta}$

29. $r = \dfrac{3}{1 + \sin\theta}$ **30.** $r = \dfrac{10}{3 + 2\cos\theta}$

31. $r = \dfrac{10}{2 + 3\sin\theta}$ **32.** $r = \dfrac{15}{4 - 4\cos\theta}$

In Exercises 33–46, find the polar equation of the conic section that has focus (0, 0) and satisfies the given conditions.

33. Parabola; vertex $(3, \pi)$ **34.** Parabola; vertex $(2, \pi/2)$

35. Ellipse; vertices $(2, \pi/2)$ and $(8, 3\pi/2)$

36. Ellipse; vertices $(2, 0)$ and $(4, \pi)$

37. Hyperbola; vertices $(1, 0)$ and $(-3, \pi)$

38. Hyperbola; vertices $(-2, \pi/2)$ and $(4, 3\pi/2)$

39. Eccentricity 4; directrix: $r = -2\sec\theta$

40. Eccentricity 2; directrix: $r = 4\csc\theta$

41. Eccentricity 1; directrix: $r = -3\csc\theta$

42. Eccentricity 1; directrix: $r = 5\sec\theta$

43. Eccentricity 1/2; directrix: $r = 2\sec\theta$

44. Eccentricity 4/5; directrix: $r = 3\csc\theta$

45. Hyperbola; vertical directrix to the left of the pole; eccentricity 2; $(1, 2\pi/3)$ is on the graph.

46. Hyperbola; horizontal directrix above the pole; eccentricity 2; $(1, 2\pi/3)$ is on the graph.

47. In Case 2A of the proof of the Conic Sections Theorem, show that $a > b$.

48. In Case 2B of the proof of the Conic Sections Theorem, show that the hyperbola has eccentricity e.

49. A comet travels in a parabolic orbit with the sun as focus. When the comet is 60 million miles from the sun, the line segment from the sun to the comet makes an angle of $\pi/3$ radians with the axis of the parabolic orbit. Using the sun as the pole and assuming the axis of the orbit lies along the polar axis, find a polar equation for the orbit.

50. Halley's comet has an elliptical orbit, with eccentricity .97 and the sun as a focus. The length of the major axis of the orbit is 3364.74 million miles. Using the sun as the pole and assuming the major axis of the orbit is perpendicular to the polar axis, find a polar equation for the orbit.

CHAPTER 10 *Review*

Important Concepts ▶

Section 10.1	Plane curve 588 Parameter 588 Parametric equations 588 Eliminating the parameter 591 Finding parameterizations 591, 592 Cycloid 596
Section 10.2	Ellipse: foci, center, vertices, major and minor axes 602, 603 Equations of ellipses centered at the origin 604 Hyperbola: foci, center, vertices, focal axis 606 Equations of hyperbolas centered at the origin 607 Parabola: focus, directrix, vertex, axis 609 Equations of parabolas with vertex at the origin 610 Parametric equations for conics 613
Section 10.3	Vertical and horizontal shifts 616 Standard equations of conic sections 619 Graphs of rotated conics 622 Discriminant 623
Section 10.4	Polar coordinates 627 Pole 627 Polar axis 627 Polar/rectangular coordinate conversion 630 Polar graphs 631–635
Section 10.5	Eccentricity 631 Polar equations for conics 639 Focus 641 Directrix 641

Important Facts and Formulas ▶

- Equation of ellipse with center (h, k) and axes on the lines $x = h$, $y = k$:

$$\frac{(x-h)^2}{a^2} + \frac{(y-k)^2}{b^2} = 1$$

- Equation of hyperbola with center (h, k) and vertices on the line $y = k$:

$$\frac{(x-h)^2}{a^2} - \frac{(y-k)^2}{b^2} = 1$$

- Equation of hyperbola with center (h, k) and vertices on the line $x = h$:

$$\frac{(y-k)^2}{a^2} - \frac{(x-h)^2}{b^2} = 1$$

- Equation of a parabola with vertex (h, k) and axis $x = h$:

$$(x - h)^2 = 4p(y - k)$$

- Equation of a parabola with vertex (h, k) and axis $y = k$:

$$(y - k)^2 = 4p(x - h)$$

- The rectangular and polar coordinates of a point are related by:

$$x = r\cos\theta \quad \text{and} \quad y = r\sin\theta;$$

$$r^2 = x^2 + y^2 \quad \text{and} \quad \tan\theta = \frac{y}{x}$$

Review Questions ▶

In Questions 1–4, find a viewing window that shows a complete graph of the curve whose parametric equations are given.

1. $x = 8\cos t + \cos 8t$ and $y = 8\sin t - \sin 8t \quad (0 \le t \le 2\pi)$

2. $x = (64\cos(\pi/6))t$ and $y = -16t^2 + (64\sin(\pi/6))t \quad (0 \le t \le \pi)$

3. $x = t^3 + t + 1$ and $y = t^2 + 2t \quad (-3 \le t \le 3)$

4. $x = t^2 - t + 3$ and $y = t^3 - 5t \quad (-3 \le t \le 3)$

In Questions 5–8, sketch the graph of the curve whose parametric equations are given and find an equation in x and y whose graph contains the given curve.

5. $x = 2t - 1, \quad y = 2 - t, \quad -3 \le t \le 3$

6. $x = 3\cos t, \quad y = 5\sin t, \quad 0 \le t \le 2\pi$

7. $x = \cos t, \quad y = 2\sin^2 t, \quad 0 \le t \le 2\pi$

8. $x = e^t, \quad y = \sqrt{t + 1}, \quad t \ge 1$

9. Which of the following is *not* a parameterization of the curve $x = y^2 + 1$?
(a) $x = t^2 + 1, \quad y = t,$ any real number t
(b) $x = \sin^2 t + 1, \quad y = \sin t,$ any real number t
(c) $x = t^4 + 1, \quad y = t^2,$ any real number t
(d) $x = t^6 + 1, \quad y = t^3,$ any real number t

10. Which of the curves in Questions 1–4 appear to be the graphs of functions of the form $y = f(x)$?

In Questions 11–14, find the foci and vertices of the conic and state whether it is an ellipse or a hyperbola.

11. $\dfrac{x^2}{16} + \dfrac{y^2}{20} = 1$

12. $\dfrac{x^2}{9} - \dfrac{y^2}{16} = 1$

13. $\dfrac{(x - 1)^2}{7} + \dfrac{(y - 3)^2}{16} = 1$

14. $3x^2 = 1 + 2y^2$

15. Find the focus and directrix of the parabola $10y = 7x^2$.

16. Find the focus and directrix of the parabola

$$3y^2 - x - 4y + 4 = 0.$$

In Questions 17–28, sketch the graph of the equation. If there are asymptotes, give their equations.

17. $\dfrac{x^2}{4} + \dfrac{y^2}{25} = 1$

18. $25x^2 + 4y^2 = 100$

19. $\dfrac{(x-3)^2}{9} + \dfrac{(y+5)^2}{4} = 1$

20. $\dfrac{x^2}{9} - \dfrac{y^2}{16} = 1$

21. $\dfrac{(y+4)^2}{25} - \dfrac{(x-1)^2}{4} = 1$

22. $4x^2 - 9y^2 = 144$

23. $x^2 + 4y^2 - 10x + 9 = 0$

24. $9x^2 - 4y^2 - 36x + 24y - 36 = 0$

25. $2y = 4(x-3)^2 + 6$

26. $3y = 6(x+1)^2 - 9$

27. $x = y^2 + 2y + 2$

28. $y = x^2 - 2x + 3$

29. What is the center of the ellipse $4x^2 + 3y^2 - 32x + 36y + 124 = 0$?

30. Find the equation of the ellipse with center at the origin, one vertex at (0, 4), passing through $(\sqrt{3}, 2\sqrt{3})$.

31. Find the equation of the ellipse with center at (3, 1), one vertex at (1, 1), passing through $(2, 1 + \sqrt{3/2})$.

32. Find the equation of the hyperbola with center at the origin, one vertex at (0, 5), passing through $(1, 3\sqrt{5})$.

33. Find the equation of the hyperbola with center at (3, 0), one vertex at (3, 2), passing through $(1, \sqrt{5})$.

34. Find the equation of the parabola with vertex (2, 5), axis $x = 2$, and passing through (3, 12).

35. Find the equation of the parabola with vertex $(3/2, -1/2)$, axis $y = -1/2$, and passing through $(-3, 1)$.

36. Find the equation of the parabola with vertex (5, 2) that passes through the points (7, 3) and (9, 6).

In Questions 37–40, assume that the graph of the equation is a nondegenerate conic. Use the discriminant to identify the graph.

37. $3x^2 + 2\sqrt{2}xy + 2y^2 - 12 = 0$

38. $x^2 + y^2 - xy - 4y = 0$

39. $4xy - 3x^2 - 20 = 0$

40. $4x^2 - 4xy + y^2 - \sqrt{5}x - 2\sqrt{5}y = 0$

In Questions 41–46, find a viewing window that shows a complete graph of the equation.

41. $x^2 - xy + y^2 - 6 = 0$

42. $x^2 + xy + y^2 - 3y - 6 = 0$

43. $x^2 + xy - 2 = 0$

44. $x^2 - 4xy + y^2 + 5 = 0$

45. $x^2 + 3xy + y^2 - 2\sqrt{2}x + 2\sqrt{2}y = 0$

46. $x^2 + 2xy + y^2 - 4\sqrt{2}y = 0$

47. Plot the points $(2, 3\pi/4)$ and $(-3, -2\pi/3)$ on a polar coordinate graph.

48. List four other pairs of polar coordinates for the point $(-2, \pi/4)$.

In Questions 49–58, sketch the graph of the equation in a polar coordinate system.

49. $r = 5$

50. $r = -2$

51. $\theta = 2\pi/3$

52. $\theta = -5\pi/6$

53. $r = 2\theta \ (\theta \leq 0)$

54. $r = 4 \cos \theta$

55. $r = 2 - 2 \sin \theta$

56. $r = \cos 3\theta$

57. $r^2 = \cos 2\theta$

58. $r = 1 + 2 \sin \theta$

59. Convert $(3, -2\pi/3)$ from polar to rectangular coordinates.

60. Convert $(3, \sqrt{3})$ from rectangular to polar coordinates.

61. What is the eccentricity of the ellipse $3x^2 + y^2 = 84$?

62. What is the eccentricity of the ellipse $24x^2 + 30y^2 = 120$?

In Questions 63–66, sketch the graph of the equation, labeling the vertices and identifying the conic.

63. $r = \dfrac{12}{2 - \sin \theta}$

64. $r = \dfrac{14}{7 + 7 \cos \theta}$

65. $r = \dfrac{-24}{3 - 9 \cos \theta}$

66. $r = \dfrac{10}{3 + 4 \sin \theta}$

In Questions 67–70, find a polar equation of the conic that has focus (0, 0) and satisfies the given conditions.

67. Ellipse; vertices $(4, 0)$ and $(6, \pi)$

68. Hyperbola; vertices $(5, \pi/2)$ and $(-3, 3\pi/2)$

69. Eccentricity 1; directrix $r = 2 \sec \theta$

70. Eccentricity .75; directrix $r = -3 \csc \theta$

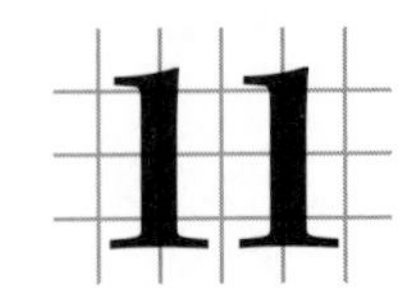

Systems of Equations

Roadmap

> Chapters 11 and 12 are independent of each other and may be read in any order.
>
> In this chapter, Section 4 is independent of the rest of the chapter and may be read at any time. So, the interdependence of sections is
>
> 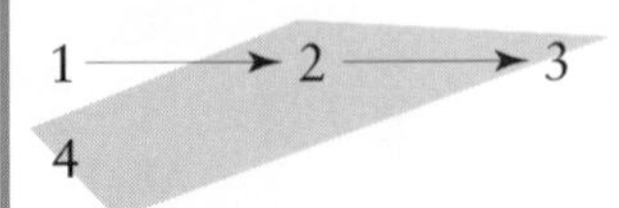
>
>
> Section 1 may be omitted by readers who are familiar with solving systems of two linear equations.

This chapter deals with *systems of equations,* such as

$\begin{aligned} 2x - 5y + 3z &= 1 \\ x + 2y - z &= 2 \\ 3x + y + 2z &= 11 \end{aligned}$	$\begin{aligned} 2x + 5y + z + w &= 0 \\ 2y - 4z + 41w &= 5 \\ 3x + 7y + 5z - 8w &= -6 \end{aligned}$	$\begin{aligned} x^2 + y^2 &= 25 \\ x^2 - y &= 7 \end{aligned}$
Three equations in three variables	Three equations in four variables	Two equations in two variables

Sections 11.1–11.3 deal with systems of linear equations (such as the first two shown above). Systems involving nonlinear equations are considered in Section 11.4.

A *solution of a system* is a solution that satisfies *all* the equations in the system. For instance, in the first system of equations above, $x = 1$, $y = 2$, $z = 3$ is a solution of all three equations (check it) and hence is a solution of the system. On the other hand, $x = 0$, $y = 7$, $z = 12$ is a solution of the first two equations, but not of the third (check it). So $x = 0$, $y = 7$, $z = 12$ is not a solution of the system.

11.1 SYSTEMS OF LINEAR EQUATIONS IN TWO VARIABLES

One method of solving a system of two linear equations in two variables is **substitution,** which is explained in Example 1.

EXAMPLE 1 Any solution of the system

$$\begin{aligned} x + 2y &= 3 \\ 5x - 4y &= -6 \end{aligned}$$

must necessarily be a solution of the first equation. Hence x must satisfy

$$x + 2y = 3, \qquad \text{or equivalently,} \qquad x = 3 - 2y.$$

Substituting this expression for x in the second equation, we have

$$\begin{aligned} 5x - 4y &= -6 \\ 5(3 - 2y) - 4y &= -6 \\ 15 - 10y - 4y &= -6 \\ -14y &= -21 \\ y &= \frac{-21}{-14} = \frac{3}{2}. \end{aligned}$$

Therefore, every solution of the original system must have $y = 3/2$. But when $y = 3/2$, we see from the first equation that:

$$\begin{aligned} x + 2y &= 3 \\ x + 2\left(\frac{3}{2}\right) &= 3 \\ x + 3 &= 3 \\ x &= 0. \end{aligned}$$

(We would also have found that $x = 0$ if we had substituted $y = 3/2$ in the second equation.) Consequently, the original system has exactly one solution: $x = 0$, $y = 3/2$. This solution could also have been found by solving the first equation for y instead of x and substituting this value in the second equation. ■

WARNING In order to guard against arithmetic mistakes, you should always *check your answers* by substituting them into *all* the equations of the original system. We have in fact checked the answers in all the examples, but these checks are omitted to save space.

The Elimination Method

The **elimination method** of solving systems of linear equations is often more convenient than substitution. It depends on this fact:

> **Multiplying both sides of an equation by a nonzero constant does not change the solutions of the equation.**

For example, the equation $x + 3 = 5$ has the same solution as $2x + 6 = 10$ (the first equation multiplied by 2). The elimination method also uses this fact from basic algebra:

$$\text{If } A = B \text{ and } C = D, \qquad \text{then } A + C = B + D \qquad \text{and} \qquad A - C = B - D.$$

EXAMPLE 2 In the system

$$\begin{aligned} x - 3y &= 4 \\ 2x + y &= 1 \end{aligned}$$

we first replace the first equation by an equivalent one (that is, one with the same solutions):

$$\begin{aligned} -2x + 6y &= -8 \qquad \textit{[First equation multiplied by } -2] \\ 2x + y &= 1. \end{aligned}$$

The multiplier -2 was chosen so that the coefficients of x in the two equations would be negatives of each other. Any solution of this last system must also be a solution of the sum of the two equations:

$$\begin{array}{rl} -2x + 6y &= -8 \\ \underline{2x + y} &\underline{= 1} \\ 7y &= -7. \end{array} \qquad \textit{[The first variable has been eliminated]}$$

Solving this last equation we see that $y = -1$. Substituting this value in the first of the original equations shows that

$$\begin{aligned} x - 3(-1) &= 4 \\ x &= 1. \end{aligned}$$

Therefore, $x = 1$, $y = -1$ is the solution of the original system. ■

EXAMPLE 3 Any solution of the system

$$\begin{aligned} 5x - 3y &= 3 \\ 3x - 2y &= 1 \end{aligned}$$

must also be a solution of this system:

$$\begin{aligned} 10x - 6y &= 6 \qquad \textit{[First equation multiplied by 2]} \\ -9x + 6y &= -3. \qquad \textit{[Second equation multiplied by } -3] \end{aligned}$$

The multipliers 2 and -3 were chosen so that the coefficients of y in the new equations would be negatives of each other. Any solution of this last system must also be a solution of the equation obtained by adding these two equations:

$$\begin{array}{rl} 10x - 6y &= 6 \\ \underline{-9x + 6y} &\underline{= -3} \\ x &= 3. \end{array} \qquad \textit{[The second variable has been eliminated]}$$

Substituting $x = 3$ in the first of the original equations shows that

$$\begin{aligned} 5(3) - 3y &= 3 \\ -3y &= -12 \\ y &= 4. \end{aligned}$$

Therefore the solution of the original system is $x = 3$, $y = 4$. ■

EXAMPLE 4 To solve the system

$$2x - 3y = 5$$
$$4x - 6y = 1$$

we multiply the first equation by -2 and add:

$$\begin{aligned} -4x + 6y &= -10 \\ 4x - 6y &= 1 \\ \hline 0 &= -9. \end{aligned}$$

Since $0 = -9$ is always false, the original system cannot possibly have any solutions. A system with no solutions is said to be **inconsistent.** ■

EXAMPLE 5 To solve the system

$$3x - y = 2$$
$$6x - 2y = 4$$

we multiply the first equation by 2 to obtain the system:

$$6x - 2y = 4$$
$$6x - 2y = 4.$$

The two equations are identical. So the solutions of this system are the same as the solutions of the single equation $6x - 2y = 4$, which can be rewritten as:

$$2y = 6x - 4$$
$$y = 3x - 2.$$

This equation, and hence the original system, has infinitely many solutions. They can be described as follows: Choose any real number for x, say $x = b$. Then $y = 3x - 2 = 3b - 2$. So the solutions of the system are all pairs of numbers of the form

$$x = b, \qquad y = 3b - 2 \qquad \text{where } b \text{ is any real number.}$$

A system such as this is said to be **dependent.** ■

Some nonlinear systems can be solved by replacing them with equivalent linear systems.

EXAMPLE 6 To solve the system

$$\frac{1}{x} + \frac{3}{y} = -1$$
$$\frac{2}{x} - \frac{1}{y} = 5$$

we let $u = 1/x$ and $v = 1/y$ so that the system becomes:

$$\begin{aligned} u + 3v &= -1 \\ 2u - v &= 5. \end{aligned}$$

We can solve this system by multiplying the first equation by -2 and adding it to the second equation:

$$\begin{aligned} -2u - 6v &= 2 \\ 2u - v &= 5 \\ \hline -7v &= 7 \\ v &= -1. \end{aligned}$$

Substituting $v = -1$ in the equation $u + 3v = -1$, we see that $u = -3(-1) - 1 = 2$. Consequently, the possible solution of the original system is

$$x = \frac{1}{u} = \frac{1}{2} \quad \text{and} \quad y = \frac{1}{v} = \frac{1}{(-1)} = -1.$$

You should substitute this possible solution in both equations of the original system to check that it is actually a solution of the system. ■

Applications

EXAMPLE 7 575 people attend a ball game and total ticket sales are \$2575. If adult tickets cost \$5 and children's tickets \$3, how many adults attended the game? How many children?

Solution Let x be the number of adults and y the number of children. Then,

Number of adults + Number of children = Total attendance

$$x + y = 575.$$

We can obtain a second equation by using the information about ticket sales:

Adult ticket sales + Children ticket sales = Total ticket sales

$$\begin{pmatrix}\text{Price}\\\text{per}\\\text{ticket}\end{pmatrix} \times \begin{pmatrix}\text{Number}\\\text{of}\\\text{adults}\end{pmatrix} + \begin{pmatrix}\text{Price}\\\text{per}\\\text{ticket}\end{pmatrix} \times \begin{pmatrix}\text{Number}\\\text{of}\\\text{children}\end{pmatrix} = 2575$$

$$5x + 3y = 2575.$$

In order to find x and y we need only solve this system of equations:

$$\begin{aligned} x + \quad y &= \ 575 \\ 5x + 3y &= 2575 \end{aligned}$$

Multiplying the first equation by -3 and adding we have:

$$\begin{aligned} -3x - 3y &= -1725 \\ \underline{5x + 3y} &= \underline{\ \ 2575} \\ 2x \qquad &= \quad 850 \\ x &= \quad 425 \end{aligned}$$

So, 425 adults attended the game. The number of children was $y = 575 - x = 575 - 425 = 150$. ■

EXAMPLE 8 How many pounds of tin and how many pounds of copper should be added to 1000 pounds of an alloy that is 10% tin and 30% copper in order to produce a new alloy that is 27.5% tin and 35% copper?

Solution Let x be the number of pounds of tin and y the number of pounds of copper to be added to the 1000 pounds of the old alloy. Then there will be $1000 + x + y$ pounds of the new alloy. We first find the *amounts* of tin and copper in the new alloy:

	Pounds in old alloy +	*Pounds added* =	*Pounds in new alloy*
Tin	10% of 1000 +	x	$= 100 + x$
Copper	30% of 1000 +	y	$= 300 + y$

Now consider the *percentages* of tin and copper in the new alloy.

	Percentage in new alloy ×	*Total weight of new alloy*	= *Pounds in new alloy*
Tin	27.5%	of $1000 + x + y$	$= .275(1000 + x + y)$
Copper	35%	of $1000 + x + y$	$= .35(1000 + x + y)$

The two ways of computing the weight of each metal in the alloy must produce the same result, that is,

$$100 + x = .275(1000 + x + y) \qquad [\textit{pounds of tin}]$$

$$300 + y = .35(1000 + x + y). \qquad [\textit{pounds of copper}]$$

Multiplying out the right sides and rearranging terms produces this system of equations:

$$\begin{aligned} .725x - .275y &= 175 \\ -.35x + .65y &= 50. \end{aligned}$$

Multiplying the first equation by .65 and the second by .275 and adding the results, we have:

$$\begin{aligned} .47125x - .17875y &= 113.75 \\ -.09625x + .17875y &= 13.75 \\ \hline .37500x &= 127.50 \\ x &= 340. \end{aligned}$$

Substituting this in the first equation above and solving for y shows that $y = 260$. Therefore, 340 pounds of tin and 260 pounds of copper should be added. ■

Geometric Interpretation

The solution of any system of linear equations in two variables can be seen geometrically by graphing all the equations in the system on the same coordinate plane. As we saw in Section 3.1 the graph of a linear equation is a straight line, and every point on the graph represents a solution of the equation. Therefore, a solution of the system will be given by the coordinates of a point that lies on *all* of the lines representing the system.

There are exactly three geometric possibilities for two lines in the plane: They are parallel, they intersect at a single point, or they coincide, as illustrated in Figure 11–1. Therefore,

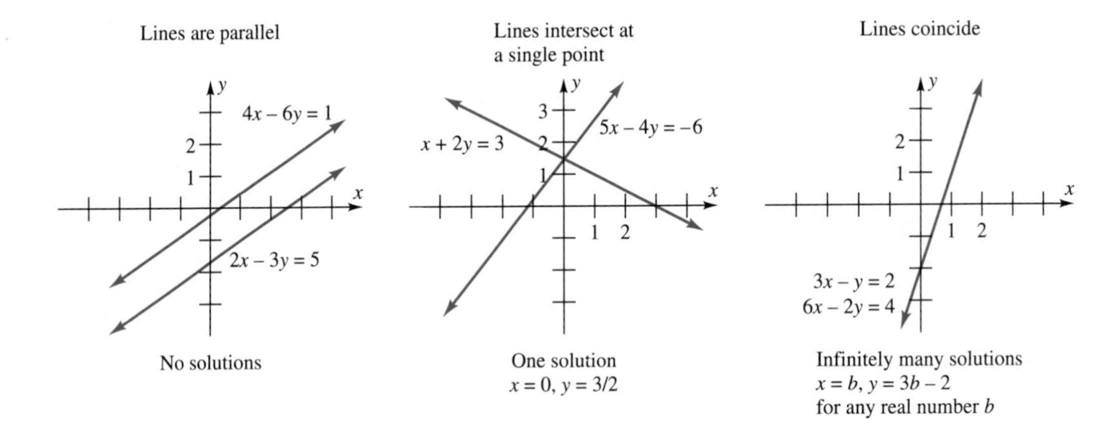

Figure 11-1

Number of Solutions of a System ▶

A system of two linear equations in two variables must have

no solutions (an inconsistent system) *or*

exactly one solution *or*

an infinite number of solutions (a dependent system).

EXERCISES 11.1

In Exercises 1–6, determine whether the given values of x, y, and z are a solution of the system of equations.

1. $x = -1, y = 3$
$$2x + y = 1$$
$$-3x + 2y = 9$$

2. $x = 3, y = 4$
$$2x + 6y = 30$$
$$x + 2y = 11$$

3. $x = 2, y = -1$
$$\frac{1}{3}x + \frac{1}{2}y = \frac{1}{6}$$
$$\frac{1}{2}x + \frac{1}{3}y = \frac{2}{3}$$

4. $x = .4, y = .7$
$$3.1x - 2y = -.16$$
$$5x - 3.5y = -.48$$

5. $x = \frac{1}{2}, y = 3, z = -1$
$$2x - y + 4z = -6$$
$$3y + 3z = 6$$
$$2z = 2$$

6. $x = 2, y = \frac{3}{2}, z = -\frac{1}{2}$
$$3x + 4y - 2z = 13$$
$$\frac{1}{2}x + 8z = -3$$
$$x - 3y + 5z = -5$$

In Exercises 7–14, use substitution to solve the system.

7. $x - 2y = 5$
$2x + y = 3$

8. $3x - y = 1$
$-x + 2y = 4$

9. $3x - 2y = 4$
$2x + y = -1$

10. $5x - 3y = -2$
$-x + 2y = 3$

11. $r + s = 0$
$r - s = 5$

12. $t = 3u + 5$
$t = u + 5$

13. $x + y = c + d$
$x - y = 2c - d$ (where c, d are constants)

14. $x + 3y = c - d$
$2x - y = c + d$ (where c, d are constants)

In Exercises 15–40, use the elimination method to solve the system.

15. $2x - 2y = 12$
$-2x + 3y = 10$

16. $3x + 2y = -4$
$4x - 2y = -10$

17. $x + 3y = -1$
$2x - y = 5$

18. $4x - 3y = -1$
$x + 2y = 19$

19. $2x + 3y = 15$
$8x + 12y = 40$

20. $2x + 5y = 8$
$6x + 15y = 18$

21. $3x - 2y = 4$
$6x - 4y = 8$

22. $2x - 8y = 2$
$3x - 12y = 3$

23. $12x - 16y = 8$
$42x - 56y = 28$

24. $\frac{1}{3}x + \frac{2}{5}y = \frac{1}{6}$
$20x + 24y = 10$

25. $9x - 3y = 1$
$6x - 2y = -5$

26. $8x + 4y = 3$
$10x + 5y = 1$

27.
$$\frac{x}{3} - \frac{y}{2} = -3$$
$$\frac{2x}{5} + \frac{y}{5} = -2$$

28.
$$\frac{x}{3} + \frac{3y}{5} = 4$$
$$\frac{x}{6} - \frac{y}{2} = -3$$

29.
$$\frac{x + y}{4} - \frac{x - y}{3} = 1$$
$$\frac{x + y}{4} + \frac{x - y}{2} = 9$$

30.
$$\frac{x - y}{4} + \frac{x + y}{3} = 1$$
$$\frac{x + 2y}{3} + \frac{3x - y}{2} = -2$$

31. $\frac{1}{x} - \frac{3}{y} = 2$
$\frac{2}{x} + \frac{1}{y} = 3$

32. $\frac{5}{x} + \frac{2}{y} = 0$
$\frac{6}{x} + \frac{4}{y} = 3$

33. $\frac{2}{x} + \frac{3}{y} = 8$
$\frac{3}{x} - \frac{1}{y} = 1$

34. $\frac{3}{x^2} + \frac{2}{y^2} = 11$ $\left[\textit{Hint}: \text{Let } u = \frac{1}{x^2} \text{ and } v = \frac{1}{y^2}.\right]$
$\frac{1}{x^2} - \frac{3}{y^2} = -11$

35. $\frac{3}{x+1} - \frac{4}{y-2} = 2$
$\frac{1}{x+1} + \frac{4}{y-2} = 5$
$\left[\textit{Hint}: \text{Let } u = \frac{1}{x+1} \text{ and } v = \frac{1}{y-2}.\right]$

36. $\frac{-5}{x^2+3} - \frac{2}{y^2-2} = -12$
$\frac{3}{x^2+3} + \frac{1}{y^2-2} = 5$

37. $3.5x - 2.18y = 2.00782$
$1.92x + 6.77y = -3.86928$

38. $463x - 80y = -13781.6$
$.0375x + .912y = 50.79624$

39. $ax + by = r$
$cx + dy = s$
(where a, b, c, d, r, s are constants and $ad - bc \neq 0$)

40. $ax + by = ab$
$bx - ay = ab$
(where a, b are nonzero constants)

41. Let c be any real number. Show that this system has exactly one solution:

$$x + 2y = c$$
$$6x - 3y = 4$$

42. **(a)** Find the values of c for which this system has an infinite number of solutions.

$$2x - 4y = 6$$
$$-3x + 6y = c$$

(b) Find the values of c for which the system in part (a) has no solutions.

In Exercises 43 and 44, find the values of c and d for which both given points lie on the given straight line.

43. $cx + dy = 2$; $(0, 4)$ and $(2, 16)$

44. $cx + dy = -6$; $(1, 3)$ and $(-2, 12)$

45. A 200-seat theater charges \$3 for adults and \$1.50 for children. If all seats were filled and the total ticket income was \$510, how many adults and how many children were in the audience?

46. A theater charges \$4 for main floor seats and \$2.50 for balcony seats. If all seats are sold, the ticket income is \$2100. At one show, 25% of the main floor seats and 40% of the balcony seats were sold and ticket income was \$600. How many seats are on the main floor and how many in the balcony?

47. The sum of two numbers is 40. The difference between twice the first number and the second is 11. What are the numbers?

48. The sum of two numbers is 50. The sum of five times one and twice the other is 136. What are the numbers?

49. An investor has part of her money in an account that pays 9% annual interest, and the rest in an account that pays 11% annual interest. If she has \$8000 less in the higher paying account than in the lower paying one and her total annual interest income is \$2010, how much does she have invested in each account?

50. Joyce has money in two investment funds. Last year the first fund paid a dividend of 8% and the second a dividend of 2% and Joyce received a total of \$780. This year the first fund paid a 10% dividend and the second only 1% and Joyce received \$810. How much money does she have invested in each fund?

51. At a certain store, cashews cost \$4.40/lb and peanuts \$1.20/lb. If you want to buy exactly 3 lb of nuts for \$6.00, how many pounds of each kind of nuts should you buy? [*Hint*: If you buy x pounds of cashews and y pounds of peanuts, then $x + y = 3$. Find a second equation by considering cost and solve the resulting system.]

52. A store sells deluxe tape recorders for \$150. The regular model costs \$120. The total tape recorder inventory would sell for \$43,800. But during a recent month the store actually sold half of its deluxe models and two thirds of the regular models and took in a total of \$26,700. How many of each kind of recorder did they have at the beginning of the month?

53. A plane flies 3000 miles from San Francisco to Boston at a constant speed in 5 hours, flying *with* the wind all the way. The return trip, against the wind, takes 6 hours. Find the speed of the plane and the speed of the wind. [*Hint*: If x is the plane's speed and y the wind speed, then on the trip to Boston (*with* the wind), the plane travels at speed $x + y$ for 5 hours. Since it goes a distance of 3000 miles, we have $5(x + y) = 3000$. Find another equation in x and y and solve the resulting system.]

54. A plane flying into a headwind travels 2000 miles in 4 hours and 24 minutes. The return flight along the same route with a tailwind takes 4 hours. Find the wind speed and the plane's speed (assuming both are constant).

55. A boat made a 4-mile trip upstream against a constant current in 15 minutes. The return trip at the same constant speed with the same current took 12 minutes. What is the speed of the boat and of the current?

56. A boat travels at a constant speed a distance of 57 km downstream in 3 hours, then turns around and travels 55 km upstream in 5 hours. What is the speed of the boat and of the current?

57. A winemaker has two large casks of wine. One wine is 8% alcohol and the other 18% alcohol. How many liters of each wine should be mixed to produce 30 liters of wine that is 12% alcohol?

58. How many cubic centimeters (cm^3) of a solution that is 20% acid and of another solution that is 45% acid should be mixed to produce 100 cm^3 of a solution that is 30% acid?

59. How many grams of a 50%-silver alloy should be mixed with a 75%-silver alloy to obtain 40 grams of a 60%-silver alloy?

60. A machine in a pottery factory takes 3 minutes to form a bowl and 2 minutes to form a plate. The material for a bowl costs .25 and the material for a plate costs .20. If the machine runs for 8 hours straight and exactly $44 is spent for material, how many bowls and plates can be produced?

11.2 LARGE SYSTEMS OF LINEAR EQUATIONS

Large systems of linear equations can be solved by using **Gaussian elimination,*** which is an extension of the elimination method used in Section 11.1. In order to understand why Gaussian elimination works, it is helpful to examine a system of two equations from a different viewpoint.

EXAMPLE 1 In Example 2 of Section 11.1, we solved the system

$$\begin{aligned} x - 3y &= 4 \\ 2x + y &= 1 \end{aligned}$$

by multiplying the first equation by -2 and adding it to the second in order to eliminate the variable x:

$$\begin{aligned} -2x + 6y &= -8 \qquad &&[-2 \text{ times first equation}] \\ 2x + y &= 1 \qquad &&[\text{Second equation}] \\ \hline 7y &= -7. \qquad &&[\text{Sum of second equation and } -2 \text{ times first equation}] \end{aligned}$$

*Named after the great German mathematician K. F. Gauss (1777–1855).

We then solved this last equation for y and substituted the answer, $y = -1$, in the original first equation to find that $x = 1$. What we did, in effect, was

> Replace the original system by the following system, in which x has been eliminated from the second equation; then solve this new system.

$$(*)\qquad \begin{aligned} x - 3y &= 4 \qquad &&\textit{[First equation]} \\ 7y &= -7. \qquad &&\textit{[Sum of second equation and } -2 \textit{ times first equation]} \end{aligned}$$

As we saw on pages 653–654, any solution of the original system must be a solution of the first equation and of the sum equation $7y = -7$, and hence of system (*). Conversely, it is easy to check that any solution of system (*) is also a solution of the original system. *Note*: We are not claiming that the second equations in the two systems have the same solutions—they don't—but only that the two *systems* have the same solution, namely, $x = 1, y = -1$. ■

Two systems of equations are said to be **equivalent** if they have the same solutions, that is, every solution of one system is a solution of the other system and vice versa. The basic technique in Gaussian elimination is to do what was done in Example 1: Replace the given system by an equivalent one (perhaps several times) until you obtain an equivalent system in which enough variables have been eliminated to make the system easy to solve. There are several ways to produce an equivalent system from a given one, the third of which was used in Example 1:

Elementary Operations ▶

Performing any of the following operations on a system of equations produces an equivalent system:

1. **Interchange any two equations in the system.**
2. **Replace an equation in the system by a nonzero constant multiple of itself.**
3. **Replace an equation in the system by the sum of itself and a constant multiple of another equation in the system.**

The reason that the first elementary operation produces an equivalent system is that rearranging the order of the equations certainly doesn't affect their solutions, and hence doesn't affect the solutions of the system. The second elementary operation produces an equivalent system because multiplying a single equation by a constant does not change the solutions of that equation, and hence does not change the solutions of any system including that equation. The third elementary operation was illustrated in Example 1. A similar argument works in the general case.

EXAMPLE 2 To solve the system

$$\begin{aligned} x + 4y - 3z &= 1 & &\text{[Equation } A\text{]} \\ -3x - 6y + z &= 3 & &\text{[Equation } B\text{]} \\ 2x + 11y - 5z &= 0 & &\text{[Equation } C\text{]} \end{aligned}$$

we first use elementary operations to produce an equivalent system in which the variable x has been eliminated from the second and third equations.

To eliminate x from equation B, replace equation B by the sum of itself and 3 times equation A:

$$\begin{aligned} &\text{[3 times } A\text{]} & 3x + 12y - 9z &= 3 \\ &[B] & -3x - 6y + z &= 3 \\ \hline & & 6y - 8z &= 6 \end{aligned}$$

$$\begin{aligned} x + 4y - 3z &= 1 & &[A] \\ 6y - 8z &= 6 & &\text{[Sum of } B \text{ and 3 times } A\text{]} \leftarrow \\ 2x + 11y - 5z &= 0 & &[C] \end{aligned}$$

To eliminate x from equation C we replace equation C by the sum of itself and -2 times equation A:

$$\begin{aligned} &[-2 \text{ times } A] & -2x - 8y + 6z &= -2 \\ &[C] & 2x + 11y - 5z &= 0 \\ \hline & & 3y + z &= -2 \end{aligned}$$

$$\begin{aligned} x + 4y - 3z &= 1 \\ 6y - 8z &= 6 \\ 3y + z &= -2 & &\text{[Sum of } C \text{ and } -2 \text{ times } A\text{]} \leftarrow \end{aligned}$$

The next step is to eliminate the y term in one of the last two equations. This can be done by replacing the second equation by the sum of itself and -2 times the third equation:

$$\begin{aligned} x + 4y - 3z &= 1 \\ -10z &= 10 & &\text{[Sum of second equation and } -2 \text{ times third equation]} \\ 3y + z &= -2 \end{aligned}$$

Finally, interchange the last two equations:

$$(*) \qquad \begin{aligned} x + 4y - 3z &= 1 \\ 3y + z &= -2 \\ -10z &= 10. \end{aligned}$$

This last system, which is equivalent to the original one, is easily solved. The last equation shows that

$$-10z = 10, \qquad \text{or equivalently,} \qquad z = -1.$$

Substituting $z = -1$ in the second equation shows that

$$\begin{aligned} 3y + \quad z &= -2 \\ 3y + (-1) &= -2 \\ 3y &= -1 \\ y &= -\frac{1}{3}. \end{aligned}$$

Substituting $y = -1/3$ and $z = -1$ in the first equation yields:

$$\begin{aligned} x + \quad 4y - \quad 3z &= 1 \\ x + 4\left(-\frac{1}{3}\right) - 3(-1) &= 1 \\ x = 1 + \frac{4}{3} - 3 &= -\frac{2}{3}. \end{aligned}$$

> ▶ **TECHNOLOGY TIP**
>
> Many systems with the same number of variables as equations can be solved directly on TI-85 (SIMULT) and Casio 9800 (SIM in equation mode).

Therefore, the original system has just one solution: $x = -2/3$, $y = -1/3$, $z = -1$. ■

The process used to solve the final system (∗) in Example 2 is called **back substitution** because you begin with the last equation and work back to the first. It works because system (∗) is in **triangular form:** The first variable in the first equation, x, does not appear in any subsequent equation; the first variable in the second equation, y, does not appear in any subsequent equation, and so on. It can be shown that the procedure in Example 2 works in every case:

Gaussian Elimination ▶

Any system of linear equations can be transformed into an equivalent system in triangular form by using a finite number of elementary operations. If the system has solutions, they can then be found by back-substitution in the triangular form system.

Matrix Methods

Once you have solved several systems of linear equations by the elimination method, one fact becomes clear. The symbols used for the variables play no real role in the solution process, and a lot of time is wasted copying the x's, y's, z's, and so on, at each stage of the process. This fact suggests a shorthand system for representing a system of equations.

EXAMPLE 3 The system of equations

$$\begin{aligned} x + 2y + 3z &= -2 \\ 2x + 6y + z &= 2 \\ 3x + 3y + 10z &= -2 \end{aligned}$$

can be represented by the following rectangular array of numbers, consisting of the coefficients of the variables and the constants on the right of the equal sign, arranged in the same order they appear in the system:

$$\left(\begin{array}{ccc|c} 1 & 2 & 3 & -2 \\ 2 & 6 & 1 & 2 \\ 3 & 3 & 10 & -2 \end{array}\right)$$

This array is called the **augmented matrix** of the system.* It has 3 horizontal **rows** and 4 vertical **columns.**

In the equation method we use elementary operations to eliminate the x terms from the last two equations and then to eliminate the y term from the last equation. As shown in the side-by-side development below, this corresponds to performing **row operations** on the augmented matrix in order to make certain entries in the first and second columns 0.

Equation Method	**Matrix Method**
Replace the second equation by the sum of itself and -2 times the first equation:	Replace the second row by the sum of itself and -2 times the first row:
$\begin{aligned} x + 2y + 3z &= -2 \\ 2y - 5z &= 6 \\ 3x + 3y + 10z &= -2 \end{aligned}$	$\left(\begin{array}{ccc\|c} 1 & 2 & 3 & -2 \\ 0 & 2 & -5 & 6 \\ 3 & 3 & 10 & -2 \end{array}\right)$
Replace the third equation by the sum of itself and -3 times the first equation:	Replace the third row by the sum of itself and -3 times the first row:
$\begin{aligned} x + 2y + 3z &= -2 \\ 2y - 5z &= 6 \\ -3y + z &= 4 \end{aligned}$	$\left(\begin{array}{ccc\|c} 1 & 2 & 3 & -2 \\ 0 & 2 & -5 & 6 \\ 0 & -3 & 1 & 4 \end{array}\right)$
Multiply the second equation by $1/2$ (so that y has coefficient 1):	Multiply the second row by $1/2$:
$\begin{aligned} x + 2y + 3z &= -2 \\ y - \frac{5}{2}z &= 3 \\ -3y + z &= 4 \end{aligned}$	$\left(\begin{array}{ccc\|c} 1 & 2 & 3 & -2 \\ 0 & 1 & -\frac{5}{2} & 3 \\ 0 & -3 & 1 & 4 \end{array}\right)$

*The plural of "matrix" is "matrices."

Replace the third equation by the sum of itself and 3 times the second equation:

$$\begin{aligned} x + 2y + 3z &= -2 \\ y - \frac{5}{2}z &= 3 \\ -\frac{13}{2}z &= 13 \end{aligned}$$

Replace the third row by the sum of itself and 3 times the second row:

$$\left(\begin{array}{ccc:c} 1 & 2 & 3 & -2 \\ 0 & 1 & -\frac{5}{2} & 3 \\ 0 & 0 & -\frac{13}{2} & 13 \end{array}\right)$$

Finally, multiply the last equation by $-2/13$:*

$$\begin{aligned} x + 2y + 3z &= -2 \\ y - \frac{5}{2}z &= 3 \\ z &= -2 \end{aligned}$$

Finally, multiply the last row by $-2/13$:

$$\left(\begin{array}{ccc:c} 1 & 2 & 3 & -2 \\ 0 & 1 & -\frac{5}{2} & 3 \\ 0 & 0 & 1 & -2 \end{array}\right)$$

The system is now easily solved. The third equations shows that $z = -2$ and substituting this in the second equation shows that

$$y - \frac{5}{2}(-2) = 3$$

$$y = 3 - 5 = -2.$$

Substituting $y = -2$ and $z = -2$ in the first equation yields

$$x + 2(-2) + 3(-2) = -2$$

$$x = -2 + 4 + 6 = 8.$$

Therefore, the only solution of the original system is $x = 8$, $y = -2$, $z = -2$. ■

▶ TECHNOLOGY TIP

Virtually all calculators have matrix capabilities. Check your instruction manual to learn how to enter and store matrices in the matrix memory and to perform row operations on them.

A matrix can be put in row echelon form with a single keystroke by using REF in the MATH or OPS submenu of the TI-83 or TI-85 MATRIX menu, or in the MATRIX submenu of the TI-92 MATH menu.

The right-hand column in Example 3 shows that matrix notation requires less writing. Furthermore, systems in matrix notation can be readily solved with a calculator (see the Tip in the margin). Consequently, we use matrix notation from now on (but omit the dotted line before the last column). Row operations replace elementary operations on equations and the solution process ends when we reach a matrix, such as the last one in Example 3, that satisfies these conditions:

All rows consisting entirely of zeros are at the bottom.

The first nonzero entry in every nonzero row is a 1 (called a *leading 1*).

Each leading 1 appears to the right of the leading 1's in any preceding rows.

Such a matrix is said to be in **row echelon form.**

*This step is not absolutely necessary, but it is often convenient to have 1 as the coefficient of the first variable in an equation.

EXAMPLE 4 To solve the system

$$\begin{aligned} x + y + 2z &= 1 \\ 2x + 4y + 5z &= 2 \\ 3x + 5y + 7z &= 2 \end{aligned}$$

we form the augmented matrix and reduce it to row echelon form via row operations:

$$\begin{pmatrix} 1 & 1 & 2 & 1 \\ 2 & 4 & 5 & 2 \\ 3 & 5 & 7 & 2 \end{pmatrix}$$

Replace second row by the sum of itself and − 2 times the first row:

$$\begin{pmatrix} 1 & 1 & 2 & 1 \\ 0 & 2 & 1 & 0 \\ 3 & 5 & 7 & 2 \end{pmatrix}$$

Replace third row by the sum of itself and − 3 times the first row:

$$\begin{pmatrix} 1 & 1 & 2 & 1 \\ 0 & 2 & 1 & 0 \\ 0 & 2 & 1 & -1 \end{pmatrix}$$

Multiply second row by 1/2:

$$\begin{pmatrix} 1 & 1 & 2 & 1 \\ 0 & 1 & \frac{1}{2} & 0 \\ 0 & 2 & 1 & -1 \end{pmatrix}$$

Replace third row by the sum of itself and − 2 times the second row:

$$\begin{pmatrix} 1 & 1 & 2 & 1 \\ 0 & 1 & \frac{1}{2} & 0 \\ 0 & 0 & 0 & -1 \end{pmatrix}$$

This matrix is almost in row echelon form. However, its last row represents the equation:

$$0x + 0y + 0z = -1.$$

Since this equation obviously has no solutions, neither does the original system. Such a system is said to be **inconsistent.** ■

EXAMPLE 5 A system such as

$$\begin{aligned} 2x + 5y + z + 3w &= 0 \\ 2y - 4z + 6w &= 0 \\ 2x + 17y - 23z + 40w &= 0 \end{aligned}$$

in which all the constants on the right side are zero, is called a **homogeneous system.** It has at least one solution, namely, $x = 0, y = 0, z = 0, w = 0$, which is called the **trivial solution.** To see if there are any nontrivial solutions, we put the system in matrix form and solve it:*

$$\begin{pmatrix} 2 & 5 & 1 & 3 & 0 \\ 0 & 2 & -4 & 6 & 0 \\ 2 & 17 & -23 & 40 & 0 \end{pmatrix}$$

Replace third row by the sum of itself and − 1 times the first row:

$$\begin{pmatrix} 2 & 5 & 1 & 3 & 0 \\ 0 & 2 & -4 & 6 & 0 \\ 0 & 12 & -24 & 37 & 0 \end{pmatrix}$$

Replace third row by the sum of itself and − 6 times the second row:

$$\begin{pmatrix} 2 & 5 & 1 & 3 & 0 \\ 0 & 2 & -4 & 6 & 0 \\ 0 & 0 & 0 & 1 & 0 \end{pmatrix}$$

Multiply first row by 1/2:

$$\begin{pmatrix} 1 & \frac{5}{2} & \frac{1}{2} & \frac{3}{2} & 0 \\ 0 & 2 & -4 & 6 & 0 \\ 0 & 0 & 0 & 1 & 0 \end{pmatrix}$$

Multiply second row by 1/2:

$$\begin{pmatrix} 1 & \frac{5}{2} & \frac{1}{2} & \frac{3}{2} & 0 \\ 0 & 1 & -2 & 3 & 0 \\ 0 & 0 & 0 & 1 & 0 \end{pmatrix} \qquad \begin{aligned} x + \frac{5}{2}y + \frac{1}{2}z + \frac{3}{2}w &= 0 \\ y - 2z + 3w &= 0 \\ w &= 0 \end{aligned}$$

The last matrix above represents the system shown to its right. Putting $w = 0$ in the second equation shows that

$$y - 2z = 0 \qquad \text{or equivalently,} \qquad y = 2z.$$

The equation $y = 2z$ has an infinite number of solutions; for instance, $z = 1$, $y = 2$ and $z = 3$, $y = -6$ are solutions. More generally, for each real number t, $z = t$, $y = 2t$ is a solution. Substituting $w = 0$, $z = t$, $y = 2t$ in the first equation shows that

$$x + \frac{5}{2}(2t) + \frac{1}{2}t = 0$$

$$x = \frac{-11t}{2}.$$

*When dealing with homogeneous systems, it really isn't necessary to rewrite the last column of zeros at every step, as we do here, since this column remains unchanged throughout the process.

Note Every system that has more variables than equations (as in Example 5) is dependent, but other systems may be dependent as well.

Therefore, this system, and hence the original one, has an infinite number of solutions, one for each real number t:

$$x = \frac{-11t}{2}, \qquad y = 2t, \qquad z = t, \qquad w = 0.$$

A system with infinitely many solutions, such as this one, is said to be **dependent.** ■

The Gauss-Jordan Method

Gaussian elimination on a calculator is an efficient method of solving systems of equations, but it may involve some messy calculations if you solve the final triangular form system by hand, as was done in the preceding examples. Hand computations can be completely avoided by using a slight variation, known as the **Gauss-Jordan Method.***

EXAMPLE 6 To solve the system

$$\begin{aligned} x - y + 5z &= -6 \\ 3x + 3y - z &= 10 \\ x - 5y + 8z &= -17 \\ x + 3y + 2z &= 5 \end{aligned}$$

we form the augmented matrix and use Gaussian elimination to reduce it to row echelon form.

$$\begin{pmatrix} 1 & -1 & 5 & -6 \\ 3 & 3 & -1 & 10 \\ 1 & -5 & 8 & -17 \\ 1 & 3 & 2 & 5 \end{pmatrix}$$

Replace second row by the sum of itself and -3 times the first row:

$$\begin{pmatrix} 1 & -1 & 5 & -6 \\ 0 & 6 & -16 & 28 \\ 1 & -5 & 8 & -17 \\ 1 & 3 & 2 & 5 \end{pmatrix}$$

Replace third row by the sum of itself and -1 times the first row:

$$\begin{pmatrix} 1 & -1 & 5 & -6 \\ 0 & 6 & -16 & 28 \\ 0 & -4 & 3 & -11 \\ 1 & 3 & 2 & 5 \end{pmatrix}$$

Replace fourth row by the sum of itself and -1 times the first row:

$$\begin{pmatrix} 1 & -1 & 5 & -6 \\ 0 & 6 & -16 & 28 \\ 0 & -4 & 3 & -11 \\ 0 & 4 & -3 & 11 \end{pmatrix}$$

*This method was developed by the German engineer Wilhelm Jordan (1842–1899).

Since the last two rows are negatives of each other we can simplify the situation by replacing the fourth row by the sum of itself and (1 times) the third row:

$$\begin{pmatrix} 1 & -1 & 5 & -6 \\ 0 & 6 & -16 & 28 \\ 0 & -4 & 3 & -11 \\ 0 & 0 & 0 & 0 \end{pmatrix}$$

Multiply second row by 1/6:

$$\begin{pmatrix} 1 & -1 & 5 & -6 \\ 0 & 1 & -\frac{8}{3} & \frac{14}{3} \\ 0 & -4 & 3 & -11 \\ 0 & 0 & 0 & 0 \end{pmatrix}$$

Replace third row by the sum of itself and 4 times the second row:

$$\begin{pmatrix} 1 & -1 & 5 & -6 \\ 0 & 1 & -\frac{8}{3} & \frac{14}{3} \\ 0 & 0 & -\frac{23}{3} & \frac{23}{3} \\ 0 & 0 & 0 & 0 \end{pmatrix}$$

Multiply third row by − 3/23:

$$\begin{pmatrix} 1 & -1 & 5 & -6 \\ 0 & 1 & -\frac{8}{3} & \frac{14}{3} \\ 0 & 0 & 1 & -1 \\ 0 & 0 & 0 & 0 \end{pmatrix}$$

At this point in Gaussian elimination we could use back substitution to solve the triangular form system represented by the last matrix. In the Gauss-Jordan method, however, additional elimination of variables replaces back substitution, as follows.

Replace second row by the sum of itself and 8/3 times the third row:

$$\begin{pmatrix} 1 & -1 & 5 & 6 \\ 0 & 1 & 0 & 2 \\ 0 & 0 & 1 & -1 \\ 0 & 0 & 0 & 0 \end{pmatrix}$$

Replace first row by the sum of itself and − 5 times the third row:

$$\begin{pmatrix} 1 & -1 & 0 & -1 \\ 0 & 1 & 0 & 2 \\ 0 & 0 & 1 & -1 \\ 0 & 0 & 0 & 0 \end{pmatrix}$$

Replace first row by the sum of itself and 1 times the second row:

$$\begin{pmatrix} 1 & 0 & 0 & 1 \\ 0 & 1 & 0 & 2 \\ 0 & 0 & 1 & -1 \\ 0 & 0 & 0 & 0 \end{pmatrix}$$

The last matrix represents the following system, whose solution is obvious:

$$\begin{aligned} x \quad\quad &= 1 \\ y \quad &= 2 \\ z &= -1 \end{aligned} \quad ■$$

In the Gauss-Jordan method, row operations can be performed in any order. For instance, instead of first setting up the system for back substitution (as in Example 6), you can work column by column to obtain columns with one entry 1 and the rest 0. All that matters is that you finish with a matrix in row echelon form in which any column containing a leading 1 has zeros in all other positions, such as the last matrix in Example 6. Such a matrix is said to be in **reduced row echelon form.**

> ▶ **TECHNOLOGY TIP**
> To put a matrix in reduced row echelon form with a single keystroke, use RREF in the MATH or OPS submenu of the TI-83 or TI-85 MATRIX menu, or in the MATRIX submenu of the MATH menu on TI-92 and HP-38.

As a general rule, Gaussian elimination is the method of choice when working by hand (the additional row operations of the Gauss-Jordan method are usually more time-consuming than back substitution). Even when using a calculator on a small system, it is often more efficient to put the matrix in row echelon form and use back substitution by hand. For larger systems or small ones in which back substitution involves significant computation, however, the Gauss-Jordan method with a calculator is usually the best way to avoid error.

Applications

In calculus it is sometimes necessary to write a complicated rational expression as the sum of simpler ones. One technique for doing this involves systems of equations.

EXAMPLE 7 Find constants A, B, and C such that

$$\frac{2x^2 + 15x + 10}{(x-1)(x+2)^2} = \frac{A}{x-1} + \frac{B}{x+2} + \frac{C}{(x+2)^2}.$$

Solution Multiply both sides of the equation by the common denominator $(x-1)(x+2)^2$ and collect like terms on the right side:

$$\begin{aligned} 2x^2 + 15x + 10 &= A(x+2)^2 + B(x-1)(x+2) + C(x-1) \\ &= A(x^2 + 4x + 4) + B(x^2 + x - 2) + C(x-1) \\ &= Ax^2 + 4Ax + 4A + Bx^2 + Bx - 2B + Cx - C \\ &= (A+B)x^2 + (4A + B + C)x + (4A - 2B - C). \end{aligned}$$

Since the polynomials on the left and right sides of the last equation are equal, their coefficients must be equal term by term, that is,

$$\begin{aligned} A + B \quad\quad &= 2 \qquad [\textit{Coefficients of } x^2] \\ 4A + B + C &= 15 \qquad [\textit{Coefficients of } x] \\ 4A - 2B - C &= 10 \qquad [\textit{Constant terms}] \end{aligned}$$

We can consider this as a system of equations with unknowns A, B, C, and solve it by Gaussian elimination:

$$\begin{pmatrix} 1 & 1 & 0 & 2 \\ 4 & 1 & 1 & 15 \\ 4 & -2 & -1 & 10 \end{pmatrix}$$

Replace second row by the sum of itself and − 4 times the first row:
Replace third row by the sum of itself and − 4 times the first row:

$$\begin{pmatrix} 1 & 1 & 0 & 2 \\ 0 & -3 & 1 & 7 \\ 0 & -6 & -1 & 2 \end{pmatrix}$$

Replace third row by the sum of itself and − 2 times the third row:

$$\begin{pmatrix} 1 & 1 & 0 & 2 \\ 0 & -3 & 1 & 7 \\ 0 & 0 & -3 & -12 \end{pmatrix}$$

Multiply second row by − 1/3:
Multiply third row by − 1/3:

$$\begin{pmatrix} 1 & 1 & 0 & 2 \\ 0 & 1 & -\frac{1}{3} & -\frac{7}{3} \\ 0 & 0 & 1 & 4 \end{pmatrix} \qquad \begin{aligned} A + B \quad &= 2 \\ B - \frac{1}{3}C &= -\frac{7}{3} \\ C &= 4 \end{aligned}$$

Verify that the solution of this last system is $C = 4, B = -1, A = 3$. Therefore,

$$\frac{2x^2 + 15x + 10}{(x-1)(x+2)^2} = \frac{3}{x-1} + \frac{-1}{x+2} + \frac{4}{(x+2)^2}.$$

The right side of this equation is called the **partial fraction decomposition** of the fraction on the left side. ■

EXAMPLE 8 Charlie is starting a small business and borrows \$10,000 on three different credit cards, with annual interest rates of 18%, 15%, and 9%, respectively. He borrows three times as much on the 15% card as on the 18% card, and his total annual interest on all three cards is \$1244.25. How much did he borrow on each credit card?

Solution Let x be the amount on the 18% card, y the amount on the 15% card, and z the amount on the 9% card. Then, $x + y + z = 10{,}000$. Furthermore,

Interest on 18% card + Interest on 15% card + Interest on 9% card = Total interest

$$.18x + .15y + .09z = 1244.25$$

Finally, we have

Amount on 15% card = 3 times amount on 18% card

$$y = 3x,$$

which is equivalent to $3x - y = 0$. Therefore, we must solve this system of equations:

$$\begin{aligned} x + \quad y + \quad z &= 10{,}000 \\ .18x + .15y + .09z &= 1{,}244.25 \\ 3x - \quad y \qquad &= 0 \end{aligned}$$

whose augmented matrix is

$$\begin{pmatrix} 1 & 1 & 1 & 10{,}000 \\ .18 & .15 & .09 & 1{,}244.25 \\ 3 & -1 & 0 & 0 \end{pmatrix}$$

GRAPHING EXPLORATION Show that this matrix can be put in the following row echelon form:

$$\begin{pmatrix} 1 & 1 & 1 & 10{,}000 \\ 0 & 1 & 3 & 18{,}525 \\ 0 & 0 & 1 & 4{,}900 \end{pmatrix}$$

Then solve the corresponding system of equations.

The Investigation shows that Charlie borrowed \$1275 on the 18% card, \$3825 on the 15% card, and \$4900 on the 9% card. ■

The preceding examples illustrate the following fact, whose proof is omitted.

Number of Solutions of a System ▶

Any system of linear equations must have

no solutions (an inconsistent system) ***or***

exactly one solution ***or***

an infinite number of solutions (a dependent system).

EXERCISES 11.2

In Exercises 1–4, write the augmented matrix of the system.

1. $$\begin{aligned} 2x - 3y + 4z &= 1 \\ x + 2y - 6z &= 0 \\ 3x - 7y + 4z &= -3 \end{aligned}$$

2. $$\begin{aligned} x + 2y - 3w + 7z &= -5 \\ 2x - y \qquad + 2z &= 4 \\ 3x \qquad + 7w - 6z &= 0 \end{aligned}$$

3. $$\begin{aligned} x - \frac{1}{2}y + \frac{7}{4}z &= 0 \\ 2x - \frac{3}{2}y + 5z &= 0 \\ -2y + \frac{1}{3}z &= 0 \end{aligned}$$

4. $$\begin{aligned} 2x - \frac{1}{2}y + \frac{7}{2}w - 6z &= 1 \\ \frac{1}{4}x - 6y + 2w - z &= 2 \\ 4y - \frac{1}{2}w + z &= 3 \\ 2x + 3y + \frac{1}{2}z &= 4 \end{aligned}$$

In Exercises 5–8, the augmented matrix of a system of equations is given. Express the system in equation notation.

5. $\begin{pmatrix} 2 & -3 & 1 \\ 4 & 7 & 2 \end{pmatrix}$ **6.** $\begin{pmatrix} 2 & 3 & 5 & 2 \\ 1 & 6 & 9 & 0 \end{pmatrix}$

7. $\begin{pmatrix} 1 & 0 & 1 & 0 & 1 \\ 1 & -1 & 4 & -2 & 3 \\ 4 & 2 & 5 & 0 & 2 \end{pmatrix}$

8. $\begin{pmatrix} 1 & 7 & 0 & 4 \\ 2 & 3 & 1 & 6 \\ -1 & 0 & 2 & 3 \end{pmatrix}$

In Exercises 9–12, the reduced row echelon form of the augmented matrix of a system of equations is given. Find the solutions of the system.

9. $\begin{pmatrix} 1 & 0 & 0 & 0 & 3/2 \\ 0 & 1 & 0 & 0 & 5 \\ 0 & 0 & 1 & 0 & -2 \\ 0 & 0 & 0 & 1 & 0 \end{pmatrix}$ **10.** $\begin{pmatrix} 1 & 0 & 0 & 0 & 0 & 5 \\ 0 & 1 & 0 & 0 & 0 & 4 \\ 0 & 0 & 1 & 0 & 0 & 3 \\ 0 & 0 & 0 & 0 & 1 & 2 \\ 0 & 0 & 0 & 0 & 0 & 1 \end{pmatrix}$

11. $\begin{pmatrix} 1 & 0 & 0 & 1 & 2 \\ 0 & 1 & 0 & 2 & -3 \\ 0 & 0 & 1 & 0 & 4 \\ 0 & 0 & 0 & 0 & 0 \end{pmatrix}$ **12.** $\begin{pmatrix} 1 & 0 & 0 & 0 & 7 \\ 0 & 1 & 0 & 0 & 1 \\ 0 & 0 & 1 & 0 & -5 \\ 0 & 0 & 0 & 1 & 4 \\ 0 & 0 & 0 & 0 & 0 \\ 0 & 0 & 0 & 0 & 0 \end{pmatrix}$

In Exercises 13–16, use Gaussian elimination to solve the system.

13. $$\begin{aligned} -x + 3y + 2z &= 0 \\ 2x - y - z &= 3 \\ x + 2y + 3z &= 0 \end{aligned}$$

14. $$\begin{aligned} 3x + 7y + 9z &= 0 \\ x + 2y + 3z &= 2 \\ x + 4y + z &= 2 \end{aligned}$$

15. $$\begin{aligned} x + y + z &= 1 \\ x - 2y + 2z &= 4 \\ 2x - y + 3z &= 5 \end{aligned}$$

16. $$\begin{aligned} 2x - y + z &= 1 \\ 3x + y + z &= 0 \\ 7x - y + 3z &= 2 \end{aligned}$$

In Exercises 17–20, use the Gauss-Jordan method to solve the system.

17. $$\begin{aligned} x - 2y + 4z &= 6 \\ x + y + 13z &= 6 \\ -2x + 6y - z &= -10 \end{aligned}$$

18. $$\begin{aligned} x - y + 5z &= -6 \\ 3x + 3y - z &= 10 \\ x + 3y + 2z &= 5 \end{aligned}$$

19. $$\begin{aligned} x + y + z &= 200 \\ x - 2y &= 0 \\ 2x + 3y + 5z &= 600 \\ 2x - y + z &= 200 \end{aligned}$$

20. $$\begin{aligned} 3x - y + z &= 6 \\ x + 2y - z &= 0 \end{aligned}$$

In Exercises 21–36, solve the system.

21. $$\begin{aligned} 11x + 10y + 9z &= 5 \\ x + 2y + 3z &= 1 \\ 3x + 2y + z &= 1 \end{aligned}$$

22. $$\begin{aligned} -x + 2y - 3z + 4w &= 8 \\ 2x - 4y + z + 2w &= -3 \\ 5x - 4y + z + 2w &= -3 \end{aligned}$$

23. $$\begin{aligned} x + y &= 3 \\ 5x - y &= 3 \\ 9x - 4y &= 1 \end{aligned}$$

24. $$\begin{aligned} 2x - y + 2z &= 3 \\ -x + 2y - z &= 0 \\ x + y - z &= 1 \end{aligned}$$

25. $$\begin{aligned} x - 4y - 13z &= 4 \\ x - 2y - 3z &= 2 \\ -3x + 5y + 4z &= 2 \end{aligned}$$

26. $$\begin{aligned} 2x - 4y + z &= 3 \\ x + 3y - 7z &= 1 \\ -2x + 4y - z &= 10 \end{aligned}$$

27. $$\begin{aligned} 4x + y + 3z &= 7 \\ x - y + 2z &= 3 \\ 3x + 2y + z &= 4 \end{aligned}$$

28. $$\begin{aligned} x + 4y + z &= 3 \\ -x + 2y + 2z &= 0 \\ 2x + 2y - z &= 3 \end{aligned}$$

29. $$\begin{aligned} x + y + z &= 0 \\ 3x - y + z &= 0 \\ -5x - y + z &= 0 \end{aligned}$$

30. $$\begin{aligned} x + y + z &= 0 \\ x - y - z &= 0 \\ x - y + z &= 0 \end{aligned}$$

31. $$\begin{aligned} 2x + y + 3z - 2w &= -6 \\ 4x + 3y + z - w &= -2 \\ x + y + z + w &= -5 \\ -2x - 2y + 2z + 2w &= -10 \end{aligned}$$

32. $$\begin{aligned} x + y + z + w &= -1 \\ -x + 4y + z - w &= 0 \\ x - 2y + z - 2w &= 11 \\ -x - 2y + z + 2w &= -3 \end{aligned}$$

33.
$$\begin{aligned} x - 2y - z - 3w &= -3 \\ -x + y + z &= 0 \\ 4y + 3z - 2w &= -1 \\ 2x - 2y + w &= 1 \end{aligned}$$

34.
$$\begin{aligned} 3x - y + 2z &= 0 \\ -x + 3y + 2z + 5w &= 0 \\ x + 2y + 5z - 4w &= 0 \\ 2x - y + 3w &= 0 \end{aligned}$$

35.
$$\begin{aligned} \frac{3}{x} - \frac{1}{y} + \frac{4}{z} &= -13 \\ \frac{1}{x} + \frac{2}{y} - \frac{1}{z} &= 12 \\ \frac{4}{x} - \frac{1}{y} + \frac{3}{z} &= -7 \end{aligned}$$
[*Hint*: Let $u = 1/x$, $v = 1/y$, $w = 1/z$.]

36.
$$\begin{aligned} \frac{1}{x+1} - \frac{2}{y-3} + \frac{3}{z-2} &= 4 \\ \frac{5}{y-3} - \frac{10}{z-2} &= -5 \\ \frac{-3}{x+1} + \frac{4}{y-3} - \frac{1}{z-2} &= -2 \end{aligned}$$
[*Hint*: Let $u = 1/(x+1)$, $v = 1/(y-3)$, $w = 1/(z-2)$.]

In Exercises 37–42, find the constants A, B, C.

37. $\dfrac{x}{(x+1)(x+2)} = \dfrac{A}{x+1} + \dfrac{B}{x+2}$

38. $\dfrac{1}{(x+1)(x-1)} = \dfrac{A}{x+1} + \dfrac{B}{x-1}$

39. $\dfrac{2x+1}{(x+2)(x-3)^2} = \dfrac{A}{x+2} + \dfrac{B}{x-3} + \dfrac{C}{(x-3)^2}$

40. $\dfrac{x^2 - x - 21}{(2x-1)(x^2+4)} = \dfrac{A}{2x-1} + \dfrac{Bx + C}{x^2+4}$

41. $\dfrac{5x^2 + 1}{(x+1)(x^2 - x + 1)} = \dfrac{A}{x+1} + \dfrac{Bx + C}{x^2 - x + 1}$

42. $\dfrac{x-2}{(x+4)(x^2 + 2x + 2)} = \dfrac{A}{x+4} + \dfrac{Bx + C}{x^2 + 2x + 2}$

Exercises 43–46 deal with systems such as this one:

$$\begin{aligned} x + y + 4z - w &= 1 \\ y - 2z + 3w &= 0 \end{aligned}$$

Verify that for each pair *of real numbers, s and t, the system has a solution:*

$$w = s, \qquad z = t, \qquad y = 2t - 3s,$$
$$x = 1 - (2t - 3s) - 4t + s = 1 - 6t + 4s$$

With this model in mind, solve these dependent systems by the elimination method.

43.
$$\begin{aligned} x - y + 2z + 3w &= 0 \\ x + z + w &= 0 \\ 3x - 2y + 5z + 7w &= 0 \end{aligned}$$

44.
$$\begin{aligned} x + 2y + z + 4w &= 1 \\ y + 3z - w &= 2 \\ x + 4y + 7z - 2w &= 5 \\ 3x + 7y + 6z + 11w &= 5 \end{aligned}$$

45.
$$\begin{aligned} x + y + z - w &= 0 \\ 2x - 4y - 4z + w &= 0 \\ 4x - 2y + 2z - 3w &= 0 \\ 7x - y - z - 3w &= 0 \end{aligned}$$

46.
$$\begin{aligned} x + 2y + 3z + 4v &= 0 \\ 2x + 4y + 6z + w + 9v &= 0 \\ x + 2y + 3z + w + 5v &= 0 \end{aligned}$$

47. A collection of nickels, dimes, and quarters totals \$6.00. If there are 52 coins altogether and twice as many dimes as nickels, how many of each kind of coin are there?

48. A collection of nickels, dimes, and quarters totals \$8.20. The number of nickels and dimes together is twice the number of quarters. The value of the nickels is one third of the value of the dimes. How many of each kind of coin are there?

49. Lillian borrows \$10,000. She borrows some from her friend at 8% annual interest, twice as much as that from her bank at 9%, and the remainder from her insurance company at 5%. She pays a total of \$830 in interest for the first year. How much did she borrow from each source?

50. An investor puts a total of \$25,000 into three very speculative stocks. She invests some of it in Crystalcomp and \$2000 more than one-half that amount in Flyboys. The remainder is invested in Zumcorp. Crystalcomp rises 16% in value, Flyboys 20%, and Zumcorp 18%. Her investment in the three stocks is now worth \$29,440. How much was originally invested in each stock?

51. An investor has \$70,000 invested in a mutual fund, bonds, and a fast food franchise. She has twice as much invested in bonds as in the mutual fund. Last year the mutual fund paid a 2% dividend, the bonds 10%, and the fast food franchise 6%; her dividend income was \$4800. How much is invested in each of the three investments?

52. Tickets to a band concert cost \$2 for children, \$3 for teenagers, and \$5 for adults. 570 people attended the concert and total ticket receipts were \$1950. Three-fourths as many teenagers as children attended. How many children, adults, and teenagers attended?

53. If Tom, Dick, and Harry work together, they can paint a large room in 4 hours. When only Dick and Harry work together, it takes 8 hours to paint the room. Tom and Dick, working together, take 6 hours to paint the room. How long would it take each of them to paint the room alone? [*Hint*: If x is the amount of the room painted in 1 hour by Tom, y the amount painted by Dick, and z the amount painted by Harry, then $x + y + z = 1/4$.]

54. Pipes R, S, T are connected to the same tank. When all three pipes are running, they can fill the tank in 2 hours. When only pipes S and T are running, they can fill the tank in 4 hours. When only R and T are running, they can fill the tank in 2.4 hours. How long would it take each pipe running alone to fill the tank?

55. Peanuts cost \$3 per pound, almonds \$4 per pound, and cashews \$8 per pound. How many pounds of each should be used to produce 140 pounds of a mixture costing \$6 per pound, in which there are twice as many peanuts as almonds?

56. A company produces three camera models, A, B, and C. Each model A requires 3 hours of lens polishing, 2 hours of assembly time, and 2 hours of finishing time. Each model B requires 2 hours of lens polishing, 2 hours of assembly time, and 1 hour of finishing time. Each model C requires 1, 3, and 1 hours of lens polishing, assembly, and finishing time. There are 100 hours available for lens polishing, 100 hours for assembly, and 65 hours for finishing each week. How many of each model should be produced if all available time is to be used?

57. A furniture manufacturer has 1950 machine hours available each week in the cutting department, 1490 hours in the assembly department, and 2160 in the finishing department. Manufacturing a chair requires 0.2 hours of cutting, 0.3 hours of assembly, and 0.1 hours of finishing. A chest requires 0.5 hours of cutting, 0.4 hours of assembly, and 0.6 hours of finishing. A table requires 0.3 hours of cutting, 0.1 hours of assembly, and 0.4 hours of finishing. How many chairs, chests, and tables should be produced in order to use all the available production capacity?

58. A stereo equipment manufacturer produces three models of speakers, R, S, and T, and has three kinds of delivery vehicles: trucks, vans, and station wagons. A truck holds 2 boxes of model R, 1 of model S, and 3 of model T. A van holds 1 box of model R, 3 of model S, and 2 of model T. A station wagon holds 1 box of model R, 3 of model S, and 1 of model T. If 15 boxes of model R, 20 boxes of model S, and 22 boxes of model T are to be delivered, how many vehicles of each type should be used so that all operate at full capacity?

11.3 OTHER MATRIX METHODS FOR SOLVING SYSTEMS OF EQUATIONS

Matrices were used in Section 11.2 as a convenient shorthand for solving systems of linear equations. We now consider matrices in a more general setting and show how the algebra of matrices provides an alternative method for solving systems of equations that are not dependent and have the same number of equations as variables. This method is very efficient when used with a calculator that has matrix capabilities, but not as efficient as the methods of Section 11.2 if you are working by hand.

Procedures for dealing with matrices vary from calculator to calculator. Check your instruction manual to find out how to enter matrices, multiply them, and find matrix inverses, as these topics are introduced below.

Let m and n be positive integers. An $\boldsymbol{m \times n}$ **matrix** (read "m by n matrix") is a rectangular array of numbers, with m horizontal **rows** (numbered from top to bottom) and n vertical **columns** (numbered from left to right). For example,

$$\begin{pmatrix} 3 & 2 & -5 \\ 6 & 1 & 7 \\ -2 & 0 & 5 \end{pmatrix} \qquad \begin{pmatrix} 3/2 \\ -5 \\ 0 \\ 12 \end{pmatrix}$$

3×3 matrix $\qquad$ 4×1 matrix

Each entry in a matrix can be located by stating the row and column in which it appears. For instance, in the 3×3 matrix above, -5 is the entry in row 1 column 3. When you enter a matrix on a calculator, the words "row" and "column" won't be displayed, but the row number will always be listed before the column number. Thus a display such as "A[3, 2]," or simply "3, 2," indicates the entry in row 3, column 2.

Two matrices are said to be **equal** if they have the same size (same number of rows and columns) and the corresponding entries are equal. For example,

$$\begin{pmatrix} 3 & (-1)^2 \\ 6 & 12 \end{pmatrix} = \begin{pmatrix} 3 & 1 \\ \sqrt{36} & 12 \end{pmatrix}, \quad \text{but} \quad \begin{pmatrix} 6 & 4 \\ 5 & 1 \end{pmatrix} \neq \begin{pmatrix} 6 & 5 \\ 4 & 1 \end{pmatrix}.$$

Matrix Multiplication

Although there is an extensive arithmetic of matrices, we shall need only matrix multiplication. The simplest case is the product of a matrix with a single row and a matrix with a single column, where the row and column have the same number of entries. This is done by multiplying corresponding entries (first by first, second by second, and so on) and then adding the results. An example is shown in Figure 11–2.

$$(3 \;\; 1 \;\; 2)\begin{pmatrix} 2 \\ 0 \\ 1 \end{pmatrix} = 3 \cdot 2 + 1 \cdot 0 + 2 \cdot 1 = 8$$

First Terms $\quad$ *Second Terms* $\quad$ *Third Terms*

Figure 11-2

Note that the product of a row and a column is a single number.

Now let A be an $m \times n$ matrix and B an $n \times p$ matrix, so that the number of columns of A is the same as the number of rows of B (namely, n). The product matrix AB is defined to be an $m \times p$ matrix (same number of rows as A and same number of columns as B). The entry in row i, column j, of AB is this number:

the product of row i of A and column j of B.

EXAMPLE 1 If

$$A = \begin{pmatrix} 3 & 1 & 2 \\ -1 & 0 & 4 \end{pmatrix} \quad \text{and} \quad B = \begin{pmatrix} 2 & -3 & 0 & 1 \\ 0 & 5 & 2 & 7 \\ 1 & 8 & -4 & 1 \end{pmatrix}$$

then A has 3 columns and B has 3 rows. So the product matrix AB is defined. AB has 2 rows (same as A) and 4 columns (same as B). Its entries are calculated as follows. The entry in row 1, column 1 of AB is the product of row 1 of A and column 1 of B, which is the number 8, as shown in Figure 11–2 on the previous page and indicated at the right here:

row 1, column 1 $$\begin{pmatrix} 3 & 1 & 2 \\ -1 & 0 & 4 \end{pmatrix}\begin{pmatrix} 2 & -3 & 0 & 1 \\ 0 & 5 & 2 & 7 \\ 1 & 8 & -4 & 1 \end{pmatrix} = \begin{pmatrix} 8 & & & \\ & & & \end{pmatrix} \qquad 3\cdot 2 + 1\cdot 0 + 2\cdot 1 = 8$$

The other entries in AB are obtained similarly:

row 1, column 2 $$\begin{pmatrix} 3 & 1 & 2 \\ -1 & 0 & 4 \end{pmatrix}\begin{pmatrix} 2 & -3 & 0 & 1 \\ 0 & 5 & 2 & 7 \\ 1 & 8 & -4 & 1 \end{pmatrix} = \begin{pmatrix} 8 & 12 & & \\ & & & \end{pmatrix} \qquad 3(-3) + 1\cdot 5 + 2\cdot 8 = 12$$

row 1, column 3 $$\begin{pmatrix} 3 & 1 & 2 \\ -1 & 0 & 4 \end{pmatrix}\begin{pmatrix} 2 & -3 & 0 & 1 \\ 0 & 5 & 2 & 7 \\ 1 & 8 & -4 & 1 \end{pmatrix} = \begin{pmatrix} 8 & 12 & -6 & \\ & & & \end{pmatrix} \qquad 3\cdot 0 + 1\cdot 2 + 2(-4) = -6$$

row 1, column 4 $$\begin{pmatrix} 3 & 1 & 2 \\ -1 & 0 & 4 \end{pmatrix}\begin{pmatrix} 2 & -3 & 0 & 1 \\ 0 & 5 & 2 & 7 \\ 1 & 8 & -4 & 1 \end{pmatrix} = \begin{pmatrix} 8 & 12 & -6 & 12 \\ & & & \end{pmatrix} \qquad 3\cdot 1 + 1\cdot 7 + 2\cdot 1 = 12$$

row 2, column 1 $$\begin{pmatrix} 3 & 1 & 2 \\ -1 & 0 & 4 \end{pmatrix}\begin{pmatrix} 2 & -3 & 0 & 1 \\ 0 & 5 & 2 & 7 \\ 1 & 8 & -4 & 1 \end{pmatrix} = \begin{pmatrix} 8 & 12 & -6 & 12 \\ 2 & & & \end{pmatrix} \qquad (-1)2 + 0\cdot 0 + 4\cdot 1 = 2$$

row 2, column 2 $$\begin{pmatrix} 3 & 1 & 2 \\ -1 & 0 & 4 \end{pmatrix}\begin{pmatrix} 2 & -3 & 0 & 1 \\ 0 & 5 & 2 & 7 \\ 1 & 8 & -4 & 1 \end{pmatrix} = \begin{pmatrix} 8 & 12 & -6 & 12 \\ 2 & 35 & & \end{pmatrix} \qquad (-1)(-3) + 0\cdot 5 + 4\cdot 8 = 35$$

row 2, column 3 $$\begin{pmatrix} 3 & 1 & 2 \\ -1 & 0 & 4 \end{pmatrix}\begin{pmatrix} 2 & -3 & 0 & 1 \\ 0 & 5 & 2 & 7 \\ 1 & 8 & -4 & 1 \end{pmatrix} = \begin{pmatrix} 8 & 12 & -6 & 12 \\ 2 & 35 & -16 & \end{pmatrix} \qquad (-1)0 + 0\cdot 2 + 4(-4) = -16$$

row 2, column 4 $$\begin{pmatrix} 3 & 1 & 2 \\ -1 & 0 & 4 \end{pmatrix}\begin{pmatrix} 2 & -3 & 0 & 1 \\ 0 & 5 & 2 & 7 \\ 1 & 8 & -4 & 1 \end{pmatrix} = \begin{pmatrix} 8 & 12 & -6 & 12 \\ 2 & 35 & -16 & 3 \end{pmatrix} \qquad (-1)1 + 0\cdot 7 + 4\cdot 1 = 3$$

The last matrix on the right is the product AB. ■

EXAMPLE 2 Use matrix multiplication to express this system of equations in matrix form:

$$\begin{aligned} x + y + z &= 2 \\ 2x + 3y \quad &= 5 \\ x + 2y + z &= -1. \end{aligned}$$

Solution Let A be the 3×3 matrix of coefficients on the left side of the equations, B the column matrix of constants on the right side, and X the column matrix of unknowns:

$$A = \begin{pmatrix} 1 & 1 & 1 \\ 2 & 3 & 0 \\ 1 & 2 & 1 \end{pmatrix}, \quad X = \begin{pmatrix} x \\ y \\ z \end{pmatrix} \quad B = \begin{pmatrix} 2 \\ 5 \\ -1 \end{pmatrix}.$$

Then AX is a matrix with 3 rows and 1 column, as is B:

$$AX = \begin{pmatrix} 1 & 1 & 1 \\ 2 & 3 & 0 \\ 1 & 2 & 1 \end{pmatrix}\begin{pmatrix} x \\ y \\ z \end{pmatrix} = \begin{pmatrix} x + y + z \\ 2x + 3y + 0z \\ x + 2y + z \end{pmatrix} \quad \text{and} \quad B = \begin{pmatrix} 2 \\ 5 \\ -1 \end{pmatrix}.$$

The entries in AX are just the left sides of the equations of the system and the entries in B are the constants on the right sides. Therefore, the system can be expressed as the matrix equation $AX = B$. ■

It can be shown that matrix multiplication is associative, that is, $A(BC) = (AB)C$ whenever all the products are defined. However, matrix multiplication is not commutative, that is, AB may not be equal to BA, even when both products are defined (Exercises 13–16).

Identity Matrices and Inverses

The $n \times n$ **identity matrix** I_n is the matrix with 1's on the diagonal from upper left to lower right and 0's everywhere else; for example,

$$I_2 = \begin{pmatrix} 1 & 0 \\ 0 & 1 \end{pmatrix} \quad I_3 = \begin{pmatrix} 1 & 0 & 0 \\ 0 & 1 & 0 \\ 0 & 0 & 1 \end{pmatrix} \quad I_4 = \begin{pmatrix} 1 & 0 & 0 & 0 \\ 0 & 1 & 0 & 0 \\ 0 & 0 & 1 & 0 \\ 0 & 0 & 0 & 1 \end{pmatrix}.$$

The number 1 is the multiplicative identity of the number system because $a \cdot 1 = a = 1 \cdot a$ for every number a. The identity matrix I_n is the multiplicative identity for $n \times n$ matrices:

Identity Matrix ▶

For any $n \times n$ matrix A, $AI_n = A = I_nA$.

For example, in the 2×2 case

$$\begin{pmatrix} a & b \\ c & d \end{pmatrix}\begin{pmatrix} 1 & 0 \\ 0 & 1 \end{pmatrix} = \begin{pmatrix} a \cdot 1 + b \cdot 0 & a \cdot 0 + b \cdot 1 \\ c \cdot 1 + d \cdot 0 & c \cdot 0 + d \cdot 1 \end{pmatrix} = \begin{pmatrix} a & b \\ c & d \end{pmatrix}.$$

Verify that the same answer results if you reverse the order of multiplication.

Every nonzero number c has a multiplicative inverse $c^{-1} = 1/c$ with the property that $cc^{-1} = 1$. The analogous statement for matrix multiplication does not always hold, and special terminology is used when it does. An $n \times n$ matrix A is said to be **invertible** (or **nonsingular**) if there is an $n \times n$ matrix B such that $AB = I_n$. In this case it can be proved that $BA = I_n$ also. The matrix B is called the **inverse** of A and is sometimes denoted A^{-1}.

EXAMPLE 3 You can readily verify that

$$\begin{pmatrix} 2 & 1 \\ 3 & 1 \end{pmatrix}\begin{pmatrix} -1 & 1 \\ 3 & -2 \end{pmatrix} = \begin{pmatrix} 1 & 0 \\ 0 & 1 \end{pmatrix} = \begin{pmatrix} -1 & 1 \\ 3 & -2 \end{pmatrix}\begin{pmatrix} 2 & 1 \\ 3 & 1 \end{pmatrix}.$$

Therefore, $A = \begin{pmatrix} 2 & 1 \\ 3 & 1 \end{pmatrix}$ is an invertible matrix with inverse $A^{-1} = \begin{pmatrix} -1 & 1 \\ 3 & -2 \end{pmatrix}$. ■

EXAMPLE 4 Find the inverse of the matrix $\begin{pmatrix} 2 & 6 \\ 1 & 4 \end{pmatrix}$.

Solution We must find numbers x, y, u, v such that

$$\begin{pmatrix} 2 & 6 \\ 1 & 4 \end{pmatrix}\begin{pmatrix} x & u \\ y & v \end{pmatrix} = \begin{pmatrix} 1 & 0 \\ 0 & 1 \end{pmatrix}$$

which is the same as

$$\begin{pmatrix} 2x + 6y & 2u + 6v \\ x + 4y & u + 4v \end{pmatrix} = \begin{pmatrix} 1 & 0 \\ 0 & 1 \end{pmatrix}.$$

Since corresponding entries in these last two matrices are equal, finding x, y, u, v amounts to solving these systems of equations:

$$\begin{aligned} 2x + 6y &= 1 \\ x + 4y &= 0 \end{aligned} \qquad \text{and} \qquad \begin{aligned} 2u + 6v &= 0 \\ u + 4v &= 1. \end{aligned}$$

We shall solve the systems by the Gauss-Jordan method of Section 11.2. Since the coefficient matrices of the two systems are the same, we shall use the following matrix, whose first three columns form the augmented matrix of the first system and whose first two and last columns form the augmented matrix of the second system:

$$\left(\begin{array}{cc|cc} 2 & 6 & 1 & 0 \\ 1 & 4 & 0 & 1 \end{array}\right).$$

Performing row operations on this matrix amounts to simultaneously doing the operations on the two augmented matrices:

Multiply row 1 by 1/2: $\left(\begin{array}{cc|cc} 1 & 3 & 1/2 & 0 \\ 1 & 4 & 0 & 1 \end{array}\right)$

Replace row 2 by the sum of itself and -1 times row 1: $\left(\begin{array}{cc|cc} 1 & 3 & 1/2 & 0 \\ 0 & 1 & -1/2 & 1 \end{array}\right)$

Replace row 1 by the sum of itself and -3 times row 2: $\left(\begin{array}{cc|cc} 1 & 0 & 2 & -3 \\ 0 & 1 & -1/2 & 1 \end{array}\right)$

The first three columns of the last matrix show that $x = 2$ and $y = -1/2$. Similarly, the first two and last columns show that $u = -3$ and $v = 1$. Therefore,

$$A^{-1} = \begin{pmatrix} 2 & -3 \\ -1/2 & 1 \end{pmatrix}.$$

Observe that A^{-1} is just the right half of the final form of the augmented matrix above and that the left half is the identity matrix I_2. ■

> ▶ **TECHNOLOGY TIP**
> The inverse A^{-1} of matrix A is found by keying in A and using the x^{-1} key. Using the ∧ key and -1 produces an error message on most calculators.

Although the technique in Example 4 can be used to find the inverse of any matrix that has one, it's quicker to use a calculator. Any calculator with matrix capabilities can find the inverse of an invertible matrix (see the Tip in the margin).

WARNING A calculator should produce an error message when asked for the inverse of a matrix A that does not have one. However, because of round-off errors, it may sometimes display a matrix which it says is A^{-1}. As an accuracy check when finding inverses, multiply A by A^{-1} to see if the product is the identity matrix. If it isn't, A does not have an inverse. See the Calculator Investigation at the end of the section.

Inverse Matrices and Systems of Equations

Suppose a system of equations is written in matrix form $AX = B$ as in Example 2 and that the matrix A has an inverse. Then we can solve $AX = B$ by multiplying both sides by A^{-1}:

$$A^{-1}(AX) = A^{-1}B$$

$$(A^{-1}A)X = A^{-1}B \quad \text{[Matrix multiplication is associative]}$$

$$I_nX = A^{-1}B \quad [A^{-1}A \text{ is the identity matrix}]$$

$$X = A^{-1}B \quad \text{[Product of identity matrix and } X \text{ is } X]$$

The next example shows how this works in practice.

EXAMPLE 5 Solve the system

$$\begin{aligned} x + y + z &= 2 \\ 2x + 3y &= 5 \\ x + 2y + z &= -1. \end{aligned}$$

Solution As we saw in Example 2, this system is equivalent to the matrix equation

$$AX = B$$

$$\begin{pmatrix} 1 & 1 & 1 \\ 2 & 3 & 0 \\ 1 & 2 & 1 \end{pmatrix}\begin{pmatrix} x \\ y \\ z \end{pmatrix} = \begin{pmatrix} 2 \\ 5 \\ -1 \end{pmatrix}.$$

Use a calculator to find the inverse of the coefficient matrix A and multiply both sides of the equation by A^{-1}:

$$A^{-1}AX = A^{-1}B$$

$$\begin{pmatrix} 1.5 & .5 & -1.5 \\ -1 & 0 & 1 \\ .5 & -.5 & .5 \end{pmatrix}\begin{pmatrix} 1 & 1 & 1 \\ 2 & 3 & 0 \\ 1 & 2 & 1 \end{pmatrix}\begin{pmatrix} x \\ y \\ z \end{pmatrix} = \begin{pmatrix} 1.5 & .5 & -1.5 \\ -1 & 0 & 1 \\ .5 & -.5 & .5 \end{pmatrix}\begin{pmatrix} 2 \\ 5 \\ -1 \end{pmatrix}$$

$$\begin{pmatrix} 1 & 0 & 0 \\ 0 & 1 & 0 \\ 0 & 0 & 1 \end{pmatrix}\begin{pmatrix} x \\ y \\ z \end{pmatrix} = \begin{pmatrix} 1.5 & .5 & -1.5 \\ -1 & 0 & 1 \\ .5 & -.5 & .5 \end{pmatrix}\begin{pmatrix} 2 \\ 5 \\ -1 \end{pmatrix} \qquad [\textit{since } A^{-1}A = I_3]$$

$$\begin{pmatrix} x \\ y \\ z \end{pmatrix} = \begin{pmatrix} 7 \\ -3 \\ -2 \end{pmatrix} \qquad [\textit{since } I_3X = X]$$

Therefore, the solution of the original system is $x = 7$, $y = -3$, $z = -2$. ■

Only a matrix with the same number of rows as columns can possibly have an inverse. Consequently, the method of Example 5 can be tried only when the system has the same number of equations as unknowns (so that the coefficient matrix has the same number of rows as columns). In this case, you should use your calculator to verify that the coefficient matrix actually has an inverse (see the Warning on page 681). If it does not, other methods must be used. Here is a summary of the possibilities:

Matrix Solution of a System of Equations ▶

Suppose the system of equations is written in matrix form as $AX = B$.

If the matrix A has an inverse, then the unique solution of the system is

$$X = A^{-1}B.$$

If A does not have an inverse (which is always the case when the number of equations differs from the number of unknowns), then the system either has no solutions or has infinitely many solutions. Its solutions (if any) may be found by using Gaussian elimination (Section 11.2).

EXAMPLE 6 Solve the system

$$\begin{aligned} 2x + y - z &= 2 \\ x + 3y + 2z &= 1 \\ x + y + z &= 2. \end{aligned}$$

Solution Since there are the same number of equations as unknowns, we can try the method of Example 5. In this case we have

$$A = \begin{pmatrix} 2 & 1 & -1 \\ 1 & 3 & 2 \\ 1 & 1 & 1 \end{pmatrix} \quad \text{and} \quad B = \begin{pmatrix} 2 \\ 1 \\ 2 \end{pmatrix}.$$

CALCULATOR EXPLORATION Verify that the matrix A does have an inverse. Show that the solutions of the system are $x = 2$, $y = -1$, $z = 1$ by computing $A^{-1}B$. ■

CALCULATOR INVESTIGATION 11.3

1. Inverse of a Matrix Put the matrix $A = \begin{pmatrix} 1 & 2/3 \\ 2 & 4/3 \end{pmatrix}$ into your calculator (enter 2/3 and 4/3 as $2 \div 3$ and $4 \div 3$). Find A^{-1}. You should get an error message because this matrix does not have an inverse (Exercise 40), but many calculators will display an ''inverse.'' If your calculator does this, check it by computing the product AA^{-1}. Do you get the identity matrix?

EXERCISES 11.3

In Exercises 1–6, determine if the product AB or BA is defined. If a product is defined, state its size (number of rows and columns). Do not calculate any products.

1. $A = \begin{pmatrix} 3 & 6 & 7 \\ 8 & 0 & 1 \end{pmatrix}$, $B = \begin{pmatrix} 2 & 5 & 9 & 1 \\ 7 & 0 & 0 & 6 \\ -1 & 3 & 8 & 7 \end{pmatrix}$

2. $A = \begin{pmatrix} -1 & -2 & -5 \\ 9 & 2 & -1 \\ 10 & 34 & 5 \end{pmatrix}$, $B = \begin{pmatrix} 17 & -9 \\ -6 & 12 \\ 3 & 5 \end{pmatrix}$

3. $A = \begin{pmatrix} 1 & 0 \\ 1 & 1 \\ 0 & 1 \end{pmatrix}$, $B = \begin{pmatrix} 5 & 6 & 11 \\ 7 & 8 & 15 \end{pmatrix}$

4. $A = \begin{pmatrix} 1 & -5 & 7 \\ 2 & 4 & 8 \\ 1 & -1 & 2 \end{pmatrix}$, $B = \begin{pmatrix} -2 & 4 & 9 \\ 13 & -2 & 1 \\ 5 & 25 & 0 \end{pmatrix}$

5. $A = \begin{pmatrix} -4 & 15 \\ 3 & -7 \\ 2 & 10 \end{pmatrix}$, $B = \begin{pmatrix} 1 & 2 \\ 3 & 4 \end{pmatrix}$

6. $A = \begin{pmatrix} 10 & 12 \\ -6 & 0 \\ 1 & 23 \\ -4 & 3 \end{pmatrix}$, $B = \begin{pmatrix} 1 & 2 & 3 \\ 3 & 2 & 1 \end{pmatrix}$

In Exercises 7–12, find AB.

7. $A = \begin{pmatrix} 3 & 2 \\ 2 & 4 \end{pmatrix}$, $B = \begin{pmatrix} 1 & -2 & 3 \\ 0 & 3 & 1 \end{pmatrix}$

8. $A = \begin{pmatrix} -1 & 2 & 3 \\ 0 & -1 & 2 \\ 1 & 2 & 0 \end{pmatrix}$, $B = \begin{pmatrix} 3 & -2 & -1 \\ 1 & 0 & 5 \\ 1 & -1 & -1 \end{pmatrix}$

9. $A = \begin{pmatrix} 1 & 0 & -4 \\ 0 & 2 & -1 \\ 2 & 3 & 4 \end{pmatrix}$, $B = \begin{pmatrix} 1 & 1 \\ 1 & 0 \\ 0 & 1 \end{pmatrix}$

10. $A = \begin{pmatrix} 1 & -2 \\ 3 & 0 \\ 0 & -1 \\ 2 & 1 \end{pmatrix}$, $B = \begin{pmatrix} -1 & 3 & -2 & 0 \\ 6 & 1 & 0 & -2 \end{pmatrix}$

11. $A = \begin{pmatrix} 2 & 0 & -1 \\ 1 & 1 & 2 \\ 0 & 2 & -3 \\ 2 & 3 & 0 \end{pmatrix}$, $B = \begin{pmatrix} 1 & 0 & 1 & 1 \\ 1 & 1 & 0 & 1 \\ 1 & 1 & 1 & 0 \end{pmatrix}$

12. $A = \begin{pmatrix} 10 & 0 & 1 & 0 \\ -1 & 1 & 0 & 1 \end{pmatrix}$, $B = \begin{pmatrix} 2 & -1 & 0 & 1 \\ -2 & 3 & 1 & -4 \\ 3 & 5 & 2 & -5 \end{pmatrix}$

In Exercises 13–16, show that AB is not equal to BA by computing both products.

13. $A = \begin{pmatrix} 3 & 2 \\ 5 & 1 \end{pmatrix}$, $B = \begin{pmatrix} 7 & -5 \\ -2 & 6 \end{pmatrix}$

14. $A = \begin{pmatrix} 3/2 & 2 \\ 4 & 7/2 \end{pmatrix}$, $B = \begin{pmatrix} 1 & -3/2 \\ 5/2 & 1 \end{pmatrix}$

15. $A = \begin{pmatrix} 4 & 2 & -1 \\ 0 & 1 & 2 \\ -3 & 0 & 1 \end{pmatrix}$, $B = \begin{pmatrix} 1 & 7 & -5 \\ 2 & -2 & 6 \\ 0 & 0 & 0 \end{pmatrix}$

16. $A = \begin{pmatrix} 1 & 1 & -1 & 1 \\ 2 & 0 & 3 & 2 \\ -3 & 0 & 0 & 1 \\ 1 & -1 & 1 & 2 \end{pmatrix}$, $B = \begin{pmatrix} 0 & 1 & 7 & 7 \\ 2 & 3 & -2 & 1 \\ 5 & 0 & 1 & 0 \\ -1 & 0 & 1 & 0 \end{pmatrix}$

In Exercises 17–24, find the inverse of the matrix, if it exists.

17. $\begin{pmatrix} 1 & 2 \\ 3 & 4 \end{pmatrix}$

18. $\begin{pmatrix} 3 & 5 \\ 1 & 4 \end{pmatrix}$

19. $\begin{pmatrix} 3 & -1 \\ -6 & 2 \end{pmatrix}$

20. $\begin{pmatrix} 1 & -1 & 0 \\ 1 & 0 & -1 \\ 6 & -2 & -3 \end{pmatrix}$

21. $\begin{pmatrix} 1 & 2 & 0 \\ 3 & -1 & 2 \\ -2 & 3 & -2 \end{pmatrix}$

22. $\begin{pmatrix} 1 & -3 & 4 \\ 2 & -5 & 7 \\ 0 & -1 & 1 \end{pmatrix}$

23. $\begin{pmatrix} 5 & 0 & 2 \\ 2 & 2 & 1 \\ -3 & 1 & -1 \end{pmatrix}$

24. $\begin{pmatrix} -1 & 3 & 1 \\ 2 & 5 & 0 \\ 3 & 1 & -2 \end{pmatrix}$

In Exercises 25–28, solve the system of equations by using the method of Example 5.

25. $\begin{aligned} -x + y \qquad &= 1 \\ -x \qquad + z &= -2 \\ 6x - 2y - 3z &= 3 \end{aligned}$

26. $\begin{aligned} x + 2y + 3z &= 1 \\ 2x + 5y + 3z &= 0 \\ x + \qquad 8z &= -1 \end{aligned}$

27. $\begin{aligned} 2x + y \qquad &= 0 \\ -4x - y - 3z &= 1 \\ 3x + y + 2z &= 2 \end{aligned}$

28. $\begin{aligned} -3x - 3y - 4z &= 2 \\ y + z &= 1 \\ 4x + 3y + 4z &= 3 \end{aligned}$

In Exercises 29–39, solve the system by any method.

29. $\begin{aligned} x + y + \qquad 2w &= 3 \\ 2x - y + z - w &= 5 \\ 3x + 3y + 2z - 2w &= 0 \\ x + 2y + z \qquad &= 2 \end{aligned}$

30. $\begin{aligned} x - 2y + 3z \qquad &= 1 \\ y - z + w &= -2 \\ -2x + 2y - 2z + 4w &= 5 \\ 2y - 3z + w &= 8 \end{aligned}$

31. $\begin{aligned} x + y + 6z + 2v \qquad &= 1.5 \\ x + \qquad 5z + 2v - 3w &= 2 \\ 3x + 2y + 17z + 6v - 4w &= 2.5 \\ 4x + 3y + 21z + 7v - 2w &= 3 \\ -6x - 5y - 36z - 12v + 3w &= 3.5 \end{aligned}$

32. $\begin{aligned} x - 1.5y + \qquad 1.5v - 4w &= 3 \\ -1.5y + \qquad .5w &= 0 \\ -x + 2y + .5z - 2v + 4.5w &= 2 \\ -.5y - 2.5z + .5v - .75w &= 8 \\ y - .5z + \qquad .5w &= 4 \end{aligned}$

33. $\begin{aligned} x + 2y + 2z - 2w &= -23 \\ 4x + 4y - z + 5w &= 7 \\ -2x + 5y + 6z + 4w &= 0 \\ 5x + 13y + 7z + 12w &= -7 \end{aligned}$

34. $\begin{aligned} x + 4y + 5z + 2w &= 0 \\ 2x + y + 4z - 2w &= 0 \\ -x + 7y + 10z + 5w &= 0 \\ -4x + 2y + z + 5w &= 0 \end{aligned}$

35. $\begin{aligned} x + 2y + 5z - 2v + 4w &= 0 \\ 2x - 4y + 6z + v + 4w &= 0 \\ 5x + 2y - 3z + 2v + 3w &= 0 \\ 6x - 5y - 2z + 5v + 3w &= 0 \\ x + 2y - z - 2v + 4w &= 0 \end{aligned}$

36. $$\begin{aligned} 4x + 2y + 3z + \quad\quad 3v + 2w &= 1 \\ 2x + \quad\quad 2z + 2u + v - w &= 2 \\ 10x + 2y + 10z + 10u + 3v + 4w &= 5 \\ 16x + 4y + 16z + 18u + 7v - 2w &= -3 \\ x + 2y + 4z - 6u + 2v + w &= -2 \\ 6x + 2y + 6z + 6u + 3v \quad\quad &= 1 \end{aligned}$$

37. $$\begin{aligned} x + 2y + 3z \quad\quad &= 1 \\ 3x + 2y + 4z \quad\quad &= -1 \\ 2x + 6y + 8z + w &= 3 \\ 2x + \quad\quad 2z - 2w &= 3 \end{aligned}$$

38. $$\begin{aligned} x + 2y + 4z &= 6 \\ y + z &= 1 \\ x + 3y + 5z &= 10 \end{aligned}$$

39. $$\begin{aligned} x \quad\quad\quad + 3w &= -2 \\ x - 4y - z + 3w &= -7 \\ 4y + z \quad\quad &= 5 \\ -x + 12y + 3z - 3w &= 17 \end{aligned}$$

40. If the matrix $A = \begin{pmatrix} 1 & 2/3 \\ 2 & 4/3 \end{pmatrix}$ had an inverse, then there would be a solution to this equation:

$$\begin{pmatrix} 1 & 2/3 \\ 2 & 4/3 \end{pmatrix}\begin{pmatrix} x & u \\ y & v \end{pmatrix} = \begin{pmatrix} 1 & 0 \\ 0 & 1 \end{pmatrix}.$$

(a) Multiply out the left side of the equation above and obtain two systems of equations, one in x, y and one in u, v (see Example 4).

(b) Use Gaussian elimination to show that neither of the systems in part (a) has a solution. Conclude that the matrix A does not have an inverse (even though your calculator may claim that it does).

In Exercises 41–46, find constants a, b, c such that the three given points lie on the parabola $y = ax^2 + bx + c$. See the hint for Exercise 41.

41. $(-3, 2)$, $(1, 1)$, $(2, -1)$ [*Hint:* Since $(-3, 2)$ is to be on the graph, we must have $a(-3)^2 + b(-3) + c = 2$, that is, $9a - 3b + c = 2$. In a similar manner, the other two points lead to *linear* equations in a, b, c. Solve this system of three equations for a, b, c.]

42. $(1, -2)$, $(3, 1)$, $(4, -1)$

43. $(1, 0)$, $(-1, 6)$, $(2, 3)$

44. $(1, 1)$, $(0, 0)$, $(-1, 2)$

45. $(-1, 6)$, $(-2, 16)$, $(1, 4)$

46. $(-1, -6)$, $(2, -3)$, $(4, -25)$

47. Find constants a, b, c such that the points $(0, -2)$, $(\ln 2, 1)$, and $(\ln 4, 4)$ lie on the graph of $f(x) = ae^x + be^{-x} + c$. [*Hint:* See Exercise 41.]

48. Find constants a, b, c such that the points $(0, -1)$, $(\ln 2, 4)$, and $(\ln 3, 7)$ lie on the graph of $f(x) = ae^x + be^{-x} + c$.

49. A candy company produces three types of gift boxes, A, B, and C. A box of variety A contains .6 lb of chocolates and .4 lb of mints. A box of variety B contains .3 lb of chocolates, .4 lb of mints, and .3 lb of caramels. A box of variety C contains .5 lb of chocolates, .3 lb of mints, and .2 lb of caramels. The company has 41,400 lb of chocolates, 29,400 lb of mints, and 16,200 lb of caramels in stock. How many boxes of each kind should be made in order to use up all their stock?

50. Certain circus animals are fed the same three food mixes, R, S, and T. Lions receive 1.1 units of R, 2.4 units of S, and 3.7 units of T each day. Horses receive 8.1 units of R, 2.9 units of S, and 5.1 units of T each day. Bears receive 1.3 units of R, 1.3 units of S, and 2.3 units of T each day. If 16,000 units of R, 28,000 units of S, and 44,000 units of T are available each day, how many of each type of animal can be supported?

11.4 SYSTEMS OF NONLINEAR EQUATIONS

Some systems that include nonlinear equations can be solved algebraically.

EXAMPLE 1 To solve the system

$$-2x + y = -1$$

$$xy = 3$$

solve the first equation for y:

$$y = 2x - 1$$

and substitute this into the second equation:

$$xy = 3$$
$$x(2x - 1) = 3$$
$$2x^2 - x = 3$$
$$2x^2 - x - 3 = 0$$
$$(2x - 3)(x + 1) = 0$$
$$2x - 3 = 0 \qquad \text{or} \qquad x + 1 = 0$$
$$x = 3/2 \qquad\qquad x = -1$$

Using the equation $y = 2x - 1$ to find the corresponding values of y, we see that:

$$\text{If } x = \frac{3}{2}, \qquad \text{then } y = 2\left(\frac{3}{2}\right) - 1 = 2.$$

$$\text{If } x = -1, \qquad \text{then } y = 2(-1) - 1 = -3.$$

Therefore, the solutions of the system are $x = 3/2$, $y = 2$ and $x = -1$, $y = -3$. ■

EXAMPLE 2 To solve the system

$$x^2 + y^2 = 8$$
$$x^2 - y = 6$$

solve the second equation for y, obtaining $y = x^2 - 6$, and substitute this into the first equation:

$$x^2 + y^2 = 8$$
$$x^2 + (x^2 - 6)^2 = 8$$
$$x^2 + x^4 - 12x^2 + 36 = 8$$
$$x^4 - 11x^2 + 28 = 0$$
$$(x^2 - 4)(x^2 - 7) = 0$$
$$x^2 - 4 = 0 \qquad \text{or} \qquad x^2 - 7 = 0$$
$$x^2 = 4 \qquad\qquad x^2 = 7$$
$$x = \pm 2 \qquad\qquad x = \pm\sqrt{7}$$

Using the equation $y = x^2 - 6$ to find the corresponding values of y, we find that the solutions of the system are

$x = 2, y = -2$; $\quad x = -2, y = -2$; $\quad x = \sqrt{7}, y = 1$; $\quad x = -\sqrt{7}, y = 1$. ■

Algebraic techniques were successful in Examples 1 and 2 because substitution led to equations whose exact solutions could be found. When this is not the case, geometric methods can be used to approximate the solutions of the system.

EXAMPLE 3 Use substitution to solve the system:

$$y = x^4 - 4x^3 + 9x - 1$$
$$y = 3x^2 - 3x - 7.$$

Solution Since the first equation is already solved for y, this expression can be substituted in the second equation:

$$y = 3x^2 - 3x - 7$$
$$x^4 - 4x^3 + 9x - 1 = 3x^2 - 3x - 7$$
$$x^4 - 4x^3 + 9x - 1 - 3x^2 + 3x + 7 = 0$$
$$(*) \qquad x^4 - 4x^3 - 3x^2 + 12x + 6 = 0.$$

Equation $(*)$ is not readily solvable algebraically, but approximate solutions can be found geometrically.

GRAPHING EXPLORATION Verify that the graph of $y = x^4 - 4x^3 - 3x^2 + 12x + 6$ has four x-intercepts in the standard viewing window. Since a polynomial of degree 4 has at most 4 roots, there are no other x-intercepts. Hence, equation $(*)$ has four solutions.

The four solutions of the equation can be found by using zoom-in or a graphical root finder. Then each x solution of equation $(*)$ can be substituted in either of the equations of the original system to find the corresponding value of y as in Examples 1 and 2, thus producing the solutions of the system.

If a graphical root finder is used to find the x solutions, then both the x and y solutions will be as accurate as the calculator can be. If zoom-in is used to find the x solutions (the x-intercepts of the graph of equation $(*)$), however, there may be a slight problem. Checking the tic marks in the zoom-in process allows you to determine the maximum error in x as usual. But the fact that a value of x has a maximum error of .01, say, does not guarantee that the corresponding value of y also has a maximum error of .01. Consequently, the alternative geometric approach presented in Example 4 below is preferable to using zoom-in here. ■

Recall that a solution of an equation in x and y can be considered as a point (x, y) on the graph of the equation. A solution of a system of equations is a point that is on the graph of every equation in the system. So solving systems geometrically amounts to finding the points of intersection of the graphs of equations in the system.

EXAMPLE 4 Solve the system of equations in Example 3 geometrically, with a maximum error of .01 in each variable.

Solution Graph both $y = x^4 - 4x^3 + 9x - 1$ and $y = 3x^2 - 3x - 7$ on the same screen. In the viewing window in Figure 11–3, the graphs intersect at three points. However, the graphs appear to be getting closer together as they run off the screen at the top right, which suggests that there may be another intersection point. The larger viewing window in Figure 11–4 shows a fourth intersection point. These four points are the only intersections because, as we saw in Example 3, the solutions of the system (intersection points) correspond to the roots of a fourth-degree polynomial.

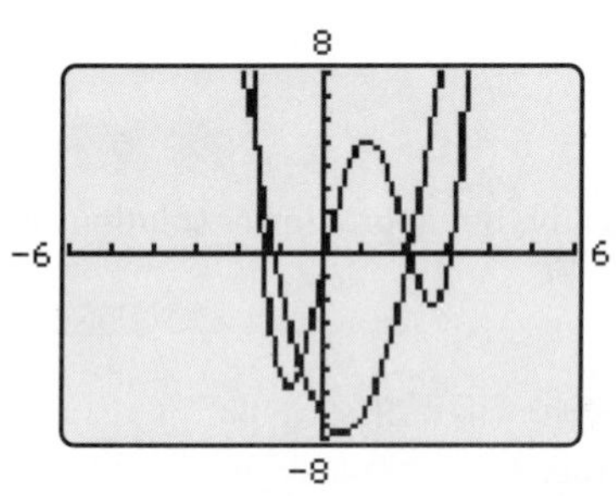

Figure 11-3

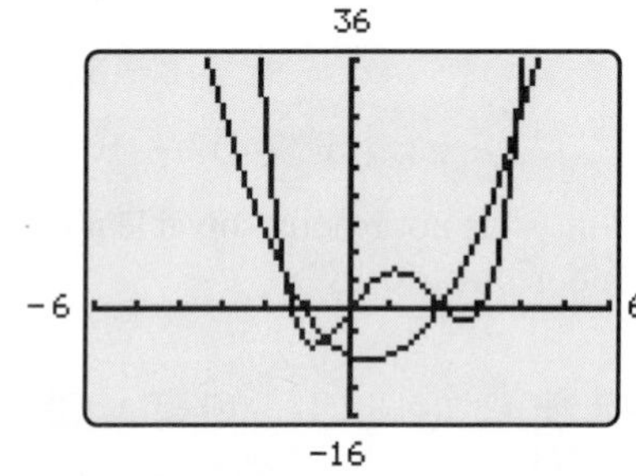

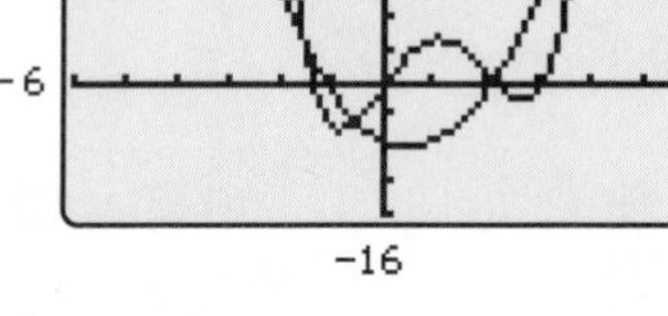

Figure 11-4

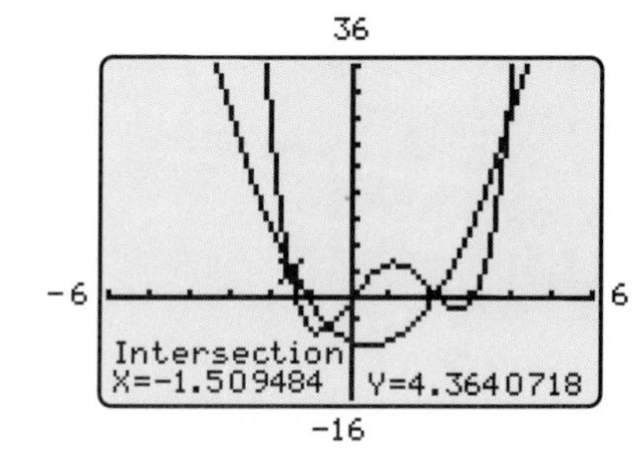

Figure 11-5

The easiest way to find these intersection points is with a graphical intersection finder. These are available on most calculators (see the Tip in the margin) and generally are extremely accurate. A typical intersection finder produced Figure 11–5, which shows that the intersection point (solution of the system) farthest to the left in Figure 11–4 is

$$x = -1.509484, \qquad y = 4.3640718.$$

▶ TECHNOLOGY TIP

The graphical intersection finder is in the CALC menu on TI-82/83, or the MATH submenu of the GRAPH menu on TI-85 and TI-92, or the FNC submenu of the graphing menu on HP-38, or the G-SOLVE menu of Casio 9800, or the JUMP menu of Sharp 9300. An intersection finder program for TI-81 is in the Program Appendix.

GRAPHING EXPLORATION If your calculator has a graphical intersection finder, use it to find the intersection point farthest to the right in Figure 11–4.

When using zoom-in to find intersection points, it helps to use the "grid on" graphing feature if your calculator has one (see the Tip on page 72). For instance, to find the first intersection point to the right of the y-axis in Figure 11–4 we zoom in the window shown in Figures 11–6 and 11–7.

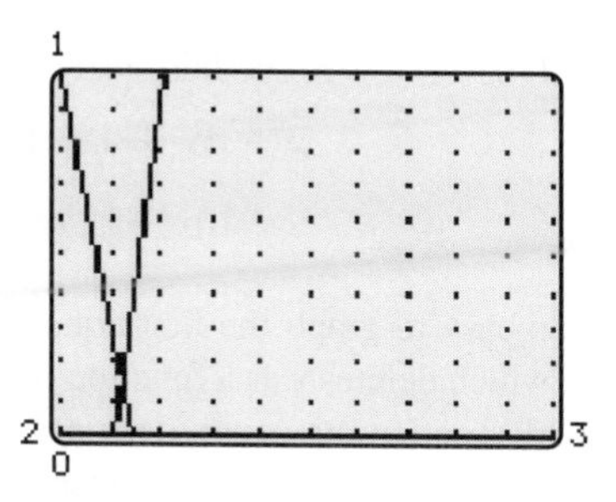

Figure 11-6

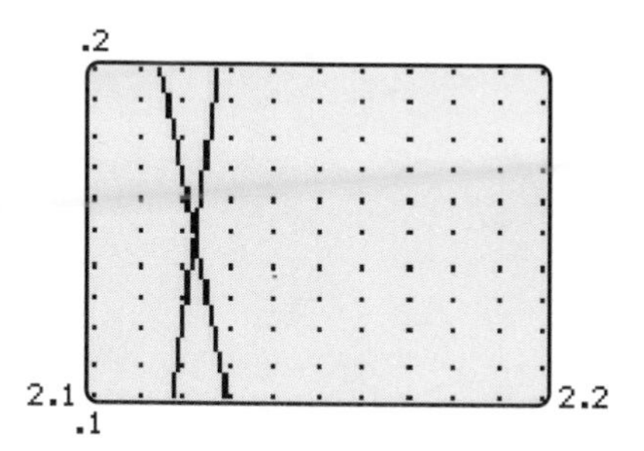

Figure 11-7

In Figure 11–7, the intersection point appears to be in the rectangle with $2.12 \le x \le 2.13$ and $.14 \le y \le .15$. Since the vertical and horizontal grid marks (scale marks) are .01 units apart, any point in this rectangle will be a solution of the system with a maximum error of .01 in each variable. We estimate this solution to be $x = 2.123$, $y = .149$.

GRAPHING EXPLORATION Use zoom-in to approximate the coordinates of the first intersection point to the left of the y-axis in Figure 11–4 with a maximum error of .01 in each coordinate.

Thus, the four solutions of the original system (intersection points of the graphs) are approximately $x = -1.509$, $y = 4.364$; $x = -.484$, $y = -4.846$; $x = 2.123$, $y = .149$; and $x = 3.871$, $y = 26.333$ with a maximum error of .01 in each variable. ■

Unless directed otherwise, use this convention when zooming in:

Accuracy Convention ▶

When finding intersection points by zoom-in (that is, solutions of systems of equations), estimate each coordinate with a maximum error of .01. It may sometimes be necessary to use a smaller maximum error in one coordinate in order to insure that both errors are at most .01.

Systems With Second-Degree Equations

Solving systems of equations graphically depends on our ability to graph each equation in the system. With some equations of higher degree, this may require special techniques.

EXAMPLE 5 Solve this system graphically:

$$x^2 - 4x - y + 1 = 0$$
$$10x^2 + 25y^2 = 100.$$

Solution It's easy to graph the first equation since it can be rewritten as $y = x^2 - 4x + 1$, which defines y as a function of x that can be entered into a calculator. The second equation, however, does not define y as a function of x (because for each value of x, there are two corresponding values of y). Since graphing calculators are designed to graph functions, we must first express this equation in terms of functions. This is done by solving for y:

$$10x^2 + 25y^2 = 100$$
$$25y^2 = 100 - 10x^2$$
$$y^2 = \frac{100 - 10x^2}{25}$$
$$y = \sqrt{\frac{100 - 10x^2}{25}} \quad \text{or} \quad y = -\sqrt{\frac{100 - 10x^2}{25}}$$

Each of these last equations does define y as a function of x and hence can be graphed on a calculator. By graphing both these functions on the same screen, we obtain the complete graph of the equation $10x^2 + 25y^2 = 100$, as shown in Figure 11–8. The graphs of both equations in the system are shown in Figure 11–9.

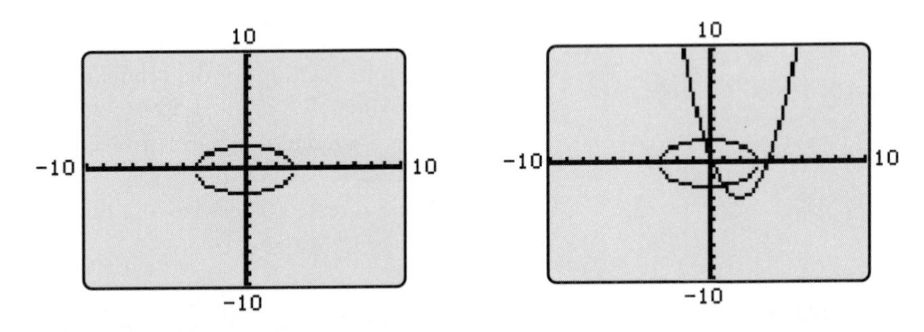

Figure 11-8

Figure 11-9

The two intersection points (solutions of the system) in Figure 11–9 can now be determined by an intersection finder or zoom-in:

$$x = -.2348, y = 1.9945 \quad \text{and} \quad x = .9544, y = -1.9067. \blacksquare$$

EXAMPLE 6 In order to solve the system

$$-4x^2 + 24xy + 3y^2 - 48 = 0$$
$$16x^2 + 24xy + 9y^2 + 100x + 50y + 100 = 0$$

we must express each equation in terms of functions in order to graph it. The first equation may be rewritten as

$$3y^2 + (24x)y + (-4x^2 - 48) = 0.$$

This is a quadratic equation of the form $ay^2 + by + c = 0$, with

$$a = 3, \qquad b = 24x, \qquad c = -4x^2 - 48$$

and hence can be solved by using the quadratic formula:

$$y = \frac{-b \pm \sqrt{b^2 - 4ac}}{2a}$$

$$y = \frac{-24x \pm \sqrt{(24x)^2 - 4\cdot 3\cdot(-4x^2 - 48)}}{2\cdot 3}$$

$$= \frac{-24x \pm \sqrt{(24x)^2 - 12(-4x^2 - 48)}}{6}.$$

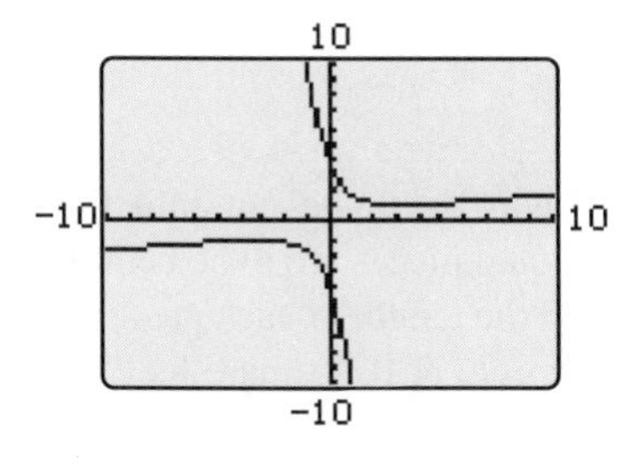

Figure 11-10

Consequently, the graph of the first equation can be obtained by graphing both of these functions on the same screen (Figure 11–10):

$$(*) \qquad y = \frac{-24x + \sqrt{(24x)^2 - 12(-4x^2 - 48)}}{6}.$$

$$y = \frac{-24x - \sqrt{(24x)^2 - 12(-4x^2 - 48)}}{6}.$$

The second equation can also be solved for y by rewriting it as follows:

$$16x^2 + 24xy + 9y^2 + 100x + 50y + 100 = 0$$

$$9y^2 + (24xy + 50y) + (16x^2 + 100x + 100) = 0$$

$$9y^2 + (24x + 50)y + (16x^2 + 100x + 100) = 0.$$

Now apply the quadratic formula with $a = 9$, $b = 24x + 50$, and $c = 16x^2 + 100x + 100$:

$$y = \frac{-b \pm \sqrt{b^2 - 4ac}}{2a}$$

$$= \frac{-(24x + 50) \pm \sqrt{(24x + 50)^2 - 4\cdot 9(16x^2 + 100x + 100)}}{2\cdot 9}.$$

Thus, the graph of the second equation (Figure 11–11 on the next page) consists of the graphs of these two functions:

$$(**) \qquad y = \frac{-(24x + 50) + \sqrt{(24x + 50)^2 - 36(16x^2 + 100x + 100)}}{18}$$

$$y = \frac{-(24x + 50) - \sqrt{(24x + 50)^2 - 36(16x^2 + 100x + 100)}}{18}.$$

By graphing both equations (that is, all four functions shown in (*) and (**) above), we obtain Figure 11–12. Then an intersection finder shows that the solutions of the system (points of intersection) are

$$x = -3.623,\ y = -1.113 \qquad \text{and} \qquad x = -.943,\ y = -1.833. \quad ■$$

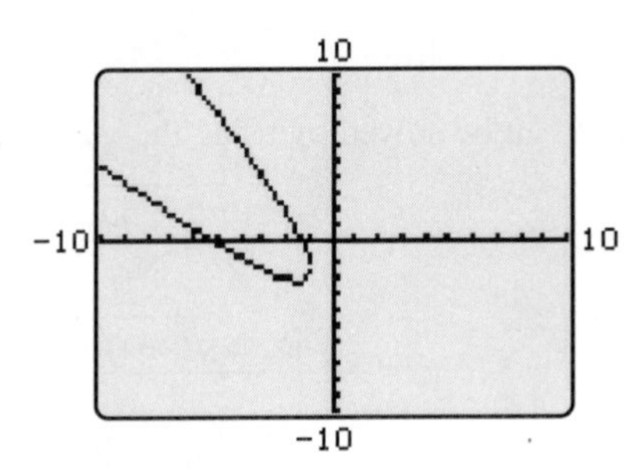

Figure 11-11

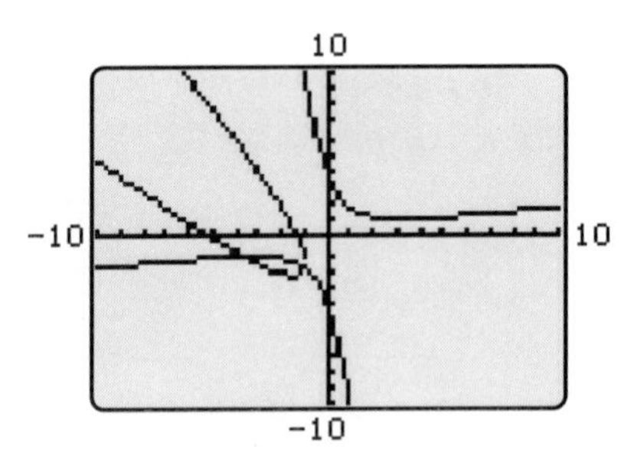

Figure 11-12

Applications

E X A M P L E 7 A 52-ft-long piece of wire is to be cut into three pieces, two of which are the same length. The two equal pieces are to be bent into squares and the third piece into a circle. What should the length of each piece be if the total area enclosed by the two squares and the circle is 100 square feet?

Solution Let x be the length of each piece of wire that is to be bent into a square and y the length of the piece that is to be bent into a circle. Since the original wire is 52 ft long,

$$x + x + y = 52, \qquad \text{or equivalently,} \qquad y = 52 - 2x.$$

If a piece of wire of length x is bent into a square, the side of the square will have length $x/4$ and hence the area of the square will be $(x/4)^2 = x^2/16$. The remaining piece of wire will be made into a circle of circumference (length) y. Since the circumference is 2π times the radius (that is, $y = 2\pi r$), the circle has radius $r = y/2\pi$. Therefore, the area of the circle is

$$\pi r^2 = \pi\left(\frac{y}{2\pi}\right)^2 = \frac{\pi y^2}{4\pi^2} = \frac{y^2}{4\pi}.$$

The sum of the areas of the two squares and the circle is 100, that is,

$$\frac{x^2}{16} + \frac{x^2}{16} + \frac{y^2}{4\pi} = 100$$

$$\frac{y^2}{4\pi} = 100 - \frac{2x^2}{16}$$

$$y^2 = 4\pi\left(100 - \frac{x^2}{8}\right).$$

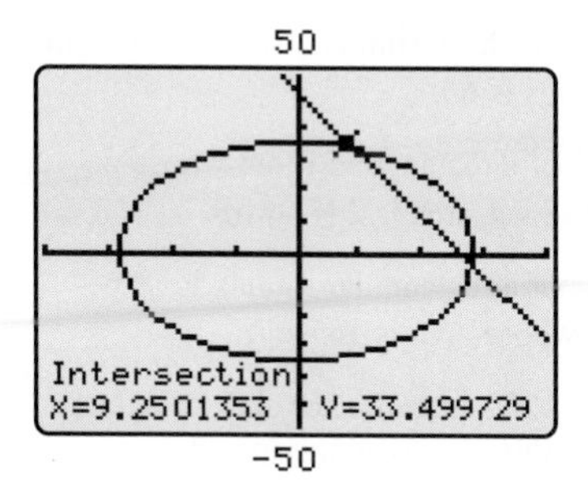

Figure 11-13

Therefore, the lengths x and y are solutions of this system of equations:

$$y = 52 - 2x$$
$$y^2 = 4\pi\left(100 - \frac{x^2}{8}\right).$$

The system may be solved either algebraically or graphically, using Figure 11–13. Since x and y are lengths, both must be positive. Consequently, we need only consider the intersection point in the first quadrant. An intersection finder shows that its coordinates are $x \approx 9.25$, $y \approx 33.50$. Therefore, the wire should be cut into two 9.25-ft pieces and one 33.5-ft piece. ■

EXERCISES 11.4

In Exercises 1–12, solve the system algebraically.

1. $\begin{aligned} x^2 - y &= 0 \\ -2x + y &= 3 \end{aligned}$

2. $\begin{aligned} x^2 - y &= 0 \\ -3x + y &= -2 \end{aligned}$

3. $\begin{aligned} x^2 - y &= 0 \\ x + 3y &= 6 \end{aligned}$

4. $\begin{aligned} x^2 - y &= 0 \\ x + 4y &= 4 \end{aligned}$

5. $\begin{aligned} x + y &= 10 \\ xy &= 21 \end{aligned}$

6. $\begin{aligned} 2x + y &= 4 \\ xy &= 2 \end{aligned}$

7. $\begin{aligned} xy + 2y^2 &= 8 \\ x - 2y &= 4 \end{aligned}$

8. $\begin{aligned} xy + 4x^2 &= 3 \\ 3x + y &= 2 \end{aligned}$

9. $\begin{aligned} x^2 + y^2 - 4x - 4y &= -4 \\ x - y &= 2 \end{aligned}$

10. $\begin{aligned} x^2 + y^2 - 4x - 2y &= -1 \\ x + 2y &= 2 \end{aligned}$

11. $\begin{aligned} x^2 + y^2 &= 25 \\ x^2 + y &= 19 \end{aligned}$

12. $\begin{aligned} x^2 + y^2 &= 1 \\ x^2 - y &= 5 \end{aligned}$

In Exercises 13–28, solve the system by any means.

13. $\begin{aligned} y &= x^3 - 3x^2 + 4 \\ y &= -.5x^2 + 3x - 2 \end{aligned}$

14. $\begin{aligned} y &= -x^3 + 3x^2 + x - 3 \\ y &= -2x^2 + 5 \end{aligned}$

15. $\begin{aligned} y &= x^3 - 3x + 2 \\ y &= \frac{3}{x^2 + 3} \end{aligned}$

16. $\begin{aligned} y &= .25x^4 - 2x^2 + 4 \\ y &= x^3 - x^2 - 2x + 1 \end{aligned}$

17. $\begin{aligned} y &= x^3 + x + 1 \\ y &= \sin x \end{aligned}$

18. $\begin{aligned} y &= x^2 - 4 \\ y &= \cos x \end{aligned}$

19. $\begin{aligned} 25x^2 - 16y^2 &= 400 \\ -9x^2 + 4y^2 &= -36 \end{aligned}$

20. $\begin{aligned} 9x^2 + 16y^2 &= 140 \\ -x^2 + 4y^2 &= -4 \end{aligned}$

21. $\begin{aligned} 5x^2 + 3y^2 - 20x + 6y &= -8 \\ x - y &= 2 \end{aligned}$

22. $\begin{aligned} 4x^2 + 9y^2 &= 36 \\ 2x - y &= -1 \end{aligned}$

23. $\begin{aligned} x^2 + 4xy + 4y^2 - 30x - 90y + 450 &= 0 \\ x^2 + x - y + 1 &= 0 \end{aligned}$

24. $\begin{aligned} 3x^2 + 4xy + 3y^2 - 12x + 2y + 7 &= 0 \\ x^2 - 10x - y + 21 &= 0 \end{aligned}$

25. $\begin{aligned} 4x^2 - 6xy + 2y^2 - 3x + 10y &= 6 \\ 4x^2 + y^2 &= 64 \end{aligned}$

26. $\begin{aligned} 5x^2 + xy + 6y^2 - 79x - 73y + 196 &= 0 \\ x^2 - 2xy + y^2 - 8x - 8y + 48 &= 0 \end{aligned}$

27. $\begin{aligned} x^2 + 3xy + y^2 &= 2 \\ 3x^2 - 5xy + 3y^2 &= 7 \end{aligned}$

28. $\begin{aligned} 2x^2 - 8xy + 8y^2 + 2x - 5 &= 0 \\ 16x^2 - 24xy + 9y^2 + 100x - 200y + 100 &= 0 \end{aligned}$

29. What would be the answer in Example 7 if the original piece of wire were 70 ft long? Explain.

30. A 52-ft-long piece of wire is to be cut into three pieces, two of which are the same length. The two equal pieces are to be bent into circles and the third piece into a square. What should the length of each piece be if the total area enclosed by the two circles and the square is 100 square feet?

31. A rectangular box (including top) with square ends and a volume of 16 cubic meters is to be constructed from 40 square meters of cardboard. What should its dimensions be?

32. A rectangular sheet of metal is to be rolled into a circular tube. If the tube is to have a surface area (excluding ends) of 210 square inches and a volume of 252 cubic inches, what size sheet of metal should be used?

33. Find two real numbers whose sum is -16 and whose product is 48.

34. Find two real numbers whose sum is 34.5 and whose product is 297.

35. Find two positive real numbers whose difference is 1 and whose product is 4.16.

36. Find two real numbers whose difference is 25.75 and whose product is 127.5.

37. Find two real numbers whose sum is 3 such that the sum of their squares is 369.

38. Find two real numbers whose sum is 2 such that the difference of their squares is 60.

39. Find the dimensions of a rectangular room whose perimeter is 58 ft and whose area is 204 sq ft.

40. Find the dimensions of a rectangular room whose perimeter is 53 ft and whose area is 165 sq ft.

41. A rectangle has area 120 square inches and a diagonal of length 17 inches. What are its dimensions?

42. A right triangle has area 225 square centimeters and a hypotenuse of length 35 centimeters. To the nearest tenth of a centimeter, how long are the legs of the triangle?

43. Find the equation of the straight line that intersects the parabola $y = x^2$ *only* at the point (3, 9). [*Hint*: What condition on the discriminant guarantees that a quadratic equation has exactly one real solution?]

CHAPTER 11 *Review*

Important Concepts ▶

Section 11.1

System of equations 652
Solution of a system 652
Substitution method 652
Elimination method 653
Inconsistent system 655
Dependent system 655
Number of solutions of a system 659

Section 11.2

Gaussian elimination 661, 664
Equivalent systems 662
Elementary operations 662
Back substitution 664
Triangular form system 664
Augmented matrix 665
Row operations 665
Row echelon form matrix 666
Inconsistent system 667
Homogeneous system 667
Dependent system 669
Gauss-Jordan method 669
Reduced row echelon form matrix 671
Partial fraction decomposition 672
Number of solutions of a system 673

Section 11.3	$m \times n$ matrix 676 Rows and columns 677 Equality of matrices 677 Multiplication of matrices 677 Matrix form of a system of equations 679 Identity matrix 679 Invertible matrix 680 Inverse of a matrix 680 Inverses and solutions of a system 681
Section 11.4	Algebraic solution of systems of nonlinear equations 685 Geometric solution of systems of nonlinear equations 688 Graphing quadratic equations 690–692

Review Questions ▶

In Questions 1–10, solve the system of linear equations by any means you want.

1. $\begin{aligned} -5x + 3y &= 4 \\ 2x - y &= -3 \end{aligned}$

2. $\begin{aligned} 3x - y &= 6 \\ 2x + 3y &= 7 \end{aligned}$

3. $\begin{aligned} 3x - 5y &= 10 \\ 4x - 3y &= 6 \end{aligned}$

4. $\begin{aligned} \frac{1}{4}x - \frac{1}{3}y &= -\frac{1}{4} \\ \frac{1}{10}x + \frac{2}{5}y &= \frac{2}{5} \end{aligned}$

5. $\begin{aligned} 3x + y - z &= 13 \\ x + 2z &= 9 \\ -3x - y + 2z &= 9 \end{aligned}$

6. $\begin{aligned} x + 2y + 3z &= 1 \\ 4x + 4y + 4z &= 2 \\ 10x + 8y + 6z &= 4 \end{aligned}$

7. $\begin{aligned} 4x + 3y - 3z &= 2 \\ 5x - 3y + 2z &= 10 \\ 2x - 2y + 3z &= 14 \end{aligned}$

8. $\begin{aligned} x + y - 4z &= 0 \\ 2x + y - 3z &= 2 \\ -3x - y + 2z &= -4 \end{aligned}$

9. $\begin{aligned} x - 2y - 3z &= 1 \\ 5y + 10z &= 0 \\ 8x - 6y - 4z &= 8 \end{aligned}$

10. $\begin{aligned} 4x - y - 2z &= 4 \\ x - y - \tfrac{1}{2}z &= 1 \\ 2x - y - z &= 8 \end{aligned}$

11. The sum of one number and three times a second number is -20. The sum of the second number and two times the first number is 55. Find the two numbers.

12. You are given \$144 in \$1, \$5, and \$10 bills. There are 35 bills. There are two more \$10 bills than \$5 bills. How many bills of each type do you have?

13. Let L be the line with equation $4x - 2y = 6$ and M the line with equation $-10x + 5y = -15$. Which of the following statements is true?

(a) L and M do not intersect.
(b) L and M intersect at a single point.
(c) L and M are the same line.
(d) All of the above are true.
(e) None of the above are true.

14. Which of the following statements about this system of equations are *false*?

$$\begin{aligned} x + z &= 2 \\ 6x + 4y + 14z &= 24 \\ 2x + y + 4z &= 7 \end{aligned}$$

(a) $x = 2, y = 3, z = 0$ is a solution.
(b) $x = 1, y = 1, z = 1$ is a solution.
(c) $x = 1, y = -3, z = 3$ is a solution.
(d) The system has an infinite number of solutions.
(e) $x = 2, y = 5, z = -1$ is not a solution.

15. Tickets to a lecture cost \$1 for students, \$1.50 for faculty, and \$2 for others. Total attendance at the lecture was 460, and the total income from tickets was \$570. Three times as many students as faculty attended. How many faculty members attended the lecture?

16. An alloy containing 40% gold and an alloy containing 70% gold are to be mixed to produce 50 pounds of an alloy containing 60% gold. How much of each alloy is needed?

17. Write the augmented matrix of the system

$$\begin{aligned} x - 2y + 3z &= 4 \\ 2x + y - 4z &= 3 \\ -3x + 4y - z &= -2 \end{aligned}$$

18. Use matrix methods to solve the system in Question 17.

19. Write the coefficient matrix of the system

$$\begin{aligned} 2x - y - 2z + 2u &= 0 \\ x + 3y - 2z + u &= 0 \\ -x + 4y + 2z - 3u &= 0 \end{aligned}$$

20. Use matrix methods to solve the system in Question 19.

In Questions 21 and 22, find the constants A, B, C that make the statement true.

21. $\dfrac{4x - 7}{x^2 - x - 6} = \dfrac{A}{x - 3} + \dfrac{B}{x + 2}$

22. $\dfrac{6x^2 + 6x - 6}{(x^2 - 1)(x + 2)} = \dfrac{A}{x + 1} + \dfrac{B}{x - 1} + \dfrac{C}{x + 2}$

In Questions 23–26, perform the indicated matrix multiplication or state that the product is not defined. Use these matrices:

$$A = \begin{pmatrix} -1 & 0 \\ 0 & -1 \end{pmatrix}, \quad B = \begin{pmatrix} 2 & -3 \\ 4 & 1 \end{pmatrix}, \quad C = \begin{pmatrix} 3 & 2 \\ 2 & 4 \end{pmatrix},$$

$$D = \begin{pmatrix} -3 & 1 & 2 \\ 1 & 0 & 4 \end{pmatrix}, \quad E = \begin{pmatrix} 1 & 2 \\ -3 & 4 \\ 0 & 5 \end{pmatrix}, \quad F = \begin{pmatrix} 2 & 3 \\ 6 & 3 \\ 6 & 1 \end{pmatrix}$$

23. AB **24.** CD **25.** AE **26.** DF

In Questions 27–30, find the inverse of the matrix, if it exists.

27. $\begin{pmatrix} 3 & -7 \\ 4 & -9 \end{pmatrix}$ **28.** $\begin{pmatrix} 2 & 6 \\ 1 & 3 \end{pmatrix}$

29. $\begin{pmatrix} 3 & 2 & 6 \\ 1 & 1 & 2 \\ 2 & 2 & 5 \end{pmatrix}$ **30.** $\begin{pmatrix} 1 & -1 & 1 \\ 2 & -3 & 2 \\ -4 & 6 & 1 \end{pmatrix}$

In Questions 31 and 32, use matrix inverses to solve the system.

31. $\begin{aligned} x + \phantom{4y - {}} 2z + 6w &= 2 \\ 3x + 4y - 2z - w &= 0 \\ 5x + \phantom{4y - {}} 2z - 5w &= -4 \\ 4x - 4y + 2z + 3w &= 1 \end{aligned}$

32. $\begin{aligned} 2x + y + 2z + u \phantom{{} + 2v} &= 2 \\ x + 3y - 4z - 2u + 2v &= -2 \\ 2x + 3y + 5z - 4u + v &= 1 \\ x \phantom{{} + 3y} - 2z \phantom{{} - 2u} + 4v &= 4 \\ 2x \phantom{{} + 3y} + 6z \phantom{{} - 2u} - 5v &= 0 \end{aligned}$

In Questions 33–38, solve the system.

33. $\begin{aligned} x^2 - y &= 0 \\ y - 2x &= 3 \end{aligned}$

34. $\begin{aligned} x^2 + y^2 &= 25 \\ x^2 + y &= 19 \end{aligned}$

35. $\begin{aligned} x^2 + y^2 &= 16 \\ x + y &= 2 \end{aligned}$

36. $\begin{aligned} 6x^2 + 4xy + 3y^2 &= 36 \\ x^2 - xy + y^2 &= 9 \end{aligned}$

37. $\begin{aligned} x^3 + y^3 &= 26 \\ x^2 + y &= 6 \end{aligned}$

38. $\begin{aligned} x^2 - 3xy + 2y^2 - y + x &= 0 \\ 5x^2 - 10xy + 5y^2 &= 8 \end{aligned}$

39. An animal feed is to be made from corn, soybeans, and meat by-products. One bag is to supply 1800 units of fiber, 2800 units of fat, and 2200 units of protein. Each pound of corn has 10 units of fiber, 30 units of fat, and 20 units of protein. Each pound of soybeans has 20 units of fiber, 20 units of fat, and 40 units of protein. Each pound of by-products has 30 units of fiber, 40 units of fat, and 25 units of cottonseed. How many pounds of corn, soybeans, and by-products should each bag contain?

40. A home supply store sells three models of dehumidifiers. The Standard model weighs 10 lb and comes in a 10-cu-ft box. The Sleek model weighs 20 lb and comes in an 8-cu-ft box. The Super model weighs 60 lb and comes in a 28-cu-ft box. The store's delivery van has 248 cu ft of space and can hold a maximum of 440 lb. In order for the van to be fully loaded, how many boxes of each model should it carry? [There are several correct answers.]

CHAPTER 12

Discrete Algebra

Roadmap

Sections 1, 4, and 5 are independent of each other and may be read in any order. The interdependence of sections is:

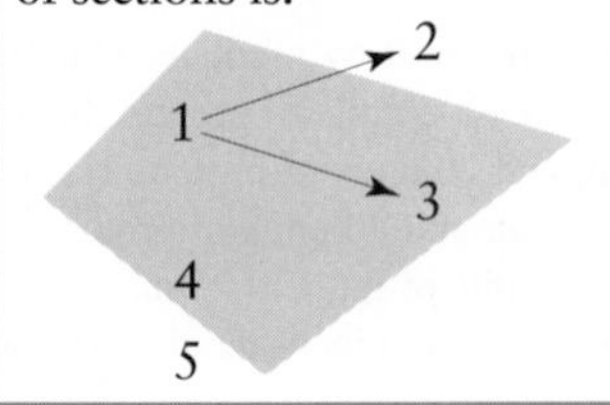

This chapter deals with a variety of subjects involving counting processes and the nonnegative integers 0, 1, 2, 3, ...

12.1 SEQUENCES AND SUMS

A **sequence** is an ordered list of numbers. We usually write them horizontally, with the ordering understood to be from left to right. The same number may appear several times on the list. Each number on the list is called a **term** of the sequence. We are primarily interested in infinite sequences, such as

$$2, 4, 6, 8, 10, 12, \ldots$$

$$1, -3, 5, -7, 9, -11, 13, \ldots$$

$$2, 1, \frac{2}{3}, \frac{2}{4}, \frac{2}{5}, \frac{2}{6}, \frac{2}{7}, \ldots$$

where the dots indicate that the same pattern continues forever.*

*Such a list defines a function f whose domain is the set of positive integers. The rule is $f(1)$ = first number on the list, $f(2)$ = second number on the list, and so on. Conversely, any function g whose domain is the set of positive integers leads to an ordered list of numbers, namely, $g(1)$, $g(2)$, $g(3)$, ... So a sequence is formally defined to be a function whose domain is the set of positive integers.

When the pattern in an ordered list of numbers isn't obvious, the sequence is usually described as follows: The first term is denoted a_1, the second term a_2, and so on. Then a formula is given for the nth term a_n.

EXAMPLE 1 Consider the sequence $a_1, a_2, a_3, \ldots, a_n, \ldots$ where a_n is given by the formula

$$a_n = \frac{n^2 - 3n + 1}{2n + 5}.$$

To find a_1 we substitute $n = 1$ in the formula for a_n; to find a_2 we substitute $n = 2$ in the formula; and so on:

$$a_1 = \frac{1^2 - 3\cdot 1 + 1}{2\cdot 1 + 5} = -\frac{1}{7}$$

$$a_2 = \frac{2^2 - 3\cdot 2 + 1}{2\cdot 2 + 5} = -\frac{1}{9}$$

$$a_3 = \frac{3^2 - 3\cdot 3 + 1}{2\cdot 3 + 5} = \frac{1}{11}.$$

Thus, the sequence begins $-1/7, -1/9, 1/11, \ldots$. The 39th term is

$$a_{39} = \frac{39^2 - 3\cdot 39 + 1}{2\cdot 39 + 5} = \frac{1405}{83}. \quad ■$$

EXAMPLE 2 It is easy to list the first few terms of the sequence

$$a_1, a_2, a_3, \ldots \quad \text{where} \quad a_n = \frac{(-1)^n}{n+2}.$$

Substituting $n = 1$, $n = 2$, and so on, in the formula for a_n shows that:

$$a_1 = \frac{(-1)^1}{1+2} = -\frac{1}{3}, \qquad a_2 = \frac{(-1)^2}{2+2} = \frac{1}{4}, \qquad a_3 = \frac{(-1)^3}{3+2} = -\frac{1}{5}.$$

Similarly,

$$a_{41} = \frac{(-1)^{41}}{41+2} = -\frac{1}{43} \quad \text{and} \quad a_{206} = \frac{(-1)^{206}}{206+2} = \frac{1}{208}. \quad ■$$

EXAMPLE 3 Here are some other sequences whose nth term can be described by a formula:

Sequence	nth Term	First 5 Terms
$a_1, a_2, a_3, \ldots$	$a_n = n^2 + 1$	$2, 5, 10, 17, 26$
$b_1, b_2, b_3, \ldots$	$b_n = \frac{1}{n}$	$1, \frac{1}{2}, \frac{1}{3}, \frac{1}{4}, \frac{1}{5}$
$c_1, c_2, c_3, \ldots$	$c_n = \frac{(-1)^{n+1}2n}{(n+1)(n+2)}$	$\frac{1}{3}, -\frac{1}{3}, \frac{3}{10}, -\frac{4}{15}, \frac{5}{21}$
$x_1, x_2, x_3, \ldots$	$x_n = 3 + \frac{1}{10^n}$	$3.1, 3.01, 3.001, 3.0001, 3.00001.$ ■

▶ TECHNOLOGY TIP

The terms of a sequence that is given by a formula can be displayed by using a table on calculators that have this feature. They can also be displayed on TI calculators (except TI-81) by using SEQ in the OPS submenu of the LIST menu or in the LIST submenu of the TI-92 MATH menu. Check your instruction manual.

TI-82/83, TI-92, and HP-38 also have a sequence graphing mode.

A **constant sequence** is a sequence in which every term is the same number, such as the sequence 7, 7, 7, 7, . . . or the sequence $a_1, a_2, a_3, \ldots$ where $a_n = -18$ for every $n \geq 1$.

The subscript notation for sequences is sometimes abbreviated by writing $\{a_n\}$ in place of $a_1, a_2, a_3, \ldots$.

EXAMPLE 4 $\{1/2^n\}$ denotes the sequence whose first four terms are

$$a_1 = \frac{1}{2^1}, \qquad a_2 = \frac{1}{2^2} = \frac{1}{4}, \qquad a_3 = \frac{1}{2^3} = \frac{1}{8}, \qquad a_4 = \frac{1}{2^4} = \frac{1}{16}.$$

Similarly, $\{(-1)^n n^2\}$ denotes the sequence with first three terms

$$a_1 = (-1)^1 \cdot 1^2 = -1, \qquad a_2 = (-1)^2 \cdot 2^2 = 4, \qquad a_3 = (-1)^3 \cdot 3^2 = -9$$

and 23rd term $a_{23} = (-1)^{23} \cdot 23^2 = -529$. ■

A sequence is said to be defined **recursively** (or **inductively**) if the first term is given (or the first several terms) and there is a method of determining the nth term by using the terms that precede it.

EXAMPLE 5 Consider the sequence whose first two terms are

$$a_1 = 1 \qquad \text{and} \qquad a_2 = 1$$

and whose nth term (for $n \geq 3$) is the sum of the two preceding terms:

$$a_3 = a_2 + a_1 = 1 + 1 = 2$$

$$a_4 = a_3 + a_2 = 2 + 1 = 3$$

$$a_5 = a_4 + a_3 = 3 + 2 = 5.$$

For each integer n, the two preceding integers are $n - 1$ and $n - 2$. So

$$a_n = a_{n-1} + a_{n-2} \qquad (n \geq 3).$$

This sequence 1, 1, 2, 3, 5, 8, 13, . . . is called the **Fibonacci sequence,** and the numbers that appear in it are called **Fibonacci numbers.** Fibonacci numbers have many surprising and interesting properties. See Exercises 54–60 for details. ■

> **▶ TECHNOLOGY TIP**
> Recursively defined sequences can be entered into the sequence memory on TI-92, TI-82/83, and HP-38, subject to some restrictions: On HP-38 and TI-83 the nth term may involve the two preceding terms (as in Example 5), but on TI-82 it can only involve the immediately preceding term (as in Example 6).

EXAMPLE 6 The sequence given by

$$a_1 = -7 \qquad \text{and} \qquad a_n = a_{n-1} + 3 \qquad \text{for } n \geq 2$$

is defined recursively. Its first three terms are

$$a_1 = -7, \qquad a_2 = a_1 + 3 = -7 + 3 = -4,$$
$$a_3 = a_2 + 3 = -4 + 3 = -1. \quad ■$$

Sometimes it is convenient or more natural to begin numbering the terms of a sequence with a number other than 1. So we may consider sequences such as

$$b_4, b_5, b_6, \ldots \qquad \text{or} \qquad c_0, c_1, c_2, \ldots .$$

EXAMPLE 7 The sequence 4, 5, 6, 7, . . . can be conveniently described by saying $b_n = n$, with $n \geq 4$. In the brackets notation, we write $\{n\}_{n \geq 4}$. Similarly, the sequence

$$2^0, 2^1, 2^2, 2^3, \ldots$$

may be described as $\{2^n\}_{n \geq 0}$ or by saying $c_n = 2^n$, with $n \geq 0$. ■

Summation Notation and Partial Sums

It is sometimes necessary to find the sum of various terms in a sequence. For instance, we might want to find the sum of the first nine terms of the sequence $\{a_n\}$. Mathematicians often use the Greek letter sigma (Σ) to abbreviate such a sum:*

$$\sum_{k=1}^{9} a_k = a_1 + a_2 + a_3 + a_4 + a_5 + a_6 + a_7 + a_8 + a_9.$$

Similarly, for any positive integer m and numbers $c_1, c_2, \ldots, c_m$

Summation Notation ▶

$$\sum_{k=1}^{m} c_k \qquad \textbf{means} \qquad c_1 + c_2 + c_3 + \cdots + c_m.$$

*Σ is the letter S in the Greek alphabet, the first letter in *Sum.*

EXAMPLE 8 $\sum_{k=1}^{5} k^2$ denotes the sum of all terms of the form k^2, as k takes values from 1 to 5, that is

$$\sum_{k=1}^{5} k^2 = 1^2 + 2^2 + 3^2 + 4^2 + 5^2 = 55. \quad \blacksquare$$

EXAMPLE 9 To find $\sum_{k=1}^{4} k^2(k-2)$, we successively substitute 1, 2, 3, 4 for k in the expression $k^2(k-2)$ and add up the result:

$$\sum_{k=1}^{4} k^2(k-2) = 1^2(1-2) + 2^2(2-2) + 3^2(3-2) + 4^2(4-2)$$
$$= 1(-1) + 4(0) + 9(1) + 16(2) = 40. \quad \blacksquare$$

EXAMPLE 10 The sum $\sum_{k=1}^{6} (-1)^k k$ is

$$(-1)^1 \cdot 1 + (-1)^2 \cdot 2 + (-1)^3 \cdot 3 + (-1)^4 \cdot 4 + (-1)^5 \cdot 5 + (-1)^6 \cdot 6$$
$$= -1 + 2 - 3 + 4 - 5 + 6 = 3. \quad \blacksquare$$

▶ TECHNOLOGY TIP

To compute sums such as those in Examples 8–10, use SUM SEQ or CUMSUM in the TI-82/83/85 LIST menu or its submenus, or use SUM SEQ in the LIST submenu of the TI-92 MATH menu, or use Σ in the LOOP submenu of the HP-38 MATH menu.

In sums such as $\sum_{k=1}^{5} k^2$ and $\sum_{k=1}^{6} (-1)^k k$, the letter k is called the **summation index.** Any letter may be used for the summation index, just as the rule of a function f may be denoted by $f(x)$ or $f(t)$ or $f(k)$, etc. For example, $\sum_{n=1}^{5} n^2$ means: Take the sum of the terms n^2 as n takes values from 1 to 5. In other words, $\sum_{n=1}^{5} n^2 = \sum_{k=1}^{5} k^2$. Similarly,

$$\sum_{k=1}^{4} k^2(k-2) = \sum_{j=1}^{4} j^2(j-2) = \sum_{n=1}^{4} n^2(n-2).$$

The Σ notation for sums can also be used for sums that don't begin with $k = 1$. For instance,

$$\sum_{k=4}^{10} k^2 = 4^2 + 5^2 + 6^2 + 7^2 + 8^2 + 9^2 + 10^2 = 371$$

$$\sum_{j=0}^{3} j^2(2j+5) = 0^2(2\cdot 0 + 5) + 1^2(2\cdot 1 + 5) + 2^2(2\cdot 2 + 5) + 3^2(2\cdot 3 + 5)$$
$$= 142.$$

Suppose $\{a_n\}$ is a sequence and k is a positive integer. The sum of the first k terms of the sequence is called the ***k*th partial sum** of the sequence. Thus,

Partial Sums ▶

$$\text{The } k\text{th partial sum of } \{a_n\} = \sum_{n=1}^{k} a_n = a_1 + a_2 + a_3 + \cdots + a_k.$$

EXAMPLE 11 Here are some partial sums of the sequence $\{n^3\}$:

First partial sum: $\sum_{n=1}^{1} n^3 = 1^3 = 1$

Second partial sum: $\sum_{n=1}^{2} n^3 = 1^3 + 2^3 = 9$

Sixth partial sum: $\sum_{n=1}^{6} n^3 = 1^3 + 2^3 + 3^3 + 4^3 + 5^3 + 6^3 = 441.$ ■

EXAMPLE 12 The sequence $\{2^n\}_{n\geq 0}$ begins with the 0th term, so the fourth partial sum (the sum of the first four terms) is

$$2^0 + 2^1 + 2^2 + 2^3 = \sum_{n=0}^{3} 2^n.$$

Similarly, the fifth partial sum of the sequence $\left\{\frac{1}{n(n-2)}\right\}_{n\geq 3}$ is the sum of the first five terms:

$$\frac{1}{3(3-2)} + \frac{1}{4(4-2)} + \frac{1}{5(5-2)} + \frac{1}{6(6-2)} + \frac{1}{7(7-2)}$$

$$= \sum_{n=3}^{7} \frac{1}{n(n-2)}.$$ ■

Certain calculations can be written very compactly in summation notation. For example, the distributive law shows that

$$ca_1 + ca_2 + ca_3 + \cdots + ca_r = c(a_1 + a_2 + a_3 + \cdots + a_r).$$

In summation notation this becomes

$$\sum_{n=1}^{r} ca_n = c\left(\sum_{n=1}^{r} a_n\right).$$

This proves the first of the following statements.

Properties of Sums ▶

1. $\sum_{n=1}^{r} ca_n = c\left(\sum_{n=1}^{r} a_n\right)$ **for any number** c**.**
2. $\sum_{n=1}^{r} (a_n + b_n) = \sum_{n=1}^{r} a_n + \sum_{n=1}^{r} b_n$
3. $\sum_{n=1}^{r} (a_n - b_n) = \sum_{n=1}^{r} a_n - \sum_{n=1}^{r} b_n$

To prove statement 2, use the commutative and associative laws repeatedly to show that:

$$(a_1 + b_1) + (a_2 + b_2) + (a_3 + b_3) + \cdots + (a_r + b_r)$$
$$= (a_1 + a_2 + a_3 + \cdots + a_r) + (b_1 + b_2 + b_3 + \cdots + b_r)$$

which can be written in summation notation as

$$\sum_{n=1}^{r} (a_n + b_n) = \sum_{n=1}^{r} a_n + \sum_{n=1}^{r} b_n.$$

The last statement is proved similarly.

EXERCISES 12.1

In Exercises 1–10, find the first five terms of the sequence $\{a_n\}$.

1. $a_n = 2n + 6$ **2.** $a_n = 2^n - 7$

3. $a_n = \dfrac{1}{n^3}$ **4.** $a_n = \dfrac{1}{(n+3)(n+1)}$

5. $a_n = (-1)^n\sqrt{n+2}$

6. $a_n = (-1)^{n+1}n(n-1)$

7. $a_n = 4 + (-.1)^n$ **8.** $a_n = 5 - (.1)^n$

9. $a_n = (-1)^n + 3n$

10. $a_n = (-1)^{n+2} - (n+1)$

In Exercises 11–14, express the sum in Σ *notation.*

11. $1 + 2 + 3 + 4 + 5 + 6 + 7 + 8 + 9 + 10 + 11$

12. $1^1 + 2^2 + 3^3 + 4^4 + 5^5$

13. $\dfrac{1}{2^7} + \dfrac{1}{2^8} + \dfrac{1}{2^9} + \dfrac{1}{2^{10}} + \dfrac{1}{2^{11}} + \dfrac{1}{2^{12}} + \dfrac{1}{2^{13}}$

14. $(-6)^{11} + (-6)^{12} + (-6)^{13} + (-6)^{14} + (-6)^{15}$

In Exercises 15–20, find the sum.

15. $\sum_{i=1}^{5} 3i$ **16.** $\sum_{i=1}^{4} \dfrac{1}{2^i}$

17. $\sum_{n=1}^{6} (2n-3)$ **18.** $\sum_{n=1}^{7} (-1)^n(3n+1)$

19. $\sum_{n=3}^{6} (n^2-8)$ **20.** $\sum_{k=0}^{5} (2k^2 - 5k + 1)$

In Exercises 21–26, find a formula for the nth term of the sequence whose first few terms are given.

21. $-1, 1, -1, 1, -1, 1, \ldots$

22. $2, -2, 2, -2, 2, -2, \ldots$

23. $\dfrac{1}{2}, \dfrac{2}{3}, \dfrac{3}{4}, \dfrac{4}{5}, \dfrac{5}{6}, \ldots$

24. $\dfrac{1}{2\cdot 3}, \dfrac{1}{3\cdot 4}, \dfrac{1}{4\cdot 5}, \dfrac{1}{5\cdot 6}, \dfrac{1}{6\cdot 7}, \ldots$

25. $2, 7, 12, 17, 22, 27, \ldots$

26. $8, -5, 2, -11, -4, -17, -10, \ldots$

In Exercises 27–34, find the first five terms of the given sequence.

27. $a_1 = 4$ and $a_n = 2a_{n-1} + 3 \quad$ for $n \geq 2$

28. $a_1 = -3$ and $a_n = (-1)^n 4a_{n-1} - 5 \quad$ for $n \geq 2$

29. $a_1 = 1, a_2 = -2, a_3 = 3$, and $a_n = a_{n-1} + a_{n-2} + a_{n-3} \quad$ for $n \geq 4$

30. $a_1 = 1, a_2 = 3$, and $a_n = 2a_{n-1} + 3a_{n-2} \quad$ for $n \geq 3$

31. $a_0 = 2, a_1 = 3$, and $a_n = (a_{n-1})\left(\frac{1}{2}a_{n-2}\right) \quad$ for $n \geq 2$

32. $a_0 = 1, a_1 = 1$, and $a_n = na_{n-1} \quad$ for $n \geq 2$

33. a_n is the nth digit in the decimal expansion of π.

34. a_n is the nth digit in the decimal expansion of $1/13$.

In Exercises 35–38, find the third and the sixth partial sums of the sequence.

35. $\{n^2 - 5n + 2\}$ **36.** $\{(2n - 3n^2)^2\}$

37. $\{(-1)^{n+1}5\}$ **38.** $\{2^n(2 - n^2)\}_{n\geq 0}$

In Exercises 39–42, express the given sum in Σ notation.

39. $\frac{1}{3} + \frac{1}{5} + \frac{1}{7} + \frac{1}{9} + \frac{1}{11} + \frac{1}{13}$

40. $2 + 1 + \frac{4}{5} + \frac{5}{7} + \frac{2}{3} + \frac{7}{11} + \frac{8}{13}$

41. $\frac{1}{8} - \frac{2}{9} + \frac{3}{10} - \frac{4}{11} + \frac{5}{12}$

42. $\frac{2}{3\cdot 5} + \frac{4}{5\cdot 7} + \frac{8}{7\cdot 9} + \frac{16}{9\cdot 11} + \frac{32}{11\cdot 13} + \frac{64}{13\cdot 15} + \frac{128}{15\cdot 17}$

In Exercises 43–48, use a calculator to approximate the required term or sum.

43. a_{12} where $a_n = \left(1 + \frac{1}{n}\right)^n$

44. a_{50} where $a_n = \frac{\ln n}{n^2}$

45. a_{102} where $a_n = \frac{n^3 - n^2 + 5n}{3n^2 + 2n - 1}$

46. a_{125} where $a_n = \sqrt[n]{n}$

47. $\sum_{k=1}^{14} \frac{1}{k^2}$ **48.** $\sum_{n=8}^{22} \frac{1}{n}$

Thinkers

Exercises 49–53 deal with prime numbers. A positive integer greater than 1 is prime *if its only positive integer factors are itself and 1. For example, 7 is prime because its only factors are 7 and 1, but 15 is not prime because it has factors other than 15 and 1 (namely, 3 and 5).*

49. (a) Let $\{a_n\}$ be the sequence of prime integers in their usual ordering. Verify that the first ten terms are 2, 3, 5, 7, 11, 13, 17, 19, 23, 29.

(b) Find $a_{17}, a_{18}, a_{19}, a_{20}$.

In Exercises 50–53, find the first five terms of the sequence.

50. a_n is the nth prime integer larger than 10. [*Hint:* $a_1 = 11$.]

51. a_n is the square of the nth prime integer.

52. a_n is the number of prime integers less than n.

53. a_n is the largest prime integer less than $5n$.

Exercises 54–60 deal with the Fibonacci sequence $\{a_n\}$ that was discussed in Example 5.

54. Leonardo Fibonacci discovered the sequence in the 13th century in connection with this problem: A rabbit colony begins with one pair of adult rabbits (one male, one female). Each adult pair produces one pair of babies (one male, one female) every month. Each pair of baby rabbits becomes adult and produces the first offspring at age two months. Assuming that no rabbits die, how many adult pairs of rabbits are in the colony at the end of n months ($n = 1, 2, 3. . . .$)? [*Hint:* It may be helpful to make up a chart listing for each month the number of adult pairs, the number of one-month-old pairs, and the number of baby pairs.]

55. (a) List the first ten terms of the Fibonacci sequence.

(b) List the first ten partial sums of the sequence.

(c) Do the partial sums follow an identifiable pattern?

56. Verify that every positive integer less than or equal to 15 can be written as a sum of Fibonacci numbers, with none used more than once.

57. Verify that $5(a_n)^2 + 4(-1)^n$ is always a perfect square for $n = 1, 2, . . . , 10$.

58. Verify that $(a_n)^2 = a_{n+1}a_{n-1} + (-1)^{n-1}$ for $n = 2, . . . , 10$.

59. Show that $\sum_{n=1}^{k} a_n = a_{k+2} - 1$. [*Hint:* $a_1 = a_3 - a_2$; $a_2 = a_4 - a_3$; etc.]

60. Show that $\sum_{n=1}^{k} a_{2n-1} = a_{2k}$, that is, the sum of the first k odd-numbered terms is the kth even-numbered term. [*Hint:* $a_3 = a_4 - a_2$; $a_5 = a_6 - a_4$; etc.]

12.2 ARITHMETIC SEQUENCES

In this section and the next we consider two types of sequences that arise frequently. Both are easy to deal with because there are simple formulas for their nth terms and various partial sums.

An **arithmetic sequence** (sometimes called an **arithmetic progression**) is a sequence in which the difference between each term and the preceding one is always the same constant.

EXAMPLE 1 In the sequence 3, 8, 13, 18, 23, 28, . . . the difference between each term and the preceding one is always 5. So this is an arithmetic sequence. ■

EXAMPLE 2 In the sequence 14, 10, 6, 2, -2, -6, -10, -14, . . . the difference between each term and the preceding one is -4 (for example, $10 - 14 = -4$ and $-6 - (-2) = -4$). Hence, this sequence is arithmetic. ■

If $\{a_n\}$ is an arithmetic sequence, then for each $n \geq 2$, the term preceding a_n is a_{n-1} and the difference $a_n - a_{n-1}$ is some constant—call it d. Therefore, $a_n - a_{n-1} = d$, or equivalently,

Arithmetic Sequences ▶

In an arithmetic sequence $\{a_n\}$

$$a_n = a_{n-1} + d$$

for some constant d and all $n \geq 2$.

The number d is called the **common difference** of the arithmetic sequence.

EXAMPLE 3 If $\{a_n\}$ is an arithmetic sequence with $a_1 = 3$ and $a_2 = 4.5$, then the common difference is $d = a_2 - a_1 = 4.5 - 3 = 1.5$. So the sequence begins 3, 4.5, 6, 7.5, 9, 10.5, 12, 13.5, ... ■

EXAMPLE 4 The sequence $\{-7 + 4n\}$ is an arithmetic sequence because for each $n \geq 2$,

$$\begin{aligned} a_n - a_{n-1} &= (-7 + 4n) - (-7 + 4(n-1)) \\ &= (-7 + 4n) - (-7 + 4n - 4) = 4. \end{aligned}$$

Therefore, the common difference is $d = 4$. ■

If $\{a_n\}$ is an arithmetic sequence with common difference d, then for each $n \geq 2$ we know that $a_n = a_{n-1} + d$. Applying this fact repeatedly shows that

$$\begin{aligned} a_2 &= a_1 + d \\ a_3 &= a_2 + d = (a_1 + d) + d = a_1 + 2d \\ a_4 &= a_3 + d = (a_1 + 2d) + d = a_1 + 3d \\ a_5 &= a_4 + d = (a_1 + 3d) + d = a_1 + 4d \end{aligned}$$

and in general,

nth Term of an Arithmetic Sequence ▶

In an arithmetic sequence $\{a_n\}$ with common difference d

$$a_n = a_1 + (n-1)d \qquad \text{for every } n \geq 1.$$

EXAMPLE 5 Find the nth term of the arithmetic sequence with first term -5 and common difference 3.

Solution Since $a_1 = -5$ and $d = 3$, the formula in the box shows that

$$a_n = a_1 + (n-1)d = -5 + (n-1)3 = 3n - 8. \quad ■$$

EXAMPLE 6 What is the 45th term of the arithmetic sequence whose first three terms are 5, 9, 13?

Solution The first three terms show that $a_1 = 5$ and that the common difference d is 4. Applying the formula in the box with $n = 45$, we have

$$a_{45} = a_1 + (45 - 1)d = 5 + (44)4 = 181. \quad ■$$

EXAMPLE 7 If $\{a_n\}$ is an arithmetic sequence with $a_6 = 57$ and $a_{10} = 93$, find a_1 and a formula for a_n.

Solution Apply the formula $a_n = a_1 + (n-1)d$ with $n = 6$ and $n = 10$:

$$\begin{aligned} a_6 &= a_1 + (6-1)d &\quad \text{and} \quad a_{10} &= a_1 + (10-1)d \\ 57 &= a_1 + 5d & 93 &= a_1 + 9d \end{aligned}$$

We can find a_1 and d by solving this system:

$$a_1 + 9d = 93$$
$$a_1 + 5d = 57.$$

Subtracting the second equation from the first shows that $4d = 36$, and hence $d = 9$. Substituting $d = 9$ in the second equation shows that $a_1 = 12$. So the formula for a_n is

$$a_n = a_1 + (n - 1)d = 12 + (n - 1)9 = 9n + 3. \blacksquare$$

Partial Sums

It's easy to compute partial sums of arithmetic sequences by using the following formulas.

Partial Sums of an Arithmetic Sequence ▶

If $\{a_n\}$ is an arithmetic sequence with common difference d, then for each positive integer k the kth partial sum can be found by using *either* of these formulas:

1. $\sum_{n=1}^{k} a_n = \frac{k}{2}(a_1 + a_k)$ **or**
2. $\sum_{n=1}^{k} a_n = ka_1 + \frac{k(k-1)}{2}d$

Proof Let S denote the kth partial sum $a_1 + a_2 + \cdots + a_k$. For reasons that will become apparent later we shall calculate the number $2S$:

$$2S = S + S = (a_1 + a_2 + \cdots + a_k) + (a_1 + a_2 + \cdots + a_k).$$

Now we rearrange the terms on the right by grouping the first and last terms together, then the first and last of the remaining terms, and so on:

$$2S = (a_1 + a_k) + (a_2 + a_{k-1}) + (a_3 + a_{k-2}) + \cdots + (a_k + a_1).$$

Since adjacent terms of the sequence differ by d we have:

$$a_2 + a_{k-1} = (a_1 + d) + (a_k - d) = a_1 + a_k.$$

Using this fact,

$$a_3 + a_{k-2} = (a_2 + d) + (a_{k-1} - d) = a_2 + a_{k-1} = a_1 + a_k.$$

Continuing in this manner we see that every pair in the sum for $2S$ is equal to $a_1 + a_k$. Therefore,

$$\begin{aligned} 2S &= (a_1 + a_k) + (a_2 + a_{k-1}) + (a_3 + a_{k-2}) + \cdots + (a_k + a_1) \\ &= (a_1 + a_k) + (a_1 + a_k) + (a_1 + a_k) + \cdots + (a_1 + a_k) \qquad (k \text{ terms}) \\ &= k(a_1 + a_k). \end{aligned}$$

Dividing both sides of this last equation by 2 shows that $S = \frac{k}{2}(a_1 + a_k)$. This proves the first formula. To obtain the second one, note that

$$a_1 + a_k = a_1 + (a_1 + (k-1)d) = 2a_1 + (k-1)d.$$

Substituting the right side of this equation in the first formula for S shows that

$$S = \frac{k}{2}(a_1 + a_k) = \frac{k}{2}(2a_1 + (k-1)d) = ka_1 + \frac{k(k-1)}{2}d.$$

This proves the second formula. ❑

EXAMPLE 8 To find the 12th partial sum of the arithmetic sequence that begins $-8, -3, 2, 7, \ldots$ we first note that the common difference d is 5. Since $a_1 = -8$ and $d = 5$, the second formula in the box with $k = 12$ shows that

$$\sum_{n=1}^{12} a_n = 12(-8) + \frac{12(11)}{2}5 = -96 + 330 = 234. \quad \blacksquare$$

EXAMPLE 9 The sum of all multiples of 3 from 3 to 333 can be found by noting that it is just a partial sum of the arithmetic sequence 3, 6, 9, 12, ... Since this sequence can be written in the form

$$3 \cdot 1, 3 \cdot 2, 3 \cdot 3, 3 \cdot 4, 3 \cdot 5, 3 \cdot 6, \ldots$$

we see that $333 = 3 \cdot 111$ is the 111th term. The 111th partial sum of this sequence can be found by using the first formula in the box with $k = 111$, $a_1 = 3$, and $a_{111} = 333$:

$$\sum_{n=1}^{111} a_n = \frac{111}{2}(3 + 333) = \frac{111}{2}(336) = 18{,}648. \quad \blacksquare$$

EXAMPLE 10 If the starting salary for a job is \$20,000 and you get a \$2000 raise at the beginning of each subsequent year, what will your salary be during the tenth year? How much will you earn during the first ten years?

Solution Your yearly salary rates form a sequence: 20000, 22000, 24000, 26000, and so on. It is an arithmetic sequence with $a_1 = 20{,}000$ and $d = 2000$. Your tenth-year salary is

$$a_{10} = a_1 + (10-1)d = 20{,}000 + 9 \cdot 2000 = \$38{,}000.$$

Your ten-year total earnings are the 10th partial sum of the sequence:

$$\frac{10}{2}(a_1 + a_{10}) + \frac{10}{2}(20{,}000 + 38{,}000) = 5(58{,}000) = \$290{,}000. \quad \blacksquare$$

EXERCISES 12.2

In Exercises 1–6, the first term a_1 and the common difference d of an arithmetic sequence are given. Find the fifth term and the formula for the nth term.

1. $a_1 = 5, d = 2$ **2.** $a_1 = -4, d = 5$

3. $a_1 = 4, d = \frac{1}{4}$ **4.** $a_1 = -6, d = \frac{2}{3}$

5. $a_1 = 10, d = -\frac{1}{2}$ **6.** $a_1 = \pi, d = \frac{1}{5}$

In Exercises 7–12, find the kth partial sum of the arithmetic sequence $\{a_n\}$ with common difference d.

7. $k = 6, a_1 = 2, d = 5$

8. $k = 8, a_1 = \frac{2}{3}, d = -\frac{4}{3}$

9. $k = 7, a_1 = \frac{3}{4}, d = -\frac{1}{2}$

10. $k = 9, a_1 = 6, a_9 = -24$

11. $k = 6, a_1 = -4, a_6 = 14$

12. $k = 10, a_1 = 0, a_{10} = 30$

In Exercises 13–18, show that the sequence is arithmetic and find its common difference.

13. $\{3 - 2n\}$ **14.** $\left\{4 + \frac{n}{3}\right\}$

15. $\left\{\frac{5 + 3n}{2}\right\}$ **16.** $\left\{\frac{\pi - n}{2}\right\}$

17. $\{c + 2n\}$ (c constant)

18. $\{2b + 3nc\}$ (b, c constants)

In Exercises 19–24, use the given information about the arithmetic sequence with common difference d to find a_5 and a formula for a_n.

19. $a_4 = 12, d = 2$ **20.** $a_7 = -8, d = 3$

21. $a_2 = 4, a_6 = 32$ **22.** $a_7 = 6, a_{12} = -4$

23. $a_5 = 0, a_9 = 6$ **24.** $a_5 = -3, a_9 = -18$

In Exercises 25–28, find the sum.

25. $\sum_{n=1}^{20} (3n + 4)$ **26.** $\sum_{n=1}^{25} \left(\frac{n}{4} + 5\right)$

27. $\sum_{n=1}^{40} \frac{n + 3}{6}$ **28.** $\sum_{n=1}^{30} \frac{4 - 6n}{3}$

29. Find the sum of all the even integers from 2 to 100.

30. Find the sum of all the integer multiples of 7 from 7 to 700.

31. Find the sum of the first 200 positive integers.

32. Find the sum of the positive integers from 101 to 200 (inclusive). [*Hint:* What's the sum from 1 to 100? Use it and Exercise 31.]

33. A business makes a $10,000 profit during its first year. If the yearly profit increases by $7500 in each subsequent year, what will the profit be in the tenth year and what will the total profit for the first ten years be?

34. If a man's starting salary is $15,000 and he receives a $1000 increase every six months, what will his salary be during the last six months of the sixth year? How much will he earn during the first six years?

35. A lecture hall has 6 seats in the first row, 8 in the second, 10 in the third, and so on, through row 12. Rows 12 through 20 (the last row) all have the same number of seats. Find the number of seats in the lecture hall.

36. A monument is constructed by laying a row of 60 bricks at ground level. A second row, with 2 fewer bricks, is centered on that; a third row, with 2 fewer bricks, is centered on the second; and so on. The top row contains 10 bricks. How many bricks are there in the monument?

37. A ladder with nine rungs is to be built, with the bottom rung 24 inches wide and the top rung 18 inches wide. If the lengths of the rungs decrease uniformly from bottom to top, how long should each of the seven intermediate rungs be?

38. Find the first eight numbers in an arithmetic sequence in which the sum of the first and seventh term is 40 and the product of the first and fourth terms is 160.

12.3 GEOMETRIC SEQUENCES

A **geometric sequence** (sometimes called a **geometric progression**) is a sequence in which the quotient of each term and the preceding one is the same constant r. This constant r is called the **common ratio** of the geometric sequence.

EXAMPLE 1 The sequence 3, 9, 27, . . . , 3^n, . . . is geometric with common ratio 3. For instance, $a_2/a_1 = 9/3 = 3$ and $a_3/a_2 = 27/9 = 3$. If 3^n is any term ($n \geq 2$), then the preceding term is 3^{n-1} and

$$\frac{3^n}{3^{n-1}} = \frac{3 \cdot 3^{n-1}}{3^{n-1}} = 3. \quad ■$$

EXAMPLE 2 The sequence $\{5/2^n\}$ which begins 5/2, 5/4, 5/8, 5/16, . . . is geometric with common ratio $r = 1/2$ because for each $n \geq 1$

$$\frac{5/2^n}{5/2^{n-1}} = \frac{5}{2^n} \cdot \frac{2^{n-1}}{5} = \frac{2^{n-1}}{2^n} = \frac{2^{n-1}}{2^{n-1} \cdot 2} = \frac{1}{2}. \quad ■$$

If $\{a_n\}$ is a geometric sequence with common ratio r, then for each $n \geq 2$ the term preceding a_n is a_{n-1} and

$$a_n/a_{n-1} = r, \qquad \text{or equivalently,} \qquad a_n = ra_{n-1}.$$

Applying this last formula for $n = 2, 3, 4, \ldots$ we have

$$\begin{aligned} a_2 &= ra_1 \\ a_3 &= ra_2 = r(ra_1) = r^2a_1 \\ a_4 &= ra_3 = r(r^2a_1) = r^3a_1 \\ a_5 &= ra_4 = r(r^3a_1) = r^4a_1 \end{aligned}$$

and in general

nth Term of a Geometric Sequence ▶

If $\{a_n\}$ is a geometric sequence with common ratio r, then for all $n \geq 1$,

$$a_n = r^{n-1}a_1.$$

EXAMPLE 3 To find a formula for the nth term of the geometric sequence $\{a_n\}$ where $a_1 = 7$ and $r = 2$, we use the equation in the box:

$$a_n = r^{n-1}a_1 = 2^{n-1} \cdot 7.$$

So the sequence is $\{7 \cdot 2^{n-1}\}$. ■

EXAMPLE 4 If the first two terms of a geometric sequence are 2 and $-2/5$, then the common ratio must be

$$r = \frac{a_2}{a_1} = \frac{-2/5}{2} = \frac{-2}{5} \cdot \frac{1}{2} = -\frac{1}{5}.$$

Using the equation in the box, we now see that the formula for the nth term is

$$a_n = r^{n-1}a_1 = \left(-\frac{1}{5}\right)^{n-1}(2) = \frac{(1)^{n-1}}{(-5)^{n-1}}(2) = \frac{2}{(-5)^{n-1}}.$$

So, the sequence begins $2, -2/5, 2/5^2, -2/5^3, 2/5^4, \ldots$. ■

EXAMPLE 5 If $\{a_n\}$ is a geometric sequence with $a_2 = 20/9$ and $a_5 = 160/243$, then by the equation in the box above,

$$\frac{160/243}{20/9} = \frac{a_5}{a_2} = \frac{r^4 a_1}{r a_1} = r^3.$$

Consequently,

$$r = \sqrt[3]{\frac{160/243}{20/9}} = \sqrt[3]{\frac{160}{243} \cdot \frac{9}{20}} = \sqrt[3]{\frac{8 \cdot 9}{243}} = \sqrt[3]{\frac{8}{27}} = \frac{2}{3}.$$

Since $a_2 = ra_1$ we see that

$$a_1 = \frac{a_2}{r} = \frac{20/9}{2/3} = \frac{20}{9} \cdot \frac{3}{2} = \frac{10}{3}.$$

Therefore,

$$a_n = r^{n-1}a_1 = \left(\frac{2}{3}\right)^{n-1} \cdot \frac{10}{3} = \frac{2^{n-1} \cdot 2 \cdot 5}{3^{n-1} \cdot 3} = \frac{2^n \cdot 5}{3^n} = 5\left(\frac{2}{3}\right)^n.$$ ■

Partial Sums

If the common ratio r of a geometric sequence is the number 1, then we have

$$a_n = 1^{n-1}a_1 \qquad \text{for every } n \geq 1.$$

Therefore, the sequence is just the constant sequence $a_1, a_1, a_1, \ldots$. For any positive integer k, the kth partial sum of this constant sequence is

$$\underbrace{a_1 + a_1 + \cdots + a_1}_{k \text{ terms}} = ka_1.$$

In other words, the kth partial sum of a constant sequence is just k times the constant. If a geometric sequence is not constant (that is, $r \neq 1$), then its partial sums are given by the following formula.

Partial Sums of a Geometric Sequence ▶

The kth partial sum of the geometric sequence $\{a_n\}$ with common ratio $r \neq 1$ is

$$\sum_{n=1}^{k} a_n = a_1\left(\frac{1-r^k}{1-r}\right).$$

Proof If S denotes the kth partial sum, then the formula for the nth term of a geometric sequence shows that

$$S = a_1 + a_2 + \cdots + a_k = a_1 + a_1r + a_1r^2 + a_1r^3 + \cdots + a_1r^{k-1}.$$

Use this equation to compute $S - rS$:

$$\begin{aligned} S &= a_1 + a_1r + a_1r^2 + a_1r^3 + \cdots + a_1r^{k-1} \\ rS &= \qquad\; a_1r + a_1r^2 + a_1r^3 + \cdots + a_1r^{k-1} + a_1r^k \\ \hline S - rS &= a_1 \qquad\qquad\qquad\qquad\qquad\qquad\qquad\quad - a_1r^k \\ (1-r)S &= a_1(1-r^k) \end{aligned}$$

Since $r \neq 1$, we can divide both sides of this last equation by $1 - r$ to complete the proof:

$$S = \frac{a_1(1-r^k)}{1-r} = a_1\left(\frac{1-r^k}{1-r}\right). \quad \square$$

EXAMPLE 6 To find the sum

$$-\frac{3}{2} + \frac{3}{4} - \frac{3}{8} + \frac{3}{16} - \frac{3}{32} + \frac{3}{64} - \frac{3}{128} + \frac{3}{256} - \frac{3}{512}$$

we note that this is the 9th partial sum of the geometric sequence $\left\{3\left(\frac{-1}{2}\right)^n\right\}$.

The common ratio is $r = -1/2$. The formula in the box shows that

$$\begin{aligned} \sum_{n=1}^{9} 3\left(\frac{-1}{2}\right)^n &= a_1\left(\frac{1-r^9}{1-r}\right) = \left(\frac{-3}{2}\right)\left(\frac{1-(-1/2)^9}{1-(-1/2)}\right) \\ &= \left(\frac{-3}{2}\right)\left(\frac{1+1/2^9}{3/2}\right) = \left(\frac{-3}{2}\right)\left(\frac{2}{3}\right)\left(1+\frac{1}{2^9}\right) \\ &= -1 - \frac{1}{2^9} = -1 - \frac{1}{512} = -\frac{513}{512}. \quad ■ \end{aligned}$$

EXAMPLE 7 A superball is dropped from a height of 9 ft. It hits the ground and bounces to a height of 6 ft. It continues to bounce up and down. On each bounce it rises to 2/3 of the height of the previous bounce. How far has the ball traveled (both up and down) when it hits the ground for the seventh time?

Solution We first consider how far the ball travels on each bounce. On the first bounce it rises 6 ft and falls 6 ft for a total of 12 ft. On the second bounce it rises and falls 2/3 of the previous height, and hence travels 2/3 of 12 ft. If a_n denotes the distance traveled on the nth bounce, then

$$a_1 = 12 \qquad a_2 = \left(\frac{2}{3}\right) a_1 \qquad a_3 = \left(\frac{2}{3}\right) a_2 = \left(\frac{2}{3}\right)^2 a_1$$

and in general

$$a_n = \left(\frac{2}{3}\right) a_{n-1} = \left(\frac{2}{3}\right)^{n-1} a_1.$$

So $\{a_n\}$ is a geometric sequence with common ratio $r = 2/3$. When the ball hits the ground for the seventh time it has completed six bounces. Therefore, the total distance it has traveled is the distance it was originally dropped (9 ft) plus the distance traveled in six bounces, namely,

$$\begin{aligned} 9 + a_1 + a_2 + a_3 + a_4 + a_5 + a_6 &= 9 + \sum_{n=1}^{6} a_n = 9 + a_1\left(\frac{1 - r^6}{1 - r}\right) \\ &= 9 + 12\left(\frac{1 - (2/3)^6}{1 - (2/3)}\right) \\ &= 9 + 36(1 - (2/3)^6) \approx 41.84 \text{ ft.} \end{aligned}$$

EXERCISES 12.3

In Exercises 1–8, determine whether the sequence is arithmetic, geometric, or neither.

1. 2, 7, 12, 17, 22, . . .

2. 2, 6, 18, 54, 162, . . .

3. 13, 13/2, 13/4, 13/8, . . .

4. $-1, -\frac{1}{2}, 0, \frac{1}{2}, \ldots$

5. 50, 48, 46, 44, . . .

6. 2, −3, 9/2, −27/4, −81/8, . . .

7. 3, −3/2, 3/4, −3/8, 3/16, . . .

8. −6, −3.7, −1.4, 9, 3.2, . . .

In Exercises 9–14, the first term a_1 and the common ratio r of a geometric sequence are given. Find the sixth term and a formula for the nth term.

9. $a_1 = 5, r = 2$

10. $a_1 = 1, r = -2$

11. $a_1 = 4, r = \frac{1}{4}$

12. $a_1 = -6, r = \frac{2}{3}$

13. $a_1 = 10, r = -\frac{1}{2}$

14. $a_1 = \pi, r = \frac{1}{5}$

In Exercises 15–18, find the kth partial sum of the geometric sequence $\{a_n\}$ with common ratio r.

15. $k = 6, a_1 = 5, r = \frac{1}{2}$

16. $k = 8, a_1 = 9, r = \frac{1}{3}$

17. $k = 7, a_2 = 6, r = 2$

18. $k = 9, a_2 = 6, r = \frac{1}{4}$

In Exercises 19–22, show that the given sequence is geometric and find the common ratio.

19. $\left\{\left(-\frac{1}{2}\right)^n\right\}$

20. $\{2^{3n}\}$

21. $\{5^{n+2}\}$

22. $\{3^{n/2}\}$

In Exercises 23–28, use the given information about the geometric sequence $\{a_n\}$ to find a_5 and a formula for a_n.

23. $a_1 = 256, a_2 = -64$ **24.** $a_1 = 1/6, a_2 = -1/18$

25. $a_2 = 4, a_5 = 1/16$ **26.** $a_3 = 4, a_6 = -32$

27. $a_4 = -4/5, r = 2/5$ **28.** $a_2 = 6, a_7 = 192$

In Exercises 29–34, find the sum.

29. $\sum_{n=1}^{7} 2^n$ **30.** $\sum_{k=1}^{6} 3\left(\frac{1}{2}\right)^k$

31. $\sum_{n=1}^{9} \left(-\frac{1}{3}\right)^n$ **32.** $\sum_{n=1}^{5} 5 \cdot 3^{n-1}$

33. $\sum_{j=1}^{6} 4\left(\frac{3}{2}\right)^{j-1}$ **34.** $\sum_{t=1}^{8} 6(.9)^{t-1}$

Thinkers

35. Suppose $\{a_n\}$ is a geometric sequence with common ratio $r > 0$ and each $a_n > 0$. Show that the sequence $\{\log a_n\}$ is an arithmetic sequence with common difference $\log r$.

36. Suppose $\{a_n\}$ is an arithmetic sequence with common difference d. Let C be any positive number. Show that the sequence $\{C^{a_n}\}$ is a geometric sequence with common ratio C^d.

37. A ball is dropped from a height of 8 ft. On each bounce it rises to half its previous height. When the ball hits the ground for the seventh time, how far has it traveled?

38. A ball is dropped from a height of 10 ft. On each bounce it rises to 45% of its previous height. When it hits the ground for the tenth time, how far has it traveled?

39. In the geometric sequence 1, 2, 4, 8, 16, . . . , show that each term is 1 plus the sum of all preceding terms.

40. In the geometric sequence 2, 6, 18, 54, . . . , show that each term is twice the sum of 1 and all preceding terms.

41. If you are paid a salary of 1¢ on the first day of March, 2¢ on the second day, and your salary continues to double each day, how much will you earn in the month of March?

42. Starting with your parents, how many ancestors do you have for the past ten generations?

43. A car that sold for $8000 depreciates in value 25% each year. What is it worth after five years?

44. A vacuum pump removes 60% of the air in a container at each stroke. What percentage of the original amount of air remains after six strokes?

45. The minimum monthly payment for a certain bank credit card is the larger of 1/25 of the outstanding balance or $5. If the balance is less than $5, the entire balance is due. If you make only the minimum payment each month, how long will it take to pay off a balance of $200 (excluding any interest that might be due)?

12.3.A *Excursion* INFINITE SERIES

We now introduce a topic that is closely related to infinite sequences and has some very useful applications. We can only give a few highlights here; complete coverage requires calculus.

Consider the sequence $\{3/10^n\}$ and let S_k denote its kth partial sum; then

$$S_1 = \frac{3}{10}$$

$$S_2 = \frac{3}{10} + \frac{3}{10^2} = \frac{33}{100}$$

$$S_3 = \frac{3}{10} + \frac{3}{10^2} + \frac{3}{10^3} = \frac{333}{1000}$$

$$S_4 = \frac{3}{10} + \frac{3}{10^2} + \frac{3}{10^3} + \frac{3}{10^4} = \frac{3333}{10{,}000}.$$

These partial sums $S_1, S_2, S_3, S_4, \ldots$ themselves form a sequence:

$$\frac{3}{10}, \frac{33}{100}, \frac{333}{1000}, \frac{3333}{10{,}000}, \ldots.$$

The terms in the sequence of partial sums appear to be getting closer and closer to 1/3. In other words, as k gets larger and larger, the corresponding partial sum S_k gets closer and closer to 1/3. Consequently, we write

$$\frac{3}{10} + \frac{3}{10^2} + \frac{3}{10^3} + \frac{3}{10^4} + \cdots = \frac{1}{3}$$

and say that 1/3 is the *sum* of the *infinite series*

$$\frac{3}{10} + \frac{3}{10^2} + \frac{3}{10^3} + \frac{3}{10^4} + \cdots.$$

In the general case, an **infinite series** (or simply **series**) is defined to be an expression of the form

$$a_1 + a_2 + a_3 + a_4 + a_5 + \cdots$$

in which each a_n is a real number. This series is also denoted by the symbol $\sum_{n=1}^{\infty} a_n$.

EXAMPLE 1

(a) $\sum_{n=1}^{\infty} 2(.6)^n$ denotes the series

$$2(.6) + 2(.6)^2 + 2(.6)^3 + 2(.6)^4 + \cdots.$$

(b) $\sum_{n=1}^{\infty} \left(\frac{-1}{2}\right)^n$ denotes the series

$$-\frac{1}{2} + \left(\frac{-1}{2}\right)^2 + \left(\frac{-1}{2}\right)^3 + \left(\frac{-1}{2}\right)^4 + \cdots$$

$$= -\frac{1}{2} + \frac{1}{4} - \frac{1}{8} + \frac{1}{16} + \cdots. \quad ■$$

The **partial sums** of the series $a_1 + a_2 + a_3 + a_4 + \cdots$ are

$$S_1 = a_1$$

$$S_2 = a_1 + a_2$$

$$S_3 = a_1 + a_2 + a_3$$

and in general, for any $k \geq 1$

$$S_k = a_1 + a_2 + a_3 + a_4 + \cdots + a_k.$$

If it happens that the terms $S_1, S_2, S_3, S_4, \ldots$ of the *sequence* of partial sums get closer and closer to a particular real number S in such a way the partial sum S_k is arbitrarily close to S when k is large enough, then we say that the series **converges** and that S is the **sum of the convergent series.** For example, we just saw that the series

$$\frac{3}{10} + \frac{3}{10^2} + \frac{3}{10^3} + \frac{3}{10^4} + \cdots$$

converges and that its sum is 1/3.

A sequence is a *list* of numbers $a_1, a_2, a_3, \ldots$. Intuitively, you can think of a convergent series $a_1 + a_2 + a_3 + \cdots$ as an "infinite sum" of numbers. But be careful: Not every series has a sum. For instance, the partial sums of the series

$$1 + 2 + 3 + 4 + \cdots$$

get larger and larger (compute some) and do not get closer and closer to a single real number. So this series is not convergent.

EXAMPLE 2 Although no proof will be given here, it is intuitively clear that every infinite decimal may be thought of as the sum of a convergent series. For instance,

$$\pi = 3.1415926 \cdots = 3 + .1 + .04 + .001 + .0005 + .00009 + \cdots.$$

Note that the third partial sum is $3 + .1 + .04 = 3.14$, which is π to 2 decimal places. Similarly, the kth partial sum of this series is just π to $k - 1$ decimal places. ■

Infinite Geometric Series

If $\{a_n\}$ is a geometric sequence with common ratio r, then the corresponding infinite series

$$a_1 + a_2 + a_3 + a_4 + a_5 + \cdots$$

is called an **infinite geometric series.** By using the formula for the nth term of a geometric sequence, we can also express the corresponding geometric series in the form

$$a_1 + ra_1 + r^2a_1 + r^3a_1 + r^4a_1 + \cdots.$$

Under certain circumstances, an infinite geometric series is convergent and has a sum:

Sum of an Infinite Geometric Series ▶

If $|r| < 1$, then the infinite geometric series

$$a_1 + ra_1 + r^2a_1 + r^3a_1 + r^4a_1 + \cdots$$

converges and its sum is

$$\frac{a_1}{1 - r}.$$

Although we cannot prove this fact rigorously here, we can make it highly plausible both geometrically and algebraically.

EXAMPLE 3 $\displaystyle\sum_{n=1}^{\infty} \frac{8}{5^n} = \frac{8}{5} + \frac{8}{5^2} + \frac{8}{5^3} + \cdots$ is an infinite geometric series with $a_1 = 8/5$ and $r = 1/5$. The kth partial sum of this series is the same as the kth partial sum of the sequence $\{8/5^n\}$ and hence from the box on page 713 we know that

$$S_k = a_1\left(\frac{1 - r^k}{1 - r}\right) = \frac{8}{5}\left(\frac{1 - \left(\frac{1}{5}\right)^k}{1 - \frac{1}{5}}\right)$$

$$= \frac{8}{5}\left(\frac{1 - \frac{1}{5^k}}{\frac{4}{5}}\right) = \frac{8}{5}\cdot\frac{5}{4}\left(1 - \frac{1}{5^k}\right)$$

$$= 2\left(1 - \frac{1}{5^k}\right) = 2 - \frac{2}{5^k}.$$

The function $f(x) = 2 - 2/5^x$ is defined for all real numbers. When $x = k$ is a positive integer, then $f(k)$ is S_k, the kth partial sum of the series. Using a calculator we obtain the graph of $f(x)$ in Figure 12–1.

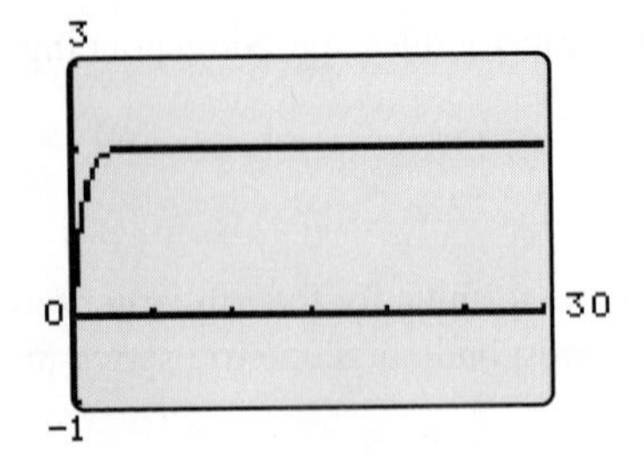

Figure 12-1

GRAPHING EXPLORATION Graph $f(x)$ in the same viewing window as in Figure 12–1. Use the trace feature to move the cursor along the graph. As x gets larger, what is the apparent value of $f(x)$ (that is, the value of the partial sum)?

Your calculator will probably tell you that every partial sum is 2, once you move beyond approximately $x = 15$. Actually, the partial sums are slightly smaller than 2, but are rounded to 2 by the calculator. In any case, the horizontal line through 2 is a horizontal asymptote of the graph (meaning that the graph gets very close to the line as x gets larger), so it is very plausible that the sequence converges to the number 2. But 2 is exactly what the box above says the sum should be:

$$\frac{a_1}{1-r} = \frac{\frac{8}{5}}{1-\frac{1}{5}} = \frac{\frac{8}{5}}{\frac{4}{5}} = \frac{8}{4} = 2. \quad \blacksquare$$

Example 3 is typical of the general case, as can be seen algebraically. Consider the geometric series $a_1 + a_2 + a_3 + \cdots$ with common ratio r such that $|r| < 1$. The kth partial sum S_k is the same as the kth partial sum of the geometric sequence $\{a_n\}$ and hence

$$S_k = a_1\left(\frac{1-r^k}{1-r}\right).$$

As k gets larger and larger, the number r^k gets very close to 0 because $|r| < 1$ (for instance, $(-.6)^{20} \approx .0000366$ and $.2^9 \approx .000000512$). Consequently, when k is very large, $1 - r^k$ is very close to $1 - 0$ so that

$$S_k = a_1\left(\frac{1-r^k}{1-r}\right) \quad \text{is very close to} \quad a_1\left(\frac{1-0}{1-r}\right) = \frac{a_1}{1-r}.$$

EXAMPLE 4 $\sum_{n=1}^{\infty}\left(\frac{-1}{2}\right)^n = -\frac{1}{2} + \frac{1}{4} - \frac{1}{8} + \frac{1}{16} + \cdots$ is an infinite geometric series with $a_1 = -1/2$ and $r = -1/2$. Since $|r| < 1$, this series converges and its sum is

$$\frac{a_1}{1-r} = \frac{-\frac{1}{2}}{1-\left(-\frac{1}{2}\right)} = \frac{-\frac{1}{2}}{\frac{3}{2}} = -\frac{1}{3}. \quad \blacksquare$$

Infinite geometric series provide another way of writing an infinite repeating decimal as a rational number.

EXAMPLE 5 To express $6.8573573573\cdots$ as a rational number, we first write it as $6.8 + .0573573573\cdots$. Consider $.0573573573\cdots$ as an infinite series:

$$.0573 + .0000573 + .0000000573 + .0000000000573 + \cdots$$

which is the same as

$$.0573 + (.001)(.0573) + (.001)^2(.0573) + (.001)^3(.0573) + \cdots.$$

This is a convergent geometric series with $a_1 = .0573$ and $r = .001$. Its sum is

$$\frac{a_1}{1-r} = \frac{.0573}{1-.001} = \frac{.0573}{.999} = \frac{573}{9990}.$$

Therefore,

$$\begin{aligned} 6.8573573573\ldots &= 6.8 + [.0573 + .0000573 + \cdots] \\ &= 6.8 + \frac{573}{9990} \\ &= \frac{68}{10} + \frac{573}{9990} \\ &= \frac{68{,}505}{9990} = \frac{4567}{666}. \end{aligned}$$ ■

EXERCISES 12.3.A

In Exercises 1–8, find the sum of the infinite series, if it has one.

1. $\sum_{n=1}^{\infty} \frac{1}{2^n}$ **2.** $\sum_{n=1}^{\infty} \left(-\frac{3}{4}\right)^n$ **3.** $\sum_{n=1}^{\infty} (.06)^n$

4. $1 - .5 + .25 - .125 + .0625 - \cdots$

5. $500 + 200 + 80 + 32 + \cdots$

6. $9 - 3\sqrt{3} + 3 - \sqrt{3} + 1 - \frac{1}{\sqrt{3}} + \cdots$

7. $2 + \sqrt{2} + 1 + \frac{1}{\sqrt{2}} + \frac{1}{2} + \cdots$

8. $\sum_{n=1}^{\infty} \left(\frac{1}{2^n} - \frac{1}{3^n}\right)$

In Exercises 9–15, express the repeating decimal as a rational number.

9. $.22222\cdots$ **10.** $.37373737\cdots$

11. $5.4272727\cdots$ **12.** $85.131313\cdots$

13. $2.1425425425\cdots$ **14.** $3.7165165165\cdots$

15. $1.74241241241\cdots$

16. If $\{a_n\}$ is an arithmetic sequence with common difference $d > 0$ and each $a_i > 0$, explain why the infinite series $a_1 + a_2 + a_3 + a_4 + \cdots$ is not convergent.

17. **(a)** Verify the $\sum_{n=1}^{\infty} 2(1.5)^n$ is a geometric series with $a_1 = 3$ and $r = 1.5$.

(b) Find the kth partial sum of the series and use this expression to define a function f, as in Example 3.

(c) Graph the function f in a viewing window with $0 \le x \le 30$. As x gets very large, what happens to the corresponding value of $f(x)$? Does the graph get closer and closer to some horizontal line, as in Example 3? What does this say about the convergence of the series?

18. Use the graphical approach illustrated in Example 3 to find the sum of the series in Example 4. Does the graph get very close to the horizontal line through $-1/3$? What's going on?

12.4 THE BINOMIAL THEOREM

The Binomial Theorem provides a formula for calculating the product $(x + y)^n$ for any positive integer n. Before we state the theorem, some preliminaries are needed.

Let n be a positive integer. The symbol $n!$ (read ***n* factorial**) denotes the product of all the integers from 1 to n. For example,

$$2! = 1\cdot 2 = 2, \qquad 3! = 1\cdot 2\cdot 3 = 6, \qquad 4! = 1\cdot 2\cdot 3\cdot 4 = 24,$$

$$5! = 1\cdot 2\cdot 3\cdot 4\cdot 5 = 120,$$

$$10! = 1\cdot 2\cdot 3\cdot 4\cdot 5\cdot 6\cdot 7\cdot 8\cdot 9\cdot 10 = 3{,}628{,}800$$

and in general,

***n* Factorial** ▶

$$n! = 1\cdot 2\cdot 3\cdot 4\cdots(n-2)(n-1)n.$$

We *define* 0! to be the number 1.

Learn to use your calculator to compute factorials. You will find ! in the MATH menu or its PROB (or PRB) submenu.

CALCULATOR EXPLORATION 15! is such a large number your calculator will switch to scientific notation to express it. What is this approximation? Many calculators cannot compute factorials larger than 69! If yours does compute larger ones, how large a one can you compute without getting an error message [or on HP-38, getting the number 9.9999 · · · E499]?

If r and n are integers with $0 \le r \le n$, then

Binomial Coefficients ▶

Either of the symbols $\binom{n}{r}$ or ${}_nC_r$ denotes the number $\dfrac{n!}{r!(n-r)!}$.
$\binom{n}{r}$ is called a *binomial coefficient.*

For example,

$${}_5C_3 = \binom{5}{3} = \frac{5!}{3!(5-3)!} = \frac{5}{3!2!} = \frac{1\cdot 2\cdot 3\cdot 4\cdot 5}{(1\cdot 2\cdot 3)(1\cdot 2)} = \frac{4\cdot 5}{2} = 10$$

$${}_4C_2 = \binom{4}{2} = \frac{4!}{2!(4-2)!} = \frac{4!}{2!2!} = \frac{1\cdot 2\cdot 3\cdot 4}{(1\cdot 2)(1\cdot 2)} = \frac{3\cdot 4}{2} = 6.$$

Binomial coefficients can be computed on a calculator by using *nCr* or Comb in the MATH menu or its PROB (or PRB) submenu.

> **CALCULATOR EXPLORATION** Compute ${}_{56}C_{47} = \binom{56}{47}$. Although calculators cannot compute 475!, they can compute many binomial coefficients, such as $\binom{475}{400}$, because most of the factors cancel out (as in the example above). Check yours. Will it also compute $\binom{475}{50}$?

The preceding examples illustrate a fact whose proof will be omitted: *Every binomial coefficient is an integer.* Furthermore, for every nonnegative integer n,

$$\binom{n}{0} = 1 \qquad \text{and} \qquad \binom{n}{n} = 1$$

because

$$\binom{n}{0} = \frac{n!}{0!(n-0)!} = \frac{n!}{0!n!} = \frac{n!}{n!} = 1 \qquad \text{and}$$

$$\binom{n}{n} = \frac{n!}{n!(n-n)!} = \frac{n!}{n!0!} = \frac{n!}{n!} = 1.$$

If we list the binomial coefficients for each value of n in this manner:

$$\begin{array}{lccccccccc}
n = 0 & & & & & \binom{0}{0} & & & & \\
n = 1 & & & & \binom{1}{0} & & \binom{1}{1} & & & \\
n = 2 & & & \binom{2}{0} & & \binom{2}{1} & & \binom{2}{2} & & \\
n = 3 & & \binom{3}{0} & & \binom{3}{1} & & \binom{3}{2} & & \binom{3}{3} & \\
n = 4 & \binom{4}{0} & & \binom{4}{1} & & \binom{4}{2} & & \binom{4}{3} & & \binom{4}{4} \\
\vdots & \iddots & & & & & & & & \ddots
\end{array}$$

and then calculate each of them, we obtain the following array of numbers:

row 0					1				
row 1				1		1			
row 2			1		2		1		
row 3		1		3		3		1	
row 4	1		4		6		4		1
$\vdots$	$\iddots$								$\ddots$

This array is called **Pascal's triangle.** Its pattern is easy to remember: Each entry (except the 1's at the beginning or end of a row) is the sum of the two closest entries in the row above it. In the fourth row, for instance, 6 is the sum of the two 3's above it, and each 4 is the sum of the 1 and 3 above it. See Exercise 47 for a proof.

In order to develop a formula for calculating $(x + y)^n$, we first calculate these products for small values of n to see if we can find some kind of pattern:

$$
\begin{array}{lll}
n = 0 & (x + y)^0 = & 1 \\
n = 1 & (x + y)^1 = & x + y \\
n = 2 & (x + y)^2 = & x^2 + 2xy + y^2 \\
n = 3 & (x + y)^3 = & x^3 + 3x^2y + 3xy^2 + y^3 \\
n = 4 & (x + y)^4 = & x^4 + 4x^3y + 6x^2y^2 + 4xy^3 + y^4
\end{array}
$$

Some parts of the pattern are already clear. For each positive n, the first term is x^n and the last term is y^n. Beginning with the second term,

The successive exponents of y are 1, 2, 3, . . . , n.

In each term before the last one, the exponent of x is 1 less than the preceding term. Suppose this pattern holds true for larger values of n as well. Then for a fixed n, the expansion of $(x + y)^n$ would have first term x^n. In the second term, the exponent of x would be 1 less than n, namely, $n - 1$, and the exponent of y would be 1. So the second term would be of the form (constant)$x^{n-1}y$. In the next term, the exponent of x would be 1 less again, namely, $n - 2$, and the exponent of y would be 2. Continuing in this fashion, we would have

$$(x + y)^n = x^n + (*)x^{n-1}y + (*)x^{n-2}y^2 + (*)x^{n-3}y^3 + \cdots + (*)xy^{n-1} + y^n$$

where the symbols $(*)$ indicate the various constant coefficients.

In order to determine the constant coefficients in the expansion of $(x + y)^n$, we return to the computations made above for $n = 0, 1, 2, 3, 4$. Each of the terms x, y, x^2, y^2, and so on, has coefficient 1. If we omit the x's and y's and just list the coefficients that appear in the computations above, we obtain this array of numbers:

$n = 0$					1				
$n = 1$				1		1			
$n = 2$			1		2		1		
$n = 3$		1		3		3		1	
$n = 4$	1		4		6		4		1

But this is just the top of Pascal's triangle. In the case $n = 4$, it means that the coefficients of the expansion of $(x + y)^4$ are just the binomial coefficients $\binom{4}{0}$, $\binom{4}{1}$, $\binom{4}{2}$, $\binom{4}{3}$, $\binom{4}{4}$; and similarly for the other small values of n. If this pattern holds true for larger n as well, then the coefficients of the expansion of $(x + y)^n$ are just the binomial coefficients

$$\binom{n}{0}, \binom{n}{1}, \binom{n}{2}, \binom{n}{3}, \ldots, \binom{n}{n-1}, \binom{n}{n}.$$

Since $\binom{n}{0} = 1$ and $\binom{n}{n} = 1$ for every n, the first and last coefficients on this list are 1. This is consistent with the fact that the first and last terms are x^n and y^n.

The preceding discussion suggests that the following result is true:

The Binomial Theorem ▶

For each positive integer n,

$$(x + y)^n = x^n + \binom{n}{1} x^{n-1}y + \binom{n}{2} x^{n-2}y^2 + \binom{n}{3} x^{n-3}y^3 + \cdots + \binom{n}{n-1} xy^{n-1} + y^n.$$

Using summation notation and the fact that $\binom{n}{0} = 1 = \binom{n}{n}$, we can write the Binomial Theorem compactly as

$$(x + y)^n = \sum_{j=0}^{n} \binom{n}{j} x^{n-j}y^j.$$

The Binomial Theorem will be proved in Section 12.5 by means of mathematical induction. We shall assume its truth for now and illustrate some of its uses.

EXAMPLE 1 In order to compute $(x+y)^8$ we apply the Binomial Theorem in the case $n=8$:

$$(x+y)^8 = x^8 + \binom{8}{1}x^7y + \binom{8}{2}x^6y^2 + \binom{8}{3}x^5y^3 + \binom{8}{4}x^4y^4 + \binom{8}{5}x^3y^5 + \binom{8}{6}x^2y^6 + \binom{8}{7}xy^7 + y^8.$$

Now verify that

$$\binom{8}{1} = \frac{8!}{1!7!} = 8, \qquad \binom{8}{2} = \frac{8!}{2!6!} = 28,$$

$$\binom{8}{3} = \frac{8!}{3!5!} = 56, \qquad \binom{8}{4} = \frac{8!}{4!4!} = 70.$$

Using these facts, we see that

$$\binom{8}{5} = \frac{8!}{5!3!} = \binom{8}{3} = 56, \qquad \binom{8}{6} = \frac{8!}{6!2!} = \binom{8}{2} = 28,$$

$$\binom{8}{7} = \frac{8!}{7!1!} = \binom{8}{1} = 8.$$

Substituting these values in the expansion above, we have

$$(x+y)^8 = x^8 + 8x^7y + 28x^6y^2 + 56x^5y^3 + 70x^4y^4 + 56x^3y^5 + 28x^2y^6 + 8xy^7 + y^8. \quad ■$$

EXAMPLE 2 To find $(1-z)^6$, we note that $1-z = 1+(-z)$ and apply the Binomial Theorem with $x=1$, $y=-z$, and $n=6$:

$$(1-z)^6 = 1^6 + \binom{6}{1}1^5(-z) + \binom{6}{2}1^4(-z)^2 + \binom{6}{3}1^3(-z)^3 + \binom{6}{4}1^2(-z)^4 + \binom{6}{5}1(-z)^5 + (-z)^6$$

$$= 1 - \binom{6}{1}z + \binom{6}{2}z^2 - \binom{6}{3}z^3 + \binom{6}{4}z^4 - \binom{6}{5}z^5 + z^6$$

$$= 1 - 6z + 15z^2 - 20z^3 + 15z^4 - 6z^5 + z^6. \quad ■$$

EXAMPLE 3 To expand $(x^2 + x^{-1})^4$, use the Binomial Theorem with x^2 in place of x and x^{-1} in place of y:

▶ TECHNOLOGY TIP

Binomial expansions, such as those in Examples 1–3, can be done on TI-92 by using EXPAND in the ALGEBRA menu.

$$(x^2 + x^{-1})^4 = (x^2)^4 + \binom{4}{1}(x^2)^3(x^{-1}) + \binom{4}{2}(x^2)^2(x^{-1})^2 + \binom{4}{3}(x^2)(x^{-1})^3 + (x^{-1})^4$$
$$= x^8 + 4x^6x^{-1} + 6x^4x^{-2} + 4x^2x^{-3} + x^{-4}$$
$$= x^8 + 4x^5 + 6x^2 + 4x^{-1} + x^{-4}. \quad ■$$

EXAMPLE 4 To show that $(1.001)^{1000} > 2$ without using a calculator we write 1.001 as $1 + .001$ and apply the Binomial Theorem with $x = 1$, $y = .001$, and $n = 1000$:

$$(1.001)^{1000} = (1 + .001)^{1000}$$
$$= 1^{1000} + \binom{1000}{1}1^{999}(.001) + \text{other positive terms}$$
$$= 1 + \binom{1000}{1}(.001) + \text{other positive terms.}$$

But $\binom{1000}{1} = \frac{1000!}{1!999!} = \frac{1000 \cdot 999!}{999!} = 1000$. Therefore, $\binom{1000}{1}(.001) = 1{,}000(.001) = 1$ and

$$(1.001)^{1000} = 1 + 1 + \text{other positive terms} = 2 + \text{other positive terms.}$$

Hence, $(1.001)^{1000} > 2$. ■

Sometimes we need to know only one term in the expansion of $(x + y)^n$. If you examine the expansion given by the Binomial Theorem, you will see that in the second term y has exponent 1, in the third term y has exponent 2, and so on. Thus,

Properties of the Binomial Expansion ▶

In the binomial expansion of $(x + y)^n$,

The exponent of y is always one less than the number of the term.

Furthermore, in each of the middle terms of the expansion,

The coefficient of the term containing y^r is $\binom{n}{r}$.

The sum of the x exponent and the y exponent is n.

For instance, the *ninth* term of the expansion of $(x + y)^{13}$, y has exponent 8, the coefficient is $\binom{13}{8}$, and x must have exponent 5 (since $8 + 5 = 13$). Thus, the ninth term is $\binom{13}{8} x^5y^8$.

EXAMPLE 5 What is the ninth term of the expansion of $\left(2x^2 + \frac{\sqrt[4]{y}}{\sqrt{6}}\right)^{13}$? We shall use the Binomial Theorem with $n = 13$ and with $2x^2$ in place of x and $\sqrt[4]{y}/\sqrt{6}$ in place of y. The remarks above show that the ninth term is

$$\binom{13}{8}(2x^2)^5\left(\frac{\sqrt[4]{y}}{\sqrt{6}}\right)^8.$$

Since $\sqrt[4]{y} = y^{1/4}$ and $\sqrt{6} = \sqrt{3}\sqrt{2} = 3^{1/2}2^{1/2}$, we can simplify as follows:

$$\begin{aligned}\binom{13}{8}(2x^2)^5\left(\frac{\sqrt[4]{y}}{\sqrt{6}}\right)^8 &= \binom{13}{8}2^5(x^2)^5\frac{(y^{1/4})^8}{(3^{1/2})^8(2^{1/2})^8} = \binom{13}{8}2^5x^{10}\frac{y^2}{3^4\cdot 2^4}\\ &= \binom{13}{8}\frac{2}{3^4}x^{10}y^2 = \frac{13\cdot 12\cdot 11\cdot 10\cdot 9}{5\cdot 4\cdot 3\cdot 2}\cdot\frac{2}{3^4}x^{10}y^2\\ &= \frac{286}{9}x^{10}y^2. \quad ■\end{aligned}$$

EXERCISES 12.4

In Exercises 1–10, evaluate the expression.

1. $6!$ **2.** $\frac{11!}{8!}$ **3.** $\frac{12!}{9!3!}$ **4.** $\frac{9! - 8!}{7!}$

5. $\binom{5}{3} + \binom{5}{2} - \binom{6}{3}$ **6.** $\binom{12}{11} - \binom{11}{10} + \binom{7}{0}$

7. $\binom{6}{0} + \binom{6}{1} + \binom{6}{2} + \binom{6}{3} + \binom{6}{4} + \binom{6}{5} + \binom{6}{6}$

8. $\binom{6}{0} - \binom{6}{1} + \binom{6}{2} - \binom{6}{3} + \binom{6}{4} - \binom{6}{5} + \binom{6}{6}$

9. $\binom{100}{96}$ **10.** $\binom{75}{72}$

In Exercises 11–16, expand the expression.

11. $(x + y)^5$ **12.** $(a + b)^7$ **13.** $(a - b)^5$

14. $(c - d)^8$ **15.** $(2x + y^2)^5$ **16.** $(3u - v^3)^6$

In Exercises 17–26, use the Binomial Theorem to expand and (where possible) simplify the expression.

17. $(\sqrt{x} + 1)^6$ **18.** $(2 - \sqrt{y})^5$

19. $(1 - c)^{10}$ **20.** $\left(\sqrt{c} + \frac{1}{\sqrt{c}}\right)^7$

21. $(x^{-3} + x)^4$ **22.** $(3x^{-2} - x^2)^6$

23. $(1 + \sqrt{3})^4 + (1 - \sqrt{3})^4$

24. $(\sqrt{3} + 1)^6 - (\sqrt{3} - 1)^6$

25. $(1 + i)^6$, where $i^2 = -1$

26. $(\sqrt{2} - i)^4$, where $i^2 = -1$

In Exercises 27–32, find the indicated term of the expansion of the given expression.

27. third, $(x + y)^5$ **28.** fourth, $(a + b)^6$

29. fifth, $(c - d)^7$ **30.** third, $(a + 2)^8$

31. fourth, $\left(u^{-2} + \frac{u}{2}\right)^7$ **32.** fifth, $(\sqrt{x} - \sqrt{2})^7$

33. Find the coefficient of x^5y^8 in the expansion of $(2x - y^2)^9$.

34. Find the coefficient of $x^{12}y^6$ in the expansion of $(x^3 - 3y)^{10}$.

35. Find the coefficient of $1/x^3$ in the expansion of $\left(2x + \frac{1}{x^2}\right)^6$.

36. Find the constant term in the expansion of $\left(y - \frac{1}{2y}\right)^{10}$.

37. **(a)** Verify that $\binom{9}{1} = 9$ and $\binom{9}{8} = 9$.

(b) Prove that for each positive integer n, $\binom{n}{1} = n$ and $\binom{n}{n-1} = n$. [*Note*: Part (a) is just the case when $n = 9$ and $n - 1 = 8$.]

38. **(a)** Verify that $\binom{7}{2} = \binom{7}{5}$.

(b) Let r and n be integers with $0 \le r \le n$. Prove that $\binom{n}{r} = \binom{n}{n-r}$. [*Note*: Part (a) is just the case when $n = 7$ and $r = 2$.]

39. Prove that for any positive integer n,

$$2^n = \binom{n}{0} + \binom{n}{1} + \binom{n}{2} + \cdots + \binom{n}{n}.$$

[*Hint*: $2 = 1 + 1$.]

40. Prove that for any positive integer n,

$$\binom{n}{0} - \binom{n}{1} + \binom{n}{2} - \binom{n}{3} + \binom{n}{4} - \cdots + (-1)^k\binom{n}{k} + \cdots + (-1)^n\binom{n}{n} = 0.$$

41. Use the Binomial Theorem with $x = \sin\theta$ and $y = \cos\theta$ to find $(\cos\theta + i\sin\theta)^4$ where $i^2 = -1$.

42. **(a)** Use DeMoivre's Theorem to find

$$(\cos\theta + i\sin\theta)^4.$$

(b) Use the fact that the two expressions obtained in part (a) and in Exercise 41 must be equal to express $\cos 4\theta$ and $\sin 4\theta$ in terms of $\sin\theta$ and $\cos\theta$.

43. **(a)** Let f be the function given by $f(x) = x^5$. Let h be a nonzero number and compute $f(x + h) - f(x)$ (but leave all binomial coefficients in the form $\binom{5}{r}$ here and below).

(b) Use part (a) to show that h is a factor of $f(x + h) - f(x)$ and find $\frac{f(x+h) - f(x)}{h}$.

(c) If h is *very* close to 0, find a simple approximation of the quantity $\frac{f(x+h) - f(x)}{h}$. (See part (b).)

44. Do Exercise 43 with $f(x) = x^8$ in place of $f(x) = x^5$.

45. Do Exercise 43 with $f(x) = x^{12}$ in place of $f(x) = x^5$.

46. Let n be a fixed positive integer. Do Exercise 43 with $f(x) = x^n$ in place of $f(x) = x^5$.

Thinkers

47. Let r and n be integers such that $0 \le r \le n$.

(a) Verify that $(n - r)! = (n - r)(n - (r + 1))!$

(b) Verify that $(n - r)! = ((n + 1) - (r + 1))!$

(c) Prove that $\binom{n}{r+1} + \binom{n}{r} = \binom{n+1}{r+1}$ for any $r \le n - 1$. [*Hint*: Write out the terms on the left side and use parts (a) and (b) to express each of them as a fraction with denominator $(r + 1)!(n - r)!$. Then add these two fractions, simplify the numerator, and compare the result with $\binom{n+1}{r+1}$.]

(d) Use part (c) to explain why each entry in Pascal's triangle (except the 1's at the beginning or end of a row) is the sum of the two closest entries in the row above it.

48. **(a)** Find these numbers and write them one *below* the next: $11^0, 11^1, 11^2, 11^3, 11^4$.

(b) Compare the list in part (a) with rows 0 to 4 of Pascal's triangle. What's the explanation?

(c) What can be said about 11^5 and row 5 of Pascal's triangle?

(d) Calculate all integer powers of 101 from 101^0 to 101^8, list the results one under the other, and compare the list with rows 0 to 8 of Pascal's triangle. What's the explanation? What happens with 101^9?

12.5 MATHEMATICAL INDUCTION

Mathematical induction is a method of proof that can be used to prove a wide variety of mathematical facts, including the Binomial Theorem, DeMoivre's Theorem, and statements such as:

The sum of the first n positive integers is the number $\frac{n(n+1)}{2}$.

$2^n > n$ for every positive integer n.

For each positive integer n, 4 is a factor of $7^n - 3^n$.

All of the statements above have a common property. For example, a statement such as

The sum of the first n positive integers is the number $\frac{n(n+1)}{2}$

or, in symbols,

$$1 + 2 + 3 + \cdots + n = \frac{n(n+1)}{2}$$

is really an infinite sequence of statements, one for each possible value of n:

$$n = 1: \qquad 1 = \frac{1(2)}{2}$$

$$n = 2: \qquad 1 + 2 = \frac{2(3)}{2}$$

$$n = 3: \qquad 1 + 2 + 3 = \frac{3(4)}{2}$$

and so on. Obviously, there isn't time enough to verify every one of the statements on this list, one at a time. But we can find a workable method of proof by examining how each statement on the list is *related* to the *next* statement on the list.

For instance, for $n = 50$, the statement is

$$1 + 2 + 3 + \cdots + 50 = \frac{50(51)}{2}.$$

At the moment, we don't know whether or not this statement is true. But just *suppose* that it were true. What could then be said about the next statement, the one for $n = 51$:

$$1 + 2 + 3 + \cdots + 50 + 51 = \frac{51(52)}{2}?$$

Well, *if* it is true that

$$1 + 2 + 3 + \cdots + 50 = \frac{50(51)}{2}$$

then adding 51 to both sides and simplifying the right side would yield these equalities:

$$1 + 2 + 3 + \cdots + 50 + 51 = \frac{50(51)}{2} + 51$$

$$1 + 2 + 3 + \cdots + 50 + 51 = \frac{50(51)}{2} + \frac{2(51)}{2} = \frac{50(51) + 2(51)}{2}$$

$$1 + 2 + 3 + \cdots + 50 + 51 = \frac{(50 + 2)51}{2}$$

$$1 + 2 + 3 + \cdots + 50 + 51 = \frac{51(52)}{2}.$$

Since this last equality is just the original statement for $n = 51$, we conclude that

If the statement is true for $n = 50$, *then* it is also true for $n = 51$.

We have *not* proved that the statement actually *is* true for $n = 50$, but only that *if* it is, then it is also true for $n = 51$.

We claim that this same conditional relationship holds for any two consecutive values of n. In other words, we claim that for any positive integer k,

① ***If* the statement is true for $n = k$, *then* it is also true for $n = k + 1$.**

The proof of this claim is the same argument used above (with k and $k + 1$ in place of 50 and 51): *If* it is true that

$$1 + 2 + 3 + \cdots + k = \frac{k(k + 1)}{2} \qquad [\textit{Original statement for } n = k]$$

then adding $k + 1$ to both sides and simplifying the right side produces these equalities:

$$1 + 2 + 3 + \cdots + k + (k + 1) = \frac{k(k + 1)}{2} + (k + 1)$$

$$1 + 2 + 3 + \cdots + k + (k + 1) = \frac{k(k + 1)}{2} + \frac{2(k + 1)}{2} = \frac{k(k + 1) + 2(k + 1)}{2}$$

$$1 + 2 + 3 + \cdots + k + (k + 1) = \frac{(k + 2)(k + 1)}{2}$$

$$1 + 2 + 3 + \cdots + k + (k + 1) = \frac{(k + 1)((k + 1) + 1)}{2}.$$

[*Original statement for* $n = k + 1$]

We have proved that claim ① is valid for each positive integer k. We have *not* proved that the original statement is true for any value of n, but only that *if* it is true for $n = k$, then it is also true for $n = k + 1$. Applying this fact when $k = 1, 2, 3, \ldots,$ we see that

② *If* the statement is true for $n = 1$, *then* it is also true for $n = 1 + 1 = 2$;
If the statement is true for $n = 2$, *then* it is also true for $n = 2 + 1 = 3$;
If the statement is true for $n = 3$, *then* it is also true for $n = 3 + 1 = 4$;
⋮
If the statement is true for $n = 50$, *then* it is also true for $n = 50 + 1 = 51$;
If the statement is true for $n = 51$, *then* it is also true for $n = 51 + 1 = 52$;
⋮

and so on.

We are finally in a position to *prove* the original statement: $1 + 2 + 3 + \cdots + n = n(n + 1)/2$. Obviously, it *is true* for $n = 1$ since $1 = 1(2)/2$. Now apply in turn each of the propositions on list ② above. Since the statement *is* true for $n = 1$, it must also be true for $n = 2$, and hence for $n = 3$, and hence for $n = 4$, and so on, for every value of n. Therefore, the original statement is true for *every* positive integer n.

The preceding proof is an illustration of the following principle:

Principle of Mathematical Induction ▶

Suppose there is given a statement involving the positive integer n and that:

(i) The statement is true for $n = 1$.

(ii) If the statement is true for $n = k$ (where k is any positive integer), then the statement is also true for $n = k + 1$.

Then the statement is true for every positive integer n.

Property (i) is simply a statement of fact. To verify that it holds, you must prove the given statement is true for $n = 1$. This is usually easy, as in the preceding example.

Property (ii) is a *conditional* property. It does not assert that the given statement *is* true for $n = k$, but only that *if* it is true for $n = k$, then it is also true for $n = k + 1$. So to verify that property (ii) holds, you need only prove this conditional proposition:

If the statement is true for $n = k$, *then* it is also true for $n = k + 1$.

In order to prove this, or any conditional proposition, you must proceed as in the example above: Assume the "if" part and use this assumption to prove the "then" part. As we saw above, the same argument will usually work for any possible k. Once this conditional proposition has been proved, you can use it *together* with property (i) to conclude that the given statement is necessarily true for every n, just as in the preceding example.

Thus proof by mathematical induction reduces to two steps:

STEP 1 Prove that the given statement is true for $n = 1$.

STEP 2 Let k be a positive integer. Assume that the given statement is true for $n = k$. Use this assumption to prove that the statement is true for $n = k + 1$.

Step 2 may be performed before step 1 if you wish. Step 2 is sometimes referred to as the **inductive step.** The assumption that the given statement is true for $n = k$ in this inductive step is called the **induction hypothesis.**

EXAMPLE 1 Prove that $2^n > n$ for every positive integer n.

Solution Here the statement involving n is $2^n > n$.

STEP 1 When $n = 1$, we have the statement $2^1 > 1$. This is obviously true.

STEP 2 Let k be any positive integer. We assume that the statement is true for $n = k$, that is, we assume that $2^k > k$. We shall use this assumption to prove that the statement is true for $n = k + 1$, that is, that $2^{k+1} > k + 1$.

We begin with the induction hypothesis:* $2^k > k$. Multiplying both sides of this inequality by 2 yields:

$$2 \cdot 2^k > 2k$$

$$2^{k+1} > 2k. \qquad (3)$$

Since k is a positive integer, we know that $k \geq 1$. Adding k to each side of the inequality $k \geq 1$, we have

$$k + k \geq k + 1$$

$$2k \geq k + 1.$$

Combining this result with inequality (3) above, we see that

$$2^{k+1} > 2k \geq k + 1.$$

The first and last terms of this inequality show that $2^{k+1} > k + 1$. Therefore, the statement is true for $n = k + 1$. This argument works for any positive integer k. Thus, we have completed the inductive step. By the Principle of Mathematical Induction, we conclude that $2^n > n$ for every positive integer n. ■

*This is the point at which you usually must do some work. Remember that what follows is the "finished proof." It does not include all the thought, scratch work, false starts, and so on, that were done before this proof was actually found.

EXAMPLE 2 Simple arithmetic shows that

$$7^2 - 3^2 = 49 - 9 = 40 = 4 \cdot 10 \quad \text{and} \quad 7^3 - 3^3 = 343 - 27 = 316 = 4 \cdot 79.$$

In each case, 4 is a factor. These examples suggest that

For each positive integer n, 4 is a factor of $7^n - 3^n$.

This conjecture can be proved by induction as follows.

STEP 1 When $n = 1$, the statement is "4 is a factor of $7^1 - 3^1$." Since $7^1 - 3^1 = 4 = 4 \cdot 1$, the statement is true for $n = 1$.

STEP 2 Let k be a positive integer and assume that the statement is true for $n = k$, that is, that 4 is a factor of $7^k - 3^k$. Let us denote the other factor by D, so that the induction hypothesis is: $7^k - 3^k = 4D$. We must use this assumption to prove that the statement is true for $n = k + 1$, that is, that 4 is a factor of $7^{k+1} - 3^{k+1}$. Here is the proof:

$$\begin{aligned}
7^{k+1} - 3^{k+1} &= 7^{k+1} - 7 \cdot 3^k + 7 \cdot 3^k - 3^{k+1} && [\textit{Since } -7 \cdot 3^k + 7 \cdot 3^k = 0] \\
&= 7(7^k - 3^k) + (7 - 3)3^k && [\textit{Factor}] \\
&= 7(4D) + (7 - 3)3^k && [\textit{Induction hypothesis}] \\
&= 7(4D) + 4 \cdot 3^k && [7 - 3 = 4] \\
&= 4(7D + 3^k). && [\textit{Factor out } 4]
\end{aligned}$$

From this last line, we see that 4 is a factor of $7^{k+1} - 3^{k+1}$. Thus, the statement is true for $n = k + 1$, and the inductive step is complete. Therefore, by the Principle of Mathematical Induction the conjecture is actually true for every positive integer n. ■

Another example of mathematical induction, the proof of the Binomial Theorem, is given at the end of this section.

Sometimes a statement involving the integer n may be false for $n = 1$ and (possibly) other small values of n, but true for all values of n beyond a particular number. For instance, the statement $2^n > n^2$ is false for $n = 1, 2, 3, 4$. But it is true for $n = 5$ and all larger values of n. A variation on the Principle of Mathematical Induction can be used to prove this fact and similar statements. See Exercise 28 for details.

A Common Mistake with Induction

It is sometimes tempting to omit step 2 of an inductive proof when the given statement can easily be verified for small values of n, especially if a clear pattern seems to be developing. As the next example shows, however, *omitting step 2 may lead to error.*

EXAMPLE 3 An integer (> 1) is said to be *prime* if its only positive integer factors are itself and 1. For instance, 11 is prime since its only positive integer factors are 11 and 1. But 15 is not prime because it has factors other than 15 and 1 (namely, 3 and 5). For each positive integer n, consider the number

$$f(n) = n^2 - n + 11.$$

You can readily verify that

$$f(1) = 11, \quad f(2) = 13, \quad f(3) = 17, \quad f(4) = 23, \quad f(5) = 31$$

and that *each of these numbers is prime.* Furthermore, there is a clear pattern: The first two numbers (11 and 13) differ by 2; the next two (13 and 17) differ by 4; the next two (17 and 23) differ by 6; and so on. On the basis of this evidence, we might conjecture:

For each positive integer n, the number $f(n) = n^2 - n + 11$ is prime.

We have seen that this conjecture is true for $n = 1, 2, 3, 4, 5$. Unfortunately, however, it is *false* for some values of n. For instance, when $n = 11$,

$$f(11) = 11^2 - 11 + 11 = 11^2 = 121.$$

But 121 is obviously *not* prime since it has a factor other than 121 and 1, namely, 11. You can verify that the statement is also false for $n = 12$ but true for $n = 13$. ■

In the preceding example, the proposition

If the statement is true for $n = k$, then it is true for $n = k + 1$

is false when $k = 10$ and $k + 1 = 11$. If you were not aware of this and tried to complete step 2 of an inductive proof, you would not have been able to find a valid proof for it. Of course, the fact that you can't find a proof of a proposition doesn't always mean that no proof exists. But when you are unable to complete step 2, you are warned that there is a possibility that the given statement may be false for some values of n. This warning should prevent you from drawing any wrong conclusions.

Proof of the Binomial Theorem

We shall use induction to prove that for every positive integer n,

$$(x + y)^n = x^n + \binom{n}{1} x^{n-1}y + \binom{n}{2} x^{n-2}y^2 + \binom{n}{3} x^{n-3}y^3 + \cdots + \binom{n}{n-1} xy^{n-1} + y^n.$$

This theorem was discussed and its notation explained in Section 12.4.

STEP 1 When $n = 1$, there are only two terms on the right side of the equation above, and the statement reads $(x + y)^1 = x^1 + y^1$. This is certainly true.

STEP 2 Let k be any positive integer and assume that the theorem is true for $n = k$, that is, that

$$(x + y)^k = x^k + \binom{k}{1} x^{k-1}y + \binom{k}{2} x^{k-2}y^2 + \cdots + \binom{k}{r} x^{k-r}y^r + \cdots + \binom{k}{k-1} xy^{k-1} + y^k.$$

[On the right side above, we have included a typical middle term $\binom{k}{r}x^{k-r}y^r$. The sum of the exponents is k, and the bottom part of the binomial coefficient is the same as the y exponent.] We shall use this assumption to prove that the theorem is true for $n = k + 1$, that is, that

$$(x + y)^{k+1} = x^{k+1} + \binom{k+1}{1} x^k y + \binom{k+1}{2} x^{k-1}y^2 + \cdots + \binom{k+1}{r+1} x^{k-r}y^{r+1} + \cdots + \binom{k+1}{k} xy^k + y^{k+1}.$$

We have simplified some of the terms on the right side; for instance, $(k + 1) - 1 = k$ and $(k + 1) - (r + 1) = k - r$. But this is the correct statement for $n = k + 1$: The coefficients of the middle terms are $\binom{k+1}{1}$, $\binom{k+1}{2}$, $\binom{k+1}{3}$, and so on; the sum of the exponents of each middle term is $k + 1$, and the bottom part of each binomial coefficient is the same as the y exponent.

In order to prove the theorem for $n = k + 1$, we shall need this fact about binomial coefficients: For any integers r and k with $0 \le r < k$,

$$④ \qquad \binom{k}{r+1} + \binom{k}{r} = \binom{k+1}{r+1}.$$

A proof of this fact is outlined in Exercise 47 on page 728.

To prove the theorem for $n = k + 1$, we first note that

$$(x + y)^{k+1} = (x + y)(x + y)^k.$$

Applying the induction hypothesis to $(x + y)^k$, we see that

$$(x+y)^{k+1} = (x+y)\left[x^k + \binom{k}{1}x^{k-1}y + \binom{k}{2}x^{k-2}y^2 + \cdots + \binom{k}{r}x^{k-r}y^r \right.$$
$$\left. + \binom{k}{r+1}x^{k-(r+1)}y^{r+1} + \cdots + \binom{k}{k-1}xy^{k-1} + y^k\right]$$
$$= x\left[x^k + \binom{k}{1}x^{k-1}y + \cdots + y^k\right] + y\left[x^k + \binom{k}{1}x^{k-1}y + \cdots + y^k\right].$$

Next we multiply out the right-hand side. Remember that multiplying by x increases the x exponent by 1 and multiplying by y increases the y exponent by 1.

$$(x+y)^{k+1} = \left[x^{k+1} + \binom{k}{1}x^k y + \binom{k}{2}x^{k-1}y^2 + \cdots + \binom{k}{r}x^{k-r+1}y^r \right.$$
$$\left. + \binom{k}{r+1}x^{k-r}y^{r+1} + \cdots + \binom{k}{k-1}x^2y^{k-1} + xy^k\right]$$
$$+ \left[x^k y + \binom{k}{1}x^{k-1}y^2 + \binom{k}{2}x^{k-2}y^3 + \cdots + \binom{k}{r}x^{k-r}y^{r+1} \right.$$
$$\left. + \binom{k}{r+1}x^{k-(r+1)}y^{r+2} + \cdots + \binom{k}{k-1}xy^k + y^{k+1}\right]$$
$$= x^{k+1} + \left[\binom{k}{1} + 1\right]x^k y + \left[\binom{k}{2} + \binom{k}{1}\right]x^{k-1}y^2 + \cdots$$
$$+ \left[\binom{k}{r+1} + \binom{k}{r}\right]x^{k-r}y^{r+1} + \cdots + \left[1 + \binom{k}{k-1}\right]xy^k + y^{k+1}.$$

Now apply statement ④ above to each of the coefficients of the middle terms. For instance, with $r = 1$, statement ④ shows that $\binom{k}{2} + \binom{k}{1} = \binom{k+1}{2}$. Similarly, with $r = 0$, $\binom{k}{1} + 1 = \binom{k}{1} + \binom{k}{0} = \binom{k+1}{1}$, and so on. Then the expression above for $(x+y)^{k+1}$ becomes

$$(x+y)^{k+1} = x^{k+1} + \binom{k+1}{1}x^k y + \binom{k+1}{2}x^{k-1}y^2 + \cdots$$
$$+ \binom{k+1}{r+1}x^{k-r}y^{r+1} + \cdots + \binom{k+1}{k}xy^k + y^{k+1}.$$

Since this last statement says the theorem is true for $n = k + 1$, the inductive step is complete. By the Principle of Mathematical Induction the theorem is true for every positive integer n.

EXERCISES 12.5

In Exercises 1–18, use mathematical induction to prove that each of the given statements is true for every positive integer n.

1. $1 + 2 + 2^2 + 2^3 + 2^4 + \cdots + 2^{n-1} = 2^n - 1$

2. $1 + 3 + 3^2 + 3^3 + 3^4 + \cdots + 3^{n-1} = \dfrac{3^n - 1}{2}$

3. $1 + 3 + 5 + 7 + \cdots + (2n - 1) = n^2$

4. $2 + 4 + 6 + 8 + \cdots + 2n = n^2 + n$

5. $1^2 + 2^2 + 3^2 + \cdots + n^2 = \dfrac{n(n + 1)(2n + 1)}{6}$

6. $\dfrac{1}{2} + \dfrac{1}{4} + \dfrac{1}{8} + \cdots + \dfrac{1}{2^n} = 1 - \dfrac{1}{2^n}$

7. $\dfrac{1}{1\cdot 2} + \dfrac{1}{2\cdot 3} + \dfrac{1}{3\cdot 4} + \cdots + \dfrac{1}{n(n + 1)} = \dfrac{n}{n + 1}$

8. $\left(1 + \dfrac{1}{1}\right)\left(1 + \dfrac{1}{2}\right)\left(1 + \dfrac{1}{3}\right)\cdots\left(1 + \dfrac{1}{n}\right) = n + 1$

9. $n + 2 > n$

10. $2n + 2 > n$

11. $3^n \geq 3n$

12. $3^n \geq 1 + 2n$

13. $3n > n + 1$

14. $\left(\dfrac{3}{2}\right)^n > n$

15. 3 is a factor of $2^{2n+1} + 1$

16. 5 is a factor of $2^{4n-2} + 1$

17. 64 is a factor of $3^{2n+2} - 8n - 9$

18. 64 is a factor of $9^n - 8n - 1$

19. Let c and d be fixed real numbers. Prove that

$$c + (c + d) + (c + 2d) + (c + 3d) + \cdots + (c + (n - 1)d) = \frac{n(2c + (n - 1)d)}{2}.$$

20. Let r be a fixed real number with $r \neq 1$. Prove that

$$1 + r + r^2 + r^3 + \cdots + r^{n-1} = \frac{r^n - 1}{r - 1}.$$

[*Remember*: $1 = r^0$; so when $n = 1$ the left side reduces to $r^0 = 1$.]

21. **(a)** Write *each* of $x^2 - y^2$, $x^3 - y^3$, and $x^4 - y^4$ as a product of $x - y$ and another factor.
(b) Make a conjecture as to how $x^n - y^n$ can be written as a product of $x - y$ and another factor. Use induction to prove your conjecture.

22. Let $x_1 = \sqrt{2}$; $x_2 = \sqrt{2 + \sqrt{2}}$; $x_3 = \sqrt{2 + \sqrt{2 + \sqrt{2}}}$; and so on. Prove that $x_n < 2$ for every positive integer n.

In Exercises 23–27, if the given statement is true, prove it. If it is false, give a counterexample.

23. Every odd positive integer is prime.

24. The number $n^2 + n + 17$ is prime for every positive integer n.

25. $(n + 1)^2 > n^2 + 1$ for every positive integer n.

26. 3 is a factor of the number $n^3 - n + 3$ for every positive integer n.

27. 4 is a factor of the number $n^4 - n + 4$ for every positive integer n.

28. Let q be a *fixed* integer. Suppose a statement involving the integer n has these two properties:

(i) The statement is true for $n = q$.
(ii) *If* the statement is true for $n = k$ (where k is any integer with $k \geq q$), then the statement is also true for $n = k + 1$.

Then we claim that the statement is true for every integer n greater than or equal to q.

(a) Give an informal explanation that shows why the claim above should be valid. Note that when $q = 1$, this claim is precisely the Principle of Mathematical Induction.
(b) The claim made above will be called the *Extended Principle of Mathematical Induction.* State the two steps necessary to use this principle to prove that a given statement is true for all $n \geq q$. (See discussion on page 731–732.)

In Exercises 29–34, use the Extended Principle of Mathematical Induction (Exercise 28) to prove the given statement.

29. $2n - 4 > n$ for every $n \geq 5$. (Use 5 for q here.)

30. Let r be a fixed real number with $r > 1$. Then $(1 + r)^n > 1 + nr$ for every integer $n \geq 2$. (Use 2 for q here.)

31. $n^2 > n$ for all $n \geq 2$

32. $2^n > n^2$ for all $n \geq 5$

33. $3^n > 2^n + 10n$ for all $n \geq 4$

34. $2n < n!$ for all $n \geq 4$

Thinkers

35. Let n be a positive integer. Suppose that there are three pegs and on one of them n rings are stacked, with each ring being smaller in diameter than the one below it (see Figure 12–2). We want to transfer the stack of rings to another peg according to these rules: (i) Only one ring may be moved at a time; (ii) a ring can be moved to any peg, provided it is never placed on top of a smaller ring; (iii) the final order of the rings on the new peg must be the same as the original order on the first peg.

(a) What is the smallest possible number of moves when $n = 2$? $n = 3$? $n = 4$?

(b) Make a conjecture as to the smallest possible number of moves required for any n. Prove your conjecture by induction.

36. The basic formula for compound interest $T(x) = P(1 + r)^x$ was discussed on page 326. Prove by induction that the formula is valid whenever x is a positive integer. [*Note*: P and r are assumed to be constant.]

37. Use induction to prove DeMoivre's Theorem: For any complex number $z = r(\cos\theta + i\sin\theta)$ and any positive integer n,

$$z^n = r^n(\cos(n\theta) + i\sin(n\theta)).$$

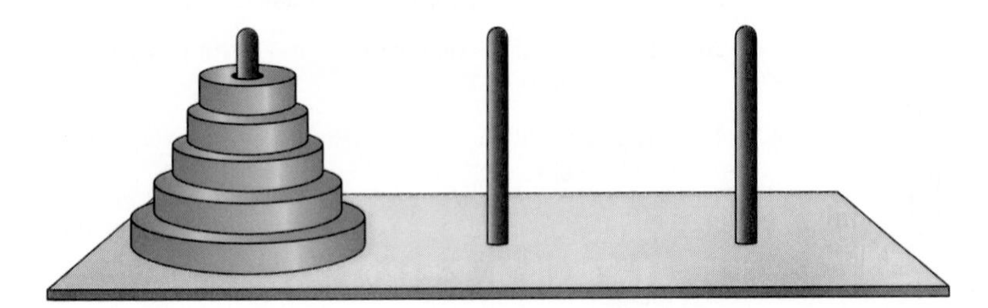

Figure 12-2

CHAPTER 12 *Review*

Important Concepts ▶

Important Facts and Formulas ▶

- In an arithmetic sequence $\{a_n\}$ with common difference d:

$$a_n = a_1 + (n-1)d \qquad \sum_{n=1}^{k} a_n = \frac{k}{2}(a_1 + a_k)$$

$$\sum_{n=1}^{k} a_n = ka_1 + \frac{k(k-1)}{2}d$$

- In a geometric sequence $\{a_n\}$ with common ratio $r \neq 1$:

$$a_n = r^{n-1}a_1 \qquad \sum_{n=1}^{k} a_n = a_1\left(\frac{1-r^k}{1-r}\right)$$

- $n! = 1 \cdot 2 \cdot 3 \cdots (n-2)(n-1)n$
- $\binom{n}{r} = \dfrac{n!}{r!(n-r)!} = {}_nC_r$
- *The Binomial Theorem:*

$$(x+y)^n = x^n + \binom{n}{1}x^{n-1}y + \binom{n}{2}x^{n-2}y^2 + \binom{n}{3}x^{n-3}y^3 + \cdots + \binom{n}{n-1}xy^{n-1} + y^n$$

$$= \sum_{j=0}^{n} \binom{n}{j} x^{n-j}y^j$$

Review Questions ▶

In Questions 1–4, find the first four terms of the sequence $\{a_n\}$.

1. $a_n = 2n - 5$ **2.** $a_n = 3^n - 27$

3. $a_n = \left(\dfrac{-1}{n}\right)^2$ **4.** $a_n = (-1)^{n+1}(n-1)$

5. Find the 5th partial sum of the sequence $\{a_n\}$, where $a_1 = -4$ and $a_n = 3a_{n-1} + 2$.

6. Find the 4th partial sum of the sequence $\{a_n\}$, where $a_1 = 1/9$ and $a_n = 3a_{n-1}$.

7. $\sum_{n=0}^{4} 2^n(n+1) = ?$ **8.** $\sum_{n=2}^{4} (3n^2 - n + 1) = ?$

In Questions 9–12, find a formula for a_n; assume that the sequence is arithmetic.

9. $a_1 = 3$ and the common difference is -6.

10. $a_2 = 4$ and the common difference is 3.

11. $a_1 = -5$ and $a_3 = 7$.

12. $a_3 = 2$ and $a_7 = -1$.

In Questions 13–16, find a formula for a_n; assume that the sequence is geometric.

13. $a_1 = 2$ and the common ratio is 3.

14. $a_1 = 5$ and the common ratio is $-1/2$.

15. $a_2 = 192$ and $a_7 = 6$.

16. $a_3 = 9/2$ and $a_6 = -243/16$.

17. Find the 11th partial sum of the arithmetic sequence with $a_1 = 5$ and common difference -2.

18. Find the 12th partial sum of the arithmetic sequence with $a_1 = -3$ and $a_{12} = 16$.

19. Find the 5th partial sum of the geometric sequence with $a_1 = 1/4$ and common ratio 3.

20. Find the 6th partial sum of the geometric sequence with $a_1 = 5$ and common ratio 1/2.

21. Find numbers b, c, d such that 4, b, c, d, 23 are the first five terms of an arithmetic sequence.

22. Find numbers c and d such that 8, c, d, 27 are the first four terms of a geometric sequence.

23. Is it better to be paid \$5 per day for 100 days or to be paid 5¢ the first day, 10¢ the second day, 20¢ the third day, and have your salary increase in this fashion every day for 100 days?

24. Tuition at Bigstate University is now \$3000 per year and will increase \$150 per year in succeeding years. If a student starts school now, spends four years as an undergraduate, three years in law school, and five years getting a PhD, how much tuition will she have paid?

Find the following sums, if they exist.

25. $\sum_{n=1}^{\infty} \frac{1}{2^{n-1}}$ 26. $\sum_{n=1}^{\infty} \left(\frac{-1}{4^n}\right)$

27. Use the Binomial Theorem to show that $(1.02)^{51} > 2.5$.

28. What is the coefficient of u^3v^2 in the expansion of $(u + 5v)^5$?

29. $\binom{15}{12} = ?$ 30. $\binom{18}{3} = ?$

31. Let n be a positive integer. Simplify $\binom{n+1}{n}$.

32. Use the Binomial Theorem to expand $(\sqrt{x} + 1)^5$. Simplify your answer.

33. $\frac{20!5!}{6!17!} = ?$ 34. $\frac{7! - 5!}{4!} = ?$

35. Find the coefficient of x^2y^4 in the expansion of $(2y + x^2)^5$.

36. Prove that for every positive integer n,

$$1^3 + 2^3 + 3^3 + \cdots + n^3 = \frac{n^2(n+1)^2}{4}.$$

37. Prove that for every positive integer n,

$$1 + 5 + 5^2 + 5^3 + \cdots + 5^{n-1} = \frac{5^n - 1}{4}.$$

38. Prove that $2^n \geq 2n$ for every positive integer n.

39. If x is a real number with $|x| < 1$, then prove that $|x^n| < 1$ for all $n \geq 1$.

40. Prove that for any positive integer n, $1 + 5 + 9 + \cdots + (4n - 3) = n(2n - 1)$.

41. Prove that for any positive integer n, $1 + 4 + 4^2 + 4^3 + \cdots + 4^{n-1} = \frac{1}{3}(4^n - 1)$.

42. Prove that $3n < n!$ for every $n \geq 4$.

43. Prove that for every positive integer n, 8 is a factor of $9^n - 8n - 1$.

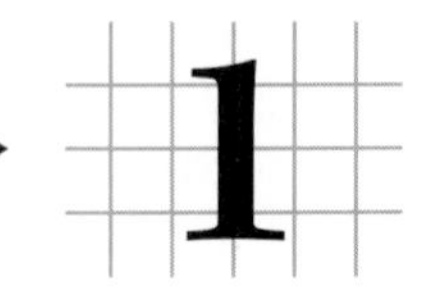

Algebra Review

This appendix reviews the fundamental algebraic facts that are used frequently in this book. You must be able to handle these algebraic manipulations in order to succeed in this course and in calculus.

1.A INTEGRAL EXPONENTS

Exponents provide a convenient shorthand for certain products. If c is a real number, then c^2 denotes cc and c^3 denotes ccc. More generally, for any positive integer n

c^n denotes the product $ccc \cdots c$ (n factors).

In this notation c^1 is just c, so we usually omit the exponent 1.

EXAMPLE 1 $3^4 = 3 \cdot 3 \cdot 3 \cdot 3 = 81$ and

$$(-2)^5 = (-2)(-2)(-2)(-2)(-2) = -32.$$

For every positive integer n, $0^n = 0 \cdots 0 = 0$. ■

EXAMPLE 2 To find $(2.4)^9$ use the [∧] (or [a^b] or [x^y]) key on your calculator:*

[2.4] [∧] [9] [ENTER]**

which produces the (approximate) answer 2641.80754. ■

Because exponents are just shorthand for multiplication, it is easy to determine the rules they obey. For instance,

*All calculator keystroke illustrations in this book are for calculators that use algebraic notation. Make the necessary changes if your calculator uses Reverse Polish Notation (RPN).

**The ENTER key is labeled EXE on Casio calculators.

$$c^3c^5 = (ccc)(ccccc) = c^8, \qquad \text{that is,} \qquad c^3c^5 = c^{3+5}.$$

$$\frac{c^7}{c^4} = \frac{ccccccc}{cccc} = \frac{\cancel{cccc}ccc}{\cancel{cccc}} = ccc = c^3, \qquad \text{that is,} \qquad \frac{c^7}{c^4} = c^{7-4}.$$

Similar arguments work in the general case:

To multiply c^m by c^n, add the exponents: $c^mc^n = c^{m+n}$.

To divide c^m by c^n, subtract the exponents: $c^m/c^n = c^{m-n}$.

EXAMPLE 3 $4^2 \cdot 4^7 = 4^{2+7} = 4^9$ and $2^8/2^3 = 2^{8-3} = 2^5$. ■

The notation c^n can be extended to the cases when n is zero or negative as follows:

If $c \neq 0$, then c^0 is defined to be the number 1.

If $c \neq 0$ and n is a positive integer, then

$$c^{-n} \text{ is defined to be the number } \frac{1}{c^n}.$$

Note that 0^0 and negative powers of 0 are *not* defined (negative powers of 0 would involve division by 0). The reason for choosing these definitions of c^{-n} for nonzero c is that the multiplication and division rules for exponents remain valid.* For instance,

$$c^5 \cdot c^0 = c^5 \cdot 1 = c^5, \qquad \text{so that} \qquad c^5c^0 = c^{5+0}.$$

$$c^7c^{-7} = c^7(1/c^7) = 1 = c^0, \qquad \text{so that} \qquad c^7c^{-7} = c^{7-7}.$$

EXAMPLE 4 $6^{-3} = 1/6^3 = 1/216$ and $(-2)^{-5} = 1/(-2)^5 = -1/32$. A calculator shows that $(.287)^{-12} \approx 3{,}201{,}969.857$.** ■

If c and d are nonzero real numbers and m and n are integers (positive, negative, or zero), then we have these

Exponent Laws ►

1. $c^mc^n = c^{m+n}$
2. $\dfrac{c^m}{c^n} = c^{m-n}$
3. $(c^m)^n = c^{mn}$
4. $(cd)^n = c^nd^n$
5. $\left(\dfrac{c}{d}\right)^n = \dfrac{c^n}{d^n}$
6. $\dfrac{1}{c^{-n}} = c^n$

*In mathematics you may define a concept any way you want, as long as it is consistent with previously established facts.

**$\approx$ means "approximately equal to."

EXAMPLE 5 Here are examples of each of the six exponent laws.

1. $\pi^{-5}\pi^{2} = \pi^{-5+2} = \pi^{-3} = 1/\pi^{3}$.
2. $x^{9}/x^{4} = x^{9-4} = x^{5}$.
3. $(5^{-3})^{2} = 5^{(-3)2} = 5^{-6}$.
4. $(2x)^{5} = 2^{5}x^{5} = 32x^{5}$.
5. $\left(\frac{7}{3}\right)^{10} = \frac{7^{10}}{3^{10}}$.
6. $1/x^{-5} = \frac{1}{\left(\frac{1}{x^{5}}\right)} = x^{5}$. ■

WARNING $(2x)^5$ is *not* the same as $2x^5$. Part 4 of Example 5 shows that $(2x)^5 = 32x^5$ and *not* $2x^5$.

The exponent laws can often be used to simplify complicated expressions.

EXAMPLE 6

(a) $(2x^{2}y^{3}z)^{4} = 2^{4}(x^{2})^{4}(y^{3})^{4}z^{4} = 16x^{8}y^{12}z^{4}$.

(first equality: *Law (4)*; second equality: *Law (3)*)

(b) $(r^{-3}s^{2})^{-2} = (r^{-3})^{-2}(s^{2})^{-2} = r^{6}s^{-4} = r^{6}/s^{4}$.

(first equality: *Law (4)*; second equality: *Law (3)*)

(c) $\frac{x^{5}(y^{2})^{3}}{(x^{2}y)^{2}} = \frac{x^{5}y^{6}}{(x^{2}y)^{2}} = \frac{x^{5}y^{6}}{(x^{2})^{2}y^{2}} = \frac{x^{5}y^{6}}{x^{4}y^{2}} = x^{5-4}y^{6-2} = xy^{4}$. ■

(*Law (3)*, *Law (4)*, *Law (3)*, *Law (2)*)

It is usually more efficient to use the exponent laws with the negative exponents rather than first converting to positive exponents. If positive exponents are required, the conversion can be made in the last step.

EXAMPLE 7 Simplify and express without negative exponents $\frac{a^{-2}(b^{2}c^{3})^{-2}}{(a^{-3}b^{-5})^{2}c}$.

Solution

$$\frac{a^{-2}(b^{2}c^{3})^{-2}}{(a^{-3}b^{-5})^{2}c} = \frac{a^{-2}(b^{2})^{-2}(c^{3})^{-2}}{(a^{-3})^{2}(b^{-5})^{2}c} = \frac{a^{-2}b^{-4}c^{-6}}{a^{-6}b^{-10}c}$$

(first equality: *Law (4)*; second equality: *Law (3)*)

$$= a^{-2-(-6)}b^{-4-(-10)}c^{-6-1} = a^{4}b^{6}c^{-7} = \frac{a^{4}b^{6}}{c^{7}}.$$ ■

(*Law (2)*)

Since $(-1)(-1) = +1$, any even power of -1, such as $(-1)^4$ or $(-1)^{12}$, will be equal to 1. Every odd power of -1 is equal to -1; for instance $(-1)^5 = (-1)^4(-1) = 1(-1) = -1$. Consequently, for every positive number c

$$(-c)^n = ((-1)c)^n = (-1)^n c^n = \begin{cases} c^n & \text{if } n \text{ is even} \\ -c^n & \text{if } n \text{ is odd} \end{cases}.$$

EXAMPLE 8 $(-3)^4 = 3^4 = 81$ and $(-5)^3 = -5^3 = -125$. ■

WARNING Be careful with negative bases. For instance, if you want to compute $(-12)^4$, which is a positive number, but you key in (−) 12 ∧ 4 ENTER the calculator will interpret this as $-(12^4)$ and produce a negative answer. To get the correct answer, you must key in the parentheses:

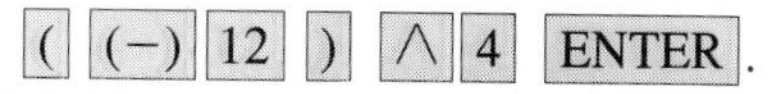

((−) 12) ∧ 4 ENTER.

EXERCISES 1.A

In Exercises 1–18, evaluate the expression.

1. $(-6)^2$
2. -6^2
3. $5 + 4(3^2 + 2^3)$
4. $(-3)2^2 + 4^2 - 1$
5. $\dfrac{(-3)^2 + (-2)^4}{-2^2 - 1}$
6. $\dfrac{(-4)^2 + 2}{(-4)^2 - 7} + 1$
7. $\left(\dfrac{-5}{4}\right)^3$
8. $-\left(\dfrac{7}{4} + \dfrac{3}{4}\right)^2$
9. $\left(\dfrac{1}{3}\right)^3 + \left(\dfrac{2}{3}\right)^3$
10. $\left(\dfrac{5}{7}\right)^2 + \left(\dfrac{2}{7}\right)^2$
11. $2^4 - 2^7$
12. $3^3 - 3^{-7}$
13. $(2^{-2} + 2)^2$
14. $(3^{-1} - 3^3)^2$
15. $2^2 \cdot 3^{-3} - 3^2 \cdot 2^{-3}$
16. $4^3 \cdot 5^{-2} + 4^2 \cdot 5^{-1}$
17. $\dfrac{1}{2^3} + \dfrac{1}{2^{-4}}$
18. $3^2\left(\dfrac{1}{3} + \dfrac{1}{3^{-2}}\right)$

In Exercises 19–38, simplify the expression. Each letter represents a nonzero real number and should appear at most once in your answer.

19. $x^2 \cdot x^3 \cdot x^5$
20. $y \cdot y^4 \cdot y^6$
21. $(.03)y^2 \cdot y^7$
22. $(1.3)z^3 \cdot z^5$
23. $(2x^2)^3 3x$
24. $(3y^3)^4 5y^2$
25. $(3x^2y)^2$
26. $(2xy^3)^3$
27. $(a^2)(7a)(-3a^3)$
28. $(b^3)(-b^2)(3b)$
29. $(2w)^3(3w)(4w)^2$
30. $(3d)^4(2d)^2(5d)$
31. $a^{-2}b^3a^3$
32. $c^4d^5c^{-3}$
33. $(2x)^{-2}(2y)^3(4x)$
34. $(3x)^{-3}(2y)^{-2}(2x)$
35. $(-3a^4)^2(9x^3)^{-1}$
36. $(2y^3)^3(3y^2)^{-2}$
37. $(2x^2y)^0(3xy)$
38. $(3x^2y^4)^0$

In Exercises 39–42, express the given number as a power of 2.

39. $(64)^2$
40. $(1/8)^3$
41. $(2^4 \cdot 16^{-2})^3$
42. $(1/2)^{-8}(1/4)^4(1/16)^{-3}$

In Exercises 43–60, simplify and write the given expression without negative exponents. All letters represent nonzero real numbers.

43. $\dfrac{x^4(x^2)^3}{x^3}$
44. $\left(\dfrac{z^2}{t^3}\right)^4 \cdot \left(\dfrac{z^3}{t}\right)^5$
45. $\left(\dfrac{e^6}{c^4}\right)^2 \cdot \left(\dfrac{c^3}{e}\right)^3$
46. $\left(\dfrac{x^7}{y^6}\right)^2 \cdot \left(\dfrac{y^2}{x}\right)^4$

47. $\left(\dfrac{ab^2c^3d^4}{abc^2d}\right)^2$

48. $\dfrac{(3x)^2(y^2)^3x^2}{(2xy^2)^3}$

49. $\left(\dfrac{a^6}{b^{-4}}\right)^2$

50. $\left(\dfrac{x^{-2}}{y^{-2}}\right)^2$

51. $\left(\dfrac{c^5}{d^{-3}}\right)^{-2}$

52. $\left(\dfrac{x^{-1}}{2y^{-1}}\right)\left(\dfrac{2y}{x}\right)^{-2}$

53. $\left(\dfrac{3x}{y^2}\right)^{-3}\left(\dfrac{-x}{2y^3}\right)^2$

54. $\left(\dfrac{5u^2v}{2uv^2}\right)^2\left(\dfrac{-3uv}{2u^2v}\right)^{-3}$

55. $\dfrac{(a^{-3}b^2c)^{-2}}{(ab^{-2}c^3)^{-1}}$

56. $\dfrac{(-2cd^2e^{-1})^3}{(5c^{-3}de)^{-2}}$

57. $(c^{-1}d^{-2})^{-3}$

58. $((x^2y^{-1})^2)^{-3}$

59. $a^2(a^{-1}+a^{-3})$

60. $\dfrac{a^{-2}}{b^{-2}}+\dfrac{b^2}{a^2}$

In Exercises 61–66, determine the sign of the given number without calculating the product.

61. $(-2.6)^3(-4.3)^{-2}$

62. $(4.1)^{-2}(2.5)^{-3}$

63. $(-1)^9(6.7)^5$

64. $(-4)^{12}6^9$

65. $(-3.1)^{-3}(4.6)^{-6}(7.2)^7$

66. $(45.8)^{-7}(-7.9)^{-9}(-8.5)^{-4}$

In Exercises 67–72, r, s, and t are positive integers and a, b, and c are nonzero real numbers. Simplify and write the given expression without negative exponents.

67. $\dfrac{3^{-r}}{3^{-s-r}}$

68. $\dfrac{4^{-(t+1)}}{4^{2-t}}$

69. $\left(\dfrac{a^6}{b^{-4}}\right)^t$

70. $\dfrac{c^{-t}}{(6b)^{-s}}$

71. $\dfrac{(c^{-r}b^s)^t}{(c^tb^{-s})^r}$

72. $\dfrac{(a^rb^{-s})^{-t}}{(b^tc^r)^{-s}}$

Errors to Avoid

In Exercises 73–80, give an example to show that the statement may be false *for some numbers.*

73. $a^r + b^r = (a+b)^r$

74. $a^ra^s = a^{rs}$

75. $a^rb^s = (ab)^{r+s}$

76. $c^{-r} = -c^r$

77. $\dfrac{c^r}{c^s} = c^{r/s}$

78. $(a+1)(b+1) = ab+1$

79. $(-a)^2 = -a^2$

80. $(-a)(-b) = -ab$

1.B ARITHMETIC OF ALGEBRAIC EXPRESSIONS

Expressions such as

$$b + 3c^2, \qquad 3x^2 - 5x + 4, \qquad \sqrt{x^3 + z}, \qquad \frac{x^3 + 4xy - \pi}{x^2 + xy}$$

are called **algebraic expressions.** Each expression represents a number that is obtained by performing various algebraic operations (such as addition or taking roots) on one or more numbers, some of which may be denoted by letters.

A letter that denotes a particular real number is called a **constant;** its value remains unchanged throughout the discussion. For example, the Greek letter π has long been used to denote the number $3.14159\cdots$. Sometimes a constant is a fixed but unspecified real number, as in "an angle of k degrees" or "a triangle with base of length b."

A letter that can represent *any* real number is called a **variable.** In the expression $2x + 5$, for example, the variable x can be any real number. If $x = 3$, then $2x + 5 = 2 \cdot 3 + 5 = 11$. If $x = \frac{1}{2}$, then $2x + 5 = 2 \cdot \frac{1}{2} + 5 = 6$, and so on.*

*We assume any conditions on the constants and variables necessary to guarantee that an algebraic expression does represent a real number. For instance, in $\sqrt{z}$ we assume $z \geq 0$ and in $1/c$ we assume $c \neq 0$.

Constants are usually denoted by letters near the beginning of the alphabet and variables by letters near the end of the alphabet. Consequently, in expressions such as $cx + d$ and $cy^2 + dy$, it is understood that c and d are constants and x and y are variables.

The usual rules of arithmetic are valid for algebraic expressions.

EXAMPLE 1 Use the distributive law to *combine like terms;* for instance,

$$3x + 5x + 4x = (3 + 5 + 4)x = 12x.$$

In practice, you do the middle part in your head and simply write $3x + 5x + 4x = 12x$. ■

EXAMPLE 2 In more complicated expressions, eliminate parentheses, use the commutative law to group like terms together, and then combine them.

$$\begin{aligned} (a^2b - 3\sqrt{c}) + (5ab + 7\sqrt{c}) + 7a^2b &= a^2b - 3\sqrt{c} + 5ab + 7\sqrt{c} + 7a^2b \\ \textit{Regroup:} \quad &= \underbrace{a^2b + 7a^2b}\underbrace{- 3\sqrt{c} + 7\sqrt{c}} + 5ab \\ \textit{Combine like terms:} \quad &= \quad 8a^2b \quad + \quad 4\sqrt{c} \quad + 5ab. \end{aligned}$$ ■

WARNING Be careful when parentheses are preceded by a minus sign: $-(b + 3) = -b - 3$ and *not* $-b + 3$. Here's the reason: $-(b + 3)$ means $(-1)(b + 3)$, so that by the distributive law,

$$-(b + 3) = (-1)(b + 3) = (-1)b + (-1)3 = -b - 3.$$

Similarly, $-(7 - y) = -7 - (-y) = -7 + y$.

The examples in the Warning Box illustrate the following.

Rules for Eliminating Parentheses ▶

Parentheses preceded by a plus sign (or no sign) may be deleted.

Parentheses preceded by a minus sign may be deleted *if* the sign of every term within the parentheses is changed.

The usual method of multiplying algebraic expressions is to use the distributive laws repeatedly, as shown in the following examples. The net result is to *multiply every term in the first sum by every term in the second sum.*

EXAMPLE 3 To compute $(y-2)(3y^2-7y+4)$, we first apply the distributive law, treating $(3y^2-7y+4)$ as a single number:

$$(y-2)(3y^2-7y+4) = y(3y^2-7y+4) - 2(3y^2-7y+4)$$

Distributive law: $= 3y^3 - 7y^2 + 4y - 6y^2 + 14y - 8$

Regroup: $= 3y^3 - \underbrace{7y^2 - 6y^2} + \underbrace{4y + 14y} - 8$

Combine like terms: $= 3y^3 - 13y^2 + 18y - 8.$ ■

EXAMPLE 4 We follow the same procedure with $(2x-5y)(3x+4y)$:

$$\begin{aligned}(2x-5y)(3x+4y) &= 2x(3x+4y) - 5y(3x+4y)\\ &= 2x\cdot 3x + 2x\cdot 4y + (-5y)\cdot 3x + (-5y)\cdot 4y\\ &= 6x^2 + \underbrace{8xy - 15xy} - 20y^2\\ &= 6x^2 - 7xy - 20y^2.\end{aligned}$$ ■

Observe the pattern in the second line of Example 4 and its relationship to the terms being multiplied:

$$(2x-5y)(3x+4y) = 2x\cdot 3x + 2x\cdot 4y + (-5y)\cdot 3x + (-5y)\cdot 4y$$

$(2x-5y)(3x+4y)$ — *First terms* ($2x\cdot 3x$)

$(2x-5y)(3x+4y)$ — *Outside terms* ($2x\cdot 4y$)

$(2x-5y)(3x+4y)$ — *Inside terms* ($(-5y)\cdot 3x$)

$(2x-5y)(3x+4y)$ — *Last terms* ($(-5y)\cdot 4y$)

This pattern is easy to remember by using the acronym FOIL (**F**irst, **O**utside, **I**nside, **L**ast). The FOIL method makes it easy to find products such as this one mentally, without the necessity of writing out the intermediate steps.

WARNING The FOIL method can be used only when multiplying two expressions that each have two terms.

EXAMPLE 5

$$(3x+2)(x+5) = 3x^2 + 15x + 2x + 10 = 3x^2 + 17x + 10.$$ ■

(*First*: $3x^2$, *Outside*: $15x$, *Inside*: $2x$, *Last*: 10)

EXERCISES 1.B

In Exercises 1–54, perform the indicated operations and simplify your answer.

1. $x + 7x$
2. $5w + 7w - 3w$
3. $6a^2b + (-8b)a^2$
4. $-6x^3\sqrt{t} + 7x^3\sqrt{t} - 15x^3\sqrt{t}$
5. $(x^2 + 2x + 1) - (x^3 - 3x^2 + 4)$
6. $\left(u^4 - (-3)u^3 + \frac{u}{2} + 1\right) + \left(u^4 - 2u^3 + 5 - \frac{u}{2}\right)$
7. $\left(u^4 - (-3)u^3 + \frac{u}{2} + 1\right) - \left(u^4 - 2u^3 + 5 - \frac{u}{2}\right)$
8. $(6a^2b + 3a\sqrt{c} - 5ab\sqrt{c}) + (-6ab^2 - 3ab + 6ab\sqrt{c})$

9. $(4z - 6z^2w - (-2)z^3w^2) + (8 - 6z^2w - zw^3 + 4z^3w^2)$

10. $(x^5y - 2x + 3xy^3) - (-2x - x^5y + 2xy^3)$

11. $(9x - x^3 + 1) - (2x^3 + (-6)x + (-7))$

12. $(x - \sqrt{y} - z) - (x + \sqrt{y} - z) - (\sqrt{y} + z - x)$

13. $(x^2 - 3xy) - (x + xy) - (x^2 + xy)$

14. $2x(x^2 + 2)$ **15.** $(-5y)(-3y^2 + 1)$

16. $x^2y(xy - 6xy^2)$

17. $3ax(4ax - 2a^2y + 2ay)$ **18.** $2x(x^2 - 3xy + 2y^2)$

19. $6z^3(2z + 5)$ **20.** $-3x^2(12x^6 - 7x^5)$

21. $3ab(4a - 6b + 2a^2b)$ **22.** $(-3ay)(4ay - 5y)$

23. $(x + 1)(x - 2)$ **24.** $(x + 2)(2x - 5)$

25. $(-2x + 4)(-x - 3)$ **26.** $(y - 6)(2y + 2)$

27. $(y + 3)(y + 4)$ **28.** $(w - 2)(3w + 1)$

29. $(3x + 7)(-2x + 5)$ **30.** $(ab + 1)(a - 2)$

31. $(y - 3)(3y^2 + 4)$ **32.** $(y + 8)(y - 8)$

33. $(x + 4)(x - 4)$ **34.** $(3x - y)(3x + y)$

35. $(4a + 5b)(4a - 5b)$ **36.** $(x + 6)^2$

37. $(y - 11)^2$ **38.** $(2x + 3y)^2$

39. $(5x - b)^2$

40. $(2s^2 - 9y)(2s^2 + 9y)$ **41.** $(4x^3 - y^4)^2$

42. $(4x^3 - 5y^2)(4x^3 + 5y^2)$ **43.** $(-3x^2 + 2y^4)^2$

44. $(c - 2)(2c^2 - 3c + 1)$

45. $(2y + 3)(y^2 + 3y - 1)$

46. $(x + 2y)(2x^2 - xy + y^2)$

47. $(5w + 6)(-3w^2 + 4w - 3)$

48. $(5x - 2y)(x^2 - 2xy + 3y^2)$

49. $2x(3x + 1)(4x - 2)$ **50.** $3y(-y + 2)(3y + 1)$

51. $(x - 1)(x - 2)(x - 3)$

52. $(y - 2)(3y + 2)(y + 2)$

53. $(x + 4y)(2y - x)(3x - y)$

54. $(2x - y)(3x + 2y)(y - x)$

In Exercises 55–64, find the coefficient of x^2 in the given product. Avoid doing any more multiplying than necessary.

55. $(x^2 + 3x + 1)(2x - 3)$ **56.** $(x^2 - 1)(x + 1)$

57. $(x^3 + 2x - 6)(x^2 + 1)$ **58.** $(\sqrt{3} + x)(\sqrt{3} - x)$

59. $(x + 2)^3$ **60.** $(x^2 + x + 1)(x - 1)$

61. $(x^2 + x + 1)(x^2 - x + 1)$ **62.** $(2x^2 + 1)(2x^2 - 1)$

63. $(2x - 1)(x^2 + 3x + 2)$

64. $(1 - 2x)(4x^2 + x - 1)$

In Exercises 65–70, perform the indicated multiplication and simplify your answer if possible.

65. $(\sqrt{x} + 5)(\sqrt{x} - 5)$

66. $(2\sqrt{x} + \sqrt{2y})(2\sqrt{x} - \sqrt{2y})$

67. $(3 + \sqrt{y})^2$ **68.** $(7w - \sqrt{2x})^2$

69. $(1 + \sqrt{3}x)(x + \sqrt{3})$ **70.** $(2y + \sqrt{3})(\sqrt{5}y - 1)$

In Exercises 71–76, compute the product and arrange the terms of your answer according to decreasing powers of x, with each power of x appearing at most once.
Example: $(ax + b)(4x - c) = 4ax^2 + (4b - ac)x - bc$.

71. $(ax + b)(3x + 2)$ **72.** $(4x - c)(dx + c)$

73. $(ax + b)(bx + a)$ **74.** $rx(3rx + 1)(4x - r)$

75. $(x - a)(x - b)(x - c)$ **76.** $(2dx - c)(3cx + d)$

In Exercises 77–82, assume that all exponents are nonnegative integers and find the product.
Example: $2x^k(3x + x^{n+1}) = (2x^k)(3x) + (2x^k)(x^{n+1}) = 6x^{k+1} + 2x^{k+n+1}$.

77. $3^r3^43^t$ **78.** $(2x^n)(8x^k)$

79. $(x^m + 2)(x^n - 3)$ **80.** $(y^r + 1)(y^s - 4)$

81. $(2x^n - 5)(x^{3n} + 4x^n + 1)$

82. $(3y^{2k} + y^k + 1)(y^k - 3)$

Errors to Avoid

In Exercises 83–92, find a numerical example to show that the given statement is false. *Then find the mistake in the statement and correct it.* Example: *The statement* $-(b + 2) = -b + 2$ *is false when* $b = 5$, *since* $-(5 + 2) = -7$ *but* $-5 + 2 = -3$. *The mistake is the sign on the 2. The correct statement is* $-(b + 2) = -b - 2$.

83. $3(y + 2) = 3y + 2$

84. $x - (3y + 4) = x - 3y + 4$

85. $(x + y)^2 = x + y^2$ **86.** $(2x)^3 = 2x^3$

87. $(7x)(7y) = 7xy$ **88.** $(x + y)^2 = x^2 + y^2$

89. $y + y + y = y^3$ **90.** $(a - b)^2 = a^2 - b^2$

91. $(x - 3)(x - 2) = x^2 - 5x - 6$

92. $(a + b)(a^2 + b^2) = a^3 + b^3$

Thinkers

In Exercises 93 and 94, explain algebraically why each of these parlor tricks always works.

93. Write down a nonzero number. Add 1 to it and square the result. Subtract 1 from the original number and square the result. Subtract this second square from the first one. Divide by the number with which you started. The answer is 4.

94. Write down a positive number. Add 4 to it. Multiply the result by the original number. Add 4 to this result and then take the square root. Subtract the number with which you started. The answer is 2.

95. Invent a similar parlor trick in which the answer is always the number with which you started.

1.C FACTORING

Factoring is the reverse of multiplication: We begin with a product and find the factors that multiply together to produce this product. Factoring skills are necessary to simplify expressions, to do arithmetic with fractional expressions, and to solve equations and inequalities.

The first general rule for factoring is

Common Factors ▶

If there is a common factor in every term of the expression, factor out the common factor of highest degree.

EXAMPLE 1 In $4x^6 - 8x$, for example, each term contains a factor of $4x$, so that $4x^6 - 8x = 4x(x^5 - 2)$. Similarly, the common factor of highest degree in $x^3y^2 + 2xy^3 - 3x^2y^4$ is xy^2 and

$$x^3y^2 + 2xy^3 - 3x^2y^4 = xy^2(x^2 + 2y - 3xy^2).$$ ■

You can greatly increase your factoring proficiency by learning to recognize multiplication patterns that appear frequently. Here are the most common ones.

Quadratic Factoring Patterns ▶

Difference of Squares	$u^2 - v^2 = (u + v)(u - v)$
Perfect Squares	$u^2 + 2uv + v^2 = (u + v)^2$
	$u^2 - 2uv + v^2 = (u - v)^2$

EXAMPLE 2

(a) $x^2 - 9y^2$ can be written $x^2 - (3y)^2$, a difference of squares. Therefore, $x^2 - 9y^2 = (x + 3y)(x - 3y)$.

(b) $y^2 - 7 = y^2 - (\sqrt{7})^2 = (y + \sqrt{7})(y - \sqrt{7})$.*

(c) $36r^2 - 64s^2 = (6r)^2 - (8s)^2 = (6r + 8s)(6r - 8s)$
$= 2(3r + 4s)2(3r - 4s) = 4(3r + 4s)(3r - 4s)$. ■

EXAMPLE 3 Since the first and last terms of $4x^2 - 36x + 81$ are perfect squares, we try to use the perfect square pattern with $u = 2x$ and $v = 9$:

$$\begin{aligned} 4x^2 - 36x + 81 &= (2x)^2 - 36x + 9^2 \\ &= (2x)^2 - 2 \cdot 2x \cdot 9 + 9^2 = (2x - 9)^2. \end{aligned}$$ ■

Cubic Factoring Patterns ▶

Difference of Cubes	$u^3 - v^3 = (u - v)(u^2 + uv + v^2)$
Sum of Cubes	$u^3 + v^3 = (u + v)(u^2 - uv + v^2)$
Perfect Cubes	$u^3 + 3u^2v + 3uv^2 + v^3 = (u + v)^3$
	$u^3 - 3u^2v + 3uv^2 - v^3 = (u - v)^3$

EXAMPLE 4

(a) $x^3 - 125 = x^3 - 5^3 = (x - 5)(x^2 + 5x + 5^2)$
$= (x - 5)(x^2 + 5x + 25)$.

(b) $x^3 + 8y^3 = x^3 + (2y)^3 = (x + 2y)(x^2 - x \cdot 2y + (2y)^2)$
$= (x + 2y)(x^2 - 2xy + 4y^2)$.

(c) $x^3 - 12x^2 + 48x - 64 = x^3 - 12x^2 + 48x - 4^3$
$= x^3 - 3x^2 \cdot 4 + 3x \cdot 4^2 - 4^3$
$= (x - 4)^3$. ■

When none of the multiplication patterns applies, use trial and error to factor quadratic polynomials. If a quadratic has two first-degree factors, then the factors must be of the form $ax + b$ and $cx + d$ for some constants a, b, c, d. The product of such factors is

$$\begin{aligned} (ax + b)(cx + d) &= acx^2 + adx + bcx + bd \\ &= acx^2 + (ad + bc)x + bd. \end{aligned}$$

Note that *ac is the coefficient of x^2* and *bd is the constant term* of the product polynomial. This pattern can be used to factor quadratics by reversing the FOIL process.

EXAMPLE 5 If $x^2 + 9x + 18$ factors as $(ax + b)(cx + d)$, then we must have $ac = 1$ (coefficient of x^2) and $bd = 18$ (constant term). Thus, $a = \pm 1$

*When a polynomial has integer coefficients, we normally look only for factors with integer coefficients. But when it is easy to find other factors, as here, we shall do so.

and $c = \pm 1$ (the only integer factors of 1). The only possibilities for b and d are

$$\pm 1, \pm 18 \qquad \text{or} \qquad \pm 2, \pm 9 \qquad \text{or} \qquad \pm 3, \pm 6.$$

We mentally try the various possibilities, using FOIL as our guide. For example, we try $b = 2$, $d = 9$ and check this factorization: $(x + 2)(x + 9)$. The sum of the outside and inside terms is $9x + 2x = 11x$, so this product can't be $x^2 + 9x + 18$. By trying other possibilities we find that $b = 3$, $d = 6$ leads to the correct factorization: $x^2 + 9x + 18 = (x + 3)(x + 6)$. ■

EXAMPLE 6 To factor $6x^2 + 11x + 4$ as $(ax + b)(cx + d)$, we must find numbers a and c whose product is 6, the coefficient of x^2, and numbers b and d whose product is the constant term 4. Some possibilities are

$ac = 6$

a	± 1	± 2	± 3	± 6
c	± 6	± 3	± 2	± 1

$bd = 4$

b	± 1	± 2	± 4
d	± 4	± 2	± 1

Trial and error shows that $(2x + 1)(3x + 4) = 6x^2 + 11x + 4$. ■

Occasionally the patterns above can be used to factor expressions involving larger exponents than 2.

EXAMPLE 7

(a) $$\begin{aligned} x^6 - y^6 &= (x^3)^2 - (y^3)^2 = (x^3 + y^3)(x^3 - y^3) \\ &= (x + y)(x^2 - xy + y^2)(x - y)(x^2 + xy + y^2). \end{aligned}$$

(b) $$\begin{aligned} x^8 - 1 = (x^4)^2 - 1 &= (x^4 + 1)(x^4 - 1) \\ &= (x^4 + 1)(x^2 + 1)(x^2 - 1) \\ &= (x^4 + 1)(x^2 + 1)(x + 1)(x - 1). \end{aligned}$$ ■

EXAMPLE 8 To factor $x^4 - 2x^2 - 3$, let $u = x^2$. Then,

$$\begin{aligned} x^4 - 2x^2 - 3 &= (x^2)^2 - 2x^2 - 3 \\ &= u^2 - 2u - 3 = (u + 1)(u - 3) \\ &= (x^2 + 1)(x^2 - 3) \\ &= (x^2 + 1)(x + \sqrt{3})(x - \sqrt{3}). \end{aligned}$$ ■

EXAMPLE 9 $3x^3 + 3x^2 + 2x + 2$ can be factored by regrouping and using the distributive law to factor out a common factor:

$$\begin{aligned} (3x^3 + 3x^2) + (2x + 2) &= 3x^2(x + 1) + 2(x + 1) \\ &= (3x^2 + 2)(x + 1). \end{aligned}$$ ■

EXERCISES 1.C

In Exercises 1–58, factor the expression.

1. $x^2 - 4$
2. $x^2 + 6x + 9$
3. $9y^2 - 25$
4. $y^2 - 4y + 4$
5. $81x^2 + 36x + 4$
6. $4x^2 - 12x + 9$
7. $5 - x^2$
8. $1 - 36u^2$
9. $49 + 28z + 4z^2$
10. $25u^2 - 20uv + 4v^2$
11. $x^4 - y^4$
12. $x^2 - 1/9$
13. $x^2 + x - 6$
14. $y^2 + 11y + 30$
15. $z^2 + 4z + 3$
16. $x^2 - 8x + 15$
17. $y^2 + 5y - 36$
18. $z^2 - 9z + 14$
19. $x^2 - 6x + 9$
20. $4y^2 - 81$
21. $x^2 + 7x + 10$
22. $w^2 - 6w - 16$
23. $x^2 + 11x + 18$
24. $x^2 + 3xy - 28y^2$
25. $3x^2 + 4x + 1$
26. $4y^2 + 4y + 1$
27. $2z^2 + 11z + 12$
28. $10x^2 - 17x + 3$
29. $9x^2 - 72x$
30. $4x^2 - 4x - 3$
31. $10x^2 - 8x - 2$
32. $7z^2 + 23z + 6$
33. $8u^2 + 6u - 9$
34. $2y^2 - 4y + 2$
35. $4x^2 + 20xy + 25y^2$
36. $63u^2 - 46uv + 8v^2$
37. $x^3 - 125$
38. $y^3 + 64$
39. $x^3 + 6x^2 + 12x + 8$
40. $y^3 - 3y^2 + 3y - 1$
41. $8 + x^3$
42. $z^3 - 9z^2 + 27z - 27$
43. $-x^3 + 15x^2 - 75x + 125$
44. $27 - t^3$
45. $x^3 + 1$
46. $x^3 - 1$
47. $8x^3 - y^3$
48. $(x - 1)^3 + 1$
49. $x^6 - 64$
50. $x^5 - 8x^2$
51. $y^4 + 7y^2 + 10$
52. $z^4 - 5z^2 + 6$
53. $81 - y^4$
54. $x^6 + 16x^3 + 64$
55. $z^6 - 1$
56. $y^6 + 26y^3 - 27$
57. $x^4 + 2x^2y - 3y^2$
58. $x^8 - 17x^4 + 16$

In Exercises 59–64, factor by regrouping and using the distributive law (as in Example 9).

59. $x^2 - yz + xz - xy$
60. $x^6 - 2x^4 - 8x^2 + 16$
61. $a^3 - 2b^2 + 2a^2b - ab$
62. $u^2v - 2w^2 - 2uvw + uw$
63. $x^3 + 4x^2 - 8x - 32$
64. $z^8 - 5z^7 + 2z - 10$

Thinker

65. Show that there do *not* exist real numbers c and d such that $x^2 + 1 = (x + c)(x + d)$.

1.D FRACTIONAL EXPRESSIONS

Quotients of algebraic expressions are called **fractional expressions.** A quotient of two polynomials is sometimes called a **rational expression.** The basic rules for dealing with fractional expressions are essentially the same as those for ordinary numerical fractions. For instance, $\frac{2}{4} = \frac{3}{6}$ and the "cross products" are equal: $2 \cdot 6 = 4 \cdot 3$. In the general case we have

Properties of Fractions ▶

1. ***Equality rule:*** $\frac{a}{b} = \frac{c}{d}$ **exactly when** $ad = bc$.*
2. ***Cancellation property:*** **If** $k \neq 0$, **then** $\frac{ka}{kb} = \frac{a}{b}$.

*Throughout this section we assume that all denominators are nonzero.

The cancellation property follows directly from the equality rule because $(ka)b = (kb)a$.

EXAMPLE 1 Here are examples of the two properties:

1. $\dfrac{x^2 + 2x}{x^2 + x - 2} = \dfrac{x}{x - 1}$ because the cross products are equal:

$$(x^2 + 2x)(x - 1) = x^3 + x^2 - 2x = (x^2 + x - 2)x.$$

2. $\dfrac{x^4 - 1}{x^2 + 1} = \dfrac{(x^2 + 1)(x^2 - 1)}{(x^2 + 1)} = \dfrac{x^2 - 1}{1} = x^2 - 1.$ ■

A fraction is in **lowest terms** if its **numerator** (top) and **denominator** (bottom) have no common factors except ± 1. To express a fraction in lowest terms, factor numerator and denominator and cancel common factors.

EXAMPLE 2 $\dfrac{x^2 + x - 6}{x^2 - 3x + 2} = \dfrac{(x - 2)(x + 3)}{(x - 2)(x - 1)} = \dfrac{x + 3}{x - 1}.$ ■

To add two fractions with the same denominator, simply add the numerators as in ordinary arithmetic: $\dfrac{a}{b} + \dfrac{c}{b} = \dfrac{a + c}{b}$. Subtraction is done similarly.

EXAMPLE 3

$$\begin{aligned}\frac{7x^2 + 2}{x^2 + 3} - \frac{4x^2 + 2x - 5}{x^2 + 3} &= \frac{(7x^2 + 2) - (4x^2 + 2x - 5)}{x^2 + 3}\\ &= \frac{7x^2 + 2 - 4x^2 - 2x + 5}{x^2 + 3}\\ &= \frac{3x^2 - 2x + 7}{x^2 + 3}.\end{aligned}$$ ■

To add or subtract fractions with different denominators, you must first find a common denominator. One common denominator for a/b and c/d is the product of the two denominators bd because both fractions can be expressed with this denominator:

$$\frac{a}{b} = \frac{ad}{bd} \quad \text{and} \quad \frac{c}{d} = \frac{bc}{bd}.$$

Consequently,

$$\frac{a}{b} + \frac{c}{d} = \frac{ad}{bd} + \frac{bc}{bd} = \frac{ad + bc}{bd} \quad \text{and} \quad \frac{a}{b} - \frac{c}{d} = \frac{ad}{bd} - \frac{bc}{bd} = \frac{ad - bc}{bd}.$$

EXAMPLE 4

$$\frac{2x+1}{3x} - \frac{x^2-2}{x-1} = \frac{(2x+1)(x-1)}{3x(x-1)} - \frac{3x(x^2-2)}{3x(x-1)}$$
$$= \frac{(2x+1)(x-1) - 3x(x^2-2)}{3x(x-1)} = \frac{2x^2 - x - 1 - 3x^3 + 6x}{3x^2 - 3x}$$
$$= \frac{-3x^3 + 2x^2 + 5x - 1}{3x^2 - 3x}. \quad ■$$

Although the product of the denominators can always be used as a common denominator, it's often more efficient to use the *least common denominator.* The least common denominator can be found by factoring each denominator completely (with integer coefficients) and then taking the product of the highest power of each of the distinct factors.

EXAMPLE 5 In the sum $\frac{1}{100} + \frac{1}{120}$, the denominators are $100 = 2^2 \cdot 5^2$ and $120 = 2^3 \cdot 3 \cdot 5$. The distinct factors are 2, 3, 5. The highest exponent of 2 is 3, the highest of 3 is 1, and the highest of 5 is 2. So the least common denominator is $2^3 \cdot 3 \cdot 5^2 = 600$. ■

EXAMPLE 6 To find the least common denominator of $\frac{1}{x^2 + 2x + 1}$, $\frac{5x}{x^2 - x}$, and $\frac{3x-7}{x^4 + x^3}$, factor each of the denominators completely:

$$(x+1)^2, \qquad x(x-1), \qquad x^3(x+1).$$

The distinct factors are x, $x + 1$, and $x - 1$. The least common denominator is $x^3(x+1)^2(x-1)$. ■

To express one of several fractions in terms of the least common denominator, multiply its numerator and denominator by those factors in the common denominator that *don't* appear in the denominator of the fraction.

EXAMPLE 7 The preceding example shows that the least common denominator of $\frac{1}{(x+1)^2}$, $\frac{5x}{x(x-1)}$, and $\frac{3x-7}{x^3(x+1)}$ is $x^3(x+1)^2(x-1)$. Therefore,

$$\frac{1}{(x+1)^2} = \frac{1}{(x+1)^2} \cdot \frac{x^3(x-1)}{x^3(x-1)} = \frac{x^3(x-1)}{x^3(x+1)^2(x-1)}$$

$$\frac{5x}{x(x-1)} = \frac{5x}{x(x-1)} \cdot \frac{x^2(x+1)^2}{x^2(x+1)^2} = \frac{5x^3(x+1)^2}{x^3(x+1)^2(x-1)}$$

$$\frac{3x-7}{x^3(x+1)} = \frac{3x-7}{x^3(x+1)} \cdot \frac{(x+1)(x-1)}{(x+1)(x-1)} = \frac{(3x-7)(x+1)(x-1)}{x^3(x+1)^2(x-1)}. \quad ■$$

EXAMPLE 8 To find $\frac{1}{z} + \frac{3z}{z+1} - \frac{z^2}{(z+1)^2}$ we use the least common denominator $z(z+1)^2$:

$$\frac{1}{z} + \frac{3z}{z+1} - \frac{z^2}{(z+1)^2} = \frac{(z+1)^2}{z(z+1)^2} + \frac{3z^2(z+1)}{z(z+1)^2} - \frac{z^3}{z(z+1)^2}$$
$$= \frac{(z+1)^2 + 3z^2(z+1) - z^3}{z(z+1)^2}$$
$$= \frac{z^2 + 2z + 1 + 3z^3 + 3z^2 - z^3}{z(z+1)^2}$$
$$= \frac{2z^3 + 4z^2 + 2z + 1}{z(z+1)^2}. \quad ■$$

Multiplication of fractions is easy: Multiply corresponding numerators and denominators, then simplify your answer:

EXAMPLE 9

$$\frac{x^2-1}{x^2+2} \cdot \frac{3x-4}{x+1} = \frac{(x^2-1)(3x-4)}{(x^2+2)(x+1)}$$
$$= \frac{(x-1)(x+1)(3x-4)}{(x^2+2)(x+1)} = \frac{(x-1)(3x-4)}{x^2+2}. \quad ■$$

Division of fractions is given by the rule:

Invert the divisor and multiply: $\frac{a}{b} \div \frac{c}{d} = \frac{a}{b} \cdot \frac{d}{c} = \frac{ad}{bc}$.

EXAMPLE 10

$$\frac{x^2+x-2}{x^2-6x+9} \div \frac{x^2-1}{x-3} = \frac{x^2+x-2}{x^2-6x+9} \cdot \frac{x-3}{x^2-1}$$
$$= \frac{(x+2)(x-1)}{(x-3)^2} \cdot \frac{x-3}{(x-1)(x+1)}$$
$$= \frac{x+2}{(x-3)(x+1)}. \quad ■$$

Division problems can also be written as fractions. For instance, $8/2$ means $8 \div 2 = 4$. Similarly, the compound fraction $\frac{a/b}{c/d}$ means $\frac{a}{b} \div \frac{c}{d}$. So, the basic rule for simplifying compound fractions is: *Invert the denominator and multiply it by the numerator.*

EXAMPLE 11

(a)
$$\frac{16y^2z/8yz^2}{yz/6y^3z^3} = \frac{16y^2z}{8yz^2}\cdot\frac{6y^3z^3}{yz} = \frac{16\cdot 6\cdot y^5z^4}{8y^2z^3}$$
$$= 2\cdot 6\cdot y^{5-2}z^{4-3} = 12y^3z.$$

(b)
$$\frac{\dfrac{y^2}{y+2}}{y^3+y} = \frac{y^2}{y+2}\cdot\frac{1}{y^3+y} = \frac{y^2}{(y+2)(y^3+y)}$$
$$= \frac{y^2}{(y+2)y(y^2+1)}$$
$$= \frac{y}{(y+2)(y^2+1)}. \quad ■$$

Rationalizing Numerators and Denominators

It is sometimes necessary to eliminate all radicals from either the numerator or denominator of a fraction. For instance, we can eliminate the radical in the denominator of $1/\sqrt{2}$ by using the fact that $\sqrt{2}/\sqrt{2} = 1$:

$$\frac{1}{\sqrt{2}} = \frac{1}{\sqrt{2}}\cdot\frac{\sqrt{2}}{\sqrt{2}} = \frac{\sqrt{2}}{2}.$$

Here are some other examples of **rationalizing the denominator:**

EXAMPLE 12

$$\sqrt{\frac{5}{2x+1}} = \frac{\sqrt{5}}{\sqrt{2x+1}} = \frac{\sqrt{5}}{\sqrt{2x+1}}\cdot\frac{\sqrt{2x+1}}{\sqrt{2x+1}} = \frac{\sqrt{10x+5}}{2x+1}. \quad ■$$

EXAMPLE 13 In order to rationalize the denominator of $\dfrac{7}{\sqrt{5}+\sqrt{3}}$, we must multiply both top and bottom by something that will eliminate the radicals in the denominator. Observe that $(\sqrt{5}+\sqrt{3})(\sqrt{5}-\sqrt{3}) = (\sqrt{5})^2 - (\sqrt{3})^2 = 5 - 3 = 2$. Thus,

$$\frac{7}{\sqrt{5}+\sqrt{3}} = \left(\frac{7}{\sqrt{5}+\sqrt{3}}\right)\left(\frac{\sqrt{5}-\sqrt{3}}{\sqrt{5}-\sqrt{3}}\right)$$
$$= \frac{7(\sqrt{5}-\sqrt{3})}{(\sqrt{5}+\sqrt{3})(\sqrt{5}-\sqrt{3})} = \frac{7(\sqrt{5}-\sqrt{3})}{2}. \quad ■$$

The same techniques can be used to **rationalize the numerator** of a fraction.

EXAMPLE 14

$$\frac{\sqrt{x+3}-\sqrt{3}}{5} = \frac{\sqrt{x+3}-\sqrt{3}}{5} \cdot \frac{\sqrt{x+3}+\sqrt{3}}{\sqrt{x+3}+\sqrt{3}}$$
$$= \frac{(\sqrt{x+3})^2-(\sqrt{3})^2}{5(\sqrt{x+3}+\sqrt{3})} = \frac{x+3-3}{5(\sqrt{x+3}+\sqrt{3})}$$
$$= \frac{x}{5\sqrt{x+3}+5\sqrt{3}}. \quad \blacksquare$$

EXERCISES 1.D

In Exercises 1–10, express the fraction in lowest terms.

1. $\frac{63}{49}$ 2. $\frac{121}{33}$ 3. $\frac{13 \cdot 27 \cdot 22 \cdot 10}{6 \cdot 4 \cdot 11 \cdot 12}$ 4. $\frac{x^2-4}{x+2}$

5. $\frac{x^2-x-2}{x^2+2x+1}$ 6. $\frac{z+1}{z^3+1}$ 7. $\frac{a^2-b^2}{a^3-b^3}$

8. $\frac{x^4-3x^2}{x^3}$ 9. $\frac{(x+c)(x^2-cx+c^2)}{x^4+c^3x}$

10. $\frac{x^4-y^4}{(x^2+y^2)(x^2-xy)}$

In Exercises 11–28, perform the indicated operations.

11. $\frac{3}{7}+\frac{2}{5}$ 12. $\frac{7}{8}-\frac{5}{6}$ 13. $\left(\frac{19}{7}+\frac{1}{2}\right)-\frac{1}{3}$

14. $\frac{1}{a}-\frac{2a}{b}$ 15. $\frac{c}{d}+\frac{3c}{e}$ 16. $\frac{r}{s}+\frac{s}{t}+\frac{t}{r}$

17. $\frac{b}{c}-\frac{c}{b}$ 18. $\frac{a}{b}+\frac{2a}{b^2}+\frac{3a}{b^3}$ 19. $\frac{1}{x+1}-\frac{1}{x}$

20. $\frac{1}{2x+1}+\frac{1}{2x-1}$

21. $\frac{1}{x+4}+\frac{2}{(x+4)^2}-\frac{3}{x^2+8x+16}$

22. $\frac{1}{x}+\frac{1}{xy}+\frac{1}{xy^2}$ 23. $\frac{1}{x}-\frac{1}{3x-4}$

24. $\frac{3}{x-1}+\frac{4}{x+1}$ 25. $\frac{1}{x+y}+\frac{x+y}{x^3+y^3}$

26. $\frac{6}{5(x-1)(x-2)^2}+\frac{x}{3(x-1)^2(x-2)}$

27. $\frac{1}{4x(x+1)(x+2)^3}-\frac{6x+2}{4(x+1)^3}$

28. $\frac{x+y}{(x^2-xy)(x-y)^2}-\frac{2}{(x^2-y^2)^2}$

In Exercises 29–42, express in lowest terms.

29. $\frac{3}{4}\cdot\frac{12}{5}\cdot\frac{10}{9}$ 30. $\frac{10}{45}\cdot\frac{6}{14}\cdot\frac{1}{2}$ 31. $\frac{3a^2c}{4ac}\cdot\frac{8ac^3}{9a^2c^4}$

32. $\frac{6x^2y}{2x}\cdot\frac{y}{21xy}$ 33. $\frac{7x}{11y}\cdot\frac{66y^2}{14x^3}$ 34. $\frac{ab}{c^2}\cdot\frac{cd}{a^2b}\cdot\frac{ad}{bc^2}$

35. $\frac{3x+9}{2x}\cdot\frac{8x^2}{x^2-9}$ 36. $\frac{4x+16}{3x+15}\cdot\frac{2x+10}{x+4}$

37. $\frac{5y-25}{3}\cdot\frac{y^2}{y^2-25}$ 38. $\frac{6x-12}{6x}\cdot\frac{8x^2}{x-2}$

39. $\frac{u}{u-1}\cdot\frac{u^2-1}{u^2}$ 40. $\frac{t^2-t-6}{t^2-6t+9}\cdot\frac{t^2+4t-5}{t^2-25}$

41. $\frac{2u^2+uv-v^2}{4u^2-4uv+v^2}\cdot\frac{8u^2+6uv-9v^2}{4u^2-9v^2}$

42. $\frac{2x^2-3xy-2y^2}{6x^2-5xy-4y^2}\cdot\frac{6x^2+6xy}{x^2-xy-2y^2}$

In Exercises 43–60, compute the quotient and express in lowest terms.

43. $\frac{5}{12}\div\frac{4}{14}$ 44. $\frac{\frac{100}{52}}{\frac{27}{26}}$ 45. $\frac{uv}{v^2w}\div\frac{uv}{u^2v}$

46. $\dfrac{3x^2y}{(xy)^2} \div \dfrac{3xyz}{x^2y}$

47. $\dfrac{\frac{x+3}{x+4}}{\frac{2x}{x+4}}$

48. $\dfrac{\frac{(x+2)^2}{(x-2)^2}}{\frac{x^2+2x}{x^2-4}}$

49. $\dfrac{x+y}{x+2y} \div \left(\dfrac{x+y}{xy}\right)^2$

50. $\dfrac{\frac{u^3+v^3}{u^2-v^2}}{\frac{u^2-uv+v^2}{u+v}}$

51. $\dfrac{\frac{(c+d)^2}{c^2-d^2}}{cd}$

52. $\dfrac{\frac{1}{x}-\frac{3}{2}}{\frac{2}{x-2}+\frac{5}{x}}$

53. $\dfrac{\frac{1}{x^2}-\frac{1}{y^2}}{\frac{1}{x}+\frac{1}{y}}$

54. $\dfrac{\frac{x}{x+1}+\frac{1}{x}}{\frac{1}{x}+\frac{1}{x+1}}$

55. $\dfrac{\frac{6}{y}-3}{1-\frac{1}{y-1}}$

56. $\dfrac{\frac{1}{3x}-\frac{1}{4y}}{\frac{5}{6x^2}+\frac{1}{y}}$

57. $\dfrac{\frac{1}{x+h}-\frac{1}{x}}{h}$

58. $\dfrac{\frac{1}{(x+h)^2}-\frac{1}{x^2}}{h}$

59. $(x^{-1}+y^{-1})^{-1}$

60. $\dfrac{(x+y)^{-1}}{x^{-1}+y^{-1}}$

In Exercises 61–70, rationalize the denominator.

61. $\dfrac{2}{\sqrt{5}}$ **62.** $\sqrt{\dfrac{16}{5}}$ **63.** $\sqrt{\dfrac{7}{10}}$ **64.** $\sqrt{\dfrac{9x^4}{23}}$

65. $\dfrac{1}{\sqrt{x}}$ **66.** $\dfrac{\sqrt{6}}{\sqrt{6}+\sqrt{2}}$ **67.** $\dfrac{x+1}{\sqrt{x}+1}$ **68.** $\dfrac{\sqrt{r}+\sqrt{s}}{\sqrt{r}-\sqrt{s}}$

69. $\dfrac{1}{\sqrt{a}-2\sqrt{b}}$ **70.** $\dfrac{u^2-v^2}{\sqrt{u+v}-\sqrt{u-v}}$

Errors to Avoid

In Exercises 71–77, find a numerical example to show that the given statement is false. *Then find the mistake in the statement and correct it.*

71. $\dfrac{1}{a}+\dfrac{1}{b}=\dfrac{1}{a+b}$

72. $\dfrac{x^2}{x^2+x^6}=1+x^3$

73. $\left(\dfrac{1}{\sqrt{a}+\sqrt{b}}\right)^2=\dfrac{1}{a+b}$

74. $\dfrac{r+s}{r+t}=1+\dfrac{s}{t}$

75. $\dfrac{u}{v}+\dfrac{v}{u}=1$

76. $\dfrac{\frac{1}{x}}{\frac{1}{y}}=\dfrac{1}{xy}$

77. $(\sqrt{x}+\sqrt{y})\dfrac{1}{\sqrt{x}+\sqrt{y}}=x+y$

APPENDIX 2

Geometry Review

An **angle** consists of two half-lines that begin at the same point P, as in Figure A–1. The point P is called the **vertex** of the angle and the half-lines the **sides** of the angle.

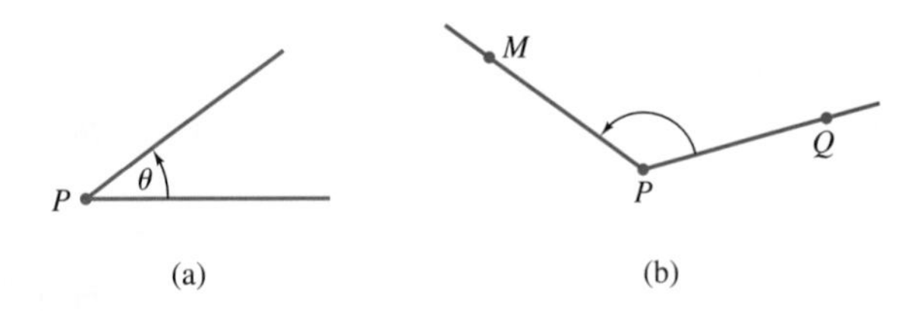

Figure A-1

An angle may be labeled by a Greek letter, such as angle θ in Figure A–1(a), or by listing three points (a point on one side, the vertex, a point on the other side), such as angle QPM in Figure A–1(b).

In order to measure the size of an angle, we must assign a number to each angle. Here is the classical method for doing this:

1. Construct a circle whose center is the vertex of the angle.
2. Divide the circumference of the circle into 360 equal parts (called **degrees**) by marking 360 points on the circumference, beginning with the point where one side of the angle intersects the circle. Label these points $0°$, $1°$, $2°$, $3°$, and so on.
3. The label of the point where the second side of the angle intersects the circle is the degree measure of the angle.

For example, Figure A–2 shows an angle θ of measure 25 degrees (in symbols, $25°$) and an angle β of measure $135°$.

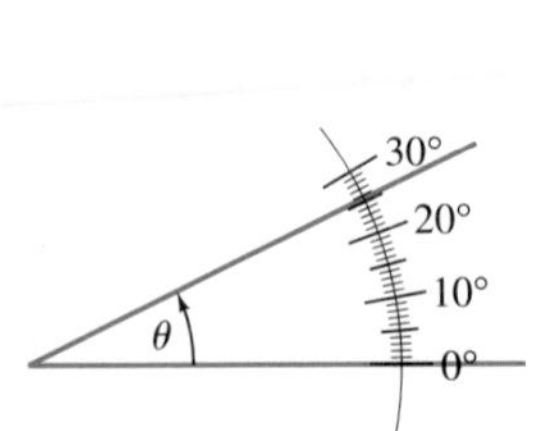

Figure A-2

An **acute angle** is an angle whose measure is strictly between 0° and 90°, such as angle θ in Figure A–2. A **right angle** is an angle that measures 90°. An **obtuse angle** is an angle whose measure is strictly between 90° and 180°, such as angle β in Figure A–2.

A **triangle** has three sides (straight line segments) and three angles, formed at the points where the various sides meet. When angles are measured in degrees,

the sum of the measures of all three angles of a triangle is *always* 180°.

For instance, see Figure A–3.

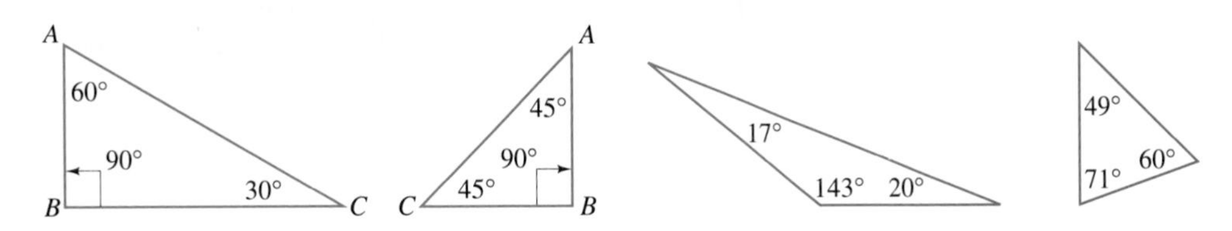

Figure A-3

A **right triangle** is a triangle, one of whose angles is a right angle, such as the first two triangles shown in Figure A–3. The side of a right triangle that lies opposite the right angle is called the **hypotenuse.** In each of the right triangles in Figure A–3, side AC is the hypotenuse.

Pythagorean Theorem ▶

If the sides of a right triangle have lengths a and b and the hypotenuse has length c, then

$$c^2 = a^2 + b^2$$

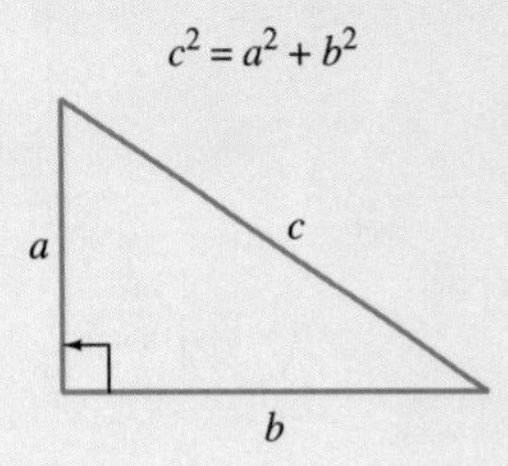

EXAMPLE 1 Consider the right triangle with sides of lengths 5 and 12, as shown in Figure A–4.

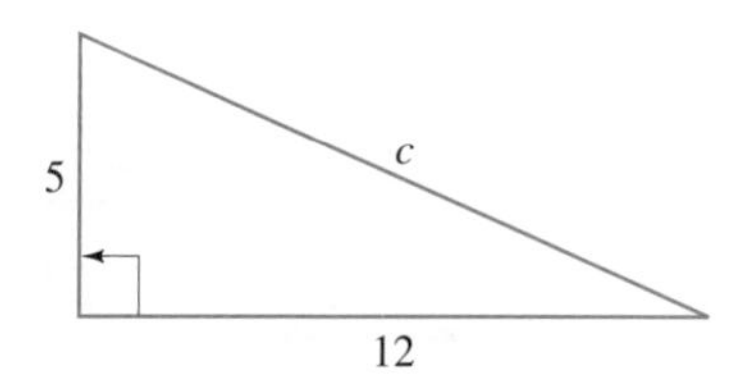

Figure A-4

According to the Pythagorean Theorem the length c of the hypotenuse satisfies the equation: $c^2 = 5^2 + 12^2 = 25 + 144 = 169$. Since $169 = 13^2$, we see that c must be 13. ■

Theorem I ▶

If two angles of a triangle are equal, then the two sides opposite these angles have the same length.

EXAMPLE 2 Suppose the hypotenuse of the right triangle shown in Figure A–5 has length 1 and that angles B and C measure 45° each.

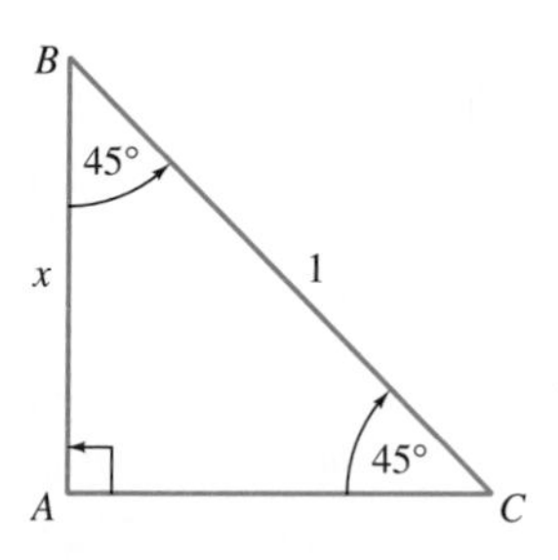

Figure A-5

Then by Theorem I, sides AB and AC have the same length. If x is the length of side AB, then by the Pythagorean Theorem:

$$x^2 + x^2 = 1^2$$

$$2x^2 = 1$$

$$x^2 = \frac{1}{2}$$

$$x = \sqrt{\frac{1}{2}} = \frac{1}{\sqrt{2}} = \frac{\sqrt{2}}{2}.$$

(We ignore the other solution of this equation, namely, $x = -\sqrt{1/2}$, since x represents a length here and thus must be nonnegative.) Therefore, the sides of a $90°-45°-45°$ triangle with hypotenuse 1 are each of length $\sqrt{2}/2$. ■

Theorem II ▶

In a right triangle that has an angle of 30°, the length of the side opposite the 30° angle is one-half the length of the hypotenuse.

EXAMPLE 3 Suppose that in the right triangle shown in Figure A–6 angle B is 30° and the length of hypotenuse BC is 2.

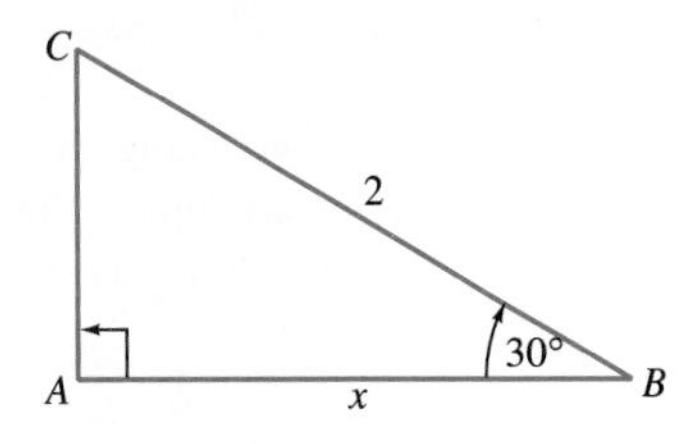

Figure A-6

By Theorem II the side opposite the 30° angle, namely, side AC, has length 1. If x denotes the length of side AB, then by the Pythagorean Theorem:

$$1^2 + x^2 = 2^2$$
$$x^2 = 3$$
$$x = \sqrt{3}. \quad ■$$

EXAMPLE 4 The right triangle shown in Figure A–7 has a 30° angle at C, and side AC has length $\sqrt{3}/2$.

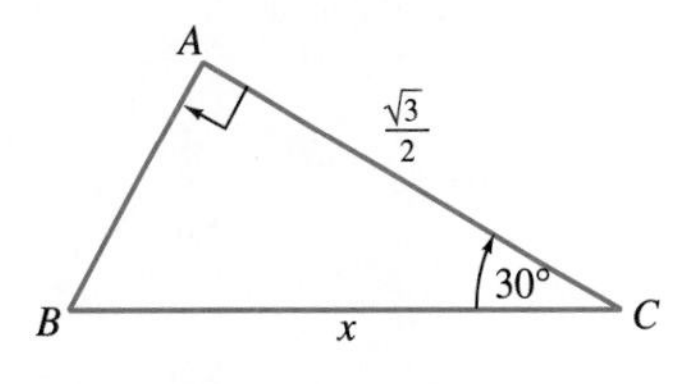

Figure A-7

Let x denote the length of the hypotenuse BC. By Theorem II, side AB has length $\frac{1}{2}x$. By the Pythagorean Theorem:

$$\left(\frac{1}{2}x\right)^2 + \left(\frac{\sqrt{3}}{2}\right)^2 = x^2$$

$$\frac{x^2}{4} + \frac{3}{4} = x^2$$

$$\frac{3}{4} = \frac{3}{4}x^2$$

$$x^2 = 1$$

$$x = 1.$$

Therefore, the triangle has hypotenuse of length 1 and sides of lengths 1/2 and $\sqrt{3}/2$. ■

Two triangles, as in Figure A–8, are said to be **similar** if their corresponding angles are equal (that is, $\measuredangle A = \measuredangle D$; $\measuredangle B = \measuredangle E$; and $\measuredangle C = \measuredangle F$). Thus, similar triangles have the same *shape* but not necessarily the same *size*.

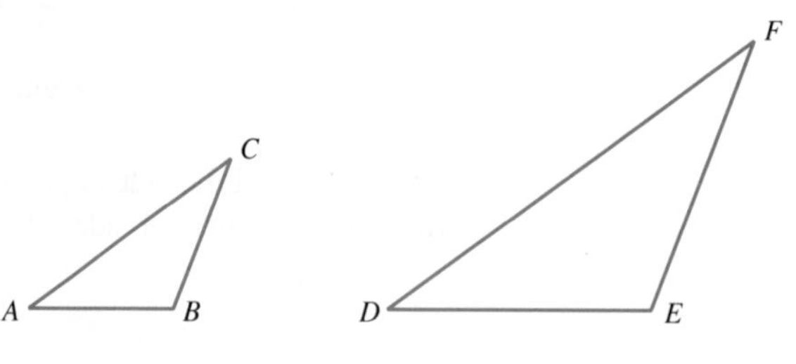

Figure A-8

Theorem III ▶

Suppose triangle *ABC* with sides *a*, *b*, *c* is similar to triangle *DEF* with sides *d*, *e*, *f* (that is, $\measuredangle A = \measuredangle D$; $\measuredangle B = \measuredangle E$; $\measuredangle C = \measuredangle F$),

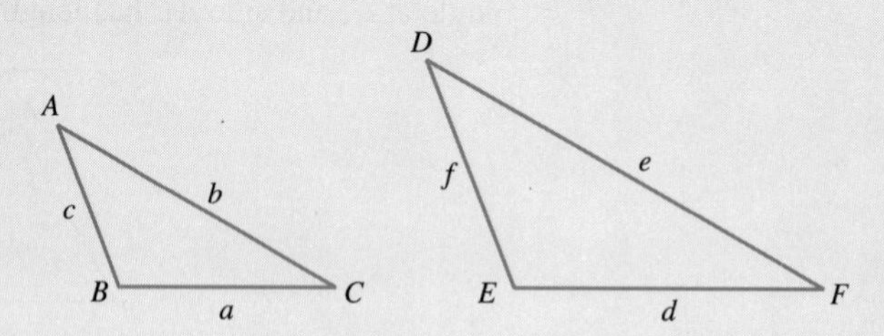

then

$$\frac{a}{d} = \frac{b}{e} = \frac{c}{f}.$$

These equalities are equivalent to:

$$\frac{a}{b} = \frac{d}{e}, \quad \frac{b}{c} = \frac{e}{f}, \quad \frac{a}{c} = \frac{d}{f}.$$

The equivalence of the equalities in the conclusion of the theorem is easily verified. For example, since

$$\frac{a}{d} = \frac{b}{e}$$

we have

$$ae = db$$

Dividing both sides of this equation by be yields:

$$\frac{ae}{be} = \frac{db}{be}$$

$$\frac{a}{b} = \frac{d}{e}.$$

The other equivalences are proved similarly.

EXAMPLE 5 Suppose the triangles in Figure A–9 are similar and that the sides have the lengths indicated.

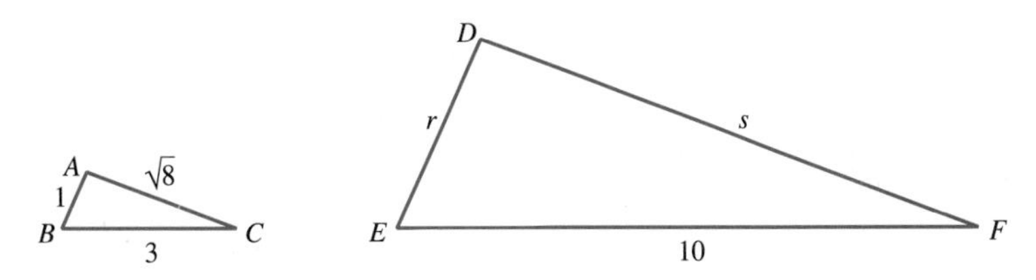

Figure A-9

Then by Theorem III,

$$\frac{\text{length } AC}{\text{length } DF} = \frac{\text{length } BC}{\text{length } EF}$$

In other words,

$$\frac{\sqrt{8}}{s} = \frac{3}{10}$$

so that

$$3s = 10\sqrt{8}$$

$$s = \left(\frac{10}{3}\right)\sqrt{8}.$$

Similarly, by Theorem III,

$$\frac{\text{length } AB}{\text{length } DE} = \frac{\text{length } BC}{\text{length } EF}$$

so that

$$\frac{1}{r} = \frac{3}{10}$$

$$3r = 10$$

$$r = \frac{10}{3}$$

Therefore the sides of triangle *DEF* are of lengths 10, $\frac{10}{3}$, and $\frac{10}{3}\sqrt{8}$. ■

APPENDIX 3

Programs

The programs listed here are of two types: programs to give older calculators some of the features that are built-in on newer ones (such as a root finder) and programs to do specific tasks discussed in this book (such as synthetic division). Each program is preceded by a *Description,* which describes, in general terms, how the program operates and what it does. Some programs require that certain things be done before the program is run (such as entering a function in the function memory); these requirements are listed as *Preliminaries.* Occasionally, italic remarks appear in brackets after a program step; they are *not* part of the program, but are intended to provide assistance when you are entering the program into your calculator. A remark such as "[*MATH NUM menu*]" means that the symbols or commands needed for that step of the program are in the NUM submenu of the MATH menu.

FRACTION CONVERSION (Built-in on TI-82/83/85 and HP-38)

Description: Enter a repeating decimal; the program converts it into a fraction. The denominator is displayed on the last line and the numerator on the line above.

TI-81

```
:ClrHome [Optional]
:Lbl 2
:Input N
:0 → D
:Lbl 1
:D + 1 → D
:If fPart round (ND, 7) ≠ 0 [MATH NUM menu]
:Goto 1
:Int (ND + .5) → N
:Disp N
:Disp D
```

Casio

```
Fix 7
Lbl 2
"N ="? → N
0 → D
Lbl 1
D + 1 → D
N × D
Rnd [MATH NUM menu]
(Frc Ans) ≠ 0 ⇒ Goto 1 [MATH NUM menu]
(Ans + .5) → N
Norm [DISP menu]
(Int N) ▲
D
```

QUADRATIC FORMULA (Built-in on TI-85/92, HP-38, and Casio 9800)

Description: Enter the coefficients of the quadratic equation $ax^2 + bx + c = 0$; the program finds all real solutions.

TI-81/82/83

```
:ClrHome [Optional]
:Disp "AX² + BX + C = 0" [Optional]
:Prompt A*
:Prompt B*
:Prompt C*
:(B² − 4AC) → S
:If S < 0
:Goto 1
:Disp (−B + √S)/2A
:Disp (−B − √S)/2A
:Stop
:Lbl 1
:Disp "NO REAL ROOTS"
```

Sharp 9300

[Gives both real and complex solutions when complex mode is selected.]

```
Print "ax² + bx + c = 0"
x₁ = (1/(2a))(−b + √(b² − 4ac))
Print x₁
x₂ = (1/(2a))(−b − √(b² − 4ac))
Print x₂
```

Casio 7700/8700

```
"AX² + BX + C = 0" [Optional]
"A ="? → A
"B ="? → B
"C ="? → C
B² − 4AC → D
D < 0 ⇒ Goto 1
(−B + √D) ÷ (2A) ◢
(−B − √D) ÷ (2A)
Goto 2
Lbl 1
"NO REAL ROOTS"
Lbl 2
```

BISECTION METHOD SOLVER

Preliminaries: Enter the function whose roots are to be found in the equation memory as Y_1 (or f_1).

Description: Enter two numbers; the program uses the bisection method to find a root (x-intercept) of the function that lies between the two numbers (if there is one).

**On TI-81, replace* Prompt A *by two lines:* :Disp "A = " *and* :Input A.

TI

```
:ClrHome [ClLCD on TI-85] [Optional]
:Disp "LOWER BOUND"
:Input A
:Disp "UPPER BOUND"
:Input B
:0 → K
:Lbl 1
:B → X [Use the variable key for X.]
:If Y1 = 0 [Y-VARS menu]
:Goto 9
:A → X
:Y1 → C
:If C = 0
:Goto 5
:(A + B)/2 → M
:M → X
:Y1 → D
:If CD < 0
:Goto 2
:M → A
:Goto 3
:Lbl 2
:1 → K
:M → B
:Lbl 3
:If abs(B − A) < .000000001
:Goto 4
:Goto 1
:Lbl 4
:(B + A)/2 → M
:If K = 0
:Goto 7
:Lbl 8
:Disp "ROOT ="
:Disp M
:Stop
:Lbl 7
:Disp "NO SIGN CHANGE"
:stop
:Lbl 5
:A → M
:Goto 8
:Lbl 9
:B → M
:Goto 8
```

Casio

```
"LOWER BOUND"? → A
"UPPER BOUND"? → B
0 → K
Lbl 1
B → X [Use the variable key for X.]
f1 → C [7700/8700 FMEM menu on Y1 on 9800 VAR menu]
C = 0 ⇒ Goto 5
((A + B) ÷ 2) → M
M → X
f1 → D
CD < 0 ⇒ Goto 2
M → A
Goto 3
Lbl 2
1 → K
M → B
Lbl 3
abs(B − A) < .000000001 ⇒ Goto 4
Goto 1
Lbl 4
(B + A) ÷ 2 → M
K = 0 ⇒ Goto 7
Lbl 8
"ROOT = "
M
Goto 6
Lbl 7
"NO SIGN CHANGE"
Goto 6
Lbl 5
A → M
Goto 8
Lbl 9
B → M
Goto 8
Lbl 6
```

TI-81 GRAPHICAL ROOT FINDER, MIN/MAX FINDER, AND INTERSECTION FINDER

Preliminaries: Enter the function whose roots or extrema are to be found in the function memory as Y_1. If an intersection point is to be found, enter the other function as Y_2. Set the RANGE so that the desired portions of the graph(s) will be displayed.

Description: Select root, intersection, or min/max. The appropriate function(s) are graphed. Use the arrow keys to move the cursor close to the desired root (x-intercept), intersection point, or extremum. Press ENTER, and its approximate coordinates are displayed. [If the calculator continues working for a long time without displaying an answer, press ON to quit the program. Then run it again, but pick a different point than before near the point you want to find.]

```
:ClrHome [Optional]
:Disp "SELECT NUMBER, PRESS ENTER"
:Disp "1 = ROOT"
:Disp "2 = INTERSECTION"
:Disp "3 = MIN/MAX"
:Input C
:If C = 1
:Goto 1
:If C = 2
:Goto 2
:Goto 3
:Lbl 2
:"Y1 − Y2" → Y4 [Y-VARS menu]
:Y1 − On [Y-VARS ON menu]
:Y2 − On
:Y3 − Off [Y-VARS OFF menu]
:Y4 − Off
:Input
:Goto 4
:Lbl 3
:"NDeriv(Y1, .000001)" → Y4 [MATH menu]
:Y1 − On
:Y2 − Off
:Y3 − Off
:Y4 − Off
:Input
:Goto 4
:Lbl 1
:Y2 − Off
:Y3 − Off
:"Y1" → Y4
:Input
:Lbl 4
:X − Y4/NDeriv(Y4, .000001) → G [Use the X|T key for X.]
:If abs (X − G) < .0000000001
:Goto 5
:G → X
:Goto 4
:Lbl 5
:ClrHome [Optional]
:G → X
:Disp "X ="
:Disp X
:Disp "Y ="
:Disp Y1
```

TABLE MAKER (Built-in on TI-82/83/92, HP-38, and Casio 9800)

Preliminaries: Enter the function to be evaluated in the function memory as Y_1.

Description: Select a starting point and an increment (the amount by which adjacent x entries differ); the program displays a table of function values. To scroll through the table a page at a time, press "down" or "up" on TI-85, and "0 ENTER" or "1 ENTER" on TI-81. Press "quit" on TI-85 or "2 ENTER" on

TI-81 to end the program. [*Note:* An error message will occur if the calculator attempts to evaluate the function at a number for which it is not defined. In this case, change the starting point or increment to avoid the undefined point.]

TI-81

```
:ClrHome [Optional]
:Disp "START AT"
:Input A
:Disp "INCREMENT"
:Input D
:6 → Arow [VARS DIM menu]
:2 → Acol [VARS DIM menu]
:A → X [Use the variable key for X.]
:1→ J
:Lbl 1
:X → [A](J,1) [Second function of 1]
:Y1 → [A](J,2) [Y-VARS menu]
:J + 1 → J
:If J > 6
:Goto 2
:X + D → X
:Goto 1
:Lbl 2
:Disp [A]
:Disp "0 = DN 1 = UP 2 = QT"
:Input K
:If K = 2
:Stop
:If K = 1
:Goto 4
:1 → J
:X + D → X
:Goto 1
:Lbl 4
:1 → J
:X − 11D → X
:Goto 1
```

TI-85

```
:Lbl SETUP
:ClLCD [Optional]
:Disp " "TABLE SETUP"
:Input "TblMin = ", tblmin
:Input "ΔTbl = ", dtbl [CHAR GREEK menu]
:tblmin → x [Use the x-var key for x.]
:Lbl CONTD
:ClLCD
:Outpt(1,1,"x") [I/O menu]
:Outpt(1,10,"y1")
:For (cnt,2,7,1)
:Outpt(cnt,1,x)
:Outpt(cnt,8," ")
:Outpt(cnt,9,y1)
:x + dtbl → x
:End
:Menu(1,"Down",CONTD, 2,"Up",CONTU,
        4,"Setup",SETUP, 5,"quit",TQUIT)
:Lbl CONTU
:x − 12*dtbl → x
:Goto CONTD
:Lbl TQUIT
:ClLCD
:Stop
```

SECANT METHOD ROOT FINDER

Preliminaries: Enter the function whose roots are to be found in the equation memory as Y_1.

Description: Enter two different guesses (estimates) for the value of the root (x-intercept); the program uses the secant method to approximate the root.

TI

```
:ClrHome [ClLCD on TI-85] [Optional]
:Disp "FIRST GUESS?"
:INPUT A
:Disp "SECOND GUESS?"
:Input B
:Lbl 1 [Lbl L1 on TI-85]
:A → X [Use the variable key for X.]
:Y1 → C [Y-VARS menu]
:B → X
:Y1 → D
:(D − C)/(B − A) → M
:B − D/M → X
:If abs(B − X) < .00000001
:Goto 2 [L2 on TI-85]
:B → A
:X → B
:Goto 1 [L1 on TI-85]
:Lbl 2 [L2 on TI-85]
:Disp "SOLUTION ="
:Disp X
```

Casio

```
"FIRST GUESS"? → A
"SECOND GUESS"? → B
Lbl 1
A → X [Use the variable key for X.]
f1 → C [7700/8700 FMEM menu or Y1 on 9800 VAR menu]
B → X
f1 → D
(D − C) ÷ (B − A) → M
B − D ÷ M → X
abs(B − X) < .00000001 ⇒ Goto 2
B → A
X → B
Goto 1
Lbl 2
"SOLUTION ="
X
```

SYNTHETIC DIVISION (Built-in on TI-92)

Description: Enter the constant term on the divisor $x - a$ and the coefficients of the dividend $F(x)$ (in order of decreasing powers of x, putting in zeros for missing coefficients); the program displays the coeffcients of the quotient $Q(x)$ (in order of decreasing powers of x) and then displays the remainder.

TI-81

[*Enter coefficients of $F(x)$ one at a time, pressing ENTER after each one. The coefficients of $Q(x)$ are displayed one at a time; press ENTER to get the next one.*]

```
:ClrHome [Optional]
:Disp "DIVISOR is X − A"
:Disp "A ="
:Input A
:Disp "DIVIDEND IS F(X)"
:DISP "DEGREE OF F(X)"
:Input N
:0 → P
:Disp "COEFFS OF F(X)"
:Lbl 1
:P + 1 → P
:Input B
:B → {x}(P) [{x} on keyboard]
:If P < N + 1
:Goto 1
:1 → P
:0 → S
:Disp "COEFFS OF Q(X)"
:Lbl 2
:{x}(P) → K
:K + S → Q
:Q → {y}(P) [{y} on keyboard]
:Disp Q
:Pause
:AQ → S
:P + 1 → P
:If P < N + 1
:Goto 2
:{x}(P) → K
:K + S → Q
:Disp "REMAINDER"
:Disp Q
```

TI-82/83/85

[*Coefficients are entered as a list (use { } on the TI-82/83 keyboard or in the TI-85 LIST menu). When the list of quotient coefficients is displayed, use the left/right arrow keys to scroll through the list and press ENTER when finished.*]

:ClrHome [ClLCD *on TI-85*] [*Optional*]
:Disp "DIVISOR IS X − A"
:Prompt A
:Disp "ENTER DIVIDEND AS"
:Disp "COEFFICIENT LIST"
:Disp "(DECREASING POWERS)" [*Optional*]
:Input L_1 [*Use L_1 on the TI-82/83 keyboard; type it in on TI-85.*]
:dim $L_1 \rightarrow$ N
:$L_1 \rightarrow L_2$ [*Use L_1, L_2 on the TI-82/83 keyboard; type them in on TI-85.*]
:For (k,2,n)
:$(L_1(K) + A*L_2(K - 1)) \rightarrow L_2(K)$
:End
:round(L_2(N),9) $\rightarrow$ R
:(N − 1) $\rightarrow$ dim L_2
:Disp "DEGREE OF QUOTIENT"
:Disp dim L_2 − 1
:Disp "COEFFICIENTS"
:Pause L_2
:Disp "REMAINDER"
:Disp R

HP-38 RECTANGULAR/POLAR CONVERSION PROGRAM

Description: Enter the rectangular coordinates of a point in the plane; the program displays the polar coordinates of the point.

Input X; "RECTANGULAR TO POLAR"; "X ="; "ENTER X"; 0:
Input Y: "RECTANGULAR TO POLAR"; "Y ="; "ENTER Y"; 0:
If X > 0
Then (ATAN(Y/X)) $\rightarrow$ C
Else (ATAN(X,Y) + π) $\rightarrow$ C:
End
MSGBOX "R =" $\sqrt{X^2 + Y^2}$ "θ =" C:

HP-38 POLAR/RECTANGULAR CONVERSION PROGRAM

Description: Enter the polar coordinates of a point in the plane; the program displays the rectangular coordinates of the point.

Input R; "POLAR TO RECTANGULAR"; "R ="; "ENTER R"; 0:
Input θ; "POLAR TO RECTANGULAR"; "θ ="; "ENTER θ"; 0: [*CHARS menu*]
MSGBOX "X =" R(cos θ) "Y =" R(sin θ):

RECTANGULAR/POLAR CONVERSION PROGRAM

[illegible]

POLAR/RECTANGULAR CONVERSION PROGRAM

[illegible]

Answers to Odd-Numbered Exercises

CHAPTER 1

Section 1.1, page 10

1.

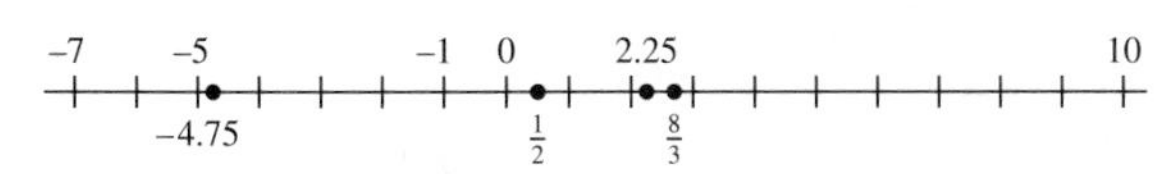

3. $-4 > -8$ **5.** $\pi < 100$ **7.** $y \le 7.5$

9. $t > 0$ **11.** $c \le 3$ **13.** $c < 4 \le d$

15. $<$ **17.** $=$ **19.** $>$

21. $b + c = a$ **23.** a lies to the right of b.

25. $a < b$ **27.** 11 **29.** 0 **31.** 10

33. 169 **35.** $\pi - \sqrt{2}$ **37.** π **39.** $<$

41. $>$ **43.** $<$

45. –2 0 2 4 6 8

47. –3 –2 –1 0 1 2 3

49. –4 –2 0 1

51. $[5, 8]$ **53.** $(-3, 14)$ **55.** $[-8, \infty)$

57. $(-\infty, 15]$

59. Many answers, including: true for $x = 0$, $y = 1$ and $x = -1$, $y = 0$; false for $x = 1$, $y = 1$ and $x = 2$, $y = -3$

61. Many answers, including: true for $x = 1$, $y = 2$ and $x = 3$, $y = 4$; false for $x = -3$, $y = 1$ and $x = -2$, $y = 5$

63. 7 **65.** 14.5 **67.** $\pi - 3$ **69.** $\sqrt{3} - \sqrt{2}$

71. t^2 **73.** $(-3 - y)^2 = (-1)^2(3 + y)^2 = (3 + y)^2$

75. $b - 3$ **77.** $-(c - d) = d - c$ **79.** 0

81. $|(c - d)^2| = (c - d)^2 = c^2 - 2cd + d^2$

83. $|x - 5| < 4$ **85.** $|x + 4| \le 17$

87. $|c| < |b|$ **89.** $|x| > |x + 6|$

91. The distance from x to 3 is less than 2 units.

93. The distance from x to -7 is at most 3 units.

95. The distance from b to 0 is less than the distance from c to 3.

97. There is no number that is within 2 units of 1 and at the same time within 3 units of 12.

99. $x = 1$ or -1 **101.** $x = 1$ or 3

103. $x = -\pi + 4$ or $-\pi - 4$ **105.** $-7 < x < 7$

107. $x \le -5$ or $x \ge 5$

109. Since $|a| \ge 0$, $|b| \ge 0$, and $|c| \ge 0$, the sum $|a| + |b| + |c|$ is positive only when one or more of $|a|$, $|b|$, $|c|$ is positive. But $|a|$ is positive only when $a \ne 0$; similarly for b, c.

111. (a) By a property of inequalities, $c \le |c|$ implies that $c + d \le |c| + d$. Similarly, $d \le |d|$ implies that $|c| + d \le |c| + |d|$. Hence, $c + d \le |c| + d$ and $|c| + d \le |c| + |d|$ imply that $c + d \le |c| + |d|$. A similar argument shows that $(-c) + (-d) \le |c| + |d|$. Since $-(c + d) = (-c) + (-d)$, we have $-(c + d) \le |c| + |d|$.

(b) If $c + d$ is nonnegative, then by the definition of absolute value and part (a), $|c + d| = c + d \le |c| + |d|$. If $c + d$ is negative, then by the definition of absolute value and part (a), $|c + d| = -(c + d) \le |c| + |d|$.

Excursion 1.1.A, page 14

1. $.7777\cdots$ **3.** $1.6428571428571\cdots$

5. $.0526315789473684210 52\cdots$

7. No; $\frac{2}{3} = .6666\cdots$ **9.** Yes; $\frac{1}{64} = .015625$

11. No **13.** Yes; $\frac{1}{.625} = 1.6$

15. $\frac{37}{99}$ **17.** $\frac{758{,}679}{9900} = \frac{252{,}893}{3300}$

19. $\frac{5}{37}$ **21.** $\frac{517{,}896}{9900} = \frac{14{,}386}{275}$

23. If $d = .74999\cdots$, then $1000d - 100d = (749.999\cdots) - (74.999\cdots) = 675$. Hence $900d = 675$ so that $d = \frac{675}{900} = \frac{3}{4}$. Also $.75000\cdots = .75 = \frac{75}{100} = \frac{3}{4}$.

25. $\frac{6}{17} = .35294117647058823529\cdots$

27. $\frac{1}{29} = .0344827586206896551724137931 0344\cdots$

29. $\frac{283}{47} = 6.02127659574468085106382978723404255319148936170212\cdots$

31. (a) One of many possible ways is to use the nonrepeating decimal expansion of π. For instance, with .75, associate $.7531415926\cdots$; with 6.593 associate $6.59331415926\cdots$, etc. Thus different terminating decimals correspond to different nonrepeating ones.

(b) As suggested in the *Hint,* associate with $.134134134\cdots$ the irrational number $.134013400134000134\cdots$. With $6.17398419841\cdots$ associate $6.173984109841009841000 9841\cdots$, etc. Thus, different repeating decimals lead to different nonrepeating ones.

Section 1.2, page 22

1. $x = 3$ or 5 **3.** $x = -2$ or 7

5. $y = \frac{1}{2}$ or -3 **7.** $t = -2$ or $-1/4$

9. $u = 1$ or $-\frac{4}{3}$ **11.** $x = \frac{1}{4}$ or $-\frac{4}{3}$

13. $x = -3$ or 5

15. $x = (1 + \sqrt{5})/2$ or $(1 - \sqrt{5})/2$

17. $x = 2 \pm \sqrt{3}$ **19.** $x = -3 \pm \sqrt{2}$

21. No real number solutions

23. $x = \frac{1}{2} \pm \sqrt{2}$ **25.** $x = \frac{2 \pm \sqrt{3}}{2}$

27. $u = \frac{-4 \pm \sqrt{6}}{5}$ **29.** 2

31. 2 **33.** 1

35. $x = -5$ or 8 **37.** $x = \frac{-5 \pm \sqrt{57}}{8}$

39. $x = -3$ or -6 **41.** $x = \frac{-1 \pm \sqrt{2}}{2}$

43. $x = 5$ or $-\frac{3}{2}$

45. No real number solutions

47. No real number solutions

49. $x \approx 1.824$ or $.470$ **51.** $x = 13.79$

53. $y = \pm 1$ or $\pm\sqrt{6}$ **55.** $x = \pm\sqrt{7}$

57. $y = \pm 2$ or $\pm 1/\sqrt{2}$ **59.** $x = \pm 1/\sqrt{5}$

61. $k = 10$ or -10 **63.** $k = 16$ **65.** $k = 4$

Excursion 1.2.A, page 23

1. $x = -6$ or 3 **3.** $x = 3/2$

5. $x = -5$ or 1 or -3 or -1

7. $x = 1$ or 4 or $\dfrac{5 + \sqrt{33}}{2}$ or $\dfrac{5 - \sqrt{33}}{2}$

Section 1.3, page 30

1. $A\,(-3, 3)$; $B\,(-1.5, 3)$; $C\,(-2.3, 0)$; $D\,(-1.5, -3)$; $E\,(0, 2)$; $F\,(0, 0)$; $G\,(2, 0)$; $H\,(3, 1)$; $I\,(3, -1)$.

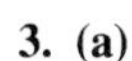
3. (a)

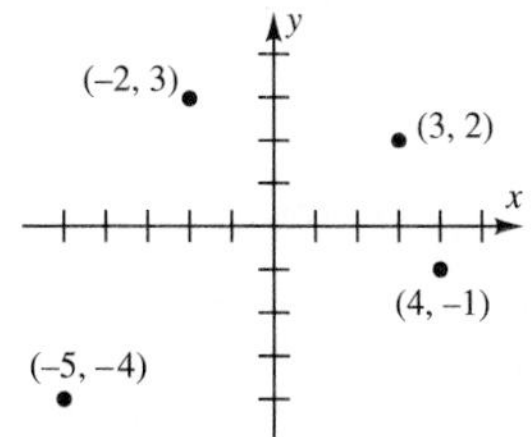

(b)

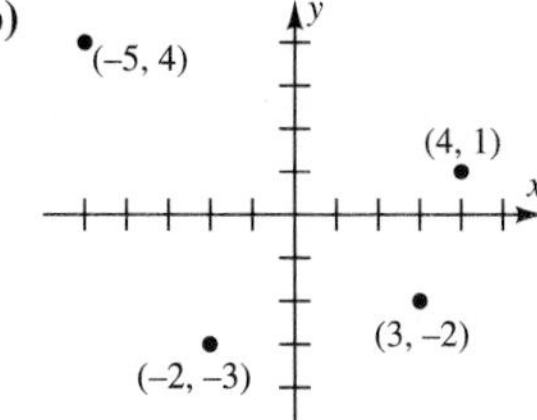

(c) They are mirror images of each other, with the x-axis being the mirror. In other words, they lie on the same vertical line, on opposite sides of the x-axis, the same distance from the axis.

5. $13; \left(-\frac{1}{2}, -1\right)$ **7.** $\sqrt{17}; \left(\frac{3}{2}, -3\right)$

9. $\sqrt{6 - 2\sqrt{6}} \approx 1.05; \left(\dfrac{\sqrt{2} + \sqrt{3}}{2}, \dfrac{3}{2}\right)$

11. $\sqrt{2}|a - b|; \left(\dfrac{a + b}{2}, \dfrac{a + b}{2}\right)$

13. Yes **15.** Yes **17.** No

19. $(x + 3)^2 + (y - 4)^2 = 4$ **21.** $x^2 + y^2 = 2$

23.

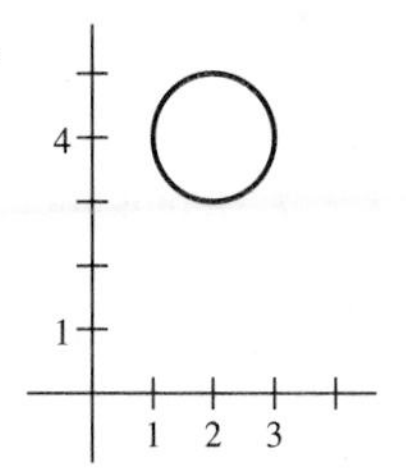

25.

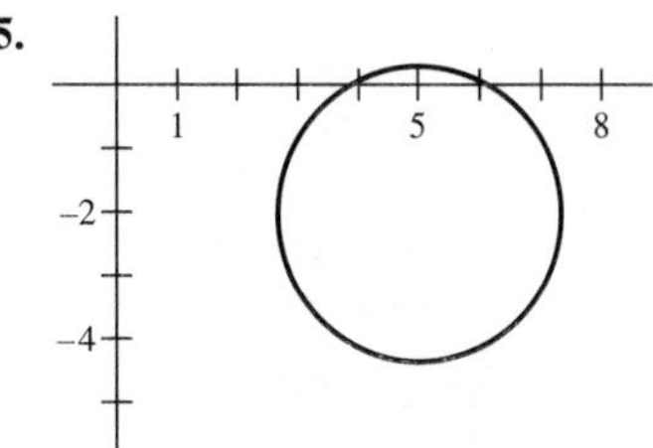

27. Center $(-4, 3)$, radius $2\sqrt{10}$

29. Center $(-3, 2)$, radius $2\sqrt{7}$

31. Center $(-12.5, -5)$, radius $\sqrt{169.25}$

33. Hypotenuse from $(1, 1)$ to $(2, -2)$ has length $\sqrt{10}$; other sides have lengths $\sqrt{2}$ and $\sqrt{8}$. Since $(\sqrt{2})^2 + (\sqrt{8})^2 = (\sqrt{10})^2$, this is a right triangle.

35. Hypotenuse from $(-2, 3)$ to $(3, -2)$ has length $\sqrt{50}$; other sides have lengths $\sqrt{5}$ and $\sqrt{45}$. Since $(\sqrt{5})^2 + (\sqrt{45})^2 = (\sqrt{50})^2$, this is a right triangle.

37. Horizontal straight line through $(0, 5)$

39. The coordinate axes

41.

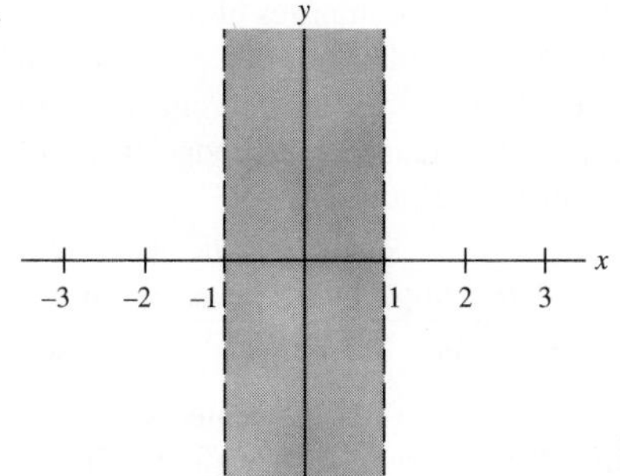

43. $(x - 2)^2 + (y - 2)^2 = 8$

45. $(x - 1)^2 + (y - 2)^2 = 8$

47. $(x + 5)^2 + (y - 4)^2 = 16$

49. $(-3, -4)$ and $(2, 1)$

51. Assume $k > d$. The other two vertices of one possible square are $(c + k - d, d)$, $(c + k - d, k)$; those of another square are $(c - (k - d), d)$, $(c - (k - d), k)$; those of a third square are $\left(c + \frac{k-d}{2}, \frac{k+d}{2}\right)$, $\left(c - \frac{k-d}{2}, \frac{k+d}{2}\right)$.

53. $(0, 0)$, $(6, 0)$ **55.** $(3, -5 + \sqrt{11})$, $(3, -5 - \sqrt{11})$

57. $x = 6$

59. M has coordinates $(s/2, r/2)$ by the midpoint formula. Hence the distance from M to $(0, 0)$ is

$$\sqrt{\left(\frac{s}{2} - 0\right)^2 + \left(\frac{r}{2} - 0\right)^2} = \sqrt{\frac{s^2}{4} + \frac{r^2}{4}},$$

and the distance from M to $(0, r)$ is the same:

$$\sqrt{\left(\frac{s}{2} - 0\right)^2 + \left(\frac{r}{2} - r\right)^2} = \sqrt{\left(\frac{s}{2}\right)^2 + \left(-\frac{r}{2}\right)^2} = \sqrt{\frac{s^2}{4} + \frac{r^2}{4}}$$

as is the distance from M to $(s, 0)$:

$$\sqrt{\left(\frac{s}{2} - s\right)^2 + \left(\frac{r}{2} - 0\right)^2} = \sqrt{\left(-\frac{s}{2}\right)^2 + \left(\frac{r}{2}\right)^2} = \sqrt{\frac{s^2}{4} + \frac{r^2}{4}}.$$

61. Place one vertex of the rectangle at the origin, with one side on the positive x-axis and another on the positive y-axis. Let $(a, 0)$ be the coordinates of the vertex on the y-axis and $(0, b)$ the coordinates of the vertex on the x-axis and $(0, b)$ the coordinates of the vertex on the y-axis. Then the fourth vertex has coordinates (a, b) (draw a picture!). One diagonal has endpoints $(0, b)$ and $(a, 0)$, so that its length is $\sqrt{(0-a)^2 + (b-0)^2} = \sqrt{a^2 + b^2}$. The other diagonal has endpoints $(0, 0)$ and (a, b) and hence has the same length: $\sqrt{(0-a)^2 + (0-b)^2} = \sqrt{a^2 + b^2}$

63. The circle $(x - k)^2 + y^2 = k^2$ has center $(k, 0)$ and radius $|k|$ (the distance from $(k, 0)$ to $(0, 0)$). So the family consists of every circle that is tangent to the y-axis *and* has center on the x-axis.

65. **(a)** $(0, -5)$ goes to $(0, 0)$; $(2, 2)$ goes to $(2, 7)$; $(5, 0)$ goes to $(5, 5)$; $(5, 5)$ goes to $(5, 10)$; $(4, 1)$ goes to $(4, 6)$

(b) $(0, -10)$ goes to $(0, -5)$; $(2, -3)$ goes to $(2, 2)$; $(5, -5)$ goes to $(5, 0)$; $(5, 0)$ goes to $(5, 5)$; $(4, -4)$ goes to $(4, 1)$

(c) $(a, b + 5)$ **(d)** (a, b)

(e) $(-4a, b - 5)$ **(f)** No points go to themselves.

67. The points are on opposite sides of the origin because one first coordinate is positive and one is negative. They are equidistant from the origin because the midpoint on the line segment joining them is $\left(\frac{c + (-c)}{2}, \frac{d + (-d)}{2}\right) = (0, 0)$.

Section 1.4, page 46

1. $P = (3, 5)$; $Q = (-6, 2)$

3. $P = (-12, 8)$; $Q = (-3, -8.5)$

5–10. Answers vary.

11. c **13.** d **15.** e **17.** (c)

19. $0 \le x \le 42$ and $0 \le y \le 4500$, with x-scl $= 5$, y-scl $= 500$. Both x (time) and y (concentration) must be nonnegative and this window shows the part of the graph where they are.

21. $0 \le x \le 4000$ and $0 \le y \le 100$, with x-scl $= 500$, y-scl $= 10$. Both x and y (numbers of barrels) must be nonnegative and this window shows the part of the graph where they are.

23. **(b)** The graphs are different.

25. False **27.** Possibly true

In Answers 29–33, the graphs shown were made on a TI-82, which has the same size screen as TI-81 and most Casios and Sharps. For wide-screen calculators such as TI-85, TI-92, and HP-38, the x-axis should be longer than this one to have a square window with the same y-axis as here.

29.

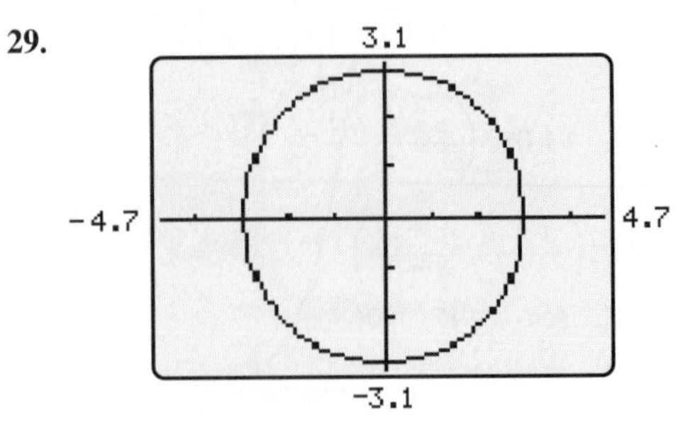

31. The two halves of the graph should be connected, but the ends of the two pieces are almost vertical so the calculator could not plot enough points near them to make the graph appear connected.

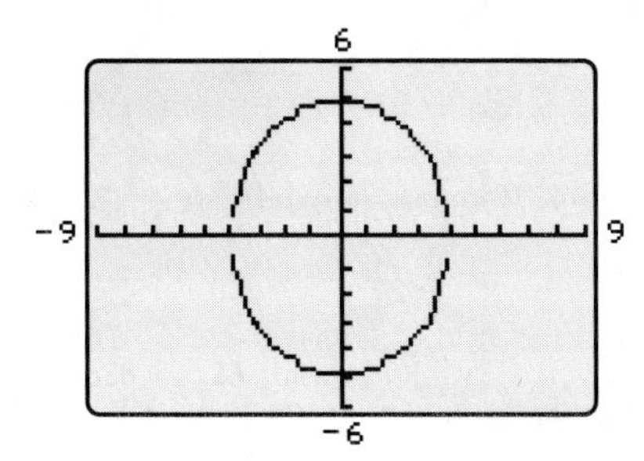

33.

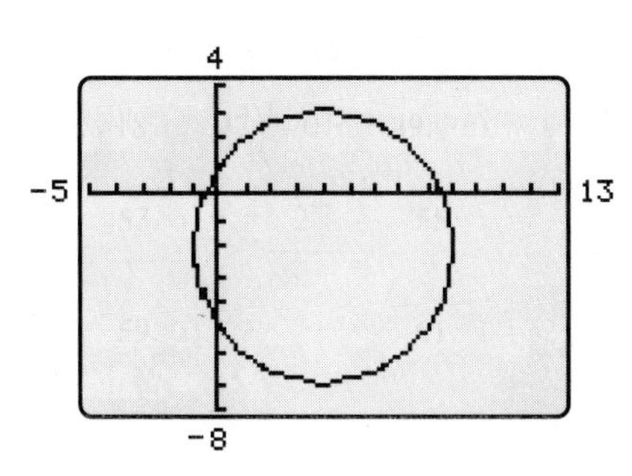

35. $-5 \le x \le 5$ and $-100 \le y \le 100$

37. $-10 \le x \le 10$ and $-2 \le y \le 20$ [Where is the right half of the graph?]

39. $-6 \le x \le 12$ and $-100 \le y \le 250$

41.

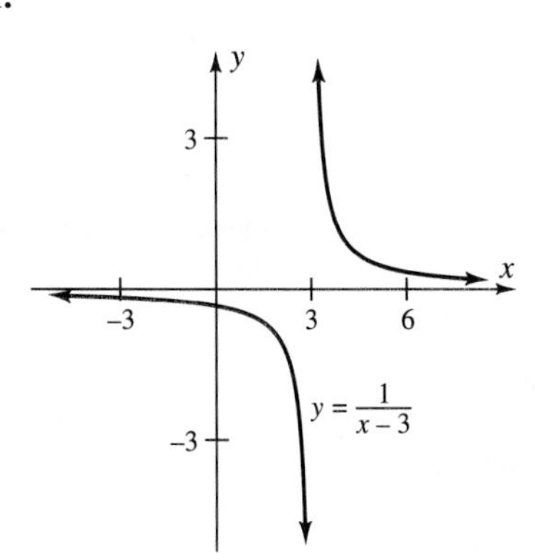

43.

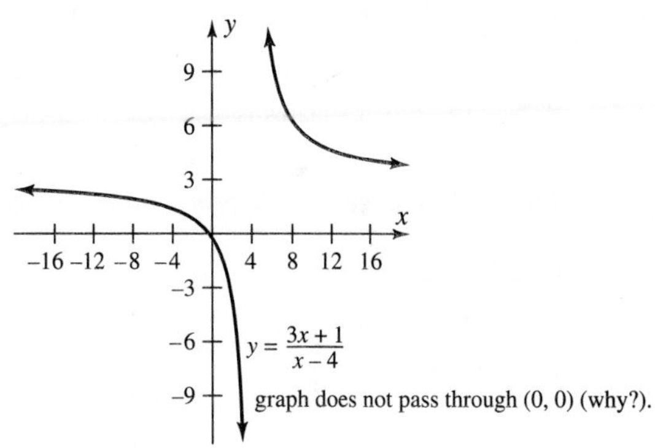

45. All four graphs have the same shape. Graph (b) is graph (a) shifted 5 units vertically upward, graph (c) is graph (a) shifted 5 units vertically downward, and graph (d) is graph (a) shifted 2 units vertically downward.

47. Graph (b) is graph (a) stretched by a factor of 2 vertically away from the x-axis; similarly for graph (c), except the stretching factor is 3. Graph (d) is graph (a) shrunk vertically toward the x-axis by a factor of 1/2.

49. The graphs are ''mirror images'' of each other, with the straight line $y = x$ being the mirror.

51. Same answer as Exercise 49.

53. Same answer as Exercise 49.

Section 1.5, page 58

1. **(a)** .1 **(b)** .05

3. **(a)** .0002 **(b)** .0001

5. 3 **7.** 3 **9.** 2

11. $x = 2.66889$ **13.** $x = -2.42645$

15. $x = 1.1640$ **17.** $x = -1.1043$

19. $x = -1.1038$ **21.** $x = 2.1017$

23. $x = -1.7521$ **25.** $x = .9505$

27. $x = 0$ or 2.2074 **29.** $x = -.4125$ or 1.1211

31. $x = 1.23725$ **33.** $x = 1.1921$

35. $x = -1.6005$ **37.** $x = -1.09128$ or 1.72699

39. $x = 2/3$ **41.** $x = 1/12$

43. $x = \sqrt{3}$ **45.** $x = 1.4528$

47. $x = 1.9097$

Section 1.6, page 68

1. The two numbers: x and y; their sum is 15: $x + y = 15$; the difference of their squares is 5: $x^2 - y^2 = 5$.

3.

English Language	*Mathematical Language*
Length of rectangle	x
Width of rectangle	y
Perimeter is 45	$x + y + x + y = 45$ or $2x + 2y = 45$
Area is 112.5	$xy = 112.5$

5. Let x be the old salary. Then the raise is 8% of x. Hence,

$$\text{old salary} + \text{raise} = \$1600$$
$$x + (8\% \text{ of } x) = 1600$$
$$x + .08x = 1600.$$

7. The circle has radius $r = 16/2 = 8$, so its area is $\pi r^2 = \pi \cdot 8^2 = 64\pi$. Let x be the amount by which the radius is to be reduced. Then $r = 8 - x$ and the new area is $\pi(8 - x)^2$, which must be 48π less than the original area, that is, $\pi(8 - x)^2 = 64\pi - 48\pi$, or equivalently, $\pi(8 - x)^2 = 16\pi$.

9. \$366.67 at 12% and \$733.33 at 6%

11. $2\frac{2}{3}$ quarts 13. 65 mph

15. 38.25 and 44 17. Approximately 1.753 ft

19. Red Riding Hood, 54 mph; wolf, 48 mph

21. (a) Approximately 6.3 sec (b) Approximately 4.9 sec

23. (a) Approximately 4.4 sec (b) After 50 sec

25. 23 cm by 24 cm by 25 cm 27. $r = 4.658$

29. $x = 2.234$ 31. 2.2 by 4.4 by 4 ft high

Section 1.7, page 77

1. $(-1, 8)$ 3. $(1.42857, -2.0625)$

5. $(-.3409, .0003222)$

7. (a) $(-1, 4)$ (b) $(-1, 4)$ and $(2, 4)$ (c) $(3, 20)$

9. 39.1487 by 39.1487 by 19.5743

11. (a) 4.4267 by 4.4267 inches. (b) 10/3 by 10/3 inches

13. (a) Approximately 206
(b) Approximately 269; approximately \$577

15. (a) 600 (b) 958 17. $x = 9.306$

19. Approximately (1.870, 1.936) 21. 12 times

Chapter 1 Review, page 81

1. (a) $>$ (b) $<$ (c) $<$ (d) $>$ (e) $=$

3. $\frac{28}{99}$ 5. (a) $-10 < y < 0$ (b) $0 \le x \le 10$

7. (a) $|x + 7| < 3$ (b) $|y| > |x - 3|$

9. $x = 2$ or 8 11. $x = -1/2$ or $-11/2$

13. $-4 \le x \le 0$ 15. (a) $7 - \pi$ (b) $\sqrt{23} - \sqrt{3}$

17. (a) $(-8, \infty)$ (b) $(-\infty, 5]$ 19. $x = 44/7$

21. No real solutions 23. $z = \dfrac{-3 \pm 2\sqrt{11}}{5}$

25. 2 27. $x = 3$ or -3 or $\sqrt{2}$ or $-\sqrt{2}$

29. $x = -1$ or $5/3$ 31. 3/11 oz gold; 8/11 oz silver

33. $2\frac{2}{9}$ hours 35. 9.6 ft 37. 4 ft

39. $\sqrt{58}$ 41. $\sqrt{c^2 + d^2}$ 43. $\left(d, \dfrac{c + 2d}{2}\right)$

45. (a) $\sqrt{17}$ (b) $(x - 2)^2 + (y + 3)^2 = 17$

47.

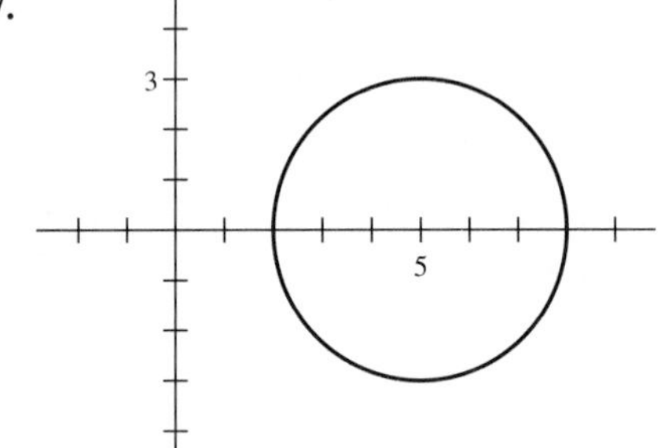

49. (b) and (d) 51. (c)

53. (a) a, d
(b) b, c do not show peaks and valleys near the origin; e crowds the graph onto the y-axis.
(c) a or $-4 \le x \le 6$ and $-10 \le y \le 10$

55. (a) None of them
(b) a, b, d do not show any peaks or valleys; c does not show the valleys; e shows only one point on the graph (which can't be distinguished because it's on the y-axis).
(c) $-7 \le x \le 11$ and $-1000 \le y \le 500$

57. **(a)** b, c
(b) a shows no peaks or valleys; d is too crowded horizontally; e shows only one point.
(c) $-10 \le x \le 10$ and $-150 \le y \le 150$

59. As x moves left or right from the origin, x^2 grows larger and larger, and hence so does $x^2 - 10$. Therefore, the graph always rises as it moves away from the y-axis and is complete.

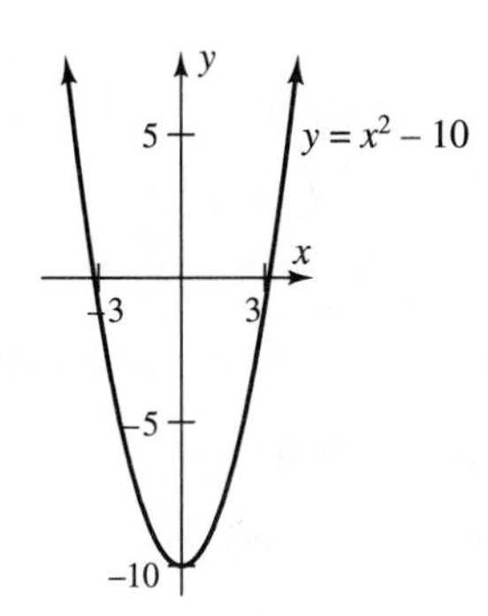

61. Note that y is defined only when $x \ge 5$ (why?). Also, $x - 5$ grows larger as x gets larger (that is, as you move to the right) and hence the same is true of $\sqrt{x-5}$. Therefore, the graph always rises as you move to the right and is complete.

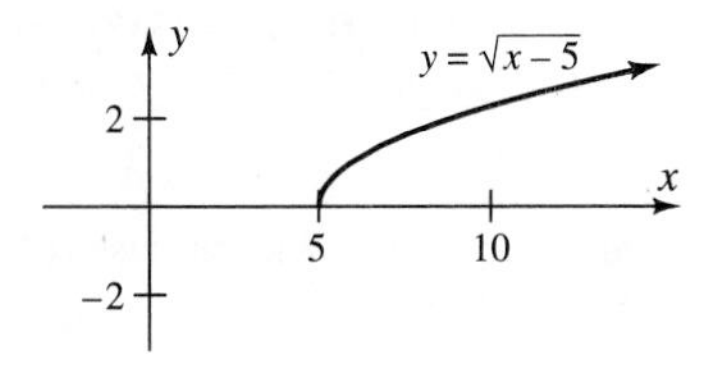

63.

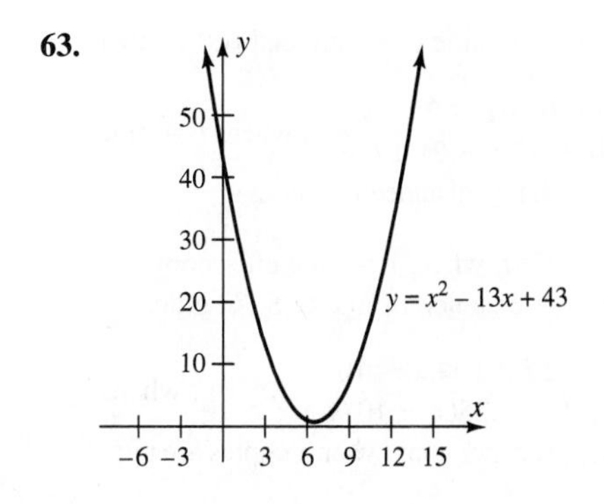

65.

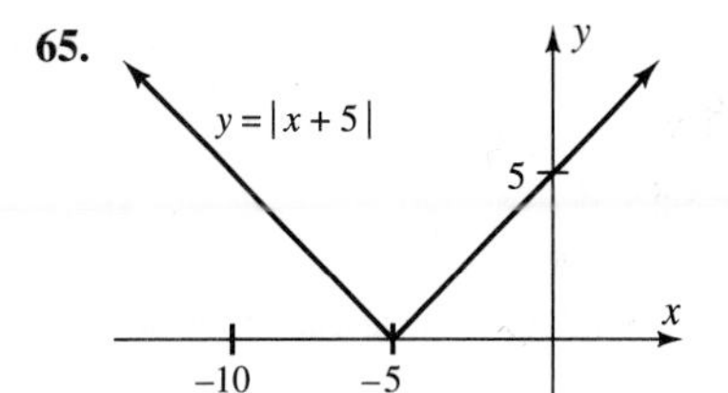

67. $x = 2.7644$ **69.** $x = 1$ or 2.8284

71. $x = -3.2843$ **73.** $x = 1.6511$ **75.** 25

77. 20 yds by 30 yds (interior fence is 20 yds)

79. $x = \sqrt{3} \approx 1.732$

CHAPTER 2

Section 2.1, page 93

1. 6 **3.** -2 **5.** 0 **7.** -17

9. Many correct answers, including: true for $u = 2$, $v = 3/4$; false for $u = 2$, $v = 1$.

11. The area is a function of the radius; domain: all radii (that is, all positive real numbers); range: all areas (all positive real numbers); rule: area equals πr^2, where r is the radius.

13. Sales are a function of (money spent on) advertising; domain: all amounts (in dollars) to be spent on advertising; range: all total sales amounts (in dollars); rule: insufficient information.

15. Tax is 0 on \$500 and \$1509. Tax is \$35.08 on \$3754. Tax is \$119.15 on \$6783; \$405 on \$12,500; and \$2547.10 on \$55,342.

17. *Each* of the different numbers 175, 560, 1120, 1800 in the domain is assigned to the *one* number 0 in the range. The definition of the rule can be contradicted only if one number in the domain is assigned to several different numbers in the range.

19. Postage is a function of weight since each weight determines one and only one postage amount. But weight is *not* a function of postage since a given postage amount may apply to several different weights. For instance, *all* letters under 1 ounce use just one first-class stamp.

21. y is a function of x.

23. x is a function of y.

25. y is a function of x and x is a function of y.

27. Neither x nor y is a function of the other.

29. Domain: approximately $[-2.8, 3]$; range: approximately $[-4, 3.8]$

31. 2 is assigned to 1/2; 0 to 5/2; and -3 to $-5/2$.

33. -0.7 (approximately) is assigned to -2; 3 to 0; 2 to 1; -2.1 (approximately) to 2.5; and 1 to -1.5.

35. 1 is assigned to -2; -3 to -1; -1 to 0; 0.5 (approximately) to 1/2; and 1.5 to 1.

37. (a) All positive numbers that can be entered in your calculator
(b) All numbers between -1 and 1 (inclusive) that can be displayed on your calculator

39. (a) The integer part of a positive number c is the integer that is closest to the number and less than or equal to the number, that is, $[c]$.
(b) All negative integers: $-1, -2, -3, \ldots$
(c) All negative numbers that are not integers

Section 2.2 page 103

1. $\sqrt{3} + 1$ **3.** $\sqrt{11/2} - \frac{3}{2}$

5. $\sqrt{\sqrt{2} + 3} - \sqrt{2} + 1$ **7.** 4 **9.** $\frac{34}{3}$

11. $\frac{59}{12}$ **13.** $(a + k)^2 + \frac{1}{a + k} + 2$

15. $(2 - x)^2 + \frac{1}{2 - x} + 2 = 6 - 4x + x^2 + \frac{1}{2 - x}$

17. 8 **19.** -1 **21.** $(s + 1)^2 - 1 = s^2 + 2s$

23. $t^2 - 1$

	$f(r)$	$f(r) - f(x)$	$\frac{f(r) - f(x)}{r - x}$
25.	r	$r - x$	1
27.	$3r + 7$	$3(r - x)$	3
29.	$r - r^2$	$r - r^2 - x + x^2$	$1 - r - x$
31.	$\sqrt{r}$	$\sqrt{r} - \sqrt{x}$	$\frac{1}{\sqrt{r} + \sqrt{x}}$

33. 1 **35.** 3 **37.** $-2x - h + 1$

39. $\frac{1}{\sqrt{x + h} + \sqrt{x}}$

41. $f(-3) = 1.1$ (approximately); $f\left(-\frac{3}{2}\right) = 1.5$ (approximately); $f(0) = -2.8$ (approximately); $f(1) = -0.2$ (approximately); $f\left(\frac{5}{2}\right) = 2$; $f(4) = 1.5$

43. $f\left(-\frac{5}{2}\right) = -1.2$ (approximately); $f\left(-\frac{3}{2}\right) = 0$; $f(0) = 1$; $f(3) = 3$; $f(4) = 1$

45. (iii) or (v) **47.** All real numbers

49. All real numbers **51.** All real numbers

53. All nonnegative real numbers

55. All nonzero real numbers

57. All real numbers **59.** All real numbers

61. All real numbers except -2 and 3 **63.** $[6, 12]$

65. Many possible answers, including $f(x) = x^2$ and $g(x) = |x|$

67. $f(x) = 0$ for all x

69. (iii) and (iv) are true; (i) is false because $f(3^2) = f(9) = 5 \cdot 9 = 45$, but $(f(3))^2 = (5 \cdot 3)^2 = 15^2 = 225$; (ii) is false because $f(-2) = 5(-2) = -10$, but $f(2) = 5 \cdot 2 = 10$.

71. (i)–(iii) are true; (iv) is false because $f(3 \cdot 2) = f(6) = 36$, but $3f(2) = 3 \cdot 2^2 = 12$.

73. $f(r) = 2\pi r$, where r = radius and $f(r)$ = circumference

75. $f(s) = s^2$, where s = side of square and $f(s)$ = area

77. $d(t) = \begin{cases} 55t \text{ if } 0 \le t \le 2 \\ 110 + 45(t - 2) \text{ if } t > 2 \end{cases}$, where t = time (in hours) and $d(t)$ = distance (in miles)

79. $d(t) = 2000 - 475t$, where t = time after noon (in hours) and $d(t)$ = distance from city S (in miles)

81. (a) $p(x) = \begin{cases} 12 \text{ for } 1 \le x \le 10 \\ 12 - .25(x - 10) \text{ for } x > 10 \end{cases}$, where $p(x)$ is the price per copy when x copies are purchased

(b) $T(x) = \begin{cases} 12x \text{ for } 1 \le x \le 10 \\ 14.5x - .25x^2 \text{ for } x > 10 \end{cases}$, where $T(x)$ is the total cost of x copies

83. $c(x) = 5.75x + (45{,}000/x)$

Section 2.3, page 113

1.

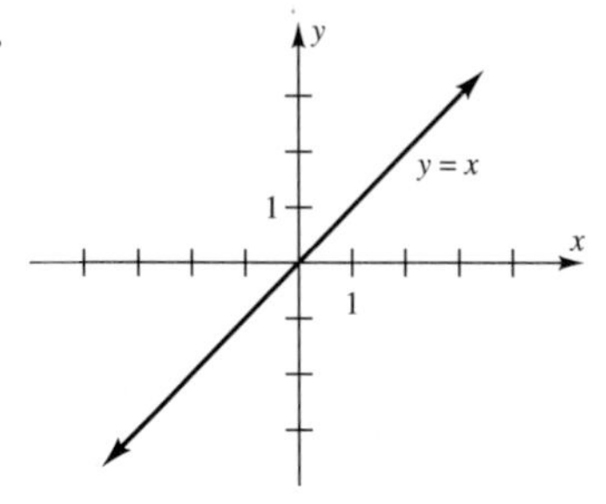

3.

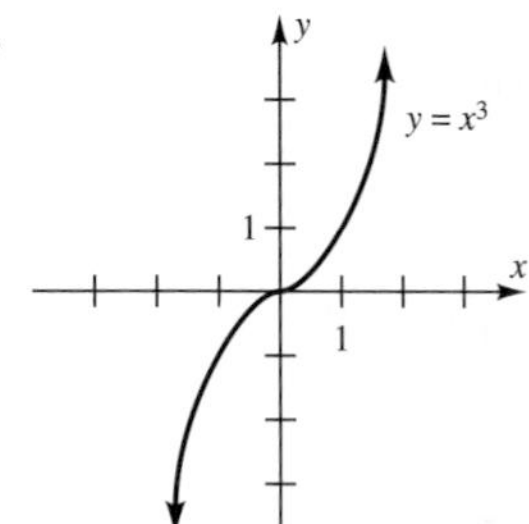

5.

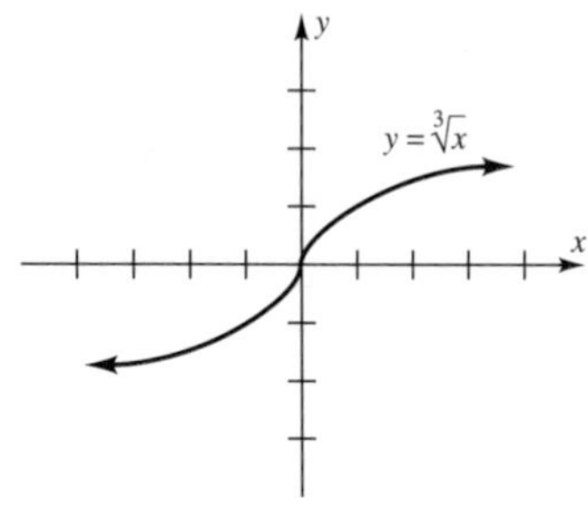

7. Increasing on $(-2.5, 0)$ and $(1.7, 4)$; decreasing on $(-6, -2.5)$ and $(0, 1.7)$

9. Increasing on $(-5.5, -3.5)$, $(-2, 0)$, and $(1, 3)$; decreasing on $(-6, -5.5)$, $(-3.5, -2)$, and $(0, 1)$.

11.

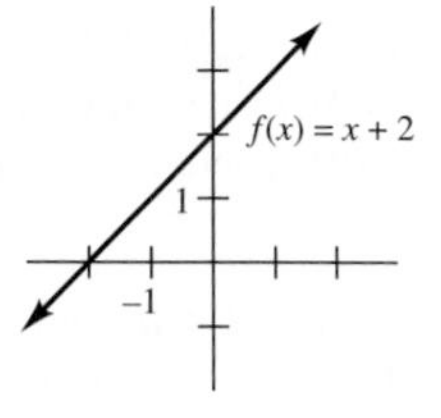

13.

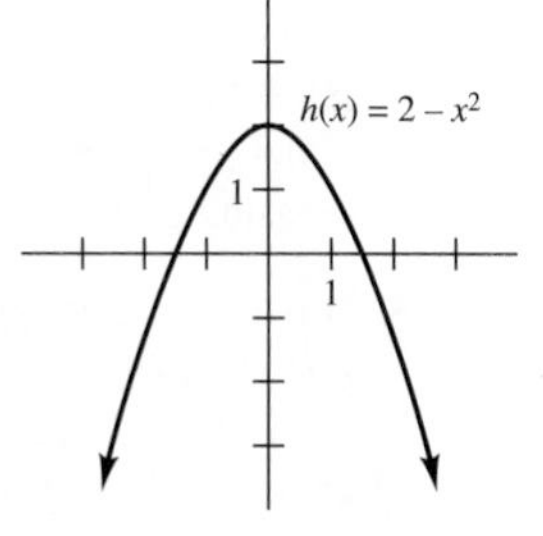

15.

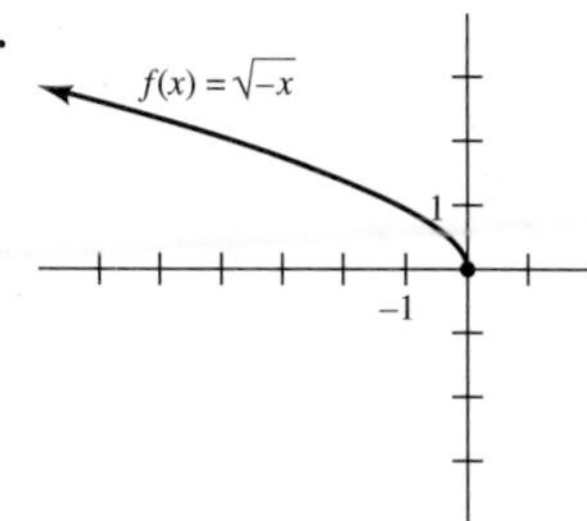

17.

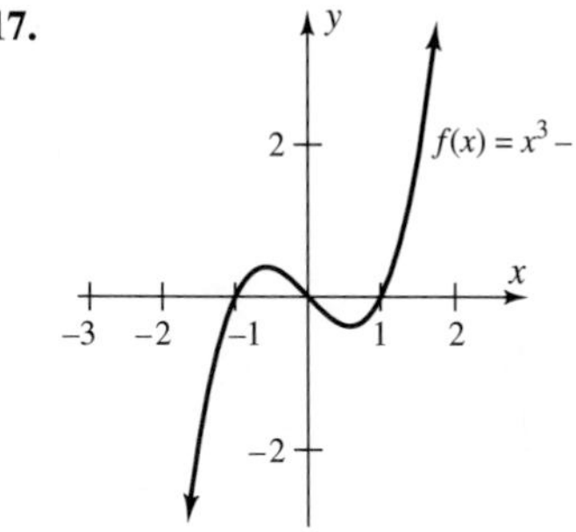

19.

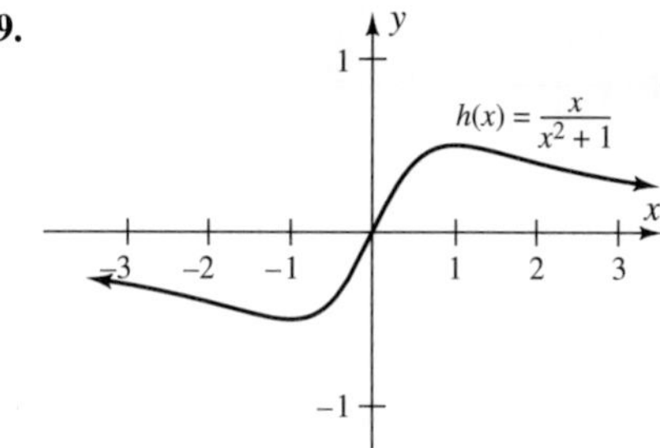

21.

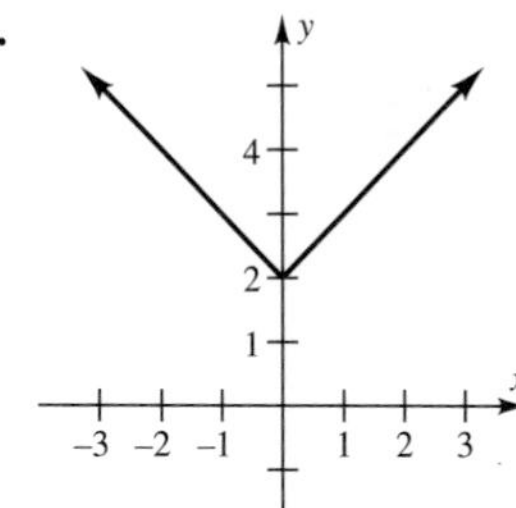

23.

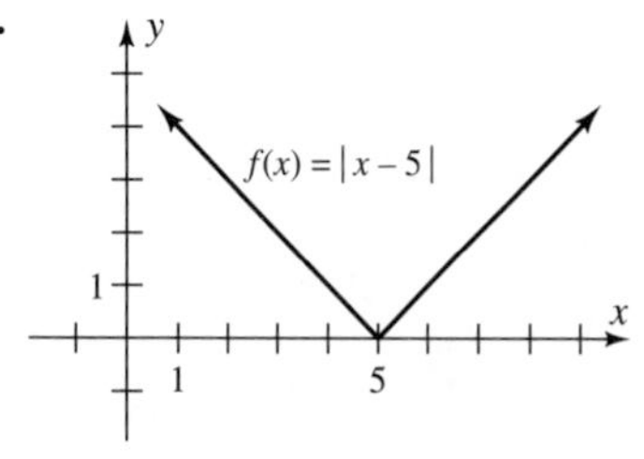

25.

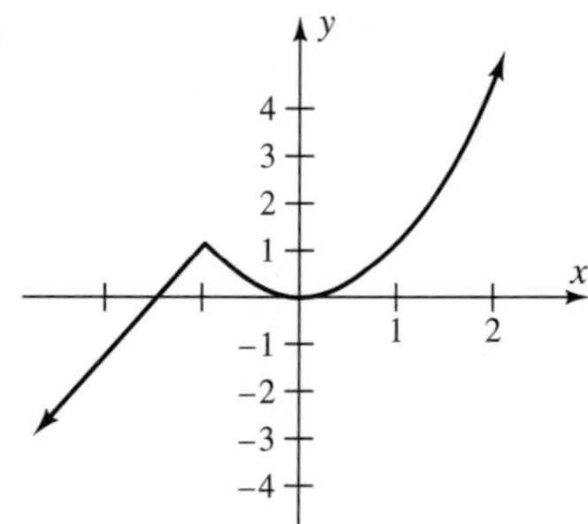

27.

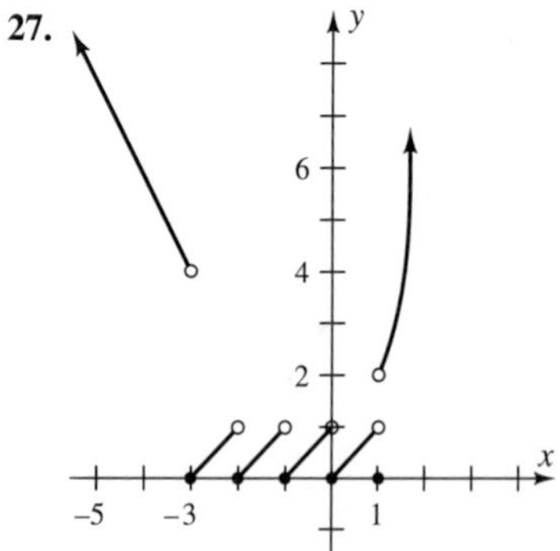

29. Graph for $-2 \le x < 3$:

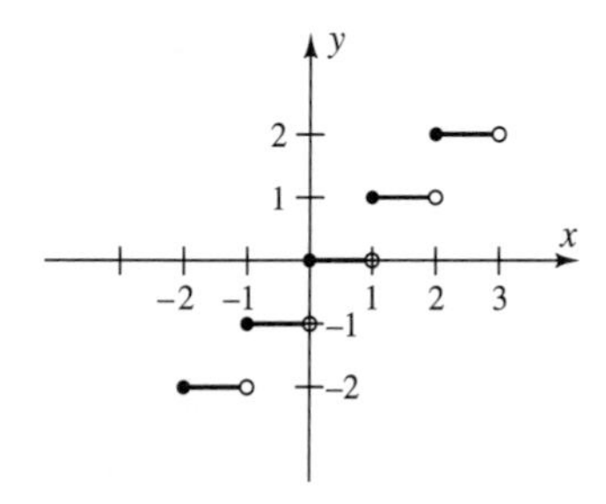

31.

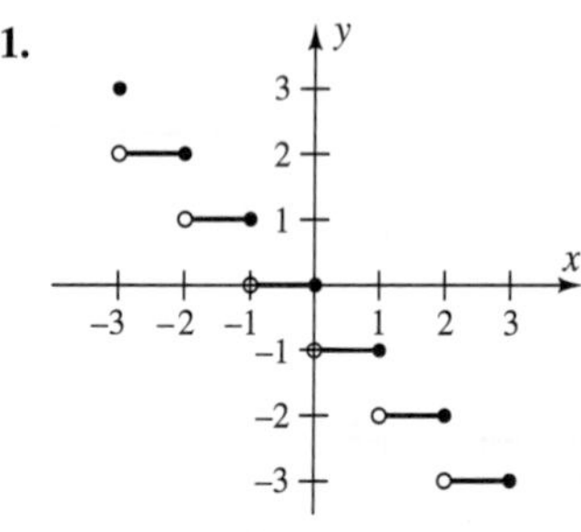

33. **(a)** Several possibilities, including

$$p(x) = \begin{cases} [x] \text{ if } x \text{ is an integer} \\ [x] + 1 \text{ if } x \text{ is not an integer} \end{cases}$$

or $p(x) = -[-x]$, with $x > 0$ in all cases, where x is the weight in ounces

(b) Graph for $0 < x \le 4$:

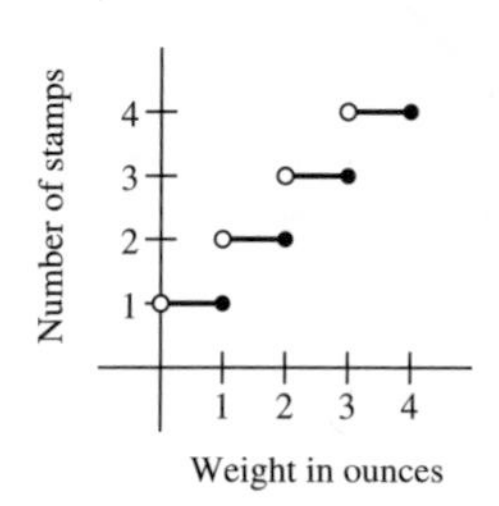

(c)

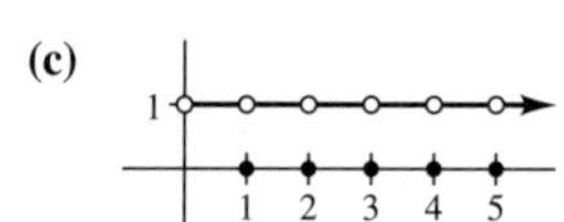

35. Many correct answers, including

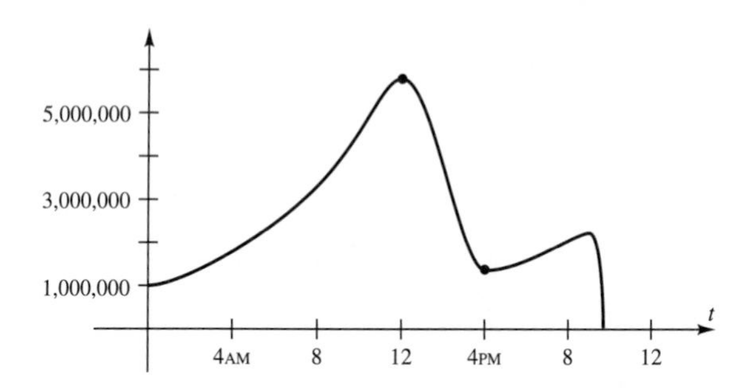

37. Decreasing on $(-\infty, \infty)$

39. Decreasing on $(-\infty, -5.7936)$ and $(.460277, \infty)$; increasing on $(-5.7936, .460277)$.

41. Decreasing on $(-\infty, 0)$ and $(.8672, 2.8828)$; increasing on $(0, .8672)$ and $(2.8828, \infty)$.

43. If $0 < c < d \le 10$, then $c^2 < d^2$. Hence, $c^2 + 3 < d^2 + 3$. But this says $f(c) < f(d)$. We have shown that if $0 < c < d \le 10$, then $f(c) < f(d)$. Therefore, f is increasing on $(0, 10]$.

45. If $0 < c < d \le 10$, then $c^2 < d^2$. But $c^2 < d^2$ and $c < d$ imply $c^2 + c < d^2 + d$. Hence, $c^2 + c + 5 < d^2 + d + 5$, that is, $h(c) < h(d)$. Therefore, h is increasing on $(0, 10]$.

47.

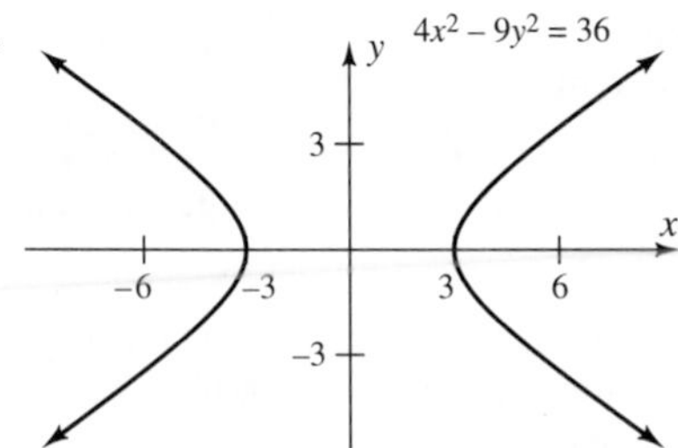

49.

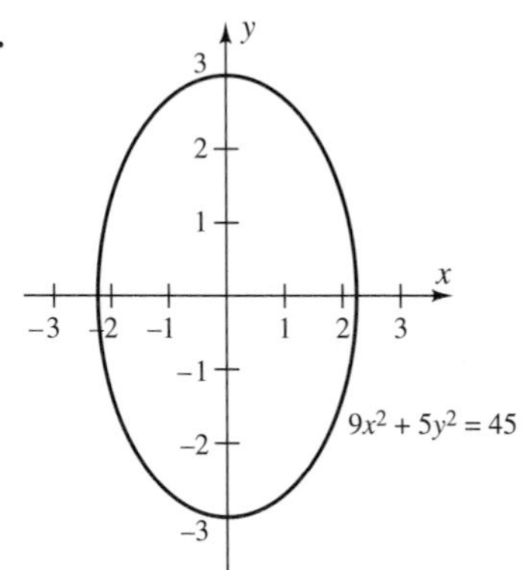

51.

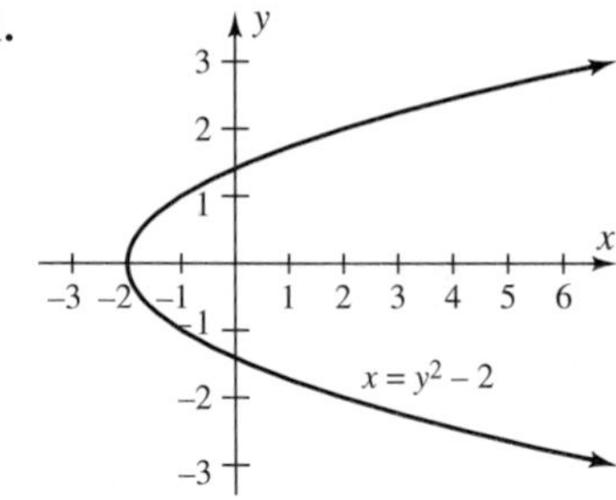

53. $-15 \le x \le 30$ and $-10 \le y \le 10$ $(-10 \le t \le 10)$

55. $-10 \le x \le 6$ and $-7 \le y \le 7$ $(-7 \le t \le 7)$

57. $-15 \le x \le 0$ and $0 \le y \le 4$ $(0 \le t \le 4)$

59. $-5 \le x \le 45$ and $-65 \le y \le 65$

61. Entire graph: $-2 \le x \le 32$ and $-10 \le y \le 75$; near the origin: $-2 \le x \le 5$ and $-10 \le y \le 10$

63. Entire graph: $-16 \le x \le 2$ and $-62 \le y \le 60$; near the origin: $-2 \le x \le 2$ and $-4 \le y \le 2$

65. 7 times

67. The graphs are ''mirror images'' of each other, with the line $y = x$ being the mirror.

69. Same as 67.

71.

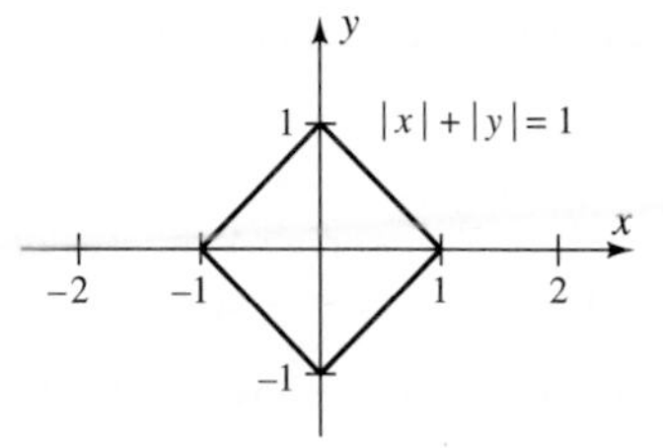

Section 2.4, page 120

1. Domain: all real numbers; range: all real numbers

3. Domain: all real numbers x such that $x \le -2$ or $x \ge 2$; range: all nonnegative real numbers

5. Domain: all nonzero real numbers; range: all real numbers y such that $y \le -2$ or $y \ge 2$

7. $[-3, 5]$ **9.** 4 **11.** 3.5 **13.** 4.5

15. 4 **17.** 1, 5 **19.** $[-3, 3]$

21. Many correct answers, including

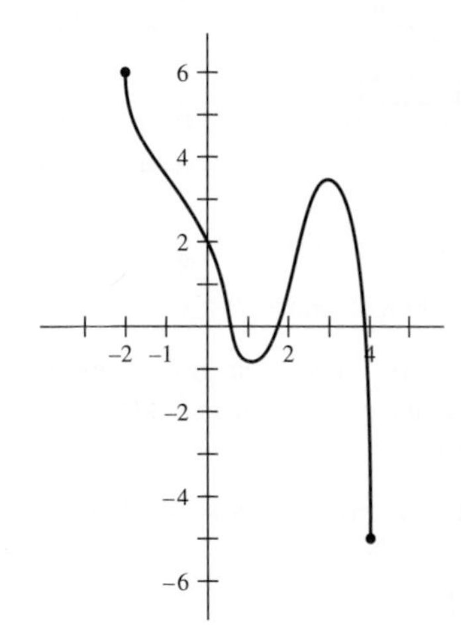

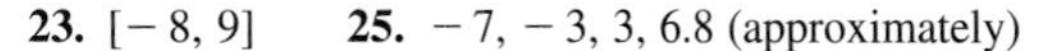

23. $[-8, 9]$ **25.** $-7, -3, 3, 6.8$ (approximately)

27. 1, 3, 5, and others **29.** 1 and -9

31. Approximately 5.5, and others

33. Domain $f = [-6, 7]$; domain $g = [-8, 9]$

35. Approximately -1.5 and $-.2$ **37.** $x = 3$

39. $[-2, -1]$ and $[3, 7]$ **41.** graph 4 **43.** graph 2

45. All five; for many values of x

47. g = graph 1 and f = graph 1 or 4; g = graph 2 and f = graph 1, 2, or 4; g = graph 3 and f = graph 1, 2, 4, or 5; g = graph 5 and f = graph 1, 2, 4, or 5

49. Any pair of distinct functions, except f = graph 3 and g = graph 5

51. f = graph 1 and g = graph 1, 2, 3, or 5; f = graph 2 and g = graph 1, 2, 3, or 5; f = graph 4 and g = graph 1, 2, 3, 4, or 5; f = graph 5 and g = graph 3

53. Approximately − \$13,000 (that is, a loss of \$13,000)

55. Approximately 12,300

7. Shift the graph of f horizontally 3 units to the right; then shift it 2 units vertically upward.

9. Reflect the graph of f in the x-axis; shrink it toward the x-axis by a factor of 1/2; shift it vertically 6 units downward.

11. Shift the graph of f horizontally 2 units to the right; stretch it away from the x-axis by a factor of 2; then reflect it in the x-axis.

57.

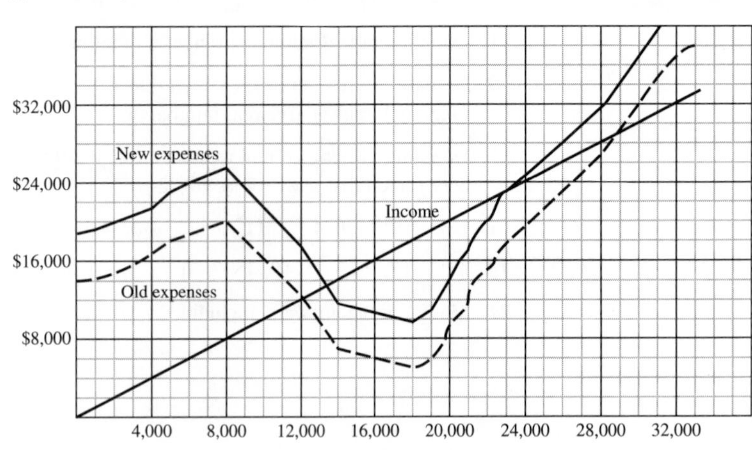

59. tem(10) ≈ 47°; tem(3 + 12) ≈ 63°

61. Approximately 10:30 A.M. and 8:30 P.M.

63. tem(10) ≈ 47°, so (10, 47) is on the graph; tem(16) ≈ 64°, so (16, 64) is on the graph. The point (10, 47) lies 17 units lower than (16, 64).

65. No **67.** Approximately 10:30 A.M. and 8:30 P.M.

13. $g(x) = f(x + 5) + 4 = (x + 5)^2 + 2 + 4 = (x + 5)^2 + 6$

15. $g(x) = 2f(x - 6) - 3 = 2\sqrt{x - 6} - 3$

17. $g(x) = 2f(x - 2) + 2$
$= 2[(x - 2)^2 + 3(x - 2) + 1] + 2$
$= 2(x - 2)^2 + 6(x - 2) + 4$

Section 2.5, page 131

1. Viewing window: $-10 \le x \le 10$ and $-36 \le y \le 42$

3. Viewing window: $-13 \le x \le 12$ and $-2 \le y \le 14$

5.

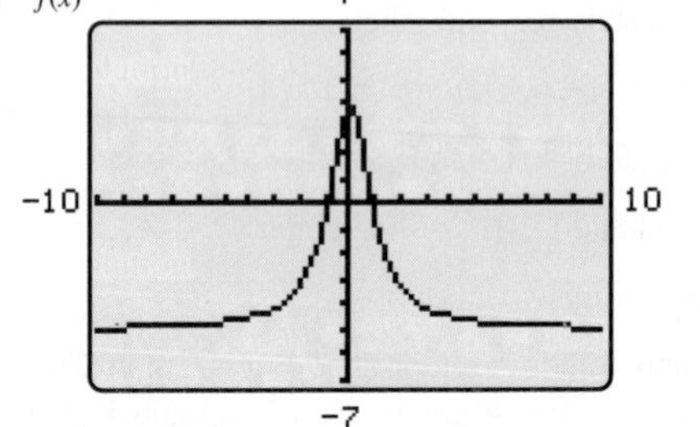

g(x)
7
-10
10
-7

19.

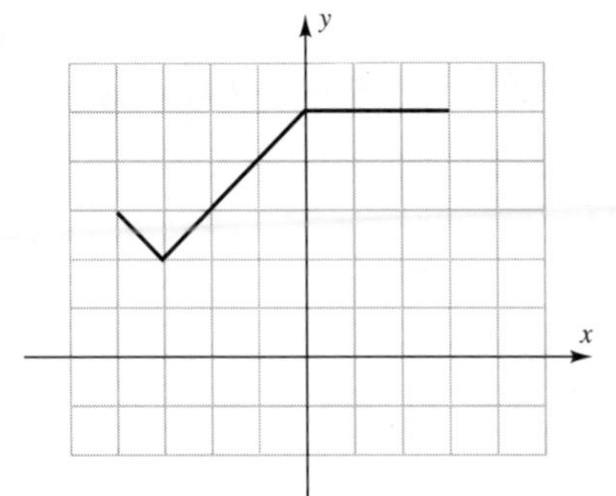

21.

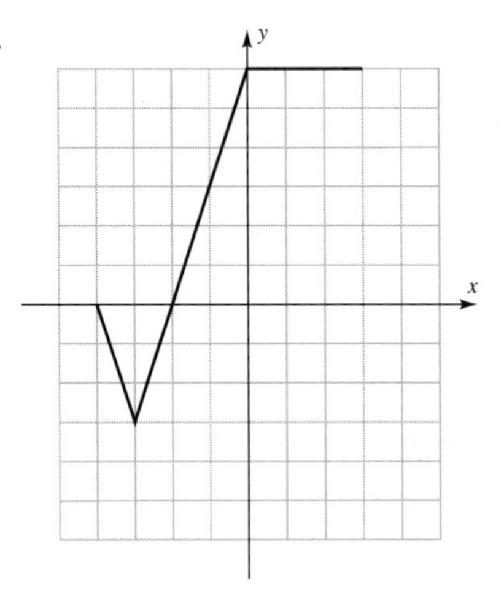

23.

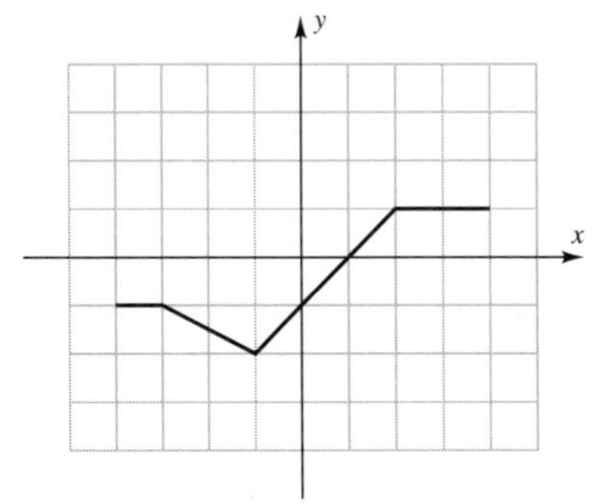

25.

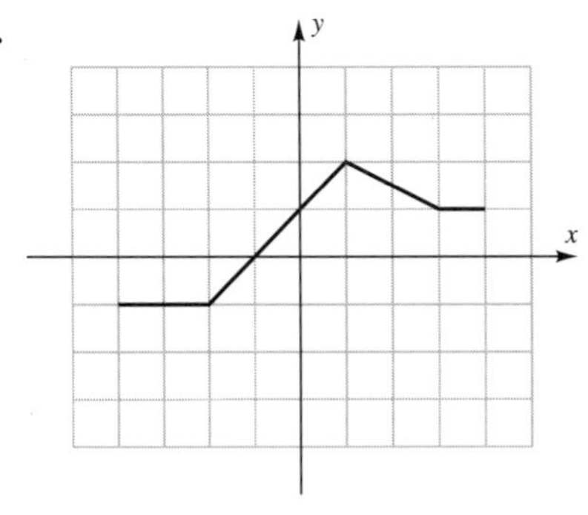

27.

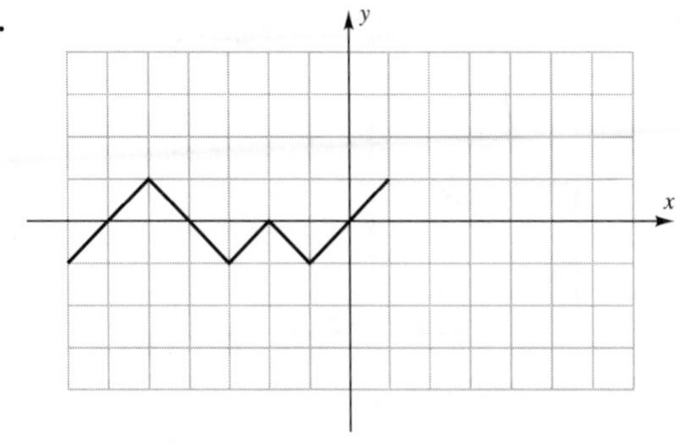

29.

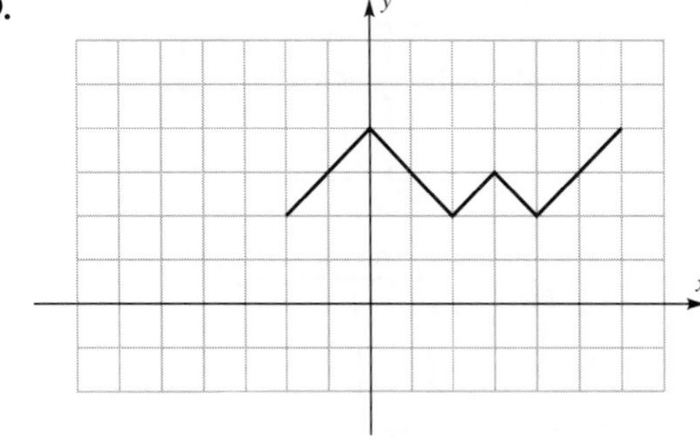

31.

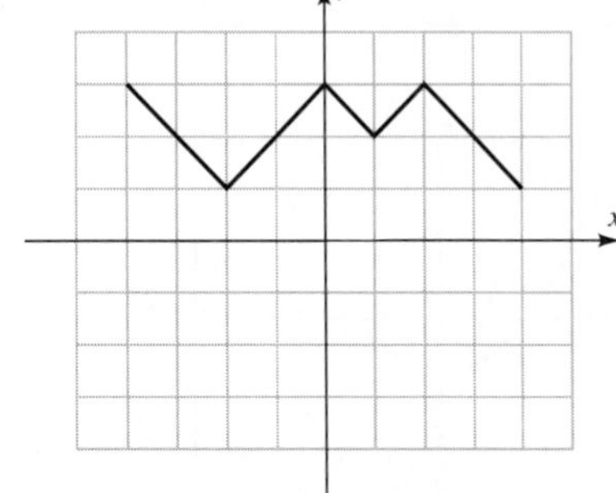

33.

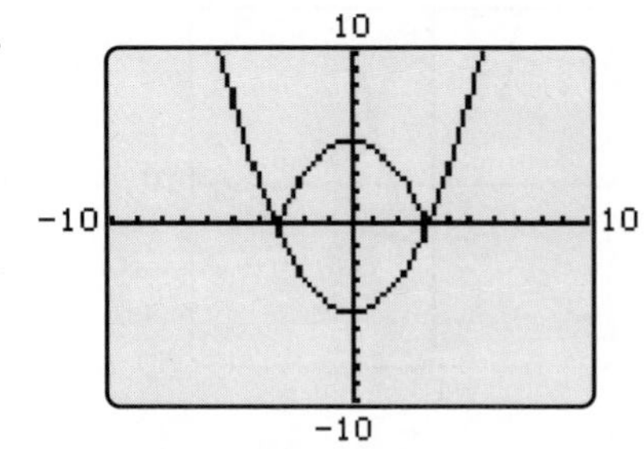

35.

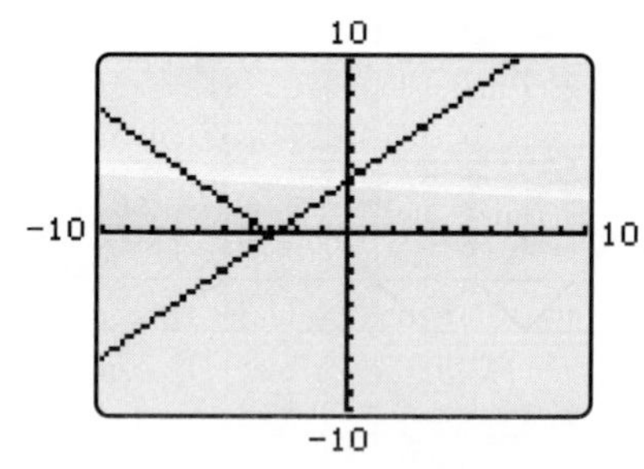

37.

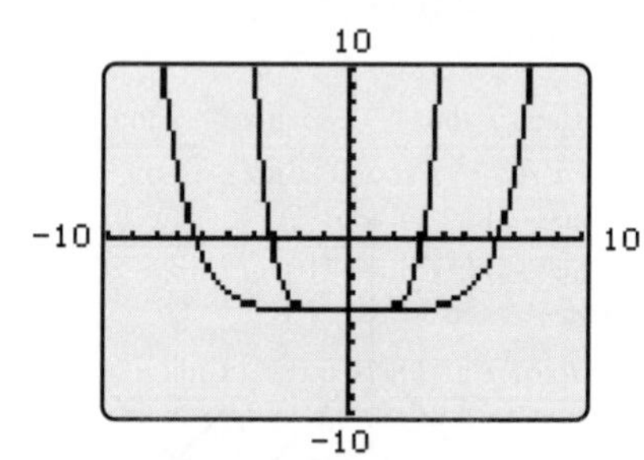

39.

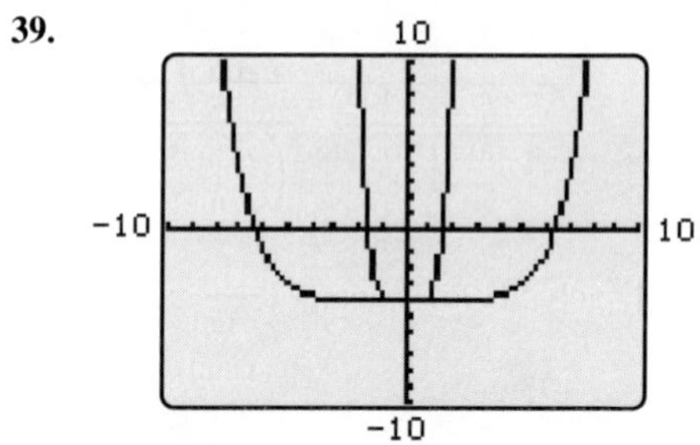

41.

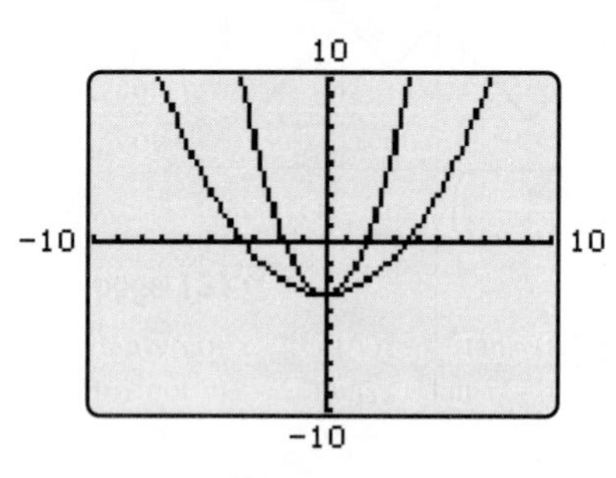

43.

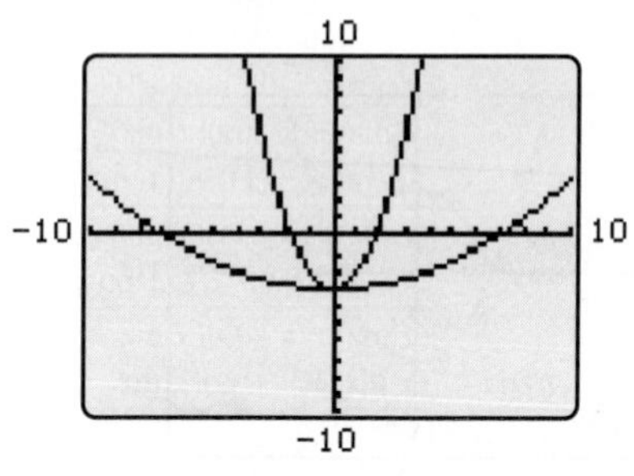

45.

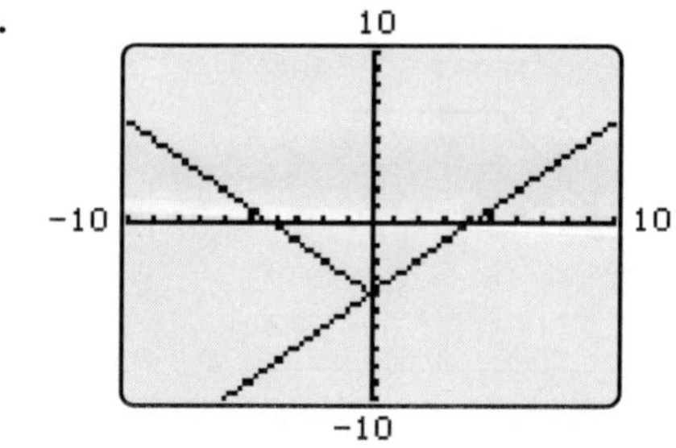

47.

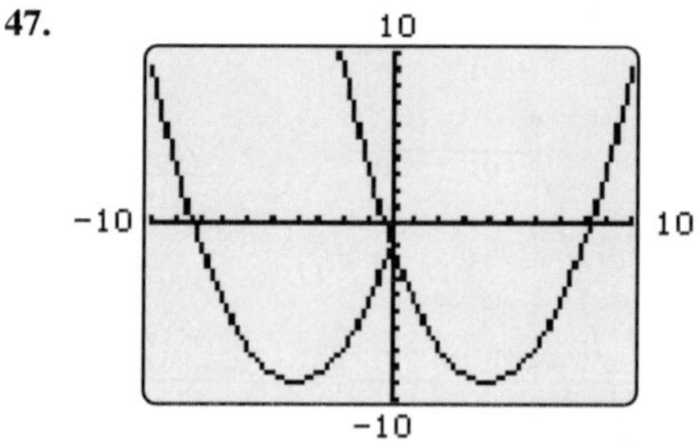

Excursion 2.5.A, page 137

1. Symmetric with respect to the vertical line $x = -2$

3. Symmetric with respect to the point (0, 2)

5. Odd **7.** Even **9.** Even **11.** Even

13. Neither **15.** Yes **17.** No **19.** Origin

21. Origin **23.** y-axis

25.

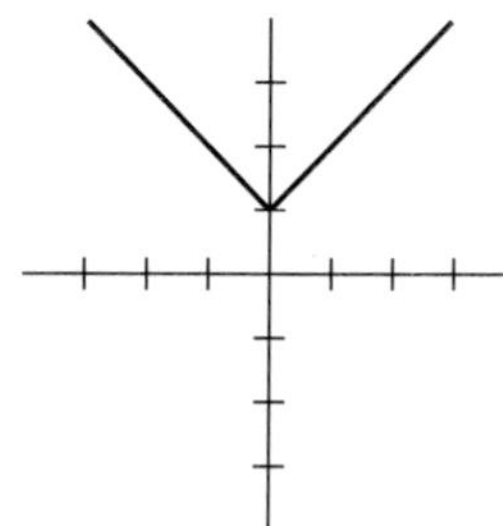

27.

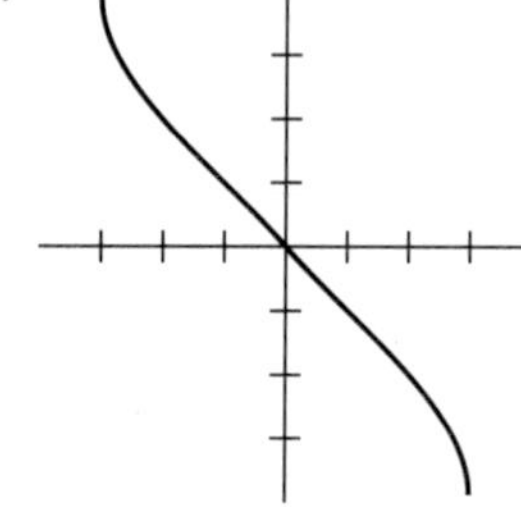

29.

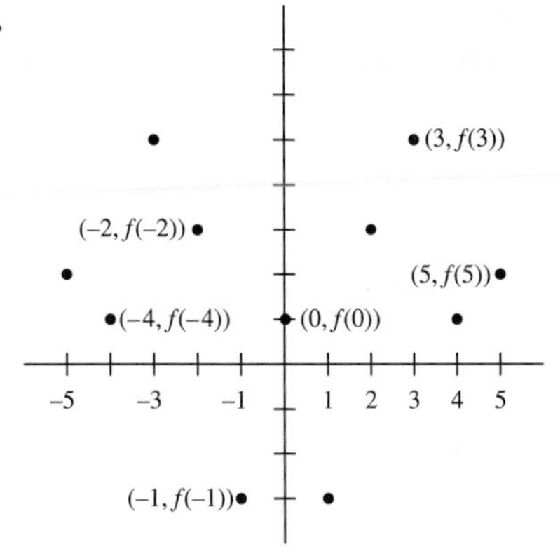

31. Many correct graphs, including the one shown here:

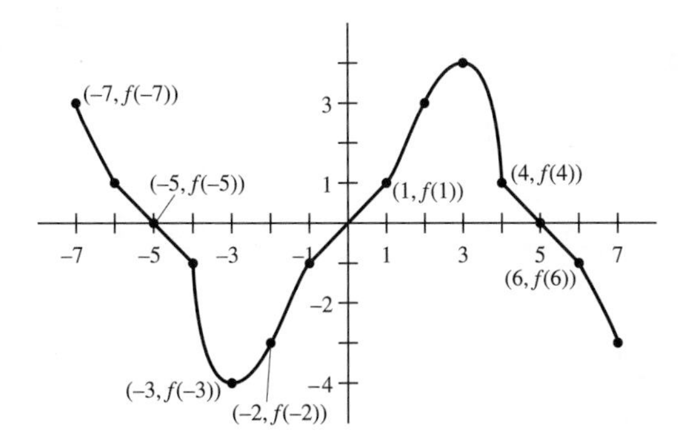

33. Suppose the graph is symmetric to the x-axis and the y-axis. If (x, y) is on the graph, then $(x, -y)$ is on the graph by x-axis symmetry. Hence, $(-x, -y)$ is on the graph by y-axis symmetry. Therefore, (x, y) on the graph implies that $(-x, -y)$ is on the graph, so the graph is symmetric with respect to the origin. Next suppose that the graph is symmetric to the y-axis and the origin. If (x, y) is on the graph, then $(-x, y)$ is on the graph by y-axis symmetry. Hence, $(-(-x), -y) = (x, -y)$ is on the graph by origin symmetry. Therefore, (x, y) on the graph implies that $(x, -y)$ is on the graph, so the graph is symmetric with respect to the x-axis. The proof of the third case is similar to that of the second case.

Section 2.6, page 145

1. 0 **3.** 30 **5.** 49; 1; −8 **7.** −3; −3; 0

9. $(f \circ g)(x) = (x + 3)^2$; $(g \circ f)(x) = x^2 + 3$

11. $(f \circ g)(x) = 1/\sqrt{x}$; $(g \circ f)(x) = 1/\sqrt{x}$

13. $(f \circ g)(x) = \sqrt[3]{x^2 + 1}$; $(g \circ f)(x) = (\sqrt[3]{x})^2 + 1$

15. $(f \circ g)(x) = f\left(\frac{x-2}{9}\right) = 9\left(\frac{x-2}{9}\right) + 2 = x$ and

$(g \circ f)(x) = g(9x + 2) = \dfrac{(9x + 2) - 2}{9} = x$

17. $(f \circ g)(x) = f((x - 2)^3) = \sqrt[3]{(x - 2)^3} + 2 = x$ and $(g \circ f)(x) = g(\sqrt[3]{x} + 2) = (\sqrt[3]{x} + 2 - 2)^3 = x$

19. $(f + g)(x) = x^3 - 3x + 2$; $(f - g)(x) = -x^3 - 3x + 2$; $(g - f)(x) = x^3 + 3x - 2$

21. $(f + g)(x) = \frac{1}{x} + x^2 + 2x - 5$; $(f - g)(x) = \frac{1}{x} - x^2 - 2x + 5$; $(g - f)(x) = x^2 + 2x - 5 - \frac{1}{x}$

23. $(fg)(x) = -3x^4 + 2x^3$; $\left(\frac{f}{g}\right)(x) = \dfrac{-3x + 2}{x^3}$; $\left(\frac{g}{f}\right)(x) = \dfrac{x^3}{-3x + 2}$

25.

x	−4	−3	−2	−1	0
$f(x)$	−2.9	−.9	0	.6	1
$g(x) = f(f(x))$	−.8	.7	1	1.2	1.3

x	1	2	3	4
$f(x)$	1.3	1	−2	−2
$g(x) = f(f(x))$	1.4	1.3	0	0

27.

x	1	2	3	4	5
$(g \circ f)(x)$	4	2	5	4	4

29.

x	1	2	3	4	5
$(f \circ f)(x)$	1	3	3	5	1

In Answers 31–35, the given function is $B \circ A$, where A and B are the functions listed here. In some cases other correct answers are possible.

31. $A(x) = x^2 + 2$, $B(x) = \sqrt[3]{x}$

33. $A(x) = 7x^3 - 10x + 17$, $B(x) = x^7$

35. $A(x) = 3x^2 + 5x - 7$, $B(x) = \dfrac{1}{x}$

37. Several possible answers, including $h(x) = x^2 + 2x$; $k(t) = t + 1$

39. $(f \circ g)(x) = (\sqrt{x})^3$, domain $[0, \infty)$; $(g \circ f)(x) = \sqrt{x^3}$, domain $[0, \infty)$

41. $(f \circ g)(x) = \sqrt{5x + 10}$, domain $[-2, \infty)$; $(g \circ f)(x) = 5\sqrt{x + 10}$, domain $[-10, \infty)$

43. **(a)** $f(x^2) = 2x^6 + 5x^2 - 1$
(b) $(f(x))^2 = (2x^3 + 5x - 1)^2 = 4x^6 + 20x^4 - 4x^3 + 25x^2 - 10x + 1$
(c) No; $f(x^2) \neq (f(x))^2$ in general

45. Not the same

47. **(a)** One day, .00012246 sq in; one week, .0000025 sq in; one 31-day month, .00000013 sq in.
(b) No; no. The model is probably reasonable until the puddle is about the size of a period, with a radius of approximately .01. This occurs after approximately 15 hours.

49. **(a)** $A = \pi\left(\dfrac{d}{2}\right)^2 = \pi \cdot \dfrac{d^2}{4} = \dfrac{\pi}{4}\left(6 - \dfrac{50}{t^2 + 10}\right)^2$
(b) $\pi/4 \approx .7854$ sq in; 22.2648 sq in
(c) In approximately 11.39 weeks

51. $V = 256\pi t^3/3$; 17,157.28 cu cm **53.** $s = 10t/3$

55. One such function is $f(x) = \dfrac{x - 1}{x}$.

Section 2.7, page 155

1. No **3.** Yes **5.** Yes **7.** No

9. $g(x) = -x$ **11.** $g(x) = \dfrac{x + 4}{5}$

13. $g(x) = \sqrt[3]{\dfrac{5 - x}{2}}$ **15.** $g(x) = \dfrac{x^2 + 7}{4}$, $(x \geq 0)$

17. $g(x) = \dfrac{1}{x}$ **19.** $g(x) = \dfrac{1}{2x} - \dfrac{1}{2}$

21. $g(x) = \sqrt[3]{\dfrac{5x + 1}{1 - x}}$

23. $(f \circ g)(x) = f(g(x)) = f(x - 1) = (x - 1) + 1 = x$ and $(g \circ f)(x) = g(f(x)) = g(x + 1) = (x + 1) - 1 = x$

25. $(f \circ g)(x) = f\left(\dfrac{1 - x}{x}\right) = \dfrac{1}{\left(\dfrac{1 - x}{x}\right) + 1} = \dfrac{1}{\dfrac{(1 - x) + x}{x}} = x$ and $(g \circ f)(x) = g\left(\dfrac{1}{x + 1}\right) = \dfrac{1 - \dfrac{1}{x + 1}}{\dfrac{1}{x + 1}} = \dfrac{\dfrac{(x + 1) - 1}{x + 1}}{\dfrac{1}{x + 1}} = x$

27. $(f \circ g)(x) = f(\sqrt[5]{x}) = (\sqrt[5]{x})^5 = x$ and $(g \circ f)(x) = g(x^5) = \sqrt[5]{x^5} = x$

29. $(f \circ f)(x) = f(f(x)) = \dfrac{2f(x) + 1}{3f(x) - 2} = \dfrac{2\left[\dfrac{2x + 1}{3x - 2}\right] + 1}{3\left[\dfrac{2x + 1}{3x - 2}\right] - 2} = \dfrac{\dfrac{2(2x + 1) + (3x - 2)}{3x - 2}}{\dfrac{3(2x + 1) - 2(3x - 2)}{3x - 2}} = \dfrac{\dfrac{7x}{3x - 2}}{\dfrac{7}{3x - 2}} = x$

31.

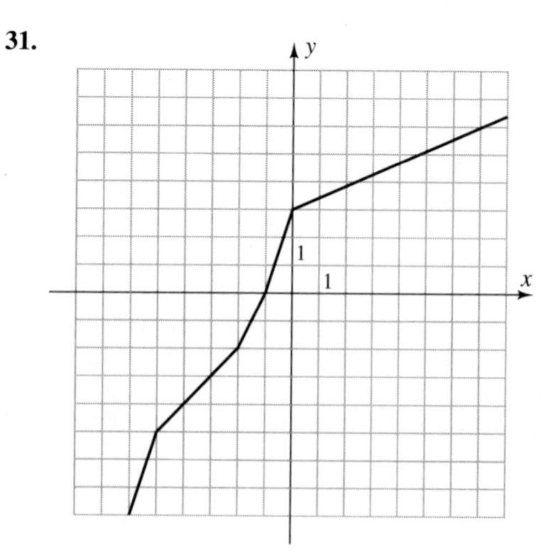

33.

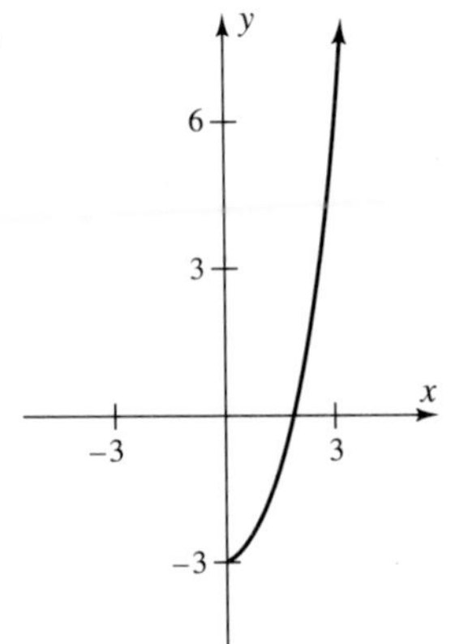

35.

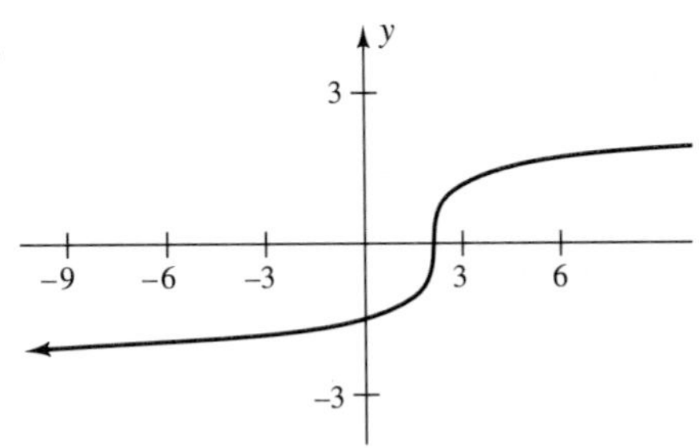

37.

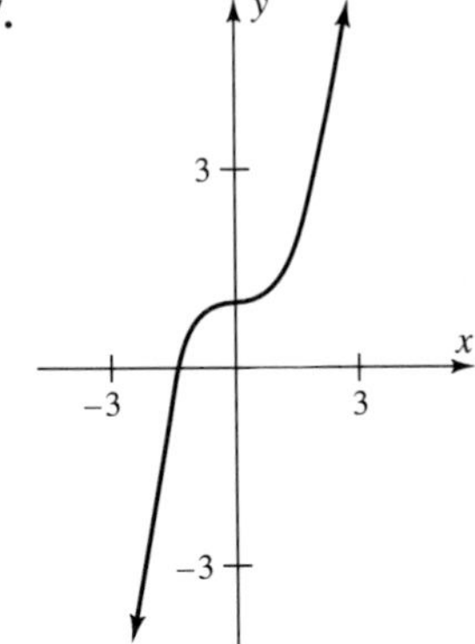

In Exercises 39–45, there are several correct answers for each, including these:

39. One restricted function is $h(x) = |x|$ with $x \geq 0$ (so that $h(x) = x$); inverse function $g(y) = y$ with $y \geq 0$.

41. One restricted function is $h(x) = -x^2$ with $x \leq 0$; inverse function $g(y) = -\sqrt{-y}$ with $y \leq 0$. Another restricted function is $h(x) = -x^2$ with $x \geq 0$; inverse function $g(y) = \sqrt{-y}$ with $y \leq 0$.

43. One restricted function is $h(x) = \dfrac{x^2 + 6}{2}$ with $x \geq 0$; inverse function $g(y) = \sqrt{2y - 6}$ with $y \geq 3$.

45. One restricted function is $f(x) = \dfrac{1}{x^2 + 1}$ with $x \leq 0$; inverse function $g(y) = -\sqrt{\dfrac{1}{y} - 1} = -\sqrt{\dfrac{1 - y}{y}}$ with $0 < y \leq 1$.

47. **(a)** $f^{-1}(x) = \dfrac{x - 2}{3}$ **(b)** $f^{-1}(1) = -1/3$ and $1/f(1) = 1/5$

49. Let $y = f(x) = mx + b$. Since $m \neq 0$, we can solve for x and obtain $x = \dfrac{y - b}{m}$. Hence, the rule of the inverse function g is $g(y) = \dfrac{y - b}{m}$, and we have: $(f \circ g)(y) = f(g(y)) = f\left(\dfrac{y - b}{m}\right) = m\left(\dfrac{y - b}{m}\right) + b = y$ and $(g \circ f)(x) = g(f(x)) = g(mx + b) = \dfrac{(mx + b) - b}{m} = x$.

51. **(a)** Length $OQ = \sqrt{(b - 0)^2 + (a - 0)^2} = \sqrt{a^2 + b^2} = \sqrt{(a - 0)^2 + (b - 0)^2} =$ length OP and length $RP = \sqrt{(c - a)^2 + (c - b)^2} = \sqrt{(c - b)^2 + (c - a)^2} =$ length RQ
(b) $OQ = OP$ and $RP = RQ$ by (a) and certainly $OR = OR$, so the triangles are congruent by side-side-side.
(c) Their sum is angle PRQ, a straight line segment, that is, a 180° angle. Since the two angles are congruent by (b), each must be 90°.
(d) OR is on the line $y = x$ and OR is perpendicular to PQ by (c). $PR = PQ$ by (b). Hence, $y = x$ is the perpendicular bisector of PQ.

53. **(a)** Suppose $a \neq b$. If $f(a) = f(b)$, then $g(f(a)) = g(f(b))$. But $a = g(f(a))$ by (1) and $b = g(f(b))$. Hence, $a = b$, contrary to our hypothesis. Therefore, it cannot happen that $f(a) = f(b)$, that is, $f(a) \neq f(b)$. Hence, f is one-to-one.
(b) If $g(y) = x$, then $f(g(y)) = f(x)$. But $f(g(y)) = y$ by (2). Hence, $y = f(g(y)) = f(x)$.
(c) If $f(x) = y$, then $g(f(x)) = g(y)$. But $g(f(x)) = x$ by (1). Hence, $x = g(f(x)) = g(y)$.

Chapter 2 Review, page 159

1. **(a)** -3
(b) 1755
(c) 2
(d) -14

3.

x	0	1	2	-4	t	k
$f(x)$	7	5	3	15	$7-2t$	$7-2k$

x	$b-1$	$1-b$	$6-2u$
$f(x)$	$9-2b$	$5+2b$	$-5+4u$

5. Many possible answers, including:
(a) $f(x)=x^2$, $a=2$, $b=3$; $f(a+b)=f(2+3)=5^2=25$, but $f(a)+f(b)=f(2)+f(3)=2^2+3^2=13$, so the statement is false.
(b) $f(x)=x+1$, $a=0$, $b=1$; $f(ab)=f(0)=1$, but $f(a)f(b)=f(0)f(1)=1\cdot 2=2$, so the statement is false.

7. $[4, \infty)$ **9.** $(t+2)^2-3(t+2)=t^2+t-2$

11. $2\left(\frac{x}{2}\right)^3+\left(\frac{x}{2}\right)+1=\frac{x^3}{4}+\frac{x}{2}+1$

13.

15. $-3\le x\le 3$ and $-6\le y\le 2$. Larger viewing windows may not show the peak and valley near the origin, but do indicate that the graph continually falls to the left of $x=-2$ and continually rises to the right of $x=1$, so that the graph in this window is complete.

17. $-2\le x\le 10$ and $-4\le y\le 4$

19.

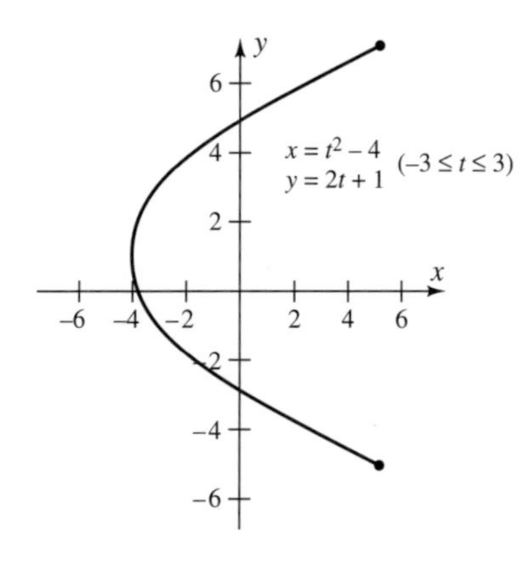

21. Many correct answers, including

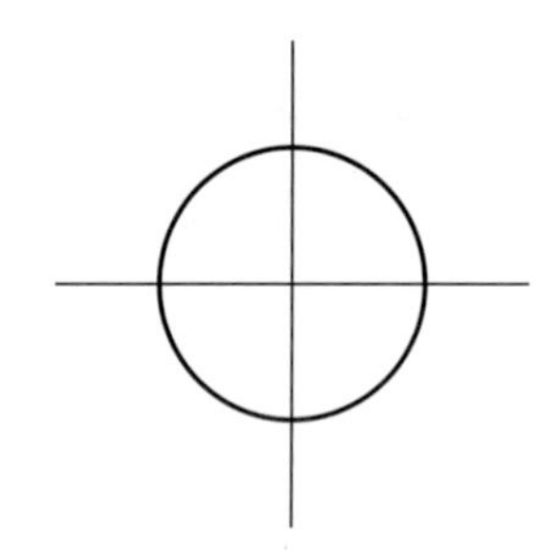

23. None **25.** None

27. x-axis, y-axis, origin **29.** Even

31. Odd **33.** y-axis

35.

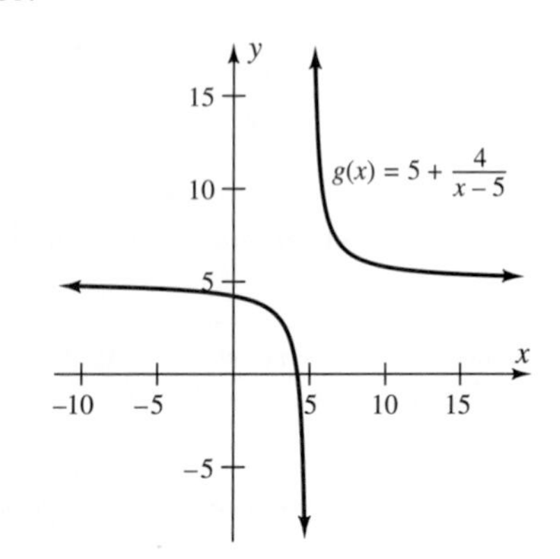

37. Approximately $[-3, 3.8]$

39. Many correct answers, including $x = -2$; all x in the interval $(2.5, 3.8)$; all x in the interval $[5, 6]$

41. 1 **43.** -3 **45.** True **47.** $x = 4$

49. $x \leq 3$ **51.** $x < 3$

53. Shift the graph of g vertically 4 units downward.

55. Shrink the graph of g toward the x-axis by a factor of .25, then shift the graph vertically 2 units upward.

57. Shift the graph of g horizontally 7 units to the right; then stretch it away from the x-axis by a factor of 3; then reflect it in the x-axis; finally, shift the graph vertically 2 units upward.

59. (e)

61. **(a)** 1/3 **(b)** $(x-1)\sqrt{x^2+5}$ $(x \neq 1)$
(c) $\dfrac{\sqrt{c^2+2c+6}}{c}$

63.

x	-4	-3	-2	-1	0
$g(x)$	1	4	3	1	-1
$h(x)$	-3	-3	-4	-3	1

x	1	2	3	4
$g(x)$	-3	-2	-4	-3
$h(x)$	4	3	1	4

65. $\dfrac{82}{27}$ **67.** $\dfrac{1}{x^3} + 3$ **69.** $\dfrac{1}{4}$

71. $(f \circ g)(x) = f(x^2 - 1) = \dfrac{1}{x^2-1}$; $(g \circ f)(x) = g\left(\dfrac{1}{x}\right) = \dfrac{1}{x^2} - 1$

73. All nonnegative numbers except 1

75.

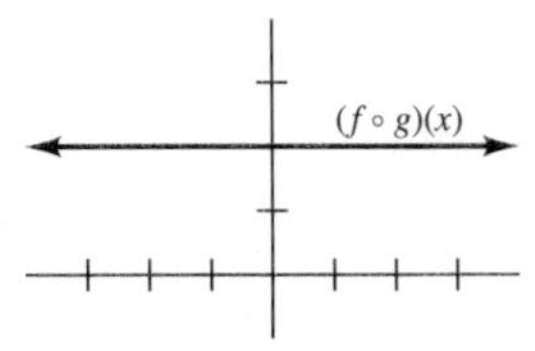

77. $g(x) = 5 - (x-7)^2 = -x^2 + 14x - 44$; $x \geq 7$

79.

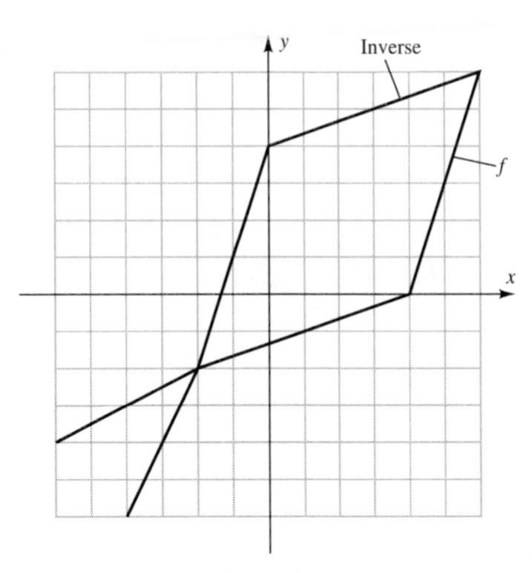

81. The graph of f passes the horizontal line test and hence has an inverse function. It is easy to verify either geometrically [by reflecting the graph of f in the line $y = x$] or algebraically [by calculating $f(f(x))$] that f is its own inverse function.

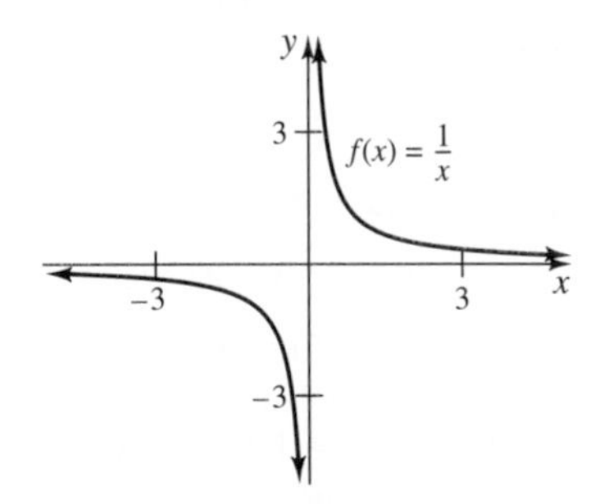

83. There is no inverse function because the graph of f fails the horizontal line test (use the viewing window with $-10 \leq x \leq 20$ and $-200 \leq y \leq 100$).

CHAPTER 3

Section 3.1, page 175

1. Many correct answers, including these:

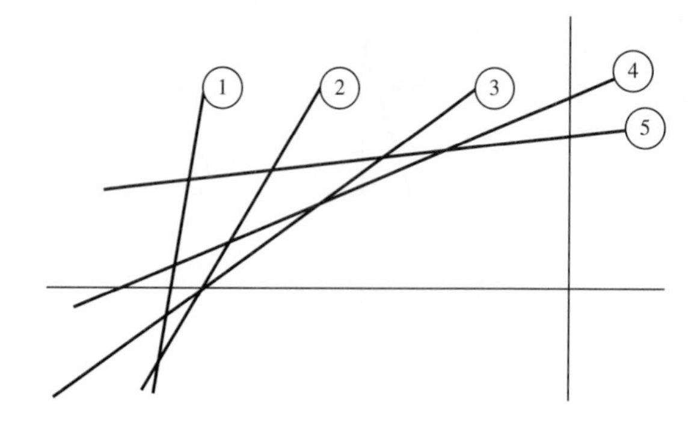

3. **(a)** C **(b)** B **(c)** B **(d)** D

5. Slope $m = 2$, y-intercept $b = 5$

7. $m = -3/7$, $b = -11/7$ **9.** 5/2 **11.** 4

13. $t = 22$ **15.** $t = 12/5$

17. P has coordinates $(1, y)$ for some y. Since $(0, 0)$ and $(1, y)$ are on L, the slope of L is $(y - 0)/(1 - 0) = y$, the second coordinate of P.

19. $y = x + 2$ **21.** $y = -x + 8$

23. $y = -x - 5$ **25.** $y = -\frac{7}{3}x + \frac{34}{9}$

27. Perpendicular **29.** Parallel **31.** Parallel

33. The side joining $(-5, -2)$ and $(3,0)$ has slope $(0 + 2)/(3 + 5) = 1/4$; the side joining $(-3, 1)$ and $(5, 3)$ has slope $(3 - 1)/(5 + 3) = 1/4$; so these two sides are parallel. Similarly, the side joining $(-5, -2)$ and $(-3, 1)$ and the side joining $(3, 0)$ and $(5, 3)$ are parallel since both have slope $3/2$.

35. Yes **37.** $y = 3x + 7$ **39.** $y = 3x/2$

41. $y = x - 5$ **43.** $y = -x + 2$ **45.** $k = -11/3$

47. $y - 4 = -\frac{3}{4}(x - 3)$ **49.** $y - 5 = -\frac{1}{2}(x - 2)$

51. The equation $Ax + By + C = 0$ is equivalent to $By = -Ax - C$ and hence to $y = (-A/B)x - (C/B)$; so this line has slope $-A/B$. Similarly, $Ax + By + D = 0$ is equivalent to $y = (-A/B)x - (D/B)$ so that this line also has slope $-A/B$. Two lines with the same slopes are parallel.

53. **(a)** $f(x) = 20{,}000x + 100{,}000$
(b) \$220,000
(c) Midway through 2001 (7.5 years from the start)

55. **(a)** $c(x) = 8.5x + 50{,}000$ **(b)** \$11; \$9.50; \$9

57. **(a)** $c(x) = 25x + 180{,}000$ **(b)** $r(x) = 40x$
(c) $p(x) = 15x - 180{,}000$ **(d)** 12,000 books

59. **(a)** Distance from P to Q is 9400 ft.
(b) B is 88 ft from the road.

61. If L and M are distinct nonvertical lines with slope m, then the equation of L is $y = mx + b$ for some b and the equation of M is $y = mx + c$ for some c. Since L and M are distinct, they cannot have the same equation, so we must have $b \neq c$. If L and M were not parallel, they would intersect at some point (x_1, y_1). Since (x_1, y_1) is on L, its coordinates satisfy the equation of L, that is, $y_1 = mx_1 + b$. Similarly, since (x_1, y_1) is on M, we have $y_1 = mx_1 + c$. Hence $mx_1 + b = y_1 = mx_1 + c$, which implies that $b = c$. This is a contradiction, which arose because we assumed that L and M had a point of intersection. Therefore that assumption must have been false; that is, L and M do not intersect and hence are parallel.

63. Denote the length of the side of the square by c and place the square in the plane with one corner at the origin and its sides on the positive axes, as shown. Then one diagonal goes through $(0, 0)$ and (c, c) and has slope $\frac{c - 0}{c - 0} = 1$. The other diagonal goes through $(0, c)$ and $(c, 0)$ and has slope $\frac{0 - c}{c - 0} = -1$. Since the product of the slopes is -1, the diagonals are perpendicular.

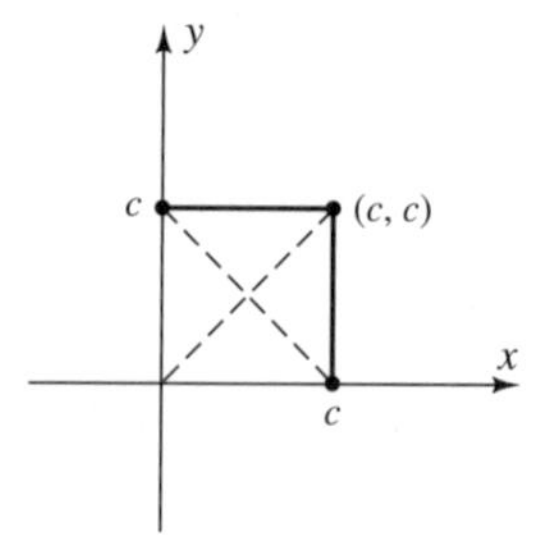

65. The line $y = x$ has slope 1 (why?). The line through (a, b) and (b, a) has slope $\frac{a - b}{b - a} = \frac{-(b - a)}{b - a} = -1$. Since the product of the slopes is -1, the lines are perpendicular.

Section 3.2, page 188

1. **(a)** 14 ft/sec **(b)** 54 ft/sec **(c)** 112 ft/sec **(d)** $93\frac{1}{3}$ ft/sec

3. **(a)** .709 gal/in **(b)** 2.036 gal/in

5. **(a)** 250 ties/mo **(b)** 438 ties/mo **(c)** 500 ties/mo **(d)** 563 ties/mo **(e)** −188 ties/mo **(f)** −792 ties/mo **(g)** −1500 ties/mo **(h)** −375 ties/mo

7. **(a)** −55.5 **(b)** −92.5 **(c)** −462.5

9. −2 **11.** −1 **13.** 1.5858

15. 1 **17.** $2x + h$ **19.** $2t + h - 8000$

21. $2\pi r + \pi h$

23. **(a)** Average rate of change is −7979.9, which means that water is leaving the tank at a rate of 7979.9 gal/min. **(b)** −7979.99 **(c)** −7980

25. **(a)** 6.5π **(b)** 6.2π **(c)** 6.1π **(d)** 6π **(e)** It's the same.

27. **(a)** C, 62.5 ft/sec; D, 75 ft/sec **(b)** Approximately $t = 4$ to $t = 9.8$ sec **(c)** The average speed of car D from $t = 4$ to $t = 10$ seconds is the slope of the secant line joining the (approximate) points (4, 100) and (10, 600), namely, $\frac{600 - 100}{10 - 4} \approx 83.33$ ft/sec. The average speed of car C is the slope of the secant line joining the (approximate) points (4, 475) and (10, 800), namely, $\frac{800 - 475}{10 - 4} \approx 54.17$ ft/sec.

29. **(a)** From day 0 until any day up to day 94, the growth rate is positive **(b)** From day 0 to day 95 **(c)** −28, meaning that the population is decreasing at a rate of 28 chipmunks per day **(d)** 20, −20, and 0 chipmunks per day

Excursion 3.2.A, page 195

1. $x = 2$ **3.** $x = 1.63735$

5. $x = -.224284$ or 3.15233 or 7.07195

7. On TI-82, these (and other) initial guesses fail to produce a solution: 1 and 2; −2 and 4; −1 and −2. These (and other) guesses produce the correct solution: 0 and 1; −1 and 1. Answers may vary for other calculators.

9. **(a)**

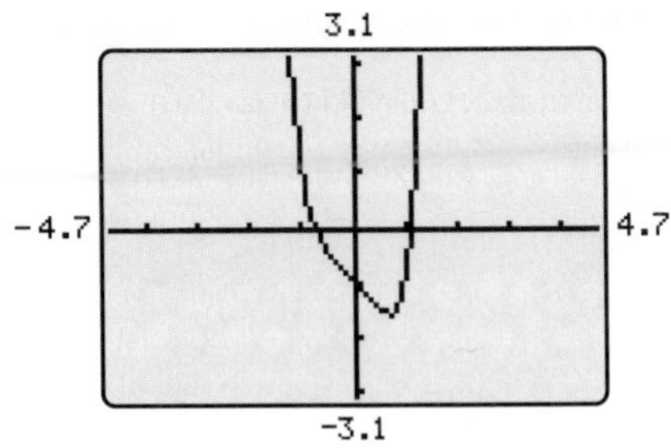

(b), (c), (d) Answers vary.

Section 3.3, page 202

1. (5, 2), upward **3.** (1, 2), upward

5. (3, −6), upward **7.** (−3/2, 15/4), upward

9. (−3, −21), upward **11.** (4, 14), downward

13. (2/3, 19/3), downward **15.** (1/2, 1/4), downward

17. $g(x) = 2x^2 - 5$; vertex $(0, -5)$

19. $h(x) = 2(x - 3)^2 + 4$; vertex (3, 4)

21. $f(x) = 3x^2$ **23.** $b = 0$ **25.** $b = -4$, $c = 8$

27. Minimum product is −4; numbers are 2 and −2

29. 9 and 18 **31.** $h = 15$, $b = 15$

33. Two 50-ft sides and one 100-ft side

35. 30 stores **37.** $3.50

39. $3.67 (but if tickets must be priced in multiples of .20, then $3.60 is best)

41. $t = 2.5$ sec, $h = 196$ ft **43.** 22 ft

45. $t = \frac{125}{8}$ sec, $h = \frac{125^2}{4} = 3906.25$ ft

Chapter 3 Review, page 205

1. **(a)** 1 **(b)** 4/5

3. $y = 3x - 7$ **5.** $y = -2x + 1$

7. $x - 5y = -29$ **9.** 25,000 ft **11.** False

13. False **15.** False **17.** False **19.** (d)

21. (e) **23.** 5/3 **25.** **(a)** −1/3 **(b)** 5/8

27. 6 **29.** 3 **31.** $2x + h$

33. **(a)** \$290/ton **(b)** \$230/ton **(c)** \$212/ton

35. **(a)** Approximately day 45 to day 50.
(b) Approximately any 10-day interval between day 20 and day 35
(c) Approximately day 30 to day 40

37. (2, 3) **39.** (4, −4) **41.** (d)

43. 30, 60, 30 ft

CHAPTER 4

Section 4.1, page 215

1. Polynomial of degree 3; leading coefficient 1; constant term 1

3. Polynomial of degree 3; leading coefficient 1; constant term −1

5. Polynomial of degree 2; leading coefficient 1; constant term −3

7. Not a polynomial

9. Quotient $3x^3 - 3x^2 + 5x - 11$; remainder 12

11. Quotient $x^2 + 2x - 6$; remainder $-7x + 7$

13. Quotient $5x^2 + 5x + 5$; remainder 0

15. No **17.** Yes **19.** 0, 2 **21.** $2\sqrt{2}, -1$

23. 2 **25.** 6 **27.** −30 **29.** 170,802

31. 5,935,832 **33.** No **35.** No **37.** Yes

39. $(x + 4)(2x - 7)(3x - 5)$

41. $(x - 3)(x + 3)(2x + 1)^2$

43. $f(x) = (x + 2)(x + 1)(x - 1)(x - 2)(x - 3) = x^5 - 3x^4 - 5x^3 + 15x^2 + 4x - 12$

45. $f(x) = x(x + 1)(x - 1)(x - 2)(x - 3) = x^5 - 5x^4 + 5x^3 + 5x^2 - 6x$

47. Many correct answers, including $(x - 1)(x - 7)(x + 4)$

49. Many correct answers, including $(x - 1)(x - 2)^2(x - \pi)^3$

51. $f(x) = \dfrac{17}{100}(x - 5)(x - 8)x$

53. $k = 1$ **55.** $k = 1$

57. If $x - c$ were a factor of $x^4 + x^2 + 1$, then c would be a solution of $x^4 + x^2 + 1 = 0$, that is, c would satisfy $c^4 + c^2 = -1$. But $c^4 \geq 0$ and $c^2 \geq 0$, so that is impossible. Hence, $x - c$ is not a factor.

59. **(a)** Many possible answers, including: if $n = 3$ and $c = 1$, then $x + 1 = x - (-1)$ is not a factor of $x^3 - 1$ since -1 is not a solution of $x^3 - 1 = 0$.
(b) Since n is odd, $(-c)^n = -c^n$ and hence $-c$ is a solution of $x^n + c^n = 0$. Thus, $x - (-c) = x + c$ is a factor of $x^n + c^n$ by the Factor Theorem.

61. $k = 5$

Excursion 4.1.A, page 220

1.
$$\begin{array}{r|rrrrr} 2 & 3 & -8 & 0 & 9 & 5 \\ & & 6 & -4 & -8 & 2 \\ \hline & 3 & -2 & -4 & 1 & \boxed{7} \end{array}$$
quotient $3x^3 - 2x^2 - 4x + 1$; remainder 7

3.
$$\begin{array}{r|rrrrr} -3 & 2 & 5 & 0 & -2 & -8 \\ & & -6 & 3 & -9 & 33 \\ \hline & 2 & -1 & 3 & -11 & \boxed{25} \end{array}$$
quotient $2x^3 - x^2 + 3x - 11$; remainder 25

5.
$$\begin{array}{r|rrrrr} 7 & 5 & 0 & -3 & -4 & 6 \\ & & 35 & 245 & 1{,}694 & 11{,}830 \\ \hline & 5 & 35 & 242 & 1{,}690 & \boxed{11{,}836} \end{array}$$
quotient $5x^3 + 35x^2 + 242x + 1690$; remainder 11,836

7.
$$\begin{array}{r|rrrrr} 2 & 1 & -6 & 4 & 2 & -7 \\ & & 2 & -8 & -8 & -12 \\ \hline & 1 & -4 & -4 & -6 & \boxed{-19} \end{array}$$
quotient $x^3 - 4x^2 - 4x - 6$; remainder −19

9. Quotient $3x^3 + \dfrac{3}{4}x^2 - \dfrac{29}{16}x - \dfrac{29}{64}$; remainder $\dfrac{483}{256}$

11. Quotient $2x^3 - 6x^2 + 2x + 2$; remainder 1

13. $g(x) = (x + 4)(3x^2 - 3x + 1)$

15. $g(x) = \left(x - \dfrac{1}{2}\right)(2x^4 - 6x^3 + 12x^2 - 10)$

17. Quotient $x^2 - 2.15x + 4$; remainder 2.25

19. $c = -4$

Section 4.2, page 227

1. $x = \pm 1$ or -3 **3.** $x = \pm 1$ or -5

5. $x = -4, 0, 1$, or $1/2$ **7.** $x = -3$ or 2

9. $x = 2$ **11.** $x = -5, 2,$ or 3

13. $(x - 2)(2x^2 + 1)$ **15.** $x^3(x^2 + 3)(x + 2)$

17. $(x - 2)(x - 1)^2(x^2 + 3)$ **19.** Lower -5; upper 2

21. Lower -7; upper 3 **23.** $x = 1, 2,$ or $-1/2$

25. $x = 1, 1/2,$ or $1/3$ **27.** $x = 2$ or $\dfrac{-5 \pm \sqrt{37}}{2}$

29. $x = 1/2$ or $\pm\sqrt{2}$ or $\pm\sqrt{3}$ **31.** $x = -1, 5,$ or $\pm\sqrt{3}$

33. $x = 1/3$ or -1.8393

35. $x = -2.2470$ or $-.5550$ or $.8019$ or 50

37. **(a)** $g(x) = 2x^4 - 2x^3 - 4x^2 - 12$
(b) $g(.5) = -13.125$

39. **(a)** The only possible rational roots of $f(x) = x^2 - 2$ are ± 1 or ± 2 (why?). But $\sqrt{2}$ is a root of $f(x)$ and $\sqrt{2} \neq \pm 1$ or ± 2. Hence, $\sqrt{2}$ is irrational.
(b) $\sqrt{3}$ is a root of $x^2 - 3$ whose only possible rational roots are ± 1 or ± 3 (why?). But $\sqrt{3} \neq \pm 1$ or ± 3.

41. 2 by 2 inches

43. **(a)** 6 deg/day at the beginning; 6.6435 deg/day at the end
(b) Day 2.0330 and day 10.7069
(c) Day 5.0768 and day 9.6126
(d) Day 7.6813

Section 4.3, page 238

1. Yes **3.** Yes **5.** No

7. Degree 3, yes; degree 4, no; degree 5, yes **9.** No

11. Degree 3, no; degree 4, no; degree 5, yes

13. The graph in the standard viewing window does not rise at the far right as does the graph of the highest degree term x^3, so it is not complete.

15. The graph in the standard viewing window does not rise at the far left and far right as does the graph of the highest degree term $.005x^4$, so it is not complete.

17. $-9 \le x \le 3$ and $-20 \le y \le 40$

19. $-6 \le x \le 6$ and $-60 \le y \le 320$

21. $-3 \le x \le 4$ and $-35 \le y \le 20$

23. Left half: $-33 \le x \le -2$ and $-50{,}000 \le y \le 250{,}000$; right half: $-2 \le x \le 3$ and $-20 \le y \le 30$

25. $-90 \le x \le 120$ and $-15{,}000 \le y \le 5000$

27. Overall: $-3 \le x \le 3$ and $-20 \le y \le 20$; near y-axis: $-.1 \le x \le .2$ and $4.997 \le y \le 5.001$

29. **(a)** Since the graph must resemble the graph of $\pm x^4$ when $|x|$ is large, either both "ends" of the graph rise or both ends fall. So if you move along the graph from left to right you must either fall, then rise or rise, then fall. There must be at least one local extrema in between, where the graph changes from rising to falling or vice versa. The graph of a fourth-degree polynomial cannot have exactly two local extrema because it would have one peak and one valley, which means that the graph falls on one end and rises on the other, unlike the graph of $\pm x^4$. The graph might have exactly three local extrema, as shown in part (b), but cannot have more since this is the maximum number for degree 4.

(b)

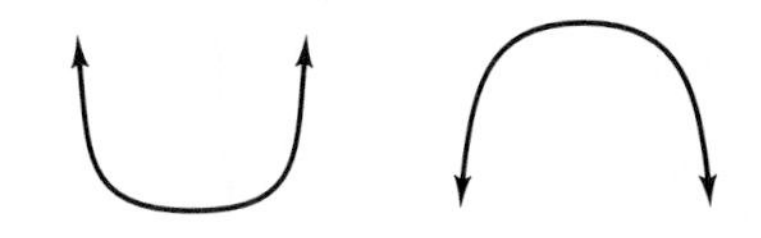

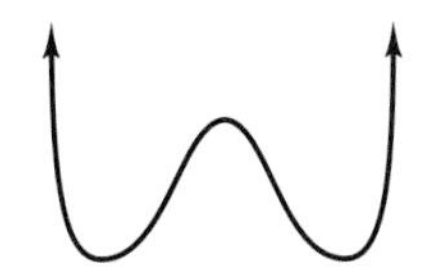

31. (d)

33.

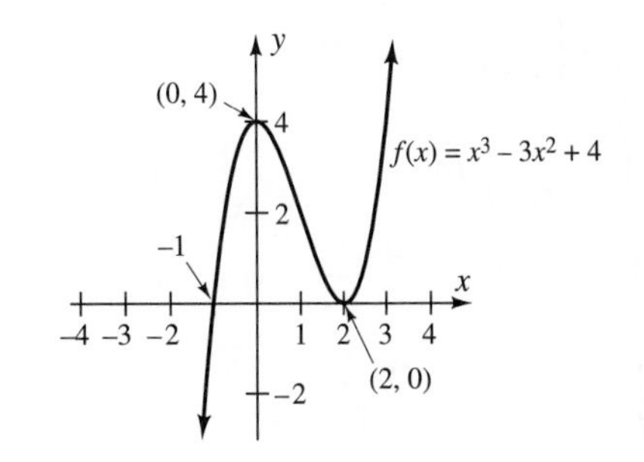

35.

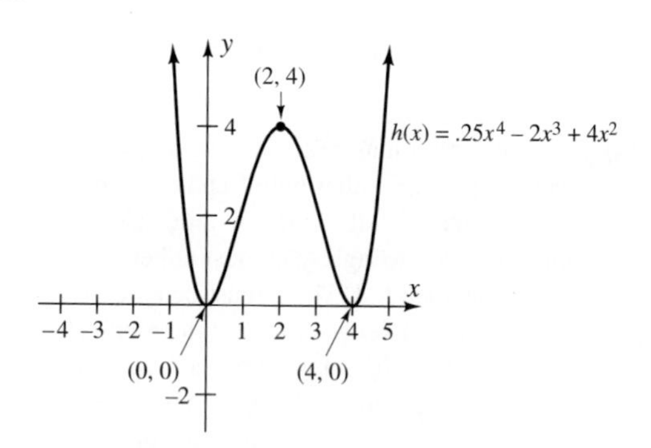

37.

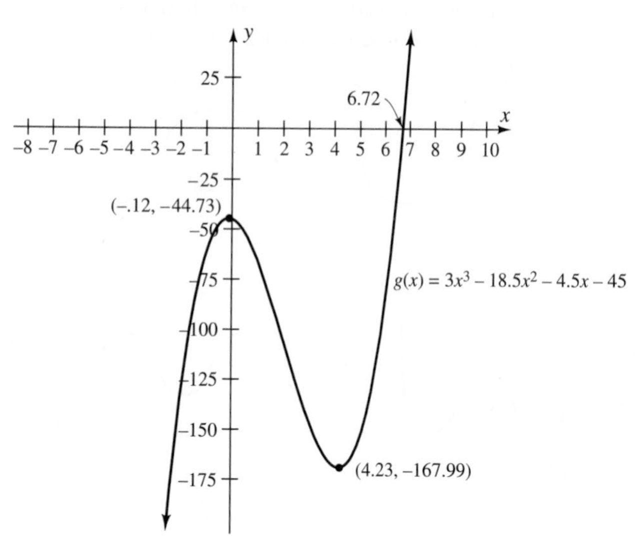

39.

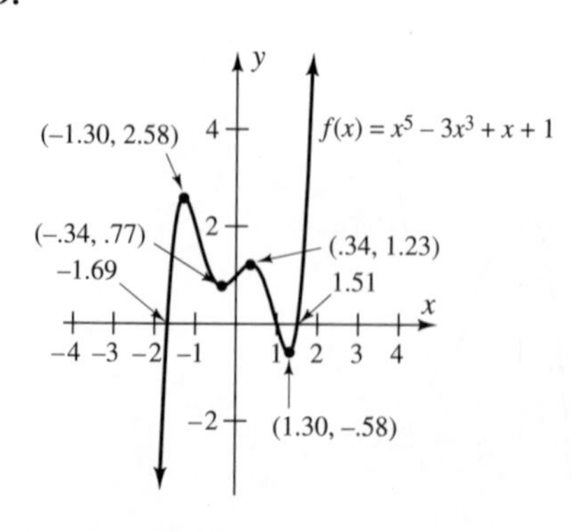

41.

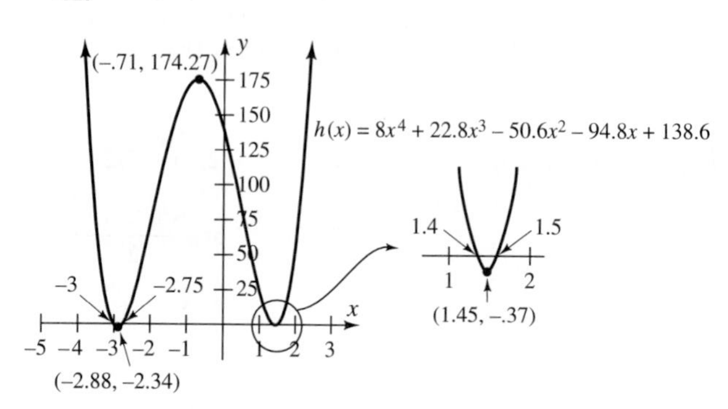

43. **(a)** 82,794; \$1546.39 **(b)** 147,200

45. **(a)** $R(x) = 350x$; $P(x) = -.01x^2 + 375x - 600{,}000$
(b) 1675 **(c)** 35,825 **(d)** 18,750

47. 2.3542 by 2.3542 inches

49. **(a)** The solutions are roots of $g(x) - 4 = .01x^3 - .06x^2 + .12x - .08$. This polynomial has degree 3 and hence has at most 3 roots.
(b) $1 \leq x \leq 3$ and $3.99 \leq y \leq 4.01$
(c) Suppose $f(x)$ has degree n. If the graph of $f(x)$ had a horizontal segment lying on the line $y = k$ for some constant k, then the equation $f(x) = k$ would have infinitely many solutions (why?). But the polynomial $f(x) - k$ has degree n (why?) and thus has at most n roots. Hence the equation $f(x) = k$ has at most n solutions, which means the graph cannot have a horizontal segment.

51. **(a)** No, except between $x = 0$ and $x = 2$
(b) Yes
(c) Approximately .0007399

53. **(a)**

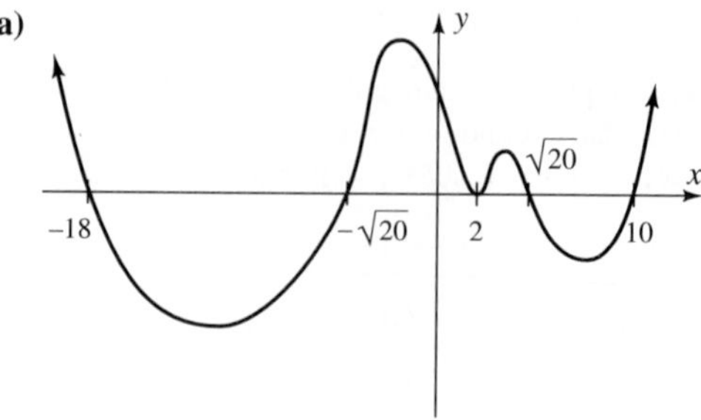

(b)–(d) Depends on the calculator.

Section 4.4, page 256

1. All real numbers except $-5/2$

3. All real numbers except $3 + \sqrt{5}$ and $3 - \sqrt{5}$

5. All real numbers except $-\sqrt{2}$, 1, and $\sqrt{2}$

7. Vertical asymptotes $x = -1$ and $x = 6$

9. Hole at $x = 0$; vertical asymptote $x = -1$

11. Vertical asymptotes: $x = -2$ and $x = 2$

13. $y = 3$; any window with $-115 \le x \le 110$

15. $y = -1$; any window with $-31 \le x \le 35$

17. $y = 5/2$; any window with $-40 \le x \le 42$

19.

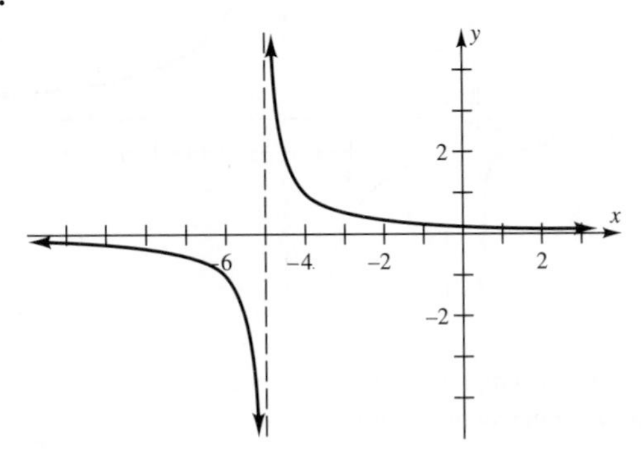

vertical asymptote $x = -5$
horizontal asymptote $y = 0$

21.

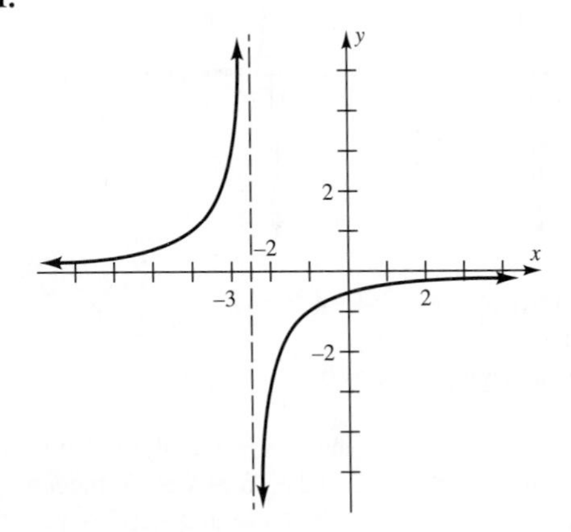

vertical asymptote $x = -2.5$
horizontal asymptote $y = 0$

23.

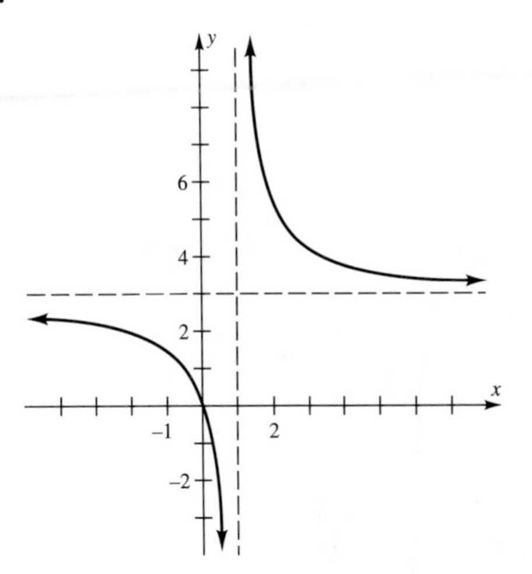

vertical asymptote $x = 1$
horizontal asymptote $y = 3$

25.

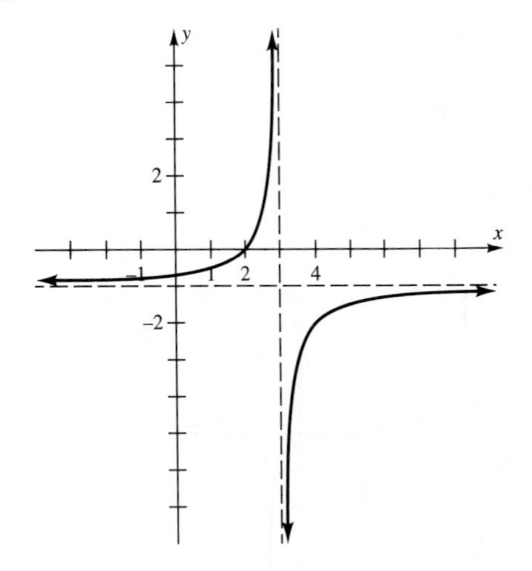

vertical asymptote $x = 3$
horizontal asymptote $y = -1$

27.

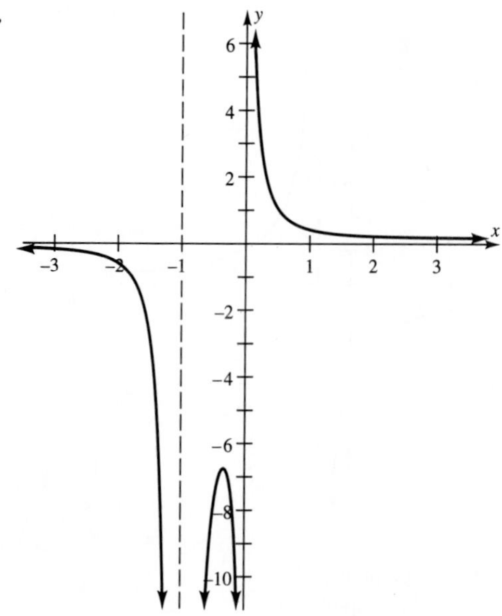

vertical asymptotes $x = -1, x = 0$
horizontal asymptote $y = 0$

29.

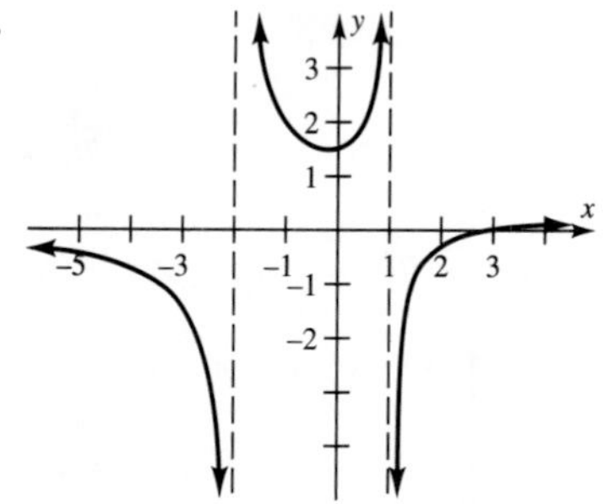

vertical asymptotes $x = -2, x = 1$
horizontal asymptote $y = 0$

31.

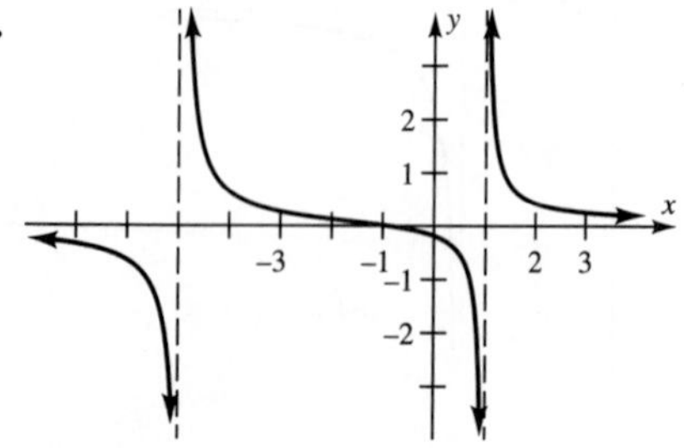

vertical asymptotes $x = -5, x = 1$
horizontal asymptote $y = 0$

33.

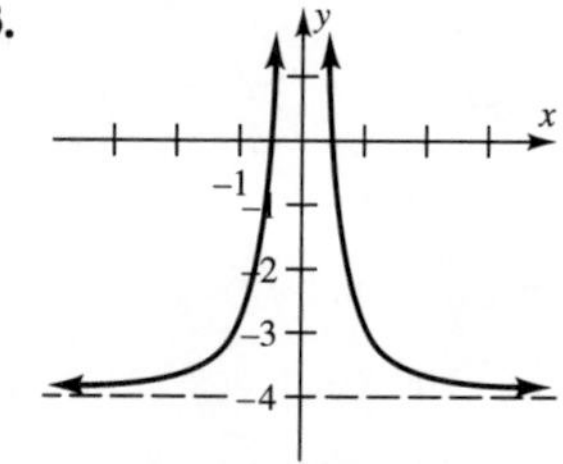

vertical asymptote $x = 0$
horizontal asymptote $y = -4$

35.

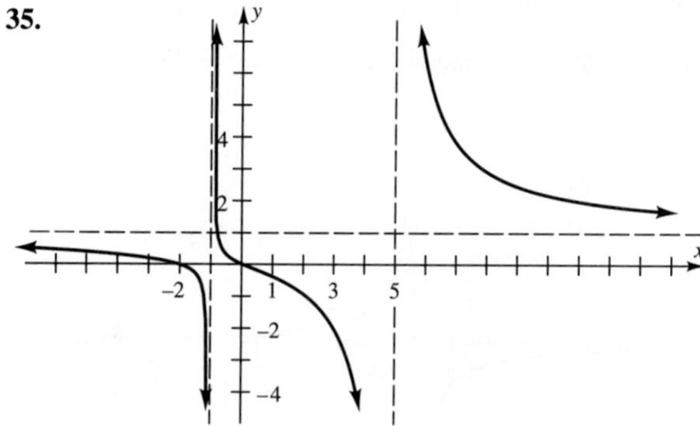

vertical asymptotes $x = -1, x = 5$
horizontal asymptote $y = 1$

37.

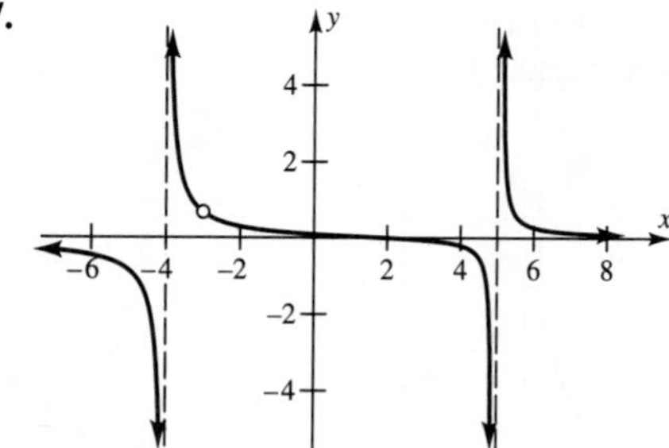

vertical asymptotes $x = -4, x = 5$
hole at $x = -3$
horizontal asymptote $x = 0$

39. Overall: $-5 \leq x \leq 4.4$ and $-8 \leq y \leq 4$; hidden area near origin: $-2 \leq x \leq 2$ and $-.5 \leq y \leq .5$; hidden area near $x = -5$: $-15 \leq x \leq -3$ and $-.07 \leq y \leq .02$

41. $-9.4 \leq x \leq 9.4$ and $-4 \leq y \leq 4$; there is a hole at $x = 2$.

43. Overall: $-4.7 \leq x \leq 4.7$ and $-2 \leq y \leq 2$; there is a hole at $x = -1$; to see the vertical asymptote, use $.65 \leq x \leq .75$ and $-3 \leq y \leq 3$.

45. For vertical asymptotes and x-intercepts: $-4.7 \leq x \leq 4.7$ and $-8 \leq y \leq 8$; to see graph get close to the horizontal asymptote: $-40 \leq x \leq 35$ and $-2 \leq y \leq 3$

47. Overall: $-4.7 \leq x \leq 4.7$ and $-3 \leq y \leq 3$; hidden area near $x = 4$: $3 \leq x \leq 15$ and $-.02 \leq y \leq .01$

49. **(b)** Stretch the graph of $f(x)$ away from the x-axis by a factor of 2.

(c) The graph of $h(x)$ is the graph of $f(x)$ shifted vertically 4 units upward; the graph of $k(x)$ is the graph of $f(x)$ shifted horizontally 3 units to the right; the graph of $t(x)$ is the graph of $f(x)$ shifted horizontally 2 units to the left.

(d) Shift the graph of $f(x)$ horizontally 3 units to the right, stretch by a factor of 2, then shift vertically 4 units upward.

(e) $p(x) = \dfrac{4x - 10}{x - 3}$

(f) Shift the graph of $f(x)$ horizontally $|s|$ units (to the left if $s > 0$; to the right if $s < 0$); stretch (or shrink) the graph by a factor of $|r|$ (away from the x-axis if $|r| > 1$, toward the x-axis if $0 < |r| < 1$); also if $r < 0$, reflect the graph in the x-axis; then shift vertically $|t|$ units (upward if $t > 0$; downward if $t < 0$).

(g) $q(x) = \dfrac{tx + (r + ts)}{x + s}$

51. **(a)** $\dfrac{-1}{x(x + h)}$

(b) $-1/4.2 \approx -.2381$; $-1/4.02 \approx -.2488$; $-1/4.002 \approx -.2499$; instantaneous rate of change $-1/4 = -.25$

(c) $-1/9.3 \approx -.1075$; $-1/9.03 \approx -.1107$; $-1/9.003 \approx -.1111$; instantaneous rate of change $-1/9 = -.1111\cdots$

(d) They are the same.

53. **(a)** $y = \dfrac{x - 1}{x - 2}$ **(b)** $y = \dfrac{x - 1}{-x - 2}$

(c) Graph (a) has a vertical asymptote at $x = 2$ and graph (b) has a vertical asymptote at $x = -2$.

55. 8.4343 in × 8.4343 in × 14.057 in

57. **(a)** $c(x) = \dfrac{2800 + 3.5x^2}{x}$ **(b)** $13.91 \leq x \leq 57.52$

(c) 28.28 mph

59. **(a)** $p(x) = \dfrac{500 + x^2}{x}$ **(b)** $10 \leq x \leq 50$

(c) $x = 22.36$; 22.36 m by 11.18 m

61. **(a)** $h_1 = h - 2$

(b) $h_1 = \dfrac{150}{\pi r^2} - 2$ (because $\pi r^2 h = 150$)

(c) $V = \pi(r - 1)^2 \left(\dfrac{150}{\pi r^2} - 2\right)$

(d) The walls are 1 ft thick. **(e)** $r \approx 2.88$ ft; $h \approx 5.76$ ft

63. **(a)** $g(0) = 9.801$ meters/sec^2

(b)

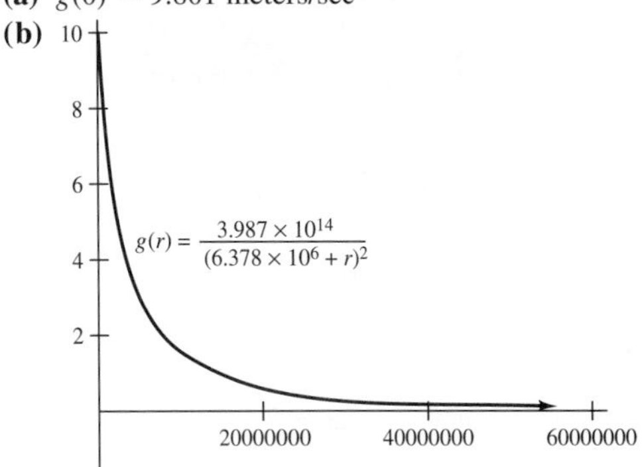

(c) There are no r-intercepts because the numerator is never zero. So you can never completely escape the pull of gravity.

Excursion 4.4.A, page 263

1. asymptote: $y = x$; window: $-14 \leq x \leq 14$ and $-15 \leq y \leq 15$

3. asymptote: $y = x^2 - x$; window: $-15 \leq x \leq 6$ and $-40 \leq y \leq 240$

5.

vertical asymptote $x = 2$
oblique asymptote $y = x + 1$

7.

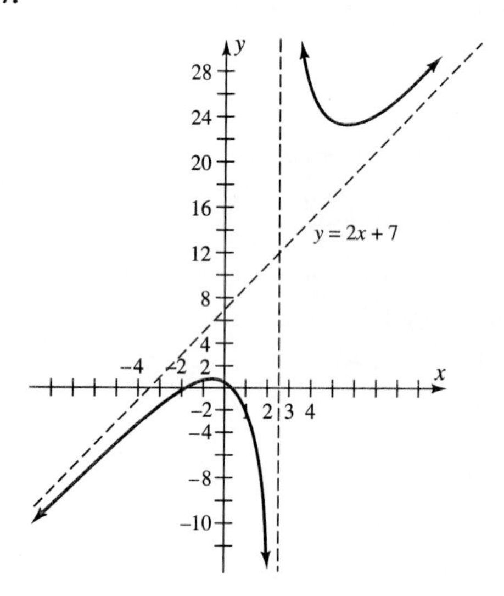

vertical asymptote $x = 5/2$
oblique asymptote $y = 2x + 7$

9.

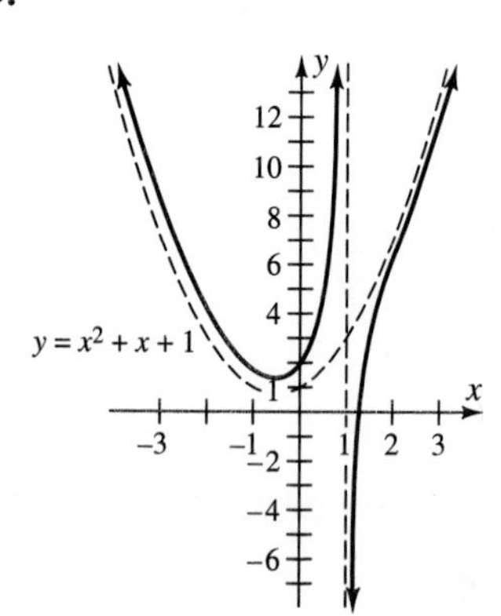

vertical asymptote $x = 1$
parabolic asymptote $y = x^2 + x + 1$

11.

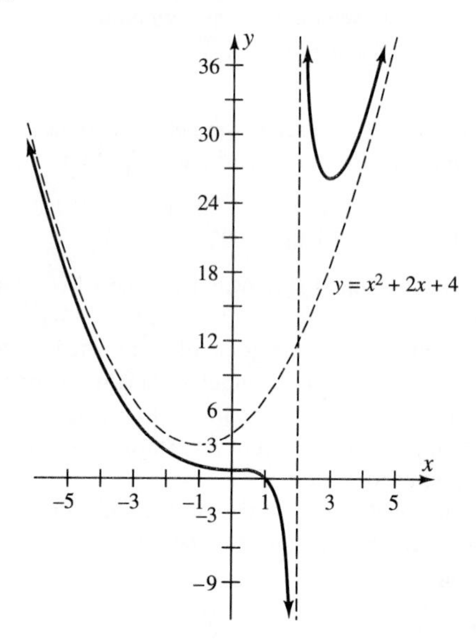

vertical asymptote $x = 2$
parabolic asymptote $y = x^2 + 2x + 4$

13. $-15.5 \le x \le 8.5$ and $-16 \le y \le 8$

15. $-4.7 \le x \le 4.7$ and $-12 \le y \le 8$

17. Overall: $-13 \le x \le 7$ and $-20 \le y \le 20$; hidden area near the origin: $-2.5 \le x \le 1$ and $-.02 \le y \le .02$

19. **(b)** Approximately $.06 \le x \le 2.78$

Section 4.5, page 273

1. $(-\infty, 3/2]$ **3.** $(-2, \infty)$ **5.** $(-\infty, -8/5]$

7. $(1, \infty)$ **9.** $(2, 4)$ **11.** $[-3, 5/2)$

13. $(-\infty, 4/7)$ **15.** $[-7/17, \infty)$ **17.** $[-1, 1/8)$

19. $[5, \infty)$ **21.** $x < \dfrac{b+c}{a}$ **23.** $c < x < a + c$

25. $1 \le x \le 3$ **27.** $x \le -7$ or $x \ge -2$

29. $x \le -2$ or $x \ge 3$ **31.** $-1 \le x \le 0$ or $x \ge 1$

33. $x < -1$ or $0 < x < 3$

35. $-2 < x < -1$ or $1 < x < 2$

37. $-2.26 \le x \le 0.76$ or $x \ge 3.51$

39. $.5 < x < .84$ **41.** $x < -1/3$ or $x > 2$

43. $-2 < x < -1$ or $1 < x < 3$ **45.** $x > 1$

47. $x \le -9/2$ or $x > -3$ **49.** $-3 < x < 1$ or $x \ge 5$

51. $-\sqrt{7} < x < \sqrt{7}$ or $x > 5.34$

53. $x < -3$ or $1/2 < x < 5$ **55.** $0 \le C \le 100$

57. Approximately 8.608 cents per kwh

59. More than \$12,500 **61.** Between \$4000 and \$5400

63. $1 < x < 19$ and $y = 20 - x$ **65.** $10 < x < 35$

67. $1 \le t \le 4$ **69.** $2 < t < 2.25$

71. **(a)** $x^2 < x$ when $0 < x < 1$ and $x^2 > x$ when $x < 0$ or $x > 1$.
(b) If c is nonzero and $|c| < 1$, then either $0 < c < 1$ or $-1 < c < 0$ (which is equivalent to $1 > -c > 0$). If $0 < c < 1$, then $|c| = c$ and c is a solution of $x^2 < x$ by part (a), so that $c^2 < c = |c|$. If $1 > -c > 0$, then $|c| = -c$, which is a solution of $x^2 < x$ by part (a), so that $c^2 = (-c)^2 < (-c) = |c|$.
(c) If $|c| > 1$, then either $c < -1$ or $c > 1$. In either case, c is a solution of $x^2 > x$ by part (a).

Excursion 4.5.A, page 277

1. $-4/3 \le x \le 0$ **3.** $7/6 < x < 11/6$

5. $x < -2$ or $x > -1$ **7.** $x \le -11/20$ or $x \ge -1/4$

9. $x < -53/40$ or $x > -43/40$

11. $x \le -7/2$ or $x \ge -5/4$

13. $x < -5$ or $-5 < x < -4/3$ or $x > 6$

15. $-1/7 < x < 3$

17. $-\sqrt{3} < x < -1$ or $1 < x < \sqrt{3}$

19. $x < -\sqrt{6}$ or $x > \sqrt{6}$

21. $x \le -2$ or $-1 \le x \le 0$ or $x \ge 1$

23. $0 < x < 2/3$ or $2 < x < 8/3$

25. $-1.43 < x < 1.24$ **27.** $x < -.89$ or $x > 1.56$

29. $x \le 2$ or $x \ge 14/3$

31. $-1.13 < x < 1.35$ or $1.35 < x < 1.67$

33. If $|x - 3| < E/5$, then multiplying both sides by 5 shows that $5|x - 3| < E$. But $5|x - 3| = |5|\cdot|x - 3| = |5(x - 3)| = |5x - 15| = |(5x - 4) - 11|$. Thus, $|(5x - 4) - 11| < E$.

Section 4.6, page 283

1. $8 + 2i$ **3.** $-2 - 10i$ **5.** $-\frac{1}{2} - 2i$

7. $\left(\frac{\sqrt{2} - \sqrt{3}}{2}\right) + 2i$ **9.** $1 + 13i$

11. $-10 + 11i$ **13.** $-21 - 20i$

15. 4 **17.** $-i$ **19.** i **21.** i

23. $\frac{5}{29} + \frac{2}{29}i$ **25.** $-\frac{1}{3}i$ **27.** $\frac{12}{41} - \frac{15}{41}i$

29. $\frac{-5}{41} - \frac{4}{41}i$ **31.** $\frac{10}{17} - \frac{11}{17}i$ **33.** $\frac{7}{10} + \frac{11}{10}i$

35. $-\frac{113}{170} + \frac{41}{170}i$ **37.** $6i$

39. $\sqrt{14}i$ **41.** $-4i$ **43.** $11i$

45. $(\sqrt{15} - 3\sqrt{2})i$ **47.** $\frac{2}{3}$

49. $-41 - i$ **51.** $(2 + 5\sqrt{2}) + (\sqrt{5} - 2\sqrt{10})i$

53. $\frac{1}{3} - \frac{\sqrt{2}}{3}i$ **55.** $x = 2, y = -2$

57. $x = -3/4, y = 3/2$ **59.** $x = \frac{1}{3} \pm \frac{\sqrt{14}}{3}i$

61. $x = -\frac{1}{2} \pm \frac{\sqrt{7}}{2}i$ **63.** $x = \frac{1}{4} \pm \frac{\sqrt{31}}{4}i$

65. $x = \frac{3 \pm \sqrt{3}}{2}$ **67.** $x = 2, -1 + \sqrt{3}i, -1 - \sqrt{3}i$

69. $x = 1, -1, i, -i$ **71.** -1

73.
$$\begin{aligned} z + w &= (a + bi) + (c + di) \\ &= (a + c) + (b + d)i \text{ and hence} \\ \overline{z + w} &= (a + c) - (b + d)i \\ &= a + c - bi - di \\ &= (a - bi) + (c - di) = \overline{z} + \overline{w} \end{aligned}$$

75. We first express z/w in standard form:
$\frac{z}{w} = \frac{a + bi}{c + di} = \frac{a + bi}{c + di}\cdot\frac{c - di}{c - di} = \frac{(ac + bd) + (bc - ad)i}{c^2 + d^2}$. Hence, $\overline{\left(\frac{z}{w}\right)} = \frac{(ac + bd) - (bc - ad)i}{c^2 + d^2} = \frac{ac + bd - bci + adi}{c^2 + d^2}$. On

the other hand, $\frac{\bar{z}}{\bar{w}} = \frac{a-bi}{c-di} = \frac{a-bi}{c-di} \cdot \frac{c+di}{c+di} = \frac{ac+bd-bci+adi}{c^2+d^2}$.

77. If $z = a + bi$, with a, b real numbers, then $z - \bar{z} = (a+bi) - (a-bi) = 2bi$. If $z = a + bi$ is real, then $b = 0$ and hence, $z - \bar{z} = 2bi = 0$. Therefore, $z = \bar{z}$. Conversely, if $z = \bar{z}$, then $0 = z - \bar{z} = 2bi$, which implies that $b = 0$. Hence, $z = a$ is real.

79. $\frac{1}{z} = \left(\frac{a}{a^2+b^2}\right) + \left(\frac{-b}{a^2+b^2}\right) i$

Section 4.7, page 290

1. 2 **3.** 6 **5.** -30

7. $x = 0$ (multiplicity 54); $x = -4/5$ (multiplicity 1)

9. $x = 0$ (multiplicity 15); $x = \pi$ (multiplicity 14); $x = \pi + 1$ (multiplicity 13)

11. $x = 1 + 2i$ or $1 - 2i$; $f(x) = (x - 1 - 2i)(x - 1 + 2i)$

13. $x = -\frac{1}{3} + \frac{2\sqrt{5}}{3} i$ or $-\frac{1}{3} - \frac{2\sqrt{5}}{3} i$;
$f(x) = \left(x + \frac{1}{3} - \frac{2\sqrt{5}}{3} i\right)\left(x + \frac{1}{3} + \frac{2\sqrt{5}}{3} i\right)$

15. $x = 3$ or $-\frac{3}{2} + \frac{3\sqrt{3}}{2} i$ or $-\frac{3}{2} - \frac{3\sqrt{3}}{2} i$;
$f(x) = (x-3)\left(x + \frac{3}{2} - \frac{3\sqrt{3}}{2} i\right)\left(x + \frac{3}{2} + \frac{3\sqrt{3}}{2} i\right)$

17. $x = -2$ or $1 + \sqrt{3}i$ or $1 - \sqrt{3}i$;
$f(x) = (x+2)(x - 1 - \sqrt{3}i)(x - 1 + \sqrt{3}i)$

19. $x = 1$ or i or -1 or $-i$;
$f(x) = (x-1)(x-i)(x+1)(x+i)$

21. $x = \sqrt{5}$ or $-\sqrt{5}$ or $\sqrt{2}i$ or $-\sqrt{2}i$;
$f(x) = (x - \sqrt{5})(x + \sqrt{5})(x - \sqrt{2}i)(x + \sqrt{2}i)$

23. Many correct answers, including $(x-1)(x-7)(x+4)$

25. Many correct answers, including $(x-1)(x-2)^2(x-\pi)^3$

27. $f(x) = 2x(x-4)(x+3)$

In Exercises 29–40, there are many correct answers, including the following:

29. $x^2 - 4x + 5$ **31.** $(x-2)(x^2 - 4x + 5)$

33. $(x+3)(x^2 - 2x + 2)(x^2 - 2x + 5)$

35. $x^2 - 2x + 5$ **37.** $(x-4)^2(x^2 - 6x + 10)$

39. $(x^4 - 3x^3)(x^2 - 2x + 2)$

41. $3x^2 - 6x + 6$ **43.** $-2x^3 + 2x^2 - 2x + 2$

45. Many correct answers, including
$x^2 - (1 - i)x + (2 + i)$

47. Many correct answers, including
$x^3 - 5x^2 + (7 + 2i)x - (3 + 6i)$

49. $3, -\frac{1}{2} + \frac{\sqrt{3}}{2} i, -\frac{1}{2} - \frac{\sqrt{3}}{2} i$ **51.** $i, -i, -1, -2$

53. $1, 2i, -2i$ **55.** $i, -i, 2 + i, 2 - i$

57. **(a)** Since $z + w = (a + c) + (b + d)i$, $\overline{z + w} = (a + c) - (b + d)i$. Since $\bar{z} = a - bi$ and $\bar{w} = c - di$, $\bar{z} + \bar{w} = (a - bi) + (c - di) = (a + c) - (b + d)i$.
(b) Since $zw = (ac - bd) + (ad + bc)i$, $\overline{zw} = (ac - bd) - (ad + bc)i$. Since $\bar{z} = a - bi$ and $\bar{w} = c - di$, $\bar{z}\,\bar{w} = (a - bi)(c - di) = (ac - bd) - (ad + bc)i$.

59. **(a)**
$$\begin{aligned}
\overline{f(z)} &= \overline{az^3 + bz^2 + cz + d} && \text{(definition of } f(z))\\
&= \overline{az^3} + \overline{bz^2} + \overline{cz} + \bar{d} && \text{(Exercise 57(a))}\\
&= \bar{a}\,\overline{z^3} + \bar{b}\,\overline{z^2} + \bar{c}\,\bar{z} + \bar{d} && \text{(Exercise 57(b))}\\
&= a\,\overline{z^3} + b\,\overline{z^2} + c\,\bar{z} + d && (r = \bar{r} \text{ for } r \text{ real})\\
&= a\bar{z}^3 + b\,\bar{z}^2 + c\,\bar{z} + d && \text{(Exercise 57(b))}\\
&= f(\bar{z}) && \text{(definition of } f)
\end{aligned}$$
(b) Since $f(z) = 0$, we have $0 = \bar{0} = \overline{f(z)} = f(\bar{z})$. Hence $\bar{z}$ is a root of $f(x)$.

61. If $f(z)$ is a polynomial with real coefficients, then $f(z)$ can be factored as $g_1(z)g_2(z)g_3(z)\cdots g_k(z)$, where each $g_i(z)$ is a polynomial with real coefficients and degree 1 or 2. The rules of polynomial multiplication show that the degree of $f(z)$ is the sum: degree $g_1(z)$ + degree $g_2(z)$ + degree $g_3(z) + \cdots +$ degree $g_k(z)$. If all of the $g_i(z)$ have degree 2, then this last sum is an even number. But $f(z)$ has odd degree, so this can't occur. Therefore, at least one of the $g_i(z)$ is a first-degree polynomial and hence must have a real root. This root is also a root of $f(z)$.

Chapter 4 Review, page 293

1. (a), (c), (e), (f) **3.** 0

5.
$$\begin{array}{r|rrrrrrr} 2 & 1 & -5 & 8 & 1 & -17 & 16 & -4 \\ & & 2 & -6 & 4 & 10 & -14 & 4 \\ \hline & 1 & -3 & 2 & 5 & -7 & 2 & \boxed{0} \end{array}$$
other factor: $x^5 - 3x^4 + 2x^3 + 5x^2 - 7x + 2$

7. Many correct answers, including $f(x) = 5(x - 1)^2(x + 1) = 5x^3 - 5x^2 - 5x + 5$

9. -1 and $5/3$ **11.** $\sqrt[3]{2}$

13. $0, \pm\sqrt{\dfrac{1 + \sqrt{21}}{2}}$

15. **(a)** $1, -1, 3, -3, \dfrac{1}{2}, -\dfrac{1}{2}, \dfrac{3}{2}, -\dfrac{3}{2}$ **(b)** 3

(c) $3, \dfrac{1 + \sqrt{3}}{2}, \dfrac{1 - \sqrt{3}}{2}$

17. 1 **19.** $3, -3, \sqrt{2}, -\sqrt{2}$

21. $(x^4 - 4x^3 + 16x - 16) \div (x - 5)$ is $x^3 + x^2 + 5x + 41$ with remainder 189. Since all coefficients and the remainder are positive, 5 is an upper bound for the roots.

23. $1, -1, 1.867, -.867$

25. $2x + h + 1$

27. Many correct answers

29. (c)

31. $-3 \le x \le 9$ and $-35 \le y \le 15$

33. $-2 \le x \le 18$ and $-500 \le y \le 1100$

35.

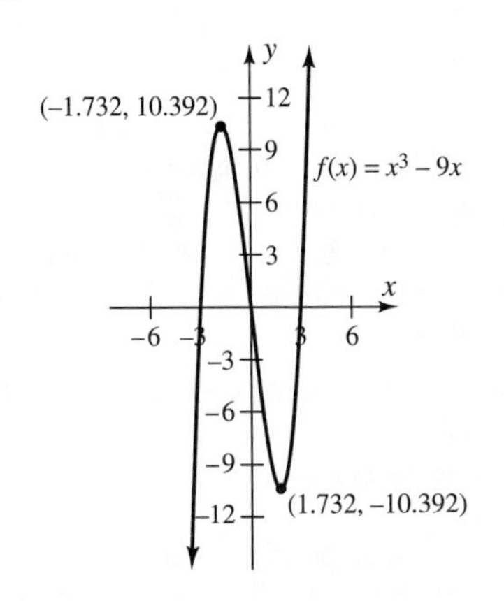

37.

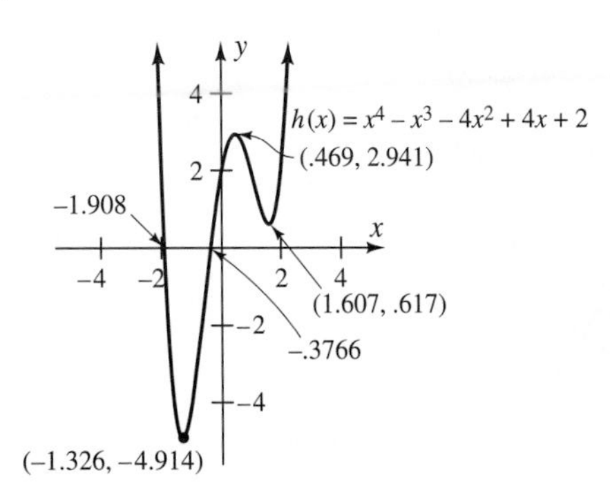

39.

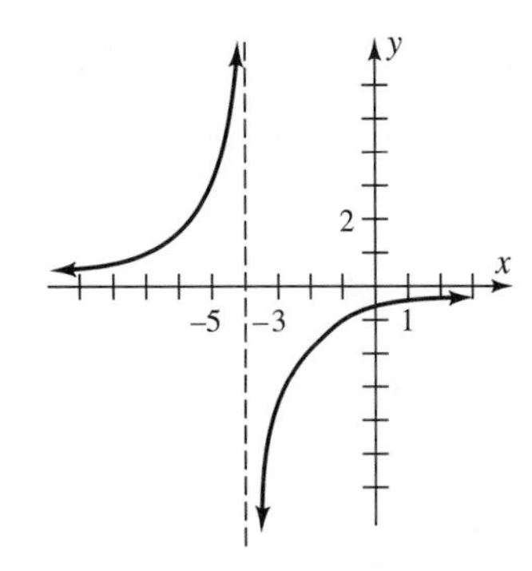

41.

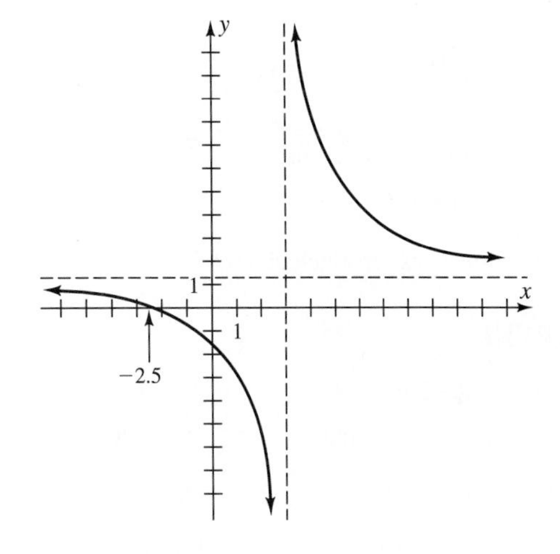

43. Vertical asymptotes $x = -2$ and $x = 3$; horizontal asymptote $y = 0$; hole at $x = 1$

45. $-4.7 \le x \le 4.7$ and $-5 \le y \le 5$

47. $-19 \le x \le 19$ and $-8 \le y \le 8$

49. At least 400 bags (priced at \$1.75 each) and at most 2500 bags (priced at \$.70 each)

51. (a) $T = \dfrac{40}{x} + \dfrac{110}{x+25}$, where x is the speed of the car and $0 < x < 55$ (speed limit)
(b) At least 44.08 mph

53. $\dfrac{1}{(x+h+1)(x+1)}$ **55.** (d)

57. $y \le -17$ or $y \ge 13$

59. $(-\infty, -2)$ and $\left(-\dfrac{1}{3}, \infty\right)$

61. $-\sqrt{3} \le x < -1$ or $-1 < x < 1$ or $1 < x \le \sqrt{3}$

63. $x \le -1$ or $0 \le x \le 1$

65. (e) **67.** $x \le -4/3$ or $x \ge 0$

69. $x \le -7$ or $x > -4$

71. $x < -2\sqrt{3}$ or $-3 < x < 2\sqrt{3}$

73. $x < \dfrac{1-\sqrt{13}}{6}$ or $x > \dfrac{1+\sqrt{13}}{6}$

75. $x = \dfrac{-3 \pm \sqrt{31}i}{2}$

77. $x = \dfrac{3 \pm \sqrt{31}i}{10}$ **79.** $x = \sqrt{2/3}$ or $-\sqrt{2/3}$ or i or $-i$

81. $x = -2$ or $1 + \sqrt{3}i$ or $1 - \sqrt{3}i$

83. $i, -i, 2, -1$

85. Many correct answers, including $x^4 - 2x^3 + 2x^2$

CHAPTER 5

Section 5.1, page 305

1. .09 **3.** $.08^6$ **5.** $6\sqrt{2}$

7. 1/2 **9.** $-1 + \sqrt{3}$ **11.** $14 + 3\sqrt{3}$

13. $(a^2 + b^2)^{1/3}$ **15.** $a^{3/16}$ **17.** $4t^{27/10}$

19. $4a^4/b$ **21.** $d^5/(2\sqrt{c})$ **23.** $15\sqrt{5}$

25. $(4x + 2y)^2$ **27.** 1 **29.** $x^{9/2}$

31. $c^{42/5}d^{10/3}$ **33.** $\dfrac{a^{1/2}}{49b^{5/2}}$ **35.** $\dfrac{2^{9/2}a^{12/5}}{3^4b^4}$

37. a^x **39.** $\dfrac{1}{x^{1/5}y^{2/5}}$ **41.** 1

43. $x^{7/6} - x^{11/6}$ **45.** $x - y$

47. $x + y - (x + y)^{3/2}$ **49.** $(x^{1/3} + 3)(x^{1/3} - 2)$

51. $(x^{1/2} + 3)(x^{1/2} + 1)$

53. $(x^{2/5} + 9)(x^{1/5} + 3)(x^{1/5} - 3)$

55. 7.9327×10^4 **57.** 2×10^{-3}

59. 5.963×10^{12} **61.** 740,000

63. .0000000000038 **65.** 34,570,000,000

67. 3. $\sqrt[n]{\sqrt[m]{c}} = \sqrt[mn]{c}$ 4. $\sqrt[m]{cd} = \sqrt[m]{c}\sqrt[m]{d}$
5. $\sqrt[m]{\dfrac{c}{d}} = \dfrac{\sqrt[m]{c}}{\sqrt[m]{d}}$

69. (a) Since its graph passes the horizontal line test, f is one-to-one and hence has an inverse.

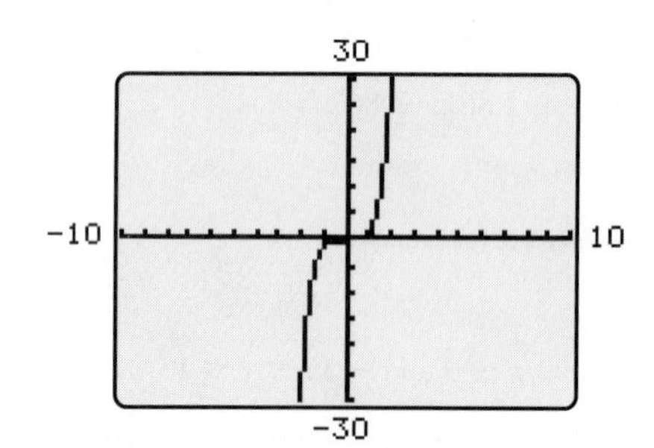

(b) $(g \circ f)(x) = g(f(x)) = (f(x))^{1/5} = (x^5)^{1/5} = x$ and $(f \circ g)(x) = f(g(x)) = (g(x))^5 = (x^{1/5})^5 = x$

71. (a) $x^{1/2} < x^{1/4} < x^{1/6}$ when $0 < x < 1$ because the graph of $y = x^{1/2}$ lies below the graph of $y = x^{1/4}$, which lies below the graph of $y = x^{1/6}$.
(b) $x^{1/2} > x^{1/4} > x^{1/6}$ when $x > 1$ because the graph of $y = x^{1/2}$ lies above the graph of $y = x^{1/4}$, which lies above the graph of $y = x^{1/6}$.

73. (a) The graph of g is the graph of f shifted horizontally 3 units to the left.
(b) The graph of h is the graph of f shifted vertically 2 units downward.
(c) The graph of k is the graph of f shifted horizontally 3 units to the left, then vertically 2 units downward.

Excursion 5.1.A, page 312

1. $x = 7$ **3.** $x = 4$ **5.** $x = -2$

7. $x = \pm 3$ **9.** $x = -1$ or 2 **11.** $x = 9$

13. $x = 1/2$ **15.** $x = 1/2$ or -4 **17.** $x \approx \pm .73$

19. $x \approx -1.17$ or 2.59 or $x = -1$

21. $x \approx 1.658$ **23.** $x = 6$

25. $x = 3$ or 7 **27.** No solutions

29. $x \approx -.457$ or 1.40 **31.** $b = \sqrt{\dfrac{a^2}{A^2 - 1}}$

33. $u = \sqrt{\dfrac{x^2}{1 - K^2}}$ **35.** $x = 4$ **37.** $x = 4$

39. $x = -1$ or -8 **41.** $x = -64$ or 8

43. $x = 16$ **45.** $x = -1/2$ or $1/3$

47. $x \approx 105.236$ **49.** $x \approx -.283$

51. **(a)** 11.47 ft or 29.91 ft **(b)** 21.00 ft

53. **(a)** $I = \dfrac{x}{(x^2 + 1024)^{3/2}}$ **(b)** 22.63 ft

Section 5.2, page 322

1.

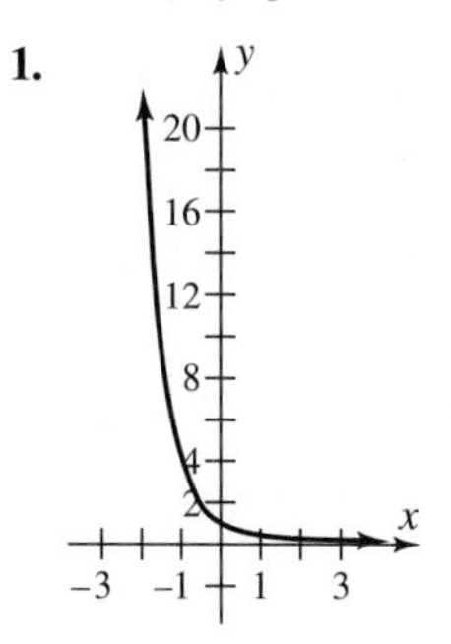

3.

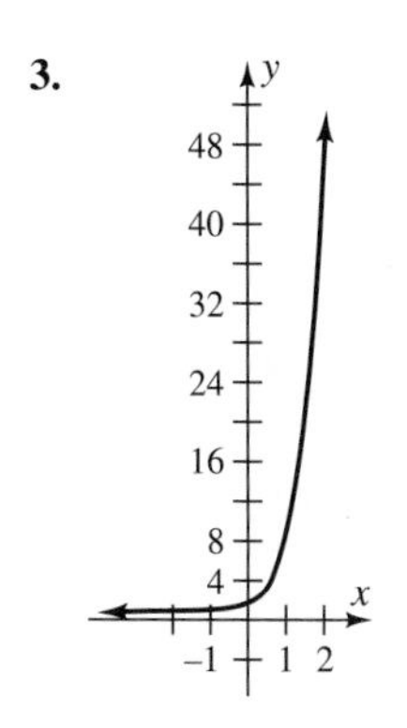

5.

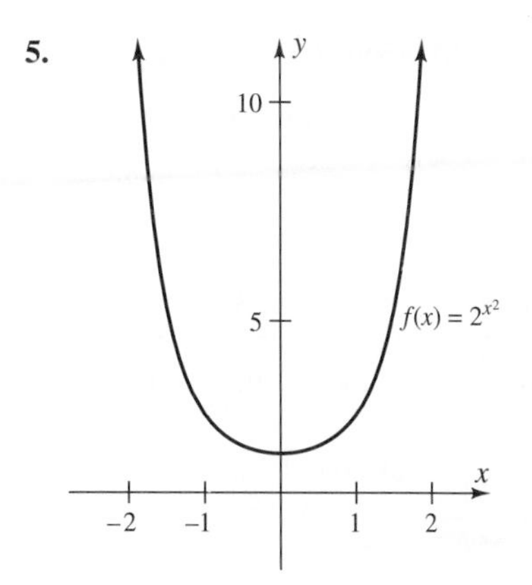

7. Shift the graph of h vertically 5 units downward.

9. Stretch the graph of h away from the x-axis by a factor of 3.

11. Shift the graph of h horizontally 2 units to the left, then vertically 5 units downward.

13. $x > 0$ **15.** $x \leq 0$ **17.** Neither **19.** Even

21. Even **23.** 11 **25.** .8

27. $\dfrac{10^{x+h} - 10^x}{h}$ **29.** $\dfrac{(2^{x+h} + 2^{-x-h}) - 2^x - 2^{-x}}{h}$

31. $-3 \leq x \leq 3$ and $0 \leq y \leq 12$

33. $-4 \leq x \leq 4$ and $0 \leq y \leq 10$

35. $-10 \leq x \leq 10$ and $0 \leq y \leq 20$

37. $-10 \leq x \leq 10$ and $0 \leq y \leq 6$

39. The x-axis is a horizontal asymptote for the left side of the graph; local minimum at $(-1.44, -.53)$.

41. No asymptotes; local minimum at (0, 1)

43. The x-axis is a horizontal asymptote; local maximum at (0, 1).

45. 100; 722.47; 2700

47. **(a)** 100 kg
(b) Approximately 139.3 kg, 163.2 kg, and 186.5 kg

49. **(a)** About 520 in 15 days; about 1559 in 25 days
(b) In 29.3 days

51. **(a)** $g(x) = 67.4(1.026)^x$ **(b)** 112.62 million

53. **(a)** $f(x) = 75(1.05)^x$ **(b)** About 164

55. **(a)** $f(x) = .75^x$ **(b)** About 8 ft

57. **(a)** $f(x) = 6(3^x)$ or $f(x) = 18(3^{x-1})$ **(b)** 3
(c) No; yes

59. **(a)**

Folds	0	1	2	3	4
Thickness	.002	.004	.008	.016	.032

(b) $f(x) = .002(2^x)$ **(c)** 2097.15 in = 174.76 ft
(c) 43

61. About 256; about 654

63. **(a)** $M(x) = 5(.5^{x/5730})$ **(b)** 3.08 gr; 1.90 gr
(c) After 13,304.65 years

65. **(a)** 100,000 now; 83,527 in 2 months; 58,275 in 6 months
(b) No. The graph continues to decrease toward zero.

67. **(a)** The current population is 10, and in 5 years it will be about 149.
(b) After about 9.55 years

69. Many correct answers: $f(x) = a^x$ for any nonnegative constant a

71. **(a)** The graph of f is the mirror image of the graph of g.
(b) $k(x) = f(x)$; see (a).

73. **(a)** Not entirely
(b) The graph of $f_8(x)$ appears to coincide with the graph of $g(x)$ on most calculator screens; when $-2.4 \le x \le 2.4$, the maximum error is at most .01.
(c) Not at the right side of the viewing window; $f_{12}(x)$

Excursion 5.2.A, page 331

1. Annually: \$1469.33; quarterly: \$1485.95; monthly: \$1489.85; weekly: \$1491.37

3. \$738.73 **5.** \$821.81 **7.** \$1065.30

9. \$661.56 **11.** \$844.30 **13.** \$568.59

15. About 9.9 years **17.** About 5.00%

19. About 5.92%

Section 5.3, page 341

1. 4 **3.** -2.5

5. $10^3 = 1000$ **7.** $10^{2.88} = 750$

9. $e^{1.0986} = 3$ **11.** $e^{6.9078} = 1000$

13. $e^{-4.6052} = .01$ **15.** $e^{z+w} = x^2 + 2y$

17. $\log .01 = -2$ **19.** $\log 3 = .4771$

21. $\ln 25.79 = 3.25$ **23.** $\ln 5.5527 = 12/7$

25. $\ln w = 2/r$ **27.** $\sqrt{43}$

29. 15 **31.** 1/2

33. 931 **35.** $\sqrt{37}$

37. $x + y$ **39.** x^2

41. $\ln(x^2 y^3)$ **43.** $\log(x - 3)$

45. $\ln(x^{-7})$ **47.** $2u + 5v$

49. $\frac{1}{2}u + 2v$ **51.** $\frac{2}{3}u + \frac{1}{6}v$

53. $(-1, \infty)$ **55.** $(-\infty, 0)$

57. **(a)** For all $x > 0$.
(b) According to the fourth property of natural logarithms on page 336, $e^{\ln x} = x$ for every $x > 0$.

59. False; the right side is not defined when $x < 0$, but the left side is.

61. True by the Power Law

63. False; the graph of the left side differs from the graph of the right side.

65. Stretch the graph of g away from the x-axis by a factor of 2.

67. Shift the graph of g horizontally 4 units to the right.

69. Shift the graph of g horizontally 3 units to the left, then shift it vertically 4 units downward.

71.

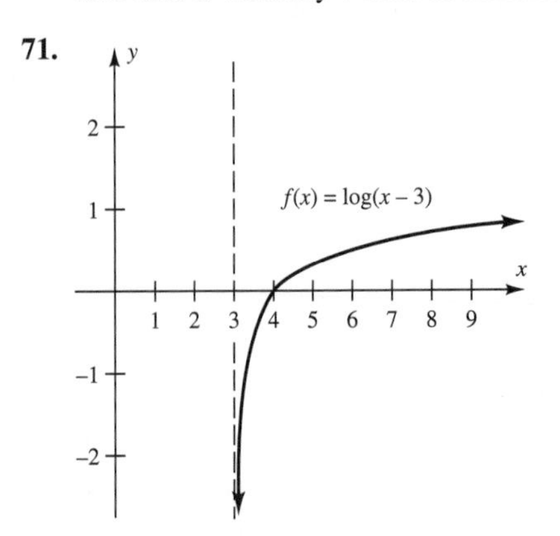

73.

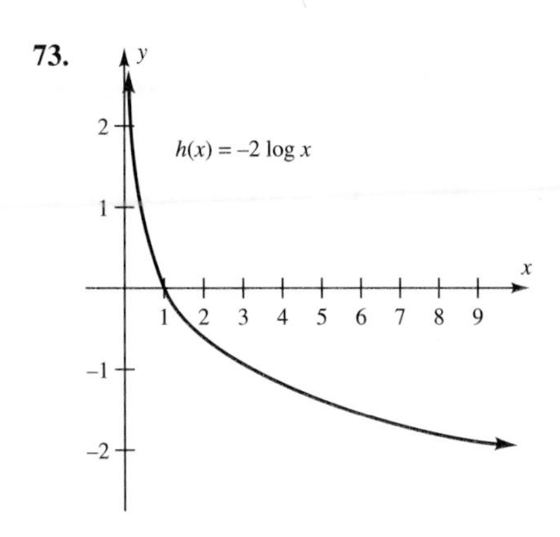

75. $0 \le x \le 9.4$ and $-6 \le y \le 6$ (vertical asymptote at $x = 1$)

77. $-10 \le x \le 10$ and $-3 \le y \le 3$

79. $0 \le x \le 20$ and $-6 \le y \le 3$

81. .5493 **83.** $-.2386$

85. **(a)** $\dfrac{\ln(3+h) - \ln 3}{h}$ **(b)** $h \approx 2.2$

87. **(a)** Both are acid. **(b)** .01 **(c)** Yes

89. $A = -9$; $B = 10$

91.
$$(f\circ g)(x) = \frac{1}{1 + e^{-\ln(x/(1-x))}} = \frac{1}{1 + \dfrac{1}{e^{\ln(x/(1-x))}}}$$
$$= \frac{1}{1 + \dfrac{1}{\dfrac{x}{1-x}}} = \frac{1}{1 + \dfrac{1-x}{x}} = \frac{1}{\dfrac{1}{x}} = x;$$
$$(g\circ f)(x) = \ln\left(\frac{\dfrac{1}{1+e^{-x}}}{1 - \dfrac{1}{1+e^{-x}}}\right)$$
$$= \ln\left(\frac{\dfrac{1}{1+e^{-x}}}{1 - \dfrac{1}{1+e^{-x}}} \cdot \frac{1+e^{-x}}{1+e^{-x}}\right)$$
$$= \ln\left(\frac{1}{1+e^{-x}-1}\right) = \ln(e^x) = x$$

93. 2 **95.** Approximately 2.54

97. 20 decibels

99. Approximately 66 decibels

101. About 4392 meters

103. **(a)** 77 **(b)** 66; 59 **(c)** 14 weeks

105. **(a)** 9.9 days **(b)** About 6986

107. $n = 30$ gives an approximation with a maximum error of .00001 when $-.7 \le x \le .7$.

109. **(a)** No advertising: about 120 bikes; \$1000: about 299 bikes; \$10,000: about 513 bikes
(b) \$1000, yes; \$10,000; no
(c) Yes; yes (but not as worthwhile as spending \$1000)

Excursion 5.3.A, page 350

1.

x	0	1	2	4
$f(x) = \log_4 x$	Not defined	0	.5	1

3.

x	1/36	1/6	1	216
$h(x) = \log_6 x$	-2	-1	0	3

5.

x	0	1/7	$\sqrt{7}$	49
$f(x) = 2\log_7 x$	Not defined	-2	1	4

7.

x	-2.75	-1	1	29
$h(x) = 3\log_2(x+3)$	-6	3	6	15

9. $\log .01 = -2$ **11.** $\log \sqrt[3]{10} = 1/3$

13. $\log r = 7k$ **15.** $\log_7 5{,}764{,}801 = 8$

17. $\log_3(1/9) = -2$ **19.** $10^4 = 10{,}000$

21. $10^{2.88} \approx 750$ **23.** $5^3 = 125$ **25.** $2^{-2} = \dfrac{1}{4}$

27. $10^{z+w} = x^2 + 2y$ **29.** $\sqrt{43}$ **31.** $\sqrt{x^2 + y^2}$

33. 1/2 **35.** 6 **37.** $b = 3$ **39.** $b = 20$

41. 5 **43.** 3 **45.** 4 **47.** .43

49. .26 **51.** .1 **53.** .7 **55.** 3.3219

57. .8271 **59.** 1.1115 **61.** 1.6199 **63.** True

65. True **67.** False **69.** 397^{398}

71. $\log_b u = \dfrac{\log_a u}{\log_a b}$ **73.** $\log_{10} u = 2 \log_{100} u$

75. $\log_b x = \dfrac{1}{2} \log_b v + 3 = \log_b\sqrt{v} + \log_b b^3 = \log_b(b^3\sqrt{v})$; hence $x = b^3\sqrt{v}$.

77. $f(x) = g(x)$ only when $x \approx .123$, so the statement is false.

79.

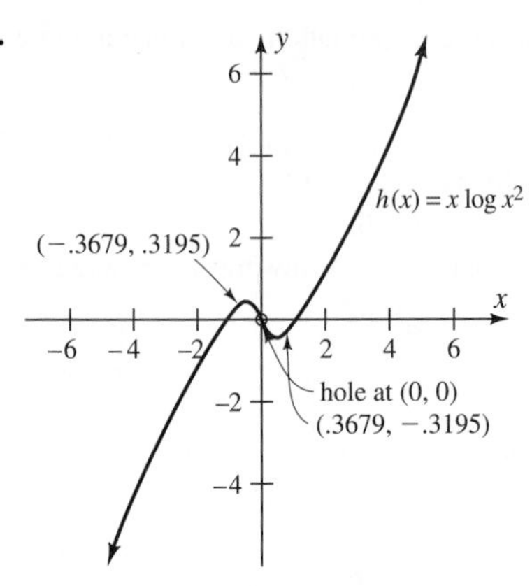

Section 5.4, page 357

1. $x = 4$ **3.** $x = 1/9$ **5.** $x = \dfrac{1}{2}$ or -3

7. $x = -2$ or $-1/2$ **9.** $x = \ln 5/\ln 3 \approx 1.465$

11. $x = \ln 3/\ln 1.5 \approx 2.7095$

13. $x = \dfrac{\ln 3 - 5 \ln 5}{\ln 5 + 2 \ln 3} \approx -1.825$

15. $x = \dfrac{\ln 2 - \ln 3}{3 \ln 2 + \ln 3} \approx -.1276$

17. $x = (\ln 5)/2 \approx .805$

19. $x = (-\ln 3.5)/1.4 \approx -.895$

21. $x = 2 \ln(5/2.1)/\ln 3 \approx 1.579$ **23.** $x = 0$ or 1

25. $x = \ln 2 \approx .693$ or $x = \ln 3 \approx 1.099$

27. $x = \ln 3 \approx 1.099$

29. $x = \ln 2/\ln 4 = 1/2$ or $x = \ln 3/\ln 4 \approx .792$

31. $x = \ln\left(t + \sqrt{t^2 + 1}\right)$

33. $x = \dfrac{1}{2} \ln\left(\dfrac{1 + t}{1 - t}\right) = \dfrac{1}{2} [\ln(1 + t) - \ln(1 - t)]$

35. $x = 9$ **37.** $x = 5$ **39.** $x = 6$ **41.** $x = 3$

43. $x = \dfrac{-5 + \sqrt{37}}{2}$ **45.** $x = 9/(e - 1)$ **47.** $x = 5$

49. $x = \pm\sqrt{10001}$ **51.** $x = \sqrt{\dfrac{e + 1}{e - 1}}$

53. Approximately 3689 years old

55. Approximately 2534 years ago

57. Approximately 444,000,000 years

59. Approximately 10.413 years

61. Approximately 9.853 days

63. Approximately 6.99%

65. **(a)** Approximately 22.5 years
(b) Approximately 22.1 years

67. \$3197.05 **69.** 79.36 years

71. **(a)** About 2.1548% **(b)** In the year 2012

73. **(a)** $k \approx 21.459$ **(b)** $t \approx .182$

75. **(a)** There are 20 bacteria at the beginning and 2500 three hours later.
(b) $2/\ln 5 \approx .43$

77. **(a)** At the outbreak: 200 people; after 3 weeks: about 2795 people
(b) In about 6.02 weeks

79. **(a)** $k \approx .229$, $c \approx 83.3$
(b) 12.43 weeks

Chapter 5 Review, page 361

1. c^2 **3.** $a^{10/3} b^{42/5}$ **5.** $u^{1/2} - v^{1/2}$

7. $c^2d^4/2$ **9.** $x = \dfrac{5 - \sqrt{5}}{2}$ **11.** No solutions

13. $x = -1.733$ or 5.521 **15.** $-3 \le x \le 3$ and $0 \le y \le 2$

17.

19.

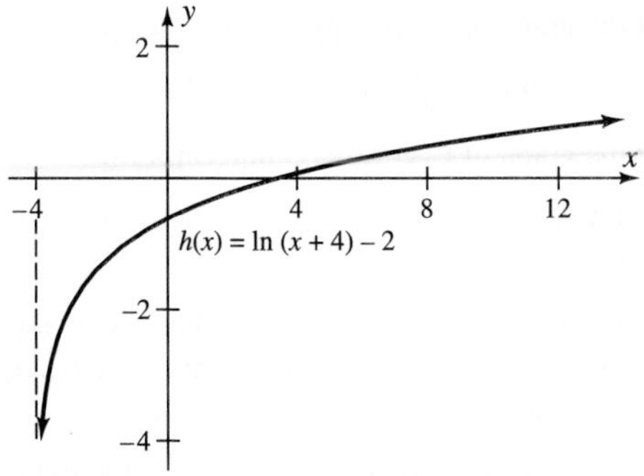

21. $\ln 756 = 6.628$ **23.** $\ln(u + v) = r^2 - 1$

25. $\log 756 = 2.8785$ **27.** $e^{7.118} = 1234$

29. $e^t = rs$ **31.** $5^u = cd - k$ **33.** 3 **35.** 3/4

37. $2 \ln x$ **39.** $\ln(9y/x^2)$ **41.** Undefined

43. 3/2 **45.** .18 **47.** .70 **49.** (c)

51. (c) **53.** $x = \pm\sqrt{2}$ **55.** $x = \frac{3 \pm \sqrt{57}}{4}$

57. $x = -\frac{1}{2}$ **59.** $x = e^{(u-c)/d}$ **61.** $x = 2$

63. $x = 101$ **65.** About 1.64 mg

67. Approximately 12 years **69.** \$452.89 **71.** 7.6

CHAPTER 6

Section 6.1, page 373

1. $2\pi/9$ **3.** $\pi/9$ **5.** $\pi/18$ **7.** $\frac{\pi}{30}$

9. $-\frac{\pi}{15}$ **11.** $\frac{5\pi}{12}$ **13.** $\frac{3\pi}{4}$ **15.** $-\frac{5\pi}{4}$

17. $\frac{31\pi}{6}$ **19.** 36° **21.** −18° **23.** 135°

25. 4° **27.** −75° **29.** 972°

31. $\frac{9\pi}{4}, \frac{17\pi}{4}, -\frac{7\pi}{4}, -\frac{15\pi}{4}$

33. $\frac{11\pi}{6}, \frac{23\pi}{6}, -\frac{13\pi}{6}, -\frac{25\pi}{6}$

35. $\frac{5\pi}{3}$ **37.** $\frac{3\pi}{4}$ **39.** $\frac{3\pi}{5}$ **41.** $7 - 2\pi$

43. $\frac{4\pi}{3}$ **45.** $\frac{7\pi}{6}$ **47.** $\frac{41\pi}{6}$ **49.** 8π cm

51. $\frac{17}{4}$ **53.** $\frac{50}{9}$ **55.** 2000 **57.** 5

59. 8.75 **61.** 3490.66 mi **63.** 942.48 mi

65. 7π **67.** 4π **69.** 42.5π **71.** $2\pi k$

73. 3 radians (≈ 171.9°)

75. **(a)** 400π radians per min
(b) 800π in per min or $200\pi/3$ ft per min

77. **(a)** 5π radians per sec **(b)** 6.69 mph

79. 15.92 ft

Section 6.2, page 381

1. −1 **3.** 0 **5.** 0 **7.** 0 **9.** 0

11. $\sin t = 1/\sqrt{5}$, $\cos t = -2/\sqrt{5}$, $\tan t = -1/2$

13. $\sin t = -4/5$, $\cos t = -3/5$, $\tan t = 4/3$

15. $(fg)(t) = 3 \sin^2 t + 6 \sin t \cos t$

17. $3 \sin^3 t + 3 \sin^2 t \tan t$ **19.** $(\cos t - 2)(\cos t + 2)$

21. $(\sin t - \cos t)(\sin t + \cos t)$ **23.** $(\tan t + 3)^2$

25. $(3 \sin t + 1)(2 \sin t - 1)$

27. $(\cos^2 t + 5)(\cos t + 1)(\cos t - 1)$

29. $(f \circ g)(t) = \cos(2t + 4)$, $(g \circ f)(t) = 2 \cos t + 4$

31. $(f \circ g)(t) = \tan(t^2 + 2)$, $(g \circ f)(t) = \tan^2(t + 3) - 1$

33. Many correct answers, including $c = \pi$ and $d = \pi/2$. In this case $\sin(c + d) = \sin\left(\pi + \frac{\pi}{2}\right) = \sin \frac{3\pi}{2} = -1$, but $\sin c + \sin d = \sin \pi + \sin \frac{\pi}{2} = 0 + 1 = 1$. Similarly, $\cos(c + d) = \cos \frac{3\pi}{2} = 0$, but $\cos c + \cos d = \cos \pi + \cos \frac{\pi}{2} = -1 + 0 = -1$.

35. Positive since $0 < 1 < \pi/2$

37. Negative since $\pi/2 < 3 < \pi$

39. Positive since $0 < 1.5 < \pi/2$

41. All t in the intervals $\left[0, \frac{\pi}{2}\right)$ and $\left(\frac{\pi}{2}, 2\pi\right)$

43. $t = \frac{\pi}{2} + 2\pi n$, n any integer

45. $t = \pi n$, n any integer

47. $t = \frac{\pi}{2} + \pi n$, n any integer

49. $\frac{\sin(t+h) - \sin t}{h}$ **51.** $\frac{\tan(t+h) - \tan t}{h}$

53. **(a)** $-.42069$; $-.41660$; $-.41619$; $-.41615$
(b) $\cos 2$

55. $\sin(\cos 0) = \sin 1$, while $\cos(\sin 0) = \cos 0 = 1$. Since $\sin 1 < 1$ (draw a picture!), $\cos(\sin 0)$ is larger than $\sin(\cos 0)$.

57. **(a)** Each horse moves through an angle of 2π radians in one minute. The angle between horses A and B is $\pi/4$ radians ($= \frac{1}{8}$ of 2π radians). It takes $\frac{1}{8}$ min for each horse to move through an angle of $\pi/4$ radians. Thus the position occupied by B at time t will be occupied by A $\frac{1}{8}$ min later, that is, at time $t + \frac{1}{8}$. Therefore, $B(t) = A(t + \frac{1}{8})$.
(b) $C(t) = A(t + \frac{1}{3})$
(c) $E(t) = D(t + \frac{1}{8})$; $F(t) = D(t + \frac{1}{3})$
(d) The triangles in Figure 6–21 are similar, so that $\frac{5}{1} = \frac{D(t)}{A(t)}$. Therefore, $D(t) = 5A(t)$.
(e) $E(t) = 5B(t) = 5A(t + \frac{1}{8})$; $F(t) = 5C(t) = 5A(t + \frac{1}{3})$
(f) Since horse A travels through an angle of 2π radians each minute and its starting angle is 0 radians, then at the end of t min horse A will be on the terminal side of an angle of $2\pi t$ radians, at the point where it intersects the unit circle. $A(t)$ is the second coordinate of this point; hence, $A(t) = \sin(2\pi t)$.
(g) $B(t) = A(t + \frac{1}{8}) = \sin(2\pi(t + \frac{1}{8})) = \sin(2\pi t + \pi/4)$; $C(t) = \sin(2\pi t + 2\pi/3)$
(h) $D(t) = 5\sin(2\pi t)$; $E(t) = 5\sin(2\pi t + \pi/4)$; $F(t) = 5\sin(2\pi t + 2\pi/3)$

Section 6.3, page 394

1. $t = \ldots, -2\pi, -\pi, 0, \pi, 2\pi, \ldots$; or $t = \pi k$, where k is any integer

3. $t = \ldots, -7\pi/2, -3\pi/2, \pi/2, 5\pi/2, 9\pi/2 \ldots$; or $t = \pi/2 + 2\pi k$, where k is any integer

5. $t = \ldots, -3\pi, -\pi, \pi, 3\pi, \ldots$; or $t = \pi + 2k\pi$, where k is any integer

7. 11 **9.** 1.4

11. Shift the graph of f vertically 3 units upward.

13. Reflect the graph of f in the horizontal axis.

15. Shift the graph of f vertically 5 units upward.

17. Stretch the graph of f away from the horizontal axis by a factor of 3.

19. Stretch the graph of f away from the horizontal axis by a factor of 3, then shift the resulting graph vertically 2 units upward.

21. Shift the graph of f horizontally 2 units to the right.

23. 2 solutions **25.** 2 solutions

27. 2 solutions **29.** 2 solutions

31. Possibly an identity **33.** Possibly an identity

35. Possibly an identity **37.** Not an identity

39. Possibly an identity **41.** Not an identity

43. **(a)** Yes if proper value of k is used; no
(b) 0, 2π, 4π, 6π, etc. So why do the graphs look identical?

45. **(a)** 80
(b) 14 or 15 on 96-pixel-wide screens; up to 40–50 on wider screens; quite different from part (a). Explain what's going on. [*Hint:* How many points have to be plotted in order to get even a rough approximation of one full wave? How many points is the calculator plotting for the entire graph?]

47. The y-coordinate of the new point is the same as the x-coordinate of the point on the unit circle. To explain what's going on, look at the definition of the cosine function.

49. **(a)** There is no such number k.
(b) If we substitute $t = 0$ in $\cos(t + k) = \cos t$, we get $\cos k = \cos 0 = 1$.
(c) If there were such a number k, then by part (b), $\cos k = 1$, which is impossible by part (a). Therefore, there is no such number k, and the period is 2π.

Section 6.4, page 401

1. Yes **3.** No **5.** No **7.** $\sin t = -\sqrt{3}/2$

9. $\sin t = \sqrt{3}/2$ **11.** $-3/5$ **13.** $-3/5$

15. 3/4 **17.** $-3/4$ **19.** $-3/5$

21. $-\sqrt{21}/5$ **23.** $-2/5$ **25.** $-\sqrt{21}/5$

27. $\sqrt{21}/2$ **29.** $2/5$ **31.** $\sqrt{2}/2$

33. $\dfrac{\sqrt{2+\sqrt{2}}}{2}$ **35.** $\sqrt{3}$ **37.** $\sqrt{2}/2$

39. $\dfrac{\sqrt{2-\sqrt{2}}}{2}$ **41.** $-\sqrt{2}/2$ **43.** $-\dfrac{\sqrt{2-\sqrt{2}}}{2}$

45. $1/2$ **47.** 0 **49.** 0

51. $\sin^2 t - \cos^2 t$ **53.** $\sin t$

55. $|\sin t \cos t|\sqrt{\sin t}$ **57.** $1/4$

59. $\cos t + 2$ **61.** $\cos t$

63. $f(t+\pi) = \sin 2(t+\pi) = \sin(2t + 2\pi) = \sin 2t = f(t)$

65. $f(t+\pi) = \sin^2(t+\pi) = (\sin(t+\pi))^2 = (-\sin t)^2 = \sin^2 t = f(t)$

67. $f(t+\pi/2) = \tan 2(t+\pi/2) = \tan(2t+\pi) = \tan 2t = f(t)$

69. $\sin(t-\pi) = \sin(-(\pi - t)) = -\sin(\pi - t) = -\sin t$

71. $\cos(\pi - t) = \cos(-t+\pi) = -\cos(-t) = -\cos t$

73. $\sin(2\pi - t) = \sin(-t + 2\pi) = \sin(-t) = -\sin t$

75. Not an identity

Section 6.5, page 409

1. $\sin t = 7/\sqrt{53}$, $\cos t = 2/\sqrt{53}$, $\tan t = 7/2$

3. $\sin t = -6/\sqrt{61}$, $\cos t = -5/\sqrt{61}$, $\tan t = 6/5$

5. $\sin t = -10/\sqrt{103}$, $\cos t = \sqrt{3}/\sqrt{103}$, $\tan t = -10/\sqrt{3}$

7. $\sin\theta = \sqrt{\dfrac{2}{11}}$, $\cos\theta = \dfrac{3}{\sqrt{11}}$, $\tan\theta = \dfrac{\sqrt{2}}{3}$

9. $\sin\theta = \sqrt{\dfrac{3}{7}}$, $\cos\theta = \dfrac{2}{\sqrt{7}}$, $\tan\theta = \dfrac{\sqrt{3}}{2}$

11. $\sin\theta = \dfrac{h}{m}$, $\cos\theta = \dfrac{d}{m}$, $\tan\theta = \dfrac{h}{d}$

13. $\sin\left(\dfrac{5\pi}{6}\right) = \dfrac{1}{2}$, $\cos\left(\dfrac{5\pi}{6}\right) = -\dfrac{\sqrt{3}}{2}$, $\tan\left(\dfrac{5\pi}{6}\right) = -\dfrac{\sqrt{3}}{3}$

15. $\sin\left(\dfrac{7\pi}{3}\right) = \dfrac{\sqrt{3}}{2}$, $\cos\left(\dfrac{7\pi}{3}\right) = \dfrac{1}{2}$, $\tan\left(\dfrac{7\pi}{3}\right) = \sqrt{3}$

17. $\sin\left(\dfrac{11\pi}{4}\right) = \dfrac{\sqrt{2}}{2}$, $\cos\left(\dfrac{11\pi}{4}\right) = -\dfrac{\sqrt{2}}{2}$, $\tan\left(\dfrac{11\pi}{4}\right) = -1$

19. $\sin\left(-\dfrac{3\pi}{2}\right) = 1$, $\cos\left(-\dfrac{3\pi}{2}\right) = 0$, $\tan\left(-\dfrac{3\pi}{2}\right)$ not defined

21. $\sin\left(-\dfrac{23\pi}{6}\right) = \dfrac{1}{2}$, $\cos\left(-\dfrac{23\pi}{6}\right) = \dfrac{\sqrt{3}}{2}$, $\tan\left(-\dfrac{23\pi}{6}\right) = \dfrac{\sqrt{3}}{3}$

23. $\sin\left(-\dfrac{19\pi}{3}\right) = -\dfrac{\sqrt{3}}{2}$, $\cos\left(-\dfrac{19\pi}{3}\right) = \dfrac{1}{2}$, $\tan\left(-\dfrac{19\pi}{3}\right) = -\sqrt{3}$

25. $\sin\left(-\dfrac{15\pi}{4}\right) = \dfrac{\sqrt{2}}{2}$, $\cos\left(-\dfrac{15\pi}{4}\right) = \dfrac{\sqrt{2}}{2}$, $\tan\left(-\dfrac{15\pi}{4}\right) = 1$

27. $\sin\left(-\dfrac{17\pi}{2}\right) = -1$, $\cos\left(-\dfrac{17\pi}{2}\right) = 0$, $\tan\left(-\dfrac{17\pi}{2}\right)$ not defined

29. $-\sqrt{3}/2$ **31.** $-\sqrt{2}/2$ **33.** $\dfrac{\sqrt{2}}{4}(1-\sqrt{3})$

35. $y = \dfrac{\sqrt{3}}{3}x$ **37.** $y = -\sqrt{3}x$ **39.** $y = \sqrt{3}x$

41. $\sin t = -3/\sqrt{10}$, $\cos t = 1/\sqrt{10}$, $\tan t = -3$

43. $\sin t = -5/\sqrt{34}$, $\cos t = 3/\sqrt{34}$, $\tan t = -5/3$

45. $\sin t = 1/\sqrt{5}$, $\cos t = -2/\sqrt{5}$, $\tan t = -1/2$

47. $a = 6.05$ **49.** $b = 18.63$

51. $b = 22.73$ **53.** $c = 19.42$

55. $t = .85$ radians **57.** $t = .84$ radians

59. $A = .65$ radians **61.** $A = .60$ radians

63. $h = 25\sqrt{2}/2$ **65.** $h = 300$

67. $h = 50\sqrt{3}$ **69.** $c = 4\sqrt{3}/3$

71. $a = 10\sqrt{3}/3$ **73.** 15.67 ft

75. .0534 radians **77.** $\pi/6$ radians

79. **(a)** $A(t) = 200 \cos t \sin t$
(b) $t \approx .7854$; approximately 14.1421 by 7.0711

Section 6.6, page 418

1. Fourth quadrant **3.** Second quadrant

5. Fourth quadrant

7. $\sin t = \frac{4}{5}, \cos t = \frac{3}{5}, \tan t = \frac{4}{3}, \cot t = \frac{3}{4}, \sec t = \frac{5}{3},$ $\csc t = \frac{5}{4}$

9. $\sin t = 12/13, \cos t = -5/13, \tan t = -12/5, \cot t = -5/12, \sec t = -13/5, \csc t = 13/12$

11. $\sin t = 5/\sqrt{26}, \cos t = -1/\sqrt{26}, \tan t = -5, \cot t = -1/5, \sec t = -\sqrt{26}, \csc t = \sqrt{26}/5$

13. $\sin t = \sqrt{3}/\sqrt{5}, \cos t = \sqrt{2}/\sqrt{5}, \tan t = \sqrt{3}/\sqrt{2}, \cot t = \sqrt{2}/\sqrt{3}, \sec t = \sqrt{5}/\sqrt{2}, \csc t = \sqrt{5}/\sqrt{3}$

15. $\sin t = \frac{3}{\sqrt{12+2\sqrt{2}}}, \cos t = \frac{1+\sqrt{2}}{\sqrt{12+2\sqrt{2}}},$

$\tan t = \frac{3}{1+\sqrt{2}}, \cot t = \frac{1+\sqrt{2}}{3},$

$\sec t = \frac{\sqrt{12+2\sqrt{2}}}{1+\sqrt{2}}, \csc t = \frac{\sqrt{12+2\sqrt{2}}}{3}$

17. $\sin t = 1/\sqrt{5}, \cos t = 2/\sqrt{5}, \tan t = 1/2, \cot t = 2, \sec t = \sqrt{5}/2, \csc t = \sqrt{5}$

19. $\sin t = 5/\sqrt{34}, \cos t = 3/\sqrt{34}, \tan t = 5/3, \cot t = 3/5, \sec t = \sqrt{34}/3, \csc t = \sqrt{34}/5$

21. -3.8287

23. **(a)** 5.6511; 5.7618; 5.7731; 5.7743
(b) $(\sec 2)^2$

25. $\cos t + \sin t$ **27.** $1 - 2\sec t + \sec^2 t$

29. $\cot^3 t - \tan^3 t$

31. $\sin\left(\frac{4\pi}{3}\right) = -\frac{\sqrt{3}}{2}, \cos\left(\frac{4\pi}{3}\right) = -\frac{1}{2},$

$\tan\left(\frac{4\pi}{3}\right) = \sqrt{3}, \cot\left(\frac{4\pi}{3}\right) = \frac{1}{\sqrt{3}},$

$\sec\left(\frac{4\pi}{3}\right) = -2, \csc\left(\frac{4\pi}{3}\right) = \frac{-2}{\sqrt{3}}$

33. $\sin\left(\frac{7\pi}{4}\right) = -\frac{\sqrt{2}}{2}, \cos\left(\frac{7\pi}{4}\right) = \frac{\sqrt{2}}{2},$

$\tan\left(\frac{7\pi}{4}\right) = -1, \cot\left(\frac{7\pi}{4}\right) = -1, \sec\left(\frac{7\pi}{4}\right) = \sqrt{2}, \csc\left(\frac{7\pi}{4}\right) = -\sqrt{2}$

35. $\csc t(\sec t - \csc t)$ **37.** $(-1)(\tan^2 t + \sec^2 t)$

39. $(\cos^2 t + 1 + \sec^2 t)(\cos t - \sec t)$ **41.** $\cot t$

43. $\frac{2\tan t + 1}{3\sin t + 1}$ **45.** $4 - \tan t$ **47.** $c = 5.49$

49. $a = 1.48$ **51.** $a = 2.52$ **53.** $A = 1.1$

55. $A = .5$ **57.** $A = 1$

59. Possibly an identity **61.** Not an identity

63. $1 + \cot^2 t = 1 + \frac{\cos^2 t}{\sin^2 t} = \frac{\sin^2 t + \cos^2 t}{\sin^2 t} = \frac{1}{\sin^2 t} = \csc^2 t$

65. $\sec(-t) = \frac{1}{\cos(-t)} = \frac{1}{\cos t} = \sec t$

67. $\sec(\pi - t) = \frac{1}{\cos(\pi - t)} = \frac{1}{-\cos t} = -\sec t$

69. $\cot(\pi - t) = \frac{\cos(\pi - t)}{\sin(\pi - t)} = \frac{-\cos t}{\sin t} = -\cot t$

71. $\csc(\pi - t) = \frac{1}{\sin(\pi - t)} = \frac{1}{\sin t} = \csc t$

73. $\sin t = 1, \cos t = 0$, $\tan t$ is undefined, $\cot t = 0$, $\sec t$ is undefined, $\csc t = 1$

75. $\sin t = \frac{12}{13}, \cos t = -\frac{5}{13}, \tan t = -\frac{12}{5}, \cot t = -\frac{5}{12},$ $\sec t = -\frac{13}{5}, \csc t = \frac{13}{12}$

77. Look at the graph of $y = \sec t$ in Figure 6–54 on page 417. If you draw in the line $y = t$, it will pass through $(-\pi/2, -\pi/2)$ and $(\pi/2, \pi/2)$, and obviously will not intersect the graph of $y = \sec t$ when $-\pi/2 \leq t \leq \pi/2$. But it will intersect each part of the graph that lies above the horizontal axis, to the right of $t = \pi/2$; it will also intersect those parts that lie below the horizontal axis, to the left of $-\pi/2$. The first coordinate of each of these infinitely many intersection points will be a solution of $\sec t = t$.

Section 6.7, page 429

1. Amplitude: 3; period: π, phase shift: $+\frac{\pi}{2}$

3. Amplitude: 7; period: $\frac{2\pi}{7}$, phase shift: $-\frac{1}{49}$

5. Amplitude: 1; period: 1; phase shift: 0

7. Amplitude: 6; period: $\frac{2}{3}$; phase shift: $-\frac{1}{3\pi}$

9. $f(t) = 3\sin\left(8t - \frac{8\pi}{5}\right)$, (other answers possible)

11. $f(t) = \frac{2}{3}\sin(2\pi t)$

13. $f(t) = 7\sin\left(\frac{6\pi}{5}t + \frac{3\pi^2}{5}\right)$

15. $f(t) = 2\sin 4t$ **17.** $f(t) = 1.5\cos\frac{t}{2}$

19. **(a)** $f(t) = -12\sin\left(10t + \frac{\pi}{2}\right)$

(b) $g(t) = -12\cos 10t$

21. **(a)** $f(t) = -\sin 2t$

(b) $g(t) = -\cos\left(2t - \frac{\pi}{2}\right)$

23. **(a)** $f(t) = \frac{1}{2}\sin 8t$

(b) $g(t) = \frac{1}{2}\cos\left(8t - \frac{\pi}{2}\right)$

25.

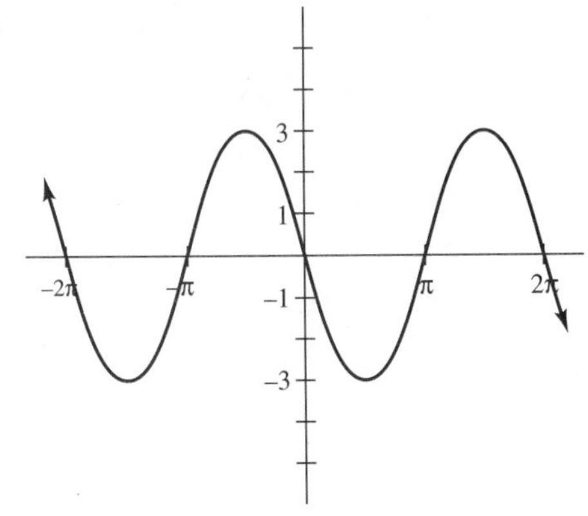

27.

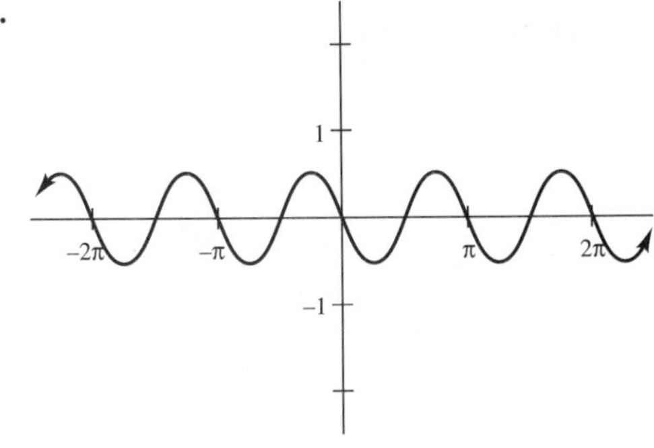

29.

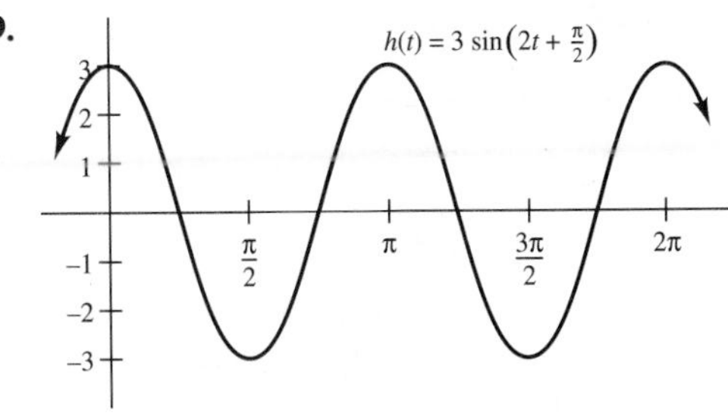

31. Local maximum at $t = 5\pi/6 \approx 2.6180$; local minimum at $t = 11\pi/6 \approx 5.7596$

33. Local maxima at $t = \pi/6 \approx .5236$, $t = 5\pi/6 \approx 2.6180$, $t = 3\pi/2 \approx 4.7124$; local minima at $t = \pi/2 \approx 1.5708$, $t = 7\pi/6 \approx 3.6652$, $t = 11\pi/6 \approx 5.7596$

35. $A \approx 5.3852$, $b = 1$, $c \approx 1.1903$

37. $A \approx 3.8332$, $b = 4$, $c \approx 1.4572$

39. All waves in the graph of g are of equal height, which is not the case with the graph of f.

41. 1/980,000; 980,000

43. $f(t) = 125\sin\left(\frac{\pi t}{5}\right)$

45. $f(t) = \cos 20\pi t + \sqrt{16 - \sin^2(20\pi t)}$

47. $h(t) = 6\sin(\pi t/2)$

49. $h(t) = 6\cos(\pi t/2)$

51. $d(t) = 10\sin(\pi t/2)$

53. **(a)** $k = 9.8/\pi^2$

(b) When k is replaced by (k + .01% of k), the value of ω changes and the period of the pendulum becomes approximately 2.000099998 seconds, meaning that the clock loses .000099998 second every 2 seconds, for a total of approximately 397.43 seconds (6.62 min) during the three months.

55. **(a)** At least four (starting point, high point, low point, ending point)

(b) At least 301 (4 points for the first wave and 3 for each of the remaining 99 waves because the starting point of one is the ending point of the preceding one)

(c) Answers vary from 95 to 239.

57. **(b)**

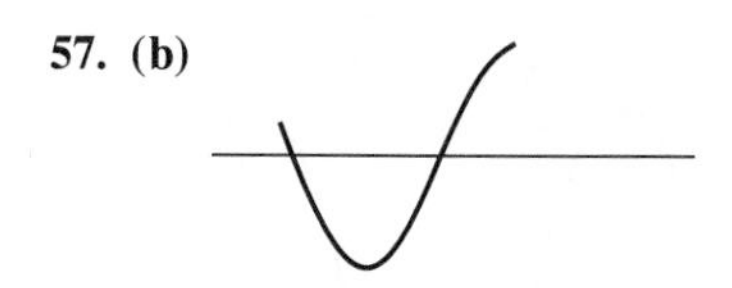

59. **(a)** $-\pi \le t \le \pi$
(b) $n = 15$; $f_{15}(2)$ and $g(2)$ are identical in the first nine decimal places and differ in the tenth, a very good approximation.

61. $r(t)/s(t)$, where $r(t) = f_{15}(t)$ in Exercise 59 and $s(t) = f_{16}(t)$ in Exercise 60

Excursion 6.7.A, page 440

1. $A \approx 2.2361$, $b = 1$, $c \approx 1.1071$

3. $A \approx 5.3852$, $b = 4$, $c \approx -1.1903$

5. $A \approx 5.1164$, $b = 3$, $c \approx -.7442$

7. $0 \le t \le 2\pi$ and $-5 \le y \le 5$ (one period)

9. $-10 \le t \le 10$ and $-10 \le y \le 10$

11. $0 \le t \le \pi/50$ and $-2 \le y \le 2$ (one period)

13. $0 \le t \le .04$ and $-7 \le y \le 7$ (one period)

15. $0 \le t \le 10$ and $-6 \le y \le 10$ (one period)

17. To the left of the y-axis, the graph lies above the t-axis, which is a horizontal asymptote of the graph. To the right of the y-axis, the graph makes waves of amplitude 1, of shorter and shorter period.
Window: $-3 \le t \le 3.2$ and $-2 \le y \le 2$

19. The graph is symmetric with respect to the y-axis and consists of waves along the t-axis, whose amplitude slowly increases as you move farther from the origin in either direction.
Window: $-30 \le t \le 30$ and $-6 \le y \le 6$

21. The graph is symmetric with respect to the y-axis and consists of waves along the t-axis whose amplitude rapidly decreases as you move farther from the origin in either direction.
Window: $-30 \le t \le 30$ and $-.3 \le y \le 1$

23. The function is periodic with period π (why?). The graph lies on or below the t-axis because the logarithmic function is negative for numbers between 0 and 1 and $|\cos t|$ is always between 0 and 1. The graph has vertical asymptotes when $t = \pm\pi/2, \pm 3\pi/2, \pm 5\pi/2, \pm 7\pi/2, \ldots$ ($\cos t = 0$ at these points and $\ln 0$ is not defined).
Window: $-2\pi \le t \le 2\pi$ and $-3 \le y \le 1$ (two periods)

Chapter 6 Review, page 443

1. $\frac{\pi}{3}$ **3.** 324° **5.** $\frac{11\pi}{9}$

7. $-495°$ **9.** $\cos v = -\frac{1}{3}$ **11.** 0

13. .809 **15.** $-\frac{\sqrt{3}}{2}$ **17.** $\frac{\sqrt{3}}{2}$

19.

t	0	$\pi/6$	$\pi/4$	$\pi/3$	$\pi/2$
$\sin t$	0	1/2	$\sqrt{2}/2$	$\sqrt{3}/2$	1
$\cos t$	1	$\sqrt{3}/2$	$\sqrt{2}/2$	1/2	0

21. 9/4 **23.** 0 **25.** $-\sqrt{\frac{2}{3}}$

27. $-\frac{3}{5}$ **29.** $-\frac{12}{13}$ **31.** $-\frac{1}{2}$

33. $\frac{\sqrt{3}}{2}$ **35.** (c) **37.** $-\sqrt{3}$

39. $\frac{-3}{\sqrt{58}}$ **41.** $y = -\sqrt{3}x$

43. **(a)** .97 **(b)** 8.3 **(c)** 5.6

45. $80\sqrt{3}$ feet **47.** $-\frac{3}{2}$

49. $-\frac{2}{\sqrt{13}}$

51. See Figure 6–55 on page 418.

53. (d) **55.** $-\frac{3}{2}$ **57.** 0 **59.** (c)

61. (d) **63.** $-\frac{1}{\sqrt{3}}$

65. **(a)** $\frac{3}{2}$ **(b)** $t = \frac{\pi}{5}$

67.

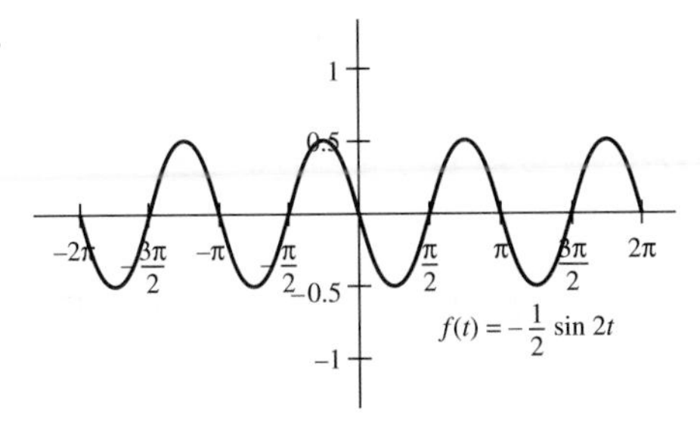

69. Not an identity **71.** Possibly an identity

73. 1/2 **75.** $2\cos(5t/2)$

77. $f(t) = 8\sin\left(\frac{2\pi t - 28\pi}{5}\right)$ is one possibility.

79. $A \approx 10.5588$, $b = 4$, $c = .4581$

81. $0 \le t \le \pi/50$ and $-5 \le y \le 5$ (one period)

CHAPTER 7

Section 7.1, page 458

	sin	cos	tan	csc	sec	cot
1.	$\frac{3\sqrt{13}}{13}$	$\frac{2\sqrt{13}}{13}$	$\frac{3}{2}$	$\frac{\sqrt{13}}{3}$	$\frac{\sqrt{13}}{2}$	$\frac{2}{3}$
3.	$\frac{6\sqrt{61}}{61}$	$\frac{-5\sqrt{61}}{61}$	$\frac{-6}{5}$	$\frac{\sqrt{61}}{6}$	$\frac{-\sqrt{61}}{5}$	$\frac{-5}{6}$
5.	$\frac{-\sqrt{22}}{11}$	$\frac{-3\sqrt{11}}{11}$	$\frac{\sqrt{2}}{3}$	$\frac{-\sqrt{22}}{2}$	$\frac{-\sqrt{11}}{3}$	$\frac{3\sqrt{2}}{2}$
7.	$\frac{-\sqrt{21}}{6}$	$\frac{-\sqrt{15}}{6}$	$\frac{\sqrt{35}}{5}$	$\frac{-2\sqrt{21}}{7}$	$\frac{-2\sqrt{15}}{5}$	$\frac{\sqrt{35}}{7}$
9.	$\frac{1}{\sqrt{4+2\sqrt{2}}}$	$\frac{1+\sqrt{2}}{\sqrt{4+2\sqrt{2}}}$	$\frac{1}{1+\sqrt{2}}$	$\sqrt{4+2\sqrt{2}}$	$\frac{\sqrt{4+2\sqrt{2}}}{1+\sqrt{2}}$	$1+\sqrt{2}$
11.	$\frac{2\sqrt{13}}{13}$	$\frac{3\sqrt{13}}{13}$	$\frac{2}{3}$	$\frac{\sqrt{13}}{2}$	$\frac{\sqrt{13}}{3}$	$\frac{3}{2}$
13.	$\frac{4}{5}$	$\frac{3}{5}$	$\frac{4}{3}$	$\frac{5}{4}$	$\frac{5}{3}$	$\frac{3}{4}$
15.	$\frac{\sqrt{7}}{4}$	$\frac{3}{4}$	$\frac{\sqrt{7}}{3}$	$\frac{4\sqrt{7}}{7}$	$\frac{4}{3}$	$\frac{3\sqrt{7}}{7}$

17. $\frac{\sqrt{3}}{2}$ **19.** $-\sqrt{3}$ **21.** $-\sqrt{2}$ **23.** $\frac{1}{2}$

25. $c = 36$ **27.** $c = 36$ **29.** $c = 8.4$

31. $\measuredangle A = 40°$, $a = 10\cos 50° = 6.4$, $c = 10\sin 50° = 7.7$

33. $\measuredangle C = 76°$, $b = 6/\sin 14° = 24.8$, $c = 6/\tan 14° = 24.1$

35. $\measuredangle C = 25°$, $a = 5\tan 65° = 10.7$, $b = 5/\cos 65° = 11.8$

37. $\measuredangle C = 18°$, $a = 3.5\sin 72° = 3.3$, $c = 3.5\cos 72° = 1.1$

39. $\measuredangle A = 33.7°$, $\measuredangle C = 56.3°$

41. $\measuredangle A = 44.4°$, $\measuredangle C = 45.6°$

43. $\measuredangle A = 48.2°$, $\measuredangle C = 41.8°$

45. $\measuredangle A = 60.8°$, $\measuredangle C = 29.2°$

47. $\measuredangle A = 30°$, $\measuredangle C = 60°$, $c = 5\sqrt{3} \approx 8.7$

49. $\measuredangle A = 70.5°$, $\measuredangle C = 19.5°$, $a = 5\sqrt{2} \approx 7.1$

51. $\measuredangle C = 55°$, $a = 10\tan 35° = 7.0$, $b = 10/\cos 35° = 12.2$

53. 75.5° **55.** 460.2 ft **57.** 8598.3 ft

59. No **61.** 27.5 ft; yes **63.** 19.25 ft; 53.97°

65. 351.1 m **67.** 10.1 ft **69.** 1.6 mi

71. **(a)** 56.7 ft **(b)** 9.7 ft

73. 205.7 ft **75.** 173.2 mi **77.** 52.5 mph

79. 449.1 ft

Section 7.2, page 467

1. $a = 4.2$, $\measuredangle B = 125.0°$, $\measuredangle C = 35.0°$

3. $c = 13.9$, $\measuredangle A = 22.5°$, $\measuredangle B = 39.5°$

5. $a = 24.4$, $\measuredangle B = 18.4°$, $\measuredangle C = 21.6°$

7. $c = 21.5$, $\measuredangle A = 33.5°$, $\measuredangle B = 67.9°$

9. $\measuredangle A = 120°$, $\measuredangle B = 21.8°$, $\measuredangle C = 38.2°$

11. $\measuredangle A = 24.1°$, $\measuredangle B = 30.8°$, $\measuredangle C = 125.1°$

13. $\measuredangle A = 38.8°$, $\measuredangle B = 34.5°$, $\measuredangle C = 106.7°$

15. $\measuredangle A = 34.1°$, $\measuredangle B = 50.5°$, $\measuredangle C = 95.4°$

17. 54.2° at vertex (0, 0); 48.4° at vertex (5, −2); 77.5° at vertex (1, −4)

19. 334.9 km **21.** 63.7 ft **23.** 84.9°

25. 8.4 km **27.** 231.9 ft **29.** 154.5 ft

31. 4.7 cm and 9.0 cm **33.** 33.44° **35.** 978.7 mi

37. $AB = 24.27$, $AC = 21.23$, $BC = 19.5$, $\measuredangle A = 50.2°$, $\measuredangle B = 56.8°$, $\measuredangle C = 73.0°$

39. 16.99 m

Section 7.3, page 479

1. $\measuredangle C = 110°$, $b = 2.5$, $c = 6.3$

3. $\measuredangle B = 14°$, $b = 2.2$, $c = 6.8$

5. $\measuredangle A = 88°$, $a = 17.3$, $c = 12.8$

7. $\measuredangle C = 41.5°$, $b = 9.7$, $c = 10.9$

9. 7.3 **11.** 32.5 **13.** 82.3 **15.** 31.4

17. No solution

19. $\measuredangle A_1 = 55.2°$, $\measuredangle C_1 = 104.8°$, $c_1 = 14.1$; $\measuredangle A_2 = 124.8°$, $\measuredangle C_2 = 35.2°$, $c_2 = 8.4$

21. No solution

23. $\measuredangle B_1 = 65.8°$, $\measuredangle A_1 = 58.2°$, $a_1 = 10.3$; $\measuredangle B_2 = 114.2°$, $\measuredangle A_2 = 9.8°$, $a_2 = 2.1$

25. $\measuredangle C = 72°$, $b = 14.7$, $c = 15.2$

27. $a = 9.8$, $\measuredangle B = 23.3°$, $\measuredangle C = 81.7°$

29. $\measuredangle A = 18.6°$, $\measuredangle B = 39.6°$, $\measuredangle C = 121.9°$

31. $c = 13.9$, $\measuredangle A = 60.1°$, $\measuredangle B = 72.9°$

33. $\measuredangle C = 39.8°$, $\measuredangle A = 77.7°$, $a = 18.9$

35. No solution **37.** 6.5 **39.** About 7691

41. 135.5 m **43.** 5.4° **45.** 5 ft

47. 5.3° **49.** 30.1 km **51.** About 9642 ft

53. **(a)** Use the Law of Cosines in triangle *ABD* to find $\measuredangle ABD$; then $\measuredangle EBA$ is 180° − $\measuredangle ABD$ (why?). Use the Law of Cosines in triangle *ABC* to find $\measuredangle CAB$; then $\measuredangle EAB$ is 180° − $\measuredangle CAB$. You now have two of the angles in triangle *EAB* and can easily find the third. Use these angles, side *AB*, and the Law of Sines to find *AE*.
(b) 94.23 ft

55. 13.36 m **57.** 5.8 gal **59.** 11.18 sq units

61. No such triangle exists because the sum of the lengths of any two sides of a triangle must be greater than the length of the third side, which is not the case here.

Chapter 7 Review, page 483

1. $b = \sqrt{313} \approx 17.7$, $A \approx 42.7°$, $C \approx 47.3°$

3. $b \approx 14.6$, $c \approx 8.40$, $A = 55°$

5. 225.9 ft **7.** 1.5°

9. $A = 52.9°$, $B = 41.6°$, $C = 85.5°$

11. $A = 20.6°$, $b = 21.8$, $C = 29.4°$

13. Approximately 301 miles

15. $A = 25°$, $a = 2.9$, $b = 5.6$

17. $A = 52.03°$, $B = 65.97°$, $b = 86.9$

19. $B = 81.8°$, $C = 38.2°$, $c = 2.5$ and $B = 98.2°$, $C = 21.8°$, $c = 1.5$

21. 147.4 **23.** 13.4 km

25. $a = 41.6$; $C = 75°$, $c = 54.1$

27. $A = 35.5°$, $b = 8.3$, $C = 68.5°$

29. Joe is 217.9 m from the pole and Alice is 240 m from the pole.

31. **(a)** 3940.65 ft **(b)** 4377.53 ft **(c)** 3967.39 ft

33. 71.89° **35.** 10 **37.** 37.95

CHAPTER 8

Section 8.1, page 493

1. Possibly an identity

3. Possibly an identity

5. B **7.** E

9. $\tan x \cos x = \left(\frac{\sin x}{\cos x}\right) \cos x = \sin x$

11. $\cos x \sec x = \cos x \left(\frac{1}{\cos x}\right) = 1$

13. $\tan x \csc x = \left(\frac{\sin x}{\cos x}\right)\left(\frac{1}{\sin x}\right) = \frac{1}{\cos x} = \sec x$

15. $\frac{\tan x}{\sec x} = \frac{\sin x/\cos x}{1/\cos x} = \sin x$

17. $(1 + \cos x)(1 - \cos x) = 1 - \cos^2 x = \sin^2 x$

19. Not an identity

21. $\frac{\sin(-x)}{\cos(-x)} = \frac{-\sin x}{\cos x} = -\tan x$

23. $\cot(-x) = \frac{\cos(-x)}{\sin(-x)} = \frac{\cos x}{-\sin x} = -\cot x$

25. Not an identity

27. $\sec^2 x - \csc^2 x = (1 + \tan^2 x) - (1 + \cot^2 x) = \tan^2 x - \cot^2 x$

29. $\sin^2 x(\cot x + 1)^2 = [\sin x(\cot x + 1)]^2 = [\cos x + \sin x]^2 = [\cos x(1 + \tan x)]^2 = \cos^2 x(\tan x + 1)^2$

31. $\sin^2 x - \tan^2 x = \sin^2 x - \frac{\sin^2 x}{\cos^2 x} = \sin^2 x\left(1 - \frac{1}{\cos^2 x}\right) = \sin^2 x(1 - \sec^2 x) = \sin^2 x(-\tan^2 x) = -\sin^2 x \tan^2 x$

33. $(\cos^2 x - 1)(\tan^2 x + 1) = (-\sin^2 x)(\sec^2 x) = (-\sin^2 x)\left(\frac{1}{\cos^2 x}\right) = -\tan^2 x$

35. $\frac{\sec x}{\csc x} = \frac{1/\cos x}{1/\sin x} = \frac{\sin x}{\cos x} = \tan x$

37. $\cos^4 x - \sin^4 x = (\cos^2 x - \sin^2 x)(\cos^2 x + \sin^2 x) = \cos^2 x - \sin^2 x$

39. Not an identity

41. $\frac{\sec x}{\csc x} + \frac{\sin x}{\cos x} = \frac{1/\cos x}{1/\sin x} + \tan x = \frac{\sin x}{\cos x} + \tan x = \tan x + \tan x = 2 \tan x$

43. $\frac{\sec x + \csc x}{1 + \tan x} = \frac{\frac{1}{\cos x} + \frac{1}{\sin x}}{1 + \frac{\sin x}{\cos x}} \cdot \frac{\sin x \cos x}{\sin x \cos x} = \frac{\sin x + \cos x}{\sin x \cos x + \sin^2 x} = \frac{\sin x + \cos x}{\sin x(\cos x + \sin x)} = \frac{1}{\sin x} = \csc x$

45. $\frac{1}{\csc x - \sin x} = \frac{1}{\frac{1}{\sin x} - \sin x} \cdot \frac{\sin x}{\sin x} = \frac{\sin x}{1 - \sin^2 x} = \frac{\sin x}{\cos^2 x} = \left(\frac{1}{\cos x}\right)\left(\frac{\sin x}{\cos x}\right) = \sec x \tan x$

47. Not an identity

49. $\left[\frac{\cot x}{1 - \tan x} + \frac{\tan x}{1 - \cot x}\right] \cdot \frac{\tan x}{\tan x} = \frac{1}{\tan x - \tan^2 x} + \frac{\tan^2 x}{\tan x - 1} = \frac{1}{\tan x(1 - \tan x)} + \frac{\tan^2 x}{\tan x - 1} = \frac{1 - \tan^3 x}{\tan x(1 - \tan x)} = \frac{(1 - \tan x)(1 + \tan x + \tan^2 x)}{\tan x(1 - \tan x)} = \frac{\tan x + \sec^2 x}{\tan x} = 1 + \frac{\sec^2 x}{\tan x} = 1 + \sec^2 x\left(\frac{1}{\tan x}\right) = 1 + \frac{1}{\cos^2 x} \cdot \frac{\cos x}{\sin x} = 1 + \frac{1}{\cos x} \cdot \frac{1}{\sin x} = 1 + \sec x \csc x$

51. Conjecture: $\cos x$. Proof: $1 - \frac{\sin^2 x}{1 + \cos x} = \frac{1 + \cos x - \sin^2 x}{1 + \cos x} = \frac{\cos x + (1 - \sin^2 x)}{1 + \cos x} = \frac{\cos x + \cos^2 x}{1 + \cos x} = \frac{\cos x(1 + \cos x)}{1 + \cos x} = \cos x.$

53. Conjecture: $\tan x$. Proof: $(\sin x + \cos x)(\sec x + \csc x) - \cot x - 2 = \sin x \sec x + \sin x \csc x + \cos x \sec x + \cos x \csc x = \sin x \cdot \frac{1}{\cos x} + \sin x \cdot \frac{1}{\sin x} + \cos x \cdot \frac{1}{\cos x} + \cos x \cdot \frac{1}{\sin x} - \cot x - 2 = \frac{\sin x}{\cos x} + 1 + 1 + \frac{\cos x}{\sin x} - \cot x - 2 = \tan x + \cot x - \cot x = \tan x.$

55. $\dfrac{1-\sin x}{\sec x} = \dfrac{1-\sin x}{\sec x}\cdot\dfrac{(1+\sin x)}{(1+\sin x)} = \dfrac{1-\sin^2 x}{\sec x(1+\sin x)} = \dfrac{\cos^2 x}{\dfrac{1}{\cos x}(1+\sin x)} = \dfrac{\cos^3 x}{1+\sin x}$

57. $\dfrac{\cos x}{1-\sin x}\cdot\dfrac{1+\sin x}{1+\sin x} = \dfrac{\cos x(1+\sin x)}{1-\sin^2 x} = \dfrac{\cos x(1+\sin x)}{\cos^2 x} = \dfrac{1+\sin x}{\cos x} = \dfrac{1}{\cos x}+\dfrac{\sin x}{\cos x} = \sec x + \tan x$

59. $\dfrac{\cos x \cot x}{\cot x - \cos x}\cdot\dfrac{\sin x}{\sin x} = \dfrac{\cos x\cos x}{\cos x - \cos x \sin x} = \dfrac{\cos^2 x}{\cos x(1-\sin x)} = \dfrac{\cos x}{1-\sin x}\cdot\dfrac{1+\sin x}{1+\sin x} = \dfrac{\cos x(1+\sin x)}{1-\sin^2 x} = \dfrac{\cos x + \sin x\cos x}{\cos^2 x}\cdot\dfrac{\dfrac{1}{\sin x}}{\dfrac{1}{\sin x}} = \dfrac{\dfrac{\cos x}{\sin x}+\cos x}{\cos x\left(\dfrac{\cos x}{\sin x}\right)} = \dfrac{\cot x + \cos x}{\cos x \cot x}$

61. $\cot x = \dfrac{1}{\tan x}$, so, $\log_{10}(\cot x) = \log_{10}\left(\dfrac{1}{\tan x}\right) = -\log_{10}(\tan x)$

63. $\csc x + \cot x = (\csc x + \cot x)\cdot\dfrac{(\csc x - \cot x)}{(\csc x - \cot x)} = \dfrac{\csc^2 x - \cot^2 x}{\csc x - \cot x} = \dfrac{1}{\csc x - \cot x}$; so, $\log_{10}(\csc x + \cot x) = \log_{10}\left(\dfrac{1}{\csc x - \cot x}\right) = -\log_{10}(\csc x - \cot x)$

65. $-\tan x \tan y(\cot x - \cot y) = -\tan y(\tan x \cot x) + \tan x(\tan y \cot y) = -\tan y + \tan x = \tan x - \tan y$

67. $\dfrac{\cos x - \sin y}{\cos y - \sin x}\cdot\dfrac{\cos x + \sin y}{\cos x + \sin y} = \dfrac{\cos^2 x - \sin^2 y}{(\cos y - \sin x)(\cos x + \sin y)} = \dfrac{(1-\sin^2 x)-(1-\cos^2 y)}{(\cos y - \sin x)(\cos x + \sin y)} = \dfrac{\cos^2 y - \sin^2 x}{(\cos y - \sin x)(\cos x + \sin y)} = \dfrac{(\cos y - \sin x)(\cos y + \sin x)}{(\cos y - \sin x)(\cos x + \sin y)} = \dfrac{\cos y + \sin x}{\cos x + \sin y}$

Section 8.2, page 502

1. $\dfrac{\sqrt{6}-\sqrt{2}}{4}$ **3.** $2-\sqrt{3}$ **5.** $2-\sqrt{3}$

7. $-2-\sqrt{3}$ **9.** $-2-\sqrt{3}$ **11.** $\dfrac{\sqrt{6}+\sqrt{2}}{4}$

13. $\cos x$ **15.** $-\sin x$ **17.** $-1/\cos x$

19. $-\sin 2$ **21.** $\cos x$

23. $-2\sin x \sin y$ **25.** $\dfrac{4+\sqrt{2}}{6}$

27. $\dfrac{2\sqrt{6}-\sqrt{3}}{10}$ **29.** $-.393$

31. .993 **33.** -2.34

35. $\dfrac{f(x+h)-f(x)}{h} = \dfrac{\cos(x+h)-\cos x}{h} = \dfrac{\cos x\cos h - \sin x \sin h - \cos x}{h} = \dfrac{\cos x \cos h - \cos x}{h} - \dfrac{\sin x \sin h}{h} = \cos x\left(\dfrac{\cos h - 1}{h}\right) - \sin x\left(\dfrac{\sin h}{h}\right)$

37. $\sin(x+y) = -44/125$; $\tan(x+y) = 44/117$; $x+y$ is in the third quadrant.

39. $\cos(x+y) = -56/65$; $\tan(x+y) = 33/56$; $x+y$ is in the third quadrant.

41. $\sin(u+v+w) = \sin u \cos v \cos w + \cos u \sin v \cos w + \cos u \cos v \sin w - \sin u \sin v \sin w$

43. Since $y = \pi/2 - x$, $\sin y = \sin(\pi/2 - x) = \cos x$. Hence, $\sin^2 x + \sin^2 y = \sin^2 x + \cos^2 x = 1$

45. $\sin(x-\pi) = \sin x \cos \pi - \cos x \sin \pi = (\sin x)(-1) - (\cos x)(0) = -\sin x$

47. $\cos(\pi - x) = \cos \pi \cos x + \sin \pi \sin x = (-1)\cos x + (0)\sin x = -\cos x$

49. $\sin(x+\pi) = \sin x \cos \pi + \cos x \sin \pi = (\sin x)(-1) + (\cos x)(0) = -\sin x$

51. By Exercises 49 and 50, $\tan(x+\pi) = \dfrac{\sin(x+\pi)}{\cos(x+\pi)} = \dfrac{-\sin x}{-\cos x} = \tan x$

53. $\frac{1}{2}(\cos(x-y) - \cos(x+y)) =$

$\frac{1}{2}(\cos x \cos y + \sin x \sin y - (\cos x \cos y - \sin x \sin y))$

$= \frac{1}{2}(2 \sin x \sin y) = \sin x \sin y$

55. $\cos(x+y)\cos(x-y) =$
$(\cos x \cos y - \sin x \sin y)(\cos x \cos y + \sin x \sin y) =$
$(\cos x \cos y)^2 - (\sin x \sin y)^2 =$
$\cos^2 x \cos^2 y - \sin^2 x \sin^2 y$

57. $\frac{\cos(x-y)}{\sin x \cos y} = \frac{\cos x \cos y + \sin x \sin y}{\sin x \cos y} =$

$\frac{\cos x}{\sin x} + \frac{\sin y}{\cos y} = \cot x + \tan y$

59. Not an identity

61. $\frac{\sin(x+y)}{\sin(x-y)} =$

$\frac{\sin x \cos y + \cos x \sin y}{\sin x \cos y - \cos x \sin y} \cdot \frac{1/\cos x \cos y}{1/\cos x \cos y} =$

$\frac{\frac{\sin x}{\cos x} + \frac{\sin y}{\cos y}}{\frac{\sin x}{\cos x} - \frac{\sin y}{\cos y}} = \frac{\tan x + \tan y}{\tan x - \tan y}$

63. Not an identity

65. Not an identity

Excursion 8.2.A, page 508

In Answers 1–11, all angles in radians.

1. .64 **3.** 2.47 **5.** $\pi/2$

7. 1.37 or 1.77 **9.** $\pi/4$ or $3\pi/4$ **11.** 1.39 or 1.75

Section 8.3, page 515

1. $\frac{\sqrt{2+\sqrt{2}}}{2}$ **3.** $\frac{\sqrt{2+\sqrt{2}}}{2}$

5. $2 - \sqrt{3}$ **7.** $\frac{\sqrt{2+\sqrt{3}}}{2}$

9. $\frac{\sqrt{2-\sqrt{2}}}{2}$ **11.** $-\sqrt{2} + 1$

13. $\frac{1}{2}\sin 10x - \frac{1}{2}\sin 2x$ **15.** $\frac{1}{2}\cos 6x + \frac{1}{2}\cos 2x$

17. $\frac{1}{2}\cos 20x - \frac{1}{2}\cos 14x$ **19.** $2 \sin 4x \cos x$

21. $2 \sin 2x \cos 7x$

23. $\sin 2x = \frac{120}{169}$, $\cos 2x = \frac{119}{169}$, $\tan 2x = \frac{120}{119}$

25. $\sin 2x = \frac{24}{25}$, $\cos 2x = -\frac{7}{25}$, $\tan 2x = -\frac{24}{7}$

27. $\sin 2x = \frac{24}{25}$, $\cos 2x = \frac{7}{25}$, $\tan 2x = \frac{24}{7}$

29. $\sin 2x = \frac{\sqrt{15}}{8}$, $\cos 2x = \frac{7}{8}$, $\tan 2x = \frac{\sqrt{15}}{7}$

31. $\sin \frac{x}{2} = .5477$, $\cos \frac{x}{2} = .8367$, $\tan \frac{x}{2} = .6547$

33. $\sin \frac{x}{2} = \frac{1}{\sqrt{10}}$, $\cos \frac{x}{2} = \frac{-3}{\sqrt{10}}$, $\tan \frac{x}{2} = \frac{-1}{3}$

35. $\sin \frac{x}{2} = \sqrt{\frac{\sqrt{5}+2}{2\sqrt{5}}}$, $\cos \frac{x}{2} = -\sqrt{\frac{\sqrt{5}-2}{2\sqrt{5}}}$,

$\tan \frac{x}{2} = -\sqrt{9 + 4\sqrt{5}} = -\sqrt{5} - 2$

37. $\sin 2x = .96$ **39.** $\cos 2x = .28$

41. $\sin \frac{x}{2} = .3162$

43. $\cos 3x = 4\cos^3 x - 3\cos x$

45. $\cos x$ **47.** $\sin 4y$ **49.** 1

51. $\sin 16x = \sin(2(8x)) = 2 \sin 8x \cos 8x$

53. $\cos^4 x - \sin^4 x = (\cos^2 x - \sin^2 x)(\cos^2 x + \sin^2 x) =$
$\cos^2 x - \sin^2 x = \cos 2x$

55. Not an identity

57. $\frac{1+\cos 2x}{\sin 2x} = \frac{1}{\frac{\sin 2x}{1+\cos 2x}} = \frac{1}{\tan\left(\frac{2x}{2}\right)} = \cot x$

59. $\sin 3x = \sin(2x + x) = \sin 2x \cos x + \cos 2x \sin x =$
$(2 \sin x \cos x)\cos x + (1 - 2\sin^2 x)\sin x =$
$2 \sin x \cos^2 x + \sin x - 2\sin^3 x = 2 \sin x(1 - \sin^2 x) +$
$\sin x - 2\sin^3 x = \sin x[2 - 2\sin^2 x + 1 - 2\sin^2 x] =$
$\sin x[3 - 4\sin^2 x]$

61. Not an identity

63. $\csc^2\left(\frac{x}{2}\right) = \frac{1}{\sin^2\left(\frac{x}{2}\right)} = \frac{1}{\frac{1-\cos x}{2}} = \frac{2}{1-\cos x}$

65. $\dfrac{\sin x - \sin 3x}{\cos x + \cos 3x} = \dfrac{-2\cos 2x \sin x}{2\cos 2x \cos x} = \dfrac{-\sin x}{\cos x} = -\tan x$

67. $\dfrac{\sin 4x + \sin 6x}{\cos 4x - \cos 6x} = \dfrac{2\sin 5x \cos x}{2\sin 5x \sin x} = \dfrac{\cos x}{\sin x} = \cot x$

69. $\dfrac{\sin x + \sin y}{\cos x - \cos y} = \dfrac{2\sin\left(\frac{x+y}{2}\right)\cos\left(\frac{x-y}{2}\right)}{-2\sin\left(\frac{x+y}{2}\right)\sin\left(\frac{x-y}{2}\right)} = \dfrac{-\cos\left(\frac{x-y}{2}\right)}{\sin\left(\frac{x-y}{2}\right)} = -\cot\left(\dfrac{x-y}{2}\right)$

71. (a) It suffices to prove the equivalent identity $(1 - \cos x)(1 + \cos x) = (\sin x)(\sin x)$. We have $(1 - \cos x)(1 + \cos x) = 1 - \cos^2 x = \sin^2 x = (\sin x)(\sin x)$.

(b) By the half-angle identity proved in the text and part (a), $\tan\dfrac{x}{2} = \dfrac{1-\cos x}{\sin x} = \dfrac{\sin x}{1 + \cos x}$

Section 8.4, page 525

In Answers 1–11, $k = 0, \pm 1, \pm 2, \pm 3, \ldots$.

1. $x = .5275$ or $1.6868 + k\pi$

3. $x = .4959$ or 1.2538 or 1.5708 or 1.8877 or 2.6457 or $4.7124 + 2k\pi$

5. $x = .1671$ or 1.8256 or 2.8867 or $4.5453 + 2k\pi$

7. $x = 1.2161$ or $5.0671 + 2k\pi$

9. $x = 2.4620$ or $3.8212 + 2k\pi$

11. $x = .5166$ or $5.6766 + 2k\pi$

13. (a) The graph of $f(x) = \sin x$ on the interval from 0 to 2π shows that $\sin x = 1$ only when $x = \pi/2$. Since $\sin x$ has period 2π, all other solutions are obtained by adding or subtracting integer multiples of 2π from $\pi/2$, that is,

$\dfrac{\pi}{2} + 2\pi = \dfrac{5\pi}{2}, \dfrac{\pi}{2} + 2(2\pi) = \dfrac{9\pi}{2},$

$\dfrac{\pi}{2} + 3(2\pi) = \dfrac{13\pi}{2}$, etc., and $\dfrac{\pi}{2} - 2\pi = \dfrac{-3\pi}{2},$

$\dfrac{\pi}{2} - 2(2\pi) = \dfrac{-7\pi}{2}, \dfrac{\pi}{2} - 3(2\pi) = \dfrac{-11\pi}{2}$, etc.

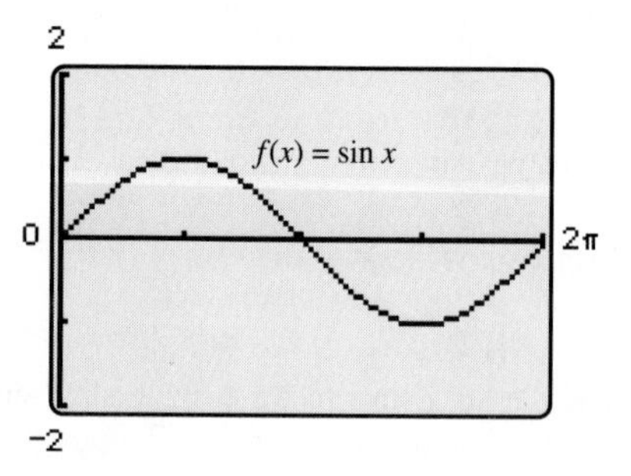

(b) Similarly, the graph shows that $\sin x = -1$ only when $x = 3\pi/2$, so that all solutions are obtained by adding or subtracting integer multiples of 2π from $3\pi/2$:

$\dfrac{3\pi}{2} + 2\pi = \dfrac{7\pi}{2}, \dfrac{3\pi}{2} + 2(2\pi) = \dfrac{11\pi}{2},$

$\dfrac{3\pi}{2} + 3(2\pi) = \dfrac{15\pi}{2}$, etc., and $\dfrac{3\pi}{2} - 2\pi = \dfrac{-\pi}{2},$

$\dfrac{3\pi}{2} - 2(2\pi) = \dfrac{-5\pi}{2}, \dfrac{3\pi}{2} - 3(2\pi) = \dfrac{-9\pi}{2}$, etc.

15. $x = \dfrac{\pi}{3}$ or $\dfrac{2\pi}{3} + 2k\pi$

17. $x = -\dfrac{\pi}{3} + k\pi$

19. $x = \pm\dfrac{5\pi}{6} + 2k\pi$

21. $x = -\dfrac{\pi}{6}$ or $\dfrac{7\pi}{6} + 2k\pi$

23. $x = .1193$ or 3.0223

25. $x = 1.3734$ or 4.5150

27. $\theta = 82.83°, 262.83°$

29. $\theta = 114.83°, 245.17°$

31. $\theta = 210°, 270°, 330°$

33. $\theta = 60°, 120°, 240°, 300°$

35. $\theta = 120°, 240°$

37. 65.38°

39. 30°

41. 27.57°

43. 14.18°

45. $x = -.4836$ or $3.6252 + 2k\pi$

47. $x = \pm 2.1700 + 2k\pi$

49. $x = -.2327 + k\pi$

51. $x = .4101 + k\pi$

53. $x = \pm 1.9577 + 2k\pi$

55. $x = -\dfrac{\pi}{6}$ or $\dfrac{2\pi}{3} + k\pi$

57. $x = \pm\dfrac{\pi}{2} + 4k\pi$

59. $x = -\dfrac{\pi}{9} + \dfrac{k\pi}{3}$

61. $x = \pm .7381 + \dfrac{2k\pi}{3}$

63. $x = 2.2143 + 2k\pi$

65. $x = 3.4814, 5.9433$

67. $x = \dfrac{3\pi}{4}, \dfrac{7\pi}{4}, 2.1588, 5.3004$

69. $x = \dfrac{\pi}{4}, \dfrac{\pi}{2}, \dfrac{5\pi}{4}, \dfrac{3\pi}{2}$

71. $x = \dfrac{\pi}{6}, \dfrac{\pi}{2}, \dfrac{5\pi}{6}, \dfrac{3\pi}{2}$

73. $x = .8481, 1.7682, 2.2935, 4.9098$

75. $x = .8213, 2.3203$

77. $x = .3649, 1.2059, 3.5065, 4.3475$

79. $x = 1.0591, 2.8679, 4.2007, 6.0095$

81. No solution

83. $x = \frac{\pi}{2}, \frac{7\pi}{6}, \frac{3\pi}{2}, \frac{11\pi}{6}$

85. $x = \frac{\pi}{6}, \frac{5\pi}{6}$

87. $x = \frac{\pi}{2}, \frac{7\pi}{6}, \frac{11\pi}{6}$

89. $x = 0, \frac{\pi}{3}, \frac{5\pi}{3}$

91. **(a)** March 1 (day 60); October 9 (day 282)
(b) June 20 (day 171)

93. .5475 or 1.0233

95. 13.25° or 76.75°

97. Sin $x = k$ and cos $x = k$ have no solutions when $k > 1$ or $k < -1$.

99. The solutions $x = 0, \pi$ are missed due to dividing by sin x.

Section 8.5, page 534

1. $\frac{\pi}{2}$ **3.** $-\frac{\pi}{4}$ **5.** 0 **7.** $\frac{\pi}{6}$

9. $-\frac{\pi}{4}$ **11.** $-\frac{\pi}{3}$ **13.** $\frac{2\pi}{3}$

15. .3576 **17.** -1.2728

19. .7168 **21.** $-.8584$

23. 2.2168

25. $\cos u = 1/2$; $\tan u = -\sqrt{3}$

27. $\frac{\pi}{2}$ **29.** $\frac{5\pi}{6}$ **31.** $-\frac{\pi}{3}$ **33.** $\frac{\pi}{3}$

35. $\frac{\pi}{6}$ **37.** $\frac{4}{5}$ **39.** $\frac{4}{5}$ **41.** $\frac{5}{12}$

43. $\cos(\sin^{-1}v) = \sqrt{1 - v^2} \quad (-1 \le v \le 1)$

45. $\tan(\sin^{-1}v) = \dfrac{v}{\sqrt{1 - v^2}} \quad (-1 < v < 1)$

47.

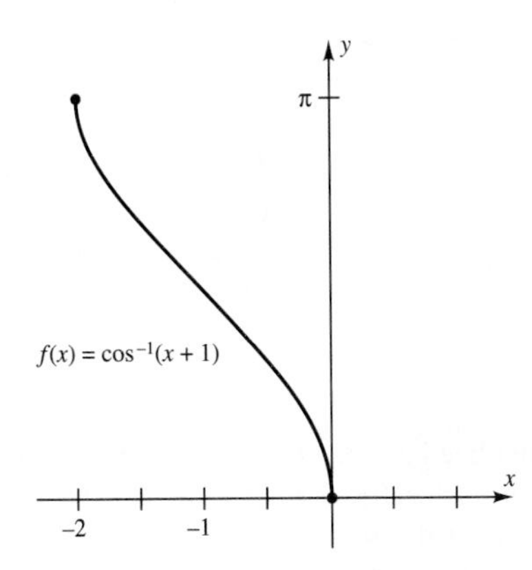

49.

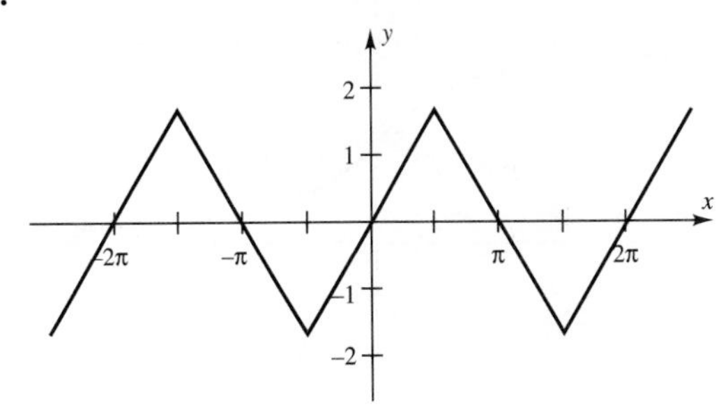

51. **(a)** $t = \dfrac{1}{2\pi f}\sin^{-1}\left(\dfrac{V}{V_{max}}\right) + \dfrac{k}{f}$ or

$t = \dfrac{1}{2\pi f}\left(\pi - \sin^{-1}\left(\dfrac{V}{V_{max}}\right)\right) + \dfrac{k}{f}$

$(k = 0, \pm 1, \pm 2, \ldots)$

(b) $t = .0005822$

53. **(a)** $\theta = \tan^{-1}\left(\dfrac{25}{x}\right) - \tan^{-1}\left(\dfrac{10}{x}\right)$ or

$\theta = \cos^{-1}\left(\dfrac{\sqrt{x^2 + 10^2} + \sqrt{x^2 + 25^2} - 15^2}{2\sqrt{x^2 + 10^2}\sqrt{x^2 + 25^2}}\right)$

(b) 15.8 ft

55. No horizontal line intersects the graph of $g(x) = \sec x$ more than once when $0 \le x \le \pi$ (see Figure 6–54).

Hence, the restricted secant function has an inverse function, as explained in Section 2.7.

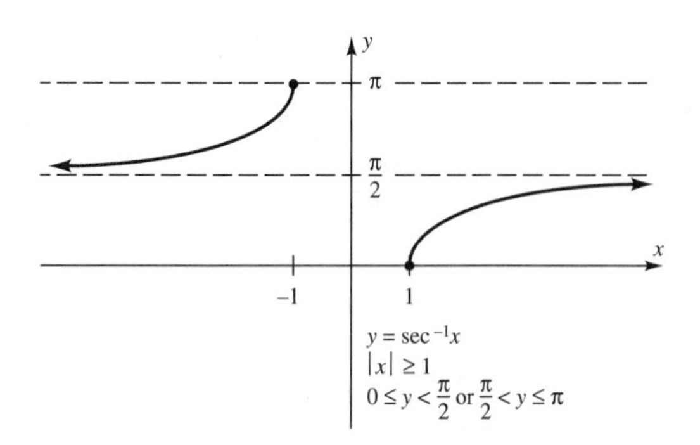

57. No horizontal line intersects the graph of $f(x) = \cot x$ more than once when $0 < x < \pi$ (see Figure 6–55). Hence the restricted cotangent function has an inverse function, as explained in Section 2.7.

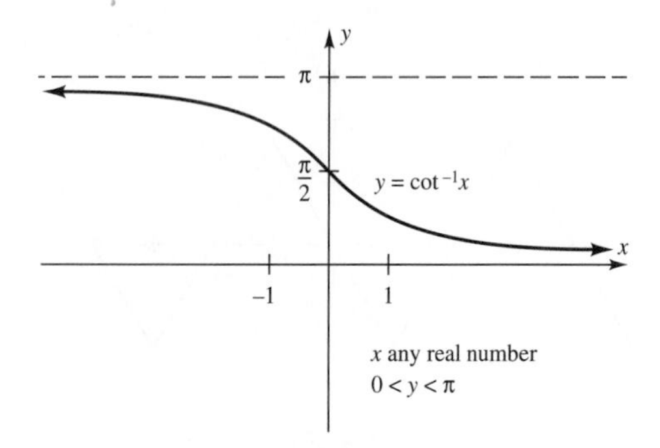

59. Let $u = \sin^{-1}(-x)$. Then $-\pi/2 \leq u \leq \pi/2$ and $\sin u = -x$; hence, $x = -\sin u = \sin(-u)$. Therefore, $\sin^{-1}x = -u$ (since $-\pi/2 \leq -u \leq \pi/2$) so that $\sin^{-1}(-x) = u = -\sin^{-1}x$.

61. Let $u = \cos^{-1}(-x)$. Then $0 \leq u \leq \pi$ and $\cos u = -x$. Since $0 \leq u \leq \pi$, we have $0 \leq \pi - u \leq \pi$. Since $\cos(\pi - u) = -\cos u = x$, $\cos^{-1}x = \pi - u = \pi - \cos^{-1}(-x)$. Therefore, $\cos^{-1}(-x) = \pi - \cos^{-1}x$.

63. If $0 < x < \pi$, then $-\frac{\pi}{2} < \frac{\pi}{2} - x < \frac{\pi}{2}$. Since $\tan\left(\frac{\pi}{2} - x\right) = \cot x$, $\tan^{-1}(\cot x) = \frac{\pi}{2} - x$.

65. Let $u = \sin^{-1}x$. Then $-\pi/2 \leq u \leq \pi/2$ and $\sin u = x$. Then $\cos u = \pm\sqrt{1 - \sin^2 u} = \pm\sqrt{1 - x^2}$. Since $-\pi/2 \leq u \leq \pi/2$, $\cos u$ is positive, so $\cos u = \sqrt{1 - x^2}$. Therefore, $\tan u = \frac{\sin u}{\cos u} = \frac{x}{\sqrt{1 - x^2}}$.
Hence, $u = \tan^{-1}\left(\frac{x}{\sqrt{1 - x^2}}\right)$, so that $\sin^{-1}x = u = \tan^{-1}\left(\frac{x}{\sqrt{1 - x^2}}\right)$.

67. Let $u = \tan^{-1}x$. Then $\tan u = x$ and $\frac{\pi}{2} - \tan^{-1}x = \frac{\pi}{2} - u$. Let $\tan^{-1}v = \frac{\pi}{2} - u$. Then $v = \tan\left(\frac{\pi}{2} - u\right) = \cot u = \frac{1}{\tan u} = \frac{1}{x}$.
Hence, $\tan^{-1}\left(\frac{1}{x}\right) = \tan^{-1}v = \frac{\pi}{2} - u = \frac{\pi}{2} - \tan^{-1}x$, so that $\tan^{-1}x + \tan^{-1}\left(\frac{1}{x}\right) = \frac{\pi}{2}$.

69. No; the graph of the left side function differs from the graph of the right side function.

Chapter 8 Review, page 538

1. $\frac{1}{3} + \cot t$ **3.** $\sin^4 x$

5. $\sin^4 t - \cos^4 t = (\sin^2 t - \cos^2 t)(\sin^2 t + \cos^2 t) = (\sin^2 t - (1 - \sin^2 t))(1) = 2\sin^2 t - 1$

7. $\frac{\sin t}{1 - \cos t} = \frac{\sin t}{1 - \cos t} \cdot \frac{(1 + \cos t)}{(1 + \cos t)} = \frac{\sin t(1 + \cos t)}{1 - \cos^2 t} = \frac{\sin t(1 + \cos t)}{\sin^2 t} = \frac{1 + \cos t}{\sin t}$

9. Not an identity

11. $(\sin x + \cos x)^2 - \sin 2x = \sin^2 x + 2\sin x\cos x + \cos^2 x - 2\sin x\cos x = \sin^2 x + \cos^2 x = 1$

13. $\frac{\tan x - \sin x}{2\tan x} \cdot \frac{\cos x}{\cos x} = \frac{\sin x - \sin x\cos x}{2\sin x} = \frac{\sin x(1 - \cos x)}{2\sin x} = \frac{1 - \cos x}{2} = \sin^2\left(\frac{x}{2}\right)$

15. $\cos(x + y)\cos(x - y) =$
$[\cos x\cos y - \sin x\sin y][\cos x\cos y + \sin x\sin y] =$
$\cos^2 x\cos^2 y - \sin^2 x\sin^2 y =$
$\cos^2 x(1 - \sin^2 y) - (1 - \cos^2 x)\sin^2 y =$
$\cos^2 x - \cos^2 x\sin^2 y - \sin^2 y + \cos^2 x\sin^2 y =$
$\cos^2 x - \sin^2 y$

17. $\dfrac{\sec x + 1}{\tan x} = \dfrac{(\sec x + 1)(\sec x - 1)}{\tan x(\sec x - 1)} = \dfrac{\sec^2 x - 1}{\tan x(\sec x - 1)} = \dfrac{\tan^2 x}{\tan x(\sec x - 1)} = \dfrac{\tan x}{\sec x - 1}$

19. $\dfrac{1 + \tan^2 x}{\tan^2 x} = \dfrac{1}{\tan^2 x} + 1 = \cot^2 x + 1 = \csc^2 x$

21. $\tan^2 x - \sec^2 x = -(\sec^2 x - \tan^2 x) = -1 = -(\csc^2 x - \cot^2 x) = \cot^2 x - \csc^2 x$

23. $\dfrac{120}{169}$ **25.** $-\dfrac{56}{65}$ **27.** $\dfrac{3\sqrt{5} + 1}{8}$

29. Yes. $\sin 2x = 2 \sin x \cos x = 2(0)\cos x = 0$

31. $\cos\left(\dfrac{\pi}{12}\right) = \cos\left(\dfrac{1}{2}\left(\dfrac{\pi}{6}\right)\right) = \sqrt{\dfrac{1 + \cos\frac{\pi}{6}}{2}} = \sqrt{\dfrac{1 + \sqrt{3}/2}{2}} = \sqrt{\dfrac{2 + \sqrt{3}}{4}} = \dfrac{\sqrt{2 + \sqrt{3}}}{2}$; $\cos\left(\dfrac{\pi}{12}\right) = \cos\left(\dfrac{\pi}{4} - \dfrac{\pi}{6}\right) = \cos\dfrac{\pi}{4}\cos\dfrac{\pi}{6} + \sin\dfrac{\pi}{4}\sin\dfrac{\pi}{6} = \dfrac{\sqrt{2}}{2}\left(\dfrac{\sqrt{3}}{2}\right) + \dfrac{\sqrt{2}}{2}\left(\dfrac{1}{2}\right) = \dfrac{\sqrt{6} + \sqrt{2}}{4}$. So, $\dfrac{\sqrt{2 + \sqrt{3}}}{2} = \dfrac{\sqrt{6} + \sqrt{2}}{4}$ or $\sqrt{2 + \sqrt{3}} = \dfrac{\sqrt{2} + \sqrt{6}}{2}$

33. $\dfrac{\sqrt{6} + \sqrt{2}}{4}$ **35.** **(a)**

37. .96 **39.** $\pi/4$

In Answers 41–59, $k = 0, \pm 1, \pm 2, \pm 3, \ldots$.

41. $x = k\pi$

43. $x = .8419$ or 2.2997 or 4.1784 or $5.2463 + 2k\pi$

45. $x = \dfrac{\pi}{6}$ or $\dfrac{5\pi}{6} + 2k\pi$

47. $x = -\dfrac{\pi}{4} + k\pi$

49. $x = .7754$ or $2.3662 + 2k\pi$

51. $x = 1.4940 + k\pi$

53. $x = \dfrac{\pi}{6}$ or $\dfrac{5\pi}{6} + 2k\pi$

55. $x = -\dfrac{\pi}{6}$ or $\dfrac{7\pi}{6} + 2k\pi$

57. $x = \pm\dfrac{\pi}{3} + k\pi$

59. $x = .8959$ or $2.2457 + 2k\pi$

61. $\theta = 225.50°, 314.50°$

63. 9.06° or 80.94°

65. $\dfrac{\pi}{4}$ **67.** $\dfrac{\pi}{3}$ **69.** $\dfrac{5\pi}{6}$

71. .75 **73.** $\dfrac{\pi}{3}$

75.

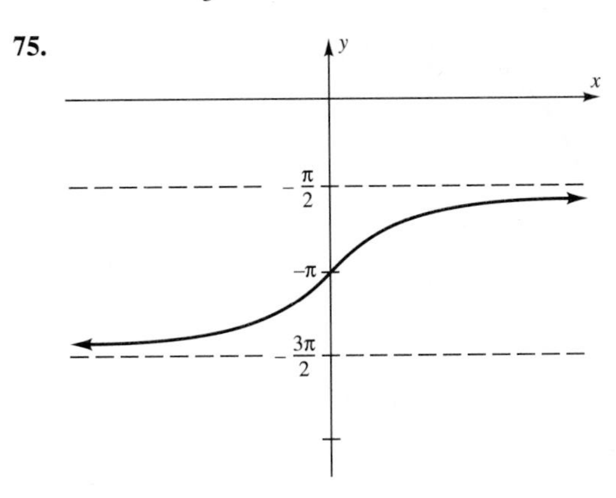

77. $\sqrt{15}/4$

CHAPTER 9

Section 9.1, page 547

1–7.

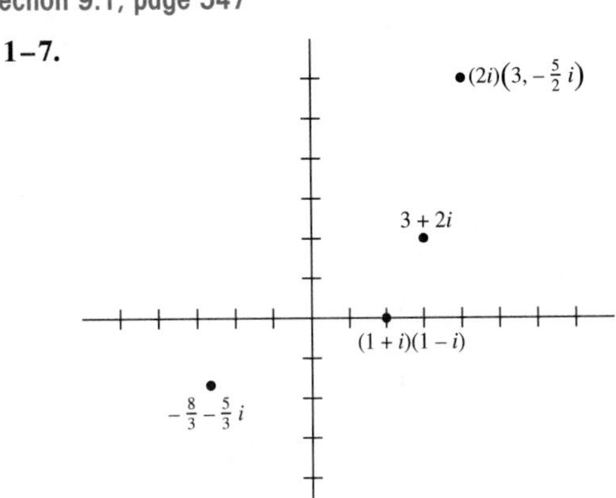

9. 13 **11.** $\sqrt{3}$ **13.** 12

15. Many correct answers, including $z = 1$, $w = i$

17.

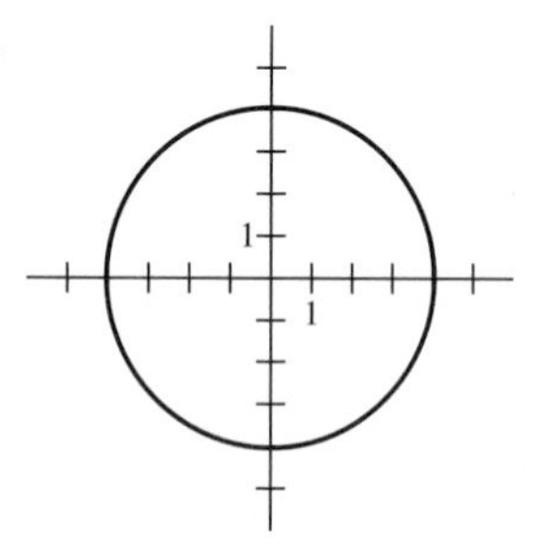

19.

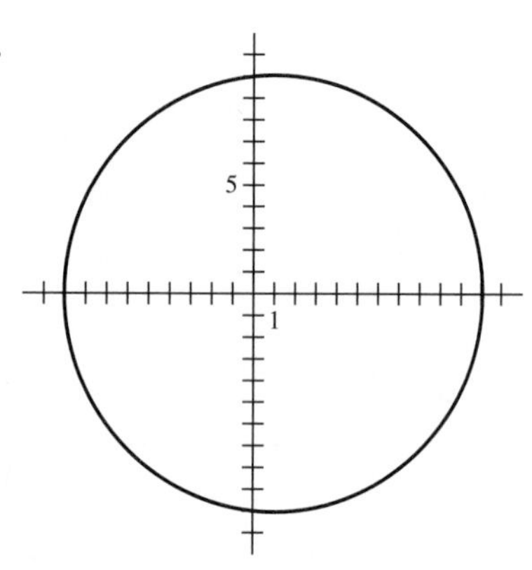

21.

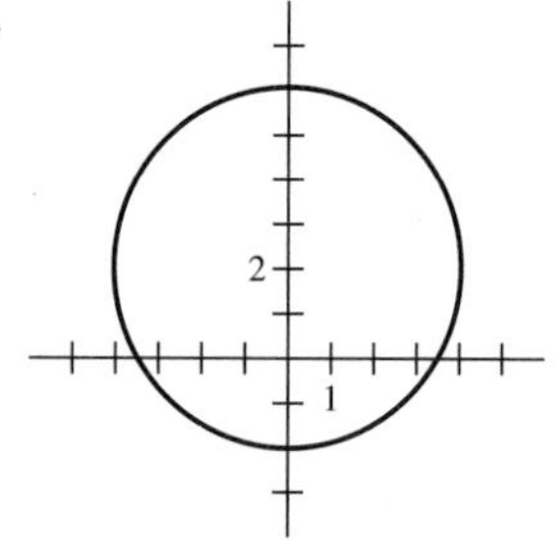

23.

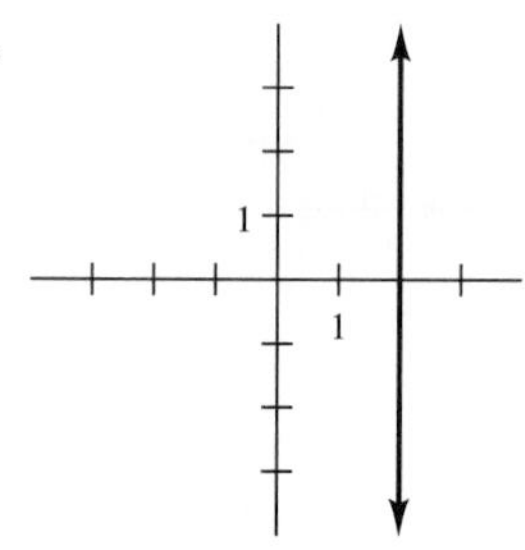

25. $5(\cos .9273 + i \sin .9273)$

27. $13(\cos 5.1072 + i \sin 5.1072)$

29. $\sqrt{5}(\cos 1.1071 + i \sin 1.1071)$

31. $\sqrt{18.5}(\cos 2.1910 + i \sin 2.1910)$

33. $6\left(\cos \frac{2\pi}{3} + i \sin \frac{2\pi}{3}\right) = -3 + 3\sqrt{3}i$

35. $42\left(\cos \frac{7\pi}{6} + i \sin \frac{7\pi}{6}\right) = -21\sqrt{3} - 21i$

37. $\frac{3}{2}\left(\cos \frac{\pi}{4} + i \sin \frac{\pi}{4}\right) = \left(\frac{3\sqrt{2}}{4}\right) + \left(\frac{3\sqrt{2}}{4}\right)i$

39. $2\sqrt{2}\left(\cos \frac{7\pi}{12} + i \sin \frac{7\pi}{12}\right)$

41. $\cos \frac{\pi}{2} + i \sin \frac{\pi}{2}$

43. $12\left(\cos \frac{2\pi}{3} + i \sin \frac{2\pi}{3}\right)$

45. $2\sqrt{2}\left(\cos \frac{19\pi}{12} + i \sin \frac{19\pi}{12}\right)$

47. The polar form of i is $1(\cos 90° + i \sin 90°)$. Hence, by the Polar Multiplication Rule

$$zi = r \cdot 1(\cos(\theta + 90°) + i \sin(\theta + 90°)).$$

You can think of z as lying on a circle with center at the origin and radius r. Then zi lies on the same circle (since it too is r units from the origin), but 90° farther around the circle (in a counterclockwise direction).

49. **(a)** $\frac{b}{a}$ **(b)** $\frac{d}{c}$ **(c)** $y - b = \left(\frac{d}{c}\right)(x - a)$

(d) $y - d = \frac{b}{a}(x - c)$

(f) $(a + c, b + d)$ lies on L since $(b + d) - b = \frac{d}{c}((a + c) - a)$ and $(a + c, b + d)$ lies on M since $(b + d) - d = \frac{b}{a}((a + c) - c)$.

51. **(a)** $r_2(\cos \theta_2 + i \sin \theta_2)(\cos \theta_2 - i \sin \theta_2) = r_2(\cos^2\theta_2 + \sin^2\theta_2) = r_2$

(b) $r_1(\cos \theta_1 + i \sin \theta_1)(\cos \theta_2 - i \sin \theta_2) =$
$r_1[(\cos \theta_1 \cos \theta_2 + \sin \theta_1 \sin \theta_2) +$
$i(\sin \theta_1 \cos \theta_2 - \cos \theta_1 \sin \theta_2)] =$
$r_1[\cos(\theta_1 - \theta_2) + i \sin(\theta_1 - \theta_2)]$

Section 9.2, page 556

1. i **3.** $\dfrac{-243\sqrt{3}}{2} - \dfrac{243}{2}i$ **5.** -64

7. $\dfrac{1}{2} - \dfrac{\sqrt{3}}{2}i$ **9.** i **11.** $1, -1, i, -i$

13. $4\left(\cos\dfrac{\pi}{15} + i\sin\dfrac{\pi}{15}\right), 4\left(\cos\dfrac{11\pi}{15} + i\sin\dfrac{11\pi}{15}\right),$
$4\left(\cos\dfrac{7\pi}{5} + i\sin\dfrac{7\pi}{5}\right)$

15. $3\left(\cos\dfrac{\pi}{48} + i\sin\dfrac{\pi}{48}\right), 3\left(\cos\dfrac{25\pi}{48} + i\sin\dfrac{25\pi}{48}\right),$
$3\left(\cos\dfrac{49\pi}{48} + i\sin\dfrac{49\pi}{48}\right),$
$3\left(\cos\dfrac{73\pi}{48} + i\sin\dfrac{73\pi}{48}\right)$

17. $\left(\cos\dfrac{\pi}{5} + i\sin\dfrac{\pi}{5}\right), \left(\cos\dfrac{3\pi}{5} + i\sin\dfrac{3\pi}{5}\right),$
$(\cos\pi + i\sin\pi), \left(\cos\dfrac{7\pi}{5} + i\sin\dfrac{7\pi}{5}\right),$
$\left(\cos\dfrac{9\pi}{5} + i\sin\dfrac{9\pi}{5}\right)$

19. $\left(\cos\dfrac{\pi}{10} + i\sin\dfrac{\pi}{10}\right), \left(\cos\dfrac{\pi}{2} + i\sin\dfrac{\pi}{2}\right),$
$\left(\cos\dfrac{9\pi}{10} + i\sin\dfrac{9\pi}{10}\right), \left(\cos\dfrac{13\pi}{10} + i\sin\dfrac{13\pi}{10}\right),$
$\left(\cos\dfrac{17\pi}{10} + i\sin\dfrac{17\pi}{10}\right)$

21. $\sqrt[4]{2}\left(\cos\dfrac{\pi}{8} + i\sin\dfrac{\pi}{8}\right), \sqrt[4]{2}\left(\cos\dfrac{9\pi}{8} + i\sin\dfrac{9\pi}{8}\right)$

23. $x = \dfrac{\sqrt{3}}{2} + \dfrac{1}{2}i$ or $\dfrac{\sqrt{3}}{2} - \dfrac{1}{2}i$ or $-\dfrac{\sqrt{3}}{2} + \dfrac{1}{2}i$ or
$-\dfrac{\sqrt{3}}{2} - \dfrac{1}{2}i$ or i or $-i$

25. $\dfrac{\sqrt{3}}{2} + \dfrac{1}{2}i$ or $-\dfrac{\sqrt{3}}{2} + \dfrac{1}{2}i$ or $-i$

27. $x = 3i$ or $\dfrac{3\sqrt{3}}{2} - \dfrac{3}{2}i$ or $-\dfrac{3\sqrt{3}}{2} - \dfrac{3}{2}i$

29. $\sqrt[4]{2}\left(\dfrac{\sqrt{3}}{2} + \dfrac{1}{2}i\right)$ or $\sqrt[4]{2}\left(-\dfrac{1}{2} + \dfrac{\sqrt{3}}{2}i\right)$ or
$\sqrt[4]{2}\left(-\dfrac{\sqrt{3}}{2} - \dfrac{1}{2}i\right)$ or $\sqrt[4]{2}\left(\dfrac{1}{2} - \dfrac{\sqrt{3}}{2}i\right)$

31. $1, .6235 \pm .7818i, -.2225 \pm .9749i, -.9010 \pm .4339i$

33. $\pm 1, \pm i, .7071 \pm .7071i, -.7071 \pm .7071i$

35. $1, .7660 \pm .6428i, .1736 \pm .9848i, -.5 \pm .8660i,$
$-.9397 \pm .3420i$

37. $x^6 - 1 = (x - 1)(x^5 + x^4 + x^3 + x^2 + x + 1)$, so the solutions of $x^5 + x^4 + x^3 + x^2 + x + 1 = 0$ are the sixth roots of unity other than 1; namely, $-1, \dfrac{1}{2} + \dfrac{\sqrt{3}}{2}i,$
$\dfrac{1}{2} - \dfrac{\sqrt{3}}{2}i, -\dfrac{1}{2} + \dfrac{\sqrt{3}}{2}i, -\dfrac{1}{2} - \dfrac{\sqrt{3}}{2}i.$

39. 12

41. For each i, u_i is an nth root of unity, so $(u_i)^n = 1$. Hence $(vu_i)^n = (v^n)(u_i)^n = v^n \cdot 1 = r(\cos\theta + i\sin\theta)$ and vu_i is a solution of the equation. If $vu_i = vu_j$, then multiplying both sides by $1/v$ shows that $u_i = u_j$. In other words, if u_i is not equal to u_j, then $vu_i \neq vu_j$. Thus, the solutions $vu_1, \ldots, vu_n$ are all distinct.

Section 9.3, page 569

1. $3\sqrt{5}$ **3.** $\sqrt{34}$ **5.** $\langle 6, 6\rangle$

7. $\langle -6, 10\rangle$ **9.** $\langle 13/5, -2/5\rangle$

11. $\mathbf{u} + \mathbf{v} = \langle 4, 5\rangle$; $\mathbf{u} - \mathbf{v} = \langle -8, 3\rangle$; $3\mathbf{u} - 2\mathbf{v} =$
$\langle -18, 10\rangle$

13. $\mathbf{u} + \mathbf{v} = \langle 3 + 4\sqrt{2}, 1 + 3\sqrt{2}\rangle$; $\mathbf{u} - \mathbf{v} =$
$\langle 3 - 4\sqrt{2}, -1 + 3\sqrt{2}\rangle$; $3\mathbf{u} - 2\mathbf{v} =$
$\langle 9 - 8\sqrt{2}, -2 + 9\sqrt{2}\rangle$

15. $\mathbf{u} + \mathbf{v} = \langle -23/4, 13\rangle$; $\mathbf{u} - \mathbf{v} = \langle -9/4, 7\rangle$; $3\mathbf{u} - 2\mathbf{v} =$
$\langle -17/2, 24\rangle$

17. $\mathbf{u} + \mathbf{v} = 14\mathbf{i} - 4\mathbf{j}$; $\mathbf{u} - \mathbf{v} = 2\mathbf{i} + 4\mathbf{j}$; $3\mathbf{u} - 2\mathbf{v} =$
$12\mathbf{i} - 8\mathbf{j}$

19. $\mathbf{u} + \mathbf{v} = -\dfrac{5}{4}\mathbf{i} - \dfrac{3}{2}\mathbf{j}$; $\mathbf{u} - \mathbf{v} = -\dfrac{11}{4}\mathbf{i} - \dfrac{3}{2}\mathbf{j}$;
$3\mathbf{u} - 2\mathbf{v} = -\dfrac{15}{2}\mathbf{i} - \dfrac{9}{2}\mathbf{j}$

21. $-7\mathbf{i}$ **23.** $-2\mathbf{i} + \dfrac{1}{2}\mathbf{j}$ **25.** $6\mathbf{i} - \dfrac{13}{4}\mathbf{j}$

27. $\mathbf{v} = \langle 4, 0\rangle$ **29.** $\mathbf{v} = \langle -5\sqrt{2}, -5\sqrt{2}\rangle$

31. $\mathbf{v} = \langle 4.5963, 3.8567\rangle$

33. $\mathbf{v} = \langle -0.1710, -0.4698\rangle$

35. $\|\mathbf{v}\| = 4\sqrt{2}, \theta = 45°$ **37.** $\|\mathbf{v}\| = 8, \theta = 180°$

39. $||\mathbf{v}|| = 6$, $\theta = 90°$

41. $||\mathbf{v}|| = 2\sqrt{17}$, $\theta = 104.04°$

43. $\left\langle \frac{4}{\sqrt{41}}, \frac{-5}{\sqrt{41}} \right\rangle$ **45.** $\frac{1}{\sqrt{5}}\mathbf{i} + \frac{2}{\sqrt{5}}\mathbf{j}$

47. Direction: 46.1°; magnitude: 108.2 lb

49. Direction: 213.4°; magnitude: 17.4 kg

51. Resultant force $= \langle -8, -2 \rangle$; $\mathbf{v} = \langle 8, 2 \rangle$

53. $\mathbf{v} + \mathbf{0} = \langle c, d \rangle + \langle 0, 0 \rangle = \langle c, d \rangle = \mathbf{v}$, and $\mathbf{0} + \mathbf{v} = \langle 0, 0 \rangle + \langle c, d \rangle = \langle c, d \rangle = \mathbf{v}$

55. $r(\mathbf{u} + \mathbf{v}) = r(\langle a, b \rangle + \langle c, d \rangle) = r\langle a + c, b + d \rangle = \langle r(a + c), r(b + d) \rangle = \langle ra + rc, rb + rd \rangle$, and $r\mathbf{u} + r\mathbf{v} = r\langle a, b \rangle + r\langle c, d \rangle = \langle ra, rb \rangle + \langle rc, rd \rangle = \langle ra + rc, rb + rd \rangle$

57. $(rs)\mathbf{v} = (rs)\langle c, d \rangle = \langle rsc, rsd \rangle$; $r(s\mathbf{v}) = r(s\langle c, d \rangle) = r\langle sc, sd \rangle = \langle rsc, rsd \rangle$; and $s(r\mathbf{v}) = s\langle rc, rd \rangle = \langle src, srd \rangle = \langle rsc, rsd \rangle$

59. 48.58 lb

61. Parallel to plane: 32.1 lb; perpendicular to plane: 38.3 lb

63. 66.4°

65. Ground speed: 253.2 mph; course: 69.1°

67. Ground speed: 304.1 mph; course: 309.5°

69. Air speed: 424.3 mph; direction: 62.4°

71. 69.08°

73. 170.32 lb on **u**; 341.77 lb on **v**

75. **(a)** $\mathbf{v} = \langle x_2 - x_1, y_2 - y_1 \rangle$; $k\mathbf{v} = \langle kx_2 - kx_1, ky_2 - ky_1 \rangle$

(b) $||\mathbf{v}|| = \sqrt{(x_2 - x_1)^2 + (y_2 - y_1)^2}$; $||k\mathbf{v}|| = \sqrt{(kx_2 - kx_1)^2 + (ky_2 - ky_1)^2}$

(c) $||k\mathbf{v}|| = \sqrt{k^2(x_2 - x_1)^2 + k^2(y_2 - y_1)^2} = \sqrt{k^2}\sqrt{(x_2 - x_1)^2 + (y_2 - y_1)^2} = |k|\,||\mathbf{v}||$

(d) $\tan\theta = \frac{y_2 - y_1}{x_2 - x_1}$; $\tan\beta = \frac{ky_2 - ky_1}{kx_2 - kx_1} = \frac{k(y_2 - y_1)}{k(x_2 - x_1)} = \frac{y_2 - y_1}{x_2 - x_1}$. Since the values of the tangent function are repeated only at intervals of π, it follows that θ and β differ by a multiple of π, and so represent either the same direction or opposite directions.

(e) If $k > 0$, the signs of the components of $k\mathbf{v}$ are the same as those of $\mathbf{v}$, so $\mathbf{v}$ and $k\mathbf{v}$ lie in the same quadrant. Therefore $\mathbf{v}$ and $k\mathbf{v}$ must have the same direction rather than opposite directions. If $k < 0$, the signs of $k\mathbf{v}$ are opposite those of $\mathbf{v}$, so $\mathbf{v}$ and $k\mathbf{v}$ do not lie in the same quadrant. They cannot have the same direction, so must have opposite directions.

77. **(a)** Since $\mathbf{u} - \mathbf{v} = \langle a - c, b - d \rangle$, the box on page 559 shows that its magnitude is $\sqrt{(a - c)^2 + (b - d)^2}$. On the other hand, the distance formula shows that the length (magnitude) of $\mathbf{w}$ is $\sqrt{(a - c)^2 + (b - d)^2}$. Hence, $||\mathbf{u} - \mathbf{v}|| = ||\mathbf{w}||$.

(b) $\mathbf{u} - \mathbf{v}$ lies on the straight line through (0, 0) and $(a - c, b - d)$ which has slope $\frac{(a - c) - 0}{(b - d) - 0} = \frac{a - c}{b - d}$. Similarly, $\mathbf{w}$ lies on the line through (a, b) and (c, d), which also has slope $\frac{a - c}{b - d}$. So, $\mathbf{u} - \mathbf{v}$ and $\mathbf{w}$ are parallel. Verify that they actually have the same direction by considering the relative positions of (a, b), (c, d), and $(a - c, b - d)$. For instance, if $\mathbf{u} - \mathbf{v}$ points upward to the right, then (a, b) lies to the right and above (c, d). Hence $c < a$ and $d < b$, so that $a - c > 0$ and $b - d > 0$, which means that the endpoint of $\mathbf{w}$ lies in the first quadrant, that is, $\mathbf{w}$ points upward to the right.

Section 9.4, page 580

1. $\mathbf{u}\cdot\mathbf{v} = -7$, $\mathbf{u}\cdot\mathbf{u} = 25$, $\mathbf{v}\cdot\mathbf{v} = 29$

3. $\mathbf{u}\cdot\mathbf{v} = 6$, $\mathbf{u}\cdot\mathbf{u} = 5$, $\mathbf{v}\cdot\mathbf{v} = 9$

5. $\mathbf{u}\cdot\mathbf{v} = 12$, $\mathbf{u}\cdot\mathbf{u} = 13$, $\mathbf{v}\cdot\mathbf{v} = 13$

7. 6 **9.** 20 **11.** -28 **13.** 1.75065 radians

15. 2.1588 radians **17.** $\pi/2$ radians

19. Orthogonal **21.** Parallel **23.** Neither

25. $k = 2$ **27.** $k = \sqrt{2}$

29. $\text{proj}_{\mathbf{u}}\mathbf{v} = \langle 12/17, -20/17 \rangle$; $\text{proj}_{\mathbf{v}}\mathbf{u} = \langle 6/5, 2/5 \rangle$

31. $\text{proj}_{\mathbf{u}}\mathbf{v} = \langle 0, 0 \rangle$; $\text{proj}_{\mathbf{v}}\mathbf{u} = \langle 0, 0 \rangle$

33. $\text{comp}_{\mathbf{v}}\mathbf{u} = 22/\sqrt{13}$ **35.** $\text{comp}_{\mathbf{v}}\mathbf{u} = 3/\sqrt{10}$

37. $\mathbf{u}\cdot(\mathbf{v} + \mathbf{w}) = \langle a, b \rangle\cdot(\langle c, d \rangle + \langle r, s \rangle) = \langle a, b \rangle\cdot\langle c + r, d + s \rangle = a(c + r) + b(d + s) = ac + ar + bd + bs$
$\mathbf{u}\cdot\mathbf{v} + \mathbf{u}\cdot\mathbf{w} = \langle a, b \rangle\cdot\langle c, d \rangle + \langle a, b \rangle\cdot\langle r, s \rangle = (ac + bd) + (ar + bs) = ac + ar + bd + bs$

39. $\mathbf{0}\cdot\mathbf{u} = \langle 0, 0\rangle\cdot\langle a, b\rangle = 0a + 0b = 0$

41. If $\theta = 0$ or π, then $\mathbf{u}$ and $\mathbf{v}$ are parallel, so $\mathbf{v} = k\mathbf{u}$ for some real number k. We know that $\|\mathbf{v}\| = |k|\,\|\mathbf{u}\|$ (Exercise 75, Section 9.3). If $\theta = 0$, then $\cos\theta = 1$ and $k > 0$. Since $k > 0$, $|k| = k$ and so $\|\mathbf{v}\| = k\,\|\mathbf{u}\|$. Therefore, $\mathbf{u}\cdot\mathbf{v} = \mathbf{u}\cdot k\mathbf{u} = k\mathbf{u}\cdot\mathbf{u} = k\,\|\mathbf{u}\|^2 = \|\mathbf{u}\|\,(k\,\|\mathbf{u}\|) = \|\mathbf{u}\|\,\|\mathbf{v}\| = \|\mathbf{u}\|\,\|\mathbf{v}\|\cos\theta$. On the other hand, if $\theta = \pi$, then $\cos\theta = -1$ and $k < 0$. Since $k < 0$, $|k| = -k$ and so $\|\mathbf{v}\| = -k\,\|\mathbf{u}\|$. Then $\mathbf{u}\cdot\mathbf{v} = \mathbf{u}\cdot k\mathbf{u} = k\,\mathbf{u}\cdot\mathbf{u} = k\,\|\mathbf{u}\|^2 = \|\mathbf{u}\|(-k\|\mathbf{u}\|) = -\|\mathbf{u}\|\,\|\mathbf{v}\| = \|\mathbf{u}\|\,\|\mathbf{v}\|\cos\theta$. In both cases we have shown $\mathbf{u}\cdot\mathbf{v} = \|\mathbf{u}\|\,\|\mathbf{v}\|\cos\theta$.

43. If $A = (1, 2)$, $B = (3, 4)$, and $C = (5, 2)$, then the vector $\overrightarrow{AB} = \langle 2, 2\rangle$, $\overrightarrow{AC} = \langle 4, 0\rangle$, and $\overrightarrow{BC} = \langle 2, -2\rangle$. Since $\overrightarrow{AB}\cdot\overrightarrow{BC} = 0$, $\overrightarrow{AB}$ and $\overrightarrow{BC}$ are perpendicular, so the angle at vertex B is a right angle.

45. Many possible answers: One is $\mathbf{u} = \langle 1, 0\rangle$, $\mathbf{v} = \langle 1, 1\rangle$, and $\mathbf{w} = \langle 1, -1\rangle$.

47. 300 pounds ($= 600 \sin 30°$)

49. 13 **51.** 24

53. The force in the direction of the lawnmower's motion is $30\cos 60° = 15$ pounds. Thus, the work done is $15(75) = 1125$ ft-lb.

55. 1368 ft-lb

Chapter 9 Review, page 585

1. $\sqrt{10} + \sqrt{20}$

3. The graph is a circle of radius 2 centered at the origin.

5. $2\left(\cos\dfrac{\pi}{3} + i\sin\dfrac{\pi}{3}\right)$ **7.** $4\sqrt{2} + 4\sqrt{2}i$

9. $2\sqrt{3} + 2i$ **11.** $\dfrac{81}{2} - \dfrac{81\sqrt{3}}{2}i$

13. $1, \cos\dfrac{\pi}{3} + i\sin\dfrac{\pi}{3}, \cos\dfrac{2\pi}{3} + i\sin\dfrac{2\pi}{3}, -1, \cos\dfrac{4\pi}{3} + i\sin\dfrac{4\pi}{3}, \cos\dfrac{5\pi}{3} + i\sin\dfrac{5\pi}{3}$

15. $\cos\dfrac{\pi}{8} + i\sin\dfrac{\pi}{8}, \cos\dfrac{5\pi}{8} + i\sin\dfrac{5\pi}{8}, \cos\dfrac{9\pi}{8} + i\sin\dfrac{9\pi}{8}, \cos\dfrac{13\pi}{8} + i\sin\dfrac{13\pi}{8}$

17. $\langle 11, -1\rangle$ **19.** $2\sqrt{29}$ **21.** $-11\mathbf{i} + 8\mathbf{j}$

23. $\sqrt{10}$ **25.** $\left\langle\dfrac{5\sqrt{2}}{2}, \dfrac{5\sqrt{2}}{2}\right\rangle$

27. $-\dfrac{1}{\sqrt{5}}\mathbf{i} + \dfrac{2}{\sqrt{5}}\mathbf{j}$

29. Ground speed: 321.87 mph; course: 126.18°

31. -26 **33.** 3 **35.** .70 radians

37. $\text{proj}_{\mathbf{v}}\mathbf{u} = \mathbf{v} = 2\mathbf{i} + \mathbf{j}$

39. $(\mathbf{u} + \mathbf{v})\cdot(\mathbf{u} - \mathbf{v}) = \mathbf{u}\cdot\mathbf{u} - \mathbf{u}\cdot\mathbf{v} + \mathbf{v}\cdot\mathbf{u} - \mathbf{v}\cdot\mathbf{v} = \mathbf{u}\cdot\mathbf{u} - \mathbf{v}\cdot\mathbf{v} = \|\mathbf{u}\| - \|\mathbf{v}\| = 0$ since $\mathbf{u}$ and $\mathbf{v}$ have the same magnitude

41. 1750 lb

CHAPTER 10

Section 10.1, page 597

1. $-5 \le x \le 6$ and $-2 \le y \le 2$

3. $-3 \le x \le 4$ and $-2 \le y \le 3$

5. $0 \le x \le 14$ and $-15 \le y \le 0$

7. $-2 \le x \le 20$ and $-11 \le y \le 11$

9. $-12 \le x \le 12$ and $-12 \le y \le 12$

11. $-2 \le x \le 20$ and $-20 \le y \le 4$

13. $-25 \le x \le 22$ and $-25 \le y \le 26$

15. $y = 2x + 7$ **17.** $y = 2x + 5$ **19.** $y = \ln x$

21. $x^2 + y^2 = 9$ **23.** $16x^2 + 9y^2 = 144$

25. Both give a straight line segment between $P = (-4, 7)$ and $Q = (2, -5)$. The parametric equations in (a) move from P to Q, and the parametric equations in (b) move from Q to P.

27. Solving $x = a + (c - a)t$ for t gives $t = \dfrac{x - a}{c - a}$ and substituting in $y = b + (d - b)t$ then gives $y = b + (d - b)\dfrac{x - a}{c - a}$, or $y = b + \dfrac{d - b}{c - a}(x - a)$. This is a linear equation and therefore gives a straight line. You can check by substitution that (a, b) and (c, d) lie on this straight line; in fact, these points correspond to $t = 0$ and $t = 1$, respectively.

29. $x = -6 + 18t$, $y = 12 - 22t$ $(0 \le t \le 1)$

31. $x = 18 - 34t$, $y = 4 + 10t$ $(0 \le t \le 1)$

33. $x = 9 + 5\cos t$, $y = 12 + 5\sin t$ $(0 \le t \le 2\pi)$

35. $x = 2 \cos t + 2,\ y = 2 \sin t + 3 \quad (0 \le t \le 2\pi)$

37. Local minimum at $(-6, 2)$

39. Local maximum at $(4, 5)$

41. (a)

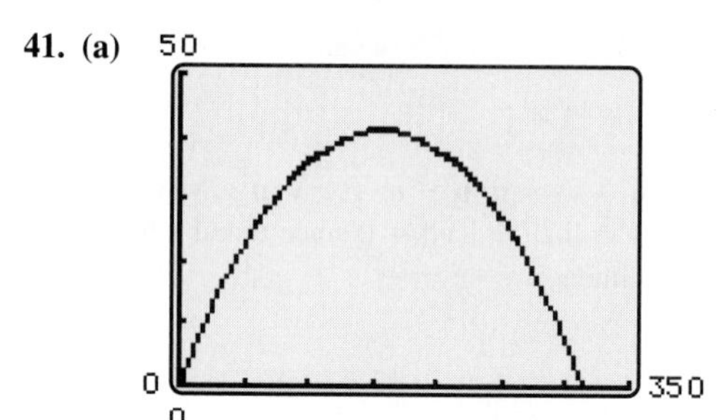

$x = (110 \cos 28°)t$
$y = (110 \sin 28°)t - 16t^2$
$0 \le t \le 3.5$

(b) About 3.2 sec
(c) 41.67 ft

43. (a)

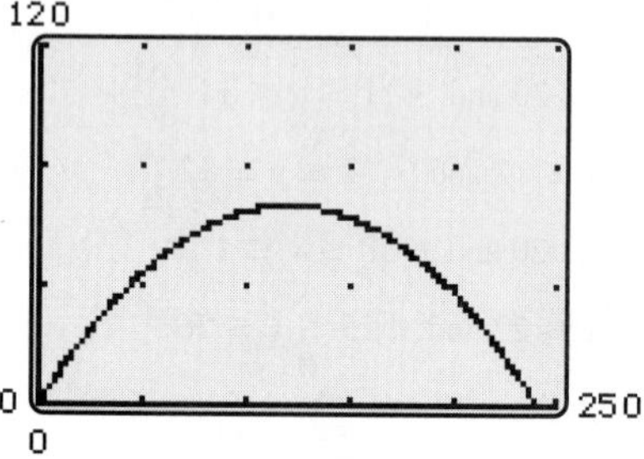

$x = (88 \cos 48°)t$
$y = (88 \sin 48°)t - 16t^2 + 4$
$0 \le t \le 4.5$

(b) Yes

45. $v = 80\sqrt[4]{3} \approx 105.29$ ft/sec

47. (b) $\cos(t - \pi/2) = \cos t \cos \pi/2 + \sin t \sin \pi/2 = (\cos t)(0) + (\sin t)(1) = \sin t$. Therefore, $3t - 3\cos(t - \pi/2) = 3t - 3 \sin t = 3(t - \sin t)$. $\sin(t - \pi/2) = \sin t \cos \pi/2 - \cos t \sin \pi/2 = (\sin t)(0) - (\cos t)(1) = -\cos t$. Therefore, $3 + 3 \sin(t - \pi/2) = 3 - 3 \cos t = 3(1 - \cos t)$.

49. (b) $\cos(t - 3\pi/2) = \cos t \cos 3\pi/2 + \sin t \sin 3\pi/2 = (\cos t)(0) + (\sin t)(-1) = -\sin t$. Therefore, $3t + 3\cos(t - 3\pi/2) = 3t - 3 \sin t = 3(t - \sin t)$ $\sin(t - 3\pi/2) = \sin t \cos 3\pi/2 - \cos t \sin 3\pi/2 = (\sin t)(0) - (\cos t)(-1) = \cos t$. Therefore, $3 - 3 \sin(t - 3\pi/2) = 3 - 3 \cos t = 3(1 - \cos t)$.

51. (a)

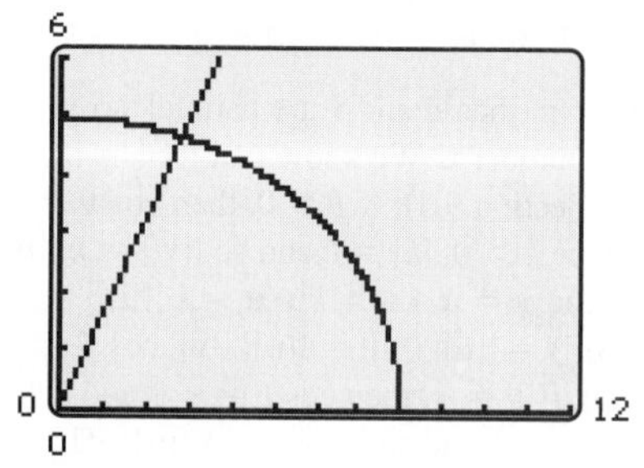

The particles do not collide.

(b) $t \approx 1.1$

(c)

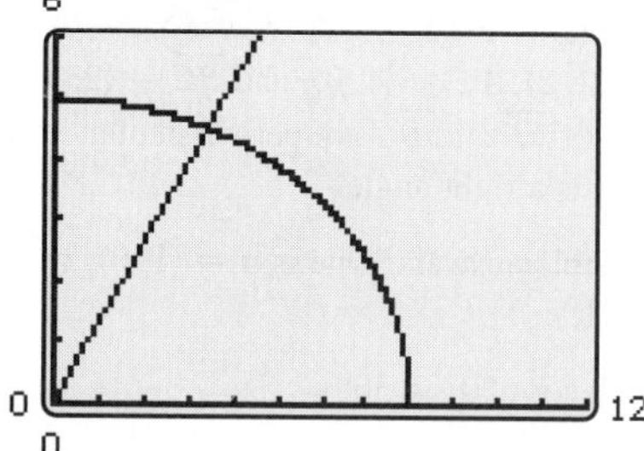

The particles do not collide; they are closest when $t \approx 1.13$.

(d)

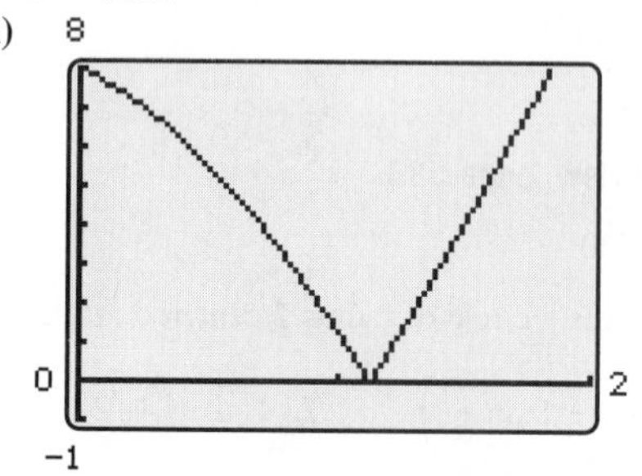

d is smallest when $t = 1.1322$.

53. (a) As shown in the diagram on page A-57, the center Q of the small circle is always at distance $a - b$ from the origin O. Suppose t is the angle that OQ makes with the x-axis. Then the coordinates of Q are $x = (a - b) \cos t,\ y = (a - b) \sin t$. Examining the smaller circle in detail, we see that the change in x-coordinate from Q to P is $b \cos u$, where u is the angle that PQ makes with the positive x-axis. Likewise, the change in y-coordinate from Q to P is $-b \sin u$. Therefore, the coordinates of P are

$$x = (a - b) \cos t + b \cos u$$
$$y = (a - b) \sin t - b \sin u.$$

The angles t and u are related by the fact that the inner circle must roll without "slipping." This means the arc length that P has moved around the inner circle must equal the arc length that the inner circle has moved along the circumference of the larger circle. In other words, the arc length from P to W must equal the arc length from S to V. Since the length of a circular arc is the radius times the angle, this means $bu = (a - b)t$, or $u = (a - b)t/b$. Substituting this for u in the above equations will give the desired parametric equations.

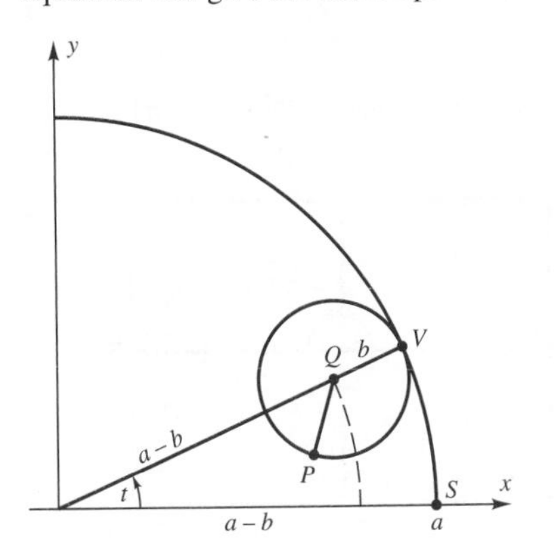

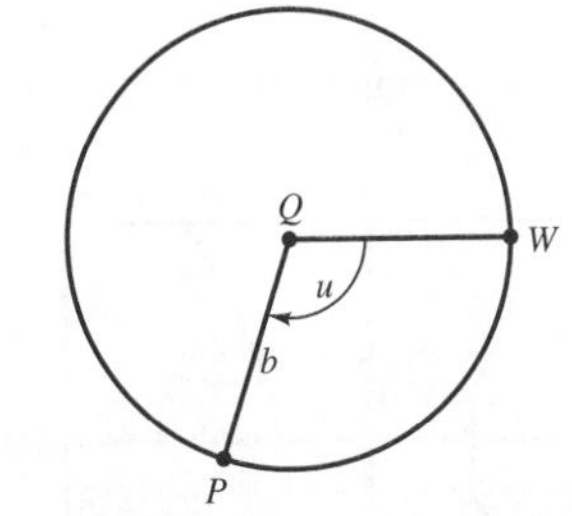

(b)

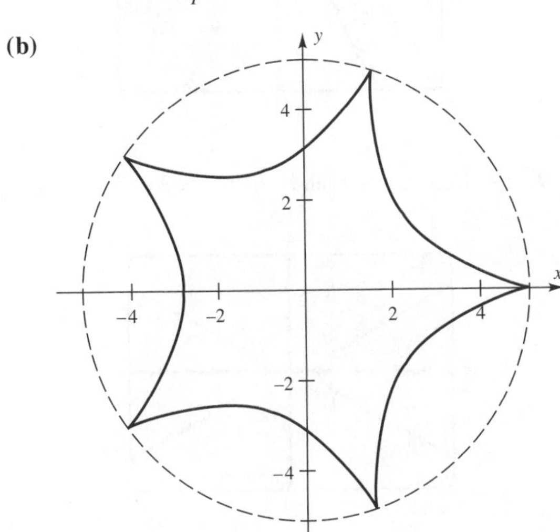

(c)

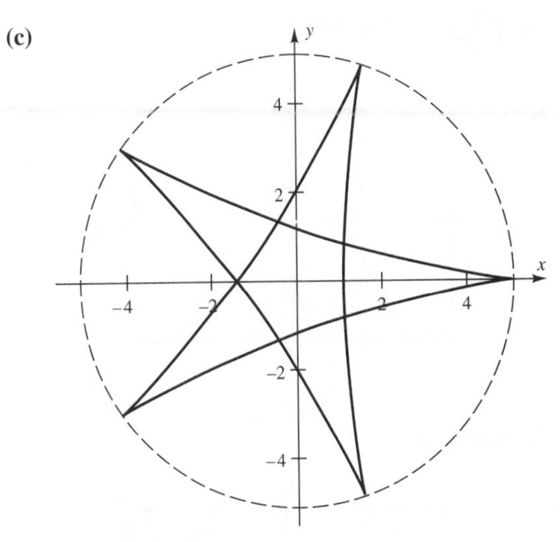

Section 10.2, page 614

1. $\dfrac{x^2}{49} + \dfrac{y^2}{4} = 1$ **3.** $\dfrac{x^2}{36} + \dfrac{y^2}{16} = 1$

5. $\dfrac{x^2}{9} - \dfrac{y^2}{36} = 1$ **7.** $\dfrac{x^2}{4} - y^2 = 1$

9. $y = 3x^2$ **11.** $y^2 = 20x$

13. $\dfrac{x^2}{9} + \dfrac{y^2}{49} = 1$ **15.** $6x = y^2$

17. $2x^2 - y^2 = 8$ **19.** $x^2 + 6y^2 = 18$

21. Ellipse

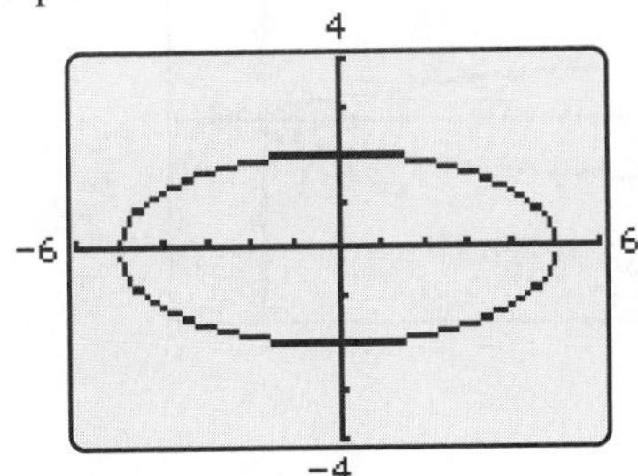

23. Ellipse

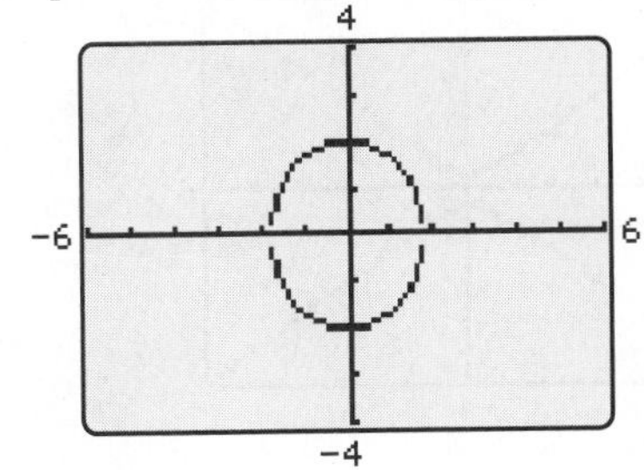

25. Hyperbola

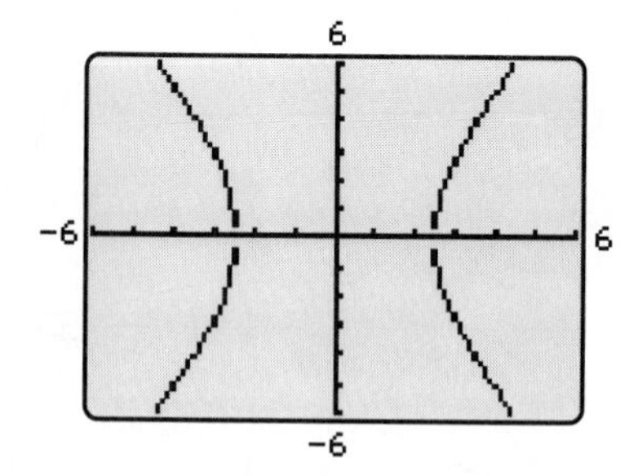

27. Hyperbola

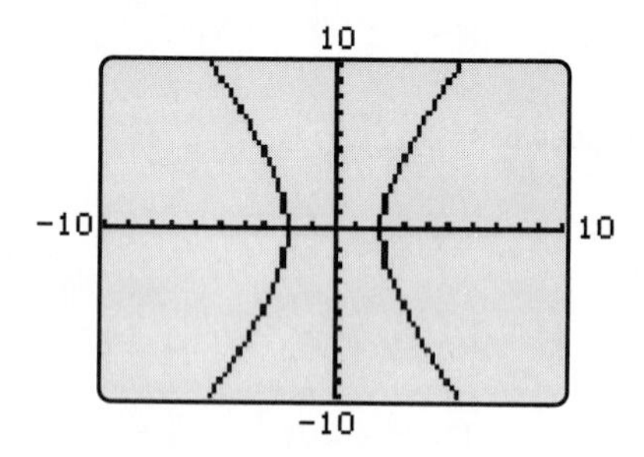

29. Parabola

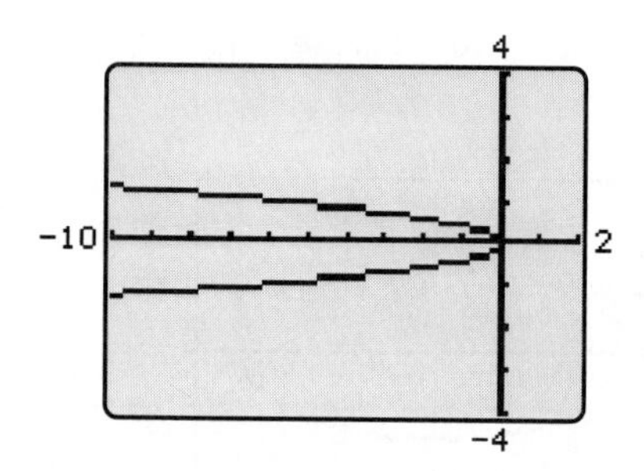

31. Hyperbola

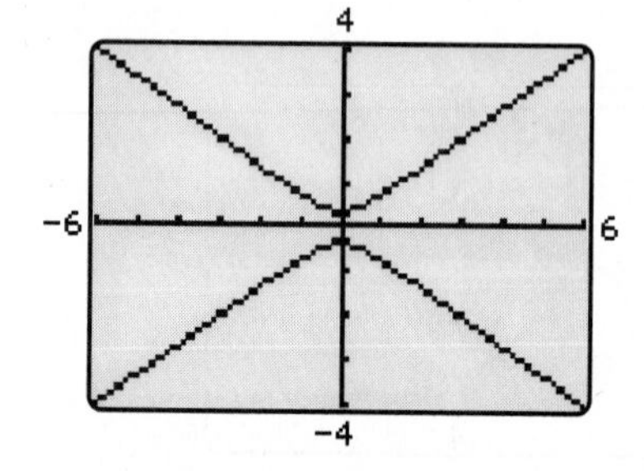

33. $x = \sqrt{10}\cos t,\ y = 6\sin t \quad (0 \le t \le 2\pi)$

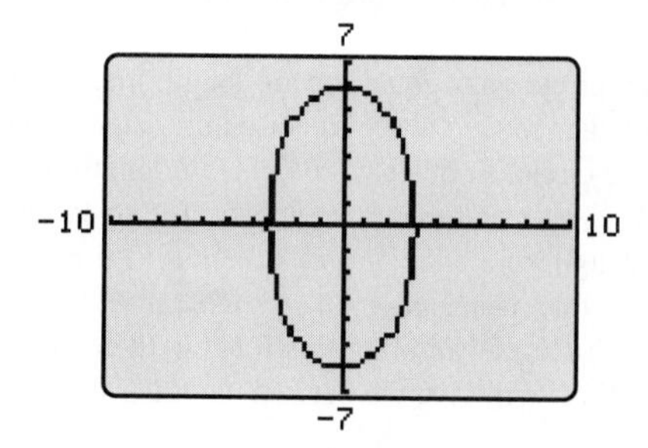

35. $x = \frac{1}{2}\cos t,\ y = \frac{1}{2}\sin t \quad (0 \le t \le 2\pi)$

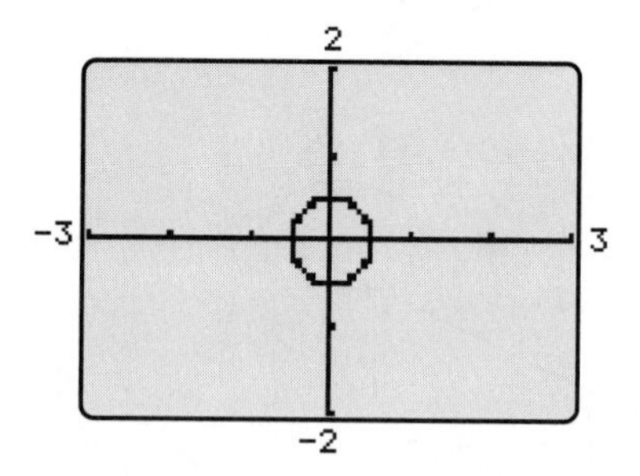

37. $x = \sqrt{10}/\cos t,\ y = 6\tan t \quad (0 \le t \le 2\pi)$

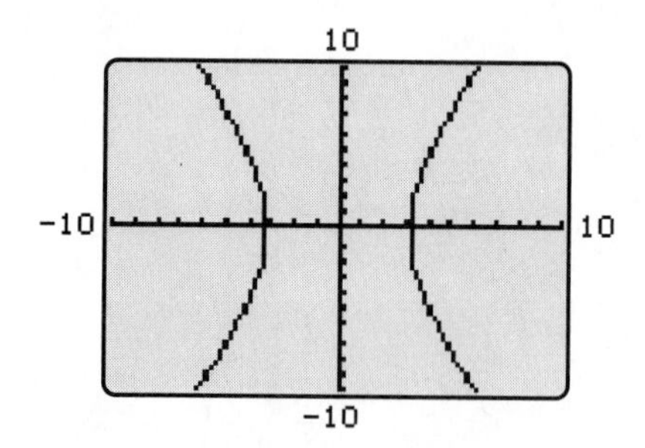

39. $x = 1/\cos t,\ y = \frac{1}{2}\tan t \quad (0 \le t \le 2\pi)$

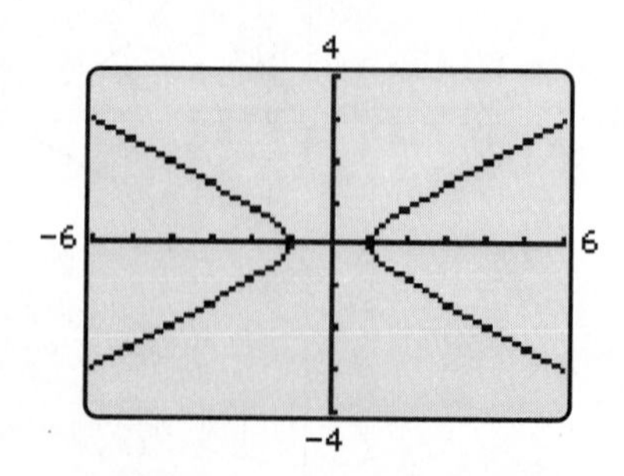

41. $x = t^2/4,\ y = t$ (t any real number)

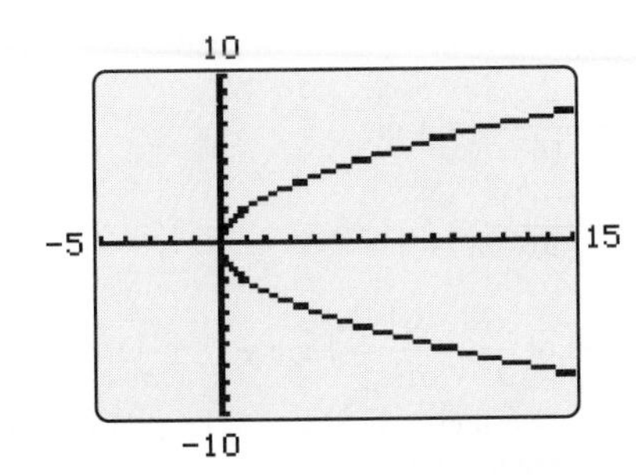

43. 8π **45.** $2\sqrt{3}\pi$ **47.** $7\pi/\sqrt{3}$

49. Focus: (0, 1/12); directrix: $y = -1/12$

51. Focus: (0, 1); directrix: $y = -1$

53. If $a = b$, then $\dfrac{x^2}{a^2} + \dfrac{y^2}{a^2} = 1$. Multiplying both sides by a^2 gives $x^2 + y^2 = a^2$, the equation of a circle of radius a with center at the origin.

55. The two branches of the hyperbola are very "flat" when b is large. With very large b and a small viewing window, the hyperbola may look like two horizontal lines, but it isn't because its asymptotes $y = \pm\dfrac{2}{b}x$ are not horizontal (their slopes, $\pm 2/b$, are close to, but not equal to, 0 when b is large).

57. $x = t^2/4p = y^2/4p$; hence, $y^2 = 4px$ and the graph is a parabola.

59. Approximately 226,335 miles and 251,401 miles

61. Let P denote the punch bowl and Q the table. In the longest possible trip starting at point X, the sum of the distance from X to Q and the distance from X to P must be 100 (since the distance from Q to P is 50). Thus, the fence should be an ellipse with foci P and Q and $r = 100$, as described on page 602 (with $c = 25$). Verify that the length of its major axis is 100 ft and the length of its minor axis is approximately 86.6 ft.

Section 10.3, page 626

1. $\dfrac{(x-2)^2}{4} + \dfrac{(y-3)^2}{16} = 1$

3. $\dfrac{4(x-7)^2}{25} + \dfrac{(y+4)^2}{36} = 1$

5. $\dfrac{(y-3)^2}{4} - \dfrac{(x+2)^2}{6} = 1$

7. $\dfrac{(x-4)^2}{9} - \dfrac{(y-2)^2}{16} = 1$

9. $y = 13(x-1)^2$ **11.** $x - 2 = 3(y-1)^2$

13. $\dfrac{(x-3)^2}{36} + \dfrac{(y+2)^2}{16} = 1$

15. $x + 3 = 4(y+2)^2$

17. Ellipse **19.** Parabola **21.** Hyperbola

23. $x = 2\cos t + 1,\ y = 3\sin t + 5$ $(0 \le t \le 2\pi)$

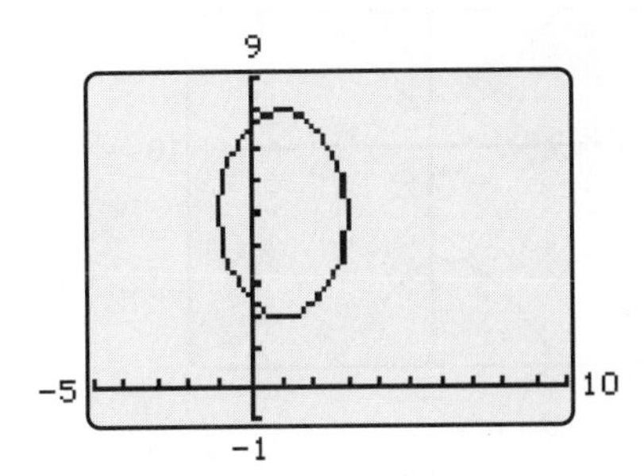

25. $x = 4\cos t - 1,\ y = \sqrt{8}\sin t + 4$ $(0 \le t \le 2\pi)$

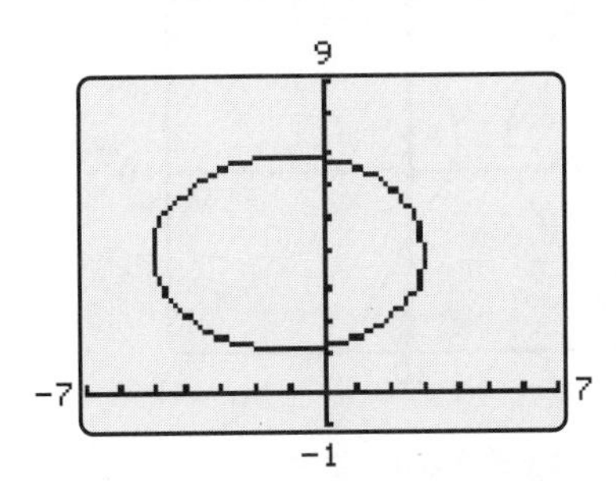

27. $x = t,\ y = 4(t-1)^2 + 2$ (any real number t)

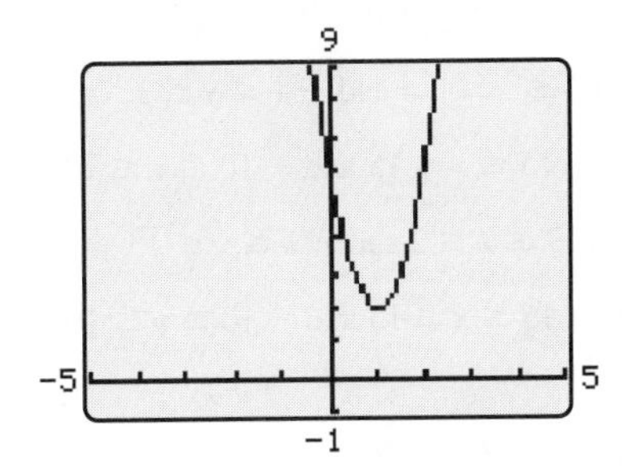

29. $x = 2(t - 2)^2, y = t$ (any real number t)

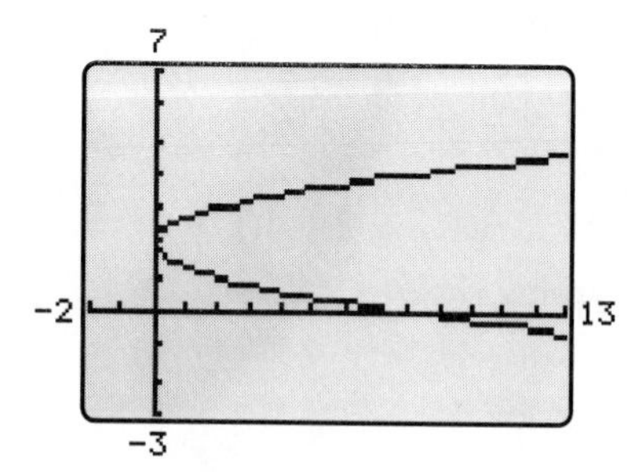

31. $x = 4 \tan t - 1, y = 5/\cos t - 3$ $(0 \leq t \leq 2\pi)$

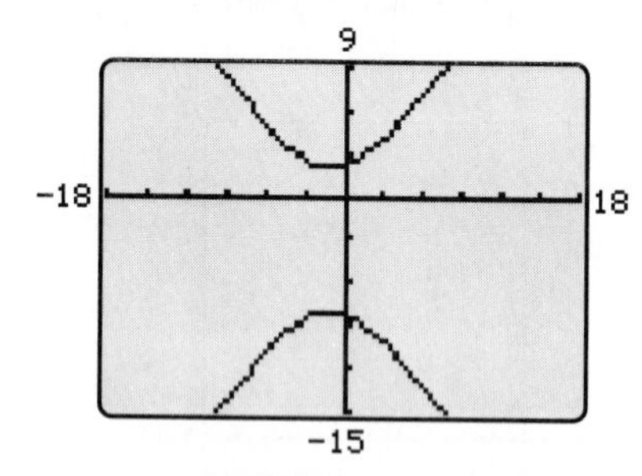

33. $x = 1/\cos t - 3, y = 2 \tan t + 2$ $(0 \leq t \leq 2\pi)$

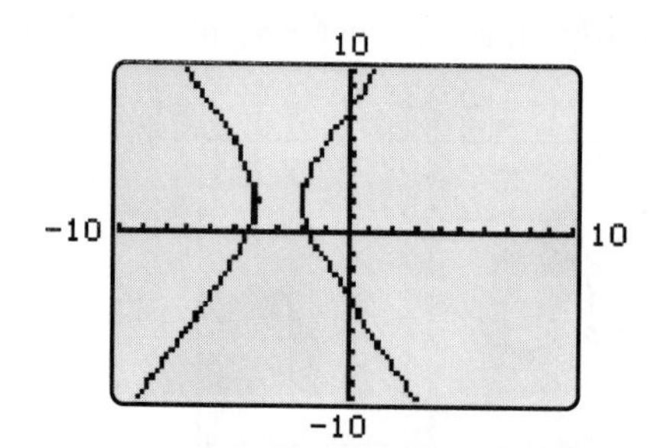

35. Ellipse; $-6 \leq x \leq 3$ and $-2 \leq y \leq 4$

37. Hyperbola; $-7 \leq x \leq 13$ and $-3 \leq y \leq 9$

39. Parabola; $-1 \leq x \leq 8$ and $-3 \leq y \leq 3$

41. Ellipse; $-1.5 \leq x \leq 1.5$ and $-1 \leq y \leq 1$

43. Hyperbola; $-15 \leq x \leq 15$ and $-10 \leq y \leq 10$

45. Parabola; $-19 \leq x \leq 2$ and $-1 \leq y \leq 13$

47. Hyperbola; $-15 \leq x \leq 15$ and $-10 \leq y \leq 10$

49. Ellipse; $-6 \leq x \leq 6$ and $-4 \leq y \leq 4$

51. Parabola; $-9 \leq x \leq 9$ and $-2 \leq y \leq 10$

53. $\dfrac{(x+5)^2}{49} + \dfrac{(y-3)^2}{16} = 1$ or $\dfrac{(x+5)^2}{16} + \dfrac{(y-3)^2}{49} = 1$

55. The asymptotes of $\dfrac{x^2}{a^2} - \dfrac{y^2}{a^2} = 1$ are $y = \pm\dfrac{a}{a}x$ or $y = \pm x$, with slopes $+1$ and -1. Since $(+1)(-1) = -1$, these lines are perpendicular.

57. $b = 0$

59. $(9, -\frac{1}{2} \pm \frac{1}{2}\sqrt{34})$

61. $\dfrac{x^2}{1{,}210{,}000} - \dfrac{y^2}{5{,}759{,}600} = 1$ (measurement in feet). The exact location cannot be determined from the given information.

Section 10.4, page 636

1. $P = (2, \pi/4)$, $Q = (3, 2\pi/3)$, $R = (5, \pi)$, $S = (7, 7\pi/6)$, $T = (4, 3\pi/2)$, $U = (6, -\pi/3)$ or $(6, 5\pi/3)$, $V = (7, 0)$

3. $(5, 2\pi)$, $(5, -2\pi)$, $(-5, 3\pi)$, $(-5, -\pi)$, and others

5. $(1, 5\pi/6)$, $(1, -7\pi/6)$, $(-1, 11\pi/6)$, $(-1, -13\pi/6)$, and others

7. $(3/2, 3\sqrt{3}/2)$

9. $(\sqrt{3}/2, -1/2)$

11. $(6, -\pi/6)$

13. $(2\sqrt{5}, 1.1071)$

15. $(\sqrt{31.25}, 2.6779)$

17.

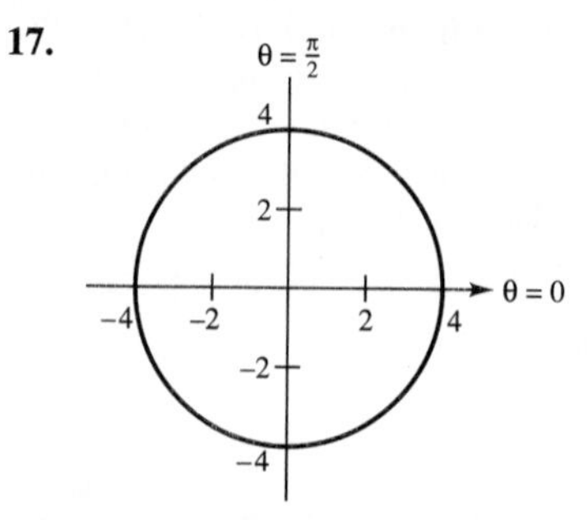

19.

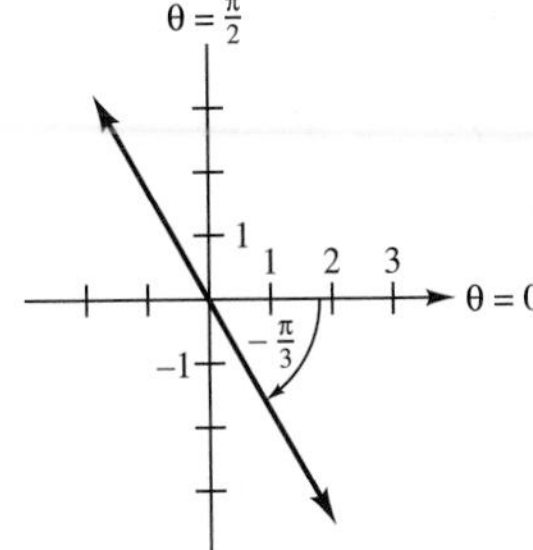

21.

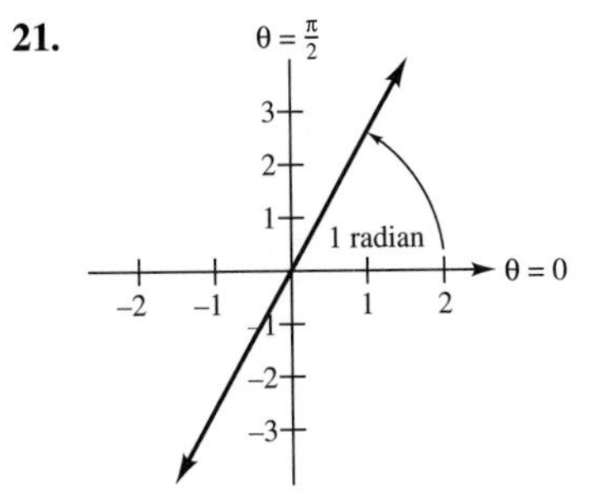

23.

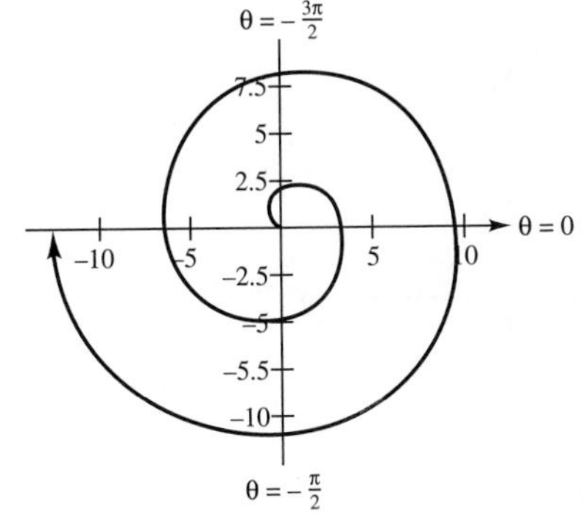

25.

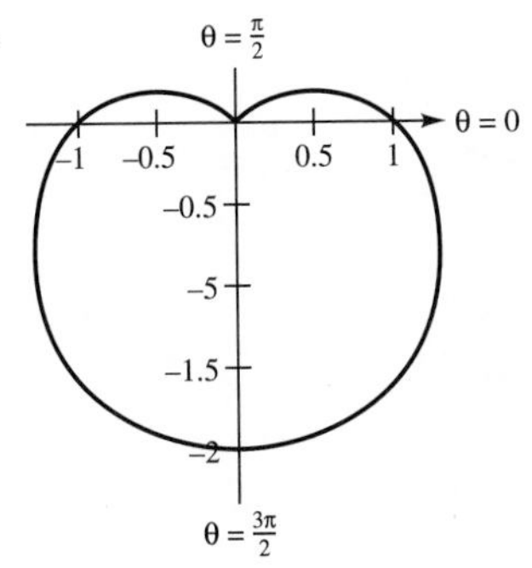

27.

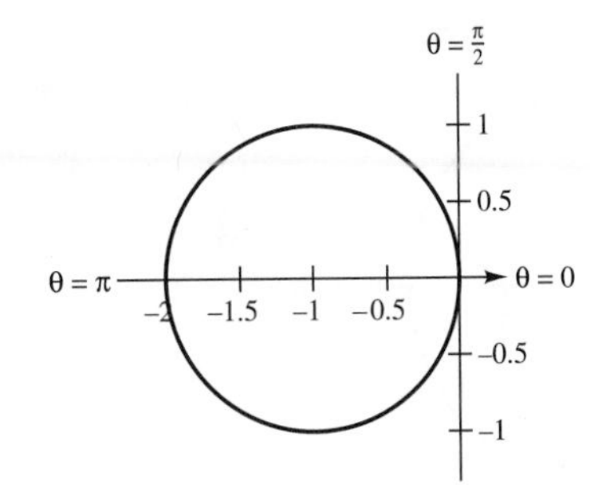

29.

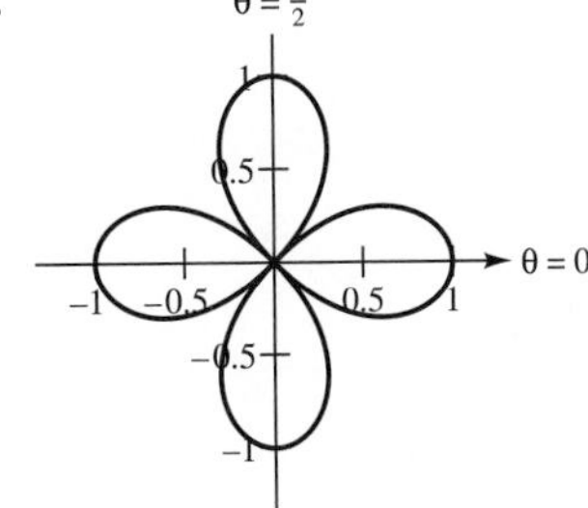

31.

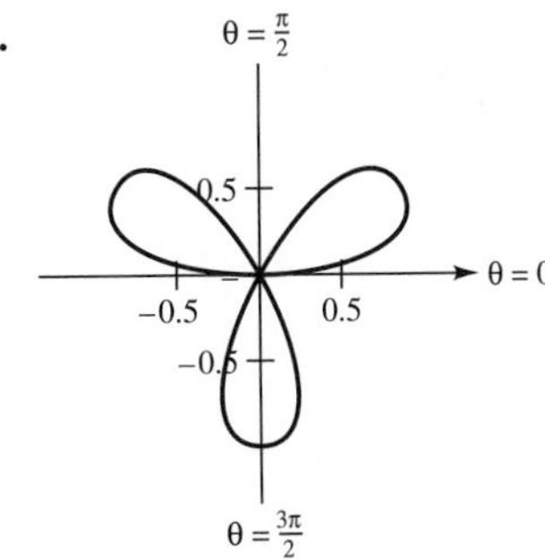

33.

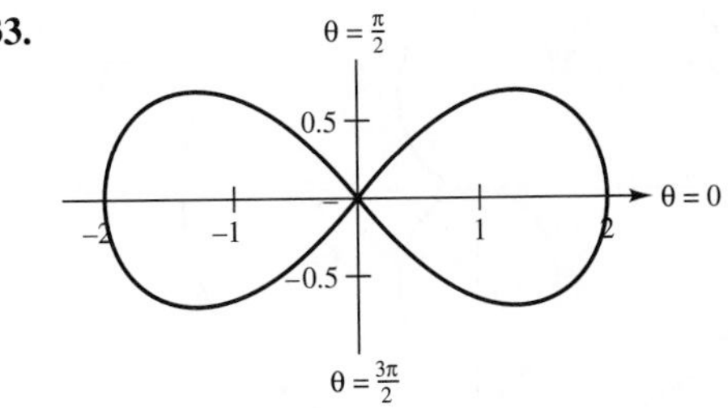

35.

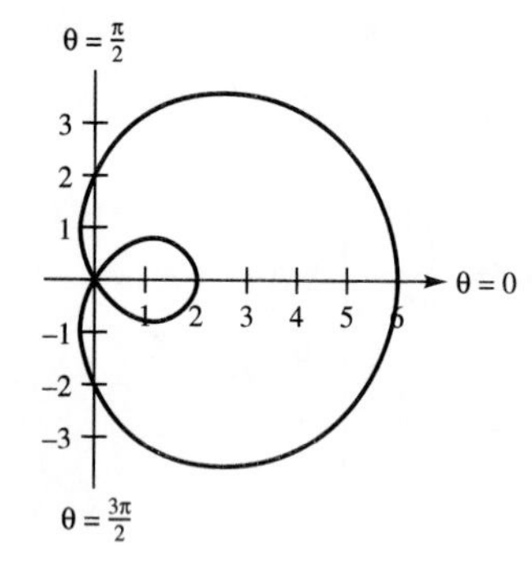

37.

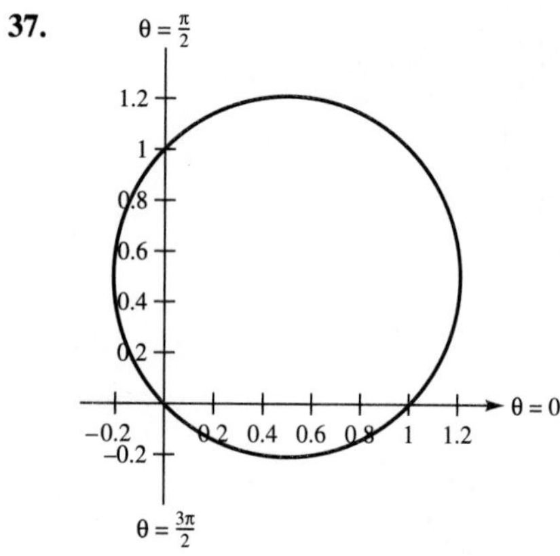

39.

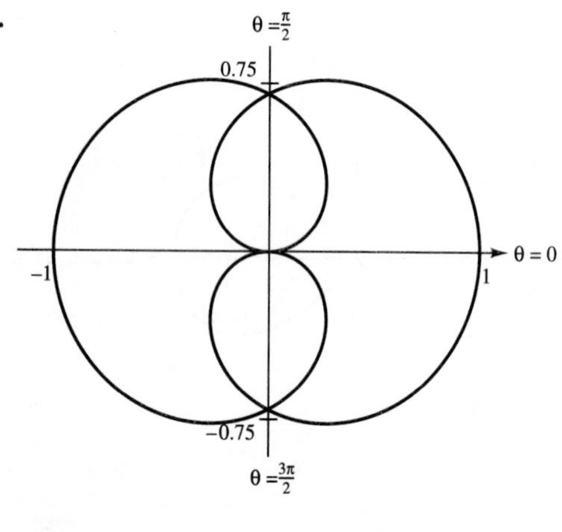

41.

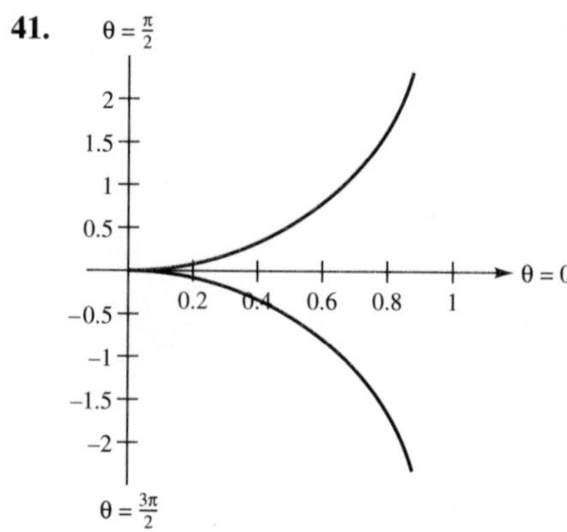

43.

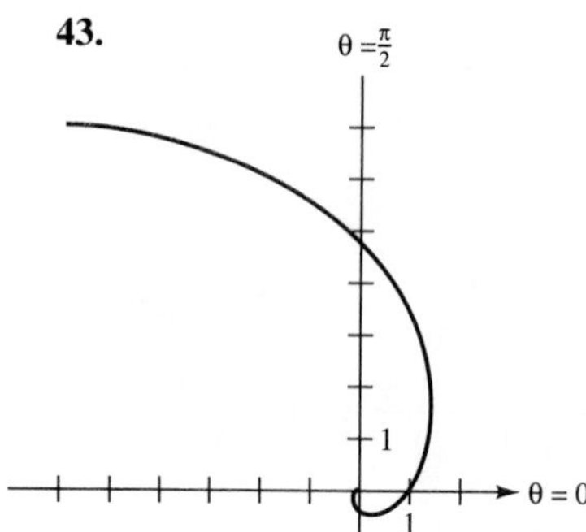

45.

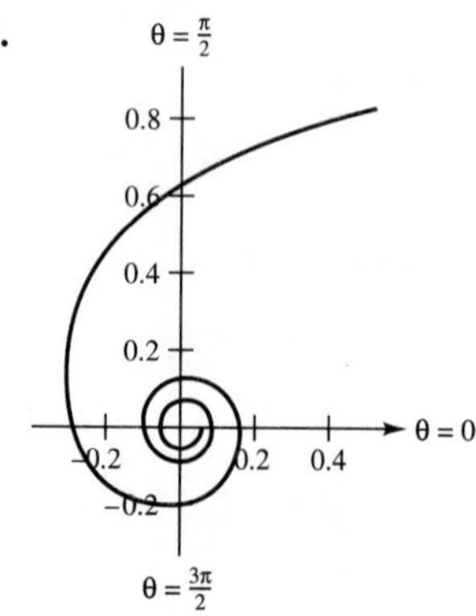

47. (a)

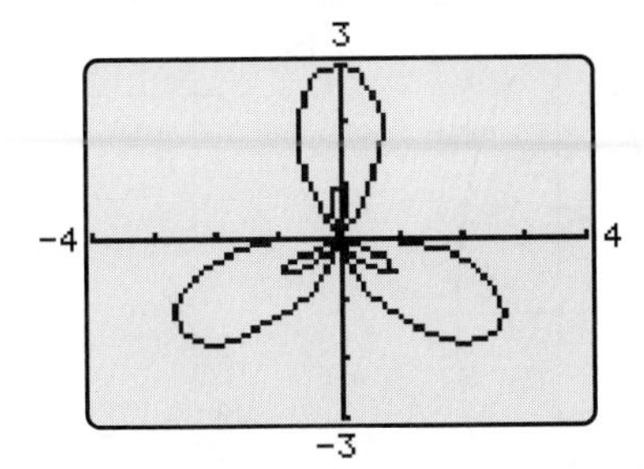

(b)

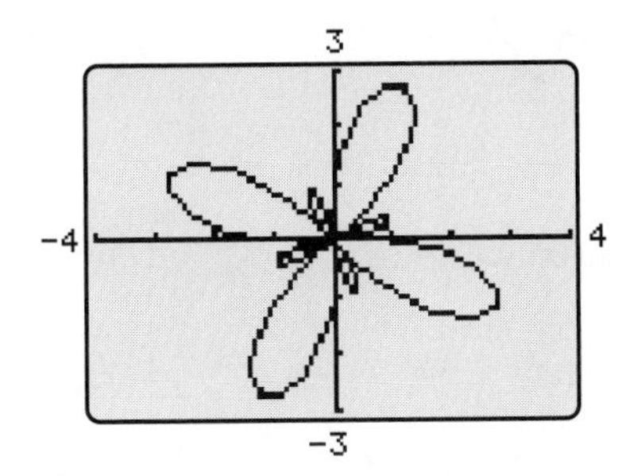

49. $r = a\sin\theta + b\cos\theta \Rightarrow r^2 = ar\sin\theta + br\cos\theta \Rightarrow$ $x^2 + y^2 = ay + bx \Rightarrow x^2 - bx + y^2 - ay = 0 \Rightarrow$ $(x^2 - bx + b^2/4) + (y^2 - ay + a^2/4) =$ $(a^2 + b^2)/4 \Rightarrow (x - b/2)^2 + (y - a/2)^2 = (a^2 + b^2)/4$, a circle with center $(b/2, a/2)$ and radius $\sqrt{a^2 + b^2}/2$.

51. Using the Law of Cosines in the following diagram, $d^2 = r^2 + s^2 - 2rs\cos(\theta - \beta)$, so $d = \sqrt{r^2 + s^2 - 2rs\cos(\theta - \beta)}$

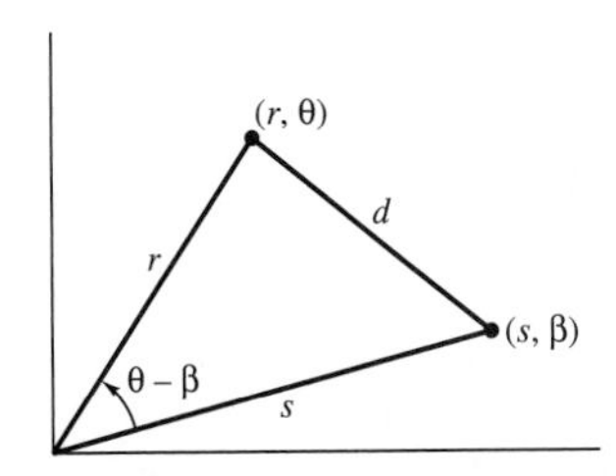

Section 10.5, page 646

1. (d) **3.** (c) **5.** (a)

7. Hyperbola, $e = 4/3$ **9.** Parabola, $e = 1$

11. Ellipse, $e = 2/3$

13. .1 **15.** $\sqrt{5}$ **17.** 5/4

19. (b) $\sqrt{15}/4, \sqrt{10}/4, \sqrt{2}/4$
(c) The smaller the eccentricity, the closer the shape is to circular.

In the graphs for Exercises 21–31, the x- and y-axes with scales are given for convenience, but coordinates of points are in polar *coordinates.*

21.

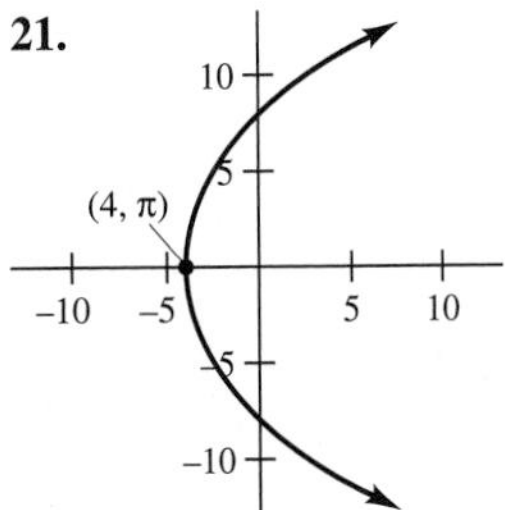

23.

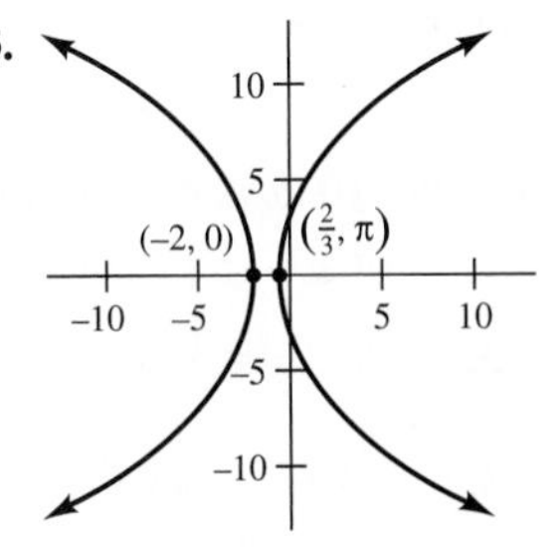

25.

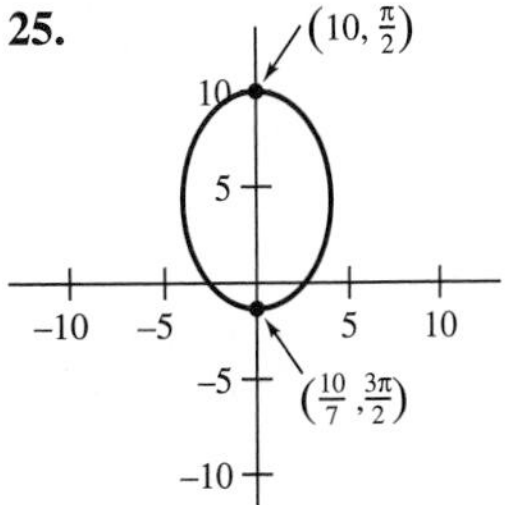

27.

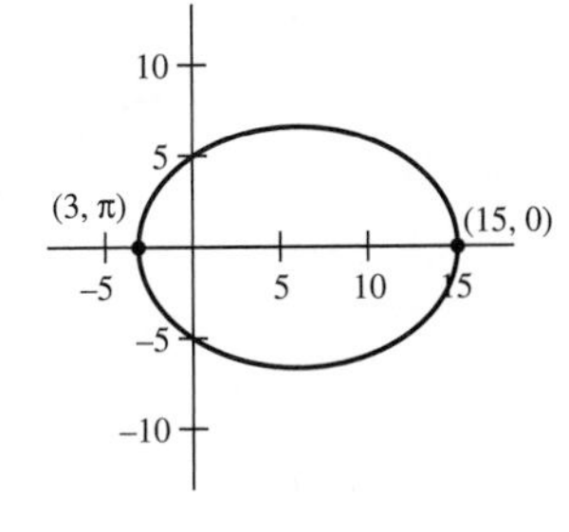

29.

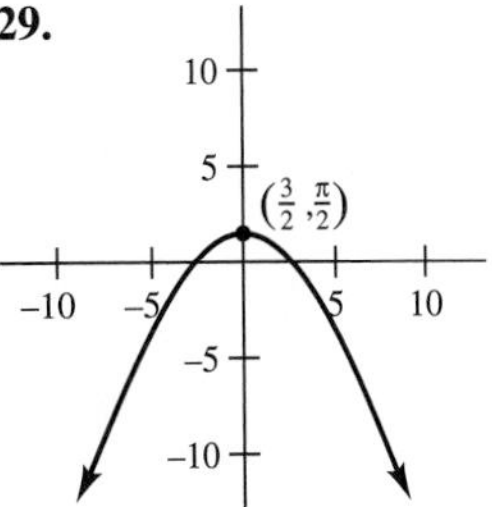

31.

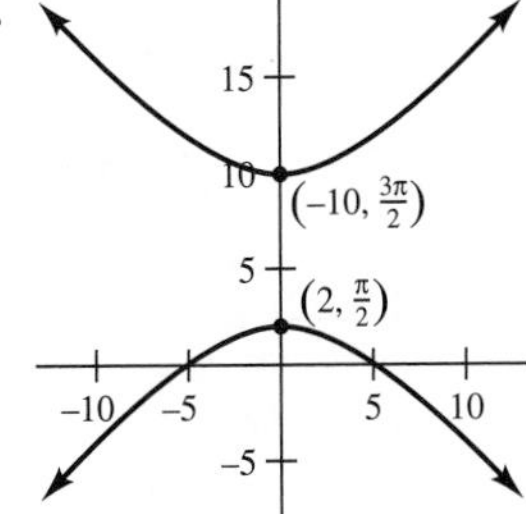

33. $r = \dfrac{6}{1 - \cos\theta}$ **35.** $r = \dfrac{16}{5 + 3\sin\theta}$

37. $r = \dfrac{3}{1 + 2\cos\theta}$ **39.** $r = \dfrac{8}{1 - 4\cos\theta}$

41. $r = \dfrac{3}{1 - \sin\theta}$ **43.** $r = \dfrac{2}{2 + \cos\theta}$

45. $r = \dfrac{2}{1 - 2\cos\theta}$

47. Since $0 < e < 1$, $0 < 1 - e^2 < 1$ as well. The formulas for a^2 and b^2 show that $a^2 = b^2/(1 - e^2)$, so $a^2 > b^2$. Since a and b are both positive, $a > b$

49. $r = \dfrac{3 \cdot 10^7}{1 - \cos\theta}$

Chapter 10 Review, page 649

1. $-15 \le x \le 15$ and $-10 \le y \le 10$

3. $-15 \le x \le 8$ and $-6 \le y \le 10$

5. $x = 3 - 2y$ or $y = -\dfrac{1}{2}x + \dfrac{3}{2}$

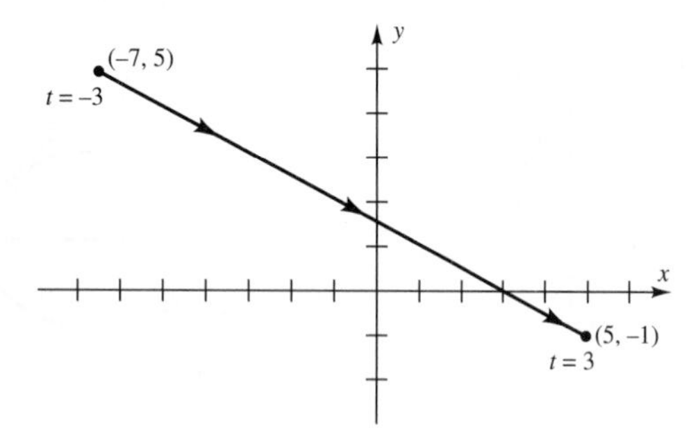

7. $y = -2x^2 + 2$

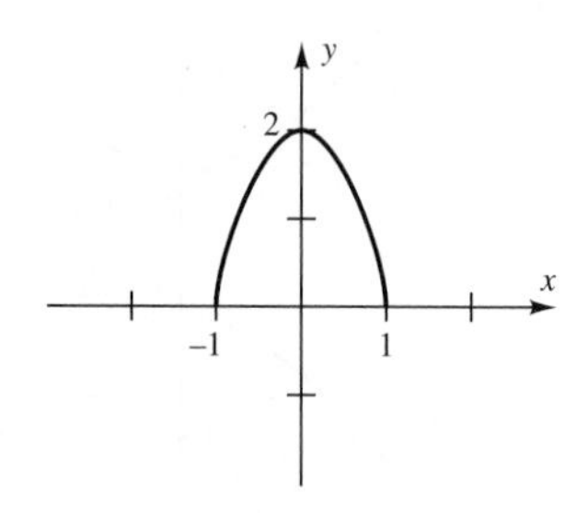

Point moves from (1, 0) to (− 1, 0) as t goes from 0 to π. Then point retraces its path, moving from (− 1, 0) to (1, 0) as t goes from π to 2π.

9. (b) and (c)

11. Ellipse, foci: (0, 2), (0, − 2), vertices: $(0, 2\sqrt{5})$, $(0, -2\sqrt{5})$

13. Ellipse, foci: (1, 6), (1, 0), vertices: (1, 7), (1, − 1)

15. Focus: (0, 5/14), directrix: $y = -5/14$

17.

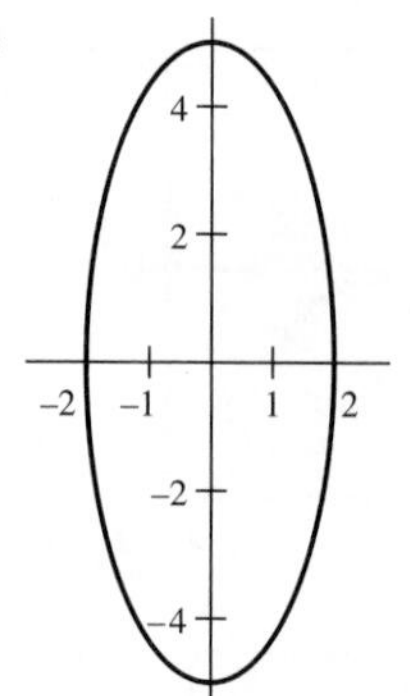

19.

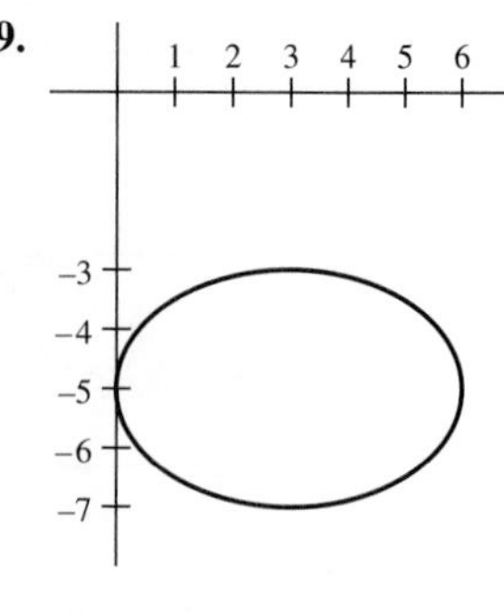

21. Asymptotes:

$y + 4 = \pm\dfrac{5}{2}(x - 1)$

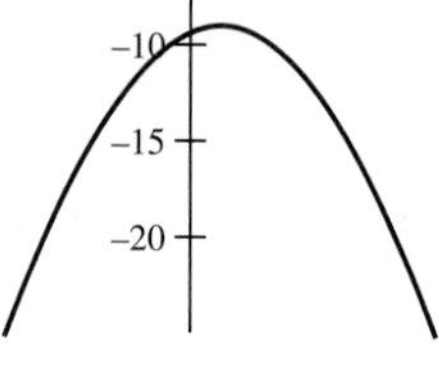
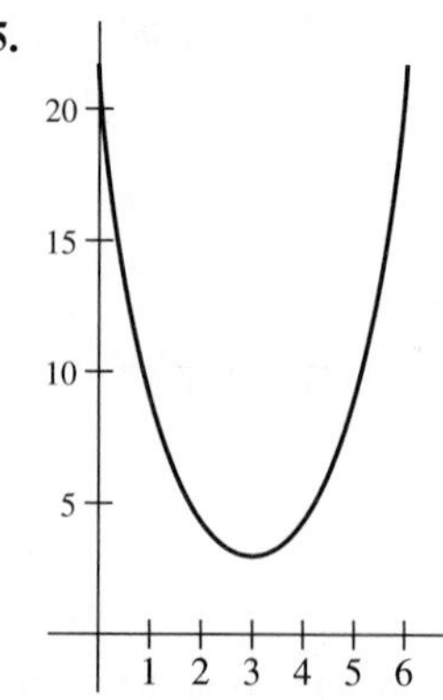

23.

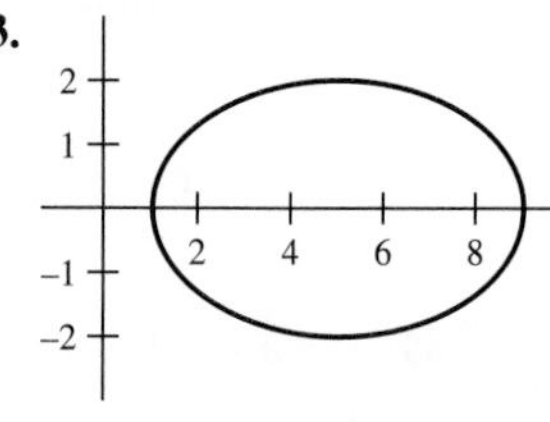

25.

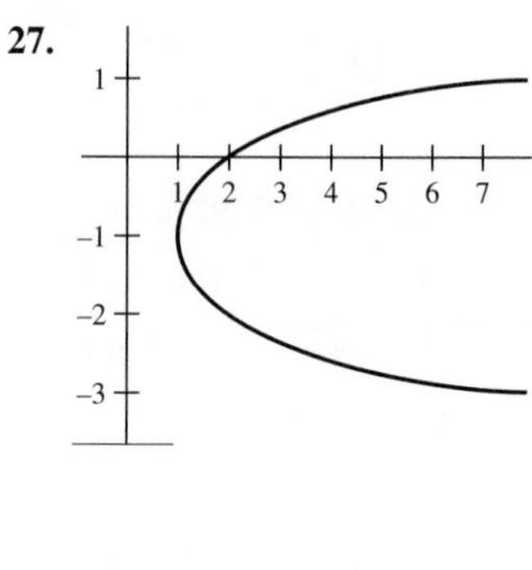

27.

29. Center: $(4, -6)$

31. $\dfrac{(x-3)^2}{4} + \dfrac{(y-1)^2}{2} = 1$

33. $\dfrac{y^2}{4} - \dfrac{(x-3)^2}{16} = 1$

35. $\left(y + \dfrac{1}{2}\right)^2 = -\dfrac{1}{2}\left(x - \dfrac{3}{2}\right)$

37. Ellipse **39.** Hyperbola

41. $-6 \le x \le 6$ and $-4 \le y \le 4$

43. $-9 \le x \le 9$ and $-6 \le y \le 6$

45. $-10 \le x \le 6$ and $-10 \le y \le 20$

47.

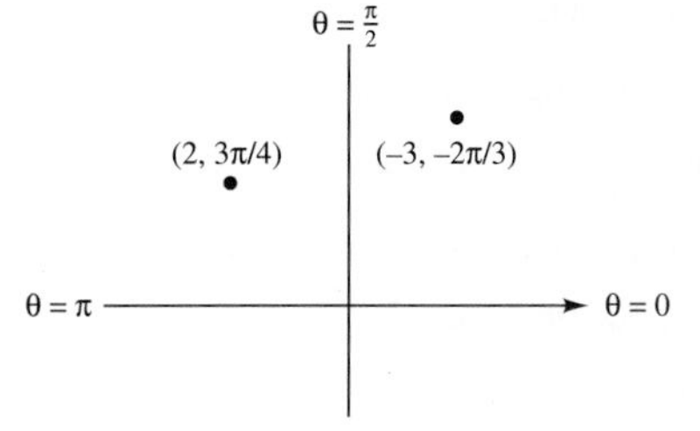

49.

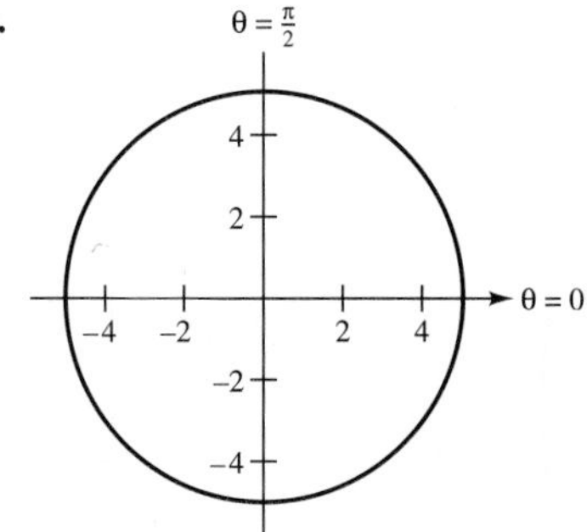

51.

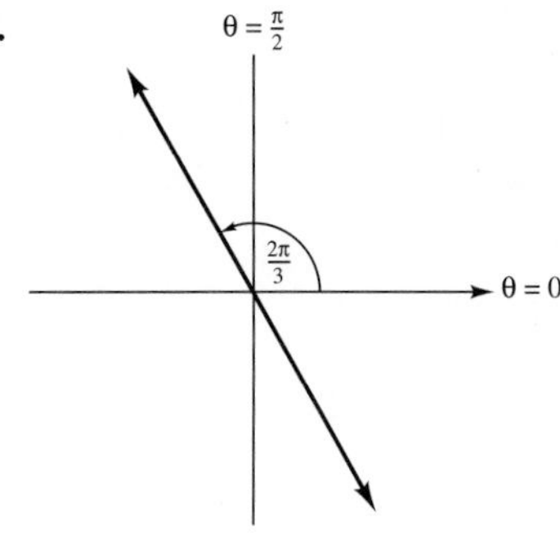

53.

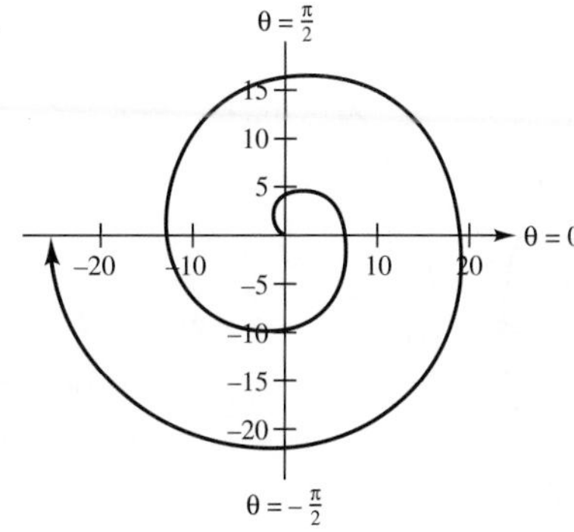

55.

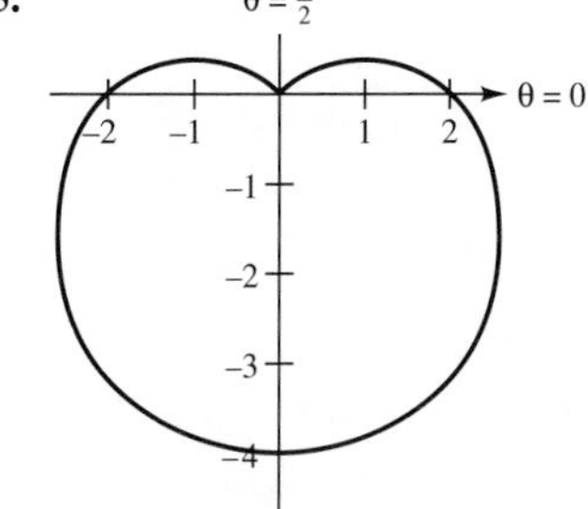

57.

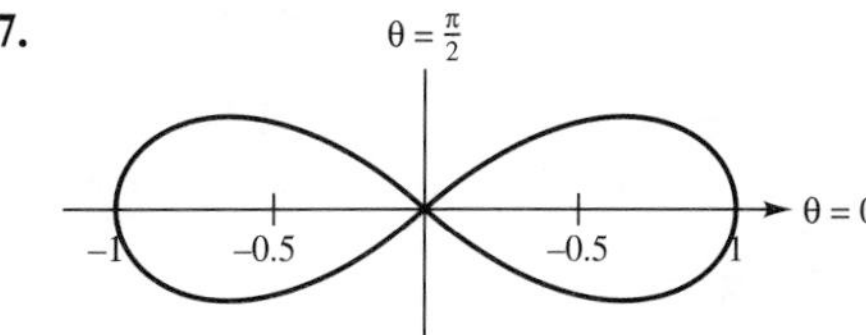

59. $\left(-\dfrac{3}{2}, -\dfrac{3\sqrt{3}}{2}\right)$

61. Eccentricity $= \sqrt{\dfrac{2}{3}} \approx 0.8165$

63. Ellipse

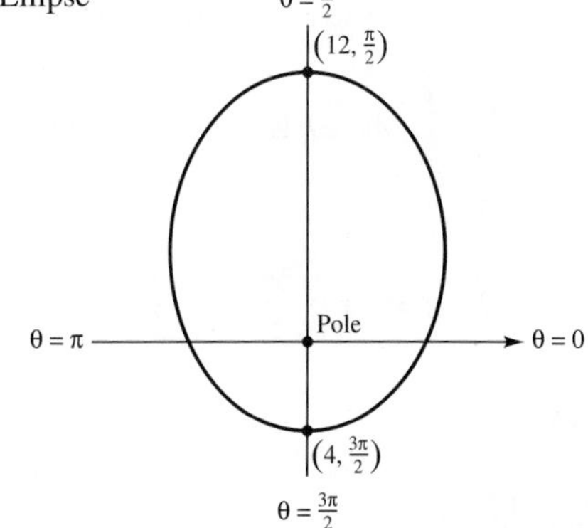

65. Hyperbola

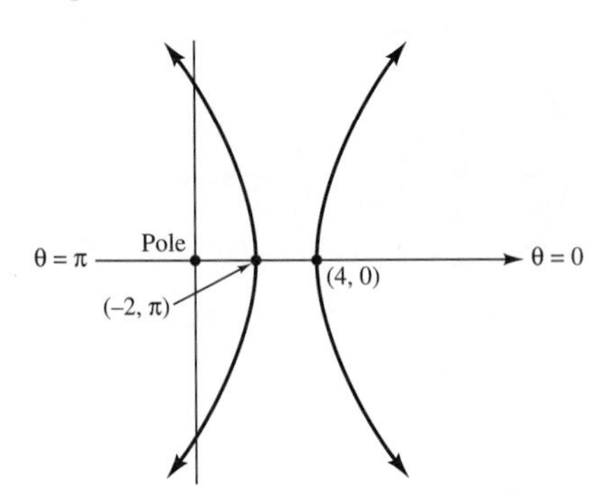

67. $r = \dfrac{24}{5 + \cos\theta}$ **69.** $r = \dfrac{2}{1 + \cos\theta}$

CHAPTER 11

Section 11.1, page 659

1. Yes **3.** Yes **5.** No

7. $x = \dfrac{11}{5}, y = -\dfrac{7}{5}$

9. $x = \dfrac{2}{7}, y = -\dfrac{11}{7}$

11. $r = \dfrac{5}{2}, s = -\dfrac{5}{2}$

13. $x = \dfrac{3c}{2}, y = \dfrac{-c + 2d}{2}$

15. $x = 28, y = 22$ **17.** $x = 2, y = -1$

19. Inconsistent

21. $x = b, y = \dfrac{3b - 4}{2}$, where b is any real number

23. $x = b, y = \dfrac{3b - 2}{4}$, where b is any real number

25. Inconsistent

27. $x = -6, y = 2$ **29.** $x = \dfrac{66}{5}, y = \dfrac{18}{5}$

31. $x = \dfrac{7}{11}, y = -7$

33. $x = 1, y = \dfrac{1}{2}$

35. $x = -\dfrac{3}{7}, y = \dfrac{42}{13}$

37. $x = .185, y = -.624$

39. $x = \dfrac{rd - sb}{ad - bc}, y = \dfrac{as - cr}{ad - bc}$

41. $x = \dfrac{3c + 8}{15}, y = \dfrac{6c - 4}{15}$ is the only solution.

43. $c = -3, d = \dfrac{1}{2}$

45. 140 adults, 60 children

47. 17 and 23

49. \$14,450 at 9% and \$6450 at 11%

51. 3/4 lb cashews and $2\frac{1}{4}$ lb peanuts

53. Plane 550 mph, wind 50 mph

55. Boat speed 18 mph; current speed 2 mph

57. 12 liters of the 18%; 18 liters of the 8%

59. 24 g of 50% alloy; 16 g of 75% alloy

Section 11.2, page 673

1. $\begin{pmatrix} 2 & -3 & 4 & 1 \\ 1 & 2 & -6 & 0 \\ 3 & -7 & 4 & -3 \end{pmatrix}$

3. $\begin{pmatrix} 1 & -\frac{1}{2} & \frac{7}{4} & 0 \\ 2 & -\frac{3}{2} & 5 & 0 \\ 0 & -2 & \frac{1}{3} & 0 \end{pmatrix}$

5. $\begin{aligned} 2x - 3y &= 1 \\ 4x + 7y &= 2 \end{aligned}$

7. $\begin{aligned} x \qquad + z \qquad &= 1 \\ x - y + 4z - 2w &= 3 \\ 4x + 2y + 5z \qquad &= 2 \end{aligned}$

9. $x = 3/2, y = 5, z = -2, w = 0$

11. $x = 2 - t, y = -3 - 2t, z = 4, w = t$, where t is any real number

13. $x = \dfrac{3}{2}, y = \dfrac{3}{2}, z = -\dfrac{3}{2}$

15. $z = t, y = -1 + \frac{1}{3}t, x = 2 - \frac{4}{3}t$, where t is any real number

17. $x = -14, y = -6, z = 2$

19. $x = 100, y = 50, z = 50$

21. $z = t, y = \frac{1}{2} - 2t, x = t$, where t is any real number

23. $x = 1, y = 2$ **25.** No solutions

27. $z = t, y = t - 1, x = -t + 2$, for any real number t

29. $x = 0, y = 0, z = 0$

31. $x = -1, y = 1, z = -3, w = -2$

33. $x = \frac{7}{31}, y = \frac{6}{31}, z = \frac{1}{31}, w = \frac{29}{31}$

35. $x = 1/2, y = 1/3, z = -1/4$

37. $A = -1, B = 2$

39. $A = -3/25, B = 3/25, C = 7/5$

41. $A = 2, B = 3, C = -1$

43. $x = -s - t, y = s + 2t, z = s, w = t$, where s and t are any real numbers

45. $x = t, y = 0, z = t, w = 2t$, where t is any real number

47. 10 quarters; 28 dimes; 14 nickels

49. \$3000 from her friend; \$6000 from the bank; \$1000 from the insurance company

51. \$15,000 in the mutual fund; \$30,000 in bonds; \$25,000 in food franchise

53. 8 hours for Tom; 24 hours for Dick; 12 hours for Harry

55. 40 lb of peanuts; 20 lb of almonds; 80 lb of cashews

57. 2000 chairs; 1600 chests; 2500 tables

Section 11.3, page 683

1. AB defined, 2×4; BA not defined

3. AB defined, 3×3; BA defined, 2×2

5. AB defined, 3×2; BA not defined

7. $\begin{pmatrix} 3 & 0 & 11 \\ 2 & 8 & 10 \end{pmatrix}$ **9.** $\begin{pmatrix} 1 & -3 \\ 2 & -1 \\ 5 & 6 \end{pmatrix}$

11. $\begin{pmatrix} 1 & -1 & 1 & 2 \\ 4 & 3 & 3 & 2 \\ -1 & -1 & -3 & 2 \\ 5 & 3 & 2 & 5 \end{pmatrix}$

13. $AB = \begin{pmatrix} 17 & -3 \\ 33 & -19 \end{pmatrix}$; $BA = \begin{pmatrix} -4 & 9 \\ 24 & 2 \end{pmatrix}$

15. $AB = \begin{pmatrix} 8 & 24 & -8 \\ 2 & -2 & 6 \\ -3 & -21 & 15 \end{pmatrix}$; $BA = \begin{pmatrix} 19 & 9 & 8 \\ -10 & 2 & 0 \\ 0 & 0 & 0 \end{pmatrix}$

17. $\begin{pmatrix} -2 & 1 \\ 3/2 & -1/2 \end{pmatrix}$ **19.** No inverse

21. No inverse **23.** $\begin{pmatrix} -3 & 2 & -4 \\ -1 & 1 & -1 \\ 8 & -5 & 10 \end{pmatrix}$

25. $x = -1, y = 0, z = -3$

27. $x = -8, y = 16, z = 5$

29. $x = -.5, y = -2.1, z = 6.7, w = 2.8$

31. $x = 10.5, y = 5, z = -13, v = 32, w = 2.5$

33. $x = -1149/161, y = 426/161, z = -1124/161, w = 579/161$

35. $x = 0, y = 0, z = 0, v = 0, w = 0$

37. Inconsistent system; no solutions

39. $x = -2 - 3w, y = 5/4 - (1/4)z$, where z and w are any real numbers.

41. $a = -\frac{7}{20}, b = -\frac{19}{20}, c = \frac{23}{10}$

43. $a = 2, b = -3, c = 1$

45. $a = 3, b = -1, c = 2$

47. $a = 1, b = -4, c = 1$

49. 15,000 of A; 18,000 of B; 54,000 of C

Section 11.4, page 693

1. $x = 3, y = 9$ or $x = -1, y = 1$

3. $x = \frac{-1 + \sqrt{73}}{6}, y = \frac{37 - \sqrt{73}}{18}$ or $x = \frac{-1 - \sqrt{73}}{6}, y = \frac{37 + \sqrt{73}}{18}$

5. $x = 7, y = 3$ or $x = 3, y = 7$

7. $x = 0, y = -2$ or $x = 6, y = 1$

9. $x = 2, y = 0$ or $x = 4, y = 2$

11. $x = 4, y = 3$ or $x = -4, y = 3$ or $x = \sqrt{21}, y = -2$ or $x = -\sqrt{21}, y = -2$

13. $x = -1.6237, y = -8.1891$ or $x = 1.3163, y = 1.0826$ or $x = 2.8073, y = 2.4814$

15. $x = -1.9493, y = .4412$ or $x = .3634, y = .9578$ or $x = 1.4184, y = .5986$

17. $x = -.9519, y = -.8145$ **19.** No solutions

21. $x = \frac{13 - \sqrt{105}}{8}, y = \frac{-3 - \sqrt{105}}{8}$ or $x = \frac{13 + \sqrt{105}}{8}, y = \frac{-3 + \sqrt{105}}{8}$

23. $x = -4.8093, y = 19.3201$ or $x = -3.1434, y = 7.7374$ or $x = 2.1407, y = 7.7230$ or $x = 2.8120, y = 11.7195$

25. $x = -3.8371, y = -2.2596$ or $x = -.9324, y = -7.7796$

27. $x = -1.4873, y = .0480$ or $x = -.0480, y = 1.4873$ or $x = .0480, y = -1.4873$ or $x = 1.4873, y = -.0480$

29. There is no solution when the wire is 70 ft long because the graphs of $y = 70 - 2x$ and $y^2 = 4\pi(100 - x^2/8)$ do not intersect.

31. Two possible boxes: one is 2 by 2 by 4 meters and the other is approximately 3.123 by 3.123 by 1.640 meters.

33. -4 and -12 **35.** 1.6 and 2.6

37. 15 and -12 **39.** 12 ft by 17 ft

41. 8×15 inches **43.** $y = 6x - 9$

Chapter 11 Review, page 695

1. $x = -5, y = -7$ **3.** $x = 0, y = -2$

5. $x = -35, y = 140, z = 22$ **7.** $x = 2, y = 4, z = 6$

9. $x = -t + 1, y = -2t, z = t$ for any real number t

11. 37 and -19 **13.** (c) **15.** 100

17. $\begin{pmatrix} 1 & -2 & 3 & 4 \\ 2 & 1 & -4 & 3 \\ -3 & 4 & -1 & -2 \end{pmatrix}$

19. $\begin{pmatrix} 2 & -1 & -2 & 2 \\ 1 & 3 & -2 & 1 \\ -1 & 4 & 2 & -3 \end{pmatrix}$ **21.** $\frac{3}{x+2} + \frac{1}{x-3}$

23. $\begin{pmatrix} -2 & 3 \\ -4 & -1 \end{pmatrix}$ **25.** Not defined

27. $\begin{pmatrix} -9 & 7 \\ -4 & 3 \end{pmatrix}$ **29.** $\begin{pmatrix} 1 & 2 & -2 \\ -1 & 3 & 0 \\ 0 & -2 & 1 \end{pmatrix}$

31. $x = -1/85, y = -14/85, z = -21/34, w = 46/85$

33. $x = 3, y = 9$ or $x = -1, y = 1$

35. $x = 1 - \sqrt{7}, y = 1 + \sqrt{7}$ or $x = 1 + \sqrt{7}, y = 1 - \sqrt{7}$

37. $x = -1.692, y = 3.136$ or $x = 1.812, y = 2.717$

39. 30 lb corn, 15 lb soybeans, 40 lb by-products

CHAPTER 12

Section 12.1, page 704

1. 8, 10, 12, 14, 16 **3.** $1, \frac{1}{8}, \frac{1}{27}, \frac{1}{64}, \frac{1}{125}$

5. $-\sqrt{3}, 2, -\sqrt{5}, \sqrt{6}, -\sqrt{7}$

7. 3.9, 4.01, 3.999, 4.0001, 3.99999

9. 2, 7, 8, 13, 14 **11.** $\sum_{i=1}^{11} i$

13. $\sum_{i=7}^{13} \frac{1}{2^i}$ (other answers possible) **15.** 45

17. 24 **19.** 54 **21.** $a_n = (-1)^n$

23. $a_n = \frac{n}{n+1}$ **25.** $a_n = 5n - 3$

27. 4, 11, 25, 53, 109 **29.** 1, -2, 3, 2, 3

31. $2, 3, 3, \frac{9}{2}, \frac{27}{4}$ **33.** 3, 1, 4, 1, 5

35. Third -10; sixth -2 **37.** Third 5; sixth 0

39. $\sum_{n=1}^{6} \frac{1}{2n+1}$ (other answers possible)

41. $\sum_{n=8}^{12} (-1)^n \left(\frac{n-7}{n}\right)$ (other answers possible)

43. 2.613035 **45.** $\frac{1051314}{31415} \approx 33.465$

47. 1.5759958 **49.** **(b)** 59, 61, 67, 71

51. 4, 9, 25, 49, 121 **53.** 3, 7, 13, 19, 23

55. **(a)** 1, 1, 2, 3, 5, 8, 13, 21, 34, 55
(b) 1, 2, 4, 7, 12, 20, 33, 54, 88, 143
(c) nth partial sum $= a_{n+2} - 1$

57. $n = 1$: $5(1)^2 + 4(-1)^1 = 1 = 1^2$; $n = 2$: $5(1)^2 + 4(-1)^2 = 9 = 3^2$; $n = 3$: $5(2)^2 + 4(-1)^3 = 16 = 4^2$; $n = 4$: $5(3)^2 + 4(-1)^4 = 49 = 7^2$; etc.

59. We have $a_1 = a_3 - a_2$; $a_2 = a_4 - a_3$; $a_3 = a_5 - a_4$; . . . $a_{k-1} = a_{k+1} - a_k$; $a_k = a_{k+2} - a_{k+1}$. If these equations are listed vertically, the sum of the left-side terms is $\sum_{n=1}^{k} a_k$. On the right side, one term in each line is the same as a term in the next line, except for sign. So the sum of the right-side terms is $a_{k+2} - a_2$. Since $a_2 = 1$, we conclude that $\sum_{n=1}^{k} a_n = a_{k+2} - 1$.

Section 12.2, page 710

1. 13; $a_n = 2n + 3$ **3.** 5; $a_n = n/4 + 15/4$

5. 8; $a_n = -n/2 + 21/2$ **7.** 87

9. $-21/4$ **11.** 30

13. $a_n - a_{n-1} = (3 - 2n) - (3 - 2(n - 1)) = -2$; arithmetic with $d = -2$

15. $a_n - a_{n-1} = \dfrac{5 + 3n}{2} - \dfrac{5 + 3(n - 1)}{2} = \dfrac{3}{2}$; arithmetic with $d = \dfrac{3}{2}$

17. $a_n - a_{n-1} = (c + 2n) - (c + 2(n - 1)) = 2$; arithmetic with $d = 2$

19. $a_5 = 14$; $a_n = 2n + 4$ **21.** $a_5 = 25$; $a_n = 7n - 10$

23. $a_5 = 0$; $a_n = -15/2 + 3n/2$ **25.** 710

27. $156\frac{2}{3}$ **29.** 2550 **31.** 20,100

33. $77,500 in tenth year; $437,500 over ten years

35. 428 **37.** 23.25, 22.5, 21.75, 21, 20.25, 19.5, 18.75

Section 12.3, page 714

1. Arithmetic **3.** Geometric **5.** Arithmetic

7. Geometric **9.** $a_6 = 160$; $a_n = 2^{n-1} \cdot 5$

11. $a_6 = \dfrac{1}{256}$; $a_n = \dfrac{1}{4^{n-2}}$

13. $a_6 = -\dfrac{5}{16}$; $a_n = \dfrac{(-1)^{n-1} \cdot 5}{2^{n-2}}$

15. 315/32 **17.** 381

19. $\dfrac{a_n}{a_{n-1}} = \dfrac{\left(-\frac{1}{2}\right)^n}{\left(-\frac{1}{2}\right)^{n-1}} = -\dfrac{1}{2}$; geometric with $r = -\dfrac{1}{2}$

21. $\dfrac{a_n}{a_{n-1}} = \dfrac{5^{n+2}}{5^{(n-1)+2}} = 5$; geometric with $r = 5$

23. $a_5 = 1$; $a_n = \dfrac{(-1)^{n-1} 64}{4^{n-2}} = \dfrac{(-1)^{n-1}}{4^{n-5}}$

25. $a_5 = \dfrac{1}{16}$; $a_n = \dfrac{1}{4^{n-3}}$

27. $a_5 = -\dfrac{8}{25}$; $a_n = -\dfrac{2^{n-2}}{5^{n-3}}$ **29.** 254

31. $-\dfrac{4921}{19683}$ **33.** $\dfrac{665}{8}$

35. $\log a_n - \log a_{n-1} = \log \dfrac{a_n}{a_{n-1}} = \log r$ **37.** 23.75 ft

39. The sequence is $\{2^{n-1}\}$ and $r = 2$. So for any k, the kth term is 2^{k-1}, and the sum of the preceding terms is the $(k - 1)$th partial sum of the sequence, $\sum_{n=1}^{k-1} 2^{n-1} = \dfrac{1 - 2^{k-1}}{1 - 2} = 2^{k-1} - 1$.

41. $\sum_{n=1}^{31} 2^{n-1} = \dfrac{1 - 2^{31}}{1 - 2} = (2^{31} - 1)$ cents = $21,474,836.47

43. $1898.44 **45.** 37 payments

Excursion 12.3.A, page 720

1. 1 **3.** $.06/.94 = 3/47$ **5.** $\dfrac{500}{.6} = 833\dfrac{1}{3}$

7. $4 + 2\sqrt{2}$ **9.** 2/9 **11.** 597/110

13. 10,702/4995 **15.** 174,067/99,900

17. **(a)** $a_1 = 2(1.5)^1 = 3$. For each $n \geq 2$, $a_{n-1} = 2(1.5)^{n-1}$ and $a_n = 2(1.5)^n$, so that the ratio $\dfrac{a_n}{a_{n-1}}$ is $\dfrac{2(1.5)^n}{2(1.5)^{n-1}} = 1.5$. Therefore, this is a geometric series.
(b) $S_k = -6 + 6(1.5)^k$ and $f(x) = -6 + 6(1.5)^x$
(c) The graph shows that as x gets large, $f(x)$ gets huge, so there is no horizontal asymptote. Hence, the series does not converge.

Section 12.4, page 727

1. 720 **3.** 220 **5.** 0 **7.** 64 **9.** 3,921,225

11. $x^5 + 5x^4y + 10x^3y^2 + 10x^2y^3 + 5xy^4 + y^5$

13. $a^5 - 5a^4b + 10a^3b^2 - 10a^2b^3 + 5ab^4 - b^5$

15. $32x^5 + 80x^4y^2 + 80x^3y^4 + 40x^2y^6 + 10xy^8 + y^{10}$

17. $x^3 + 6x^2\sqrt{x} + 15x^2 + 20x\sqrt{x} + 15x + 6\sqrt{x} + 1$

19. $1 - 10c + 45c^2 - 120c^3 + 210c^4 - 252c^5 + 210c^6 - 120c^7 + 45c^8 - 10c^9 + c^{10}$

21. $x^{-12} + 4x^{-8} + 6x^{-4} + 4 + x^4$ **23.** 56

25. $-8i$ **27.** $10x^3y^2$ **29.** $35c^3d^4$

31. $\frac{35}{8}u^{-5}$ **33.** 4032 **35.** 160

37. (a) $\binom{9}{1} = \frac{9!}{1!8!} = 9; \binom{9}{8} = \frac{9!}{8!1!} = 9$

(b) $\binom{n}{1} = \binom{n}{n-1} = \frac{n!}{1!(n-1)!} = \frac{n(n-1)!}{(n-1)!} = n$

39. $2^n = (1+1)^n = 1^n + \binom{n}{1}1^{n-1}\cdot 1 + \binom{n}{2}1^{n-2}\cdot 1^2 + \binom{n}{3}1^{n-3}\cdot 1^3 + \cdots + \binom{n}{n-1}1^1\cdot 1^{n-1} + 1^n = \binom{n}{0} + \binom{n}{1} + \binom{n}{2} + \binom{n}{3} + \cdots + \binom{n}{n-1} + \binom{n}{n}$

41. $\cos^4\theta + 4i\cos^3\theta\sin\theta - 6\cos^2\theta\sin^2\theta - 4i\cos\theta\sin^3\theta + \sin^4\theta$

43. (a) $f(x+h) - f(x) = (x+h)^5 - x^5 = \left(x^5 + \binom{5}{1}x^4h + \binom{5}{2}x^3h^2 + \binom{5}{3}x^2h^3 + \binom{5}{4}xh^4 + h^5\right) - x^5 = \binom{5}{1}x^4h + \binom{5}{2}x^3h^2 + \binom{5}{3}x^2h^3 + \binom{5}{4}xh^4 + h^5$

(b) $\frac{f(x+h) - f(x)}{h} = \binom{5}{1}x^4 + \binom{5}{2}x^3h + \binom{5}{3}x^2h^2 + \binom{5}{4}xh^3 + h^4$

(c) When h is *very* close to 0, so are the last four terms in part (b), so $\frac{f(x+h) - f(x)}{h} \approx \binom{5}{1}x^4 = 5x^4$.

45. $\frac{f(x+h) - f(x)}{h} = \frac{(x+h)^{12} - x^{12}}{h} = \binom{12}{1}x^{11} + \binom{12}{2}x^{10}h + \binom{12}{3}x^9h^2 + \binom{12}{4}x^8h^3 + \cdots + \binom{12}{10}x^2h^9 + \binom{12}{11}xh^{10} + h^{11} \approx \binom{12}{1}x^{11} = 12x^{11}$, when h is very close to 0.

47. (a) $(n-r)! = (n-r)(n-r-1)(n-r-2)(n-r-3)\cdots 2\cdot 1 = (n-r)(n-(r+1))(n-(r+1)-1)\cdots 2\cdot 1 = (n-r)(n-(r+1))!$

(b) Since $(n+1) - (r+1) = n - r$, $((n+1)-(r+1))! = (n-r)!$

(c) $\binom{n}{r+1} + \binom{n}{r} = \frac{n!}{(r+1)!(n-(r+1))!} + \frac{n!}{r!(n-r)!} = \frac{n!(n-r) + n!(r+1)}{(r+1)!(n-r)!} = \frac{n!(n+1)}{(r+1)!(n-r)!} = \frac{(n+1)!}{(r+1)!((n+1)-(r+1))!} = \binom{n+1}{r+1}$

(d) For example, rows 2 and 3 of Pascal's triangle are

```
   1   2   1
1  (3)  3   1
```

that is,

$$\binom{2}{0} \quad \binom{2}{1} \quad \binom{2}{2}$$

$$\binom{3}{0} \quad \binom{3}{1} \quad \binom{3}{2} \quad \binom{3}{3}$$

The circled 3 is the sum of the two closest entries in the row above: $1 + 2$. But this just says that $\binom{3}{1} = \binom{2}{0} + \binom{2}{1}$, which is part (c) with $n = 2$ and $r = 0$. Similarly, in the general case, verify that the two closest entries in the row above $\binom{n+1}{r+1}$ are $\binom{n}{r}$ and $\binom{n}{r+1}$ and use part (c).

Section 12.5, page 737

1. *Step 1:* For $n = 1$ the statement is $1 = 2^1 - 1$, which is true. *Step 2:* Assume that the statement is true for $n = k$: that is, $1 + 2 + 2^2 + 2^3 + \cdots + 2^{k-1} = 2^k - 1$. Add 2^k to both sides, and rearrange terms:

$$1 + 2 + 2^2 + 2^3 + \cdots + 2^{k-1} + 2^k = 2^k - 1 + 2^k$$

$$1 + 2 + 2^2 + 2^3 + \cdots + 2^{k-1} + 2^{(k+1)-1} = 2(2^k) - 1$$

$$1 + 2 + 2^2 + 2^3 + \cdots + 2^{k-1} + 2^{(k+1)-1} = 2^{k+1} - 1$$

But this last line says that the statement is true for $n = k + 1$. Therefore, by the Principle of Mathematical Induction the statement is true for every positive integer n.

Note: **Hereafter, in these answers, step 1 will be omitted if it is trivial (as in Exercise 1), and only the essential parts of step 2 will be given.**

3. Assume that the statement is true for $n = k$:
$1 + 3 + 5 + \cdots + (2k - 1) = k^2$.
Add $2(k + 1) - 1$ to both sides:
$1 + 3 + 5 + \cdots + (2k - 1) + (2(k + 1) - 1) = k^2 + 2(k + 1) - 1 = k^2 + 2k + 1 = (k + 1)^2$.
The first and last parts of this equation say that the statement is true for $n = k + 1$.

5. Assume that the statement is true for $n = k$:

$$1^2 + 2^2 + 3^2 + \cdots + k^2 = \frac{k(k + 1)(2k + 1)}{6}$$

Add $(k + 1)^2$ to both sides:

$$\begin{aligned}
&1^2 + 2^2 + 3^2 + \cdots + k^2 + (k + 1)^2 \\
&= \frac{k(k + 1)(2k + 1)}{6} + (k + 1)^2 \\
&= \frac{k(k + 1)(2k + 1) + 6(k + 1)^2}{6} \\
&= \frac{(k + 1)[k(2k + 1) + 6(k + 1)]}{6} \\
&= \frac{(k + 1)[2k^2 + 7k + 6]}{6} \\
&= \frac{(k + 1)(k + 2)(2k + 3)}{6} \\
&= \frac{(k + 1)[(k + 1) + 1][2(k + 1) + 1]}{6}
\end{aligned}$$

The first and last parts of this equation say that the statement is true for $n = k + 1$.

7. Assume that the statement is true for $n = k$:

$$\frac{1}{1\cdot 2} + \frac{1}{2\cdot 3} + \cdots + \frac{1}{k(k + 1)} = \frac{k}{k + 1}.$$

Adding $\dfrac{1}{(k + 1)((k + 1) + 1)} = \dfrac{1}{(k + 1)(k + 2)}$ to both sides yields:

$$\begin{aligned}
&\frac{1}{1\cdot 2} + \frac{1}{2\cdot 3} + \cdots + \frac{1}{k(k + 1)} + \frac{1}{(k + 1)(k + 2)} \\
&= \frac{k}{k + 1} + \frac{1}{(k + 1)(k + 2)} \\
&= \frac{k(k + 2) + 1}{(k + 1)(k + 2)} = \frac{k^2 + 2k + 1}{(k + 1)(k + 2)} \\
&= \frac{(k + 1)^2}{(k + 1)(k + 2)} = \frac{k + 1}{k + 2} = \frac{k + 1}{(k + 1) + 1}
\end{aligned}$$

The first and last parts of this equation show that the statement is true for $n = k + 1$.

9. Assume the statement is true for $n = k$: $k + 2 > k$. Adding 1 to both sides, we have: $k + 2 + 1 > k + 1$, or equivalently, $(k + 1) + 2 > (k + 1)$. Therefore, the statement is true for $n = k + 1$.

11. Assume the statement is true for $n = k$: $3^k \geq 3k$. Multiplying both sides by 3 yields: $3\cdot 3^k \geq 3\cdot 3k$, or equivalently, $3^{k+1} \geq 3\cdot 3k$. Now since $k \geq 1$, we know that $3k \geq 3$ and hence that $2\cdot 3k \geq 3$. Therefore, $2\cdot 3k + 3k \geq 3 + 3k$, or equivalently, $3\cdot 3k \geq 3k + 3$. Combining this last inequality with the fact that $3^{k+1} \geq 3\cdot 3k$, we see that $3^{k+1} \geq 3k + 3$, or equivalently, $3^{k+1} \geq 3(k + 1)$. Therefore, the statement is true for $n = k + 1$.

13. Assume the statement is true for $n = k$: $3k > k + 1$. Adding 3 to both sides yields: $3k + 3 > k + 1 + 3$, or equivalently, $3(k + 1) > (k + 1) + 3$. Since $(k + 1) + 3$ is certainly greater than $(k + 1) + 1$, we conclude that $3(k + 1) > (k + 1) + 1$. Therefore, the statement is true for $n = k + 1$.

15. Assume the statement is true for $n = k$; then 3 is a factor of $2^{2k+1} + 1$; that is, $2^{2k+1} + 1 = 3M$ for some integer M. Thus, $2^{2k+1} = 3M - 1$. Now $2^{2(k+1)+1} = 2^{2k+2+1} = 2^{2+2k+1} = 2^2\cdot 2^{2k+1} = 4(3M - 1) = 12M - 4 = 3(4M) - 3 - 1 = 3(4M - 1) - 1$. From the first and last terms of this equation we see that $2^{2(k+1)+1} + 1 = 3(4M - 1)$. Hence, 3 is a factor of $2^{2(k+1)+1} + 1$. Therefore, the statement is true for $n = k + 1$.

17. Assume the statement is true for $n = k$: 64 is a factor of $3^{2k+2} - 8k - 9$. Then $3^{2k+2} - 8k - 9 = 64N$ for some integer N so that $3^{2k+2} = 8k + 9 + 64N$. Now $3^{2(k+1)+2} = 3^{2k+2+2} = 3^{2+(2k+2)} = 3^2 \cdot 3^{2k+2} = 9(8k + 9 + 64N)$. Consequently,

$$\begin{aligned} 3^{2(k+1)+2} - 8(k+1) - 9 &= 3^{2(k+1)+2} - 8k - 8 - 9 \\ &= 3^{2(k+1)+2} - 8k - 17 \\ &= [9(8k + 9 + 64N)] - 8k - 17 \\ &= 72k + 81 + 9 \cdot 64N - 8k - 17 \\ &= 64k + 64 + 9 \cdot 64N = 64(k + 1 + 9N). \end{aligned}$$

From the first and last parts of this equation we see that 64 is a factor of $3^{2(k+1)+2} - 8(k+1) - 9$. Therefore, the statement is true for $n = k + 1$.

19. Assume that the statement is true for $n = k$: $c + (c + d) + (c + 2d) + \cdots + (c + (k-1)d) = \frac{k(2c + (k-1)d)}{2}$. Adding $c + kd$ to both sides, we have

$$\begin{aligned} c + (c+d) + (c+2d) + \cdots + {} & (c + (k-1)d) + (c + kd) \\ &= \frac{k(2c + (k-1)d)}{2} + c + kd \\ &= \frac{k(2c + (k-1)d) + 2(c + kd)}{2} \\ &= \frac{2ck + k(k-1)d + 2c + 2kd}{2} \\ &= \frac{2ck + 2c + kd(k-1) + 2kd}{2} \\ &= \frac{(k+1)2c + kd(k-1+2)}{2} \\ &= \frac{(k+1)2c + kd(k+1)}{2} = \frac{(k+1)(2c + kd)}{2} \\ &= \frac{(k+1)(2c + ((k+1) - 1)d)}{2} \end{aligned}$$

Therefore, the statement is true for $n = k + 1$.

21. **(a)** $x^2 - y^2 = (x - y)(x + y)$;
$x^3 - y^3 = (x - y)(x^2 + xy + y^2)$;
$x^4 - y^4 = (x - y)(x^3 + x^2y + xy^2 + y^3)$
(b) *Conjecture*: $x^n - y^n = (x - y)(x^{n-1} + x^{n-2}y + x^{n-3}y^2 + \cdots + x^2y^{n-3} + xy^{n-2} + y^{n-1})$.
Proof: The statement is true for $n = 2, 3, 4$, by part (a). Assume that the statement is true for $n = k$:

$$x^k - y^k = (x - y)(x^{k-1} + x^{k-2}y + \cdots + xy^{k-2} + y^{k-1}).$$

Now use the fact that $-yx^k + yx^k = 0$ to write $x^{k+1} - y^{k+1}$ as follows:

$$\begin{aligned} x^{k+1} - y^{k+1} &= x^{k+1} - yx^k + yx^k - y^{k+1} \\ &= (x^{k+1} - yx^k) + (yx^k - y^{k+1}) \\ &= (x - y)x^k + y(x^k - y^k) \\ &= (x - y)x^k + y(x - y)(x^{k-1} + x^{k-2}y \\ &\qquad + x^{k-3}y^2 + \cdots + xy^{k-2} + y^{k-1}) \\ &= (x - y)x^k + (x - y)(x^{k-1}y + \\ &\qquad x^{k-2}y^2 + x^{k-3}y^3 + \cdots + xy^{k-1} + y^k) \\ &= (x - y)[x^k + x^{k-1}y + x^{k-2}y^2 + \\ &\qquad x^{k-3}y^3 + \cdots + xy^{k-1} + y^k] \end{aligned}$$

The first and last parts of this equation show that the conjecture is true for $n = k + 1$. Therefore, by mathematical induction, the conjecture is true for every integer $n \geq 2$.

23. False; counterexample: $n = 9$

25. True: *Proof:* Since $(1 + 1)^2 > 1^2 + 1$, the statement is true for $n = 1$. Assume the statement is true for $n = k$: $(k + 1)^2 > k^2 + 1$. Then $[(k + 1) + 1]^2 = (k + 1)^2 + 2(k + 1) + 1 > k^2 + 1 + 2(k + 1) + 1 = k^2 + 2k + 2 + 2 > k^2 + 2k + 2 = k^2 + 2k + 1 + 1 = (k + 1)^2 + 1$. The first and last terms of this inequality say that the statement is true for $n = k + 1$. Therefore, by induction the statement is true for every positive integer n.

27. False; counterexample: $n = 3$

29. Since $2 \cdot 5 - 4 > 5$, the statement is true for $n = 5$. Assume the statement is true for $n = k$ (with $k \geq 5$): $2k - 4 > k$. Adding 2 to both sides shows that $2k - 4 + 2 > k + 2$, or equivalently, $2(k + 1) - 4 > k + 2$. Since $k + 2 > k + 1$, we see that $2(k + 1) - 4 > k + 1$. So the statement is true for $n = k + 1$. Therefore, by the Extended Principle of Mathematical Induction, the statement is true for all $n \geq 5$.

31. Since $2^2 > 2$, the statement is true for $n = 2$. Assume that $k \geq 2$ and that the statement is true for $n = k$: $k^2 > k$. Then $(k + 1)^2 = k^2 + 2k + 1 > k^2 + 1 > k + 1$. The first and last terms of this inequality show that the statement is true for $n = k + 1$. Therefore, by induction, the statement is true for all $n \geq 2$.

33. Since $3^4 = 81$ and $2^4 + 10 \cdot 4 = 16 + 40 = 56$, we see that $3^4 > 2^4 + 10 \cdot 4$. So the statement is true for $n = 4$. Assume that $k \geq 4$ and that the statement is true for $n = k$: $3^k > 2^k + 10k$. Multiplying both sides by 3 yields: $3 \cdot 3^k > 3(2^k + 10k)$, or equivalently, $3^{k+1} > 3 \cdot 2^k + 30k$. But

$$3 \cdot 2^k + 30k > 2 \cdot 2^k + 30k = 2^{k+1} + 30k.$$

Therefore, $3^{k+1} > 2^{k+1} + 30k$. Now we shall show that $30k > 10(k + 1)$. Since $k \geq 4$, we have $20k \geq 20 \cdot 4$, so that $20k > 80 > 10$. Adding $10k$ to both sides of $20k > 10$ yields: $30k > 10k + 10$, or equivalently, $30k > 10(k + 1)$. Consequently,

$$3^{k+1} > 2^{k+1} + 30k > 2^{k+1} + 10(k + 1).$$

The first and last terms of this inequality show that the statement is true for $n = k + 1$. Therefore, the statement is true for all $n \geq 4$ by induction.

35. (a) 3 (that is, $2^2 - 1$) for $n = 2$; 7 (that is, $2^3 - 1$) for $n = 3$; 15 (that is, $2^4 - 1$) for $n = 4$.

(b) *Conjecture:* The smallest possible number of moves for n rings is $2^n - 1$. *Proof:* This conjecture is easily seen to be true for $n = 1$ or $n = 2$. Assume it is true for $n = k$ and that we have $k + 1$ rings to move. In order to move the *bottom* ring from the first peg to another peg (say, the second one), it is first necessary to move the top k rings off the first peg *and* leave the second peg vacant at the end (the second peg will have to be used *during* this moving process). If this is to be done according to the rules, we will end up with the top k rings on the third peg in the *same* order they were on the first peg. According to the induction assumption, the least possible number of moves needed to do this is $2^k - 1$. It now takes one move to transfer the bottom ring (the $(k + 1)$st) from the first to the second peg. Finally, the top k rings now on the third peg must be moved to the second peg. Once again by the induction hypothesis, the least number of moves for doing this is $2^k - 1$. Therefore, the smallest total number of moves needed to transfer all $k + 1$ rings from the first to the second peg is $(2^k - 1) + 1 + (2^k - 1) = (2^k + 2^k) - 1 = 2 \cdot 2^k - 1 = 2^{k+1} - 1$. Hence, the conjecture is true for $n = k + 1$. Therefore, by induction it is true for all positive integers n.

37. *De Moivre's Theorem:* For any complex number $z = r(\cos\theta + i \sin\theta)$ and any positive integer n, $z^n = r^n(\cos(n\theta) + i \sin(n\theta))$. *Proof:* The theorem is obviously true when $n = 1$. Assume that the theorem is true for $n = k$, that is, $z^k = r^k(\cos(k\theta) + i \sin(k\theta))$. Then

$$z^{k+1} = z \cdot z^k = [r(\cos\theta + i \sin\theta)]\,[r^k(\cos(k\theta) + i \sin(k\theta))].$$

According to the multiplication rule for complex numbers in polar form (multiply the moduli and add the arguments) we have:

$$z^{k+1} = r \cdot r^k(\cos(\theta + k\theta) + i \sin(\theta + k\theta)) = r^{k+1}[\cos((k + 1)\theta) + i \sin((k + 1)\theta)].$$

This statement says the theorem is true for $n = k + 1$. Therefore, by induction, the theorem is true for every positive integer n.

Chapter 12 Review, page 740

1. $-3, -1, 1, 3$ **3.** $1, \frac{1}{4}, \frac{1}{9}, \frac{1}{16}$ **5.** -368

7. 129 **9.** $a_n = 9 - 6n$

11. $a_n = 6n - 11$ **13.** $a_n = 2 \cdot 3^{n-1}$

15. $a_n = \frac{3}{2^{n-8}}$ **17.** -55

19. $\frac{121}{4}$ **21.** 8.75, 13.5, 18.25

23. Second method is better. **25.** 2

27. $(1.02)^{51} = (1 + .02)^{51} = 1^{51} + \binom{51}{1} 1^{50}(.02) + \binom{51}{2} 1^{49}(.02)^2 +$ other positive terms $= 2.53 +$ other positive terms > 2.5

29. 455 **31.** $n + 1$ **33.** 1140 **35.** 80

37. True for $n = 1$. If the statement is true for $n = k$, then $1 + 5 + \cdots + 5^{k-1} = \frac{5^k - 1}{4}$ so that

$$1 + 5 + \cdots + 5^{k-1} + 5^k = \frac{5^k - 1}{4} + 5^k = \frac{5^k - 1 + 4 \cdot 5^k}{4} = \frac{5 \cdot 5^k - 1}{4} = \frac{5^{k+1} - 1}{4}.$$

Hence, the statement is true for $n = k + 1$ and therefore true for all n by induction.

39. Since the statement is obviously true for $x = 0$, assume $x \neq 0$. Then the statement is true for $n = 1$. If the statement is true for $n = k$, then $|x^k| < 1$. Then $|x^k| \cdot |x| < |x|$. Thus, $|x^{k+1}| = |x^k| \cdot |x| < |x| < 1$. Hence, the statement is true for $n = k + 1$ and therefore true for all n by induction.

41. True for $n = 1$. If the statement is true for $n = k$, then $1 + 4 + \cdots + 4^{k-1} = \frac{1}{3}(4^k - 1)$. Hence,
$1 + 4 + \cdots + 4^{k-1} + 4^k = \frac{1}{3}(4^k - 1) + 4^k =$
$\frac{1}{3}(4^k - 1) + \frac{3 \cdot 4^k}{3} = \frac{4^k - 1 + 3 \cdot 4^k}{3} = \frac{4 \cdot 4^k - 1}{3}$
$= \frac{1}{3}(4^{k+1} - 1)$. Hence, the statement is true for $n = k + 1$ and therefore for all n by induction.

43. If $n = 1$, then $9^n - 8n - 1 = 0$. Since $0 = 0 \cdot 8$, the statement is true for $n = 1$. If the statement is true for $n = k$, then $9^k - 8k - 1 = 8D$, so that $9^k - 1 = 8k + 8D = 8(k + D)$. Consequently,
$9^{k+1} - 8(k + 1) - 1 = 9^{k+1} - 8k - 8 - 1 =$
$9^{k+1} - 9 - 8k = 9(9^k - 1) - 8k =$
$9[8(k + D)] - 8k = 8[9(k + D) - k]$. Thus, 8 is a factor of $9^{k+1} - 8(k + 1) - 1$ and the statement is true for $n = k + 1$. Therefore it is true for all n by induction.

ALGEBRA REVIEW

Section 1.A, page 745

1. 36 **3.** 73 **5.** -5 **7.** $-125/64$

9. 1/3 **11.** -112 **13.** 81/16 **15.** $-211/216$

17. 129/8 **19.** x^{10} **21.** $.03y^9$

23. $24x^7$ **25.** $9x^4y^2$ **27.** $-21a^6$

29. $384w^6$ **31.** ab^3 **33.** $8x^{-1}y^3$

35. a^8x^{-3} **37.** $3xy$ **39.** 2^{12} **41.** 2^{-12}

43. x^7 **45.** ce^9 **47.** $b^2c^2d^6$ **49.** $a^{12}b^8$

51. $1/(c^{10}d^6)$ **53.** $1/(108x)$ **55.** a^7c/b^6

57. c^3d^6 **59.** $a + \frac{1}{a}$ **61.** Negative

63. Negative **65.** Negative **67.** 3^s

69. $a^{6t}b^{4t}$ **71.** b^{rs+st}/c^{2rt}

73. Many possible examples, including $3^2 + 4^2 = 9 + 16 = 25$, but $(3 + 4)^2 = 7^2 = 49$

75. Many possible examples, including $3^2 \cdot 2^3 = 9 \cdot 8 = 72$; but $(3 \cdot 2)^{2+3} = 6^5 = 7776$

77. Many possible examples, including $2^6/2^3 = 64/8 = 8$, but $2^{6/3} = 2^2 = 4$

79. False for all nonzero a; for instance, $(-3)^2 = (-3)(-3) = 9$, but $-3^2 = -9$

Section 1.B, page 748

1. $8x$ **3.** $-2a^2b$

5. $-x^3 + 4x^2 + 2x - 3$

7. $5u^3 + u - 4$

9. $4z - 12z^2w + 6z^3w^2 - zw^3 + 8$

11. $-3x^3 + 15x + 8$

13. $-5xy - x$ **15.** $15y^3 - 5y$

17. $12a^2x^2 - 6a^3xy + 6a^2xy$

19. $12z^4 + 30z^3$

21. $12a^2b - 18ab^2 + 6a^3b^2$ **23.** $x^2 - x - 2$

25. $2x^2 + 2x - 12$ **27.** $y^2 + 7y + 12$

29. $-6x^2 + x + 35$

31. $3y^3 - 9y^2 + 4y - 12$

33. $x^2 - 16$ **35.** $16a^2 - 25b^2$

37. $y^2 - 22y + 121$

39. $25x^2 - 10bx + b^2$

41. $16x^6 - 8x^3y^4 + y^8$

43. $9x^4 - 12x^2y^4 + 4y^8$

45. $2y^3 + 9y^2 + 7y - 3$

47. $-15w^3 + 2w^2 + 9w - 18$

49. $24x^3 - 4x^2 - 4x$

51. $x^3 - 6x^2 + 11x - 6$

53. $-3x^3 - 5x^2y + 26xy^2 - 8y^3$

55. 3 **57.** -6 **59.** 6 **61.** 1

63. 5 **65.** $x - 25$ **67.** $9 + 6\sqrt{y} + y$

69. $\sqrt{3}x^2 + 4x + \sqrt{3}$

71. $3ax^2 + (3b + 2a)x + 2b$

73. $abx^2 + (a^2 + b^2)x + ab$

75. $x^3 - (a + b + c)x^2 + (ab + ac + bc)x - abc$

77. 3^{4+r+t}

79. $x^{m+n} + 2x^n - 3x^m - 6$

81. $2x^{4n} - 5x^{3n} + 8x^{2n} - 18x^n - 5$

83. Example: if $y = 4$, then $3(4 + 2) \neq (3 \cdot 4) + 2$; correct statement: $3(y + 2) = 3y + 6$

85. Example: if $x = 2$, $y = 3$, then $(2 + 3)^2 \neq 2 + 3^2$; correct statement: $(x + y)^2 = x^2 + 2xy + y^2$

87. Example: if $x = 2$, $y = 3$, then $(7 \cdot 2)(7 \cdot 3) \neq 7 \cdot 2 \cdot 3$; correct statement: $(7x)(7y) = 49xy$

89. Example: if $y = 2$, then $2 + 2 + 2 \neq 2^3$; correct statement: $y + y + y = 3y$

91. Example: if $x = 4$, then $(4 - 3)(4 - 2) \neq 4^2 - 5 \cdot 4 - 6$; correct statement: $(x - 3)(x - 2) = x^2 - 5x + 6$

93. If x is the chosen number, then adding 1 and squaring the result gives $(x + 1)^2$. Subtracting 1 from the original number x and squaring the result gives $(x - 1)^2$. Subtracting the second of these squares from the first yields: $(x + 1)^2 - (x - 1)^2 = (x^2 + 2x + 1) - (x^2 - 2x + 1) = 4x$. Dividing by the original number x now gives $\frac{4x}{x} = 4$. So the answer is always 4, no matter what number x is chosen.

95. Many correct answers

Section 1.C, page 753

1. $(x + 2)(x - 2)$ **3.** $(3y + 5)(3y - 5)$

5. $(9x + 2)^2$ **7.** $(\sqrt{5} + x)(\sqrt{5} - x)$

9. $(7 + 2z)^2$

11. $(x^2 + y^2)(x + y)(x - y)$

13. $(x + 3)(x - 2)$ **15.** $(z + 3)(z + 1)$

17. $(y + 9)(y - 4)$ **19.** $(x - 3)^2$

21. $(x + 5)(x + 2)$ **23.** $(x + 9)(x + 2)$

25. $(3x + 1)(x + 1)$ **27.** $(2z + 3)(z + 4)$

29. $9x(x - 8)$ **31.** $2(x - 1)(5x + 1)$

33. $(4u - 3)(2u + 3)$ **35.** $(2x + 5y)^2$

37. $(x - 5)(x^2 + 5x + 25)$

39. $(x + 2)^3$ **41.** $(2 + x)(4 - 2x + x^2)$

43. $(-x + 5)^3$ **45.** $(x + 1)(x^2 - x + 1)$

47. $(2x - y)(4x^2 + 2xy + y^2)$

49. $(x^3 + 2^3)(x^3 - 2^3) = (x + 2)(x^2 - 2x + 4)(x - 2)(x^2 + 2x + 4)$

51. $(y^2 + 5)(y^2 + 2)$ **53.** $(9 + y^2)(3 + y)(3 - y)$

55. $(z + 1)(z^2 - z + 1)(z - 1)(z^2 + z + 1)$

57. $(x^2 + 3y)(x^2 - y)$ **59.** $(x + z)(x - y)$

61. $(a + 2b)(a^2 - b)$

63. $(x^2 - 8)(x + 4) = (x + \sqrt{8})(x - \sqrt{8})(x + 4)$

65. If $x^2 + 1 = (x + c)(x + d) = x^2 + (c + d)x + cd$, then $c + d = 0$ and $cd = 1$. But $c + d = 0$ implies that $c = -d$ and hence that $1 = cd = (-d)d = -d^2$, or equivalently, that $d^2 = -1$. Since there is no real number with this property, $x^2 + 1$ cannot possibly factor in this way.

Section 1.D, page 758

1. $\frac{9}{7}$ **3.** $\frac{195}{8}$ **5.** $\frac{x - 2}{x + 1}$

7. $\frac{a + b}{a^2 + ab + b^2}$ **9.** $1/x$ **11.** $\frac{29}{35}$

13. $\frac{121}{42}$ **15.** $\frac{ce + 3cd}{de}$ **17.** $\frac{b^2 - c^2}{bc}$

19. $\frac{-1}{x(x + 1)}$ **21.** $\frac{x + 3}{(x + 4)^2}$ **23.** $\frac{2x - 4}{x(3x - 4)}$

25. $\frac{x^2 - xy + y^2 + x + y}{x^3 + y^3}$

27. $\frac{-6x^5 - 38x^4 - 84x^3 - 71x^2 - 14x + 1}{4x(x + 1)^3(x + 2)^3}$

29. 2 **31.** $2/(3c)$ **33.** $3y/x^2$ **35.** $\frac{12x}{x - 3}$

37. $\frac{5y^2}{3(y + 5)}$ **39.** $\frac{u + 1}{u}$

41. $\frac{(u + v)(4u - 3v)}{(2u - v)(2u - 3v)}$ **43.** $\frac{35}{24}$

45. $u^2/(vw)$ **47.** $\frac{x + 3}{2x}$

49. $\frac{x^2y^2}{(x + y)(x + 2y)}$ **51.** $\frac{cd(c + d)}{c - d}$

53. $\frac{y - x}{xy}$ **55.** $\frac{-3y + 3}{y}$

57. $\dfrac{-1}{x(x+h)}$ **59.** $\dfrac{xy}{x+y}$

61. $2\sqrt{5}/5$ **63.** $\sqrt{70}/10$

65. $\sqrt{x}/x$ **67.** $\dfrac{(x+1)(\sqrt{x}-1)}{x-1}$

69. $\dfrac{\sqrt{a}+2\sqrt{b}}{a-4b}$

71. Example: if $a = 1$, $b = 2$, then $\dfrac{1}{1} + \dfrac{1}{2} \neq \dfrac{1}{1+2}$; correct statement: $\dfrac{1}{a} + \dfrac{1}{b} = \dfrac{b+a}{ab}$

73. Example: if $a = 4$, $b = 9$, then $\left(\dfrac{1}{\sqrt{4}+\sqrt{9}}\right)^2 \neq \dfrac{1}{4+9}$; correct statement: $\left(\dfrac{1}{\sqrt{a}+\sqrt{b}}\right)^2 = \dfrac{1}{a+2\sqrt{ab}+b}$

75. Example: if $u = 1$, $v = 2$, then $\dfrac{1}{2} + \dfrac{2}{1} \neq 1$; correct statement: $\dfrac{u}{v} + \dfrac{v}{u} = \dfrac{u^2+v^2}{vu}$

77. Example: if $x = 4$, $y = 9$, then $(\sqrt{4}+\sqrt{9})\cdot\dfrac{1}{\sqrt{4}+\sqrt{9}} \neq 4 + 9$; correct statement: $(\sqrt{x}+\sqrt{y})\cdot\dfrac{1}{\sqrt{x}+\sqrt{y}} = 1$

Index of Applications

Biology and Life Science Applications

Business and Manufacturing Applications

Business and Manufacturing Applications *(Continued)*

Business and Manufacturing Applications *(Continued)*

Chemistry and Physics Applications

Chemistry and Physics Applications *(Continued)*

Construction Applications

Consumer Applications

Consumer Applications *(Continued)*

Geometric Applications

Geometric Applications *(Continued)*

Interest Rate Applications

Miscellaneous Applications

Time and Distance Applications

Time and Distance Applications *(Continued)*

Time and Distance Applications *(Continued)*

Time and Distance Applications *(Continued)*

Index

TRIGONOMETRY

Trigonometric Functions of Real Numbers

If t is a real number and P is the point where the terminal side of an angle of t radians in standard position meets the unit circle, then

$\cos t = x$-coordinate of P $\qquad$ $\sin t = y$-coordinate of P

$$\tan t = \frac{\sin t}{\cos t} \qquad \csc t = \frac{1}{\sin t} \qquad \sec t = \frac{1}{\cos t} \qquad \cot t = \frac{\cos t}{\sin t}$$

Point-in-the-Plane Description

For any real number t and point (x, y) on the terminal side of an angle of t radians in standard position:

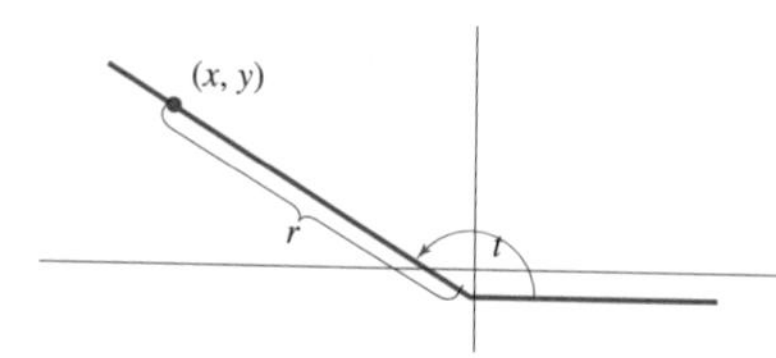

$$\sin t = \frac{y}{r} \qquad \cos t = \frac{x}{r} \qquad \tan t = \frac{y}{x} \quad (x \neq 0)$$

$$\csc t = \frac{r}{y} \quad (y \neq 0) \qquad \sec t = \frac{r}{x} \quad (x \neq 0) \qquad \cot t = \frac{x}{y} \quad (y \neq 0)$$

Periodic Graphs

If $A \neq 0$ and $b > 0$, then each of $f(t) = A \sin(bt + c)$ and $g(t) = A \sin(bt + c)$ has

amplitude $|A|$, $\qquad$ period $2\pi/b$, $\qquad$ phase shift $-c/b$.

Right Triangle Trigonometry

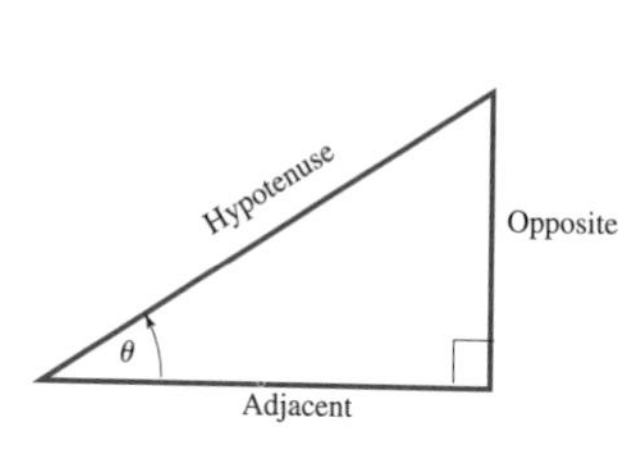

$$\sin \theta = \frac{\text{opposite}}{\text{hypotenuse}}$$

$$\tan \theta = \frac{\text{opposite}}{\text{adjacent}}$$

$$\cos \theta = \frac{\text{adjacent}}{\text{hypotenuse}}$$

Special Right Triangles

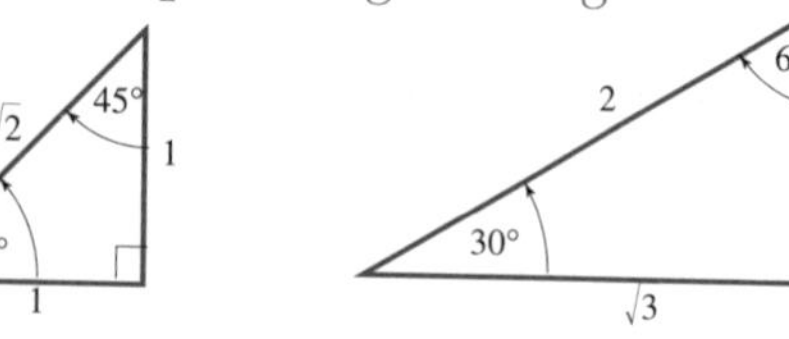

Special Values

θ Degrees	θ Radians	$\sin \theta$	$\cos \theta$	$\tan \theta$
0°	0	0	1	0
30°	$\frac{\pi}{6}$	$\frac{1}{2}$	$\frac{\sqrt{3}}{2}$	$\frac{\sqrt{3}}{3}$
45°	$\frac{\pi}{4}$	$\frac{\sqrt{2}}{2}$	$\frac{\sqrt{2}}{2}$	1
60°	$\frac{\pi}{3}$	$\frac{\sqrt{3}}{2}$	$\frac{1}{2}$	$\sqrt{3}$
90°	$\frac{\pi}{2}$	1	0	undefined

Law of Cosines

$$a^2 = b^2 + c^2 - 2bc \cos A$$

$$b^2 = a^2 + c^2 - 2ac \cos B$$

$$c^2 = a^2 + b^2 - 2ab \cos C$$

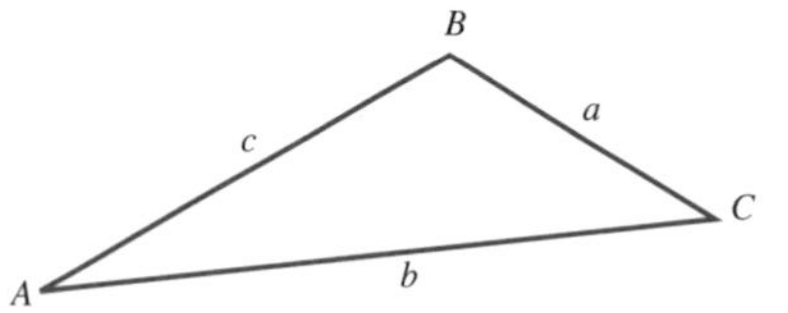

Law of Sines

$$\frac{a}{\sin A} = \frac{b}{\sin B} = \frac{c}{\sin C}$$

Area Formulas

Herron's Formula: Area $= \sqrt{s(s-a)(s-b)(s-c)}$, where $s = \frac{1}{2}(a + b + c)$ $\qquad$ Area $= \frac{1}{2} ab \sin C$

Distance Formula

Length of segment $PQ =$

$$\sqrt{(x_1 - x_2)^2 + (y_1 - y_2)^2}$$

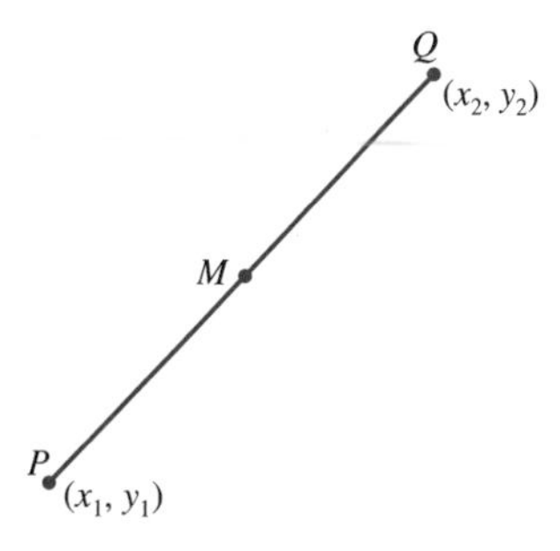

Midpoint Formula

Midpoint M of segment $PQ =$

$$\left(\frac{x_1 + x_2}{2}, \frac{y_1 + y_2}{2}\right)$$

Slope

If $x_1 \neq x_2$, slope of line $PQ =$

$$\frac{y_2 - y_1}{x_2 - x_1}$$

The equation of the straight line through (x_1, y_1) with slope m is $y - y_1 = m(x - x_1)$.

The equation of line with slope m and y-intercept b is $y = mx + b$.

Rectangular and Parametric Equations for Conic Sections

Circles

Center (h, k), radius r

$$(x - h)^2 + (y - k)^2 = r^2$$

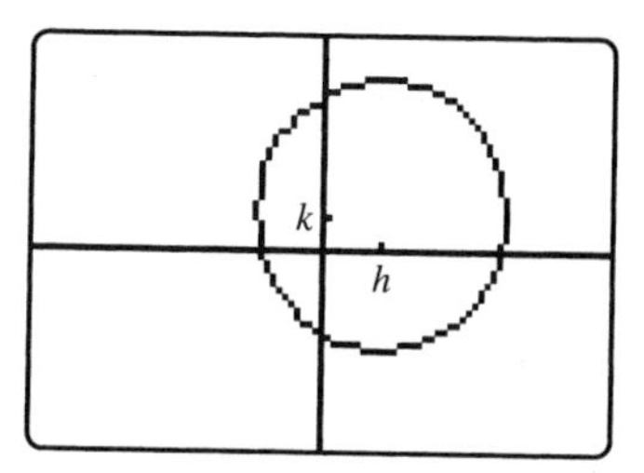

$x = r \cos t + h$
$y = r \sin t + k$ $\quad (0 \leq t \leq 2\pi)$

Ellipse

Center (h, k)

$$\frac{(x - h)^2}{a^2} + \frac{(y - k)^2}{b^2} = 1$$

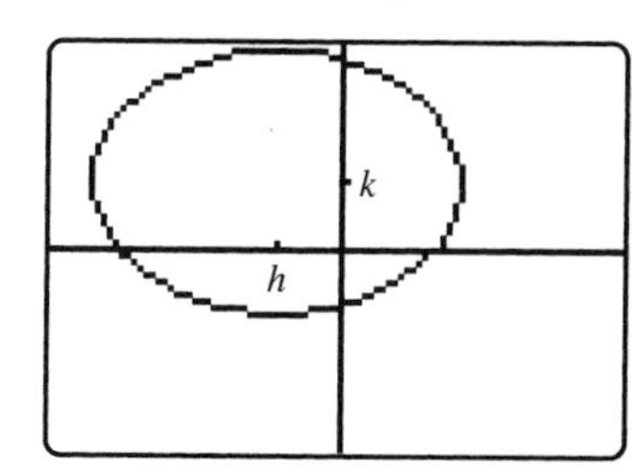

$x = a \cos t + h$
$y = b \sin t + k$ $\quad (0 \leq t \leq 2\pi)$

Parabola

Vertex (h, k)

$$(x - h)^2 = 4p(y - k)$$

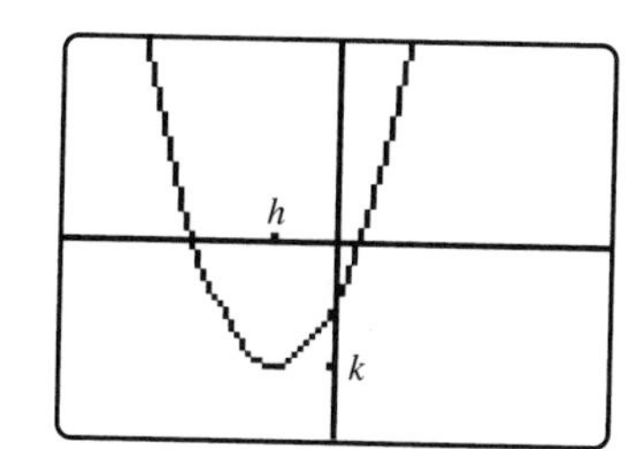

$x = t$
$y = \dfrac{(t - h)^2}{4p} + k$ $\quad$ (t any real)

Parabola

Vertex (h, k)

$$(y - k)^2 = 4p(x - h)$$

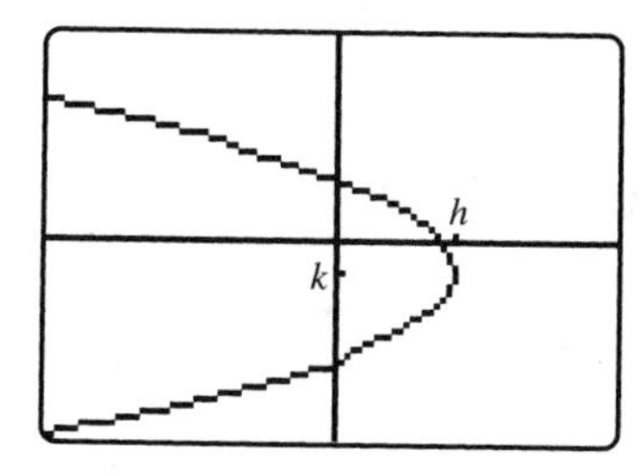

$x = \dfrac{(t - k)^2}{4p} + h$
$y = t$ $\quad$ (t any real)

Hyperbola

Center (h, k)

$$\frac{(x - h)^2}{a^2} - \frac{(y - k)^2}{b^2} = 1$$

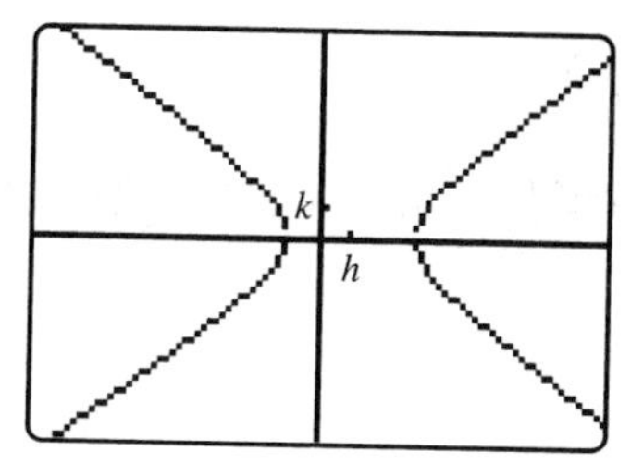

$x = \dfrac{a}{\cos t} + h$
$y = b \tan t + k$ $\quad (0 \leq t \leq 2\pi)$

Hyperbola

Center (h, k)

$$\frac{(y - k)^2}{a^2} - \frac{(x - h)^2}{b^2} = 1$$

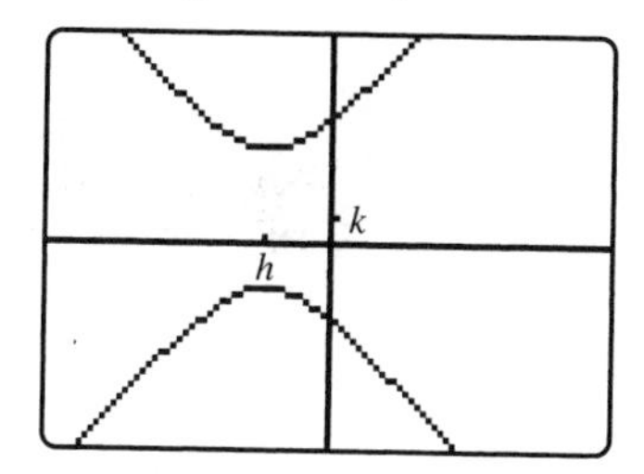

$x = b \tan t + h$
$y = \dfrac{a}{\cos t} + k$ $\quad (0 \leq t \leq 2\pi)$